The Ecology of Seashores

The CRC Marine Science Series is dedicated to providing state-of-the-art coverage of important topics in marine biology, marine chemistry, marine geology, and physical oceanography. The series includes volumes that focus on the synthesis of recent advances in marine science.

CRC MARINE SCIENCE SERIES

SERIES EDITOR

Michael J. Kennish, Ph.D.

PUBLISHED TITLES

Artificial Reef Evaluation with Application to Natural Marine Habitats, William Seaman, Jr.
Chemical Oceanography, Second Edition, Frank J. Millero
Coastal Ecosystem Processes, Daniel M. Alongi
Ecology of Estuaries: Anthropogenic Effects, Michael J. Kennish
Ecology of Marine Bivalves: An Ecosystem Approach, Richard F. Dame
Ecology of Marine Invertebrate Larvae, Larry McEdward
Environmental Oceanography, Second Edition, Tom Beer
Estuary Restoration and Maintenance: The National Estuary Program, Michael J. Kennish
Eutrophication Processes in Coastal Systems: Origin and Succession of Plankton Blooms and Effects on Secondary Production in Gulf Coast Estuaries, Robert J. Livingston
Handbook of Marine Mineral Deposits, David S. Cronan
Handbook for Restoring Tidal Wetlands, Joy B. Zedler
Intertidal Deposits: River Mouths, Tidal Flats, and Coastal Lagoons, Doeke Eisma
Morphodynamics of Inner Continental Shelves, L. Donelson Wright
Ocean Pollution: Effects on Living Resources and Humans, Carl J. Sindermann
Physical Oceanographic Processes of the Great Barrier Reef, Eric Wolanski
The Physiology of Fishes, *Second Edition*, David H. Evans
Pollution Impacts on Marine Biotic Communities, Michael J. Kennish
Practical Handbook of Estuarine and Marine Pollution, Michael J. Kennish
Seagrasses: Monitoring, Ecology, Physiology, and Management, Stephen A. Bortone

The Ecology of Seashores

George A. Knox, M.B.E., F.R.S.N.Z.

CRC Press
Boca Raton London New York Washington, D.C.

Library of Congress Cataloging-in-Publication Data

Knox, G. A.
The ecology of seashores / by George A. Knox.
p. cm. — (Marine science series)
Includes bibliographical references (p.).
ISBN 0-8493-0008-8
1. Seashore ecology. I. Title. II. Series.

QH541.5.S35 K66 2000
577.69′9—dc21 00-058573
CIP

Direct all inquiries to CRC Press LLC, 2000 N.W. Corporate Blvd., Boca Raton, Florida 33431.

International Standard Book Number 0-8493-0008-8
Library of Congress Card Number 00-058573
Printed in the United States of America 1 2 3 4 5 6 7 8 9 0
Printed on acid-free paper

Preface

Because of its accessibility, the intertidal zone has offered excellent opportunities to study the adaptations of individual organisms and populations to their environment, and the factors controlling community composition. Early work on seashores concentrated on the problems of life in an environment characterized by steep gradients in physical conditions, but in more recent years, the focus of research on the fascinating shore ecosystems has been on understanding the processes controlling their productivity and dynamic functioning. The emphasis has been on processes such as primary production, grazing, predation, competition, impact of disturbance, secondary production, detritus formation, decomposition, and the role of microorganisms.

My own involvement in seashore research began when I embarked on a M.Sc. thesis in zoology on the ecology of the rocky shores at Taylors Mistake, Banks Peninsula, with special reference to the serpulid polychaete *Pomatoceros cariniferus* (Knox, 1948; 1953). In the late 1950s, I became involved in research on a local estuary, the Avon-Heathcote Estuary. This research evolved into a comprehensive interdisciplinary research program that has continued until the present time. It has involved a large number of research associates, research assistants, and research students and culminated in two major reports (Knox and Kilner, 1973; Knox, 1992). A multiauthor book bringing together the results of 50 years of research on this estuarine ecosystem is in an advanced stage (Knox and Robb, in preparation). Over the period of 1959 to 1983, I directed the activities of the Estuarine Research Unit, Zoology Department, University of Canterbury, Christchurch. This unit carried out research on estuarine and coastal ecosystems throughout New Zealand and compiled some 28 major reports. The research aimed at understanding the interaction of estuarine plants, microorganisms, animals, and man with each other and their environment. Such research, while contributing to basic ecological principles, also provided information necessary for the management of New Zealand coastal ecosystems (Knox, 1983c). I was scientific coordinator of a multidisciplinary study of the Ahuriri Estuary, Hawke Bay (Knox and Fenwick, 1978; Knox, 1979b), and coordinator of the biological aspects of the Upper Waitemata Harbour Catchment Study, a comprehensive five-year interdisciplinary study of the land and water resources of the mangrove-fringed Upper Waitemata Harbour, Auckland (Knox, 1983a; 1983b). A fellowship at the East-West Center, Honolulu, enabled me to undertake a study of coastal zone resource development and conservation in Southeast Asia with special reference to Indonesia (Knox and Miyabara, 1983). This estuarine research culminated in 1983 with the publication of a two-volume work on estuarine ecosystems (Knox, 1983a,b).

Over the years I have also been involved in research on the intertidal ecology of rocky shores (Knox, 1963a; 1968; 1969b; 1988; in preparation; Knox and Duncan, in preparation) and sand beaches (Knox, 1969a). Field work has been carried out on rocky shores throughout New Zealand, the Subantarctic Islands (Snares, Auckland, Campbell, and Chatham Islands), Antarctica, the west coast of Chile, and the southern coasts of Australia. Shores in Peru, Argentina, the eastern and western coasts of Canada and the United States, England and Scotland, and tropical shores in Guam, the Palau Islands, Fiji, and Hawaii have been briefly examined. A book on New Zealand seashore ecology will be published shortly (Knox, in preparation). This work has also involved an examination of the evolution and biogeography of the Southern Hemisphere, especially the Pacific Ocean, intertidal and shallow water biotas (Knox, 1960; 1963a; 1975; 1988).

The approach used in this book is somewhat different from that used in many texts in marine ecology in that the emphasis is on ecological processes and the use of systems analysis in understanding such processes. This book is an attempt to bring some order to some of the most complex of ecosystem types, those of the seashores of the world. Wherever possible the energy circuit language of symbols and diagrams developed by H. T. Odum has been used as a basis of understanding (Odum, 1983).

The book is divided into seven chapters. The first provides an introduction and sets the scene for the succeeding chapters. The second deals with zonation patterns on hard shores, the basic causes of zonation, the biology of the major divisions of the biota, the biology of some key faunal components (mussels, limpets, and barnacles), and the flora and fauna of special habitats. The third chapter covers the physicochemical environment of soft shores, and the general ecology of the biota from producers through microbes to macrofauna. Two large sections deal with microbial ecology and detritus and nutrient cycles in representative ecosystems. The fourth chapter gives an account of ecological niches on the shore and the establishment and maintenance of zonation patterns. Chapter 5 attempts to synthesize the various factors such as grazing, competition, predation, disturbance, and succession that determine the distribution, diversity, biomass, and production of the various categories of consumers. Energy budgets, patterns of energy flow, trophic structure, and food webs are discussed in Chapter 6, while in Chapter 7, examples of ecosystem models for the various ecosystem types on both hard and soft shores are detailed. Finally, the application of the relatively new technique of network analysis to gain greater insight into ecosystem processes and enable the comparison of different ecosystems is outlined.

As will be seen by the reference list at the end of this book, there is a considerable volume of recent literature on seashore ecosystems, although the list contains only a fraction of the published work. Extensive literature citations have been included so that the book might serve as a resource for those engaged in research on and management of the coastal zone. Because of this growing volume of literature, I have had to be selective in the material included. Thus, of necessity, I have concentrated on work published since 1970, and in particular in the last decade. Regretfully, except in a few instances, it has not been possible to develop the history of the concepts considered. Examples have been carefully chosen from the pool of published research to illustrate these concepts. There are doubtless other examples that could have been equally used, and I apologize to authors whose work has not been included. I have attempted to include as wide a geographic range of examples as possible, and in particular have included Southern Hemisphere examples that often do not appear in texts originating in the Northern Hemisphere.

The book has been designed for use by upper level undergraduate and graduate students and professionals engaged in coastal zone research and management. I hope that they find it useful.

Author

George A. Knox, M.B.E., F.R.S.N.Z., was head of the Department of Zoology, University of Canterbury, Christchurch, New Zealand, from 1959 to 1976. He is now professor-emeritus in zoology.

Professor Knox was born in New Zealand and received his education at the University of Canterbury where he was appointed a staff member in 1948. He has been a visiting fellow at the East-West Center, Honolulu, and a visiting professor at the Department of Oceanography, Texas A&M University and the Department of Environmental Engineering, University of Florida, Gainesville. He has visited and worked in laboratories in the U.S., Canada, Chile, Japan, Australia, Western Europe, the USSR, and China.

Professor Knox's research has been wide ranging and includes: (1) the systematics and distribution of polychaeta with special reference to New Zealand and Antarctica; (2) rocky shore intertidal ecology and biogeography; (3) the ecology and conservation of islands; (4) studies on the pelagic and benthic ecosystems beneath the sea ice in McMurdo Sound, Antarctica; and (5) estuarine and coastal ecology and management. He established and directed the Estuarine Research Unit in the Department of Zoology and the University of Canterbury Antarctic Research Unit. He has participated in many field expeditions, including the Chatham Islands 1954 Expedition (leader); the Royal Society of London Darwin Centennial Expedition to southern Chile (marine biologist and deputy leader); thirteen summer expeditions to McMurdo Sound, Antarctica; the establishment of the Snares Islands Research Programme (participated in three field expeditions); and participation in field expeditions to Campbell and Auckland Islands. He has published over 100 scientific papers and 28 environmental reports, written five books, and edited and co-authored three other volumes.

Professor Knox has received a number of awards and fellowships for his contributions to science, including Fellow of the Royal Society of New Zealand (FRSNZ), 1963; Hutton Medal, Royal Society of New Zealand, 1978; Conservation Trophy, New Zealand Antarctic Society; Member of the Most Excellent Order of the British Empire (MBE), 1985; New Zealand Marine Sciences Society Award for Outstanding Contribution to Marine Science in New Zealand, 1985; and the New Zealand Association of Scientists' Sir Ernest Marsden Medal for Service to Science, 1985.

Throughout his career, professor Knox has been active in international scientific organizations. He has been a member of the Scientific Committee for Oceanic Research (SCOR) and the Special Committee for the International Biological Programme (SCIBP). He has been a delegate to the Scientific Committee for Antarctic Research (SCAR) since 1969, serving 4-year terms as secretary and president, and a member of the governing board of the International Association for Ecology from 1965 to 1990, also serving 4-year terms as secretary-general and president.

Acknowledgments

I would like to express my indebtedness to the late Professor Edward Percival, who by his enthusiasm and teaching skills started me on my career as a marine biologist, to Professor Howard Odum, who inspired by interest in the energy analysis approach to ecosystem modeling, and to my colleagues in New Zealand and various parts of the world, with whom I have discussed many of the ideas in this book.

My thanks are also due to all those who gave me permission to reproduce original figures and tables. Finally, I am indebted to John Sulzycki and the staff of CRC Press, in particular Amy Rodriguez and Pat Roberson, for their patience and support during the preparation of this volume and for seeing the project through to the completion of such a high quality product.

George A. Knox
November, 2000
Christchurch, New Zealand

Contents

1 The Environment

CONTENTS

1.1 INTRODUCTION

The relatively narrow strip where the land meets the sea provides the most diverse range of habitats for living organisms anywhere on the earth. It offers a unique blend of habitats not found elsewhere. Animals and plants living there experience marked daily changes in environmental conditions as the tides ebb and flow. These variations create stresses to which the species have adapted by organic evolution over long periods of time.

The seashore is best defined as that part of the coastline extending from the lowest level uncovered by the spring tides to the highest point washed or splashed by the waves. This intertidal or littoral zone does not, however, merge abruptly with the land above or the permanently submerged sublittoral zone below. Above the highest point reached by the waves, salt spray influences a zone which, on very exposed coasts, may extend several hundred meters up the shore. Below the lowest point reached by the tides, a gradient in the quantity and quality of light is reflected in the depth distribution of the plants and animals found there. Furthermore, some sublittoral species extend up into the intertidal zone proper. In this book, both of the transitional zones between the intertidal zone and the land above and the sublittoral below will be considered.

The extent of the intertidal zone depends on a variety of factors, the most important of which are the angle of slope of the shore and the amplitude of the tides. This may vary from less than a meter on vertical rock faces where the tidal amplitude is small to several kilometers on soft shores where the tidal amplitude is high. The physical features of this zone are of immense complexity, varying not only geographically, but from coast to coast in the same region and within comparatively short distances on any one shore. Whether narrow or wide, this region is astonishingly rich in both the number of species of plants and animals and the densities of the individual species. In the following pages we shall explore the complex shore environment, the ways in which representative species are adapted to life in a constantly changing environment, how they interact with each other, how the structure of the various communities is controlled, and how energy and materials cycle in the different ecosystems.

Three interlocking factors determine the type of community on the seashore: (1) the amount and intensity of wave action; (2) the type of substrate (whether rock, sand, mud, or some combination of these); and (3) the amplitude of the tides. To consider each of these in turn:

1. Waves are caused by wind, and their size is primarily determined by the uninterrupted distance or "fetch" over which the wind can blow, the velocity and direction of the wind, and to a lesser extent the depth of the water. Thus, the severity of wave action in any given locality is

determined by its geographical position in relation to the factors listed above. As a consequence, there is every gradation in wave shock, from the pounding surf of exposed shores on the western coasts of the main land masses to the quiet waters of deep inlets, narrow estuaries, and fjords.
2. From a biological point of view, there are two major types of shores, "hard" and "soft," the former often referred to as rocky shores and the latter as sand and mud shores. The environmental features and modes of life differ so much between these two types that little overlap, if any, exists between the species populations inhabiting them. There is, however, a gradation from rocky shores through boulder beaches, pebble beaches to sand beaches, and the latter grade through muddy sand to mudflats. Nevertheless, it is convenient to preserve the distinction between "hard" (rock and boulders) and "soft" (sand and mud) shores.
3. The third important factor, tidal exposure, will be considered in detail later. It is sufficient here to note that the tidal range, or vertical distance, between high and low water not only varies with the lunar cycle, but also geographically. The ebb and flow of the semidiurnal rise and fall of the tide result in periodic emersion and submersion of the intertidal zone, the extent of which depends on a variety of factors (see Section 1.3).

1.2 ENVIRONMENTAL GRADIENTS AND STRESSES ON THE SHORE

Seashores are characterized by three main environmental gradients:

1. The vertical (intertidal) gradient from sea to land.
2. The horizontal gradient of exposure to wave action.
3. The particle size gradient from solid rock through boulders, pebbles, and coarse and fine sands, to silts and clays.

Across the intertidal gradient, there is a change from the highly stable, buffered subtidal environment to an unstable and increasingly stressed environment with increasing tidal height. Upshore the relatively stable environment changes until at the top of the intertidal zone (including the area influenced by wave splash and spray) it merges with the unstable, terrestrial environment with diurnal and seasonal changes in temperature, light, humidity, wind, and variable salinity due to rainfall and evaporation. Thus, across the intertidal gradient shore, organisms are subjected to increasing stress.

As wave action increases along the horizontal exposure gradient, the upper shore is increasingly subject to wetting by surging waves and spray, more food is transported to filter-feeders, and better supplies of oxygen and nutrients are available to the plants and animals. On the other hand, increased stresses may be imposed on some organisms, e.g., problems of settlement, adhesion, and propensity for dislodgement and abrasive action of sand.

The type of substrate, as discussed in Chapters 2 and 3, determines the type of community that can survive.

These gradients interact in very complex ways with the biological communities living on and in the substrate. In the succeeding chapters we shall explore these interactions.

1.3 SALIENT FEATURES OF THE SHORE ENVIRONMENT

1.3.1 SEAWATER

When intertidal organisms are covered by the tides, they are subjected to the same physiological conditions of temperature and salinity as those permanently submerged in the sublittoral below. Seawater from a physiological point of view is a very complex solution. The aspects of this complexity that are of biological significance are ion concentration, density, and osmotic pressure. The osmotic properties of seawater result from the total amount of dissolved salts. Seawater is composed of a number of different compounds that can be divided into the following phases (Millero and Sohn, 1991):

1. Solids (material that does not pass through a 0.45 μm filter)
   a. Particulate organic matter (plant and animal detritus)
   b. Particulate inorganic material (minerals)
2. Gases
   a. Conservative (N_2, Ar, Xe)
   b. Nonconservative (O_2 and CO_2)
3. Colloids (pass through 0.45 μm filter, but are not dissolved)
   a. Organic
   b. Inorganic
4. Dissolved solutes
   a. Inorganic solutes
      • Major (> 1 ppm)
      • Minor (< 1 ppm)
   b. Organic solutes

On average, seawater is composed of 96.52% water and 3.49% dissolved substances, mostly salts. The latter

TABLE 1.1
Concentrations of the Principal Ions in Seawater in Moles per Kilogram, in Parts per Thousand by Weight, and in Percent of Total Salts

Ion	Moles kg^{-1}	Percent by Weight	Percent of Total Salts	
Chloride (Cl^-)	0.549	18.980	55.2	
Sulfate (SO_4^{2-})	0.0762	2.649	7.71	
Bicarbonate (HCO_3^-)	0.0023	0.140	0.35	negative ions
Bromide (Br^-)	0.008	0.065	0.19	(anions) = 21.861%
Borate ($H_2BO_3^-$)	0.0004	0.026	0.07	
Fluoride (F^-)	0.000004	0.001	0.0001	
Sodium (Na^+)	0.468	10.556	30.40	
Magnesium (Mg^{++})	0.532	3.70	3.70	
Calcium (Ca^{++})	0.0103	0.400	1.16	positive ions
Potassium (K^{++})	0.0099	0.400	1.10	(cations) = 12.621%
Strontium (Sr^{++})	0.0002	0.013	0.035	
Overall total salinity = 34.482%				

can be conveniently grouped into major and minor constituents. The major constituents, which are found everywhere in the ocean in virtually the same relative proportions, are termed *conservative elements*. The minor constituents, on the other hand, show marked variations in their relative concentrations due to selective removal from the water by living organisms, and are termed *nonconservative elements.*

The major constituents, comprising 99.9% of all dissolved salts, are sodium, magnesium, calcium, potassium, and strontium cations and the chloride, sulfate, carbonate, bicarbonate, and bromide anions, together with boric acid mostly in the undissociated state (Table 1.1). The most important nonconservative constituents are the major plant nutrients, phosphates and nitrates, together with silicon, which is required by diatoms for the construction of their frustules and by radiolarians for their skeletons.

In addition to the dissolved inorganic substances, seawater, especially at inshore locations, contains appreciable amounts of organic material, both particulate (detritus) and dissolved. The role of organic matter in material cycling and in shore food webs will be dealt with in later sections of this book.

The concentrations of the dissolved substances in seawater provide a means of determining the salt content of seawater. The total amount of inorganic material dissolved in seawater expressed as weight in grams per kilogram (or parts per thousand) is termed the *salinity,* and is usually around 35 gram kg^{-1}, or 35 parts per mille. Until the early 1980s, salinity values were expressed in parts per thousand, or per mille, for which the symbol is "‰," with the average salinity being 35‰. However, it is now standard practice to dispense with the symbol "‰," because as detailed below, salinity is now defined in terms of ratio.

The international standard method of estimating salinity was based on determination by titration of the chlorine content (chlorinity) of the water (see Strickland and Parsons, 1972), and the salinity was calculated from the relationship: S‰ = 1.8065 Cl‰ (Sharp and Cuthbertson, 1982). This definition assumes, among other things, that all organic matter is oxidized, the carbonates are converted to oxides, and the bromide and iodide have been replaced by chloride. Titration is time consuming, and various types of conductivity salinometers are now used for the determination of salinity. Thus, the practical salinity of a sample of seawater is defined in terms of the conductivity ratio, K_{15}, which is defined by:

$$K_{15} = \frac{\text{Conductivity of the seawater sample}}{\text{Conductivity of standard KCl solution}}$$

at 15°C and 1 atmosphere pressure, the concentration of the standard KCl being 32.3456 gram kg^{-1}. The practical salinity is related to the ratio K_{15} by a complicated equation.

Other methods of determining salinity include the refractometer, which measures light refraction and the hydrometer, which measures density.

In the open sea salinities commonly range from 32.00 to 37.00. The differences reflect local effects of evaporation, rain, freezing and melting of ice, or the influx of river water. Certain seas have markedly higher or lower salinities: the Mediterranean, due to high evaporation and little freshwater influx, has salinities ranging from 38.40 to 39.00; the Baltic, which is a large brackish water body, ranges from 10.00 at the mouth to 3.00 at the northern extremity.

Regions where freshwaters mix with seawater are termed estuarine. In such regions, salinities undergo great variations depending on the state of the tide, the amount

TABLE 1.2
Salinities of Various Types of Water

Type of Water	Salinity (%)
Freshwater	0–0.5
Oligohaline brackish water	0.5–3.0
Mesohaline brackish water	3.0–10.0
Polyhaline brackish water	10.0–17.0
Oligohaline seawater	17–30
Mesohaline seawater	30–34
Polyhaline seawater	34–38
Brine	> 38

of freshwater input from the inflowing rivers and streams, and in some land-locked systems, the balance between rainfall and evaporation. On the basis of the range of salinities encountered, the various types of water can be classified in the categories listed in Table 1.2. The variations in salinity that occur in estuaries and lagoons will be dealt with in Section 4.4.

The important biological effects of salinity variations are the consequences of movement of water molecules along water potential gradients and the flow of ions along electrochemical gradients (Lobban et al., 1985). These processes are regulated in part by semipermeable membranes that surround cells, chloroplasts, vacuoles, etc. Estuarine organisms and those living in tide pools on rocky coasts, especially those high on the shore, are subject to salinity fluctuations, and they must cope with the osmotic stress that this entails. When exposed to the air by receding tides, evaporation causes an increase in the salinity of surface films on seaweeds and the soft part of animals. Rainfall, on the other hand, results in a decrease in the salinity of these surface films.

1.3.2 Tides

Tides are caused by the gravitational pull of the Sun and Moon on the Earth, and their regular variations in height and time result from the regular differences in the positions of the Sun and Moon to each other (Figure 1.1). The Earth and the Moon together rotate around a common point (C) called the center of rotation. Any point on the surface of the Earth (or Moon) is subjected to two forces: a centrifugal force tending to displace it away from C, and the gravitational acceleration of the Moon tending to displace it toward the Moon. If two points A and B on the opposite sides of the Earth are considered, both the magnitude and vectorial sum of these forces will be different, since A lies further from the center of rotation than B. The vectorial sum of the forces at both A and B will be directed perpendicularly away from the Earth's center. This causes the seawater to be displaced to form two high tides on the opposite sides of the Earth (Figure 1.2a, b). These two tides appear to pass around the Earth due to the Earth's rotation about its axis. The period of rotation of the Earth-Moon couple, however, is 29.53 D; thus there is an average retardation of the time of each tide by 24.5 min.

The same process applies to the couple formed by the Earth and the Sun. However, the center of rotation is closer to the Sun because of the latter's greater mass, although the tide-generating forces are approximately 46% of those produced by the Moon, due to the vastly greater distance of the Sun from the Earth. Thus, there are two pairs of tides with different periodicities. Those caused by the Sun occur at intervals of 12 h, while those caused by the Moon occur at intervals of 12.4 h. These two tides drift in and out of phase over a period of c. 14.7 days. When the tides caused by the Sun coincide with those caused by the Moon, they result in maximal high and low tides, called spring tides. When the two tides are out of phase, tidal rise and fall are at a minimum; these are called neap tides. Thus, spring and neap tides are related to the phases of the Moon as shown in Figures 1.2c and d. Theoretically this results in a semidiurnal tidal pattern, with two tidal maxima and two minima in each lunar day, with a lag of a little under an hour between the corresponding tides on successive days (Figure 1.2). However, the actual situation is not as simple as this, as tides with unequal highs and unequal lows occur, and in some cases there is only one tide per day.

The inclination of the Earth's axis at 23.5° off the vertical (relative to the plane of the Earth's orbit about the Sun), and the 5° inclination of the Moon's orbital plane to the orbital plane of the Earth are the causes of inequalities in the magnitude of the two tides each day at certain latitudes and at different times of the year.

1.3.3 Tidal Range and Proportions

In general, tidal cycles are of three types (Figure 1.3). Diurnal tides have one high and one low per day; this is an unusual type, occurring in parts of the Gulf of Mexico. Semidiurnal tides rise and fall twice a day, with successive highs and lows more or less equal in height; this type is common along open Atlantic Ocean and New Zealand coasts (Figure 1.3). Mixed tides occur twice a day with unequal highs and lows (Figure 1.4). They are characteristic of Pacific and Indian Ocean coasts, as well as in smaller basins such as the Caribbean Sea and the Gulf of St. Lawrence (Gross, 1982). In addition, there are storm tides with irregular periods, usually of several days, caused by barometric changes and winds.

Spring tides occur on average about every 14 days. However, this period is not quite constant. During the year, the spring tides narrow their range while the neap tides widen their range until there is a long period with indeterminate tides, and then the tides that were spring become

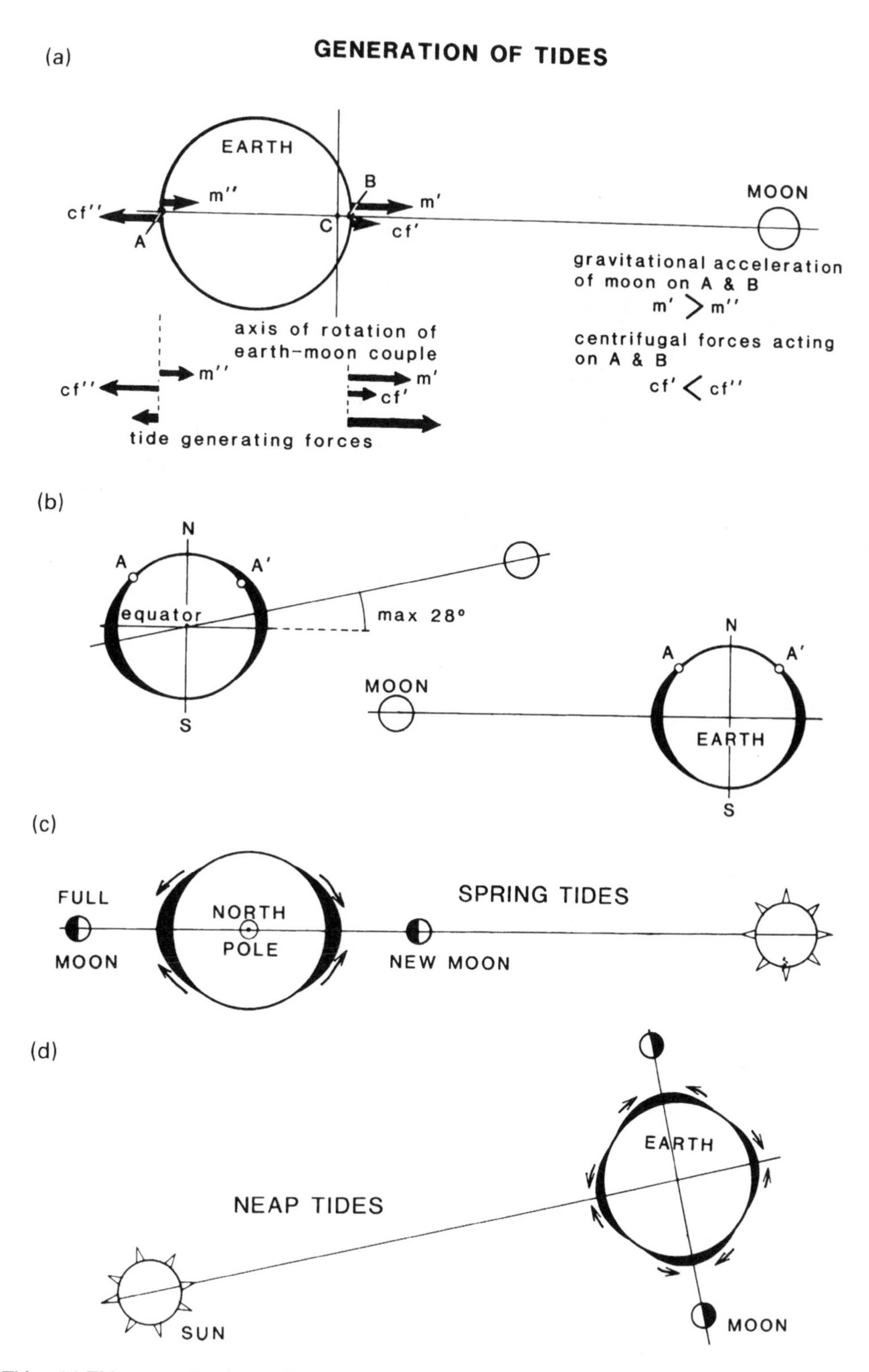

FIGURE 1.1 Tides. (a) Tide-generating forces. C, center of rotation of the Earth-Moon couple (vertical line axis of rotation). A, B, diametrically opposite points on the Earth's surface. The gravitational acceleration of the Moon on point B(m') is greater than on A(m''). The centrifugal force acting on B(cf'') is less than on A(c''). Tide-generating forces (arrows below Earth) are the vectorial sum of gravitational and centrifugal forces at A and B (broad arrows). (b) Inequality of tides. (c), (d) spring and neap tides, respectively. See text for explanation. (From Harris, V.A., *Sessile Animals of the Seashore,* Chapman & Hall, London, 1990, 6. With permission.).

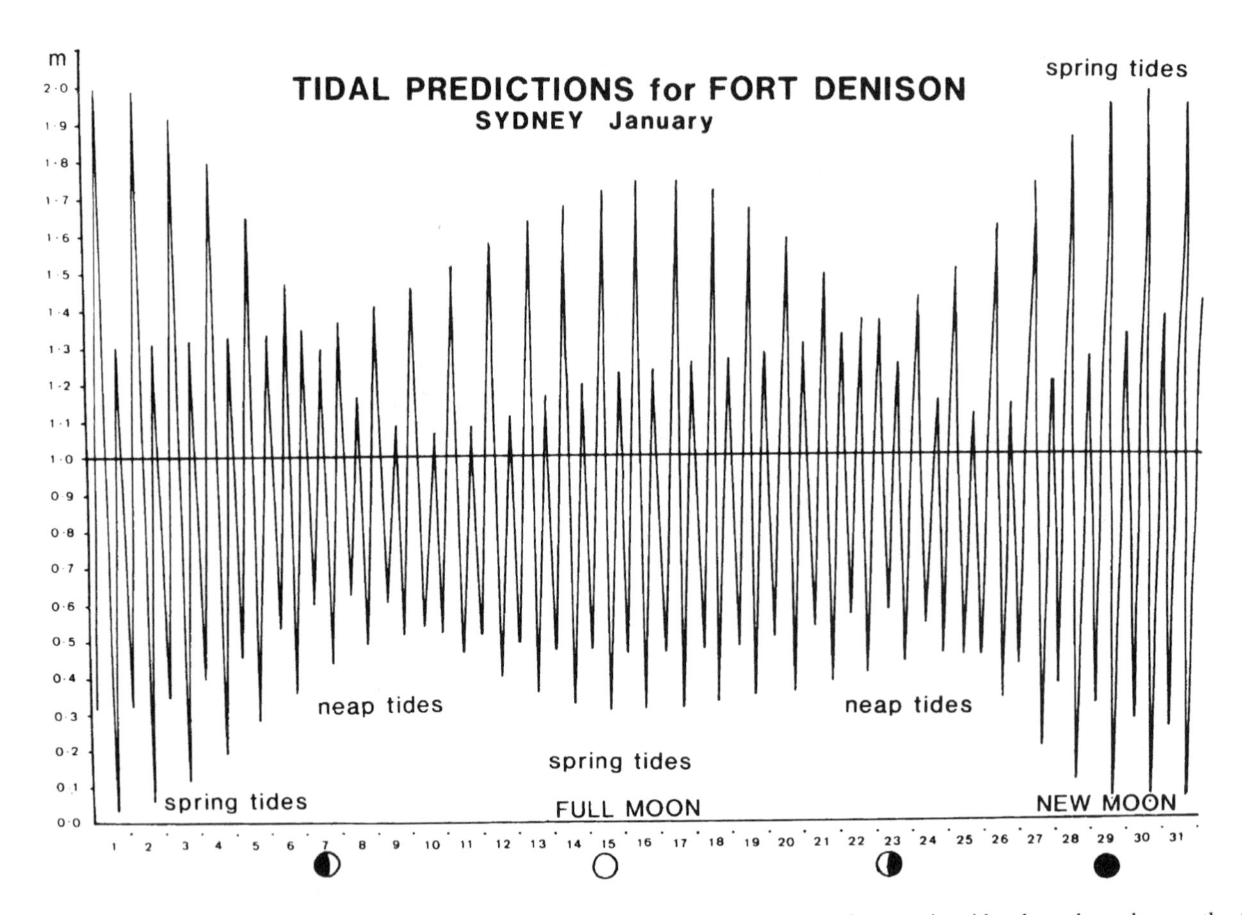

FIGURE 1.2 Tide chart for Fort Denison, Sydney, Australia, showing predicted height of successive tides throughout the month of January and the corresponding phases of the moon (based on tidal predictions for 1987). (From Harris, V.A., *Sessile Animals of the Seashore*, Chapman & Hall, London, 1990, 9. With permission.)

neaps and the neaps become springs. This change takes place about every 12 to 14 months. Spring/neap tide differences reach a maximum twice a year near the vernal and autumnal equinoxes.

The height of the waterline on the shore is measured with reference to a standard level, the Chart Datum, or zero tide. The definition of this level varies from country to country and even from coast to coast in the same country. Predicted tides are given in tide tables published by the appropriate agency in each country. However, the actual height reached at high and low water may depart from these predictions due to difference in barometric

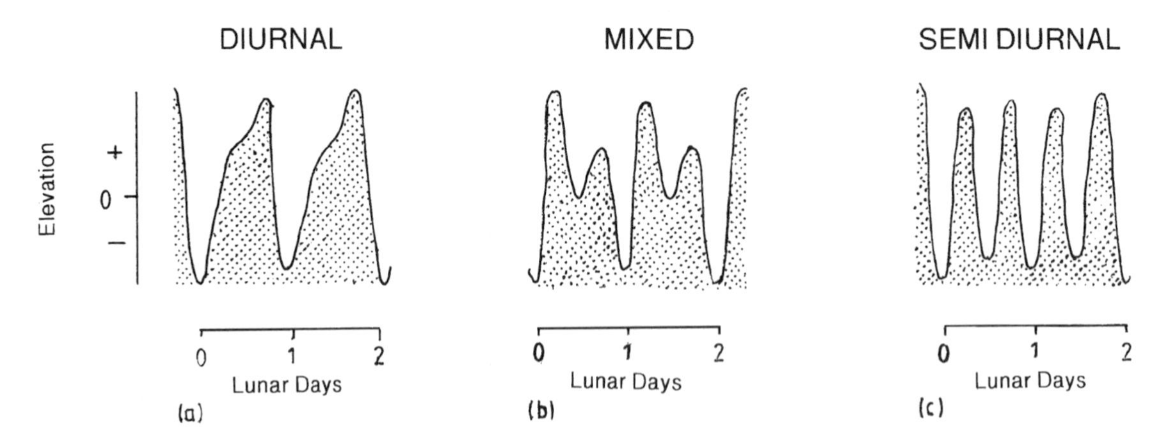

FIGURE 1.3 Diagram illustrating the three main tidal regimes: (a) diurnal (one high and one low tide each lunar day; (b) mixed tides (with highs and lows of different heights each lunar day); (c) semidiurnal (with only a slight difference between the two high tides on each lunar day). HHW = High high water; LHW = Low high water; HLW = High low water; LLW = Low low water.

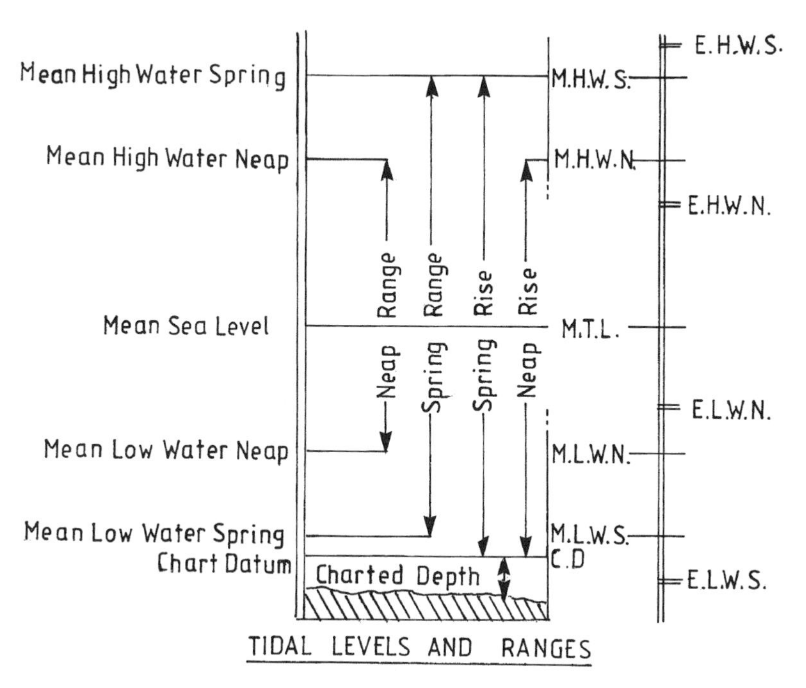

FIGURE 1.4 The nomenclature of tidal levels for springs and neaps, showing the extent of the range and tidal rise and fall as related to "Chart Datum."

pressure and wind (see Figure 2.28). A rise in barometric pressure equivalent to 2.4 cm of mercury is sufficient to depress the water level by approximately 29 cm. Strong winds tend to hasten or retard the arrival of the tidal crest as it progresses along a coastline.

The configuration of the land masses may impede water movement and markedly delay the tide in different areas of a coastline. In deep narrow inlets and estuaries, the high and low tides may be several hours later at the head than at the entrance. In the open sea, the effects of the Moon's gravitational pull is spread over a wide area producing a tidal bulge no more than 0.5 cm high. This explains why some small oceanic islands have small tidal ranges seldom exceeding 1.0 m. This contrasts with 17 m spring tides in the Bay of Fundy, Canada, the 13 m tides in the Severn Estuary, England, and the 12+ m tides in Northern Australia.

By averaging the height of the tidal oscillations, horizontal lines can be imagined on the shore representing various tidal levels as depicted in Figure 1.4. These tidal levels are used when referring to tidal ranges and are useful when defining the positions of plants and animals on the shore. Table 1.3 gives definitions of the terms and the common abbreviations used when referring to the tidal levels.

1.3.4 Submersion and Emersion

If the percentage monthly submergence at high tide and the percentage monthly exposure at low tide for a given locality are plotted (Figure 1.5), it can be seen that there is a significant change in the percentage of submergences between EHWN and MHWN. From EHWN up, the plants and animals on the shore will be subject to increasing periods of continuous exposure to air. Those above MHWN will generally experience several days each fortnight without tidal submergence, and the higher they are on the shore, the longer will be the period of continuous emergence until above MHWN; this period may last several weeks.

A significant change in the percentage of exposure also occurs between MLWN and MLWS. From levels below ELWS, where submersion and emersion occur twice daily, the percentage of emergences decreases rapidly, until below MLWS the plants and animals are uncovered only a few times a year at the ELWS tides. The annual percentage of exposure to air for various levels on the shore can be plotted from tidal records to produce an emersion curve. Two such curves are shown in Figure 1.6, one for Taylors Mistake, New Zealand (Knox, 1953) and one for Holyhead in Great Britain (Lewis, 1964). The difference in the rate of change of percentage exposure to air is greatest in the vicinity of high and low water neaps. Such levels are often referred to as "critical levels" (Doty, 1946), and the concept will be dealt with in detail in Section 2.3.8.

It should be pointed out that the impact of changes in the percentage exposure to air as deduced from tide tables is subject to modification by wave action, climatic conditions, and the times at which low water occurs. This is illustrated in Figure 1.6, where the rise and fall of three consecutive spring tides during the midsummer on the

TABLE 1.3
Tidal Terms and Definitions

Term	Abbreviation	Definition
Extreme high-water spring	EHWS	The highest level reached by the sea on the greatest spring tides of the year.
Mean high-water spring	MHWS	The average of the higher levels reached by the sea on the fortnightly spring tides.
Mean high-water neap	MHWN	The average of the lowest high waters reached by the sea on the fortnightly neap tides.
Extreme high-water neap	EHWN	The lowest of the high waters reached by the neap tides during the year.
Mean tide level	MTL	The average of the high and low waters throughout the year.
Extreme low-water neap	ELWN	The highest of the low waters reached by the neap tides during the year.
Mean low-water neap	MLWN	The average of highest low waters reached by the sea on the fortnightly neap tides.
Mean low-water spring	MLWS	The average of lowest levels reached by the sea on the fortnightly spring tides.
Extreme low-water spring	ELWS	The lowest level reached by the sea on the greatest spring tides of the year.

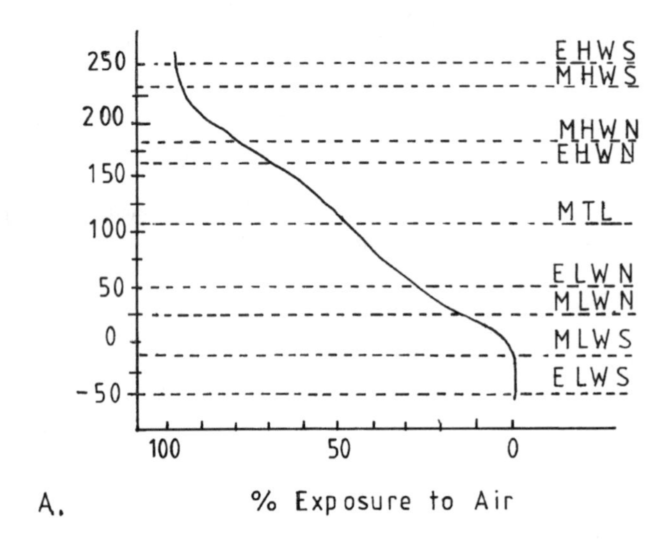

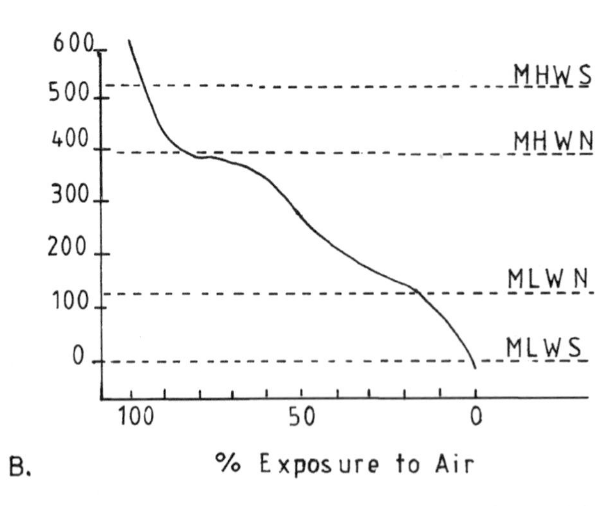

FIGURE 1.5 Annual percentage exposure to air at various levels on the shore. A. Taylors Mistake, South Island, New Zealand. (Redrawn from Knox, G.A., *Trans. Roy. Soc. N.Z.*, 83, 193, 1953. With permission.) B. Holyhead, Great Britain. (Redrawn from Lewis, J.R., *The Ecology of Rocky Shores*, English University Press, London, 1964, 25. With permission.)

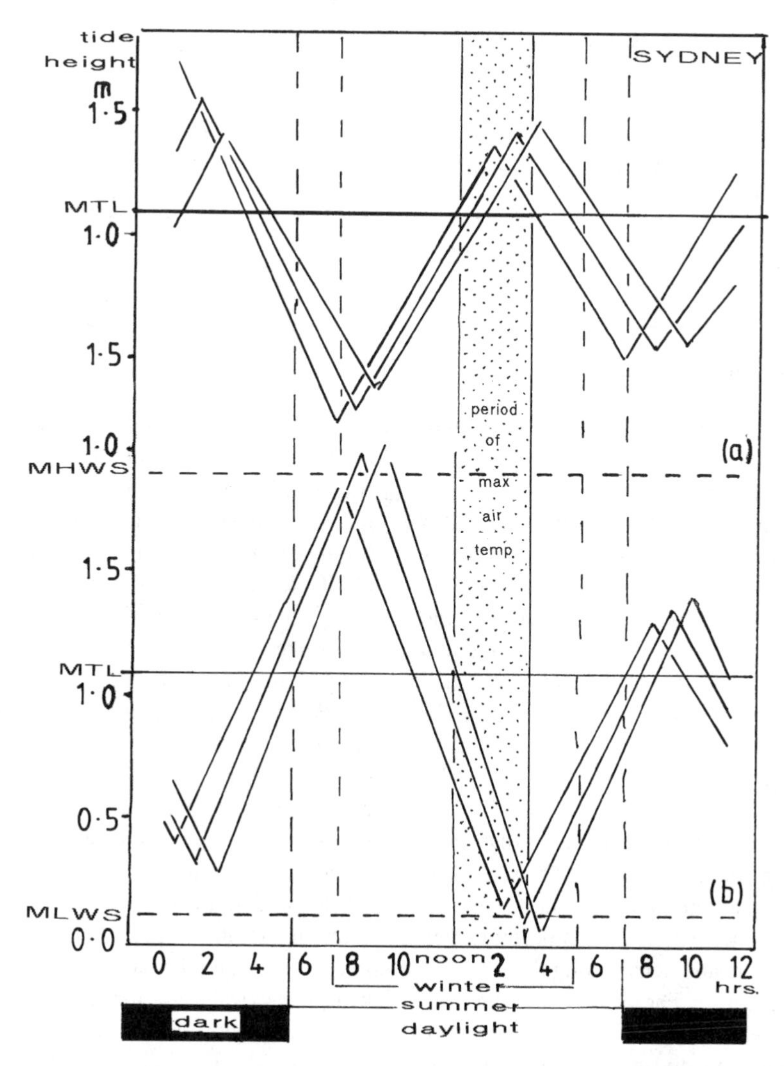

FIGURE 1.6 Comparison of exposure to sun during daytime tidal cycles near Sydney, Australia. (a) During neap tides, organisms on the middle shore are immersed during the period of maximum insolation (stippled). (b) Maximum periods of exposure to the sun occur *c*. 2 days before maximum spring tides, when low tide occurs between 13:30 and 15:30 hours. MTL, mean tide level; MHWS, mean high water spring tides; MLWS, mean low water spring tides. (Redrawn from Harris, V.A., *Sessile Animals of the Seashore,* Chapman & Hall, London, 1990, 10. With permission.)

coast of Sydney, New South Wales, Australia, has been plotted (Figure 1.6a), together with three neap tides occurring 7 days later (Figure 1.6b). Solar radiation at these times would have been most intense during the middle of the day (10.00 to 15.00 local time), but the impact of the sun at high temperatures would have been very different on the two occasions. During spring tides, the middle shore would be exposed to the sun throughout the low tide period, subjecting both the plants and animals on the shore to severe stress. On the other hand, during neap tides the shore would be covered during the middle of the day, thus the plants and animals would not be subject to stress. During the hours of darkness, there would be no harmful impact of sunlight and increased temperatures, and it would be expected that the activity of the animals would increase at such times. During neap tides, organisms on the upper shore may not be submerged for several days in succession, in contrast to the spring tides, during which at least one period of submergence occurs each day.

1.3.5 Modifying Factors

So far our discussion of tidal rise and fall has been concerned with the predicted levels of rise and fall giving average tidal heights and exposure times. As already noted, there are departures from predicted levels for a variety of reasons, and it should be emphasized that one single period of extreme conditions may be critical for many species. In addition, the operation of the tidal factor varies throughout the year, from month to month as well as from week to week and day to day. It also operates with different intensity at different stages in the life cycle of many intertidal

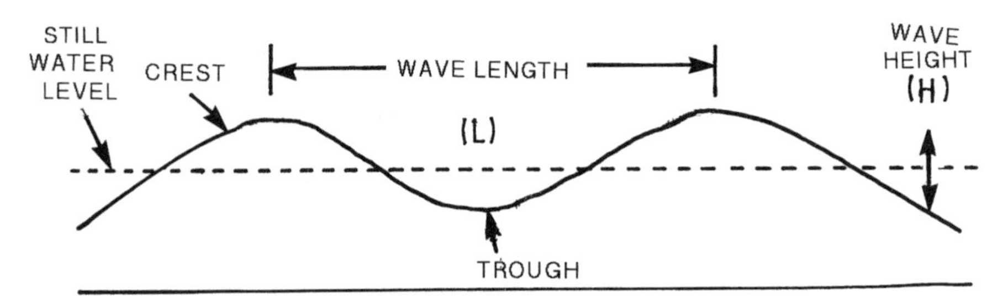

FIGURE 1.7 Features of a progressive wave.

organisms. A period of continuous exposure to air that may be tolerated by an adult plant or animal might well be fatal to a newly settled sporeling or larval stage.

The effect of the tidal factor is also subject to alteration by a range of modifying factors, as they are sometimes called. These have previously been listed as wave action, topography (including aspect and angle of slope), the nature of the substratum, climatic factors (climate), and a variety of biotic interactions of which competition and predation are the most important.

1.3.5.1 Wave Action

This is perhaps the most important single factor that modifies the vertical range of organisms on the shore and the species composition of the communities found there. In addition, on soft shores, wave action is an important agent in molding the shore profile and determining the type and distribution of the sediment; on hard shores it is the principal erosive agent. Wavecut platforms are prominent features of hard shore topography and shore slope, crevicing, channeling, and undercutting. The presence of boulders and pebbles are consequences of wave action combined with the weathering of rocks of different geological composition and hardness.

Waves — Waves are generated by wind stress on the water surface, with friction between air and water causing a viscous drag which stretches the surface like an elastic membrane. This distortion by the wind and its restoration by surface tension generates the undulations that we call waves. Thus, waves transfer energy from winds at sea to the coastal zone.

Basic wave features are illustrated in Figure 1.7. Wave length (L) is the horizontal distance between successive crests, and wave height (H) is the vertical height from trough to crest. The time required for successive crests to pass a fixed point is the wave period (T). The wave steepness is H/L and the wave speed C = L/T. The height and period of a wave, and hence its energy, are related to the velocity, time, and fetch (the unobstructed distance over which it has been blowing) of the wind that generated it. Ocean swell can transfer prodigious amounts of energy, generated by storms at sea, over vast distances with little loss before it is expended against some distant coast. In the Southern Hemisphere, huge swells generate very large waves that crash on exposed western coasts. When a wave breaks against a vertical cliff, its velocity is abruptly brought to zero, causing sudden pressure changes. Harger (1970) found that the mean wave impact ranged from 1.4×10^3 N m^{-2} to 9×10^5 N m^{-2} (i.e., from c. 0.01 to 9 atmospheres) on the coast of California. Pressures during storms could rise considerably above these values.

Water particles within a wave oscillate clockwise in a circular motion that is equal to the height of the wave (Figure 1.8). This orbital motion decreases with depth until at the depth of half the wavelength, the orbital motion becomes negligible. As the wave approaches shallow water, its advance becomes slowed, and the wave form changes when the depth becomes less than half the wave length. Friction with the seabed causes the water particles to move in an ellipse, and the particles over the seafloor move more or less backward and forward. As a result of this friction, the deeper water in the wave lags behind the wave crest, the wave shortens, and the crest tends to overhang the trough until a point is reached where it topples over and breaks. This occurs when H/L = 1/7 and the water depth = 1.3 H. The wave then surges up the shore as the "wash."

Breakers are of three types: spilling, plunging, and surging. Plunging breakers occur when circumstances cause the water particles under the crest to travel faster than the wave crest itself and the crest then plunges into the wave trough ahead as a water jet. Spilling breakers (Figure 1.9) occur when the maximum vertical acceleration in the wave motion increases until it exceeds the downward gravitational motion. The water particles then start popping out of the wave surface.

The type of breaker is determined by two factors, the deepwater wave steepness (H/L) and the beach slope (Figure 1.10). Spilling breakers occur when steep waves reach a gently sloping beach, while plunging breakers occur on all slopes, but with a lower wave steepness. The third type of breaker, the surging breaker, occurs with very low wave steepness and steep beach slope. Here the wave does not break, but surges up the beach face and then is reflected back to the sea. There is not a sharp transition between these types and they grade into each other.

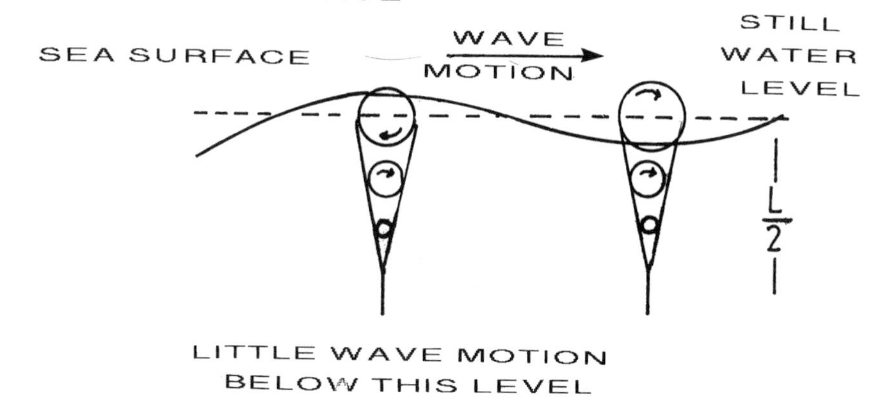

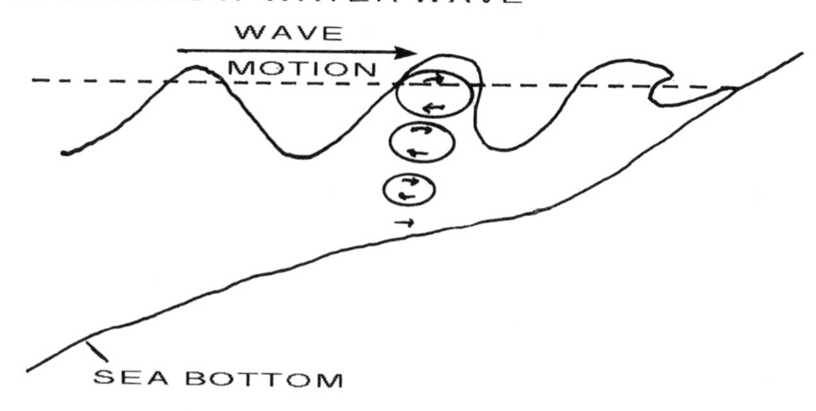

FIGURE 1.8 Particle motion in waves.

Wave energy is dissipated in the breaker zone. Spilling breakers dissipate their energy gradually, while plunging breakers dissipate it forcibly. The dissipated energy sustains a "set-up" (or rise) of the mean water level within the breaker zone. At the surface, within the breaker zone, there is a mass flow of water shoreward. Water accumulated on the beach by the waves is discharged out of the surf zone by rip currents (see Chapter 3).

When the wave motion reaches the bottom, the wave decelerates. Such a change in one section of the wave

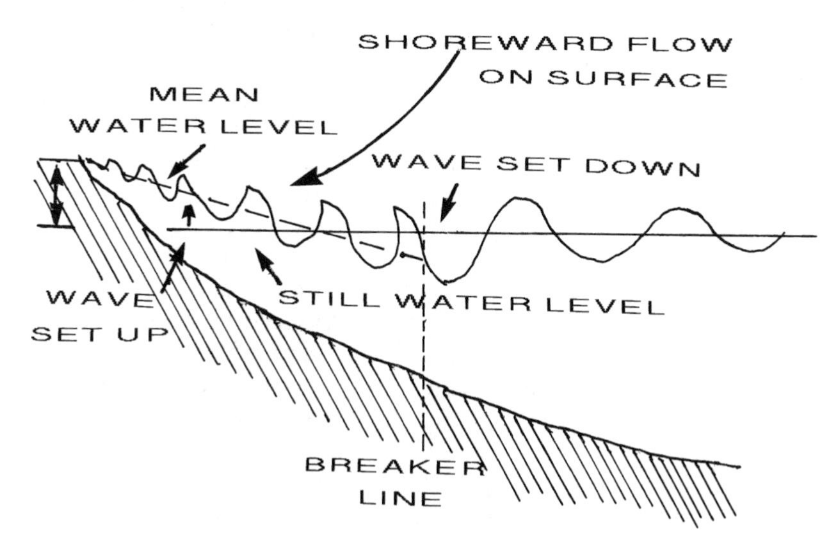

FIGURE 1.9 Wave setup for spilling breakers. (After Swart, D.H., in *Sandy Beaches as Ecosystems,* McLachlan, A. and Erasmus, T., Eds., Dr. W. Junk Publishers, The Hague, 1983, 19. With permission.)

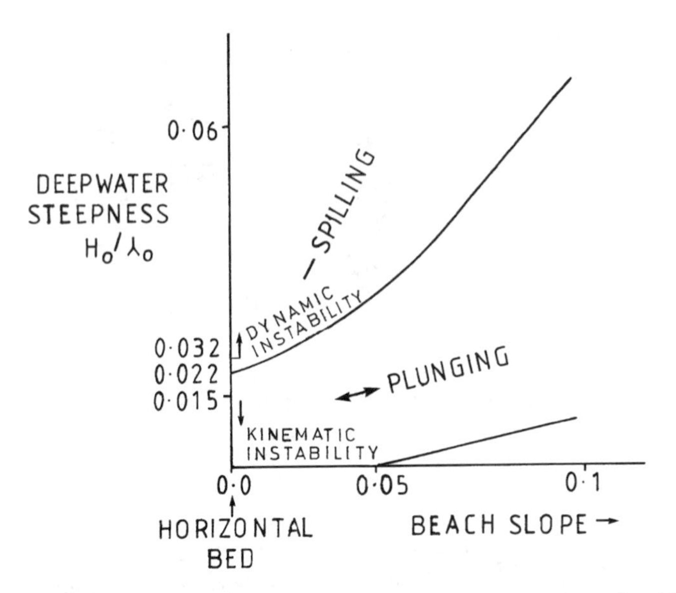

FIGURE 1.10 Types of wave breaking as a function of wave steepness and beach slope. (After Kjeldsen, S.P. and Olsen, G.B. (1972). Redrawn from McLachlan, A., Erasmus, T., Eds., *Sandy Beaches as Ecosystems*, Dr. W. Junk, 1983, 17. With permission.)

causes a change in direction. This refraction, or bending of waves as they approach the shore, tends to align them to the contours of the coastline. Waves approaching a headland meet the shallower water sooner. Consequently, they converge on the headland and diverge and spread out in the bays. This tends to focus the wave energy on the headlands and dissipate it in the bays (Figure 1.11). Convergence of wave energy also occurs over raised areas of the seabed, such as reefs and bars. In Chapter 3 we shall consider in detail the constructive and destructive effects of waves on beaches.

A major factor complicating the action of waves on rocky shores in particular is the topography of the shore, which can completely upset expectation of the effect of waves based solely on considerations of wind strength and direction and length of fetch. In general, the more gentle the slope and the more extensive the shore, the less the wave effect, as the energy is dissipated low on the shore with the result that the upper parts are little affected by splash and spray. To the leeward side of promontories, or in sheltered bays and estuaries, the force of the waves is greatly diminished, and in landlocked fjords or lochs the effect of the waves is reduced to virtually zero. In these quiet waters, only locally generated waves of low energy lap the shore. On hard shores, the impact of the waves is complicated by local conditions, with wave action varying greatly on the seaward and landward sides of stacks, rocks, boulders and reefs, and the angle of slope of the shore. With increasing steepness of slope, wave action tends to be more violent, and the level reached by splash and spray extends higher up the shore. Because of these variations, it is very difficult to establish specific categories of exposure to wave action based solely on physical data. However, developments in instrumentation using dynamometers to measure wave forces (Jones and Demetropoulos, 1968) can provide quantitative data.

It is customary in discussing the impact of wave action to use the term "exposed" to describe its severity. This is difficult to quantify, but a number of attempts have been made to categorize shores from the viewpoint of wave exposure (Guiler, 1959; Ballantyne, 1961; Lewis, 1964; Rasmussen, 1965; Dalby, 1980). Table 1.4 lists the categories and the bases upon which they have been established for several of these attempts. The categories of Guiler and Rasmussen are based on physical data, whereas those of Ballantyne and Lewis are based on the use of biological indicator organisms, with each category established on the abundance of a series of such indicator organisms. In the succeeding chapters in this book, the categories defined by Lewis (1964) are used. "Very exposed" shores are those found on headlands, stacks, and offshore islands subject to strong swell and violent wave action, while "very sheltered" shores are found at the headwaters of deep inlets where wave action is virtually absent.

Where waves and swell are constantly experienced, their general effect is to raise the effective tidal levels (see Section 2.3.1 for a detailed discussion). This leads to a clear quantitative difference in the vertical distribution of the plants and animals on the shore. In addition, there is a qualitative difference in the kinds and numbers of species present between sheltered and wave-exposed shores. Many species cannot tolerate wave exposure and are only found on sheltered shores, while other species cannot live on shores where wave action is absent or slight.

Rapidly moving water caused by waves also exerts a strong lateral force (drag) on any plant or animal that projects above the substratum. However, the direction from which this force comes is unpredictable and varies

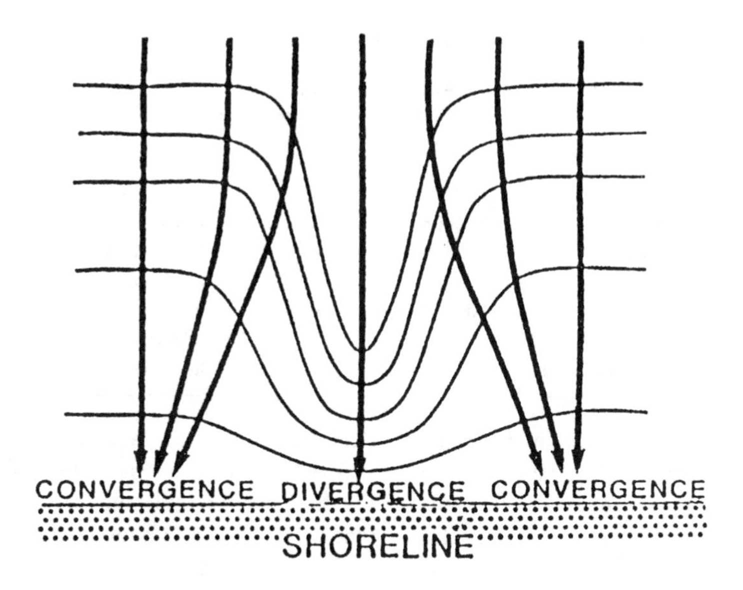

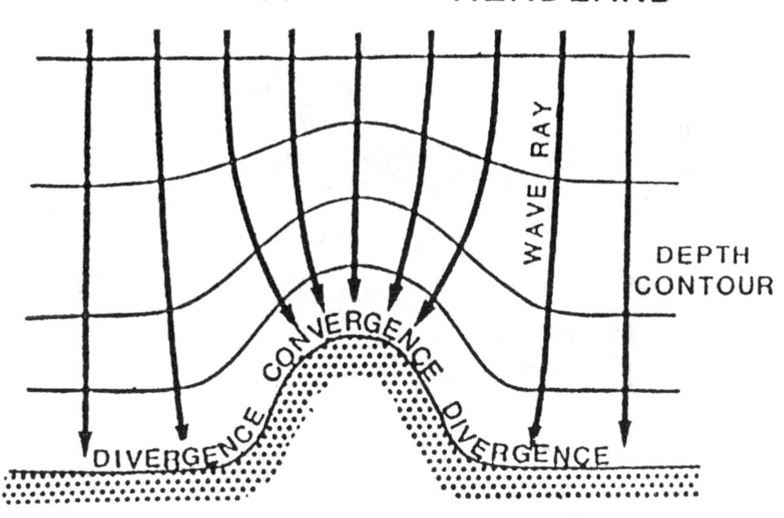

FIGURE 1.11 Wave refraction.

from moment to moment, but nevertheless the species regularly exposed to such conditions have evolved many adaptations to cope with them.

In order to understand processes such as exchange of gases, uptake of minerals, and settlement of sporelings by plants, and the processes of respiration, filter feeding, and larval settlement in animals, there is a need to know something about water flow close to surfaces (Lobban et al., 1985). As water flows over a surface, the velocity decreases toward the surface because of the drag that is created. Water flow may be laminar or turbulent, and the slow-moving velocity boundary layer is much thinner if the water is turbulent. Figure 1.12 illustrates the turbulent flow over an irregular shore at four stages during a tidal cycle. It can be seen that the intensity and direction of the water flow vary not only with the state of the tide but with the contours of the surface. Such variations will impact on the processes discussed above.

1.3.5.2 Topography and Aspect

The effects of topography on the distribution of intertidal organisms is very complicated due to the enormous variety of shores ranging from mudflats through sand-mud

TABLE 1.4
Wave Exposure Scales for Hard Shores

Ballantyne (1961)	Lewis (1964)	Rasmussen (1965)	Coast Type
1. Extremely exposed	1. Very exposed		Open
2. Very exposed		Class A	Open
3. Exposed	2. Exposed		Open
4. Semi-exposed	3. Semi-exposed	Class B	Protected outer
5. Fairly sheltered		Class D	Protected outer
6. Sheltered	4. Sheltered	Class E	Inlets and estuaries
7. Very sheltered	5. Very sheltered		Inlets and estuaries
8. Extremely sheltered			Inlets and estuaries

beaches, sand beaches, pebble beaches, and boulder beaches to wide rock platforms and steep vertical cliffs. In addition, the type of rock, whether soft and friable, hard and resistant to erosion, or smooth or rough, is important in determining the kinds and numbers of species present. Where the rock shores are steep and smooth, they will drain rapidly as the tide falls or the waves recede after impact. On the other hand, undulating platforms and gentle slopes will retain water and dry out more slowly, while the presence of tide pools can complicate distribution patterns considerably. The desiccating effects of emersion are mitigated on the latter type of shore, and consequently

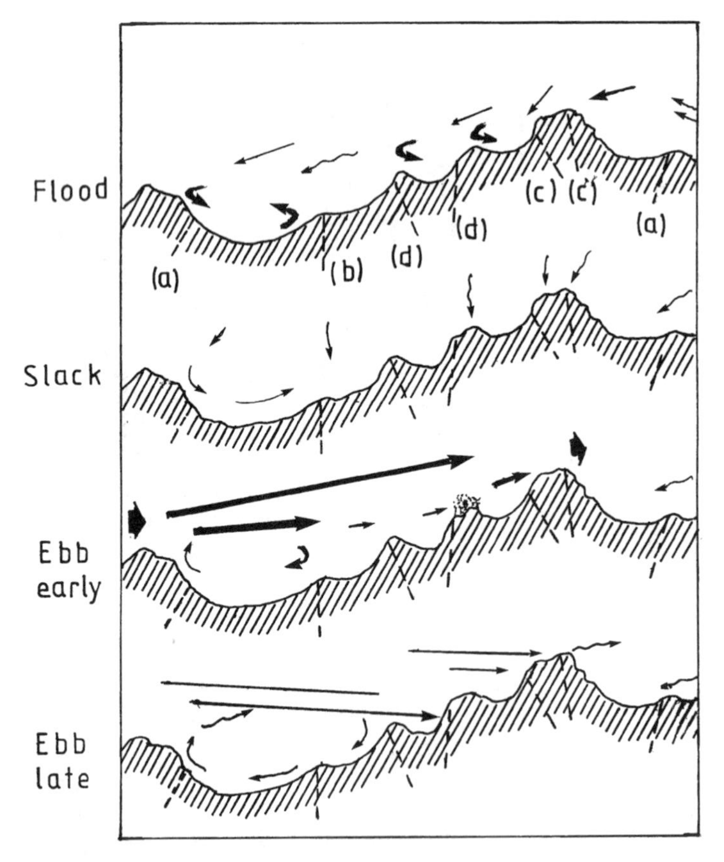

FIGURE 1.12 Water currents along an irregular shore at four stages during a tidal cycle, showing the complexities of water flow. The sizes of the arrows indicate relative current strengths. The seven transects marked by dashed lines were categorized into five current regimes (a–e): (a) sheltered back eddy with no measurable currents; (b) moderate ebb (0.20–0.25 m sec^{-1}) and very reduced flood; (c) moderate ebb (0.35 m sec^{-1}) and flood (0.20 m sec^{-1}); (d) strong ebb (0.40–0.60 m sec^{-1}) and very reduced flood; (e) strong ebb (0.80 m sec^{-1}) and moderate flood (0.20–0.25 m sec^{-1}). The most exposed transects (d and e) exhibited erratic pulsations of water motion during ebb tide. (Redrawn from Matheson, A.C., Tveter, E., Daly, M., and Howard, J., *Botanica Marina*, 20, 282, 1977. With permission.)

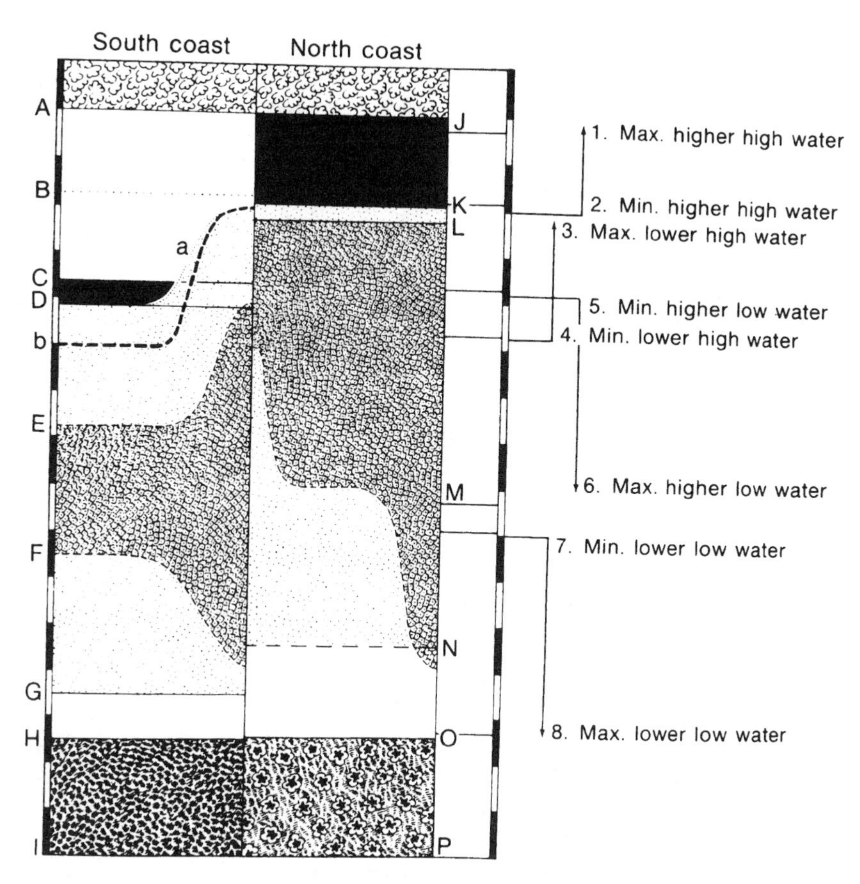

FIGURE 1.13 A diagrammatic section through Brandon Island, British Columbia, with the coast of Departure Bay in the background: *A–J*, lower limit of terrestrial lichens; *D–K*, upper limit of main barnacle population; *b*, lower limit of *Chthalamus* subzone; *E–L*, upper limit of main *Fucus* population; *G–N*, lower limit of most barnacles; *H*, upper limit of 'beard' zone; *L*, low water of a low spring tide. (From Stephenson, T.A. and Stephenson, A., *Life between the Tidemarks on Rocky Shores*, W.H. Freeman & Co., San Francisco, 1972, 211. With permission.)

there is an upward extension of the upper limits of many species. Dipping strata and rocks that weather easily provide overhangs and caves under or in which light is reduced and humidity increased. The slope of the shore plays an important role in modifying the effects of wave action; on angled slopes the waves sweep higher up the shore and steep faces increase the effects of spray from strong wave action. This is reflected in the upward extension of the vertical ranges of many species. Thus, one of the important effects of variations in topography is to modify the effects of wave action.

Another related factor is the shore aspect or a particular region of a shore. Typically, organisms extend higher up the shore on the shaded sides of boulders and gullies than on the sides exposed to the sun. Shaded positions mitigate the problems of desiccation and thermal stress, and when coupled with strong wave action allow a significant upshore extension of sublittoral organisms (for examples see Lewis, 1954a,b). While organisms that live high on the shore are more resistant to desiccation and thermal stress, they are nevertheless affected by aspect. Evans (1947) demonstrated that the British limpet *Patella vulgata* occurred at higher levels on the shore on shaded surfaces than on illuminated ones, and the effect was more pronounced when the shaded surface coincided with exposure to intense wave action.

Stephenson and Stephenson (1961; 1972) carried out a detailed survey of the intertidal flora and fauna of Brandon Island in Departure Bay, British Columbia. The island is largely sheltered from wave action, and while the northern side consists of steep shaded cliffs, the southern side slopes gently and is exposed to the sun (Figure 1.13). The zonation patterns on each of the two coasts are markedly different, despite the similarity of the tidal range and wave action on each side of the island. Figure 1.14 illustrates the relation of the zone boundaries to some eight tidal levels. Line A marks the lower limit of terrestrial lichens on both coasts and lies very close to level I (the maximum higher high water). Lines D and K show the upper boundaries of barnacles on the south and north coasts, respectively. Line D lies between levels 4 and 5, the minimum of higher low waters, while line K corresponds with levels 2 and 3, which coincide as the maximum higher low waters and the minimum higher high waters. Thus, the

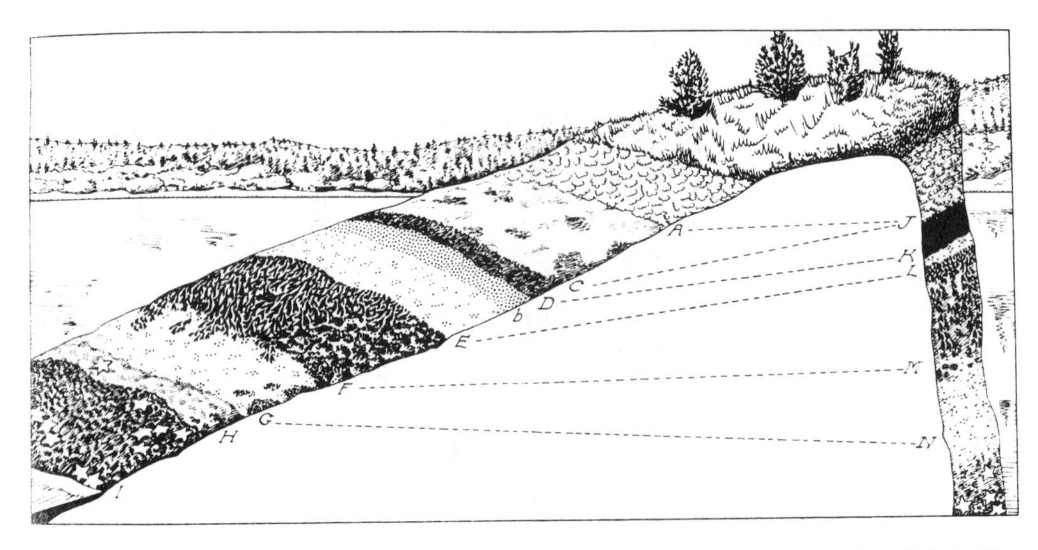

FIGURE 1.14 The relation of some zone boundaries to tide levels at Brandon Island, Departure Bay, British Columbia. (From Stephenson, T.A. and Stephenson, A., *Life between the Tidemarks on Rocky Shores,* W.H. Freeman & Co., San Francisco, 1972, 13. With permission.)

barnacle limit differs in relation to tidal levels on the two shores. Line H indicates the beard (brown birdlike and mosslike algae) zone on the south coast, which is almost the same as that of the large anemone, or *Metridium* (*M. senile fimbricatum*) zone on the north coast, both at tide level 8, the maximum of lower low waters. Other zonal limits also differ on the two shores. Stephenson and Stephenson (1972) point out that such differences in the zonation patterns must be largely due to the slope and aspect of the respective shores, although tidal level must also play a part in controlling the distribution patterns.

1.3.5.3 Climatic Factors

Climatic factors such as temperature, humidity, degree of insolation, rainfall, and amount of ice (in high latitudes) can play an important role in the distribution of plants and animals on the shore. These climatic factors interact with the type of substrate, aspect, topography, tidal rise and fall, and wave action. The shore climate will be discussed in detail in Chapter 5.

1.4 PATTERNS OF ZONATION ON THE SHORE

Few features of the shore are more obvious than zonation, or the vertical banding of the organisms living there. While such banding is most clearly evident on hard shores, it nevertheless occurs to a greater or lesser extent on soft shores. In estuaries, as we shall see later, the marsh vegetation characteristically exhibits a zonal pattern of the distribution of the component species. Such patterns are one response to the impact of tidal rise and fall. Generally, where the range of the tides is small or the slope of the beach is steep, the bands are narrow; where the range of the tides is great, or where the slope of the beach is gentle, the zones are wide. Heavy wave action widens the zones and the upper and lower boundaries tend to become less distinct.

Superimposed on this vertical zonation pattern is a gradient in species diversity at various levels on the shore. Higher up the shore, environmental conditions become more extreme, and the abundance and distribution of the organisms there will be controlled by physical factors such as extremes of temperature and desiccation. Thus, there is a reduction in the number of species higher up the shore (Knox, 1953). This pattern of species abundance where species diversity is high at the lower tidal levels and decreases up shore occurs in both plants and animals, although other factors such as competition for space and food (see Dayton, 1971; Connell, 1972; Harger, 1970) and predation (Connell, 1961a,b; Paine, 1966; 1971) can have important modifying effects on the distribution limits of many species. Figure 1.15 depicts the distribution of the algae and the major faunal components on a rocky shore at Dalebrook, on the East Cape Peninsula, South Africa. A number of features are evident in this diagram. First, the total number of species present declines with increasing height on the shore, similar to what occurs on other shores (see Colman, 1933). Secondly, filter-feeding organisms are dominant lower on the shore, where they are immersed longer and are thus able to feed for longer periods of time. At higher levels their biomass is drastically reduced, reflecting the reduced time available for feeding. Herbivores also predominate on the lower shore, where they are responsible for the reduction that occurs in algal biomass. In contrast, carnivores, which are able to exploit food resources over much of the intertidal zone, are present in smaller densities than their prey organisms.

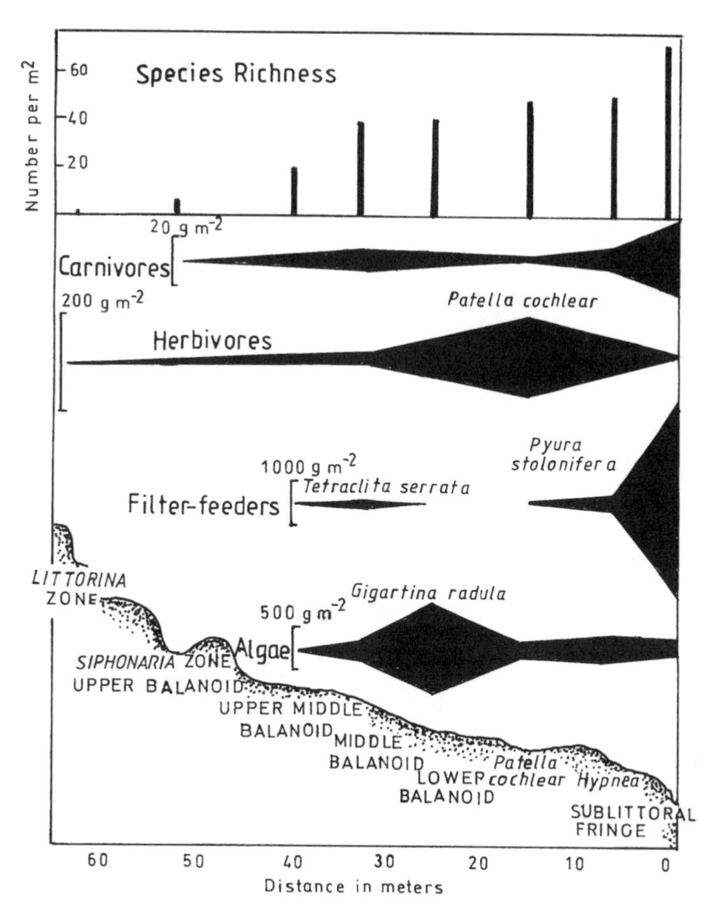

FIGURE 1.15 Transect of a rocky shore at Dalebrook, on the East Cape Peninsula, South Africa, showing the number of species of macrofauna per 2 × 0.5 m^2 samples (above) and the distribution of the macrophytes and principal components of the macrofauna. Note that (a) the number of species decreases toward the unpredictable conditions of the upper shore; (b) the standing stock of algae is reduced on the lower shore by the effects of grazing by herbivores (principally *Patella*); (c) the filter feeders decline dramatically with height on the shore, reflecting the decrease in available feeding time; and (d) the biomass of predators is small but moderately uniformly spread over the intertidal zone. Biomass data are expressed as grams (ash-free dry wgt m^{-2}). (Redrawn from Newell, R.C., *The Biology of Intertidal Animals,* 3rd ed., Marine Ecological Surveys, Fasherman, Kent, 1979, 3. With permission.)

2 Hard Shores

CONTENTS

2.1 ZONATION PATTERNS ON HARD SHORES

2.2.1 The Shore Environment and Zonation Patterns

The vertical distribution of plants and animals on the shore is rarely, if ever, random. On most shores, as the tide recedes, conspicuous bands appear on the shore as a result of the color of the organisms dominating a particular level roughly parallel to the water line (Figure 2.1). In other places, while the bands or zones are less conspicuous and less readily distinguishable, they are rarely, if ever, completely absent. Stephenson and Stephenson (1949; 1972) and Southward (1958) and Lewis (1955; 1961; 1964) have summarized much of the earlier information on zonation distribution patterns of intertidal organisms, and have shown that such zones are of universal occurrence on rocky shores, although their tidal level and width is dependent on a number of factors, of which exposure to wave action is the most important. More recent reviews of zonation patterns are to be found in Knox (1960; 1963a; 1975), Newell (1979), Lobban et al. (1985), Peres (1982a,b), Norton (1985), and Russell (1991).

2.2.2 Zonation Terminology

A variety of schemes have been proposed to delineate the various zones found on rocky shores, and I do not propose to review them here. Details of these schemes can be found in Southward (1958), Hedgpeth (1962), Hodgkin (1960), and Lewis (1964). Based on the work of Lewis (1964) and

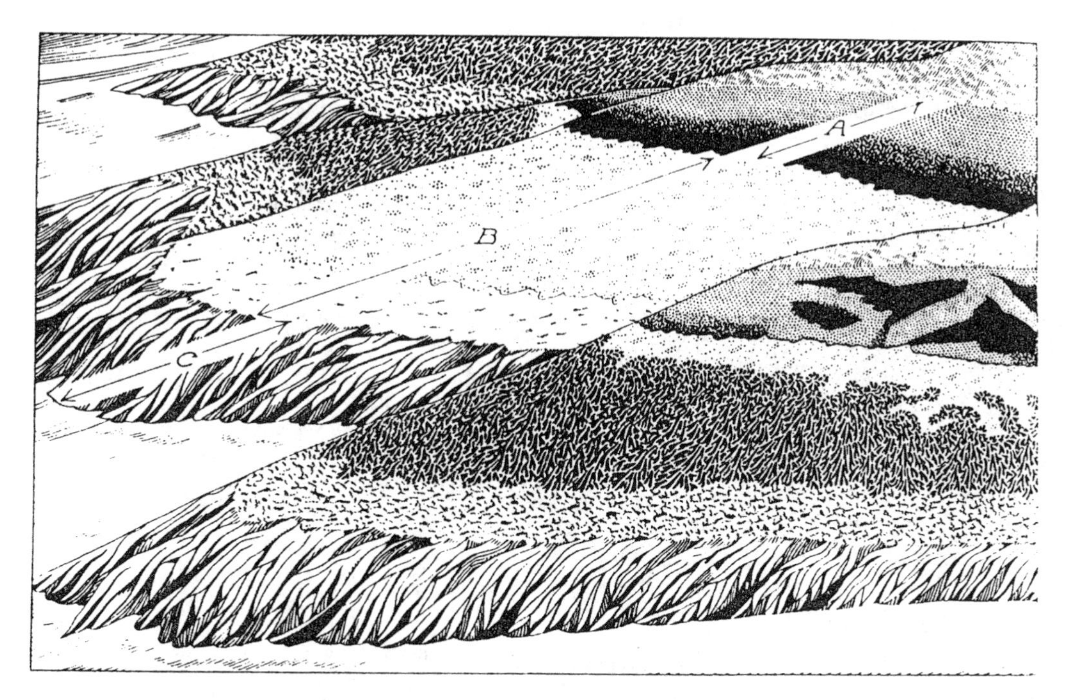

FIGURE 2.1 A comparison of the widespread features of zonation with an example that complicates them. A coast is shown on which smooth granite spurs are exposed to considerable wave action. On the middle spur, some of the widespread features are summarized and the following succession is shown. *A*, littoral fringe (= *Littorina zone*), blackened below by myxophyceans; *B*, eulittoral (balanoid zone), occupied by barnacles above and lithothamnia below; *C*, sublittoral fringe, dominated in this case by laminarians growing over lithothamnia. On the other spurs (*foreground and background*) the actual zonation from the Atlantic coast of Nova Scotia is shown. Here the simplicity of the basic plan is complicated by maplike black patches in the littoral fringe, consisting of *Codiolum*, *Calothrix*, and *Plectonema*; the existence of a strongly developed belt of *Fucus* (mostly *F. vesciculosus* and *F. endentatus*, in this example) occupying a large part of the eulittoral zone and overgrowing the uppermost barnacles; and a distinct belt of *Chondrus crispus* growing over the lower part of the eulittoral zone and largely obliterating the belt of lithothamnia, which, on the middle spur, extends over the laminarians. (From Stephenson, T.A. and Stephenson, A., *Life Between Tidemarks on Rocky Shores*, W.H. Freeman, San Francisco, 1972, 386. With permission.)

Stephenson and Stephenson (1972), who recognized three primary zones on marine rocky shores, each characterized by particular kinds of organisms, the scheme given in Table 2.1 and shown in Figure 2.2 will be used in the following discussion.

In the Stephenson and Stephenson scheme, the intertidal zone is called the *littoral zone* extending from the extreme high water of spring tides (EHWS) to the extreme low water of spring tides (ELWS). A *midlittoral zone* extends from the upper limit of the barnacles down to the lower limit of large brown algae (e.g., laminarians). A *supralittoral fringe* straddles EHWS extending from the upper limit of the barnacles to the lower limit of the terrestrial vegetation of the *supralittoral zone*. Its upper limit often coincides with the upper limit of littorinid snails. Below ELWS is the *infralittoral zone*, which is the upper part of the permanently submerged subtidal or sublittoral zone. Between the upper limits of the infralittoral zone a fringing zone, between the midlittoral and the infralittoral, the *infralittoral fringe*, is often distinguished.

The principal difference between the Stephensons' scheme and that proposed by Lewis is that the latter accounts for the impact of wave action in broadening and extending the vertical height of the zones. This takes into account the actual exposure time and not the theoretical time as determined from tide tables. In his scheme, Lewis extended the term *littoral* to include the Stephensons' supralittoral fringe and called the latter the *littoral fringe*. The rest of the littoral zone down to the upper limit of the laminarians is called the *eulittoral zone*. Lewis did not distinguish a zone equivalent to the Stephensons' infralittoral fringe. In this book, cases where a fringing zone between the eulittoral and the sublittoral is recognized will be called the *sublittoral fringe*.

As Russell (1991) points out, identification of the primary zones by inspection of a shore is necessarily influenced by the species composition of the topmost layer of the communities. He illustrates this in the diagram reproduced in Figure 2.3 of the stratification of the algal vegetation of the eulittoral zone on a Netherlands dyke as described by Den Hartog (1959). At the rock face surface, the entire extent of the zone is covered by the crustose red alga *Hildenbrandia rubra*. The middle stratum, also of red algae, has an upper band of *Catenella caespitosa* and a

TABLE 2.1
Table Showing the Principal Zones of Universal Occurrence on Hard Shores

Tidal Level		Zone	Indicator Organisms
		MARITIME ZONE	Terrestrial vegetation, orange and green lichens
Extreme high water of spring tides	LITTORAL ZONE	LITTORAL FRINGE	Upper limit of littorinids *Melaraphe (=Littorina) neritoides* *Ligia, Petrobius,* *Verrucaria etc.*
		EULITTORAL ZONE	*Upper limit of barnacles* Barnacles Mussels Limpets Fucoids (plus many other organisms)
Extreme low water of spring tides		SUBLITTORAL ZONE	*Upper limit of laminarians* Rhodophyceae Ascidians (plus many other organisms)

lower band of *Mastocarpus stellatus.* Finally the outer canopy layer consists of large brown (fucoid) algae in four conspicuous belts, with *Pelvetia caniculata* at the top, followed successively by *Fucus spiralis, Ascophyllum nodosum,* and *Fucus serratus.* This demonstrates that zonation is a three-dimensional phenomenon and that the zones defined by the uppermost stratum may conceal a number of other patterns.

As Lobban et al. (1985) point out, there are difficulties in defining zones on the shore in terms of the organisms

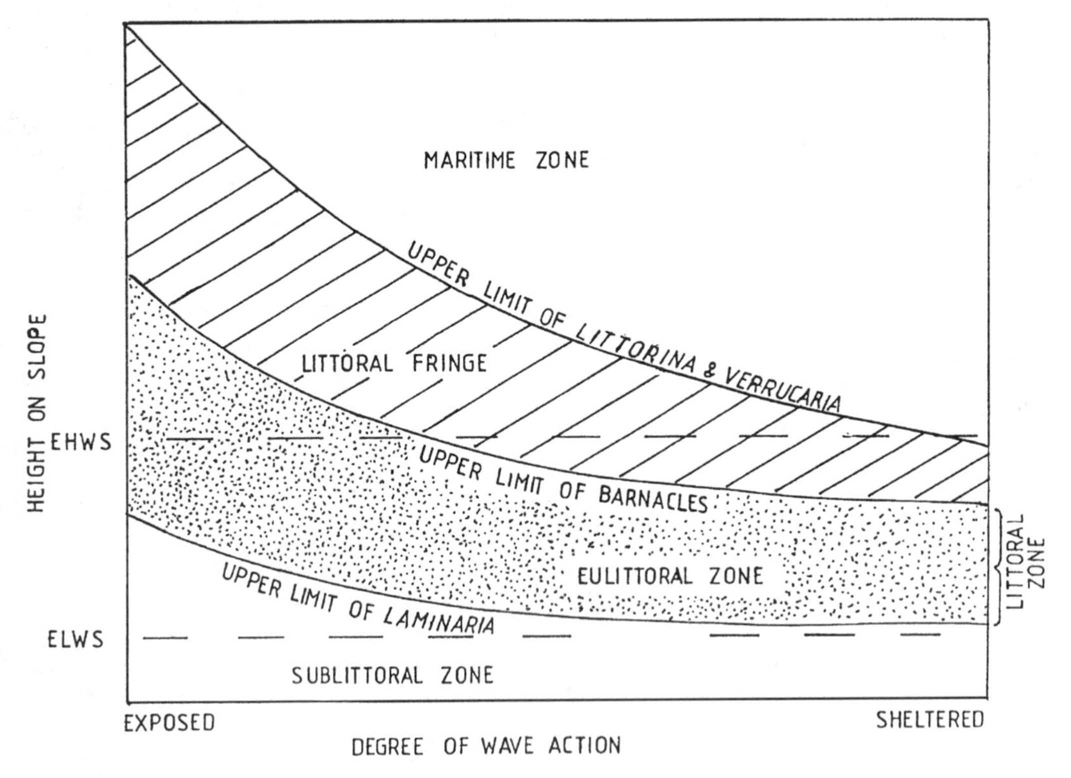

FIGURE 2.2 Diagram showing the effect of exposure to wave action on the intertidal zones of shore in the British Isles. (Modified from Lewis, J.R., *The Ecology of Rocky Shores*, English University Press, London, 1964, 49. With permission.)

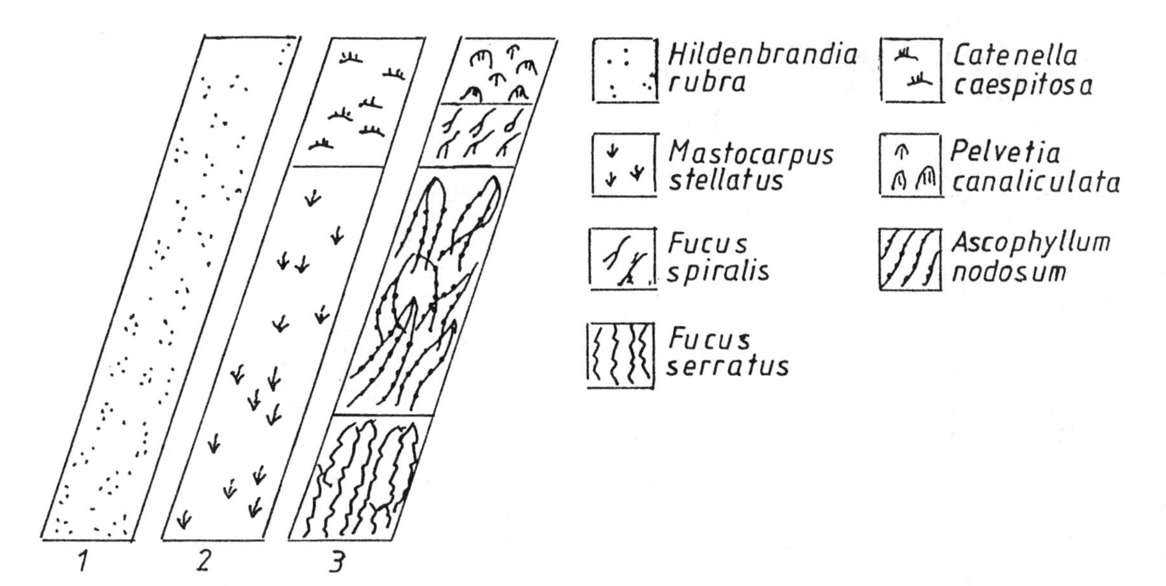

FIGURE 2.3 Stratification of vegetation in the eulittoral zone of a dyke in The Netherlands. The rock surface (1) bears the encrusting red alga *Hildenbrandia rubra*, the second stratum (2) consists of *Cantenella caespitosa* and *Mastocarpus stellatus,* and the canopy (3) comprises, in descending order, *Pelvetia canaliculatus, Fucus spiralis, Ascophyllum nodosum,* and *Fucus serratus*. Based on a diagram in den Hartog (1959). (Redrawn from Russell, G. in *Intertidal and Littoral Ecosystems, Ecosystems of the World* 24, Mathieson, A.C. and Nienhuis, P.H., Eds., Elsevier, Amsterdam, 1991, 44. With permission.)

found on them. Floras and faunas change geographically, and while a topographically uniform shore may have a uniform zonal distribution pattern, a broken shoreline of varying exposure to wave action and/or a broken substratum of irregular rocks and boulders can present a confusing pattern, with the zones breaking down into patches. However, if comparisons of surfaces with the same exposure, slope, and aspect are compared, then like patterns emerge.

In addition to variability in space, there is also variability in time. There are seasonal and successional changes in the vegetation and in the timing of disturbance that make space available for settlement (Dethier, 1984). The net result is a changing mosaic pattern of distribution. Vertical limits of many species can vary from year to year (Figure 2.4), perhaps dependent on variations in emersion-submersion histories. Relative abundances and distribution of species which may be nearly equal competitors change over time (Lewis, 1982). Among the algae, the presence or absence of a particular species at a given locality can be interpreted to mean that conditions there have been suitable for its growth since it settled (Lobban et al., 1985). Absence, on the other hand, only indicates that at a particular time conditions were unfavorable for the settlement of the reproductive bodies of that species, such as unfavorable currents or extreme desiccating conditions.

2.1.3 Widespread Features of Zonation Patterns

A consideration of the zonation patterns discussed above and in the next section (2.2) of this chapter reveals a number of widespread features or tendencies (Stephenson and Stephenson, 1972) as follows:

1. Near the high water mark there is a zone that is wetted by waves only in heavy weather, but affected by spray to a greater or lesser extent. The number of species is relatively small, and includes particular species adapted to semiarid conditions, and belonging to the gastropod genus *Littorina*, and related genera, or to genera of snails containing similarly adapted species. Semi-terrestrial crustaceans, such as isopods of the genus *Ligia,* are also characteristic of this zone.
2. The surface of the rock in the zone described above, especially in the lower part, is blackened by encrustations of blue-green algae, or lichens of the *Verrucaria* type, or both. This is a most persistent feature of the zone. Depending on the latitude and geographic location, other grey, green blue-green, and orange lichens (the latter belonging to the genus *Caloplaca*) paint splashes of color on the rocks.
3. The middle part of the shore typically includes numerous balanoid barnacles belonging to genera such as *Balamus, Semibalanus, Chthalamus,* and *Tetraclita*. The upper limit of the zone is marked by the disappearance of barnacles in quantity. Herbivorous and carnivorous gastropods, especially limpets, whelks, and chitons are often abundant. On some shores, algae, especially fucoids, may form conspicuous bands.

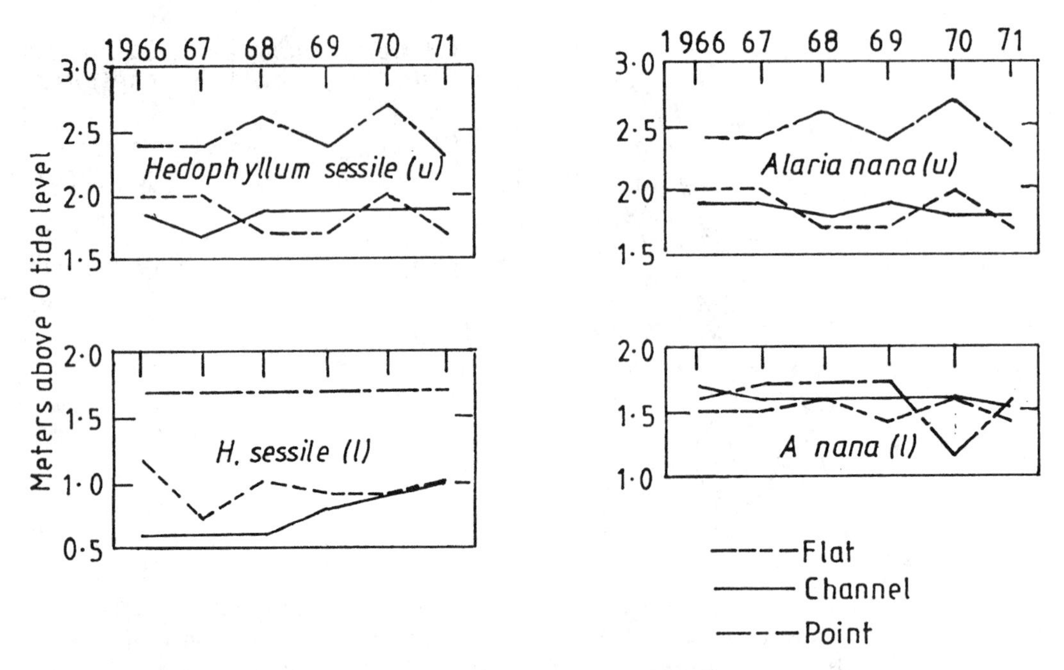

FIGURE 2.4 Year-to-year changes in upper and lower limits of two intertidal kelp species on three transects at an exposed site on the west coast of Vancouver Island, British Columbia. A gently shelving platform, a rocky point, and a narrow channel are compared. (Redrawn from Druehl, L.D. and Green, J.M., *Mar. Ecol. Prog. Ser.*, 9, 168, 1982. With permission.)

4. The lowest part of the shore is uncovered only by spring tides and is characterized by a diverse assemblage of species. In cold-temperate regions, it consists of a forest of brown algae (e.g., laminarians) with an undergrowth of smaller algae, especially reds, between the holdfasts. In warm-temperate regions, it may support (a) a dense covering of simple ascidians (e.g., *Pyura*), (b) a dense mat of small mixed algae, primarily reds, or (c) other communities.

2.2 ZONATION PATTERNS ON REPRESENTATIVE SHORES

In this section we will briefly detail the principal zonation patterns on a range of shore from both the southern and northern hemispheres. From this survey it will be seen that although there are similarities between the patterns, and while some taxa (e.g., barnacles, mussels, herbivorous and carnivorous gastropods, limpets, and some algal species) are found on most shores, there are considerable differences in the distribution patterns related to the latitude of the shore (affecting seasonal ranges in temperature and other climatic variables), the patterns of the tides, and in the species composition of the shore communities.

2.2.1 The British Isles

The British Isles are approximately 1,125 km long and are subject to cool-temperate climatic conditions. The seasonal variation in sea temperatures is roughly 7°C in northern parts and up to 12°C in part of the Irish Sea and the southeastern coasts. The range of spring tides varies from 0.6 m to 12 m, although ranges of between 7 and 12 m are more common. Detailed accounts of the zonation patterns on the shores of the British Isles are to be found in Lewis (1964) and Stephenson and Stephenson (1972).

The general pattern of zonation is as follows: (1) a littoral fringe dominated by "black" lichens, dark microphytes, and littorinid snails, (2) a eulittoral zone dominated by various combinations of barnacles, mussels, limpets, snails, and brown (fucoid) and red algae; and (3) a sublittoral fringe dominated by laminarian algae.

Littoral Fringe: The upper limit of the littoral fringe is placed at the junction between the black lichens and the band of orange and/or grey lichens above, although on other shores this latter zone is regarded as the upper littoral fringe. Two species of lichens dominate much of this black zone, *Verrucaria* throughout and *Lichina confinis* toward the upper limit. In wave-swept places, algal growth superimposed on the lichens takes the form of a very fine layer dominated by cyanophyceans (*Calothrix* spp. in particular), and, more locally, filamentous green and red algae (*Ulothrix, Urospora,* and *Bangia*). Superimposed on this are the larger red alga, *Porphyra umbilicalis,* and species of the green algal genus, *Enteromorpha.* Most of these algae are seasonal in occurrence.

The lower limit of the littoral fringe is taken as the upper limit of barnacles in quantity. Where *Chthalamus stellatus* predominates (in southwestern areas generally and exposed situations in the west and northwest), the "barnacle line" is higher than in areas where *Balanus*

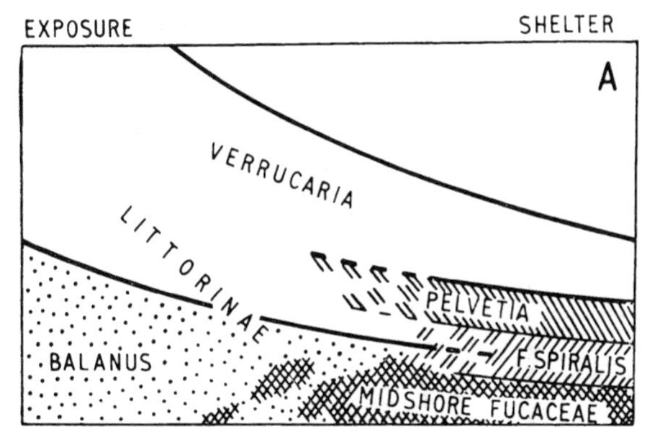

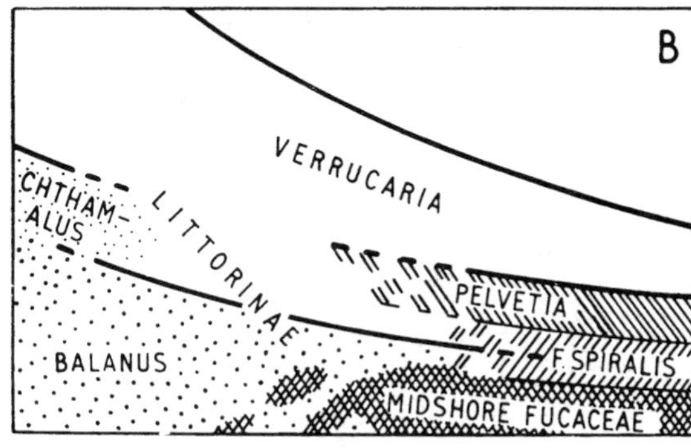

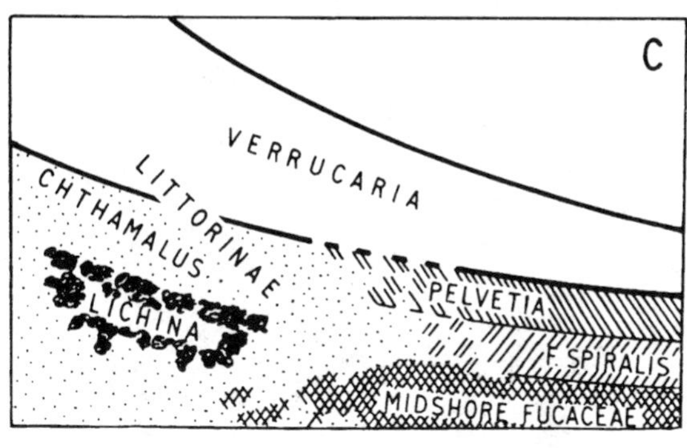

FIGURE 2.5 Simplified diagram showing the littoral fringe on: A. *Balanus* shore; B. situations on the north-west coasts where *Chthalamus* is confined to exposure; and C. *Chthalamus* shores in the British Isles. (From Lewis, J.R., *The Ecology of Rocky Shores*, English University Press, London, 1964, 54. With permission.)

balanoides is present alone (on the north and east coasts, and in sheltered areas of the west and northwest). Consequently, some conspicuous zone-forming plants of narrow vertical range (the lichen *Lichina pygmaea* and *Fucus spiralis*) lie largely within the eulittoral zone on "*Chthalamus* shores" and partly, or completely in the littoral fringe on "*Balanus* shores" (Figure 2.5). The characteristic animals in the littoral fringe are littorinid snails, *Littorina neritoides* and *L. saxatilus*. Other animals are mites, the thysanuran, *Pterobius maritimus*, and the eulittoral molluscs, *Patella vulgata* and *Littorina littorea*.

Eulittoral Zone: At one extreme this zone is dominated by (1) barnacles or mussels (or both), and (2) at the other by exceptionally heavy growths of long-fronded fucoid algae.

1. *Barnacle-dominated shores* (Figure 2.6): Where they are abundant, barnacles can extend from their sharp upper limit to within a few centimeters of the topmost laminarians. *Balanus balanoides* is the most ubiquitous, while *Chthalamus stellatus* predominates in the southwest but is absent from North Sea coasts and the entire eastern half of the English Channel. On moderately exposed sites in southwest England and Wales a third larger species, *Balanus perforatus*, occupies a belt 60 to 90 cm high immediately above the laminarians, or forms isolated patches at higher levels. Since the late 1940s, the Australasian barnacle, *Elminius modestus*, has established itself in harbors and estuaries and along the less exposed coasts, mainly at the expense of *Balanus balanoides*. Associated animals include limpets (*Patella depressa*, *P. vulgata*, and *P. aspersa*) and whelks (*Gibbula cineraria*, *G. umbilicalis*, and *Nucella lapillus*).
2. *Fucus-dominated shores* (Figure 2.7): As exposure decreases, there is a progressive replacement of barnacle- and mussel-dominated communities by fucoids, beginning with the appearance of *Fucus vesiculosus* f. *linearis*. *Pelvetia* gradually appears in the littoral fringe and *F. serratus* begins to mingle with the low level *Himanthalia*. As the larger and sheltered shore form of *F. vesiculosus* replaces *F. vesiculosus* f. *linearis*, *F. spiralis* appears and *F. serratus* displaces *Himanthalia*. Next, *Ascophyllum nodosum* starts to appear in the flatter and more protected places among the *F. vesiculosus*. This process culminates in very sheltered bays and locks with luxuriant narrow belts of *Pelvetia* and *F. spiralis* surrounding a midshore belt of long-fronded *Ascophyllum*, with a narrow belt of *F. spiralis* just above the laminarians (Figure 2.7). The relative proportions of the eulittoral zone occupied by *Ascophyllum*, *F. vesiculosus*, and *F. serratus* vary greatly.

The shade of the fucoids enables *Laurencia*, *Leathesia*, and other members of the red algal belt to extend upshore, but under the dense growths of *Ascophyllum* and *F. serratus* they are replaced by lithothamnion. As the fucoids develop there is a loss of such open-coast species as *Littorina littorea*, *Patella aspersa*, *P. depressa*, *Balanus perforatus*, and *Mytilus edulis*, with its associated fauna. The topshell, *Gibbula umbilicalis*, becomes plentiful throughout the middle zone and is joined by *G. cineraria*

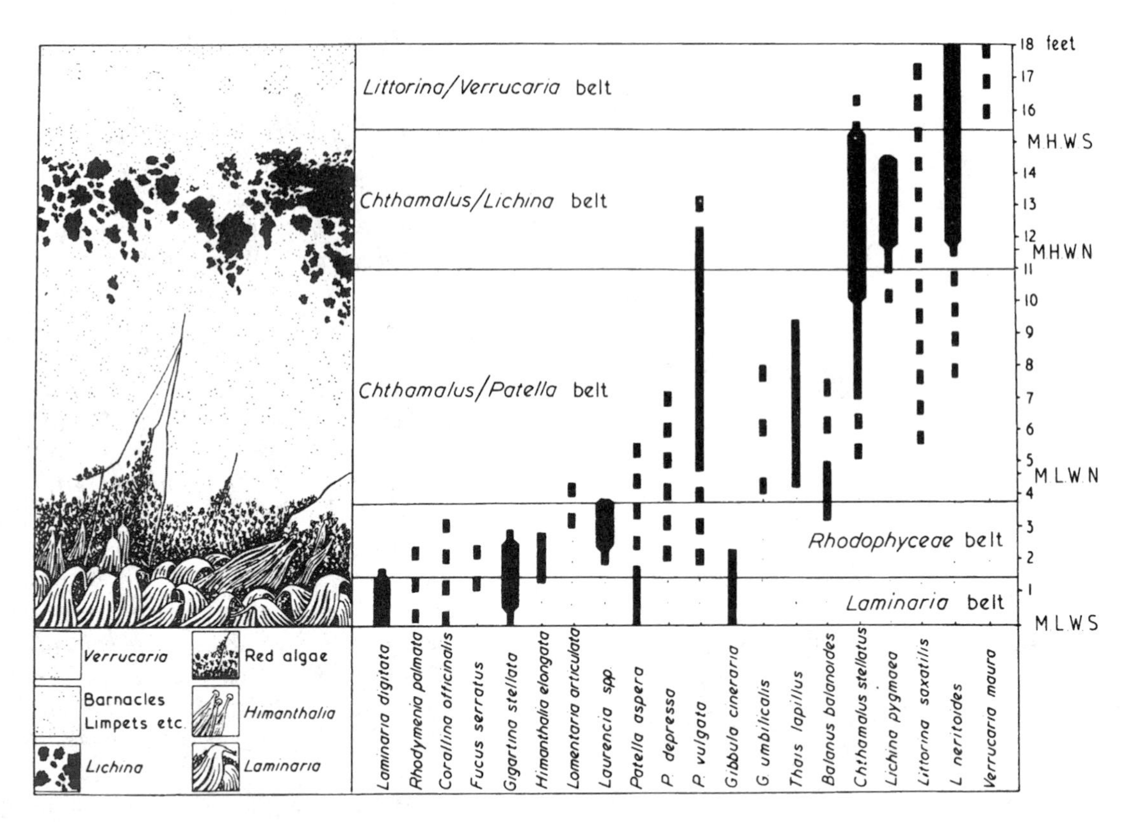

FIGURE 2.6 A barnacle-dominated face near Hope Cove, South Devon, typical of many exposed and south-facing areas of the English Channel coast. (From Lewis, J.R., *The Ecology of Rocky Shores*, English University Press, London, 1964, 78. With permission.)

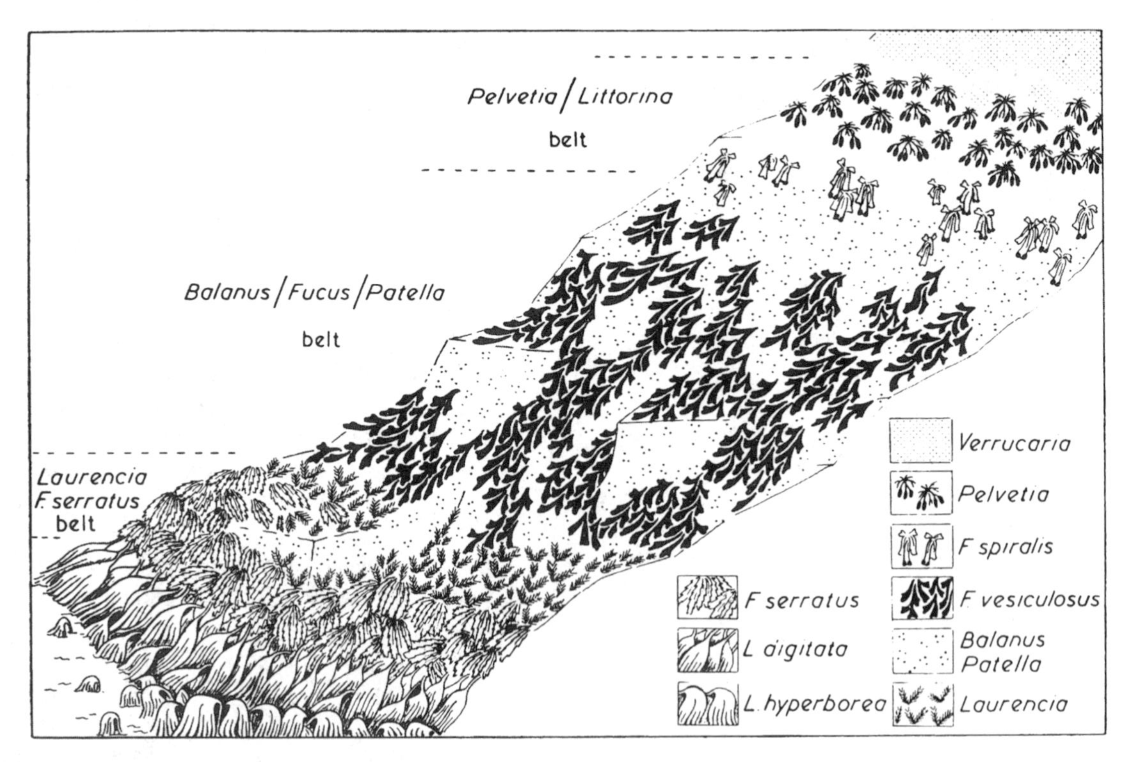

FIGURE 2.7 Representation of a moderately sheltered *Fucus*-dominated shore. (From Lewis, J.R., *The Ecology of Rocky Shores*, English University Press, London, 1964, 119. With permission.)

and *Monodonta lineata* in the lower and upper levels, respectively. *Littorina saxatilus* is joined by *L. obtusata*, mainly in the fucoids, and by large number of *L. littorea*.

Sublittoral Fringe and Upper Sublittoral: The flora of the sublittoral fringe is characteristically dominated by laminarians. Most of the permanently submerged "forest" consists of *Laminaria hyperborea*. Above this species on open coasts, two species predominate — *Alaria esculenta* in very exposed situations, and *Laminaria digitata* elsewhere. They form a continuous narrow belt, typically not more than 30 to 60 cm deep. As *L. digitata* replaces *Alaria* the undercanopy algal growth becomes more variable and luxuriant, and commonly includes species such as *Ceramium* spp., *Chondrus crispus, Cladophora rupestris, Cystoclonium purpureum, Delessaria sanguinea, Dictyota dichotoma, Membranoptera alata, Plocamium coccineum, Plumaria elegans, Polysiphonia* spp., and *Rhodymenia palmata*.

The fauna of this zone changes from one of relatively large numbers of a few species to one of small numbers of very many species. A few eulittoral species extend down into this zone such as *Patella aspersa, Gibbula cineraria, Mytilus edulis*, and barnacle (*Verrucaria stroemia* and *Balanus crenatus*). Sublittoral species that occur, depending on the degree of wave exposure, include sponges, hydroids, anemones, tubiculous polychaetes, bryozoans, and ascidians.

2.2.2 The Northwest Atlantic Shores

This encompasses the North American coastline between Cape Cod/Nantucket Shoals and Newfoundland, and exhibits conspicuous regional differences in temperature, tidal fluctuation, ice scouring, wave exposure, and nutrient enrichment. This area has been extensively studied by a number of investigators (see references by A.R.O. Chapman, 1981, 1984, 1990; C.R. Johnson, 1985; C.S. Lobban; J. Lubchenco; K.H. Mann; A.C. Mathieson; B.A. Menge; J.L. Menge; J.D. Pringle, 1987; and R.S. Steneck, 1982, 1983, 1986. Stephenson and Stephenson (1954a,b; 1972) have given accounts of the zonation patterns in Nova Scotia and Prince Edward Island, and Mathieson et al. (1991) have recently reviewed northwest Atlantic shores.

Tidal ranges within the area vary considerably. Average tidal ranges within the Gulf of Maine vary from 2.5 to 6.5 m (mean spring tides = 2.9 to 6.4 m), while those elsewhere vary from 2.7 to 11.7 m (mean spring tides = 3.1 to 13.3 m) in the Bay of Fundy, 0.7 to 2.2 m on the Atlantic coast of Nova Scotia, and 0.8 to 1.9 on the coast of Newfoundland. In the Bay of Fundy the annual temperature range is moderate, the maximum being 1.8°C in February to a maximum of 11.4°C in September. Salinities of the surface waters vary from 30 to 33. For the Atlantic coast of Nova Scotia, intertidal populations are subjected to very cold waters (sometimes below 0°C) in the winter, and relatively warm water (often near 20°C, or locally even higher) in the summer.

Descriptions of **zonation patterns on New England** coasts can be found in J.L. Menge (1974; 1975), B.A. Menge (1976), B.A. Menge and Sutherland (1976), Lubchenco and Menge (1978), Menge and Lubchenco (1981), Mathieson et al. (1991), and Vadas and Elner (1992). The basic zonation patterns of the New England coasts are depicted in Figure 2.8.

Littoral Fringe: The littoral fringe is characterized by blue-green algae (*Calothrix, Lyngbya, Rivularia*, etc.) and ephemeral macrophytes (such as *Bangia, Blidingia, Coliolum, Porphyra, Prasiola, Ulothrix*, and *Urospora*, lichens (such as *Verrucaria maura*), and a periwinkle (*Littorina saxatilis)*.

Eulittoral Zone: On a typical semi-exposed rocky shore, three major zones occur (Lubchenco, 1980): (1) an upper barnacle zone with *Semibalanus balanoides* dominating; (2) a mid-shore brown algal zone with *Ascophyllum nodosum* and/or *Fucus* spp.; and (3) a lower red algal zone with *Chondrus crispus* and *Mastocarpus stellatus*.

The *S. balanoides* zone exhibits a conspicuous uplifting with increasing wave action, while the brown and red algal zones are compressed and displaced downwards. Barnacles may also extend down into the lower eulittoral zone, particularly in extremely exposed habitats. Other species include the predatory dogwhelk, *Nucella lapillus*, and the periwinkle, *Littorina littorea*. On some exposed shores the dwarf fucoid, *Fucus distichus* ssp. *uncaps*, grows on the barnacles. Depending upon wave action and other associated physical and biological factors, either *A. nodosum* or *Fucus* spp. will dominate the mid-shore (Lubchenco, 1980). As in Europe, *A. nodosum* is most abundant in sheltered sites and is replaced by *F. vesiculosus* and *F. distichus* ssp. *dentatus* with increasing wave exposure. Under extreme wave action the fucoids are limited and *Mytilus edulis* becomes the major occupier of space in the mid-shore. In the lower eulittoral zone, *C. crispus* and/or *Mastocarpus stellatus* dominate at all but the most exposed sites, where mussels are the most abundant macroorganism. *C. crispus* is found mainly on shelving and horizontal surfaces, whereas *M. stellatus* dominates the vertical ones (Pringle and Mathieson, 1987). Substrata with intermediate slopes are populated by a mixture of both algae.

In the mid-eulittoral, competition between *Mytilus edulis* and *Semibalanus balanoides* is the dominant biological interaction. Predation and herbivory are the main factors affecting space utilization (Menge and Sutherland, 1976; Menge, 1978a,b; Lubchenco, 1983; 1986). By clearing space, *Nucella lapillus* and other predators of *Mytilus edulis* allow the persistence of *Fucus vesciculosus* and *Ascophyllum nodosum* on semiprotected and protected sites, respectively (Keser and Larson, 1984a,b). Both fucoids are competitively inferior to many ephemeral

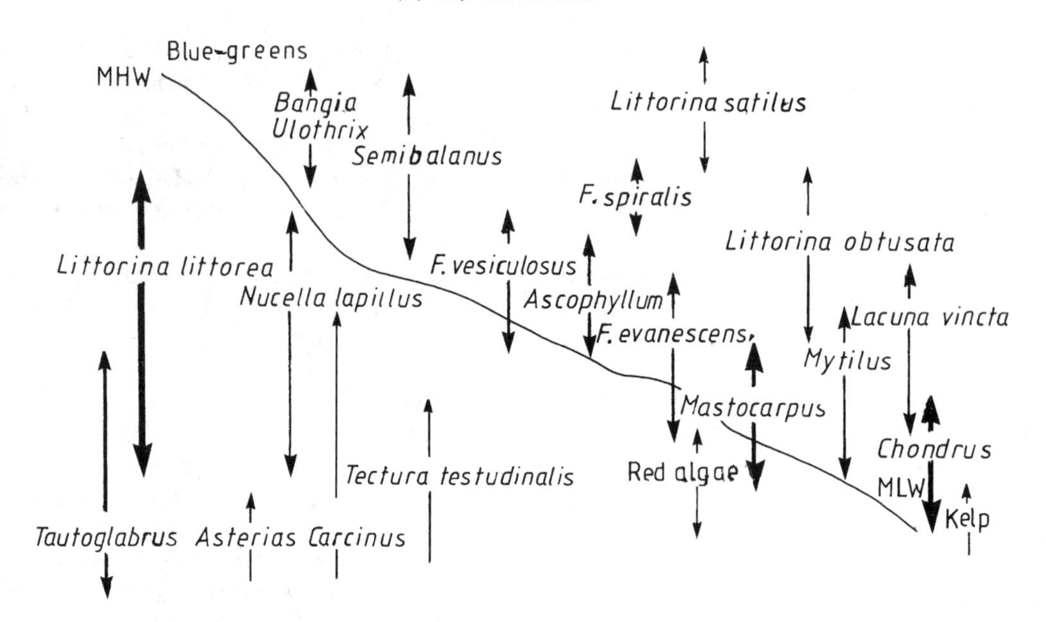

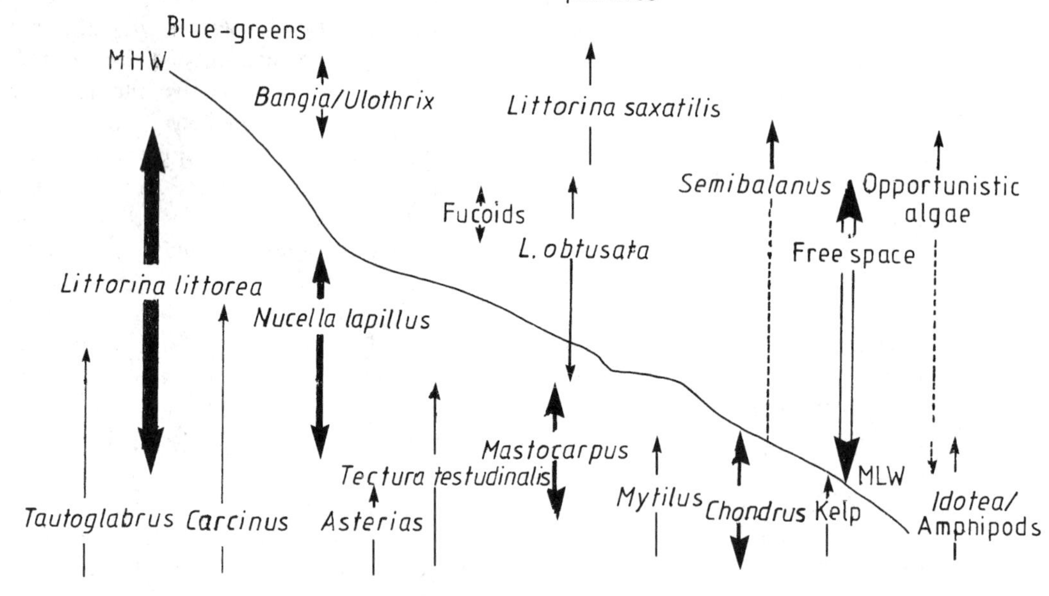

FIGURE 2.8 Schematic diagram showing the vertical distribution patterns of major taxa on northwest Atlantic shores. (a) A relatively exposed shore. (b) A moderately sheltered shore. The vertical distribution is shown by the length of the arrows, while the width depicts the relative abundance or functional importance. A dashed line indicates a changing or ephemeral, seasonal pattern. (Redrawn from Vadas, R.L. and Elner, R.W., in *Plant-Animal Interactions in the Marine Benthos,* John, D.M. and Hawkins, S.J., Eds., Clarendon Press, Oxford, 1992, 36, 37. With permission.)

algae (such as *Enteromorpha* spp., *Porphyra* spp., and *Ulva lactuca*.)

In addition to *Mytilus edulis* and *Semibalanus balanoides*, numerous other invertebrate species, both sessile and motile, characterize the eulittoral zone. Several herbivorous crustaceans and gastropods are common (Vadas, 1985), including amphipods (such as *Hyale nilssoni*), herbivorous snails (*Littorina littorea, L. obtusata, L. saxatilis,* and *Lacuna vincta*), and limpets (*Acmaea testudinalis*). The chiton, *Tonicella ruber*, and the sea urchin, *Strongylocentrotus droebachiensis,* graze within the lower eulittoral and sublittoral zones. The whelk, *Nucella lapillus*, and two crab species, *Carcinus maenas* and *Cancer irrotatus*, and a starfish, *Asterias vulgaris,* are important predators in both the lower eulittoral and sublittoral zones. The abundance of these species decreases with increasing wave exposure. This allows *M. edulis* to achieve dominance over *Chondrus crispus* and *Semibalanus balanoides*

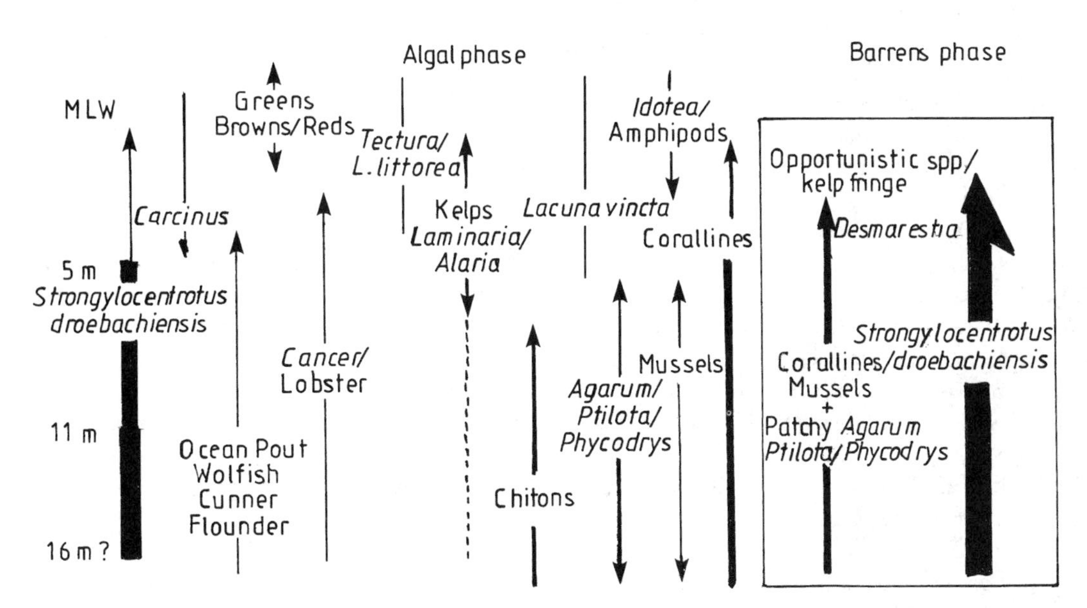

FIGURE 2.9 Schematic diagram showing the patterns of distribution of the New England sublittoral. (Redrawn from Vadas, R.L. and Elner, R.W., in *Plant-Animal Interactions in the Marine Benthos*, John, D.M. and Hawkins, S.J., Eds., Clarendon Press, Oxford, 1992, 44. With permission.)

(Menge, 1976; 1983; Lubchenco and Menge, 1978; Lubchenco, 1980). Conversely on sheltered shores, heavy predation on mussels by the starfish, *Asterias* spp., the crab, *Carcinus maenas*, the lobster, *Homarus americanus*, and sea ducks, allow its replacement by the alga *C. crispus*.

Sublittoral Zone (Figure 2.9): The composition and ecology of this zone is discussed in detail in Section 2.9.5. It supports a diverse epiflora and epifauna on rocky substrates. Dominant large macroalgae are the kelps *Laminaria longicirrus, L. digitata, L. saccharina*, and *Agarum cribosum*. However, other algae such as *Desmarestia* can form extensive beds, and species such as *Chondrus* dominate the understory layer. Crustose corallines such as species of *Lithothamnion, Clathromorphum*, and *Phymatolithon* are ubiquitous (Steneck, 1983; 1986). The principal predators of sublittoral grazers are lobsters, *Homarus americanus*, *Cancer* spp., green crabs, seastars, *Asterias* spp., and numerous fish species (see Keats et al., 1987).

Two alternative community states of the sublittoral community exist, depending on the population density of the sea urchin, *Strongylocentrotus droebachiensis*. When sea urchins are rare, communities of kelp and other macrophytes flourish. Where corallines dominate, the system ("barrens phase") is maintained through intense grazing by sea urchins (see Figure 2.9).

2.2.3 The Pacific Coast of North America

The northeast Pacific coastline stretches nearly 2000 km from Alaska (53°N) to the tip of Baja California (23°N). In the northeast Pacific, temperatures range from 5°C near the Aleutian Islands to 20°C near Baja California. Inshore temperatures are affected by coastal upwelling, a seasonal feature most prevalent off California and Baja California. Tides along the northeast Pacific coast are mixed semidiurnal: two highs and two lows occur in each lunar day, with successive high and low waters and successive low waters each having different heights. The difference between high and low tides range from 1.6 to 6.0 m, generally increasing with higher latitude. The coastline from Alaska to Baja California includes three main zones: cold-temperate, warm-temperate, and tropical.

There is considerable literature on the ecology of the shores of this region. There are a number of recent reviews of northeast Pacific shores in general (Carefoot, 1977; Ricketts et al., 1985), rocky shores (Moore and Seed, 1986; Foster et al., 1988; 1991), and shallow subtidal rocky reefs (Dayton, 1985b; Foster and Schiel, 1985; Schiel and Foster, 1986).

There is considerable variation in the **zonation patterns** and species composition both geographically and locally. Zonation patterns will be described with special reference to central California.

Littoral Fringe: This zone is only infrequently wetted by storm waves and spray. It is mainly bare rock or covered with small green and blue-green algae. Larger green algae (*Enteromorpha* spp., *Ulva* spp.) or red (*Porphyra* spp., *Bangia vermicularis)* algae and masses of benthic diatoms may be present, especially in winter and spring (Cubit, 1975). The few animals that occupy this zone include the limpet, *Collisella digitalis*, other gastropods such as *Littorina keenae*, and isopods (*Ligia* spp.).

Upper Eulittoral: This zone is characterized by dense populations of barnacles (*Balanus glandula*). In addition to the barnacles, the algae *Endocladia muricata, Mastocarpus papillatus,* and *Pelvetia fastigiata* are conspicuous and characteristic members of this zone. The small periwinkle, *Littorina scutulata*, the turban snail, *Tegula funebralis*, and several species of limpets also occur in this zone.

Mid-eulittoral: On moderate to fully exposed shores, the most conspicuous species are mussels (predominantly *Mytilus californianus*) and gooseneck barnacles (*Pollicipes polymerus*). Other characteristic species, especially on the more protected shores, are the predatory snail, *Nucella emarginata*, the chitons, *Katharina tunicata* and *Nuttallina californica*, and the red alga, *Iridaea flaccida*.

Lower Eulittoral: This zone is typically covered by carpets of the surf-grass (*Phyllospadix* spp.), various kelps (particularly *Laminaria setchellii*), and a variety of red algae. This zone grades into the sublittoral, with its upper margin forming the sublittoral fringe.

Sublittoral Zone: This zone has a marked three-dimensional structure provided by large stipate and float-bearing kelps. Surface-canopy kelps such as *Macrocystis pyrifera* occupy the entire water column; stipate kelps such as *Pterogophora californica* and *Laminaria* spp. may form an additional vegetation layer within two meters of the bottom of the kelp forests. A third layer of foliose red and brown algae, as well as articulated corallines, is common beneath the understory kelps, with a final layer of filamentous and encrusting species on the bottom. There is considerable variation in species composition over the length of the coastline. Common surface-canopy algae are *Alaria fistulosa* in southwestern Alaska, *Macrocystis integrifolia* and *Nereocystis leutkeana* from eastern Alaska to Pt. Conception, California, and *Macrocystis pyrifera* from near Santa Cruz, California, to Baja California.

Stephenson and Stephenson (1972) have described in detail the zonation patterns on the coasts of Pacific Grove on the Monterey Peninsula (36° 30'N to 36° 38"N) in central California. Figure 2.10 depicts these patterns on steep slopes with reference to the influence of wave action.

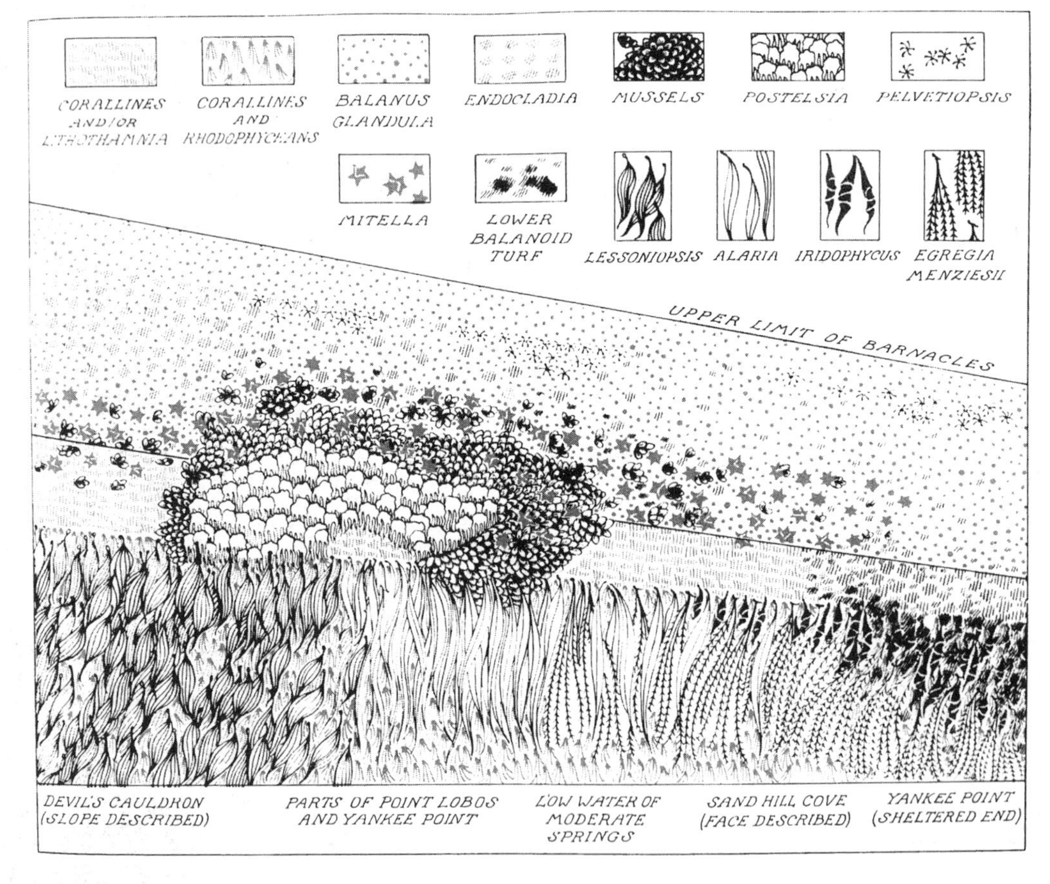

FIGURE 2.10 A comparison of the zonation of steep slopes at different localities in the Pacific Grove region of California. The various slopes are subject to different types and degrees of wave action. The line across the center of the figure indicates the boundary between the upper and lower balanoid zones. (From Stephenson, T.A. and Stephenson, A., *Life between Tidemarks on Rocky Shores,* W.H. Freeman, San Francisco, 1972, Pl. 16. With permission.)

Littoral Fringe: Two periwinkles, *Littorina planaxis* and *L. sculata* occur in vast numbers. The former continues down into the upper eulittoral, while the latter extends further down the shore. Other animals characteristic of this zone are the isopod, *Ligia occidentalis*, the crab, *Pachygrapsus crassipes*, and the limpet, *Acmaea digitalis*. Algae present are encrusting forms, especially *Hildenbrandia occidentalis*. Blue-green algae cause a blackening of the rocks (species include *Entophysalis granulosa, Calothrix crustacea, Rivularia battersii*).

Upper Eulittoral: The dominant barnacle is *Balanus glandula*, often occurring in dense sheets. The smaller *Chthalamus fissus* is scattered among the *Balanus*, while a third species, *Tetraclita squamosa*, is most abundant in a belt overlapping the junction between the upper and lower eulittoral. A fourth species, *Balanus cariosus*, is present but never abundant. Other animals are two small whelks, *Acanthina lapilloides* and *Thais emarginata*, and the trochid, *Tegula funebralis*. The two commonest limpets are *Acmaea scabra* and *A. digitalis*.

Algal growths consist mainly of irregular patches or tufts of turf algae (a mixture of species including *Gigatina papillata, Endocladia muricata, Cladophora trichotoma*, sporelings of *Fucus*, and small plants of *Pelvetia, Porphyra*, and *Rhodoglossum affine*).

Lower Eulittoral: All three barnacles that characterize the upper eulittoral extend somewhat into the lower eulittoral, where they are in competition with algal turfs and only flourish in clearings. *Balanus glandula* may be plentiful at the top of this zone, *Tetraclita* is often common, and *Chthalamus dalli* may form dense sheets at the bottom of the zone. As exposure increases, dense beds of mussels (*Mytilus californianus*) occur and may extend up into the lower part of the upper eulittoral. Another common clump-forming species is the goose-necked barnacle, *Pollicipes polymerus*. The trochid, *Tegula finebralis*, may occur in vast quantities in this zone. The limpets, *Acmaea pelta, A. limatula, A. scutum*, and *A. mitra*, all occur here, while the giant limpet, *Lottia gigantea*, occupies a restricted band overlapping the upper part of this zone and the lower part of the upper eulittoral. Common chitons include *Nuttallina californica, Lepidochiton hartwegii, Mopalia mucosa*, and the large *Katharina tunicata*. Anemones are also a conspicuous feature of the lower eulittoral. Three species occur, and the smallest, *Anthopleura elegantissima*, often forms extensive sheets in sheltered places. The other two larger species are *A. xanthogrammica* and *A. artemisia*.

A conspicuous feature of the sheltered rocks is a blackish-brown turf of short algae. The species involved are red algae comprising four species of *Gigartina* (*G. agardhii, G. canaliculata, G. leptorhynchus*, and *G. papillosa*), *Rhodoglossum affine, Porphyra perforata*, and *Endocladia muricata*. Larger plants that occur among the turf include species of *Iridaea*. As exposure increases, this turf gives way to a coarser coralline turf composed of *Corallina gracilis, C. chilensis, Bossea dichotoma*, and *Calliarthron setchelliae*. A distinctive feature of all but the more exposed shores in the area where *M. californianus* beds occur is the appearance of the characteristic Pacific coast palmlike laminarian, *Postelsia palmaeformis*.

Sublittoral Fringe: A number of laminarians are characteristic of the sublittoral fringe, especially on steep slopes. Three species, *Egregia menziesii, Alaria marginata*, and *Lessoniopsis littoralis*, form a sequence with *Lessoniopsis* on the most exposed shores. Other fringe species are the laminarians, *Costaria costata*, and *Laminaria setchellii* and *Cystoseira osmundacea* (Sargassacea). Animals found in this zone include abalones (*Haliotis rufescens* and *H. cracherodii*), seastars (*Pisaster ochraceus, Patiria minuata*), and sea urchins (*Strongylocentrotus purpuratus*). The sublittoral is dominated by the laminarians *Nereocystis* and *Macrocystic pyrifera*.

2.2.4 New Zealand

Zonation patterns on New Zealand coasts will be described with reference to the central South Island east coast (Figure 2.11). Descriptions of zonation patterns on New Zealand shores will be found in Batham (1956; 1958), Knox (1953; 1960; 1963; 1969b; 1975), and Morton and Miller (1968).

Littoral Fringe: This is subdivided into two subzones: (1) an upper subzone with dense lichen cover comprising the black *Verrucaria*, yellow-orange species of *Caloplaca* and *Xanthoria*, and white, grey, or grey-green species of *Ramalina, Physicia, Lecanora, Placopsis, Parmelia*, and *Pertusaria*; and (2) a lower or black zone characterized by blue-green algae and the lichen *Verrucaria*. The mosslike red alga *Stichosiphonia arbuscula* often straddle the margin between the littoral fringe and the barnacle zone below. Two species of littorinids, *Littorina unifasciata* and *L. cincta* are codominant. The former species extends down to MLWN while the latter extends down to MTL. Several seasonal alga species can extend into the lower zone, including *Porphyra* spp., *Pylarella littoralis, Prasiola crispa*, and species of *Ectocarpus, Ulothrix, Rhizoclonium, Lola*, and *Enteromorpha*.

Eulittoral Zone: The upper and mid-eulittoral zones are a barnacle-mussel-tubeworm zone. On exposed coasts, the small barnacle *Chamaesipho columna* may dominate the zone and form an almost complete cover. Where the surface is broken the small black mussel, *Xenostrobus pulex*, is scattered throughout, while the serpulid tubeworm, *Pomatoceros cariniferus*, and the blue mussel *Mytilus galloprovincialis* form aggregations in crevices and on ledges. Where surf action is stronger, the larger barnacle *Epopella plicata* joins *C. columna* to form a mixed community. *E. plicata*, however, does not extend as high as *C. columna*, but in some places it may dominate to the almost complete exclusion of the latter species below

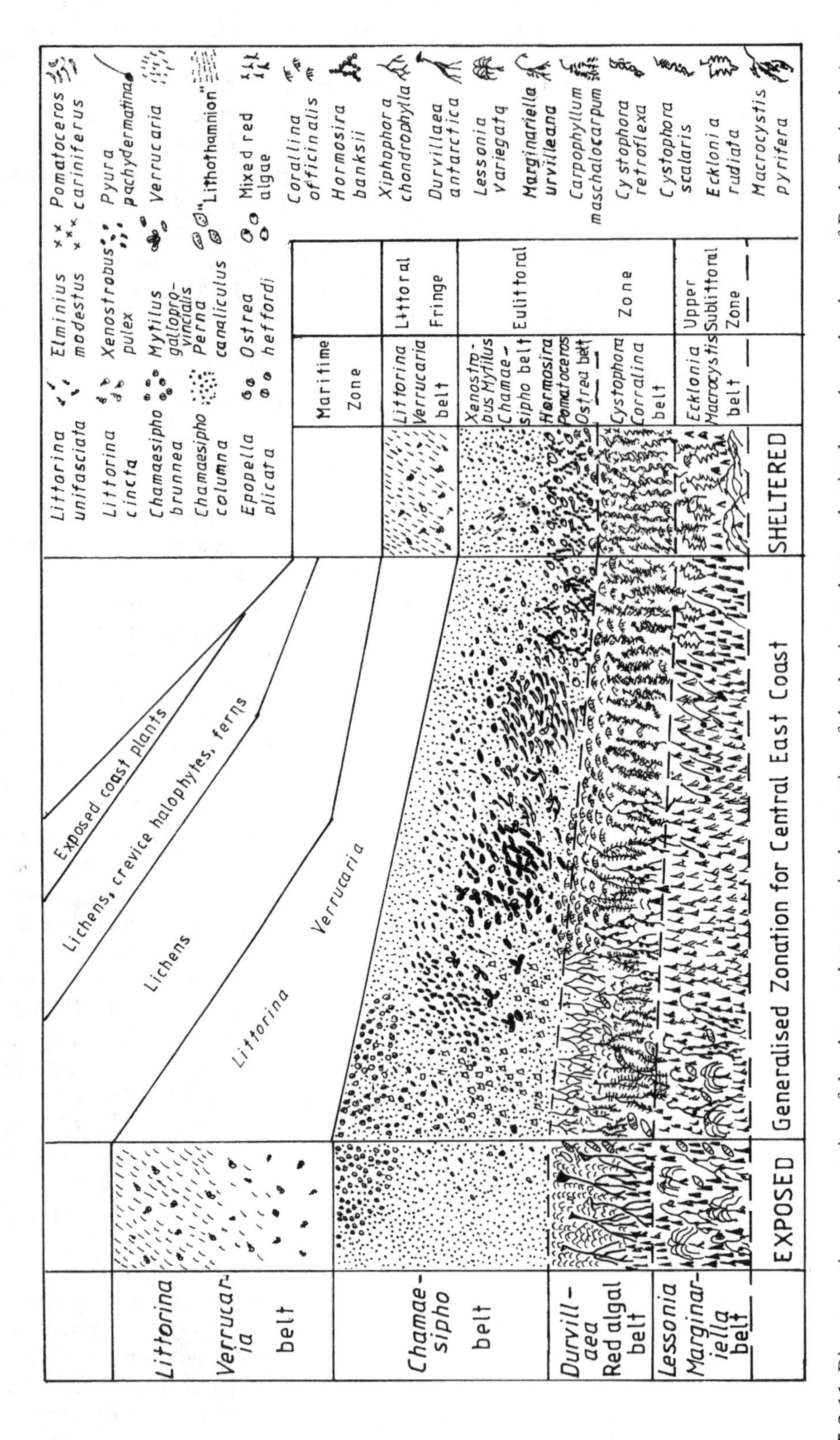

FIGURE 2.11 Diagrammatic representation of the changes that occur in the zonation of the dominant plants and animals on the rocky shores of Banks Peninsula (east coast, South Island, New Zealand) with increase in shelter from wave exposure. (Modified from Knox, G.A., in *The Natural History of Canterbury*, Knox, G.A., Ed., A.H. & A.W. Reed, Wellington, 1969, 551. With permission.)

MLWN. The barnacles may be replaced by a broad band of mussels from EHWN down to MLWN. This "mussel band" may consist only of the blue mussel, but often the ribbed mussel, *Aulacomya ater maoriana*, may be codominant or subordinate, and the green-lipped mussel, *Perna canaliculus*, may penetrate the lower portion of the band from below. With increase in shelter, the mussels may be replaced from ELWN down by a thick encrustation of the serpulid tubeworm, *Pomatoceros cariniferus*. Throughout both the upper and mid-eulittoral, *X. pulex* may form subzones at any level. A number of other barnacle species, *Elminius modestus, Tetraclita purpurascens*, and the stalked barnacle, *Pollicipes spinosus*, may be present, depending on the exposure to wave action.

Two limpets are characteristic of the "barnacle zone," *Cellana ornata* occurring throughout its vertical extent, while *C. radians* is more characteristic of the mid-eulittoral. The chiton, *Chiton pelliserpentis*, is highly characteristic of coasts of all degrees of exposure. Other species that occur throughout are the limpets *Notoacmea parviconoidea, Patelloida corticata*, and *Siphonaria zelandica*. Herbivorous gastropods include *Melagraphia aethiops* (above MLW to ELWS), *Risellopsis varia, Turbo smaragada*, and *Zediloma digna*, while carnivorous whelks include *Lepsiella albomarginata, L. scobina, Neothais scalaris*, and *Lepsithais lacunosa*.

Superimposed on the barnacle are a series of bands of algae, many of which are seasonal in occurrence. The most constant species are browns, *Scytothamnus australis*, replaced by *S. fasciculatus* on more sheltered shores, from MHWN to MTL, and *Splachnidium rugosum* from MTL to MLWN. Of the reds, *Porphyra umbilicalis* may be locally abundant in winter and spring throughout the upper eulittoral. Other seasonal growths in the lower part of the mid-eulittoral are the browns *Ilea fascia, Scytosiphon lomentaria, Myriogloia, Colponemia sinuosa, Leathesia difformis, Adenocystis utricularis*, and the greens, especially species of *Ulva, Enteromorpha*, and *Bryopsis*.

With an increase in local shelter, the barnacle, mussels, and tubeworms are replaced in the lower portion of the mid-eulittoral by a *Corallina-Hormosira banksii* band. This association may carpet a wide area where wavecut platforms occur at this level. Below the *Hormosira*, the development of the *Corallina* turf is rather variable. It ranges from a pink paint formed by the basal portions of *Corallina officinalis* to a mixed turf of *Corallina, Gigartina* spp., *Echinothamnion* spp., *Polysiphonia* spp., *Champia novaezealandiae*, and *Halopteris spicigera*.

Lower Eulittoral: While animals are generally the dominant forms throughout the upper and mid-eulittoral zones, a sharp change takes place at the boundary of the lower eulittoral. Except for the large mussel, *Perna canaliculis*, the catseye topshell, *Turbo smargada*, and the stalked ascidian, *Pyura pachydermatina*, the dominant species are algae. On wave-beaten coasts the salient feature of the lower eulittoral from about MLWN to MLWS is a well-defined band of large bull kelps, *Durvillaea antarctica* and *D. willana*, with the former generally higher on the shore than the latter. In some places *D. antarctica* may be the only species present, while in others the two species intermingle.

Depending on the substrate and degree of wave exposure, the rock surface between the holdfasts may be covered with calcareous "lithothamnion" or with encrusting growths of species of *Lithophyllum, Melobesia, Crodelia*, and other calcareous red algae, or with dense mats of *Corallina, Jania*, or *Amphiura*, or may bear a rich underflora of predominantly red algae.

Large animals extending into the lower eulittoral from the sublittoral fringe are the large chitons *Guildingia obtecta, Diaphoroplax biramosa*, the pauas *Haliotis iris* and *H. auatralis*, the large herbivorous gastropod *Cookia sulcata*, the whelks *Haustrum haustorium* and *Lepsithais lacunosus*, the starfish *Patierella regularis, Calvasterias suteri, Astrostole scabra*, and *Coscinasterias calamaria*, and the sea urchin *Evechinus chloroticus*.

With a decrease in wave action the *Durvillaea* is replaced by a narrow band of the brown alga *Xiphophora chondrophylla* along the upper part of the zone, with *Carpophyllum maschalocarpum* below. On some shores *Xiphophora* may be replaced by a "mixed" or a *Carpophyllum* turf. The mixed turf is composed of *Halopteris* spp., *Glossophora kunthii, Zonaria subarticulata*, dwarfed *Cystophora*, and species of *Polysiphonia, Lophurella*, and *Laurencia*. With further increase in shelter, the brown alga *Cystophora scalaris* joins *Carpophyllum, Xiphophora* disappears, and these two species extend upward to replace it. Two other species of *Cystophora* may be codominant, *C. retrofexa* on the more exposed parts of the range and *C. torulosa* in the more sheltered parts.

Sublittoral Fringe: While the dominant *Durvillaea, Cystophora*, and *Carpophyllum* species of the lower eulittoral may extend varying distances through the sublittoral fringe into the sublittoral, the dominant algae change. The dominant species are the brown algae *Lessonia variegata* and *Marginarella boryana*. Often a band of red algae characterizes the sublittoral fringe below the *Durvillaea* or *Carpophyllum* bands. Many of the species are the same as those found beneath the *Durvillaea* in the lower eulittoral. Other large algae characteristic of the sublittoral zone are *Cystophora platylobium, Sargassum sinclairii, Ecklonia radiata*, and *Macrocystic pyrifera*.

2.2.5 South Africa

The southern African region extends from the northern border of Namibia (17°S) to the southern border of Mozambique (21°S) (Figure 2.12). The overall length of the coastline is about 4,000 km. The eastern coast is influenced by the warm Agulhus Current, while the west-

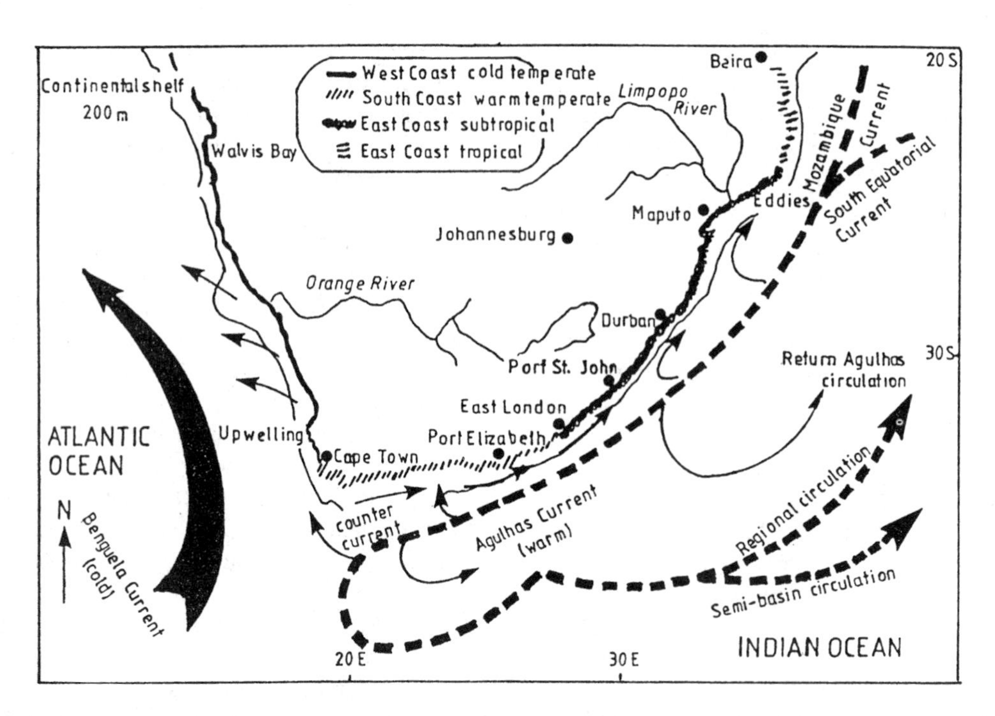

FIGURE 2.12 Major oceanographic features around the coast of southern Africa and associated shore communities. (Adapted from Branch, G.M. and Branch, M.L., *The Living Shores of Southern Africa,* C. Struik, Cape Town, 1984, 14. With permission.)

ern coast is bathed by the cold Benguela Current. Surface water in the Agulhus Current ranges from 21 to 26°C and the salinity is 35.4. Along the west coast there are regions of upwelling of cold water (8 to 14°C). On the east coast, mean monthly sea surface temperatures range from 22°C in the winter to 27°C in the summer, while on the south coast they range from 15 to 22°C, respectively. The entire region is subject to a simple diurnal tidal regime, with a spring-tide amplitude of some 2 to 2.5 m and a neap-tidal range of about 1 m. Three biogeographic provinces can be distinguished: East Coast Subtropical, South Coast Warm Temperate, and West Coast Cold Temperate.

Vertical zonation (Figure 2.13) has been studied in southern African rocky shores since the 1930s when T.A. Stephenson and his team conducted surveys round the coast, summarized in Stephenson (1948) and Stephenson and Stephenson (1972). More recent accounts are given by Brown and Jarman (1978), Branch and Branch (1981), and Field and Griffiths (1991).

Littoral Fringe — East Coast: Littorinid snails (*Littorina kraussii, L. africana,* and *Nodilittorina natalensis*) are the most abundant animals. Algae usually form moss-like patches that include species of *Bostrychia, Rhizoclonium, Gelidium,* and *Herposiphonia.*

Littoral Fringe — South Coast: *Littorina africana* var. *knysnaensis* is incredibly abundant in this zone. Snails that invade the zone from the eulittoral below are *Oxystele variegata* and *Thais dubia* together with the limpet *Helcion pectunculus*. Algae are few, apart from the patches of *Bostrychia mixta* and a variable amount of *Porphyra capensis*.

Littoral Fringe — West Coast: The dominant species is *Littorina africana* var. *knysnaensis*. As on the south coasts it is invaded by species from below, the three species listed for the south coast plus outliers of the limpet, *Patella granularis*. Patches of *Bostrychia mixta* and clumps of *Porphyra capensis* are locally abundant.

Eulittoral Zone — East Coast: The upper eulittoral is populated primarily by barnacles and limpets. The principal barnacle species are *Chthalamus dentatus, Tetraclita serrata,* and *Octomeris angulosa*. The limpets are *Patella concolor, P. granularis,* and *Cellana capensis* and species of *Siphonaria*. Another characteristic species is the oyster, *Saccostrea cucullata*. A very common snail is *Oxystele tabularis*. The algae present are mostly small, primarily *Gelidium reptans* and *Caulacanthus ustulatus*.

The lower eulittoral in many areas is dominated by algae. Typical constituents of the algal turf are *Gelidium reptans* and *Caulacanthus ustulatus* at higher levels and *Gigartina minima, Hypnea arenaria, Centroceros clavulatum,* and *Herposiphonia heringii* at lower levels. Barnacles and serpulid tubeworms, *Pomatoleios kraussii,* compete with the algal turf. With increasing exposure, the barnacles and *Pomatoleios* are replaced by zooanthids (*Zoanthus natalensis*). In very exposed conditions, the alga *H. spicifera* displaces the zooanthids and the brown mussel, *Perna perna*, forms extensive clumps.

Eulittoral Zone — South Coast: The upper eulittoral supports vigorous populations of the barnacles *Chthalamus dentalus, Tetraclita serrata,* and *Octomeris angulosa*. Limpets, especially *Patella granularis,* are common, and in many places it is associated with *Helcion pentunctulus*

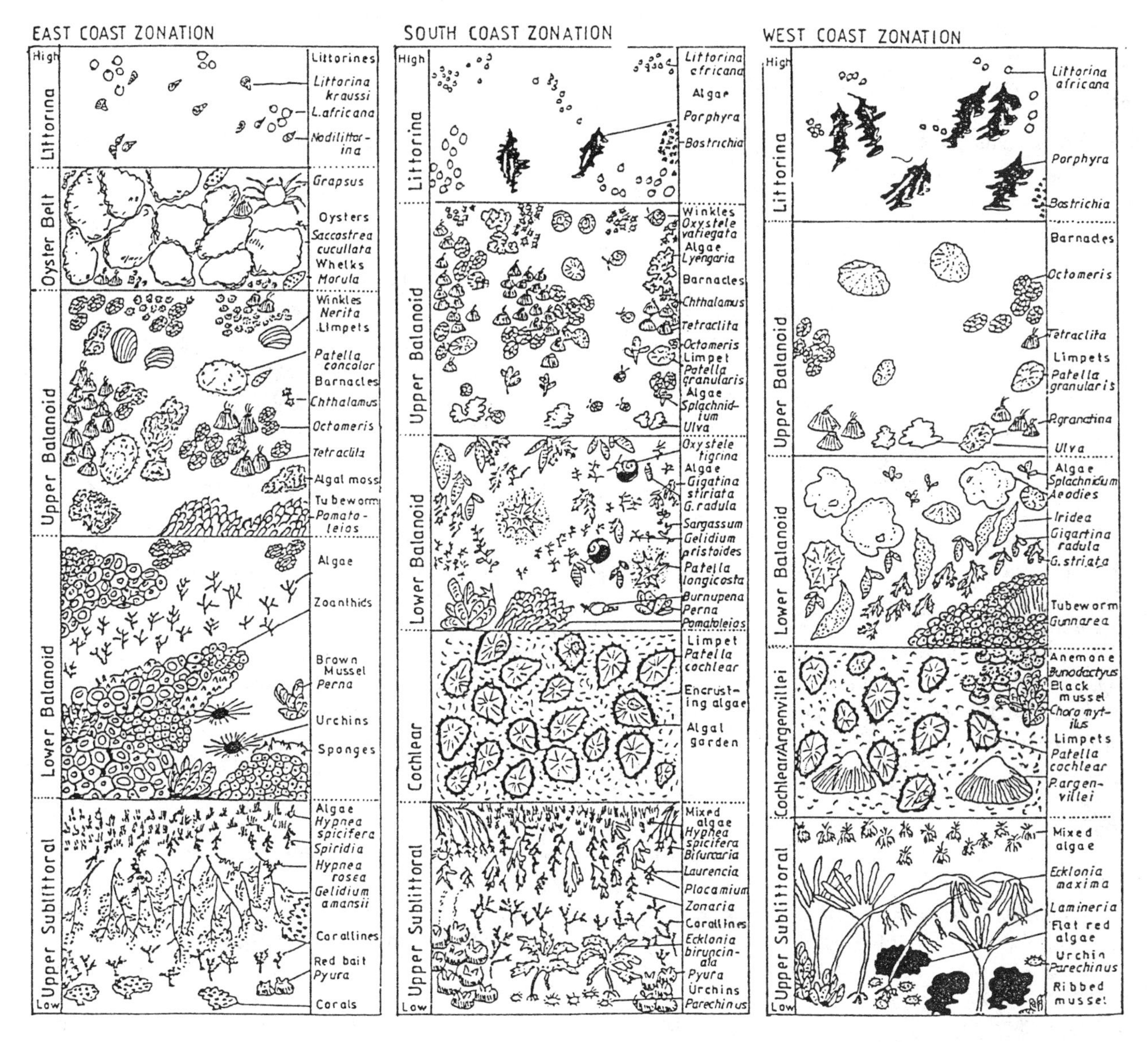

FIGURE 2.13 The principal features of the zonation patterns on the west, south, and east coasts of South Africa. Only a few species characteristic of the zones are shown, and perhaps overlaps in vertical distribution are ignored for the sake of simplicity. (Redrawn from Branch, G.M. and Branch, M.L., *The Living Shores of Southern Africa,* C. Struik, Capetown, 1984, 27, 26, 29. With permission.)

and species of *Siphonaria.* The periwinkle *Oxystele variegata* replaces the east coast *O. tubularis.* Algae are scarce, but there is often *Porphyra, Colpomenia capensis, Splachnidium rugosum, Ulva* sp., and mosslike *Caulacanthus ustulatua* and *Bostrychia mixta.*

The lower eulittoral has extensive growths of the serpulid tubeworm *Pomatoleios kraussii* and the sandy tube of *Gunnarea capensis.* A third polychaete, *Dodecaceria pulchra,* is a common feature of this subzone. In the lower part there are extensive beds of the mussel *Perna perna.* At the bottom of this subzone, a well-developed mosaic of the large limpet *Patella cochlear* and lithothamnion covers the rock surface with ephemeral algae forming typical algal gardens. Where wave action is strong, the *P. cochlear* is replaced by *Perna perna* and algae such as *Gelidium cartiligineum.* With increasing shelter there is a short turf of mosslike algae including corallines with epiphytes and *Hypnea spicifera, Gigartina radula,* and *G. papillosa.* Larger algae include species such as *Sargassum heterophyllum, Caulerpa ligulata, Dictyota dichotoma, Colpomenia capensis, Laurencia flexuosa, L. glomerata,* and *L. natalensis.*

Eulittoral Zone — West Coast: The upper eulittoral has a sparse barnacle cover (*Chthalamus dentatus, Tetraclita serrata,* and *Octomerus angulosa*) and *Patella granularis* is the most conspicuous animal. There is a belt of high-growing *Porphyra.* Below the *Porphyra* are mixed algal growths of *Chatangeum saccatum, C. ornatum,* and *Ulva lactuca*; these algae in turn become mixed with and largely replaced by *Iridopsis capensis, Aeodes orbitosa,* and two brown algae, *Splachnidium rugosum* and *Chordaria capensis*.

In the lower eulittoral, lithothamnion covers the rocks. The most common larger algae in the uppermost part of the lithothamnia are *Aeodes orbitosa, Splachnidium rugosum* and *Chordaria capensis*. The sandy tubes of *Gunnarea capensis* may cover the rocks over extensive areas. As shelter increases, the tubes of this species form a narrow band or ridge between the *Patella cochlear* belt and the main *P. granularis* population. *P. cochlear* is dominant on the southern part of the west coast, but to the north it is replaced by *P. argenvillei*. Four important algae of the *P. cochlear* — *P. argenvillei* belt are *Champia lumbricalis, Plocamium cornutum, Gigartina striata,* and *G. radula*. The dominant mussel of the south coast, *Perna perna*, disappears on the west coast to be replaced by the blue-black *Chloromytilus meridionalis* and the ribbed *Aulacomya ater.*

Sublittoral Fringe — East Coast: The population of the sublittoral fringe varies greatly from place to place. In some regions it is occupied by the ascidian, *Pyura stolonifera*, in dense concentrations; in other places it is replaced by limpets and lithothamnia, but in most places algae dominate. The latter takes the form of a dense sward of small species of varying composition; in places it is dominated by *Hypnea rosea* and *Rhodymenia natalensis*, and elsewhere by *Gelidium rigidum* and *Galaxaura natalensis*. Larger species such as the brown algae *Sargassum longifolium, Ecklonia radiata,* and *Dictyopteris dichotoma,* and the green alga *Caulerpa ligulata* may be locally important.

Sublittoral Fringe — South Coast: In most places the sharply defined lower limit of *Patella cochlear* marks the beginning of the sublittoral fringe. Here the ascidian *Pyura stolonifera* usually forms a continuous cover. The associated fauna is varied — anemones, compound ascidians, and *Alcyonium falax*. Where ascidians do not dominate, algae, especially the larger corallines, dominate. Associated species are *Gelidium cartilagineum, G. rigidum, Caulerpa ligulata, Plocamium corallorhiza,* and *Hypnea spicifera*. Among the larger species are *Sargassum heterophyllum, S. longifolium,* and *Zonaria interrupta*.

Sublittoral Fringe — West Coast: Here the sublittoral region proper is occupied by giant laminarians, *Laminaria schinzli, L. pallida,* and *Macrocystis pyrifera*. While the lower limit of the *P. cochlear* — *P. argenvillea* belt marks the top of the sublittoral fringe, the latter does not form a distinct zone. The upper fringe of kelp beds is inhabited by species characteristic of the lower part of the *P. cochlear* — *P. argenvillea* belt (*P. argenvillea, Bunodactis reynaudi, Gunnerea capensis, Champia lumbricalis, Plocamium cornutum,* species of *Gigartina*, corallines, and lithothamnion). The kelp bed community structure has been described by Field et al. (1980). It includes a mosaic patchwork of understory algae, mussels (*Aulacomya*), sea urchins (*Parechinus*), holothurians, and the spiny lobster, *Jasus lalandii*.

2.3 CAUSES OF ZONATION

2.3.1 Wave Action and Zonation

2.3.1.1 Introduction

Wave action is the most important factor that causes variation in the patterns of distribution of organisms on the shore, modifying the height of a particular zone and determining the kinds of species present. Wave-exposed shores are characterized by large water forces due to the action of breaking and surging waves, little or no sediment settling on the shore (although in some situations sand scouring), and thorough mixing of the inshore waters, resulting in variations in temperature, salinity, and nutrients. Sheltered habitats, in contrast, are characterized by little hydrodynamic stress, siltation, and inshore water stratification, causing marked daily or seasonal changes in temperature salinity and nutrient concentrations. The effects of wave action on a small scale are shown in the differences in zonation patterns between the landward and seaward sides of rocks and large boulders. On a larger scale these differences can be seen in progression from exposed headlands into sheltered bays and harbors. This can be seen in Figure 2.14, which illustrates the distribution of seaweeds in relation to wave exposure in North Wales. Here seaweeds thrive best on sheltered or semi-sheltered shores where luxuriant stands of *Fucus* spp., and *Ascophyllum nodosum* thrive with individual plants often of large size, reaching, in the case of *Ascophyllum*, a length of several meters. With increasing wave exposure, fucoid algae become progressively sparser and the plants stunted. As exposure increases, the fucoids are usually replaced by red algae, *Porphyra* and *Mastocarpus*, and on the lower shore the laminarians *Laminaria saccharina* and *L. digitata* are replaced by the kelp *Alaria esculenta*.

2.3.1.2 The Problem of Defining Wave Exposure

How to define wave exposure? It has proved to be difficult to precisely define the degree of wave exposure that any particular shore experiences, and it is usually taken to be an integrated index of the severity of the hydrodynamic environment to which the plants and animals are exposed. Thus it has tended to be defined on the basis of the type of community of the plants and animals on the shore and the presence or absence of so-called indicator species.

Denny (1995) has recently developed a method for predicting physical disturbance of wave action. The steps involved in this method are depicted in Figure 2.15. In his paper Denny (1995) outlines the theoretical basis for calculating the various steps depicted in the figure. Step 5 is

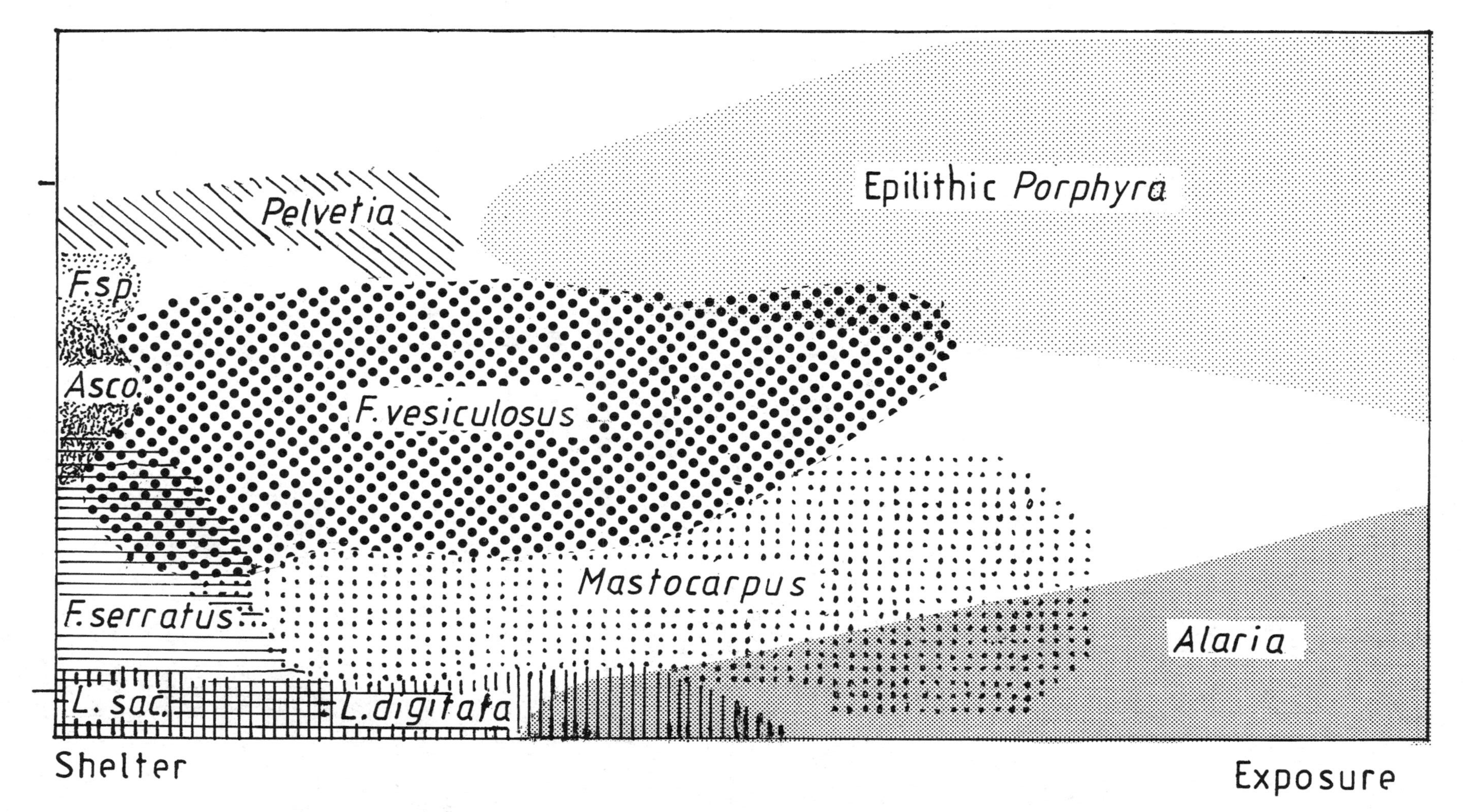

FIGURE 2.14 The distribution of littoral seaweeds in relation to wave exposure in North Wales. *F.sp.* = *Fucus spiralis; Asco* = *Ascophyllum nodosum*; *L. sac.* = *Laminaria saccharina*; EHWS = Extreme High Water of Spring Tides; ELWS = Extreme Low Water of Spring Tides; *Mastocarpus* = *Gigartina stellata*. (From Norton, T.A., in *The Ecology of Rocky Coasts,* Moore, P.G. and Seed, R., Eds., Hodder & Straughton, London, 1985, 8. After Jones and Demetropoulos, 1968. With permission.)

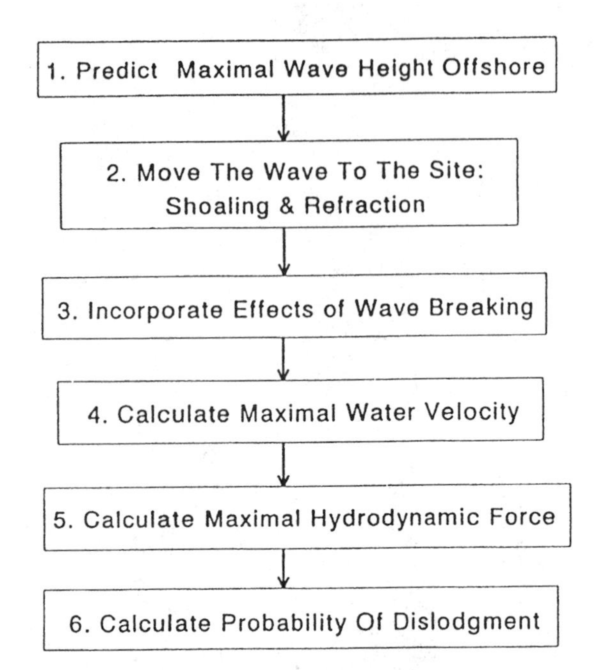

FIGURE 2.15 A flow chart of the steps involved in calculating the probability of dislodgement for an individual plant or animal at a given site on a rocky shore. (Redrawn from Denny, M.W., *Ecol. Monogr.*, 65, 374, 1995. With permission.)

an important one in which three forces are involved: drag, acceleration force, and lift. Drag is the force that tends to push objects in the direction of the flow. This force increases with the square of the water velocity relative to an organism, and is proportional to an organism's projected area. The second force, the acceleration force, acts along the direction of the flow (Denny et al., 1985; Denny, 1988; 1989; 1993; Gaylord et al., 1994). It scales linearly with the water's acceleration and is proportional to the volume of the organism. The third force, lift, acts perpendicular to the direction of the flow. Denny and his coworkers (Denny, 1988; 1989; 1991; 1993; Denny and Gaines, 1990) have developed methologies for estimating these three forces and thus determining the forces required to dislodge plants and animals of different sizes.

The powerful forces discussed above scale with size and consequently set mechanical limits as to the size of organisms in wave-swept environments. Water motion along wave-swept rocky shores produce some of the most powerful hydrodynamic forces on earth, and since such forces scale with size, they may exert selective pressures for small size. The first theoretical and quantitative attempt to explore the possibility that wave forces could set mechanical limits on size was undertaken by Denny et al. (1985). They hypothesized that hydrodynamic forces acting on organisms along wave-swept shores tend to increase with increasing body size, faster than the ability of the organism to maintain its attachment to the rock surface. Hydrodynamic forces depend on an organism's area and volume, as well as on the velocity and acceleration of the fluid past the organism.

As originally noted by Denny et al. (1985), attachment strength tends to scale with area; thus at large size, isometrically growing organisms (whose volumes increase faster than their area) will feel increasingly large acceleration forces relative to their attachment strengths. This means that acceleration forces (acting in conjunction with drag) have the potential to set upper limits on size in wave-exposed organisms. Blanchette (1997) tested this prediction in the field by reciprocally transplanting individuals of the brown alga *Fucus gardneri* between wave-exposed and wave-protected sites. Mean sizes of wave-exposed plants transplanted to protected sites increased significantly relative to exposed control transplants. Mean sizes of wave-protected plants transplanted to exposed sites decreased significantly in size relative to protected control transplants.

Denny (1995) tested his predictive model by predicting the rate at which patches of bare substratum are formed in the beds of the mussel *Mytilus californianus*, a dominant competitor for space on the rocky shore of the Pacific Northwest. Predicted rates were very similar to those measured in the field. Thus the model has the potential to provide useful input into models of intertidal patch dynamics. An analysis of data from several sites round the world suggested that the yearly average "waviness" of the ocean at any particular site can (over the course of decades) vary by as much as 80% of the long-term mean. Denny predicted that an increase of 1 m in yearly average significant wave height would result in a fourfold increase in the rate of patch formation in a mussel bed.

2.3.1.3 The General Effects of Wave Action

The general effects of wave action can be summarized as follows:

1. A general shift of the concentration center of most of the species of the eulittoral and the littoral fringe.
2. An expansion of the vertical range of each species.
3. A relative lowering of the concentration centers of some species in the upper sublittoral and the lower eulittoral.
4. The disappearance of a many species that are intolerant of wave action.
5. The appearance of a few species that appear to tolerate or require wave action.
6. A marked increase in the filter-feeding biomass on wave-exposed shores.
7. A higher overall biomass on the more exposed shores.

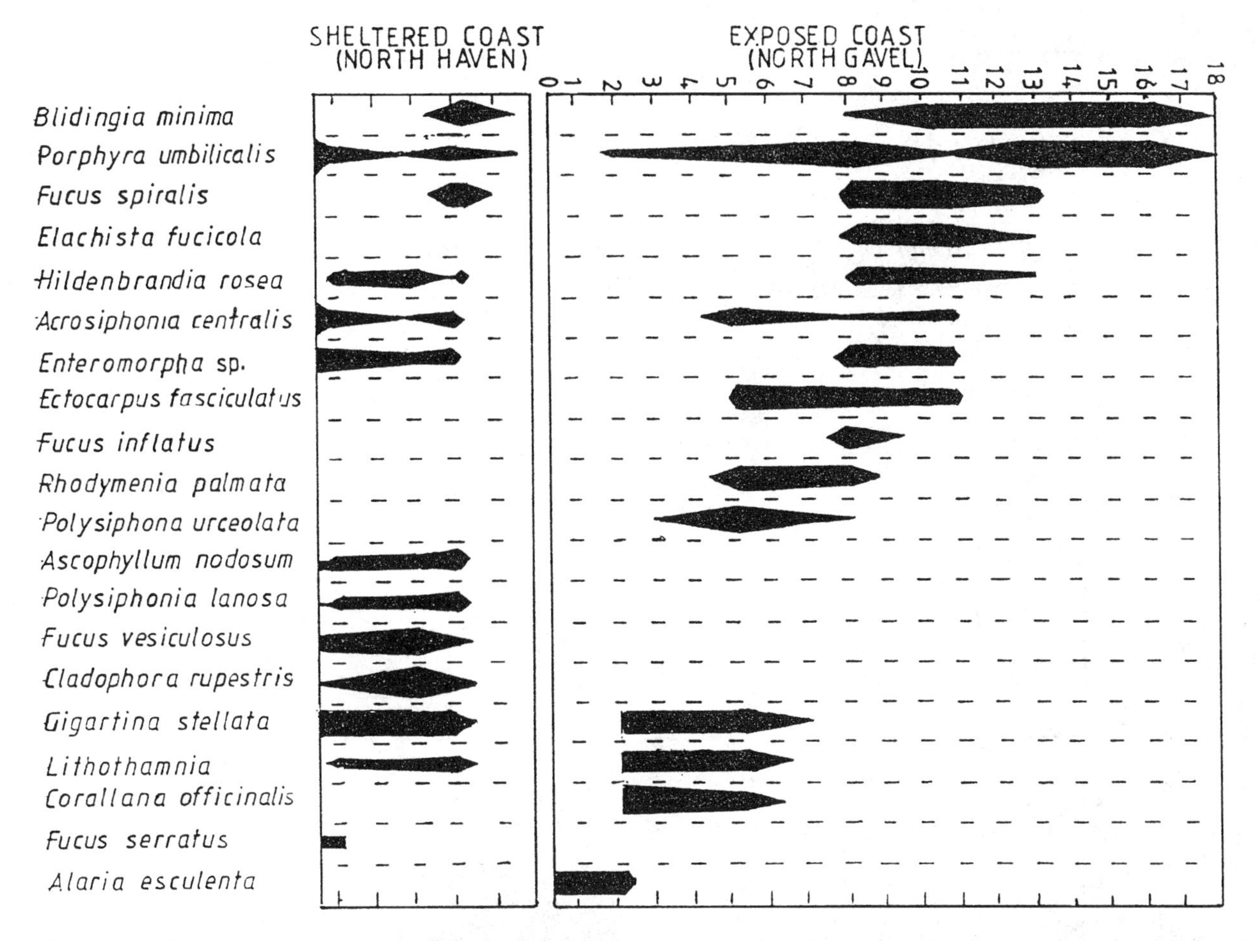

FIGURE 2.16 Diagram showing the vertical distribution (feet above chartum datum) of intertidal algae on an exposed coast (North Gravel) and sheltered coast (North Haven) on Fair Isle, Scotland. (Modified from Burrows, E.M., Conway, E., Lodge, S.M., and Powell, H.T., *J. Ecol.*, 42, 286, 1954. With permission.)

8. A change in trophic structure. On sheltered shores, attached algae are the primary producers at the base of the food web, whereas on exposed shores, water column primary production is at the base of the food web in shore communities in which the standing crop of consumers is higher than that of the primary producers.

Selected examples illustrating the above effects follow. Burrows et al. (1954) compared the distribution of a number of intertidal algal species on Fair Isle, Scotland. They found that not only were the species different in exposed and sheltered areas, but that the corresponding algal zones were displaced upward by as much as 12 ft (3.05 m) on shores exposed to strong wave action. From Figure 2.16 it can be seen that *Ectocarpus fasciculatus, Fucus inflatus* f. *disticles, Rhodomenia palmata, Polysiphonia urceolata, Corallina officinalis,* and *Alalia esculenta* were all absent on the sheltered coast of North Haven. On the other hand, species such as *Ascophyllum nodosum, Polysipohonia vesiculosus,* and *Cladophora rupestris* were present on sheltered shores at North Haven, but absent on the exposed coast at North Gravel. It can also be seen that the vertical zones of those species that occurred at both localities were considerably elevated on the exposed coast, e.g., *Porphyra umbilicalis* had a vertical range of 1.5 ft on the sheltered coast compared with 16 ft on the exposed coast.

Figure 2.17 depicts the distribution of two periwinkle species *Littorina unifasciata* and *Littorina cincta* with reference to exposure and shelter at three New Zealand localities from north to south. *L. unifasciata* is rare or absent on the northern coasts, but increases in density to the south. The reverse trend is evident for *L. cincta*. The vertical distribution of both species increases with wave exposure. The density of *L. unifasciata* tends to increase with wave exposure, whereas that of *L. cincta* is maintained especially at the southernmost locality.

Ohgaki (1989) investigated the daily vertical movement if the littoral fringe periwinkle *Nodolittorina exigua* in relation to wave height on the Japanese coast. The position of the snails on a cliff shore were high when the wave-reach was high and ascended with increasing height of the wave-reach. The snails moved a long distance

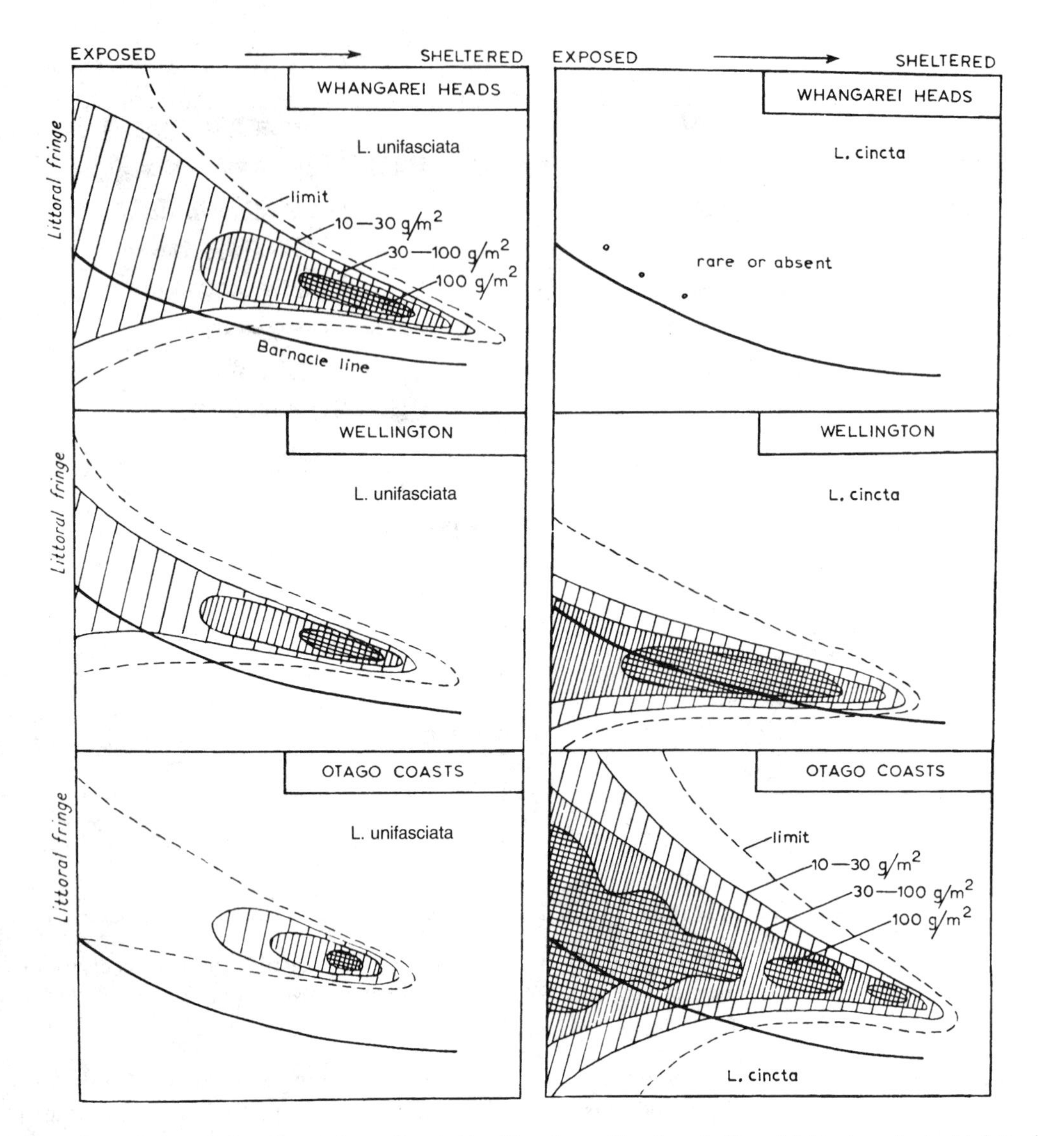

FIGURE 2.17 Distribution and abundance of two littorinids, *Littorins unifasciata* (left) and *Littorina cincta* (right) with reference to exposure and shelter at three New Zealand localities from North to South. Density expressed as grams per m^{-2}. (From Morton, J. and Miller, M., *The New Zealand Seashore*, Collins, Auckland, 1968, 350. Courtesy of W.J. Ballantyne.)

upward in the late summer, when typhoon swells occur, and moved gradually downward again in the autumn, parallel to decreasing wave-reach.

McQuaid and Branch (1985) have investigated the trophic structure of rocky intertidal communities in the Cape of Good Hope, South Africa, in relation to wave action, and discussed the implications for energy flow through the communities. Figures 2.18A and 2.18B compare the vertical distribution of the total and trophic compartment biomass on sheltered and exposed shores. Exposure influenced both the vertical distribution and the trophic composition of the total biomass. Total biomass showed a simple decrease upshore on sheltered shores, but the pattern was more complex with greater exposure. Filter feeders, carnivores, and omnivores all exhibited significantly higher biomass under exposed conditions. Grazing by the high densities of the limpet *Patella cochlear* (up to 1000 m^{-2}; Branch 1975b) in the cochlear zone on exposed shores resulted in a dramatic decrease in algal cover in this zone. Filter-feeding biomass in the sublittoral fringe on exposed shores was high and there was a decrease in algal biomass relative to that on sheltered shores. On exposed shores the filter-feeding biomass was high in the upper balanoid zone. Among the minor trophic components, trends of vertical zonation were less obvious, but the biomass of carnivores did correlate positively with that of filter feeders and was greatest in the sublittoral fringe. The essential differences between the two shore types is the addition of a very large filter-feeding component on exposed shores. Filter-feeding biomass is generally low on sheltered shores; on exposed shores it is very much higher, up to 6,533 g m^{-2} shell-free dry weight.

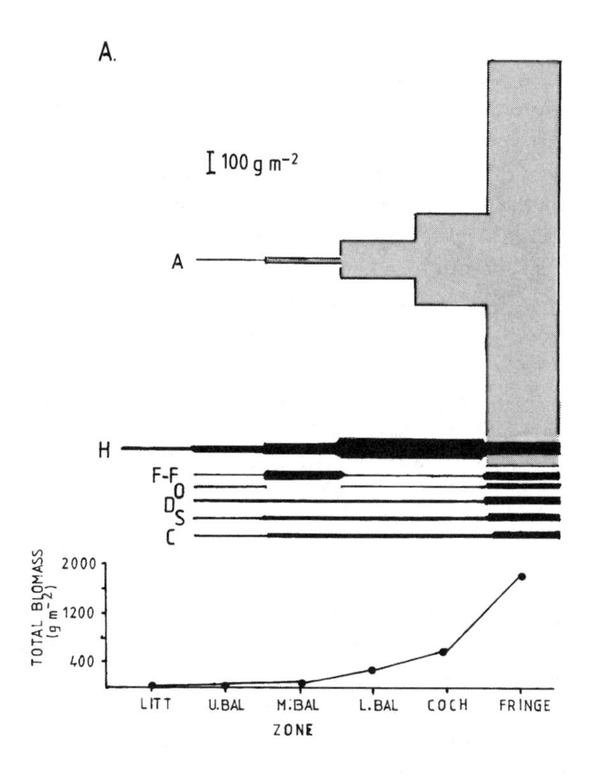

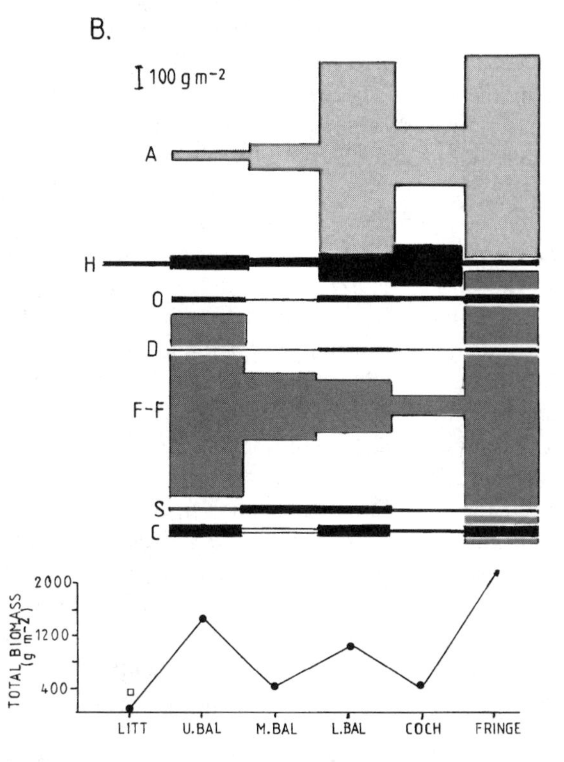

FIGURE 2.18 Vertical distribution of total and trophic compartment biomass on: *A*, sheltered shores; *B*, exposed shores. A: algae; H: herbivores; F.F: filter-feeders; O: omnivores; D: detritovores; S: scavengers; C: carnivores. (Redrawn for McQuaid, C.D. and Branch, G.M., *Mar. Ecol. Progr. Ser.*, 22, 158, 1985. With permission.)

Thus the balance between consumers and primary producers is considerably different on the two shore types, implying alterations in the net balance between import and export of production between these two communities and the inshore marine system. The high filter-feeding biomass on the exposed shores results from the importation of primary production from the water column to the shore community in which the standing crop of consumers is considerably higher than that of the primary producers.

2.3.2 Tidal Currents and Zonation

Swift tidal currents are developed where there are narrow inlets to lochs, fjords, and enclosed embayments. Where such inlets are lined with rocky shores, a distinctive flora and fauna and vertical zonation is found. Such tidal rapids provide conditions intermediate between sheltered and exposed coasts. Swift water currents maintain an ample supply of plankton for filter feeders and intertidal animals such as hydroids and anemones that feed on small plankters, a plentiful supply of nutrients for plant growth, prevent deposition of silt, and provide protection from wave action.

Many of the algae growing in the tidal rapids have morphological adaptations to withstand strong currents. For example, *Macrocystis integrifolia* blades from tidal rapids are intermediate in size and shape between sheltered and exposed plants (Druehl, 1978). Plants in tidal rapids frequently grow to immense size, perhaps due to the ample nutrient supply.

Some of the most thoroughly studied tidal rapids in the world are those at the entrance to Lough Ine, County Cork, Ireland (Kitching, 1987). Although these rapids are somewhat atypical in having a large population of the brown alga *Saccorhiza polygchides* (Lewis, 1964), other features are typical of loch and fjord channels of western Ireland and Scotland. As water velocity increases from the inside of the loch toward the channel, calm water plants such as *Halidrys siliquosa* and *Laminaria saccharina* gradually give way to plants characteristic of moderately exposed shores, such as *Himanthalia elongata*. In the fastest currents, such as over the sill in the middle of Lough Ine Rapids, where current velocity reaches 2.6 m s^{-1}, the water becomes turbulent and exposed coast plants such as *Laminaria digitata* and *L. hyperborea* generally appear. In some rapids *Halidrys* may persist into rapid currents, flourishing side by side with *L. digitata,* an unusual combination of sheltered and exposed coast plants (Lewis, 1964).

The fauna of tidal rapids is invariably dominated by five groups of animals: sponges, hydroids, anemones, polyzoans, and ascidians. The distribution of these species shows a similar trend to that discussed above for the algae. Thus there is a tendency for the more robust, less easily damaged types of colonies to appear when the current becomes strongest.

2.3.3 Substrate, Topography, Aspect, and Zonation

The nature of the substratum can influence the kinds of plants and animals that may be present on hard shores. Rock surfaces may be smooth and polished or pitted and rugose. This surface texture influences the settling of the larval stages of many species (see Section 5.4.3). Moore and Kitching (1939) have shown that minor variations in the abundance of algae, and the barnacles *Balanus balanoides* and *Chthalamus stellatus* on rocky shores are often associated with variations in the roughness of the rock surface. The hardness of the rock also determines whether rock-boring species such as bivalve molluscs of the family Pholaridae are present.

Wave action, topography, and aspect need to be considered together, as the two latter features may modify the effects of the first. The effects of topography are very complicated, arising from the great variety of rocky shores ranging from boulder-strewn shores to wide rock platforms and steep cliffs. On broken shores, elevation of vertical zone with strong wave action may be evident on the seaward side of rock masses, while on the sheltered side, zonation patterns and species characteristic of sheltered shores may be found. The angle of slope is important in modifying the zonation patterns and species composition on shores of comparable wave exposure. Under conditions of shelter from wave action, zonation patterns are determined primarily by emersion/submersion factors. A shore with a gentle slope will have a wide littoral zone where poor drainage and extensive tidal pools permit an upward extension of sublittoral fringe species. Conversely, where the shore is steep, such as on cliff faces, jetties, and piers, the whole of the littoral zone is condensed into a narrow band corresponding to tidal rise and fall.

Under exposed conditions, the angle of the slope plays an important role in modifying the effect of wave action. Where the slope is gentle there may be little uplift of the higher intertidal zones. Gently sloping surfaces generally remain damper than vertical ones, and this can influence the vertical distribution of many species. Local topography also affects the presence of many species that find suitable conditions in depressions, drainage channels, tide pools, and crevices. Here conditions of humidity and shade may enable them to penetrate higher on the shore than they can on open surfaces.

The aspect of a shore, or a particular region of a shore, is important in determining the upper limits of many intertidal species. Broken and gullied shores provide many examples of the upward extension of both plants and animals on shaded surfaces. Such surfaces, to some extent, offset the rigors of desiccation and thermal stress and, for example, allow an upward extension of sublittoral organisms. The example discussed earlier of zonation patterns on Brandon Island (see Section 1.3.5.2) is a good example of the modifying effect of aspect and angle of slope on tidal-dependent zonation patterns. Figure 2.19 after Batham (1958) shows the vertical zonation of a number of species at Portobello, Otago Harbour, New Zealand, on sun-facing and shaded surfaces. Various species such as the red algae *Stichosiphonia arbuscula* and littorinid snails have elevated ranges and greater densities on shaded surfaces; encrusting and tufted coralline algae up to well above MTL. Aspect also affects the abundance of many eulittoral species. On the one hand, the black lichen *Lichina pygmaea,* which is dense on sun-facing rocks, is almost absent on shaded ones; while on the other the tunicate *Pyura suteri* is practically confined to shaded sites. In general, aspect affects vertical zoning more on the upper parts of the shore where desiccation is more pronounced.

2.3.4 Sand and Zonation

Most rocky shores include considerable sand intermixed with the biota attached to rock substrates, and fluctuations in the degree of sand deposition and coverage are common (Littler, 1980a; Littler and Littler, 1981; Littler et al., 1983; 1991). Devinny and Volse (1978) postulated the following three mechanisms of sediment damage to attached algae: (1) smothering due to reduced light, nutrients, or dissolved gases; (2) physical injury due to scouring; and (3) detrimental changes of the surrounding interstitial microenvironment. These three mechanisms also apply to sessile and motile animal species. Taylor and Littler (1982) distinguish different effects due to sand impacts, stress (smothering), and disturbance (scouring), with the greater effect due to the former. In addition, opportunities for feeding, both for filter feeders, grazers, and predators, are reduced with sand cover. Sand has been reported to physically scour the underlying substratum, thus making bare space available for colonization when the substrate reemerges from the sand cover (Climberg et al., 1973).

Littler et al. (1983) studied over a 3-year period the impact of variable sand deposition on a Southern California rocky intertidal system, ranging from about zero to total inundation over different portions of the study area. An apparent subclimax association of delicate high-producing macrophytes (*Chaetomorpha linum, Cladophora columbina, Ulva lobata,* and *Enteromorpha intestinalis*) and highly productive macroinvertebrates (*Tetraclita rubescens, Chthalamus fissus, C. dalli, Phragmatopoma californica*) that corresponds to opportunistic strategists (*sensu* Grime, 1977) dominated the low-lying intertidal areas routinely buried by sand and exhibited zonational patterns reflecting both tidal height and degree of sand coverage. A number of characteristics (Odum, 1971) distinguish those species subjected to recurrent mortalities due to sand stress including: (1) high productivity, (2) low biomass, (3) opportunistic life histories, and (4) emphasis on the herbivore trophic level. For example, high produc-

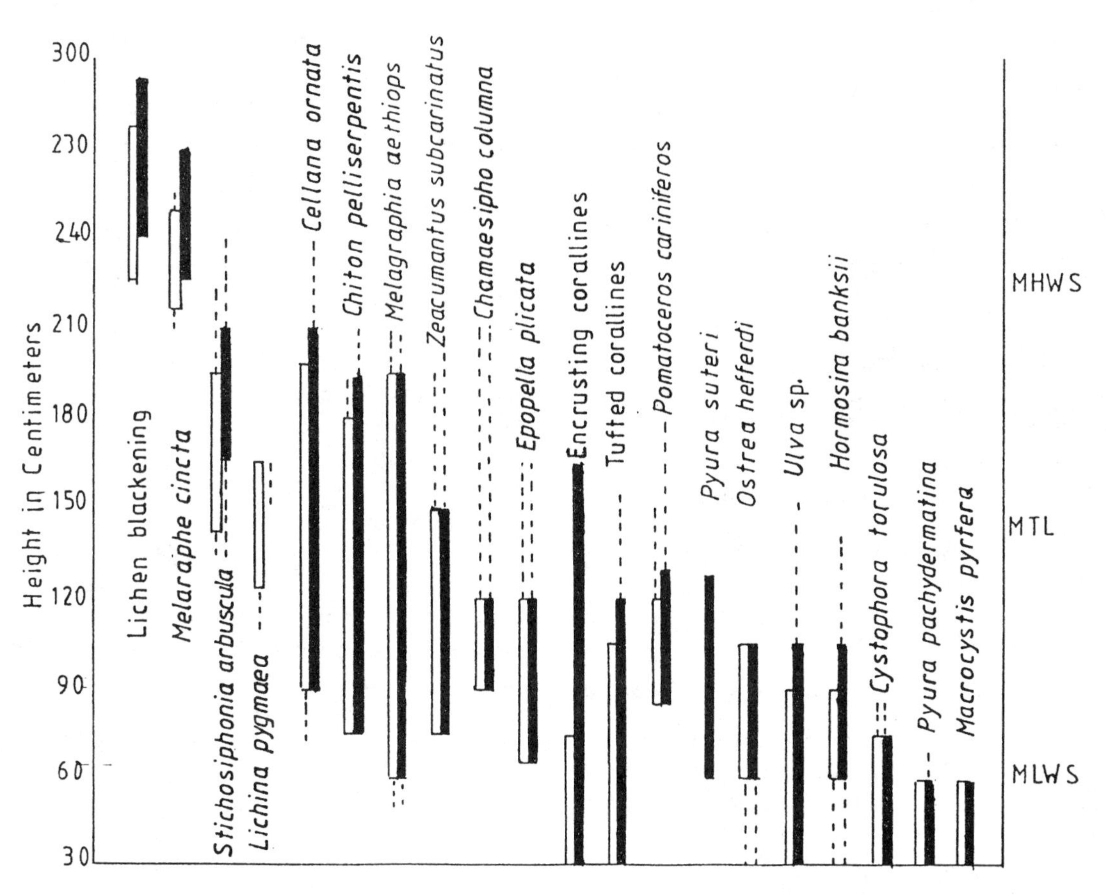

FIGURE 2.19 Vertical zonation of selected species at Portobello, South Island, New Zealand, on shaded (■) and sun-facing surfaces (□). Redrawn from Batham, E.J., *Trans. R. Soc. N.Z.*, 85, 459, 1956. With permission.)

tivity has been reported for the green algae *Ulva lobata, Enteromprpha intestinalis, Cladophora columbina,* and *Chaetomorpha lineum* (Littler, 1980b; Littler and Arnold, 1982), and they are all of low biomass. Opportunistic reproductive strategies have been suggested for *Enteromorpha* sp. (Fahey, 1953) and *Ulva* sp. (Littler and Murray, 1975). It is well documented that these two species are rapid colonizers (Littler and Murray, 1975; Sousa, 1979b; Littler, 1980b).

Refuge habitats on rock pinnacles (sand free) were dominated by long-lived molluscs such as *Mytilus californianus, Haliotis cracherodii,* and *Lottia gigantea.* The lower limits of these biologically competent taxa (*sensu* Vermeij, 1978) appear to be determined by the physical smothering action of the sand. The stress-tolerant anemone *Anthropleura elegantissima* dominated the upper intertidal macroinvertebrate cover because of reproductive, behavioral, and physiological adaptations to the stresses of aerial exposure and sand burial. The dominant plant in the lower intertidal pools was the biotically competent surf grass *Phyllospadix scouleri,* because of its large size and rhizomatous root system, which traps and binds sediments. The most numerous of the mobile macroinvertebrates, the snail *Tegula funebralis,* is able to migrate away from the winter sand inundation to refuge habitats.

Sand inundation thus resulted in subclimax and mature intertidal communities being intermixed in a mosaic-like pattern, and this augmented the within-habitat diversity, contrary to the belief that periodic inundation by sand would reduce species diversity by eliminating organisms intolerant of sand scour and sand smothering (e.g., Daly and Mathieson, 1977; Littler and Littler, 1980). Levin and Paine (1974) predicted, and others (Sousa, 1979a; Littler and Littler, 1981; McQuaid and Dower, 1990) found that disturbances such as sand scour and inundation, when localized, may induce diversity as a result of mixed patches undergoing different stages of succession. McQuaid and Dower (1990) recently studied faunal richness on 10 regularly sand-inundated shores on the Cape region of South Africa. They confirmed that inundation promoted richness by increasing habitat heterogeneity. Table 2.2 compares total faunal species richness for sandy beaches, rocky shores, and sand-inundated rocky shores. It shows that the number of "rocky shore" species recorded was remarkably similar to results for noninundated shores in the Western Cape of South Africa (McQuaid, 1980). In addition, there is a component of psammophilic or "sandy shore" species found in the sand deposits themselves. Thus, inundation of these shores clearly caused enrichment, rather than impoverishment, of the biota.

TABLE 2.2
Total Faunal Species Richness for Sandy Beaches, Rocky Shores, and Sand-Inundated Shores of South Africa

	Rocky Shore Species	Psammophilic Species	Total
Sandy beaches (East Cape)	0	21	21
Rocky shores (West Cape)	254	0	254
Inundated rocky shores (East Cape)	255	26	281

Note: Data for sandy beaches (East Cape) from McLachlan (1977a,b), McLachlan et al. (1981), and Woolridge et al. (1981). Data for rocky shores (West Cape) from McQuaid (1980). Data for sand-inundated shores from McQuaid & Dower (1990).

Source: From McQuaid, C.D. and Dower, K.M., *Oecologia*, 143, 1990. With permission.

Several macroalgal species have adapted to resist sand scour and even months of burial. These species include *Gymnogongrus linearis, Laminaria sinclairii, Phaeostrophion irregulare,* and *Ahnfeltia* spp. from the west coast of North America and *Sphacelaria radicans* on the east coast of North America (Daly and Mathieson, 1977). Characteristics of these algae (Lobban et al., 1985) include tough, usually cylindrical thalli with thick cell walls; great ability to regenerate, or an asexual reproductive cycle functionally equivalent to regeneration (Norton, 1985); reproduction timed to occur when the plants are uncovered; and physiological adaptations to withstand nutrient deprivation, anaerobic conditions, and H_2S.

2.3.5 Climatic Factors and Zonation

2.3.5.1 Solar Radiation

Light is a key factor affecting both plants and animals on the shore, but it is also very complex. The ebb and flow of the tides has a profound effect on the quantity and quality of the light reaching the photosynthetic plants on the shore. The primary importance of light to the plants is in providing the initial energy for photosynthesis, and ultimately for all biological processes. It is also the signal for many events throughout the life cycles of the algae, including reproduction, growth, and distribution. Light also influences the behavior and activity of most animal species. Many more animals are active on the shore during nighttime low tides than in the daytime. The barnacle *Semibalanus balanoides* is often larger when growing in the shade than in direct sunlight. Wethey (1985) found that this barnacle could survive higher on the shore when shielded from direct sunlight, although this may be a response to reducing the impact of the infrared end of the spectrum. In contrast, calcification, and hence growth in corals takes place more rapidly during daylight (Goreau and Goreau, 1959). Reef-building corals have symbiotic algae in their tissues that carry out photosynthesis during daylight hours.

Sunlight influences the behavior of marine animals in many ways apart from inhibition of activity. The direction of light is used by the larvae of many intertidal animals as a cue for orientation. The transition from daylight to darkness activates diurnal rhythms of activity and many physiological processes, and day length (photoperiod) may determine the onset of breeding, or the timing of events such as spawning.

Radiant energy from the sun's rays encompasses the electromagnetic spectrum from long-wave, low-energy to short-wave, high-energy rays. "Light" refers to the narrow region of the spectrum visible to the human eye, plus the ultraviolet and infrared wavelengths. One of the most important variables controlling plant photosynthesis is the "photosynthetically active radiation," PAR, or light in the range of wavelengths from 400 to 700 nm. There is, however, some evidence that photosynthetic absorbance extends down to 300 nm in the green alga *Ulva lactuca* and the tetrasporangial stage (*Tralliella intricata*) of the red alga *Bonnemaisonia hamifera* (Halldal, 1964).

Light hitting the surface of the water is reduced by two processes, refraction and absorption. The percentage of reflected light depends on the angle of the sun to the water and also on the state, or roughness of the water. As solar energy penetrates the water it is attenuated in both quantity and quality. The attenuation results from absorption and scattering by dissolved and suspended substances in the water. Water itself absorbs maximally in the infrared and far red above 700 nm. Other wavelengths are screened out as the light passes through the water. The quality of light is important in determining the distribution of seaweeds with a number of pigments that absorb various portions of the visible wavelengths for photosynthesis and reflect others (Figure 2.20). The reflected wavelengths impart the distinctive colors to the plants.

Green seaweeds use mainly *chlorophyll* pigments, which absorb light in the red and blue portions of the spectrum. Because they rely heavily on red light for photosynthesis, they are found in shallow habitats. Red algae also have chlorophyll, but this pigment is masked by *phycoerythrin* and *phycocyanin* pigments, which absorb in the green and orange portions of the spectrum. Because they use most of the visible spectrum for photosynthesis, and can use light from the middle, or green, portion of the spectrum more effectively than light from the blue or red regions, they can live at all depths but generally prefer the low intertidal or upper sublittoral. Brown algae have both chlorophyll and *fucxanthin* pigments, the latter absorbing in the blue-green wavelengths. Brown algae are

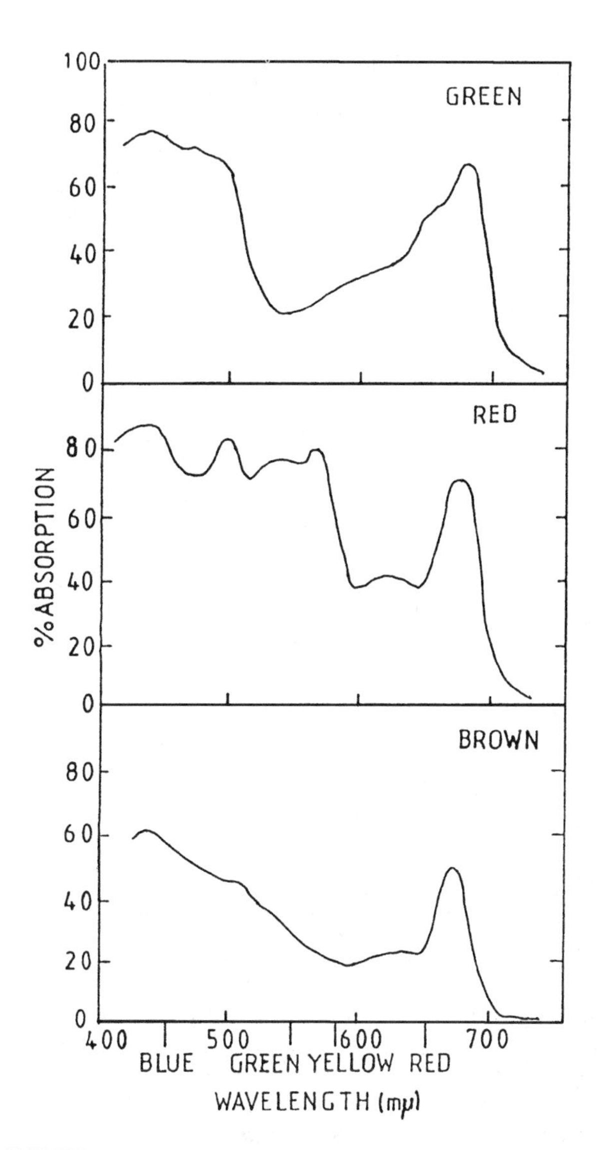

FIGURE 2.20 Absorption of light by green, red, and brown seaweeds. Green seaweeds absorb maximally in the blue and red portions of the spectrum; hence, they appear green in color. The brown color of seaweeds results from absorption near the middle of the spectrum, which removes more of the green. Red algae absorb light in the green portion of the spectrum and thus appear red in color. (After Blinks, L.R., *J. Mar. Res.*, 14, 366, 367, 384, 1955.)

found in the mid- and lower eulittoral and extend to depths of 10 to 15 m or more in the upper sublittoral.

Light is an important factor in determining the upper limits of some species. Algae such as *Ulva, Cladophora,* and *Porphyra* living on the middle and upper shores are unaffected by strong sunlight, but the kelp *Laminaria,* which is adapted to low light levels in the sublittoral zone, reduces its rate of photosynthesis in strong sunlight and during emersion. Long exposure can cause destruction of photosynthetic pigments. Many seaweeds from the sublittoral zone are killed by 2 hours of exposure to direct sunlight when out of the water. The ultraviolet part of the spectrum can have deleterious effects on both plant and animal tissues. This can cause bleaching in some seaweeds.

The role of light in photosynthesis and primary production will be dealt with later in the relevant sections.

2.3.5.2 Temperature

The intertidal zone experiences varying degrees of exposure to atmospheric conditions depending on the level on the shore, and hence exposure to solar radiation. Depending on the latitude and climate air temperatures, this zone may have a daily fluctuation that exceeds the annual fluctuation of the sea. Temperatures on the shore in the winter due to freezing conditions may fall 10°C or more below the sea temperatures, and in the summer may rise 15 to 20° or more above them. Rock surfaces receiving direct insolation will have surface temperatures much higher than the air temperatures. Black basaltic rock heats up more rapidly and reaches higher temperatures than does light-colored coral limestone, mudstones, and chalk. Basaltic reefs on the New South Wales coast of Australia frequently exceed 45°C on summer days.

Temperature is a most fundamental factor for all organisms because of its effects on molecular activities and properties, and hence, on virtually all aspects of metabolism. Atmospheric temperatures are subject to extensive modification by a suite of microenvironmental situations. Factors such as shading affect the influx of heat to an organism, whereas other factors such as evaporation may reduce body temperatures. Irradiance (heating) may be reduced by shading, clouds, water, algal growth, and shore topography (including overhangs, crevices, and direction of slope). Small-scale topographic features also give shelter from wind (air movement over the surface), and hence from evaporative cooling.

Figure 2.21 gives two examples of measurements of intertidal temperatures recorded during emersion on a hot day. In Figure 2.21 the alga *Enterocladia muricata* is a stiff tufted plant; the temperature of the interior of the clump, which is shaded yet open to the air, remains considerably cooler than the air or open rock surface (Glynn, 1965). The red alga *Porphyra fasicola*, in contrast, is flattened against the rock surface when the tide is out, and on a calm day it heats up to a much higher temperature than the air (Biebel, 1970). The graphs also show the sharp drop that occurs in the temperature when the tide covers the plants. Most notable is the drop in surface temperature of the *Porphyra* thallus from 33°C to 13°C in a matter of minutes as the water reaches it.

2.3.6 Desiccation and Zonation

Many observations have demonstrated that desiccation effects are particularly important in setting the upper limits of the intertidal distributions of many species, such as: (1) the elevation of zones in areas of wave splash (see Figure

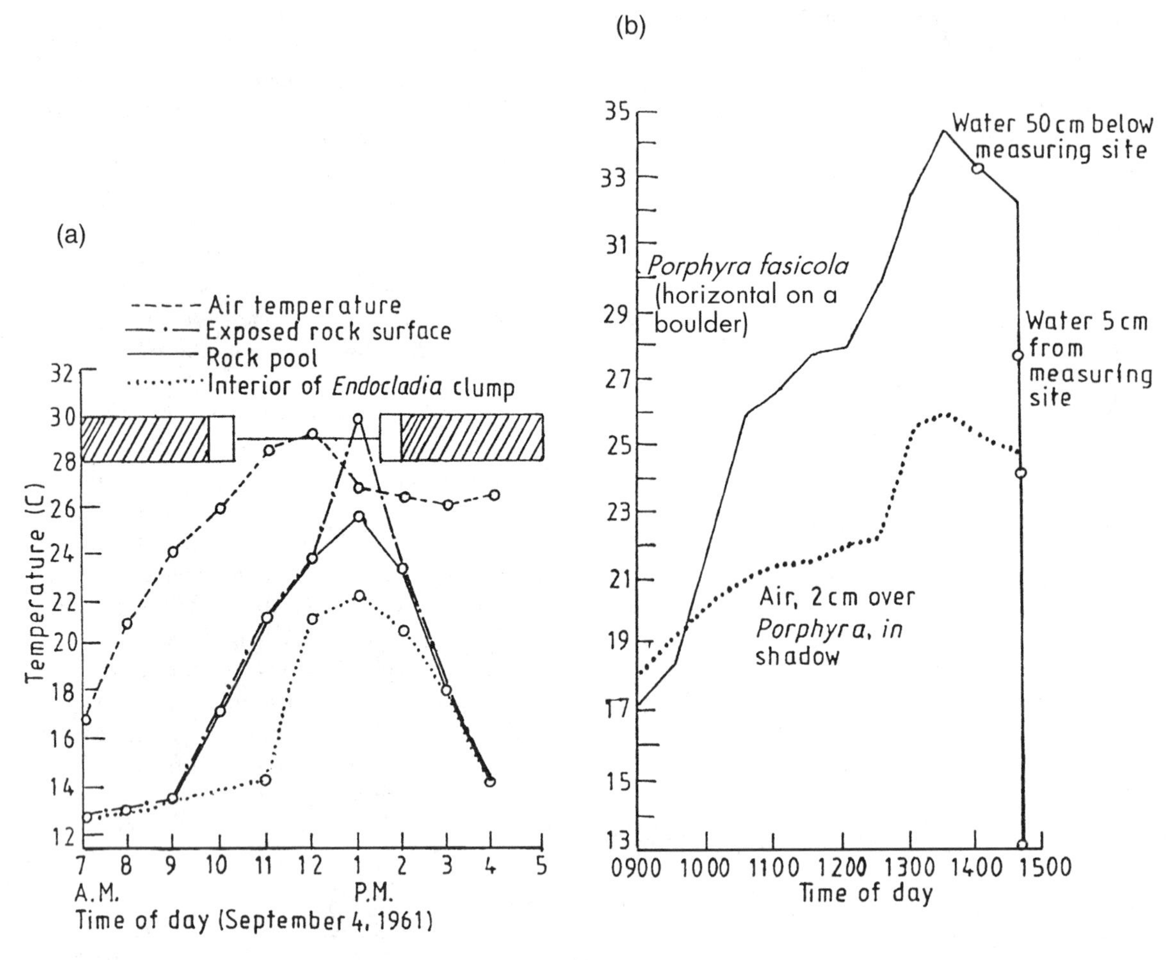

FIGURE 2.21 (Left) Temperature observations of three microhabitats in the high intertidal *Endocladia-Balanus* association at Monterey, California, as related to low water exposure. Also shown is the air temperature at a nearby weather station during the observation period. The horizontal bar and line at the top of the graph show, for the level observed, the approximate duration of submerged (cross-hatching), awash (clear), and exposed (line) periods. (Right) *Porphyra fasicola* thallus temperature during ebb tide on a calm day. Left from Glynn (1965); Right from Biebel (1970). (Redrawn from Lobban, C.S., Harrison, P.T., and Duncan, M.T., *The Physiological Ecology of Seaweeds*, Cambridge University Press, New York, 1985, 37. With permission.)

2.14), and on shaded slopes; (2) the enhanced growth of certain organisms, e.g., the green alga *Enteromorpha* in areas of freshwater seepage; (3) the higher distribution of some species, such as mussels, where seawater seeps from high tidal pools. Direct evidence of the effects of desiccation on intertidal organisms is well documented in the scientific literature and will be discussed further in Chapter 6.

2.3.7 Biotic Factors and Zonation

Plants and animals may influence the zonation of other species in a variety of ways. Firstly, the presence of algae at a certain level on the shore may reduce the problems of desiccation for other species by providing a microclimate when the shore is exposed at low tide, where suitable humidity conditions enable an animal species to extend higher on the shore than they would in the absence of the algae. Mussel beds also provide a similar favorable microclimate for many species. Secondly, the algae themselves provide a substrate for a great variety of epiphytic algae, sessile animals such as anemones, hydroids, tube-building polychaetes, colonial ascidians, etc., and motile species such as amphipods, isopods, and polychaetes.

Other biological interactions involved in determining the vertical distributions are grazing, predation, and competition. These interactions will be considered fully in Chapter 6.

2.3.8 Factor Interactions

The environment of an organism comprises many factors, each almost constantly varying and interacting with each other to determine the vertical zone occupied by a particular species on the shore. According to Lobban et al. (1985) factor interactions can be grouped as follows: (1) multifaceted factors; (2) interactions between environmental variables; (3) interaction between environmental variables and biological factors; and (4) sequential effects. These will not be discussed in detail here, but will be examined further in Chapters 5 and 6.

Many environmental variables are complex, e.g., light quality and quantity not only change with depth, but also change with turbidity and the nature of the particles in suspension in the water. Emersion usually involves desiccation, heating or chilling, removal of most of the nutrients, and frequently changes in the salinity of the water in the surface films on plants, and in the respiratory surfaces of animals. There are also complex interactions among environmental variables. Water motion can affect turbidity and siltation as well as nutrient availability. Interactions between environmental variables and biological factors include both the ways in which biological parameters such as age, phenotype, and genotype affect an organism's response to an environmental variable, and also the effects that an organism has on the environment. The environment of a given species includes other organisms, with which it interacts through intra- and interspecific competition, predator–prey relationships, and basiphyte–epiphyte relationships. Other organisms may greatly modify the environment of a particular individual or population. Protection from strong irradiance and desiccation by canopy algae is important to the survival of newly settled sporelings and the larvae of many species, and the survival of understory algae and many sessile (e.g., sponges, bryozoans, colonial ascidians) and motile (e.g., gastropod molluscs) animal species. Grazing damage may destroy plants, yet some species depend on grazers for their own survival. Finally, there are factor interactions through sequential effects. In general, any factor that alters the growth, form, reproductive, or physiological condition of an organism is apt to change the response of that organism to other factors, both at the same time and in the future.

2.3.9 Critical Levels

Early hypotheses to account for the patterns of zonation and vertical distribution of species on rocky shores involved the concept of critical tidal levels (CTLs) (Colman, 1933; Southward, 1958; Lewis, 1964; Newell, 1979). The critical level concept was originally developed by Colman (1933) and elaborated by Doty (Doty, 1946; Doty and Archer, 1950). They advanced the view that at certain levels on the shore (*critical levels*), the rate of change of tidal emersion and submersion was greater than at other levels. At these levels a disproportionate number of species reached their limit of tolerance to the periods of emersion or submersion during the tidal cycle. While the critical tidal level hypothesis has been widely supported as a major factor in explaining zonation on rocky shore intertidal communities (Beveridge and Chapman, 1950; Doty and Archer, 1950; Knox, 1953; Lewis, 1964; Townsend and Lawson, 1972; Carefoot, 1977; Druehel and Green, 1982; Swithenbanks, 1982), its validity has been frequently challenged (e.g., Connell, 1972; Stephenson and Stephenson, 1972; Edwards, 1972; Wolcott, 1973; Underwood, 1978a; Chaloupka and Hall, 1984). Critical reviews of the CTLs are to be found in Underwood (1978a) and Chaloupka and Hall (1984).

As Underwood (1978a) points out, if CTLs actually exist, then the boundaries of distribution of intertidal organisms should be dispersed nonrandomly, i.e., they should be underdispersed along the intertidal gradient. Knox (1953) tested this for a rocky shore at Taylors Mistake, Banks Peninsula, New Zealand. From Figure 2.22 it can be seen that the greatest number of upper and lower limits were grouped in the vicinity of MLWN and EHWN. The former marks the lower limits of the main mid-eulittoral populations such as the barnacles *Epopella plicata* and *Chamaesipho columna*, the tubeworm *Pomatoceros cariniferus*, and the mussel *Mytilus galloprovincialis*, and the upper limits of the principal lower eulittoral algae such as *Durvillaea willana, Carpophyllum maschalocarpum,* and *Cystophora scalari*. The EHWN level marks the lower limits of a few high intertidal species such as the alga *Stichosiphonia arbuscula* and the upper limit of a number of filter feeders (barnacles, bivalves, and *Pomatoceros*).

Generally, earlier studies have calculated the submersion/emersion history of a shore from predicted or recorded tidal levels. However, as Druehl and Green (1982) point out, such approaches fail to describe the submersion/emersion history accurately, insofar as they do not account for wave conditions over extended periods, or for the influence of local topographic conditions. Figure 2.23 from Druehl and Green (1982) compares submergence/emergence data for a rocky point on Vancouver Island, British Columbia derived from predicted tide heights over one lunar cycle with the actual tidal heights that take into account wave conditions. Their data demonstrate that wave action causes substantial changes in actual submergence/submergence events from those predicted from tidal data, and that the extent of these changes is dependent upon topography and season. This is demonstrated in Figure 2.23, which compares the measured accumulated time submerged as a function of elevation above zero tide level for a rocky point, a channel, and a gently sloping rock face, all within 50 m of each other, compared with data from tidal predictions. Druehl and Green (1982) found that in some instances, limits of vertical distribution of algae over their 6-year study period ranged over 1 m (or over 1/3 of the maximum tidal amplitude). Other studies have demonstrated a seasonal change in the vertical limits of intertidal plants (Druehel and Hsiao, 1977; Schonbeck and Norton, 1978). Over the period of their study, Druehel and Green (1982) found that while vertical floristic patterns changed from year to year, as well as among the three topographies (rocky point, channel, and gently shelving rock face), there was a general tendency

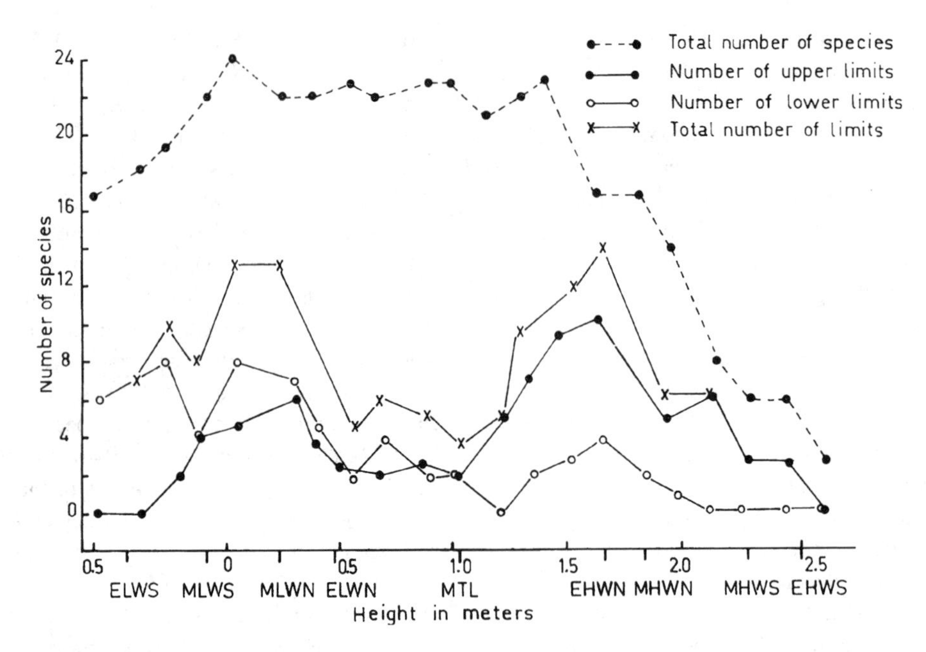

FIGURE 2.22 Total number of species, number of upper limits, number of lower limits, and total number of limits at various levels on the shore at Taylors Mistake, South Island, New Zealand. (Redrawn from Knox, G.A., *Trans. R. Soc. N.Z.*, 82, 192, 1953. With permission.)

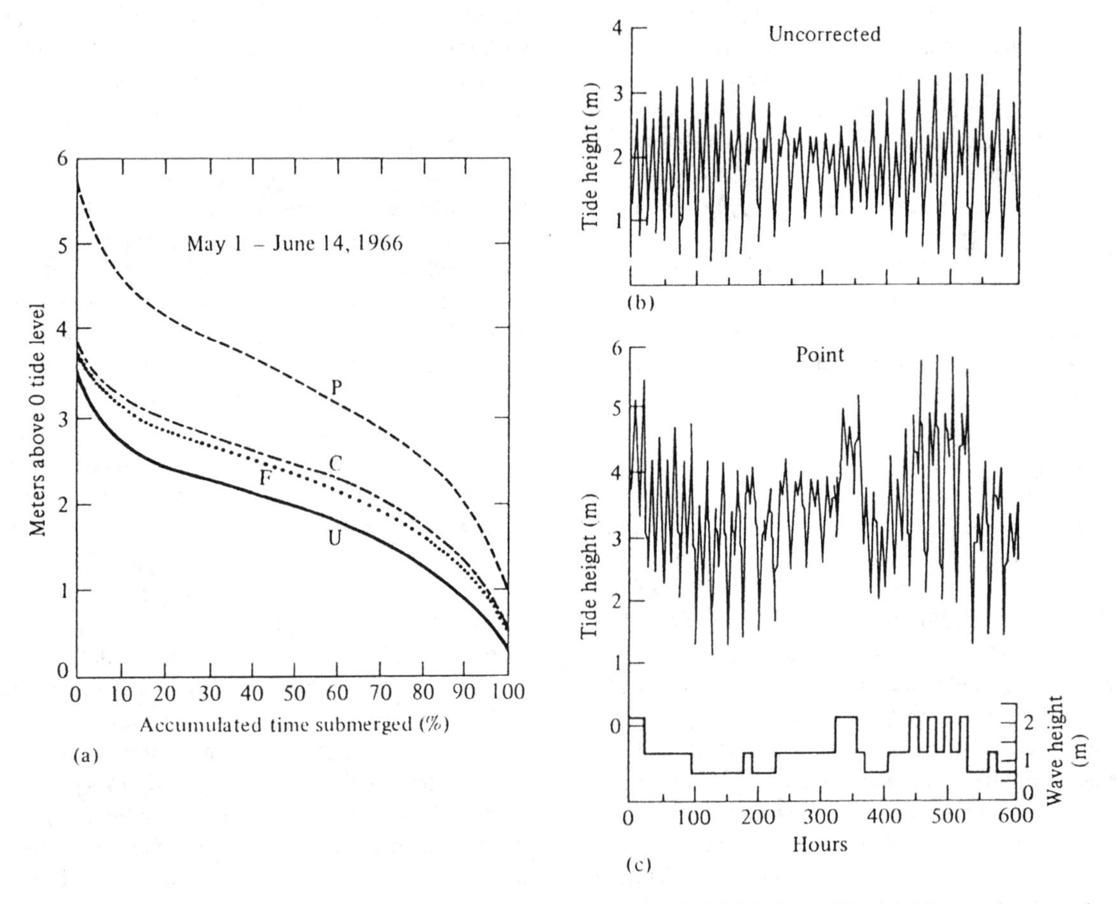

FIGURE 2.23 Submergence-emergence data from a site on Vancouver Island, British Columbia. (a) Measured accumulated time submerged as a function of elevation above zero tide for a rocky point (P), a channel (C), and a gently shelving rock face (F), all within 50 m of one another, compared to data from 6-min tidal predictions (U). (b) Predicted tide heights over a lunar cycle. (c) Actual tide heights and wave heights at the rocky point. The wave height data are derived form twice-daily observations. (Redrawn from Druehl, L.D. and Green, M., *Mar. Ecol. Progr. Ser.*, 9, 165, 166, 1982. With permission.)

for the majority of species to alter their limits in a common mode from year to year. These results suggest the presence of common factor(s) where stress/benefit effects on the vertical limits of the plants vary from year to year, but that for any one year, they have a more or less uniform effect.

Underwood (1978a), in a study of the upper and lower boundaries on five shores in different parts of Great Britain, found no evidence that critical tide levels exist, because there was no evidence that the upper or lower boundaries of the distribution of vertical distribution were in any way aggregated. He points out that if CTLs actually exist, then the boundaries of distribution of intertidal species should be dispersed nonrandomly, i.e., they should be underdispersed, or clumped, along an intertidal gradient. Underwood (1978b) described a method for detecting nonrandom patterns of distribution of species along a gradient. He developed an occupancy model based on the method of Pielou (1975), but correcting for biases in the method as applied by Pielou and Routledge (1976). Chaloupka and Hall (1984) later elaborated a restricted occupancy model, as an alternative model to the unrestricted occupancy model used by Underwood to test the null hypothesis that the upper and lower limits of intertidal species are dispersed randomly with respect to the intertidal gradient. They used their model to analyze the distribution of species boundaries on the intertidal rocky shores of sub-Antarctic Macquarie Island. They found that the observed species were randomly dispersed along the intertidal gradient, concluded that there was no evidence to support the CTL hypothesis, and suggested that tidal emersion was not a significant factor in structuring intertidal communities on Macquarie Island.

Knox and Duncan (in preparation) tested the CTL hypothesis in an investigation of species vertical zonation patterns along 13 transects ranging from sheltered to exposed on the sub-Antarctic Snares Islands to the south of New Zealand. Figure 2.24 plots the upper and lower limits for the dominant species on the Snares Islands shores. From the plots it can be seen that there are clusters of vertical limits at a number of levels on the shore. The major ones coincide with the upper limits of the bull kelp *Durvillaea antarctica* and the red alga *Pachymenia lusoria* and the lower limits of the lichens of the littoral fringe. Possible causes of these clumpings are discussed below.

The upper limit of the kelp *D. antarctica* is a conspicuous feature of southern shores, and it coincides approximately with the mean low water of neap tides. On steep exposed cliff faces, the attachment zone of the *Durvillaea* holdfasts occupies a vertical zone of no more than 1 meter, and often less, while on broken exposed coasts with reefs and boulders it may extend over a vertical height of 4 to 5 meters. Hay (1982) found that on the New Zealand mainland, the upper limit of *Durvillaea* extended to higher levels in the southern part of the South Island than in the northern part. Chapalouka and Hall (1984) state that *D. antarctica*, a prominant species on Macquarie Island shores, has been shown to be greatly affected by the grazing activities of limpets on newly settled sporelings (Hay, 1982) and speculate that the limpet *Nacella macquariensis* on Macquarie Island probably plays an important role in determining the upper limit of the kelp. However, they fail to mention that the removal of both limpets and barnacles on southern South Island shores did not result in kelp sporelings colonizing levels above the normal limit of *Durvillaea.* Furthermore, the sporelings that settled and grew above the normal upper limit of *Durvillaea* at Kaikoura further north on the South Island coast were unable to survive hot, dry conditions during the summer, and Hay (1982) considered that while the removal of limpets did raise the level colonized by sporelings, the upper margin of *D. antarctica* is determined mainly by the physiological tolerance of the sporelings to desiccation. On the Snares Islands, while numerous small plants were found under the *D. antarctica* canopy, only a few scattered, stunted plants occured above the the upper limit of the holdfast attachments. Limpets were not abundant in the zone above the *Durvillaea,* and it may be inferred that the upper limit of *D. antarctica* on the Snares Islands shores is probably determined mainly by physical factors.

A second major concentration of upper and lower limits occurs at the upper limit of the *Pachymenia* zone. This zone is dominated by the red algae *Pachymenis lusoria, Gigartina* spp., *Notogenia fastigista,* and *Haliptilon roseum.* Other green, brown, and red algae are minor components. Grazing by the limpet, *Cellana strigilis,* and the siphonariids, *Kerguelenella strwartiana* and *Siphonaria zelandica,* which are common in the zone above, probably plays a role in combination with desiccation in determining the upper limits of the algae.

Field observations, however, lend support to the view that desiccation plays an important role in determining the upper limits of intertidal algae. Many algae extend further upshore wherever they are protected from desiccation by repeated wave splash (Burrows et al., 1954; Lewis, 1964), by an overlying canopy of larger algae (Menge, 1976), or by inhabiting shady places (Norton et al., 1981). Year-round observations on the condition of the intertidal algae growing *in situ* on the shores of the Isle of Cumbrae, Scotland, showed that the upper limit of the zones occupied by the fucoid algae *Pelvetia canaliculata, Fucus spiralis,* and *Ascophyllum nodosum* were periodically pruned back by environmental factors when drying conditions coincided with neap tides, which exposed the plants to aerial conditions for long periods. Laboratory experiments also demonstrated that the ability to tolerate desiccation and then resume growth when resubmerged was greatest in *P. canaliculata*, the species found highest on the shore, and was progressively less in the species inhabiting successively lower levels. Similar correlations between vertical distribution and drought tolerance have been reported

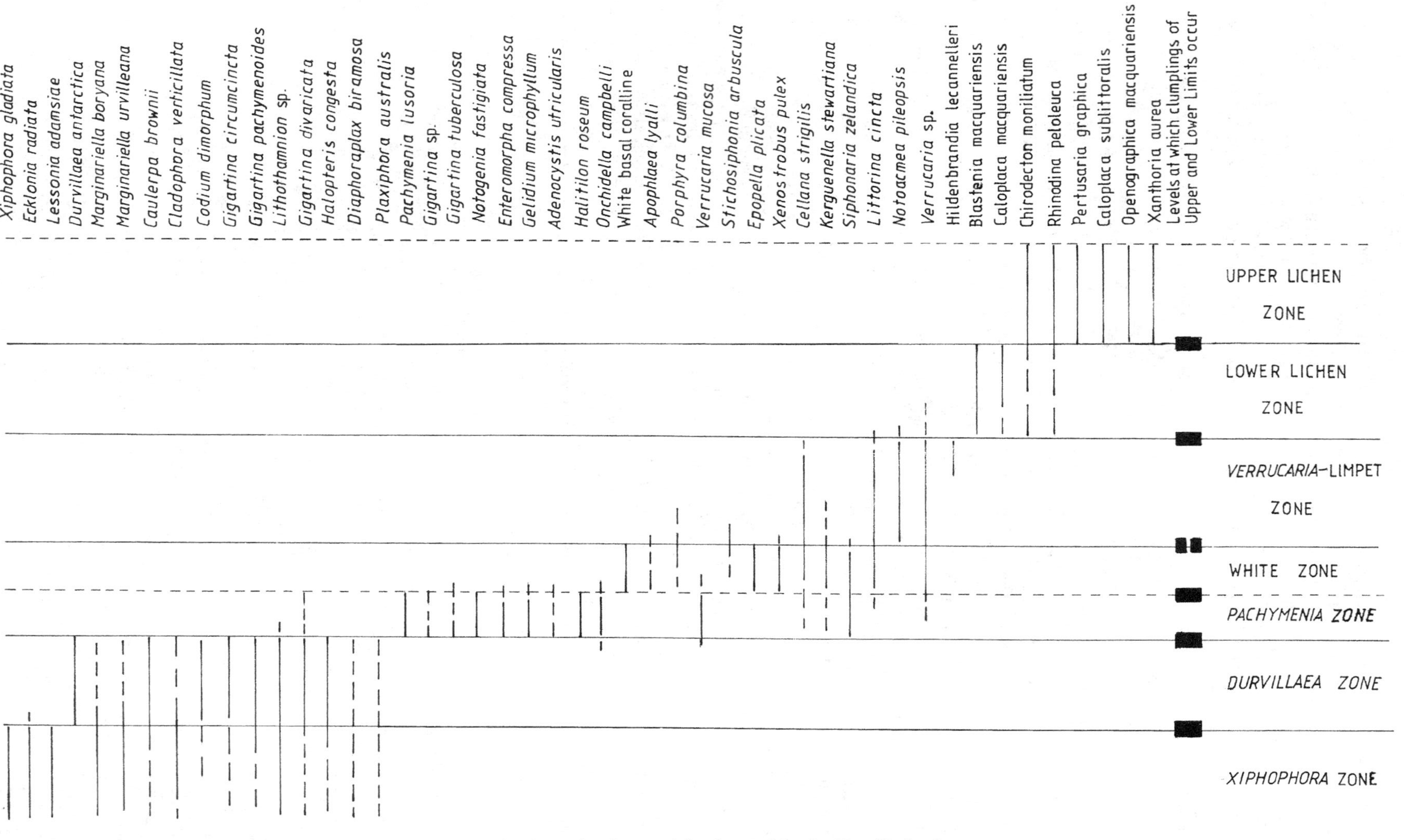

FIGURE 2.24 Upper and lower limits of intertidal plants and animals on the shores of the Snares Islands, New Zealand.

for other species of intertidal algae. Other workers, however, (e.g., Edwards, 1972; Dromgoole, 1980) have found no direct relationship between rate of dehydration and location on the shore.

On a number of occasions in the field I have observed the pruning back effect discussed by Schonbeck and Norton (1978). In January 1986 on the Snares Islands, when a period of calm weather coincided with exceptional neap tides with a small range and relatively high temperatures (16°C+), there was a dramatic die-off of the alga at the top of the *Pachymenia* zone. In particular, the red algae *Pachymenis lusoria* and *Halipton roseum* were bleached white.

A third major grouping of limits occurs near the lower boundary of the lichen zone, which is a conspinuous feature of the littoral zonation patterns of the Subantarctic islands (Knox, 1968, 1975, 1988c). This boundary is marked by the lower limit of the white lichen, *Pertusaria graphica*. On the east coast of the Snares Islands, this is several meters above low water spring tide level and progressively extends lower down the shore as shelter from wave action increases. It appears that the high shore lichens cannot withstand prolonged immersion in seawater.

It has thus been established that on some shores, the upper and lower limits of species distributions are concentrated at particular levels on the shore. However, the relationship of these levels (CTLs) to the tidal factor is not clear-cut, and there is a dearth of experimental evidence that has tested growth, survival, and reproduction just above and below a CTL. A variety of studies have shown that abiotic factors associated with an intertidal position can limit the vertical distribution of marine algae. Observations on postearthquake shores of Alaska demonstrated that moderately uplifted intertidal communities could not survive at a higher intertidal level, but that communities shifted downward could survive at their new position (Haven, 1971). Further, transplant studies on intertidal fucoids have demonstrated that the upper limits of two *Fucus* species are determined by physical factors (Schonbeck and Norton, 1978).

A number of factors complicate possible correlations between distributional limits and CTLs, notably wave action and the ability of organisms to become acclimated during periods of subcritical conditions. Further, the critical period may apply to reproduction and settlement, or survival after settlement, rather than the ability of the adults to survive and grow at a particular level. Also, as discussed above, it may not be the average continuous emersion that is critical, but a period of unusual continuous emersion that coincides with abnormally high summer temperatures. As will be discussed in Chapter 6, physical factors and their impact on the physiological functioning of intertidal species are only part of the explanation of zonation patterns. There are also a number of biological controls that determine the upper and lower limits: grazing, predation, competition, and other intra- and interspecific interactions.

2.4 HARD SHORE MICROALGAE

A film of organic material and microorganisms coats the littoral rock surfaces, the shells of animals such as bivalve molluscs, and the fronds of algae. In addition to diatoms and blue-green algae, bacteria and protozoa are abundant in such films, which are the first site of attachment and early growth of all settling macroalgae and sessile animals (Wahl, 1989). This microbial film is the main food resource of microphagous herbivores whose grazing activities regulate and even prevent macroalgal recruitment and growth (for reviews see Underwood, 1979; Lubchenco and Gaines, 1981; Hawkins and Hartnoll, 1983a). Although the importance of this microbial film has long been recognized, its study has received much less attention than that of the macrobiota (MacLulich, 1983; Hill and Hawkins, 1990).

Temporal variation in the microalgal community has been investigated at a number of different geographic localities (North America, Castenholz, 1961; Australia, MacLulich, 1983; Underwood, 1984a,b,c; Great Britain, Hill and Hawkins, 1990), but very little work has been carried out on the patchiness of these benthic microalgal communities (Hill and Hawkins, 1990). The importance of microalgae in the diet of microphagous grazers has been demonstrated many times (Medlin, 1981; Raffaelli, 1985; Hill and Hawkins, 1990).

Hill and Hawkins (1990) studied the seasonal and spatial variation in the distribution and abundance of rocky shore microalgae on moderately exposed shores on the Isle of Man. They found that the microbial biomass increased during the late autumn, peaked in the winter, and declined to relatively low levels during late spring and summer. This result was generally consistent with previously reported studies (Castenholz, 1961; 1963; MacLulich, 1983; Underwood, 1984a). Both MacLulich (1983) and Hill and Hawkins (1990) found that algal diversity increased in the late spring and summer. Hill and Hawkins (1990) also found considerable spatial differences in microalgal abundance and composition. Diatom abundance, dominated by the firmly attached stalked species, *Acanthes*, was greatest on barnacles, while filamentous algal cover was greatest on open rock. In his investigation of the microalgal flora on an intertidal rock platform near Sydney, Australia, MacLulich (1983) found that the community was dominated by a blue-green alga, *Anacystis* sp., a situation not previously described in any similar system. Also present were diatoms and various red, green, and brown algal sporelings. Both density and variety were greater lower on the shore and at more exposed sites. He suggested that the great variety of the microalgal assemblages that he found may be due to: (1) the density of *Anacystis* sp. spores and microscopic red, green, and

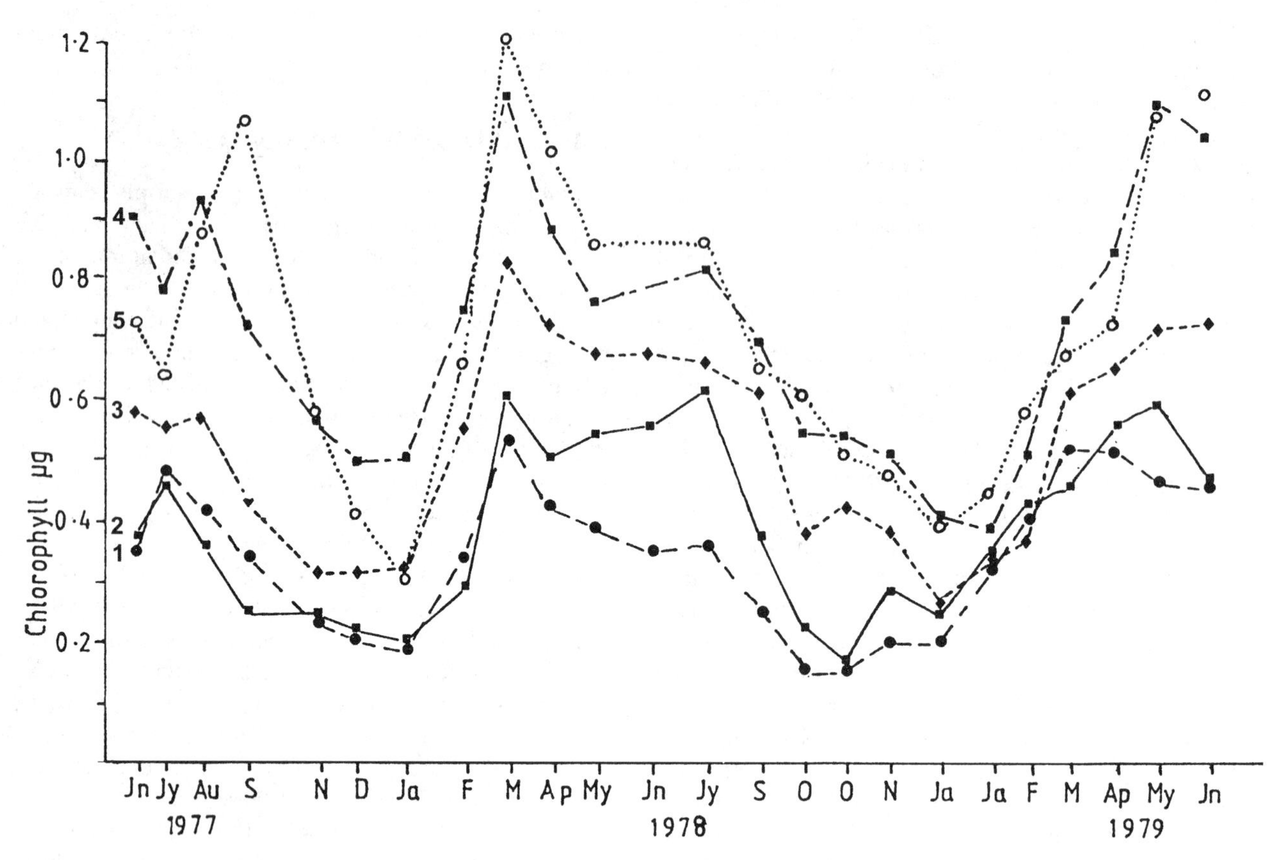

FIGURE 2.25 Seasonal patterns of vertical distribution of microalgae measured by chlorophyll concentrations in samples at five levels of the shore: $n = 8$ in each case; standard. Errors are not shown, to avoid confusion, but ranged from 0.04–0.65 (with 85% of samples < 0.32); levels sampled were ●, 1 (highest); ■, 2; ♦, 3; ⬓, 4; ○, 5 (lowest). (Redrawn from Underwood, A.J., *J. Exp. Mar. Biol. Ecol., 78,* 199, 1984a. With permission.)

brown algae was greatly reduced during the summer, thereby increasing the observed variety; (2) diatoms (as a group) increased both in variety and number during the summer; and (3) gastropods increased in number and activity during the summer, which may have resulted in a reduction in the density of certain preferred species (MacLulich, 1983).

Gastropod molluscs, especially limpets, are the principal grazers on these microalgal communities, and the question as to whether there is preferential grazing has been investigated in a number of studies (see Castenholz, 1961; 1963; Foster, 1975; Underwood, 1984a,b). Gastropod grazing will be further explored in Chapter 6.

Underwood (1984a) estimated the abundance of microalgae on the shores of Cape Banks, Botany Bay, New South Wales, Australia, by assays of the concentration of chlorophyll on and in the surface of sandstone rock at five levels on the shore at approximately monthly intervals. This (Figure 2.25) revealed a vertical gradient of increasing abundance of microalgae toward the bottom of the shore, except in the summer. This trend was not correlated with decreasing abundance of microalgal grazers. During the summer the abundance of microalgae declined abruptly at all levels. This decrease was greatest at the lower levels, resulting in little or no difference among the abundances of the microalgae at the five heights during the summer. These seasonal changes were unrelated to any major changes in the densities of the grazers. Other workers have also described a greater concentration of chlorophyll (Nicotri, 1977), or number of diatoms (Round, 1971) toward the lower levels on the shore. These investigators explained this as a response to periods of emersion and the inability of the microalgae to withstand prolonged periods of desiccation, increased temperature, and increased light intensity at the higher levels on the shore. Castenholz (1961), however, pointed out that the increased abundance of diatoms toward the lower levels on Oregon shores may have been due to the fact that grazing pressure due to limpets and species of *Littorina* decreased at lower levels. In later studies Castenholz (1963) revealed that the pattern of distribution on various surfaces depended to a very great extent on the aspect of the substratum. Thus, on balance it seems that much of the vertical distribution of intertidal microalgae may be a direct result of physiological stresses during low tide. Superimposed on this is the effect of grazers.

2.5 HARD SHORE MICRO- AND MEIOFAUNA

Mare (1942) divided the benthos into three categories: the *macrobenthos*, including all organisms too large to pass through a 1-mm sieve; the *meiobenthos*, including those that do pass through a 1-mm mesh but are retained by a 0.1-mm mesh; and the *microbenthos*, those forms that pass through a 0.1-mm mesh. Coull and Wells (1983), on the other hand, define meiofauna as micrometazoans that pass through a 0.5-mm mesh. The microbenthos comprises mainly bacteria, small protozoans, especially flagellates, ciliates, and amoebae. The meiobenthos includes many small copepods, nematodes, flatworms, the larval stages of various molluscs, polychaetes, and other animals, rotifers, gastrotrichs, gnathostomulids, and the larger foraminiferans. Of these, the copepods and the nematodes are numerically the most important. They are metabolically important members of the benthic ecosystem (Gerlach, 1971) and are known to be prey items for a variety of larger animals (Coull and Bell, 1979), particularly juvenile fishes (Hicks and Coull, 1983).

Most of the investigations on marine meiobenthic ecology have been carried out on soft shores, and there is a considerable body of literature concerning this. In contrast, there is very limited knowledge of the meiofauna of rocky shores, apart from the harpacticoid copepods (Hicks, 1980; 1985a; Hicks and Coull, 1983). In 1980, Platt and Warwick reviewed the significance of free-living nematodes in littoral ecosystems and cited only two references (Moore, 1978; Warwick, 1977) that were studies on the nematodes of rocky shores.

The exceptions to the paucity of studies of rocky shore meiofauna include those of Gibbons and Griffiths (1986) and Gibbons (1989) in South Africa, Coull et al. (1983) on the phytal meiofauna in South Carolina, U.S.A., and the extensive study of the phytal meiofauna of the rocky intertidal at Island Bay, Wellington, New Zealand (Hicks, 1977a,b,c; 1980; Coull and Wells, 1983). Gibbons (1989) investigated the impact of organic sediments on a rocky shore in relation to tidal elevation, using artificial "algal" mats of differing complexity. Sediment accumulation was correlated with habitat structure and increased at higher elevations. The meiofaunal communities were similar to those previously reported from intertidal algae (Hicks, 1977a; Edgar, 1983a,b,c,d; Johnson and Scheibling, 1987; Coull, 1988). The meiofauna was dominated by small interstitial forms, notably harpacticoid copepods, copepod nauplii, amphipods, and nematodes. Densities recorded on the high shore were as great as those on the low intertidal, although the diversity (especially of the larger forms) and biomass were markedly lower. As a result of the increasingly favorable environment toward the bottom of the shore, meiofaunal density and biomass increased (Figure 2.26).

In 1985 Heip et al. reviewed the ecology of marine nematodes, noting that when macrophytes are present in the littoral zone, an enormous increase in food availability, habitat complexity, and shelter is created. In sheltered areas the algae accumulate sediment and detritus, and such plants have a richer nematode fauna. Moore (1978) found that the nematodes inhabiting holdfasts of kelp (*Laminaria*) were mostly epigrowth feeders and omnivores. Kito (1982) studied the nematodes from the thalli of the brown alga *Sargassum confusum* in Japan and found that the nematodes were second in abundance to the harpacticoid copepods.

Harpacticoid copepods are usually the second most abundant meifaunal taxon in marine sediments, with free-living nematodes regularly ranking first in terms of the total number of individuals. However, harpacticoids are reported to comprise from 11 to 60% of the phytal (seaweed-associated) meiofauna (Hicks 1977b). Phytal harpacticoids have been investigated in South Africa (Bechley and McLachlan, 1980), New Zealand (Hicks, 1977a,b,c; 1985a; Coull and Wells, 1983). They comprise two groups: (1) those characteristic of sediments trapped by the algae, and (2) the true phytal dwelling forms. The highest diversity for harpacticoid assemblages is recorded from algal holdfast communities (Moore, 1979), while Hicks (1977c) has also recorded high densities from the fronds of the brown alga *Ecklonia radiata*.

Often when algal fronds are heavily loaded with silt-clay or detritus, a fauna typical of most sediments, i.e., a nematode-dominated fauna, is found. If only the fronds or blades of the algae are considered, then copepods are the dominant taxon. Increased surface area (complexity) of the plant substratum generally leads to a concomitant increase in copepod numbers and/or species (Hicks, 1977c; 1980). Hicks (1977c) distinguishes two general subassociations of phytal harpacticoids, those characteristic of the sediments trapped by the algae, and the true phytal dwelling forms. Many of the latter are ovoid to subovoid in shape and often dorsoventrally flattened or laterally compressed (amphipod shaped). Phytal assemblages are remarkably similar in different parts of the world (Hicks and Coull, 1983). The highest known species diversity for harpacticoid assemblages has been recorded from algal holdfast communities (Moore, 1979) and shallow water algal frond assemblages also have high diversity (Hicks, 1977c).

There is a considerable volume of literature demonstrating the importance of harpacticoids as food for larval, juvenile, and small fishes (for references see Hicks and Coull, 1983). Meiobenthic harpacticoids ingest either epipelic (benthic) or epiphytic diatoms, phytoflagellates, bacteria, either as aggregated cells or as detritus associates, blue-green algae, ciliates, and mucoid substances

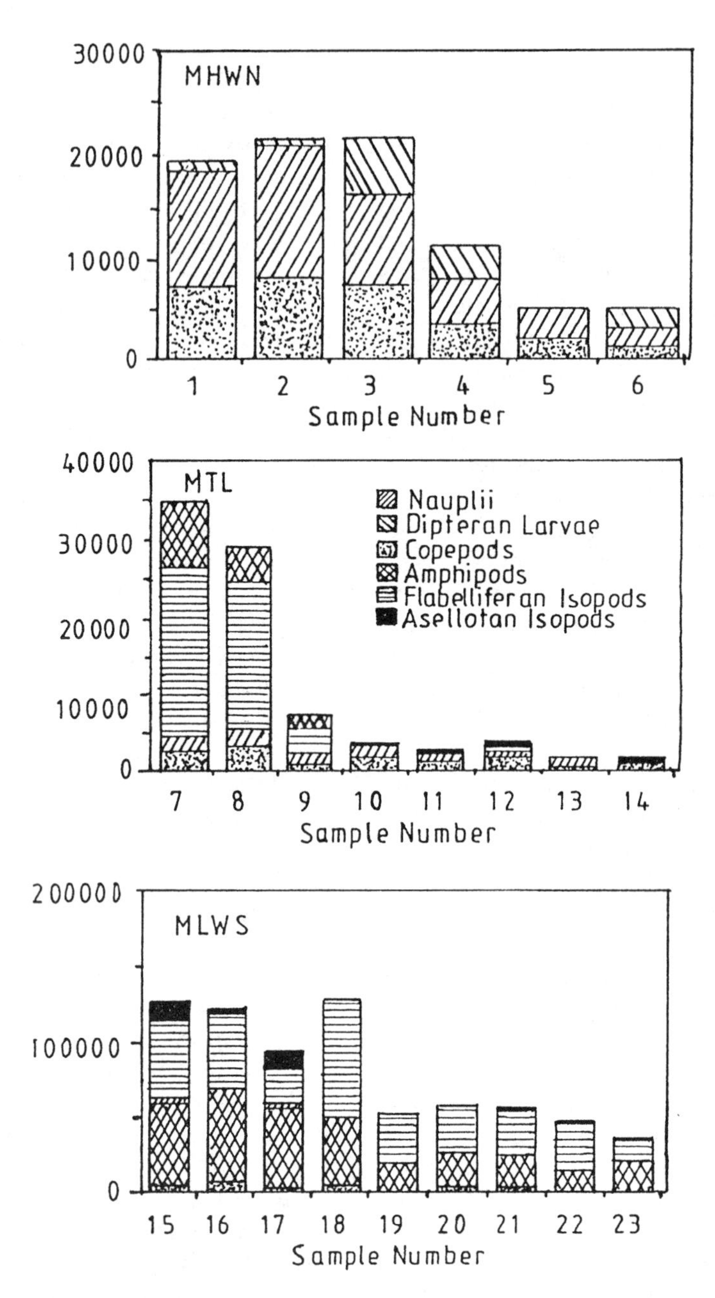

FIGURE 2.26 The distribution of major (5% total biomass) meiofaunal groups among artificial mats at three studied elevations above Chart Datum. A: MHWN; B: MTL; C: MLWS. Complex mats are sample numbers 1, 3, 7, 8, 15–17; mats of medium complexity size are numbers 4, 5, 10–12, 18–20; and simple mats are numbers 6, 13, 14, 21-23. (Redrawn from Gibbons, M.J., *Est. Coastal Shelf Sci.*, 27, 585, 1989. With permission.)

(Hicks and Coull, 1983). A number of species such as the frond-mining species *Diarthrodes cystoecus* feed on the medullary tissues of the macroalgae. Coull and Wells (1983) in a series of experiments on the rocky intertidal at Island Bay, Wellington, New Zealand, demonstrated that substrate complexity was an important factor in reducing fish predation on phytal harpacticoids. They found that the red alga *Corallina officinalis*, the most complex structure tested, was the only refuge for harpacticoids from blenny (*Helcogramma medium*) predation for the total meiofaua and the dominant taxon, copepods.

2.6 ROCKY SHORE LICHENS

2.6.1 Species Composition and Distribution Patterns

Although lichens are a common feature of most rocky shores, they are more often than not neglected in accounts of shore zonation patterns. As Fletcher (1980) points out, although almost 450 species of lichens inhabit and can dominate the eulittoral, littoral fringe, and supralittoral rocks of seashores around the British Isles, there is little published information on their distribution, biological interactions, and physiological properties. Of these 450 species, the vast majority occur in the supralittoral.

On most temperate shores the region above high water mark is blackened by an overgrowth of blue-green algae and lichens called the "black zone" by Stephenson and Stephenson (1972). The dominant plant associations are: (1) Cyanophyta (*Calothrix, Plectonema, Gloecapsa,* etc.); (2) *Lichina pygmaea* (Lecanorales, Lichenes); and (3) *Lichina maura* (Verrucariales, Lichenes). Lichen distribution on rocky shores will be illustrated with reference to the British Isles and the southern sub-Antarctic region.

2.6.1.1 British Isles

Eulittoral Zone: This zone is rarely dominated by lichens. *Arthropyvenia sublittoralis* may be locally frequent on chalk and limestone shores where it penetrates down to the uppermost limit of *Laminaria. Lichina pygmaea* on most shores straddles the margin between the littoral fringe and the upper eulittoral, but on *Chthalamus* shores is confined to the barnacle zone.

Littoral Fringe: This zone, in contrast to the one below, has an extensive lichen cover. It is dominated by *Verrucaria mucosa* and *V. halizoa* in the lower part, above which is *V. striatula*, followed by *V. amphibia* and finally *V. maurea,* which is by far the most important species. *Lichina pygmaea* appears among *V. mucosa* and the uppermost barnacle on sunny shores, provided that wave action is moderate, while *Lichina confinis* occupies the middle and upper parts of the *Verrucaria* zone. *V. mucosa* similarly becomes rare on wave-exposed shores. *Lichina confinis* is common among the uppermost patches of *V. maura* on sunny, sheltered shores. The three British marine *Atrhopyrenia* spp. intermingle with marine *Verrucaria* spp.

On some shores in the upper part of the littoral fringe and extending up into the supralittoral, there occurs what has been termed the "Orange Belt." The dominant species are orange *Caloplaca marina* and the white-to-grey *Lecanora helicopis, L. actophila,* and *Catillaria chalybeia. Arthonia phaeobaea, Lecania erysibe,* and *Caloplaca thallincola* are frequent under suitable conditions. Often *Caloplaca marina* forms a lower zone, with *Lacanora actophila* above, but much local intermingling can occur. In addition, the species composition of the belt can vary. For example, on shaded shores, *Caloplaca marina* is replaced by a sparser cover of orange *Caloplaca thallincola*, while on these same shore *Arthronia phaeobaea, Lacania erysibe, Catillaria chalybeia,* and *Lacanora helicopis* become more frequent. Consequently this subzone is seldom orange on shaded shores, and becomes leaden-grey, with only occasional patches of orange *Caloplaca* spp. On the other hand, sunny shores are dominated by creamy-white *Lacanora actophila* and bright orange *Caloplaca marina*, with a reduced cover of grey species.

2.6.1.2 The sub-Antarctic Region

One of the characteristic features of the littoral zones of the sub-Antarctic region is the prominent role played by lichens of various species, especially in the littoral fringe (Skottsberg, 1941; Knox, 1960, 1968). Anyone who has visited the islands of the sub-Antarctic must be impressed with the broad band of white and grey lichens, often with bands of orange, in the transition zone between the eulittoral and the maritime terrestrial vegetation. This band, depending on the degree of exposure to wave action, may be many meters in vertical height.

The distribution of the lichen flora has only been examined in detail on two of the island groups, the Auckland Islands (Knox, 1968) and the Snares Islands (Sainsbury, 1972). Personal observations have also been made on Campbell Island and the sub-Antarctic coasts of South America. In the Auckland Islands, a total of 21 species has been recorded. This list will no doubt be extended considerably with more intensive collecting. The zonation is dominated by four species of *Verrucaria* that form a series from the upper littoral fringe to the lower eulittoral. Species of *Verrucaria* are also prominent in other areas of the sub-Antarctic, e.g., Marion and Prince Edward (De Villiers, 1976) and Kerguelen (Delepine and Hureau, 1968; Arnaud, 1974), and in the Antarctic (Dodge, 1962; Knox, 1968).

Sainsbury (1972) subdivided the lichen zone on the Snares Islands into an upper portion (upper littoral fringe) terminating at the lower range of *Caloplaca sublobulata.* This upper portion receives a constant low level of spray and often comparatively high levels of desiccation. The lower portion (lower littoral fringe) is not as strongly desiccated, and receives spray and wave wash that changes greatly with weather conditions.

Sainsbury (1972) has investigated the faunal community associated with the lichens and other plants of the littoral fringe on the Snares Islands. He recorded 60+ species including 16 Diptera (4 breeding), 1 Hymenoptera, 4 Coleoptera (all of which had larvae associated with the lichens), 5 Collembola, 1 Tardigrada, 1 Pseudoscorpionidae, 1 Araneae, 26 Acarina, and 2 Amphipoda. Rotifers and nematodes were also present. He found that the various lichen species had their own specific assemblage and he was able to develop a preliminary food web for the system. The primary consumers could be divided into four groups, feeding either on the lichens themselves, on free-living microalgae, on free-living fungi, or on detritus.

Kromberg (1988) investigated the fauna of the "black zone" in the littoral fringe of the rocky shores of northern Europe. Here the encrusting lichen *Verrucaria maura* grows preferentially along dry and wave-exposed shores.

Where crevices are available, they are inhabited by a terrestrial fauna composed of *Nanorchestes amphibius* and *Bdella septemtrionalis* (Acarina), *Anurida maritima* (Collembola), *Strigamia maritima* (Chilopoda), *Pterobius brevistylis* (Archaeognatha), and by aquatic forms such as *Ligia oceanica* (Isopoda) and *Littorina neritoides* (Gastropoda). Small shrub-like lichens (*Lichina pygmaea*) are inhabited by species such as the isopod *Campecopea hirsuta*, the molluscs *Lasaea rubra*, *Mytilus edulis* (juvenile), *Littorina neritoides*, *L. saxatilis*, the mites *Hyadesia fusca*, *Ameronothrus* spp., and Limoniidae instars (*Geranomyia unicolor*).

Cyanophyta prefer sheltered shores and show a high tolerance to freshwater influences. The fauna here is composed of a few aquatic animals: *Hyadesia fusca* (Acarina), *Mniobia symbotica* (Rotatoria), *Telmatogeton japonicus* (Chironomidae), *Echiniscoides sigismundi* (Tardigrada), *Littorina saxatilis* (Gastropoda), and several eulittoral and terrestrial invaders.

Kromberg (1988) recorded a total of 67 species including 17 Nematoda, 13 Acarina, 11 Insecta, and 8 Copepoda-Harpacticoida. The most abundant species were Rotatoria (*Mniobia symbiotica*), Tardigrada (*Echiniscoides sigismundi*), Acarina (*Hyadesia fusca*), Collembola (*Anurida maritima*, *Hypogastrura viatica*), and Chironomidae (*Telematogeton japonicus*). Terrestrial animals enter this zone from the supralittoral for feeding, e.g., *Anurida maritima*, *Hypogastrura viatica*, *Bdella septemtrionalis*, and *Abrolophus rubipes*.

2.7 HARD SHORE MACROALGAE

2.7.1 ZONATION PATTERNS

Aspects of the distribution of rocky shore macroalgae have already been dealt with in previous sections of this chapter. Here we will describe and compare algal zonation patterns on representative shores from a range of geographic areas. As a starting point, we shall illustrate such patterns with those on an intermediately exposed shore in the British Isles, a shore dominated by fucoid and laminarian algae. On a shore where there is a complete coverage of fucoids, the vertical sequence from the top of the shore is *Pelvetia canaliculata*, *Fucus spiralis*, *Ascophyllum nodosum*, and/or *Fucus vesiculosus*, *F. serratus* (with *Laurencia pinnatifida*), *Laminaria digitata*, and *L. hyperborea*. Figure 2.27 compares this zonation pattern with that of shores on the North American east coast, the North American west coast, Chile, New Zealand, temperate Australia, and South Africa.

2.7.2 FACTORS CONTROLLING THE LOWER LIMITS OF INTERTIDAL MACROALGAE

It has been demonstrated that a number of algal species of the upper shore exhibit their most rapid growth when kept permanently submerged rather than repeatedly being exposed to air (Norton, 1985). Therefore it can be assumed that they could live lower on the shore and that other factors must restrict their lower limits. The only seaweed that may require periodic emersion is the high-level species *Pelvetia canaliculata*. When transplanted to the midshore, it thrived throughout the summer, but during the winter it declined and eventually decayed (Schonbeck and Norton, 1980). Thus most species of eulittoral macroalgae can survive at lower levels on the shore than those they normally inhabit.

A number of investigations have tested the role of interspecific competition in limiting the distribution of macroalgae. Removal of the plants of one species usually allows other species to extend their ranges. For example, on North American shores, the removal of *Chondrus crispus* allowed *Fucus* to colonize the *Chondrus* zone (Menge, 1976), and *Fucus serratus* occasionally appeared in cleared areas in the sublittoral on British shores (Kain, 1975). In another experiment, the clearing of *Ascophyllum nodosum* resulted in the colonization of the cleared areas by *Fucus spiralis* from above (Kain, 1975). Schonbeck and Norton (1980) denuded areas of the shore in which *Pelvetia* and *F. spiralis* grew when both species were fertile. They allowed the areas to become colonized by both species and then weeded out *F. spiralis* plants from parts of the cleared areas. In the unweeded areas, *Pelvetia* came to dominate only within its usual zone, but when *F. spiralis* was removed, *Pelvetia* ranged well down into the *Fucus* zone. *F. spiralis* outcompeted *Pelvetia* primarily due to its vastly greater growth rate.

On a gradient of environmental conditions, competition between species can result in zonal distribution patterns. It has been shown that several species of fucoids grow more slowly toward their upper limits than they do lower on the shore (Schonbeck and Norton, 1979). Such growth inhibition is partially due to the fact that the plants grow faster when submerged longer, but also because sublethal aerial emersion causes intermittent hiatuses in growth, the effects of which may persist after resubmergence. As a consequence, the uppermost plants of a zone may be stunted, fail to form a substantial canopy, and exhibit reduced fertility (Norton et al., 1981).

2.7.3 FACTORS CONTROLLING THE UPPER LIMITS OF INTERTIDAL MACROALGAE

For intertidal macroalgae, aerial emersion should constitute an adverse stress since it has been demonstrated that they grow best when submerged (Schonbeck and Norton, 1980), and prolonged emersion high on the shore can prove fatal. Protection from the worst effects of aerial emersion (such as in shaded situations, in rockpools, or beneath an overlying canopy of other macroalgae) enables many species to survive higher on the shore than normal. Thus, it

Zone	ENGLAND	PACIFIC GROVE CALIFORNIA	SOUTH COAST SOUTH AFRICA	CENTRAL CHILE	BANKS PEN. NEW ZEALAND
LITTORAL FRINGE	*Porphyra umbilicalis* *Pelvetia canaliculata* *Fucus spiralis* Filamentous algae (*Enteromorpha*, *Bangia*, *Eurospora*)	*Hildenbrandia occidentalis* Blue-green algae (*Enterophysalis*, *Calothrix*, *Plectonema*)	*Porphyra capensis* *Bostrychia mixta*	*Porphyra columbina*	*Porphyra umbilicalis* *Pylarella littoralis* *Prasiola crispa* *Ectocarpus* *Enteromorpha*
EULITTORAL	Upper: *Fucus spiralis* Mid: *Ascophyllum nodosum* *Fucus vesiculosus* *Laurencia pinnatifolia* *Leathesia difformis* Algal turf (*Laurencia*, *Leathesia*, *Chondrus*, *Cladophora*) Lower: *Fucus serratus* *Rhodomenia palmata* *Laurencia pinnatifolia* *Lomentaria articulata* *Plumaria elegans*	Upper: Algal turf (*Gigartina* spp. *Entocladia muricata* *Cladophora trichotoma*· *Pelvetia*, *Porphyra*, *Rhodoglossum*) Lower: Algal turf (*Gigartina* spp. *Porphyra perforata* *Entocladia muricata* *Iridea*, *Plocamium*, *Chondria*) Coralline turf (*Corallina* spp. *Calliarthron setchelliae* *Bossea dichotomon*) *Postelsia palmaeformis*	Upper Balanoid: *Porphyra capensis* *Bostrychia mixta* *Splachnidium rugosum* *Caulacanthus ustulatus* Lower Balanoid: *Gigartina striata* *Gigatina radula* *Gelidium pristoides* *Colpomenia capensis* Cochlear: *Gelidium cartilagineum* *Hypnea spicifera* *Gigartina radula* *Gigartina papillata* Algal turf (*Laurencia* spp. *Sargassum heterophyllum* *Caulerpa ligulata* *Dictyota dichotoma*	Upper: *Centroceros clavulatum* *Colpomenia sinuosa* *Gelidium filicinum* *Iridea laminarioides* *Ulva rigida*, *Enteromorpha* Mid: *Codium dimorphum* *Colpomenia sinuosa* *Adenosystis utricularis* *Scytosiphon lomentaria* *Halopteris hardacea* *Glossophora kunthii* *Corallina chilensis* *Notogenia fastigiata* *Iridaea laminarioides* *Griffithsia chilensis* *Plocamium pacificum* Lower: *Lessonia nigrescens* *Durvillaea antarctica* *Hildenbrandia* spp. *Ahnfeltia plicata* *Gymnogongrus furcellatus*	*Porphyra umbilicalis* *Ilea fascia* *Scytosiphon lomentaria* *Colpomenia sinuosa* *Leathesia difformis* *Adenocystis utricularis* *Ulva* *Bryopsis* *Enteromorpha* - - - - - - *Durvillaea antarctica* *Durvillaea williana* *Xiphophora chondrophylla* *Cystophora maschalocarpum* *C. scalaris* *C. retroflexa* Algal turf (*Halopteris* spp. *Glossophora kunthii*, *Zonaria subarticulata* *Polysiphonia*, *Laurencia* Encrusting corallines
UPPER SUBLITTORAL	*Alaria esculenta* *Laminaria digitata* *Laminaria hyperborea* "Lithothamnia" *Corallina officinalis* Mixed red and green algal turf	*Egregaria menziesii* *Alaria marginata* *Lessoniopsis littoralis* *Cystoseira osmundacea* *Iridophycus* Coralline turf "Lithothamnion"	*Hypnea spicifera* *Plocamium* spp. *Gelidium cartilagineum* *Bifurcaria brassicaeformis* *Sargassum heterophyllum* *Zonaria* spp. *Ecklonia radiata*	*Lessonia trabeculata* Calcareous crusts *Ralfsia*, *Hildenbrandia*, *Peysonella*, *Glossophora kunthii*, *Halopteris funicularis*, *Plocamium secundatum* *Macrocystis pyrifera*	*Lessonia variegata* *Marginariella urvillina* *M. boryana* *Cystophora platylobium* *Sargassum sinclairii* *Ecklonia radiata* Algal turf "Lithothamnion"

FIGURE 2.27 Comparison of algal zonation patterns on the shores of Great Britain, North American east coast, North American west coast, Chile, New Zealand, and South Africa.

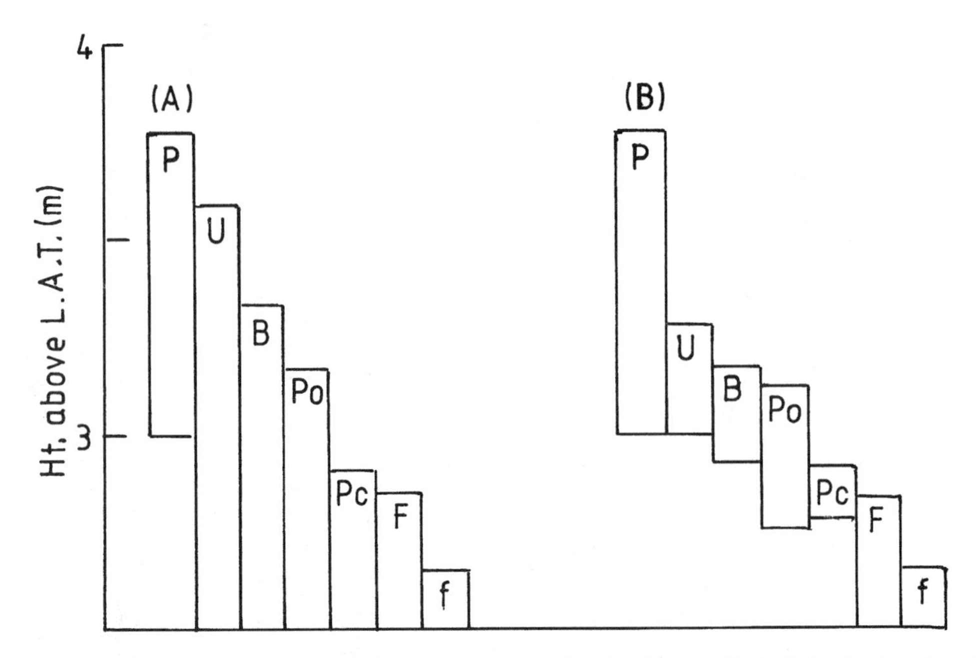

FIGURE 2.28 Zone sharpening in the littoral fringe at Hunterston, Ayreshire, Scotland. *A*: Vertical distribution of settled propagules, from summer, 1975, until winter, 1978–79. *B*: Vertical distribution of macroscopic plants, February, 1979. P = *Prasiola stipata;* U = *Urospora bangio*ides; B = *Bildingia minima*; Po = *Porphyra umbilicalis;* Pc = *Pelvetia canaliculata;* F = *Fucus spiralis*; f = *Fucus* plants of indeterminate species. L.A.T = Lowest Astronomical Tide. (Modified from Clokie, J.J.P. and Boney, A.D., in *The Shore Environment*, Vol. 2. *Ecosystems*, Price, J.H., Irvine, D.E.G., and Farnham, W.F., Eds., Academic Press, London, 1980b, 645. With permission.)

would appear that desiccation and overheating are of importance in limiting the upward extension of macroalgae on the shore. This was confirmed by Schonbeck and Norton (1978), who studied the zonation of littoral fucoid algae on the west coast of Scotland where there were five species of fucoids, each forming a well-marked zone.

Schonbeck and Norton (1978) correlated the condition of algae on the shore with climatic conditions over a period of two and a half years. They found that the upper limits of the zones highest on the shore, i.e., those dominated by *Pelvetia canaliculata, Fucus spiralis,* and *Ascophyllum nodosum*, were periodically "pruned back" during the summer. The picture that emerged from this study was of plants that could survive for long periods at the upper margin of their zone, but that a few critical days in the summer were sufficient to damage and often kill individuals that had settled above the level where they could survive such a critical period. In species such as *F. spiralis*, in which growth is rapid, damaged plants were quickly replaced as those at the upper edge of the zone could be killed off and recolonized during a single summer. However, if the slower growing *Pelvetia* plants were killed, it took much longer for them to recolonize the area.

Propogules of rocky shore algae generally settle over a wider vertical range on the shore than occupied by the adult plants (Figure 2.28). They therefore settle in positions where survival to adulthood is extremely unlikely. Considerable truncation of the settlement vertical range takes place due to the death of the sporelings, and although some may grow for a period, they are subsequently killed off so that the zone of the adult plants is "sharpened." This sharpening is illustrated by the study of Hay (1982) on factors affecting the upper limits of the southern bull kelp, *Durvillaea antarctica*, on southern New Zealand shores.

2.8 KEY FAUNAL COMPONENTS

Three groups of animals dominate the eulittoral zone: mussels, limpets, and barnacles. In this section we shall examine the dynamics of these groups and the role they play in structuring intertidal ecosystems.

2.8.1 Mussels

2.8.1.1 Introduction

Mussels (Family Mytilidae) are dominant space occupiers on exposed or semiexposed rocky shores in temperate habitats throughout the world, especially on horizontal or gently sloping surfaces. They are one of the most productive species on earth, rivaling the productivity of tropical rainforests and kelp beds (Whittaker, 1975), with a standing crop of up to 6.5 kg m^{-2} and a productivity of ca. 2.0 kg m^{-2} yr^{-1} for low littoral *Mytilus californianus* beds at Tatoosh Island, Washington (U.S.A.).

Although most mytilid populations are found in the mid- and low eulittoral zones, all appear to have the ability to live sublittorally. However, species such as *Aulacomya ater* and *Choromytilus meridionalis* in South Africa and *Modiolus modiolus* in New England (U.S.A.), have dense shallow sublittoral populations. *Mytilus edulis* and its species complex are most the widely distributed species living in circumpolar temperate habitats in both the Northern and Southern Hemispheres (Seed, 1976). Suchanek (1985) provides a detailed summary of the vertical and geographical ranges of the principal mytilid species found throughout the world.

2.8.1.2 Factors Limiting Mussel Zonation

Upper Limits — Mytilid upper limits are often very constant over long periods of time (Lewis, 1964; 1977; Paine, 1974; 1984). At Tatoosh Island, Washington, Paine (1974; 1976; 1994) has shown a constancy in the upper limits of *Mytilus californianus* over 12 years (1971–83). Data by Suchanek (provided by Paine) show moderate stability for the upper limits of both *M. edulis* and *M. californianus*.

While reduced food intake at high shore levels may result in slower growth rates and smaller-sized adults (Seed, 1976; Griffiths, 1981a,b), it appears that physiological intolerance of desiccation is the single most significant factor determining the upper limit of mussels. Mortality from high temperatures and/or desiccation has been well documented (Suchanek, 1979; Griffiths, 1981a; Tsuchiya, 1983). In a well-planned series of field and laboratory studies, Kennedy (1976) determined the influence of temperature and desiccation on three New Zealand mussel species, *M. galloprovincialis, Perna canaliculus,* and *Aulacomya mater.* His results indicated that their relative positions on the shore (*M. galloprovincialis* highest and *P. canaliculus* lowest) were consistent with their respective tolerances to temperature, and especially desiccation. He also found that there were combined and/or synergistic effects of temperature, wind, relative humidity, and species-specific or age-specific behavioral differences on desiccation-related mortality. In high latitudes, freezing can be equally important as desiccation.

Lower Limits — A number of observational and/or experimental studies have shown that biological factors (i.e., predation and competition) set the lower limits at which mussels grow on the shore. The pioneer work on this was carried out by Newcombe (1935) in the Bay of Fundy. He found that a suite of predators, a sea urchin (*Strongylocentrotus*), a dogwhelk (*Nucella*), and especially two species of seastar (*Asterias*) determined the lower limits of *Mytilus edulis* populations. In Ireland, Ebling et al. (1964) and Kitching and Ebling (1967) have described a somewhat similar suite of predators that feed on *M. edulis* (the crabs *Carcinus* and *Liocarcinus*, the dogwhelk *Nucella,* and the seastar *Marthasterias*), and suggested that they controlled its littoral distribution. In New England the distribution of *M. edulis* is also controlled by a guild of predators: the dogwhelk *Nucella*, two seastars *Asterias* spp., and three crabs *Cancer* spp. and *Carcinus* sp. (Menge, 1976; 1978a,b; 1979; 1983; Lubchenco and Menge, 1978; Menge and Lubchenco, 1981).

Predator-controlled lower limits of three other mytilid species have been shown experimentally by R.T. Paine in which he removed predatory seastars: *Pisaster* in Washington, U.S.A. (Paine, 1974), *Stichaster* in New Zealand (Paine, 1971), and *Heliaster* in Chile (Paine et al., 1985). In each case when predators were removed, the dominant mussels at each site (*Mytilus californianus, Perna canaliculus,* and *Perimytilus purpuratus,* respectively) extended their vertical ranges downward. When the predators were allowed to return to the sites, the *Perimytilus* system returned to its original condition. However, in some instances *Perna* and *Mytilus* were able to grow beyond the size capable of being consumed by the seastars (Paine, 1976).

Competition has also been shown to control the lower limits of mussel species. Experimental manipulations of populations of *Perna* in New Zealand by Paine (1971) have shown that in addition to being controlled at its lower limit by the predatory seastar *Stichaster*, it can also be outcompeted by the bull kelp, *Durvillaea antactica*, which occur at the same level on the shore. When *Durvillaea* is removed, *Perna* settles and grows in the vacated space. On Washington coasts, *M. edulis* occurs on (1) a high littoral band occupying about 0.3 m in vertical extent, (2) in mid- to low-shore gaps formed in the *M. californianus* band, and (3) in holdfasts of the kelp *Lessoniopsis*. Whenever *M. californianus* was removed, either by winter storms or experimentally, *M. edulis* colonized the lower littoral sites.

2.8.1.3 Mussels as a Habitat for Associated Organisms

Mussel beds are often multilayered and as such they provide a habitat and refuge for a wide variety of associated organisms. The mussel bed habitat consists of three major components: the mussel matrix (mussel shells and byssal attachments), a diverse assemblage of associated organisms, and the accumulated sediment and detritus within the bed, especially at the base. The mussel matrix is more complicated than the surrounding substratum and as such provides a greater diversity of niches, thus leading to increased species richness. As the matrix thickness increases, it: (1) decreases the influence of wave action, temperature, and sunlight at the base of the bed; and (2) increases relative humidity and sedimentation. For *M. edulis* the matrix may reach a thickness of 10 cm (Nixon et al., 1971), while for *M. californianus* the matrix is often five or six mussel layers deep, reaching a thickness of about 35 cm (Suchanek, 1985).

The macroflora and fauna found in the mussel beds fall into three categories: (1) epibiota (the fauna and flora that grow on, or bore into mussel shells); (2) mobile fauna (organisms that move freely through the mussel matrix); and (3) infauna (those organisms that are restricted to and often dependent upon the organic detritus, sediment, and shell fragments that collect within the matrix).

1. *Epibiota*: A number of algae grow on the mussel shells, especially encrusting corallines and seasonal ephemerals such as *Enteromorpha* and *Ulva*. Sessile invertebrates include barnacles, serpulid tubeworms, other tube-building polychaetes, bryozoans, hydroids, anemones, and ascidians. Limpets and chitons are often found attached to the mussel shells. A number of species including polychaetes and barnacles bore into the calcarious mussel shells.
2. *Mobile fauna*: This includes a great variety of species, especially crustaceans (isopods, amphipods, tanaids, shrimps, and crabs). Other species include a variety of errant polychaetes (e.g., hesionids, syllids, nereids, polynoids), asteroids, ophiuroids, herbivorous and carnivorous gastropods, opisthobranchs, and some small fishes. Some of these species prey on the mussels, while others use the mussel bed as shelter from desiccation when the tide is out or as protection from predators. A number of small bivalves (such as *Lasaea* spp.) can also be found in the mussel matrix. Many of the species found in the beds are the same as the epifauna of the macroalgae at the same level on the shore.
3. *Infauna*: The accumulated detritus comprises feces, pseudofeces, other organic detritus, and inorganic components such as sediment, shell fragments, sponge spicules, and sea urchin spines. In a 22-cm thick *Mytilus californianus* bed in Washington, the dry weight of detritus may reach 65 kg m^{-2} (Suchanek, 1985). A comparable value for a 10-cm thick *M. edulis* bed in Rhode Island is 14.4 kg m^{-2}, with an organic content of 3.86% (Nixon et al., 1971)

Species richness of the associated macrobenthic community has been determined for a number of mussel communities: 303 taxa for a *Mytilus californianus* community (Suchanek, 1979; 1980), 90 for a *Modiolus modiolus* community (Brown and Seed, 1977), and 69 for a *M. edulis* community (Tsuchiya and Nishihara, 1985). Both species richness and associated diversity indices are correlated with age and the structural complexity of the mussel matrix (Suchanek, 1979). Species richness is also correlated with height on the shore; the higher on the shore, the fewer associated species.

2.8.1.4 Role of Mussel Beds in Coastal Ecosystems

Mussels as filter feeders are an important component of rocky shore ecosystems in most geographic areas. They often represent a considerable biomass, and their filtration activities can have a considerable impact on the properties of the water column in their vicinity. It is not unusual for the filtration capacity of these suspension feeders to be on the order of one fifth to one third of the total system water volume, thus imposing on the phytoplankton a mortality rate on the order of 0.20 to 0.33 day^{-1} (Dame et al., 1991; Hily, 1991).

Many studies have shown that mussel suspension feeders play a significant role in coupling the pelagic and benthic components of coastal ecosystems by:

1. Filtration of large quantities of material from the water columns.
2. Reduction and possible local depletion of phytoplankton concentrations.
3. The biodeposition of large quantities of high-quality organic material.
4. Remineralization of biodeposits.
5. Release of inorganic nutrients to the water column.
6. Increasing the availability of dissolved nutrients.
7. Affecting the nutrient ratios in the water column.

In addition, the mussel beds themselves provide: (1) a rich food source for a variety of predators, especially gastropods, starfish, crabs, lobsters, fishes, and birds (see Sections 6.2.3.2 and 6.3.5.7); (2) a substrate for the attachment of a range of sessile animals and plants; and (3) a habitat and food resource for a rich community of meiofauna and invertebrate macrofauna.

Tidal currents transport water masses from a large surrounding area to the sites where mussel beds occur, and thus enable a continuous input of fresh phytoplankton. Phytoplankton, however, constitutes only a minor proportion of the particulate organic matter in the water column seston (Hickel, 1984). The seston will also contain detritus derived from a variety of sources, in particular kelp and other algal detritus, resuspended phytoplankton, protozoa, and bacteria. While the non-phytoplankton food sources may contribute to the energy requirements of the mussels, laboratory experiments have shown that *Mytilus edulis* exhibits poor growth when supplied only with non-phytoplankton food (Williams, 1981). Maximal clearance rates for mussel beds range between 6 and 12 m^3 m^{-2} hr^{-1} (for a summary see Jorgensen, 1990). Feeding by the mussels gives rise to the biodeposition of large quantities of feces and pseudofeces. Some of these will be retained within the mussel bed or among the algal growths in the vicinity. Depending on wave action and the intensity of tidal currents, some will be carried away to be deposited in adjacent bottom areas, where they provide food for detritivores and deposit feeders. Through their metabolic activities, the mussels release nutrients to the water column and remineralize deposited material, either in the mussel bed matrix or on adjacent hard or soft sediments releasing nutrients for further algal growth.

2.8.2 Limpets

Limpets are a major component of rocky shore intertidal and shallow sublittoral ecosystems. Recent reviews of their biology and role in the dynamics of littoral and sublittoral community dynamics include those of Branch (1981; 1986), while their role as herbivores has been reviewed by Lubchenco and Gaines (1981) and Hawkins and Hartnoll (1983a).

2.8.2.1 Adaptations to Intertidal Living

Homing to a fixed site on the shore after feeding excursions is characteristic of many limpet species, and has long been assumed to be a behavior that protects against desiccation (see Figure 5.11). Homing ensures that the shell fits flush with the substratum. Limpets can thus clamp down on the rock surface and protect the soft parts from drying out and from extreme external salinities. Temperatures, which may approach lethal limits on hot summer days, can be countered by lifting the shell, thus cooling the body by evaporation (Lowell, 1984). Several littoral limpets migrate upshore during cooler conditions and then retreat downshore in the summer (Lewis, 1954a; Breen, 1972). Shell shape and texture can be correlated with physical conditions. High shore limpets tend to be tall, thus decreasing the area exposed to solar radiation (Vermeij, 1973). Shade-dwelling individuals often have smooth shells, while those exposed to direct sunlight are often sculptured.

The relative tolerance of limpets to wave action is influenced partly by shell shape and partly by adhesion (Branch, 1981). Kelp-dwelling species have streamlined shells, while in rock-dwelling species, shell shape profoundly influences the drag imposed by wave action. Branch and Marsh (1978) demonstrated fourfold differences in drag between tall-domed species and those with flat shells. Despite these patterns they could find no correlations between shell shape and relative intensity of wave action that each species normally experienced. There was, however, a correlation between wave action and the power of attachment, which was found to vary considerably between species from 69,990 g m^{-2} in *Patella argenvillei* to 484 g m^{-2} in *Acmaea dorsuosa* (see Branch, 1981). Limpets could be divided into two groups: those with a high ratio of drag to tenacity (0.80), all of which live where wave action is severe, and those with a much lower ratio (0.36), which live at less wave-exposed sites. The attachment of limpets is largely due to the pedal mucus (Grenon and Walker, 1980).

2.8.2.2 Factors Controlling Vertical Distribution

While there is evidence that physical conditions (extreme high temperatures or extreme cold) may at times be lethal, whether this controls the upper limits of limpet distribution is open to question. We do have many reported examples of limpet death due to extremely cold winters, or when high temperatures coincide with prolonged low tides. Wolcott (1973) studied the ecology and distribution of five species of acmaeid limpets over three years. He found that field temperatures never became lethal and that salinity variations seemed unlikely to control zonation. However, high shore species were periodically affected by desiccation. Wolcott developed a hypothesis that high-shore limpets fringe on a habitat that is physically demanding (and potentially lethal), but relatively unexploited. He proposed that it would profit high-shore limpets to expand their range upward to the limit of their tolerance to benefit from unexploited food resources, even if this involved the risk of death. The increased reproductive effort due to the higher food supply would offset losses due to mortality. In contrast, mid- to low-shore limpets do not have access to such unexploited resources and consequently remain well within their limits of physical tolerance, their zonation not set by physical conditions, but rather by behavior and interaction with other organisms. Choat (1977) found that when he experimentally removed a high-shore limpet (*Collisella digitalis*), a mid-shore species (*C. strigatella*) expanded its range upward in the absence of competition.

Juveniles of many limpets are limited to the lower shore, while adults move upshore as they grow, thus establishing size gradients on the shore (Vermeij, 1973). Juveniles of other species are restricted to cryptic habitats until they are large enough to tolerate the stresses of open rock faces. Recruitment, coupled with early mortality, are major factors determining the population density of limpets.

2.8.2.3 Algal–Limpet Interactions

Many studies (e.g., Underwood, 1980; Hay, 1982; Jernakoff, 1983; and see reviews by Branch, 1981; 1986; Lubchenco and Gaines, 1981; Hawkins and Hartnoll, 1983a) have established that limpets can regulate the abundance and distribution of littoral algae. Other authors have described how limpets eliminate or control the growth of microflora on hard surfaces (Castenholz, 1961; Nicotri, 1977). Patallacean limpets with their strong radular teeth are capable of excavating the rock surface and removing all of the microalgae and algal sporelings, while Siphonarian limpets with their relatively fragile teeth can only rasp at the surface, and usually feed by breaking off pieces of macroalgae (Creese and Underwood, 1982), thus seldom affecting the abundance of the microalgae. The regulation of algae by limpets depends on the rate of algal growth. On the mid- and high shore where physical stresses reduce algal growth, limpets dominate and little algal growth is evident. However, lower on the shore the growth of algal sporelings may be rapid enough for some plants to escape grazing and develop.

On many shores, limpets are abundant on the high and mid-shore, but are replaced by dense algal growths on the low shore. A clear exception to this pattern is the large limpet *Patella cochlear,* which dominates a narrow band at the level of low water spring tides on African shores (see Figure 2.12), reaching densities of 1600 m^{-2} (Branch, 1975a). Algae flourish both above and below the *cochlear* zone, and experimental removal of *P. cochlear* results in substantial algal growth. Once macroalgae establish themselves, they may prevent further settling of *P. cochlear* for extended periods. *P. cochlear* reaches its greatest density under moderate, but not severe wave action, being replaced by algae on sheltered shores and by mussels on very exposed shores. The success of *P. cochlear* is related to its specialized habit of developing "gardens" of fine red algae that are highly productive and apparently essential to sustain the high densities (Branch, 1981). If isolated limpets are removed, their scars are occupied by juvenile limpets that abandon their normal position on the shells of larger adults to occupy the vacant scars. This is the reason for the stable populations of *P. cochlear.*

The limpet *Acmaea tessulata* is regularly associated with the coralline alga *Clathromorphum circumscriptum,* which depends on its grazing to remove epiphytes that would otherwise smother it (Steneck, 1982). In tropical Panama the abundance of a blue-green algal crust (? *Schizothrix calicula*) is inversely related to the dominant limpet *Siphonaria gigas.* Experimental removal of *S. gigas* increased the cover of *Schizothrix* but decreased other encrusting species (e.g., *Ralfsia*) that appeared to be outcompeted by *Schizothrix* in the absence of *S. gigas* (Levings and Garrity, 1984).

Various authors have suggested that a heteromorphic algal life cycle — an alternation of an encrusting stage with an upright foliose stage — is an adaptation to variations in the intensity of grazing. For example, Slocum (1980) demonstrated that the alga *Gigatina papillosa* has a flat crustose phase that is dependent on grazers to prevent its overgrowth and an upright foliose phase that flourishes when grazing is reduced, and is a superior competitor. Dethier (1981) came to similar conclusions, and Lubchenco and Cubit (1980) have linked the two phases with seasonal changes in herbivory with grazing gastropods less active in the winter when the upright phase is prevalent.

Limpet grazing can also influence the species composition and diversity of the algae (Lubchenco and Gaines, 1981). Jara and Moreno (1984) have shown that limpets and chitons virtually prevent the growth of the upright foliose form of the red alga *Iridaea boryana* in southern Chile. This then permits the development of short-lived algae such as *Petalonia* and *Scytosiphon,* which are normally competitively excluded by *I. boryana.*

2.8.2.4 Limpet–Barnacle Interactions

Barnacles and limpets may have both positive and negative effects on each other. Limpets often hinder or prevent the establishment of barnacle by grazing on newly settled cyprids or "bulldozing" juvenile barnacles from the rocks (Dayton, 1975; Denley and Underwood, 1979; Underwood et al., 1983). Barnacles reduce the growth rate and reproductive output of some limpets (Choat, 1977; Underwood et al., 1983). At the same time, densities of juvenile limpets are often positively correlated with the density of barnacles (Choat, 1977). Barnacles appear to hinder the grazing of at least three limpets that are moderate or large in size (Choat, 1977). The relationship between barnacle, and limpets is further complicated when interactions with algae are taken into consideration. Underwood et al. (1983) have shown that at very high densities, the limpet *Cellana tramoserica* reduces recruitment and survival of the barnacle *Tesseropora rosea*, but in the absence of this limpet, and even when densities are low, algae develop and smother the barnacles, or prevent further settlement.

A number of authors have discussed cyclic changes that occur in limpet and algal abundance on British shores (Southward and Southward, 1978; Hartnoll and Hawkins, 1980; Hawkins, 1981a,b; 1983). Large algae such as *Fucus* species reduce barnacle settlement. The limpet *Patella vulgata* immigrates onto strands of *Fucus* and prevents the development of further algal sporelings. Eventually the adult *Fucus* are eliminated by wave action, assisted by the limpets, which weaken their stipes. As the fucoids decline, barnacle settlement increases. The barnacles retard colonization by some algae, such as diatoms and *Enteromorpha*, but they seem to enhance the settlement of fucoids because their sporelings settle on and between the barnacles where they are relatively inaccessible to the limpets. As a result, these sporelings can escape the limpet grazing until they are large enough to be comparatively immune to grazing (Hawkins, 1981b). In Chile, Jara and Moreno (1984) have shown that the removal of limpets and chitons allows the red alga *Iridaea boryana* to form a foliose canopy largely excluding barnacles. However, in the presence of the grazers, the growth of *I. boryana* is restricted to a crustose sheet and barnacles become abundant.

2.8.2.5 Intra- and Interspecific Competition

In a range of studies, intraspecific competition has been shown to influence growth, survival, recruitment, body weight, and reproductive output (Branch, 1975a,b; Creese, 1980; Fletcher, 1984a,b; and see Branch, 1981; 1984 for reviews). There are several possible mechanisms for reducing intraspecific competition. Breen (1972) and Branch (1975b) have described how the limpets *Colisella digitalis* and *Patella granularis* migrate upshore as they age, leading to size gradients, with the largest limpets highest on the shore. Workman (1983) analyzed the energetics of high- and low-shore populations of *Patella vulgata* and showed that although consumption and growth are higher on the low shore, longevity is greater on the high shore, with the result that lifetime reproductive effort is greater on the upper shore.

Interspecific competition between limpet species has been reported in widely separated geographic localities. Haven (1973) experimentally removed either *Collisella digitalis* or *C. scabra* from shores where the two species coexisted, and found that each of the species grew faster in the absence of the other. Similarly, by manipulating the densities of three gastropods in experimental cages, Underwood (1978b) demonstrated that the snail *Nerita atramentosa* outcompeted both the snail *Bembicium nanum* and the limpet *Cellana tramoserica,* with the body weight of the two latter species decreasing in proportion to the density of *C. tramentosa,* which moves farther and feeds faster than the other species. *C tramoserica*, in an experiment where it was caged with two pulmonate limpets, *Siphonaria denticulata* and *S. virgulata,* was found to outcompete the two siphonarians, which grew more slowly and experienced a higher mortality rate (Creese and Underwood, 1982).

It has been demonstrated that the zonation of certain limpets is influenced by interspecific competition. Choat (1977) demonstrated that the removal of the high-shore *Collisella digitalis* allowed another species, *C. strigatella,* to move upshore. Black (1979) found that the siphonarian *Siphonaria kurrachaensis* had a bimodal pattern, being more abundant above and below the mid-shore zone dominated by *Notoacmaea onychitis*. After experimental removal of all the limpets from the shore, Black recorded that *S. kurrachaensis* established itself and developed its highest densities in the zone normally occupied by *N. onychitis*. On South African shores, the large limpet, *Patella cochlear,* dominates the low-shore zone (Stephenson, 1936; Stephenson and Stephenson, 1972) and excludes practically all other limpet species (see Figure 2.29). One species, *Patella longicosta,* is common above and below this "cochlear zone," but only occurs in this zone after experimental removal of *P. cochlear* (Branch, (1976).

Interference and territoriality also play important roles in the distribution of limpets. A dramatic example of territorial behavior is that of *Lottia gigantea*, a large limpet that occupies patches of algal film on the western North American coast, which it defends by thrusting against intruders (Stimpson, 1970). Removal of *L. gigantea* results in invasion of its territory by numerous small acmaeid limpets that eliminate the algal film. *Lottia* generally excludes these acmaeids from its territory and also ousts herbivorous gastropods and some space-occupying organisms, and retards the encroachment of mussels. The growth rate of *L. gigantea* is correlated with the size of its territory, and it expands its territory when algae are scarce so that its density is higher when food is abundant (Stimpson, 1973). Territoriality is well known in a number of other limpet species, e.g., *Patella compressa* (Branch, 1975a), *P. cochlear* (Branch, 1975b), *P. longicosta,* and *P. tabularis* (Branch, 1976; 1981). Most species of limpets exhibit evasive action when they intrude on other limpets' territory (Branch, 1986).

2.8.2.6 Limpet–Predator Interactions

Many animals are important predators on limpets, including whelks (Black, 1978), crabs (Chaplin, 1968), starfish (Dayton et al., 1977; Blankley and Branch, 1985), octopus (Wells, 1980), gulls (Blankley, 1981; Lindberg and Chu, 1983), oystercatchers (Hockey, 1983; Hockey and Underhill, 1984), and fish (Stobbs, 1980).

Blankley and Branch (1985) have estimated the annual consumption of the limpet *Nacella delesserti*, the dominant macroinvertebrate on the shores of the Subantarctic Marion Island. They found that the starfish *Anasteria rupicola* accounted for 40% of the mortality of the limpets in the size range 25 to 60 mm; in shallow water the kelp gull *Larus dominicanus* consumed 20% of the limpets over 35 mm in length; and the Antarctic cod *Notothenia corriceps* accounted for an estimated 15 to 30% of the limpets. Parry (1982) estimated that each year oystercatchers (*Haematopus fuliginosus*) ate 9% of the population of *Cellana tramoserica* at Victoria, Australia, and that wrasses (*Pseudolabrus* spp.) annually remove 52 to 80% of the population of *Patellate alticostata* and about 100% of *Patella peroni.* Thus it is clear that for at least some limpets, predation is a major source of mortality.

In South Africa there have been extensive studies of the influence of seabirds on rocky shore communities comparing mainland sites lacking large bird populations with offshore seabird islands. The African black oystercatcher, *Haematopus moquini,* reaches a density of 75 birds km^{-1} on the islands, whereas on the mainland the

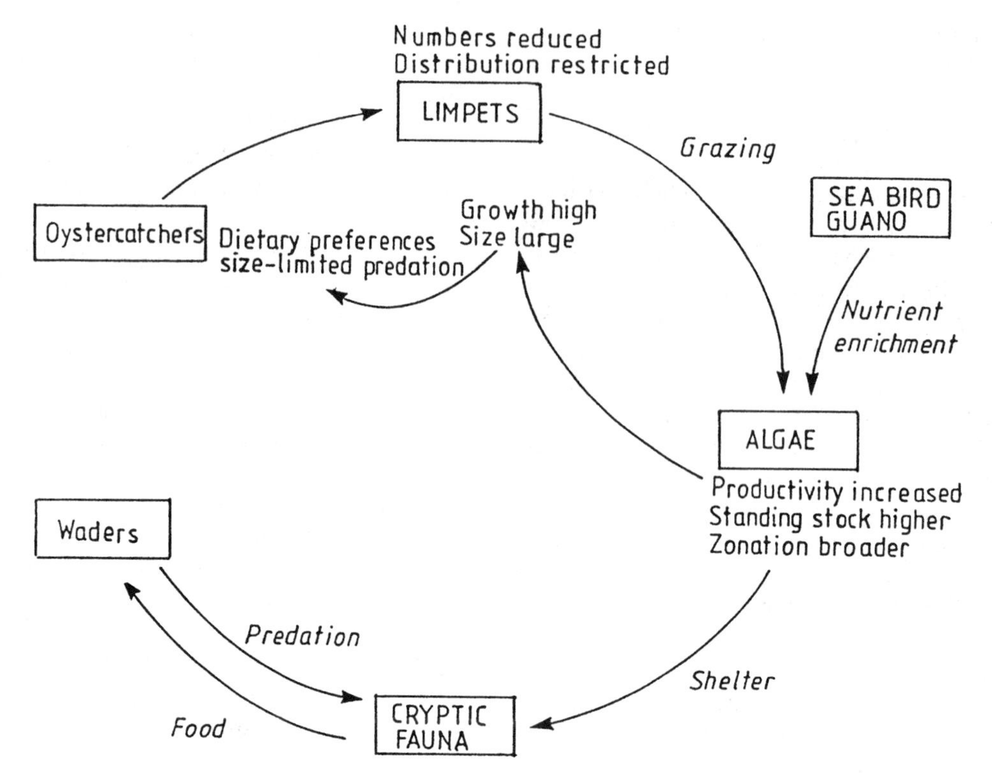

FIGURE 2.29 Conceptual model of the interactions between birds, limpets, and macroalgae on seabird islands off the west coast of South Africa. (Redrawn from Branch, G.M., *The Ecology of Rocky Coasts,* Moore, P.C. and Seed, R., Eds., Hodder & Straughton, London, 1986, 116. With permission.)

corresponding density is about 3 birds km^{-1}. On the islands the oystercatcher predation reduces the limpet population, and the resulting reduced limpet grazing enables large beds of macroalgae to develop in conditions of nutrient enrichment from seabird guano. This favors waders, which feed on the abundant cryptic fauna associated with the macroalgae. Figure 2.29 illustrates the interactions between birds, limpets, and macroalgae on the islands (Branch, 1986).

Limpets have been shown to detect predators at a distance and they have evolved species-specific responses that allow them to escape. In most cases this involves elevation of the shell and rapid flight away from the predator (Branch, 1978). Large individuals of *Patella oculus* and *P. grenatina* on South African shores lift the shell and smash it down on predators — a behavior that is size specific and predator specific (Branch, 1979). Both the habit of homing to a scar and the timing of activity rhythms may be important in reducing predation.

2.8.3 Barnacles

Barnacles are one of the most conspicuous features of the eulittoral zone of temperate shores throughout the world where they may cover the rock surface to the exclusion of other attached or almost stationary animals, except limpets and chitons. Depending on the type of shore and the degree of wave action, competition for the space occupied by the barnacles occurs with mussels and algae. While a single barnacle species may dominate the shore, more often two or more species of barnacle compete for space. On most shores, limpets and chitons coexist with the barnacles, as well as littorinid snails, other herbivorous gastropods, and whelks. Spaces among the barnacles harbor other smaller animals, including the small bivalve *Lasaea* spp, small crustaceans (harpacticoid copepods, amphipods, isopods), and often considerable numbers of mites and dipteran larvae.

2.8.3.1 Adaptations to Intertidal Life

Differential barnacle zonation on rocky shores implies varying adaptation to the tidal regime and the modifying factors discussed above. It is well established that the upper limits are largely determined by physical factors associated with tidal emersion and that the lower limits are set by a complex of biological relationships. Distribution and density is also strongly influenced by recruitment (see below).

The relatively impermeable shell of the barnacle is both protection against predators and desiccation loss. In order to survive high on the shore, barnacles must be able to tolerate stresses such as reduced feeding time, high temperatures, and desiccation. Those species that

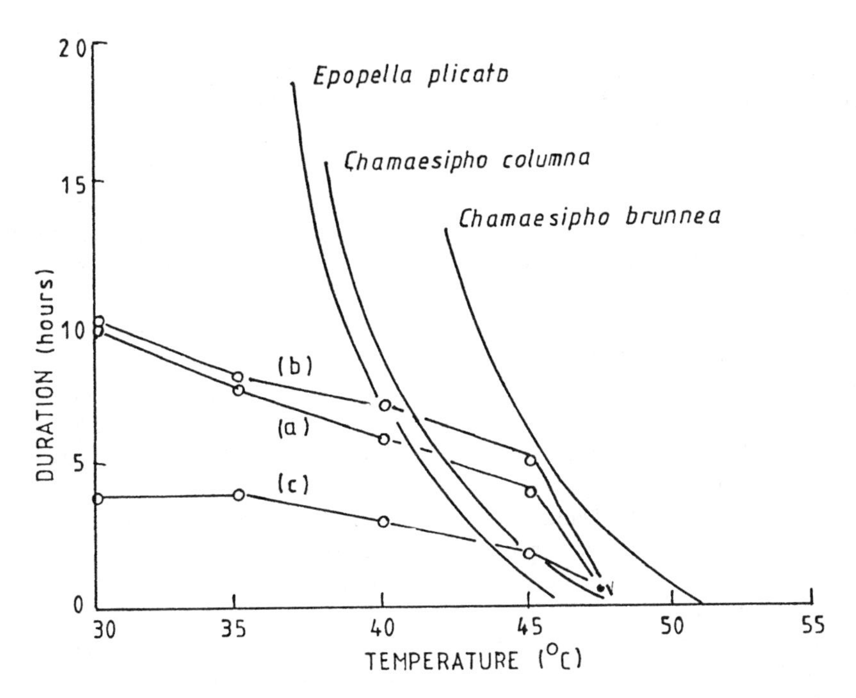

FIGURE 2.30 Time–temperature relationships of shore and barnacles at Leigh, New Zealand. Maximum duration of temperature at three levels (a = highest; c = lowest) compared to time–temperature survival curves in adults of three barnacle species from the same shore. Adult *Epopella plicata* and *Chamaesipho columna* are never exposed for more than one tidal cycle; *Chamaesipho brunnea* can be emersed for two weeks between spring tides during calm weather. (Redrawn from Forster, B.A., in *Barnacle Biology*, Southward, A.J., Ed., A.A. Balkema, Rotterdam, 1987, 121. With permission.)

occur high on the shore are better able to cope with these stresses, as has been demonstrated for barnacles on British shores (Forster, 1969; 1971a). High temperatures accelerate evaporation and desiccation during emersion. Newly settled cyprids are particularly susceptible to desiccation (Connell, 1961b; Forster, 1971b), but with growth they can survive the effects of desiccation longer (Forster, 1971a). High tidal chthamalid barnacles have particularly good desiccation tolerance and frequently have to survive several days of continuous emersion on calm days. Figure 2.30 depicts the maximum duration of temperatures measured at three levels on the shore at Leigh, New Zealand, compared with the time-temperature survival curves in adults of three species of barnacles on the same shore (Forster, 1969). The vertical distribution of these three species is shown in Figure 2.30. It can be seen that the temperature tolerance correlates with their vertical distribution. Adult *Epopella plicata* and *Chamaesipho columna* are never exposed to air for more than part of a tidal cycle, whereas *Chamaesipho brunnea* can be emersed for two weeks between spring tides during calm weather. Many shore barnacles are living close to their tolerance limits. The adaptations that enable them to do this are a combination of physiological and behavioral ones, the former for enhanced temperature tolerance, the latter for keeping the opercular closed tightly during emersion to reduce water loss (Forster, 1971a).

2.8.3.2 Settlement

Adult barnacles as sessile organisms require a suitable area of substratum to which each individual becomes permanently attached. Connell (1985) defines settlement as the point when an individual first takes up permanent residence on the substratum. Recruitment, on the other hand, is a measure of the recently settled individuals that have survived for a period after settlement. Numerous studies have addressed variations in the distribution and abundance of barnacles on rocky shores (e.g., Lewis, 1964; Hawkins and Hartnoll, 1983b; Caffey, 1985; Connell, 1985; Gaines and Roughgarden, 1985; Pineda, 1994). Many factors influence the settlement of intertidal barnacles. These include light quality, orientation and intensity, surface reflection, color and background illumination (Smith, 1948), surface angle (Pomerat and Renier, 1942), water movement (Crisp, 1955), surface texture (Crisp and Barnes, 1954; Yule and Walker, 1985), water agitation and exposure to waves (Smith, 1946), water depth (Bousefield, 1954), shore level (Strathman and Branscomb, 1979), the presence and abundance of conspecifics, and fragments or extracts of conspecifics or related species (see Crisp, 1974; Larman et al., 1982; Rittschoff et al., 1984), algal primary films (Strathman et al,, 1981; Hudson et al., 1983), and the nature of molecular films (Rittschoff et al., 1984).

Settlement patterns in barnacles have recently been reviewed by Hui and Moyse (1987). Settlement refers to all the behavioral events in the barnacle life cycle leading to and including attachment. These events can be divided into two phases: planktonic, or swimming, and an epibenthic phase (Chabot and Bouget, 1988) as detailed below.

1. *The swimming phase.* In the swimming phase, upward movement concentrates the cyprids near the surface of the water. This ensures contact with the intertidal substrate given onshore winds and currents.
2. *The epibenthic phase.* Contact with the substratum initiates exploratory crawling behavior. At this point cyprids are able to make active comparisons and choices between various prospective settlement positions near their original point of contact (Yule and Walker, 1985). During this phase, the cyprids use benthic diatoms on the substrate as cues. If the conditions are suitable, the cyprids more closely explore the substratum (see Newell, 1979). Three epibenthic phases are involved in examining the substratum (Crisp, 1974): (a) a "broad exploration phase" during which the cyprid takes a more or less straight path on the substratum, turning infrequently; then (b) after a long enough stay on the substratum, another behavior is adopted, that of "close exploration," characterized by short movements with frequent changes of direction; and lastly, (c) as attachment nears, the cyprid enters the "inspection phase" during which there are at most to-and-fro movements within its own length. Finally, the cyprid becomes attached to the substratum by a small disc on each of the antennules, secretes the basal attachment from the cement glands, which occupy much of the preoral region, and metamorphoses into the adult barnacle.

2.8.3.3 Factors Affecting Settlement

A number of factors have been shown to affect the settlement of barnacles, including rock type, the presence of adults of the same or related species, and the presence of substrate previously occupied by a mobile predator. The different composition of the rock on the shore can potentially affect the settlement and establishment of barnacles. Caffey (1982) showed that there were no consistent differences in the recruitment and survivorship of the intertidal barnacle *Tessopora rosea* on four different types of rock experimentally transplanted to widely separated shores. However, Moore and Kitching (1939) suggested that rock type strongly influenced the distribution of the barnacle *Chthalamus stellatus* on British shores. Raimondi (1988a) recently investigated the settlement, recruitment, and zonation of the barnacle *Chthalamus anisopoma* on two shores in the northern Gulf of California. He found that its vertical upper limit was, on average, 25 cm lower on a basalt shore than on a nearby granite shore, probably because of greater postsettlement mortality on basalt than on granite near the upper limits of their distribution; this difference may be related to the thermal capacity of each rock type, as the basalt gets much hotter than the granite.

In a series of experiments, Johnson and Strathman (1989) found that settlement of the larvae of the barnacle *Balanus glandula* was decreased on slate tiles that had previously been occupied by the postmetamorphic stages of the whelk *Nucella lamellosa,* which is an important predator on barnacles (Connell, 1970; Dayton, 1971; Palmer, 1984). Settlement of another barnacle species, *Semibalanus carinosus,* was not as greatly affected (Figure 2.31). A similar but lesser effect occurred after the treatment of the tiles with traces of a potential disturber of settled larvae, the limpet *Tectura scutum* and mucous of the brown alga *Fucus dischus.*

Most barnacle cyprids tested in the laboratory settle more readily on surfaces bearing settled conspecifics or on surfaces that have been treated with extracts of conspecifics than on bare, untreated controls (Knight-Jones, 1953; Gabbot and Larman, 1987). This effect, termed gregariousness, has been interpreted as an attraction that increases settlement density on surfaces suitable for subsequent survival in the field. Barnett and Crisp (1979) have suggested that gregariousness leads to reductions in local diversity, tending to aid the establishment of monospecific zones on the shore.

What, then, is the adaptive advantage of gregariousness? Although adult *Semibalanus balanoides* are rarely found above HWNT, its cyprids sometimes settle densely above this level, but die before metamorphosis, or when neap tides coincide with hot weather (Connell, 1961a). A recent field study (Wethey, 1984) found that *S. balanoides* cyprids are more responsive to other recently settled conspecifics than to recently metamorphosed spat, or to the bases of detached adults. Any settlement outside the zone occupied by adults would, therefore, lead to increased settlement there. Thus, gregariousness cannot restrict intertidal barnacles to specific zones on the shore. Hui and Moyse (1987) postulate that gregariousness is an effect that contributes mainly to the rejection of unfavorable substratum as opposed to one that increases settlement on favorable substratum.

There is extensive literature on the cues that induce the settlement of barnacle cyprids. Water movement (Crisp, 1955), surface texture (Barnes, 1956), light (Barnes et al., 1951), the presence of conspecifics (Knight-Jones, 1953; Wethey, 1984), and other indicator species (Strathman et al., 1981) have all been found to be important determinants of local spatial patterns in intertidal barnacles. It is important for the larvae to settle between the upper and lower limits at which the species can survive

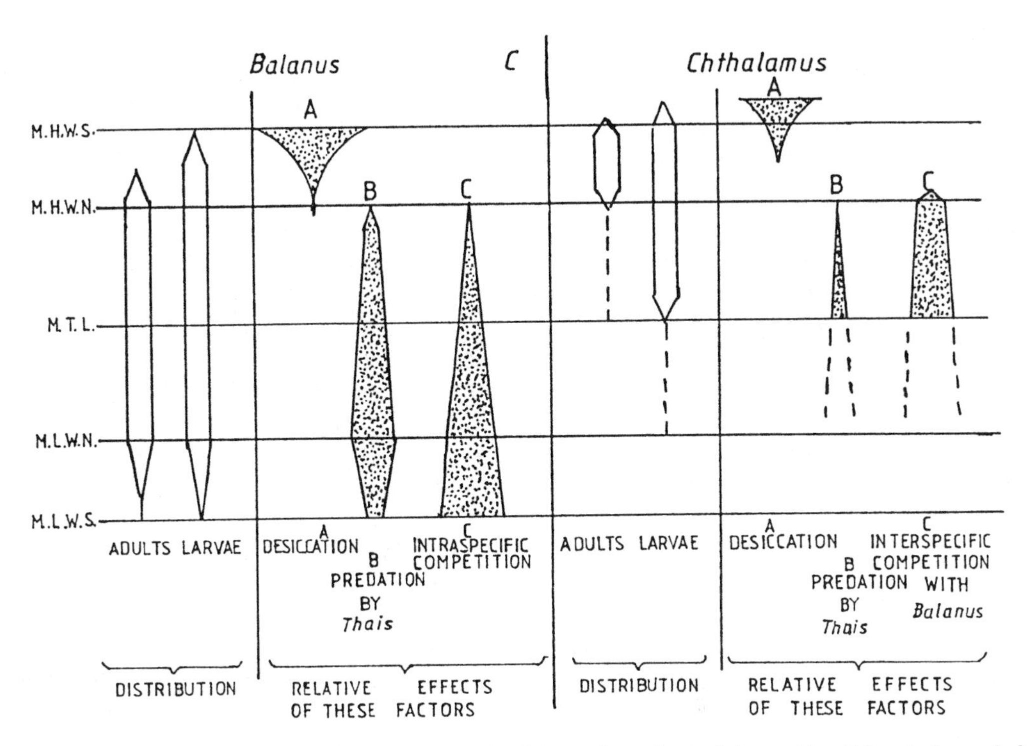

FIGURE 2.31 Comparison of the patterns of distribution of adults and newly settled cyprids of *Semibalanus balanoides* and *Chthalamus montagui* at Millport, Scotland, with diagrammatic representation of factors that result in adult pattern, and the influence of interspecific competition and other factors on the distribution of the barnacle *Chthalamus stellatus*. (Based on data from Connell, J.H., *Ecol. Monogr.*, 31, 710, 1961b. With permission.)

and reproduce. Strathman and Branscomb (1979) have argued that cues indicating the upper limits of a species distribution are more reliable guides for settlement than those that indicate a lower limit, because the physical stress that defines upper limits is less variable than biological stress, which largely determines the lower limit.

Raimondi (1988b) found that the upper limit of *Chthalamus anisopoma* in the northern Gulf of California was extremely variable, ranging from 1.5 to 2.6 m above MTL. *Chthalamus* showed increased settlement in response to chemical cues produced by several species whose vertical distribution overlapped with its vertical distribution. He found that *Chthalamus* did not recruit to areas with the blue-green bacterium *Calothrix,* regardless of whether barnacle extract had been applied, but that where *Calothrix* was removed, there were significantly more recruits to sites where barnacle extracts were applied. There was significantly more settlement in response to cues associated with the gastropod *Nerita funiculata* that occurred with *Chthalamus* than to *N. funiculata* that occurred above the upper limit of *Chthalamus*. Another species that induced settlement was the whelk, *Acanthina angelica*, a major predator of adult *Chthalamus*. *Acanthina* may be a good settlement cue because it is a reliable indicator of the vertical distribution of *Chthalamus*, and because (1) the clusters of *Acanthina* that induce settlement may move away before the new barnacle become edible, and (2) *Chthalamus* can develop a predator-resistant morphology in response to *Acanthina*. The variability of the upper limits noted above may be the result of the variability of the settlement cues at the time of settlement. In a further study, Raimondi (1991) found that settlement of *C. anisopoma* was largely restricted to areas within the adult distribution. This is support for the hypothesis that the larvae are induced to settle by a biochemical attribute of *Chthalamus*, or some related factor (e.g., a bacterial film associated with the barnacle).

2.8.3.4 Variability in Settlement and Recruitment

Variability in settlement and recruitment of intertidal barnacles has been widely documented (Connell, 1961a; 1985; Denley and Underwood, 1979; Caffey, 1982; Grosberg, 1982; Hawkins and Hartnoll, 1982; Kendall et al., 1982; Gaines et al., 1985; Gaines and Roughgarden, 1985; Wethey, 1985; Raimondi, 1990; Bertness et al., 1992). A variety of mechanisms have been proposed to account for this settlement variability. Factors affecting the delivery of cyprids from offshore to nearshore include reproductive output of adults (Barnes, 1956; 1962; Connell, 1961a; Geraci and Romairone, 1982), winds and current (Barnes, 1956; Hawkins and Hartnoll, 1982), tides, and tidally generated internal waves (De Wolf, 1973; Shanks, 1983). Mechanisms affecting delivery from nearshore waters to the rock surface are inadequately studied, but nearshore turbulence (Denny and Shibata, 1989), stratification of the larvae in the water column (Grosberg, 1982), and site-specific variation in abundance of competent larvae (Gaines et al., 1985) are all possibilities. Factors affecting attachment to the rock surface and subsequent metamorphosis have been the subject of a number of studies. Certain physical characteristics of surfaces, such as texture or contour (Crisp and Barnes, 1954; Barnes, 1956; Wethey, 1986) and composition (Rittschof et al., 1984; Raimondi, 1988b) have been shown to influence settlement. The presence of conspecifics can increase (Knight-Jones, 1953), or in high densities decrease (Connell, 1961b; 1985) settlement. Other species may act as settlement inducers (Strathman et al., 1981; Raimondi, 1988b) or inhibitors (Rittschof et al., 1984).

Bertness et al. (1992) examined the roles of variation in larval supply, settlement rates, and early settlement mortality in recruitment success in two populations of the barnacle *Semibalanus balanoides*. Water column larval supplies differed fourfold between populations. In the high larval supply population, settlement sites were rapidly saturated with settlers. As a consequence of the saturation of settlement sites, settlement in the remaining free space was intensified to 15 times higher than in the population with lower larval abundances. Within-population patterns in settlement were also conspicuous, but of lower magnitude than interpopulation differences. Settlement decreased with increasing tidal height but increased both with wave action and proximity of conspecifics. Both Bertness et al. (1992) and Pineda (1994) consider the available settling area important in understanding spatial variation between sites and temporal variation within sites. Settlement intensification takes place where available settlement area becomes limiting.

2.8.3.5 Barnacle Distribution Patterns

In order to breed, the adults of cross-fertilizing individuals must be close to each other. However, the relationships among neighboring conspecific adults may range from intense competition to obligate parasitism. Spatial competition among barnacles is greatest in the intertidal species (Stanley and Newman, 1980). The spatfall of abundant shore barnacles often occurs in such density that as they grow, contact between individuals is common. Connell (1961a,b) has termed the effects of continued growth of contiguous barnacles "crowding." Growth in such situations can lead to physical crushing of an individual, undercutting and uplifting, and smothering, all leading to the death of an individual. Sometimes when diametric growth is restricted by crowding, growth occurs in an upward direction, leading to the formation of hummocks (Barnes and Powell, 1950). Such hummocks are more susceptible to dislodgement.

Where two or more species of barnacle densely populate the same substratum, competition for space can occur. Where *Semibalanus balanoides* and *Chthalamus montagui* are found on the same shore in the British Isles, the latter species is zoned above the former. In a study of the ecology of these species at Millport, Isle of Cumbrae, Scotland, Connell (1961a,b) found that the upper limit was determined by the desiccation tolerance of the species, while the lower limit of *C. montagui* was attributable to its inability to survive the spatial competition in the *S. balanoides* zone. Although the settlement of *C. montagui* cyprids was not limited to the zone occupied by the adults, nearly all of the *C. montagui* that settled in the *S. balanoides* zone were crushed, lifted off, or smothered before the end of the first year. This mortality seemed to be due almost entirely to crowding by the fast-growing *S. balanoides*, while little intraspecific crowding was observed in *C. montagui*. The net result was a very narrow overlap between the adults of the two species (Figure 2.31).

On other shores in Great Britain, a third barnacle species is present, the immigrant species from New Zealand, *Elminius modestus*. This species has a wide vertical distribution, ranging from the sublittoral almost up to HWST, and while it dominates on estuarine shores, it is rare, though present, on exposed clear water shores. The spread of *E. modestus* in the British Isles and in northwest Europe has been well documented. Hui (1983) and Hui and Moyse (1987) have studied the settlement and survival of this species at Munbles, South Wales, and arrived at the following conclusions: (1) *E. modestus* settled readily on established adult barnacles, as well as on bare rock; (2) crowding by both newly settled and established *S. balanoides* was the major cause of mortality in *E. modestus*; (3) *E. modestus*, which had settled epizoically on *S. balanoides*, suffered crowding mortality comparable to those settled on bare rock; (4) *E. modestus* suffered mortality due to interspecific crowding of the same order as *C. montagui* when the aggressor was newly settled *S. balanoides*, but fared somewhat better than did *C. montagui* when the mortality was caused by established adult *S. balanoides*; and (5) neither epizoics nor nearby settlement of *E. modestus* caused appreciable mortality to *S. balanoides* adults. While *S. balanoides* has a minimum 1-year generation time, *E. modestus*, with a maximum generation time of 14 weeks (Barnes and Barnes, 1962), can ensure an abundant continuous spatfall during the summer that will enable it to occupy all the available space at the time of settlement of *S. balanoides*.

Luckens (1970; 1975a,b; 1976) studied the relationship between the barnacles *Epopella plicata*, *Chamaesipho columna*, and *C. brunnea* at Leigh, New Zealand (Figure 2.32). She found that *C. brunnea* seemed to be restricted to the upper shore by *C. columna* in the same way as described above for the exclusion of *S. balanoides* by *E. modestus* on sheltered shores, i.e., the prolonged settlement season of *C. columna* reduced the availability of substrata for *C. brunnea*, which has a short settlement season and avoids settlement on top of established barna-

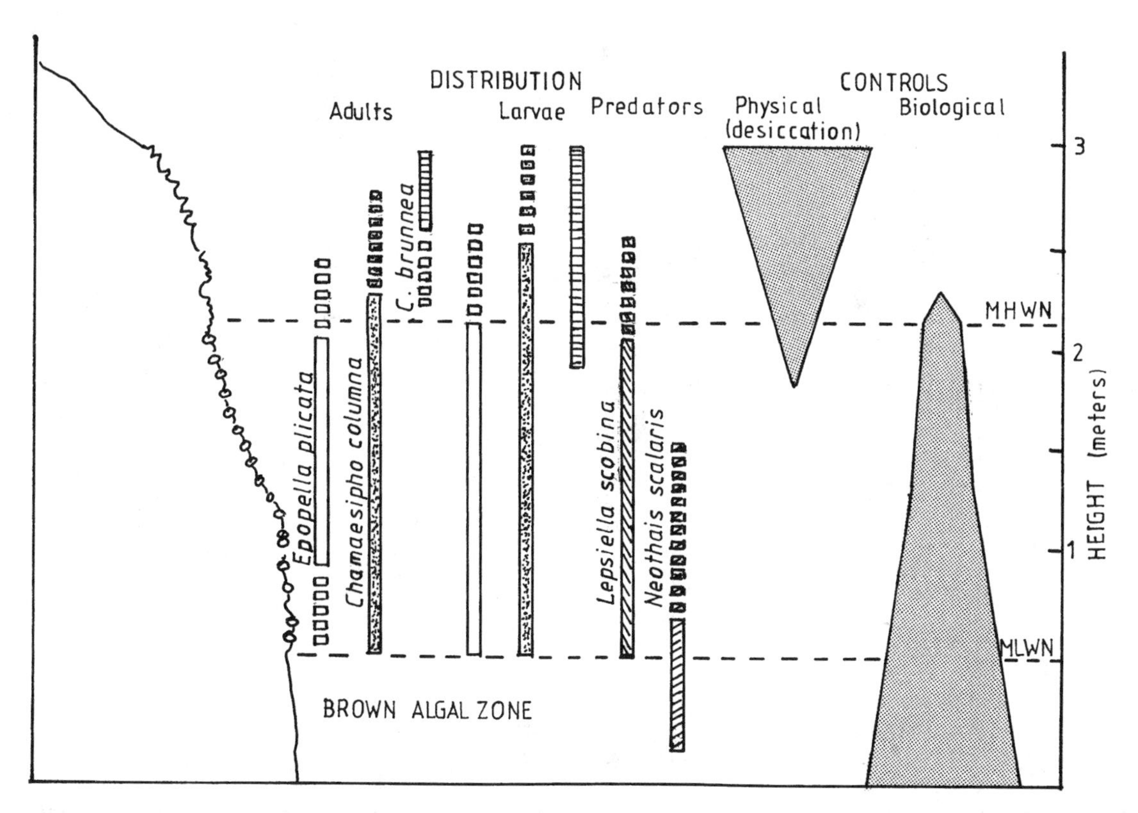

FIGURE 2.32 Vertical distribution of the adults and larvae of the barnacles *Epopella plicata, Chamaesipho columna,* and *C. brunnea* and the predatory gastropods *Lepsiella scobina* and *Thais orbita* on the shore at Leigh, New Zealand, and the relative importance of physical and biological controls that operate to determine the distributions. (After Luckens, P.A., *N.Z. J. Mar. Freshw. Res.*, 9, 38, 1975b. With permission.)

cles. *E. plicata* and *C. columna*, on the other hand, have completely overlapping distributions on the same shore. Their coexistence seems to be due, as in *E. modestus* and *S. balanoides* on British shores, to their different settlement patterns, growth rates, and shell strength. Whereas both breed throughout the year, *C. columna* settled mainly on sheets of bare rock or on the top of other barnacles, whereas *E. plicata* frequently settled on the distal parts of the shell wall of adult conspecifics to form clumps. *C. columna* is short lived but fast growing, whereas *E. plicata* is long lived and slow growing. As a result, where *E. plicata* become large and well established, it becomes difficult for *C. columna* to displace them, while where *C. columna* was densely settled, *E. plicata* could only settle on the top of sheets of *C. columna* where they are poorly attached and easily dislodged. Thus, the outcome of interspecific barnacle interaction is clearly dependent on the timing and density of the first colonists, the influence of the first colonizers on the settlement of others, and the relative growth rates of the competing species in effecting mechanical smothering or dislodging.

2.8.3.6 Predation and Other Biotic Pressures

The principal predators on barnacles are muricid gastropods (see Figure 2.33) (Connell, 1961a,b; Luckens, 1970; 1975a; Menge, 1976; 1983; Palmer, 1983; 1984; and others), flatworms (Luckens, 1970; Hurley, 1975), starfish (Paine, 1966), crabs (Menge, 1983), and fish (Newman, 1960).

In northern New Zealand, the muricid gastropod *Lepsiella scobina* does not occur with *Chamaesipho brunnea* (Figure 2.33). The slower growth rate of *C. brunnea*, reaching reproductive maturity in 3 years compared with 3 months for *C. columna*, means that it would be threatened with extinction if it were in the range of muricid predators, and the breeding populations of *C. brunnea* are restricted to the upper shore. Thus, the high shore is a refuge for slow-growing barnacles. Others attain a size refuge in that their shells become too thick for the muricids to complete drilling through them during a tidal period. The mid-shore *Epopella plicata* grow to about 2 cm diameter, and at about 1 cm, are immune to predation from *L. scobina* (Luckens, 1975a; 1976). However, they are still vulnerable to a different and larger muricid, *Thais orbita*, which is restricted to the lower shore and crevices. Its predation results in a mid-shore band of *E. plicata* in a tidal refuge from *T. orbita* and a size refuge from *L. scobina.* Sebens and Lewis (1985) found that populations of the large barnacle *Semibalanus cariosus* in the San Juan Islands, Washington, U.S.A., were dominated by small (2 to 9 mm diameter) individuals, but there was

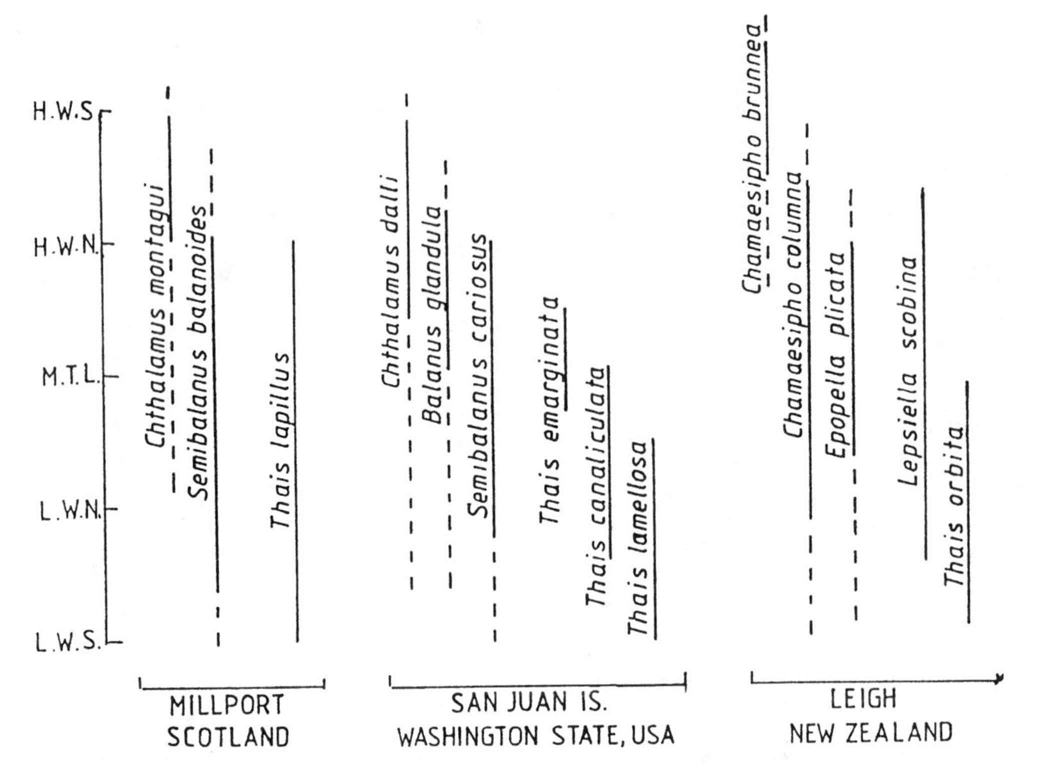

FIGURE 2.33 Distribution of barnacles and muricid gastropod predators on three shores. Barnacle settlements range to limits of dashed lines; adult survival to solid lines. HWS — high water spring tides; HWN — high water neap tides; MTL — mid tide level; LWN — low water neap tides; LWS — low water spring tides. Millport (after Connell, 1961a), Leigh (after Luckens, 1975b) with semidiurnal tides, San Juan Islands (after Connell, 1970) with mixed tides. (Redrawn from Forster, B.A., in *Barnacle Biology*, Southward, A.J., Ed., A.A. Balkema, Rotterdam, 1987, 120. With permission.)

often a distinct peak of larger barnacle (>20 mm) in the population size–frequency distribution at a particular site. During the winter of 1968–69, a massive death of predatory gastropods (*Thais* spp.) occurred. After that event, an extraordinarily large fraction of the *S. cariosis* population survived and grew to a large size. It was these survivors that produced the distinct peak of large barnacles present in 1973–77.

Predation is not the only cause of barnacle mortality. Small barnacles can be bulldozed by grazing limpets (Dayton, 1971; Creese, 1982), whiplashed by macroalgal fronds (Menge, 1976; Grant, 1977), or encroached by mussels (Paine, 1974; Grant, 1977).

2.9 SPECIAL HABITATS

2.9.1 Boulder Beaches

2.9.1.1 Boulder Types

Rocks that are present between the tides may be arranged in a variety of ways. Large irregular boulders may cover a shore and be so lodged that they resist movement by waves. In a like manner on tidal flats, boulders may be immobilized by being embedded in sediment (sand or silt). On some shores, small boulders or large stones and pebbles under the influence of strong wave action may build up a mobile shore in which the stones may be smoothly rounded by constant abrasion against each other. Such beaches develop shelving ramps and the component rocks become roughly graded, with the larger rocks lower on the shore and the smallest on the upper shore. The most inhospitable of shore habitats are those beaches where small pebbles and gravel, often several meters thick, are constantly in motion due to wave action. Thus, there is a great variety of boulder types and consequently a great variety of microclimates and microhabitats.

Recent investigation of boulder communities include those of Sousa (1979b), McGuiness (1984; 1987a,b), and McGuiness and Underwood (1986). A comprehensive account of the flora and fauna of New Zealand beaches is to be found in Morton and Miller (1968).

2.9.1.2 The Boulder Environment

Boulder beaches are subject to the same physical factors that are important on rocky shores, such as tidal rise and fall and wave action. However, they have a number of special features:

1. *The size range of the boulders.* This is important for two reasons: (a) small boulders are prone to

overturning and casting about by wave action (Osman, 1977; Sousa, 1979a,b), whereas large boulders are more stable, and depending on the specific conditions of exposure to wave action and the degree to which they may be embedded in sediment, they may not be subject to disturbance; (b) the top of a small boulder at the same level of the shore as a large boulder will be immersed considerably longer on both the ebb and flood tides (Underwood, 1980; Underwood and Jernakoff, 1984); and (c) the landward side of large boulders provides greater shelter and shade than small boulders.

2. *The amount of sediment present.* Depending on the coarseness of the sediment, the boulders are subject to sand scour and burial by sand.
3. *The nature of the boulder beach interstitial environment.* This varies with the sizes of the boulders, large stones, or pebbles that comprise the beach. Where the sizes are small enough and the beach comes under strong wave action, they build up a mobile shore of smoothly rounded or flatly abraded stones. On these shores, water drains rapidly through the interstices between the stones and pebbles, and while the surface is mobile, the deeper portions are subject to little movement. The nature and diversity of the community living between the interstices of such beaches is largely determined by the input of organic matter largely derived from macroalgae on nearby littoral and sublittoral rocky habitats and soft sediment sea grass beds. Environmental stresses such as desiccation are lessened in this interstitial environment where temperature extremes are dampened and humidity is high.

Where the constituent boulders or stones are stable, a distinctive assemblage develops on their undersides. Here, light is reduced, temperature changes are dampened, and humidity is high. The outer margins of the undersurfaces often have a fauna of encrusting organisms, coralline algae, barnacles, tubeworms, limpets, and chitons that are absent from the upper surfaces. Also under the boulder, the assemblage often includes encrusting coralline algae and bryozoans, vermetid molluscs, sponges, anemones, colonial and solitary ascidians, flatworms, tube-building polychaetes, rock oysters, brachiopods, and opisthobranchs. On the substrate beneath the stones, a number of scavengers or detritovores can often be found in dense aggregations. The dominant taxa include amphipods, crabs (especially species belonging to the family Grapsidae), porcellanid half-crabs (on New Zealand shore the half-crab, *Petrolithses elongatus*, occurs in thousands of individuals m^{-2}), and various carnivorous gastropods.

On the more sheltered boulder beaches where sediment and shell fragments collect, an assemblage of polychaetes belonging to the families Nereidae, Cirratulidae, Capitellidae, and Terebellidae are to be found. The deeper sediments are anaerobic and the species present resemble the faunal associations of mudflats.

Figure 2.34 gives a profile of a boulder beach in northern New Zealand with the intertidal distribution of the dominant species. On such a beach, the surface may remain bare of plants and animals down to ELWN and beyond to where the boulders are larger and cemented together by coralline and lithothamnion encrusting algae. Down to MLWN, the common mobile snails are the periwinkle *Littorina unifasciata,* and the topshells *Melagraphia aethiops* and *Nerita melanotragus*. An abbreviated barnacle zone develops at LWN, beyond which scattered rock oysters and crusts of the red alga, *Apophloea sinclarii,* are found. Further down the shore there may be a narrow belt of *Corallina,* with the brown kelp *Carpophyllum maschalocarpum,* the seasonal brown alga *Petalonia fascia*, and patches of the encrusting alga *Ralfsia verrucosa.*

The interstitial fauna toward high tide level is well developed, and zoned by depth as well as by distance down the shore. The chief source of food is terrestrial plant debris. Several lizards (skinks and geckos) are found here. Other common species are the wingless beach earwig *Anisolabis littorea*, amphipods (*Talorchestia* sp.), beetles (especially wingless staphylinids), collembolans, spiders, mites, and the beach centipede, *Scolioplanes* sp. Further below the surface is the air-breathing snail, *Suterilla neozelanica.* Further down the shore *Suterilla* is replaced by another snail, *Marinula filholi.* Where algal wrack collects, abundant larvae of the kelp-flies *Coelopa littoralis* and *Borborus empiricus* are found. On the upper shore the highshore crab *Cyclograpsus lavauxii,* and the isopod *Ligia novaezealandiae,* scavenge between the boulders. Under the boulders in the middle shore, two common crabs are found, *Leptograpsus variegatus* and *Heterozius rotundifrons*, together with the half-crab *Petrolithses elongatus* and the shrimps *Betaeus aequimanus* and *Alope spinifrons.*

Clean boulders fully submerged at each tide have a distinctive fauna beneath, including the anemones *Isactinia tenebrosa* and *Isactinia olivacea.* Other sessile animals beneath the stones include the serpulid tubeworms *Pomatoceros cariniferus* and *Hydroides norvegica.* A variety of gastropods may be abundant, including *Nerita melanotragus, Zediloma atrovirens, Z. arida, Z. digna,* and *Anisodiloma lugubri,* and the fragile limpets, *Notoacmea daedala* and *Atalacmea fragilis.* Common chitons on these boulders include *Chiton pelliserpentis* and *Amaurochiton glaucus.* Toward the low tidal coralline fringe, the catseye, *Turbo smaragda*, and the larger limpets *Cellana radians* and *C. stellifera,* are found on and beneath the boulders. At and beyond mean low water, two abalone species, *Haliotis iris* and *H. australis,* are found.

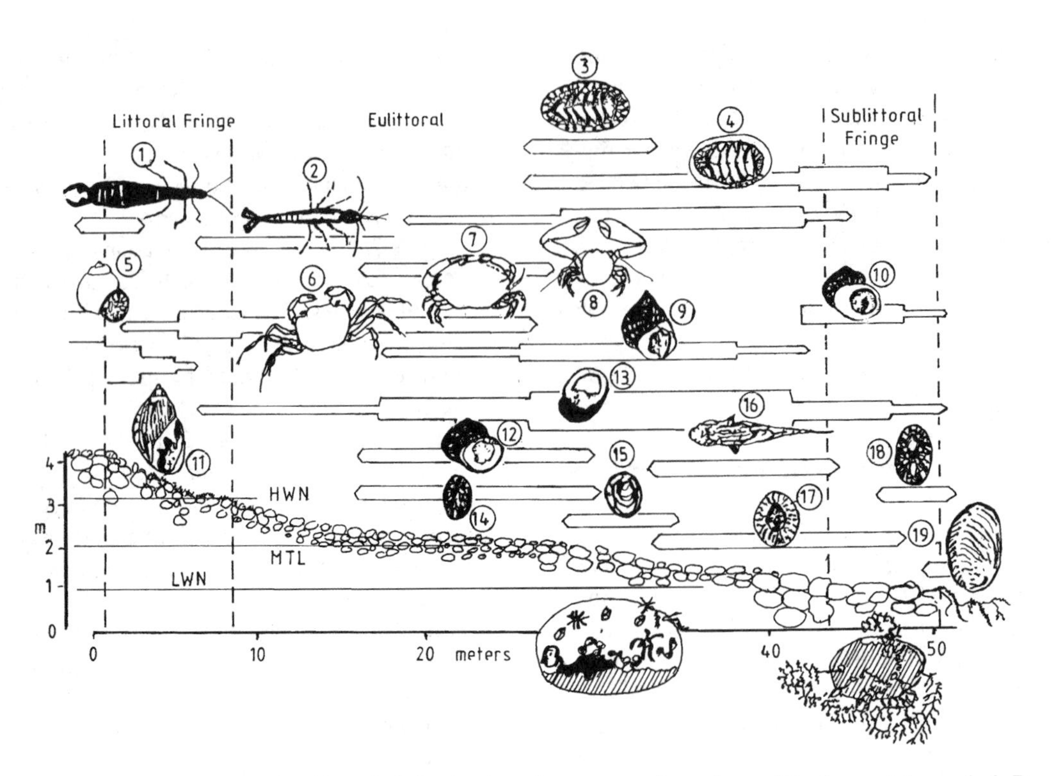

FIGURE 2.34 Profile of a boulder beach with typical distribution of some typical members of the fauna at Smuggler's Bay, Northern New Zealand. Species: 1. *Anisolabis littorea*; 2. *Beatus aequimanus*: 3. *Chiton pellesserpentis*; 4. *Amaurochiton glaucus*; 5. *Suterilla neozelanica*; 6. *Cyclograpsis lavauxi*; 7. *Heterozius rotundifrons*; 8. *Petrolosthes elongatus*; 9. *Melagraphia aethiops*; 10. *Turbo smargada*; 11. *Marinula filholi*; 12. *Zediloma atrovirens*; 13. *Nerita melanotragus;* 14. *Notoacmea daedala*; 15. *Atalacmea fragilis*; 16. *Trachelochismus pinnulatus*; 17. *Cellana radians*; 18. *Cellana stellifera*; 19. *Haliotis iris*. (From Morton, J. and Miller, M., *The New Zealand Seashore,* Collins, London, 1964, 103. With permission.)

2.9.1.3 Disturbance and Boulder Community Structure

Boulder communities are subject to disturbance on a greater scale than rocky shore communities. The most common physical disturbances in boulder fields are due to the movement of the boulders themselves and the movement of the surrounding sediment resulting in coverage or scouring by sand. Movement of the boulders is dependent on the intensity of the waves. Such movement is greater in the intertidal than the sublittoral (Osman, 1977). Sousa (1979a,b; 1980) noted that one type of disturbance to algae in intertidal boulders resulted when a boulder was overturned, and another when a boulder was cast against another. In the latter case, abrasion can denude the boulder of its biota.

Sousa (1979a,b; 1980) studied algal assemblages on boulders and found that the total space occupied increased with rock size, with nearly 100% cover on large rocks. Diversity, in contrast, was greatest on rocks of intermediate size. He explained these results by the Intermediate Disturbance Model (Connell, 1978). Diversity was small, and cover great, on the largest rocks because one foliose alga, resistant to physical stress and grazing, was able to outlast other species and dominate space. The greater rates of disturbance on somewhat smaller rocks prevented dominance and allowed more species to coexist. The very small rocks were so disturbed, however, that only the few species able to colonize and grow quickly were present.

In contrast to Sousa's findings, McGuiness (1987a,b), in a study of intertidal boulders on rock platforms on the New South Wales coasts, found that the percentage of the tops of rocks occupied by sessile species, predominantly algae, was never related to the size of the rock. Further, while some algae were less abundant on the smaller rocks, consistent with the effects of disturbance, others were most abundant on these rocks, or occasionally on rocks in the middle size classes. The assemblages on the largest rocks were more diverse (McGuiness, 1984), but this could result from their larger surface-area sampling more of the available colonists (McGuiness 1984; 1987a,b).

On the basis of his experiments, McGuiness (1987a) concluded that few of the patterns of abundance that he observed could result from domination of space in the absence of disturbance. An alternative explanation for these patterns involves the interaction of the effects of exposure during low tide and grazing, both known to be important on open platforms (Underwood, 1980; 1985; Underwood and Jernakoff, 1981; 1984). The time for which the top of the rock was exposed above water increased with rock size (McGuiness and Underwood, 1986). The difference was especially pronounced in the low shore area where, during the average low tide, small rocks were not exposed at all, but the tops of large rocks were out of the water for up to two hours (McGuiness and Underwood, 1986). The decline in the abundance of crusts with rock size on the shore studied may have been due to this increase in the period of exposure.

Disturbance by sand and waves did, however, appear to be of major importance to sessile species on the undersides of boulders. Patterns of diversity consistent with the Intermediate Disturbance Model were observed by McGuiness (1987b) in low shore areas. Further, with very few exceptions, species were most abundant on the medium to large rocks. Indeed, most space was typically occupied on the largest rocks, and total cover often approached 100%. Sponges, ascidians, and bryozoans often overgrew more solitary forms, notably polychaetes and barnacles, and experiments revealed that the abundance of the latter forms was limited by a lack of space for settlement. High on the shore, these interactions were not observed, and competition for space could not reduce abundance on larger rocks.

2.9.2 The Fauna Inhabiting Littoral Seaweeds

2.9.2.1 Introduction

Marine algae provide a number of resources for both invertebrates and fishes. Due to their abundance, especially on the lower shore and the sublittoral, they supply a greatly increased surface area for the attachment of sessile or nonsessile sedentary species. They also provide shelter, offering a refuge for motile species during periods of low tide, where desiccation effects are mitigated, and where they have protection from predators. They also act as sediment traps, thus providing an additional habitat for burrowing species, especially the meiofauna. Finally, the algae are an important food source, either directly, in the case of algal browsing species, or indirectly for species grazing on microalgal films and epiphytes on the fronds, or by providing detritus eventually fed upon by detritovores, especially in nearby soft sediments.

The animals that utilize one or more of these resources are generally termed epiphytes. Species found on the algae may be confined to one or a group of species, e.g., the anemone *Crichophorus nutrix*, which is only found on the fronds of the New Zealand brown algae *Carpophyllum* and *Cystophora*, or may occur on a variable range of species depending on their morphology. Others only temporarily utilize the algae for food and/or shelter and do not have a permanent association with the algae upon which they may be found. Hayward (1980) considers that the term *epiphyte* should apply only to a species that is directly dependent on an algal resource for at least the most significant stages of its life history. Such species might be termed primary, or obligate epiphytes. Secondary epiphytes are purely opportunistic, eurytrophic species that are able to utilize one or more of the algae over part of their habitat range, but are neither restricted to nor dependent upon them. Primary epiphytes would include sessile species whose larvae display a faculty for discrimination at settlement, leading to a restricted distribution on one or a few algal species. Numerous examples are to be found among tubeworms, bryozoans, and hydroids. The definition would also include specialized algal grazers and browsers. The secondary epiphytes of importance in the dynamics of algal communities include: (1) nonselective algal feeders; (2) species that depend on epiphytic algae as a food source (many algal species are host to a wide range of smaller epiphytic algae, especially filamentous reds and greens); and (3) predators that depend on the primary epiphytes as a food source.

2.9.2.2 Community Composition

The study of the fauna of coastal seaweeds started with Colman (1940), who studied the fauna of eight algal species along intertidal transects at Wembury, England, and identified 177 species of invertebrates. The numbers of each taxonomic group per 100 g of algae are given in Table 2.3. Numerically, the dominant groups were the Copepoda, Isopoda, and Amphipoda. Following this pioneer investigation, other studies were carried out by Wieser (1953) and Hagerman (1966). Broad-based reviews of algal invertebrate communities are those of Meadows and Campbell (1972) and Scheltema (1974). Hayward (1980) has provided an excellent recent review of such communities. The most comprehensive study of phytal communities is that of Edgar (1983a,b,c,d), who carried out an investigation on the Tasmanian coast from low water to a depth of 5 m. A list of the number of species in each taxonomic group and their total abundance is given in Table 2.4. Gammaridean amphipods had the largest number of species (70) and the greatest numerical abundance. Polychaetes (57 species) and Gastropoda (26 species) were the next most diverse taxa. In terms of total abundance, caprellid amphipods and polychaetes ranked second and third. In the bulk of studies to date on the phytal fauna, Crustacea, especially Amphipoda and Isopoda, dominate.

Morton and Miller (1968) described the fauna of the coralline alga, *Corallina officinalis*, on northern New Zealand shores. *Corallina* carries a wealth of small epiphytic algae, and silt and shell debris collects between its tangled fronds. Especially in rock pools, *Corallina* forms a multilayered habitat. Two molluscs that use it directly for food are *Turbo smaragda* and *Zeacumantus subcarinatus*, as well as a host of smaller gastropods, including juvenile *T. smaragda*, *Zebittum exile*, *Eatoniella olivacea*,

TABLE 2.3
Abundances of the Fauna Inhabiting Seaweeds on British Shores (In no. 100 g^{-1})

Invertebrate Fauna	*Pelvetia canaliculata*	*Fucus spiralis*	*Lichina pygmnea*	*Fucus resiculosus*	*Ascophyllum nodosum* and *Polycriphorics lanosa*	*Fucus serratus*	*Gigartina stellata*	*Laminaria digitata* holdfrots
Porifera	—	—	—	—	—	—	—	8.3
Coelenterata	—	—	—	—	3.0	0.2	216.0	31.0
Turbellana and Nemertinea	—	0.3	—	5.3	63.6	0.5	2.8	40.2
Nematoda	3.4	3.2	—	3.3	76.1	21.2	16.5	247.8
Polychaeta	—	0.8	—	—	72.6	0.6	73.8	2056.0
Oligochaeta	0.2	3.2	11.0	0.3	39.2	0.2	—	9.7
Sipunculoidea	—	—	—	2.0	—	0.3	10.8	1.5
Ostricoda	—	0.5	—	16.0	353.3	3.0	0.5	7.5
Copepoda	—	25.8	—	221.0	272.2	178.0	1676.2	54.0
Cirripedia	—	—	287.0	—	—	—	1.2	51.0
Tanaidacea	—	—	—	—	0.2	—	—	13.7
Isopoda	15.2	0.5	2886.0	4.3	30.0	8.1	32.2	6.3
Amphipoda	15.4	46.3	35.0	1.3	48.1	4.3	83.8	125.3
Decapoda	—	0.2	—	—	0.2	0.1	—	4.0
Pycnogonida	—	—	—	—	0.2	—	—	3.7
Acarina	—	3.2	436.0	135.7	222.4	75.6	758.8	7.3
Inscita	0.8	1.3	161.0	2.3	58.3	2.0	—	—
Bivalvia	—	—	9447.0	—	14.8	0.8	42.0	96.2
Gastropoda	8.8	13.5	435.0	67.0	163.3	23.5	72.2	19.5
Byozoa	—	—	—	—	0.2	6.1	158.2	73.5
Turicata	—	—	—	—	—	—	—	7.8
TOTAL	43.8	98.8	13716.0	458.5	1417.7	324.6	3145.0	2864.3

Source: Data from Colman, J., *J. Mar. Biol. Assoc. U.K.*, 24, 172, 1940. With permission.

E. varicolor, E. delli, E. huttoni, Rissoa hamiltoni, and *Estea zosterophila*. Many amphipod species are to be found on the fronds, including representatives of the families Lysianassidae, Stegocephalidae, Talitridae, Gammaridae, Amphithoidae, and Ampeliscidae. The isopod fauna includes representatives of the genera *Cerceis, Cilicaea,* and *Isocladus*. Small bivalves are usually very numerous, including *Notolepton citrinum, Lasaea hinemoa,* and *Nucula hartvigiana*. Other species include tanaids, the opisthobranch, *Alaja cylindrica,* small ophiuroids, *Amphiura* sp., the small crab, *Halicarcinus innominata*, larvae of the marine caddis, *Philanisus plebejus*, and the sipunculid, *Dendrosromum aeneum*. Where sediment collects at the base of the plants, a host of polychaete species, including deposit and filter feeders and carnivores (especially syllids) are found. Common species are the nereid, *Perinereis camiguinoides*, the hesionid, *Podarke angustifrons*, and the polynoid, *Lepidonotus polychroma*.

2.9.2.3 Seasonal Changes in Species Composition

Seasonal changes in species composition and abundance are correlated with seasonal changes in the species composition and abundance of the algal hosts. Tropical and temperate marine plants are generally considered to show opposing phenologies (Conover, 1964). The standing crops of temperate plants are greatest in summer while those of the tropical species reach maxima during periods of low water temperatures. (De Wneede, 1976). In his study of algal growth in temperate Tasmanian waters, Edgar (1983b) found that there were two major algal growth pulses: one in spring-summer and one in winter-spring. *Sargassum bracteolosum, Cystophora retroflexa, Caulocystis cephalornithos,* and *Zonaria turneriana* had maximum growth rates in winter-spring, while *Sargassum verruculosum* grew most rapidly over summer. Epiphytic algae generally respond to the same environmental conditions as the host algae. *Sargassum muticum* in the British Isles was found to be colonized by epiphytes shortly after the onset of fertility in summer (Jephson and Gray, 1977). The biomass of the epiphytes and the epiphytic fauna was greatly reduced in the winter.

In Tasmania, Edgar (1983b) found almost identical patterns in the monthly variation in epiphytic weight and animal abundance. Close relationships between epiphytic algae and animal abundance were also found previously by Wieser (1959) and Hagerman (1966), while Jannson

TABLE 2.4
Abundances and Number of Species of Invertebrates Inhabiting Macroalgae on Tasmanian Shores

Animal Group	Number of Species	Total Abundance
Gammaridean Amphipoda	70	22,262
Caprellid Amphipoda	5	7,772
Tanaidacea	6	918
Sphaeromatid Isopoda	6	530
Asellote Isopoda	11	520
Anthurid Isopoda	4	49
Gnathid Isopoda	1	32
Idotheid Isopoda	1	24
Cumacea	2	22
Mysidacea	1	9
Caridea	4	28
Anomura	1	1
Brachyura	5	17
Insecta	1	28
Pycnogonida	4	221
Chaetognatha	2	9
Anthozoa	1	18
Platyhelminthes	4	38
Polychaeta	51	3,559
Oligochaeta	3	33
Nemertea	9	96
Gastropoda	26	531
Amphiura	1	2
Holothuridea	3	13
Echinoidea	1	4
Ophiuroidea	1	34
Asteroidea	2	32
Pisces	4	24
Total	230	36,826

Source: Data from Edgar, G.J., *J. Exp. Mar. Biol. Ecol.*, 70, 137, 1983c. With permission.

(1974) showed that fluctuations in the density of meiofauna corresponds directly with changes in the standing crop of filamentous *Cladophora* and *Ceramium*. As well as being comparatively inaccessible to external predators, phytal grazers and detritovores living among the dense growth of filamentous algae would also benefit from the considerable abundance of food resources. Many of the animal groups in the Tasmanian study showed monthly doubling of population sizes in late spring and early summer. The decrease in animal numbers in autumn was equally rapid, indicating that recruitment was low and mortality was high at this time of the year. It was unlikely that mass emigration of animals occurred, because of limited algal growth in adjacent habitats, or that the fauna was reduced by extreme physical factors. Thus, predators must have been responsible for this decline in population levels. The gut contents of two carnivorous fish (*Neodax balteatus* and *Nesogobius pulchellus*) and one omnivorous fish (*Acanthaluteres spilomelanurus*) contained many of the phytal species, and these predators were extremely abundant within the study area. Predation pressure would be maximal in autumn due to the increased metabolic rates of the fish at higher water temperatures, considerable prey densities, and declining structural complexity of the filamentous algae.

Edgar (1983a) concluded that the most likely hypothesis to explain the observed seasonal flux in the phytal fauna involves epiphytic growth regulating the dynamics of the phytal community. The low number of grazers present during the winter months cannot control a burst of productivity of filamentous algae as light levels increase in spring and early summer. Animal numbers rise during the summer months in response to the considerable habitat heterogeneity and production of epiphytes, but, as these resources are removed by grazers in autumn, predation and lack of recruitment reduce the number of invertebrates to comparatively low winter and spring levels.

2.9.2.4 Factors Influencing Community Diversity and Abundance

Phytal communities are controlled by a variety of physical and chemical factors, including algal shape (Hicks, 1977b), water depth (Dahl, 1948), wave exposure and water movement (Norton, 1971; Fenwick, 1976), season (Mukai, 1971), turbidity and detrital load (Dahl, 1948; Moore, 1978), and salinity (Dahl, 1948). Determining the relative importance of each of these factors has proved difficult due to interactions among them.

1. *Water depth and wave action*: The zonation of phytal communities with depth was initially shown by the investigations of Kitching et al. (1934) and Dahl (1948), and is generally recognized to be caused by diminishing wave turbulence with depth (Hagerman, 1966; Fenwick, 1976). Edgar (1983a) found that in Tasmania, water depth had the greatest influence on the distribution of the phytal fauna. Amphipods peaked in abundance in shallow water (<2 m depth), while isopods and molluscs were most common in the deepest (>5 m) areas investigated. Podocerid and caprellid amphipods, which are filter feeders at one site, were poorly represented in sheltered bays at any depth, but comprised >90% of the motile fauna on an exposed site. At another site, caprellids and podocerids dominated the fauna down to 2 m, but were virtually absent below that level. At that depth, there was a major dichotomy in the fauna. Water movement differed significantly

between the two fauna zones, with turbulent flow providing a large suspension-feeding niche in the upper zone and greater settling of particles in the lower zone allowing for increased numbers of detritovores.

2. *Algal structure*: Edgar (1983a) found that within a depth zone there were quantitative differences in the abundance of animal species on dissimilar algae. These differences were partly caused by a close correspondence between algal shape and faunal size structure. Small animals, particularly amphipods, were more likely to be present on filamentous algae than on plants with wide thalli, while larger animals showed the opposite response. Animals retained on a 0.5-mm sieve were most abundant on plants with branch widths considerably less than their body sizes. These animals would be more likely to respond to individual branches than to total surface area. Larger amphipods (2.8-mm sieve size) had peaks of abundance on blades 4.7 mm wide and possibly may distribute themselves with reference to surface area. The large amphipod, *Erichthonius braziliensis,* was found by Connell (1963) to use behavioral mechanisms to space individuals evenly over a flat surface, and Stoner (1980) concluded after a series of experiments that the amphipods, *Cymadusa compta* and *Grandidierella bonnieroides,* were equally distributed among differently shaped sea grasses with similar surface areas, but that a third species, *Melita elongata*, showed a much greater preference for filamentous algae than sea grasses.
3. *Habitat complexity*: Edgar (1983c) found that simple, flat-thalloid algae and finely filamentous plants had very low species richness and heterogeneity. Both number of species and abundance were positively related to the "diversity" (sensu heterogeneity) of physical structures such as branch widths within the plant, but not to the surface area/weight ratio (the degree of dissection). These relationships are probably dependent on an increase in the number of habitats with increasing algal complexity and/or increasing animal abundance with increasing surface area. The size structures within the habitat also seem to be of considerable importance, possibly because additional species may be able to survive within an environment if sufficient refuges within a given size range are available from predators.
4. *Environmental stability:* In Edgar's (1983c) study, a bell-shaped relationship similar to that predicted by Huston (1979) was deduced between environmental stability (as measured by wave exposure) and species density. Animal heterogeneity was also maximal at intermediate levels of wave exposure. The reduced values of diversity indices in conditions of extreme exposure certainly occurred because few phytal species are capable of surviving the disturbances that elongate algae underwent at wave-swept sites. The reasons for the reduced species richness in calm habitats are more complex and are possibly the result of the aversion of many phytal species to the high detritus levels present on the sheltered algae, and also to the slightly greater seasonal variation in water temperatures. Alternatively, superiorly competitive phytal species may have excluded less competitive species.
5. *Food resources*: Both the biomass of filamentous epiphytes and water depth (which was considered to be directly related to water movement and the flux of suspended food particles) were found by Edgar (1983c) to be strongly correlated with the abundance and dominance of the phytal assemblage. Thus, the availability of food is an important factor affecting community diversity and abundance.

2.9.3 Rock Pools

2.9.3.1 Introduction

Rock pools (or tidepools) form an unusual and patchily distributed habitat with markedly different characteristics from other coastal habitats. During low tide they are isolated bodies of water providing a kind of subtidal refuge without the effects of waves or current flow. Conditions in the pools are, however, highly regulated by the tidal cycle. The degree of fluctuation in the physical conditions within the pools will vary greatly with tidal height, with the lower pools being less variable than the higher ones. In addition, depending on their depth and shape, each pool can be differentially affected by disturbance. Some pools located higher on the shore above the influence of tides and waves (except in exceptional conditions) harbor what are basically freshwater or brackish water (due often to salt input from spray) communities.

Studies on intertidal rock pools have been mostly qualitative (e.g., Sze, 1980; Femino and Mathieson, 1980). Most studies have focused on macroalgae, with little attention being given to sessile or motile animals (Goss-Custard et al., 1979). The studies have generally concentrated on three aspects of rock pool ecology: seasonal cycle of algae in the pools, zonation within the pools, and the role of disturbance in pool. Variability in the biological communities in the pools has been attributed to differences in

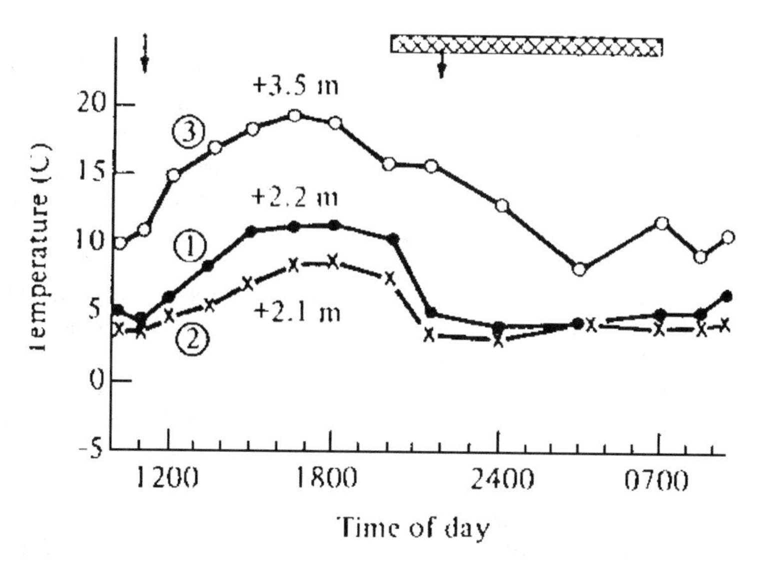

FIGURE 2.35 Temperature changes in three pools at different heights in the intertidal zone near Halifax, Nova Scotia, May 8–9, 1970. Times of high tide indicated by arrows, darkness by the cross-hatched bars. Pools 1 and 2, between neap and spring higher high waters, were generaly flushed twice daily, except for pool 1 during periods of calm or neap tides. Pool 3, 0.4 m above extreme high water, was washed only during severe storms and contained no perennial macroscopic algae. (Redrawn from Edelstein, T. and McLachlan, J., *Mar. Biol.*, 30, 310, 311, 1975. With permission.)

their physical characteristics (e.g., area, volume, and depth) (Astles, 1993; Metaxis et al., 1994). Also, because a number of factors determine the extent of tidal exchange into the pools (e.g., orientation, wave exposure, height of the surrounding rocks, and drainage patterns), height above Chart Datum does not accurately describe the tidal position of a pool. Tidepools separated by a very small vertical distance may receive very different tidal inputs and thus harbor different biological communities.

2.9.3.2 The Physicochemical Environment

Physical conditions within pools vary greatly, but the larger and deeper a pool and the lower its position on the shore, the more it will correspond to a sublittoral habitat in terms of the stability of temperature, salinity, etc. Conversely, small or very shallow pools are more prone to temperature and salinity changes. This is especially so for pools at the highest tidal levels where they may be isolated from the sea over several tidal cycles.

Temperature and salinity within the pools depend largely upon weather conditions, although shade and land drainage can have an overriding influence. Temperatures can vary daily by up to 15° depending on the height of the pool along the intertidal gradient, wave exposure, the volume of the pool, and the degree of shading (Goss-Custard et al., 1979; Morris and Taylor, 1983; Huggett and Griffiths, 1986). In general, depending on the level of the pool on the shore, temperatures follow that of the air to a much greater extent than the temperature of the sea, so that in the summer the shallow pools, or the surface water of the deeper ones, are warmer by day but colder at night (Figure 2.35), while in the winter they may be appreciably colder than the sea. Daily changes in temperature can often result in thermal stratification of the water in the deeper supralittoral pool (McGregor, 1965).

High temperatures and evaporation can lead to an increase in salinity, especially in pools not flooded for several days by the tide. Heavy rainfall followed by tidal flooding can lead to particularly violent fluctuations in salinity. Salinity stratification can occur seasonally because of freezing in the winter, evaporation in the summer, and rainfall (Morris and Taylor, 1983). Daily fluctuations in oxygen saturation, alkalinity, and pH have also been recorded, which are due to biological processes in the pools (McGregor, 1965; Morris and Taylor, 1983). Thus the physical environment of tidepools fluctuates vertically, horizontally, diurnally, and seasonally, although not as much as the adjacent emergent rock surfaces. The fluctuations, in turn, will vary with the volume, surface area, and depth of the pool, as well as its height on the shore, degree of shading, drainage pattern (which depends on the aspect), and exposure to waves and splash. Thus it is virtually impossible for two natural pools to be similar in all these characteristics; individual tidepools are unique in their physical regimes.

2.9.3.3 Temporal and Spatial Patterns in the Tidepool Biota

In general, the biological assemblages found in tidepools are similar to those described for emergent surfaces. Differences between the two types of habitat arise because of the smaller fluctuations in physical conditions and/or more intense biological interactions in the pools. Two main trends have been observed:

1. The uplift of species to higher levels on the shore in response to the absence of desiccation.
2. The increased specialization and/or replacement of fully marine species toward higher levels.

Several taxa are more abundant in pools than on emergent substrata. These include algae, e.g., the genera *Ceramium, Spongoclonium, Corallina,* and *Rhizoclonium* in Maine, U.S.A. (Johnson and Skutch, 1928), *Prionitis* in Washington, U.S.A. (Dethier, 1982), and *Fucus distiches* in Nova Scotia, Canada (Chapman and Johnson, 1990); and gastropods, e.g., the genus *Cellana* in New South Wales, Australia (Underwood, 1976), and *Littorina littorea* in Massachusetts, U.S.A. (Lubchenco, 1982). Macroalgae such as fucoids, *Scytosiphon, Spongomorpha,* and *Ulva,* occur at higher intertidal levels in tidepools than on emergent surfaces on the northeast coast of North America (Johnson and Skutch, 1928; Femino and Mathieson, 1980; Chapman and Johnson, 1990). Similar observations have been made for mussels, chitons, limpets, and sea urchins in tidepools in British Columbia, Canada (Green, 1971) and for the surfgrass, *Phyllospadix scouleri*, in Washington (Dethier, 1984).

Many species of benthic invertebrates and fish also show zonation along the intertidal gradient. The periwinkle, *Littorina rudis,* is mainly found in high pools, whereas *L. littorea*, whelks, mussels, and sea urchins are found in low pools (Goss-Custard et al., 1979; Femino and Mathieson, 1980). Sze (1980) found that the abundance of *L. littorea* increased from low to high pools. Zonation has also been observed for various meiofaunal groups: flatworms, rotifers, oligochaetes, cladocerans, cyclopoid copepods, ostracods, barnacles, amphipods, isopods, and chironomid larvae. Fish zonation in tidepools has been documented extensively but the results are not quantitative (Green, 1971; Gibson, 1982; Bennett and Griffiths, 1984; Mgaya, 1992). Bennett and Griffiths (1984) detected a decrease in the number of fish species with increasing height above low water, which they attributed to intolerance to extreme physical conditions.

2.9.3.4 Factors Affecting Community Organization

Herbivory: Herbivory (grazing mainly by limpets and littorinids) has been found to limit the distribution and abundance of algae in tidepools. Vadas (1977) showed that the removal of sea urchins resulted in increased macroalgal abundance and diversity in shallow tidepools in Washington. In Massachusetts, Lubchenco (1978) observed the effect of herbivory in two mid-zone pools. In one, littorinid snails were absent and the dominant alga was *Enteromorpha* sp., and in the other, snails were present and the dominant alga was *Chondrus crispus*. Lubchenco (1978) added snails to the first pool and observed a decrease in the cover of the dominant *Enteromorpha* sp. On the other hand, when she removed snails from the second pool she observed a decrease in the cover of the dominant *Chondrus crispus*. The cover of *Fucus vesiculosus* and ephemerals increased in a number of tidepools in the mid-intertidal zone of a protected and a semiexposed shore in Maine and Massachusetts, when littorinids were excluded (Lubchenco, 1982).

A number of other studies have shown that grazers in tidepools (mainly littorinids) have a negative effect on the abundance of some algal species, especially fucoids, a positive effect on the abundance of ephemeral algae, and no effect on the encrusting genus *Hildenbrandia* and corallines (Chapman, 1990; Dethier, 1982; Underwood and Jernakoff, 1984).

Predation: A limited number of studies have demonstrated the importance of predation in tidepools. Fairweather (1987) found that whelks introduced into shallow (1 to 4 cm deep) tidepools in New South Wales reduced the abundance of barnacles, tubeworms, and limpets. Lubchenco (1978) suggested that littorinid populations in tidepools in Massachusetts may be controlled by predation by the green crab, *Carcinus maenas*. In Washington, Dethier (1980) showed that fish, and to a lesser extent sea anemones, can reduce the abundance of an harpacticoid copepod, *Tigriopus californicus*, in tidepools in the high zone of rocky shores. In Island Bay, New Zealand, Coull and Wells (1983) observed high meiofauna mortality due to fish predation in tidepools in the absence of *Corallina officinalis,* which acts as a refuge.

Competition: Interspecific competition may be important in regulating tidepool community structure, but the evidence is sparse. Lubchenco (1982) and Chapman (1990) have documented decreases in fucoid canopy due to competition with ephemeral algae and *Chondrus cripsis* in tidepools in Massachusetts and Nova Scotia, respectively. Chapman and Johnson (1990) suggested that the absence of a canopy of *Ascophyllum nodosum* can enhance recruitment of *Fucus spiralis* in tidepools in Nova Scotia. Cecchi and Cinelli (1992) found that canopy removal in *Cystoseira* spp. dominated tidepools on the west coast of Italy had no effect on either encrusting or articulated corallines (e.g., *Corallina* spp.) or on coarsely branched algae (e.g., *Gelidium pulchellum*), but enhanced the abundance of delicately branched (e.g., *Ceramium, Cladophora* spp.) and thickly branched (e.g., *Padina pavonica*) algal species.

Recruitment: Recruitment is potentially an important factor in the organization of tidepool communities, although no studies have addressed this directly. The variability in the response of the tidepool community to grazer removal (Paine and Vadas, 1969) and in recovery from disturbance (Dethier, 1984) have been partially attributed to seasonal availability and "vagaries of recruitment" of algal spores and invertebrate larvae from the surrounding seawater. Chapman and Johnson (1990) suggested that differential recruitment success in high tidepools can lead to competitive displacement between *Fucus evanescens* and *F. vesiculosus*.

Physical Factors: The abundance of tidepool algae has been correlated with pool elevation (which determines the length of emergence and the extent of temperature fluctuations), and topography and shading by surrounding rocks (Johnson and Skutch, 1928). The number of species present is also correlated with tidepool depth and volume. Several studies have shown that deeper pools may support more plant and invertebrate species (Fairweather and Underwood, 1991). Other studies have shown that fish biomass, species number, and abundance may show significant correlations with pool area, depth, or volume (Mgaya, 1992).

As on emergent substrata (Menge, 1976; 1978b; 1983; Lubchenco and Menge, 1978; Underwood and Jernakoff, 1981), algal cover and abundance of littorinids and fish in tidepools are correlated with wave exposure. Sze (1982) showed that some algae such as *Enteromorpha, Spongomorpha,* and *Scytosiphon* are more abundant in tidepools on exposed shores where littorinids are absent, whereas fucoids are more abundant in pools on protected shores. Dethier (1984) also found that the cover of the dominant algal species varied among pools of different wave exposure. For example, the green alga, *Collonsiella tuberculata*, and the red alga, *Rhodomelia larix*, were found in pools higher on the shore in more exposed habitats (Dethier, 1982). Grossman (1982) found that the abundance of fish in tidepools decreased with increased wave action, possibly because few species can adapt to higher turbulence in exposed pools (Gibson, 1982). Green (1971) also found that the vertical distribution of cottid fish was related to the degree of wave exposure in tidepools, with fewer fish found in pools on lower, more exposed shores.

Physical Disturbance: There is little information on the effect of disturbance in determining the structure of communities in tidepools. In tidepools in Washington, Dethier (1984) used an operational definition of disturbance as the destruction of biomass over a period of less than six months, which she subjectively categorized as severe, moderate, or minimal (affecting most, some, or one or two species of a pool, respectively). Freezing and heat stress were types of physical disturbance for the surfgrass, *Phyllospadiz scouleri*; bashing by logs and rocks were types of disturbance for mussels, anemones, and *Cladophora* spp. More disturbances were recorded in low than in high zone pools, and the frequency of disturbance was the same in wave-exposed pools as in more protected sites. The rate of recovery from disturbance varied with species and depended upon the magnitude of the disturbance.

2.9.3.5 Conclusions

A number of similarities and differences exist in community organization between tidepools and emergent substrata of rocky shores. Since many of the same species are common to both habitats, they are biologically similar. However, certain differences in the physical regime can result in differences in the species composition of the two habitats. On the other hand, the amplitude of the fluctuations in the physical regime tends to be smaller in some tidepools, particularly those located lower on the shore, making them more benign habitats. However, grazing and predation may be more intense in tidepools where both food and favorable foraging conditions (due to the ability of herbivores to graze irrespective of the state of the tide) are provided. In addition, tidepools that are high on the shore and infrequently flushed by wave action can become stagnant, resulting in harsh conditions due to lack of nutrients and food, and pronounced changes in physical conditions such as pH, salinity, and temperature. The decrease in species diversity with increasing intertidal height is due to the inability of many species to tolerate such conditions.

The variability in community structure among pools is larger than that on emergent substrata, with pools at the same height on the same shore showing large variability in species composition and abundance. Despite the large variability, some general patterns of species distribution in tidepools along an intertidal gradient have emerged from the various studies. Most have shown that the dominant space occupiers in lower tidepools in the northern temperate region are fucoid and coralline algae and mussels, whereas the higher pools are dominated by green algae.

2.9.4 Kelp Beds

2.9.4.1 Introduction

One of the most common sublittoral habitats throughout the temperate regions of the world are large "forests" or "beds" of kelp plants (large assemblages of brown algae of the order Laminariales). These communities are important because of their high productivity and the diverse associated invertebrate and fish communities they support. They also give rise to large quantities of drift algae, especially after storm events. These detached plants are often cast up on nearby beaches where they provide an energy source for the beach communities.

Since the kelp forests can extend up to 30 m or more above the substratum, they give vertical structure to the sublittoral rocky communities. The sometimes extensive canopies can modify light levels to the substrate below, reduce the impact of waves and turbulence, impact on the nutrient dynamics of the surrounding water, and limit the space available for other species to colonize. Recent reviews of the distribution and ecology of kelp beds include those of North (1971a,b), Rosenthal et al. (1974),

Dayton (1985a,b), Foster and Schiel (1985), Schiel and Foster (1986), Schiel (1990; 1994), and Kennelly (1995).

2.9.4.2 Species Composition, Distribution, and Zonation

Species of the order Laminariales occur in both hemispheres, but most are found in northern temperate and boreal areas (Moe and Silva, 1977). They are normally confined to depths of 50 m or less in the sublittoral zone (Foster and Schiel, 1985). Representatives of another order, the Fucales, are most important in the Southern Hemisphere, usually occurring in the shallow upper sublittoral and the sublittoral fringe. Laminarians, but not fucaleans, have alternate generations, the haploid, microscopic gametophyte, and the macroscopic sporophyte plant (see Figure 2.36).

Along the west coast of North America, three broad depth zones can be categorized in the upper sublittoral (e.g., Druehl, 1970; Foster and Schiel, 1985) (Figure 2.36). Inshore (0 to 5 m depth) there is a mixture of fucalians such as *Cystoseira* and *Sargassum,* and laminarians such as *Egregia* and *Eisenia*. Middle depths (5 to 20 m) are dominated by the larger kelps *Nereocystis* and *Macrocystis*. Here there is a lush understory of a number of stipate laminarians. On the rock surfaces between the kelp holdfasts there is a complex assemblage of smaller brown , red, and green algae, as well as sponges, ascidians, bryozoans, and mobile invertebrates, especially the sea urchin *Stronglocentrotus*.

The larger kelps with their surface canopies do not occur on the North Atlantic coasts. There, *Fuscus* species occupy the shallow areas, and stipate laminarians such as *Alaria* are dominant (Mann, 1972a). On British coasts (Figure 2.36b), *Laminaria digitata* occurs in the shallow subtidal, with *L. hyperborea, L. saccaharina,* and *Saccorhiza polyschides* occupying progressivley deeper areas.

In contrast, southern Chile has few laminarians, with *Lessonia* species in the shallow sublittoral and a dominant canopy of the bladder kelp *Macrocystis pyrifera,* with a subcanopy of *Lessonia* species and other algae below (Figure 2.36c). In South Africa (Figure 2.36d), kelp forests are dominated by two species of laminarians. At all depths to 30 m the smaller *Laminaria pallida* can form dense beds (Diechmann, 1980), while in the shallower sublittoral, *Ecklonia maxima,* which reaches several meters in length and is held above the substratum by a bouyant stipe, is very abundant.

Australian and New Zealand sublittoral zones have a more diverse flora with many fucalean species in the shallower waters and laminarians deeper down (Choat and Schiel, 1982; Kennelly, 1995). In northern New Zealand (Figure 2.36e), the shallowest areas (0 to 3 m) are dominated by *Carpophyllum angustifolia* and *C. maschalocarpum*. At 3 to 5 m there are dense stands of *Ecklonia radiata* and *Lessonis variegata*, as well as patches of *Carpophyllum* spp. In many areas, depths of 7 to 10 m are devoid of large brown algae and are dominated by the sea urchin *Evechinus chloroticus*. Deeper areas (11 to 20 m) are characterized by stands of *E. radiata* and *C. flexuosum*. On southern New Zealand shores (Figure 2.36f), the upper fringe of the sublittoral is occupied by the bull kelp *Durvillaea antarctica* (Schiel, 1990) with *D. willana* below, extending on some shores to 3 m (Hay, 1982). *Cystophora* spp. and *Marginarella boryana* often form dense stands at various depths, interspersed with stands of *Carpophyllum* and the laminarians *E. radiata* and *L. variegata*. In semiexposed and sheltered waters, the bladder kelp *Macrocystis pyrifera* often forms extensive beds. Between the holdfasts there often is a lush covering of red and brown algae.

Australian shores have fewer laminarians but more fucalean species than in New Zealand. Along the New South Wales coasts (Figure 2.36g), a suite of *Sargassum* and *Cystophora* spp. are found in the upper sublittoral. Below, the kelp community is dominated by *Phyllospora comosa* in intermediate depths, and *E. radiata* below (Kennelly and Larkum, 1984; Larkum, 1986). On southern shores (Figure 2.36h), the upper fringe of the sublittoral is dominated by the bull kelp *Durvillaea potatorum*, with *Cystophora, Sargassum,* and *Scyothalia* spp. below. *Ecklonia radiata* dominates from 5 to 40 m depths. In the southernmost regions, especially in Tasmania, the bladder kelps (*Macrocystis angustifolia* in the north and *M. pyrifera* in the south) may be abundant, depending on the degree of wave action.

The largest and most studied laminarian is the "giant kelp," *Macrocystis pyrifera*, which is found along the east and west coasts of South America, the southern part of the North American west coast, South Africa, southern Australia, New Zealand, and some sub-Antarctic islands. Another species, *M. integrifolia,* replaces *M. pyrifera* on the North American west coast.

2.9.4.3 Kelp Bed Fauna

A wide variety of sessile invertebrates, including sponges, ascidians, bryozoans, hydroids, sedentary polychaetes, molluscs, and brachiopods are found on the kelp fronds and the understory plants, in the kelp holdfasts, and on the rock surface. There is also a diverse fauna of motile invertebrates, including herbivorous and carnivorous gastropods, and especially crustaceans that live on the algal fronds. In a study of the fauna in the holdfasts of *Macrocystis pyrifera* in southern Chile, Ojeda and Santelices (1984) found that it was about half as diversified as in California, where Ghelardi (1971) found 100+ species. In Chile the most abundant taxa were echinoids and decapod crustaceans, whereas in California they were amphipods, polychaetes, and isopods.

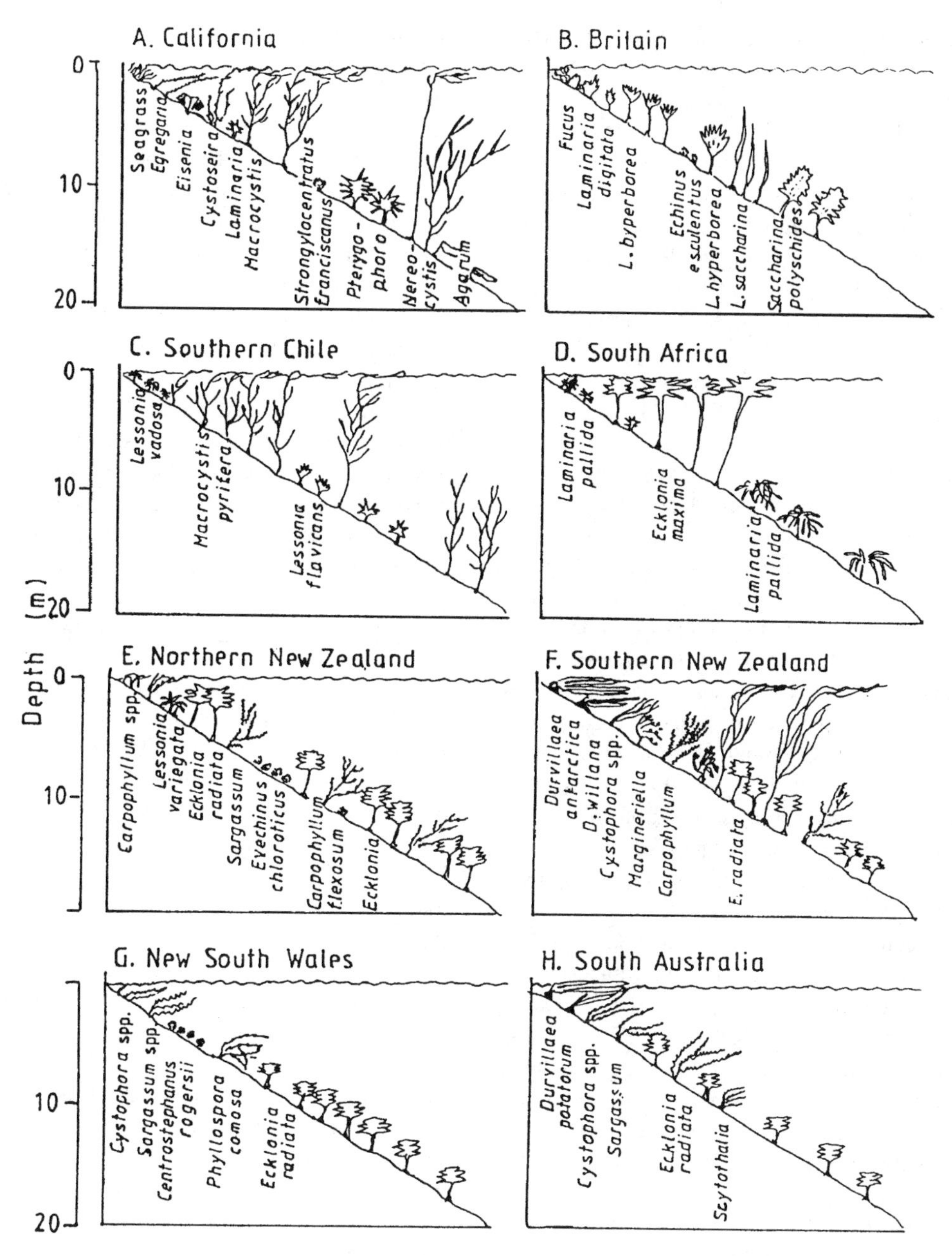

FIGURE 2.36 Stylized depth profiles of kelp forests from several areas round the world. All of the species represented in a single diagram may not be represented at a single site. A: Central California — several surface canopy kelps and understory kelps occur; B: United Kingdom — stipate laminarians predominate; C: Southern Chile — *Lessonia* spp. form the upper boundary of *Macrocystis* forests and occur in the understory; D: South Africa — *Ecklonia maxima* forms a surface canopy and *Laminaria pallida* occurs at all depths to 20 m; E: Northern New Zealand — fucalean algae occur in shallow water, middle depths are dominated by sea urchins, and *Ecklonia radiata* occurs in both shallow and deep water; F: Southern New Zealand — *Durvillaea* occupies shallow water and *Lessonia, Ecklonia,* and several fucaleans occupy most depths; G: New South Wales — *Cystophora* spp. and *Sargassum* spp. occur in shallow water above a zone of sea urchins, *Phyllospora* occurs at intermediate depths, and *Ecklonia* occurs at most depths; H: South Australia — where *Durvillaea* occurs in shallow water and *Ecklonia* and several fucaleans occupy intermediate and greater depths. (Redrawn from Scheil, D.R., in *Marine Biology,* Hammond, L.S. and Synnot, R.N., Eds., Longman Chesire, Sydney, 1994, 350. With permission.)

Typically, the major herbivorous invertebrates of the kelp beds are sea urchins and gastropods. Sea urchins, as discussed below, in particular exert a considerable influence on the species composition and abundance of the algal communities. Abalone (*Haliotis* spp.) are very abundant in exposed habitats, where they feed on drift algae.

The most conspicuous motile components are the fish. Nearly 100 species of fish are known to inhabit the southern Californian kelp forests (Feder et al., 1974). In a study comparing fish densities in *Macrocystis* forests and the structurally less complex, temporarily unstable, *Nereocystis* beds, Bodkin (1986) found the biomass densities of the dominant water-column fish to be >50 % lower in the *Nereocystis* beds. Experimental clearing of *Macrocystis* from one 1-ha rock reef resulted in a 63% decline in the standing stock of fishes (Bodkin, 1988). Similar fish diversities and densities are found in other kelp forests, e.g., those of New Zealand (Choat and Ayling, 1987).

2.9.4.4 Reproduction, Recruitment, and Dispersal

Laminarian and fucalean algae have discrete annual reproductive periods, and recruits usually appear within a few months of these (Schiel, 1988). Most of the species have limited means of propagule dispersal from the fertile parent plants, and recruits appear within a few meters of the reproductive adults (Paine, 1979). In northern New Zealand, 70% of viable recruits appeared within 8 m of fertile adults for *Ecklonia* and three fucalean species (Schiel, 1988). Vast numbers of propagules may be necessary to produce successful recruitment, e.g., Chapman (1984) found that for *Laminaria longicirurus* in Nova Scotia, 9×10^9 spores m^{-2} yr^{-1} produced 9×10^6 microscopic stages, from which only a single sporophyte developed. The question, then, is how the species become widely distributed. This probably takes place via drifting fertile plants.

Sea urchins can vary enormously in abundance on kelp beds. The question is posed, do the differences in sea urchin density, especially between the so-called barren grounds where densities are high and the kelp forests where densities are low, result from differential larval settlement, postsettlement mortality, or movement? The relationship between the alternative communities of sea urchin-dominated barren grounds (barrens) and kelp forests have been investigated throughout the world. Several of these studies have noted high densities of small urchins in barrens relative to kelp bed habitats. These observations have involved several species in different geographical areas: *Strongylocentrotus purpuratus* and *S. franciscanus* in California, U.S.A. (Pearse et al., 1970), *S. droebachiensis* in Nova Scotia, Canada (Lang and Mann, 1976), and *Evechinus chloroticus* in northeast New Zealand (Andrew and Choat, 1982; 1985).

Rowley (1989) has investigated settlement and recruitment in the kelp beds of southern California. He found that newly settled individuals of both purple sea urchins (*Strongylocentrotus purpuratus*) and red sea urchins (*S. franciscanus*) were present in similar, high densities (1,000 *S. purpuratus* m^{-2}) on foliose red algal turf, a dominant substratum of the kelp bed, and on crustose coralline algae, the dominant substratum of adjacent barrens. Larvae of *S. putpuratus* reared and tested in the laboratory showed high rates of settlement on both red algal turf and on crustose coralline algae, but significantly lower rates on rock. Larvae also settled in response to a partially purified extract of coralline algae. The reduced settlement on natural rock surfaces relative to either algal treatment and the significant settlement in response to the extract of coralline algae indicate that larvae discriminate between natural substrata and that they respond to a chemical settlement cue. The similar densities of young recruits of *S. purpuratus* on dominant substrata of barrens and kelp beds show that differential settlement cannot explain the high densities of sea urchins in the barrens habitat. Movement between barrens and kelp beds is unlikely given the small sizes of the newly recruited sea urchins relative to the large distances often involved. Reduced postsettlement mortality of newly settlement individuals in the barrens is the likely mechanism leading to higher recruitment densities of the sea urchins in the barrens relative to the kelp bed habitats.

2.9.4.5 Impact of Grazers on Kelp Communities

Grazing by sea urchins can have devastating effects on the populations of marine algae, especially kelps of the order Laminariales (Lawrence, 1975; Harrold and Pearse, 1987). Such effects have been noted in algal communities over a broad geographical range (see reviews of Lawrence, 1975; Duggins, 1980; and Schiel, 1990; 1994). Grazing is most pronounced when the urchins form moving aggregations, or "fronts," that eat almost all macroalgae in their pathway producing "barren grounds" (e.g., Pearse et al., 1970; Leighton, 1971; Lawrence, 1975; Breen and Mann, 1976). One such front reported by Leighton (1971) had densities of red and purple urchins exceeding 60 m^{-2}, and moved at a rate of 10 m $month^{-1}$. Almost all the macroalgae disappeared in the wake of the front. These areas can remain barren for several years because some urchins remain behind the front and graze on newly recruited plants. Urchins in barren grounds appear to persist primarily on drift algae (Mathieson et al., 1977), but may also gain nourishment by grazing on microalgae (Chapman, 1981) or by absorbing organic matter.

Dean et al. (1984) studied the effects of grazing by two species of sea urchins on two species of kelp (*Macrocystis pyrifera* and *Pterygophora californica*) in southern California from 1978 through 1981. Both red sea urchins, *Strongylocentrotus franciscanus*, and white sea urchins, *Lytechinus anamesus*, were abundant and lived in aggregations. The aggregation of red urchins were either relatively small and stationary, or relatively large and motile (advancing at about 2 m $month^{-1}$). Both stationary and moving aggregations were observed at the same time, within 100 m of one another. Stationary aggregations largely subsisted on drift kelp and had no effect on kelp recruitment or on the adult kelp abundance. In contrast, red sea urchins in large, motile aggregations or "fronts" ate all the macroalgae in their path. Dean et al. proposed that a scarcity of drift algae for food resulted in a change in the behavior pattern of the red urchins, and that this led to the formation of large, motile aggregations.

The removal of sea urchins usually results in a large recruitment of algae. This has occurred in experimental areas (Andrew and Choat, 1982), and on a larger scale along the coastline of Nova Scotia following a mass mortality of sea urchins (Miller and Colodey, 1983). The subsequent survival of the algae in these denuded areas depends on many factors, including the impact of other grazers such as gastropods and fish, the availability of algal propagules, and the seasonal timing of the clearance.

In addition to sea urchins, other grazers are gastropod molluscs. In California, three trochids of the genus *Tegula* live and feed on the fronds of *Macrocystis*, but appear to have no major impact. In northern New Zealand, high numbers of gastropods, including the limpet, *Cellana stellifera*, and the turbinid, *Cookia sulcata*, invade areas that have been heavily grazed by sea urchins (Ayling, 1981). These gastropods consume algal spores and germlings, and thus help to maintain these areas free from kelp (Choat and Andrew, 1986). Abalone are the largest grazers found in kelp beds, and while small individuals graze on encrusting algae and algal turfs, the larger animals feed almost exclusively on drift algae.

Some fish species are important herbivores in kelp beds. In California the scorpidid, *Medialuna californiensis*, and the kyposid, *Girella nigricans*, have been found to graze 90% of the exposed recruits of *Macrocystis* (>10 cm tall) (Harris et al., 1984). In New Zealand, the fish *Odax pullus* sometimes feeds on the reproductive branchlets of large brown algae, and at other times feeds on the fronds of *Ecklonia radiata*. In New South Wales, grazing by *O. cyanomelas* causes the death of *E. radiata* in patches of up to 100 m^{-2} (Andrew and Jones, 1990).

2.9.4.6 Predation

Since sea urchins can have significant impact on kelp beds, there has been considerable interest in the roles of sea urchin predators. In eastern Canada, the fishes, *Anarhichas lupus* and *Hippoglossoides platessoides*, can break up large aggregations of sea urchins, thereby modifying their grazing effects (Bernstein et al., 1981). The sheepshead wrasse, *Semicossyphus pulcher*, also feeds extensively on sea urchins (Cowen, 1983), affecting their abundance. Similar predation impacts of fish have been noted in New Zealand (Andrew and Choat, 1982) and southern Australia (Hutchins and Thompson, 1983). Lobsters can also have an impact on sea urchin numbers and behavior. Increases in sea urchin abundances in both California and Nova Scotia may be due in part to the decline in lobster abundances following intensive fishing (Tegner and Dayton, 1981; Wharton and Mann, 1981).

At some sites along the coasts of central California and the Aleutian Islands in the Bering Sea, sea otters play a key role in kelp community dynamics, since they prey heavily on sea urchins (Estes et al., 1978; Estes and Duggins, 1995). Where sea otters are abundant, the algal beds are well developed and the sea urchins tend to be small in size and confined to cryptic habitats (Duggins, 1980).

Predatory seastars can also impact on kelp communities by influencing the distribution and abundance of sea urchins and gastropods. In southern California, the seastar *Patiria miniata*, by feeding on the urchin *Lytechinus abamesus*, lessens its grazing impact on juvenile kelp plants (Schroeter et al., 1983; Dean et al. 1984). In the northern waters of the west coast of North America, the large seastar *Pycnopodia helianthoides* feeds on sea urchins, reducing their densities and enabling algal recruitment to take place (Duggins, 1983).

Of all the factors affecting the population dynamics of sea urchins, disease causes the greatest reduction in numbers. Mass mortalities of sea urchins due to disease such as those that have occurred in California (Pearse et al., 1977) and Nova Scotia (Miller and Colodey, 1983) can bring about a sudden switch from echinoid to algal dominance (Miller, 1983).

2.9.4.7 Kelp Production

Productivity of kelps is very high, and the genus *Macrocystis* has been extensively studied in the Californian and British Columbian kelp beds where its growth, production, and standing crop have been measured (North, 1971a,b; Wheeler and Druehl, 1984; Coon, 1982). Productivity varies considerably and in typical beds ranges from about 400 to 1,500 g C m^{-2} yr^{-1} for *Macrocystis pyrifera*, 1,200 to 1,900 g C m^{-2} yr^{-1} for *Laminaria* in the North Atlantic, and between 600 and 1,000 g C m^{-2} yr^{-1} for *Ecklonia* in Australia and South Africa (Mann, 1973; 1982; Kirkman, 1984; Foster and Schiel, 1985). On an aerial basis, this is up to 40 times that of the open ocean. The turnover of biomass ranges from 3.8 to 20 times yr^{-1} for *Laminaria* in Nova Scotia (Mann, 1973) and is at least 3.5 times yr^{-1} for *Ecklonia* in western Australia (Kirkman, 1984), and 7 times yr^{-1} for *Macrocystis* in California.

Growth rates of *Macrocystis* species are high. In Otago Harbour, New Zealand, Brown et al. (1997) measured elongation rates of canopy 2 cm day^{-1} for canopy and 1.3 cm day^{-1} for submerged fronds. These were higher than those obtained by von Tussenbrock (1989) from a shallow water site in the Falklands (1.3 to 2 cm day^{-1} for canopy and 0.5 to 0.8 cm day^{-1} for submerged fronds). However, the growth rates listed above are considerably lower than those found by Zimmerman and Kremer (1986) at a deep-water location in southern California (6 to 14 cm day^{-1} for submerged fronds). For *M. intgrifolia* in British Columbia, the mean growth rates varied from a high of 4.3% day^{-1} to a low of –3.6% day^{-1} (Wheeler and Druehl, 1986). Net mean assimilation rates of carbon varied from a high of 0.65 g C m^{-2} of foliage day^{-1} to a low of –0.87 g C m^{-2} day^{-1}. Annual carbon input on a foliage area basis was calculated at 250 g C m^{-2} yr^{-1}. Annual carbon input to the kelp forest was estimated at 1300 g C m^{-2} of ocean bottom yr^{-1}. Wheeler and Druehl calculated that the forest production was 52 kg wet wgt m^{-2} yr^{-1}. This compares with estimates of 75 to 125 kg wet wgt m^{-2} yr^{-1} calculated by Coon (1982) for Californian kelp forests.

3 Soft Shores

CONTENTS

3.1 SOFT SHORES AS A HABITAT

As mentioned previously, there is a gradation in shore type from rock through pebble and sand to mud, although mixed shores of sand or mud with rocky outcrops are common. The characteristics of the flora and fauna of hard (rocky) shores has already been discussed in detail. We now turn our attention to the other shore types, which are characterized by their relative instability. They are composed of particles of various sizes ranging from pebbles through coarse sands, fine sands to muds (silt and clay). In this chapter we shall consider the ways in which soft shores are formed, their physical and chemical characteristics, the nature of the communities found in the various shore types, and their dynamic functioning.

3.1.1 Beach Formation

Hard shores are erosion shores cut by wave action. Soft shores, on the other hand, are depositing shores formed from particles that have been carried by water currents from other areas. The material that forms these depositing shores is in part derived from the erosion shores, but the bulk of the material, especially the silts and clays, is derived from the land and transported down the rivers to the sea.

Beaches generally consist of a veneer of beach material covering a beach platform of underlying rock formed by wave erosion. On sand beaches the two main types of beach material are quartz (or silica) sands of terrestrial origin and carbonate sands of marine origin (particles weathered from mollusc shells and the skeletons of other animals). Other materials that may contribute to beach sands include heavy minerals, basalt (of volcanic origin), and feldspar. On sand-mud and mudflats, silts and clays of terrestrial origin and organic material derived from river input and from the remains of dead animals and plants contribute to the composition of the sediment.

3.1.2 Sediment Characteristics

The most important feature of beach material particles is their size. Particle size is generally classified according to the Wentworth scale in phi units, where $\phi = -\log_2$ diameter (mm). The Wenthworth classification is summarized in Table 3.1. A classification scheme (Figure 3.1) is generally used to describe differences in sediment texture by reference to the proportion of sand, silt, and clay. Such classifications are essentially arbitrary and many such gradings can be found in the engineering and geological literature.

TABLE 3.1
Wentworth Scale for Sediments

	Generic Name	Wentworth Scale Size Range	Particle Diameter (mm)
Gravel	Boulder	< –8	> 256
	Cobble	–6 to –8	64 to 256
	Pebble	–2 to –6	4 to 64
	Granule	–1 to –2	2 to 4
Sand	Very coarse	0 to –1	1.0 to 2.0
	Coarse	1 to 0	0.50 to 2.0
	Medium	2 to 1	0.25 to 0.50
	Fine	3 to 2	0.125 to 0.50
	Very fine	4 to 3	0.0625 to 0.125
Mud	Silt	8 to 4	0.0039 to 0.0625
	Clay	> 8	< 0.0039

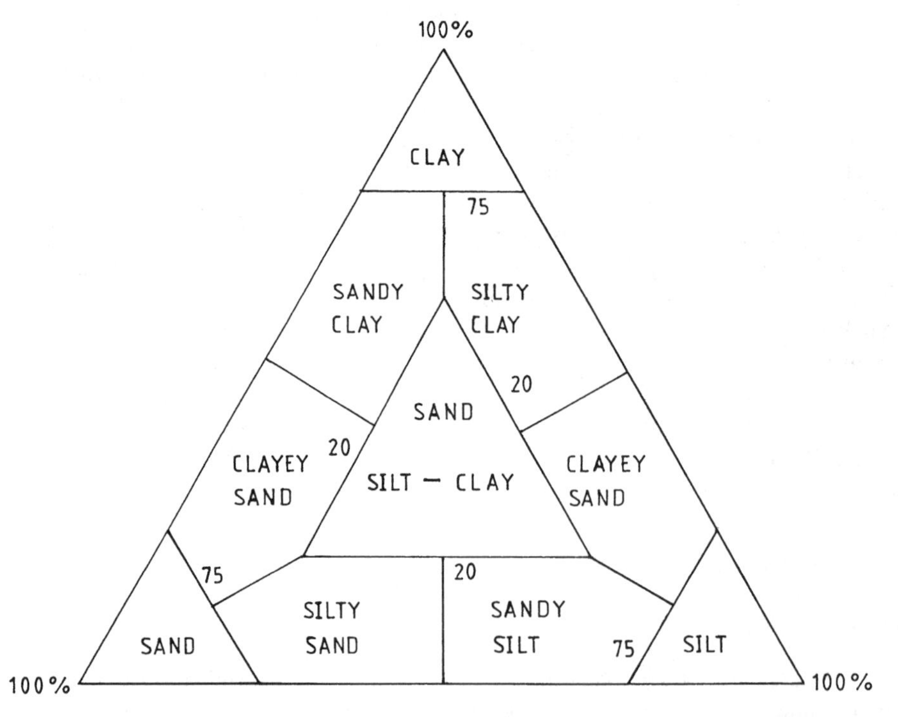

FIGURE 3.1 Classification scheme for sediment texture according to percentage composition of silt, clay, and sand. (Redrawn from Parsons, T.R., Takashi, M., and Hargrave, B.T., *Biological Oceanographic Processes*, 2nd. ed., Pergamon Press, Oxford, 1977, 193. With permission.)

TABLE 3.2
Measures of Sediment Parameters

1. *Measures of average size*
   (a) Median particle diameter (Mdϕ) is the diameter corresponding to the 50% mark on the cumulative curve (ϕ50).
   (b) Graphic mean particle curve (ϕ50).

$$M_Z = (\phi 16 + \phi 50 + \phi 84)/3.$$

2. *Measures of uniformity of sorting*
   (a) Phi quartile deviation (QDϕ), where QDϕ = (ϕ75 – ϕ25)/2
   (b) Inclusive graphic standard deviation (ϕI), where

$$I = \frac{(\phi 84 - \phi 16)}{4} + \frac{(\phi 95 - \phi 5)}{6.6}$$

3. *Measures of skewness* (Skqϕ), where
   (a) Phi quartile skewness (Skqϕ), where
Skq ϕ = (ϕ25 + ϕ75 – 2 ϕ 50)/2
   (b) Inclusive graphic skewness (Sk_1), where

$$Sk_1 = \frac{(\phi 16 + \phi 84 - \phi 50)}{(2(\phi 84 - \phi 16))} + \frac{(\phi 5 + \phi 95 - \phi 50)}{(2(\phi 95 - \phi 5))}$$

4. *Measures of kurtosis or peakedness:*
   Graphic kurtosis (K_i), where

$$K_G = \frac{\phi 95 - \phi 5}{2.44(\phi 75 - \phi 25)}$$

Analysis of sediment size fractions is generally carried out by passing the dried sediment through a set of sieves of varying sizes and weighing the fractions retained on the sieves. Following this, further graphical analysis is generally carried out by plotting cumulative curves on probability paper and calculating the parameters listed in Table 3.2 (Folk, 1966). An important property is the degree to which the sediments are sorted, i.e., of uniform particle size or varying mixtures of different sized particles. Unless the sands are badly skewed, median and mean particle diameters are very similar, and for most ocean beaches are in the range of fine to coarse sand. The inclusive graphic standard deviation is the best index of the sorting of the sediments. Values below 0.5 indicate good sorting, values between 0.5 and 1.0 moderate sorting, and values above 1.0 poor sorting, with a wide range of particle sizes present. Skewness measures the asymmetry of the cumulative curve, and plus or minus values indicate excess amounts of fine or coarse material, respectively. The inclusive graphic skewness is the best measure of this. Values between –0.1 and +0.1 indicate near symmetry, values above +0.1 indicate fine skewed sediments, while values below –0.1 indicate coarse skewed sediments. For normal curves, K_G (Kurtosis) is 1.0, while leptokurtic curves with a wide spread have values over 1.0, and platykurtic curves, with little spread and much peakedness, have values below 1.0.

TABLE 3.3
Settling Velocities of Sediments

Material	Median Diameter (μm)	Settling Velocity (m day^{-1})
Fine sand	125–250	1040
Very fine sand	62–125	301
Silt	31.2	75.2
	15.6	18.8
	7.8	4.7
	3.9	1.2
Clay	1.95	0.3
	0.98	0.074
	0.49	0.018
	0.25	0.004
	0.12	0.001

Source: After King, C.M., *Introduction to Marine Geology and Geomorphology,* Edward Arnold, London, 1975, 196. With permission.

The type of beach developed in any particular locality is dependent on the velocity of the water currents and the particle sizes of the available sediments. This is due to the fact that particles carried in suspension fall out when the current velocity falls below a certain level. This relationship is shown in Table 3.3. From this table it can be seen that sands and coarse material settle rapidly, and any sediment coarser than 15 μm will settle within a tidal cycle. For finer particles, the settling velocities are much lower. Consequently, the waters in estuaries and enclosed inlets tend to be turbid as silt and clay are carried in suspension until they settle on mudflats, as they will not be deposited unless the water is very still.

Thus, pebble beaches are formed only in areas of strong wave action, with sand beaches where wave action is moderate, and muddy shores are characteristic of quiet waters of semienclosed bays, deep inlets, and estuaries. The relationship between current speed and the erosion, transportation, and deposition of sediments is shown in Figure 3.2. From this figure it can be seen that for pebbles 10^4 μm (1 cm) in diameter, erosion of the sediments will take place at current speeds over 150 cm sec^{-1}. At current speeds between 150 and 90 cm sec^{-1}, the pebbles will be transported by the current, while at speeds of less than 90 cm sec^{-1} they will be deposited. Similarly, for a fine sand of 10^2 μm (0.1 mm) diameter, erosion will occur at speeds greater than 30 cm sec^{-1} and deposition will occur at speeds less than 15 cm sec^{-1}. For silts and clays, a similar relationship exists. However, erosion velocities are affected by the degree of consolidation of the sediment, which is a function of its water content.

Throughout the intertidal area of a beach there is a gradient in substratum texture of finer particles at low tidal

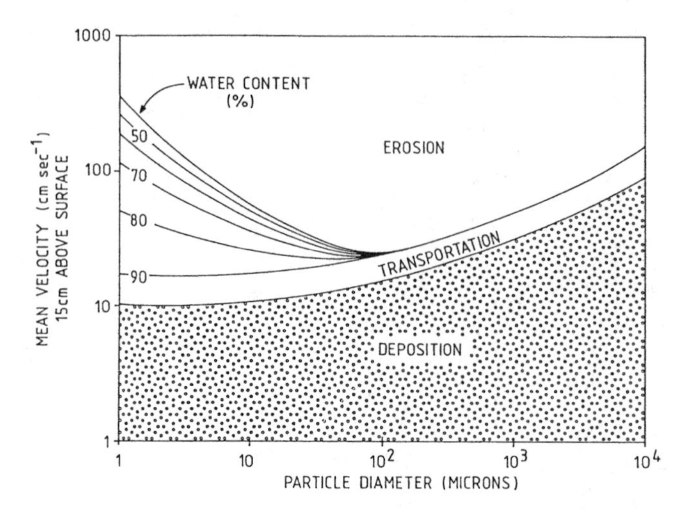

FIGURE 3.2 Erosion, transportation, and depositional velocities for different particle sizes. Also illustrated is the effect of the water content of the sediment on the degree of consolidation, which in turn modifies the erosion velocities. (Redrawn from Postma, H., in *Estuaries*, Lauff, G., Ed., Publication No. 83, American Association for the Advancement of Science, Washington, D.C., 1967, 158. With permission.)

levels to coarse particles higher up the shore. Figure 3.3 illustrates these processes on a sand beach at Howick in the Upper Waitemata Harbour, New Zealand. It is interesting to note that on the *Zostera* (eelgrass) flat there are finer deposits due to the reduction of water velocity by the leaves of the plants.

3.1.3 Currents, Wave Action, and Beach Formation

Figure 3.4 is a profile of a typical sandy beach environment on a temperate shore. Beyond the highest point reached by waves on spring tides is the "Dune Zone." The "Beach Zone" extends from the upper limit of the drift line to the extreme low water level. It is subdivided into the "Bachshore Zone" above high water, which is covered only on exceptional tides, and the "Foreshore Zone" extending from low water up to the limit of high water wave swash. The "Nearshore Zone" extends from low water to the deepest limit of wave erosion. It is subdivided into an "Inner Turbulent Zone" covering the region of breaking waves and an "Outer Turbulent Zone." The profile of a sand beach may exhibit structures such as "berms" and "ridges." A berm is a flat-topped terrace on the backshore, while a ridge is a bar running along the beach near low water. Below low water the corresponding features are a "bar" and a "trough."

The physical feature of beaches of importance to ecologists can be found in King (1975), and McLachlan and Erasmus (1983). Water movement results in shear stress on the sea bed. This may move sediment off the bed, whereupon it may be transported by currents and waves. Cyclic water movement leads to the formation of ripples on the sand. Sand can be transported in two modes — as bed load and as suspended load. Bed load is defined as that part of the total volume of transported material moving close to the bed, and not much above the ripple height. Suspended load is that part transported above the bed.

Movement of material up and down beaches varies with the nature of the waves and shore level. As waves approach the beach and as the water becomes shallower and the breakpoint approached, more and more sand is caught up and transported. Inside the breakpoint the direction of transport depends chiefly on the slope of the waves. Steep waves are destructive, tending to move material seaward, while flatter waves are constructive, tending to move particles up the beach. The slope of the beach face depends on the interaction of the swash/backwash processes planing it. Swash running up a beach carries sand with it and therefore tends to cause accretion and a steep beach face. Backwash has the opposite effect. If a beach consists of very coarse material such as pebbles, the uprunning swash tends to drain into the beach face, thus eliminating the backwash. The sediments are thus carried up the beach but not back again, resulting in a steep beach face. Fine sand and sand-mud beaches, on the other hand, stay waterlogged because of their low permeability, so that each swash is flattened by a full backwash, which flattens the beach by removing sand. Thus the beach slope is a function of the relationships between particle size and wave action (Figure 3.5). Each grade of beach material has a characteristic angle of slope; a gravel or shingle beach has a depositional slope of about 12°, and a rubble beach may have a slope of about 20°. Fine sand and mud beaches may have a slope of under 2°.

Material removed from a beach is carried out to sea in the undertow. Waves that break obliquely on the shore carry material up the shore at an angle, while the backwash, with its contained sediment, runs directly down the beach. This means that with each breaking wave, some

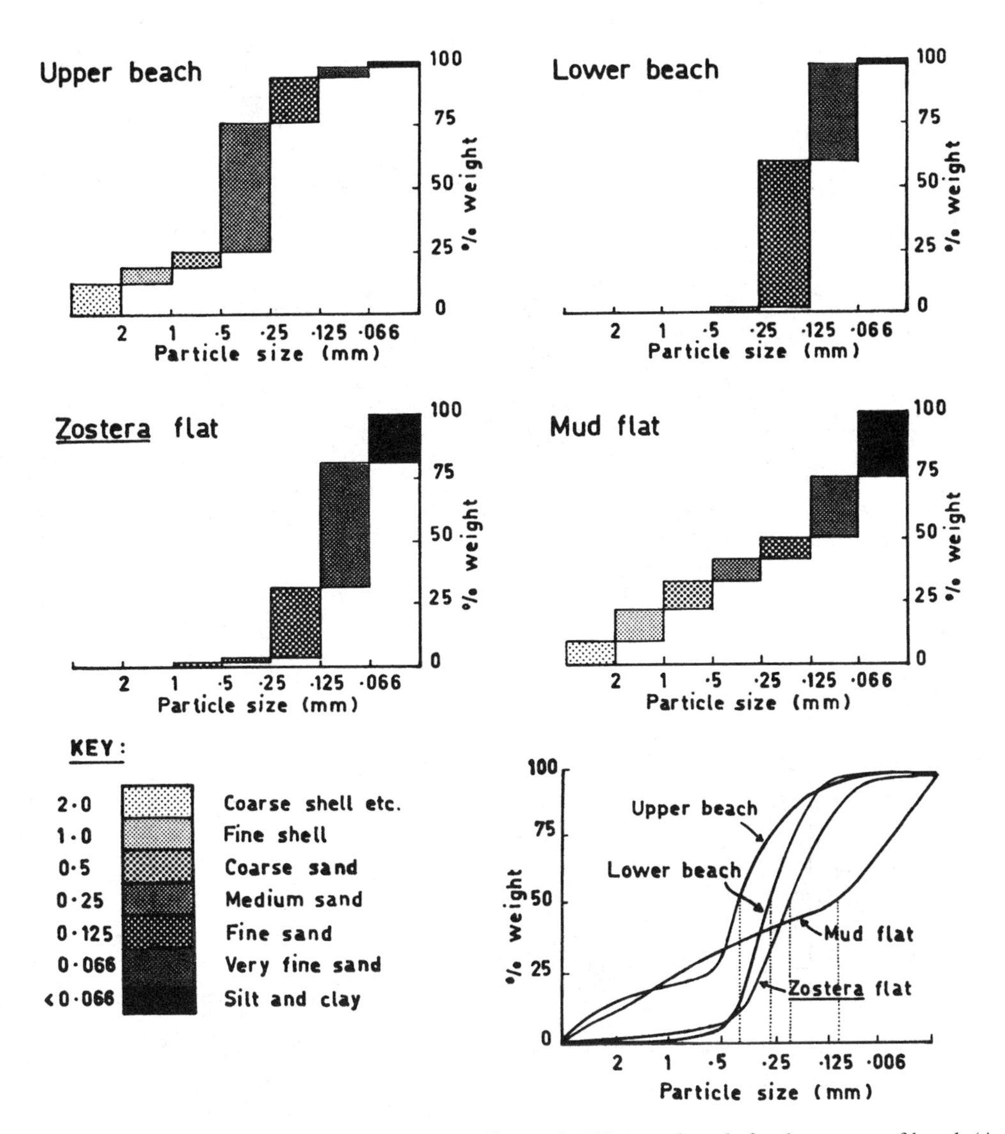

FIGURE 3.3 Distribution of sediment particles analyzed according to the Wentworth scale for three types of beach (A, upper beach; B, lower beach; C, *Zostera* flat at Howick, Upper Waitemata Harbour, North Island, New Zealand), and a tidal mudflat in Lyttleton Harbour, South Island, New Zealand (D). In E the same information is presented as cumulative curves. (A to D redrawn from Morton, J. and Miller, M., *The New Zealand Seashore,* Collins, London, 1968, 441. With permission.)

material is carried a short distance along the beach by a process known as beach drifting. Beach material can thus be transported considerable distances along the shore.

The longshore transport of sediments is assisted by longshore currents, which run parallel to the shore (Figure 3.6). Such currents are found when more water is brought ashore than can escape in the undertow; this leads to a piling up of water that escapes by running parallel to the beach. The interaction of surface gravity waves moving toward the beach, and edge waves moving along shore, produces alternating zones of high and low waves, which determines the position of rip currents. The classical pattern that results from this is the horizontal eddy or cell known as the nearshore circulation pool.

Water filtration by the sediments: Large volumes of seawater are filtered by the intertidal and subtidal sediments. In the intertidal this occurs by the swash flushing unsaturated sediments and in the subtidal by wave pumping, that is, by pressure changes associated with wave crests and troughs (Riedl, 1971; Riedl et al., 1972). Most filtration occurs on the upper beach around high tide. Water seeps out of the beach slowly by gravity drainage, mostly below the mean tide level. The volume of water filtered increases with coarser sands and steeper beaches (McLachlan, 1982). Tidal range also has an influence, with maximum filtration volumes associated with small to moderate tidal ranges.

3.1.4 Exposure Rating

In the literature on beach ecology, the terms "exposed" and "sheltered," or "high" and "low," are used in a very

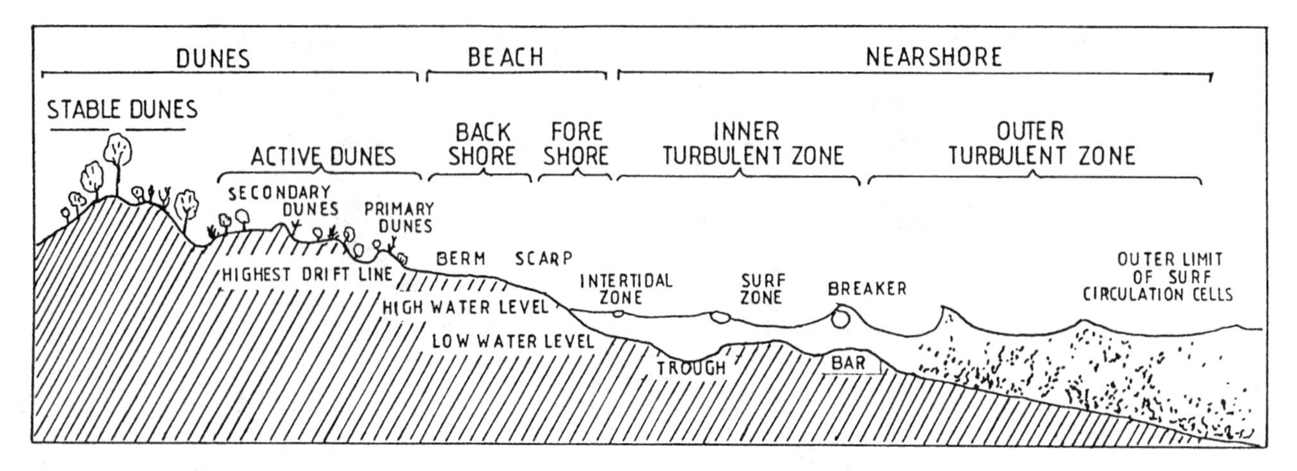

FIGURE 3.4 Profiles of a typical sandy beach environment, showing the areas referred to in the text. (After McLachlan, A., in *Sandy Beaches as Ecosystems*, McLachlan, A. and Erasmus, T., Eds., Dr. W. Junk Publishers, The Hague, 1983, 332. With permission.)

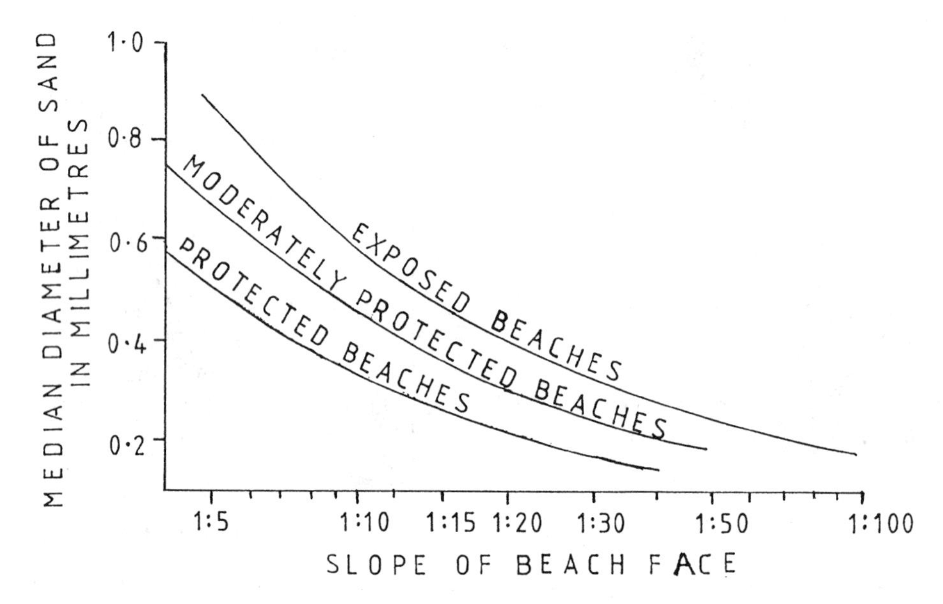

FIGURE 3.5 The relationship between beach particle size, exposure to wave action, and beach face angle in the western U.S.A. (Redrawn from Brown, A.C. and McLachlan, A., *Ecology of Sandy Shores,* Elsevier, Amsterdam, 1990, 21. After Wiegel, 1964. With permission.)

subjective way. To one worker, a beach may be "exposed," while to another it may be "moderately sheltered." McLachlan (1980a) developed a more objective exposure rating for beaches on a 20 point scale. The parameters considered included wave action, surf zone width, percent of very fine sand, median particle diameter, depth of the reduced layer, and the presence or absence of animals with stable burrows (Table 3.4). On the basis of the total score, beaches were rated as shown in Table 3.5.

3.2 THE PHYSICOCHEMICAL ENVIRONMENT

Differences in particle size and the degree of sorting result in important changes in the physicochemical properties of the sediments, which are reflected in the density and kinds of plants and animals that characterize the deposits. Among the most important of these are the interstitial pore space, water content, mobility, and depth to which the deposits are disturbed by wave action, the salinity and temperature of the interstitial water, the oxygen content, the organic content, and the depth of the reducing layer.

3.2.1 Interstitial Pore Space and Water Content

The interstitial water of a beach is either retained in the interstices between the sand grains as the tide falls, or is replenished from below by capillary action. The quantity of water that is retained within the sediment is a function of the available pore space, which in turn is dependent on the degree of packing and the degree of sorting of the

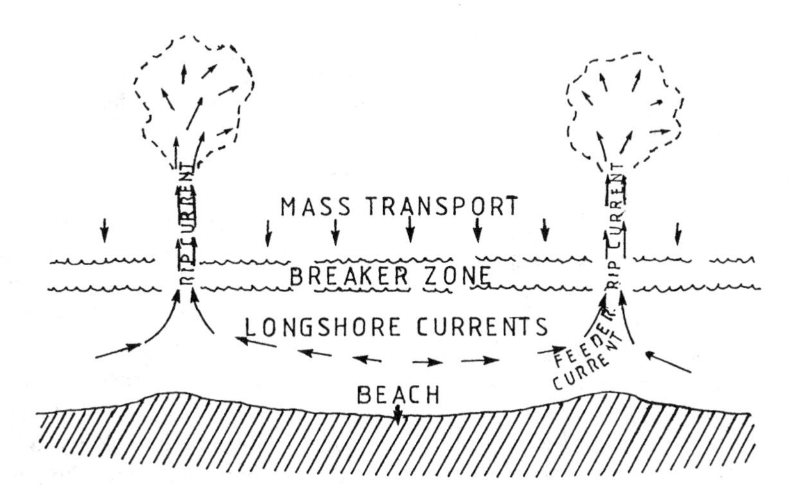

FIGURE 3.6 Nearshore cell circulation consisting of feeder offshore currents, rip currents, and a slow mass transport returning water to the surf zone. (Redrawn from Brown, A.C. and McLachlan, A., *Ecology of Sandy Shores,* Elsevier, Amsterdam, 1990, 32. After Sheppard and Inman, 1950. With permission.)

TABLE 3.4
Rating Scheme for Assessing the Degree of Exposure of Sandy Beaches

	Parameter	Rating					Score
1.	Wave action	Practically absent					0
		Variable, slight to moderate, wave height seldom exceeds 1 m					1
		Continuous, moderate, wave height seldom exceeds 1 m					2
		Continuous, heavy, wave height mostly exceeds 1 m					3
		Continuous, extreme, wave height less than 1.5 m					4
2.	Surf zone width (if wave score exceeds 1)	Very wide, waves first break on bars					0
		Moderate, waves usually break 50 to 150 m from shore					1
		Narrow, large waves break on the beach					2
3.	% very fine sand (62 to 125 μm)	5%					0
		1 to 5%					1
		1%					2
4.	Median particle diameter (μm)	**Intertidal slope:**					
		1/10	**1/10–1/15**	**1/15–1/25**	**1/25–1/50**	**1/50**	
	>710	5	6	7	7	7	
	500–710	4	5	6	7	7	
	350–500	3	4	5	6	7	
	250–350	2	3	4	5	6	
	180–250	1	2	3	4	5	
	<180	0	0	1	2	3	
5.	Depth of reduced layers (cm)	0 to 10					0
		10 to 25					1
		25 to 50					2
		50 to 80					3
		80 or more					4
6.	Animals with stable burrows	Present					0
		Absent					1
Maximum score							20
Minimum score							0

Source: From Brown A.C. and McLachlan, A., *Ecology of Sandy Shores,* Elsevier, Amsterdam, 1990, 38. With permission.

TABLE 3.5
Beach Types and Descriptions (Scores as Awarded in Table 3.4)

Score	Beach Type	Description
1 to 5	Very sheltered	Virtually no wave action, shallow reduced layers, abundant macrofaunal burrows
6 to 10	Sheltered	Little wave action, reduced layers present, usually some macrofaunal burrows
11 to 15	Exposed	Moderate to heavy wave action, reduced layers deep, usually no macrofaunal burrows
16 to 20	Very exposed	Heavy wave action, no reduced layers, macrofauna of highly mobile forms only

Source: From Brown, A.C. and McLachlan, A., *Ecology of Sandy Shores*, Elsevier, Amsterdam, 1990, 31. With permission.

sediment. In poorly sorted sediments the smaller particles pack into the interstices between the larger particles and thus reduce the percentage pore space. Coarse, ill-sorted sandy beaches have a relatively low porosity (approximately 20%), whereas in more sheltered localities where the deposits are well-sorted, the water retention may approach 45%.

The rate of replacement of water lost by evaporation from the surface of the deposits is dependent upon the diameter of the channels between the sand grains. These channels decrease in size with a decrease in grain size so that capillary rise is greatest in fine deposits. Thus, on beaches with fine deposits where the slope of the shore is low and the water retention (porosity) is high, the sediment is permanently damp, whereas on coarse, ill-sorted beaches where the slope is steep and the water retention low, the sediment contains less water and dries out quickly.

Related to the above characteristic of the sediments are the properties of "thixotrophy" and "dilatancy," which affect the ease with which burrowing animals can penetrate into the substratum (Chapman, 1949). Visitors to the seaside will have noticed the whitening of the sand that occurs underfoot. This is caused by water being driven from the interstices by the pressure applied until the sand becomes hard packed and dry. This property is called dilatancy, and such sands are called dilatant. These sands are difficult to penetrate because the application of pressure causes them to harden. Dilatant sands usually have a water content of less than 22% by weight. When the water content of the sand is greater than 25% the sands become thixotrophic, and consequently softer and easier to penetrate. Thixotrophic sediments become less viscous upon agitation and show a reduction in resistance with increased rate of shear in contrast to dilitant sediments, which show an increase in resistance. The most notorious examples of thixotrophic sands are quicksands, which liquify when pressure is applied. In experiments with burrowing worms, e.g., *Arenicola*, it has been shown that the speed of burrowing is dependent on the water content of the sediments and their resistance to shear.

3.2.2 Temperature

The temperature within the sediments is determined by insolation, evaporation, wind, rain, tidal inundation, and the amount of pore water. In general there is a gradient across the intertidal zone with maximum and minimum values occurring at the high water mark and low water mark, respectively. Marked vertical temperature gradients can develop in the upper 10 cm of the sediments, below which the temperature is fairly uniform, approaching that of the overlying seawater. The vertical gradient is much steeper in the summer than in the winter in temperate regions. Thus, animals living in the sediment are buffered against the temperature extremes that can occur when the tide is out.

3.2.3 Salinity

The salinity of the interstitial water of the sediments represents an equilibrium between the overlying seawater and the fresh water seeping out from the land. In estuaries there is a horizontal salinity gradient from low water to high water. The nature of this gradient depends on the pattern of estuarine circulation and salinity stratification. There may be considerable differences between the interstitial salinities and those of the overlying water (see Figure 3.7 for some data from the Avon-Heathcote Estuary, New Zealand). It can be seen that the interstitial salinity is considerably dampened when compared to that of the overlying water. Tube-building invertebrates that irrigate their burrows can play a significant role in maintaining the interstitial water salinity and other chemical properties so that it approximates that of the overlying water (Aller, 1980; Montague, 1982). Many such species cease irrigation when the salinity of the overlying water falls below a certain level.

Interstitial salinity variations are greatest on intertidal flats. During exposure to air, the salinities of the surface sediment are subject to dilution by rain and concentration by evaporation. In a two-month study (September to October) of salinity in a *Salicornia- Spartina* marsh at Mission Bay, San Diego, California, the water retained on the marsh had a higher salinity than that of the bay (ca. 34) for 75% of the time, exceeding 40 for 37% of the time, exceeding 45 for 10% of the time, and had a recorded maximum value of 50 (Bradshaw, 1968).

3.2.4 Oxygen Content

The oxygen content depends to a large extent on the drainage of water through the sediments. Porosity and drainage time increase sharply when there is 20% or more of fine

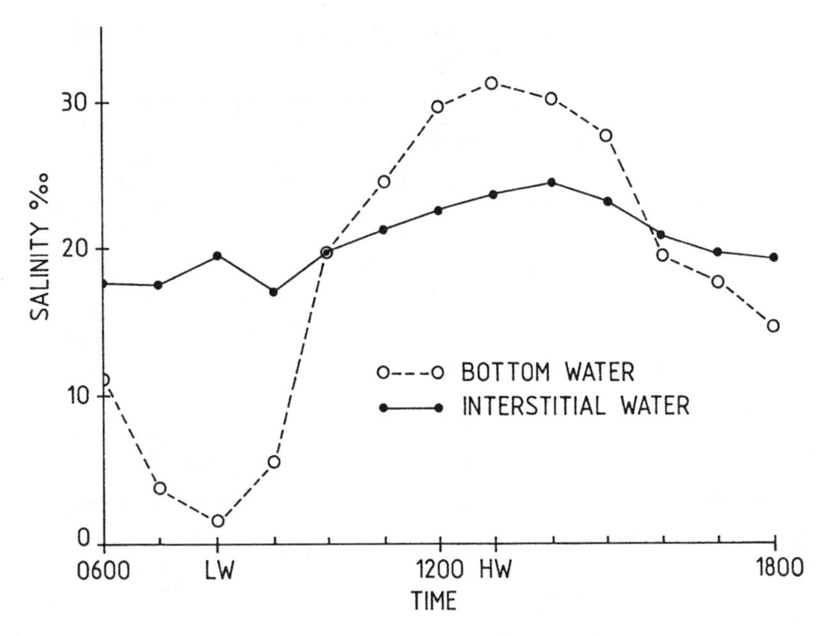

FIGURE 3.7 Salinity of bottom water at a low tide in the upper part of the Avon-Heathcote Estuary, New Zealand. (After Voller, R.W., Salinity, sediment, exposure and invertebrate macrofaunal distributions on the mudflats of the Avon-Heathcote estuary, Christchurch, M.Sc. thesis, Universiry of Canterbury, Christchurch, New Zealand, 1975. With permission.)

sand in a deposit, and in a similar manner, the oxygen concentration in the interstitial water varies with the percentage of fine sand. In general, coarse sands have more oxygen than do fine sands and muds. In poorly drained mudflats there is a pronounced vertical gradient in oxygen concentrations, high in the oxygenated layer and decreasing rapidly below (Bradfield, 1964). In one study the values varied from saturation at the surface (due to the photosynthetic activities of the microalgae) to 1.4 ml O_2 l^{-1} at 2 cm and 0.3 ml O_2 l^{-1} at 5 cm.

As discussed below, the vertical gradient in oxygen content is related to the amount of organic matter in the sediment and the depth of the reducing layer. Animals that live in the deeper layer of the sediment where oxygen levels are low must either be tolerant of anaerobic conditions or must retain a connection with the oxygenated surface layers and the overlying water, and maintain a flow of water over their respiratory surfaces. Burrowing polychaetes either live in U-shaped burrows through which water is circulated (e.g., *Arenicola*) by a pumping action of the body, or they maintain water circulation through simple burrows or tubes by ciliary and/or muscular action. Bivalves that live in anaerobic sediments maintain a connection to the surface via their siphons and thus maintain a circulation of oxygenated water across their gills.

3.2.5 Organic Content

Sediments are profoundly modified by the input of organic detritus that becomes incorporated in the surface sediments and through the activities of burrowing invertebrates if mixed to the deeper part of the sediment column. This detritus is derived from the variety of sources discussed in Section 3.8.2. Since organic particles are light and only settle out where the water is quiet, there is an inverse relationship between the organic content of the sediment and the turbulence of the water and hence grain size. Organic detritus tends to clog the interstices between the sediment grains and bind them together.

The sediments of mudflats that are found in sheltered bays and inlets and in the quiet lateral and upper parts of estuaries are composed of fine material (clays and silts) with ample organic material. These sediments have a characteristic vertical layering (Figure 3.8) in bands of color (Fenchel, 1969; Fenchel and Riedl, 1970) due to the one-way supply of light and oxygen and the biological activity of the burrowing invertebrates, the meiofauna, and the microflora and fauna. At the sediment surface there is a layer of often semifluid yellow or brown (due to the presence of ferric iron) oxidized sediment that is readily resuspended by turbulent currents. In this layer the redox potential as measured by the Eh is around +400 mV close to the surface and around +200 mV deeper in this layer. Below this layer is the "grey zone" or redox potential discontinuity (RPD) layer, a layer where oxidizing processes become replaced by reducing processes. According to Fenchel and Riedl (1970): "Food availability higher than oxygen input sufficient for the oxidization of food causes anaerobic conditions; hence the steepness and depth of the RPD layer depend basically on the equilibrium 'food: oxygen flow into the interstices'." The depth of the RPD layer depends upon the organic content and grain size composition (mean size, sorting, % clay) of the sediment; an increase in the

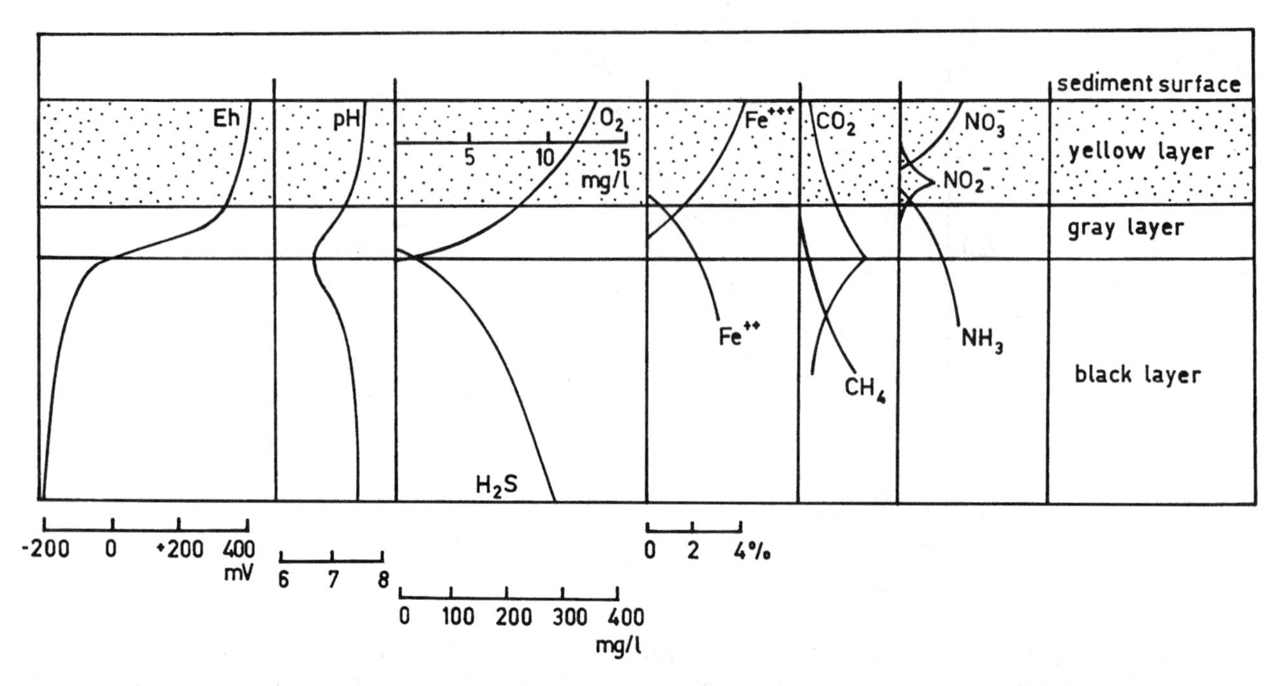

FIGURE 3.8 Schematic representation of Eh and pH profiles and the vertical distribution of some compounds and ions in estuarine sediments. The fully oxidized layer is dotted. (Redrawn from Fenchel, T., *Ophelia*, 6, 61, 1969. With permission.)

clay fraction and the amount of organic matter sharpens and raises the RPD layer toward the surface, while an increase in mean grain size and sorting causes the RPD layer to sink deeper into the sediment. In this RPD layer, oxygen, as well as reduced compounds such as hydrogen sulfide, are present in small amounts, while the Eh decreases quickly from positive to negative values. The third and deepest layer is the "black zone" or "sulfide zone." This layer is totally anaerobic, and within it, H_2S occurs in substantial amounts, up to 700 ml l^{-1} in the interstitial water of muddy sediments, while values around 300 ml l^{-1} are common (Fenchel and Reidl, 1970). Considerable amounts of H_2S are found as ferrous sulfides, giving the sediment its characteristic black color. The sediment layers can undergo vertical migrations correlated with changes in the following parameters (Fenchel, 1969; Fenchel and Reidl, 1970): (1) an increase in protection from water movement reduces permeability and brings the RPD layer closer to the surface; (2) higher temperatures bring about a higher position of the RPD layer, thus giving rise to seasonal changes in its level; (3) input of organic matter cause the RPD layer to rise; even a circadian rhythm has been observed where sunlight, due to the oxidizing activity of phototropic bacteria and microalgae, keeps the RPD layer down during daylight, but it rises toward the surface for a major part of the night. The biochemical processes mediated by the abundant microorganisms that occur within these layers are of considerable importance for the functioning of the sediment ecosystem and they will be considered in detail in Section 3.8.

The interstitial system on sand beaches is subject to cyclic changes related to storm/calm, tidal, diel, and seasonal cycles. In physically dynamic beaches, this results in fluctuations in the water table, pore moisture content, and surface temperature and it may result in sharp changes in the sediment chemical gradients. During warm conditions, the reduced layer may rise toward the surface, while storms or photosynthetic activity by surface diatoms can drive this layer deeper.

3.2.6 Stratification of the Interstitial System

Particularly on exposed beaches toward the physical extreme, the great vertical extent of the system and the drainage it experiences at low tide permit the subdivision of the intertidal beach into layers or strata. Various schemes have been proposed to describe this, and one such scheme is shown in Figure 3.9 (Salvat, 1964; Pollock and Hammon, 1971; McLachlan, 1980b). The layers range from dry surface sand at the top of the shore to permanently saturated sand lower down. The permanently saturated layers have little circulation and tend to become stagnated, while the resurgence zone has gravitational water drainage through it during ebb tide; the retention zone loses gravitational water but retains capillary moisture at low tide; and the zones of dry sand and drying sand lose even capillary movement. The zone of retention represented optimum conditions for interstitial fauna, since there is a good balance between water, oxygen and food input, physical stability, and lack of stagnation (Brown and McLachlan, 1990).

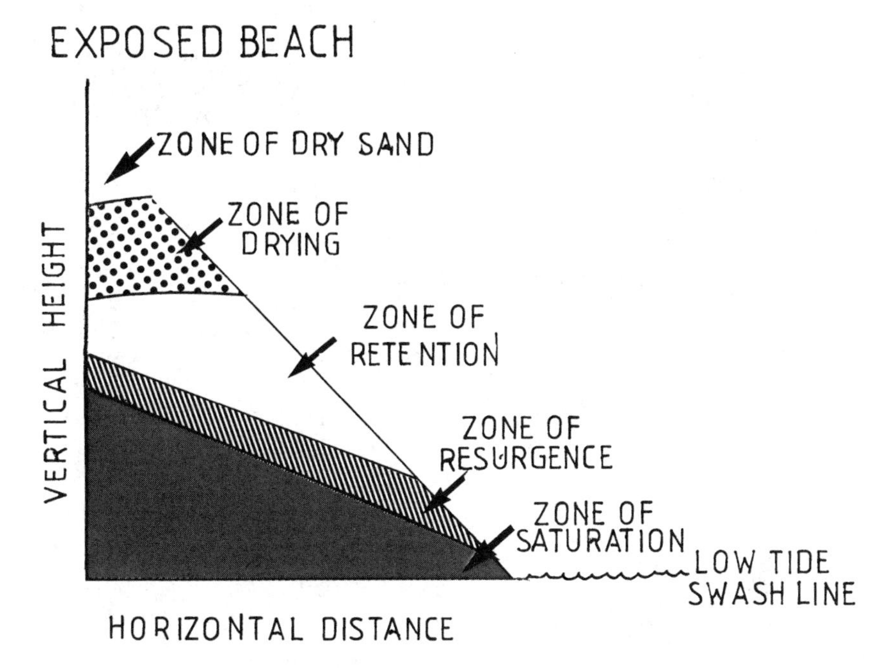

FIGURE 3.9 Stratification of the interstitial system on an exposed sandy beach. (Redrawn from Brown, A.C. and McLachlan, A., *Ecology of Sandy Shores,* Elsevier, Amsterdam, 1990, 148. After Salvat (1964) and Pollock & Hummon (1971). With permission.)

3.3 SOFT SHORE TYPES

On most coasts four main types of soft shores can be recognized (Figure 3.10):

1. *Shingle and pebble beaches*: These occur where wave action is strong, and they are characterized by their steep slope.
2. *Open sand beaches*: These occur along straightish exposed coasts as smooth beaches with moderately steep slopes. They are frequently backed by wind-blown dunes. Sediments are coarse to fine sands that are often freely disturbed by wave action, and the beach profile is frequently modified by storm events.
3. *Protected sand beaches*: These occur within bays where there is some protection from strong wave action. Destructive wave action is slight, due to the low pitch of the beach and the reduced fetch and steepness of the waves. The sediments have a large proportion of fine sand and very fine sand and often some of the silt/clay fraction.
4. *Protected mudflats*: These occur at the upper ends of deep inlets and harbors or on the inner side of barrier islands. Wave action is slight, enabling the deposition of fine sediments and organic matter, and consequently are characterized by a high proportion of the silt/clay fraction. The beach slope is gentle.

Graduations between the above types also occur. Special types of both protected sand beaches and protected mudflats occur in estuaries where wave action is slight and the main forces shaping the beaches are tidal rise and fall, river flow, and wind-generated turbulence. In many estuaries the upper shore intertidal areas are colonized by marsh macrophytes or mangroves (on tropical shores).

Short and Wright (1983) and Wright and Short (1983) have proposed a classification of beaches into three main types as shown in Figure 3.11, with their special features listed in Table 3.6. The two extremes of this system are the dissipative beach/surf zone and the reflective beach/surf zone, with a series of intermediate states. The reflective end of the scale occurs when the beach is exposed to strong wave action (surging breakers) and the sediments are coarse. Shingle beaches are of this type. On such beaches all the sediment is stored on the subaerial beach; there is no surf zone, and waves surge directly up the beach face. Cusps caused by edge waves are typical of such beaches. The beach face is characterized by a step on the lower shore and a berm above the intertidal slope.

As larger waves cut back a beach and spread out sediments to form a surf zone, the reflective beach is replaced by a series of intermediate forms. At the dissipative end of the scale, the beach is flat and maximally eroded, and the sediment is stored in a board surf zone that may have multiple bars parallel to the beach. Waves on such beaches tend to be spilling, and break a long way from the beach, often reforming and breaking again.

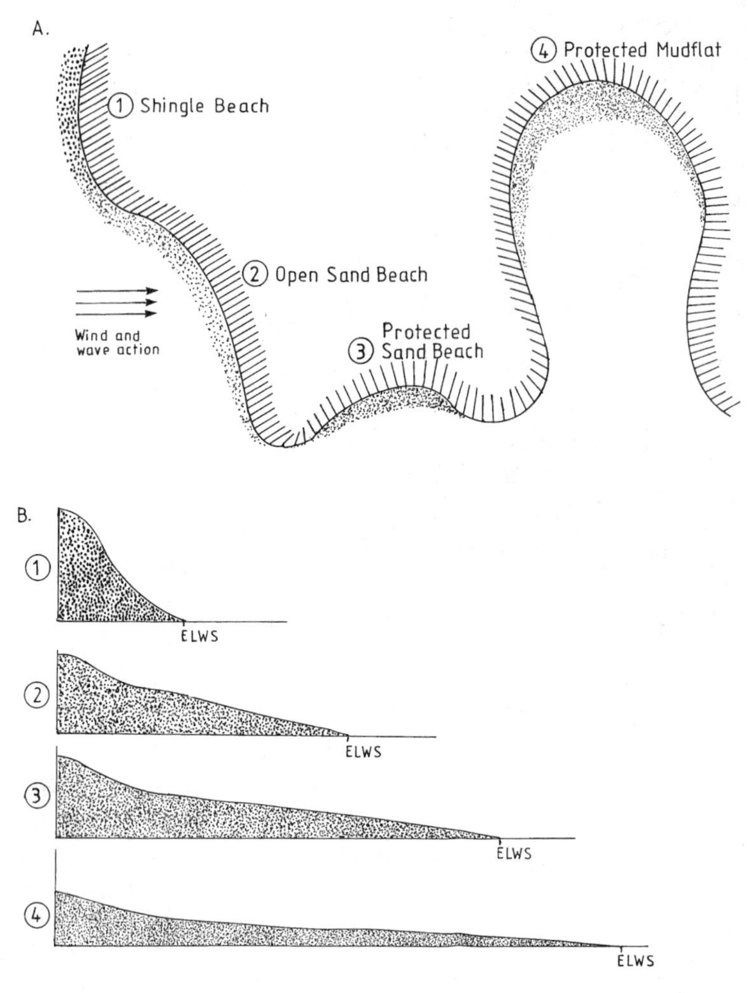

FIGURE 3.10 A. The four main types of soft shore. B. Beach profiles of the four types.

Between the two extremes, four intermediate states (Table 3.7) can be recognized. Intermediate beaches are characterized by high temporal variability, sand storage both on the beach and in the surf zone, and bars and troughs in a surf zone, usually supporting well-developed rip currents. Reflective beaches usually occur where waves exceed 2 m and sands are finer than 200 μm, whereas reflective beaches are found where waves are less than 0.5 m and sands are coarser than 400 μm. Short and Wright (1984) have shown that the morphodynamic state of a beach can be described by the parameter $Hb/W \cdot T$, where Hb is the breaker height, T the wave period, and W the fall velocity (Stokes law) of the sand. This generally resolves into a scale of 1 to 6, which agrees closely with the six states from reflective to dissipative. Swart (1983) has demonstrated that this correlates well with breaker types (Figure 3.12); spilling breakers being characteristic of dissipative beaches, plunging waves of most intermediate beaches, and surging breakers of reflective beaches. On fine-grained, macrotidal beaches, the lower tidal zones are often flat and highly dissipative, while the high tidal zones are reflective. This is the condition of many North European and British beaches.

3.4 ESTUARIES

3.4.1 What is an Estuary?

Estuaries constitute the transition zone between the freshwater and marine environments. As such they have some characteristics of both environments, but they also have unique properties of their own. Estuaries have been the preferred sites of human settlement, and many of the world's major cities are situated on estuaries. From a purely anthropomorphic point of view, estuaries have multiple values. Their values lie in the biological resources of fish and shellfish, which are of prime economic importance; their function as a fundamental link in the development of many species of fish and crustaceans (including economically important species such as flatfish, mullet, and prawns), and in the migration of important species such as salmon; their provision of feeding and breeding sites and

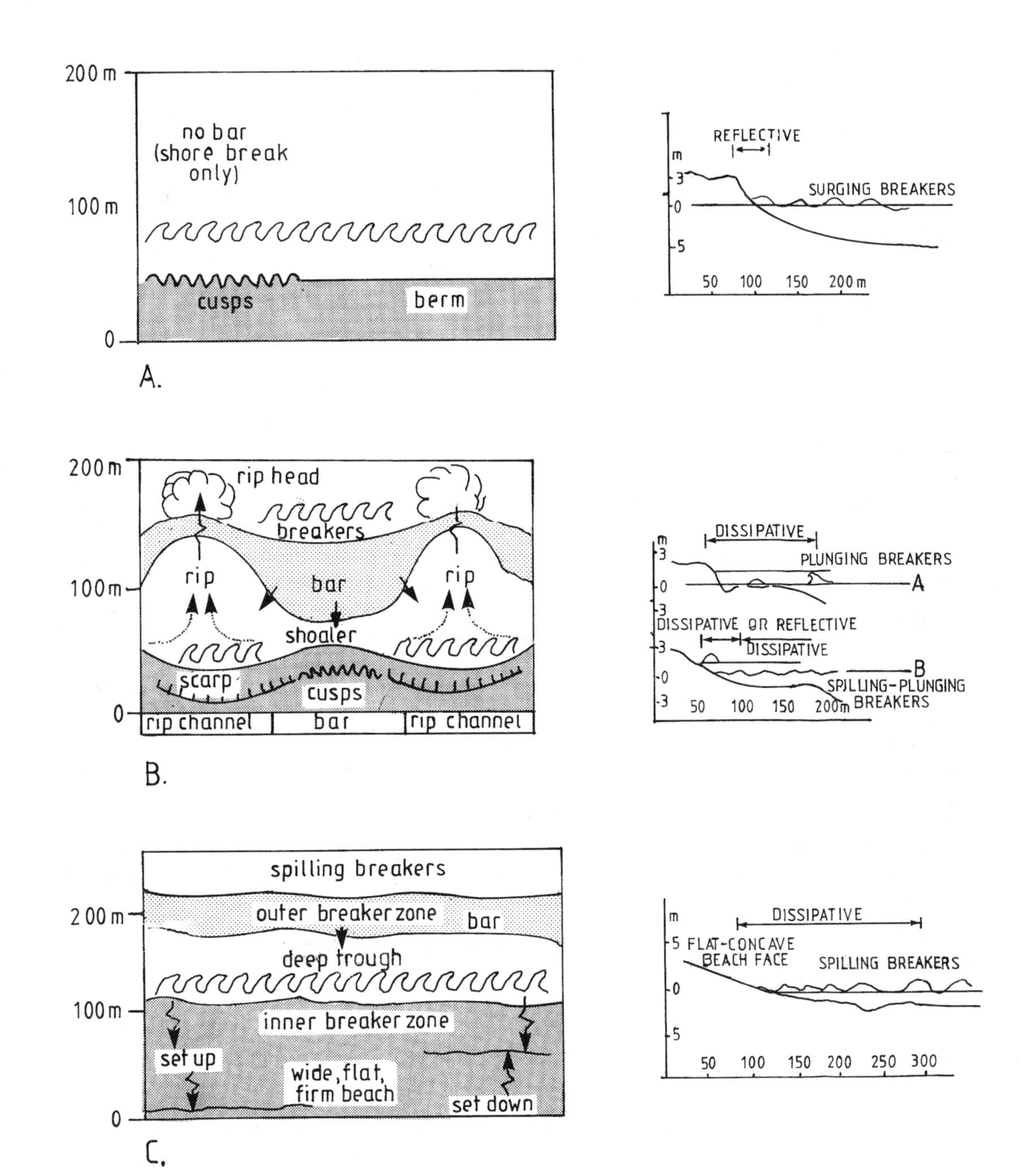

FIGURE 3.11 The three major morphodynamic states of beaches and surf zones. (After Jones, A.R and Short, A.D., *Coastal Marine Ecology of Temperate Australia*, University of New South Wales Press, Sydney. 1995, 139 and 140; and Short, A.D. and Wright, I.D., *Sandy Beaches as Ecosystems*, Mclachlan, A. and Erasmus, T., Dr. W. Junk Publishers, The Hague, 1983, 135. With permission.)

stopover points on migration routes for many species of ducks, geese, swans, and a great variety of wading birds; their mineral resources of sand, gravel, and sometimes oil; their provision of harbors and transportation routes for commerce; their provision of locations for housing and industrial plants; the recreational opportunities they provide for hunting, fishing, boating, swimming, and aesthetic enjoyment; and the opportunities they provide for education and scientific research (Allen, 1964). However, estuaries not only have multipurpose values, they are extremely vulnerable to impact from human activities.

The term estuary is derived from the Latin *aestuarium*, meaning "tidal inlet of the sea." *Aestuarium* in turn, is derived from the term *aestus*, meaning "tidal." There are many definitions of an estuary that have been used in the literature reflecting the particular scientific discipline of the proposer, whether physical or chemical oceanographer, geologist, geomorphologist, or biologist. At the First International Conference on Estuaries held in Georgia in 1964, a confusing array of definitions was proposed. There was wide agreement that variable salinity is an essential feature of all estuarine systems, and the defini-

TABLE 3.6
General Features of Different Beach Types

	Modal Beach Type		
	Dissipative	Intermediate	Reflective
Energy source	Infragravity standing waves and bores	Gravity and infragravity waves, rips	Gravity and edge waves
Morphology	Flat, with multiple bars	Variable bars	Deep water inshore
Slope	0.01	0.10–0.20	0.10–0.20
Sand storage	Stores in surf zone	Shifts between	Stores on
Width of the surf zone (m)	300–500	100–200	None
Dunes	Usually large	Intermediate	Usually small
Filtered volume[a]	Small	Intermediate	Large
Residence time[a]	About 24 h	6 to 24 h	About 6 h
Surf circulation	Vertical bores on surface, undertow below	Horizontal cells	No surf zone. Minicirculation within cusps
Surf zone diatoms	Rich	Variable	None
Intertidal fauna	Rich	Variable	Poor

[a] Filtered volume is the volume of seawater flushed daily through the intertidal sand; residence time is the time it takes to percolate.

Source: After Brown, A.C. and McLachlan, A., *Ecology of Sandy Shores,* Elsevier, Amsterdam, 1990, 31. With permission.

TABLE 3.7
Major Features of the Intermediate Beach Types

Beach Type and Beach Morphology	Width of Surf Zone (m)	Intertidal Beach Slope (m)	Subtidal Trough	Bar	Rip Current	Beach Cusps[3]
(a) Longshore bar-trough	ca. 200	0.05–0.20	Deep	Straight or crescentic	Weak	+
(b) Rhythmic bar and beach	150–200	0.05–0.10	Variable	Crescentic	Medium	+
(c) Transverse[1] bar and rip	150–200	0.05	—	Transverse	Strong	+
(d) Edge-runnel[2]	100–150	0.01	—	Terrance	Weak	+

Key: 1 — transverse = bar perpendicular to beach; 2 — runnel = small, shallow and narrow trough; and 3 — cusp = regular crescentic sand formation in beach.

Source: After Short, A.D. and Wright, L.D., in *Coastal Geomorphology in Australia*, Thom, G.B., Ed., Acadenic Press, Sydney, 1984. With permission.

tion proposed by Pritchard (1967a) was generally accepted. According to Pritchard, "an estuary is a semi-enclosed coastal body of water which has a free connection with the sea and within which seawater is measurably diluted with freshwater derived from land drainage." However, as Day (1980; 1981b) points out, such a definition excludes saline lakes with a salinity composition different than that of the sea, as well as marine inlets and lagoons without freshwater inflow, and inlets on arid coasts whose salinity is the same as that of the seas. Fjords, which exhibit many of the characteristics of estuaries, are also excluded.

However, the existing definitions are, as Day et al. (1980; 1987) point out, neither satisfying nor useful to those concerned with estuarine ecology and dynamic functioning of such systems, or to the diversity of environments that are involved. Consequently, Day et al. (1987) proposed the following functional definition:

> An estuarine system is a coastal indentation that has a restricted connection to the ocean and remains open at least intermittently. The estuarine system can be subdivided into three regions:
>
> (a) *A tidal river zone,* a fluvial zone characterized by lack of ocean salinity but subject to tidal rise and fall.
> (b) *A mixing zone* (the estuary proper) characterized by water mass mixing and existence of strong gradients of physical, chemical, and biotic quantities reaching

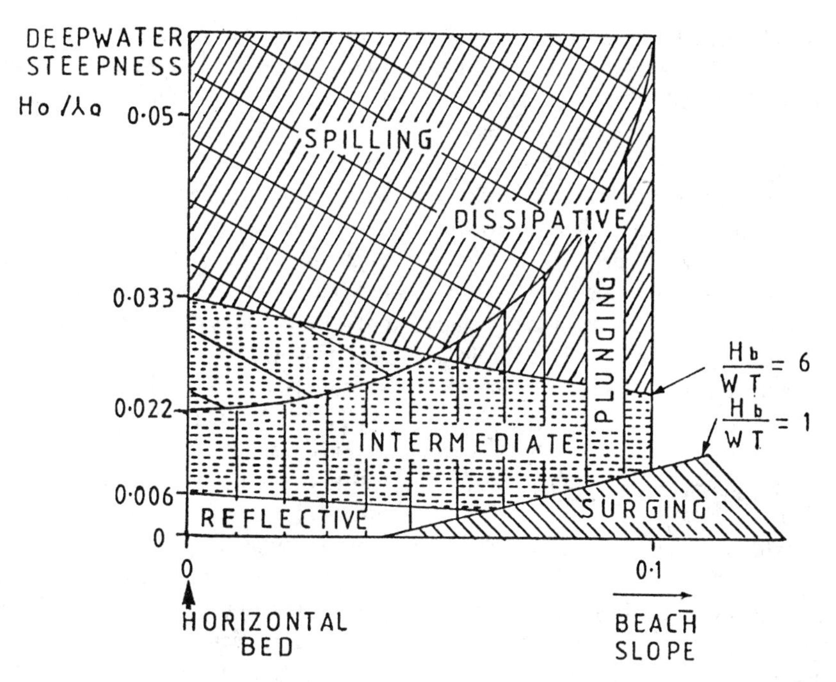

FIGURE 3.12 Correlation between beach states and breaker types. (Redrawn from Brown, A.C. and McLachlan, A., *Ecology of Sandy Shores*, Elsevier, Amsterdam, 1990, 30. After Schwart, 1983. With permission.)

from the tidal river zone to the seaward location of a river-mouth bar or ebb-tidal delta.

(c) *A nearshore turbid zone* in the open ocean between the mixing zone and the seaward edge of the tidal plume at full ebb tide.

This definition differs from other proposed definitions in that it recognizes and includes a nearshore component that is estuarine in character. As such the definition applies to what can be considered as typical estuaries with all three zones present. However, as Day et al. point out, all three zones may not be present. For example, lagoons in arid regions with a small tidal range may not have a tidal river zone and they may also lack a nearshore zone if the freshwater input is nonexistent. If the river discharge is large, e.g., the Amazon, the mixing zone may be absent, with the mixing taking place in the nearshore zone. The three zones are not static in any given system but are dynamic, with their positions changing on a variety of time scales from a tidal cycle to annual cycles to geological time scale.

3.4.2 Special Features of Estuarine Ecosystems

The salient feature of estuarine ecosystems is that they are highly productive; in fact they have been considered to be among the most productive ecosystems of the world (Scheleske and Odum, 1961; Knox, 1986a). This arises from the unique set of geomorphological, physical, chemical, and biological factors that are characteristic of such systems. While no two estuaries are identical, they nevertheless share a number of essential features in common, and it is this combination of characteristics that makes them unique. Above all, it is the fluctuating salinity that gives rise to their unique biological characteristics. Research over the past few decades has led to the development of a number of generalizations concerning their structure and function. These are discussed below.

1. *Circulation patterns and salinity distributions*: Among the most distinctive characteristics of estuaries are their circulation patterns and salinity distributions (Dyer, 1973). The particular patterns of water movement that occur within estuaries are the result of the combined influences of freshwater inflow, density distributions, wind, waves, and tidal action. This results in patterns that vary from stratified two-layered systems, characteristic of drowned river valleys and fjords, to well-mixed vertically homogeneous systems characteristic of shallow estuaries constantly well mixed by waves and tides. Circulation patterns are important since they transport nutrients, distribute plankton and larval stages of fish and invertebrates, control salinity patterns, transport sediments, mix water masses, dilute pollutants, and generally do useful work (Carricker, 1967).

2. *Sediment distributions*: Estuarine sediments have their own characteristics and complexity. In general, there is an upstream-downstream gradient from fine muddy sediments to sand, due to the interplay of different physical factors. In addition there is a gradient in substratum particle size across the intertidal areas from finer particles at low tidal levels to coarser particles at high tidal level. The nature of the sediment is profoundly modified by the activity of burrowing invertebrates, especially the deposit feed-

ers, as well as by the deposition of fecal pellets and pseudo-feces from deposit and suspension feeders, and the input of organic matter from inflowing rivers and from marsh plants that line the margins and submerged macrophytes. They are also the sites of intense microbial activity which decomposes the complex organic detritus derived from plants and animals and makes available nutrients such as ammonium nitrate and phosphate.

3. *Estuarine productivity*: As mentioned above, estuaries are among the most productive ecosystems in the world. Primary production in estuaries is complex, involving various combinations of the following groups of plants:

a. Macrophytes (sea grasses, sedges, cordgrasses, mangroves, etc.)
b. Epiphytic microalgae
c. Epiphytic macroalgae
d. Benthic microalgae (diatoms, flagellates, blue-green algae)
e. Benthic macroalgae
f. Phytoplankton

While no single study has simultaneously measured all these components of organic matter production, their relative importance in different estuarine ecosystems is reasonably well known. Many of the components are highly dynamic and productive, e.g., a dense sea grass meadow may have more than 4,000 plants m^{-2}, and have a standing stock of 1 to 2 kg dry wgt m^{-2}. Reported productivity of the sea grasses alone ranges from at least 5 to 15 g C m^{-2} day^{-1}, and when other associated primary producers such as the epiphytes are taken into account, the daily production can be well over 20 g C m^{-2} day^{-1} (McRoy and McMillan, 1977).

Westlake (1963) reviewed plant productivity on a global scale and concluded that when agricultural systems were excluded, tropical rainforests appeared to be the most productive (5 to 8 kg m^{-2} organic dry wgt $year^{-1}$), while salt marshes, reed swamps, and submerged macrophytes were the next most productive (in the ranges of 2.9 to 7.5 kg m^{-2} $year^{-1}$). Mean net primary production for estuaries as a whole is about 2 kg m^{-2} $year^{-1}$; this compares with 0.75 for the total land and 0.1555 for the total ocean (Ryther, 1969; Mann, 1972a).

4. *Estuarine food webs and energy flow*: The trophic dynamics of estuaries are complex. As we have seen, estuaries differ from the open ocean, which has phytoplankton as the sole producer in that there are always several different primary producers present (see 3 above). Direct grazing by herbivores in general consumes only a very small proportion of the macrophyte and macroalgal production. The great bulk of the organic matter produced (something over 90%) is processed through the detrital food web. Annual plant growth and decay provides a continuous source of large quantities of organic detritus (Teal, 1962; Darnell, 1967b; Fenchel, 1970). In addition there is also a considerable input of detritus from the inflowing rivers, especially during storm events (Naimen and Sibert, 1978). Nevertheless, the grazing food web based on phytoplankton plays an important role in many estuaries, with the phytoplankton being consumed by filter feeders such as bivalve molluscs, zooplankton, and small planktivorous fish. In addition, the epiphytic microalgae are eaten by crustacean and fish grazers, and the benthic microalgae are consumed by deposit-feeding invertebrates, fishes such as mullets, and the sediment micro- and meiofauna. However, a substantial proportion of the microalgal production is not consumed and is added to the detrital pool. The detritus becomes colonized by bacteria, fungi, and other microorganisms (Fenchel, 1970). The detrital particles and their associated microorganisms provide a basic food source for primary consumers such as zooplankton, most benthic invertebrates, and some fishes.

Estuaries can range from those in which phytoplankton dominates as the principal primary producer to those in which macrophytes (marsh grasses or mangroves) dominate, with every possible gradation in between. Many of the estuarine consumers are selective or indiscriminate feeders on particles in suspension in the water column or in the sediments they ingest. Thus, most of the biota of estuaries are best described as particle producers (microalgae and organic particles derived primarily from plant production) and particle consumers (Correll, 1978), and it is difficult to relate these to the traditional primary producer-primary consumer categories. The first trophic level in the estuarine ecosystem is therefore best described as a mixed trophic level, which in varying degrees is composed of herbivores, omnivores, and primary carnivores.

5. *Factors that determine the specific nature of estuaries and estuarine productivity*: The following features are important in this context:

a. *Protection from oceanic forces*: The degree to which they are protected and hence buffered from direct oceanic forces.
b. *Freshwater inflow*: The amount of freshwater inflow, together with the input of nutrients and organic matter, both dissolved and particulate, plays an important role in the dynamics of estuarine ecosystems. It also plays a major role in the nutrient trap effect detailed below.
c. *Water circulation patterns and tidal mixing*: Water circulation patterns are determined by riverine and tidal currents, density distributions, and geomorphology. The rise and fall of the tide is important in promoting the mixing of nutrient-rich water from the bottom (Mann, 1982). When the volume of the tidal exchange is large compared with river input, vertical salinity gra-

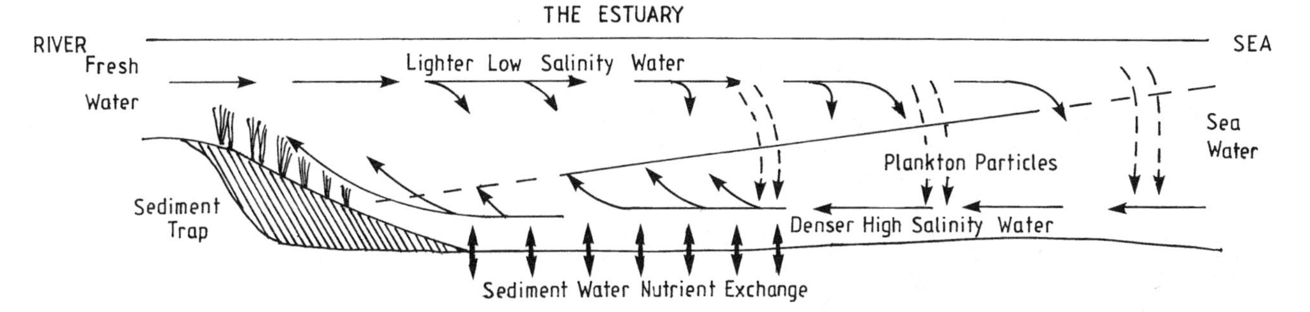

FIGURE 3.13 Schematic diagram of the nutrient conserving and modulating processes in estuaries, including the two-layered salt wedge, plankton circulation pattern, sediment trap, the tidal marsh, vascular plant "nutrient pump," and sediment-water nutrient exchange.

dients may be broken down so that the salinity is uniform from top to bottom. The most notable salinity gradient, then, is that from the river to the open sea. The sharpness of this gradient has a profound impact on water circulation and on many biological properties. Under such conditions, nutrients regenerated at the surface of the sediments are rapidly carried to the surface where they become available to the phytoplankton, e.g., in Narragansett Bay, Rhode Island (Kremer and Nixon, 1978).

d. *The depth of the estuary*: When estuaries are shallow the interaction between the water column and the bottom is strong. This allows nutrients released from the bottom to be rapidly mixed through the water column and made available to the phytoplankton.

e. *Tidal marsh nutrient modulation*: At times when nutrients are high in the upper estuary surface waters, they tend to be taken up rapidly by the tidal marshes, mudflats, and bottom sediments (Correll et al., 1975). Conversely, at times of low nutrient concentrations in the surface waters, a net release of nutrients occurs (Gardener, 1975). From a study of nutrient fluxes across the sediment-water interface in the turbid area of the Patuxent Estuary in Chesapeake Bay, Boynton et al. (1980) concluded that:

> In general it appears that nutrient fluxes across the sediment water interface represent an important source to the water column in summer when photosynthesis demand is high and water column stocks are low and, conversely, serve as a sink in winter when demand is low and water column stocks are high, thereby serving as a "buffering" function between supply and demand.

Overall, in the long term, the reservoir of nutrients in the sediments remain relatively constant; in the short term, however, they act as nutrient filters or modulators (Axelrod et al., 1976). The marshes also tend to trap particulate nitrogen and phosphorus and microbial action converts them into orthophosphate, ammonia, and dissolved organic phosphorus and nitrogen, which are then exported back to the open waters of the estuary (Axelrod et al., 1976).

f. *Sediment trapping*: Rivers deliver to estuaries large quantities of particulate mineral matter derived from land erosion, e.g., the Rhode River estuary, U.S.A., receives about 1.2 t ha^{-1} of estuary per year from land runoff (Correll et al., 1976). As detailed in Section 3.5, the fine particulate matter in suspension in river water flocculates and is deposited in a portion of the estuary known as the sediment trap (Figure 3.13). In the Rhode River estuary, Correll (1978) found that sediments were deposited in this zone at an average rate of about 11 t ha^{-1} $year^{-1}$. These deposited sediments are rich in nutrients and organic matter depending on tidal level support large populations of sea grasses and marsh plants.

g. *Vascular plant "nutrient pump"*: Eelgrasses and marsh plants have the capability to act as "nutrient pumps" between the surface waters and the bottom sediments. On the one hand, they take up nutrients from the sediments, and on the other, lose them to the water via death and decomposition, leaching from the leavers and perhaps by direct excretion.

h. *Rate of geomorphological change*: The rate of geomorphological change as determined by the various physical energies that move sediments is important in determining the nature of the physical and biological characteristics of an estuary. Present-day estuaries were formed during the last interglacial stage as sea level rose 120 m from 15,000 years ago up to the present level, which was reached approximately 5,000 years ago. Any future changes in sea level as a consequence of global warming will have a

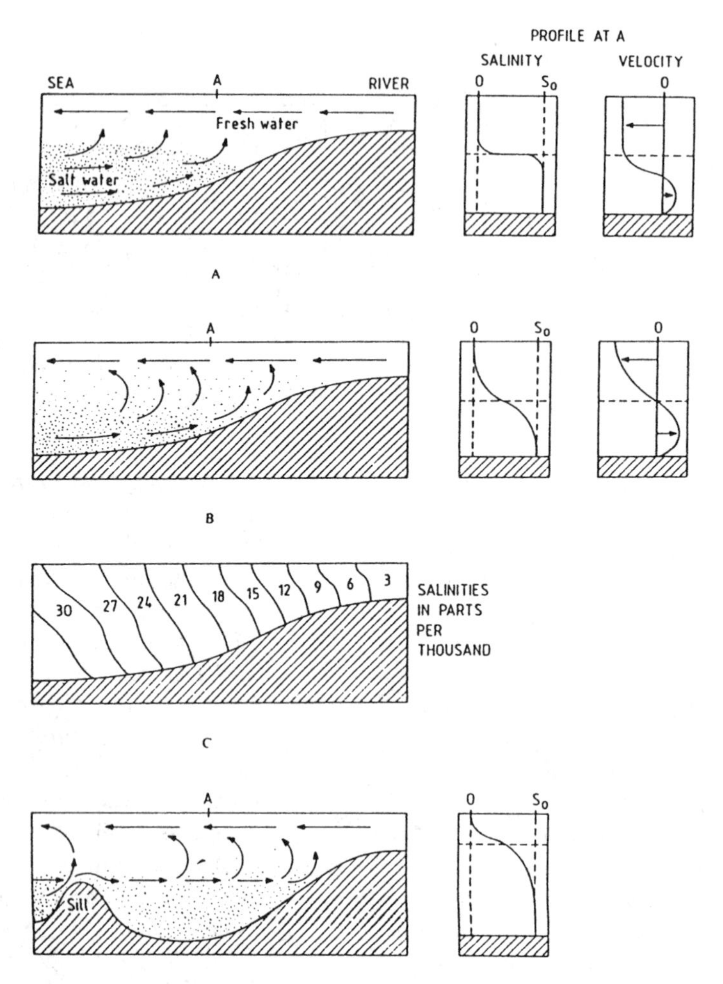

FIGURE 3.14 Diagrams illustrating the main types of estuarine circulation as seen in longitudinal section. A. Salt wedge estuary, salt water is stippled. B. Partially mixed estuary; salt and fresh water partially mixed by tidal movements and internal waves. C. Vertically homogeneous estuary isohalines for salinity are shown. D. Fjord; saline water is trapped by a sill. (From Knox, G.A., *Estuarine Ecosystems: A Systems Approach,* CRC Press, Boca Raton, Florida, 1986a, 31. With permission.)

major impact on the morphology of the present estuaries. Much will depend on the future rates of sedimentation resulting in the infilling of the estuaries. According to the scheme of Davies (1973) there is a continuum of estuarine types. At one end of the spectrum there are lagoons, which are produced by marine (wave action), while at the opposite end there are deltas, which are produced by river processes rather than by marine activities. The intermediate types are produced largely as a function of the interaction between wave energy and sediment transport by the rivers entering the system.

3.4.3 Estuarine Geomorphology

There are various classification of estuaries based on parameters such as topography, salinity structure, patterns of stratification, and circulation. From a geomorphological point of view, estuaries can be divided into four main groups (Dyer, 1973): (1) coastal plain estuaries; (2) lagoons (or bar-built estuaries); (3) fjords; and (4) tectonically produced estuaries. The distinction between lagoons and bar-built estuaries is not clear-cut. True lagoons have limited freshwater input compared with typical bar-built estuaries (Figure 3.14).

3.4.3.1 Coastal Plain Estuaries (Drowned River Valleys)

These are estuaries formed during the Flandarian Transgression, which ended about 3,000 B.C. In such estuaries sedimentation has not kept pace with inundation and their topography remains similar to that of river valleys. Sediments grade from silty muds at the top to coarse sands at the mouth. Such estuaries are usually restricted to, and are common in temperate latitudes. Examples of this type are the Thames and Mersey in the U.K., the Chesapeake Bay system in the eastern U.S., the Knysna in South Africa, and the Fitzroy River in Western Australia. In many of these estuaries the tidal range may be considerably less than half the mean depth, with a residence time of the water flowing into the estuary of only several days. Such estuaries are commonly either partially mixed or highly stratified.

3.4.3.2 Coastal Plain Salt Marsh Estuaries

Day et al. (1987) recognize another type of coastal plain estuary, the salt marsh estuary or the salt marsh creek. These are commonly found along much of the U.S. east coast, particularly from Cape Fear, North Carolina, to Cape Canaveral, Florida. They also occur on other temperate-tropical shores. They are characterized by the lack of a major river inflow, but have a well-defined tidal drainage network dendritically intersecting the extensive coastal salt marshes. Water and material exchange between the salt marsh system and adjacent waters through narrow tidal inlets. The estuary proper consists of the drainage channels, which typically occupy less than 20% of the system area.

3.4.3.3 Lagoon Type Bar-Built Estuaries

These are usually shallow estuarine basins separated from the sea by barrier sand islands and sand spits, broken by one or more inlets. In such estuaries recent sedimentation has kept pace with inundation, and they usually have a characteristic bar across the mouth. They are often shallow, only a few meters deep, and have extensive shallow waterways inside the mouth. Such estuaries are especially common in tropical regions, or in areas where there is active coastal deposition of sediments. Classic examples of such estuaries are the extensive network of marine bays in Texas and the Gippsland Lakes in Victoria, Australia. Other examples are many estuaries along the east coast of the U.S. and the Avon-Heathcote Estuary discussed in Section 6.3.2.

3.4.3.4 Lagoons

The distinction between lagoons and bar-built estuaries is not clear-cut. True lagoons have limited freshwater input compared with typical bar-built estuaries. True lagoons are common on all continents. The physical characteristics and dynamics of the extensive system of lagoons along the Gulf of Mexico have been described by Hedgpeth (1967) and Langford (1976). Other studies of lagoons are those of the Coorang Lagoon in South Australia (Noye, 1973), the lagoons on the southeastern Australian coast (Kench, 1999), and the St. Lucia Lagoon in South Africa (Day, 1981c). Lagoons characteristically have large expanses of open water and are uniformly shallow, often less than 2 m deep over large expanses. The physical processes of lagoons are mostly wind dominated. They are usually oriented parallel to the coast in contrast to coastal plain estuaries, which are usually oriented normal to the coast (Fairbridge, 1980).

3.4.3.5 Fjords

Fjords occur in regions that were covered by Pleistocene ice sheets, which deepened and widened existing river valleys to a typical U-shape, leaving rock bars or sills of glacial deposits at their mouths. They normally have rocky floors with a thin covering of sediments. Fjords are common in Norway, British Columbia, the Fjordland region of the South Island of New Zealand, and southern Chile. In regions of high rainfall such as the New Zealand Fjordland region, the waters are highly stratified with a layer of freshwater on the surface.

3.4.3.6 Estuaries Produced by Tectonic Processes

This is a catchall classification for estuaries that do not fit into the above categories. Coastal indentations formed by faulting or by land subsidence, such as San Francisco Bay, are included in this category.

3.4.4 Estuarine Circulation and Salinity Patterns (Figure 3.14)

3.4.4.1 Circulation Patterns

Circulation patterns in estuaries are complex due to the intermixing of fresh and salt water, the configuration of the estuarine basin (extent, depth, and size of opening to the ocean), and the amount of tidal rise and fall. As river water is less dense than seawater, it tends to flow seaward as a surface current, while the seawater tends to flow up-estuary as a bottom current. Depending on the degree of mixing between the two layers, the seawater forms a *salt wedge* extending toward the head of the estuary. As the tide rises and falls, the salt wedge advances and retreats (Figure 3.14A) The volume of water between the high and low water levels is known as the *tidal prism* and as it increases in volume from neap to spring tides, so does the velocity of the tidal currents.

Generally there is some degree of eddy diffusion and turbulent mixing at the interface between the surface layer of freshwater and the seawater of the salt wedge. Depending on the degree of mixing, the water column may become vertically stratified. This can vary from a gradual increase in salinity with depth, to a situation where the salinity becomes homogenous from the surface to the bottom if there is a high degree of turbulent mixing (as may occur in a shallow estuary due to wind and waves).

Estuarine circulation patterns affect or control many of the ecological processes within estuaries. An important characteristic is the residence time of a given parcel of water in the estuary, which is a function of the circulation pattern. Circulation patterns and residence time control the fluxes of dissolved constituents such as nutrients, dissolved organic matter, pollutants and salts, as well as of particulate material such as sediment, detritus, and plankton. Three main driving forces of the circulation patterns are: (1) gravitational circulation; (2) tidal currents; and (3) wind-driven circulation (Day et al., 1987).

3.4.4.2 Classification of Circulation and Salinity Patterns

A classification of estuaries based on current systems and salinity distributions has been developed by Pritchard (1967b).

1. Positive or Normal Estuaries
   a. *Salt wedge estuaries*: These are characterized by a dominant freshwater inflow, a small tidal range, and a large depth-to-width ratio (Figure 3.14A). The salt wedge penetrates up the estuary depending on river flow and the state of the tide. Saltwater mixes into the outward flowing freshwater, but the mixture of freshwater downward into the saltwater is minimal. The degree of mixing is largely dependent on the on the volume of freshwater inflow. A layer of mixed water of varying depth develops between the freshwater and the seawater with marked haloclines between them. Such estuaries are often referred to a highly stratified estuaries. The estuary of the Mississippi River and some fjords are of this type.
   b. *Partially mixed estuaries*: These are estuaries in which there are varying degrees of mixing between the outward-flowing surface freshwater and the inward-flowing bottom seawater so that the distinct boundaries that occur in salt wedge estuaries do not occur (Figure 3.14B). Turbulent mixing between the two layers is enhanced by a number of factors: it increases (a) when the ratio between tidal inflow and freshwater outflow approaches 1:1, (b) when there are irregularities in the channel bed, and (c) in shallow estuaries where the volume of the tidal prism is large compared with the total volume of the estuary basin. The velocity of the upper, freshwater layer is greatest at the surface and decreases with depth until the interface with the bottom saltwater layer is reached when the velocity is zero. In contrast, the velocity of the lower layer increases below this interface until it is retarded by friction with the bottom. The salinity of the upper layer increases down-estuary while the salinity of the lower layer decreases up-estuary until the tip of the salt wedge is reached. Examples of this type of estuary are the James River in the U.S.A., the Mersey and the Thames in the U.K., and the Hawkesbury River in Australia.
   c. *Vertically homogeneous estuaries*: In these estuaries (Figure 3.14C), the salinity decreases from the mouth toward the head without pronounced vertical gradients in salinity. This is the result of intense turbulent mixing and is characteristic of shallow estuaries with a large tidal range where the ratio of tidal inflow to river flow is on the order of 10:1. Estuaries of this type are the Solway Firth in the U.K. and Netarts Bay and Coos Bay, Oregon, U.S.
2. Hypersaline or Negative Estuaries

   Such estuaries have a reversed or negative salinity gradient where the salinity increases from seawater values at the mouth to hypersaline in the upper reaches where the water level is below sea level, so that the net flow is landward. These estuaries are found in regions subject to periodic drought. A classic example of a hypersaline estuary is the Lagoon Madre in Texas as described by Hedgpeth (1967).
3. Periodically Closed Estuaries

   These are coastal water bodies referred to by Day (1951) as *blind estuaries* and termed *estuarine lagoons* by Jennings and Bird (1967). The timing and period of their opening to the sea may vary considerably. They may be closed for periods of a year or more or they may be open to the sea at frequent intervals. When they are closed there is no tidal rise and fall and no tidal currents. According to the ratio of freshwater inflow plus precipitation and seepage through the bar separating the estuary from the sea, the salinity may vary considerably (Day, 1981b). There are many such estuaries in southern Africa such as the Umgababa estuary, in Western Australia, and other arid coasts.

3.4.5 Estuarine Sediments

Aspects of the characteristics of soft bottom sediments have already been dealt with in Section 3.1.2. Here we will relate these to estuarine sediments in particular. In terms of sedimentation, estuaries are very complex environments. One principal reason for this is that the sediments can arise from a number of sources; these include sediments of terrestrial origin transported by rivers (fluvial sediments) and sediments from the sea (marine sediments). On high energy coasts where littoral drift is strong, a flood-tide delta of marine sand may largely occlude an inlet, and if the tides have a high amplitude, the marine sand is carried rapidly to the head of the estuary. Conversely on low energy coasts where littoral drift is minimal, little marine sand enters the estuary. The estuary basin may remain deep, or if the drainage basin of the estuary is prone to erosion and the estuary is river dominated, the estuary may become largely filled by fluvial sediments. These are the extreme cases and generally there is an upstream-downstream gradient of fine silts and clays of mainly fluvial origin, to medium or coarse sands of marine origin at the mouth (J.H. Day, 1981b). The sedimentological properties of estuaries have been reviewed by Postma (1967), Dyer (1973), and J.H. Day (1981b).

For cohesive suspended particles, mainly in the clay (<2 μm) and colloidal ranges, their behavior is modified by processes causing coagulation or flocculation. Silt and clay particles bear negative charges due to the adsorption of anions, particularly OH^-, cation substitution in the crystal lattice, and broken bonds at the edge of the particles (Neihof and Loeb, 1972). These negative charges are balanced by a double layer of hydrated cations. The thickness of this double layer depends mainly on the ionic concentration of the water in which the particles are suspended. River water usually has a low electrolyte content and the charges on suspended particles repel each other. Estuarine water on the other hand, has a high electrolyte content so that the repulsive charges diminish and when the particles collide they unite to form a large spongy network or floccule. The flocculation of silt particles, most types of humus, and the clay minerals *illite* and *koalinite* mainly occurs at salinities of 1 to 4. Flocculation starts at the head of the estuary and as the floccules grow they drift downstream with the water becoming increasingly turbid. In many estuaries this produces a so-called "turbidity maximum." The presence and magnitude of this turbidity maximum is controlled by a number of factors, including the amount of suspended material in the water, the estuarine circulation pattern, and the settling velocity of the available material (Postma, 1967). It moves downstream with high river flow and upstream with low river flow. High turbidity cuts down light penetration and consequently influences primary production.

3.5 SOFT SHORE PRIMARY PRODUCERS

3.5.1 The Microflora

3.5.1.1 Sand Beaches

The benthic microflora of beach sands includes bacteria, blue-green algae, autotrophic flagellates, and diatoms. These may be attached to sand grains (episammon), or they may be free living in the interstices between the sand grains. On beaches with vigorous wave action, living diatoms may be mixed to considerable depths in the sediment; such mixing is enhanced by the activities of bioturbating macroinvertebrates. The photic zone within the sediments decreases with depth and is deeper on beaches with coarse sediment particles, but normally does not exceed 5 mm. The surface area of the sand grains with decreasing particle size provides increased area for the attachment of the microflora, but also decreased pore space.

Some species such as the pennate diatom *Hantzchia* undergo rhythmic vertical migrations associated with tidal and light cycles. On sheltered sand beaches in Massachusetts, during daytime low tides, the cells move to the surface of the sand, returning to the subsurface interstitial spaces before the incoming tide reaches them.

There have only been a limited number of estimates of microfloral densities and production in beach sands, but values are known to reach 10^3 cells cm^{-3} under optimal conditions (Brown and McLachan, 1990).

3.5.1.2 Mudflats and Estuaries

Sandy-mud and mud shores harbor a more diverse assemblage of pennate diatoms, blue-green algae, and flagellates (Fenchel, 1978; Admiraal, 1984). In the salt marshes of Georgia the benthic microflora includes a species-rich assemblage of pennate diatoms that comprise 75 to 93% of the total microalgal biomass (Williams, 1962). Williams found that on average 90% of the cells belonged to four genera: *Cylindrotheca, Gyrosigma, Navicula,* and *Nitzschia.* Filamentous blue-green algae (*Anabaena oscillarioides, Micocoleus lyngbyaceous,* and *Schizothrix calicola*) and a single species of *Euglena* constitute most of the microalgal community.

On an intertidal mudflat in the Avon-Heathcote Estuary, New Zealand, McClatchie et al. (1982) identified 64 diatom species. The dominant genera were *Nitzschia* (11 species), *Navicula* (10 species), *Acanthes* (6 species), and *Amphira* (4 species). The number of diatom taxa found were comparable with the range found in North American estuaries. Sullivan (1975) found between 57 and 62 species in edaphic communities associated with vegetated areas in a Delaware salt marsh, whereas a bare beach lacking macroscopic vegetation supported only 43 species, and a salt pan only 30 species.

The large nonflagellated euglenoid *Euglena obtusa* is a cosmopolitan component of estuarine and mudflat microalgal assemblages, being abundant in fine muds and especially in areas of high organic and nutrient inputs (Steffensen,1969). Palmer and Round (1965) in a study in the Avon River estuary, England, found that *E. obtusa* had a pattern of migration in the muddy sediment, moving down 2 mm prior to being covered by the tide, or in response to reduced light. Some diatoms' vertical migrations appear to have the characteristics of an endogenous circadian rhythm (Palmer and Round, 1965; Brown et al., 1972).

3.5.1.3 Benthic Microalgal Biomass and Production

Biomass is generally estimated by determining the chlorophyll *a* content of a measured volume or weight of sediment. There are difficulties with chlorophyll determinations as the sediments contain high concentrations of chlorophyll degredation products. There are also many technical problems that are encountered in estimating production of benthic microalgae. This stems from the characteristics of the environment in which they live. Pomeroy et al. (1981) describe it in the following terms:

> The microalgae live in and on the top few millimeters of sediment, a habitat whose microenvironment is very difficult to describe. The interface represents a boundary between a dark, nutrient-rich, anaerobic sediment and either an illuminated, aerobic, comparatively nutrient-poor water column or, at ebb tide, the atmosphere. This microenvironment is extremely patchy and is subject to rapid and extreme variation, being directly affected by many factors. It is influenced by variatons in tidal exposure, sedimentation, higher plant cover, and surface and subsurface herbivores and detritivores. These factors in turn affect light intensity, temperature, pH, salinity, levels of organic and inorganic nutrients, intensity of grazing, and the stability of the sediment surface. The habitat of epibenthic algae is virtually impossible to define or reproduce adequately; thus when attempting to measure the performance of the algae in their native habitat, we must maintain the integrity of the natural relationships of the surface layer of the sediment.

Biomass — Joint (1978) investigated microalgal production on a mudflat in the River Lynher estuary, Cornwall, England. The seasonal cycle of chlorophyll *a* content of the surface sediment is shown in Figure 3.15. The increase in chlorophyll *a* in April coincided with an increased rate of photosynthesis. The increase in the standing stock of the chlorophyll *a* between March and April was 39 g^{-1} dry sediment, equivalent to an increase in biomass of 12.5 g C m^{-2}, if a carbon to chlorophyll *a* ratio of 50 is assumed; the calculated photosynthetic production for the same period was 20 g C m^{-2}. There was a decrease in the chlorophyll *a* content of the surface sediment during May, but this increased again in July. The decrease in the chlorophyll *a* content was assumed to be due to bioturbation by benthic invertebrates and consumption by heterotrophs as they were at a maximum at this time. This was supported by studies of depth profiles of chlorophyll *a* and phaeopigments; while the maximum chlorophyll *a* values were in the top 2 cm, appreciable levels were recorded down to 14+ cm and larger quantities of phaeopigments were found at the same depth.

Underwood and Paterson (1993) recently investigated seasonal changes in benthic microalgal biomass in the Severn Estuary, England. Chlorophyll *a* concentrations at three sites along a gradient of decreasing salinity varied both spatially and seasonally. There were significantly higher concentrations of chlorophyll *a* at the upper shore stations than the lower ones. Chlorophyll *a* concentrations also varied seasonally, and were generally higher in the warmer months, being positively correlated with temperature.

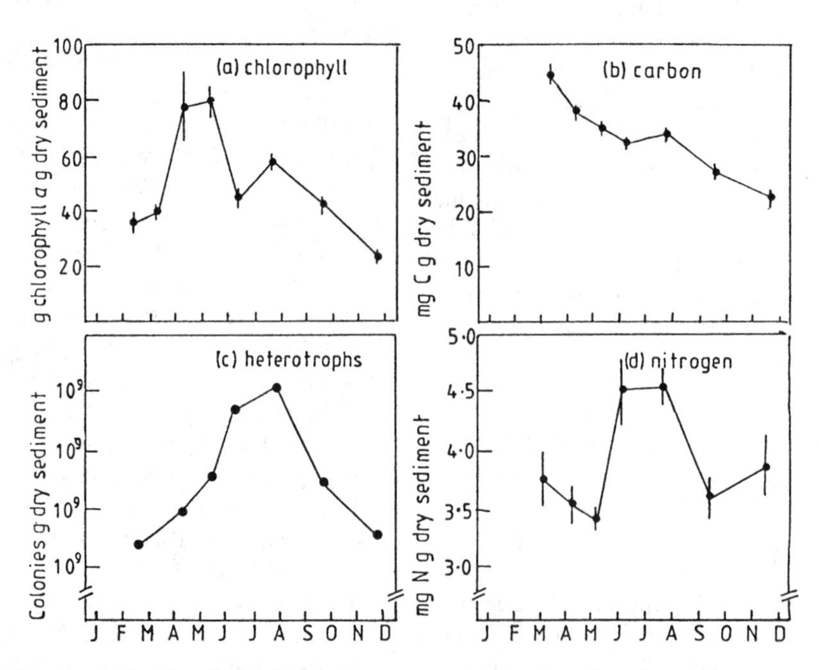

FIGURE 3.15 Seasonal cycle of (a) chlorophyll *a*, (b) carbon content, (c) number of aerobic heterotrophs, and (d) nitrogen content in the top 5 cm of the sediments in the River Lynher estuary, England. Vertical bars are two standard errors. (Redrawn from Joint, I.R., *Est. Coastal Mar. Sci.*, 7, 190, 1978. With permission.)

TABLE 3.8
Benthic Microalgal Chlorophyll *a* Levels Measured in a Range of Estuaries

Locality	Chlorophyll *a* (mg ch *a* m^{-2})	Biomass	Reference
Upper Waitemata Harbour, Auckland	0.6–6.7	0.072–8.04	Briggs (1982); Knox (1983a)
Avon-Heathcote Estuary, Christchurch	9.3–109.8	1.12–13.8	Juniper (1982)
Delaware Inlet, Nelson	12.5–30.5	1.5–3.66	Juniper (1982)
Nanaimo River Estuary, British Columbia	2.6–10.6	0.32–1.27	Naimen and Sibert (1978)
Netarts Bay, Oregon	Means		Davis and McIntire (1983)
Sand	46.2	—	
Fine sand	74.7		
Silt	93.7	—	

Note: Most of the available data for other estuaries is expressed in terms of g chlorophyll *a* g^{-1} (dry sediment) and thus are not easily translated into mg ch *a* m^{-2}. Chlorophyll *a* values have been converted into biomass estimates assuming a carbon:chlorophyll *a* ratio of 120:1.

However, increases in biomass also occurred during the autumn and winter months when temperatures were low.

A number of estimates of benthic microalgal chlorophyll *a* levels are listed in Table 3.8. A wide range of values have been recorded dependent upon tidal level, season, geographic locality, and type of beach (sheltered bay or inlet, estuary, or lagoon). For the New Zealand estuaries, the ranges differ but are in accord with the trophic status of the three systems: Delaware Inlet (range 0.6 to 30.5) is oligotrophic, the upper Waitemata Harbour (range 0.6 to 67) is mesotrophic, whereas the Avon-Heathcote Estuary (range 9.3 to 109.8) is eutrophic. Mean concentrations of chlorophyll *a* in the top cm of sediment in the entire study period were 46.2 mg m^{-2} (sand site), 749 mg m^{-2} (fine sand site), and 93.7 mg m^{-2} (silt site). Other studies have reported increases in chlo-

TABLE 3.9
Estimates of Benthic Microalgal Primary Production from a Range of Soft Shores

Estuary	Method	Production (g C m^{-2} yr^{-1}) Range:Mean	Chlorophyll (mg m^{-2})	Primary Production Rates (g C m^{-2} yr^{-1})	References
Ythan Estuary, Scotland	^{14}C	116	25–34 μg g^{-1} sediment	4–26: 10	Leach (1970)[c]
South New England, USA	^{14}C	81	100	8–31: 20	Marshall et al. (1971)[c]
Falmouth Bay, USA	^{14}C	106	n.d.	5–80	van Raalte et al. (1976)[c]
Wadden Sea, The Netherlands	^{14}C	60–140	40–400: 100	50–1100	Cadee and Hegemann (1974)[c]
Balgzand, Wadden Sea	^{14}C	29–188: 85	3–13 μg g^{-1} sediment	0–900 mg C m^{-2} day^{-1}	Cadee and Hegemann (1977)
River Lynher Estuary, England	^{14}C	143	30–80 μg g^{-1}	5–115	Joint (1978)
Ems-Dollard Estuary, The Netherlands	^{14}C	62–276	33–184	1–120: 37	Colijn and De Jonge (1983)
Ems-Dollard Estuary, The Netherlands	O_2	69–314	n.d.	0–1900 mg C m^{-2} day^{-1}	van Es (1977)[c]
False Bay, USA	O_2	143–226	30–70 μg g^{-1}	0–100	Pamatmat (1968)[c]
Georgia salt marches, USA	O_2	200	n.d.	5–140	Pomeroy (1959)[c]
Bay of Fundy, Canada	O_2	47–83	10–500	47–83	Hargrave et al. (1983)[c]
North Inlet, SC, USA	O	93	60–120	72–180	Pinckney and Zingmark (1991)
Oosterschelde, The Netherlands	^{14}C[a]	105–210	99–212	<1024[b]	Neinhuis et al. (1985)

[a] Based on annual regression of chlorophyll to ^{14}C-production as described by Colijn and De Jonge (1984).
[b] Taken from oxygen microelectrode measurements.
[c] References in Colijn and De Jonge (1984).

rophyll *a* concentrations with increasing silt/clay content of the sediment (Leach, 1970; Coles, 1979; Colijn and Dijkema, 1981).

Production — Some estimates of benthic microalgal production are given in Table 3.9. In the Sapelo Island marshes, Pomeroy (1959) estimated gross microalgal production at 200 g C m^{-2} and net production at not less than 90% of this value. Gallagher and Daiber (1974) estimated gross production for a Delaware salt marsh at 80 g C m^{-2}, which was about one third of the net angiosperm aboveground production in that particular marsh. Van Raalte et al. (1976) found that brief spring and autumn (fall) peaks in microalgal primary productivity coincided with blooms of filamentous green algae. Algal poduction was estimated at 106 g C m^{-2}, or about 25% of the aboveground macrophyte production for the marsh.

In Netarts Bay, Oregon, Davis and McIntire (1983) found that the maximum hourly rate of microalgal production occurred when *Enteromorpha prolifera* sporlings were abundant in the sediment. They found that a *fine sand* site had the highest mean hourly rates of gross primary production (47 mg C m^{-2} hr^{-1}), followed by a *sand* site (37 mg C m^{-2} hr^{-1}), and a *silt* site (25 mg C m^{-2} hr^{-1}). More recent estimates of benthic microalgal production at Sapelo Island (Pomeroy et al., 1981) gave a value ot 190 g C m^{-2}, which was close to Pomeroy's (1959) estimate. More than 75% of this microalgal production occurs when the sediment is exposed at ebb tide and at this time the bare creek bank was the most productive. Conversely, it is least productive when submerged on the flood tide. In the River Lynher Estuary, England, Joint (1978) found that the rate of photosynthesis decreases rapidly as the mudflat was submerged and was not detectable only 30 minutes after flooding.

Pickney and Zingmark (1993a,b) measured benthic microalgal production over a period from April to October in the North Inlet Estuary, South Carolina, in five different habitat types. The average annual production (in g C m^{-2} yr^{-1}) was highest in the short *Spartina* habitats (234.2), followed by intertidal mudflat (190.9), tall *Spartina* (97.9), sand flat (92.8), and subtidal (55.9) habitats. There was a unimodal peak in biomass during the winter–early spring period. Productivity was found to vary according to the tidal stage and the time of day which is related to the vertical migration of the microalgae within the sediments (Brown et al., 1972; Gallagher and Daiber, 1974, Pickney and Zingmark, 1991).

From the data in Table 3.9 it can be seen that there is a general trend of increasing production in warmer waters, e.g., mean production was 31 g C m^{-2} in the Ythan Estuary, Scotland, 99 in the North Wadden Sea, and 143 in the River Lynher Estuary. Values recorded in the New Zealand estuaries are comparable to those recorded in southern U.S. estuaries.

3.5.1.4 Factors Regulating Benthic Microalgal Distribution, Abundance, and Production

Environmental factors: The are many potential factors limiting the standing crop and production of benthic microalgae. Some of these such as sediment type, intertidal height, and seasonal changes in light intensity have been dealt with in Section 3.5. In all investigations to date, production of benthic microalgae occurred predominately when the sediments were exposed to the air with low values when the sediments were submerged. The tidal regime may also affect productivity indirectly through its influence on other parameters, including salinity, pH, temperature, light intensity, and nutrients.

The salinity of the surface sediment in estuaries varies from that of the overlying water at flood tide, to increasing salinity following evaporation, or decreasing salinity following high rainfall at low ride. In the River Lynher Estuary, England, Joint (1981) recorded salinity changes greater than 20% over a tidal cycle. However, estuarine benthic diatoms appear to be particularly tolerant of such salinity changes (McIntire and Reimer, 1974). Fourteen species of diatoms isolated from the Sapelo Island marshes grew well in salinities of 10 to 30, and several grew well over a range of 1 to 68 (Williams, 1964).

In the Sapelo Island marshes, the pH of the marsh surface sediments was found to be between 7 and 8, but during low tide it was found that algal photosynthesis could increase it to 9 (Pomeroy, 1959). It is possible that inadequate supplies of CO_2 and HCO_3 could limit photosynthesis under such conditions (Pomeroy et al., 1981).

Seasonal variations in temperature do not appear to have a marked effect on microalgal photosynthesis and production. Pomeroy (1959) and Van Raalte et al. (1976) noted that photosynthesis rates were independent of temperature at suboptimal temperatures. Temperature can also indirectly affect benthic microalgal production and biomass by influencing the activity of grazers.

Benthic microalgae growing in estuarine intertidal sediments are exposed to considerable variations in light intensity. However, they are much less sensitive to high light intensities and have the capacity to photosynthesize at lower light intensities than phytoplankton. Taylor (1964) found very little photoinhibition at "full sunlight" in experiments with diatoms from a Massachusetts intertidal sand flat, and that the photosynthesis was saturated at about 16% full sunlight. Cells receiving only 1% incident solar radiation were able to fix carbon at 35% of their maximum rate. These results have been confirmed by other studies (Williams, 1964; Cadee and Hedgemann, 1974; Colijn and van Buurt, 1975). Thus it is clear that estuarine benthic microalgae are adapted to photosynthesize at very low light intensities.

Limitation by nutrients has been investigated in a number of studies. Van Raalte et al. (1976) found that nutrient enrichment in a vegetated portion of a Massachusetts salt marsh stimulated the productivity of the benthic microalgae. One problem with experiments such as this is that fertilization increases the growth of vascular plants and thus reduces the light intensity reaching the sediment surface. Darley et al. (1981) incubated Sapalo Island fertilized sediment cores in the field and found that in cores from the short *Spartina* marsh the algal standing crop and productivity both increased. However, on bare creek banks, similar experiments indicated that the benthic microalgae were limited by the grazing activities of snails and fiddler crabs.

Biotic factors: From a number of studies it is clear that the species composition of the benthic microalgal community plays an important role in determining the overall primary productivity. In an Oregon marsh the presence of sporelings of the green alga, *Enteromotpha prolifera*, increased production rates (Davis and McIntire, 1983). In the Delaware Inlet, New Zealand, Gillespie and MacKenzie (1981) found that the highest rates of benthic microalgal $^{14}CO_2$ fixation occurred at sandy sites colonized primarily by the flagellate, *Euglena obtusa*, sometimes with occasional blooms of the blue-green alga, *Oscillatoria ornata*. Rates of fixation at these sites were generally 10 to 20 times greater than at sites inhabited by other microalgal populations. The highest rate of fixation observed (216 mg C m^{-2} hr^{-1}) occurred under bloom conditions of *Euglena*. Rates observed at other sites ranged from 1 to 5 mg C m^{-2} hr^{-1}.

Many infaunal and epifaunal deposit feeders ingest and assimilate benthic microalgae (Fenchel and Kofoed, 1976; Levinton, 1980; Juniper, 1982). Experimental manipulations of the populations of the estuarine gastropods *Hydrobia* spp. (Fenchel and Kofoed, 1976; Levinton, 1980), *Nassarius obsoletus* (Wetzel, 1977; Pace et al., 1979), *Benbicium auratum* (Branch and Branch, 1980), *Illyanassa obsoleta* (Levinton and Bianka, 1981; Connor and Edgar, 1982; Edwards and Welsh, 1982), and *Amphibola crenata* (Juniper, 1981; 1982; McClatchie et al., 1982) have demonstrated significant impacts on benthic microalgal populations. The roles of these snails are discussed in Section 7.3.5. In an experiment on estuarine sand flats of the Wash, England, Coles (1979) killed populations of invertebrates (especially those of the amphipod, *Corophium*) and found that this was followed by a dramatic explosion of diatom numbers, which reached an average of 95 × 10^4 cm^{-2} within a week compared to only 5 × 10^4 cell cm^{-2} on the surrounding sand flats.

Davis and Lee (1983) carried out a series of experiments in Yaquina Bay, Oregon, to determine the rates of recolonization of benthic microalgae and the effects of the infauna on microalgal biomass and production. Estuarine sediments were defaunated and transplanted in the field or kept in the laboratory. Microalgal colonization in the field was rapid, with chlorophyll *a* levels returning to control levels by day 10, while infaunal densities returned to control levels within 40 days. Removal of the infauna in the laboratory, primarily tanaids, increased benthic microalgal growth. After 40 days, chlorophyll *a* was four times greater and gross primary production two times greater in the defaunated sediment than in the controls. These and other experiments have demonstrated that natural densities of infauna as well as epifauna can control both microalgal biomass and production.

3.5.1.5 A Model of Estuarine Benthic Microalgal Production

Figure 3.16 is a simplified model illustrating the forcing functions and environmental variables that influence estuarine microalgal biomass and production. The photosynthetically available radiation (PAR) at any point in time or position on the shore is determined by a complex of factors including intertidal height, depth in the sediment, sediment type (fine or coarse), wave action, season (day length), and latitude (affecting the angle of incidence of the sun).

Net production (P_n) is determined by the interaction of PAR, the physiological state of the algae, including the degree to which they are shade adapted, temperature, salinity, and the availability of nutrients (nitrogen, phosphorus, and silicate). The benthic microalgae (diatoms, flagellates, and seasonally in some estuaries the sporelings of macroalgae, especially *Ulva* and *Enteromorpha*) are subject to grazing by sediment protozoa, meiofauna, infaunal deposit feeders, and epifuanal deposit feeders (especially gastropods).

3.5.1.6 Surf-Zone Phytoplankton

Many exposed beaches are characterized by persistently dense populations of phytoplankton species, clearly visible as patches of colored water. The species concerned are diatoms and collectively they are known as "surf-zone diatoms." When present they develop dense localized accumulations known as cell patches, which are clearly visible as dark brown stains composed of large numbers of cells floating on the surface where they are maintained by relatively stable foam. They have been recorded from most continents, and are characteristic of beaches with broard dissipative surf zones exposed to strong wave action. Thus they are typical of extensive beaches, not being found along short stretches of sandy coastline or pocket beaches.

Apart from the study by Cassie and Cassie (1968) on the primary productivity of *Chaetoceros armatum* and *Asterionella glacialis* (= *japonica*) on the west coast of the North Island, New Zealand, little attention was paid to the

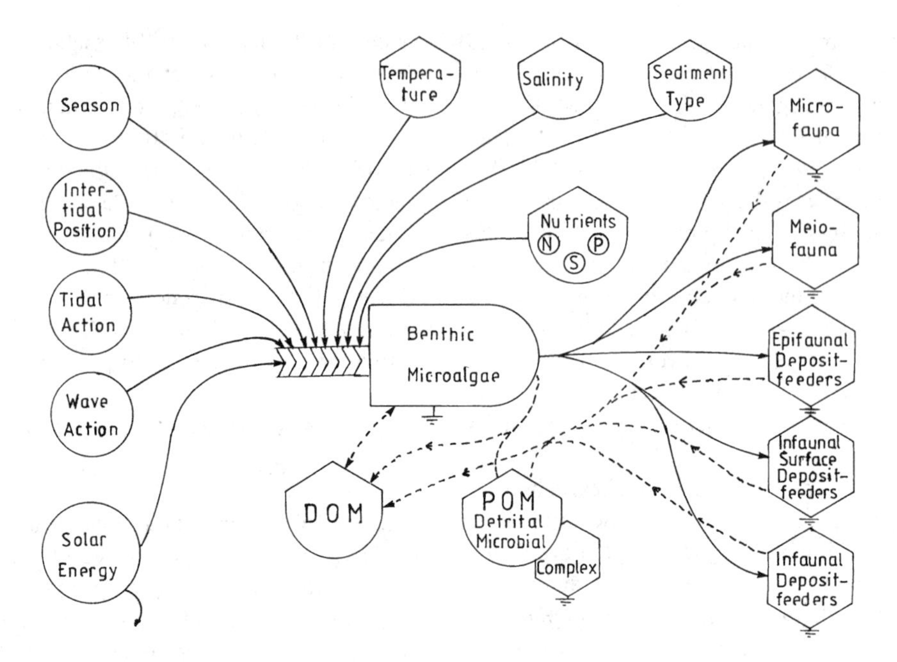

FIGURE 3.16 An energy flow model of the interstitial system.

study of surf-zone diatoms until the late 1970s with the initiation of extensive studies along the Washington coast (Oregon, U.S.A.) (reviewed by Lewin and Schaefer, 1983). Other comprehensive investigations have been concentrated on South African beaches (McLachlan and Lewin, 1981; Campbell and Bate, 1987; 1988; Talbot and Bate, 1986; 1987a,b; 1988,a,b,c,d; 1989). Talbot et al. (1989) have recently reviewed the ecology of surf-zone diatoms.

Species diversity: The dominant taxonomic feature of surf-zone diatoms is their low species diversity; all reported occurrences of large accumulations of the diatoms involve a single, or at the most two species. Six species have been reported belonging to four genera: *Anaulus, Asterionella, Aulacodiscus,* and *Chaetoceros.* At Sundays River Beach, South Africa, Campbell (1987) showed that within the surf-zone *Anaulus australis* made up 96.8% of the phytoplankton numbers in the surface layer. The remainder of the species assemblage consisted of *Asterionella glacialis* (1.3%), *Navicula* spp. (0.7%), and *Aulacodiscus johnstoni* (0.3%), with the bulk of the remaining 0.9% comprising species of *Campylosira, Hemicaulis, Leptocylindricus, Nitzschia,* and *Rhizosolenia.*

Frequency of patch occurrence: Cell patches are not a permanent feature of the surf zone and four major temporal features in their occurrence can be identified (Talbot and Bate, 1988b):

1. A diel periodicity, whereby cell patches form in the morning and are rare at dusk before disappearing by nightfall (Figure 3.17). This pattern has been reported for *Chaetoceros armatum, Asterionella socialis* (Lewin and Rao, 1975), *Aulacodiscus kittonii* (Kindley, 1983), and *Anaulus australis* (McLachlan and Lewin, 1981).
2. Superimposed on the regular periodicity of appearance-disappearance is a mesoscale variability comprising a sequence of presence-absence over a scale of days, or even weeks (Talbot and Bate, 1988c).
3. The third temporal feature is seasonality, e.g., Gianucia (1983) in southern Brazil found that blooms of *Asterionella glacialis* increased from late summer, throughout autumn and winter and tended to disappear in spring.
4. A change in species composition has also been reported, both for the Washington and South African coasts.

From numerous observations Talbot and Bate (1988a) have proposed the following model of patch formation and decay. Coupled with a pattern of vertical migration between sand and foam, there are diel changes in cell division and the production of a mucous coat. In the early morning, cells begin to divide. At this time they lose their mucous sheath. The loss of mucous (by an unknown process) causes the cells to be released from the sediment by the scouring action of the waves. The cells then enter the water column and briefly become planktonic by attachment to air bubbles that have been produced by wave action. Cells then concentrate at the air-water interface and complete the process of division (Figure 3.18). By the late afternoon, the recently divided cells once again develop the mucous sheath, which provides them with an active surface and enables them to switch their attention from air bubbles to sand grains. Thus the diatom popula-

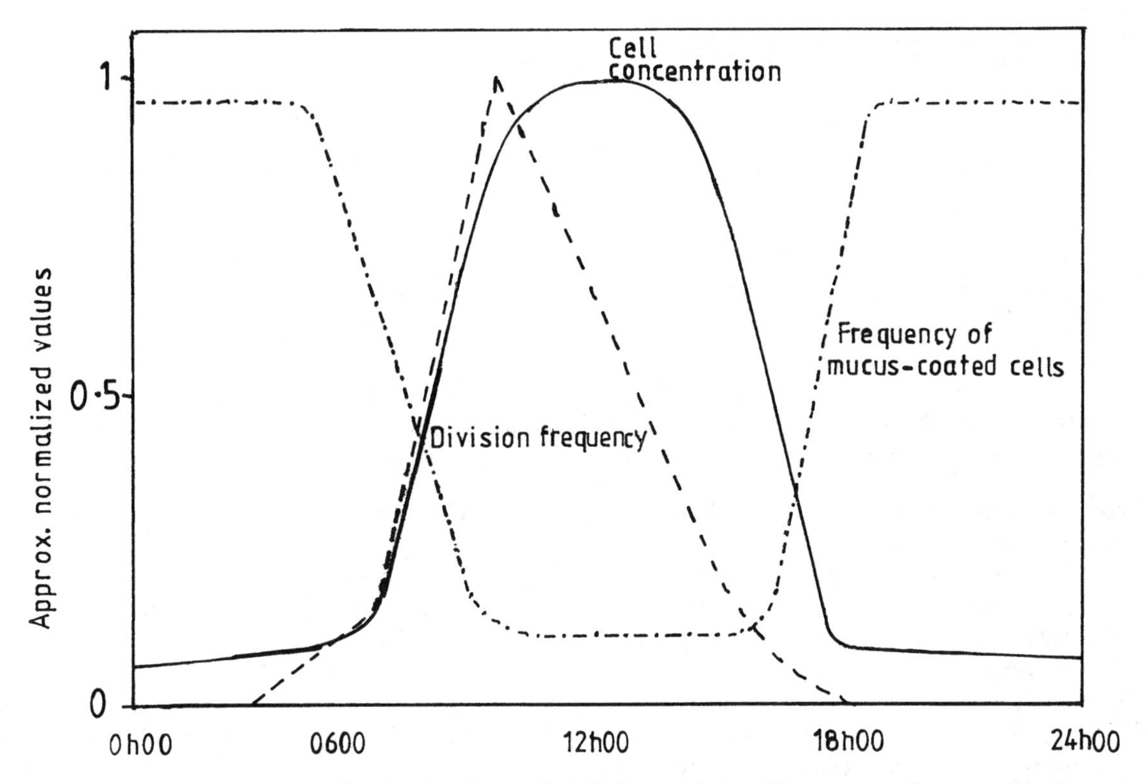

FIGURE 3.17 Stylized diagram illustrating the diel changes in cell characteristics of numerous surf-zone species. Data compiled from Lewin and Hruby (1973), Lewin and Rao (1975), and Talbot and Bate (1986; 1988c). (Redrawn from Brown, A.C. and McLachlan, A., *Ecology of Sandy Shores,* Elsevier, Amsterdam, 1990, 160. With permission.)

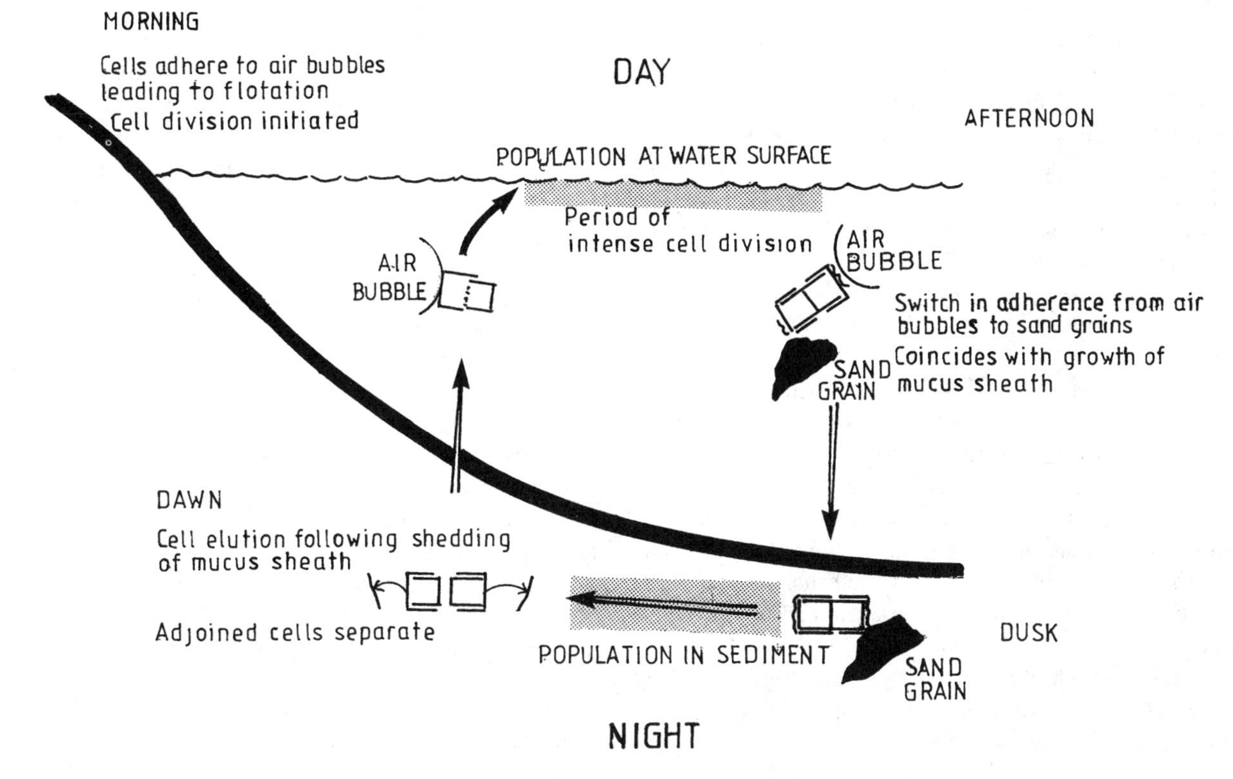

FIGURE 3.18 Model of diel patch formation and decay. (Redrawn from Talbot, M.M.B, Bate, G.C., and Campbell, E.E., *Oceanogr. Mar. Biol. Annu. Rev.*, 28, 163, 1989. With permission.)

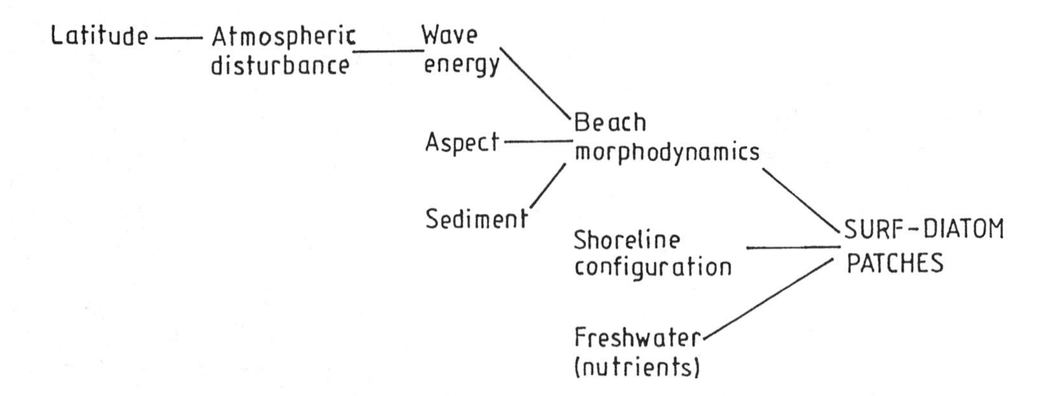

FIGURE 3.19 Model of environmental requirements of surf-zone diatom accumulations. (Redrawn from Talbot, M.M.B., Bate, G.C., and Campbell, E.E., *Oceangr. Mar. Biol. Annu. Rev.,* 28, 168, 1989. With permission.)

tion becomes nocturnally episammic until cell division is again initiated early in the day.

Hypotheses concerning geographical distribution: Why do some beaches support surf-zone diatoms and others do not? In reviewing the work of the 1970s, Lewin and Schaefer (1983) listed beach topography, wind, nutrient supply, and rainfall as important determinants. Other factors postulated as a result of subsequent research are the requirement for beach of a certain minimum length (Garner and Lewin, 1981; Campbell and Bate, 1988), and a rip-current system related to the surf circulation (Talbot and Bate, 1988a). Several investigators have suggested a direct wind influence on patch dynamics (Talbot and Bate, 1988b).

Talbot et al. (1989) have proposed the model of environmental requirements for the development of surf-zone diatoms depicted in Figure 3.19. Active areas occur, at least within the Southern Hemisphere, strictly between the latitudes 29°S and 34°S. This implies some overriding climatic requirement. They propose that the meteorological requirement is for periodic high wave energy, which accompanies the passage of atmospheric disturbances, or east-moving low pressure cyclonic systems that develop in the circumpolar westerlies of the Southern Ocean. Within the active areas, beach aspect relative to the direction of the wave approach and the sediment particle size if conducive, will result in beach morphodynamics that range from transverse bar and rip to incipiently dissipative. Within this given morphodynamic state, two more important requirements are an ample supply of freshwater runoff (enhanced nutrient supply), and an uninterrupted beach length of more than 10 km with some headland by which the cells are trapped on a large scale and not swept away from the area.

Brown and McLachlan (1990) have suggested that three cycles can explain the dynamics of surf-zone patch formation and decay:

1. The diurnal vertical movement between the foam during the day and the sediment at night. However, the bulk of the diatom population remains buried in the sediment with only a portion of it entering the water column each day.
2. An onshore/offshore migration, diatoms rising toward the surface during the day and being advected shoreward by wave bores. In the afternoon, when the cells start sinking out of the foam, they may be transported beyond the surf-zone by rip currents and be deposited on the seabed outside the breaker zone. Here they become available for entry into the surface water the next morning if they are stirred into the water column.
3. The involvement of a storm/calm cycle of events that is irregular and largely predictable. Increased wave action during storms causes greater disturbance and turnover of the sediments in which the bulk of the diatoms reside. Consequently the richest diatom accumulations occur during and immediately following conditions of high wave energy.

Spatial features in relation to rip currents: Longshore concentrations of diatoms occur at certain points along the shore, often adjacent to rip currents (McLachlan and Lewin, 1981). The opposing forces of incoming waves and outgoing rips create a bottle neck, or eddy effect, where the diatom foam accumulates. In Angola Bay, South Africa, Talbot and Bate (1987b) found that at low tide 94% of the 176 cell patches studies were found within a few meters of a rip current. In fully dissipative beaches without rip currents, such cell patches are not as discrete.

3.5.1.7 Epiphytic Microalgae

Distribution Patterns — Epiphytic microalgae grow on hard substrates (rocks, piles and other structures, and mollusc shells), on the stems and leaves of marsh plants, the leaves of sea grasses, the prop roots of mangroves, and on the fronds of macroalgae. The great diversity of sea grass species, morphology, and habitats in particular provide a substrate for the growth of a rich epiphytic community (Adams, 1976a; Stoner, 1980; Jernakoff et al., 1996). Such epiphytic communities comprise small macroalgae, microalgae, bacteria, protozoa, and meiofauna, and they are variously referred to as epiphytic, periphyton, or aufwuch associations or communities. Typically they contribute 10 to 50% of the combined sea grass-epiphyte production and standing crop (e.g., Penhale, 1977). The epiphytes can be closely connected to the sea grasses so that they exchange both nutrients and carbon (McRoy and Goering, 1974; Penhale and Smith, 1977). However, it has been demonstrated by a number of investigators that dense epiphytic colonization can reduce the light available to the sea grass leaves (Brix and Lyngby, 1985), and can directly inhibit the saturated photosynthetic response to light by up to 25% or more (Sand-Jensen, 1975; Penhale and Smith, 1977).

The epiphytic flora can be extremely diverse and can include as many as 100 species of microalgae and small macroalgae. The bulk of the flora, however, is dominated by a few species. McRoy and Helferich (1977) found that the diatom, *Isthmia nervosa*, on the leaves of *Zostera* in Alaska could contribute as much as 50% of the total leaf plus epiphyte dry weight. Kita and Harada (1962) compared the composition of phytoplankton in a *Zostera* bed near Seto, Japan, with the microalgae on the blades of the plants. They found that the two populations were distinct with very little overlap. The overwhelming majority of the epiphytes were diatoms, generally *Cocconeis scutellum* and *Nitzchia longissima*. The standing crop increased toward the tip of the blade, averaging 0.1 mg dry wgt cm^{-2}. In a study in Yaquina Estuary, Oregon, Main and McIntire (1974) identified 221 diatom taxa on the blades of *Zostera marina*.

Primary Production — Epiphytic algal production has been measured by a number of investigators who have shown that it can be significant when compared with that of the host plant and the total ecosystem. Marshall (1970) estimated the epiphytic algal productivity to be 20 g C m^{-2} yr^{-1}. In a detailed study of the epiphytes of the sea grass *Thalassia* in Florida, Jones (1968) found considerable seasonal variation in epiphytic productivity; peak rates occurred in February and March, and July to October, with very low and sometimes indictable rates of net production in the intervening months. He estimated the net epiphytic production in the summer to be 0.9 g C m^{-2} day^{-1} and 0.2 g C m^{-2} day^{-1} in the winter. The total production of the epiphytes was estimated at 200 g C m^{-2} yr^{-1}; this value was 20% of the estimated net production of the *Thalassia* bed of the area. Thayer et al. (1975) showed that a *Zostera marina* bed in North Carolina produced on average 350 g C m^{-2} yr^{-1}, while the associated epiphytic algae (both microalgae and fine macroalgae contributed a further 300 g C m^{-2} yr^{-1}. In Beaufort, North Carolina, Penhale and Smith (1977) measured epiphytic production at 73 g C m^{-2} yr^{-1} which, averaged over the total estuary, contributed 13 g C m^{-2} yr^{-1} (8.5% of the total estuarine primary production). In Flax Pond, New York, Woodwell et al. (1979) recorded an epiphytic production which, averaged over the total area, gave a value of 20 g C m^{-2} yr^{-1} (3.7% of the total annual primary production). It is thus clear that epiphytes can contribute one-fifth to one-third of the total estuarine community primary production.

3.5.2 Estuarine Phytoplankton

3.5.2.1 Introduction

Organisms in the plankton are generally assigned to three compartmental groups: bacterioplankton, phytoplankton, and zooplankton. These groups are further subdivided into trophic levels on the basis of taxonomic categories well above the species level. Unfortunately this results in the grouping together of organisms with different modes of nutrition, e.g., nonphotosynthetic flagellates are grouped together with algae and are considered to be phytoplankton. In contrast, other protozoan groups, such as the ciliates and sarcodinians, are assigned to the zooplankton as microzooplankton. In order to overcome these and other problems, Siebruth et al. (1978) proposed a scheme (Figure 3.20) based on the level of organization (ultrastructure) and mode of nutrition.

The heterotrophic organisms fall into five major categories: viroplankton, (viruses), bacterioplankton (free-living bacteria), mycoplankton (fungi), protozooplankton (apochlorite flagellates, amoeboid forms, and ciliates), and the metazooplankton (the multicellular ingesting forms). The metazooplankton span the size range from the mesoplankton through the macroplankton to the megaplankton. The mesoplankton consist mainly of copepods, while the macrozooplankton comprises mainly the larger crustacea such as mysids and euphausiids. Juvenile stages of the latter, however, fall within the mesoplankton size range. The megazooplankton comprise the larger drifting forms such as the coelenterates and appendicularians.

The protozooplankton, mycoplankton, and the phytoplankton are unicellular eucaryotes and fall into three size groupings: picoplankton (<2.0 μm), nanoplankton (2.0 to 20 μm), and microplankton (20 to 200 μm). The bacterioplankton compartment consists of unattached unicellular bacteria: these can be selectively filtered with 0.1 to 1.0-μm, porosity filters. In the scheme depicted in Figure 3.20 the heterotrophic components of the plankton have been redefined into more discrete taxonomic groupings and in an expanded range of redefined size groups. The size groupings are indicative of the growth and metabolic rates of the organisms involved, generally a function of size (Sheldon et al., 1972). Figure 3.20 shows that there is little overlap between the size categories and compartmental groups of plankton organisms, apart from the phytoplankton and protozooplankton, which occupy the same size

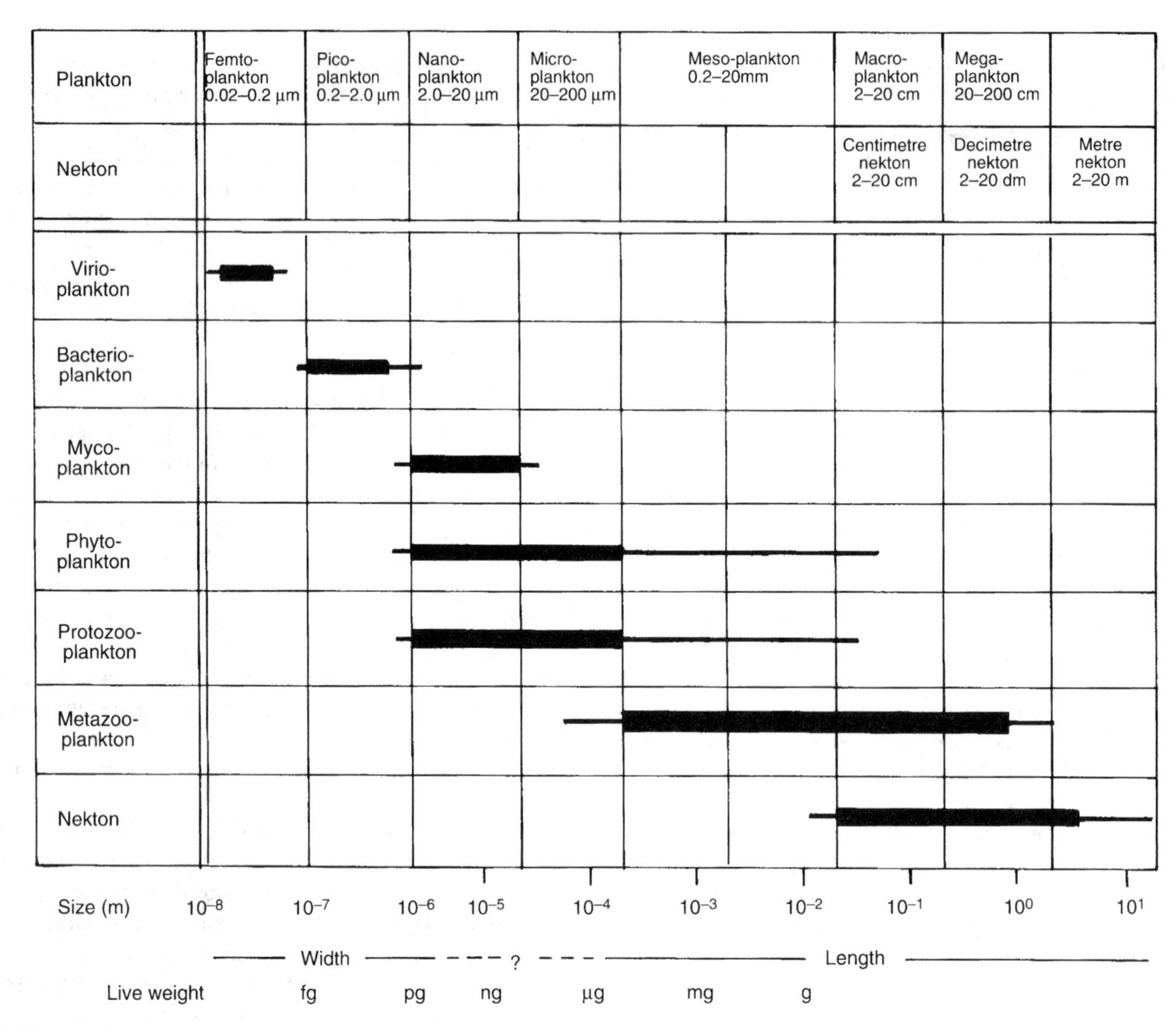

FIGURE 3.20 Distribution of different taxononic-trophic compartments of plankton in a spectrum of size fractions, in comparison with a size range of nekton. (Redrawn from Siebruth, J. McN., Smetacek, V., and Lenza, J., *Limnol. Oceanogr.*, 23, 1259, 1978. With permission.)

categories. However, they are distinguished by the presence or absence of chlorophyll, although it has now been shown that many chlorophyll-containing flagellates and ciliates are mixotrophic, being capable of ingesting phytoplankton.

In this section we will be concerned with the phytoplankton component; the bacterioplankton will be considered in Section 3.8.4 while the zooplankton will be dealt with in Section 3.6.1. The contribution of phytoplankton to the overall production of coastal waters and estuaries in particular is dependent on a number of factors of which salinity, temperature, availability of light (as influenced by turbidity) and nutrients, and the configuration of the estuarine basin are important. In estuaries that drain on the ebb tide to a system of low tide channels with large areas of exposed mud and sand flats, the phytoplankton make a much smaller contribution than in deep-water estuaries in which the exposed mud and sand flats form a small percentage of the total area.

3.5.2.2 Composition of the Phytoplankton

Generally the dominant species groups are diatoms and dinoflagellates, while other important groups include cryptophytes, chlorophytes (green microalgae), and chrysophytes (blue-green microalgae). Mention has already been made of the separation of the phytoplankton into net- or microphytoplankton, nanophytoplankton, and picophytoplankton (Table 3.10). The nanophytoplankton are numerically dominant in most estuaries, although the major part of the biomass consists of net-phytoplankton.

3.5.2.3 Distribution and Seasonal Variation in Species Composition

Due to the fluctuating temperatures and salinities that are found in estuaries, the phytoplankton tends to be both euryhaline and eurythermal. The phytoplankton of the lower reaches are generally dominated by diatoms, and

TABLE 3.10
Composition of the Macrobial Community by Size Class

Size Class	Heterotrophs	Autotrophs
Picoplankton 0.2–2.0 μm	Bacteria Microflagellates	Cyanobacteria Chemolithotrophic bacteria Eucaryote microalgae
Nanoplankton 2–20 μm	Microflagellates Naked ciliates	Phytoflagellates Nonflagellate microalgae Smaller diatoms
Microplankton 29–200 μm	Naked ciliates Tintinnids Larger dinoflagellates Amoeboid protozoa Rotifers Other metazoa	Larger diatoms Larger dinoflagellates

dinoflagellates are less abundant, although they may be important at certain seasons. Small nanoflagellates are usually abundant in the upper reaches. Neritic species from adjacent coastal waters penetrate varying distances onto the estuary depending on their euryhalinity and the number of neretic species is further reduced up an estuary. In the upper reaches, the phytoplankton community may include characteristic estuarine species not normally found in the adjacent neritic waters or in the freshwaters of the rivers. Few freshwater species can tolerate even very low salinities, and consequently die as they are carried by river flows into the higher salinities of the estuary.

An example of the changes that occur in the phytoplankton species composition while moving from the lower to higher salinities along the estuarine axis is depicted in Figure 3.21 for the Palmico River estuary, North Carolina, U.S.A. (Kuenzler et al., 1979). It can be seen that dinoflagellates dominate the samples from the low salinity areas in the upper and middle estuary, with diatoms being of secondary importance. In the lower, higher salinity reaches of the estuary, diatoms and dinoflagellates are of about equal importance, with the former dominating in the winter and spring, and the latter becoming more abundant during the summer. This latter trend appears to be common in many temperate estuaries, including the lower Chesapeake Bay (Patten et al. 1963), Long Island Sound (Riley and Conover, 1967), the lower Hudson River (Malone, 1977), and Doboy Sound, Georgia (Ragotzkie, 1959). These distribution patterns may reflect changes in temperature, insolation, water column stability, nutrient availability, and the adaptations of the species to exploit the changing conditions. Summer species (dinoflagellates) are known to have higher light optima and shorter generation times than winter forms (diatoms), which have lower light optima, longer generation times, and a greater capacity for energy storage, but being non-motile they require turbulent mixing to remain suspended in the water (Smayada, 1983). As the diatom blooms decline the cells tend to sink to the bottom (often en masse) (Smetacek, 1981). In some shallow estuarine systems, such as the Dutch Wadden Sea, blooms of diatoms and dinoflagellates alternate throughout the year, with the peaks of both groups occurring between April and September (Cadee, 1986).

The relative importance of the net phytoplankton and the nanophytoplankton varies greatly depending on the environmental conditions; competition for nutrients appears to determine the species succession. Nanophytoplankton, principally small flagellates and dinoflagellates, may play an important role particularly in the upper part of estuaries.

3.5.2.4 Biomass and Production

Measurements of the biomass and production of estuarine phytoplankton are necessary in order to understand the contribution that phytoplankton makes to the total primary productivity of the estuarine ecosystem, their role in estuarine food webs, and their contribution to the POM and DOM pools.

Biomass: Early estimates of standing crop were made by counting phytoplankton (principally diatom) cell numbers. However, this does not take into account the contribution made by the nano- and picophytoplankton. Estimates of phytoplankton standing crop are now generally made from chlorophyll *a* concentrations. The values vary widely from below 0.5 mg ch *a* m^{-3} to as high as 100 mg ch *a* m^{-3}. There are wide variations in individual estuaries in such estimates depending on the season. In addition there are geographic variations with the levels tending to be lower in estuaries at higher latitudes.

Primary productivity: The absolute fixation rate of inorganic carbon into organic molecules is the gross primary production (P_g). When corrected for the respiration of the autotrophs (R), P_g reduces to primary net productivity (P_n):

$$P_g - R = P_n$$

A major complication is that the microheterotrophs coexist with and share the same size range as the autotrophs, and in attempting to measure biomass or metabolism of one it is extremely difficult to discriminate the biomass or metabolism of the other (Li, 1986). If the respiration of all heterotrophs (both macroscopic and microscopic) is subtracted from P_n the residual is termed net community production P_c.

Estimates of primary production are dependent on the measurement techniques employed (Eppley, 1980; Peterson, 1980). The method most commonly used is the ^{14}C uptake method of Steeman-Nielsen (1952). While there has

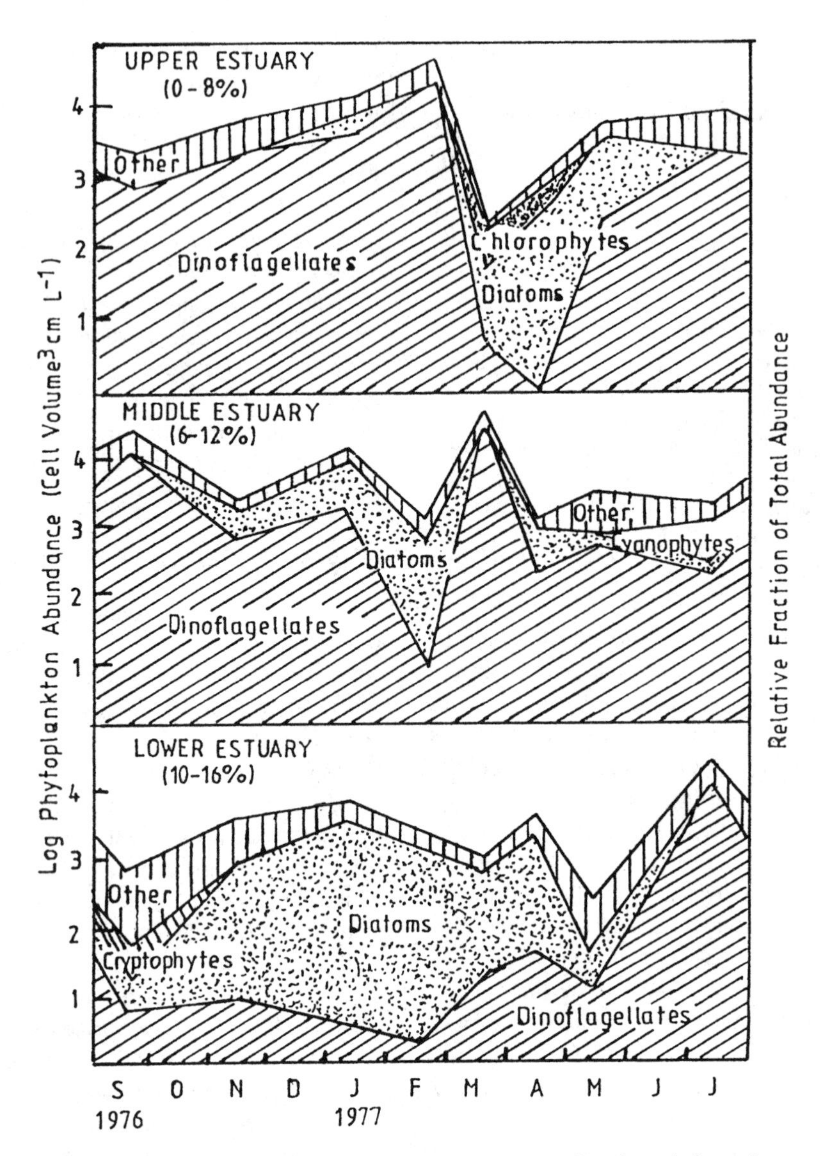

FIGURE 3.21 Seasonal changes in the total net phytoplankton abundance (as cell volumes) from the upper to the lower estuary (salinity range given) in the Pamlico River estuary, North Carolina. Total cell volume (uppermost line in each panel) is given as a logarithm, while relative fraction of total comprised by each taxonomic group is depicted as an arithmetic percent of the total. A. Upper estuary, salinity range 0–8; B. Middle estuary, 6–12; C. Lower estuary, 10–16. (Redrawn from Kuenzler, E.J., Staley, P.W., and Koenings, J.P., Water Resources Institute, Raleigh, NC, Report No. UNC-WRRI-79-139, 1979. With permission.)

been much debate as to exactly what is measured by this method (Dring and Jewson, 1979), it is currently the only technique sensitive enough to measure the low rates of production that are frequently encountered. However, the results of ^{14}C experiments are often difficult to interpret and a number of workers consider that the method in general underestimates primary production (see Gieskes et al., 1979). Bearing this in mind the production values that have been determined by this method will be discussed.

Estimates of gross and net primary phytoplankton production from various estuaries are given in Table 3.11, while the mean daily rates of primary production for selected estuarine ecosystems from a range of geographic localities are shown in Figure 3.22. Rates range from near zero to 4.8 g C m^{-2} day^{-1}. The average for all the systems is about 0.70 g C m^{-2} day^{-1} (256 g C m^{-2} yr^{-1}), a value substantially higher than the 100 g C m^{-2} yr^{-1} reported by Ryther (1963) for coastal areas and is of the same order as Ryther's value of 300 g C m^{-2} yr^{-1} estimated for upwelling areas. In high latitudes, light intensity may be critical while in tropical areas other factors such as seasonal nutrient or salinity fluctuations may be more important. Furnas et al. (1976) found that the annual carbon production in temperate Naragansett Bay, U.S.A., was 308 g C m^{-2} of which 42% occurred in July and August. In the Sapelo Island marshes (Pomeroy et al., 1981), investigators found that the highest photosynthetic rates for

TABLE 3.11
Estimates of Gross and Net Phytoplankton Production from Various Estuaries

Location	Production ($g\ C\ m^{-2}\ year^{-1}$) Gross	Net	Reference
Narragansett Bay	—	300	Furnas et al. (1976)
Duplin River, Georgia	248	—	Ragotzkie (1959)
Beaufort Channel, North Carolina	255	—	Williams & Murdock (1966)
Bogue Sound, Newport, North Carolina	100	—	Williams (1966)
Cove Sound, North Carolina	—	67	Thayer (1971)
North Inlet, South Carolina	—	346	Sellner & Zingmark (1976)
Doboy Sound-Duplin River, Georgia	—	375	Pomeroy et al. (1981)
Barataria Bay, Louisiana	598	412	Day et al. (1973)
San Francisco Bay, California	—	5–318	Cole & Cloern (1988)
Nanaimo River estuary, British Columbia	—	7.5	Naimen & Siebert (1979)
Langebaan Lagoon, South Africa	—	56–314	Christie (1981)
Fala Lagoon, Natal	—	2.8–65.7	Oliff (1976)
Cocjin Backwater, India	14–575.4	—	Quasim (1970)
Bristol Channel, England	—	7–165	Joint & Pomeroy (1981)
Westerchelde, The Netherlands	—	122–212	Van Spaendonk et al. (1993)
Oosterschelde, The Netherlands	—	301–382	Wetsteyn & Kromkamp (1994)
River Lynher estuary, England	—	81.7	Joint (1978)
Sydney Harbour, Australia	—	11–127	Relevante & Gillmartin (1978)
Upper Waitemata Harbour, New Zealand	200	140	Briggs (1982)

phytoplankton occurred in the water over the marsh on high spring tides. In this study the annual production in the Duplin River and the adjacent Doboy Sound, Sapelo Island, was estimated to be 375 g C m^{-2} yr^{-1}.

3.5.2.5 Factors Regulating Estuarine Primary Production

Numerous factors regulate the magnitude, seasonal pattern, and species composition of phytoplankton photosynthesis, including light, macronutrients, macronutrients, temperature, grazing by protozoa, zooplankton and other filter feeders, tidal mixing, and river flow effects. The relative significance of these factors will vary with the type of estuarine system and geographic locality and we will discuss their relative importance.

Light: Light is one of the most important variables controlling phytoplankton photosynthesis. Total incident radiation is a function of latitude and this is reflected in the seasonal pattern of production. In Arctic locations there is a single strong seasonal peak in production, while temperate systems often show a strong spring and a lesser autumn peak. The initiation of the winter-spring phytoplankton bloom has been demonstrated to be keyed to increasing light intensity in Long Island Sound and Narragansett Bay (Riley, 1967; Nixon et al., 1979). In tropical systems there is little predictable seasonality in phytoplankton production. These overall patterns are, however, considerably modified by other factors such as periodicity in river flows and turbidity patterns.

Of prime importance to photosynthesis is the amount of photosynthetically available radiation (PAR), or the light in the range of wavelengths from 400 to 700 nm, which can be utilized by chlorophyll-bearing algae. Light is reflected, absorbed, and refracted by dissolved and suspended particles in the water. The extent to which light is attenuated at a given depth is determined by the clarity of the water. Estuaries in general are turbid with much material, both organic and inorganic, in suspension. In addition, as the phytoplankton blooms, the cells themselves diminish the amount of light that penetrates the water column.

In deeper estuarine water bodies, extreme vertical mixing of the water column can cause reduced photosynthesis by transporting phytoplankton below the depth at which there is sufficient light to maintain growth. The depth at which gross photosynthesis (P_g) is just equal to the algal respiration rate (R) is referred to the as the *compensation depth* (D_c). Below this depth cells cannot survive because there is insufficient light for photosynthesis to produce the necessary energy for base respiration. This D_c is often equated with the depth at which 1% of the surface irradiation is available, or 2.5 times the Secchi depth. However, since phytoplankton cells are being continuously mixed throughout the water column, another critical point is *critical depth* (D_{cr}), the depth to which an entire phytoplankton population or assemblage can be mixed while still maintaining photosynthesis (P_l) (over

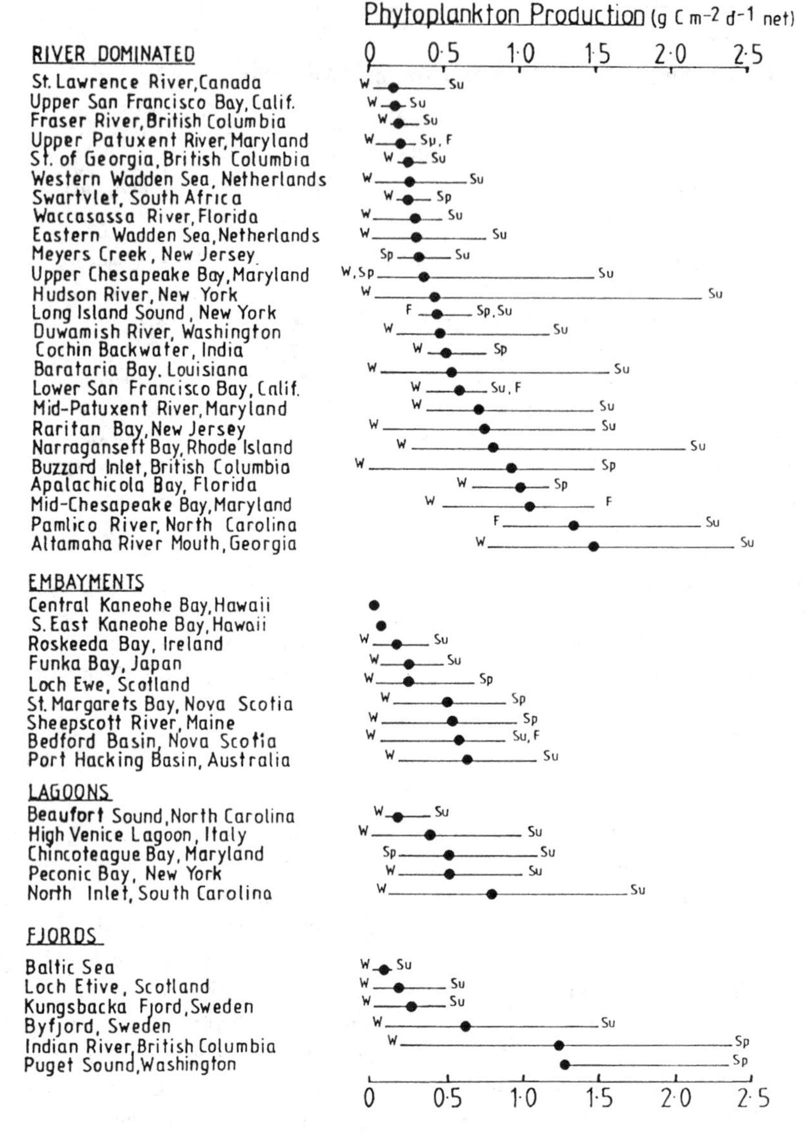

FIGURE 3.22 Summary of the average daily phytoplankton rates (solid circles) in 45 estuarine systems. Horizontal bars indicate annual ranges. Season in which maximum and minimum rates occurred is also indicated. (W, winter; Sp, spring; Su, summer, F, fall (winter). (Redrawn from Boynton, W.R., Kemp, W.M., and Keefe, C.W., in *Estuarine Comparisons*, Kennedy, V.S., Ed., Academic Press, New York, 1982, 75. With permission.)

time and depth) and integrated respiration (R_l), i.e., an average cell can be mixed from surface to D_{cr} and still maintain a positive energy balance. At D_{cr} the vertically integrated photosynthesis equals the vertically integrated respiration. When the depth of the mixed water column (D_m) exceeds D_{cr}, the assemblages of phytoplankton will not be able to develop net photosynthesis. Ragotskie (1959) showed that no net productivity occurred in estuarine water near Sapelo Island, Georgia, when D_{cr} was less than the mixed depth (which was identical to the mean depth of the water column in this shallow, well-mixed estuarine system).

Macronutrients: Algal primary production requires continuing availability of the macronutrients, nitrogen, phosphorus, and silicon. If they are in low supply they can be limiting to phytoplankton growth. The ratios of dissolved inorganic nitrogen (DIN = $NO_3 + NO_2 + NH_4$) to dissolved inorganic phosphorus (DIP) in studied estuarine systems are low during periods of high phytoplankton production, except in highly eutrophic systems. The significance of this ratio lies in the fact that algal production is constrained (among other things) by the requirement for nitrogen and phosphorus in proportion to the atomic ratio of 16:1. Redfield et al. (1963) demonstrated what is now termed the "Redfield ratio," i.e., the atomic weights of the elemental composition of microalgae, C:N:Si:P, are on the order of 106:16:15:1. Water column nutrient concentrations of DIN:DIP less than the Redfield ratio would indicate that nitrogen is less abundant than phosphorus in terms of phytoplankton demand, while values in excess of the ratio would indicate that phosphorus is less abundant.

Boynton et al. (1982) analyzed seasonal N:P ratios from 28 estuarine ecosystems and found that the ratios ranged from <1.0 to over 200. At peak production, nitrogen was consistently less abundant than phosphorus in nearly all systems. Exceptions are those systems that are heavily enriched by diffuse or point sources of nutrient enrichment such as sewage inflows. Actual concentrations vary considerably between various estuaries. In addition there may be substantial annual excursions in the N:P ratios, particularly in the river-dominated group. In Chesapeake Bay, U.S.A., during periods of high river flow, ratios of >60:1 have been recorded. These were due to very high concentrations of nitrogen rather than low concentrations of phosphorus. Thus, because the estuarine environment generally has less nitrogen available per unit of phosphorus than algal cells require, it is often concluded that nitrogen is the more limiting nutrient for estuarine photosynthesis.

While nitrogen is the macronutrient that most often limits the production of phytoplankton under most circumstances, Smith (1984) and Smith et al. (1987) have provided geochemical and mass balance data to suggest that, ultimately, P will limit production in coastal marine systems. Furthermore, diatom production may often be limited (both seasonally and annually) by the availability of silicon in the form of silicates (Officer and Ryther, 1980). Officer and Ryther (1980) suggest that diatoms utilize the plentiful nutrient supplies (including silicon) built up over the winter, and this is followed by depletion of silicon, which is regenerated more slowly than nitrogen or phosphorus.

Temperature: Most phytoplankton species exhibit a relatively narrow optimal temperature range for their photosynthesis and growth (Eppley, 1972). Goldman (1979) showed that the temperature optima for five coastal phytoplankton species, as well as for a mixed assemblage of these species, fell in the range of 20 to 25°C. Temperature seems to exert a selective force for phytoplankton populations so that the temperature optima coincide with the prevailing local conditions. However, Karentz and Smayada (1984) reported that the maximum abundances for 30 algal species in Narragansett Bay occurred at temperatures 3 to 14°C lower than their respective optimum growth temperatures. This suggests that other overriding factors and not temperature determine the seasonal phytoplankton succession patterns.

Grazing: The impact of grazing, a mechanism regulating phytoplankton productivity, has been the subject of much debate. Steeman-Nielsen (1958), for example, argued that the commonly observed seasonal patterns of more or less coincidental peaks in phytoplankton and zooplankton abundance supported the hypothesis that grazing by zooplankton maintained the algal populations in a steady-state balance, the level of which was determined by other environmental conditions (e.g., nutrients, light, temperature). In contrast, Cushing (1959), following the analysis of a simple predator-prey model, concluded that grazing did affect the magnitude and timing of phytoplankton blooms, and that a lag time between the peak abundances of phytoplankton and zooplankton populations was observed consistently. Other studies have demonstrated that zooplankton grazing can alter the species composition of phytoplankton communities by selective grazing on the larger diatoms, thus shifting the size composition of the assemblage toward smaller-sized species.

Protozoa are significant grazers on the ultraphytoplankton (cells between 0.5 and 5 μm) (Turner et al., 1986). Laboratory studies of the grazing capabilities of pelagic ciliates, including tintinnids and aloricate species have shown that they are adapted to graze 2 to 10 μm sized algal cells at high rates (Heinbokel and Beers, 1979; Verity, 1985; Jonsson, 1986). Sherr et al. (1991) found clearance rates of <6 μm sized algal prey by heterotrophic flagellates in a salt marsh tidal creek ranging from 0.004 to 0.83 l $cell^{-1}$ hr^{-1} while those for ciliates ranged from 0.24 to 8.3 l $cell^{-1}$ hr^{-1}. Estimated daily clearance rates by algivorous protozoa over a four-month period averaged 45% of the water volume. The average grazing impact of the flagellates was about 33% of that of the ciliates for 2 μm prey, and 50 and 85% that of the ciliates for 5.4 and 3.4 μm sized prey, respectively. Reports of tintinnid consumption of phytoplankton production in coastal waters range from 4 to 60% (Sherr et al., 1986). Microzooplankton, which is often reported as being dominated by aloricate ciliates, has been estimated to graze between 10 and 80% of the primary production in many marine environments (Capriulo and Carpenter, 1980; Burkill et al., 1987; Rassoulzadegan et al., 1988). Evidence for the importance of heterotrophic dinoflagellates as grazers of algae in marine systems is also accumulating (Lessard and Swift, 1985). It is thus clear that in addition to the macrozooplankton (especially copepods) and benthic invertebrate filter feeders (especially bivalve molluscs), the planktonic protozoa are very important grazers on the phytoplankton.

Mallin and Paerl (1994) investigated the impact of zooplankton grazing on the phytoplankton in the Neuse River Estuary, North Carolina. Zooplankton community grazing rates were generally lowest in the winter and highest in the spring through late summer, ranging from 0.1 to 310 ml^{-1} hr^{-1}. Community grazing was positively correlated with primary productivity and the abundance of total phytoplankton, centric diatoms, dinoflagellates, and the small centric diatom *Thalassiosira*. On an annual basis the zooplankton community grazed approximately 38 to 45% of the daily phytoplankton production. Table 3.12 gives estimates of zooplankton grazing in various estuaries. They range from 17 to 69%, demonstrating that

TABLE 3.12
Effect of Zooplankton Grazing on Primary Production from a Range of Estuarine Systems

System	Taxon Group	Period	Primary Production Grazed (%)		
			Method	Mean	Range
Long Island Sound (Riley, 1956)	Community all sizes	Annual	EST	69	—
Solent Estuary, U.K. (Burkill, 1982)	Tintinnids	Annual	EST	4	0–20
Long Island Sound, U.S. (Capriulo & Carpenter, 1989)	Tintinnida	Annual	EST	27	—
	Copepods			44	—
Gunpowder River, Maryland (Sellner,1983)	Comunity	Annual	EST	17	1–>100
Beaufort Estuary, North Carolina (Fulton, 1984)	Copepods	Annual	EST	45	0–>100
Narragansett Bay, Rhode Island (Verity, 1987)	Tintinnids	Annual	EST	26	—
Halifax Harbour, Nova Scotia (Gifford and Dagg, 1988)	Community >102 m	Annual	EXP	49	0–100
Chesapeake Bay, U.S. (White & Roman, 1992)	Community >64 m	Mar–Oct	EXP	51	0–100
Neuse Estuary, North Carolina (Mallin and Paerl, 1994)	Community	Annual	EXP	38	2–>100

Note: EXP = experimentally derived, EST = estimated by other means. Percent grazed given as means and range, if available.

the amount of phytoplankton production grazed by estuarine zooplankton is considerable.

Tidal mixing: In coastal environments, phytoplankton cells are subjected to vertical tidal mixing, in which the intensity varies on a semidiurnal cycle. The vertically mixed cells may therefore experience light variations in a combination of both circadian and semidiurnal tidal cycles. Legendre et al. (1985) in tank experiments with water samples from the St. Lawrence Estuary, Canada, observed semidiurnal cycles in photosynthetic efficiency (a^B) and intracellular chlorophyll *a*. They suggested that such variations are possibly endogenous, phased on semidiurnal variations in vertical tidal mixing (associated with variations in the light conditions).

River flow effects: Seasonal and interannual variations in river flow can influence phytoplankton production and the species composition of the phytoplankton communities in a number of ways, including (1) changes in the input of nutrients from the watershed; (2) changing the rates of dilution or advection of algal cells out of the estuary; and (3) changing the availability of light to the phytoplankton cells through stratification of the water column, gravitational circulation, and changes in the location of the turbidity maximum. Boynton et al. (1982) noted that the nutrient input as a result of a major tropical storm (Agnes) in June 1972 resulted in inputs of nitrogen 2 to 3 times higher than in the preceding or subsequent years. While production in that year was high, peak production occurred in the following year when nutrient inputs were more typical. Boynton et al. (1982) suggested that the high summer rates of production appear to be supported primarily by recycled nutrients, some fraction of which was introduced into the estuary during the spring runoff period. In contrast to the Chesapeake Bay situation, high river flow can lead to low phytoplankton abundance and production due to the washout of algal cells from the estuary as documented for the Duwamish River estuary, Oregon, during high flow years (Welsh et al., 1972).

3.5.2.6 A Model of Estuarine Phytoplankton Productivity

Boynton et al. (1982) developed a conceptual model (Figure 3.23) of the factors influencing estuarine phytoplankton production. Inputs of energy and materials or physical characteristics (morphology) common to all estuarine systems are shown as circles. Rectangles represent the mechanisms through which the inputs affect phytoplankton primary production, e.g., turbidity is depicted as influencing primary production and its impact in turn is enhanced by algal biomass sediment from both external (riverine and others) and internal (resuspension due to winds and tides) sources. Data was collected from 63 estuarine systems covering latitude, insolation, temperature, extinction coefficient, mean depth, stratification depth, mixed depth, critical depth, surface area, drainage area, freshwater input, tidal range, salinity, nutrient concentrations, nutrient loading rate, chlorophyll *a* concentrations, and phytoplankton production rate (Keefe et al., 1981).These data were analyzed statistically to test hypotheses concerning the factors regulating temporal patterns.

Prior to the statistical analysis, Boynton et al. (1982) classified the estuarine systems into four groups (fjord, lagoon, embayment, or river dominated). *Fjords* were defined as having a shallow sill and a deep basin with slow exchanges with the adjacent sea. *Lagoons* were con-

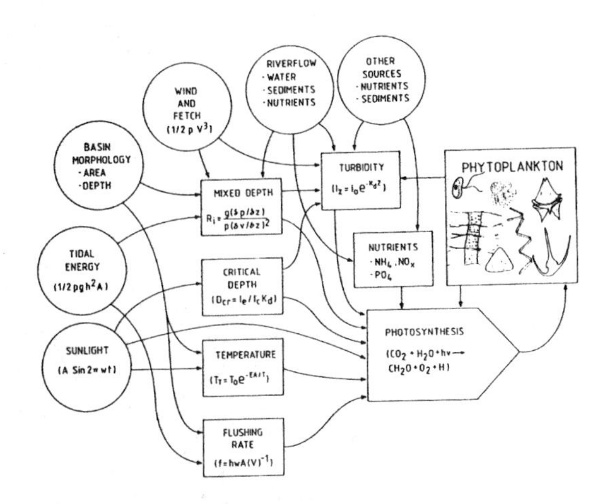

FIGURE 3.23 Mean monthly phytoplankton production rates at six stations in central Chesapeake Bay, 1972 to 1977. Values below peaks represent estimates of annual phytoplankton production (g C m^{-3} $year^{-1}$). (Redrawn from Boynton, W.R., Kemp, W.M., and Keefe, C.W., in *Estuarine Comparisons*, Kennedy, V.S., Academic Press, New York, 1982, 71. With permission.)

sidered to be those systems that were shallow, well mixed, slowly flushed, and only slightly influenced by riverine inputs. *Embayments* were considered to be deeper than lagoons, often stratified, only slightly influenced by freshwater input, and having good exchange with the ocean. The category *river dominated* contained a more diverse group of systems, but all members of the group were characterized by seasonally depressed salinities due to riverine inputs and variable degrees of stratification.

Annual phytoplankton production means for river-dominated estuaries, embayment, lagoons, and fjords were 0.58 ± 0.37, 0.36 ± 0.23, 0.49 ± 0.23, and 0.62 ± 0.53 g C m^{-2} day^{-1}, respectively. This indicated that estuaries with different physical characteristics commonly have comparable rates of primary production, suggesting there are system-specific biotic and physical mechanisms operating in different estuarine systems. Boynton et al. (1982) found that, in general, phytoplankton production and biomass exhibited weak correlations with a variety of physical and state variables in the estuarine systems that they considered. They concluded that this perhaps indicated the significance of rate processes as opposed to standing stocks in regulating phytoplankton production.

3.5.3 Estuarine Macroalgae

3.5.3.1 Composition and Distribution

Macroscopic algae in general are not as well represented in estuaries as the other producers (marsh plants, sea grasses, and mangroves). The species that occur are restricted to a small number of genera that can tolerate turbidity, silt deposition, and changing salinity patterns. There are, however, some exceptions. Species of brown algae belonging to the genera *Fucus, Pelvetia,* and *Ascophyllum* are abundant on rocky substrates in North Atlantic estuaries, while smaller plants of *Fucus* and *Pelvetia* grow among the vascular marsh plants. In the Nauset Marsh, Massachusetts (a back-barrier estuary), patches of the green alga, *Cladophora gracilis*, occur throughout (Roman et al., 1990). On vertical creek banks there is an intertidal zone of fucoid algae (*Ascophyllum nodosum* ecad. *scorpiodes* and *Fucus vesciculosus*) usually associated with tall *Spartina alterniflora*, an intertidal zone of filamentous algae attached to the substrate of exposed creek-bank peat, and a subtidal zone of assorted macroalgae. In the tropics there are smaller amounts of brown algae of the genus *Sargassum* growing on hard substrates, while other brown genera such as *Colpomenia* and *Dictyota* occur as epiphytes on sea grasses such as *Posidonia* and *Thalassodendron*. In some Australasian estuaries a free-floating form of the normally attached brown alga, *Hormosira banksii*, reproducing by vegetative division is common.

Red algae are represented mainly by small species such as *Polysiphonia, Ceramium,* and *Laurencia* growing as epiphytes on sea grasses and marsh plants, while species of *Bostrychia* and *Calloglossa* grow attached to salt marsh plants and the boles and roots of mangroves. Common estuarine algae are species of the genus *Gracilaria*, which originally grow on mudflats attached to pebbles, living bivalves, and dead shells, but as they develop they may become free floating. *G. verrucosa* is abundant in South African estuaries where it forms the basis of an agar industry (Day, 1981d). Hedgpeth (1967) records *Gracilaria* in the Laguna Madre of Texas, while species of *Gracilaria secundata* are widespread in New Zealand estuaries (Henriques, 1978).

The most common estuarine algae are green algae belonging to the genera *Enteromorpha, Ulva, Ulothrix, Cladophora, Rhizoclonium, Chaetomorpha,* and *Codium*. Filamentous species of *Enteromorpha* and *Cladophora* grow as epiphytes on sea grasses and salt marsh plants. Larger green algae, especially *Ulva lactuca* and *Enteromorpha* spp., often form extensive mats on estuarine mudflats worldwide.

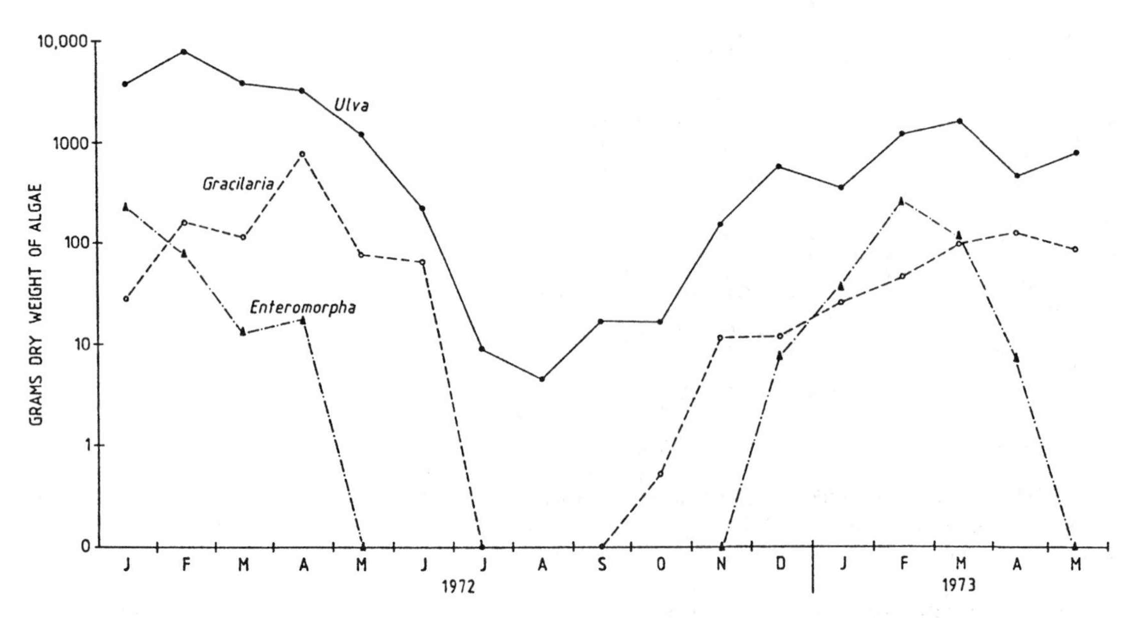

FIGURE 3.24 Total dry weight at a series of 37 representative stations (each 1 m^2 in the Avon-Heathcote Estuary, New Zealand, over the period of January 1972 to May 1973. (Redrawn from Knox, G.A., *Estuarine Ecosystems: A Systems Approach*, CRC Press, Boca Raton, Florida, 1986a, 79. With permission.)

3.5.3.2 Biomass and Production

Studies on the photosynthesis and respiration rates of estuarine algae include studies on *Polysiphonia* from Gent Bay Estuary (Fralick and Mathieson, 1975), *Hypnea* from a mangrove fringed estuary in Florida (Dawes et al., 1976), six species of algae from an estuary in Oregon (Kjeldsen and Pinney, 1972), and six species of algae from a mangrove and salt marsh estuary in Florida (Dawes et al., 1976). In addition Christie (1981) has investigated the standing crop and production of *Gracilaria verrucosa* in Langebaan Lagoon in South Africa, and Knox and Kilner (1973) and Steffensen (1974a) have studied the annual changes in the standing crop of the green algae, *Ulva lactuca* and *Enteromorpha ramulosa*, in the Avon-Heathcote Estuary, New Zealand.

In the Avon-Heathcote Estuary, New Zealand, large populations of the sea lettuce, *Ulva lactuca,* have become established over the past 40 years. Since 1960 a series of studies (Knox and Kilner, 1973; Steffensen, 1974) have documented the increase in algal density. From these studies it became evident that the increase in algal biomass was associated with the discharge of increasing amounts of treated nutrient-rich sewage effluent to the estuary.

The link between luxuriant growth of *Ulva* and available nutrients was examined at the beginning of the century (Cotton, 1911) and has been confirmed by numerous reports since (Wilkinson, 1981). As early as 1914, Forster (1914) demonstrated that growth of *U. lactuca* was stimulated by additions of urea, acetamide, and ammonium nitrate. Waite and Mitchell (1972) examined the effect of adding NH_3-N and PO_4-P in varying combinations and found that the growth in *U. lactuca* was stimulated by increasing the amount of either nutrient.

In the Avon-Heathcote Estuary where *U. lactuca* was generally the dominant species, a second species of green alga, *Enteromorpha ramulosa,* was twice as abundant as *Ulva* in the summer of 1969, but since then it has not occurred in the same abundance. *Enteromorpha* is more sensitive to temperature fluctuations than *Ulva,* and a mild preceding winter may be the explanation for its dominance. Laboratory experiments (Steffensen, 1974) showed that between 15 and 18°C, winter plants of *U. lactuca* grew 15 to 20 cm in length in 4 to 6 weeks, while below 15°C the rates of growth were very much slower. No growth of *Enteromorpha* was detected below 12°C while some growth in *Ulva* was detected at 10°C.

Figure 3.24 depicts the seasonal variation in total organic dry weight of the three dominant algal species in the estuary from 37 stations in 1972–1973. The seasonal pattern in the growth of all three species is evident. However, both *Enteromorpha* and *Gracilaria* die down in the winter, while substantial *Ulva* biomass persisted over the winter. *Ulva* plants usually develop attached to cockle shells, but as they grow the thalli reach a size and buoyancy that uproots the shell leaving the thalli to drift with the currents. These drifting thalli, and fragments broken from them, continue to grow over the summer and autumn, forming large drifts of unattached plants. These form the bulk of the winter biomass. Summer biomass values of up to 130 g m^{-2} (dry weight) have been recorded.

Table 3.13 gives macroalgal biomass and production estimates for a range of estuaries. Highest biomass den-

TABLE 3.13
Biomass and Production Estimates of Macroalgae from Various Estuaries

Estuary	Genus	Max. Biomass g DW m^{-2}	Max. Production mg C DW^{-1} h^{-1}	Annual Production g C m^{-2} yr^{-1}	References
Mai Po Pond (HK)	*Enteromorpha*	200	1.3[a]	50.6 (2.3%)	Lee (1989)
Nanaimo Estuary (BC)				0.9 (0.1%)[b]	Naiman and Sibert (1978)
Nauset Marsh (MA)	*Ascophyllum*	1500[c]		390[f]	Roman et al. (1990)
	Fucus	400[d]	110[f]		
Cladophora		640[e]		380[f] (13%)	
Wilson Cove (CA)			0.4–3.1[g]		Littler and Murray (1974)
Tagus Estuary (P)	*Ascophyllum*	375[d]	0.4	405[k]	Ferreira and Ramos (1989)
Ulva		160[h]	1.0	213[k]	
Gracilaria		80[i]	1.2	92[k]	
Coos Bays (OR)	*Enteromorpha*	1050	13.3[m]	1060[l]	Pregnall and Rudy (1985)
Gray Harbor Estuary (WA)	*Enteromorpha*		31.4[g]	48 × 10^6 (22)[n,g]	Thom (1984)

Note: In the column giving annual production values, the contribution of macroalgal production is given. HK, Hong Kong; BC, British Columbia; MA, Massachusetts; CA, California; P, Portugal; OR, Oregon; WA, Washington.

[a] $P_Q = 1$. Net primary production, estimated from O_2-method.
[b] Percentage of total production on intertidal mudflat (including intertidal phytoplankton production, 12.1%).
[c] *Ascophylum nodosum.*
[d] *Fucus vesiculosus.*
[e] *Cladophora gracilis.*
[f] Net production estimates from changes in biomass.
[g] From 3 to 4-h O_2-exchange measurements. $P_Q = 1.2$. Several macroalgae.
[h] *Ulva lactuca.*
[i] *Gracilaria verrucosa.*
[k] Net production from 0.5 to 1-h O_2-exchange measurements. $P_Q = 1$.
[l] From 3-h $^{14}CO_2$-fixation measurements.
[m] g C m^{-2} day^{-1}.
[n] k g C yr^{-1}.

Source: From Heip, H.R., Gossen, N.K., Herman, P.M.J., Kromkomp, J., Middleberg, J.J., and Stoetaert, K., *Oceanogr. Mar. Biol. Annu. Rev.*, 33, 27, 1995. With permission.

sities and production seem to occur when hydrodynamic energy is relatively low, such as lagoons and tidal inlets. In open river-dominated estuaries, especially in funnel-shaped estuaries, biomass seems to be lower, although on hard substrates in the intertidal, local densities of macroalgae are often found, especially in northern temperate estuaries. High biomass and production often occur where there is the free-floating form of the green alga *Ulva*.

3.5.3.3 Impact of Macroalgal Mats on the Benthic Fauna

Mat-forming algae, principally *Enteromorpha* spp. and *Ulva* sp., as discussed above, develop rapidly on temperate estuaries during the spring, especially where there are large inputs of nutrients from sewage discharges (Knox and Kilner, 1973; Steffenson, 1974; Reise, 1984). Such algal mats provide food for grazing molluscs (e.g., *Zeacumantus subcarinatus* and *Diloma subrostrata* in the Avon-Heathcote Estuary, Christchurch; Knox and Kilner, 1973; Knox, 1992), and on decay provide an abundant source of detritus for detritus feeders. The physical presence of the mats may provide shelter from predation for other animals, e.g., shrimps and larval fishes, although they may also act as refuges for predators. Mats also reduce oxygen exchange with the sediments, accumulate silt, and interfere with the feeding activities of suspension feeders. They also can act as a physical barrier to mudflat feeding fish and birds. When the mats decay they cause the surface sediments to become anoxic with the RPD layer reaching the surface. Foolad (1983) showed that concentrations of the ammonium ion (NH_4^+) reaches toxic levels in the mats producing sublethal effects on the bivalve *Cerastoderma*. In extreme conditions they can cause the death of infaunal macroinvertebrates.

Possible effects of algal mats are summarized in the flow diagram in Figure 3.25. Hull (1987) examined the impact of algal mats (principally *Enteromorpha* spp., especially *E. intestinalis*, with *Ulva lactuca*, *Chaetomor-*

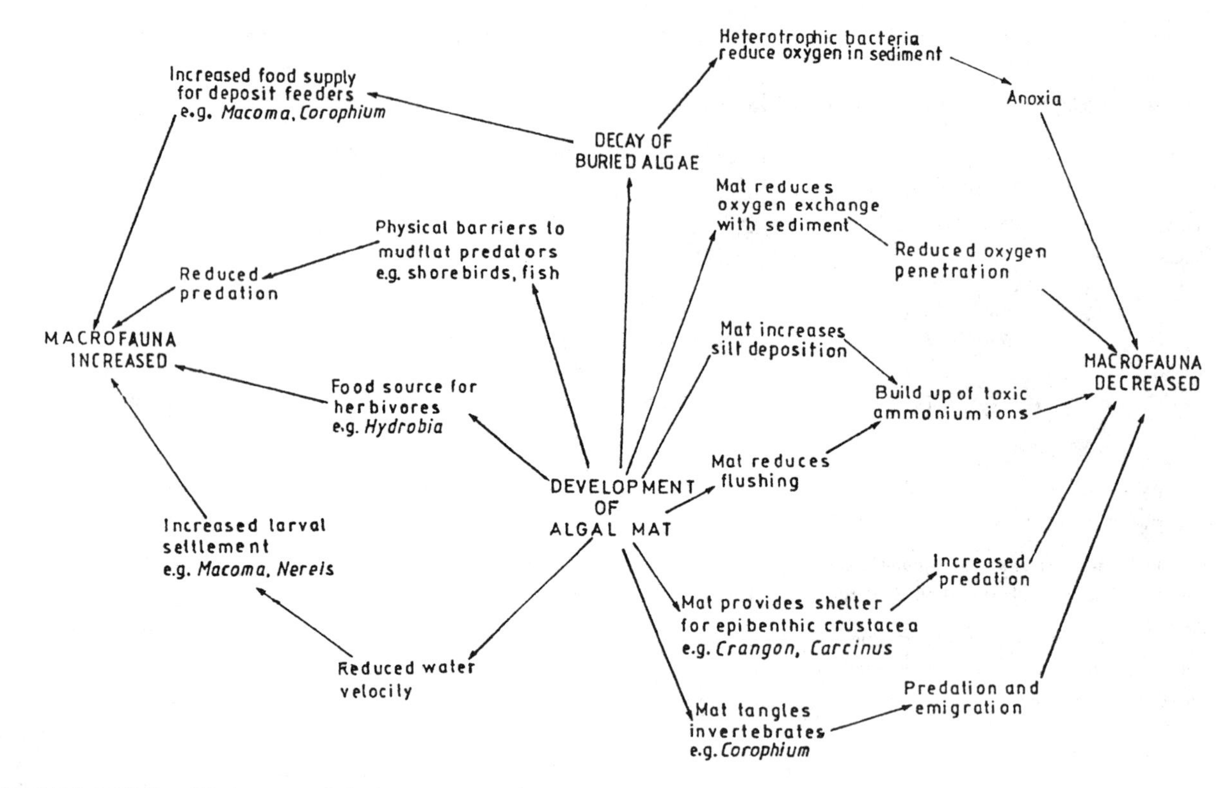

FIGURE 3.25 Possible impacts of algal mats on estuarine benthic ecosystems. (Redrawn from Hall, S.J., *Oceangr. Mar. Biol. Annu. Rev.*, 32, 210, 1994. With permission.)

pha linum, and *Cladophora* sp.) in the Ythan Estuary, Scotland, by manipulating algal density at 0, 0.3, 1, and 3 kg fresh weight m^{-2} (corresponding to control, low-, medium, and high-density). Many of the effects observed were similar to those occurring during organic enrichment, notably large numbers of the opportunistic polychaete, *Capitella capitata*, the bivalve, *Cerastoderma edule*, and the snail, *Hydrobia ulvae*, increased in abundance, while the polychaete, *Pygiospio elegans,* and the oligochaetes, *Tubifcoides benedini* and *Tubifex castatus,* showed no consistent response. The abundance of the tube-dwelling amphipod, *Corophium volutator*, was markedly affected by the algal cover. Initially densities were low but increased to high densities following the burial of the mats in October.

Nothko and Bonsdorff (1996) studied the population responses of benthic invertebrates to drifting mats of algae (*Cladophora, Enteromorpha, Pilayella,* and *Fucus*) in the northern Baltic Sea. Massive die-offs of benthic invertebrates were recorded. The community dominants in abundance (*Hydrobia* spp.) and biomass (*Macoma balthica*) experienced strong population reductions after 9 days of algal stress. Within 5 weeks population crashes were recorded for the sedentary polychaetes, *Manayurikia aestuarensis* and *Pygospio elegans*, while populations of the errant polychaete, *Nereis diversicolor*, and tubificid oligochaetes remained stable.

3.5.4 Sea Grass Systems

3.5.4.1 Introduction

Sea grasses are characteristic features of shallow coastal seas, coral reef lagoons, and estuaries. According to den Hartog (1967) there are 49 species in 12 genera of marine aquatic angiosperms with the ability to function normally and complete their reproductive cycle when fully submerged in the marine environment. Many genera have a worldwide distribution with *Thalassia, Thalassodendron, Cymodocea, Posidonia, Halodule*, and several genera in warm and tropical waters, and *Zostera, Halophila, Ruppia, Potamogeton*, and *Zannichellia* extending into temperate estuaries. The common eelgrass, *Zostera*, and its subgenus, *Zosterellia,* are widely distributed in the temperate zones of North America, Europe, Asia, and the temperate coasts of Australia and New Zealand. Many of these genera are euryhaline but their extension into regions of lowered salinity is variable. *Posidonia, Thalassia, Cymodocea*, and *Thallasodendron* prefer salinities above 20 while *Zostera* and *Halodule* tolerate salinities as low as 10; *Ruppia* and *Potamogeton* prefer low salinities while *Zannichella* extends into freshwater and is seldom found in salinities above 10.

Sea grass systems vary from a few plants to extensive meadows composed of a single species. Such meadows are highly productive ecosystems with a rich associated fauna. Excellent reviews of sea grass ecosystems are those edited by McRoy and Helfferich (1977) and Phillips and McRoy (1980). Stevenson (1988) has more recently reviewed the comparative ecology of submerged grass beds in fresh and marine waters.

3.5.4.2 Distribution and Zonation

Important factors affecting the colonization and distribution of sea grasses include the morphology and character of the bottom, the sources, routes and rates of sediment transport, and the rates of sediment accumulation. Other factors include light, temperature, salinity, waves and currents, and the availability of seed. Once they have become established, sea grasses may significantly impact on the prevailing sedimentary processes depending on the species composition and the density of the plants. Sea grasses alter the sedimentation processes in a number of ways, but the major effects are to increase the sedimentation rates by preferentially concentrating the finer particle sizes (see Figure 3.3), and to stabilize the deposited sediments. This is due to the entrapment of waterborne particles by the grass blades which slow the currents passing over the sediment surface, the formation and retention of organic particles produced locally within the grass beds, and the binding and stabilizing of the substrate by the grass rhizome and root systems. Alteration of the sediments may lead to an increase or decrease in the sea grass colonization pattern with the possibility of partial or complete replacement of the pioneering species by another, e.g., the *Halodule* to *Thalassia* succession documented by Phillips (1960). Sediments in sea grass beds are often anoxic below the surface layer.

Table 3.14 gives the depth distribution of the principal sea grass genera. The main factors limiting the depth to which the species grow is light availability and abrasion by sand-laden water in the more turbulent habitats. In many estuaries *Zostera* extends no more than a meter below low water, but in clear estuaries it may extend down to 6 m, and it grows to a depth of 30 m off California. *Halophila* has been recorded from depths as great as 90 m, although many species have restricted depth distributions. In deeper tropical waters, where wave action is low and water clarity high, species such as *Halophila* have ovate leaves with prostrate stems unlike the linear leaves of species that occur where wave action is higher. *Posidonia* descends down to at least 60 m. *Phyllospadix* extends down from the lower eulittoral to 30 m under favorable conditions. In southern Mozambique *Halodule uninervis* may extend from MLWN to below MLWS. *Thalassodendron ciliatum* grows among the lower *Halodule* and extends to 5 to 10 m below MLWS. *Cymodocea rotundata, C. serrulata, Zostera capensis,* and *Halophila ovale* grow near MLWS, while *Syringodium isoetifolium* and *Thalassia hemprichii* both grow subtidally (Day, 1981d). The depth to which a particular species grows is a complex function of water turbidity, sedimentation regimes, wave action, and salinity.

3.5.4.3 Biomass and Production

Biomass: While sea grasses have a large biomass of leaves, the majority of the biomass of the plants is in the roots and rhizomes in the sediment and this is very difficult to sample because of the deep penetration of the roots (Jones, 1968; Zieman, 1975). However a number of estimates of the root (including rhizomes):shoot ratio can be calculated from Table 3.15. It can be seen that this ratio is usually >1 and in some cases it is very much higher. A characteristic of many sea grasses is their rhizomatous growth. This may help protect loss of perennial meristems from wave action and conserve energy and nutrients within the beds (Kenworthy and Thayer, 1984). In addition, since the roots of aquatic vascular plants are functional in terms of nutrient uptake, large root biomass (and surface area) provides for increased nutrition from sediments in oligotrophic coastal waters (McRoy and Barsdate, 1970). Although the rhizomes themselves may not be as important in the uptake, they provide "safe storage" for nutrients and photosynthetic products protected from both wave action and from most grazers.

There are a large number of literature values for sea grass plant biomass (see Table 3.15 for some representative values). The bulk of the these measurements are concentrated in two species, *Zostera marina* and *Thalassia testudinium.* For *Z. marina,* biomass estimates for aboveground biomass range from 75 to 2,060 g dry wgt m^{-2}. The wide diversity of values that have been estimated is dependent upon the habitat sampled, the number of replicate samples taken, and the methods used.

Primary production — Methods of estimation: Sea grasses present a number of problems in the estimation of their primary production and a variety of techniques have been used to overcome this. One approach has been to measure the changes occurring in the amount of standing stock over the growing season (Pomeroy et al., 1981). However, one of the problems encountered is to account for the loss of plant material over the growing season. In addition, many sea grasses, including tropical and subarctic ones, are perennial and maintain significant winter populations. Consequently, this method, which was used in early studies, has been superseded by other techniques.

Increasingly, production estimates derived from marking techniques have been used and these have proved to be a considerable improvement over the measurement of changes in biomass. The marking technique fixes the amount of standing stock present at a particular point in time and allows the more exact measurement of plant growth over a defined time interval. The technique involves the marking of grass blades with a small stapler

TABLE 3.14
Depth Distribution of Sea Grass Species

	Halodule	*Zostera* subgen *Zosterella*	*Zostera* subgen *Zostera*	*Cymadocea*	*Thalassia*	*Syringodium*	*Phyllospadix*	*Enhalus*	*Posidonia*	*Halophila*	*Heterozostera*	*Amphibolis*	*Thalassodendron*
Mideulittoral	+	+	–	–	–	–	–	–	–	+	–	–	–
Belt between MLWN and MLWS	+	+	+	+	+	+	+	+	–	+	+	–	–
Upper sublittoral	+	+	+	+	+	+	+	+	+	+	+	+	+
Lower sublittoral	+	–	+	–	–	–	+	–	+	+	–	+	+

Source: From den Hartog, C., *Helgol. Wiss. Meeresunters,* 15, 96, 1967. With permission.

TABLE 3.15
Comparison of the Biomass and Productivity of Selected Sea Grass Species

Species /Location	Peak Biomass (g dry wgt m^{-2})			Rate of Productivity			Reference
	Above	Below	Total	(mg O_2 hr^{-1} g^{-1})	(g C m^{-2} day^{-1})	NPP (g m^{-2} day^{-1})	
Poisodonia oceanica							
Mediterranean, Italy	600–1,700	—	—			0.8–4.4	Ott and Mauer (1977)
Poisodonia australis							
Port Hacking, Aust.	80	195	275			1.9	Kirkman and Reid (1979)
Spencer Gulf, Aust.	616	—	2,457			5.5	West and Larkum (1979)
N.S.W., Aust.	137–453		847–25,550			1.9–5.2	West and Larkum (1979)
Syringodium filiforme							
St. Croix, Florida	15–200					0.3	Zieman and Wetzel (1980)
Thalassia testudinium							
Cuba	76–517			8.0			Buesa (1975)
Puerto Rico		2,800					Odum et al. (1974)
Bibimi Harbour, Bahamas	200				1.8–2.4	2.6–3.9	Capone et al. (1979)
Zostera marina							
Lake Grevelingen, The Neth.	75–94	35–50	110–144			1.7	Nienhuis and De Bree (1980)
Oresund, Den.	290	160	450			2.9	Wium-Andersen and Borum (1980)
Vellerup Vig, The Neth.	226	217	443			5.9	Sand-Jensen (1975)
Nabeta Bay, Japan	222	71	292			7.2	Mukai 918710
Izembek Lagoon, Alaska	500	900	1,400			2.8	Kentula and McIntire (1986)
Netarts Bay, Oregon	84–256	56–214	140–470		4.8	7.2–10.3	McRoy (1974)
Cymodoce nodosa							
Mediterranean	15–340				5.5–18.5		Gelner (1959)
Halodule wrightii							
North Carolina					0.5–2.0		Dillon (1971)
Amphibolis antarctica							
Shark Bay, Aust.	300–850	300–1,150	600–2,000			2–17	Walker (1985)
Heterozostera tasmanica							
Australia	83–218					1.3–3.6	Bulthuis and Woelkerling (1983)

and measuring the growth that has occurred after a specified time interval, as well as the new leaves produced during the time that has elapsed. Details of the technique will be found in Zieman (1975) and Zieman and Wetzel (1980). The technique suffers from the inability to measure short-term (hours) rates for total plant production (Zieman and Wetzel, 1980), since the experiments must be carried our for periods of a least 8 to 12 days in order to allow for the significant emergence of new leaves. In addition, production estimates by the marking technique do not measure losses associated with excretion, sloughing, and herbivore consumption.

Changes in the dissolved oxygen and pH concentrations in the water have been used by several workers to measure the products of photosynthesis and respiration. Increasingly the ^{14}C uptake technique has been used to estimate productivity in sea grass (McRoy, 1974). This technique, however, may underestimate productivity because it assumes that inorganic carbon is taken up exclusively from the water column, and $^{14}CO_2$ may be stored and recycled in the lacunar gas spaces (Penhale and Thayer, 1980; Sondergaard, 1990).

Production estimates: Reviews by McRoy and Helfferich (1977), McRoy and McMillan (1977), Phillips and McRoy (1980), and Zieman and Wetzel (1980) concluded that sea grasses potentially have both high standing crop and productivity (up to 10 to 20 g C m^{-2} day^{-1}). Table 3.15 lists some representative estimates of sea grass production. From the data in the table, two trends emerge. The values demonstrate that sea grass productivity can rival the most productive agricultural areas (Westlake, 1963). Reliable estimates have shown that *Zostera marina* can attain productivities of 8 g C m^{-2} day^{-1} and estimated yearly production of 600 g C m^{-2}. Due to greater overall radiation input and longer growing season, tropical sea grasses can have still greater productivities. Beds of *Thalassia testudinium* have been reported to produce up to 16 g C m^{-2} day^{-1}. Using a marking technique, Greenaway (1976) estimated that the yearly production of *Thalassia* was 825 g C m^{-2}. It is thus clear that sea grasses rank among the most productive communities existing.

Productivity estimates recorded in the literature have been concerned primarily with aboveground production. If the belowground production is taken into account, the production estimates would be very much higher. Root:shoot ratios range from 0:24 for *Zoztera marina* in Japan to 3:46 for the same species in Denmark. Peak aboveground biomass estimates range from 75 to 500 g dry wgt m^{-2} for *Zostera marina,* and 76 to 500 g dry wgt m^{-2} for *Thalassia testudinum.* Zieman (1975) found that *Thalassia* leaves usually constituted 15 to 22% of the total dry weight of the plant, while Patriquin (1973) estimated that the short shoot and rhizomes accounted for only 10 to 13% of the net production. Sand-Jensen (1975) found that for *Zostera* growing in Denmark, the living rhizomes increased from 100 to 200 g dry wgt m^{-2} throughout the growing season, while the leaves and flowering shoots increased from 50 to 230 g dry wgt m^{-2}.

3.5.4.4 Factors Affecting Sea Grass Production

Light: Solar radiation is often regarded as the ultimate limiting factor in any autotrophic system. Photosynthesis versus irradiance (*P-I*) curves typically exhibit an initial linear response with increasing irradiance, with a gradually decreasing slope until saturation is achieved, after which a plateau may be reached with no photosynthetic response to increased irradiance. Williams and McRoy (1982) have demonstrated similar light responses for six sea grass species growing in different regions from 10 to 50° latitude.

In most estuaries light attenuation is often severe due to the high turbidity associated with suspended seston (sediment, organic matter, and phytoplankton), which limits the availability of light to submerged sea grasses (Wetzel and Penhale, 1982). In tropical lagoons the water clarity is higher and consequently sea grass can grow to greater depths.

Nutrients: Sea grasses are capable of deriving their nitrogen and phosphorus from two sources: (1) the water column via uptake by the leaves, and (2) the sediment interstitial water via roots and rhizomes (Izumi and Hattori, 1982; Thursby and Harlin,1984; Short and McRoy, 1984). These two methods of uptake of nitrogen are geared to the ambient nutrient conditions the plants experience. When there is sufficiently high concentration in the water, leaf uptake predominates; however, when nitrogen is depletes in the water, root uptake takes over.

Phosphorus is assimilated predominately through the roots. McRoy and Barsdate (1970) have estimated the daily movement of phosphorus in an eelgrass (*Zostera marina*) stand and have shown that it is excreted into the overlying water via the leaves. Orth (1977) demonstrated a rapid a positive growth response by *Z. marina* to the application of a commercial fertilizer (containing both nitrogen and phosphorus) indicating that the plants were nutrient limited. Similar responses have been observed in Naragansett Bay (Harlin and Thorne-Miller, 1981) and in Western Australia (Baluthius and Woelkerling, 1983). The nitrogen requirement of sea grasses may in part be supplied via nitrogen fixation by bacteria and blue-green algae on the sea grass leaves and in their sediments. McRoy and Goering (1974) reported that the nitrogen fixed by the epiphytes can be transferred to and absorbed by the sea grass host. Table 3.16 compares measured nitrogen fixation rates in a range of sea grass communities. From this table it can be seen that the proportion of the nitrogen requirements that can be met by nitrogen fixation range from 1 to about 100%.

TABLE 3.16
Comparison of Nitrogen-Fixation Rates Associated with Sea Grass Communities

Sea Grass or Macrophyte	Location	N_2-Fixation (mg N m^{-2} day^{-1})	Fraction (%) N Required (for plant growth)
Thalassia testudinium	Florida		
	Sediments	0–13	0–23
	Epiphytes	11–28	20–50
	Barbados		
	Sediments	27–137	100+
Zostera marina	Long Island Sound		
	Sediments	2–17	3–28
	Epiphytes	0	0
Myriophullum spicatum	Chesapeake Bay		
	Epiphytes	3	2

Source: From Day, J.W., Jr., Kemp, W.M., and Yanez-Aranciba, A., *Estuarine Ecology*, John Wiley & Sons, New York, 15, 1989, 246. With permission.

3.5.4.5 Fate of Sea Grass Primary Production

There is a paucity of information on the fate of sea grass production. These systems, as we have seen, are highly productive. In temperate regions, direct grazing on the sea grass leaves is low (Mann, 1988), but more substantial in the tropics. The current paradigm covering sea grass grazing contends that it is small, usually less than 10% of the annual aboveground primary production, with the bulk of the sea grass leaf production entering the marine food web as detritus (Mann, 1988). Information on the fate of this detritus is scarce (Hemmiga et al., 1991). However, Valentine and Heck (1999) have recently questioned these assumptions. They analyzed evidence from over 100 publications and concluded that grazing on sea grasses has been underestimated and that it is more widespread than previously thought.

Mateo and Romero (1997) have recently investigated the detritus and nutrient dynamics of the sea grass *Posidonia oceanica* in the Mediterranean (Table 3.17). At 5 m, the amount of exported leaf litter represented carbon, nitrogen, and phosphorus losses of 7, 9, and 6% of plant primary production. About 26% of the carbon produced by the plants in one year was immobilized by burial in the underground compartment, i.e., as roots and rhizomes. Annual nitrogen and phosphorus burial in the sediment

TABLE 3.17
***Poisodonia oceanica*. Dry Weight, Carbon, and Nutrient Budget in Medes Islands, N.W. Mediterranean (Percentage Values Relative to Annual Production are Given in Parentheses)**

	Dry Weight	C	N	P
Production	826 (100%)	326 (100%)	13.9 (100%)	1.168 (100%)
Fate				
Export	112 (14%)	27 (%)	1.2 (9%)	0.070 (6%)
Burial	206 (25%)	84 (26%)	1.1 (8%)	0.060 (5%)
Nutrient recycling				
Leaves			6.2 (45%)	0.525 (46%)
Belowground			0.9 (6%)	0.372 (10%)
Respiratory consumption				
Aerobic	143 (17%)	57 (1%)	1.9 (14%)	0.114 (10%)
Anaerobic	9 (1%)	4 (1%)	2.2 (16%)	
Fine particulate detritus	356 (48%)	159 (49%)	0.4 (3%)	0.027 (2%)

Source: From Mateo, M.A. and Romero, J., *Mar. Ecol. Prog. Ser.*, 151, 60, 1997. With permission.

was 8 and 5% of the total plant N and P needs, respectively. Respiration consumption (aerobic and anaerobic) of carbon leaf detritus represented 18% of the annual consumption. An additional, but very substantial loss of carbon as very fine particulate organic matter was estimated at approximately 48%. The fate of this fine material was undetermined. It would either be exported or enter the surface sediments.

3.5.5 Salt Marshes

3.5.5.1 Introduction

Salt marshes are a common feature of most estuaries and they have a wide distribution throughout the world, especially in temperate zones. Numerous studies have shown that they are among the most productive plant communities anywhere. In addition they play a very important role in estuarine ecosystem dynamics since they: (1) provide a food source through their production of organic detritus to both estuarine and coastal water consumers; (2) serve as a habitat and nursery for large numbers of both juvenile and adult estuarine animals; and (3) regulate important components of estuarine chemical cycle (Day et al., 1987). In this section the composition, distribution, and productivity, as well as the factors affecting production will be discussed. Recently there have been a number of reviews of the ecology of salt marshes including H.T. Odum et al. (1974), Nixon (1980), Pomeroy and Wiegert (1981), Zedler (1982), Josselyn (1983), Mitsch and Gosselink (1986), Teal (1986), Dijkema (1987), and Adam (1990).

3.5.5.2 Development, Distribution, and Zonation

Salt marshes are intertidal ecosystems dominated by rooted plants in areas regularly inundated and drained by the tides. They occur primarily on temperate coasts but occasionally form in the tropics on salt flats not occupied by mangroves, and they develop in sheltered situations where silt and mud accumulate. They are shaped by the interaction of freshwater, seawater, sediments, and vegetation. For their development they require protection from high energy waves and they therefore are found in lagoons, in inlets behind barrier islands, or in the protection of an estuary. Here the slowing down of water currents permits the deposition of fine sediments and the building up of extensive, gently sloping beaches. The growth of the plants can usually keep pace with sediment deposition so that over long periods of time considerable accumulation of sediments and peat can occur, especially on shores where sea level is rising relative to the land.

The vegetation patterns in marshes is influenced by a complex of environmental factors including the frequency and range of the tides, salinity, microrelief, nature of the sediment, ice-scouring, and storms. In addition, historical factors, and more recently anthropic factors such as fires, cutting, dyking, grazing, and ditching have in many areas profoundly influenced the distribution of salt marsh species.

One of the characteristics of a mature salt marsh is the presence of creeks and drainage channels. They form a network over the marsh, becoming progressively narrower and shallower as they subdivide in a more or less regular pattern (Wiegert et al., 1981). Tidal waters enter and leave the marsh through this network (Figure 3.26).

A feature of salt marshes is the widespread occurrence of dominant genera such as *Salicornia, Suaeda, Spartina, Juncus, Arthrocnemum,* and *Plantogo*. In most marshes a characteristic banding or zonation pattern occurs. These zones are the result of differences in reproduction and growth, and differential response to environmental gradients encountered from low to high water. The plants respond to factors such as elevation, drainage, sediment type, salinity, and the oxidation-reduction state of the sediments. Some examples of typical patterns are shown in Figure 3.27. On the Atlantic coast of the United States where over 6,000,000 ha of salt marshes occur, the smooth cordgrass *Spartina alterniflora* dominates the marsh between mean sea level and mean high water. Above its upper limit, the tufted hairgrass (*Deschampsia caespitosa*) occurs. On the European side of the North Atlantic the flora is much more diverse and heterogeneous; for example, in the U.K., there are marked differences between the marshes bordering the North Sea, the English Channel, and the Atlantic Ocean. The North Sea marshes frequently have as codominants the seapink (*Ameria*), sea lavender (*Limonium*), sea plantago (*Plantago maritima*), and species of *Spergularia* and *Triglochin*. The Atlantic marshes tend to be used for cattle and sheep grazing and are dominated by the grasses *Puccinellia* and *Festuca*. On the south coast of England, the introduced cordgrasses *Spartina x townsendii* and *S. anglica* have been spreading rapidly and replacing the original more diverse flora. In South Africa the ricegrass, *Spartina maritima*, dominates a band above mean high water. Then follows a band of glasswort, *Sarcocornia*, with mixed plants of *Chenolea diffusa* and *Limonium linifolium*. This in turn on the higher shore is replaced by a band of *Sporobolus virginicus* with the rush, *Juncus krausii*, and occasional plants of *Disphyma crassifolium*.

Until the 19th century introduction of *Spartina alterniflora* into Europe from North America, only *S. maritima*, a small species rarely exceeding 0.2 m in height, was found in northwest Europe. However, *Spartina* has only become a major component of the salt marsh vegetation since the appearance of the *S. maritima*—*S. alterniflora* infertile hybrid, *S. x townsendii*, and the fertile amphidiploid derived from it, *S. anglica* (Gray et al., 1990). Since the first recorded occurrence of *S. x townsendii* in 1870 (Marchant, 1967), the species has spread both as a result of deliberate plantings and natural dispersal to

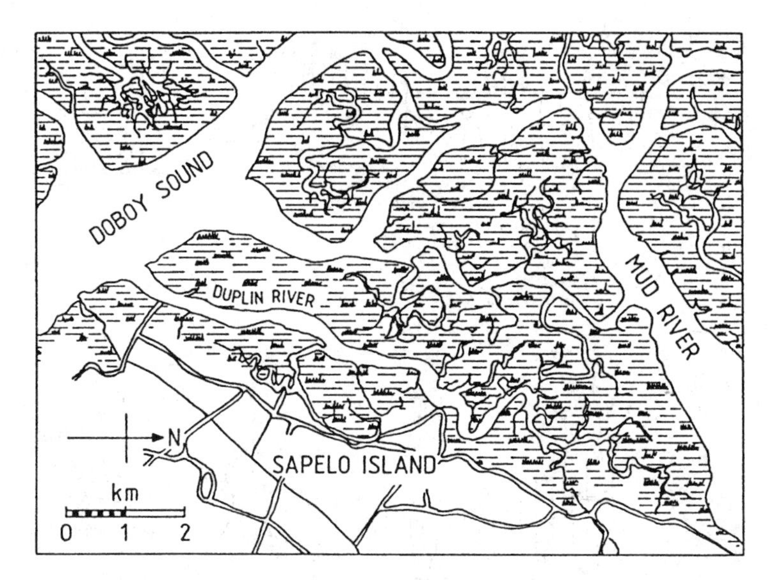

FIGURE 3.26 Diagram of the marshes of Sapelo Island, Georgia. The classic work of the *Spartina* marshes was carried out in the watershed of the Duplin River, a brackish tidal stream that is tributary to Doboy Sound. (Redrawn from Wiegert, R.G., Pomeroy, L.R., and Wiebe, W.T., *Ecology of a Salt Marsh*, Pomeroy, L.R. and Wiegert, R.G., Eds., Springer-Verlag, Berlin, 1981. 9. With permission.)

occupy an area of about 25,000 ha in northwest Europe. *S. x townsendii* has been introduced into New Zealand where the fertile *S. anglica* appears to have also arisen. In many parts of New Zealand, *Spartina* is now well established as a result of deliberate plantings and natural spread (Odum et al., 1983). In Pelorus Sound on the northern coast of the South Island, *S. anglica* can establish itself on stony beaches where it accumulates sediment and spreads to cover the original beach.

Spartina anglica generally occupies a zone on the shore between MHWN and MHWS, with occasional clumps above and below these limits. The frequency and periodicity of tidal submergence appears to be a important factor in its distribution. On both the British (Gray et al., 1990) and New Zealand shores (Odum et al., 1983), *S. anglica* invades mudflat zones not occupied by native macrophyte species. The species composition of the benthic infauna is different within the *S. anglica* from that on the open mudflats. In particular many polychaete and bivalve species are eliminated. This is believed to have limited feeding opportunities for wading birds. However, documented evidence of an impact on bird populations is rare, and the overwintering populations of most species of waders have remained constant or increased in size on British shores since nationwide counts began in the early 1970s (Goss-Custard and Moser, 1988). An exception is the Dunlin, *Calidris alpina*, the overwintering numbers of which had declined to almost half by 1988 since 1973–74 (Salmon and Moser, 1985). However, the decline may be due to a factor, or factors, unrelated to the spread of *Spartina*.

A characteristic feature of *Spartina* marshes is the development of natural levees along the banks of tidal creeks (Figure 3.28). As a result there are extensive areas that are reached only by fortnightly spring tides. In such areas increased salinity of the interstitial water due to evaporation from the marsh surface profoundly influences the size and distribution of the dominant plants. The development of the levees also influences the tidal movement of water over the marsh. A perched water table behind the levees holds interstitial water that is more saline than tidal water and largely anaerobic (Odum and Heald, 1975; Nestler, 1977). Depending on the conditions, different growth forms of *S. alterniflora* occur. Adjacent to the creeks are found tall (>2 m), robust plants with low densities (30 to 50 stems m^{-2}). Behind the levees and at the head of the creeks, where sediment salinities average 35 to 40, are found short (25 cm), high density plants (up to 300+ stems m^{-2}).

3.5.5.3 Primary Production

The primary productivity of salt marsh plants and the factors affecting it have been extensively studied. The bulk of the studies are for *Spartina alterniflora*, although there is a fair amount of information available for other species (see Turner, 1976; Long and Mason, 1983; Mitsch and Gosselink, 1986). Most of the production values have been calculated from changes in live and dead plant biomass over an annual cycle. In temperate regions the live biomass increases during the growing season (spring to autumn) and the plants then flower and die. As the grass dies the live biomass decreases and the dead organic matter increases. In spring the biomass of the dead grass decreases as it decomposes. However, in subtropical and tropical regions there is some growth year-round. In addition not all salt marsh plants have distinct seasonal patterns

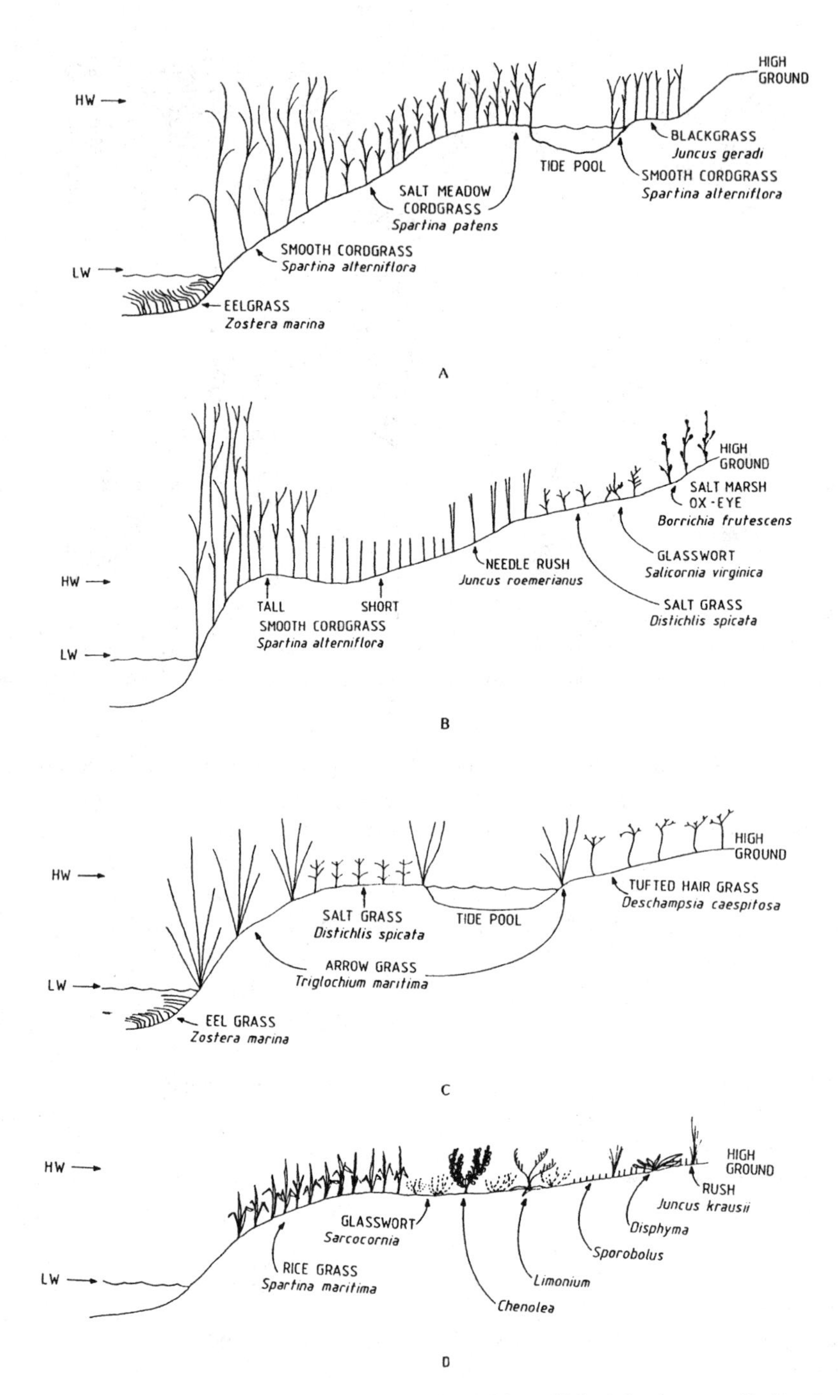

FIGURE 3.27 Some characteristic salt marsh vegetation patterns. A. Northern U.S. Atlantic coast. B. Southern U.S. Atlantic coast. C. Northern U.S. Pacific coast. D. Ashmead Knysna Estuary, South Africa. (A, B, and C redrawn from Gallagher, J.L., in *Coastal Ecosystem Management*, Clark, J.R., Ed., John Wiley & Sons, New York. 1977, 754 and 755. With permission. D redrawn from Day, J.H., in *Estuarine Ecology with Special Reference to South Africa*, Day, J.H., Ed., A.A. Balkema, Rotterdam, 1981, 92. With permission.)

of growth, and there are also distinct differences in both live and dead biomass between streamside and inland marsh areas due to a complex of factors such as tidal flooding, and the nutrient chemistry, salinity, and redox potential of the sediment.

Long and Mason (1983) have reviewed the various methodologies used for the estimation of marsh plant production using destructive (harvesting) techniques. These are: (1) estimation of live dry weight; (2) estimation of maximum standing crop; and (3) maximum-minimum (i.e., the maximum biomass reached during the year — the minimum biomass recorded during the year). These techniques assume that no material is lost before maximum biomass has been reached, but this clearly is not the

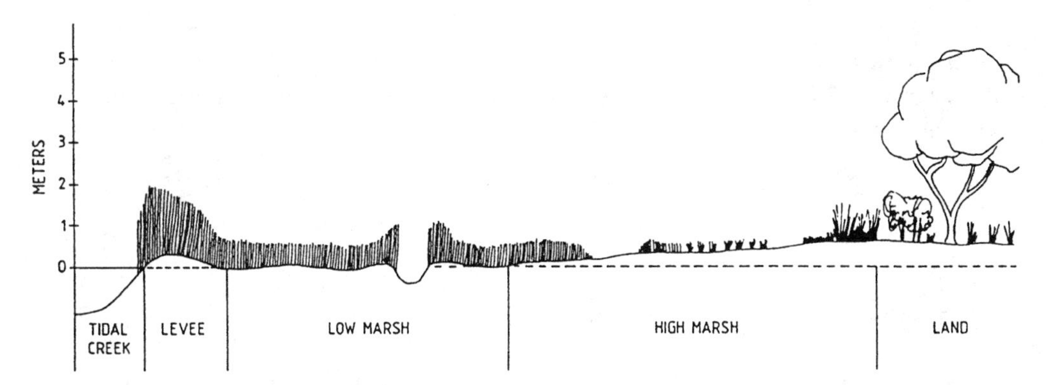

FIGURE 3.28 Schematic cross-section through the Sapelo Island salt marsh, showing levee and low marsh with tall *Spartina alterniflora*, and high marsh with short *S. alterniflora*. The level reference line is approximately at mean high water. (Redrawn from Wiegert, R.G., Pomeroy, L.R., and Wiebe, W.J., *The Ecology of a Salt Marsh,* Pomeroy, L.R. and Wiegert, R.G., Eds., Springer-Verlag, Berlin, 1981. With permission.)

case. Thus most of the estimates of production using these techniques are underestimates of the overall production. Consequently a number of methods have been developed to overcome this (e.g., Smalley's method; Smalley, 1959). One technique determines the changes in both live and dead standing crop at regular intervals and the use of techniques to estimate the loss of both live and dead standing crop at regular intervals between sampling dates (Wiegert and Evans, 1964). Other techniques include different types of tagging to measure increase in height, diameter, and number of leaves, as well as the disappearance of individual plants and leaves (Hopkinson et al., 1978). Hopkinson et al. (1978) found that the different techniques for measuring annual net production in Lousiana gave highly variable results. Gas-exchange measurements have generally given results which are in close agreement with the harvest techniques (Houghton and Woodwell, 1980). A comparison of the various methods can be found in Kirby and Gosselink (1976), Hopkinson et al. (1978), Linthurst and Reimold (1978), Shew et al. (1981), and Long and Mason (1983).

Aboveground production: At least 90% of the work carried out on salt marsh productivity has been carried out in North America and the majority of this is concentrated on *Spartina alterniflora*. Most of the values in the literature are for net production calculated from changes in the aboveground live and dead biomass over an annual cycle. Table 3.18 lists organic production estimates for emergent macrophyte species from a range of localities. From this table it can be seen that not only are there wide variations in productivity between species, but that there is also variation within individual species. Species with the highest productivity are *Distichlis spicata* (4,214 g C m^{-2} yr^{-1}), *Spartina cynoduroides* (5,996 g C m^{-2} yr^{-1}), and *Spartina patens* (3,824 g C m^{-2} yr^{-1}). However, in terms of total contribution to overall marsh production along the east coast of North America, these species make a relatively minor contribution when the huge area occupied by *S. alterniflora* is considered.

Hatcher and Mann (1975) have compared the production of various salt marshes along the Atlantic and Gulf coasts of the U.S. They record a net production of 289 g C m^{-2} yr^{-1} in Nova Scotia, a minimum of 133 g C m^{-2} yr^{-1} in New Jersey, and a maximum of 1,153 g C m^{-2} yr^{-1} in Georgia. In spite of local variations it is apparent that production increases toward the tropics.

Long and Woolhouse (1983) and Knox (1983c) give data for the production of *Spartina anglica* in England and New Zealand. Production for this species in the Humber estuary, England, was 900 g C m^{-2} yr^{-1}, while the same species in Pelorus Sound, New Zealand, had a very similar production of 955 g C m^{-2} yr^{-1}.

Production varies widely over a given marsh depending on the duration of tidal inundation, drainage, substratum elevation gradient, age of the marsh, sediment nutrient status, and salinity. In *Spartina* marshes, as mentioned above, the plants tend to be taller and more productive along the edges of the creeks (low marsh), and shorter and less productive as distance from the creek increases (high marsh). Gallagher et al. (1986) found an almost 2.5-fold difference in the production of tall and short *S. alterniflora*. The tall *Spartina* has a mean biomass of 1,966 g dry wgt m^{-2}, while the shortest stands back from the creeks had only 397 g dry wgt m^{-2}. In the Pelorus marsh, New Zealand, tall *S. anglica* had a biomasses ranging from 932 to 1,368 g dry wgt m^{-2}, whereas the short form had biomasses ranging from 215 to 766 g dry wgt m^{-2}.

Belowground production: The estimation of belowground production presents a number of difficulties. Sampling is a problem, especially in species where much of the belowground material consists of large, irregularly distributed components such as tubers and rhizomes. Table 3.19 lists data for belowground productivity for ten marsh species along the east coast of North America. Estimates

TABLE 3.18
Aboveground Primary Production Estimates for Estuarine and Salt Marsh Plants ($g\ C\ m^{-2}\ year^{-1}$)

Locality	Species	Production	Reference
North Carolina	*Juncus roemerianus*	280–612	Waits (1967), Foster (1968, 1982), Stroud & Cooper (1968)
Florida	*Juncus roemerianus*	425	Heald (1969)
East and Gulf coasts, U.S.	*Spartina alterniflora*	220–2,000	Smalley (1959), Williams & Murdock (1969), Kirby (1972), Odum & Fanning (1973)
Long Island, North Carolina	*Spartina patens*	650	Odum (1983)
	Spartina patens with *Scitpus robustus*	670	Waits (1967)
California	*Spartina foliosa*	108–276	Marshall & Park (1976)
Sapelo Island, Georgia	*Spartina alterniflora*		
	short	520	Gallagher et al. (1973),
	tall	1,480	Linthurst & Reimhold (1978b)
	Distichus spicata	4,214	
	Spartina cynosauroides	5,996	
	Spartina patens	3,824	
	Sporobolus virginicus	1,372	
England	*Spartina anglica*	900	Long & Woolhouse (1983)
New Zealand	*Spartina anglica*	950	Odum et al. (1983)
Oregon	Mixed communities (*Carex, Jucus, Scirpus*)	92–1,130	Eilers (1979)
California	*Salicornia virginica*	3,844	Packham & Liddle (1970)

are variable, ranging from 0.22 to 4.78 for the tall form of S. alterniflora, to 0.45 to 6.2 for short form. In general belowground production is higher in the short *Spartina*.

Pomeroy et al. (1981) list estimates for the underground production of seven marsh plants on the Sapelo Island marsh. While the belowground production was the same for both short and tall *S. alterniflora*, the belowground production of the former was 59% of the total production compared to only 34% for the latter. For the other species, apart from *Juncus roemerianus*, the belowground production was small compared to that of the aboveground.

Recently Schubauer and Hopkinson (1984) estimated above- and belowground emergent macrophyte production and turnover in a coastal marsh system in Georgia, U.S.A. Peak live aboveground biomass was 1.6 times higher for *Spartina cynosuroides* than for *S. alterniflora*. Live biomass was 2.4 times more belowground for *S. cynosuroides* and 1.7 times for *S. alterniflora*. They found that rhizomes made up 76 and 87% of the live belowground biomass during the year. Belowground production was measured with a technique that partially accounted for midseason decomposition. Total plant production was estimated to be 7,620 g dry wgt $m^{-2}\ yr^{-1}$ for *S. alterniflora* and 7,708 g dry wgt $m^{-2}\ yr^{-1}$ for *S. cynosuroides*. Belowground production was roughly 1.6 times that of aboveground production. Turnover rates for belowground live material were 1.42 yr^{-1} for *S. cynosuroides* and 3.22 yr^{-1} for *S. alterniflora*.

Root-to-shoot ratios: Root:shoot ratios in marsh species reflect differences between species and difference in plants of the same species growing in different environments. Table 3.20 gives root:shoot ratios for four species of marsh plants along the western Atlantic coast of the U.S., along with data from other localities. Most estimates range from 1.2 to 5.4 for the short form of *S. alterniflora*, and 0.5 to 8.25 for the tall form, although ratios as high as 50 have been reported for the short form (Gallagher, 1974). Generally in the same marsh, the ratios are much higher for the short form than for the tall one. Values for the rush, *Juncus roemerianus*, range from 0.80 to 8.7. In the Pelorus marsh, New Zealand, the ratios for the short and tall forms of *S. angelica* were 11.6 and 1.7, respectively.

3.5.5.4 Factors Affecting Production

Within salt marshes there is considerable spatial and temporal environmental heterogeneity, much of which is related to the frequency and duration of tidal cover. E.P. Odum (1969) has described salt mashes as pulse-stabilized systems due to the tidal regime that limits the type of vegetation at particular levels. In addition to tidal rise and fall, a variety of other factors influence the growth and productivity of salt marsh plants. Changes in the physical

TABLE 3.19
Belowground Production Estimates for Salt Marsh Species at Several Locations along the Atlantic Coast of the U.S.

Species	Production (kg m^{-2} year^{-1})	Location	Reference
Borrichia frutescens	0.28	Georgia	Gallagher and Plumley (1979)
Distichlis spicata	1.07	Georgia	Gallagher and Plumley (1979)
	3.20	Delaware	Gallagher and Plumley (1979)
	2.78	New Jersey	Good and Frasco (1979)
Juncus gerardia	4.29	Delaware	Gallagher and Plumley (1979)
	1.62	Maine	Gallagher and Plumley (1979)
Jumcus roemerianus	1.36	Mississippi	de la Cruz and Hackney (1977)
	4.4–7.6	Alabama	Stout (1978)
	3.36	Georgia	Gallagher and Plumley (1979)
Phragmites communis	3.65	Delaware	Gallagher and Plumley (1079)
	2.81	New Jersey	Good and Walker (1977)
Salicornia virginus	0.41	Georgia	Gallagher and Plumley (1979)
	1.43	Delaware	Gallagher and Plumley (1979)
Spartina alterniflora			
Tall form	2.1	Georgia	Gallagher and Plumley (1979)
	0.5	North Carolina	Stroud (1976)
	2.9	New Jersey	Good (1977)
	3.3	New Jersey	Good and Frasco (1979)
	2.4	New Jersey	Good and Walker (1977)
	3.5	Massachusetts	Valiela et al. (1976)
	0.22	Maine	Gallagher and Plumley (1979)
Short form	2.02	Georgia	Gallagher and Plumley (1979)
	0.56	North Carolina	Smith and Odum (1981)
	0.46	North Carolina	Stroud (1976)
	3.2	New Jersey	Good (1977)
	2.4	New Jersey	Good and Frasco (1979)
	2.3	New Jersey	Smith et al. (1979)
	3.6–6.2	Alabama	Stout (1978)
Spartina cynosuroides	2.2	Mississippi	de la Cruz and Hackney (1977)
	3.56	Georgia	Gallagher and Plumley (1979)
Spartina patens	0.31	Georgia	Gallagher and Plumley (1979)
	0.47	Delaware	Gallagher and Plumley (1979)
	2.5	Massachusetts	Valiela et al. (1979)
	0.54	Maine	Gallagher and Plumley (1979)
Sporobolis virginus	0.58	Georgia	Gallagher and Plumley (1979)

or chemical environment include solar radiation, temperature, nutrient concentrations, sediment type, drainage, oxygen concentration, and pH. The individual plant species in a marsh differ in their reaction to these factors. Here we will consider the ways in which these factors interact to determine the nature of the salt marsh plant community (Figure 3.29).

1. *Solar radiation, temperature, and evapotranspiration*: The levels of solar radiation impact directly on photosynthetic rates. They also affect plants indirectly because changes in radiation lead to changes in temperature, and temperature directly affects the rate of metabolic processes. Heat energy controls the process of evaporation that brings about the transport of nutrients and essential elements from the sediment to the leaves in the replacement of the water lost by evapotranspiration. Solar radiation, temperature, and evaporations act together to produce differences in salt marsh production over a latitudinal gradient as discussed above for salt marsh production along the U.S. Atlantic coast.
2. *Salinity*: There is considerable literature on the salinity relationships and physiological responses of halophytes to salinity gradients, and I do not propose to discuss them in detail here. Readers are referred to Waisel (1972) and Chap-

TABLE 3.20
Some Published Root:Shoot Ratios for Salt Marsh Plants

Species	Ratio	Location	Reference
Distichlis spicata	7.2	Georgia	Gallagher et al. (1978)
	4.5	New Jersey	Good and Frasco (1982)
Juncus roemerianus	0.80	Mississippi	de la Cruz and Hackney (1977)
	3.26	Alabama	Stout (1978)
	3.7–8.7	Florida	Kruczyzsko et al. (1978)
	8.2	Georgia	Gallagher (1974)
Spartina alterniflora			
Tall	1.43	Georgia	Gallagher (1974)
	0.3–0.4	North Carolina	Stroud (1976)
	4.53	New Jersey	Good and Frasco (1979)
	8.25	Massachusetts	Valiela et al. (1978)
Short	3.72	Alabama	Stout (1978)
	48.9	Georgia	Gallagher (1974)
	1.2–1.3	North Carolina	Stroud (1976)
	4.7	New Jersey	Smith et al. (1979)
	5.24	New Jersey	Good and Frasco (1979)
Sportina anglica			
Tall	1.7–1.9	New Zealand	Knox (1983d)
Short	8.6–20.0	New Zealand	Knox (1983d)
Juncus maritimus	4.1	New Zealand	Knox (1983d)
Leptocarpus simplex	3.4	New Zealand	Knox (1983d)

man (1974) for details. At high salinities, osmotic stress (resulting in reduced water uptake) and cell membrane damage are likely to be limited. Halophytes seem to deal with osmotic stress by selectively concentrating preferred ions, while they have evolved metabolic adaptations to deal with high salt concentrations. In addition they have the capacity to remove salt via salt glands (Waisel, 1972), hence the salt deposits often found on the tips of *S. alterniflora* leaves.

The effect of salinity on the growth and distribution of *Spartina alterniflora* has been intensively studied. Mooring et al. (1971) found that *S. alterniflora* grows best at a salinity of 10. Nestler (1977) found that the interstitial water in the sediments of the Sapelo Island marshes formed salinity clines across the marsh, with the lowest average values in the creek beds with increasing values with increasing distance from the creeks. Biomass, height, and leaf area (but not shoot number) were all negatively correlated with salinity. Linthurst (1980a), working in North Carolina, found that salinity increases of 15 decreased biomass, density, and mean height of *S. alterniflora* and sediment MOM (macroorganic matter, i.e., > 2 mm) an average of 42, 32, 22, and 37%, respectively. An increase in salinity from 15 to 45 decreased biomass, density, mean height, and MOM 66, 53, 38, and 61%, respectively.

3. *Aeration:* The amount of oxygen present in the sediment has been found to have a positive effect on the growth of *S. alterniflora* (Linthurst, 1979, 1980a) and other marsh plants (Linthurst and Seneca, 1981). Thus, maximum growth occurs in oxygenated sediments, and the H_2S produced in anaerobic sediments inhibits respiration and nutrient uptake. Sediment drainage is an important factor affecting oxygen concentrations and the growth of marsh plants. Well-drained sediments have higher oxygen levels because the air spaces can hold atmospheric oxygen, and streamside marshes generally have the best drainage. Field studies have shown that reduced drainage conditions result in a decrease in the total biomass of *S. alterniflora* (Mendelssohn and Seneca, 1980; Linthurst and Seneca, 1981). Wiegert et al. (1981) have found that increasing sediment drainage in an intermediate-height *S. alterniflora* marsh increased plant biomass beyond that of an adjacent undisturbed plot, and shifted a number of other characteristics toward typical tall *S. alterniflora*.
4. *Tidal inundation*: In Long Island Sound, where the tidal range changes from 0.7 near the mouth to 2.26 m near the head of the inlet, Steever et al. (1976) found that marsh productivity was correlated with tidal range, the correlation coefficient being better than 0.96. They also found

that there was a 26% reduction in the productivity of a gated marsh compared to an adjacent ungated one. An analysis of a variety of data from the North American Atlantic coast fitted the trend of increasing productivity with increasing tidal height. Odum (1969; 1974) refers to tidal energy as an "energy subsidy," performing the work of mineral cycling, food transport, waste removal, etc. He claims that: "It is clear that the energy subsidy provided by the tidal flow more than compensates for the energy drain of osmoregulation required by a high salinity environment." As Mann (1982) points out, this energy subsidy theory is an attractive one, but the evidence is mainly correlational, and as we have seen there are other factors involved in determining marsh productivity.

5. *pH:* In laboratory experiments, Linthurst (1980b) found that *S. alterniflora* growth was optimal at pH 6 as compared to pH 4 and 8. Short *S. alterniflora* with its shallow root system may be subjected to changing pH when periods of high temperature, low tides, and rainfall prevail. *S. alterniflora* was found to be inhibited at pH 8, a pH observed in areas subject to dieback in a North Carolina marsh (Linthurst and Seneca, 1981).
6. *Nutrients*: There have been numerous investigations of the impact of nutrients, especially nitrogen, on the growth of marsh plants (e.g., Sullivan and Daiber, 1974; Broome et al., 1975; Haines and Dunn, 1976; Mendelssohn, 1979; Valiela and Teal, 1979b; Buresh et al., 1980). In laboratory experiments, it was found that an increase in nitrogen from a natural level to 168 kg ha^{-1} increased biomass, density, and mean height 2.02, 1.46, and 1.26 times (Linthurst, 1980b; Linthurst and Seneca, 1981). High levels of N and aeratioin together, in comparison with low levels of both if these, produced increases of 4.53, 2.71, 1.88, and 2.24 times in biomass, density, mean height, and macroorganic matter. High levels of N and aeration together, in comparison with low levels of both of these, produced increases of 4.53, 2.71, 1.88, and 2.24 times in biomass, density, mean height, and macroorganic matter.
7. *Photosynthetic pathways:* A major factor that is partly responsible for the high productivity of salt marshes is that many species of marsh plants have C_4 biochemical pathways of photosynthesis. C_4 plants have higher levels of production than the other group of plants, the C_3 plants. The distinction refers to the number of carbon atoms in the initial product of photosynthesis. In C_3 plants it is phosphoglyceric acid and for C_4 plants it is oxalocetic acid. C_4 plants have much higher light and temperature saturation levels than C_3 plants. For example the summer temperature optimum for *S. alterniflora* (a C_4 plant) is 30 to 35°C while that of *Juncus roemerianus* (a C_3 plant) is 25°C. As a consequence, as temperature or light rises, the photosynthetic rate for C_3 plants levels off earlier than that of C_4 plants. There is also less water transpired per unit of photosynthesis in C_4 plants. Since C_4 plants (especially *S. alterniflora*) constitute the great bulk of the salt marsh plants, this is an important factor in their high productivity of salt marshes.
8. *Synthesis*: Chambers (1982) has advanced a model (Figure 3.29) of the interactions between the factors which affect *S. alterniflora* growth, either directly or indirectly. Edaphic factors control the heterogeneity in height, biomass, and productivity. Field and laboratory studies have shown that salinity is an important factor that can influence *S. alterniflora* growth, although there are some marshes where both tall and short forms occur in the absence of salinity gradients.

Fertilization experiments have shown that the growth of tall *S. alterniflora* is not nitrogen limited, but that the productivity of the short form can be increased by nitrogen additions. Other studies have shown that the apparent nitrogen limitation in the short form is not due to the shortage of available nitrogen, but to an attenuation in nitrogen uptake kinetics. Salinity stress-caused diversion of nitrogen to the production of osmotica can also reduce the amount of nitrogen available for growth.

High sulfide concentrations and consequent low oxidation-reduction potentials in the rhizosphere can affect nitrogen-uptake kinetics. High sulfide and anoxia can also cause structural damage or changes in the roots, which could affect nutrient uptake. Sediment drainage, iron concentrations, oxygen diffusion from *S. alterniflora* roots, and plant productivity itself can all affect sulfide concentrations and redox potentials.

3.5.5.5 Marsh Estuarine Carbon Fluxes

Marshes are inundated on each tide and the tidal ebb and flow have the potential to exchange dissolved and particulate organic matter between the marshes and the adjoining estuarine waters. These exchanges have been the focus of much research over the past few decades (early studies reviewed by Nixon, 1980; later studies include Roman and Daiber, 1989; Dame et al., 1991b, and Williams et al., 1992). The early studies suggested that there could be

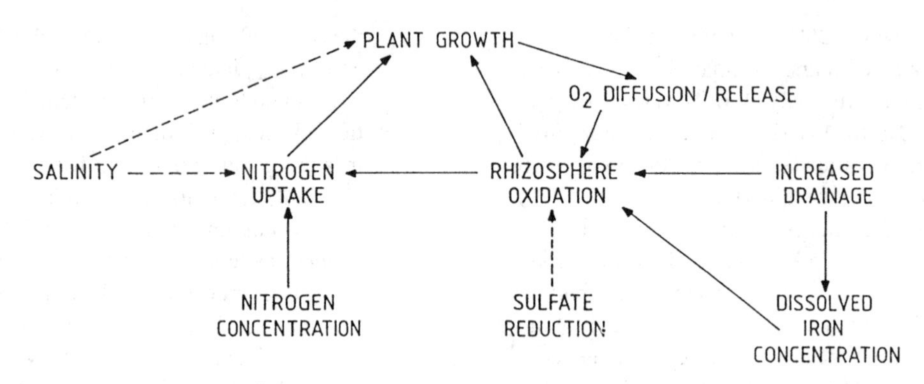

FIGURE 3.29 Relationship between factors directly and indirectly affecting the growth of *Spartina alterniflora*. Solid lines are positive effects, dashed lines are negative effects. (Redrawn from Chambers, A.G., in *Estuarine Comparisons*, Kennedy, V.S., Ed., Academic Press, New York, 1982, 239. With permission.)

large-scale exportation of organic material from the marshes to adjacent waters (Odum and de la Cruz, 1967), a view that became widely referred to as the "outwelling hypothesis" (Odum, 1980; Dame et al., 1986). However, controversy has arisen as to the correctness of the original hypothesis, and while a number of studies (e.g., Nixon and Ovaitt, 1973; Axelrod, 1974; Knox, 1983a: Dame et al., 1991b) have documented exports of organic carbon, others (e.g., Nadeau, 1972; Heinle and Flemer, 1976b; Woodwell et al., 1977; Valiela et al., 1978; Haines, 1979) were unable to measure net export from tidal wetlands. As Odum et al. (1979) point out, the export/import debate has been confused by the varied use of the term "export." To some, export is simply the transfer of marsh-produced organic carbon and nutrients to nearby water bodies such as tidal creek systems. To others this same term applies to the long-range movement of particulate and organic carbon and nutrients out of the estuaries into adjacent coastal waters. This latter view will be considered in Section 4.11.

The bulk of the studies on export/import have been conducted on the east coast of the U.S.A. and most of these, in turn, have focused on low elevation *Spartina alterniflora* marshes (exceptions include Kokkinn and Allanson, 1985; Abd Aziz and Nedwell, 1986; Baird and Winter, 1989). Only a limited number of investigations have been conducted on high elevation marshes (exceptions include Borey et al., 1983; Jordon and Correll, 1991; Taylor and Allanson, 1993). Along the east coast of the United States, high elevation marshes are characterized by the macrophytes *Spartina patens, Distichilis spicata,* and *Juncus* spp., while low marshes are typically vegetated solely by *S. alterniflora*. High marshes occur above mean high water (MHW), while low marshes can extend from MHW to mean low water (MLW) (McKee and Patrick, 1988). High marshes are flooded less frequently, and for shorter periods than low marshes.

Dame et al. (1991b) investigated the processing of carbon by the Bly Creek salt marsh-estuarine ecosystem in North Inlet, South Carolina. For the purposes of the study, the area was subdivided into the following regions: creek, oyster reef, tall marsh, mid-marsh, and short marsh. The areas of each subdivision and their annual primary production are given in Table 3.21. Total annual primary production was 1.61×10^6 kg C, with 6.7×10^5 kg C of this contributed by the aboveground production. Figure 3.30 depicts the carbon fluxes between the marsh, oyster reef, the creek, sediments, and the North Inlet. Only about 10% of the total carbon produced within the system was exported (1.6×10^5 kg C/1.61×10^6 kg C), which represents about 14% of the aboveground production. Since there is very little peat formation in this system most of the remaining organic carbon must be decomposed *in situ*. From Figure 3.30 it can be seen that only a small amount of the carbon exported in the tidal water was from the salt marsh. Most of the carbon export (89%) from the basin can be accounted for by primary production of phytoplankton, macroalgae, and benthic microalgae within the creek.

Table 3.22 summarizes data on organic carbon fluxes from salt marshes and comparisons of TOC fluxes with net aerial production. From this table it can be seen that the TOC export as a percentage of the net primary production varies from as low as 3% to as high as 63%. Values for TOC fluxes range from an import of 123 g C m^{-2} yr^{-1} to an export of 486 g C m^{-2} yr^{-1}. However, these global export/import values do not reflect the true export picture. When time series plots of the data are made (Figure 3.31) it can be seen that in most systems there is a varied pattern over a year with both import and export showing pronounced seasonal peaks. Overall, the carbon exports are a function of the relative sizes of the marsh and the adjacent open water systems.

Taylor and Allanson (1995) investigated tidal fluxes of total (TOC), dissolved (DOC), and suspended particulate (POC) organic carbon between a high *Sarcocornia perennis* — *Chenolea diffusa* salt marsh and the Karienga estuary, South Africa. The marsh showed an annual export of TOC of +16 g C m^{-2} yr^{-1}, with 80% of this occurring in the dissolved form. These fluxes were similar to those reported for high elevation *Spartina patens* and *Distichlis spicata*

TABLE 3.21
Net Primary Production in the Bly Creek Basin

Subsystem	Area (m^2)	Phytoplankton[a] ($gC\ m^{-2}\ yr^{-1}$)	Microbenthic[b] ($gC\ m^{-2}\ yr^{-1}$)	Macrobenthic[c] ($gC\ m^{-2}\ yr^{-1}$)	Grass[d] ($gC\ m^{-2}\ yr^{-1}$)	Total ($gC\ yr^{-1}$)
Creek	1.28×10^5	265	400	450	0	1.43×10^8
Oyster reef	1.00×10^3	0	400	790	0	1.19×10^6
Tall marsh	3.99×10^4	0	400	290	2078	1.10×10^8
Mid-marsh	8.46×10^4	0	400	20	666	1.19×10^7
Short marsh	4.06×10^5	0	400	10	2888	1.34×10^9
Total system	6.61×10^5					1.61×10^9

[a] Sellner et al. (1976).
[b] Zingmark unpubl.
[c] Coutinho (1987).
[d] Dame & Kenny (1986).

Source: From Dame, R.F., Spurrier, J.D., Williams, T.M., Kjerfve, B., Zingmark, R.G., Wolaver, T.G., Chzanowski, T.H., McKellar, H.N., and Vernberg, F.J., *Mar. Ecol. Prog. Ser.*, 72, 160, 1991. With permission.

marshes on the east coast of the U.S.A. Borey et al. (1983) documented TOC exports of +25 g C m^{-2} yr^{-1} from Coon Creek, a high marsh in Texas, United States, while Jordan and Correll (1991) reported a similar export of +49 g C m^{-2} yr^{-1} from a high marsh in the Rhode River, Maryland, U.S.A. These fluxes, however, are an order of magnitude smaller than for most low *S. alterniflora* marshes.

The TOC exports from the Karienga marsh amounted to about 6% of its macrophyte production, and is the lowest percentage reported in Table 3.22. It is similar to the 13% for Coon Creek (Borey et al., 1983), but is much smaller than the 21 to 24% (Dame et al., 1991b; Williams et al., 1992) and 43% (Dame et al., 1986) reported for low marshes. This suggests that for high marshes a relatively large proportion of the macrophyte production is "consumed" by the marsh by respiration or burial processes. A carbon budget for the Karienga marsh indicated that the respiration was the larger of the two pathways of carbon consumption. Respiration by marsh sediments (+189 g C m^{-2} yr^{-1}) and to a lesser extent by marsh crabs (+27 g C

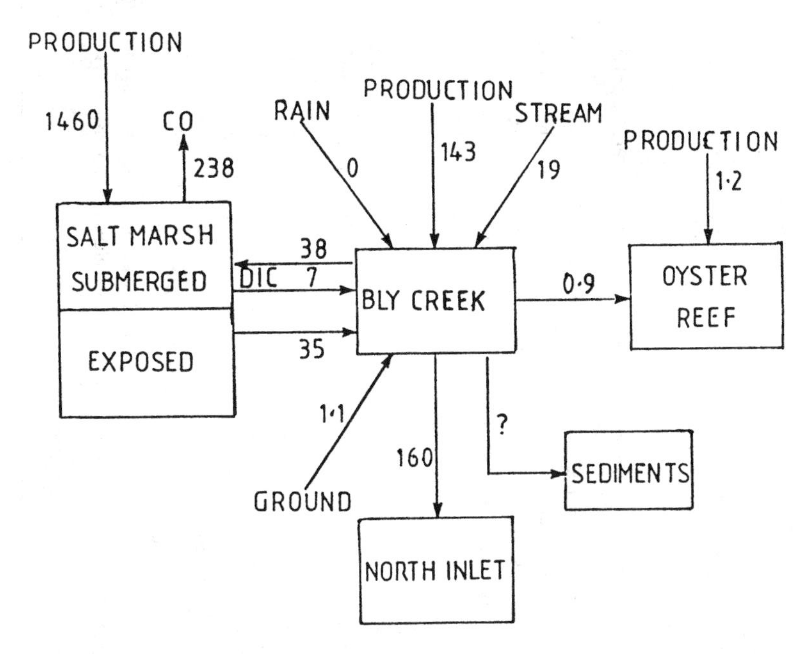

FIGURE 3.30 Carbon fluxes for the Bly Creek system, North Inlet. All production values are for net primary production adjusted for area. Total system area is 6.612×10^5 m^{-2}. Redrawn from Dame, R.F., Spurrier, T.R., Williams, T.M., Kjerfve, T.H., McKellar, H.N., and Vernberg, F.J., *Mar. Ecol. Progr. Ser.*, 72, 160, 1991. With permission.

TABLE 3.22
Summary of Organic Carbon Fluxes from Salt Marshes, and Comparison of TOC Fluxes with Net Aerial Primary Production

System	Organic Carbon Flux			Net Aerial Primary Production ($g\ C\ m^{-2}\ yr^{-1}$)	TOC Export as % of Net Aerial Primary Production	Form of Marsh	Source
	DOC	POC	TOC				
Ware Creek, Virginia, USA	+80	+35	+115	599	19	*Spartina alterniflora*	Axelrad et al. (1976)
Carter Creek, Virginia, USA	+25	+116	+142	599	24	*Spartina alterniflora*	Axelrad et al. (1976)
Flax Pond, New York, USA	+8	–61	–53	372	na	*Spartina alterniflora*	Woodwell et al. (1979)
Canary Creek, Delaware, USA	+104	+55	+159	252	63	*Spartina alterniflora*	Roman and Daiber (1989)
Coon Creek, Texas, USA	+21	+4	+25	550–900	3–5	*Spartina patens* *Distichlis spicata*	Borey et al. (1983)
High Marsh, Rhode River, Maryland, USA	+43	+14	+57	nd	nd	*Spartina patens*	Jordan et al. (1983)
			+49			*Distichlis spicata*	Jordan and Correll (1991)
Ems-Dollard, Marsh, The Netherlands	+15	–140	–125	500	na	*Puccinellietum maritima* *Spartina anglica*	Dankers et al. (1984)
North Inlet, South Carolina, USA	+328	+128	+456	1059	na	*Spartina alterniflora*	Dame et al. (1986)
Bly Creek, South Carolina, USA	+250	–31	+219	1028[a]	21	*Spartina alterniflora*	Williams et al. (1992)
	+272	–30	+242	1028[a]	24		Dame et al. (1991b)
Kariega Marsh, South Africa	+13[a]	+3[a]	+16[a]	200–300	5–8	*Spartina perennis* *Chenolea diffusa*	Taylor and Allanson (1995)

Note: Because patterns of fluxes of DOC and POC can be different, and certain studies measured only one, only the data from those studies that measured both components are included, nd: not applicable; – values net imports: + values net exports.

[a] Area-weighted aerial production values for tall, medium, and short forms of *Spartina;* areas from Dame et al. (1991).

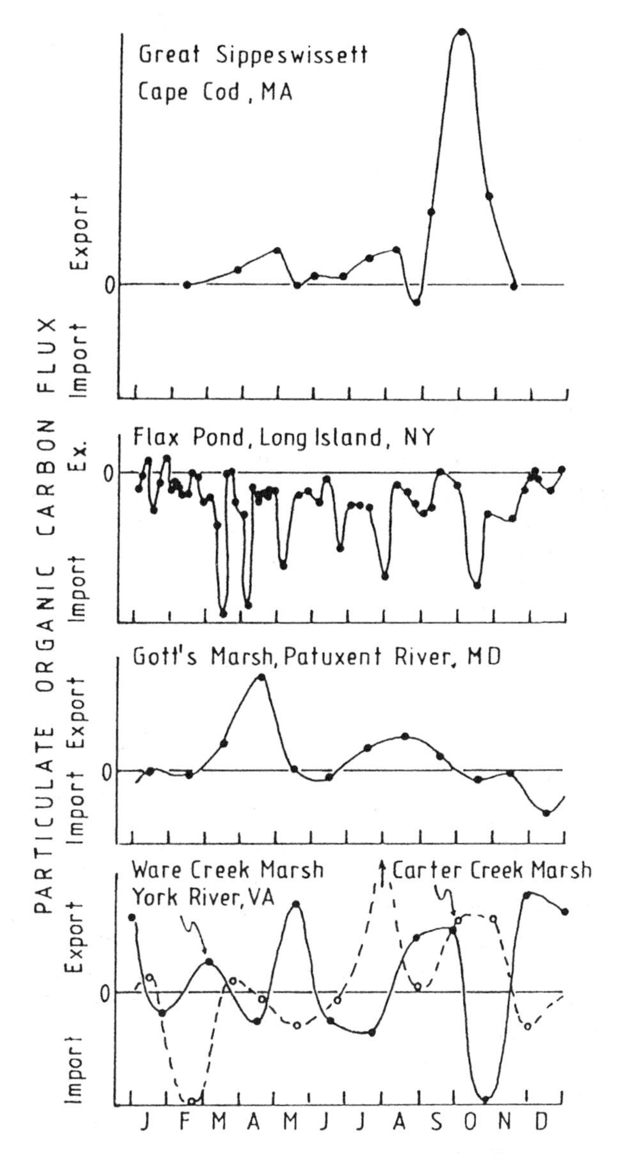

FIGURE 3.31 Time-series plots of the relative magnitude (within each data set) of the export and import of particulate organic carbon over an annual cycle reported for several marshes along the Atlantic coast of the U.S. The measurements do not include nekton, large "rafts" of *Spartina*, or bed load transport. In some cases the original paper did not present a plot of the data, and in others the data were displayed in different forms (linear interpolations or polynomial regressions). Data for Great Sippewissett from Valiela et al. (1978); for Flax Pond from Woodwell and Whitney (1977), Woodwell et al. (1977), and Woodwell et al. (1979); for Gott's Marsh from Heinle et al. (1976); for York River marshes from Moore (1974). (Redrawn from Nixon, S.W., in *Estuarine and Wetland Processes*, Hamilton, P. and MacDonald, K., Eds., Plenum Press, New York, 1980, 487. With permission.)

m^{-2} yr^{-1}) accounted for 70% of the macrophyte production. This value is less than half of the approximately 76% calculated to be available for export from the low elevation Bly Creek marsh (Dame et al., 1986; 1991b). The difference between the two marsh types is probably a function of their different elevations and degree of water exchange with adjoining systems. The high marshes are inundated only at spring tides. This infrequent inundation means that they do not support high benthic microalgal populations. Primary production of benthic microalgae has been shown to be correlated with sediment water content (Sullivan and Moncreif, 1980). In addition the plants on the high marshes form dense canopies that have a greater shading effect than that of the low marsh *S. alterniflora.* They also trap detached material in contrast to the vertical canopies of the low marsh (Borey et al., 1983) facilitating decomposition within the marsh rather than in the adjoining estuary.

Chambers et al. (1985) summarizes data on the concentrations of particulate and dissolved organic carbon in the Duplin River and Sapelo Island marshes, the tidal exchange of POC and DOC in the marsh, the standing stock and movement of *S. alterniflora* wrack, and the removal of carbon from the surface of the marsh by rain. Figure 3.32 is a diagramatic representation of the carbon cycle in the Sapelo Island marsh. Carbon enters the system as CO_2 fixed by *S. alterniflora* (and a smaller amount fixed by algae). Anaerobic degradation releases both CO_2 and methane. Material is eroded from the surface of the marsh and washed into the creeks and the upper Duplin River by rain. Some of this material diffuses downstream or is flushed out by severe rainstorms, leaving as exported POC, DOC, and some algae and other microorganisms. The bulk of the material, however, is picked up on flood tides and deposited on the marsh. In the course of this cyclic deposition and erosion, aerobic degradation can release CO_2 from the water and allow incorporation of carbon into mobile migrant organisms that can leave the system. Organisms that feed in the marsh during high tide and then return to tidal creeks and rivers during ebb tide may move substantial amounts of carbon off the marsh in their guts, and then release much of it into the water as feces. In addition, shrimp that feed in the marsh and tidal creeks as juveniles move out of the system as adults, accounting for the removal of some carbon. Female blue crabs that leave the creeks and rivers to spawn seldom return. This is another source of carbon removal.

3.5.6 Mangrove Systems

3.5.6.1 Introduction

Soft sediment tropical and subtropical coasts are characterized by dense growth of shrubs and trees usually referred to as mangrove swamps or mangrove forests. The term "mangrove" refers to two different concepts. First it describes an ecological group of halophytic shrubs and tree species belonging to some 12 genera in eight different families of plants (Waisel, 1972). In another sense the term refers to the complex of plant communities fringing trop-

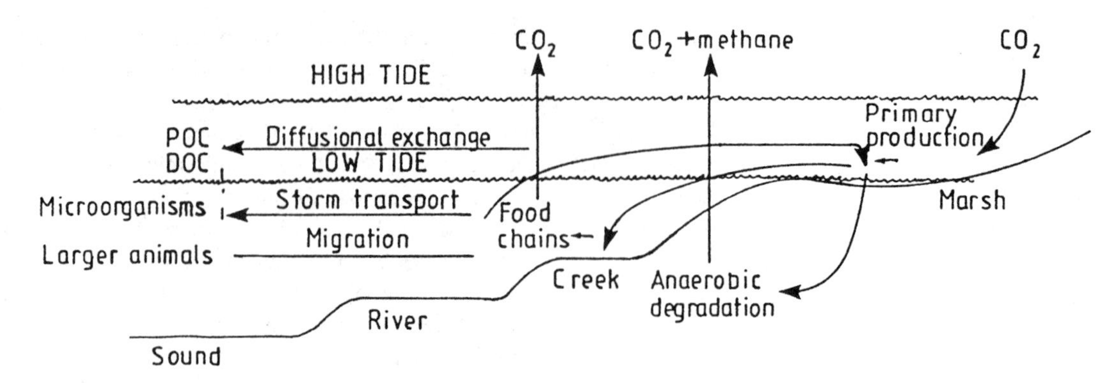

FIGURE 3.32 Diagrammatic representation of the carbon cycle in the Sapelo Island, Georgia, salt marshes.

ical and subtropical shores. Chapman (1977) prefers to refer to these communities or formations as "mangals" (MacNae, 1968; Chapman, 1977), reserving the term mangrove for the individual species or genera. Here the term mangrove system or mangrove ecosystem will be used.

Mangrove systems occur generally between 25°N and 25°S. On the east coasts of Africa, and in Australia and New Zealand they extend 10 to 15° further south. Mangroves (*Avicennia marina*) grow as far south as Ohiwa Harbour (30° 01'S) in New Zealand and Corner Inlet, Victoria (38° 55'S) in Australia. In the United States black mangrove shrubs occur along the northern coast of Mexico at 29°N. The most luxuriant and diverse mangrove forests occur in Southeast Asia and northern Australia where no fewer than 36 species of true mangroves occur. Biogeographers recognize two main group of mangroves: the Old World Group with about 60 species and the New World Group with only about 10 species. It is believed that the center of evolution of mangroves was in the Indo-Malayasian area from which they spread to other regions.

For reviews of the ecology of mangroves, readers are referred to MacNae (1968), Lugo and Snedaker (1974), Walsh (1974) Chapman (1976b; 1977), Clough (1982), Teas (1983), Robertson and Alongi (1992), and Twilley et al. (1996).

3.5.6.2 Distribution and Zonation

Mangrove systems range from the complex, species-rich assemblages such as those found on the west coast of Malaysia, which may contain as many as 20 mangrove species, to ones composed of a single species such as the New Zealand *Avicennia marina* community. They grow from the highest level of spring tides down to about mean sea level and are characteristically found on accreting shores. MacNae (1968) and Saenger et al. (1977) give many examples of the zonation patterns of mangrove forests. On some gently sloping tropical shores, the mangrove system may extend up to 5 km from the seaward edge and may cover many thousands of hectares in a single estuary.

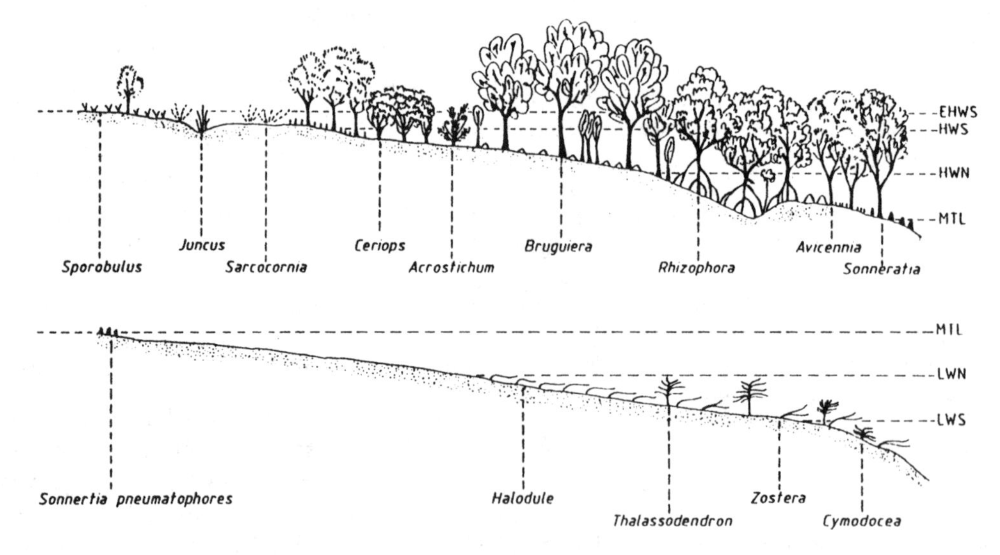

FIGURE 3.33 Zonation on a mangrove shore in Morrumbene Estuary, Mozambique, east coast of southern Africa. (Redrawn from Day, J.H., in *Estuarine Ecology with Special Reference to Southern Africa,* Day, J.H., Ed., A.A. Balkema, Rotterdam, 1981d, 90. With permission.)

A typical zonation pattern is illustrated by the mangrove shore on Morrumbene Estuary in Mozambique on the east coast of southern Africa (Figure 3.33). From extreme high water spring tide (EHWS) to mean high water spring tide (MHWS), there is a diverse and variable assemblage of halophytes dominated by *Sporobolus virginicus*, the glasswort, *Sarcocornia,* and the rush, *Juncus krausii.* The region between MHWS to mean tide level (MTL) is covered by a mature forest of mangroves. It includes an upper fringe of *Avicennia marina*, thickets of *Ceriops tagal*, then *Bruguiera* and *Rhizophora mucronata* in seepage channels, and a lower fringe of *Avicennia* and *Sonneratia alba.* The genera *Bruguieria* and *Rhizophora* can be distinguished by their prop roots, especially the latter species in which the roots emerge from the trunk high above the ground and arch downwards, forming a dense, almost impenetrable tangle that traps sediments. The genus *Avicennia* has numerous breathing roots or pneumatophores, which grow vertically upwards from the underground root system. The pneumatophores, prop roots, and the boles of the trees are covered with moss-like growths of algae such as *Bostrychia* and *Calloglossa.*

Lugo and Snedaker (1974) classified the mangrove systems of south Florida into six types (Figure 3.34), depending on the specific situation in which they grow. Four species of mangrove (*Rhizophora mangle, Avicennia germinans, Laguncularia racemosa,* and *Conocarpus errecta*) occur in these in varying mixtures. The formation and physiognomy of these six types are apparently strongly controlled by the local pattern of tides and terrestrial surface drainage. This classification, with local variations, seems to apply the mangrove systems generally.

In the more diverse Malaysian mangrove systems, Watson (1928) recognized five zones based on the frequency of inundation. Beginning at the lowest level these are: (1) species growing on sediments *flooded at all times*:

FIGURE 3.34 The six mangrove community types. (Redrawn from Day, J.W., Jr., Hall, C.A.S., Kemp, W.M., and Yanez-Aranciba, A., *Estuarine Ecology,* John Wiley & Sons, New York, 1987, 194. As modified from Lugo and Snedaker, 1974. With permission.)

no species normally thrive under these conditions but *Rhizophora mucronata* may; (2) species on sediments *flooded at medium tides*: species of *Avicennia, Sonneratia griffithii*, and bordering rivers, *Rhizophora mucronata*; (3) species on sediments flooded at normal high tides: most mangroves, but species of *Rhizophora* tend to be dominant; (4) species on sediments flooded by spring tides only: *Bruguiera gymnorphiza*, and *B. cylindrica*; and (5) species on land flooded by equinoxal or other exceptional tides only: *B. gymnorphiza* dominant, but *Rhizophora apiculata* and *Xylocarpus granatum* may coexist.

Most of the ecological literature on mangroves is concerned with descriptions of species composition and plant zonation patterns and is heavily weighted toward the classical successional view (Chapman, 1970l; 1974; 1976a,b; Walsh, 1974). This model emphasizes biotic processes including sediment accumulation and plant community changes from a pioneer through to a climax stage. The schemes proposed were based on the frequency of inundation, and salinity of the dominant tree species.

3.5.6.3 Environmental Factors

In recent years, in contrast to the older successional model, an alternative one has been advanced that seems to better fit the observed facts (Thom, 1967; 1974; 1981; 1982). This alternative model views mangroves as opportunistic species, colonizing available substrates. In this context mangroves zonation patterns are primarily seen as the result of the ecological responses of the species to external conditions of sedimentation, microtopography, estuarine hydrology, and geochemistry (Thom, 1982).

According to Thom (1982) there are three major components of the environmental setting of any locality in which mangroves occur: geophysical, geomorphic, and biologic. The first of these includes a variety of physical forces such as sea level change, climate, and tidal factors. The second component is essentially the product of the physical forces. It includes factors such as the depositional environment, the extent to which this is dominated by wave, tidal, or river processes, and the impact of the microtopography of particular landforms (e.g., river levees, beach-ridge swales) on plant establishment, growth and regeneration. Land surface elevation, drainage, and stability, in combination with substratum or sediment properties (texture, composition, structure, etc.), nutrient inputs, and the salinity regime, combine to produce environmental gradients within the coastal region. Different species, or different ecotypes within a species, according to their physiological responses to the above factors, will establish themselves where they find the combination of conditions favorable. Thus there are established species composition and distribution patterns as well as gross and net productivity along environmental gradients. For example, Lugo et al. (1975) reported differential responses in the *in situ* gas exchange characteristics of three mangrove species distributed along a salinity gradient in Florida. Thus "mangrove distribution (zonation?) can be viewed as an opportunistic response of certain species to more or less favored changing environmental conditions whose characteristics within a region are primarily controlled by past and present geomorphic processes" (Thom, 1982).

Oliver (1982) has discussed the role of environmental factors, with special reference to climate on the development of mangrove systems. Figure 3.35 from Oliver summarizes the interrelationships of the various factors. It can be seen that the system is characterized by a number of feedback loops, e.g., the establishment of a mature, well-developed mangrove community has a modifying influence on the very climate and other factors that brought about its initial development. From the diagram it can also be seen that climate has many indirect effects on mangrove communities, particularly through its influence on the nature and scale of the geomorphic process outlined above, as well as on the complex relationships of the sediment biogeochemistry.

3.5.6.4 Adaptations

Mangroves live rooted in a saline anaerobic substrate; abundant salt is usually toxic to woody plants, and oxygen is necessary for root respiration, so the question is, how do these plants persist, grow, and reproduce successfully? In addition, the plants must cope with periodic fluctuations and extremes of the physicochemical parameters of their environment. Saenger (1982) has reviewed the morphological, anatomical, and reproductive adaptations of mangroves, while Clough et al. (1982) have done likewise for their physiological processes.

1. *Leaf adaptations*: The leaves of most mangroves exhibit a range of xeromorphic features, such as a thick-walled epidermis, thick waxy cuticles, a tomentum of variously shaped hairs, sunken stomata, and the distribution of cutinized and sclerenchymatous cells throughout the leaf. These are xeric characters that have developed in response to the physiological dryness of the environment.
2. *Salt regulation*: As a group, mangroves do not appear to be obligate halophytes, as many species grow well in freshwater. However, most species grow best at salinities over the range from freshwater to normal seawater (Clough et al., 1982). As they normally grow in saline environments they absorb sodium and chloride ions, and they have evolved various physiological adaptations to control the uptake and concentration of the ions in their tissues (Walsh, 1974). The wide range of adaptations include the

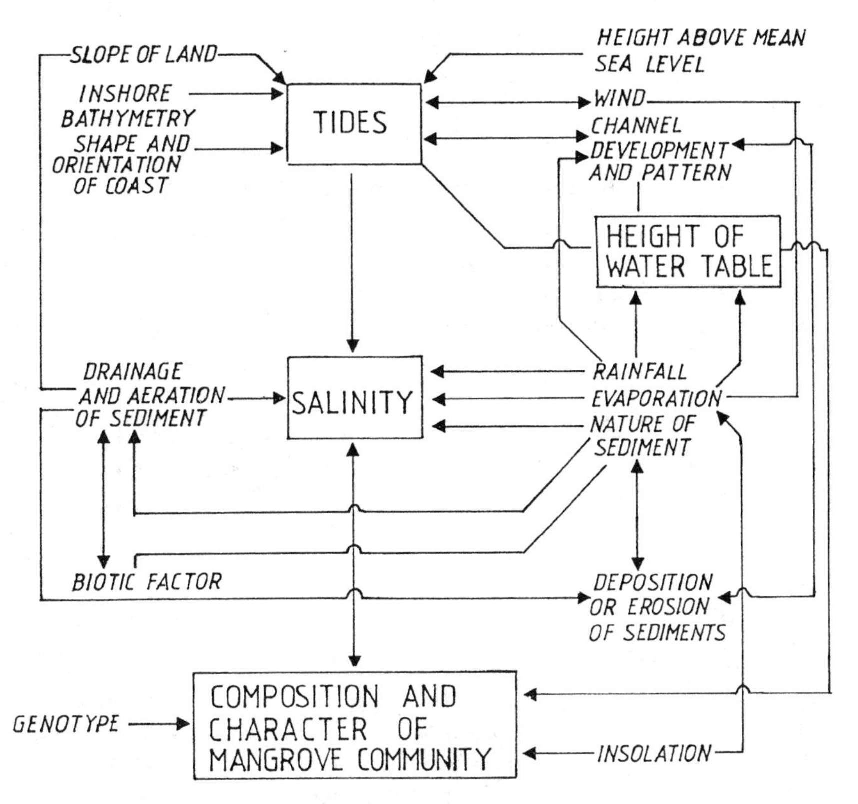

FIGURE 3.35 Interrelationships of environmental factors in mangrove ecosystems. (Redrawn from Oliver, J., *Mangrove Ecosystems in Australia,* Clough, B.F., Ed., Australian National University Press, Canberra, 1982, 29. With permission.)

capacity of the roots to discriminate against NaCl (Chapman, 1976b), the possession by some species of salt-secreting glands in the leaves (Clough et al., 1982; Saenger, 1982), the accumulation of salt in the leaves and bark (Chapman, 1976b), and the loss of salt when leaves and other organs are shed.

3. *Sediment adaptations*: Adaptations that enable mangroves to overcome the problems of anaerobic sediments and of anchoring the plants in often semifluid sediments include the sometimes complicated root systems with their great diversity of form and function and the almost universal presence of aerenchyma and lenticels. Below the sediment surface, all mangroves possess a system of laterally spreading cable roots with smaller descending anchor roots. The root system generally penetrates less than 2 m and tap roots have not been reported (Walsh, 1974). Despite the shallowness of the root system, the belowground to aboveground biomass (root:shoot ratio) is higher than that of other vegetation types. Most species of mangroves possess, in varying combinations, the following array of aboveground root types (Percival and Womersley, 1975): (1) surface cable roots (*Aegialtis, Excoecaria*); (2) pneumatophores, negatively geotrophic, unbranched (*Avicennia, Xylocarpus*), or branched (*Sonneratia*) arising from the cable root system; (3) knee roots, modified sections of the cable roots with a period of negative-geotrophic followed by a period of positive-geotrophic growth (*Bruguiera, Ceriops, Lumnitzera*); and (4) stilt roots, positively geotrophic arching (*Rhizophora*), or straight (*Ceriops*), generally branched roots arising from the trunk and growing into the substrate. The evidence that these structures are adaptations for subterranean root aeration comes from a variety of sources. Firstly, those mangroves growing at lower tidal levels, and consequently the most frequently inundated, tend to posses the greatest array of root types. Secondly, the presence within the aboveground roots of parenchymatous tissue, and the widespread occurrence of numerous lenticels. The mechanism of air intake by these roots through the development of a negative gas pressure has been investigated by Scholander et al. (1962).
4. *Reproductive adaptations*: In several genera the fruits contain seeds that develop precociously while still attached to the tree. Some species, e.g., *Bruguiera, Ceriops*, and *Rhizophora*, are viviparous in that the embryo ruptures the peri-

TABLE 3.23
Mangrove Forest Biomass

Locality	Latitude	Mangrove Species	Aboveground Biomass g dry wgt m^{-2}	g Cm^{-2}	t ha^{-1}	Reference
Panama	9°N	*Rhizophora brevistyla*	27,921	11,168	279.2	Golley et al. (1975)
Thailand	8°N	*Rhizophora apiculata*	—	—	159.0	Christensen (1978)
Puerto Rico	18°N	*Rhizophora mangle*	6258	2514	62.9	Golley et al. (1962)
Phillipines	12°N	Mixed mangrove species	—	—	45.9	de la Cruz and Banaag (1967)
Florida	26°N					
Overwash		*Rhizophora mangle*	11,958	4783	124.6	Lugo and Snedaker (1974)
Riverine		*Rhizophora mangle*	9281	3928	118.9	Lugo and Snedaker (1974)
Riverine		*Rhizophora mangle*	17,390	6956	136.0	Lugo and Snedaker (1974)
Succession		*Laguricularia racemusa*	812	324	49.0	Lugo and Snedaker (1974)
Japan	24°N	*Rhizophora mucroriata*	—	—	108.1	Suzuki and Tagawa (1983)
		Briguiera gymnorhiza	-	—	97.6	Suzuki and Tagawa (1983)
		Rhizophora + *Briguiera*	—	—	78.6	Suzuki and Tagawa (1983)
Australia						
Sydney	34°N	*Avicennia marina*	—	—	128.4	Briggs (1977)
Westernport Bay		*Avicennia marina*	8600	3440	86.0	Clough and Attiwell (1975)
New Zealand	37°S					
Upper Waitemata Harbour		*Avicennia marina*	3887	1556	—	Knox (1983b)
Tuff Crater		*Avicennia marina* (tall)	—	—	104.1	Woodroffe (1982a)
Tuff Crater		*Avicennia marina* (short)	—	—	6.8	
Lucas Creek		*Avicennia marina*	5619	2248	—	Knox (1983b)

carp, and grows beyond it, sometimes to a considerable extent (MacNae, 1968). In other species, e.g., *Aegialtis, Avicennia*, and *Langvincularia*, the embryo, while developing within the fruit, does not enlarge sufficiently to rupture the pericarp; these genera are termed cryptoviviparous. Saenger (1982) has reviewed the significance of vivparity in mangroves. Its adaptive significance could include rapid rooting, salt regulation, ionic balance, the development of buoyancy, and nutritional parasitism. However, many apparently successful mangrove species (e.g., *Osbornia, Sonneratia, Lumnitzera, Xylocarpus*, and *Excoecaria*) do not possess viviparous fruits. The propagules of all mangroves are buoyant and are adapted to water dispersal.

3.5.6.5 Biomass and Production

Biomass: In 1974 Lugo and Snedaker (1974) reviewed the then available data on the standing stocks of mangroves from a range of localities. In Table 3.23 their data has been augmented by more recent studies. The biomass data show considerable variability, but nevertheless some general trends can be seen, e.g., tropical forests, with the greatest range of species and tree height, have the highest aboveground biomass. The variability in the data is a reflection of the age, stand history, and structural differences, as can be seen in the Florida forests (riverine, overwash, succession).

Belowground biomass data are available from only a limited number of localities. Unlike temperate forests where roots seldom make up more than 20% of the total biomass, mangroves have relatively high root:shoot ratios varying from –.20 to 1.73 (Lugo and Snedaker, 1975; Clough and Attiwell, 1975). The reasons for such high ratios has been discussed above.

Primary production: With few exceptions (Bunt et al., 1979) net primary production of mangroves has been estimated from measurements of photosynthesis and respiration of individual leaves or small branches (Golley, et al., 1962; Lugo and Snedaker, 1975). The estimates derived from this method generally fall within the range that might be expected for a woody plant with the C_3-pathway of carbon fixation. The studies, however, usually do not include measurement of the respiration of parts of the plants other than leaves. Lugo et al. (1975) suggest that 4 to 10% of gross primary production might be lost via respiration from stems and surface roots. Values for gross primary productivity range from 1.4 g C m^{-2} day^{-1} to 13.9 g C m^{-2} day^{-1}, or 10.7 t ha^{-1} yr^{-1}.

3.5.6.6 Litterfall

Mangrove litter comprises leaves, twigs, inflorescences and fruits, and forms the primary source of organic detritus in tropical estuaries. Gill and Tomlinson (1969) in their studies of red mangrove in southern Florida, found that flowering, fruit formation, and leaf fall occurred at measurable rates in all seasons. However, peak rates of leaf fall occurred during the summer months when air temperatures and incident light were at their annual peaks. The following are some case history studies of litterfall.

Tuff Crater, New Zealand: Woodroffe (1982a,b; 1985) investigated litterfall in Tuff Crater near the southern latitudinal limit of mangroves in New Zealand. Here the sole species, *Avicennia marina* var. *resinifera*, adopts two distinct growth forms, taller tree-like mangroves up to 4 m tall along the banks of the tidal creek, and low stunted shrub mangroves generally less than 1 m tall on the mudflats. Average aboveground biomass for the taller mangroves was estimated to be 104 t ha^{-1} and for the lower 6.8 t ha^{-1}. While the value for the taller mangroves was similar to those reported for more complex tropical mangroves, the fact that 94% of the basin was covered by low, generally sparse mangroves resulted in an estimate for the 21.6 ha crater of 153 tons, an average of 7.6 t ha^{-1}. Litterfall beneath the tall mangroves was estimated at 7.6 ± 2.5 t ha^{-1} yr^{-1}, and 3.3 ± 0.5 t ha^{-1} yr^{-1} beneath the short mangroves. The value for the taller mangroves was similar to that reported for mangroves in many other parts of the world, but because of the extensive low sparse mangroves, the total for the basin was estimated at 2.7 t ha^{-1} yr^{-1}. Woodroffe (1985) also found that 78% of the total litterfall beneath the tall mangroves (2 to 4 m) and 69.2% beneath the low mangroves (less than 1.5 m tall) were recorded during the months of December to April (Figure 3.36). Storm events can have a marked impact on mangrove litter

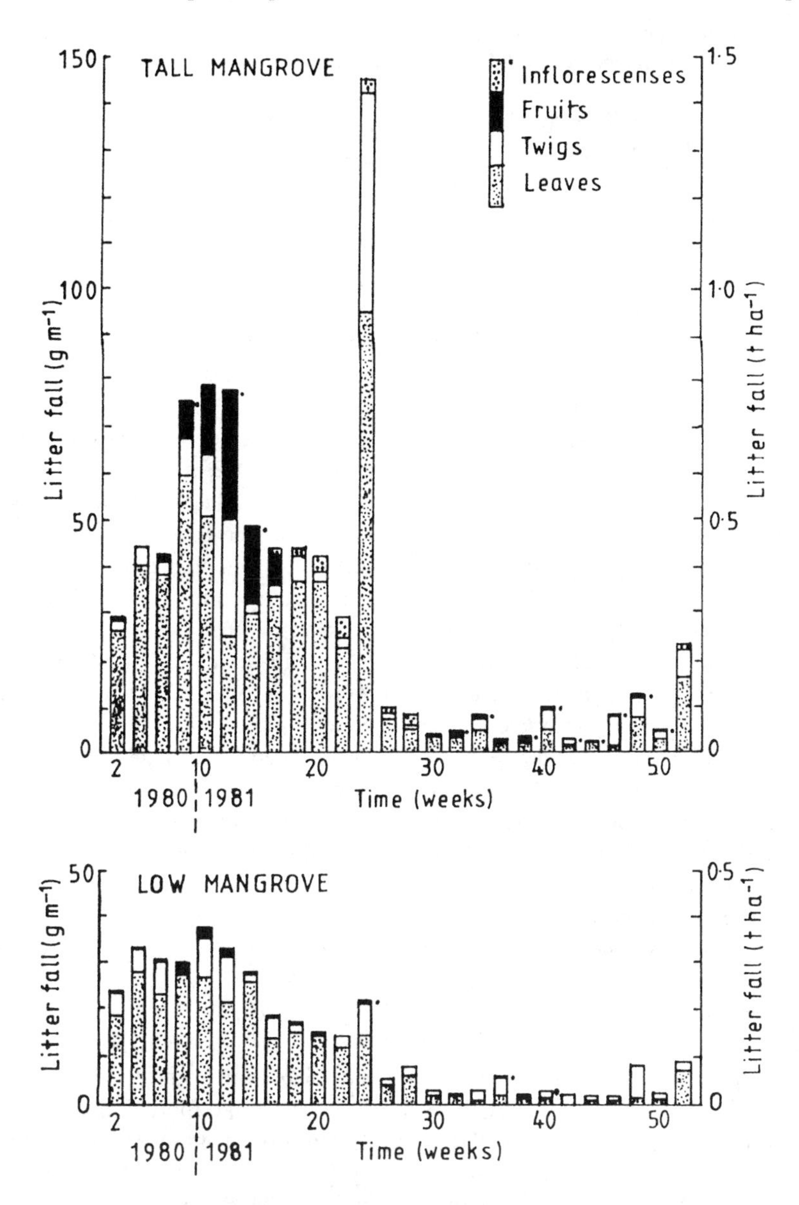

FIGURE 3.36 Litterfall by components beneath tall and short mangroves in Tuff Crater, New Zealand (4 November 1980–2 November 1982). (Redrawn from Woodroffe, C.D., *N.Z. J. Mar. Freshw. Res.*, 16, 182, 1982b. With permission.)

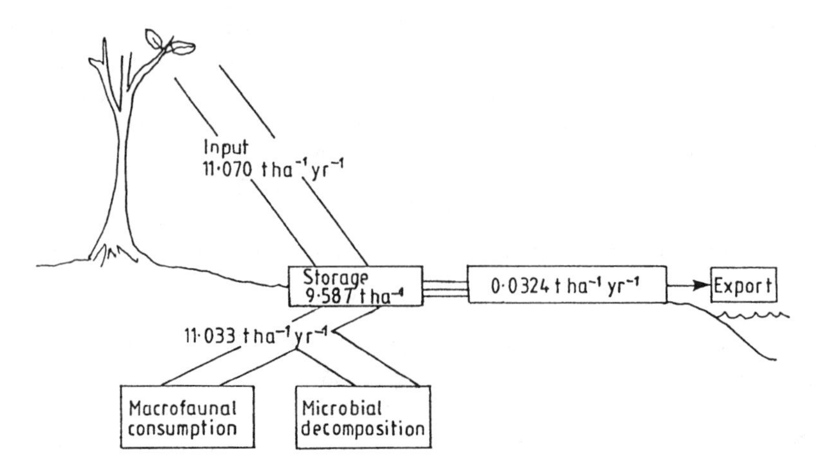

FIGURE 3.37 A schematic representation of the cycling of litter from the *K. candel* mangroves of tidal shrimp pond at Mai Po. (Redrawn from Lee, S.J., *Est. Coastal Shelf Sci.*, 29, 85, 1989. With permission.)

production (Goulter and Alloway, 1979). A storm in Tuff Crater in April 1981 with wind speeds reaching 113 km hr^{-1} accounted for a particularly large litterfall when over a period of a week 18% of the annual total was recorded.

Hong Kong tidal shrimp pond: Lee (1989) investigated production and turnover of the mangrove, *Kandelia candel*, in a tidal shrimp pond (10 ha) at Mai Po Marshes, northwest Hong Kong for a period of two years (Figure 3.37). Litter production averaged 11,070 t ha^{-1} (equivalent to 4.880×10^7 kcal ha^{-1}), with wood and leaf materials contributing, respectively, 56.5 and 53.94% of the total. The reproductive plus frass component contributed 40.69%. The litterfall pattern was bimodal, with high inputs in spring (February to June) and late summer (August to November). The mean standing litter biomass recorded during the study was 9.587 ± 0.556 t ha^{-1} (4.969×10^7 kcal ha^{-1}), with respective contributions by the three components of 19.07, 432.26, and 38.65%. Residence times (standing biomass/production) of the three components were extraordinarily long, and were estimated to be, respectively, 980, 300, and 252 days. Residence times were related to both inundation frequency and crab (*Chiromanthes* spp.) consumption. As the *Kandelia* stands are largely located above the mean high-water mark, there was little export of litter. The litter production was largely decomposed or consumed by the macrofauna *in situ* creating a large energy sink, which was not coupled to the adjacent waters.

Figure 3.37 is a schematic representation of the cycling of litter within the system. This represents the energy flow pattern typical of landward mangrove communities. Due to the marked reduced inundation frequency and duration in such forests, and *in situ* litter accumulation and turnover, low export is characteristic of such systems, resulting in internal cycling energy pathways. In addition such mangrove stands are subject to continuous accretion due to the dominance of a mainly depositional sedimentary regime. This contrasts with riverine and fringing mangrove systems, which are more open systems with much greater tidal exchange and higher rates of export.

Frequent inundation hastens litter turnover in two important ways. First, tidal inundation facilitates the export of litter, especially the freshly fallen components (Boto and Bunt, 1981; Twilley, 1985). Second, a moist environment accelerates disappearance rates by enhancing microbial decomposition and promoting consumption by amphipods, gastropods, and brachyurans (Twilley et al., 1986). In the Mai Pond system the daily export is very low, estimated to be 1.024×10^{-2} g dry wgt m^{-2}, equivalent to about 0.34% of the daily litter production.

Litter production estimates for a range of mangrove systems are given in Table 3.24. They range from 32 g m^{-2} for succession forest in Florida to 10,210 g m^{-2} for a Panamian forest. The contribution of leaf litter to total litterfall has been calculated to be 6.7 t ha^{-1} yr^{-1} in Thailand (8°) (Christensen, 1978), 4.75 t ha^{-1} yr^{-1} in Puerto Rico (Golley et al., 1962), and 5.62 t ha^{-1} yr^{-1} in Tuff Crater, New Zealand (Woodroffe, 1985). Total litterfall has been investigated for a range of mangrove species on Hinchinbrook Island, Great Barrier Reef, Australia (18°S) (Bunt, 1982). Rates of total annual litterfall on a yearly basis ranged from 365 to 2,810 g m^{-2} yr^{-1}, with a mean of 2,280 g m^{-2} yr^{-1}. Rates also varied between the different species.

3.5.6.7 The Fate of Mangrove Leaf Litter

Early models of mangrove ecology built on findings from Florida mangrove ecosystems were strongly influenced by the salt marsh research along the Atlantic coast of North America. In the Florida model, the production of organic detritus from mangroves was considered to be little affected by biotic processes such as herbivory (Heald, 1971): the bulk of the mangrove detritus being exported via tidal action to support consumers in the adjacent estuarine waters (Odum and Heald, 1972; 1975). However, recent research on non-Florida mangroves (Robertson and Alongi,

TABLE 3.24
Mangrove Forest Litter Production

Locality	Latitude	Mangrove Species	Litter Production g dry wgt m^{-2}	Litter Production g Cm^{-2}	Reference
Thailand	8°N	*Rhizophora apiculata*	670	268	Christiansen (1978)
		Rhizophora apiculata	340	—	Poorvachiranan and Chansang (1982)
Panama	9°N	*Rhizophora brevistyla*	10,210	4084	Golley et al. (1975)
Malaysia	12°N	*Rhizophora apiculata*	1390	—	Sasekumar and Loi (1983)
India	15°N	*Rhizophora officiralia*	1020	—	Wafar et al. (1997)
		Rhizophora apiculata	1180	—	Wafar et al. (1997)
		Rhizophora macronala			
		Sonnerata alba	1700	—	Wafar et al. (1997)
Puerto Rico	18°N	*Rhizophora mangle*	1399	559	Golley et al. (1962)
Florida	26°N				
Overwash		*Rhizophora mangle*	4295	1718	Lugo and Snedaker (1974)
Riverine		*Rhizophora mangle*	3393	1357	Lugo and Snedaker (1974)
Succession		*Rhizophora mangle*	32	13	Lugo and Snedaker (1974)
Equator	2.25°N	*Rhizophora mangle*	1054	—	Twilley et al. (1997)
		Rhizophora hurrisonii			
Australia					
Queensland	20°S	*Avicennia marina*	960	384	Bunt (1982)
Sydney	341S	*Avicennia marina*	580	332	Briggs (1977)
Westernport Bay	38°S	*Avicennia marina*	260	80	Clough and Attiwell (1975)
New Zealand	37°S	*Avicennia marina* (tail)	810	324	Woodroffe (1982b)
		Avicennia marina (short)	365	146	Woodroffe (1982b)

1992; Robertson et al., 1992; Lee, 1995; 1997; 1999) has led to a reappraisal of the nature and diversity of forces shaping mangrove ecosystem structure and function.

As we have seen, mangrove systems produce considerable quantities of litter. The question then is what is the fate of this litter? As discussed above a percentage of the litter will be exported (Table 3.25). Recent studies on tropical mangroves have demonstrated that sesarmine crabs play a critical role in the fate of mangrove litter. Robertson (1986) estimated that these crabs (*Sesarma* spp.) could remove >28% of the litter production in mixed *Rhizophora* forests. This finding was later backed by similar findings from southeast Asia (e.g., Lee, 1989; 1997), Africa (e.g., Emmerson and McGwynne, 1992; Steinke et al., 1993), and Australia (Micheli, 1993). Recent studies in Latin America also suggested that crabs consume much of the mangrove leaf litter (Twilley et al., 1986; 1997). Jaspar (1989) conducted a detailed study on the fate of leaf litter in a Malaysian mangrove area of tidal inundation classes III and IV (Watson, 1928), that is, areas that are inundated by normal high and spring tides, respectively. The "disappearance" of leaf litter was partitioned as follows: tidal export, which accounted for 36 to 78% (22 to 26% being leached out from the leaves and the rest by particulate export); macrofeeders (mainly sesarmid crabs) activity, depending on the population density, accounted for 10 to 48%; and microbial activity accounted for 0 to 20%. Of the 10 to 48% biomass consumed by crabs, only 14% was converted to crab biomass, with the rest being used in metabolic processes or returned to the system through crab feces (67%). Some of this fecal material would eventually be exported from the mangroves. The above values are comparable to those obtained in the Coral Creek mangroves in Queensland, which have similar inundation characteristics, that is, inundation at most high tides (Robertson, 1987). Robertson (1986) found that 28% (range 22 to 42%) of the annual leaf litter was removed by crabs. These figures are within the range of 10 to 48% reported by Jasper (1989). Boto and Bunt (1981), working on the same site, estimated the export of macroscopic litter to be 19.5 kg ha^{-1} day^{-1} without taking into consideration crab activity. Robertson (1986) modified Boto and Bunt's (1981) figures to 15.3 kg ha^{-1} day^{-1}, or 63% of the litter fall after taking into consideration removal by crabs. The export of leaf litter was 7.7 kg ha^{-1} day^{-1} or 52% of leaf litter production. From Table 3.26 it can be seen that the percentage consumption of leaf litter by crabs in a number of studies ranged from 9 to 79% with a mean in the order of 50%. Crabs are not the only invertebrates involved in the consumption of mangrove leaf litter. Slim et al. (1997) found that in the relatively elevated *Ceriops tagal* forest in Kenya, which is only flooded during spring tides, crabs were absent and the detritivorous snail *Terebralia palustris* was the major benthic organism responsible for litter removal.

Camilleri (1992) investigated leaf litter processing by invertebrates in a Queensland mangrove forest. Figure 3.38 depicts the leaf litter processing by the guild of detritovores. At least 50 species of invertebrates depend on the

TABLE 3.25
Estimates of Litter Export from Selected Mangrove Ecosystems

Locality	Litter Export (kg ha^{-1} day^{-1})	Reference
Malaysia	12.4	Gong & Ong (1990)
Northeastern Australia	15.3	Boto & Boto (1981) (amended by Robertson, 1986)
Florida	2.5	Lugo & Snedaker (1973)
Florida	8.0	Odum & Heald (1972)
Puerto Rico	11.0	Golley et al. (1962)

senescent leaves that fall from the mangrove trees as a food source. The leaf shredders consist of three feeding groups: (1) large crabs represented by *Sesarme erythrodactyla*, which eats leaves where they fall or drags them into their burrows to be consumed later; (2) amphipods, isopods, and smaller crabs; and (3) polychaetes, especially *Capitellides* sp., and chironomid larvae. These groups feed successively on smaller particles as the leaves are broken down. The role of species that shred whole leaves into small particles in mangrove forests is significant for five reasons (Camilleri, 1992):

1. They prevent mangrove leaf material from being washed out of the forest.
2. They make POM available as a food source to detritovores that feed on fine POM, e.g., a comparison of the length of particles in the rectum of *Sesarme erythrodactyla* (32 to 1 117 μm) with that in the proventriculus of sediment-eating crabs such as *Heloecius cordiformes* (30 to 600 μm), *Australopus tridentata* (53 to 888 μm), and molluscs (<50 to 1,200 μm), suggests that these latter groups feed on detrital particles egested by *S. erythrodactyla*.
3. They regulate the size of POM in the environment.
4. Since the size of POM influences the rate at which it is colonized by microfauna and microorganisms, growth of the microbial community is also influenced by the leaf shredders, since microorganisms enhance the mineralization rate of plant detritus (Fenchel and Harrison, 1976), the processes that enhance rapid leaf breakdown, such as the rate of leaf shredding into small POM, may be significant in making nutrients available to the trees.
5. The plant tissues are broken down to the level of organelles, thus simplifying the structure and chemical composition resulting in the plant cell contents (cellulose and hemicellulose) being freed from degradation-resistant material such as lignin and thus facilitating degradation by microbial organisms.

Lee (1997) investigated the value of crab-consumed and processed mangrove material to coprophagous macrofauna in a tropical mangrove ecosystem and found that crab fecal material supported faster growth and lower mortality in the amphipod *Parayallela* sp. compared to fresh mangrove detritus. The crab fecal material, now finely fragmented and low in deterrent chemical such as tannins, is ideal for microbial colonization. It forms the basis of a coprophagous food chain both in the mangrove benthos and among pelagic consumers (e.g., copepods) upon resuspension and export to adjacent waters as "micro-particulate organic carbon" (micro-POC). Robertson et al. (1992) estimated that "micro-POM" made up about 17.0% of all mangrove export from their study site in Missionary Bay, northeastern Australia.

A number of studies (e.g., Boto and Bunt, 1981; Twilley, 1985; Robertson, 1988; Robertson et al., 1988; Lee, 1995) have determined net export of detritus from mangrove forests with the contribution to adjacent sediments dependent on the ratio of the size of the vegetated area to open ocean area, the geomorphology of the tidal basin, tidal amplitude, and water motion. Robertson et al. (1988) have recently estimated direct export of particulate organic matter from the mangroves on Hinchinbrook Island and its channel on the Queensland coast to adjacent Missionary Bay and the Great Barrier Reef (GRB) lagoon to be

TABLE 3.26
Consumption of Mangrove Litter by Crabs

Locality	Crab Species	% Consumption	Reference
Northeastern Australia	Sesarmides	22–43	Robertson (1986)
Southern Africa	*Sesarma meinerii*	44	Emmerson and McGwynne (1992)
Australia	*Sesarma meinerii*	71–79	Robertson and Daniel (1989)
Malaysia	*Sesarma meinerii*	9–30	Leh and Sasekumer (1985)
Hong Kong	*Chiromanthes* sp.	>57	Lee (1989)
East Africa	Sesarmides	21.7–40.3	Slim et al. (1997)

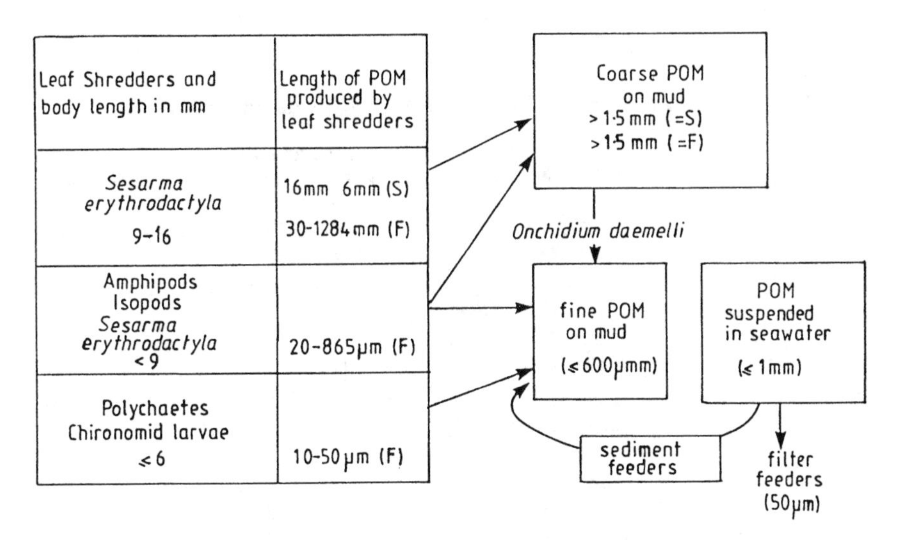

FIGURE 3.38 Leaf Litter processing by a guild of detritovores at Myora Springs, Stradbroke Island, Queensland. S = POM from "sloppy" feeding; F = fecal POM. (Redrawn from Camilleri, J.C., *Mar. Biol.*, 114, 143, 1992. With permission.)

on the order of 25,00 tons C yr^{-1}. However, Robertson et al. (1988) estimate that most of this detritus is deposited over a wide area of the GBR lagoon (260 km^2) giving an annual flux of litter to the adjacent waters of 52 g C m^{-2} yr^{-1}. Alongi et al. (1989) have determined that the mangrove litter exported to the adjacent nearshore waters is of poor nutritional quality (e.g., high C:N ratio, refractory nature of the constituents, and high tannin content), with enhancement of bacterial activity and DOC fluxes apparent only in a semienclosed area of highest litter deposition. Alongi et al. (1989) and Alongi (1990a,b) demonstrated from detritus enrichment experiments that addition of unprocessed mangrove detritus did not promote either bacteria growth, DOC flux, or microfauna abundance in laboratory microcosms.

3.5.7 Relative Contributions of the Various Producers

In Table 3.27 information is listed for eight estuarine studies for which data is available on the major primary producers. The different systems vary widely in their relative contributions. Phytoplankton contribution, where measured, varied from 43.3% in Beaufort to 2.2% in Flax Pond. Epiphyte production has been measured in only two of the studies where they were estimated to contribute 8.5% (Beaufort) and 3.7% (Flax Pond), respectively. It is likely that most systems would have the same relatively low contributions. In Flax Pond fucoid macroalgae contributed 20.5%, while in Grays Harbor macroalgae contributed 12.4%. In the other studies they were a minor contributor. Epibenthic microalgae were significant contributors in the Upper Waitemata Harbour (35.9%), while in the Nanaimo River estuary they contributed a similar 39.5%. In the other studies they contributed between 5 and 10%. While *Spartina* contributed only 9.8% in Beaufort, the emergent macrophytes were major contributors in Flax Pond (74.7%, *Spartina*), Sapelo Island (84%, *Spartina*), the Upper Waitemata Harbour (31.4%, mangroves, *Avicennia*), Grays Harbor (20.7%), and Mai Po (96.7%, *Phragmites* 46% and *Kandelia*, 50.7%). Submerged macrophytes were major contributors in the Nanaimo River estuary (42.1%, eelgrass, *Zostera*), Grays Harbor (54.7%, eelgrass, *Zostera*), and Bot River estuary (72%).

From the table it can be seen that each system has its own characteristics, depending on the mix of primary producers. In some of the studies, not all of the producers were measured, especially the epibenthic microalgae. If they had been measured, the relative contributions would have been different.

3.6 SOFT SHORE FAUNA

3.6.1 Estuarine Zooplankton

3.6.1.1 Introduction

Zooplankton are the small or weakly swimming animals that are found in estuarine waters. Together with the phytoplankton, bacterioplankton (free-living bacteria), mycoplankton (fungi), and protozooplankton (apochloritic flagellates, amoeboid forms, and ciliates) they constitute the plankton community.

Estuarine zooplankton in contrast to that of the open ocean is limited by two features. First, the turbidity can limit phytoplankton production and thus limit the food available for the zooplankton, although many estuarine zooplankters have been shown to feed on bacterial aggregates and detritus, and thus the phytoplankton concentrations may not be as limiting as has been assumed. Second, and often more importantly, currents, particularly in small

TABLE 3.27
Total System Net Primary Productivity in Selected Estuarine Ecosystems

Site	Producer	Production (g C m^{-2} year^{-1})	Reference
Beaufort, North Carolina	Phytoplankton	66.0 (43.3%)	Penhale & Smith (1977)
	Zostera marina	58.0 (38.8%)	
	Epiphytes	73.0 (8.5%)	
	Spartina alterniflora	15.0 (9.8%)	
	Total	152.6	
Flax Pond, New York	Phytoplankton	11.7 (2.2%)	Woodwell et al. (1977)
	Epiphytes	20.0 (3.7%)	
	Epibenthic macroalgae	30.0 (5.6%)	
	Fucoid algae	75.0 (20.5%)	
	Spartina alterniflora		
	Aboveground	292.0 (54.5%)	
	Belowground	108.0 (20.2%)	
	Total	535.0	
Sapelo Island, Georgia	Phytoplankton	79.0 (6.0%)	Pomeroy et al. (1981)
	Epibenthic microalgae	150.0 (10.0%)	
	Emergent macrophytes		
	Aboveground	608.0 (42.0%)	
	Belowground	608.0 (42.0%)	
	Total	1,445.0	
Grays Harbor, Washington	Phytoplankton	64.5 (2.3%)	Thom (1984)
	Epibenthic microalgae	280.5 (9.9%)	
	Macroalgae	348.7 (12.4%)	
	Zostera	1,541.5 (54.7%)	
	Emergent macrophytes	582.4 (20.7%)	
	Total	2,817.3	
Barataria Bay, Louisiana	Phytoplankton	167.2 (18.4%)	Day et al. (1973)
	Epibenthic microalgae	195.2 (21.8%)	
	Epibenthic macroalgae	10.3 (1.2%)	
	Spartina alterniflora	607.2 (69.0%)	
	Total	879.9	
Bot River Estuary, South Africa	Phytoplankton	58.0 (7.0%)	Bally et al. (1985)
	Epibenthic algae	58.0 (&.0%)	
	Phrahmites australis	111.0 (14.0%)	
	Submerged macrophytes	558.0 (72.0%)	
	Total	815.0	
Tidal Pond, Mai Po, Hong Kong	Phytoplankton	20.3 (0.9%)	Lee (1989)
	Macroalgae	50.6 (2.3%)	
	Kandelia kandel	1,108.2 (50.7%)	
	Phragmites communis	1,004.9 (46.0%)	
Upper Waitemata, Harbour, New Zealand	Phytoplankton	120.0 (25.3%)	Knox (1983a)
	Epibenthic microalgae	145.6 (30.8%)	
	Macroalgae	1.4 (0.3%)	
	Mangroves (*Avicennia*)	148.49 (31.4%)	
	Total	473.49	

shallow estuaries and those dominated by river flow, can carry some of the zooplankton out to the sea. Grindley (1981) has recently reviewed the zooplankton of estuaries with special reference to South African estuarine systems, while Miller (1983) has compiled a general review of estuarine zooplankton.

3.6.1.2 Composition and Distribution

As in the adjacent coastal waters, the zooplankton of estuaries can be subdivided into the *holoplankton* species, which are planktonic throughout their life cycle, and the *meroplankton* species, which are planktonic for only part

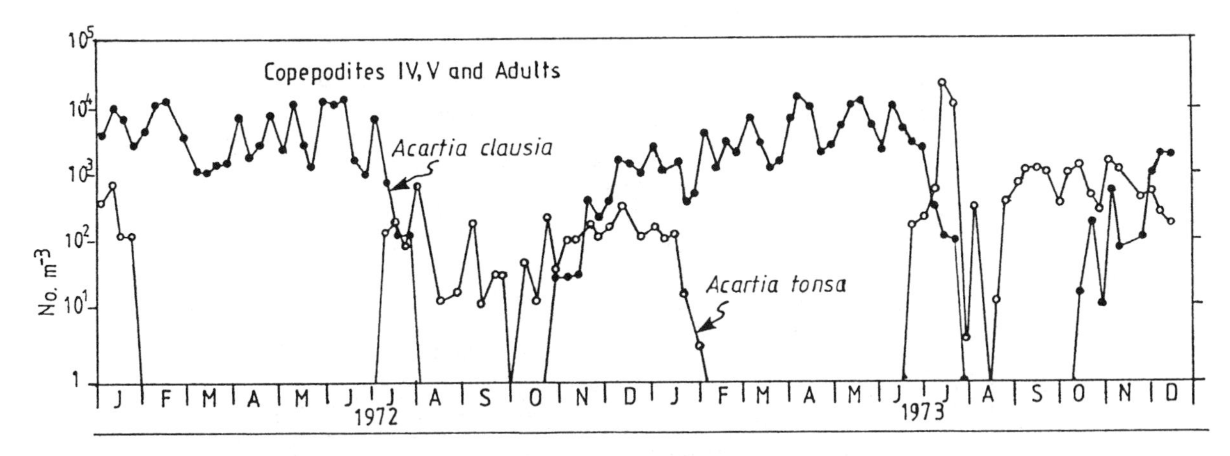

FIGURE 3.39 Abundance cycles of *Acartia clausi* and *A. tonsa* in central Narragansett Bay. Numerical densities for adults and older copepodites (summed) are shown for each species. (Redrawn from Miller, C.B., in *Estuaries and Enclosed Seas, Ecosystems of the World* 26, Ketchum, B.H., Ed., Elsevier, Amsterdam, 1983, 108. With permission.)

of the time — for example at night only, such a mysids. The mesoplankton also includes the larval stages of benthic invertebrates and fishes that spend part of their life cycle in the plankton.

Holoplankton: The holoplanktonic fauna is dominated by small species of copepods, although most of the taxa found in neritic seas are also found in estuaries, particularly if the salinities are high. Chaetognaths, ctenophores, and ciliates, especially Tintinnidae may be common, and large rhizostomid jellyfish may be seasonally abundant. However, foraminifera, hydroid medusae, euphausiids, salps, and larvaceans are usually scarce.

A dominant copepod in many estuaries is *Acartia tonsa*, a euryhaline (salinity range 0.3 to 30+) and eurythermal (5 to 35°C) species. In studies of the zooplankton of Barataria Bay, Louisiana, *A. tonsa* accounted for 60% of the zooplankton countered by Gillespie (1971) and 83% of those countered by Cuzon du Rest (1963). In Biscayne Bay and Card Sound, Florida, *A. tonsa* and *Paracalanus parvus* were the dominant species (Reeve, 1975). Mean numbers in Card Sound and south and central Biscayne Bay were 300, 187, and 2,833 m^3 for *A. tonsa* and 524, 87, and 516 m^3 for *P. parvus*, respectively. In the Damariscota River estuary, Maine, Lee and McAlice (1979) estimated that there were 28,882 copepods m^{-3}, with 8,571 *A. tonsa* m^{-3}, 7,360 *Eurytemora herdmanni* m^{-3}, and 3,753 *Acartia clausi* m^{-3}.

Many surveys of the zooplankton of Narrangasett Bay, Massachusetts, (reviewed in Miller, 1983) have shown that copepods are the dominant group, accounting for 80% or more of the individual animals on an annual average. Among the dominants are *Acartia clausi, A. tonsa, Pseudocalanus minutus,* and *Oithona* spp. Rotifers are abundant in the late winter, and cladocerans are abundant in the early summer. The two species of *Acartia* alternate seasonally (Figure 3.39). *A. clausi* copepodites IV, V, and VI are present in densities on the order of 5×10^3 m^{-3} from January through June. In July they drop quickly to zero and are rapidly replaced by densities of *A. tonsa* (IV, V, and VI) on the order of 10^2 m^{-3}. *A clausi* reappears and begins to increase in about January. *P. minutus* has the same sort of cycle as *A. clausi* with winter and spring densities around 1,400 m^{-3}. *Oithona* spp. have no clear seasonal cycle. Typical densities are 150 m^{-3}.

Meroplankton: The meroplankton are a very diverse group comprising representatives of many phyla. Most common are the larval stages of benthic invertebrates (especially polychaetes, barnacles, mussels, and gastropods), and the eggs, larvae, and juveniles of adult nekton (shrimps, crabs, and fishes), and the sexual stages of hydrozoan and scyphozoan coelenterates. Meroplanktonic mysids, cumaceans, tanaidaceans, amphipods, and isopods are abundant at night, but spend their daytime on the bottom or in the surface sediments.

Microzooplankton: It is only comparatively recently that the importance of microzooplankton in estuarine pelagic waters has been recognized. Together with the heterotrophic pico- and nanoplankton, they constitute the water column microbial community. Protozoa dominate the nano- (2 to 20 μm) and microzooplankton assemblages in pelagic waters and collectively they comprise what is called the protozooplankton. They function as predators of bacteria and small phytoplankton, as prey of larger zooplankton, and as agents for the remineralization and recycling of elements essential for phytoplankton and microbial growth (Sherr and Sherr, 1984; Porter et al., 1985). The picoheterotrophs are small heterotrophic flagellates and they are important grazers of bacteria. The nanoheterotrophs comprise two groups, microflagellates and nonloricate, or naked, ciliates (i.e., those that do not possess a lorica or shell), and they also are major consumers of bacteria. Frequently observed microflagellates include monads, euglenoids, choanoflagellates, chrysomonads, and dinoflagellates. In addition to bacteria, the

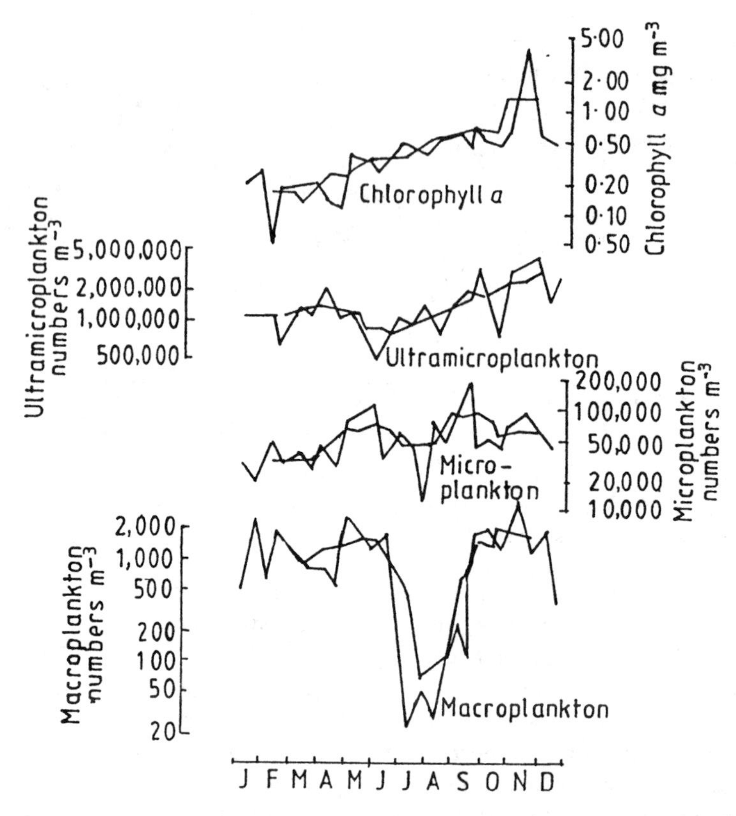

FIGURE 3.40 Seasonal changes in chlorophyll *a*, ultramicroplankton (20–64 μm), microzooplankton (64–200 μm) in Card Sound, Florida. (Redrawn from Reeve, M.R., *Estuarine Research*, Vol. 1, Cronin, L.E., Ed., Academic Press, New York, 1975, 364. Based on Reeve and Cosper, 1973.)

ciliates feed on autotrophs in both the pico- and nano-planktonic size ranges. Spirotrichous ciliates are the dominant microheterotrophs, with most species falling into two orders Oligotrichina and Tintinnida. Tintinnids feed mostly on small flagellated phytoplankton. A second group of protozoans of importance in the microzooplankton are the large, unpigmented dinoflagellates. Amoeboid protozoans including naked amoebae, foraminiferans, and radiolarians may be transiently abundant.

The relative densities from macrozooplankton, microzooplankton (64 to 200 μm), and ultrazooplankton (20 to 64 μm) in Card Sound, Florida, (Figure 3.40) indicate that the microzooplankton can be two orders of magnitude more numerous than the macrozooplankton. For example, tintinnids have a annual mean density of 121,000 individuals m^{-3} as compared with 2,933 m^{-3} for *Acartia tonsa*, the dominant zooplankter. In Long Island Sound, Capriulo and Carpenter (1980) noted that tintinnids and total microzooplankton numbers varied widely, with tintinnids ranging from 1,000 to 9,600 cell l^{-1} and total numbers ranging from 1,000 to 10,500 l^{-1}. In Terebonne Bay, Louisiana, Gifford and Dagg (1988) reported the following densities (in cells l^{-1}): tintinnids, 540^{-1} to 1,400; nonloricate ciliates, 3,160 to 20,360; other ciliates, 420 to 3,500; and zooflagellates up to 18,000. From these and other studies it is evident that microzooplankton densities are generally high, that numbers vary widely, and that the relative importance of the various groups of microzooplankton not only vary widely in any one estuary, but also from estuary, to estuary.

3.6.1.3 Temporal and Spatial Patterns

Seasonal Changes — Seasonal zooplankton patterns in estuaries are much more variable than those in the open ocean due to several factors. Meroplankton are much more important in estuaries and the recruitment of the larvae of different benthic species into the water column occurs at many different times of the year. In addition there are a variety of different food sources in estuaries and the estuarine zooplankton are less dependent than the open ocean zooplankton on phytoplankton. The comparative lack of water column stratification in shallow estuaries is also a factor. The changing salinity patterns due to seasonal high freshwater input from the inflowing rivers can have a marked impact on the species composition and abundance of the zooplankton. The closing of the estuarine mouth, which occurs seasonally in some estuaries, can bring about hypersaline conditions and this can markedly affect the zooplankton.

Vertical Migration — In many estuaries there is a scarcity of zooplankton in the waters during the hours of

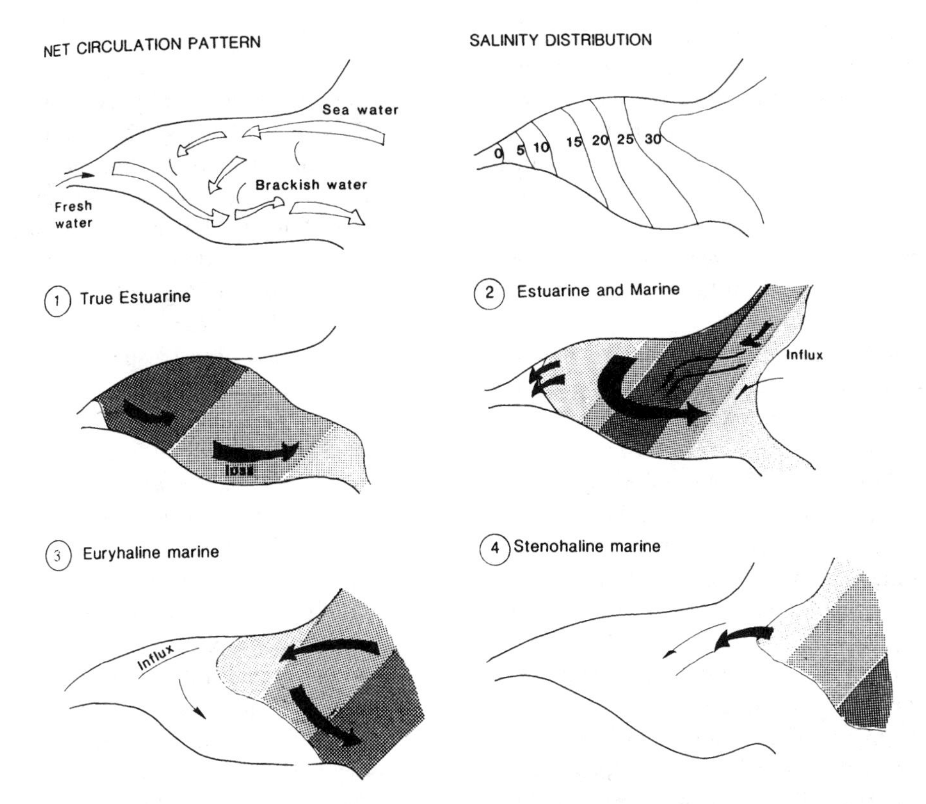

FIGURE 3.41 Stylized distributions of four categories of holoplanktonic copepods in a hypothetical estuary. Darkened arrows indicate the drift of animals produced in areas shown by diagonal lines; the lines are closely spaced in the center of propagation. Relative development in an estuary is a function of salinity distribution and net circulation. Redrawn from Jeffries, H.P., in *Estuaries*, Am. Assoc. Adv. Sci. Spec. Publ. 83, 1982, 502. With permission.)

daylight and the settled volumes of zooplankton obtained in surface hauls in daytime is commonly less than one tenth of that found at night. Many estuarine zooplankters exhibit diurnal vertical migration patterns, reacting positively to low light intensities and ascending toward the surface at sunset. In addition, meroplanktonic species such as amphipods, isopods, mysids, and tanaidacean, which spend the day in the surface sediment, ascend into the water column at night.

According to Grindley (1964) the survival value to estuarine zooplankton of inhibition by low surface salinity is that it probably helps them to maintain their position within the estuary. In estuaries where there is usually a net surface outflow of low salinity water, and a compensating influx of saline water along the bottom, the vertical migration behavior enables the zooplankton to drift alternately up and down the estuary and thus maintain their position in the estuary. When floods occur, however, they would be in danger of being swept out to sea if they rose to the surface. The observed inhibition of vertical migration produced by a strong salinity discontinuity would prevent this.

Diversity and Spatial Variability — Due to a variety of factors including water mass movements, vertical migration, influx of larvae, and predation the distribution of estuarine zooplankton tends to be very "patchy," with high variability in species composition and abundance between stations in sampling programs. In almost all estuaries, the greatest species diversity occurs near the mouth where there is a wide range of neretic species. If species diversity is plotted against salinity for a particular estuary, it is found that peak diversity occurs near 35, and diversity decreases with greater variation in salinity. Below a salinity of 2 an increase in diversity may occur as a result of penetration of freshwater species into the upper estuary. On the basis of the observed patterns, the zooplankton of estuaries may be divided into four components on the basis of salinity tolerance: (1) a stenohaline marine component penetrating only into the mouth area; (2) a euryhaline marine component penetrating further up the estuary; (3) a true estuarine component comprising species confined to the estuary; and (4) a freshwater component comprising species normally found in fresh water (Figure 3.41).

3.6.1.4 Biomass and Production

Peak abundance and peak production are not always closely connected in time, principally due to predation, which removes much of the production. Heinle (1966) studied production of *Acartia tonsa* in the Patuxent River estuary, Maryland. Growth of this copepod from egg to the next generation required 7, 9, and 13 days at 29.5,

22.4, and 15.6°C, respectively, indicating a strong temperature dependence of growth. Heinle estimated that production for *A. tonsa* during a month period in summer was 0.19 g dry wgt m^{-2} day^{-1}. The number of generations per year is dependent on the temperature. In Arctic estuaries *Acartia clausi* generally has only one generation in a year (Evans and Grainger, 1980), while in the English Channel there may be as many as five or six generations. Estimates of zooplankton biomass range from 1.0 to 1 014.5 mg m^{-3}.

Values for annual production like those for biomass vary widely. For Richards Bay, South Africa, Grindley and Woolridge (1974) estimated net secondary zooplankton production at 12 mg (dry wgt) m^{-3} day^{-1}, or approximately 4.4 g m^{-3} yr^{-1}. This gave a P:B ratio of 0.04 over 24 hours, or a *P:B* ratio of 13 over a year. Reeve (1975), for central Biscayne Bay and Card Sound, Florida, calculated a mean daily production of 97.2 and 9.2 mg (dry wgt) m^{-3}, respectively, corresponding to 46.1 and 4.6 g C m^{-3} (carbon as 50% of organic dry wgt). This can be compared with values obtained by Riley and Conover (1967) for zooplankton production in Long Island Sound, which he described as a "somewhat estuarine environment of moderately high productivity," of 27 mm m^{-2} day^{-1} in a 20-m deep water column. In the case of Long Island Sound, the P:B ratio was only 0.027 (i.e., one tenth of that for Biscayne Bay). On the other hand, in Barataria Bay, Day et al. (1973) calculated a net annual production of 25 g dry wgt m^{-2} yr^{-1} (12.5 g C m^{-2} yr^{-1}).

3.6.1.5 Factors Influencing Distribution and Production

1. *Type of estuary*: The type of estuary plays an important role in determining both the distribution and production of the zooplankton. Especially important is the ratio of estuarine water volume to tidal prism volume, and the consequent resident time of the water in the estuary. If this is high, zooplankton will have a longer time to develop and grow within the estuary and less risk of being flushed out of the system. In most estuaries there is a point above which the residence time of the water is sufficient for the zooplankton to survive, while nearer the sea the rapid tidal exchange results in this area being dominated by neritic zooplankton. For each species there is a balance between the exchange ratio of the water and the coefficient of reproduction required to maintain the population. Population increase can occur upstream, while nearer the mouth, tidal dispersion will produce a decrease (Grindley, 1977).
2. *The volume of freshwater inflow*: This is related to the tidal prism and water volume. When the tidal prism is small, large volumes of freshwater inflow can flush the zooplankton out of the estuary. The volume of the freshwater inflow also impacts on the salinity patterns in the estuary.
3. *Salinity*: Salinity tolerance is one of the most important factors limiting the distribution and abundance of estuarine zooplankton. Salinity, in addition to affecting the overall composition of the zooplankton, may affect individual species at different stages of their life cycle. Grindley (1981) has carried out experiments on the salinity tolerance of species of *Pseudodiaptomus*, which are known to inhabit estuaries with a wide range of salinities. *P. hessei* has been found in water from less than a salinity of 1 to 74, while *P. stuhlmanni* occupies a similar range from less than 1 to 75. Survival experiments for both species indicated peak survival at a salinity of 35, but a wide tolerance reaching above a salinity of 70. However, like the Australian species, *Gladioferens imparipes*, which also has a wide salinity tolerance (Hodgkin and Rippingdale, 1971), their distribution is limited by other factors since they dominate both in the hypo- and hypersaline regions of the estuary. In the more normal salinities near the mouths of estuaries they cannot compete with the marine neritic species. Grindley (1981) lists the recorded salinity ranges of 31 estuarine copepods from South African estuaries. Of these, five belonging to the genera *Arcartia, Halicyclops, Oithona*, and *Pseudodiaptomus* have wide salinity tolerances. Six other species have tolerances ranging from a salinity of about 20 to 35, while the rest are restricted to waters around 30 to 35.
4. *Temperature*: Information discussed earlier showed that the growth rates of zooplankton are temperature dependent, resulting in higher numbers of generation per year in warmer waters. Seasonal changes in abundance result from changes in larval recruitment, which is also temperature dependent.
5. *Predation*: Zooplankton predators may appear in large numbers when their prey is abundant. The impact of predators has been documented by a number of authors. It has been demonstrated that an increase in tentaculate ctenophores (and sometimes jellyfish) is often accompanied by decreases in the numbers of copepods. This has been documented in the shallow bays of Long Island Sound (Barlow, 1955), Narragansett Bay (Deason and Smayada, 1982), Delaware Bay (Cronin et al., 1962), the Patuxent River (Herman et al., 1968), the Mississippi Sound (Phillips et al., 1969), and

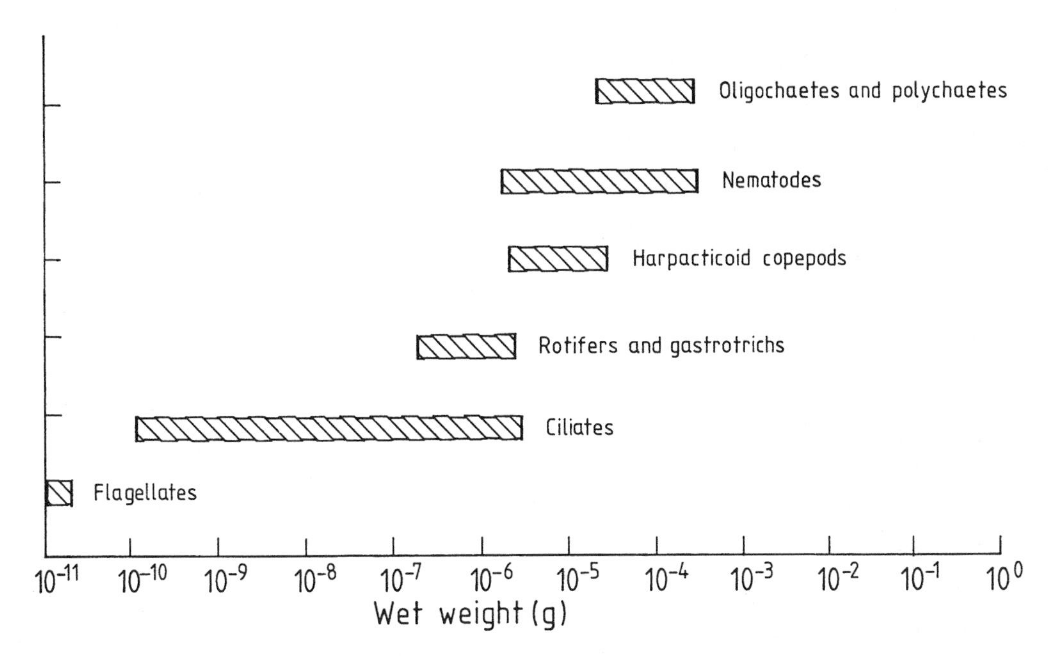

FIGURE 3.42 Approximate weight ranges of meiobenthic faunal groups. (Redrawn from Mann, K.H., *The Ecology of Coastal Waters*, Blackwell Scientific, Oxford, 1982, 186. With permission.)

Barataria Bay (Gillespie, 1971). Phillips et al. (1969) concluded that ctenophores and jellyfishes constituted the most important group of predators on zooplankton because of their periodic extreme local abundance and voracious feeding habits. Deason and Smyada (1982) studied ctenophore-zooplankton-phytoplankton interactions in Narragansett Bay, Rhode Island between 1972 and 1977. They found that in four of the years, the beginning of the pulse by the ctenophore, *Mnemiopsis leidyi*, was accompanied by a rapid decline in zooplankton density and a summer phytoplankton bloom.

3.6.2 Interstitial Fauna

3.6.2.1 The Interstitial Environment

The physicochemical characteristics of the interstitial environment have already been considered in Section 3.2. The interstitial environment spans a continuum of conditions ranging between chemical and physically controlled extremes. The chemical extreme is represented by low energy, sheltered, dissipative beaches of fine sand to mud, where water filtration through the sediment is negligible and the organic input is high. Here, apart from a thin surface layer and the oxygenated zones around the tubes and burrows of macroinvertebrates, oxygen is in limited supply. Consequently, the sediment becomes largely deoxygenated and steep gradients in the interstitial chemistry develop. Since the oxygen is concentrated at the surface, it is here that the bulk of the interstitial fauna is found.

The physical extreme occurs on coarse-grained, steep, high energy, reflective beaches subject to strong wave action. Here large volumes of water are flushed through the beach face and drain rapidly back to sea due to the high permeability of the sand. Thus on such beaches the interstitial biota is well developed and no matter how large the organic input, oxygen demand never exceeds the supply. Here the interstitial fauna has a deep vertical distribution of up to several meters. This fauna is adapted primarily to the physical environment, i.e., high interstitial water flow rates, desiccation during low tide, and disturbance of the sediment by wave action. Most beaches are intermediate between these two extremes, with a combination of both chemical and physical gradients shaping the interstitial climate.

3.6.2.2 The Interstitial Biota

It is often convenient to separate the benthic biota of soft shores according to size into macrobenthos, meiobenthos, and microbenthos with dividing lines at about 1.0 mm and 0.1 mm and 10^{-4} and 10^{-10} g wet weight (Mann, 1982) (Figure 3.42). The latter two categories, which occupy the interstices between the sediment grains include bacteria, fungi, microalgae, protozoans, and metazoans. The roles of the bacteria and protozoans will be considered in detail later. Here details will be given of one study of their dynamics.

Alongi (1985b), in a series of laboratory studies, investigated the impact of physical disturbance on selected microbiota (bacteria, the hypotrich ciliate *Aspidisca* sp., and zooflagellates of the suborder Bodina). He found that both numbers and growth rates of the bacteria were

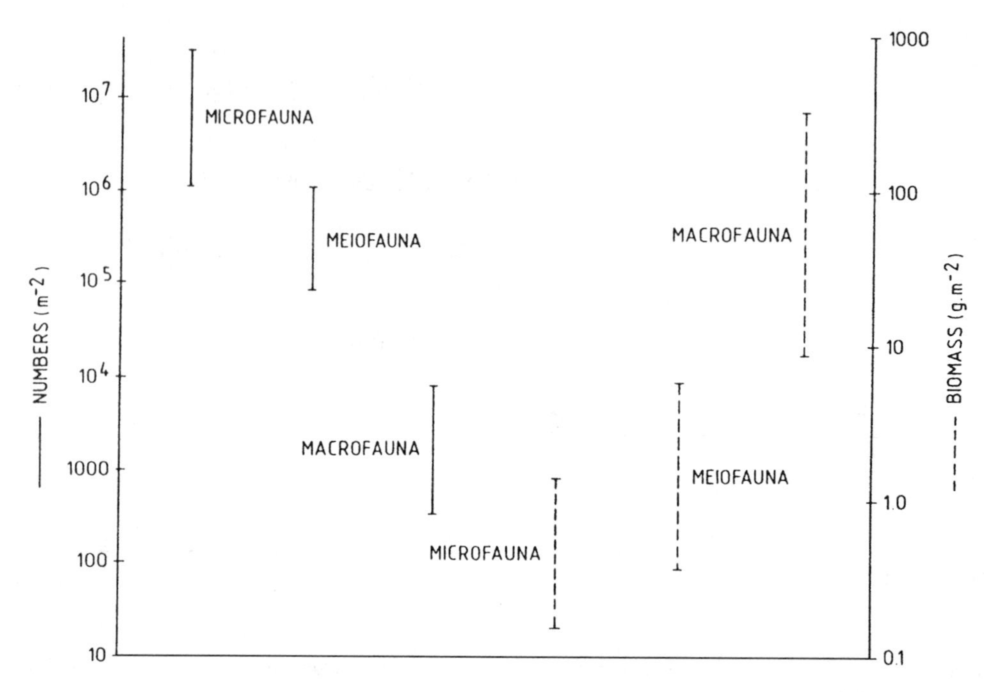

FIGURE 3.43 Abundance and biomass ranges of macro-, micro-, and meiofauna from sublittoral sandy sediments. (Redrawn from Fenchel, T., *Annu. Rev. Ecol. Syst.*, 9, 99, 1978. With permission.)

affected little by sediment disturbance, or by the presence of meiofauna. The dynamics of the bacterial populations presumably reflect a balance between nutrient supply (organic enrichment) and the effects due to physical disturbance and grazing (White, 1983). Using literature values for ingestion rates of bacteria by flagellates (Sherr et al., 1983; 1991), ciliates (Fenchel, 1975c), nematodes (Tietjen, 1980), and harpacticoid copepods (Hicks and Coull, 1983), and averaging protozoan and meiofaunal standing stocks and bacterial productivity over the experimental period, on average only 9 to 11% of the bacterial production was consumed by the protozoans and only 2 to 4% of the production was ingested by the meiofauna. Populations of the ciliate *Aspidisca* sp. were generally more abundant in disturbed cultures that received both low and high detritus rations, especially in the presence of meiofauna, but no persistent responses were detected for the flagellates.

There are a limited number of studies in which all components, macro-, meio-, and microfauna, have been measured. Figure 3.43 depicts data from a sand beach where the smaller animals (microfauna) dominate numerically, but the macrofauna dominates in terms of biomass (Fenchel, 1978). The actual ratios between the three groups depend largely on the sediment type with, for example, the microfauna being very common in fine sand but scarcer in muddy sediments.

3.6.2.3 The Meiofauna

Swedmark (1964), McIntyre (1969), Coull (1973; 1988; 1990), Fenchel (1978), Coull and Bell (1979), Higgens and Thiel (1988), Hicks and Coull (1983), Heip et al. (1985), and Coull and Chandler (1992) have all reviewed various aspects of meiofaunal ecology. The meiofauna (small metazoans) can be subdivided into two groups:

1. *Temporary* meiofauna comprising the larval stages of benthic macroinvertebrates. At times, after settlement following pelagic larval stages, they can become very dense, but they are only temporary in the sense that if they survive long enough they grow out of the meiofaunal size range.
2. *Permanent* meiofauna include representatives of the Rotifera, Gastotricha, Kinorhyncha, Nematoda, Archiannelida, Tardigrada, Copepoda, Ostracoda, Mystacocarida, Halicarina, many groups of Turbellaria and Oligochaeta, some Polychaeta, and a few specialized species of the Hydrozoa, Nemertina, Bryozoa, Gastropoda, Aplacophora, Holothuroidea, and Tunicata (Figure 3.44). These species pass through their complete life cycle as members of the interstitial fauna living in the interstices between the sediment grains. There is an overlap at the lower end of the meiofaunal size range with the protozoa of the microfauna, particularly the ciliates.

The meiofauna occupy a position of considerable significance in biodegradable processes in sediments, especially in the finer sediments of mudflats and estuaries (Fenchel, 1969; 1972; McIntyre, 1969; Gerlach, 1971).

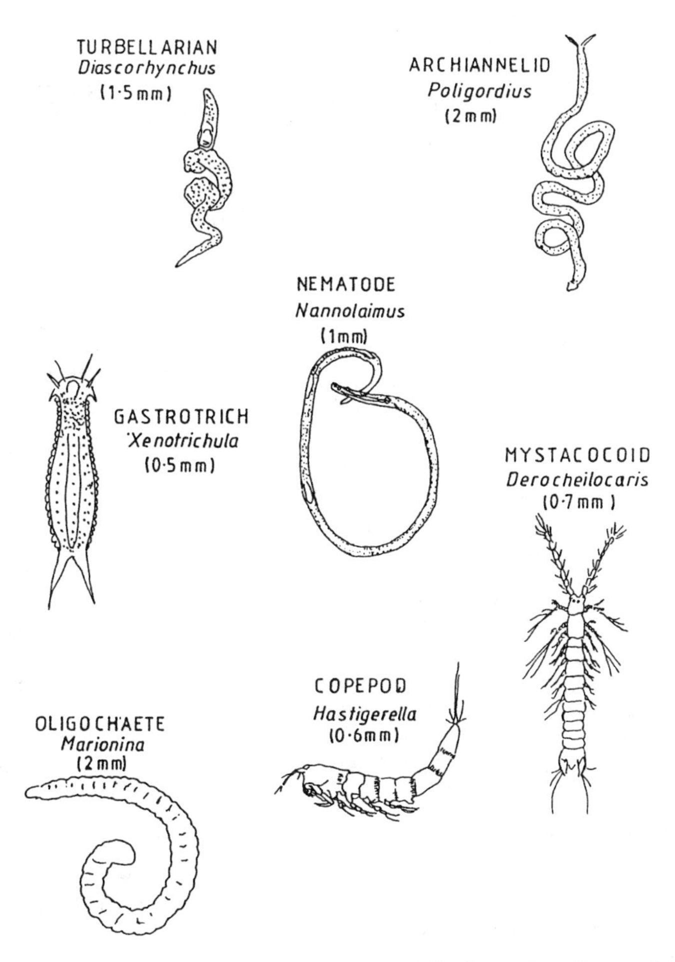

FIGURE 3.44 Some characteristic meiofaunal genera from soft sediments. (Redrawn from Brown, A.C. and Mclachlan, A., *Ecology of Sandy Shores*, Elsevier, Amsterdam, 1990, 182. With permission.)

These processes involve the breakdown of plant material to smaller detrital particles and its subsequent mineralization by microorganisms (see Section 3.6.4). A wide range of feeding types enables the meiofauna to occupy several trophic levels and this, together with their relatively high densities, greatly enhances the flow of energy to the detrital subsystem.

Two groups of meiofauna have been distinguished: burrowing muddy sediment forms and interstitial sandy sediment forms. The former group burrows through the sediment, displacing particles as they move, and their body form tends to be streamlined (Figure 3.44) making burrowing easier. Interstitial meiofaunal species reside within the spaces between the sand grains. In general nematodes are the dominant meiofaunal taxon, although they may be exceeded in some sediment by harpacticoid copepods.

In sands coarser than 200 μm median particle diameter, virtually all the meiofauna are interstitial forms (Wieser, 1959). Body size tends to decrease as grain size and consequently pore space decreases (Swedmark, 1964). Harpacticoid copepods can remain interstitial down to a particle size of between 160 and 170 μm (McLachlan, 1978), and nematodes down to 100 to 125 μm. Below 100 to 125 μm mean particle diameter, the interstitial fauna is absent. Purely interstitial groups (e.g., Gastotricha) are thus excluded from very fine sands and muds, while exclusively burrowing groups (e.g., Kinorhyncha) are excluded from medium to coarse sands (Coull, 1988). In sands of 200 to 300 μm, nematodes are usually dominant, whereas in sands coarser than 350 μm copepods are usually more abundant, while in sands of 300 to 350 μm these two groups are equally abundant (McLachlan et al., 1981). Most exposed beaches have sands coarser than 200 μm and consequently have an almost exclusive interstitial fauna.

In contrast to the robust burrowing forms, the interstitial meiofauna are often slender and vermiform (Figure 3.44) . Because of their small size they have reduced cell size and a simple organization, with a body length as little as 0.2 mm. Body walls are usually reinforced by cuticles, spines, or scales, or protection against abrasion is afforded by an ability to contract. Locomotion can be by ciliary gliding, writhing (nematodes), or crawling (in crustaceans). Due to their reduced cell numbers, gamete production is low, fertilization is internal, brood protection common, and pelagic larvae absent.

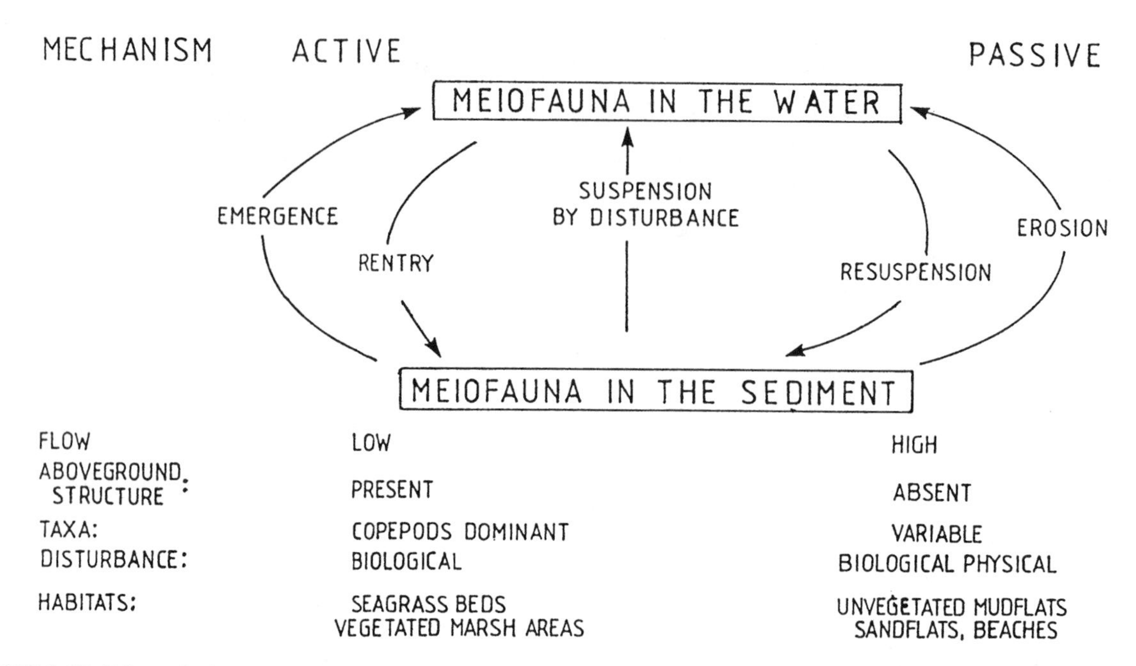

FIGURE 3.45 Schematic illustration of the mechanisms (active, passive) by which meiofauna move between the sediments and water. Four factors influencing entry are identified: flow, structure, taxonomic composition, and disturbance. Habitats are listed as generalized examples where active vs. passive recruitment of meiofauna is expected if the conditions listed in the factor column above the respective habitats prevail. (Redrawn from Palmer, M.A., *Mar. Ecol. Prog. Ser.*, 40, 48, 1988b. With permission.)

3.6.2.4 Meiofaunal Recruitment and Colonization

Early studies assumed that meiofauna had low rates of dispersal and were basically sediment bound, and that colonization of new areas occurred via migration through the sediments (Swedmark, 1964; McIntyre, 1969). A key factor here is the size of the new area and the distance the meiofauna must travel to colonize an area (Hockin and Ollason, 1981). However, studies have shown that when large areas are defaunated colonization can occur within hours (e.g., Sherman and Coull, 1980). It is highly unlikely that animals 200 to 88 μm in length could burrow distances of meters in a matter of hours. Thus it must be concluded that meiofaunal recruitment must take place primarily via the water column. This conclusion was recently confirmed by the experiments conducted by Savidge and Taghon (1988).

Palmer (1988b) has recently reviewed dispersion mechanisms in marine meiofauna. Two distinct patterns exist for recruitment via water column pathways: active entry of the meiofauna into the water and passive erosion of the meiofauna from the sediments. A number of studies have shown that both adult and juvenile meiofauna are regularly found in the water column, e.g., Kern and Bell (1984), Palmer and Gust (1985). Palmer (1988b) has advanced a conceptual model (Figure 3.45) in which four factors interact to determine whether active vs. passive mechanisms are most important for a given community, namely taxonomic composition, hydrodynamics, aboveground structure, and disturbance.

For the meiofauna of areas that are hydrodynamically benign and dominated by active swimmers (e.g., sea grass beds), water column recruitment would involve substrate choice through active swimming. Recent studies have shown that meiofaunal dispersal in sea grass beds is primarily an active process (Hicks, 1986; Walters and Bell, 1994). Hicks (1986) estimated that 3 to 13% of the available harpacticoid copepods (sediment and blade species) migrated into the water column. Walters and Bell (1994) found that of 28 species of copepods found regularly in the sediment, 19 species are also found to migrate vertically into the water column. Such migrations occurred primarily at night.

In areas free of aboveground vegetation and more rigorous hydrodynamically (e.g., tidal flats, beaches), passive recruitment processes dominate and are modified by behaviors that may influence transport and settlement. Disturbance events such as sediment resuspension by strong wave action and tidal currents, and the feeding activities of various fish species such as flatfishes and rays, can cause suspension of the meiofauna in any habitat. Passive suspension may be augmented by active swimming once the disturbance has taken place.

3.6.2.5 Meiofaunal Population Density, Composition, and Distribution

On average one can expect to find 10^6 m^{-2} meiofaunal organisms and a standing crop dry weight biomass of 1 to 2 g m^{-2}. These values, however, will vary according to season, latitude, water depth, substrate, etc. (Gerlach,

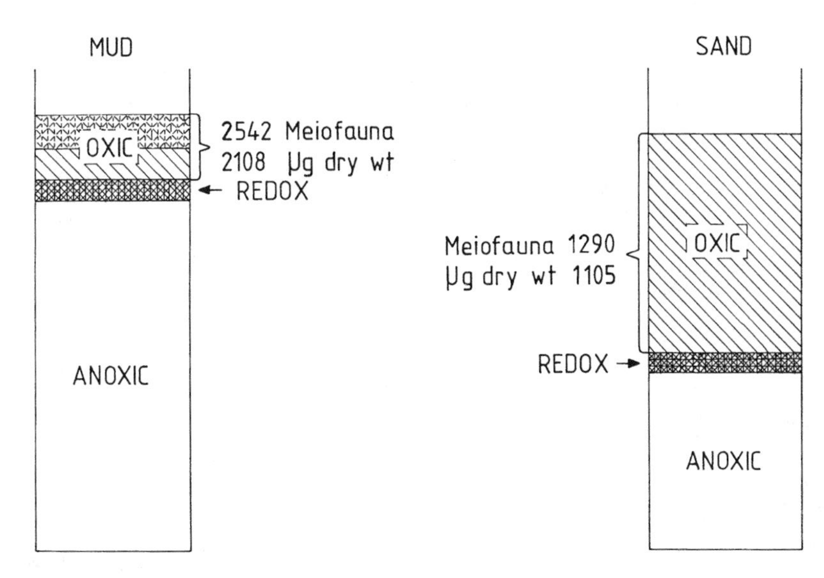

FIGURE 3.46 A schematic diagram of the sediment at a muddy and sandy site in North Inlet, South Carolina, U.S.A., with numbers and biomass at the two sites. The oxidized layer at the muddy site is limited to about 1 cm, while at the sandy site it varies between 10 and 15 cm. (Redrawn from Coull, B.C. and Bell, S.S., *Ecological Processes in Coastal and Marine Systems,* Livingston, R.J., Ed., Plenum Press, New York, 1979, 199. With permission.)

1971). According to McLachlan et al. (1977), average macrofaunal to meiofaunal ratios are 1:10^5 for numbers, 5:1 for biomass, and 1:1 for production. Thus while the meiofauna are quantitatively important, they still account for only approximately 23 μg organic carbon ml^{-1} sediment, compared with a values of approximately 90 μg organic carbon ml^{-1} for the combined macrofauna and bacteria (Gerlach, 1978). However, they may be quantitatively more important in estuaries that are considerably richer than in other sediments. The average meiofaunal density in sand flats appears to be about 1,000 10 cm^{-2}, and some areas may achieve this in nematodes alone (Fenchel, 1969). An increase in meiofaunal density is usually found on mudflats where the average is in the region of 3,000 10 cm^{-2}, although densities of up to 6,000 10 cm^{-2} have been recorded (Rees, 1940). Salt marshes are characterized by very high meiofaunal numbers. Teal and Wieser (1966) recorded 12,000 10 cm^{-2} in a Georgia salt marsh, while Wieser and Kaniwasher (1961) found 1,785 10 cm^{-2} in a Massachuetts salt marsh. In Knysna Estuary, South Africa, Dye (1977) recorded densities between 493 and 19,682 10 cm^{-2}. Possibly the highest recorded meiofaunal density is that of 30,000 to 65,000 10 cm^{-2} found in the salt marshes of the Swartkops Estuary, South Africa (Dye and Furstenberg, 1981).

There are also differences in the composition of the meiofauna according to sediment type. Fenchel (1978) distinguishes three main meiofaunal assemblages. In well-sorted sands with particle diameters greater than 100 μm with the interstices filled with water rather than clay or silt, there is a rich interstitial fauna comprising ciliates, tardigrades, turbellarians, gastrotrichs, oligochaetes, archianellids, harpacticoids, ostracods, and others. In silty and clayey sediments the meiofauna changes completely and is dominated by nematodes capable of burrowing. However, there is richer fauna inhabiting the upper 1 mm or so of these sediments, and harpacticoids, ostracods, foraminifera, and various annelids, along with nematodes, are found in abundance. Figure 3.46 schematically illustrates mud and sand cores and the redox layers with the mean number and biomass of the meiofauna from 5 years of monthly data in the North Inlet, South Carolina. The muddy core has twice the meiofaunal biomass of the sand core. The former has twice the biomass of the latter, and in it the fauna is concentrated in the top 1 cm, whereas in the sand core the fauna is distributed to a depth of 10 to 15 cm. Another meiofaunal community is that inhabiting the anoxic sediments below the redox discontinuity layer. Here occur anaerobic species of ciliates, together with a few species of flagellates, nematodes, turbellarians, rotifers, gnathostomulids, and gastrotrichs.

Nematodes are usually the dominant group accounting for more than 80% of the total meiofauna. In a study of an estuarine salt marsh in Barataria Bay, Louisiana, it was found that the nematodes comprised 87% of the total meiofauna (DeLaune et al., 1981). Total numbers of meiofauna were highest in early March and lowest in October with a mean of 2.9 × 10^6 m^{-2}. This is comparable to those of a South Carolina estuarine marsh where the nematode numbers averaged 3 × 10^6 m^{-2} (Sikora, 1977) (Figure 3.47). Sikora and Sikora (1982) investigated the distribution of nematodes in salt marsh sediments in North Inlet, South Carolina, along a transect through low intertidal, midtidal, and high intertidal creek bank to the low marsh. The annual mean number of subtidal nematodes was 45% of the mean numbers of the high intertidal creek bank nematodes. In

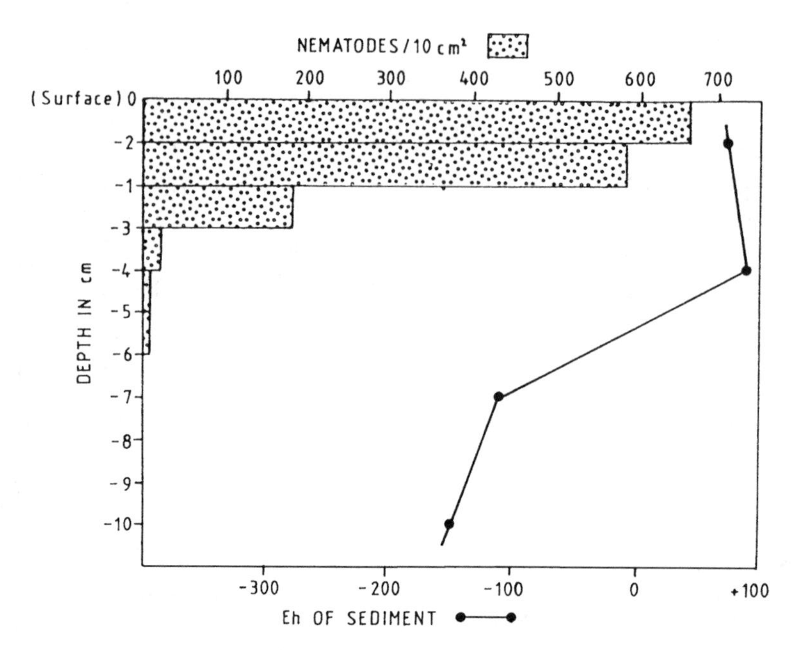

FIGURE 3.47 Distribution of nematodes with depth and Eh profile if the sediments in Barataria Bay, Louisiana, U.S.A. (Redrawn from Sikora, W.B. and Sikora, J.P., *Estuarine Comparisons,* Kennedy, V.S., Ed., Academic Press, New York, 1982, 272. With permission.)

the high marsh, or short *Spartina alterniflora* zone, the number of nematodes were lower than in the high intertidal tall *S. alterniflora* zone (Bell, 1980). Total microbial biomass, as calculated from sediment ATP determinations, was higher in the tall *S. alterniflora* zone just above the high intertidal creek bank than in the short *S. alterniflora* zone. The subdominant meiofaunal group is usually the harpacticoid copepods. Their distribution is strongly influenced by factors such as desiccation and oxygen availability. Thus, they occur near the surface of the sediment and are more abundant toward the lower tidal levels. Population density decreases markedly in muddy areas.

Dye and Furstenberg (1981) reviewed the literature on meiobenthic densities through the mid-1970s and reported an average of about 3 million individuals m^{-2} in mudflats and 5 million m^{-2} in salt marshes, but noted densities of up to 30 and 60 million m^{-2} in the Swartkops Estuary in South Africa. In general, meiobenthic densities of between 100,000 and 10 million individuals m^{-2} are reported and biomass values vary from by four orders of magnitude between approximately 0.01 and 10 g C m^{-2}.

3.6.2.6 Role of Meiofauna in Benthic Systems

Meiofauna affect the availability of detritus and its utilization by detritus-feeding macrofauna by:

1. Enhancing the mineralization of detritus (Findlay and Tenore, 1982).
2. Grazing on bacteria, fungi, and microalgae (Alongi and Tietjen, 1980; Montagna et al., 1983; Montagna, 1984); bacterial grazing, however, rarely results in a reduction in bacterial productivity and more often has a stabilizing effect, removing senescent cells and maintaining the cell growth in an exponential phase (Montagna, 1984).
3. The "gardening" effects of meiofauna enhance bacterial stocks; excreted mucus is colonized by microorganisms, and they are in turn cropped by the meiofauna.
4. Secretion of N- and P-containing dissolved metabolites by the meiofauna supports the growth of bacteria.
5. Meiofaunal bioturbation enhances geochemical fluxes (Aller and Aller, 1992), especially the diffusion rates of oxygen.
6. The mechanical breakdown detrital particles by meiofauna supports bacterial decomposition (Tenore et al., 1977).
7. Competing with infaunal macrobenthos for food and space (McIntyre, 1969; Marshall, 1970; Tenore et al., 1982).

The role of meiofauna in energy flow in benthic ecosystems will be considered in Chapter 6. Here we will deal with meiofaunal food resources and the extent to which the meiofauna form part of the diet of other animals. Individual meiofaunal species may be specialized to feed on diatoms, cyanophytes, flagellates, heterotrophic bacteria, fungi, yeasts, detritus, and other meiofauna, while some are filter feeders. Many species select prey by size and many partition food size niches by one standard deviation (Fenchel, 1968). Most turbellarians are predatory, oligochaetes feed on detritus and bacteria, nematodes feed on a variety of foods, including bacteria, diatoms, and detritus, and copepods consume bacteria (Coull, 1988). Feeding categories for nematodes include nonselective deposit feeders, epigrowth feeders, and omnivore/predators (Wieser, 1953).

Food inputs to the interstitial system include (Brown and McLachlan, 1990):

1. Primary production by interstitial benthic microalgae (photoautrophs).
2. Dissolved and particulate organics flushed through or pumped into sand beaches by wave action and reworked by burrowing invertebrates.
3. Particulate detritus (derived from macrophytes, macroalgae, and phytoplankton) deposited on the surface sediments of the more sheltered shores and sand-mudflats, and subsequently buried by the action of burrowing invertebrates.
4. Chemoautotrophic synthesis in sheltered beaches with reduced layers.

The primary consumers of the organic matter are the bacteria that mostly coat the sediment grains, but also occur in the interstitial waters. Autotrophs (cyanobacteria, diatoms, and flagellates) as well as the heterotrophic bacteria are consumed by the heterotrophic flagellates and ciliates, as well as by the meiofauna.

The belief had been that nematodes and other meiofauna in general were a trophic dead end, either acting in competition with macrofauna for food resources or simply functioning as nutrient recyclers (McIntyre, 1969; Marshall, 1970; Heip and Smol, 1976). Recent research (e.g., Elmgren, 1978) has suggested that nematodes, which were thought to have little nutritive value, are indeed consumed and their production passed up the food chain. Nonselective deposit feeders such as some polychaetes, echinoids, holothurians, and sipunculids in swallowing sediment, must also ingest the nematodes and other meiofauna. Gerlach (1978) points out, "bacterial biomass is not very much higher than meiofaunal biomass; therefore meiofauna automatically has to be considered as an important source of food, if the concept of non-selective feeding is valid." He further suggested that meiofauna (including Foraminifera) in subtidal sand bottoms contributed about 20% to the energy webs of deposit-feeding macrofauna. Platt and Warwick (1980) consider that the reason why nematodes are not more frequently reported in the guts of consumers is that, once macerated, they are difficult to identify.

Fish, especially in their larval stages, are now known to consume large quantities of meiofauna, especially harpacticoid copepods (Feller and Kaczynski, 1975; Grossman et al., 1980; Godin, 1981; Alheit and Scheibel, 1982; Coull, 1990; Feller and Coull, 1995). Feller and Kaczynski (1975) and Siebert et al. (1977) demonstrated that juvenile salmon in British Columbian estuaries feed almost exclusively on harpacticoid copepods, while W.E. Odum and Heald (1975) reported that meiobenthic copepods comprised 45% of the gut contents of North American grey mullet. In addition, Sikora (1977) has reported that nematodes provide a significant proportion of the food of the grazing glass shrimp *Palaemoneted pugio*.

Hicks (1985) has investigated the biomass and production of the meiobenthic harpacticoid copepod *Parastenhelia megarostrum* in Pauatahanui Inlet, New Zealand, and its exploitation as a food resource by juvenile flatfish. He found that *P. megarostrum* reached a density of 263,000 m^{-2}, the highest recorded density for a meiobenthic harpacticoid copepod. The mean sediment density in January, when 60% of the population was contributed to by nauplii, was 441 ± 179.9 10 cm^{-2}. Biomass was estimated at 0.605 g ash-free dry wgt m^{-2} $year^{-1}$, or 0.42 g C m^{-2} $year^{-1}$. Annual production based on a P:B ratio of 15 and a 30% correction for nauplii production gave an upper estimate of 9.074 g ash-free dry wgt m^{-2} $year^{-1}$, or 3.630 g C m^{-2} $year^{-1}$. O-group flatfish within a size range of about 8.0 to 35.0 mm standard length were found to feed predominantly on the harpacticoid copepods with *P. megarostrum* forming 95% of the intake. The mean number of *P. megarostrum* found in the juvenile flatfish guts was 264.8 ± 143.3. If a daily gut turnover rate of three is assumed, then it was estimated that the daily removal rate of *P. megarostrum* was 0.00377% and the annual removal rate 1.38%. This low removal rate of harpacticoid copepods by fish predators has been confirmed in a number of other studies; e.g., 0.59% of the standing crop each day by the spotted dragnet *Calliomymus paucirradiatus* in sea grass beds in Biscayne Bay, Florida (Sogard, 1982; 1984); 1.2% of the standing crop per day by juvenile plaice (Bregnballe, 1961); while Alheit and Scheibel (1982) estimated that the daily predation rate by juvenile tomates *Haemulon auriolineatum* would have little effect on the harpacticoid populations. Thus there is general agreement: although the harpacticoid copepods are an important food resource for juvenile fishes, they exert little regulatory control on their populations.

Meiofauna, and in particular the nematodes, play important roles in facilitating the decomposition of organic matter and in influencing sediment stability. Evidence has accumulated to show that the breakdown and mineralization of organic matter by bacteria is stimulated significantly by the meiofauna, especially by the nematodes (Tenore et al., 1977; Gerlach, 1978). The burrowing and feeding activities of the meiofauna not only improve the exchange of metabolites, but in feeding on the bacteria they maintain them in a "youthful" condition by maintaining their populations at the point of maximum growth (the log phase) or sustainable yield. Studies of impact of the burrowing activities of marine nematodes on the sediment

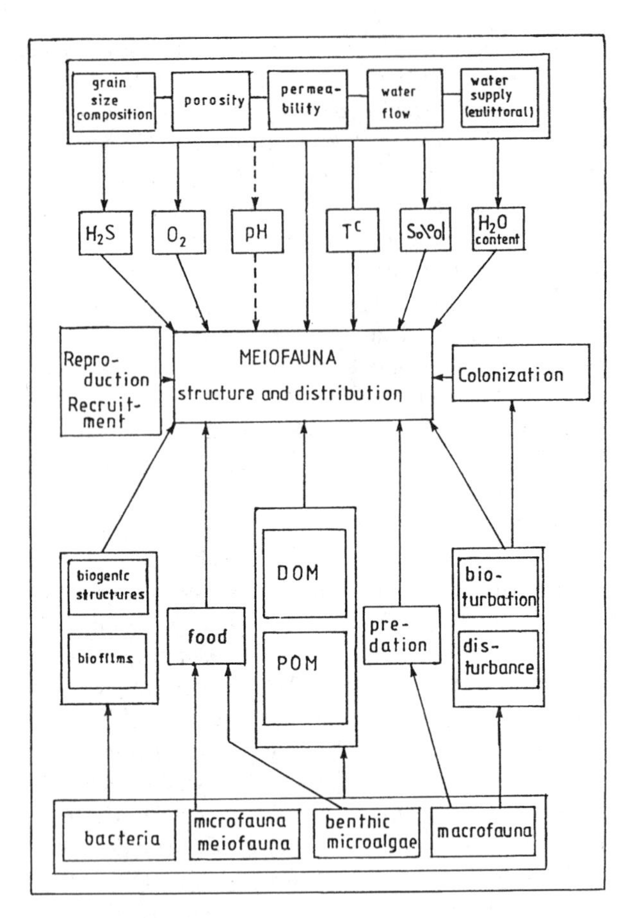

FIGURE 3.48 A schematic diagram of the factors affecting meiobenthic food web structure.

microenvironments has shown that they play two opposing roles with respect to sediment stability. First, they assist in bioturbation and second, by producing a network of mucus-lined burrows (Cullen, 1973; Rieman and Schrage, 1978), they assist sediment stabilization.

3.6.2.7 Factors Involved in the Structuring of Meiofaunal Communities

Figure 3.48 depicts the web of abiotic paramaters and ecological interactions that together act to determine the composition of meiofaunal communities. The sedimentary complex interacts with the physicochemical parameters to set the limits over which the various species can exist. Reproduction, recruitment, and colonization of new or disturbed habitats interact with the complex of environmental and biotic factors to determine species diversity and abundance. Food (bacteria and benthic microalgae), dissolved and particulate organic matter, predation by the macrobenthos, biogenic structures such as the tubes of tube-building invertebrates, sediment biofilms, and bioturbation by the benthic infauna and disturbances, such as those due to the feeding activities of some fishes, constitute a biogenic complex acting on the meiobenthic communities.

Coull (1999) has discussed the relative roles of biological controls on estuarine meiobenthos, i.e., the relative roles of top-down or bottom-up controls. *Top-down (predation) control:* As discussed above, predators appear to have minimal impact on meiofaunal populations. The meiofaunal prey have the ability to reproduce rapidly and thus they can "out-reproduce" the removal due to predation. *Bottom-up (food) control:* Montagna (1995) stated that meiofauna graze an average of 1% of the microbial biomass per hour. This rate and the rate of turnover of the bacteria and benthic microalgae led Montagna et al. (1983) to state: (i) "it is certain that meiofauna will never run out of food"; (ii) "bacterial production and turnover is more than sufficient to satisfy the demands of the meiofaunal population," and (iii) "... meiofauna are not growing so fast that they are limited by microalgal food." It is thus clear that biological controls (predation and food limitation) do not act in the ways they do on macrobenthic populations.

3.6.3 Soft Shore Benthic Macrofauna

3.6.3.1 Introduction

There is an enormous literature on the distribution and abundance of benthic macroinvertebrate animals. Wolff (1983) has reviewed estuarine benthic ecology and Reise (1985) tidal flat ecology. Excellent broad, process-orientated discussions of benthic community dynamics in coastal waters are found in Livingston (1979) and Mann (1982), while Parsons et al. (1979), Valiela (1995), and Alongi (1998) present overviews of coastal ecology. Detailed accounts of the ecology of sandy beaches are to be found in McLachlan and Erasmus (1983) and Brown and McLachlan (1990). McCall and Tevesz (1982) provide detailed accounts of animal-sediment relationships, while Lopez and Levinton (1987) discuss the ecology of marine deposit feeders. Recent accounts of trophic relationships in the marine benthos are to be found in the volumes edited by Barnes and Gibson (1990) and John et al. (1992).

Four types of sedimentary coasts providing habitats for benthic macroinvertebrates can be recognized. The habitats with the lowest species diversity are the exposed sand beaches. As shelter increases, the habitats become more complex and the diversity of the fauna increases. In addition, rooted plants (macrophytes) become a feature of the shore. In temperate regions salt marshes composed of terrestrial halophytes are either restricted around high tide level (e.g., Europe) or they extend to about mid-tide level (e.g., Atlantic coast of North America, South Africa). Muddy sediment accumulates within these marshes. Toward the tropics, mangroves replace the marshes on the upper shore, but both may occur together. On all sheltered shores, sea grasses occur at various levels on the shore.

The benthic fauna can be subdivided into the *epifauna* (living on the surface of the sediments) and the *infauna* (burrowing or tube-building species). The epifauna comprises motile species such as surface deposit-feeding and carnivorous or scavenging snails, starfishes, and crustaceans, crabs, and shrimps. In addition some sessile bivalves (mussels and oysters) build up reefs that provide a habitat for a wide variety of sessile and motile species. However, in many systems the bulk of the epifaunal biomass is composed of fish species. The dominant infaunal species comprise polychaetes, bivalves, amphipods, and callianassid shrimps.

Traditional approaches to the study of the macrobenthos have involved the use of community concepts such as classification of species associations, diversity, succession, and stability/time relationships often involving sophisticated mathematical analyses. However, such approaches have not proved to be particularly useful in understanding the dynamics of benthic communities (Wildish, 1977).

3.6.3.2 Macrofaunal Zonation Patterns

In this section the two extreme beach types, exposed sand beaches and sheltered mudflats, will be considered. However, it should be borne in mind that these two extremes grade into each other through fine sands and muddy sands. On exposed sand beaches the animals present are characterized by a high degree of mobility, including an ability to burrow rapidly. They include representatives of the major taxa, although polychaete worms, molluscs, and crustaceans (isopods, amphipods, crabs, and callianassid shrimps) predominate. Fishes are a very important macrofaunal group on such shores. In contrast, mudflats are dominated by tube-dwelling species (many polychaetes and amphipod crustaceans), relatively sedentary bivalves living at various depths within the sediment, burrowing species (especially polychaetes and crabs), and representatives of many other taxa including nemertines, anthozoans, platyhemninthes, sipunculids, echiuroids, echinoids, holothurians, ophiuroids, and insects.

3.6.3.2.1 Sand beaches

The distribution of macrobenthic animals on sandy beaches exhibit patchiness, zonation, and fluctuation due to tidal and other migrations. Patchiness results chiefly from passive sorting by waves and swash, and from localized food concentrations (McLachlan and Hesp, 1984). Depending on the morphodynamic beach state, the scale of the patches may vary from 10 m on reflective beaches with cusps (McLachlan and Hesp, 1984) to 100 m on high-energy intermediate/dissipative beaches (Bally, 1981).

The differential distribution of the biota across a beach from high tidal to subtidal levels has been the subject of a considerable number of studies. A number of schemes have been proposed to describe such zonation patterns (Figure 3.49). Dahl (1952) introduced a scheme of zonation consisting of three zones similar to those for rocky shores. These zones were based on the distribution of crustaceans and comprised a "subterrestrial fringe," char-

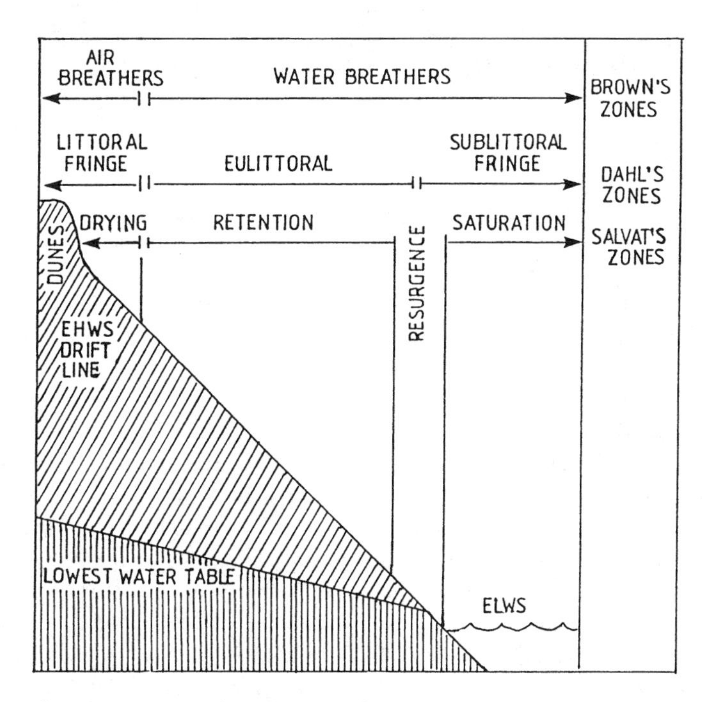

FIGURE 3.49 Three schemes of sandy beach zonation. (Redrawn from McLachlan, A., in *Sandy Beaches as Ecosystems*, McLachlan, A. and Erasmus, T., Eds., Dr. W. Junk Publishers, The Hague, 1983, 337. With permission.)

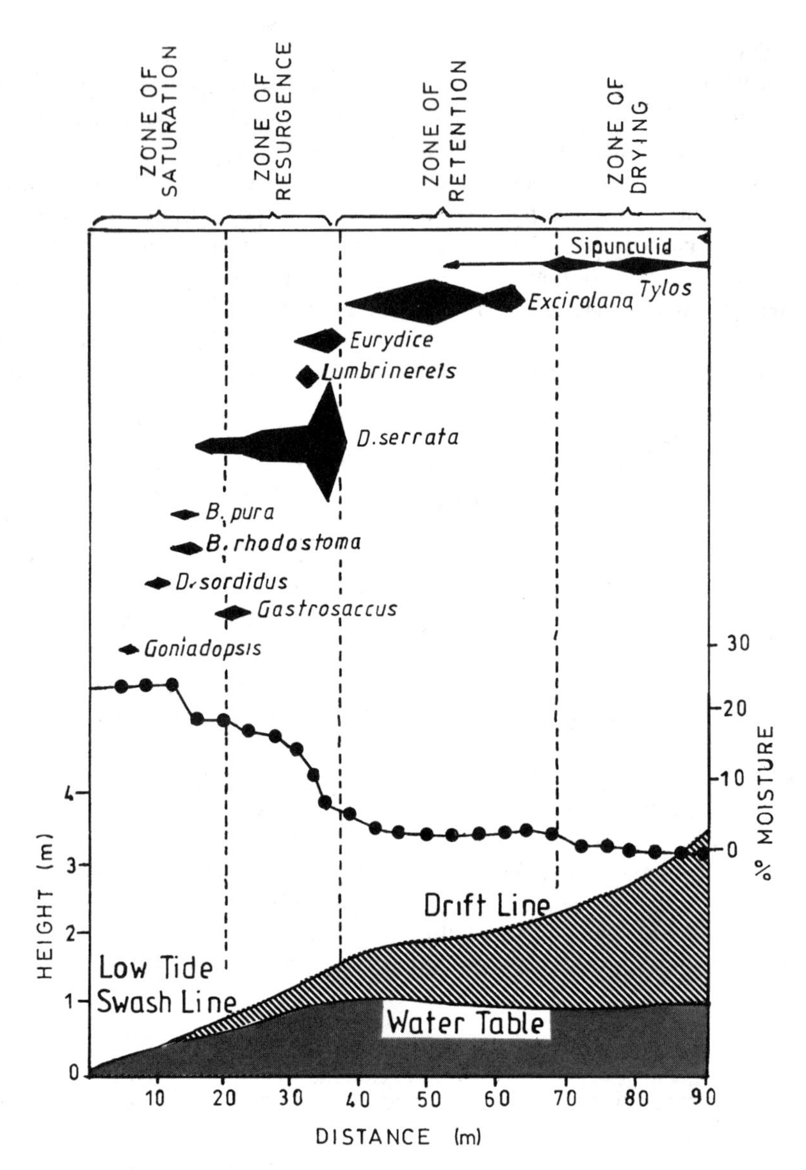

FIGURE 3.50 Zonation of intertidal macrofauna in the East Cape, South Africa, including a moisture profile for the surface sand (after Wendt and McLachlan, 1985). (Redrawn from McLachlan, A., *NATO Advanced Workshop on Behavioral Adaptation to Intertidal Life,* Chelazzi, C. and Vannani, M., Eds., Plenum Press, New York, 1988, 463. With permission.)

acterized by ocypodid crabs (in the tropics and subtropics or talitrid amphipods (on temperate and subpolar shores), a "mid-littoral zone," characterized by cirolanid isopods (although these may be absent in cold regions), and a "sublittoral fringe" with a mixed fauna characterized by hippid mole crabs in the tropics and haustoriid and other amphipods on temperate shores.

Salvat (1964; 1966; 1967) proposed an alternative scheme based on physical factors. He defined four, not three, zones: (1) a zone of drying, or of dry sand, normally wetted only by spray; (2) a zone of retention, reached by all tides, where some moisture always remains around the sand grains even after gravitational water loss; (3) a zone of resurgence, reached by all tides, which is subject to considerable water movement during ebb and flow; and (4) a zone of saturation, permanently saturated with water but with little interstitial water flow (see Figure 3.49). Pollack and Hammon (1971) further developed this scheme by subdividing the zone of drying into zones of dry sand and drying sand.

Various authors have used these different schemes with somewhat conflicting results (Knox, 1969a; Bally, 1983; Went and McLachlan, 1985). Brown (in McLachlan, 1983) considers that only two "indisputable" zones occur universally on sandy shores — a zone of air-breathing animals at the top of the shore and a zone of aquatic breathers lower down. McLachlan (1988), on the beaches of the East Cape, South Africa, related the distribution of the intertidal macrofauna to the Salvat zonation scheme (Figure 3.50). A similar pattern was found by Knox

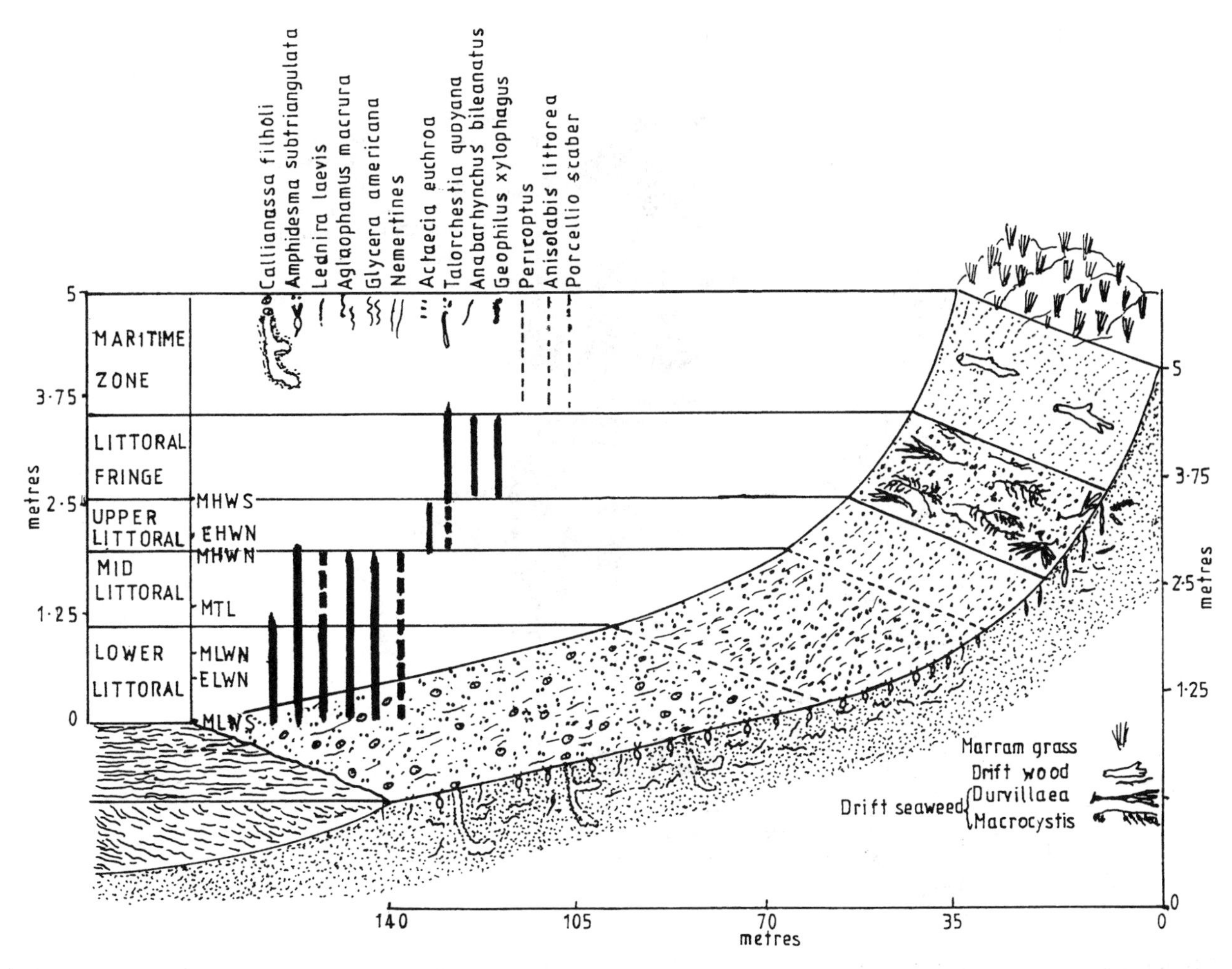

FIGURE 3.51 Schematic diagram showing the distribution of the fauna on the New Brighton Beach, South Island, New Zealand. (From Knox, G.A., in *The Natural History of Canterbury,* Knox, G.A., Ed., A.H. & A.W. Reed, Wellington, New Zealand, 1969a, 530. With permission.)

(1969a) on a New Zealand sand beach (Figure 3.51). A recent detailed analysis of zonation on Oregon beaches has led to the conclusion that four zones do indeed exist there, even if these zones are difficult to define accurately (McLachlan, 1990a,b).

From the studies detailed above it is clear that, depending on the type of beach, three major zones can be distinguished with possibly the subdivision of the lower zone into two subzones on dissipative beaches. However, other workers (e.g., Raffaelli et al., 1991) consider that apart from a high shore zone, biological zones are difficult to define on sandy shores.

Brazeiro and Defeo (1996) carried out a yearly study of the macroinfauna of a dissipative beach on the Atlantic coast of Uruguay. An average zonation pattern with three main belts was recognized between the sand dunes and the lower levels of the swash zone. (Figure 3.52). This pattern roughly matched traditional zonation schemes, whereas monthly patterns did not always fit them. Important spatial variability of the macroinfauna was observed, with aperiodic and seasonal components. They concluded that a yearly study is needed to give a representative picture of the zonation patterns in microtidal beaches.

McLachlan and Jamarillo (1995) have recently reviewed zonation patterns on sandy beaches. They emphasize that, "There is no relationship between macrofaunal zones and long-term or mean tidal levels: zones rather adjust each day to the limits of the beach as defined by the excursions of the swash zone; thus, "boundaries" that may be recognized during the low tide period include the drift zone or the highest swash line, the low-tide swash zone (which may be anywhere from 1 m to >100 m wide) and the effluent line." (Figure 3.53). Each species responds independently to the shore physical gradient and distributions vary from day to day. They concluded that virtually all sandy shores studied to date display three zones as detailed below, which may be distinguished by the presence of characteristic faunal elements.

1. *Supralittoral zone*: This zone is situated at and above the drift line in sand dry at the surface. It is inhabited by the following taxa in any

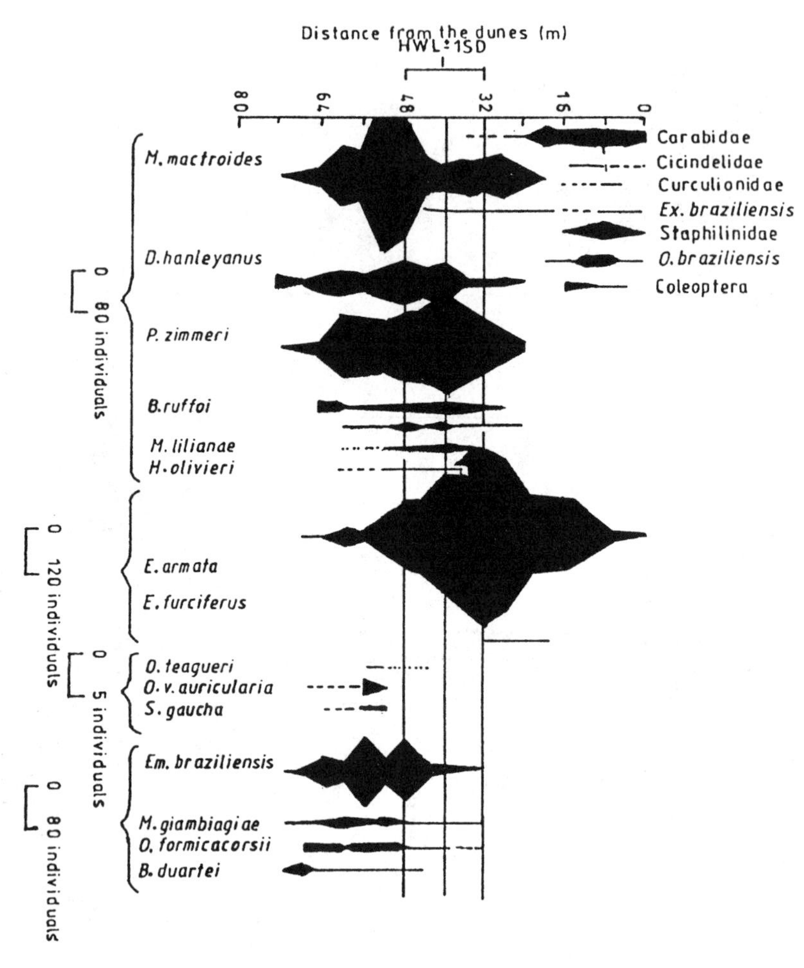

FIGURE 3.52 Intertidal distribution of macroinfauna at Abarra del Chuy beach, Uruguay. The average position of the upper limit of the swash zone (HWL) (+ 1 SD) throughout the study period is indicated. The different scales concerning the number of individuals refers to densities (individuals m^{-2}). (Redrawn from Braziero, A. and Defeo, O., *Est. Coastal Shelf Sci.*, 42, 526, 1996. With permission.)

combination: talitrid amphipods (*Tylos, Talorchestia*), oxypodid crabs, less commonly cirolanid isopods of the genus *Excirolana*, and insect larvae and adults, especially where there is macroalgal litter.

2. *Littoral zone*: This spans the intertidal zone proper. The taxa characteristic of this zone are true intertidal species and may include cirolanid amphipods, other isopods, haustoriid and other amphipods, spionids (such as *Scololepis*), opheliids (such as *Euzonus*) and other polychaetes, bivalve molluscs (especially Donocids), and calianassid shrimps.
3. *Sublittoral zone*: This zone has its upper margin near the effluent line. It is characterized by a greater species diversity including, hippid crabs, mysids, idoteid, oedicerotid and haustoriid amphipods, calianassid shrimps, nephtyid, glycerid and other polychaetes, and holothurians. On some beaches subdivisions of both the littoral and sublittoral zones can be distinguished.

McLachlan et al. (1984) investigated the distribution of the fauna across three high-energy surf zones in front of well-studied intertidal beaches and concluded that the patterns that they found could be integrated with the intertidal distribution zones. They distinguished a "surf zone," or "inner turbulent zone" benthic fauna, which included the same species as those found in the sublittoral fringe, or zone of saturation, and extended beyond the breakers. Some distance outside the breaker zone, this gave way to the "outer turbulent zone," which included species colonizing stable bottoms. Similar associations have been found by other workers off high energy beaches (Knox, 1969a; Christie, 1976). Brown and McLachlan (1990) have combined the findings of this study with those of the intertidal zone to yield a scheme for zonation patterns on open sandy shores (Figure 3.54).

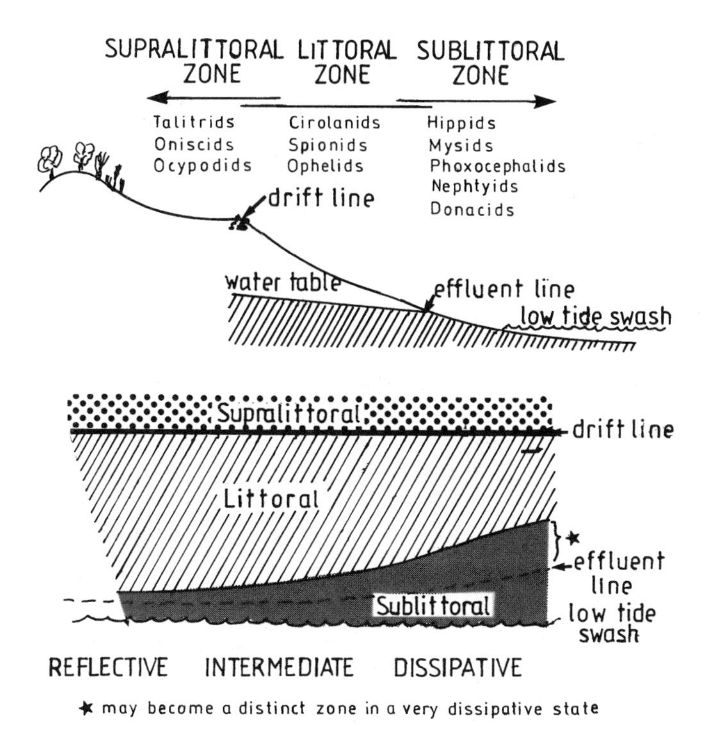

FIGURE 3.53 McLachlan and Jaramillos (1995) scheme for describing zonation patterns on sandy beaches. (a) Zonation on an intermediate ocean beach, and (b) variation in zones across a range of beach types. (Redrawn from McLachlan, A. and Jaramillo, E., *Oceanogr. Mar. Biol. Annu. Rev.*, 33, 327, 1995. With permission.)

The prime causes of zonation across a sandy beach and its surf zone are exposure, changing wave energy levels, and sediment water content and stability. All of the zones and their boundaries shift with tides and storms, as does the fauna itself. Thus the macrofaunal populations do not occupy fixed discrete area or time periods, except on the more sheltered shores and are thus difficult to define in terms of tidal levels.

3.6.3.2.2 Mudflats

Mudflats are characteristic of the upper ends of sheltered bays, inlets, and estuaries. These flats have gentle slopes and the sediments vary from muddy sands (fine sands with a high percentage of the silt/clay fraction) to muddy deposits predominantly composed of the silt/clay fraction). The redox potential discontinuity layer (RPD layer) is close to the surface with only a thin oxidized surface layer. The organic content of the sediments is high and increases with the fineness of the sediments. The upper margin of these mudflats often has a fringe of salt marsh vegetation, while lower down, eelgrass (e.g., *Zostera*) often forms extensive beds. The ecology of salt marshes and eelgrass beds will be discussed in the next chapter.

Accounts of the distribution of the macrofauna on mudflats will be found in J.H. Day (1981a) (South Africa), Morton and Miller (1968) (New Zealand), Inglis (1989) (Australia), Perkins (1974), Reise (1985) (Europe), Ricketts et al. (1985) (North American west coast), and Raffaelli and Hawkins (1996) (Northern Europe).

There are a number of characteristics common to mudflat areas worldwide. They include:

1. The presence of large numbers of deposit-feeding gastropods. Common genera are *Hydrobia, Littorina, Ilyanasssa, Potamopyrgus, Amphibola,* and *Ulva.*
2. Burrowing crabs are a common feature with large numbers of species in the tropics decreasing toward the higher latitudes. Crabs of the Families Oxopodidae, Grapsidae, Portunidae, and Hymenodomatidae dominate the epifauna. J.H. Day (1981b), for southern African estuarine mudflats, recorded 67 species in the Morrumbene Estuary, 18 species in southern Cape estuaries, and 4 species in Atlantic coast estuaries. Common Oxopodidae, burrowing, deposit-feeding species worldwide include species of the genera *Uca, Macrophthalamus, Helice, Cleistostum, Hymensoma.* Fiddler crabs (*Uca* spp.) often dominate in the mid and upper intertidal regions on subtropical and tropical shores.
3. Amphipod detritovores are often abundant, especially tube-dwelling *Corophium* spp.
4. Burrowing shrimps, such as species of *Upogebia* and *Callianassa* are a feature of many mudflats, especially on tropical shores.
5. Deposit-feeding polychaetes are usually dominant (along with bivalve molluscs), especially

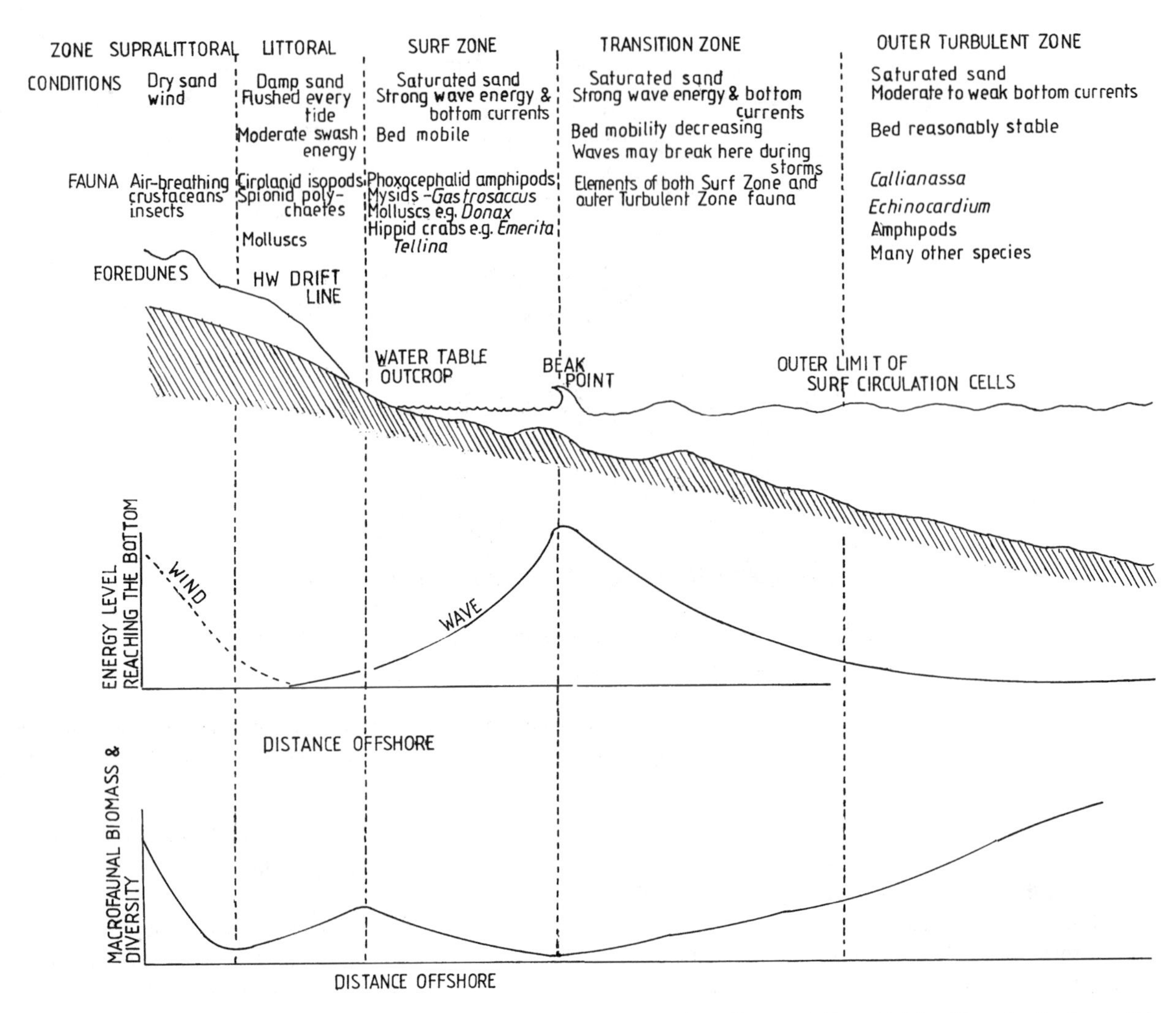

FIGURE 3.54 Generalized scheme of zonation on sandy shores. (Redrawn from Brown, A.C. and McLachlan, A., *Ecology of Sandy Shores*, Elsevier, Amsterdam, 1990, 132. With permission.)

in the mid and lower eulittoral. Species of lugworm (*Arenicola* or related genera) are a universal feature. Some species, e.g., *Capitella capitata* and *Heteromastus filiformis*, are cosmopolitan. Common genera are *Aonides, Boccardia, Polydora, Scolecolepides, Haploscoloplos, Spio, Prionospio,* and *Tharyx.* Many of these attain very high densities.

6. Both bivalve suspension feeders (common genera are *Cerastoderma, Mytilus, Mya, Austrovenus,* and *Panope*) and deposit feeders (common genera are *Macoma, Dosinia,* and *Tellina*) are widespread. Suspension feeders also feed on phytoplankton, particulate organic matter, and resuspended living and nonliving particles.
7. Scavenging and carnivorous gastropods include species belonging to the genera *Nassarius, Natacia, Polynices,* and *Cominella.*
8. Species diversity and total abundance is higher than on sand beaches. The number of species present may exceed 100, and the densities, 100,000 m^{-2} (and may approach 200,000 m^{-2}).

3.6.3.2.3 Sea grass beds

A distinguishing feature of sea grass beds is the significantly greater abundance of animals and the number of species present as compared with adjacent unvegetated areas (Kikuchi and Peres, 1977; Orth et al., 1984; Orth, 1992). Hypotheses suggested for this abundance and diversity include habitat complexity, refuge from predation, increased food supplies, hydrodynamic effects on larval supply, and stable substrates. Manipulative laboratory and field experiments have shown that this high diversity may result from complex, and highly interrelated plant-plant, plant-animal, plant-physical environmental, and animal-animal interactions. These studies have shown that:

1. Foraging success of predators is reduced by macrophyte complexity, whether measured as density of shoots or biomass.
2. There is a nonlinear relationship between habitat complexity and predation success, i.e., a threshold level of complexity was necessary before foraging success was significantly reduced.

Current research on sea grass biota centers around the role of structural complexity, initial settlement densities of recruiting individuals, the physical location of a sea grass meadow in relation to larval supply or source of adults or juveniles, and the ability of individuals to move within or between sea grass areas once settled, in determining population abundances in a given sea grass area. Since sea grasses modify hydrodynamic processes, they also play an important role in determining larval supply rates and larval settlement.

3.6.3.3 Diversity and Abundance

The distribution, diversity, and abundance of animals on beaches has been related to many factors, including sediment particle size and/or organic content (McIntyre, 1970; Bally, 1981; McLachlan et al., 1981; Lopez-Cotelo et al., 1982), beach slope (McLachlan et al., 1981), sediment moisture (Hayes, 1977; Salvat, 1964; 1966; 1967; Bally, 1981), food in the surf water (Brown, 1964; McLusky et al., 1975; McLachlan et al., 1981; Hutchings et al., 1983), and dynamic changes such as those due to storms (Brown, 1971).

Bally (1981; 1983) summarized the results of 105 beach macrofaunal surveys up to 1981. He concluded that, whereas diversity (species richness) and abundance decrease with increasing exposure, individual size increases, often yielding a high biomass. The highest reported biomass values have all been recorded from very exposed beaches of the intermediate to dissipative types, with large populations of filter feeders in the form of bivalves (*Donax* and other species) and mole crabs (*Emerita*). Examples are the Eastern Cape, South Africa, with 7,000 g m^{-2} dry biomass (McLachlan, 1987), Brazil with roughly the same biomass (Gianuca, 1983), and Peru with a dry biomass of no less than 25,700 m^{-1}.

McLachlan (1983) demonstrated a relationship between species diversity on the one hand and slope and particle size on the other, for a range of medium- to high-energy micro/mesotidal beaches around the South African coast. Both diversity and abundance increased as the sands became finer. In a more complete analysis combining data from 23 beaches in Australia, South Africa, and Oregon, U.S.A. (McLachlan, 1990a), diversity (species richness), total abundance, biomass, and mean individual biomass were all regressed against Dean's parameter (Ω), a measure of the morphodynamic state, beach slope, particle size, and wave energy (Hb modal breaker height). Several significant trends were evident. Species diversity showed a linear increase from reflective to dissipative (Figure 3.55), and from steep to flat shapes, whereas abundance showed a logarithmic increase in response to the same changes. Biomass was, however, best correlated with wave energy, increasing logarithmically with an increase in modal breaker height. The probable explanation is that wave energy controls surf-zone productivity and food inputs to the sediment community. Finally, mean individual size of an organism on a beach was correlated with Dean's parameter, slope and particle size with the latter giving the best fit. Thus the average size of organisms on a beach becomes smaller as the sand gets finer and the beach flattens and tends to move to the dissipative state. A recent study by Jaramillo and McLachlan (1993) sampled 10 exposed sandy sites covering a range from reflective to dissipative in south-central Chile and their findings confirmed the conclusions discussed above. The number of species, abundance, and biomass per beach in general decreased with increasing particle size and beach face slope (steeper beaches), and increased for reflective to dissipative conditions. The best fit for the number of species was with Dean's parameter, whereas for abundance and biomass the best fits found were with particle size.

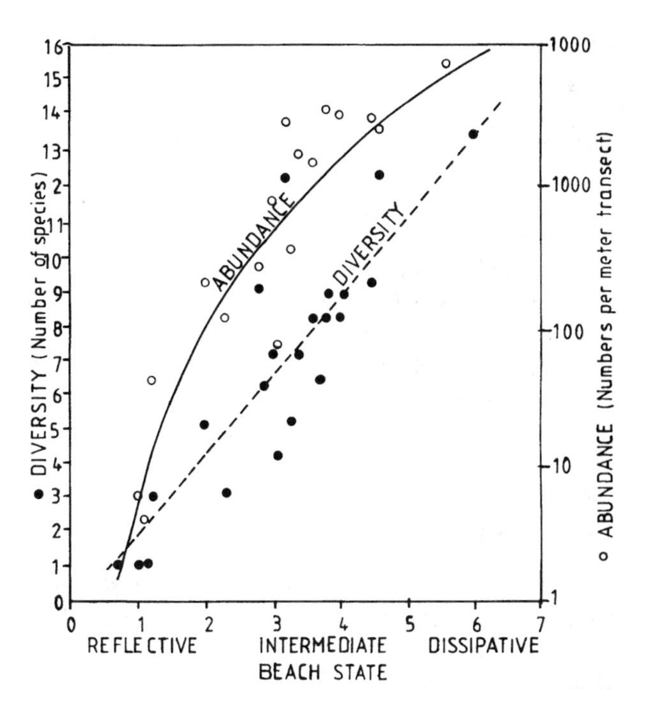

FIGURE 3.55 The relationship between beach state and macrobenthic abundance and diversity. (Redrawn from McLachlan, A., in *NATO Advanced Research Workshop on Behavioral Adaptation to Intertidal Life*, Chelazzi, C. and Vannani, M., Eds., Plenum Press, New York, 1988, 383. With permission.)

McLachlan et al. (1981) has calculated meiofaunal and macrofaunal biomass of two high energy surf-zones in southern Africa. In the three zones, surf, transition, and outer turbulent zone, the meiofaunal biomass was 32 g m^{-1}, 48 g m^{-1}, and 148 g m^{-1}, at Kings Beach, and 160 g m^{-1}, 103 g m^{-1}, and 254 g m^{-1} at Sunday River. Macrofaunal biomasses were generally 0 to 0.1 g m^{-2} in the surf zone, 1 to 10 g m^{-2} in the transition zone, and up to 150 g m^{-2} in the outer turbulent zone. Thus biomasses within the surf zone are reduced and increase offshore. Parallel to the increase in biomass is an increase in species diversity. The ratios of macrofaunal biomass/meiofaunal biomass for the three zones were 0.2, 6.0, and 28.4 at Kings Beach and 0.2, 3.3, and 46.8 at Sunday River. Thus, passing down the turbulent gradient the macrofauna increases considerably in importance in relation to the meiofauna.

3.6.3.4 Distribution Patterns of Estuarine Macrofauna

There is a very considerable literature on the distribution patterns of estuarine plants and animals. Details will be found in the reviews of Lauff (1967), Wolff (1980; 1983), Odum et al. (1974), Perkins (1974), Day (1981a), McLusky (1981), Knox (1986a,b), and Day et al. (1987). In this section we will concentrate on the distribution patterns of estuarine macrobenthic animals. The distribution of estuarine macroalgae, macrophytes, phytoplankton, and zooplankton have already been considered, and nekton will be dealt with in a later section in this chapter. From the literature, there emerges a picture of the biota responding in an often complex manner to gradients of environmental factors such as sediment grain size and organic content, salinity and nutrient distributions (both water column and interstitial), and tidal height. The biotic changes along this estuarine complex gradient can perhaps best be referred to as "coenocline," or community gradient, and the estuarine ecosystem gradient as an "ecocline" (Whittaker, 1967).

Early estuarine benthic ecologists found it useful to classify segments of the estuarine ecocline into zones of similar biotic composition and to relate these zones to the distribution of salinity. A variety of classification schemes have been proposed, especially by researchers working in the large homeohaline brackish systems such as the Baltic and Zuiderzee (reviewed by Sergerstraale (1959) and Remane (1971). Boesch (1972, 1973) investigated the zonation of the macrobenthos along a homeohaline estuarine gradient in the Chesapeake Bay-York River estuary, and a seasonally poiklohaline estuarine in the Brisbane River, Australia (Boesch, 1977). From these studies he concluded that the designation of portions of the estuary as polyhaline, mesohaline, oligohaline, etc. was futile. He found, however, that both estuaries had what can be termed euryhaline marine species, euryhaline opportunists, and estuarine endemics (see Figure 3.56). In a homiohaline estuary the dominant and characteristic species of the macrobenthos progresses from stenohaline marine species on the continental shelf to a diverse assemblage of euryhaline marine species in the lower reaches of the estuary. Many euryhaline species, however, present in fully marine habitats, are more common and abundant in the lower estuary. The coenocline further grades into domination by opportunistic euryhaline species and estuarine endemics broadly around a salinity of 18. The euryhaline opportunists are mostly small polychaetes and include several well-known opportunistic species. Euryhaline opportunists decline in importance and number up-estuary as their salinity tolerance limits are reached and give way

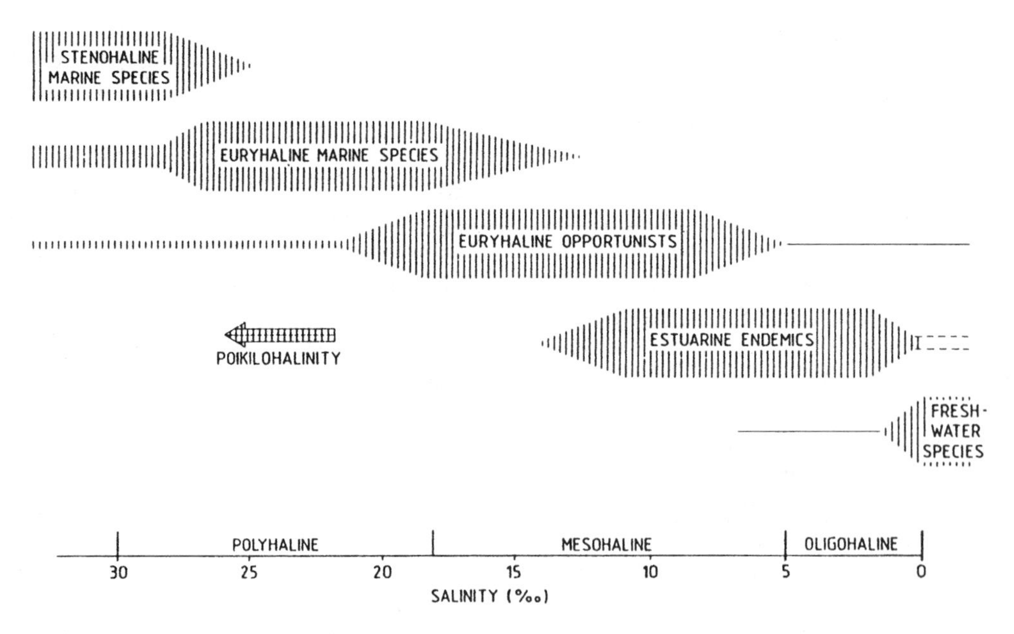

FIGURE 3.56 Distributional classes of species in a homeohaline estuary. (Redrawn from Boesch, D.F., in *Ecology of Marine Benthos*, Coull, B.C., Ed., University of South Carolina Press, Columbia, 1977, 260. With permission.)

broadly at a salinity of 5, to virtually complete dominance by estuarine endemics. The estuarine endemics decline in importance and number as the water becomes fresher.

In general, the estuarine limits of marine and estuarine organisms are set by tolerance of low salinities. Similarly, the down-estuary limits of freshwater species are generally set by tolerance of high salinity. In addition to the salinity tolerances of adult individuals, larval and juvenile tolerances and the effect of salinity on reproduction may also determine distributional limits along the estuarine coenocline (Kinne, 1971; Schleiper, 1971). There is no simple explanation of the distribution patterns within estuarine ecosystems. While salinity gradients are perhaps the dominating factor, important differences between different estuaries exist in geomorphology, sedimentology, hydrography, anthropogenic modification, and biogeography. The last factor can be important. For example, in the Avon-Heathcote Estuary, New Zealand, there are only six species of crabs (Knox and Kilner, 1973), whereas in the Brisbane River estuary there are 23 species (Snelling, 1959). Increased interspecific competition in the latter estuary will be an important factor affecting distribution patterns.

3.6.3.5 Epifauna

Animals and plants that live on the surface of beaches and mudflats are termed the epifauna. Such species are absent on high energy beaches where the animals present burrow into the sediments, although they may emerge to the surface when the tide is in or during the hours of darkness. On mudflats and in estuaries, on the other hand, epifaunal communities are developed to a greater or lesser extent. On mudflats the most conspicuous epifaunal species are surface deposit-feeding snails belonging to a variety of genera (including *Amphibola, Hydrobia, Potamopyrgus*), carnivorous gastropods, and some crab species.

In estuarine areas in particular, reef builders, oysters, and mussels form conspicuous assemblages at particular levels on the shore. These aggregations of suspension feeders can form structures of considerable vertical extent with living bivalves in the surface layer and dead shells below. The eastern oyster (*Crassostrea virginica*), which ranges along the American Atlantic coast from Nova Scotia to Venezuela, builds massive intertidal reefs, especially in estuaries. These reefs form islands of hard substrates in otherwise soft sediments. Where present they help to prevent erosion, influence the velocity and direction of water currents, regenerate nutrients, deposit large quantities of organically rich fecal pellets and pseudofeces, provide a substrate for the attachment of an array of sessile invertebrates such as barnacles, hydroids, ascidians, sponges, bryozoans, and tube-building polychaetes, and shelter and food for many motile species such as flatworms, polychaetes, gastropods, amphipods, isopods, and fishes. Detailed discussions of oyster biology and oyster reef ecology can be found in Dame (1972), Bahr (1976), and Bahr and Lanier (1981). Mussel beds on the surface of low energy beach flats are a common feature of European coasts, e.g., the Dutch Wadden Sea (Reise, 1985). Accounts of the ecology of such mussel beds are to be found in Beukema (1981).

Epibenthic crustaceans, especially penaeid shrimps and blue crabs, are abundant in sea grass beds. Densities of macrocrustaceans in Florida sea grass beds averaged 90 m^{-2}. In Texas salt marsh habitats (Zimmerman and Minello, 1984), total densities of penaeid shrimps and blue crabs (*Callinectes sapidus*) were 35.5 to 80.2 m^{-2} in vegetated areas in contrast to 1.2 to 19.9 m^{-2} in nonvegetated areas.

3.6.3.6 The Hyperbenthos

The term hyperbenthos is a term applied to the association of small animals living in the water layer close to the seabed (the boundary layer) (Mees and Jones, 1997). The fauna is composed of two groups (Wang and Dauin, 1994; Wang et al., 1994): (1) near-bottom zooplankton (subdivided into meroplankton (copepods, crustacean larvae, chaetognaths, polychaete larvae) and macrozooplankton (postlarval fish); and (2) the "permanent hyperbenthos" (mysids, cumaceans, decapods, amphipods, isopods, and pycnogonids). Many of the latter group occur in the mobile sediment surface layer and they make periodic movements into the water column. Such movements may be erratic, at specific times of the day or year, or at certain life-history stages (Day et al., 1987; Jones et al., 1989). Nocturnal vertical movements into the water column have been reported for amphipods, isopods, cumaceans, mysids, copepods, and decapod crab larvae.

The vertical migrations are probably related to the availability of food in the superficial layer (sedimented phytoplankton and detritus, resuspended phytoplankton, benthic microalgae and detritus, as well as meiobenthos). The hyperbenthos plays an important role in the coupling of the benthic and pelagic food webs. Many hyperbenthic species are omnivores, feeding on detritus, microalgae, and zooplankton. In addition the hyperbenthos, especially the crustaceans, contribute significantly to the diet of many species of postlarval and juvenile fish (McCall and Fleeger, 1995).

3.6.3.7 Soft Shore Macrofaunal Feeding Types

While invertebrates living in soft sediments are usually classified as carnivores, deposit feeders, or suspension feeders, such divisions can be misleading and may not apply to the same species in different locations or under different environmental conditions. Many infaunal species can switch between deposit feeding and suspension feeding, depending on the flow regime (Brafield and Newell, 1961; Dauer et al., 1981; Taghorn and Greene, 1992).

At low flow speeds, feeding is almost exclusively on particles on the sediment surface. At higher flow speeds, distinctive shifts in feeding behavior are now known to occur in several species. These behavioral changes are often characterized by reorientation of feeding appendages to feed on particles in the overlying water. Even in the absence of behavioral shifts, many invertebrates that feed at the sediment-water interface (e.g., some bivalves with short siphons) make little distinction between recently deposited particles and particles in suspension. Dauer et al. (1981) used the term "interface feeders" to refer to such species, which are not obligatory deposit or suspension feeders.

3.6.3.7.1 Deposit feeders

Deposit feeders meet their nutritional requirements from the organic fraction of ingested sediment. They are a dominant component of mudflat and estuarine benthic communities where they are responsible for extensive sediment reworking and microbial grazing. There is a general tendency for deposit feeders to predominate in fine sediments with a high clay and organic content. Suspension feeders, on the other hand, reach their optimum abundance in coarser sediments with a mean grain size of 0.18 mm. In their study of the intertidal benthic fauna of a Florida estuary, Bloom et al. (1972) identified species assemblages in which the number of deposit- and suspension-feeding species were inversely related. Hughes and Thomas (1971) demonstrated that 46% of the variance in the occurrence of polychaetes and echinoderms in a coastal bay was associated with an area of soft mud.

Levinton (1979), Levinton and Lopez (1987), and Lopez et al. (1989) have reviewed the role of deposit-feeding invertebrates in soft-bottom benthic communities. Levinton (1979) lists the impact that dense populations of deposit feeders have on the benthic system as: (1) altering the texture and resuspension properties of the sediment (Rhoads, 1973; 1974); (2) influencing the vertical distribution of chemical pore water properties; (3) changing bacterial and microalgal standing stocks (Fenchel and Kofoed, 1976; Lopez and Levinton, 1987); (4) influencing microbial population dynamics (Lopez et al., 1977; Juniper, 1981); and (5) influencing the patterns of larval recruitment (Rhoads, 1974; Woodin, 1974). The high densities of deposit feeders, which may exceed 100,000 m^{-2} (Knox, 1977), suggest that they play an important role in accelerating detrital breakdown and the regulation of deposit-feeding populations by influencing resource availability. Levinton (1979a) considers that three lines of evidence suggest that resources are limiting to deposit-feeding benthic populations: (1) correlations of population density with food-related parameters, such as the organic content of the sediment, percent clay, and microbial abundance; (2) patterns of niche structure of coexisting deposit-feeding species that show little niche overlap; and (3) experimental demonstrations of intra- and interspecific interactions and resource depression.

Lopez and Levinton (1987) point out that because most sediments consist primarily of mineral grains (normally in excess of 95%), and because much of the organic matter present is typically refractory humic material, deposit feeders subsist on a remarkably poor food resource. Consequently they must process large quantities of sediment to obtain their food and their feeding rates are considerable, e.g., the New Zealand marine pulmonate, *Amphibola crenata,* ingests sediment hourly at a rate of up to four times its dry tissue weight (Juniper, 1981). Potential food for deposit feeders includes microbes (bacteria, fungi, protozoa, and microalgae), meiofauna, and nonliving organic matter. While the sedimentary microbes are an important source of protein, it is clear that the majority of deposit feeders require an additional carbon source in order to meet their nutritional requirements. The only deposit feeders that appear to be able to meet most of their metabolic demands from microbial food sources are surface deposit feeders such as tellinid bivalves, deposit-feeding gastropods such as *Hydrobia* spp., *Ilyanassa obsoleta*, and *Amphibola crenata,* and some tentaculate polychaetes.

Coexistence of species requiring similar limiting resources implies a mechanism of niche partitioning. In recent years there have been a number of investigations of the differences in resource exploitation among coexisting deposit-feeding species (Fenchel, 1975a, b; Fenchel et al ., 1975; Levinton, 1979; Lopez and Kofoed, 1980; Whitlach, 1980; Levinton and De Witt, 1989). Whitlach (1980) investigated the patterns of foraging and distribution of 19 species of infaunal polychaetes inhabiting sand and mudflats in Barnstable Harbor, Massachusetts. Two major feeding groups were recognized: surface-feeding species, including all species living in vertically positioned tubes or burrows and collecting food from the sediment-water interface with palps or tentacles (e.g., the spionids, *Polydora ligni*, *Scolecolepides viridis*, and *Prionospio heterobranchiata*) and burrowing species, constituting a rather loosely defined group containing species feeding below the sediment surface (e.g., the capitellids *Capitella capitata* and *Heteromastus filiformis*, the orbiniid *Scoloplos acutus*, and the cirratulid *Chaetozone* sp.). In both groups, between-habitat species richness was positively correlated with food-resource supplies in the sediment. Food resources were allocated among the deposit feeders in two primary ways: with respect to particle size and among subsurface feeders, according to vertical space utilization. For surface feeders the median particle size ingested for the surface feeding polychaetes *Scolecolepides viridis* was about 10 μm, whereas that of *Spiophanes bombax* was 90+ μm, while for subsurface feeders it was about 15 μm for *Tharyx* sp. and about 90 μm for *Scoloplos acutus*. Many other species showed similar differences.

The majority of the species selected the smaller and more abundant size-fractions present in the sediments. While all of the surface-feeding polychaetes collected particulate material from the sediment-water interface, burrowing species were found at different levels within the sediment. Estimates of overlap in food size and vertical position among species suggested that interspecific competition influenced community structure. When organic content of the sediments was high, species that were similar in resource use were able to coexist and species richness was high, while in areas with lower organic content, ecologically similar species were excluded and species richness reduced. The excluded species generally exhibited the highest overlap in resource use when compared to other species in the community. Species having lesser amounts of resource overlap with other species in the system tended to be found ubiquitously throughout the study areas.

A convenient framework for the examination of niche partitioning is the organization of the distribution of a species into niche axes, or dimensions, with each species occupying an ecological hypervolume (Vandermeer, 1972). Levinton (1979) envisages niche separation of benthic deposit feeders along the following resource axes:

1. Sediment type.
2. Living or feeding depth below the sediment-water interface.
3. Particle size (as discussed above, coexisting species may consume particles of differing diameters).
4. Food type. Coexisting deposit feeders might potentially feed on different types of food. Lopez and Kofoed (1980) have studied the feeding behavior of four species of mudflat snails, *Hydrobia ulvae, H. ventrosa, H. neglecta,* and *Potamopyrgus jenkensi. P. jenkensi* had a strikingly different feeding behavior than *Hydrobia* spp. When two similarly sized *H. ventrosa* and *H. neglecta* are compared it was found that the former was more specialized for feeding on fine particles. All species were found to be capable of feeding on the microbial community and organic matter adhering to sediment particles by swallowing small particles and by browsing on the particle surfaces, a feeding method that Lopez and Kofoed (1980) call "epipsammic browsing." This is accomplished by taking the particles into the buccal cavity, scraping off the attached material, and then spitting out the particles. Of all the species, *H. ulvae* was the poorest browser on the larger particles. In addition, some species consume primarily the benthic microalgae in contrast to others that browse on the microbial community attached to the sediment particles.

3.6.3.7.2 Suspension feeders

Species feeding on organic particles suspended in the water are termed suspension feeders (also referred to as filter feeders); their potential food sources are phytoplankton, zooplankton, suspended bacteria, organic aggregates (detritus), and dissolved organic matter in the water above the sediment. Resuspension of benthic microalgae and bacteria enhances planktonic microbial biomass available to the suspension feeders (Blanchard et al., 1997). For instance, the contribution of resuspended benthic microalgae to total phytoplankton can make up about 60% of the total phytoplankton in the Dollard estuary (de Jonge and van Beusekon, 1992); Baillie and Welsh (1980) even calculated, in the Branford Harbor Estuary, that if only 10 to 15% of the mudflat sediments were resuspended by tidal currents to a depth of 1 mm it could account for the chlorophyll levels in the estuary. Stable isotope studies have been used to trace the flows of particulate organic matter to suspension feeders. Such studies have concluded that many bivalves are opportunistic as regards diet (Stephenson and Lyon, 1982; Langdon and Newell, 1990). Peterson et al. (1985; 1986) examined a number of bivalve species within a salt marsh and concluded that "... the bivalves ingest a mixture of bacteria, algae and detritus, with the relative proportions varying according to location and season." These and other studies (e.g., Lucas et al., 1987; Fielding and Davis, 1989) have concluded that phytoplankton and organically rich detrital particles provide the dominant proportion of the diet, although bacteria, which are relatively enriched in nitrogen, may in some circumstances be a significant nitrogen source. Zooplankton is relatively unimportant when compared to phytoplankton on a weight basis, but some suspension feeders, such as the edible oyster, *Ostrea edulis*, may consume considerable amounts of zooplankton. Dissolved organic matter is quantitatively unimportant. In muddy substrates resuspended sediments, with their ample supplies of benthic microalgae, micro- and meiofauna, and organic matter, can provide a nutritious food source. However, the physical instability of muddy sediments that are reworked by deposit feeders tends to inhibit or exclude suspension feeders from such areas (Young and Rhoads, 1971; Rhoads, 1974).

Suspension feeders include sponges, ascidians, barnacles, sabellid and serpulid polychaetes, and especially bivalve molluscs. The latter are usually the dominant species on many sandy beaches and sand flats in estuaries. Bivalve suspension feeders are important components of coastal ecosystems because they remove large quantities of suspended material from the water column and excrete considerable amounts of nutrients. They are also major prey for numerous predators including fish, mammals, and many invertebrates, especially crabs, lobsters, and gastropods. Bayne and Hawkins (1992) have recently reviewed the feeding physiology of bivalves. It is clear that mussels and other bivalves are able to select among particle types trapped by the gills, and to vary their rates of feeding in response to the composition of the seston, i.e., the relative amounts of phytoplankton and organic detritus. There is also evidence that they can vary the production of digestive enzymes in response to changes in their diet (Brock, 1989).

Levinton (1972) suggested that suspension-feeders rely largely on sedimenting phytoplankton for their food, since the gut contents of many species reflect what is available in the phytoplankton. However, phytoplankton is extremely variable, both in space and time, in terms of quantity and the kinds of species present. Thus, as suspension feeders are fixed to the substratum and cannot move to favorable feeding areas, they must be typical opportunists, rapidly exploiting favorable conditions by building up large populations. A consequence is that they are subject to equally dramatic population crashes (Levinton, 1970), as is well known in bivalve populations. Levinton (1970) argued that it is unlikely that two such species will compete for long enough to evolve niche specializations or competitively exclude each other, and he predicted that suspension feeders will have broad, overlapping niches with few feeding specializations. However, there is evidence of interference interaction between bivalve suspension-feeding species living in the same sediment.

3.6.4 Estuarine Nekton

3.6.4.1 Introduction

In contrast to the zooplankton, larger animals that can control the direction and speed of their own movements rather than drifting with the water currents constitute the *nekton*. In the open seas the nekton are largely independent of the bottom, whereas in estuaries few nektonic species can be considered to be truly independent of the bottom. Most estuarine nektonic species live near or just off the bottom, but they periodically swim or feed in the water column. The bulk of the estuarine nekton is composed of fishes, but squids, scallops, and natant crustaceans (including crabs, lobsters, and shrimp) are often important components. In tropical estuaries in particular, crocodiles, turtles, and dugons perform a role that is parallel to that of the true nekton.

The relationships of the nekton in estuaries are very complex and this section is an attempt to elucidate some of this complexity. The section commences with a description of the taxonomic composition of the dominant group, the fishes. Next the interrelated topics of nektonic life histories and food webs are discussed, emphasizing the importance of estuarine migrations. Finally the concept of the estuary as a nursery will be evaluated. A number of reviews of estuarine nekton have been published and include those of McHugh (1967), Day et al. (1981a), Haedrick (1983), Yanez-Arancibia (1975; 1978; 1985), and Yanez-Arancibia and Sanchez-Gil (1988).

3.6.4.2 Taxonomic Composition

Estuarine fishes: McHugh (1967) and Perkins (1974) have divided the estuarine fishes according to their breeding and migratory habits.

1. The largest group may be termed *marine migrants*. The great majority of fishes found in estuaries are the juveniles of species that breed in the sea and use the estuary as a nursery area for food and shelter until they mature and their gonads develop. Common examples are mullets, croakers, anchovies, menhaden, flatfishes, and soles. Species in this category increase dramatically from high latitude to tropical estuaries.
2. The *anadromous fishes* include various species of salmon (*Salmo* and *Oncorhynchus*), sturgeons (*Acipenser*), the northern smelts (Osmeridae), the shad (*Alosa*), the white perch (*Roccus americanus*), the striped bass (*R. saxatilis*), and the lamphrey (Pteromyzontidae). These fishes breed in freshwater and feed and grow in the sea. In their migrations to and from freshwater these fishes dominate the northern cold estuaries. According to Korringa (1967), anadromous fishes were equally important in Europe at one time but now, due to pollution and dam construction, they are scarce or absent except in Norway and the western coasts of Scotland and Ireland. There are no endemic anadromous fishes in the Southern Hemisphere apart from the lamphrey, *Geotria australis*, in New Zealand, Australia, and South America.
3. The *catadromous fishes* include the freshwater eels (*Anguilla* spp.), which breed in the deep ocean and return to the estuaries as elvers before entering the rivers.
4. *Estuarine residents* are species that pass through their entire life cycle in the estuary. The majority of estuarine residents are small species such as gobies, sygnathids, ambassids, antherinids, stoleophorids, and some clupeids. Of the 56 common species listed for South African estuaries, 12 breed there (Day et al., 1981a).
5. Finally there is an anomalous group whose breeding habits and migrations do not fit any of the above groups. The American garfish, *Lepidosterus spatula* and *Dorosoma spidianum*, breed in freshwater but live in estuaries. The menhaden (*Brevoortia* sp.) which is one of the most important commercial species in the Atlantic and Gulf coasts of the U.S.A, differs from one species to another. *B. tyrannus* on the Atlantic coast actually breeds in the sea but migrates at an early stage into low salinities or

even freshwater (Gunter, 1967). The Gulf species, *B. patronus*, and *B. gunteri*, both breed in the upper reaches of estuaries, but the adults migrate to the sea.

Other nektonic species: The other dominant group comprises a diverse assemblage of invertebrates of which the crustacea are the most important. Decapod crustacea are the dominant group and are divided into two suborders: the Reptantia (creeping forms) and Natantia (swimmers). Most Reptantia are nektonic but one group, the paddle crabs, are good swimmers (e.g., the blue crab, *Callinectes sapidus*, from the Atlantic and Gulf coasts of the U.S.A. and Mexico). Many members of the Natantia are active swimmers but they also crawl or bury in the substrate. Important members of this group are the shrimps, which are of great commercial importance.

Apart from the estuarine resident species, the fish found in estuaries are mainly marine species that enter these systems either infrequently and usually in small numbers, or regularly and in large numbers (Blaber and Blaber, 1980; Blaber et al., 1989; Haedrick, 1983; Claridge et al., 1986; Potter et al., 1986). These two groups have been termed *marine stragglers* or *estuarine opportunists*.

3.6.4.3 Nektonic Food Webs

Nekton play a dominant role in estuarine food webs since they occupy the role of top- and middle-level carnivores and thus regulate, through predation pressure, lower trophic levels. A great variety of taxa are eaten by the fish in particular and there is considerable spatial and temporal variation in the range of items eaten even for the same species (de Sylva, 1975; Livingstone, 1984; 1985;Yanez-Aranchiba, 1985). The food items taken change as the fish grow (ontogenetic changes) and they are usually opportunistic feeders.

A number of studies have investigated trophic interactions in coastal and estuarine fishes with particular emphasis on intra- and interspecific competition and resource partitioning. Ovaitt and Nixon (1973) found little overlap in feeding habits of the dominant demersal fishes in a Rhode Island estuary, and they considered trophic resource partitioning as a mechanism that reduced competition. Stickney et al. (1974) found that the estuarine flatfishes in Georgia estuaries showed temporal partitioning of trophic resources with morphological adaptation (relative mouth size) as a functional determinant of feeding relationships.

Mullets are dominant estuarine fishes, especially in tropical regions. Ten species of mullets occur in the St. Lucia estuarine lakes in Zululand, South Africa, and Blaber (1976) has analyzed their feeding habits. Many of the mullets feed on the small gastropod, *Assiminea bifasciata*. This gastropod is abundant and competition between the mullets only becomes apparent when it is in short supply. However, the various species of mullets also feed on foraminifera, the large centric diatom *Aptinoptychus splendens*, small centric and pennate diatoms, filamentous algae, blue-green algae, plant detritus in varying quantities, and sediment particles of different sizes. Blaber (1976) found that the different species of mullet ingest different size ranges of particles and suggested "that interspecific competition is reduced by substrate particle selection, and perhaps by differences in feeding periodicity." Yanez-Aranciba (1976) studied the food habits of the mullet, *Mugil curema,* in a number of coastal lagoons on the west coast of Mexico. Detritus usually constituted 50% of the stomach contents of the adults at all seasons studied. However, the portion of the diet that comprised filamentous algae varied from 0 to 28% depending on the season and location.

Ontogenetic changes in diet are typical of most estuarine fishes. In Apalachicola Bay, Florida, Sheridan and Livingstone (1979) found that the stomach contents of the croaker, *Micropogonias undulatus*, showed a great variation with size of the fish. Fish between 10 and 70 mm ate mostly (>60%) polychaetes, detritus, and insect larvae, as well as a diversity of other food items such as amphipods, mysids, harpacticoid copepods, and calanoid copepods. In larger fish between 100 and 150 mm, the number of food items decreased and included polychaetes, shrimps, chironomid larvae, juvenile fishes, and a small but variable amount of detritus.

Stomach contents of the seabream (*Archosargus rhomboidalis*) in Terminos lagoon, Mexico, showed considerable variation and changed with the growth of the fish (Chavance et al., 1986). Young fish (30 to 80 mm) ate mostly small crustaceans (35% of dry wgt), and occasionally polychaetes. Plant material (fragments of sea grass *Thalassia*, and filamentous algae) formed a minor component of the diet. Fishes from 80 to 160 mm fed mostly on plants (unicellular and filamentous algae and *Thalassia*); animals items were scarce but diverse (microfauna, polychaetes, and small fishes). Large fish (160 to 200 mm) ate predominantly crustaceans (e.g., shrimps and Tanaids). The proportion of detritus in the diet was variable.

De Sylva (1975) has reviewed nektonic food webs and the environmental variables affecting them. The major environmental variables are salinity, temperature, water transparency, tidal streams, and food availability. Generalized food webs (Table 3.28) are basically fueled by either phytoplankton or detritus, and its associated microbial community as the *energy* source. Secondary trophic levels are benthic invertebrates (either infaunal or epifaunal) or zooplankton and micronekton, or both, and include primary carnivores, omnivores, or benthic herbivores. Middle carnivores include planktivores, benthophagous fishes and invertebrates, pelagic fishes, and invertebrates such as squids.

TABLE 3.28
Classification of Nektonic Food Webs in Estuaries

Phytoplankton-Generated Food Webs

Phytoplankton → zooplankton → planktivorous pelagic and benthopelagic fish
Phytoplankton → zooplankton → planktivorous fish → large fish predators
Phytoplankton → phytoplanktonic fishes such as menhaden (*Brevoortia*): summer
Phytoplankton → zooplankton → menhaden: winter
Phytoplankton → zooplankton → large carnivores (*Manta*)
Phytoplankton (dinoflagellates) → mullet (Mugilidae): alteration of usual feeding habits

Detritus-Generated Food Webs

Detritus → benthos (epifauna) → benthophagous fishes
Detritus → benthos (infauna) → benthophagous fishes
Detritus → benthos → benthophagous fishes → large fish predators (sharks)
Detritus → small benthos → larger invertebrates and small benthic fishes → large fishes
Detritus → large detritivorous fishes (mullet): "telescoping" of food chain
Detritus → benthos → large predators
Detritus → micronekton → intermediate predators such as snappers (Lutjanidae) and croaker (Sciaenidae)
Detritus → zooplankton → small fishes and invertebrates
Detritus → zooplankton → small fishes and invertebrates → larger fishes

Source: From De Sylva, D., in *Estuarine Research,* Vol. 1, Cronin, L. E., Ed., Academic Press, New York, 1975, 428. With permission.

3.6.4.4 Patterns of Migration

Different species of nekton have evolved various "strategies" for utilizing estuaries at various stages of their life history. As a consequence, particular species spend different stages of their life cycle in different areas of the estuary. The classification of fishes according to their life cycle and migrations was discussed in Section 4.6.4.2. There are three major migration patterns by which nekton use estuaries for reproduction and feeding:

1. Saltwater spawning, followed by immigration of the larvae into estuaries.
2. Estuarine spawning, in which the larvae remain for the most part in the estuary.
3. Freshwater spawning, followed by the downstream drift or swimming of the larvae and juvenile nekton into the estuary.

Chambers (1980) figured a diagram of the pattern of use of the Barataria Basin by four different nekton groups:

1. Euryhaline larvae, postlarvae, and juveniles of marine nekton that spawn offshore, migrate far up the basin in the late winter and spring, and then gradually move down bay as they grow, eventually emigrating to the Gulf in late summer and winter.
2. Juvenile and adult freshwater species that move southward in the autumn into oligohaline areas as they become fresher and replace the emigrating marine species, returning to freshwater areas in the late winter.
3. During the warmer months, mesohaline juveniles of certain species move up into the mid-Basin during periods of high salinity, and later return to the lower bays and Gulf in the late autumn and winter as salinities decrease.
4. Some euryhaline species spend their entire life cycles in the estuary and often may be found anywhere from the freshwater swamps to the lower bays and barrier islands bordering the Gulf.

The data of both Wagner (1973) and Chambers (1980) suggest that euryhaline marine-spawned individuals migrate into waters of low salinity and slowly move into waters of higher salinity as they grow.

The available data suggest that nekton species, especially larval and juvenile forms, preferentially seek out shallow waters adjacent to wetlands, such as salt marsh ponds, tidal creeks, and the marsh edge in general. Data from Barataria Bay, Camiinada Bay, and Lake Pontchartrain show that nekton biomass is 7 to 10 times higher in shallow-water marsh areas as compared with open waters. These shallow-water areas seem to satisfy the three requirements outlined by Joseph (1973) for a nursery area: (1) physiological suitable in temperature, salinity, and other physicochemical parameters; (2) abundant suitable food with a minimum of competition at critical trophic levels; and (3) a degree of protection from predators.

Shrimps (Penaeidae) often follow a similar pattern to that of many fish species, spawning offshore with the juveniles after passing through a series of larval stages migrating into the estuary to feed and grow. Two to four months are spent in the estuary during which the shrimps grow rapidly, after which they migrate to the sea to complete their life cycle.

3.6.4.5 The Estuary as a Nursery

As Boesch and Turner (1984) have pointed out, it is almost an article of faith among estuarine scientists that coastal wetlands (salt marshes and mangrove forests) and sea grass beds are important nursery areas for juvenile fishes and crustaceans. This stems from the widespread use of estuaries by the larvae and juveniles of a large number of species. From this the concept of "estuarine dependence" developed, implying that the estuary is required for some part of the life cycle of such species. Within estuaries there are three primary nursery areas: wetlands (salt marshes and mangroves), including the shallow marsh fringing areas, tidal creeks, and mudflats; the low salinity areas at the head of the estuary; and sea grass beds.

A number of workers have shown correlations among estuaries, wetlands, and fisheries (Boesch and Turner, 1984; Robertson and Duke, 1987). Turner (1976) correlated shrimp yield (kg ha^{-1}) on a worldwide basis. In the northern Gulf of Mexico he found that yields of inshore shrimp were directly related to the area of estuarine vegetation, whereas they were not correlated with area, average depth, or volume of estuarine water. Moore et al. (1970), in their analysis of the distribution of demersal fish off Louisiana and Texas, suggested that the greatest fish populations occur offshore from extensive wetlands with a high freshwater input. Evidence discussed above from the Barataria Basin also suggests that wetlands enhance fisheries productivity.

In the Phillipines, Primavera (1998) over a period of 13 months, collected penaid prawns belonging to nine species from mangrove and non-mangrove sites. Of the three dominant species, *Metapenaeus ensis* and *Peneaus merguiensis* were restricted to the brackish water riverine mangroves, while *M. anchistus* was restricted to the high-salinity mangroves and tidal flats. Abundance and size composition of these species suggested a strong nursery role for the riverine mangrove, limited nursery use of island mangroves, and a non-nursery use of the tidal flats.

Robertson and Duke (1987) have shown that in tropical Australia the mangrove habitat is a major nursery site for juveniles of the banana prawn, *Penaeus merguelensis*, one of the most important commercial species of prawn in Australia. Staples et al. (1985) have shown that in the Gulf of Carpentaria, Australia, postlarvae and juveniles are restricted to mangrove habitats. Adults of the family Leiognathidae (pony fish) are the most abundant fish in several Southeast Asian trawl fisheries (Pauley, 1979) and juveniles of the two most heavily exploited species, *Leiognathus equulus* and *L. splendens*, are mangrove dependent as juveniles.

3.7 BIOLOGICAL MODIFICATION OF THE SEDIMENT

3.7.1 Bioturbation and Biodeposition

Studies of the interactions between benthic organisms and the sedimentary environment have consistently demonstrated that bioturbation has a major impact on the mass properties of the sediments (see reviews by Gray, 1974; Rhoads, 1974; Lee and Swartz, 1980; Rhoads and Boyer, 1982, and Thayer, 1983). Bioturbation involves both infauna and benthic feeding fishes; these organisms exhibit three major modes of sediment reworking (see Rhoads, 1967): ingestion (sediment passed through the gut), manipulation (handling of sediment by mouthparts or other appendages during feeding, e.g., particle scraping, or tube building), and displacement (removal of sediment by digging or burrowing) (Myers, 1977; Boucot, 1981). The mechanisms of disruption have been classified by Thayer (1983) who concluded that, on the basis of maximum recorded volumetric rates, "biological bulldozing" is the most effective form of bioturbation. This bulldozing results from plowing through the medium, manipulation of the sediment while burrowing, or manipulation while feeding. Investigations of bioturbation have involved two main approaches. The first of these are studies of the manner in which irrigation, palletization, tube structure, mucus, etc. affect flow near the seabed and the erodability of surface sediments (Rhoads et al., 1977; 1978; Grant et al., 1982; Nowell et al., 1984). The second avenue of approach involves measuring the rates at which organisms rework sediment through various activities such as burrowing and feeding (see review by Carney in Boucot, 1981).

Among other effects, bioturbation increases oxygenation and mineralization (Hines and Jones, 1985; Kristensen et al., 1985), and alters the sediment geochemistry (Aller et al. 1983; Marinelli, 1992). Bioturbation has important effects on the exchange of sediment pore water with overlying waters; it can enhance sediment-seawater exchange by as much as an order of magnitude (e.g., Blackburn and Henriksen, 1983; Rutgers van der Loeff et al., 1984; Murphy and Kremer, 1992). These exchange processes alter the chemistry of the surrounding sediments, pore water, and the overlying water (e.g., Aller, 1980; 1982; Aller and Yingst, 1985, Kristensen and Blackburn, 1987; Kristensen, 1988; Doering, 1989; Huttel, 1990; Murphy and Kremer, 1992; and others). As a result, important constituents such as carbon, nitrogen, sulfur, phosphate, and silica are increased, and productivity in the sedimentary environment and water may be stimulated (McCaffrey et al., 1980; Matisoff, 1982; Waslenchuk et al., 1983; Marinelli, 1992). A number of the characteristics of the fauna and their habitats that affect pore water composition and bioadvection pore water transport include: the size, depth, and density of infaunal dwellings (Aller, 1982; Aller and Yingst, 1985; Takeda and Kurihara, 1987), the sediment type (Kristensen et al., 1985), and the physicochemical composition of the infaunal dwelling (Hines et al., 1982; Aller, 1983; Aller et al., 1983). Bioturbation has many diverse effects on the microflora, microfauna, and meiofauna (e.g., Hanson and Tenore, 1981; Reise, 1981; 1983a,b; Bell and Woodin, 1984; Reichart, 1984; 1988; Alongi, 1985a,b; Moriarty et al., 1985; Branch and Pringle, 1987). It may result in the exclusion of certain groups of organisms (Rhoads and Young, 1970; Flint and Kaike, 1986), or alternatively increase the density and diversity of the infaunal benthos.

In a New England salt marsh, experimental reductions in densities of the fiddler crab *Uca* led to a 35% increase in root biomass and a 47% decrease in aboveground biomass of *Spartina alterniflora* over a single growing season (Bertness, 1985). Similarly, in a mixed *Rhizophora* mangrove forest in northern Australia, Smith et al. (1991) found that in plots without crabs (mostly grapsids, *Sesarma messa* and *S. semperi longicristatum*), pore water sulfide and ammonium concentrations increased compared to control sediments, while forest growth was significantly reduced in the absence of crabs. It is thus clear that burrowing crabs can significantly affect the production of marsh grasses and mangroves.

Motile deposit feeders move laterally and vertically within the sediments, causing mixing and transport of sediment particles, as well as interstitial water and dissolved gases. The fecal pellets that they produce are deposited within the sediments. Surface deposit feeders generally deposit their fecal pellets on the surface. Other deposit feeders (mainly polychaetes, but including enteropneusts, holothurians, and callianassid shrimps) can form dense populations with vertically orientated tubes. If these species feed at the lower ends of their tubes, a massive transfer of sediment may take place from depth to the surface. The tube mats may also physically bind and stabilize the sediments. Suspension feeders trap suspended seston, ingest it and produce considerable quantities of fecal pellets and pseudofeces that become incorporated in the surface sediments.

The various methods whereby deposit feeders mix and recycle sediments is illustrated in Figure 3.57. The maldanid and holothurian are termed conveyor belt species by Rhoads and Young (1970). The feeding activities of these animals form a cone of fecal pellets around the opening of the tube or burrow, and a ring of unconsolidated sediments in a depression around the cone. Filter feeders are inhibited from colonizing the depressions, but colonize the sides of the cone in great numbers. These cones confer a spatial heterogeneity on an otherwise uniform environment. The rates of reworking of sediments by deposit feeders is considerable (See Table 3.29). Comprehensive tabular summaries of available bioturbation rate data are provided by Lee and Swartz (1980) and Thayer (1983). These rates become even more impressive when multiplying the individual rates by the standing stocks found on mud bottoms and flats, which commonly reach densities of 10^2 to 10^4 individuals m^{-2}. Many such areas have sedimentation rates of between 1 and 2 m $year^{-1}$. Given this rate of accumulation and the rates of reworking shown in Table 3.29, it is not surprising to find that the sediments are passed through the benthos at least once, and in many cases, several times a year (Rhoads, 1967; 1974). In the Waddensee mudflats, the polychaetes, *Heteromastus filiformis* and *Arenicola marina,* and the bivalve, *Macoma balthica*, collectively reworked a sediment layer equivalent to 35 cm per annum (Cadee, 1990). Takeda and Kurihara (1987) in an investigation of the effects of burrowing by the mud crab *Helice tridens* in Japan found that ca. 2.9% (11,500 cm^{-3} m^{-2}) of the sediment from the surface to a depth of 40 cm was turned over each year. Numerous leaf and stem fragments were buried in the sediment by these burrowing activities.

Sediments are also profoundly modified by the activities of suspension feeders. Phytoplankton, organic matter, and sediment filtered from the water is combined into aggregates and voided as feces or pseudofeces that settle to the sediment. At high densities they are able to deplete the water mass of phytoplankton (Wildish and Kristmanson, 1984; Fretchette and Bourget, 1985). The quantitative importance of biodeposition by a variety of suspension-feeding invertebrates is given in Table 3.30. Haven and Morales-Alamo (1966) estimated that a population of the estuarine oyster *Crassostrea virginica* on 0.405 ha may produce up to 981 kg (dry weight) of feces and pseudofeces weekly. They found that the material deposited contained 77 to 91% inorganic matter, 4 to 12% organic carbon, and 1.0 kg of phosphorus.

Haven and Morales-Alamo (1966), and Samuel et al. (1986) have all emphasized the importance of mussels in the removal and deposition of suspended particulate matter. Kauysky and Evans (1987) recently investigated the role of biodeposition by *Mytilus edulis* in the circulation of matter and nutrients in the northern Baltic. Annual deposition by *M. edulis* was 1.7 g dry weight, 0.33 g ash-free dry weight, 0.13 g C, 1.7×10^{-3} g nitrogen, and 2.6×10^{-4} g phosphorus per g mussel (dry weight, including shell). At the study site, the benthos annually receives 81 g C, 10 g N, and 2 g P m^{-2} as a result of the feeding activities of the mussels. This can be compared with 245 g C, 30 g N, and 5 g P received due to natural sedimentation.

The physical stability of the sediments is closely tied to the activities of the benthos. Mobile deposit feeders produce pelletal surfaces and sediments with an open fabric, high near-surface porosity, and low compaction and

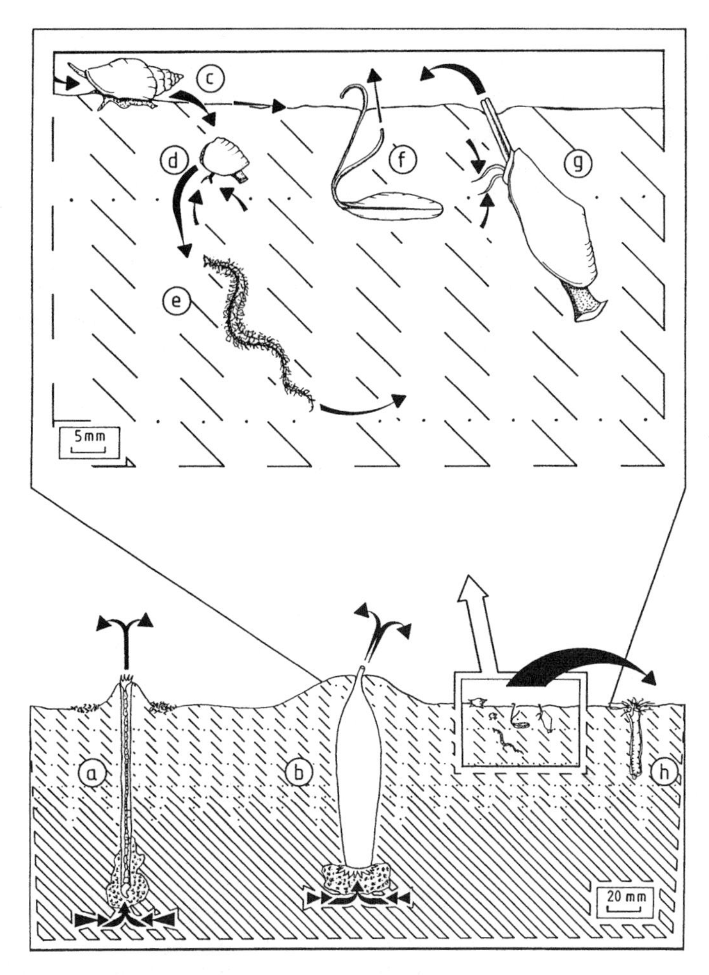

FIGURE 3.57 Methods of mixing and recycling sediments by deposit feeders: (a) maldanid polychaete; (b) holothurian; (c) gastropod (*Nassarius*); (d) nuculid bivalve (*Nucula* sp.); (e) errant polychaete; (f) tellinid bivalve (*Macoma* sp.); (g) nuculanid bivalve (*Yoldia* sp.); and (h) anemone (*Cerianthis* sp.). Oxidized sediment lightly hatched; reduced sediment densely hatched. (Redrawn from Rhoads, D.C., *Oceanogr. Mar. Biol. Annu. Rev.*, 12, 275, 1974. With permission.)

adhesion. As a consequence such sediments are readily resuspended. Sedentary deposit and suspension feeders (especially tube dwellers), on the other hand, bind the bottom by the concentration of sediment into tubes and by entrapping particles between the tubes, so that a "tighter" fabric results in greater compaction and adhesion (Rhoads, 1974). Bioturbation also substantially increases sediment water content. Intensely bioturbated sediments are frequently resuspended by wave and tidal currents (Young and Rhoads, 1971).

3.7.2 Impact of Bioturbation on Benthic Infauna

One of the best studied bioturbators is the lugworm, *Arenicola*. Reise (1985) showed that, at high worm densities, the activities of *Arenicola marina* disturbed the tube-building spionid polychaete, *Pygospio elegans*, and reduced its numbers. Similarly, in field and laboratory experiments, DeWitt and Levinton (1985) showed that the disturbance generated by the plowing action of the mud snail, *Ilyanassa obsoleta*, led to the migration of the tube-building amphipod, *Microdeutopus gryllotalpa*. Building on the earlier observations of Reise, Flach (1992) added a range of *Arenicola marina* densities to azoic experimental plots and followed the subsequent recolonization. The experiments showed that *Arenicola* had a strong negative effect on the amphipod, *Corophium volutator*, and on the juveniles of a variety of other bivalve and polychaete species. In the case of *Corophium*, densities were halved when the average density of *Arenicola* for the area was added to the experimental plots. Laboratory observations suggested that the sediment reworking by the lugworms stimulated *Corophium* to emigrate.

Other intertidal species for which the effects of bioturbation have been demonstrated include the predatory naticid gastropods that plow through the sediments. Wiltse (1980) studied one such species, *Polinices duplicatus*, and conducted a year-long caging experiment to study both the effects of predation on molluscs and disturbance to non-

TABLE 3.29
Rates of Reworking of Sediments by Deposit Feeders (ml wet sediment individ^{-1} yr^{-1})

Species	Location	Rate	Reference
		Polychaeta	
Clymenells torquata	Cape Cod, MA	274	Rhoads (1963)
Clymenells torquata	Beaufort, NC	246	Magnum (1964)
Clymenella torquata	Passamaquodduy, New Brunswick	96	Magnum (1964)
Pectinaria goiuldii	Cape Cod, MA	400	Gordon (1966)
		Bivalvia	
Yoldia limatula (pseudofeces)	Cape Cod, MA	257	Rhoads (1967)
Nucula annulata (fecal pellets)	Cape Cod, MA	365	Young and Rhoads (1971)
		Arthropoda	
Calianassa	Mugu Lagoon, CA	75-cm thick layer trasnsport to surface each year	Warne (1967)

TABLE 3.30
Rates of Biodeposition by Suspension Feeding Invertebrates

Organism	Location	Rate	Reference
		Bivalves	
Cerastoderma edule	France	648 mg wet wgt individ^{-1} day^{-1}	Dame (1986)
Cerastoderme edule	Wadden Sea	1 × 10 m-tons day total pop.$^{-1}$ yr^{-1}	Verwey (1952)
Mytilus edulis	Wadden Sea	25–175 × 10 m-tons dry total pop.$^{-1}$ yr^{-1}	Verwey (1952)
Crassostrea virginica	Japan	3 mg wet individ^{-1} day^{-1}	Ito and Imai (1955)
Crassostrea virginica	Texas	0.7 ton dry acre^{-1} day^{-1}	Lund (1957)
Crassostrea virginica	Chesapeake Bay	18.4 mg dry individ^{-1} day^{-1}	Haven and Morales-Alamo (1966)
Mya arenaria	Chesapeake Bay	25 mg dry individ^{-1} day^{-1}	Haven and Morales-Alamo (1966)
Geukensia demissa	Chesapeake Bay	125 mg dry individ^{-1} day^{-1}	Haven and Morales-Alamo (1966)
		Barnacle	
Balanus eburneus	Chesapeake Bay	18.4 mg dry individ^{-1} dry^{-1}	Haven and Morales-Alamo (1966)
		Tunicate	
Molgula manhattensis	Chesapeake Bay	60 mg dry individ^{-1} day^{-1}	Haven and Morales-Alamo (1966)

mollusc species. For both mollusc and non-mollusc species, community attributes such as diversity, evenness, the number of species, and the density of total individuals all decreased with increasing *Polinices* density. Samples taken inside and outside the trails made by *Polinices* showed that disturbance to the surface sediment layers decreased the abundance of spionid polychaetes.

Flint and Kaike (1986) who studied the infaunal benthos of Corpus Christie Bay, Texas, over a 3.5-year period found that bioturbation by larger burrowing infauna such as enteropneusts, ophiuroids, and echiurians had a major impact on community composition. These species oxygenate the sediments and make them less compact due to their movements and thus allow some species to occupy otherwise uninhabitable deeper sediments. These impacts also lead to colonization by new infaunal species. Their data (Figure 3.58) show a marked change in the depth distribution and species richness of the fauna associated with the colonization of the area by the enteropneust worm (acorn worm), *Schizocardium* sp. The community reverted to its original state two years later when *Schizocardium* disappeared.

3.7.3 Influence of Macrofaunal Activity on the Chemistry of the Sediments

Intensive biogenic mixing and irrigation of the sediment, especially in the top 10 to 30 cm, takes place depending on the mix of species present. Rhoads (1974) lists the

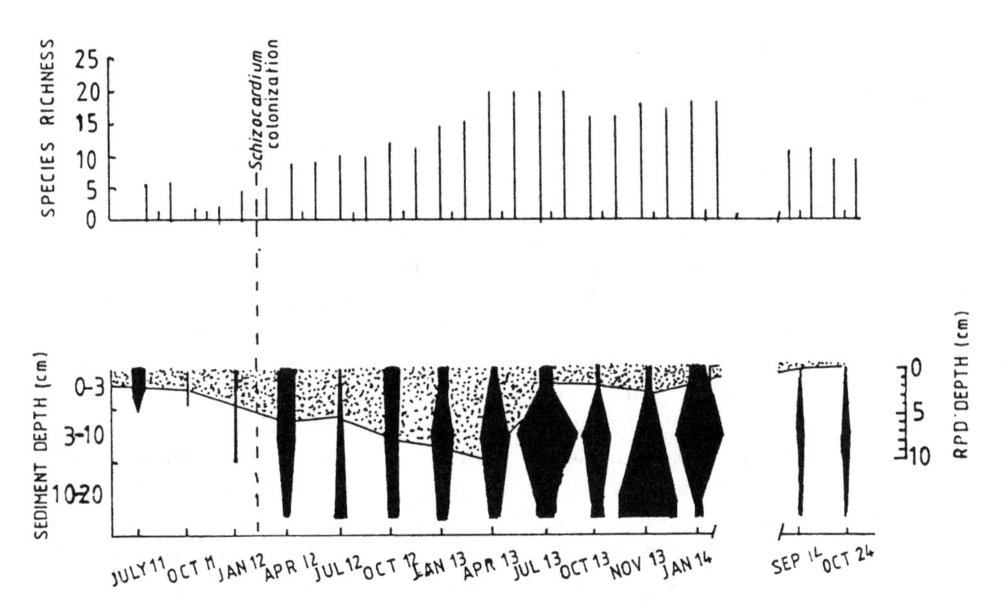

FIGURE 3.58 Vertical distribution of benthic macroinfaunal densities and total species richness at a Station in Corpus Christie Bay over 14 sampling intervals from 1981 to 1984. The time of the establishment of the bioturbating species, *Schizocardium* sp. is marked with a dashed line. (Redrawn from Flint, R.W. and Kaike, R.D., *Mar. Ecol. Progr. Ser.*, 31, 26, 1986. With permission.)

chemical processes and factors modified by macrofaunal activity as:

1. Rate of exchange of dissolved and adsorbed ions, compounds, and gases across the sediment-water interface.
2. Magnitude and form of the vertical gradients of Eh, pH, and dissolved O_2.
3. Transfer of reduced compounds from below the RPD layer to the aerated surface.
4. Cycling of carbon, nitrogen, sulfur, and phosphorus.
5. Concentration of elements in tissues and sediments that are dilute at ambient concentrations.

Space does not permit a discussion of these points in detail and only a few selected examples will be given. Montague (1980) has discussed the influence of fiddler crab burrows on the metabolic process in salt marsh sediments. On the Georgia marshes, fiddler crab (*Uca* sp.) burrow densities may range from 50 to 700 m^{-2} with the highest density near creek banks (Katz, 1980). Through their activities they increase the surface area of the salt marshes by 20 to 60% or more (Katz, 1980). Water within the burrows is oxygenated and the walls of the burrows are aerobic, similar to that of the sediment surface layer. Improved water circulation through the subsurface sediments dilute the salts that accumulate from cordgrass transpiration (Smart and Barko, 1980). Lowering of the sediment salinity can improve nutrient uptake and growth in the cordgrass *Spartina alterniflora* (Smart and Barko, 1980; Linthurst and Seneca, 1981). In their excavation of their burrows, fiddler crabs also transport organic carbon from the subsurface anaerobic sediments to the surface where more rapid aerobic decomposition occurs (Hackney and de la Cruz, 1980). Montague (1982) found that fiddler crabs in the Sapelo Island salt marsh transported 26 g organic C m^{-2} from below ground during the month of July 1979. The annual amount of such transport was estimated at 157 g C m^{-2}, equivalent to 20% of the belowground production of *Spartina alterniflora*.

Montague (1982) found that the mean burrow respiration rate was 2.1 mg O_2 hr^{-1}, accounting for 20 to 30% of the salt marsh respiration. Compared to unburrowed plots, burrows increased *S. alterniflora* standing stocks by 23% on the high marsh. He found that the chemistry of the burrow water was different from that of the interstitial water. Salinity of the burrow water was 20 to 23, while that of the interstitial water was 37 to 45. Phosphate and ammonia concentrations were significantly different from that of the interstitial water. In September, burrow water contained nearly three times the phosphate of the interstitial water. Fiddler crabs excrete ammonia, and it was estimated that 10 g of fiddler crabs m^{-2} probably regenerate more that the 4.2 to 5.7 g N m^{-2} $year^{-1}$ required to supply 19% of the production of *S. alterniflora* (Whitney et al., 1981).

Montague (1982) consider that fiddler crab burrowing activity provided five of the six factors postulated by Morris (1980) that improve growth of short *S. alterniflora*. These factors are: (1) more oxygen, even though *S. alterniflora* supplies some oxygen internally to its roots ; (2) less hydrogen sulfide; (3) lower salinity (Nestler, 1977; Smart and Barko, 1980); (4) increased rates of nutrient diffusion through pore water; (5) added nitrogen (Linthurst and Seneca, 1981;); and (6) greater exchange capacity with the sediments.

Hoffman et al. (1984) investigated the impact of fiddler crab (*Uca pugnax*) deposit feeding on the abundance of meiofauna in a tall *Spartina alterniflora* salt marsh in New England. Total removal of *U. pugnax* from experimental enclosures resulted a tenfold increase in the abundance of nematodes and meiofaunal crustaceans in comparison to experimental enclosures with *U. pugnax* at natural densities. It is clear that fiddler crab deposit feeding plays an important role in regulating the abundance and distribution of meiofauna in slat marshes.

Aller and Yingst (1985) in a series of tank experiments with three deposit feeders (the polychaete *Heteromastus filiformis*, and the bivalves *Tellina texana* and *Macoma balthica*) found that at natural population abundances, each species had major effects on sediment overlying water solute transport, bulk sediment transport rates, and microbial distributions. One-dimensional transport models of Cl^- pore water profiles demonstrated that, when macrofauna are present, effective diffusion coefficients were 2 to 5 times the molecular diffusion rates in the upper 8 to 12 cm of the sediment. Net sedimentary production rates of NH_4^+ were increased by 20 to 30% when macrofauna were present.

Deposit-feeding polychaetes have also been shown to impact on the silicate dynamics of the sediments. Marinelli (1992), in a series of laboratory experiments, tested the effects of two polychaetes, a surface deposit feeder *Eupolymenia heterobranchia*, and a dead-down deposit feeder *Abarenicola pacifica*, on silicate dynamics in sediment pore water and the overlying waters. Silicate fluxes were positively correlated with the frequency of new burrows or tube construction, and were strongly related to the activity of the organisms. Benthic diatoms were found to exert a significant effect on silicate removal from the sediments and the water column. It is thus clear that burrowers and deposit feeders can profoundly modify the nutrient and carbon dynamics of the sediments in which they live as well as that of the overlying water.

An additional chemical impact is biogenic contamination of the sediments with halogenated aromatic compounds, such as bromophenol (Woodin, 1999). A number of hemichordates and at least five plychaete families are now known to contain such compounds and to release them into the sediments (Woodin, 1991; Steward et al., 1992; Woodin et al., 1993). Sediments containing such compounds are rejected by new recruits (Woodin et al., 1993) and also appear to deter disturbance agents.

3.7.4 Influence of Macrofaunal Activity on Microbial Parameters in Intertidal Sediments

It is widely assumed that macrofaunal burrowing activity in marine sediments has a stimulating effect on microbial growth and metabolic activities (Hargrave, 1970b; Hylleberg and Henricksen, 1980; Henricksen et al., 1980; Morrison and White, 1980; Yingst and Rhoads, 1980; Hines and Jones, 1985; Kikuchi, 1986; Reichardt, 1986). Such biological interactions between the interstitial microbiota and meiofauna and the macrofauna are more prevalent toward the sheltered end of beach wave exposure and apparently are absent from coarse-grained exposed beaches.

The most extensive work on such interactions is that of Reise (1981; 1983a,b; 1985) on the sand flats of the Dutch Wadden Sea. In his 1983 study he assessed the impact of the biotic enrichment of intertidal sediments by experimental aggregates of the tellinid deposit-feeding bivalve *Macoma balthica*. The tellinids stay buried at a depth of 3 to 6 cm while their long inhalant siphons pick up deposits from the surface of the sediment. Figure 3.59 compares the vertical distribution of small zoobenthos (primarily meiofaunal species and including ciliates > 0.2 mm length) in sediment cores with and without *Macoma balthica*. From the figure it can be seen that there was a

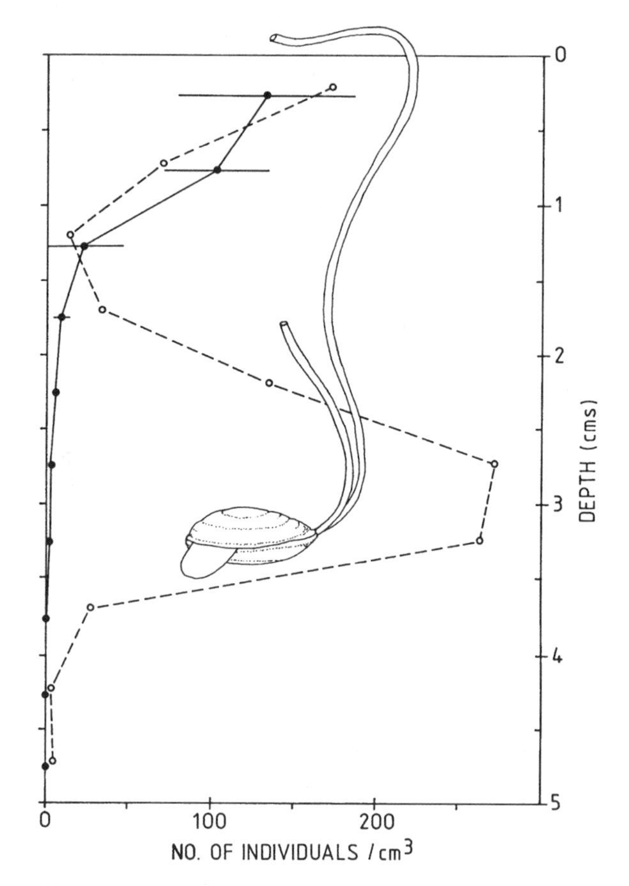

FIGURE 3.59 Vertical distribution at 0.5 cm intervals of meiofauna (including Ciliata > 0.2 mm length) in the upper 5 cm of normal sediment (mean and standard deviation of 5 cores of 2 cm^{-2}) and in a single core (dashed line) which happened to include an individual of *Macoma balthica* positioned at 3.5 cm depth. (Redrawn from Reise, E., *Mar. Ecol. Progr. Ser.*, 12, 231, 1983a. With permission.)

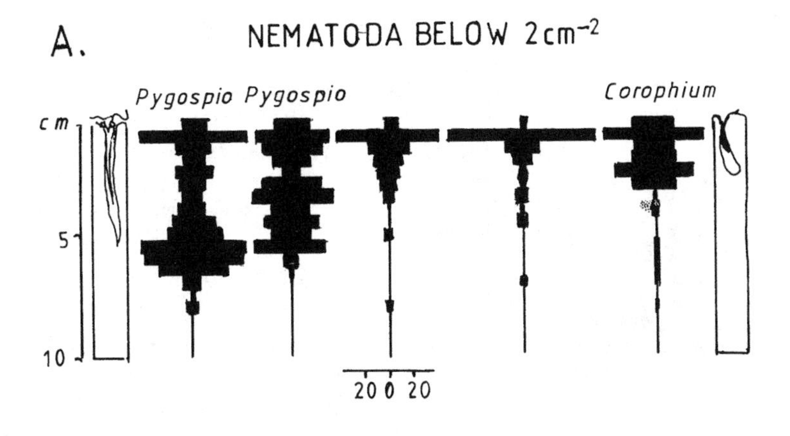

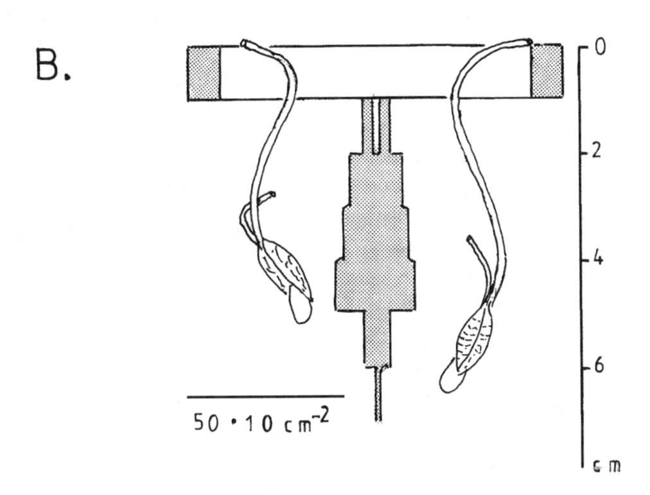

FIGURE 3.60 Effects of macrofauna on the vertical distribution of meiofauna. (a) Distribution of nematodes (individuals below 2 cm^{-2} at intervals of 0.5 cm) in the presence of the polychaete *Pygiospio elegans*, the amphipod *Corophium arenarium*, and in the absence of macrofauna. (b) Distribution of platyhelminth abundance in normal sediment (white) and with experimental aggregates of *Macoma balthica* (shaded). (Redrawn from Reise, K., *Tidal Flat Ecology: An Experimental Approach to Species Interactions*, Springer-Verlag, Berlin, 1985, 119 and 123. With permission.)

dramatic increase in their abundance at the level of the *Macoma* individual.

Reise (1983) also found that Turbellarians also showed a preference for the tellinid aggregates (Figure 3.60a). Figure 3.60b illustrates the impact of the polychaete *Pygiospio elegans* and the amphipod *Corophium arenarium* on the vertical distribution of nematodes. Small polychaetes also responded to the presence of the tellinids. The interstitial annelid *Macrophthalamus sczekowskii* populated the transition zone between the surface and subsurface when *Macoma* was present. This hesionid polychaete feeds on bacteria and microalgae.

High abundance of meiofauna in the micro-oxic zones generated by the macrofauna in the subsurface anoxic sediments is a general phenomenon, due to the increased oxygen supply and the increased microbial community, which provides food for the meiofauna (Aller and Yingst, 1978; Reise, 1981). In *Macoma* the exhalant siphon terminates below the surface and it is responsible for the micro-oxic zones that surround the level of the shell. The exhalant current probably also produces local concentrations of nutrients that stimulate the growth of the microalgae. Thus an increased food supply is provided for the diatom- and bacteria-feeding meiofauna. The meiofauna in turn provide food for the predatory Turbellarians.

Hoffman et al. (1984) investigated the impact of deposit feeding by the fiddler crab, *Uca pugnax*, on the abundance of meiofauna in the tall-form *Spartina alterniflora* zone of a sheltered New England salt marsh. Total removal of *U. pugnax* from experimental enclosures resulted in a tenfold increase in the abundance of nematodes and meiofaunal crustaceans and a fourfold increase in the abundance of segmented worms in comparison to experimental enclosures with *U. pugnax* at natural densities. These results suggest that fiddler crab deposit feeding plays an important role in regulating the abundance and distribution of meiofauna in salt marsh sediments. This fiddler crab regulation of meiofaunal abundance may be

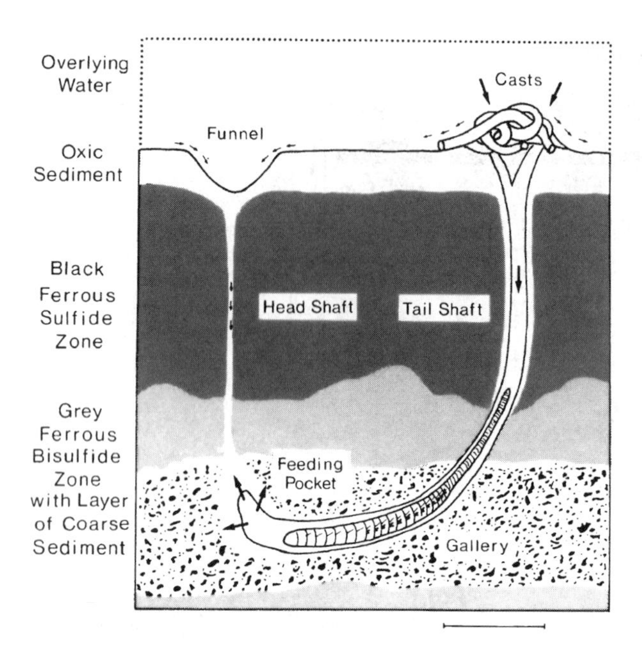

FIGURE 3.61 The lugworm *Arenicola marina* in its burrow on a sandy tidal flat of Konigshafen, Island of Sylt, Netherlands. *Large arrows* indicate flow of overlying water pumped into the burrows. *Small arrows* indicate sediment movement. (From Reise, K., *Tidal Flat Ecology: An Experimental Approach to Species Interactions*, Springer-Verlag, Berlin, 1985, 127. With permission.)

a critical process in dictating the distribution of meiofauna in salt marsh habitats and this may secondarily influence sediment decompositional processes, which are directly mediated by the meiofauna.

Reise (1985) also demonstrated significant effects of the burrows of the lugworm, *Arenicola marina*, on the sediment meiofauna. *Arenicola* dominates the macrofaunal biomass of the tidal flats of the Dutch Wadden Sea. It is also a major bioturbator and can ingest 1 cm^{-3} hr^{-1} with an annual sediment reworking of a layer up to 33 cm. Nematodes avoid the unstable funnels and fecal mounds (see Figure 3.61). However, all the meiofaunal taxa investigated were more abundant in samples taken from the burrow structures when compared to the normal anoxic sediment at the same depth (Table 3.31). Reichardt (1988) in an investigation of the impact of bioturbation by *Arenicola marina* on microbial parameters in intertidal sediments found that the burrow walls contained twice as much organic matter as the sediment surface. Viable counts of aerobic proteolytic and chitonolytic bacteria peaked at the external and internal boundary layers (sediment surface and burrow walls). Burrow walls showed maximal bacterial production, maximal microheterotrophic activity, and highest levels of certain hydrolytic enzymes. The enhanced densities of meiofauna in the vicinity of burrow walls reported by Reise (1985) is indicative of their grazing on the elevated bacterial levels. Reichardt (1986) found evidence of intensive grazing on bacteria in the burrow walls of the nereid polychaete *Nereis diversicolor*.

In the perizoic niches and microhabitats provided by the polychaete burrow systems positive feedback of macrofaunal deposit feeding on bacterial growth may occur. This has been termed "gardening" (e.g., Hylleberg, 1975; Dobbs and Whitlach, 1982; Plante et al., 1990). In sediments bioturbated by *Arenicola* or other subsurface deposit feeders, certain biochemical processes are intensified (Hines et al., 1982; Holdren and Armstrong, 1982; Henricksen et al., 1983; Reichardt, 1988). There is evidence of selective survival of bacteria in the passage of sediment through the guts of bioturbating polychaetes and other deposit feeders. In addition the growth of bacteria surviving passage through an animal's gut may be stimulated by an increased supply of nutrients in the feces, as well as by the removal of predators, inhibitors, and competitors during selective digestion (Hargrave, 1970b; Juniper, 1981; 1982; Plante et al., 1989).

3.8 MICROBIAL ECOLOGY AND ORGANIC DETRITUS

3.8.1 INTRODUCTION

Shore ecosystems are broadly open systems exchanging nutrients, producers, and consumers with adjacent coastal systems as well as with adjacent marshes and swamps (Darnell, 1967b). The inherent feature of community metabolism on soft shores is the role played by organic detritus in its various forms. In such communities, especially those fringed with salt marshes or mangroves, current views emphasize the importance of the decomposer-based food web. As Newell (1979) states: "It has become increasingly apparent during the past decade that although phytoplankton is the basis for production in oceanic waters, food chains in the intertidal zone and shallow coastal waters are based primarily on the production of attached plants." These may be utilized directly in the form of the diatoms and photosynthetic flagellates of the benthic microflora, small epiphytes attached to sand grains, epiphytic algae growing on larger macroalgae or on marsh plants, or direct consumption of the macroalgae by herbivores. Grazing by herbivores, especially in salt marshes, ingest only a small fraction of the macroalgal or macrophyte crop, while the remainder decays to form dissolved or particulate organic matter. Much of the organic matter thus produced in estuarine and coastal marine areas is eventually decomposed by aerobic and anaerobic microbial processes either on the surface or in the sediments.

The dissolved organic matter released during decomposition, together with that released during the metabolism of the phytoplankton, benthic microalgae, epiphytic

TABLE 3.31
Abundance of Meiofauna (individuals cm^{-3}) at Three Depths of Normal Sediment and in Three Regions beside Lugworm Burrows, Sampled during Summer 1978 on a Sandy Flat in Konigshaften, Island of Sylt

	Habitat and Sediment Depth					
	Sandy Flat			Beside Lugworm Burrows		
Meiofaunal Taxa	Oxic 0–1 cm	Anoxic 1.5 cm	Anoxic 5–15 cm	Head Shaft 5–15 cm	Pocket 18 cm	Tail Shaft 5–15 cm
Gnathostomulida	4	3	0	0	0	22
Platyhelminthes	111	11	1	11	115	27
Gastrotricha	8	2	1	2	49	4
Nematoda	950	270	39	462	118	357
Annelida	63	6	0	51	120	3
Copepoda	509	6	0	22	165	40
Ostracoda	78	0	0	4	1	1
Sediment sampled (cm^3)	22	64	50	30	15	60

Note: All sampling was done at low tide with a corer of 1 cm^2 cross-section. When shafts of the burrows were sampled, these were positioned at the center of the corer.

Source: From Reise, K., *Tidal Flat Rcology: An Experimental Approach to Species Interactions*, Springer-Verlag, Berlin, 1985, 131. With permission.

microalgae, macroalgae, and the macrophytes, may either be absorbed directly by animals, or together with the particulate organic matter act as a substrate for heterotrophic metabolism. Bacteria and fungi rapidly colonize the particulate detritus as it becomes available and through their activities convert the high-fiber, low-nitrogen dead plant material into variously sized particles of high nutritional value. These microbe-laden particles are either incorporated into the surface sediments, or transported to coastal waters by tidal action where they may become available to filter-feeding animals. Some settle to the bottom and become incorporated in the sediment where they may be consumed by surface and burrowing deposit feeders (Teal, 1962).

It has been generally accepted that the consumers of particulate detritus receive most of their nutrition from the attached microorganisms rather than directly from the decomposing plant particles. Egestion of stripped detrital particles returns them to the aquatic environment and the processes is repeated over and over again. Abrasion, maceration by consumers, and the continued action of the decomposers results in smaller and smaller particles. At some point in the sequence the most minute of the fractions are considered to be unidentifiable and are often referred to as amorphous aggregates or organic material of undetermined origin.

Heterotrophic organisms are the major link in the mineralization and transformation of organic matter, thus recycling inorganic nutrients and making them available to the primary producers. Two groups of organisms are involved: aerobes and anaerobes (Figure 3.62). The left side of the figure is divided into aerobic and anaerobic environments on the basis of the "oxidizing potential," or Eh. Aerobic environments are in the water column and the oxidized surface layer of the sediments (see Section 3.2), while the anaerobic environment comprises the sediment below the RPD (redox potential discontinuity) layer. The important aspect of the anaerobic zone is that it represents a storehouse of chemical energy that may have some ability to feed back into the food chain (Sorokin, 1978; 1981). In this zone different groups of bacteria exist that can decompose organic matter using sulfate and nitrate as a source of oxygen, and in the process produce reduced substances such as CH_4, H_2S, and NH_4. The latter compounds can be used by other bacteria, some of which may be strictly chemoautotrophic since they are able to use CO_2 as a source of carbon, and inorganic compounds as a source of energy. The top of Figure 3.62 is divided into energy-dependent reactions and organic carbon reactions.

3.8.2 Organic Matter

3.8.2.1 Sources of Organic Matter

Darnell (1967a) defines organic detritus (Particulate Organic Matter, POM) as "all types of biogenic material in various stages of microbial decomposition which represent potential energy sources for consumer species." Thus defined, organic detritus includes all dead organisms as well as secretions, regurgitations, and egestions of living organisms, together with all subsequent products of decomposition that still represent potential sources of

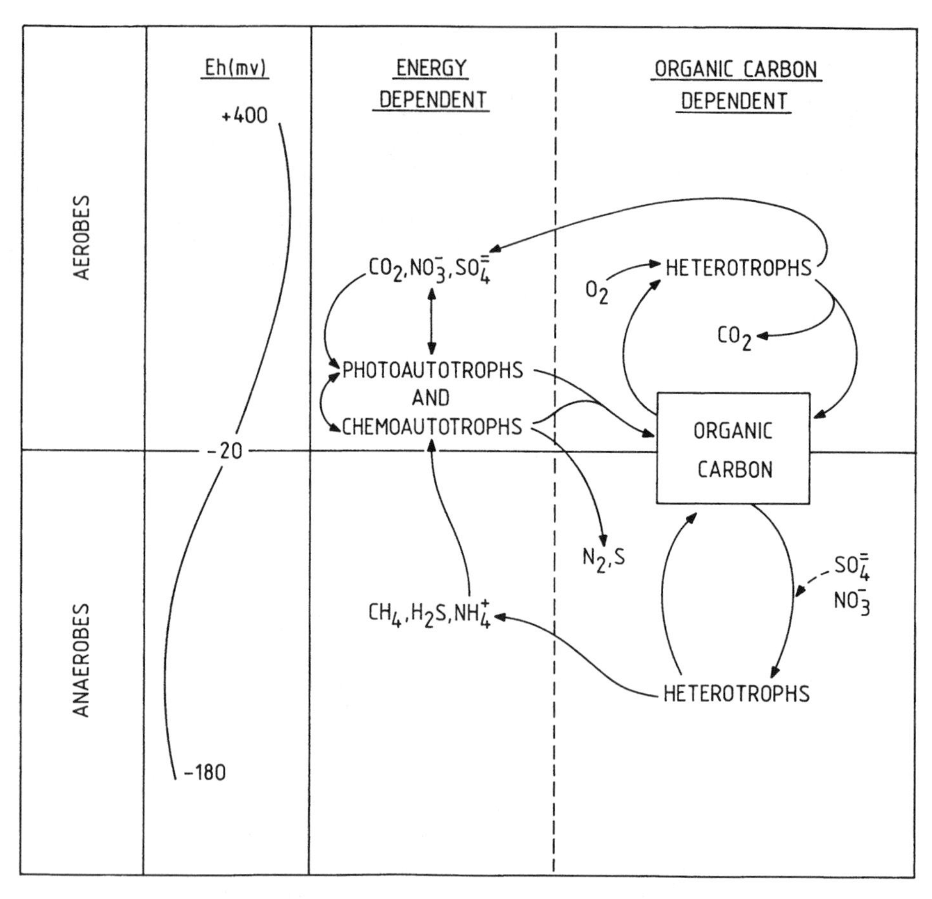

FIGURE 3.62 Organic carbon and energy-dependent cycles in marine aerobic and anaerobic environments. (Redrawn from Parsons, T.R., Takashi, M., and Hargrave, B.T., *Biological Oceanography Processes,* 2nd. ed., Pergamon Press, New York, 1979, 169. With permission.)

energy such as proteins or amino acids. Organic detritus consists of a range of particles varying from very large to very small and it is often convenient to distinguish coarse POM (that material retained by filters with apertures of 1 μm in diameter) and fine or subparticulate (sometimes called nanodetritus) POM (material that passes through such filters). Recent studies (e.g., Joint and Morris, 1982) have demonstrated that the bulk of organic matter in the coastal waters and estuaries consists of amorphous aggregates not identifiable as decomposition products of macrophytes such as macroalgae, marsh grasses, and mangroves. Odum and de la Cruz (1967) separated suspended particles in a Georgia marsh (U.S.A.) into three size fractions: "coarse detritus," retained by a netting of 0.239 μm; "fine detritus," passing through the 0.239 μm net but retained by a net of 0.064 μm; and "nanodetritus," passing through the 0.064 μm aperture but retained by a millipore filter with 0.045 μm pore size. The coarse and fine fractions comprising 1 and 4% of the total particulate, respectively, were identified as fragments of vascular plants, while the nanofraction (95% of the total) comprised the amorphous aggregates mentioned above. The nanodetritus includes colloidal micelles as well as chemically reduced organic molecules. Colloids may include molecular aggregates or large molecules, such as proteins, carbohydrates, lipids, etc. Smaller molecules may exist as dissolved liquids (bichromes, vitamins, amino acids, sugars, urea, nitrites, nitrates, etc.), or as dissolved gases (methane, hydrogen sulfide, etc.). Thus there is a continuous gradient from small molecules to large organic particles, most of which represent a potential energy source.

Figure 3.63 depicts a generalized scheme of biological decomposition and organic detritus formation. Biological decomposition involves both mechanical and chemical breakdown. The latter is brought about primarily through the processes of hydrolysis and oxidation. According to Darnell (1967b) three agents are likely to be involved: autolysis, where there is the breakdown of tissues by their own enzymes; chemical breakdown during the passage through the guts of consumers; and the chemical activities of heterotrophic organisms (bacteria and fungi).

Particulate organic matter: The great bulk of the POM is derived from primary production. However, there are a multiplicity of other sources. POM in coastal ecosystems can be classified as autochthonous (derived from inside the system), and allochthonous (from outside the system).

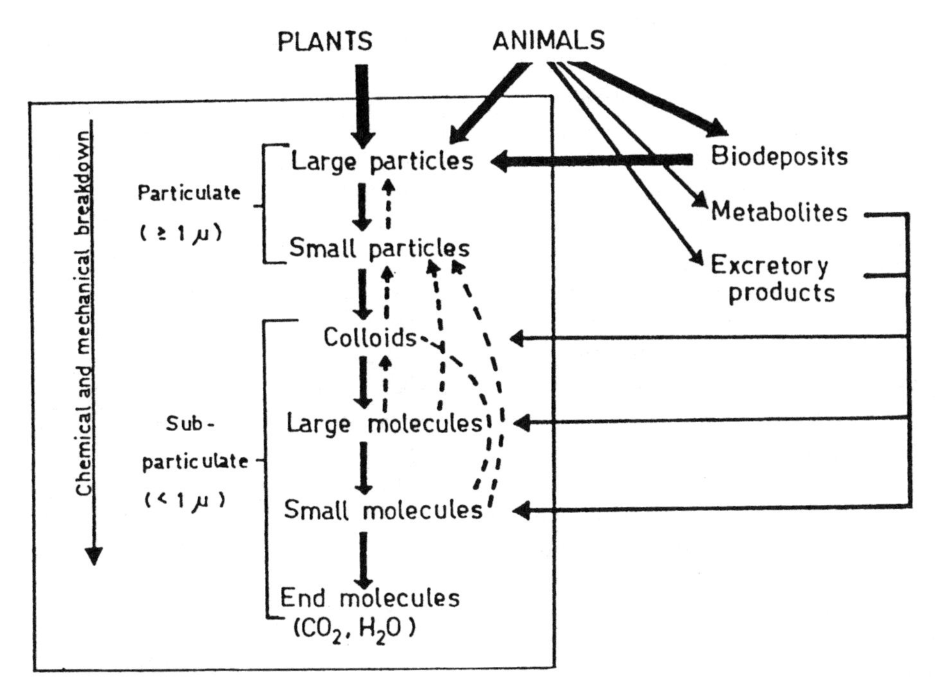

FIGURE 3.63 Schematic view of detritus formation and decomposition. (After Darnell, R.M., in *Estuaries*, Lauff, G.H., Ed., American Association for the Advancement of Science, Publ. No. 83, Washington, D.C., 1967b, 377. With permission.)

1. Autochthonous sources:
   a. Plankton (including bacteria, phytoplankton, and zooplankton)
   b. Marginal emergent vegetation (salt marsh plants and mangroves)
   c. Submerged macrophytes (sea grasses)
   d. Macroalgae
   e. Periphyton (epiphytic micro- and small macroalgae growing on the stems and leaves of emergent and submerged macrophytes and other surfaces)
   f. Benthic microalgae (diatoms, flagellates, and blue-green algae)
   g. Sediment bacteria
   h. Sediment micro- and meiofauna
   i. Sediment macrofauna
   j. Fecal pellets and pseudofeces
2. Allochthonous sources:
   a. Freshwater swamp vegetation
   b. River-borne phytoplankton and organic detritus, including sewage
   c. Beach and shore material washed into estuaries and inlets by the tides and particularly during storms
   d. Wind-blown terrigenous material, especially leaves and pollen grains

Examination of organic particles in marine ecosystems has revealed two main types: particles that clearly show remnants of previous cellular structure, and smaller particles that are clearly organic precipitates (Goldman, 1984b). There are a variety of mechanisms that bring about the conversion of dissolved organic matter to particulate form, all of them involving the collection of surface-active material at a gas-liquid interface and compression if the interface. The surfaces of bubbles and the surface of the sea are sites of the aggregation of surface-active molecules. Wave action is an important agent in the conversion process. In addition inorganic particles frequently act as a focus for the conversion of DOM to POM. Bacteria often form large quantities of extracellular POM, which often brings about the adhesion of bacteria into clumps.

Dissolved organic matter (DOM): There are two pools of DOM within coastal ecosystems: the water column pool and the sediment pool. The chemical composition of these pools is not well known. The vast majority of the pools is composed of refractory material (Bada and Lee, 1977), while a small, but significant proportion is composed of biologically active organic compounds. Of these biologically active compounds, vitamins, amino acids, peptides, proteins, mono- and polysaccharides, fatty acids, and nucleotides have been detected in seawater (see review by P.J.Le B.Williams, 1975). The DOC to POC ratio in seawater is on average 10:1 (Wetzel, 1984).

DOM is derived from a variety of sources, including:

1. Microalgal (phytoplankton, benthic microalgae, epiphytic microalgae) exudation. Excretion of a proportion of the photoassimilated carbon from microalgae is now generally accepted, although the quantities and rates of excretion are still debated (Wiebe and Smith,

1977; Fogg, 1983; Lancelot, 1984; Lignell, 1990). Fogg et al. (1965) found that 0 to 50% of photoassimilated carbon in a range of microalgae was released during *in situ* studies conducted in marine and fresh waters. Since that time Berman and Holm-Hansen (1974) and Lignell (1990) have recorded amounts within this range in marine phytoplankton. Commonly accepted values for percentage exudate release (PER) range from 5% of primary productivity (eutrophic waters) to about 40% (oligotrophic waters; Fogg, 1983). In Baltic Sea coastal waters, Lignell (1990) found that annual excreted organic carbon (EOC) and total exudation (EOC plus bacterial uptake of exudates) values amounted to 4.6 and 7.1 of primary production; at the same time, the net and total exudation averaged 7.5 and 10.8% of the current phytoplankton biomass per day. Baines and Pace (1991) analyzed 16 studies of extracellular carbon release and found that the average was 13% of total exudation. A figure of 10% is often used in calculating carbon budgets.

Epibenthic microalgae are another potential source of DOM. Their excretion has proved difficult to measure since sediment bacteria would be expected to assimilate any labile organic material as fast as it released (Pomeroy et al., 1981). Darley et al. (1976) suggested that the amount lost is very small, approximately 1% of total fixation, but this only represents the nonassimilated excreted matter that remained in the sediment after 15 min.

2. Macroalgal exudation. Siebruth (1969), Moebus and Johnson (1974), and Khailov and Burkova (1969) have reported 10 to 40% of carbon excreted from various macroalgae. However, more recent measurements (e.g., Pregnall, 1983; Arnold and Manley, 1985; Sondergaard, 1990) have given a range of 2 to 15%.
3. Zooplankton secretion and excretion.
4. Macrofaunal secretion and excretion.
5. Microalgal autolysis.
6. Macrophyte exudation.
7. Decomposition of macroalgae and macrophytes.
8. Decomposition of dead animals.
9. Release of DOM by microflagellates and ciliates.
10. Sloppy feeding.

3.8.2.2 Quantities of Particulate Organic Matter (POM)

The level of particulate organic matter in coastal waters varies considerably depending on the input of detrital material from the various sources listed above. Measured concentrations of POM in suspension in marine areas ranges from 0.01 to 2.0 g C m^{-2} with the highest concentrations occurring in inshore waters (Mann, 1982). In surf-zone waters on exposed sand beaches, McLachlan and Bate (1984) estimated that there was 1 to 5 g C m^{-3}; higher values have been recorded in estuaries. In the Sapelo Island marsh area in Georgia, U.S.A., the annual recorded average is ca. 9 g C m^{-2} (Pomeroy and Imberger, 1981). In most coastal areas much of the input of dead plant material settles fairly rapidly to the bottom where it becomes incorporated in the sediments, where much higher levels have been recorded than those found in the water. Estimates of sediment particulate organic matter vary widely ranging from 100 g C m^{-2} for a New England salt marsh to 18,200 g C m^{-2} for the Sapelo Island salt marshes. Other data available are expressed in terms of mg C g^{-1} (dry weight) of sediment and are not readily comparable with those listed above. Nixon (1980) lists values ranging from 16 mg C g^{-1} for six North Carolina marshes Broome et al. (1975) to 240 mg C g^{-1} for a marsh at Lees, Delaware (Lord, 1980). Nixon 1980) concluded that marsh sediments were richer in organic carbon than those found in the sediments of open estuarine areas and offshore areas. The distribution of organic carbon varies considerably with depth in marshes and in subtidal sediments. On the basis of the available data it appears that much of the organic matter buried in sediments is decomposed. However, a small amount of the more refractory carbon is effectively trapped within the sediment and removed from recycling within the system. The amount of carbon so removed is a function of the carbon content of the deep sediment, the accretion rate, and the density of the sediment. Nixon (1980) concluded that the quantity buried appears to lie between 150 and 400 g C m^{-2} $year^{-1}$, He further calculated that the annual marsh sediment input is likely to somewhere between 750 and 1500 g m^{-2}, and that the organic carbon content of this material is at least 50 mg C g^{-1}, giving an organic carbon input associated with sediment trapping of 37 to 75 g C m^{-2} $year^{-1}$. He also calculated that of 500 g C m^{-2} $year^{-1}$ fixed on a marsh in plant biomass, 150 g (30%) would be buried.

Input of macrophytes to sandy cove adjacent rocky shores, or beaches fringing sea grass meadows may be very high. In such situations values of between 20,000 and 30,000 g C m^{-1} (of beach) have been recorded (Robertson and Hansen, 1981; Griffiths et al., 1983; McLachlan and McGwynne, 1986). Following breakdown much dissolved and particulate material may leach into the sand. Estimates of the input of organic matter to coastal sediments vary considerably. Steele and Baird (1968) estimated that the input of organic carbon to a Scottish inlet was 30 g C^{-2} $year^{-1}$, Stephens et al. (1967) estimated 200 g C m^{-2} $year^{-1}$ in a bay on the west coast of Canada, while Webster et al. (1975) estimated 164 g C m^{-2} $year^{-1}$ in St. Margarets Bay, Nova Scotia.

3.8.2.3 Quantities of Dissolved Organic Matter (DOM)

The mean concentration of DOC in rivers flowing into coastal waters is about 10 mg l^{-1}, but may reach 50 mg l^{-1} in waters draining swamps and bogs (Carricker, 1967). Odum and de la Cruz (1967) reported DOC values of up to 20 mg l^{-1} in waters draining salt marshes. Levels of DOC in coast waters usually fall between 1 and 5 mg l^{-1} (Biddanda, 1985), but may reach 20 mg l^{-1}. High concentrations of DOC occur in coastal waters receiving domestic sewage. The DOC concentrations at any one time are dependent upon the rate at which it is being recycled by the microbial populations. The labile portion of the DOC is recycled rapidly, whereas the refractory portion is recycled slowly.

In a study of DOC-bacteria interactions in a Queensland mangrove swamp, Boto et al. (1989) found that the range of sediment interstitial DOC concentrations (4 to 50 mg C l^{-1}) was similar to that reported for temperate marine sediments. Soluble tannins constituted 2 to 51% of the interstitial DOC pools, which varied considerably with tidal height (high > mid > low) and showed significant variation with sediment depth. Surface (1 cm depth) sediment DOC concentrations varied with elevation in the intertidal zone.

Thus primary production in coastal ecosystems produces large quantities of organic matter that is not utilized directly by herbivores but is converted into detritus, which is decomposed into finer POM particles and DOM. While some of the breakdown processes occur in the water column, most occur in the surface sediments. One of the central questions concerning the functioning of marsh-estuarine ecosystems in particular is whether the plant organic matter is processed *in situ*, or whether a major proportion is exported to adjacent waters (Pomeroy et al., 1981). Another question that is basic to our understanding of the functioning of coastal ecosystems is the role of detritus and its associated microbial community in coastal food webs. Both of these questions will be discussed below.

3.8.3 River Input of Organic Carbon

River transport can be an important source of carbon to coastal ecosystems (Stephens et al., 1967; Parsons and Seki, 1970; Sutcliffe, 1973; Head, 1976; Naimen and Sibert, 1978). Naimen and Sibert (1978) examined the timing and magnitude of carbon movements from the Nanaimo River in British Columbia to the mudflats of its estuary. Dissolved organic carbon concentrations in the river ranged from 6 to 14 g m^{-3} dependent on the season with a mean of 6.44 g m^{-3}. When compared to that in many streams and rivers, the concentrations of organic carbon in the Nanaimo River were not especially large, but due to the large annual discharge, the river contributed 76.5% of the total carbon input to the estuary (Figure 3.64). Most of the annual carbon input occurred during autumn freshlets. From October to December the intertidal mudflat received 70% of the annual dissolved organic carbon (DOC: >1 mm) river export, 73% of the annual fine particulate organic carbon (FPOC: >0.5 M, <1 mm) export, and 93% of the annual larger particulate organic carbon (LPOC: >1 mm) export. DOC contributes by far the greatest amount of carbon each year, nearly 2,000 g C m^{-2}. FPOC contributes 56 g C m^{-2} $year^{-1}$, while LPOC contributes only 0.7 g C m^{-2} $year^{-1}$. The total POM concentrations were only 2.8% of the DOM concentrations. On the other hand, Quasim and Sankaranayanan (1972) found

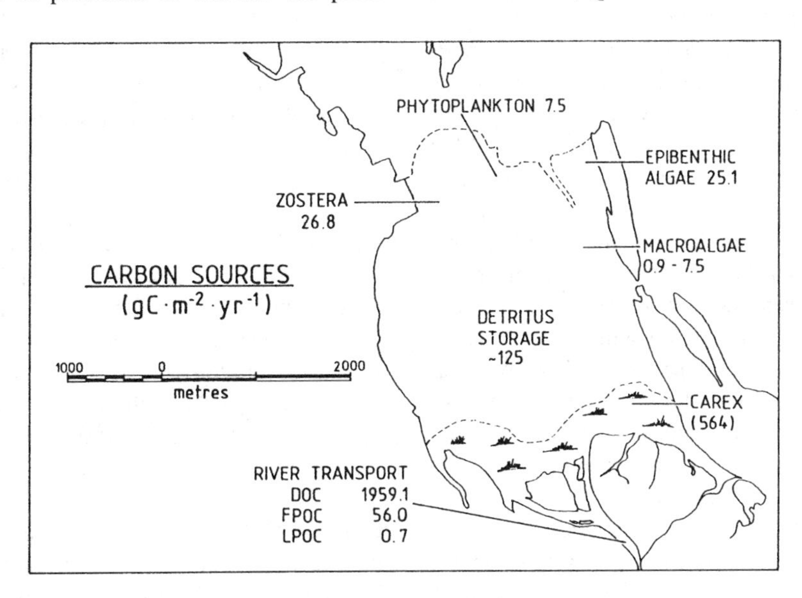

FIGURE 3.64 Sources of carbon (g C m^{-2} yr^{-1}) available to the Nanaimo Estuary, British Columbia, mudflats. (Redrawn from Naimen, R.J. and Sibert, J.R., *J. Fish Res. Bd. Canada,* 36, 517, 1979. With permission.)

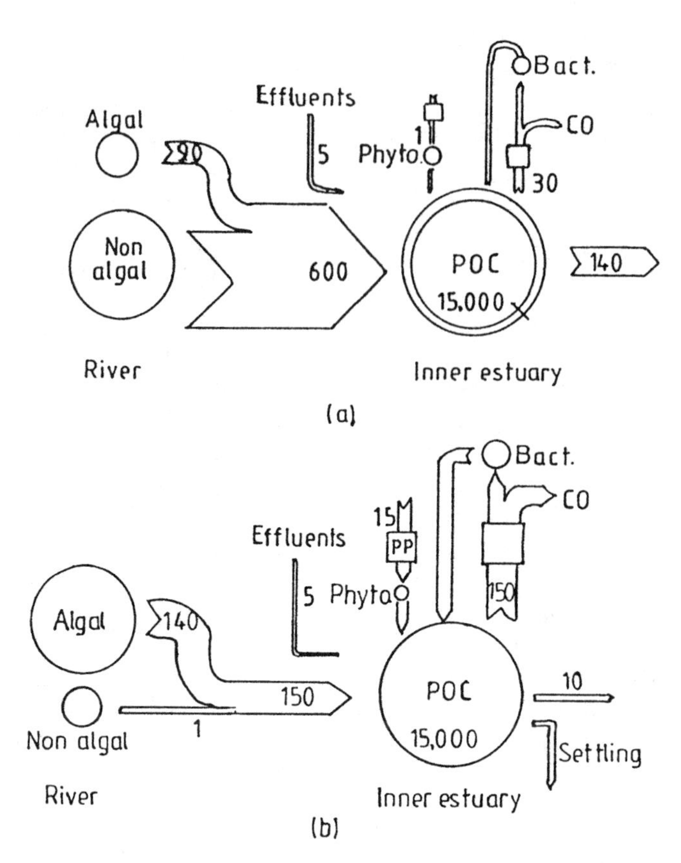

FIGURE 3.65 Approximate budget of POC in the Loire River (France) estuary in (a) a typical high discharge spring situation (Q = 2,000 m^3 sec^{-1}) and (b) a typical low discharge summer situation (Q = 300 m^3 sec^{-1}). Fluxes are expressed in tons C day^{-1}; stocks are in tons C, for the whole inner estuarine zone, i.e., an area of 30 km^2 with a mean depth of 4 m. (Redrawn from Rexellans, J.C., Meybsck, M., Billen, G., Brueaville, M., Etchieber, H., and Somville, M., *Est. Coastal Shelf Sci.*, 27, 681, 1988. With permission.)

that POM derived from terrestrial and aquatic macrophytes constituted the greater part of the POC in suspension in a tropical estuary, with the contribution of the phytoplankton being less than 1%.

There are several primary sources of DOC in rivers: leaves that leach once they fall into the river, periphyton that release significant amounts of DOC, forest soils from which carbon is leached, and atmospheric input via rain. POM in river inflow is derived from leaves and other plant litter washed or blown into the rivers, organic matter derived from marginal macrophyte vegetation or submerged macrophytes, and river phytoplankton. The input of organic matter from these sources is dependent upon the amount of fresh water inflow in relation to the tidal volume of the estuary, the nature of the catchment, and the type of river. It is clear that in some estuaries it is the source of the major input of carbon to the estuarine ecosystem, e.g., in the Nanaimo River estuary it represents 75% of the carbon input.

The input of river discharge on the POC budgets of the Loire Estuary in France is illustrated in Figure 3.65. When the river discharge is high (above 1,000 m^3 sec^{-1}), no phytoplankton develops in the river and the POC input is made up of 80% nonliving terrestrial and detrital material. At a typical discharge of 2,000 m^3 sec^{-1}, the POC flux from the river is about 600 t C day^{-1}. This means that the average load of POC in the inner estuary is 15,000 t C. When the discharge drops below 1,000 m^3 day^{-1}, phytoplankton blooms in the river and primary production reach very high values. This results in a dramatic change in the POC composition where living algae become entirely dominant (about 85% of the total). When this material reaches the inner estuary, the high turbidity strongly inhibits primary production. Phytoplankton cells die rapidly and are quickly degraded by bacterial activity. Under these conditions the amount of particulate material brought into the inner estuary by the river is about 150 t C day^{-1}. Most of this is degraded within the high turbidity zone and consequently the heterotrophic activity is high (5 times that of the high flow situation).

As colloidal and particulate matter carried by freshwater enters the saline environment, the surface charge on the particles approaches zero (Edzwald et al., 1974; Krank, 1975). As a result most of the particulate matter in suspension, sediments within the estuary. This process is effective even for such major rivers as the Mississippi and the Amazon (Milliman et al., 1975).

3.8.4 Microbial Processes in Coastal Waters

3.8.4.1 Microbial Standing Stocks

It has proved difficult to obtain accurate measurements of either the biomass or the activity of microorganisms in the water column. Several authors have addressed methods for the determination of the characteristics of microbial communities (e.g., Jones, 1979; Litchfield and Seyfried, 1979). These involve either direct observations of the organisms in question or an indirect approach involving measures of the chemical components of the microorganisms, especially the analysis of total living biomass using firefly lantern extracts for the determination of adenosine triphosphate (ATP) and other cellular nucleotide triphosphates (Holm-Hansen and Booth, 1966; Rublee, 1982b). The ATP method is rapid, sensitive, relatively inexpensive, but fraught with problems of both analysis and interpretation, particularly when used in sediments (Rublee, 1982a).

Siebruth (1976) and Morita (1977) both give estimates of the mean number of bacteria in the euphotic zone of the sea as 10^{-3} to 10^{-5} ml^{-1}. Siebruth (1976) estimates that this is equivalent to a biomass of 1.5 to 8 mg C m^{-3}. Ferguson and Rublee (1976) found concentrations of bacteria averaging 6.6 × 10^5 cells ml^{-1} in coastal waters off California and commented that several other workers had found 10^5 ml^{-1} to be characteristic of coastal waters in general. Hoppe (1978) used autoradiography to distinguish active from inactive bacteria in the Baltic Sea and found that the annual average number of active bacteria ranged from 4.5 × 10^5 ml^{-1} in offshore waters to 9.5 × 10^5 ml^{-1} in polluted coastal waters. The percentage of active bacteria ranged from 10% in the winter to 56% in the summer. Numerous other studies have indicated that heterotrophic activity, on a per cell basis, is higher in estuarine waters than in connecting fresh or marine waters (Wright, 1978; Goulder et al., 1980).

One of the most comprehensive studies of the distribution of planktonic bacteria is that of Wright and Coffin (1984) in the Essex, Parker, and Ipswich River estuaries, and Ipswich Bay in northern Massachusetts. In winter, bacterial counts and activity were uniformly low throughout the Essex River estuary. During spring there gradually developed a peak of bacteria in mid-estuarine waters, which was sustained throughout the summer and subsided in the autumn. The seasonal range in these waters was tenfold (0.7 to 70 × 10^6 ml^{-1}), while the range in coastal waters was only fourfold (Wright, 1978). Activity and counts appear to be related and follow the same pattern, with highest values in the mid-estuary and lower values in the upper and lower estuary, and still lower values in the coastal waters. The seasonal values of the estuarine bacteria appear to be strongly linked to temperature, a relationship also reported for a salt marsh coastal system in South Carolina (Wilson and Stevenson, 1980).

3.8.4.2 Role of Microorganisms in Coastal Food Webs

Traditionally bacteria have been regarded as remineralizers, responsible for converting organic matter into inorganic, and recycling nutrients to primary producers. While this is an important role in the food web, evidence has accumulated that bacterial production is much greater than previously assumed (Joris, 1977; Joint and Morris, 1982). The role of DOM is also increasingly being recognized. Watson (1978) has argued that the efficiency of utilization of ingested food at any trophic level was only 50 to 60%, and that 10 to 20% was required for growth with another 30 to 40% for maintenance. Therefore, at each trophic level, 40 to 50% of the ingested food is either excreted as DOC or released as POC in feces, and hence a very large proportion of the primary production would be available for bacterial utilization. There is also evidence that DOC is utilized efficiently by the microorganisms. Wiebe and Smith (1977) showed that in kinetic tracer experiments, of about 10-hr duration, DOM was converted to microbial biomass with an efficiency of about 97 to 99%.

Pomeroy (1974) was one of the first to draw attention to the possibility that the Protista are major consumers of bacteria in the sea. Support for this view comes from several sources. Eriksson et al. (1977) found that ciliates, microrotifers, and nauplii constituted 15% of the biomass of the zooplankton in a shallow coastal area off Sweden, but estimated that their energy requirements were 65% of the total energy flow. Beers and Stewart (1971) reached similar conclusions for two sites in the Pacific Ocean and Margalef (1967) suggested that ciliates contribute as much, or more, to organic production as do the net zooplankton. Fenchel (1982a,b) has shown that heterotrophic microflagellates in the size range of 3 to 10 μm are effective bacteriovores in marine waters, capable of filtering 12 to 67% of the water column per day (see also Siebruth, 1976; Sorokin, 1981). Sorokin (1981) has also suggested that bacteria in the sea can occur in aggregates large enough for net zooplankton to feed on them directly.

The classical paradigm concerning marine planktonic food webs, which persisted until the mid-1970s, has been summarized by Steele (1974) in his book *The Structure of Marine Ecosystems*. "The phytoplankton of the open sea is eaten nearly as fast as it is produced so that effectively all plant production goes through herbivores." Coincidentally, at the same time, Pomeroy (1974) proposed a new paradigm that included an alternative energy-flow pathway, the "microbial loop." There was emerging evidence that much of the organic matter synthesized by the primary producers entered an extracellular pool as algal exudates and losses during feeding and excretion by metazoans and was utilized by heterotrophic microorganisms, principally bacteria. As discussed above, other organisms, such as microflagellates and ciliates, also contribute to this extracellular pool.

In 1979 in a development of his views, Pomeroy (1979) presented a compartmental model of energy flow through a continental shelf ecosystem, demonstrating the various pathways by which energy may pass from primary producers to terminal consumers, and showing that pathways involving dead organic matter (detritus), bacteria, and their predators (especially Protozoa) could play a major role. Pomeroy's ideas have been expanded and modified in a series of recent papers:

1. In a landmark review P.J. Le B.Williams (1981) attempted to reconcile the classical view of an herbivore-dominated food chain with the observation of high growth yields (50 to 80%) for bacteria utilizing glucose and amino acid substrates (Crawford et al., 1974; P.J.Le B.Williams, 1975), and concluded that at least 50 to 60% of the primary production should pass through planktonic heterotrophs before it is mineralized. However, he stressed that calculations of the proportion of the primary production entering the bacterioplankon are very sensitive to estimates of the proportion of primary production exuded as DOC, as it has been shown that while net growth yield on labile soluble substrates may be high (up to 85%), the net growth yields on particulate matter is much lower (19 to 15%) (Linley and Newell, 1981).
2. Azam et al. (1983) further elaborated the concept of the "microbial loop." Their hypothesis envisaged primarily phytoplankton-derived DOM supporting bacterioplankton production, a part of which may be transferred to the traditional grazing food chain via bacterioplankton, nanoflagellate, and ciliate links. Bacteria (0.3 to 1 μm) utilize DOM. When sufficient DOM is available for bacterial growth their population generally does not exceed 3×10^{-3} cells ml^{-1} as they are preyed upon primarily by heterotrophic nanoflagellates and small ciliates (Wright and Coffin, 1984; Rassoulzadegan et al., 1988). The heterotrophic microflagellates may reach densities of 5×10^6 cell ml^{-1}.
3. Newell and Linley (1984), from a study of standing stocks off the western approaches to the English Channel during late August, concluded that the microheterotrophic community of bacteria and protozoans comprised as much as 46% of the total consumer carbon at a mixed water station, 60% at a frontal station, and 21% at a deeper station on stratified water. Thus, in a late stage of a phytoplankton bloom, 50 to 60% of the carbon flow was estimated to enter the bacteria, 34 to 41% being subsequently respired by the bacteria, and a further 9 to 12 % eaten by the heterotrophic microflagellates.
4. Pace et al. (1984) further developed Pomeroy's 1979 model and this has been further modified for a coastal water body in Figure 3.66. This model involves abandoning the classical ideas on trophic levels and regarding food webs as anatomizing structures that defy classification under trophic levels. Pomeroy (1979) demonstrated that it was possible for energy to flow through either the grazers or alternate pathways to support all the major trophic groups at a reasonable level, and to maintain fish production at about the levels where they commonly occur. In their model simulations, Pace et al. (1984) found that benthic production appeared to be related to both the quantity of primary production and the sinking rates of the phytoplankton. From these and other studies (Joris et al., 1982), it is clear that the heterotrophs become increasingly important as one progresses from the open ocean across the continental shelf to coastal and estuarine waters.

Experiments by Thingstad (Azam et al., 1983) showed that as a phytoplankton bloom declined, bacterial populations built up, and these in turn declined when flagellates became abundant. Newell (1979) summarizes the results of microcosm experiments on the degradation of DOM, kelp debris, animal feces, phytoplankton, and *Spartina* debris, which all showed the same successional pattern in natural seawater. In all such studies there is a remarkably similar pattern, with heterotrophic microflagellates controlling bacterial numbers with a lag of some 3 to 4 days between bacterial and flagellate peaks. Biddanda and Pomeroy (1988) have demonstrated that regardless of its source, organic detritus suspended in seawater develops a remarkably similar and well-defined sequence of microbial succession. They proposed that there is a regular process of aggregation of organic material, microbial colonization, and utilization, followed by a process of aggregate disaggregation. Depending on their density and the vertical density structure of the water column, the aggregates sink at varying rates, and some will sediment to the bottom. Turbidity currents and wave action will result in resuspension of the aggregates. This model is driven by inputs from primary and secondary production. Thus the fate of detritus in the water column is seen as aggregation-disaggregation sequences in time and space.

In his experiments on phytoplankton (diatom)-derived detritus, Biddanda (1985) found that about 30 to 35% of the carbon in the detritus is mineralized (34 to 39% is utilized) by the microbial community in 4 days, whereas 63% is mineralized within 16 days. The rate of detritus

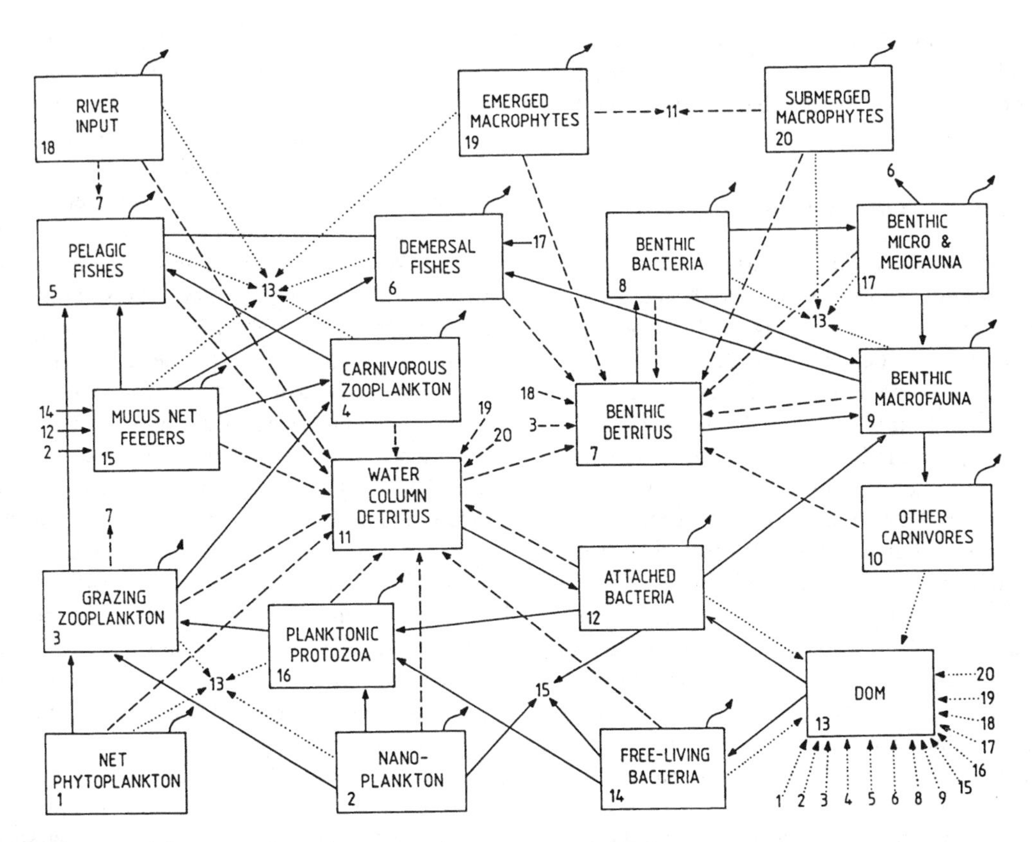

FIGURE 3.66 Diagram of the energy fluxes in an estuarine ecosystem. Solid lines = trophic flows; dotted lines = dissolved flows; broken lines = detrital flows; and wavy lines = respiratory flows. (From Knox, G.A., *Estuarine Ecosystems: A Systems Approach,* Vol. 1, CRC Press, Boca Raton, Florida, 1986a, 135. Adapted from Pace et al., 1984. With permission.)

utilization decreases steadily as the more refractive compounds in the particulate and dissolved pools are gradually mineralized or incorporated by the microbial community.

The cycle involving phytoplankton, bacteria, protozoa, and zooplankton via grazing, excretion, and remineralization, known as the microbial loop (Figure 3.67), is at the center of pelagic food-web dynamics in estuaries, bays, and lagoons. Here organic matter derived from phytoplankton, macroalgal and macrophypte detritus, dead zooplankton, fecal pellets, exuvia, and organic detritus derived from the land and the rivers are colonized by bacteria, heterotrophic flagellates, and ciliates to form

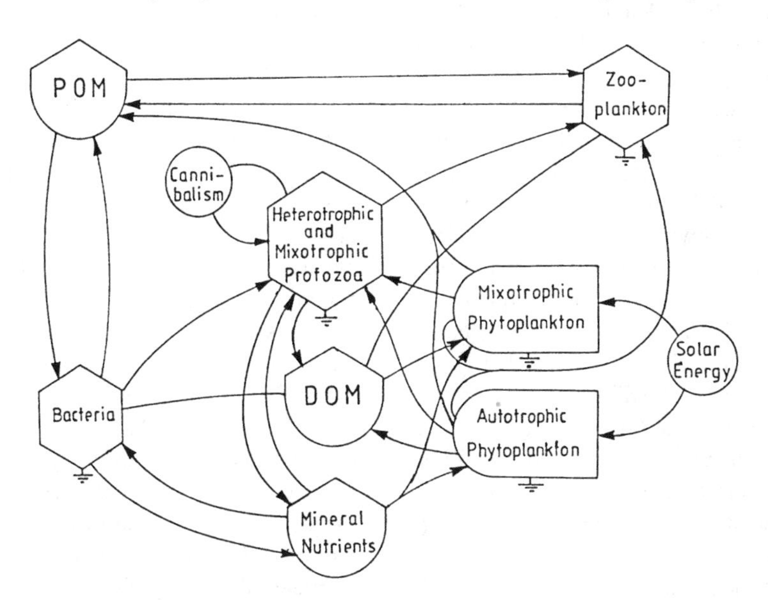

FIGURE 3.67 Model of energy flow in the microbial loop.

detrital-microbial complexes. The organic material is broken down with the release of nutrients. At the same time quantities of DOM are released to be utilized by the bacterial and other organisms.

3.8.5 Aerobic Detrital Decomposition

As we have seen, most of the primary production within coastal ecosystems (especially that of the macroalgae, marsh plants, sea grasses, and mangroves) is not utilized directly by herbivores, but enters the detrital food chain. The most detailed studies on the initial phases of the biodegradation of dead plant material have been made on marine grasses, particularly *Spartina, Zostera* and *Thalassia*. There is also a fair amount of data available for other plants and detritus at various stages of decomposition. There is often an initial decrease in the nitrogen (protein) content as soluble organics rich in nitrogen are leached from the leaves following death (Swift et al., 1979; Andersen and Hargrave, 1984). Colonization by fungi and bacteria soon begins to weaken the fibrous tissues of the dead leaves. As the plant structural material weakens due to a combination of biological and physical processes, the plant material is broken down into smaller pieces. Mechanical grinding of this tissue by amphipods, grass shrimps, crabs, and other macrofauna further reduces the particle size and increases the surface area available for further microbial transformation.

Newell et al. (1983) investigated microorganism colonization and succession in powdered leaf debris of *Spartina alterniflora* incubated in estuarine water. The natural assemblage of free-living bacteria in estuarine water (bacterial rods, coccobacilli, cocci, and filamentous forms ranging from 1.8 to 4.3 × 10^6 cells ml^{-1}) rapidly entered logarithmic growth attaining a peak density of 1.39 × 10^6 cell ml^{-1} in the presence of powdered *Spartina*, and 9.98 × 10^4 cells ml^{-1} in the presence of powdered *Juncus*. The free bacteria subsequently declined with an increase in bactivorous microflagellates. These were replaced by a mixed population of ciliates, choanoflagellates, amoeboid forms, and attached bacteria that form part of the complex community associated with particulate detritus. At this stage the bacteria were mainly attached forms associated with the particulate detritus, similar to the microbial communities associated with detritus (detrital microbial complex) from both kelp (Linley and Newell, 1981; Stuart et al., 1982) and phytoplankton (Newell et al., 1983).

E.P. Odum and de la Cruz (1967) studied the rate of decomposition of a variety of salt marsh plants including *Juncus, Distichlis, Spartina, Salicornia,* and the fiddler crab *Uca*. They found that *Uca* decomposed completely after 180 days in the field, while the plants showed variable rates of breakdown; after 300 days, the residues expressed as a percentage of the initial dry weight of the original material were: *Salicornia* 6%, *Spartina* 42%, *Distichlis* 47%, and *Juncus* 65%. These residues were then carried off the marsh into the tidal creek where the organic particulate fraction ranged from 2 mg l^{-1} at the mid-flood tide to 20 mg l^{-1} at mid-ebb tide, and was found to comprise 93.9% of fragments of *Spartina*, 4.8% of macrophytic algae, and only 1% of animal detritus.

They also found that as the detritus derived from *Spartina* decayed progressively into smaller particles it became richer in protein (Figure 3.68). The small suspended particles were 70 to 80% ash but the organic proportion comprised as much as 24% on an ash-free dry weight basis, compared with 10% in living *Spartina* and 6% in dead grass as it entered the water. Oxygen consumption per gram of detritus was found to be five times greater in this "nano-detritus" as compared with coarse detritus. Odum and de la Cruz (1967) suggested that the increase in the organic portion (protein) was due to the microbial population, which was utilizing the carbohydrate component of the detritus, while the crude fiber remained as an attachment substrate, and that the increased protein of the decomposing particles made them a potentially richer food resource than the original *Spartina* grass from which they were derived.

Valiela et al. (1985) working on the east coast of the U.S. found that the decay of salt marsh grasses occurred in three phases. First there was an early phase lasting less than a month, with fast rates of weight loss, during which 5 to 40% of the litter was lost, mainly by the leaching of soluble compounds. A second, slower phase lasted up to a year. In this second phase, microbial degradation of organic matter and subsequent leaching of hydrolyzed substances removed an additional 40 to 70% of the original material. The third phase may last an additional year. In this phase the decay is very slow, because only refractory material remains. By this third stage, as little as 10% of the original material may remain.

Investigations of detritus formation in the turtlegrass *Thalassia testudinum* (Zieman, 1968; Fenchel, 1970, 1972; Knauer and Ayers, 1977) and the eelgrass *Zostera marina* (Burkholder and Doheney, 1968; Harrison and Mann, 1975a,b) have included observations on the composition and biomass of the microbial community. Fenchel (1970) investigated the microbial community living on detrital particles derived from *Thalassia testudinum*. He found that the microbial community was a complex one whose population density and rate of oxygen consumption were similar in all fragments expressed on a unit area basis. Field samples of detrital particles had 3 × 10^6 bacteria, 5 × 10^7 flagellates , 5 × 10^4 ciliates, and 2 × 10^7 diatoms, and consumed 0.7 to 1.4 mg O_2 g^{-1} hr^{-1}. The total surface area per gram of particulate detritus depends on the diameter of the particles and thus is greater in fine particles than in coarse ones. Particles with a median diameter of 0.1 mm will have numbers of microorganisms about one order of magnitude higher than particles with

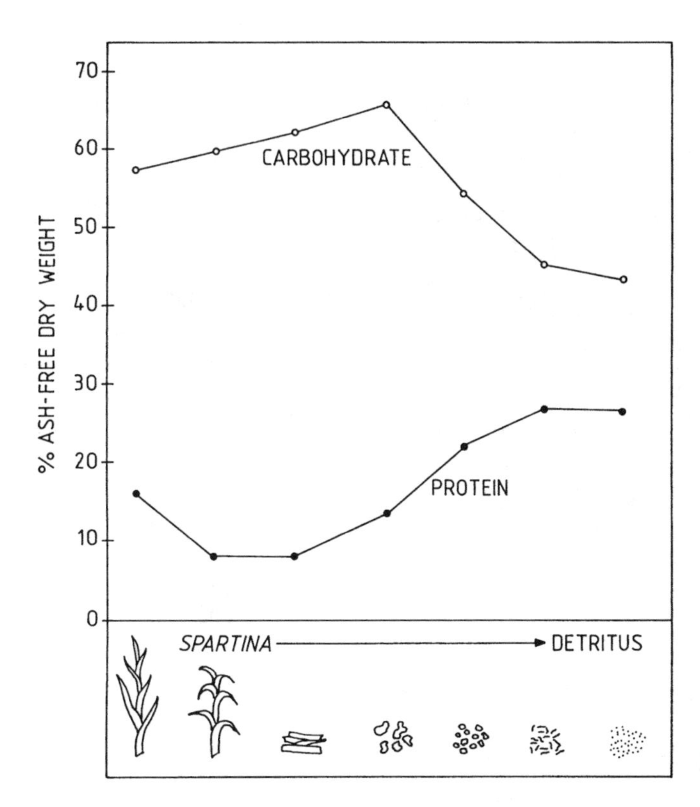

FIGURE 3.68 Diagram of the successive stages in the disintegration and subsequent degradation of organic debris from *Spartina*. The increase in protein from 10% in the fresh plant to 22% in the detritus associated with the establishment of a community of microorganisms on the surface of the finely divided organic detritus. (Redrawn from Odum, E.P. and de la Cruz, A.A., in *Estuaries*, Lauff, G.H., Ed., American Association for the Advancement of Science, Publ. No. 83, Washington, D.C., 1967b, 388. With permission.)

a median diameter of 1 mm. Figure 3.69 depicts the relationship between the numbers of microheterotrophs and particle diameter and also the oxygen composition of the detritus particles at 24°C in the dark.

Experiments on marsh grass litter decay have involved the placing of weighed samples of plant material in mesh litter bags in the natural environment. Zieman (1968) found that *Thalassia* leaves in litter bags lost weight at a rate three times as fast as those incubated in experimental tanks. Figure 3.70 depicts the rates of loss determined for a range of emergent macrophytes. Losses were lowest in *Juncus* and highest in *Thalassia*. He found that one of the factors influencing the rate of loss was the activity of invertebrate "shredders," which gained entry to the litter bags and accelerated the reduction of the particle size. Fenchel (1970) showed that the amphipod *Parhyalella whelpleyi* greatly accelerated particle size reduction in *Thalassia* detritus and, in his experiments, almost doubled the rate of oxygen uptake of the detritus over a period of four days due to the increase in microbial populations, which accompanied reduced particle size.

Woodroffe (1982b), using the litter bag technique, studied *in situ* decomposition of mangrove (*Avicennia marina*) litter in Tuft Crater, Waitemata Harbour, New Zealand. The mangrove leaves degraded rapidly at first losing approximately half their weight in 42 to 56 days and more slowly thereafter. Within the initial period, amphipods were observed to be abundant in the bags and they played an important role in the breakdown process. Other workers at Puket Island, Thailand (Boonruang, 1978) and in Florida (Heald, 1969) have found that amphipods play a similar role in detrital breakdown. In Thailand, leaves of *Avicennia* decomposed much more rapidly, losing half their weight in 20 days (Boonruang, 1978). At Roseville, Sydney, Australia, in winter, leaves of *Avicennia* were found to lose weight more slowly (Goulter and Alloway, 1979).

One of the most interesting aspects of the breakdown of detritus concerns the demonstrated increase occurring over time in the nitrogen content. The increase in protein shown in Figure 3.68 reflects nitrogen uptake from the water and its incorporation into microbial biomass.

3.8.6 Microbial Processes in the Sediments

3.8.6.1 Sediment Stratification and Microbial Processes

Sediment systems are complex and receive energy from two sources: light and organic matter, the latter either DOM produced *in situ* (from benthic microalgae and dead roots, and rhizomes of macrophytes), or POM produced

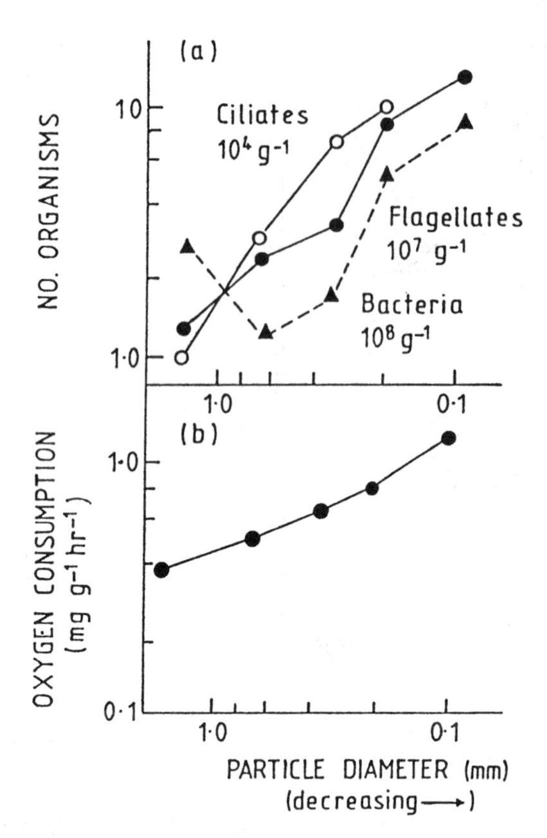

FIGURE 3.69 A. The relationship between particle size and the number of organisms on their surfaces; and B. the oxygen uptake of the detritus. (Redrawn from Mann, K.H., *The Ecology of Coastal Waters*, Blackwell Scientific, Oxford, 1982, 141. With permission.)

in situ (dead microalgae, dead roots and rhizomes), or imported in the form of detritus (dead phytoplankton, plant litter, dead animals). The processes of synthesis through oxidative and reductive pathways driven by energy sources are interrelated and their relative importance depends on the availability of light, oxygen, and various organic and inorganic hydrogen acceptors (Figure 3.71). The sediment stratification that results was discussed in Section 3.2.6. The processes and the organisms involved have been reviewed by Fenchel (1969), Fenchel and Riedl (1970) and Reise (1985).

In the upper part of the aerobic zone, as discussed in Section 3.8.5 and shown in Figure 3.71, photosynthetic unicellular algae (diatoms, coccolithophorids, euglenoids, dinoflagellates, phytomonids, and other small flagellates) as primary producers, contribute significantly to the overall primary productivity of soft bottom ecosystems. In this aerobic layer, organic detritus derived from the multiplicity of sources outlined previously is decomposed aerobically by bacteria and probably by fungi, resulting in the formation of microbial biomass on the one hand, and oxidized inorganic compounds (CO_2, N_2, NO_3^-, etc.) with oxygen as the terminal hydrogen acceptor on the other.

Below the redox potential discontinuity layer, however, conditions are anaerobic and the limiting factor for the utilization of the potential energy of the organic material is the availability of hydrogen acceptors. Heterotrophic bacteria, by various types of fermentation processes, utilize organic compounds (fermentation) as hydrogen acceptors, while at the same time oxidizing other organic substances and producing, in addition to CO_2 and H_2O, reduced compounds such as lactate, alcohols, H_2S, NH_3, etc. Other anaerobic bacteria utilize certain inorganic compounds as hydrogen acceptors for the oxidization of organic material. In this process reduced inorganic compounds (CH_4, NH_3, H_2S) which diffuse upward, are produced.

The reduced compounds thus formed, both inorganic and organic, still contain potential energy that may be utilized by chemoautotrophic bacteria in the presence of oxygen and by photoautotrophic bacteria in the presence of light. The chemoautotrophs obtain their energy from the oxidation of inorganic substances (H_2S, Fe^{++}, H_2, NH_3, NO_2, etc.) and the energy thus obtained is used to reduce CO_2 to carbohydrates. The photoautotrophs (purple and green sulfur bacteria and some protophytes) utilize reduced compounds for the reduction of CO_2 to carbohydrates in a photosynthetic process in the presence of light, instead of using H_2O as in normal photosynthesis (photoreduction). The purple and green sulfur bacteria use H_2S, SO_3, and S, and some forms may use H_2 or reduced organic compounds as hydrogen donors in this process. Sulfate reduction, carried out by the bacterium *Desulfovibrio* sp., is a dominating process in the chemistry of the anaerobic sediment layer (Fenchel, 1969; 1971). On the other hand, the chemoautotrophic white sulfur bacterial species of *Thiobacillus, Thiovolum, Macromonas, Boggiatoa*, etc. oxidize H_2S to elemental S or sulfate. When the anaerobic zone reaches the surface of the sediment, these bacteria become visible to the naked eye as a white scum on the surface.

3.8.6.2 Sediment Microbial Standing Stocks and Activity

Rublee (1982a) has summarized data on direct counts of total bacteria in estuarine intertidal and subtidal sediments in a range of estuaries in North America. Total numbers ranged from about 4 to 17 × 10^9 cells cm^{-3} in surface sediments and decreased with increasing depth in the sediment. Seasonal distribution of bacteria numbers have been studied in a North Carolina marsh (Rublee, 1982b), and an intertidal muddy beach in the sub-Antarctic Kerguelen Archipelago (Bouvay, 1988). Bouvy (1988) found obvious seasonal changes in bacterial numbers and carbon biomass with maximal values in spring (November and December). On the North Carolina marsh, numbers in the surface sediments ranged from a high of 14 × 10^9 cells cm^{-3} in the

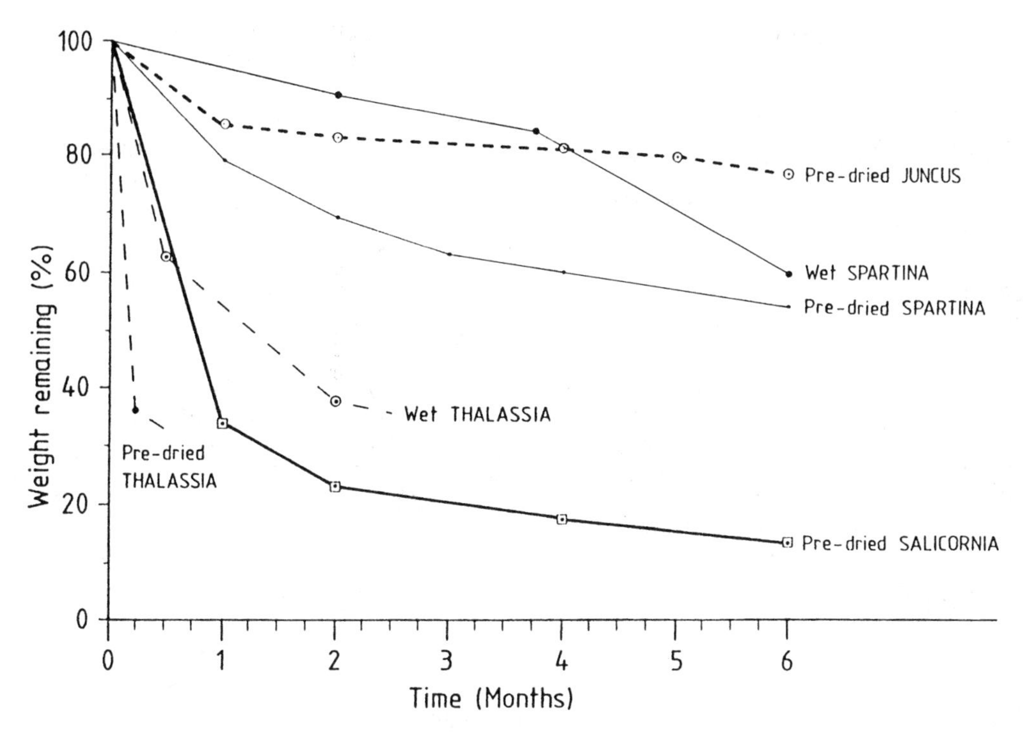

FIGURE 3.70 Rates of disappearance of named marine macrophytes from litter bags of 2.5-m mesh. (Redrawn from Wood, J.E.F., Odum, W.E., and Zieman, J.C., in *United Nations Symposio Internat. Lagunas Costeras, Nov. 1967, Mexico,* UNAM:UNESCO, 1969, 560. With permission.)

late autumn (fall) to a low of 5×10^9 cm^{-3} in the late summer. At 5 cm depth there was the same pattern but with lower maximum and minimum numbers (7 and 3×10^9 cell cm^{-3}), but at 10 and 20 cm there was no seasonality. In the Rhode River, Maryland, similar results were obtained at six stations, including four vegetated marsh sites, a tidal creek site, and a subtidal mudflat (Rublee, 1982a).

Dale (1974) also showed that bacterial numbers, estimated by direct counts, which varied from 1×10^8 to almost 1×10^{10} g^{-1}, showed a highly significant negative correlation between numbers and mean grain size. This suggests that, as for POM, the surface area of the sediment grains controls the density of the bacteria. Rublee (1982a) found that the sediment interstitial water contained only 0.1 to 1% of the total sediment bacteria, and that when filtered water was flushed through short sediment cores only a small fraction (<1%) of the total sediment bacteria was washed out. This result suggested that the majority of the bacterial cells in sediment are attached to the particles.

Rublee (1982b) monitored the number and size of bacteria at four depths (0 to 1, 5 to 6, 10 to 11, and 20 to 21) in the North Newport Estuary *Spartina* marsh for a period of 13 months. The number of bacteria reached a maximum of about 1.4×10^{10} cells cm^{-3} at the sediment surface in October, corresponding to the period of *Spartina* dieback. Estimates for microbial, bacterial, and sediment carbon and nitrogen at the various depths are given in Table 3.32. Cell numbers were lowest and most consistent throughout the year at the 20 to 21 cm depth. The mean standing crop of bacteria to a depth of 20 cm was about 14 g C m^{-2}. In the surface layer, bacteria contributed up to 15% and benthic microalgae up to 10% of the total living microbial biomass as estimated from ATP assays. Bacteria were the major biomass components at all other depths. The contribution of the bacteria in the surface layer was estimated at about 0.6% of the total sediment organic carbon and less than 2% of the nitrogen. These values compare well with those reported from a Georgia salt marsh (Christian et al., 1975), and for shallow estuaries (Ferguson and Murdock, 1975), where the contribution of bacteria to sediment organic carbon and nitrogen was in the range 0.05 to 2.98%. However, standing crop measures are not a good indicator of the metabolic activity of bacteria, which have high metabolic rates and turnover times.

Rublee (1982a) has summarized estimates of the algal and microbial biomass in estuarine marsh systems (Table 3.33). These data suggest that benthic microalgae are the dominant contributors to total microbial biomass in the surface sediments where they contribute up to 50% of the total. At greater depths, the bacteria generally comprise the bulk of the biomass. Of the estimated total microbial biomass the bacteria diatoms comprise about 60%. Rublee (1982a) points out that the estimates in Table 3.34 are generalizations at best, and that there will be site-specific variations depending on the physical, chemical, and biological conditions.

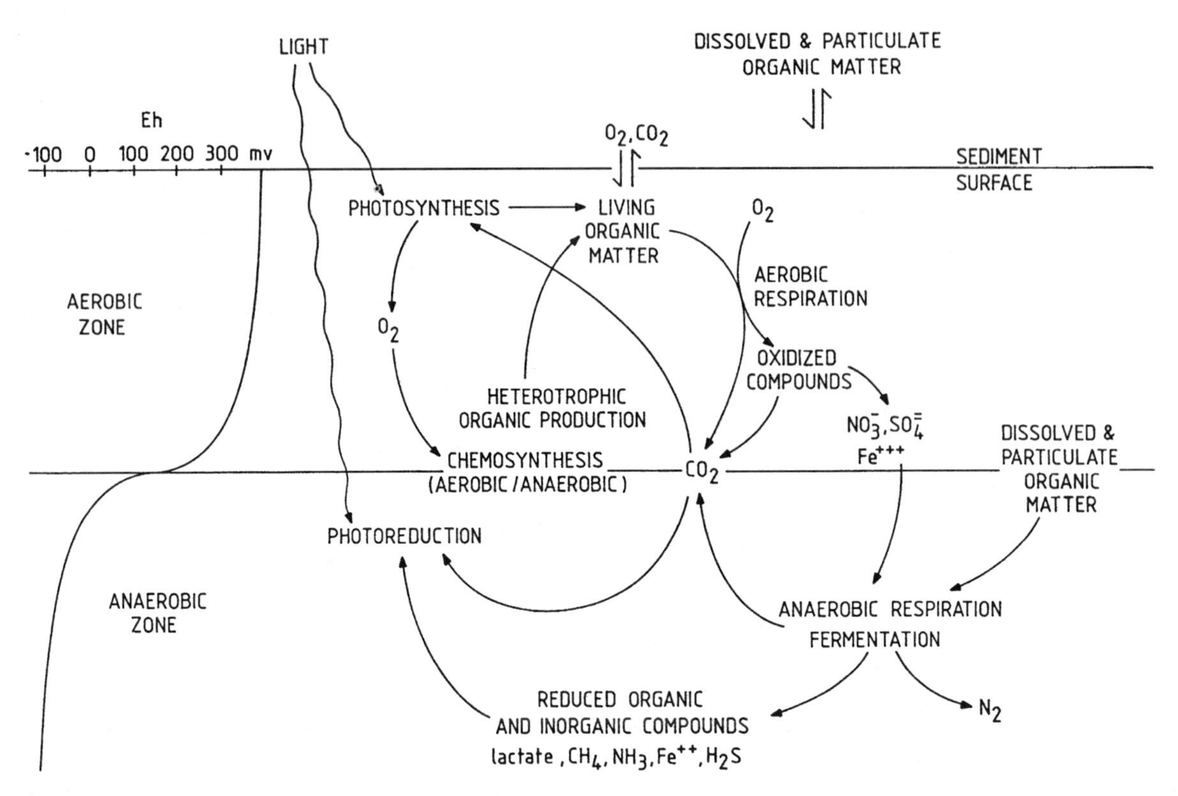

FIGURE 3.71 Interrelations between photosynthetic, heterotrophic, and chemosynthetic processes that occur in sediments. Photosynthesis and photoreduction only occur in the presence of light. *Aerobic metabolic processes*: heterotrophic production (oxidation of reduced simple organic compounds with possible reduction of external CO_2); photosynthesis (reduction of CO_2 to carbohydrates using H_2O and light); aerobic respiration (reduction of oxygen to water with organic compounds as electron doners); aerobic chemosynthesis (oxidation of CH_4, H_2S, NH_4, Fe^{++}, and H_2 to form organic carbon compounds by fixation of external CO_2). *Anaerobic metabolic processes*: anaerobic respiration (oxidized inorganic end products of aerobic decomposition used as hydrogen acceptors for the oxidation of organic matter: fermentation (organic compounds used a hydrogen acceptors to produce CO_2, H_2O and reduced organic compounds, lactate, glycollic acid, H_2S, and NH_3); photoreduction (reduced compounds in the presence of light used to reduce CO_2 to carbohydrates with H_2S, SO_3, S, H_2, or reduced organics serving as a hydrogen donor); anaerobic chemosynthesis (oxidize organic compounds H_2, H_2S, Fe^{++}, NO_2, and use energy to reduce CO_2 to carbohydrates). (Redrawn from Parsons, T.R., Takahashi, M., and Hargrave, B.T., *Biological Oceanographic Processes*, 2nd ed., Pregamon Press, New York, 1979, 225. With permission.)

TABLE 3.32
Estimated Contribution of Microbial and Bacterial Standing Crop to Total Sediment Organic Carbon and Nitrogen in a North Carolina Salt Marsh

Depth (cm)	Microbial (mg C cm^{-3})	Bacterial mg C cm^{-3})	Sediment C (mg C cm^{-3})	Microbial N (mg N cm^{-3})	Bacterial N (mg N cm^{-3})	Sediment N (mg N cm^{-1})
0–1	1.013	0.147	25.40	0.170	0.044	2.19
5–6	0.080	0.088	31.90			
10–11	0.023	0.047	33.80			
20–21	0.010	0.014	13.20			

Source: From Rublee, P.A., *Est. Coastal Shelf Sci.*, 15, 72, 1982. With permission.

TABLE 3.33
Generalized Biomass Estimates for Microbial Groups in Estuarine Marsh Systems

Depth (cm)	Total Microbial Biomass	Bacterial Biomass	Fungal Biomass	Algal Biomass	Protozoan Biomass
0–1	775 (175–1 400)	198 (92–354)	130 (63–228)	1170 (87–3420)	—
5–6	140 (80–258)	127 (57–226)	—	21	4–46[a]
10–11	56 (23–95)	80 (29–162)	—	3	—
20–21	25 (10–35)	49 (17–88)	—	3	—
0–20	3183 (1061–5678)	1975 (919–3670)	130 (63–228)	1125 (198–5756)	4–46

Note: Ranges are given in parentheses, where multiple determinants were available 0 to 20 cm values for total biomass were determined by summation of linearly interpolated values from surface to 20 cm depth. Values are given in g C m^{-3}.

[a] 0 to 5 cm estimate based on the range given by Fenchel (1969) assuming dry wgt = 20% wet wgt, and carbon = 50% dry wgt.

Source: From Rublee, P.A., in *Estuarine Comparisons,* Kennedy, V.C., Ed., Academic Press, New York, 1982a. With permission.

TABLE 3.34
Studies Measuring Direct Assimilation of Detrital Carbon

Species	Detritus Type	Assimilation Efficiency %	References
Hyalella azteca	Cellulose + lignin	0	Hargrave (1970a)
Planorbis contortus	Refractory detritus	<10	Carlow (1975)
Hydrobia ventrosa	Sterile fresh hay	34	Kofoed (1975)
Mysis stenopsis	Sterile cellulose	30–50	Foulds and Mann (1978)
Nereis succinea	Sterile *Spartina*		Cammen (1980)
	Water-extracted	10	
	Chemically extracted	Detected	
Strongylocentrotus droebachiensis	Sterile cellulose	Detected	Fong and Mann (1980)
Aulocomya ater	Kelp detritus	50	Stuart et al. (1982)
Hydrobia totteni	Detritus	low	Bianchi and Levinton (1981)
Acantholeberis curvirostris	Sterile cellulose	11	Schoenberg et al. (1984)
Pseudosida bidentata	Sterile cellulose	9	Schoenberg et al. (1984)
Daphnia magna	Sterile cellulose	2	Schoenberg et al. (1984)
Pteronarcys proteus	Cellulose	11	Sinsabaugh et al. (1985)
Tipula abdominalis	Cellulose	18	Sinsabaugh et al. (1985)
Pycnopsyche luculenta	Cellulose	12	Sinsabaugh et al. (1985)

Source: From Kemp, P.F., *J. Exp. Mar. Biol. Ecol.*, 99, 56, 1986.

3.8.6.3 Anaerobic Processes in the Sediments

The overall transformation of particulate organic matter deposited on the sediments, and that produced within the sediment (e.g., the roots and rhizomes of macrophytes) is often referred to as diagenesis. The major anaerobic processes in the sediments are fermentation, dissimilatory nitrogenous oxygen reduction (DNOR), dissimilatory sulfate reduction (sulfate reduction), and methanogenesis (Wiebe et al., 1981). Organic matter from primary producers and aerobic heterotrophs enters the anaerobic cycle through fermenters and nitrogenous oxide reducers (Figure 3.72). Sulfate reducers and methanogens utilize relatively few substrates, most of which are the end products of fermentation. Dissimilatory nitrogenous oxide reduction (DNOR) is a general term encompassing a number of pathways including denitrification, dissimilatory reduction, and dissimilatory ammonia production. In the course of these processes, plant and animal matter is broken down into simpler substances, and eventually reduced to its mineral constituents.

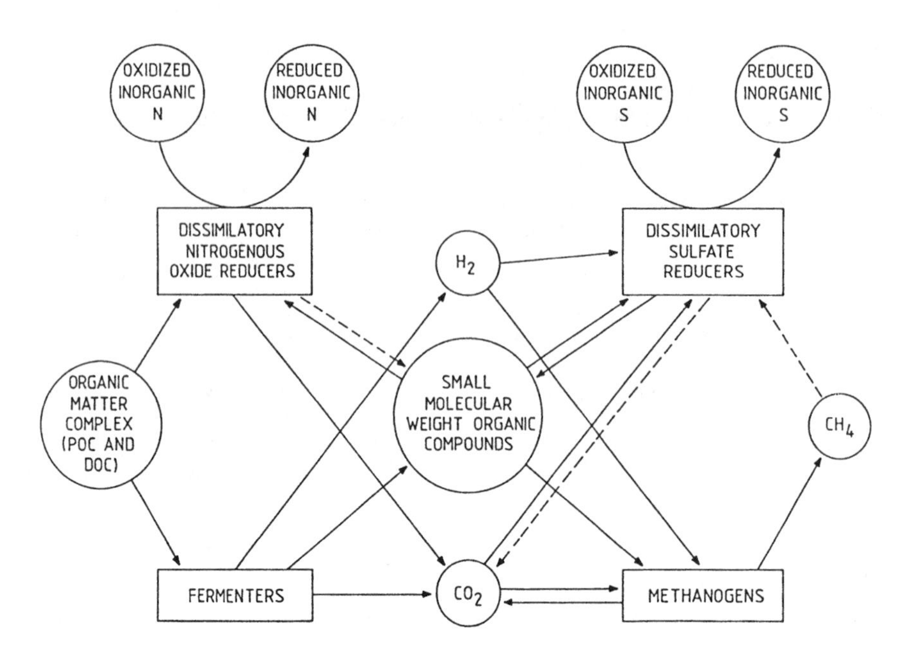

FIGURE 3.72 Conceptual model of interactions of anaerobic microbial processes in salt marsh sediments. Solid lines are confirmed fluxes; dashed lines are possible fluxes. (Redrawn from Wiebe, W.J., Christian, R.R., Hansen, J.A., King, G., Sherr, B., and Skyring, G., in *The Ecology of a Salt Marsh*, Pomeroy, L.R. and Wiegert, R.C., Eds., Springer-Verlag, Berlin, 1981, 138. With permission.)

With respect to marsh sediments, there has been debate as to the relative contributions to heterotrophic activity of the breakdown of roots and rhizomes and exudates from the roots. Christian et al. (1978) concluded that the long-term decomposition of roots and rhizomes was the main resource available to the bacteria, while Howarth and Hobbie (1982) considered that rapid release of organic matter from living roots and rhizomes, and possibly other carbon sources as well, contributed significantly. The inputs are roughly divided into two pools: those considered to have a short turnover time and those likely to have a long turnover time. Valiela et al. (1976) estimated the belowground production of the roots and rhizomes of marsh macrophytes to be approximately 1,680 g C m^{-2}, while Valiela et al. (1984) reported that about 20% of the biomass of such roots and rhizomes was rapidly leached in buried litter bag experiments. It can therefore be estimated that approximately 340 g C m^{-2} $year^{-1}$ of readily available organic matter is leached from roots and rhizomes as they die. Living roots and rhizomes contribute additional organic matter by the sloughing of material or by the excretion of organic matter. The production of chemosynthetic bacteria represents another input of organic matter to sediments. Howarth and Hobbie (1982) have estimated the inputs of readily decomposable carbon to the Great Sippewissett Marsh to be in the range of 300 to 600 g C m^{-2} $year^{-1}$, and even more if excretions of ethanol and other end products of aerobic decomposition are large. Inputs of slowly decomposable matter are in the range of 1,200 to 1,500 g C m^{-2} $year^{-1}$ or more. This compares with an estimate of carbon consumption by marsh heterotrophic activity on the order of 2,600 g C m^{-2} $year^{-1}$.

Fermentation: Fermentation is the anaerobic dissimilation of organic matter in which the terminal electron acceptor is an organic compound. As a result, organic compounds of low molecular weight are produced and these can serve as substrates for other anaerobic processes. Fermentation reactions are produced by a variety of both prokaryotic and eucaryotic organisms that may be facultative or obligatory anaerobic; the former may be capable of either aerobic respiration or fermentation, while the latter are inhibited or killed by the presence of oxygen.

Fermenters utilize many substrates including simple sugars, as well as cellulose, starch, pectin, alcohols, amino acids, and purines. The end products include organic compounds, such as short-chain fatty acids and alcohols, as well as CO_2, H_2, and NH_3. Although a wide variety of organisms are capable of fermentation reactions, bacteria and yeasts are responsible for most of the fermentation activity in sediments.

Measurement of fermentation activity in nature is difficult because of the diversity of organisms and inorganic compounds involved. A number of investigators have added ^{14}C-glucose to sediment samples and followed its incorporation into CO_2, particulate matter, and the ether-soluble end products of fermentation (Christian and Wiebe, 1978; 1979). The results of such experiments yield an estimated *potential* rather than *actual* fermentation rates, since the natural substrate concentrations are not determined in such experiments.

Dissimilatory sulfate reduction and the sulfur cycle: Extensive studies of metabolic processes in sediments have shown that much of the energy flowing through the system is modulated through the sulfur cycle.

1. Dissimilatory sulfate reduction (*Desulfovibrio*)

   $SO_4^= \longrightarrow S^-$

   (using $SO_4^=$ as the terminal electron acceptor for anaerobic respiration)

2. Assimilatory sulfate reduction

   $SO_4^= \longrightarrow R{\cdot}SH$

   (incorporation of S into organic matter)

3. Sulfer mineralization
   a. Sulfide mineralization (putrefaction)

   $R{\cdot}SH \longrightarrow H_2S$

   b Oxidized sulfur mineralization

   R·oxidized S (e.g. $SO_4^=$) $\longrightarrow SO_4^=$

4. Reduced sulfur oxidation (photosynthetic and non-photosynthetic chemoautotrophic)

   $S^= \longrightarrow S \longrightarrow S_2O_3^= \longrightarrow SO_4^=$ (*Thiobacillus*, *Thiothrix*, *Beggiatoa*)

5. Sulfur immobilization
   a reduced sulfur immobilization

   $S \longrightarrow R{\cdot}SH$ or $R_1{\cdot}S{\cdot}R_2$

   b oxidized sulfur immobilization

   $SO_4^= \longrightarrow R\cdot$ (oxidized sulfur)

6. Volatile organic sulfide formation

   $R_1{\cdot}S{\cdot}R_2 \longrightarrow$ Small molecular weight organic S compounds

FIGURE 3.73 Transformations within the sulfur cycle in estuarine systems. Names in parentheses are organisms that carry out particular reactions. (Redrawn from Day, J.W., Jr., Hall, C.A.S., Kemp, W.M., and Yanez-Arancibia, A., *Estuarine Ecology,* John Wiley & Sons, New York, 1987, 276. With permission.)

By far the most abundant electron acceptor (or oxidizing agent) in most sediments (especially muddy ones) is sulfate (SO_4^{2-}), which is reduced to sulfide (S^{2-}). This is accomplished through two main pathways, assimilatory and dissimilatory (Figure 3.73). *Assimilatory* reduction involves the uptake and biochemical incorporation of SO_4^{2-} into organic molecules (e.g., amino acids, methionine, and cystine). Sulfide is released through subsequent aerobic mineralization of these compounds (desulfuration). However, below the RPD layer *dissimilatory* reduction is more important. It is a complex cycle involving many possible steps (see Wiebe et al., 1981). Dissimilatory sulfate reduction occurs as sulfate diffuses down into the anoxic zone, where organisms such as *Desulfovibrio* sp. use it to oxidize organic substances producing elemental sulfur as in the following reaction:

$$2\ \text{lactate} + SO_4^{2-} = 2\ \text{acetate} + CO_2 + H_2O + S^{2-}$$

Some of the sulfide is trapped in the sediments by precipitation with metal ions such as iron (Fe), but much remains dissolved and diffuses to the aerobic surface layer. There it is oxidized back to SO_4^{2-}, either spontaneously or through catalysis by chemoautotrophic (in oxic sediments) or photoautotrophic (in anoxic sediments) bacteria, refixing organic carbon.

Due to the abundance of SO_4^{2-} in seawater, sulfate reduction is a dominant process in muddy sediments, and the liberation of sulfide affects the chemical and biological environment in many ways. The presence of H_2S (or depending on the pH, HS^-) provides a sink for heavy metals, especially iron, which is precipitated as the black ferrous sulfide to which reducing sediments owe their color.

The sulfur cycle has been well documented in shallow sea grass (*Zostera*) beds and in a model system with *Zostera* detritus as the sole carbon input (Johnson, 1973). In an attempt to verify the results of the model system in a natural habitat, Jorgensen (1977a) constructed a budget of the sulfur cycle at nine stations in shallow coastal sediments in Denmark. He found (Figure 3.74) that oxygen uptake and sulfate reduction changed in parallel through the seasons. The weighted daily average rate of sulfate reduction was 9.5 mmol SO_4^{2-} m^{-2}, made up of 6.2 mmol in the upper 10 cm and 3.3 mmol in the deeper sediments. However, 1 mol of sulfate oxidizes twice as much organic carbon as CO_2, as does 1 mol of oxygen. Hence, it can be concluded that a total of 34 mmol of organic carbon m^{-2} would be oxidized daily with 9.5 × 2

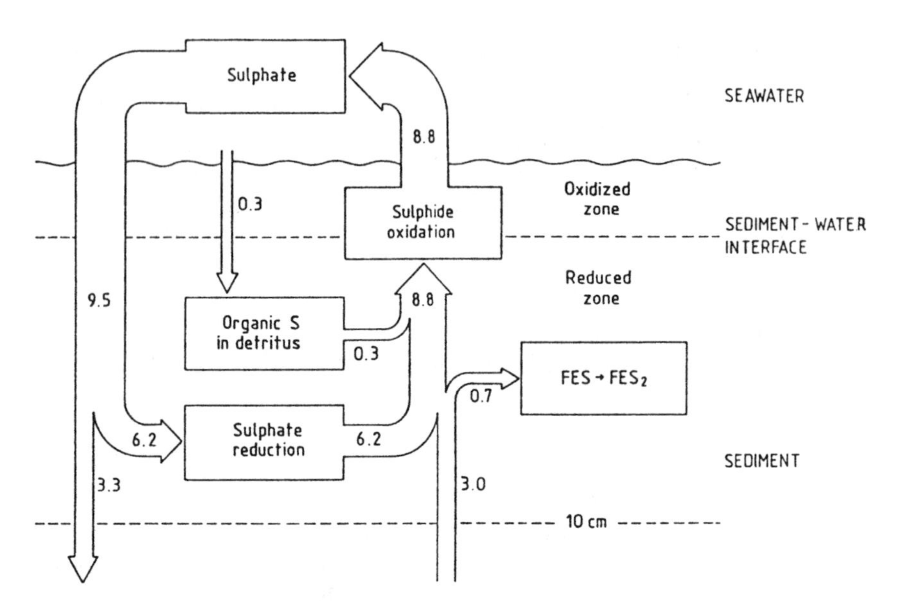

FIGURE 3.74 Transfer rates of sulfur in a marine sediment (Limfjorden, Denmark). (Redrawn from Jorgensen, B.B., *Limnol. Oceanogr.*, 22, 827, 1977a. With permission.)

= 19 mmol being oxidized by sulfate-reducing bacteria and the remainder by processes in the aerobic zone.

Dissimilatory nitrogenous oxide reduction: Dissimilatory nitrogenous oxide reduction (DONOR) encompasses four specific pathways (see Figure 3.72): (1) denitrification: (2) dissimilatory reduction (terminates at NO_2); (3) dissimilatory ammonia production; and (4) "nitrification" (N_2O pathway = ammonia to nitrous oxide). It is an obligatory anaerobic process mediated by bacteria since oxygen concentrations as low as 0.2 ml l^{-1} inhibit enzyme activity. The nitrogenous oxides serve as terminal electron acceptors during cytochrome-catalyzed oxidation of organic matter. When the end products are N_2O, or N_2, the process is called denitrification.

Methanogenesis: Methane is produced under anaerobic conditions by bacteria using CO_2 or a methyl group as an electron acceptor. This is the least studied of the sediment anaerobic processes. In the Sapelo Island marshes, methane released over a year varied substantially in the different marsh zones: 53.1 g m^{-2} for short *Spartina* (SS) sediments, 14 g m^{-2} for intermediate *Spartina* (MS) sediments, and 0.41 g m^{-2} for tall *Spartina* (TS) sediments (King and Wiebe, 1978). This represents a loss of 8.8% of the net carbon fixed in SS sediments, 0.002% in TS sediments, and about 5% for the entire marsh.

3.8.7 Microorganisms and Detritus as Food Resources

As Mann (1988) points out, there are five different pathways by which materials and energy from primary production may be transferred to invertebrates and vertebrates in aquatic environments (Figure 3.75). The processes of fragmentation, colonization by microbes, and enrichment by nitrogen to form detrital-microbial complexes proceed at varying rates depending on the nature of the original plant material and the environmental conditions. In addition dissolved organic matter precipitates to form amorphous particles, which become colonized by a rich microbial community. Numerous studies have shown that the bacteria and other microbes living on the detrital particles are assimilated efficiently by detritovores. However, several studies have found that the amount of microbial carbon that can be obtained from the bacteria is less than that required to meet the carbon requirements of detritovores (Wetzel, 1977; Cammen, 1980; Jensen and Siegsmund, 1980).

Lopez and Levinton (1987) in a critical review of the literature on whether bacterial biomass could be the main food source for deposit feeders concluded that, although bacterial carbon is more efficiently digested by the animals, bacterial biomass in the sediments is too low to satisfy the energetic needs of most deposit feeders. Thus feeding directly on detritus or the benthic microalgae is necessary to close the energy budgets. Cammen (1989) in a literature review of studies in which ingestion and respiration have been determined simultaneously, concluded that assimilation of between 3.2 and 5.5% of ingested detritus would be sufficient to meet the energetic needs for respiration. To allow for growth, this percentage should be raised to 4 to 13%.

There has been debate as to the extent to which detritovores are reliant on the microbial community that colonizes the detrital particles for their energy requirements, or whether they can utilize the organic material in the detrital particles directly. Mann (1982, Chapter 7) reviewed the evidence in detail and concluded that many prominent macrofaunal species such as gastropods, amphipods, and even tube-dwelling infauna are capable

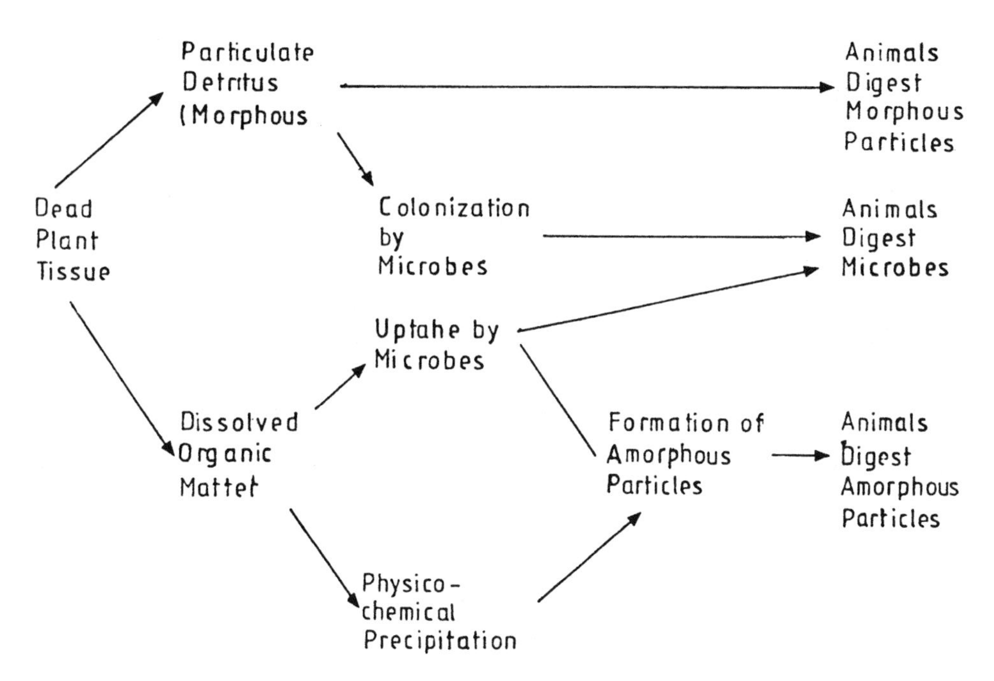

FIGURE 3.75 Pathways of utilization of plant detritus on coastal ecosystems. (Redrawn from Mann, K.H., *Limnol. Oceanogr.*, 33, 913, 1988. With permission.)

of intercepting newly settled detritus at the sediment surface and using it directly with an efficiency of about 40%. Many detritovores have been found to lack (or have low activity of) enzymes capable of hydrolyzing structural carbohydrates (Kristensen, 1988), which comprise the major part of aged detritus. These studies led to the related hypothesis that detrital material is unavailable to detritovores until it is converted into microbial biomass. On the other hand, it has been demonstrated by Foulds and Mann (1978), Wainwright and Mann (1982), and Friesen et al. (1986) that the epibenthic mysid, *Mysis stenolepis*, can digest cellulose with endogenously produced enzymes, and by Fong and Mann (1980) that sea urchins have a gut fora that can digest cellulose. The assimilability of detrital material appears to depend on its source (Tenore and Harrison, 1980; Tenore and Rice, 1980; Tenore et al., 1982).

Angiosperm tissues are composed of a number of components such as cellulose, waxes, and lignins that are both difficult to digest and resistant to bacterial degradation (Tenore, 1983; Valiela et al., 1984). Experiments have shown that vascular plant detritus is a relatively poor-quality food. Algal tissues, in contrast, are both high in nitrogen and readily available consumers and microorganisms (Findlay and Tenore, 1982). It has been shown that a significant fraction of kelp standing stock degrades *in situ* into particulate and dissolved fraction, available to a diverse assemblage of nearshore (pelagic and benthic) suspension feeders (Newell et al., 1983; Bustamante and Branch, 1996). In addition, some kelp species, as we have seen, are a continuous source of particulate organic carbon (POC) and dissolved organic carbon (DOC) as they grow releasing DOC and sloughing off of particulate material, resulting in a 20-fold turnover of the tissues per year. Stable carbon isotope analyses and *in situ* growth experiments in the North Pacific (Duggans et al., 1989) and the west coast of South Africa (Bustamante and Branch, 1996) have confirmed that kelp production provides a food resource to a broad range of secondary consumers, especially filter feeders. Laboratory feeding experiments with a serpulid polychaete and a mussel carried out by Duggans and Eckman (1997) found that fresh particles from the kelp *Laminaria groenlandica* and aged particle from the kelps *Agarum fimbriatum* and *Alaria marginata* and mixed phytoplankton promoted the highest growth in both consumers. They found that growth was inversely related to total polyphenolic concentrations in the food particles. With aging there was a rapid loss of such compounds in *Agarum* and *Alaria*.

Kemp (1986) tested the hypothesis that benthic detritovores are able to use detrital carbon directly as a food resource by feeding radiolabeled detritus to the deposit-feeding polychaete, *Euzonus mucronata*, an abundant inhabitant of coastal beaches of the Pacific coast of the U.S. The assimilation of carbon from sterile detritus at a concentration of 0.3 mg ash-free dry wgt of detritus per g dry sand (0.03% organic content) was sufficient to provide about one-half of the carbon requirements of *E. mucronata*. There was no significant difference in assimilation from chemically extracted or water-extracted sterile detritus (representing relatively refractory and available material, respectively), or from the same materials after microbial colonization. Thus, the major source of carbon may have been the supposedly refractory component of the detritus. Despite a low estimated assimilation efficiency

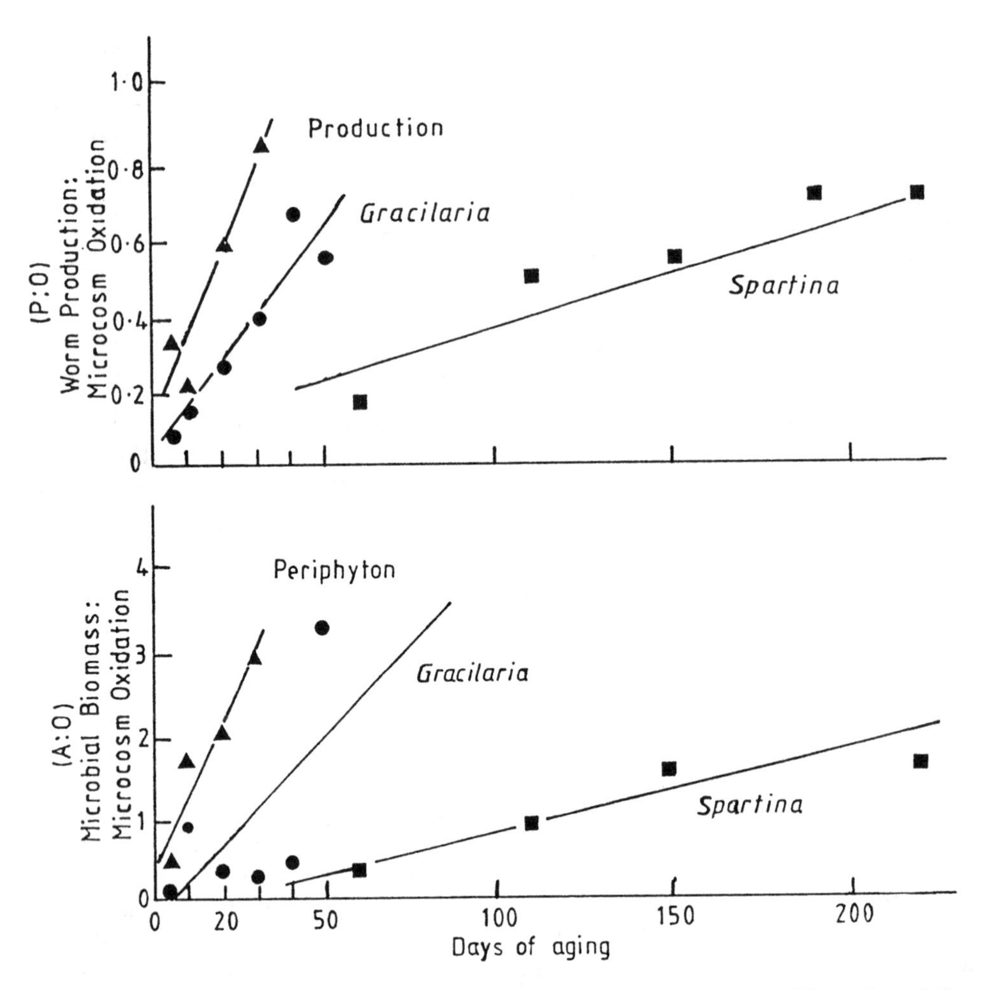

FIGURE 3.76 Results of culturing the polychaete worm, *Capitella capitata*, on detritus prepared from three different sources and aged for different times. Above: ratio of worm production to total oxidation of the culture. Below: ratio of microbial biomass to total oxidation. (Redrawn from Tenore, K.R. and Hanson, R.B., *Limnol. Oceanogr.*, 25, 555, 1980. With permission.)

(0.76 to 1.72%), the carbon requirement of *E. mucronata* could be met entirely by direct assimilation of detrital carbon if only 10% of the total organic carbon in the sand is as available as the chemically extracted detritus in the experiment. Thus, low assimilation efficiency does not preclude the acquisition of nutritionally important quantities of carbon by direct assimilation of detritus.

Tenore and Hanson (1980) prepared detritus from three sources: marsh grass (*Spartina*), seaweed (*Gracilaria*), and natural mixed periphyton. Some detritus was frozen soon after preparation; other subsamples were aged for different lengths of time. In laboratory cultures the polychaete, *Capitella capitata*, was fed detritus from different plants aged for differing lengths of time (Figure 3.76). Worm production was measured as well as microcosm oxidation (the CO_2 of the worm respiration plus the amount of detrital carbon converted to CO_2 in the process of microbial metabolism). The upper plot in Figure 3.76 shows that the ratio of worm production to microbial oxidation exceeded 90% for the periphyton aged for only 30 days. The seaweed *Gracilaria* needed aging for at least 40 days and the *Spartina* 200 days before a similar ratio was reached. Sea grass tissue would have required an even longer time (Harrison and Mann, 1975b). The lower part of Figure 3.76 shows microbial biomass as a function of microcosm oxidation and the time of aging. The results strongly suggest that it is the rapid buildup of microbial biomass on algal tissue that makes possible the efficient utilization of algal detritus by *Capitella*.

Table 3.34 lists studies measuring direct assimilation of detrital carbon. It demonstrates that detrital carbon is directly available to some detritovores although the assimilation efficiency varies widely. Cammen (1980) calculated that the polychaete, *Nereis succinea*, would obtain over two-thirds of its carbon requirements by feeding directly on detrital carbon at an assimilation efficiency of only 5.2%. Thus both the detritus itself and the associated microbial community can contribute in varying amounts to the nutrition of detritovores. Marine plants vary in the proportion of refractory material in the detritus derived from them, and consequently the rate at which detritus is decomposed. In addition the composition and biomass of the associated microbial community are also highly variable.

Thus the contributions of detritus and the associated microbial community to the nutrition of detritivores will vary depending on a variety of factors. These include the origin of the detrital material, the proportion of refractory material in the detritus, the time over which it has aged, the composition of the associated microbial community, the ability of the detritovore to manufacture enzymes capable of breaking down the detrital material, and the degree to which the detritivore possesses a gut flora capable of facilitating the breakdown of the refractory material.

3.9 NUTRIENT CYCLING

3.9.1 INTRODUCTION

Over the past few decades a considerable amount of research has been carried out on nutrient cycling in estuarine and coastal waters. Nixon (1980; 1981; 1982) has summarized the available data and given an interesting account of the changing paradigms concerning nutrient cycling, particularly concerning the importance of the various mechanisms involved in nutrient remineralization.

Shallow coastal waters, and estuaries in particular, are highly productive systems and it is clear that nutrients are a major limiting factor for the producers, consumers, and decomposers. Among the most important are the autotrophic nutrients, which serve as raw materials for the primary production of organic matter. The minerals required for plant production in the coastal zone are numerous, including C, N, P, Si, S, K, Mg, Na, Ca, Fe, Mn, Zn, Cu, B, Mo, Co, V, and the vitamins thiamin, cyanocobalamin, and biotin. Of these the first four are the ones most heavily utilized for algal growth, although silicon is required only by diatoms. The others (sulfur to biotin) are generally present in nonlimiting quantities. In estuaries in particular, nutrient input from freshwater inflow is important in maintaining the high rates of primary production. The sediments function as important sinks and transformers of nutrients, thus altering the quantity and quality of nutrients transported from the land to the sea. The processes that govern the fates of N and P in these waters differ. Consequently, the ratios of inorganic N to P in coastal waters, unlike those in the ocean (Redfield, 1934), may vary widely with time and space and may deviate greatly from the ratio of N to P in phytoplankton. The relative abundance of inorganic N in coastal waters may be decreased by denitrification (Seitsinger, 1988), or by inhibition of N fixation (Howarth and Cole, 1985), or by hydrologic factors (Smith, 1984). In contrast, the abundance of inorganic P may be increased by enhanced release from sediments in the presence of high salinity and SO_4^{2-} (Caraco et al., 1987). These processes may explain why N is the nutrient that most often limits primary production in coastal waters (Boynton et al., 1982). In some estuaries, however, high inputs of NO_3 may increase the ratios of inorganic N to P and lead to P limitation.

As Nixon (1980) has pointed out, our perspective on the importance of the various components and mechanisms of the nutrient cycling models have changed markedly over the past few decades. One of the most striking changes has been the advances in our understanding of benthic remineralization as a principal source of recycled nutrients available to photosynthesis. This runs counter to the previously held views that water column processes dominated remineralization.

In this section we shall first examine the general transformation, exchanges, and cycling of the two major nutrients, nitrogen and phosphorus. Then nutrient cycling in representative coastal ecosystems (sand beaches, estuaries, salt marshes, sea grass beds, mangroves) will be discussed, illustrating how the nutrients interact with the biota, sediments, and water. Models of mangrove-nutrient interactions will then be considered, followed by detailed discussion of nitrogen fixation, nitrification, and denitrification, and sediment-water interactions in nutrient dynamics.

3.9.2 THE NITROGEN CYCLE

3.9.2.1 Transformations of Nitrogen

Nutrients occur in many forms which can be described in terms of their chemical structure, oxidation state, and solid-liquid-gas phase. The nitrogen cycle is dominated by the gaseous phase and the microbial transformations involving changes in the oxidation state. The oxidation state ranges from nitrate (NO_3^-, +5), the most oxidized form, to ammonium (NH_4^+, –3) (Figure 3.77), and the compounds that exist in all intermediate states (Fenchel and Blackburn, 1979; Webb, 1981).

Nitrate (NO_3^-) is taken up under aerobic conditions by algae, bacteria, and macrophytes. It is also used as a terminal electron acceptor and is reduced by dissimilatory processes in the following sequence: nitrate to nitrite (NO_2^-), then NO and N_2O, and finally N_2 gas. This pathway, called denitrification, occurs under anaerobic conditions and requires a supply of organic compounds. Denitrifiers may also reduce nitrate to ammonium.

Ammonium is the preferred plant nutrient. It may also be oxidized by nitrifying bacteria. The cycle of nitrogen also includes a pathway by which nitrogen in the gaseous form can be fixed into organic form by certain bacteria ("nitrogen fixation"). Both of these processes will be discussed in Section 3.9.

Large amounts of organic compounds are released after the death of animals. These are ultimately decomposed by heterotrophs, ultimately to ammonium. Excretion by animals also releases dissolved organic nitrogen compounds.

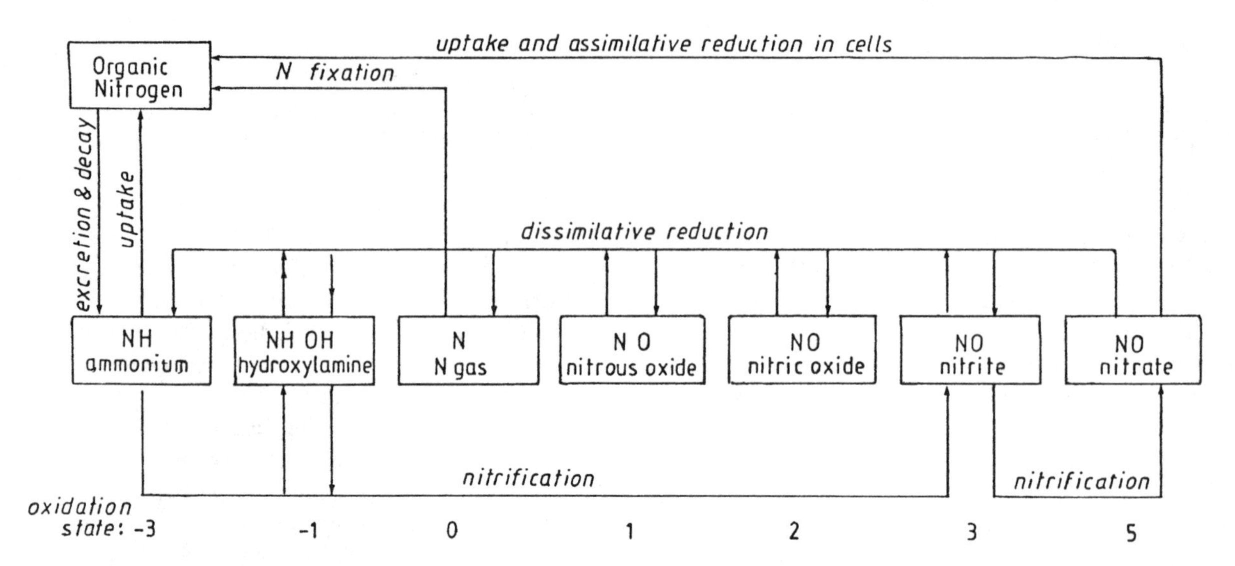

FIGURE 3.77 Transformations of nitrogen. The boxes show the various nitrogen species and arrows indicate the processes of transformations. (Adapted from Fenchel, T. and Blackburn, T.H., *Bacteria and Mineral Cycling*, Academic Press, London, 1979, 101. With permission.)

3.9.2.2 The Coastal Nitrogen Cycle

The coastal zone nitrogen cycle is complex (Figure 3.78). There are three important dissolved organic salts, NO_3^-, NO_2^-, and NH_4^+ (which are interconvertible through redox reactions), two major gaseous forms (N_2 and N_2O), and a number of soluble organic salts, including urea, uric acid, and all amino acids. In addition there is particulate nitrogen in the detrital particles, both in the water column and the sediments.

Sources of nitrogen: Nitrogen enters coastal zone waters from river inflow, sewage discharges, rain, atmospheric diffusion, transport via currents of nitrogen upwelled from deeper offshore waters, and biological fixation. The amount of nitrogen entering the coastal system via precipitation is generally small relative to other inputs. Nitrogen from this source comprises only 0.5% of the nitrogen that enters Narrangansett Bay, Rhode Island, but is greater than 12% in the Baltic where shores are heavily industrialized.

Dissolved inorganic and organic nitrogen are assimilated by the phytoplankton and bacteria, which preferentially take up NH_4^+. Small amounts of NH_4^+ enter the coastal system via nitrogen fixation. The compound NO_3^- enters via freshwater runoff and is produced in both the water column and the sediment by nitrification. Particulate nitrogen enters the coastal system in detritus in inflowing freshwater, detritus derived from the death and decay of the producers (phytoplankton, sea grasses, marsh grasses, mangroves, benthic microalgae, and macroalgae), and consumers. Some of this particulate nitrogen will eventually be made available to the producers and bacteria via mineralization, while some will be buried in the deeper sediments.

Regeneration of nitrogen: Since the inputs of new nitrogen are not sufficient to meet the needs of autotrophic production, the regeneration of nitrogen is a key process in coastal ecosystems. The principal processes involved are nutrient regeneration via decomposition by bacteria and excretion by living organisms.

Organic matter resulting from the excretion, defecation, and death of living organisms is subjected to decomposition by microorganisms (bacteria and fungi) with the eventual release of nutrients. Most organisms in the water column contribute to the regeneration of nitrogen as organic matter they produce sinks through the water column. However, due to the shallow water depths and often rapid sinking rates in the coastal zone, the residence time in the water column is short, and most of the regeneration takes place in the sediments. The processes involved will be discussed in detail in Section 3.9.5.

While bacterial decomposition of the organic matter in the water column contributes to the pool of regenerated nitrogen, a much greater contribution arises from excretion by microzooplankton (flagellates and ciliates) and the larger zooplankton (especially copepods). Nitrogen is excreted in the form of ammonium and dissolved organic nitrogen compounds (urea, uric acid, and amino acids). Excretion rates per unit weight of zooplankton are inversely proportional to the size of the individual zooplankton. Thus the smaller nano- and picozooplankton contribute more to the total excretion than would be assumed from their biomass. Recent measurements of NH_4^+ excretion by zooplankton size classes indicated that those <35 μm accounted for the bulk of the release (Harrison, 1980; Gilbert, 1982).

Losses of nitrogen: Nitrogen is lost from coastal ecosystems by burial in deep sediment, transport by water

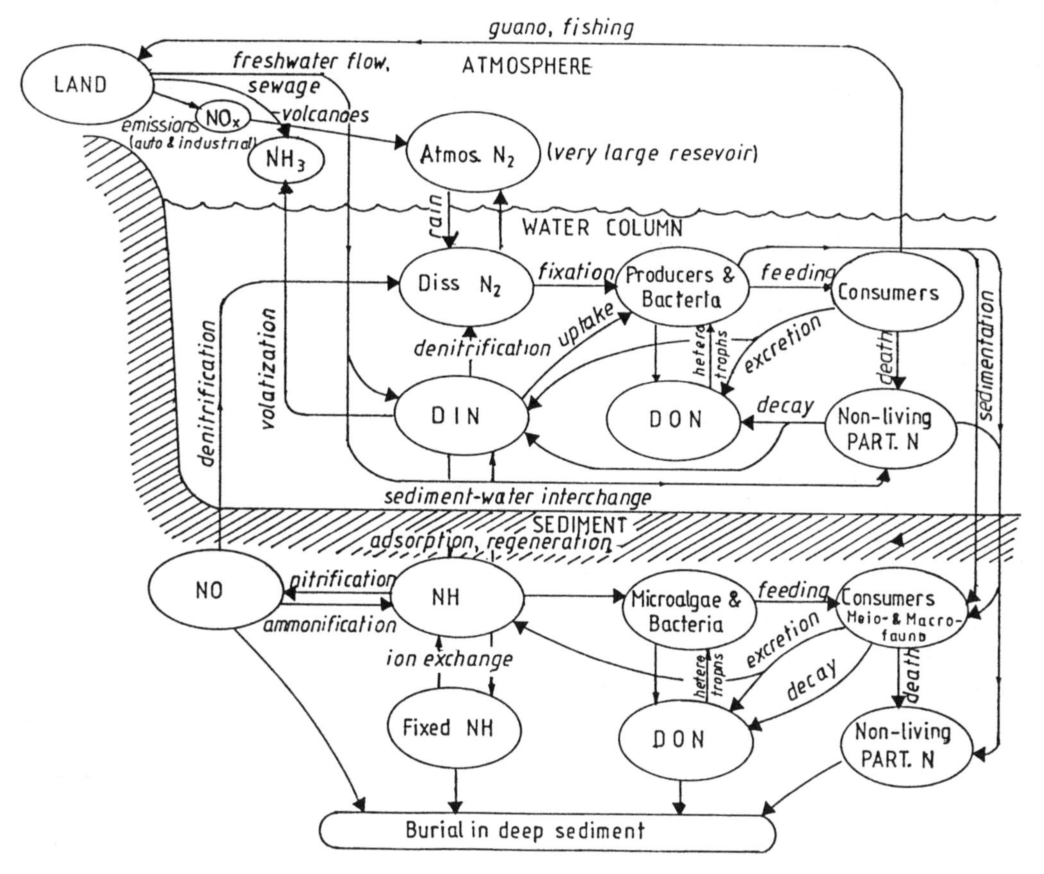

FIGURE 3.78 The coastal zone nitrogen cycle.

currents, or denitrification. Burial of organic matter was discussed in Section 3.8, while denitrification will be discussed in the succeeding section.

3.9.2.3 Nitrogen Fixation, Nitrification, and Denitrification

Nitrification: The nitrogen cycle includes a pathway in which nitrogen in the gaseous form (N_2) can be "fixed" into organic compounds by a number of procaryotic organisms (bacteria and blue-green algae). These can be free living or symbiotic with primary producers (Capone et al., 1979).

On a *Pucinellia* and *Festuca* salt marsh in northwest England, Jones (1974) showed nitrogen fixation by blue-green algae of 20 g N m^{-2} $year^{-1}$ on the mud surface and more than twice that level in pools. Whitney et al. (1975) for the marshes of Flax Pond, Long Island, calculated an average daily fixation of 4.6 mg N m^{-2} day^{-1}, which can be extrapolated to 0.5 to 1.0 g N m^{-2} $year^{-1}$. For the Sapelo Island marshes, Hanson (1977a,b) calculated that nitrogen fixation provides a new source of nitrogen to salt marsh sediments of Sapelo Island at the rate of 15 g N m^{-2} $year^{-1}$. Table 3.35 presents some comparative data on marsh nitrogen fixation.

Work by a number of investigators in different marsh systems (Hanson, 1977a,b; Patriquin and Knowles, 1972; Teal et al., 1979; Whitney et al., 1981) has demonstrated that the bulk of this fixation takes place in the rhizosphere of the plant roots. Patriquin and McLung (1978) identified a number of types of nitrogen-fixing bacteria distributed over the surface of the stems and roots, and also within the cells of the cortical layer.

However, as can be seen from Table 3.35, N_2 fixation tends to be relatively unimportant in the overall nitrogen budget, except in some specialized environments (such as sea grass beds and some salt marshes). In general, nitrogen fixation rarely supplies more than 5% of the autotrophic nitrogen demand.

Two additional processes in the nitrogen cycle are nitrification and denitrification. Nitrification occurs under anaerobic conditions, and is the oxidation of ammonia to nitrate. It is a bacteria-mediated process in which nitrifying bacteria use NH_4^+ as an energy source to fix carbon dioxide. The process occurs in two major steps; the first is mediated by bacteria of the genus *Nitrosomonas* and the second by species of *Nitrobacter:*

$$NH_3 + 3/2O_2 \rightarrow HNO_2 + H_2O \text{ (\textit{Nitrosomonas})}$$

$$HNO_2 + 1/2O_2 \rightarrow HNO_3 \text{ (\textit{Nitrobacter} spp.)}$$

TABLE 3.35
Summary of Nitrogen Fixation Rates and Associated Effect on Assimilative Demand for Selected Estuarine and Coastal Ecosystems

Ecosystem Dominant Sp. (N_2-Fixing Organisms)	N_2 Fixation (mg N m^{-2} d^{-1})	Assimilative Demand Satisfied (%)	References
Salt Marsh			
Spartina alterniflora			
(Blue-green algae)	5–45	1–5	Carpenter et al. (1978)
(Sediment bacteria)	80	15	Jones (1974)
			Whitney et al. (1975)
Sea Grass			
Thalassia testudinum			
(Epiphytes)	4–150	3–100	Capone (1982)
(Sediment bacteria)	20–80	15–50	
Zostera marina			
(Epiphytes)	0–0.1	0–0.5	Capone (1983)
(Sediment bacteria)	0.2–6	0.5–5	
Myriophyllum spicatum			
(Epiphytes and sediments)	3	2	Lipschultz et al. (1979)
Estuarine Shallows			
(Sediment bacteria)	0.5–10	1–5	Brooks et al. (1971)
Coastal Shelf			
(Planktonic blue-greens)	0.01–1.3	1–5	Carpenter and Price (1977)
(Bacteria on *Sargassum*)	0.01	1	Carpenter (1973)

In the coastal zone, nitrification rates tend to be low or negligible in the water column but occur at moderate rates in the surface layer (oxic layer). Table 3.36 gives data on nitrification and denitrification rates of various coastal ecosystems.

Denitrifying bacteria use NO_3^- as an electron acceptor to oxidize organic matter anaerobically releasing N_2 gas by the following reaction:

$$5\ C_6H_2O_6 + 24\ HNO_3 \rightarrow 30\ CO_2 + 42\ H_2O + 12\ N_2$$

Denitrification can also produce nitrous oxide (N_2O) from NO_3^- via the following reaction which is the first step in the one above:

$$C_6H_{12}O_6 + 6\ HNO_3 \rightarrow 3(CO_2 + 9\ H_2) + 12\ N_2$$

As can be seen from Table 3.36, denitrifying rates are similar to or perhaps slightly higher than those for nitrification rates.

3.9.3 Phosphorus Cycle

Phosphorus in seawater is found in living organisms or as dissolved inorganic phosphorus (DIP), dissolved organic phosphorus (DOP), and particulate organic phosphorus (POP) Figure 3.79 depicts the transformations within the phosphorus cycle on coastal water. Phosphorus enters coastal zone waters either from weathering of soil and rock and subsequent runoff, or from point sources as sewage discharges. Dissolved inorganic phosphorus (DIP) is taken up by algae and bacteria and incorporated into cellular organic matter. Some of this particulate organic phosphorus (POP) is excreted or released after death, either as DIP or as DOP, which can then be decomposed by bacteria with the release of DIP. These processes also occur in the sediments.

Two aspects of the chemistry of phosphate that are important in maintaining the generally low concentrations of the dissolved forms in the environment are facility of adsorption on particles and the propensity to form insoluble compounds with certain metals (such as Fe and Mn). Phosphate absorbs readily under aerobic conditions onto amorphous oxyhydroxides, calcium carbonate, and clay mineral particles (Krom and Berner, 1980). This sorption takes place at high concentrations but the process is reversed at low concentrations. The sorption-desorption interaction provides a buffering mechanism for DIP. Phosphate ions also precipitate with cations such as Ca^{2+}, Al^{3+}, and Fe^{3+} (ferric iron) to form insoluble minerals.

The uptake of phosphorus by primary producers and bacteria is responsible for the generally low concentrations

TABLE 3.36
Summary of Nitrification and Denitrification Rates in Various Estuarine Ecosystems

Ecosystem	Nitrification (g atoms m^{-2} hr^{-1})	Denitrification (g atoms m^{-2} hr^{-1})	Reference
Salt marsh	60	110	Kaplan et al. (1979)
Spartina alterniflora			
Sea grass	20–120	20–90	Izumi and Hattori (1982)[a]
Zostera marina			Shenton (1982)
Potamogenton perfoliatus			
Estuarine water	0–5,000	0	Billen (1975)
Estuarine sediment	10–115	10–300	Billen (1978)
			Sorenson (1978)
			Oren and Blackburn (1979)
			Nishio et al. (1982)
			Seitzinger et al. (1980)
			Henrikson et al. (1981)
			Hansen et al. (1981)

[a] Assumes bulk density of sediments at 1 gm cm^{-1}, and rates limited to upper 20 cm of sediment.

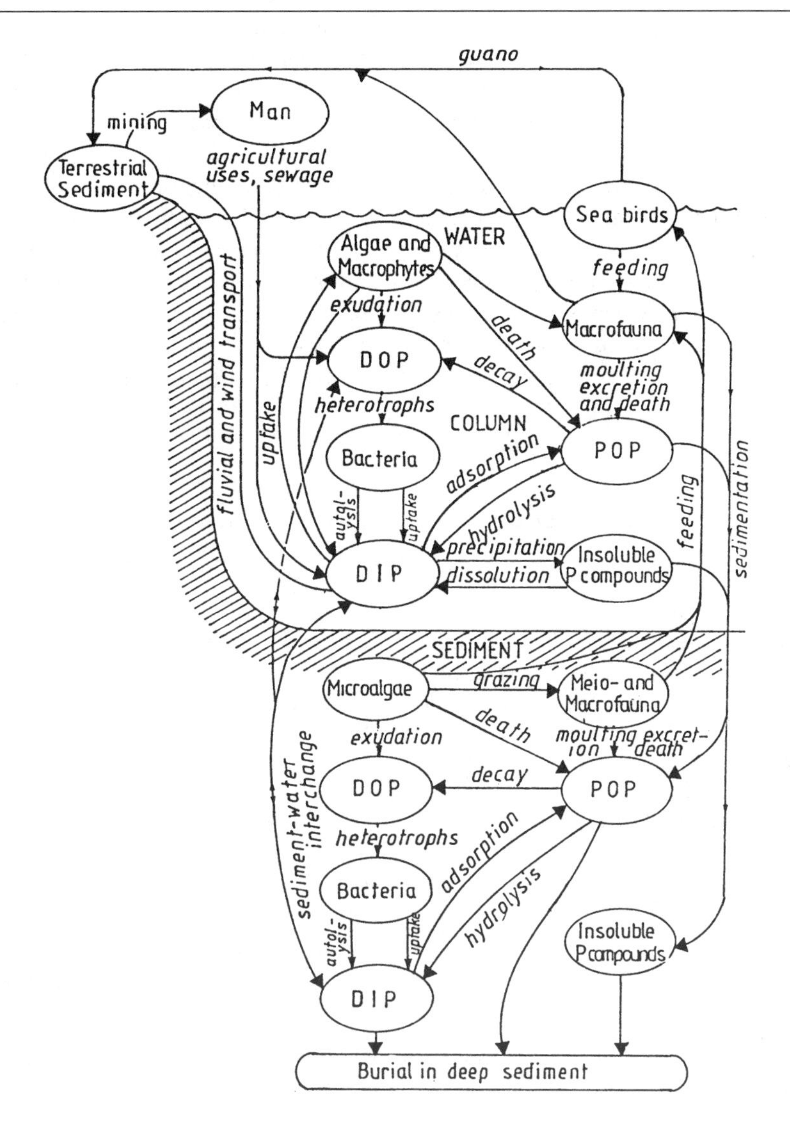

FIGURE 3.79 The coastal zone phosphorus cycle.

in coastal zone surface waters. Some coastal macrophytes such as salt marsh plants, eelgrasses, and mangroves take up phosphorus from the sediments and in turn some of this uptake may be released into the surrounding water.

3.9.4 Sediment-Water Interactions in Nutrient Dynamics

Fluxes across the sediment-water interface may be grouped into two categories: (1) the flux downwards due to deposition and burial of particulate matter in the sediments, and (2) diffusional fluxes that may be in any direction depending on the chemical and physical conditions of the reacting system (Zeitzchel, 1980). The exchange of nutrients between the bottom sediments and the overlying waters has long been recognized as one of the principal factors regulating chemical and biological cycles in the coastal zone.

3.9.4.1 Nutrient Fluxes across the Sediment-Water Interface

There have been numerous investigations of nutrient fluxes across the sediment-water interface. Some selected examples will be discussed below.

Hopkinson and Wetzel (1982) investigated nutrient and oxygen fluxes in nearshore coastal water in Georgia Bight, U.S.A. Dissolved nutrient concentrations were always greater in the top 25 cm of the sediment than in the overlying waters. For example, ammonium values were about 110 μmol in the top 25 cm of the sediment compared with 1.1 μmol in the overlying water. NO_2^- + NO_3^- values ranged from 20 to 0 μmol in the pore water and were less than 1 μmol in bottom waters; PO_4 was 25 μmol in pore water and 1.4 mol in the overlying water. The percentages of organic carbon and total nitrogen were low and averaged about 0.3 and 0.03% of dry sediment weight, respectively. The basic patterns were regeneration and release to the water column of ammonium, nitrite, nitrate, and phosphate, and uptake by the sediment of DON and no net exchange of DOP. The flux of nutrients across the sediment-water interface was overwhelmingly dominated by NH_4^+. Although there as a substantial flux of orthophosphate, it was only one-fifth of the NH_4^+.

Nixon et al. (1980) investigated phosphorus regeneration and metabolism in Narragansett Bay, Rhode Island. They found that there was almost always a net flux out of the sediments, with the magnitude ranging from near zero in winter to almost 60 mmol m^{-2} hr^{-1} in the summer. It was calculated that some 120 mg-at of inorganic P m^{-2} $year^{-1}$ were released annually to the overlying water, or enough phosphorus to support about 50% of the annual phytoplankton production. The flux of dissolved organic phosphorus was lower, and appreciable uptake, as well as release, was observed.

From Table 3.37, which compares estuarine, coastal, and offshore environments, it can be seen that with the exception of the sandy sediments of La Jolla Bight and the Newport River estuary, average values of inorganic nitrogen release from the sediments of the other eight environments were similar, ranging from 1 to 10 mmol m^{-2} day^{-1}. Phosphate exchanges were more variable, ranging from 0.03 to 1.2 mmol m^{-2} day^{-1}.

TABLE 3.37
Values of Sediment-Water Column Exchange Rates (mmol m^{-2} day^{-1})

Environment (Region)	NO_3	NO_2	NH_4	∑DIN	PO_4	O_2	Ref.
South River estuary (North Carolina)	0.03	—	2.7	2.7	0.15	42	—
Neuse River estuary (North Carolina)	0.07	—	5.4	5.5	0.34	36	—
Newport River estuary (North Carolina)	0.00	—	0.0	0.0	0.00	0	—
La Jolla Bight (California)	0.08	0.03	0.87	0.86	0.077	11	206
Narragansett Bay (Rhode Island)	–0.03	0.04	2.1	2.1	0.23	28	185
Buzzards Bay (Massachusetts)	0.05	0.02	1.3	1.4	0.03	23	460
Coastal North Sea (Belgium)	2.3	—	1.8	4.1	—	—	40
Offshore North Sea (Belgium)	1.2	—	0.90	3.1	—	—	40
Loch Thurnaig (Scotland)	0.0	0.0	0.96	0.96	—	11	100
Cap Blanc (Africa)	3.8	0.34	5.6	9.7	1.2	20	459
Average	0.73	0.09	2.2	3.0	0.28	21	—
± S.E.	0.42	0.06	0.60	0.90	0.16	4.9	—

Note: Positive values are net release to the water column, and negative values represent loss from the water column. ∑DIN (dissolved inorganic nitrogen) = NO_3 + NO_2 + NH_4.

Source: From Fisher, T.R., Carlson, P.R., and Barber, R.T., *Est. Coastal Shelf Sci.*, 14, 111, 1982a. With permission.

From this table it can be seen that the proportion of the nutrient demand by the producers that was supplied by release from the sediments ranged from 0 to 79% (average 35%) for nitrogen and 0 to 75% (average 28%) for phosphorus. Thus on average for these systems, about one quarter of the water-column nitrogen demand is supplied by benthic regeneration and flux across the sediment-water interface. One of the most comprehensive budgets is that of Valiela and Teal (1977a) for the Great Sippwissett Marsh (Table 3.38) of the total input precipitation accounts for 0.53%, groundwater flow 16.3%, N_2 fixation 4.5%, and tidal water exchange 75%.

Boynton et al. (1980) have summarized comparative benthic fluxes of ammonium and dissolved inorganic phosphate from estuarine and coastal ecosystems (Figure 3.80). The highest fluxes were recorded in the Patuxent Estuary. The annual mean flux of NH_4^+ and DIP was 295 and 43 μg-at m^{-2} hr^{-1}, respectively. Fluxes of NH_4^+ and DIP in Narragansett Bay (Nixon et al., 1976) were estimated at 100 and 20 μg-at m^{-2} hr^{-1}, whereas in Buzzards Bay (Rowe et al., 1975) these fluxes were 68 and –3 μg-at m^{-2} hr^{-1}, respectively. Boynton et al. (1980) considered that the high fluxes they observed in the Patuxent Estuary were related to the high seasonal organic matter supply to the sediment and the physical dynamics of the sediment-water interface. They concluded that the benthic fluxes of NH_4^+ and DIP could satisfy 0 to 190% and 52 to 330% of the estimated daily demand for these nutrients by the phytoplankton, values somewhat higher than those listed in Table 3.36.

Nixon et al. (1980) have discussed the nitrogen to phosphorus ratios (N:P) released from the sediments. In oceanic waters, this ratio remains remarkably constant at 16 N:1 P. In estuarine and nearshore waters, however, this ratio is usually much lower, as previously discussed, and nitrogen generally is the most limiting nutrient. In Narragansett Bay, the N:P ratio ranged from 10 to 15 in the winter, but it declined sharply in the spring and remained below 5 during the summer. For the Patuxent Estuary, Boynton et al. (1980) found that the N:P ratio was considerably lower (3:1) than the expected 16:1, indicating that the flux of nitrogen back to the water column was only about 19% of the expected phosphorus release. Explanations for the low N:P flux ratios are: (1) regeneration of organic matter in the sediments results in a return of inorganic nutrients to the water column that is low in fixed nitrogen relative to phosphorus; (2) nitrogen is being stored in the sediments as particulate nitrogen or as one of the dissolved forms; and (3) denitrification takes place in all seasons, particularly in the water column. Boynton et al. (1980) developed a simple nitrogen budget for the Patuxent Estuary (Figure 3.81) and concluded that the low N:P benthic flux ratios resulted from both incomplete remineralization of particulate nitrogen and denitrification. Annual sedimentation rates (2 cm $year^{-1}$), coupled with the concentration of particulate nitrogen in the water column yielded an estimate of about 7.0 g-at N m^{-2} $year^{-1}$. Vertical profiles of particulate N concentration in the sediment were relatively constant, suggesting that after some initial remineralization, a considerable amount of nitrogen (2.9 g-at N m^{-2}) remained to a depth of 2.0 cm. Flux of NH_4^+ from the sediment accounted for virtually all dissolved nitrogen exchange from the sediments to the water column and amounted to 2.4 g-at m^{-2} $year^{-1}$.

TABLE 3.38
Nitrogen Budget for the Great Sippwissett Salt Marsh (kg N $year^{-1}$)

Process	Input	Output	Net Exchange
Precipitation			179
NO_3 — N	52		
NO_2 — N	0.2		
NH_4 — N	31		
DON[a]	89		
Particulate N	7		
Subtotal	179		
Groundwater flow			5,471
NO_3 — N	2,495		
NO_2 — N	29		
NH_4 — N	492		
DON[a]	2,455		
Subtotal	5,471		
N_2 fixation			1,592
Algal	145		
Rhizosphere bacteria	1,273		
Nonrhizosphere bacteria	174		
Subtotal	1,592		
Tidal water exchange			–5,352
NO_3 — N	386	1,215	
NO_2 — N	154	166	
NH_4 — N	2,623	3,539	
DON[a]	16,346	18,479	
Particulate N	6,743	8,205	
Subtotal	26,252	31,604	
Denitrification		1,558 (41,273)	–2,831
Sedimentation		25	–25
Volatilization of NH_4		8	8
Deposition of bird feces	9		9
Shellfish harvest		9	9
Total	33,503	34,477	–974

[a] Dissolved organic nitrogen.

Source: From Valiela, I. and Teal, H.M., *Nature (London)*, 286, 653, 1979a. With permission.

Thus, of the annual particulate nitrogen input to the sediments of the Patuxent Estuary, about 14% remains as particulate N, 34% is returned to the water column as NH_4^+, and 24% is not accounted for and may represent that por-

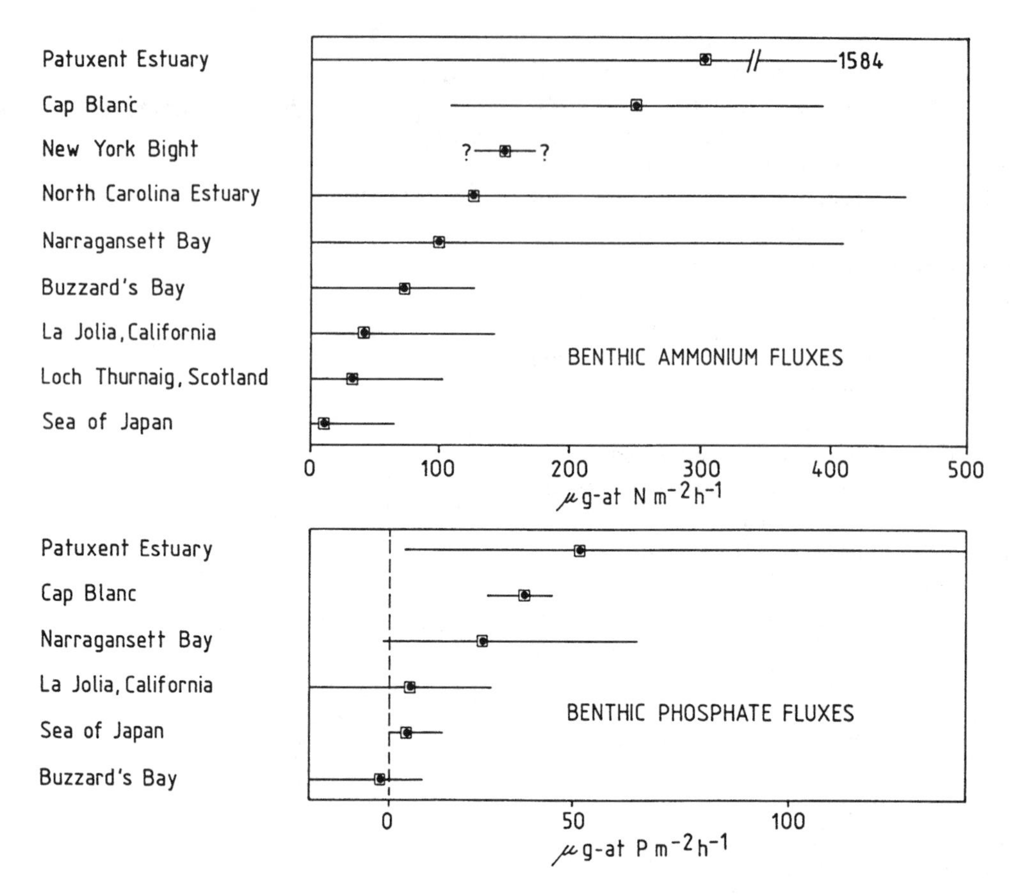

FIGURE 3.80 Comparative benthic fluxes (annual mean and range) of ammonium and dissolved inorganic phosphate from selected estuarine and coastal ecosystems. Data sources are as follows: Patuxent (Boynton et al. (1980), Cap Blank (Rowe et al., 1977), New York Bight (Rowe et al., 1975), North Carolina Estuary (Fisher and Carlson, 1979), Narranganset Bay (Nixon et al., 1976), Buzzards Bay (Rowe et al., 1975), La Jolla, California (Hartwig, 1978), Loch Turnaig (Davies, 1975), and Sea of Japan (Propp et al., 1980). (Redrawn from Boynton, W.R., Kemp, W.M., and Osborne, C.G., in *Estuarine Perspectives*, Kennedy, V.S., Ed., Academic Press, New York, 1980, 102. With permission.)

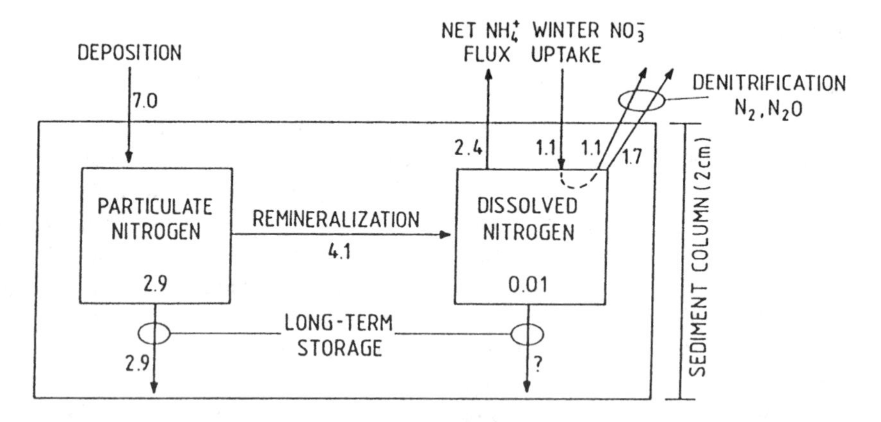

FIGURE 3.81 Model of annual nitrogen exchanges and storages between the water column and sediments in the turbid portion of the Patuxent Estuary. Units are g-at N m^{-2} $year^{-1}$ for fluxes and g-at N m^{-2} for storages. (Redrawn from Boynton, W.R., Kemp, W.M., and Osborne, C.G., in *Estuarine Perspectives,* Kennedy, V.S., Ed., Academic Press, New York, 1980, 105. With permission.)

tion of the annual particulate N input that enters the denitrification process. Since the NO_3^- flux into the sediments during winter amounts to 1.1 g-at m^{-2}, and if it is assumed that most of this nitrate is denitrified, then about 35% of the N input to the sediments undergoes denitrification.

3.9.4.2 Causes and Mechanisms of the Migration of Nutrients at the Sediment-Water Interface

According to Berner et al. (1976) the migration of nutrients and other materials across the benthic boundary layer is caused by: (1) physical processes in the water column immediately above the benthic boundary layer, and (2) biological, physical, and chemical processes within the sediment. Kemp et al. (1982) point out that it has been assumed that gradient-supported diffusive (and adjective) flux accounts for the mass transport of nutrients across the sediment-water interface (Berner et al., 1976). When measurements of flux, as calculated from pore water concentration gradients using a diffusion model, are compared with direct *in situ* measurements of flux into benthic chambers, the ratios of the latter to the former can vary from 1.5:1 to 15:1. The fact that the actual transport of pore water ions can exceed that predicted from molecular diffusion is due to a number of factors, including:

1. Turbulence and resuspension of sediments — mechanical energy transferred from the water to the sediments results in sediment resuspension and dispersion of pore water in the upper part of the sediment layer. Turbulent stresses affect the physical structure of the sediment-water interface and this can enhance the transfer of dissolved substances from the upper part of the sediment column.
2. Bioturbation of the sediment by deposit feeders and burrowing organisms may result in homogenization of the chemical composition of the upper layers of the sediment, and in addition, animals that build permanent burrows circulate water from the overlying water resulting in the flushing of interstitial water (Rhoads, 1974).
3. Diffusion of dissolved constituents can be a migrational mechanism of prime importance (Lerman, 1977).
4. Advection, or bulk flow, is a migrational mechanism causing net displacement of mass, relative to the sediment-water interface.
5. Vertical gravity displacement of interstitial water (flushing), due to changes in the density of bottom water. This process results in the provision of oxygen to the microbenthos and nutrients to the phytoplankton.
6. Interstitial water is drawn to the surface by capillary action caused by evaporation when the intertidal flats are exposed by the ebbing tide. The nutrient salts precipitate at the surface and are dissolved by the flooding tide.
7. Eubillation from methane bubbles.

The rates at which these fluxes occur are determined by the input of organic matter, bioturbation by the benthos, and the effects of temperature and water flow in the benthic boundary layer. The overriding factor is the input of organic matter. This input of organic matter as it is deposited on the bottom is subject to rapid decomposition in the aerobic surface layer, which is unconsolidated and flocculent. The sharp discontinuities that occur in profiles of nitrogen distribution in sediment suggest that rapid decomposition and remineralization occurs in the flocculent surface layer. Phosphorus, on the other hand, exhibits a local maximum concentration at the sediment surface, probably due to sorption or precipitation occurring in the oxidized surface sediments.

Fisher et al. (1982a) postulate that there are probably three potential fates of newly sedimented organic matter (Figure 3.82): (1) rapid and at least partly aerobic degradation in the upper layer of the sediments (0 to 10 cm) by micro- and macrobenthos, and return to the water column in less than a year of the remineralized nutrients through the combined effects of bioturbation and diffusion; (2) temporary burial in the top meter of the sediments, anaerobic decomposition by microorganisms, and return to the water column over decades via diffusion and bulk flow of the interstitial waters; and (3) effective removal from the aquatic system by deeper (>1 m) burial on much longer time scales. The relative importance of these three possibilities will depend on the composition of the sedimented material, sedimentation rates, oxygen availability, resuspension, bioturbation, etc. As we have seen, Nixon (1980) showed that one-quarter to one-half of the organic input to shallow marine sediments is oxidized, suggesting that one-quarter of the material is returned to the water column (fates 1 and 2), and one-half to three-quarters is buried (fate 3) or is oxidized in the water column.

The diffusion of material such as ammonium and phosphate, normally a slow process, is the result of the large concentration gradients that exist between interstitial pore waters, typically in the mmol l^{-1} range, and the overlying water column, typically in the mol l^{-1} range. A 100- to 1,000-fold concentration gradient thus drives the fluxes of phosphate and ammonia into the overlying water. Other important processes that play a role in this process are sorption and mineral formation.

Zeitzchel (1980) concluded that in shallow-water coastal ecosystems such as estuaries, a succession of major events take place:

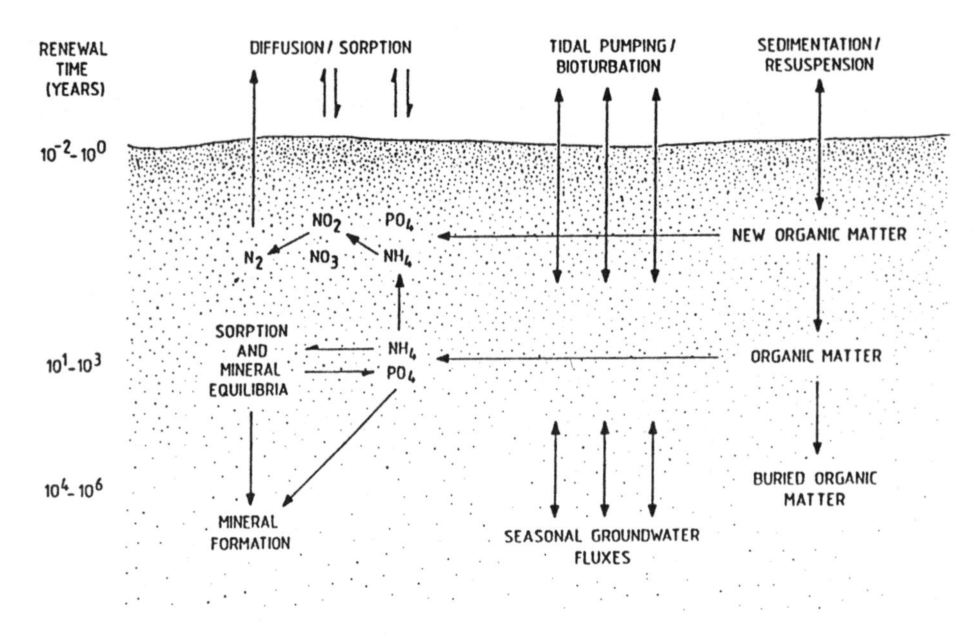

FIGURE 3.82 Summary of processes responsible for nutrient exchange at the sediment-water column interface. (Redrawn from Fisher, T.R., Carlson, P.R., and Barber, R.T., *Est. Coastal Shelf Sci.*, 14, 109, 1982a. With permission.)

1. Phytoplankton blooms.
2. Sedimentation of this bloom (directly or via fecal pellets of herbivores).
3. Decomposition of the organic matter (mainly at the sediment-water interface).
4. Release of the nutrients from the sediment, the recycling of these nutrients to the euphotic zone by turbulent mixing.

3.9.5 Nutrient Cycling in a High-Energy Surf-Zone Beach

McLachlan et al. (1981) suggested that the phenomenon of phytoplankton blooms that occur on high-energy beaches were associated with beach regeneration of nutrients. Lewin and Rau (1975) monitored populations of surf diatoms at Copalis Beach, Washington, U.S.A., over a 2-year period and found that during those times when nitrogen was absent, the nitrogen source for algal growth was apparently supplied by ammonia, which was always present in high concentrations. Subsequently, Lewin et al. (1979) showed that the surf clam, *Siliqua patula*, regenerated sufficient quantities of ammonia in the surf zones. McLachlan et al. (1981) studied open sandy beach ecosystems along the southeastern coast of South Africa and noted that the surf zones in this area were sites of intensive phytoplankton growth and that dense blooms of surf diatoms were common (see Section 3.5.1.6). They also concluded that a broad, shallow surf zone best develops circulation cells that allow a buildup of phytoplankton blooms. As a result of these circulation cells the residence times of beach-generated nutrients are prolonged in the surf zone, thereby stimulating phytoplankton growth. Furthermore, they suggested that abundant filter-feeding populations will generally only develop on beaches where surf-zone circulation patterns are such that nutrients, produced by beach mineralization of organic matter, can be trapped long enough to allow the development of phytoplankton blooms. They suggested that on the beaches they studied the macrofauna may contribute as much as 37%, and the interstitial fauna 63% of the total inorganic compounds generated by the mineralization of organic material in the beach.

Prosch and McLachlan (1984) studied the significance of ammonia excretion by the sand mussel, *Donax serra,* on a high-energy South African beach. They calculated a figure of 300 g NH_4-N per m of shoreline per year on that beach, which is enough to generate the total inorganic nitrogen pool of the surf zone approximately 15 times every year. These estimates are conservative as they do not include amino-N, and do not take into account the removal of NH_4-N, possibly by bacteria. As noted above, the interstitial fauna will generate about twice as much inorganic nitrogen as the *D. serra.*

Cockroft and McLachlan (1993) developed a nitrogen budget for an exposed beach and surf zone at Sundays Beach, South Africa. Nitrogen budgets show that the major contributors to nitrogen recycling are the subtidal interstitial fauna (4,610 g N m^{-2} $year^{-1}$) and the microbial loop (4,607 g N m^{-2} $year^{-1}$). Figure 3.83 illustrates the nitrogen cycle on this beach. Inputs of nitrogen that occur via groundwater inflow and from an estuary with rain and airborne detritus are of minor importance. Losses of nitrogen to the sediment as a result of beach accretion or to the atmosphere by denitrification are considered negligible because of sediment reworking and the highly oxidized

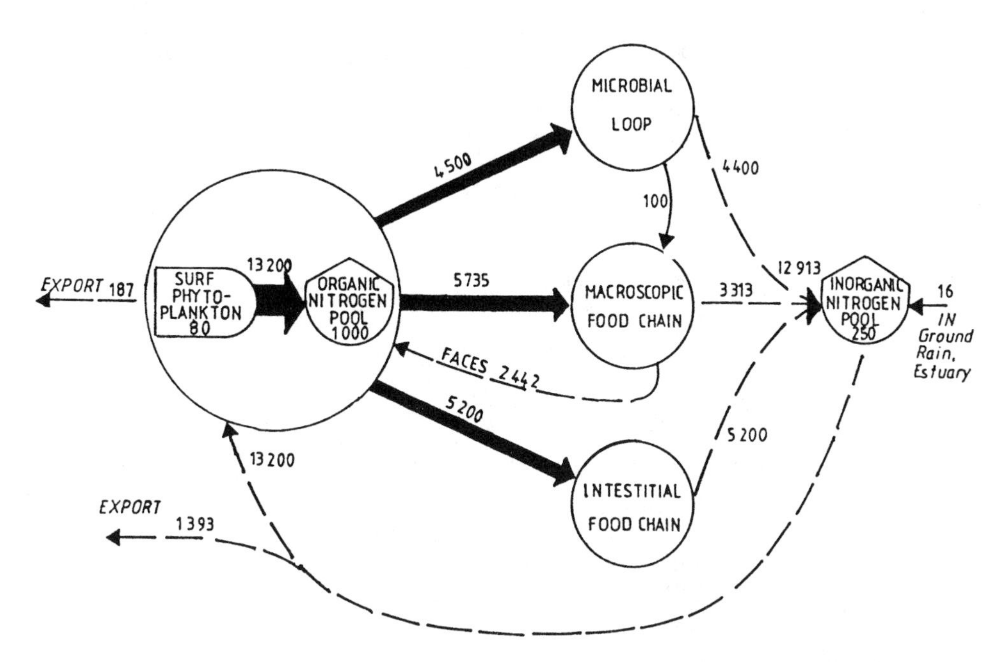

FIGURE 3.83 Nitrogen budget and recycling in the Sundays River beach/surf zone ecosystem (South Africa). Solid lines indicate grazing, broken lines recycling or nutrient flux. All values in g N m^{-1} $year^{-1}$. (Redrawn from Cockroft and McLachlan, *Mar. Ecol. Prog. Ser.*, 100, 294, 1993. With permission.)

state of the high-energy surf zone. The budget indicates that: (1) inorganic nitrogen inputs supply 13% of the phytoplankton requirements for primary production; (2) the 3 trophic assemblages together recycle 99% of the nitrogen required for total primary production; and (3) the dissolved and particulate organic nitrogen recycled by the macrofauna can supply 24% of the nitrogen requirements estimated for the microbial loop. Thus the surf zone can be considered to be a self-sustaining, semiclosed ecosystem.

3.9.6. Nutrient Cycling in Estuaries

Nutrient budgets have been developed in various levels of detail and completeness for a number of estuaries, including the Avon-Heathcote Estuary (Knox and Kilner, 1973) and the Upper Waitemata Harbour (Van Roon, 1983) in New Zealand, the Duplin River estuary, Sapelo Island, Georgia (Pomeroy and Wiegert, 1981), the Great Sippeswissett Marsh, Massachusetts (Valiela and Teal, 1979a,), The Hudson River Estuary (Simpson et al., 1975), San Francisco Bay (Peterson et al., 1975), Narragansett Bay (Nixon, 1981), portions of Chesapeake Bay (Taft et al., 1978), and the Rhode River Estuary (Jordon et al., 1983).

Jordon et al. (1991) investigated the fates of nutrients entering the Rhode River Estuary from its watershed and from atmospheric deposition. Figure 3.84 depicts the average flux rates for the various forms of phosphorus: dissolved phosphate (DPO_4^{3-}), particulate phosphate (PPO_4^{3-}), dissolved ortho-phosphate (DOP), and particulate ortho-phosphate (POP). Watershed inputs were greatest for the particulate forms of phosphorus and of this the bulk was exported down-estuary. Dissolved phosphate was released from the sediments while particulate phosphate was taken up. Figure 3.85 depicts the average flux rates of nitrogen: nitrate (NO_3^-), dissolved ammonia (DNH_4^+), particulate ammonia (PNH_4^+), dissolved organic nitrogen (DON), and particulate organic nitrogen (PON). The upper estuary consumed NO_3^-, DNH_4^+, PNH_4^+, and DON. Nearly all of the NO_3^- entering in bulk precipitation and watershed discharge was consumed in the upper estuary. The marshes, unlike the subtidal areas, released DNH_4^+ and DON and took up PON. The release of DNH_4^+ from the marshes was about one-third the uptake of the subtidal areas, and the exchanges of DON and PON by the marshes were about half those by the subtidal areas.

The consumption of nitrate due to phytoplankton production and the production of DPO_4^{3-} due to releases from particulate P after deposition in the sediments resulted in low inorganic N:P ratios, contrasting sharply with the lower estuary and Chesapeake Bay. Dissolved inorganic N and P entered the upper estuary from the watershed at an atomic ratio of 26 but left the estuary at a ratio of 2.7, thus creating the potential for nitrogen limitation. Although the marshes had relatively little effect on NO_3^-, they contributed substantially to the release of DPO_4^{3-}. The release of DOP_4^{3-} from terrigenous sediments may be a feature common to many estuaries and may partly explain why N, rather than P, is usually limiting in coastal waters.

3.9.7 Nutrient Cycling in Salt Marsh Ecosystems

Nutrient cycling in salt marsh ecosystems is complex. Childers et al. (1993) have presented a conceptual model

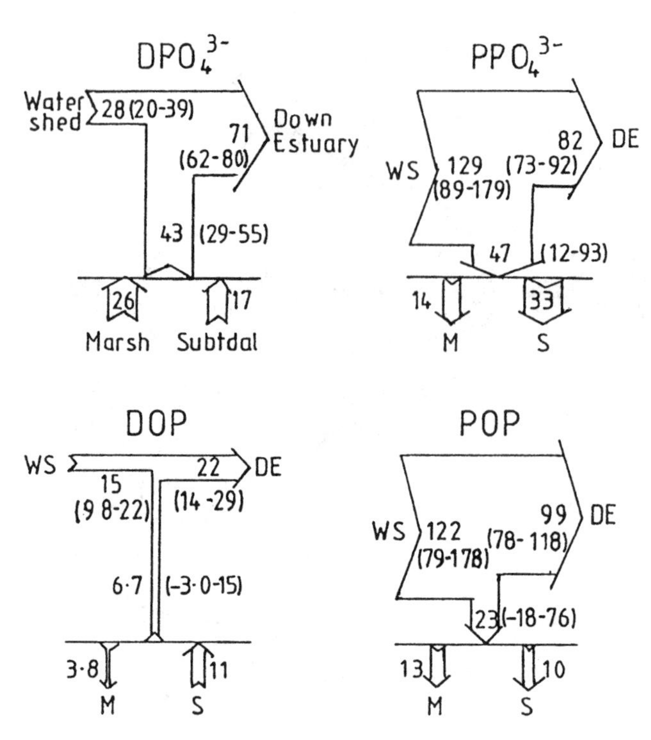

FIGURE 3.84 Average flux rates (g-atoms day^{-1}) for the forms of phosphorus in the upper Rhode River Estuary from March to November (95% C.L. in parentheses). Marsh fluxes are from Jordon et al. (1983). (Redrawn from Jordan, T.E., Correll, D.L. Miklas, J., and Weller, D.E., *Limnol. Oceanogr.*, 36, 261 and 262, 1991. With permission.)

of the six subsystems of the North Inlet Estuary, South Carolina, U.S.A.; forested uplands, salt marsh, tidal creek water column, subtidal benthic, and oyster reefs (Figure 3.86). Oyster reefs are a feature of some estuaries, especially those on the North American east coast but are absent elsewhere. The tidal water creek column is the subsystem that links all other habitats of the ecosystem. The figure also illustrates the important nutrient and material fluxes. Nutrient cycling is influenced by nutrient input via freshwater inflow, precipitation, and groundwater, sedimentation, input of organic matter to the sediment, nutrient exchanges across the sediment-water interface, nitrification and denitrification, and the activities of the benthic fauna.

Correll and Ford (1982) have compared data for precipitation and land runoff as sources of nitrogen for the Rhode River, a subestuary of Chesapeake Bay. The Rhode River Estuary has a surface area of 485 ha, while the watershed has an area of 3,330 ha. They found that during an average year about the same amount of readily available nitrogen entered the system via precipitation as entered via land runoff; during summer and autumn, precipitation was the largest source. Nitrogen inputs via precipitation were calculated to provide about 40% of the annual total nitrogen inputs, and about 50% of the readily available nitrogen inputs to the estuary. However, from a short-term viewpoint, the loadings from precipitation and the watershed are probably less important than benthic regeneration or recycling of nitrogen in the water column by the plankton. Phytoplankton produce about 1 kg organic carbon m^{-2} year^{-1} (Correll, 1978). To do this they need 320 g N m^{-2} year^{-1}. This is equivalent to 3.2 t N ha^{-1} year^{-1}, or 1,550 t year^{-1} for the whole estuary. Thus the approximately 12 t of total N derived by the Rhode River Estuary from bulk precipitation plus land runoff in an average year would provide less than 1% of the required amount. Other systems will vary in the total input depending on the relative areas of the estuary and the drainage basin, the amount of anthropogenic input (e.g., sewage), and the total rainfall.

The Great Sippwissett Marsh — One of the most complete studies carried out is that on the nitrogen cycle of the Great Sippwissett Marsh, Cape Cod, Massachusetts (Valiela and Teal, 1974; 1979a,b). This marsh comprises a mozaic of habitats, the major ones being creek sandy bottoms (42,500 m^{-2}), muddy creek bottoms (17,100 m^{-2}), salt pans (2,190 m^{-2}), algal mats on mudflats (6,200 m^{-2}), low marsh covered by short *Spartina alterniflora* (90,900 m^{-2}), tall *S. alterniflora* (7,350 m^{-2}), and high marsh consisting of the grasses *S. patens* and *Distichlis spinosa* (28,000 m^{-2}). Inputs are via precipitation, groundwater, and fixation of atmospheric N_2 by microorganisms. Tidal exchanges result in both input and losses of nitrogen. Nitrogen also leaves the marsh by denitrification, volatization of NH_3 from the marsh water, and movement of nitrogen into the sediments below the root zone of the marsh vegetation.

Table 3.38 presents the nitrogen budget for the marsh ecosystem while Figure 3.87 depicts the standing stocks

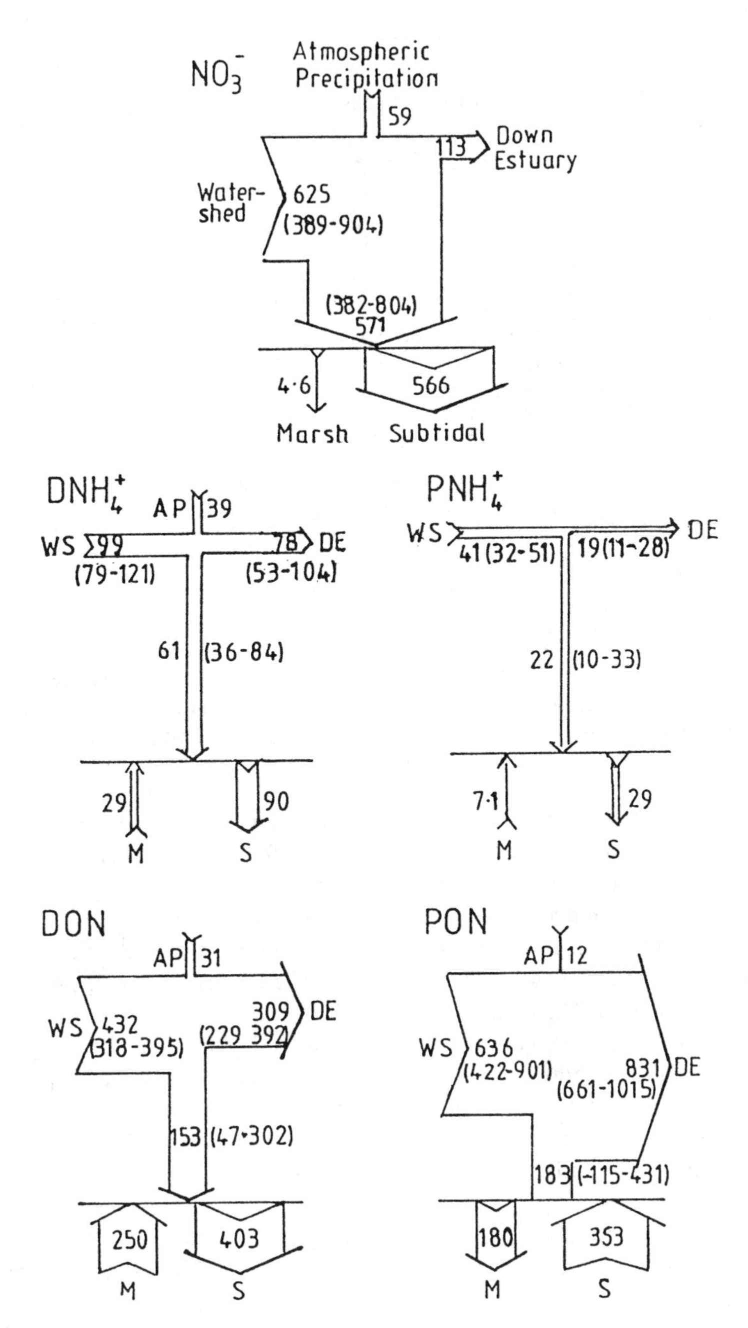

FIGURE 3.85 Average flux rates (g-atoms day^{-1}) of the forms of N in the upper Rhode River Estuary from March to December. (Redrawn from Jordan, T.E., Correll, D.L., Miklas, J., and Weller, D.E., *Limnol. Oceanogr.*, 36, 240, 1991. With permission.)

and fluxes of nitrogen among the major components of the Great Sippeswissett marsh ecosystem. The major sources of nitrogen are flooding of tidal water (16,252 kg N year^{-1}), flow of groundwater (5,471 kg N year^{-1}), and nitrogen fixation (1,592 kg N year^{-1}). The losses, principally via the ebbing tide and denitrification (1,558 kg N year^{-1}), exceeds gains by 974 kg year^{-1}. This amounts to only 2.9% of the inputs, primarily due to the overwhelmingly magnitude of the dissolved organic nitrogen component.

The sediments contained by far the greatest amount of nitrogen. As Valiela and Teal (1974) note, the large pool of nitrogen in the sediments raises an apparent paradox since there is evidence that the marsh vegetation is nitrogen limited (Valiela et al, 1973; Valiela and Teal, 1974). Furthermore the dissolved nitrogen in the tidal water is high (Valiela et al., 1978). There is evidence that nitrogen fixation is among the highest recorded. It is apparent that much of the nitrogen in the sediment and tidal water is not immediately available to the plants.

Duplin River Marshes, Sapelo Island — Whitney et al. (1981) have summarized the large amount of research that has been carried out on the nitrogen and

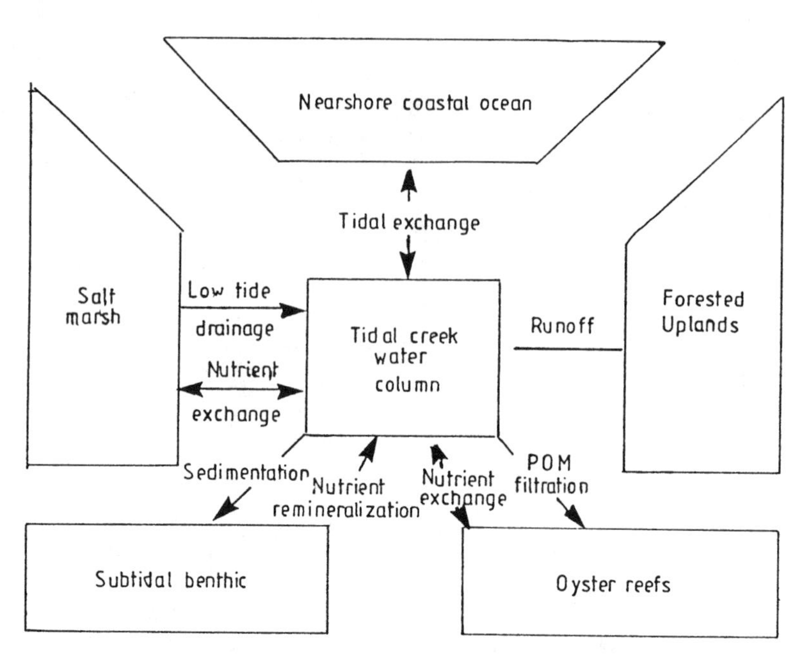

FIGURE 3.86 Conceptual diagram of the six subsystems, or habitats, of the salt marsh ecosystem including the important nutrient and materials fluxes that have been measured in the North Inlet estuary. Note the central position of the tidal creek water column as the integrator of these subsystem exchanges. (After Childers and McKellar, 1987. Redrawn from Childers, D.L., McKellar, H.V., Dame, R.F., Sklar, E.H., and Blood, E.R., *Est. Coastal Shelf. Sci.*, 36, 107, 1993. With permission.)

phosphorus cycles of the salt marshes of the Duplin River, Sapelo Island, Georgia. The amount of phosphorus that is present in the various compartments of the salt marsh system is in some instances quite large but highly variable. An estimate of the average standing stocks is given in Table 3.39. The chief sink and the largest standing stock of phosphorus is that of the sediments. The biota, particularly *Spartina* and the microorganisms, are the only other large stocks. Whitney et al. (1981) point out, however, that smaller pools of phosphorus, along with the sediment bacteria, may contribute disproportionately to the total flux within the system since they have been shown to have high turnover rates.

Depending on their relative concentrations, phosphorus is exchanged between the water and the sediment. The plants obtain the bulk of their phosphorus from the sediments. This phosphorus is recycled into the water and circulated through the system as the grass dies and decomposes and the resulting detritus is consumed by detritovores. While Reimhold (1972) found that a substantial amount of phosphate is lost from *Spartina* plants by leaching when they are covered by the tidal waters, this has been disputed by subsequent workers (Gardner, 1975).

At least half of the annual production of *Spartina* is degraded in the marsh resulting in release on the order of 3 g P m^{-2}. While Reimhold (1972) and Gardner (1975) considered that the marsh was a source of phosphorus for the estuary and coastal waters, Whitney et al. (1981) consider that on balance the marsh receives organic phosphorus and releases inorganic phosphorus, with the net flux of phosphorus being into the marsh. This pattern is similar to that found in the Flax Pond salt marsh (Woodwell and Whitney, 1977) where inorganic phosphate is exported during the warm months and organic phosphate is imported throughout the year.

The residence time of phosphate in the water is brief (from hours to days) depending on the rates of photosynthesis and respiration (Pomeroy et al., 1972). Since *Spartina* forms a link between the phosphorus in the subsurface sediments and that in the latter, the growth and microbial breakdown of dead *Spartina* are key biochemical processes. Other processes that complete the recycling are the deposition of sediment and detritus on the surface of the marsh and the seaward transport of particulate material during storms. The large reservoir of phosphorus in the sediments, and its continued recycling through *Spartina* would thus appear to confer stability to the salt marsh ecosystem (Pomeroy, 1975). The Duplin River marsh system is naturally eutrophic with a high capacity to store and assimilate phosphorus.

Studies on the nitrogen cycling in the Sapelo Island marshes have involved investigations of the microbial transformation of nitrogen, such as nitrogen fixation and denitrification, and changes in the standing stocks of the various forms of nitrogen: ammonia, nitrite, and nitrate. The quantity of nitrogen in the salt marsh sediments is determined by three interrelated factors: tides, physical and chemical exchanges with water and air, and biological fluxes. Similar processes determine the quantity of nitrogen in the estuarine water that participates in the exchange

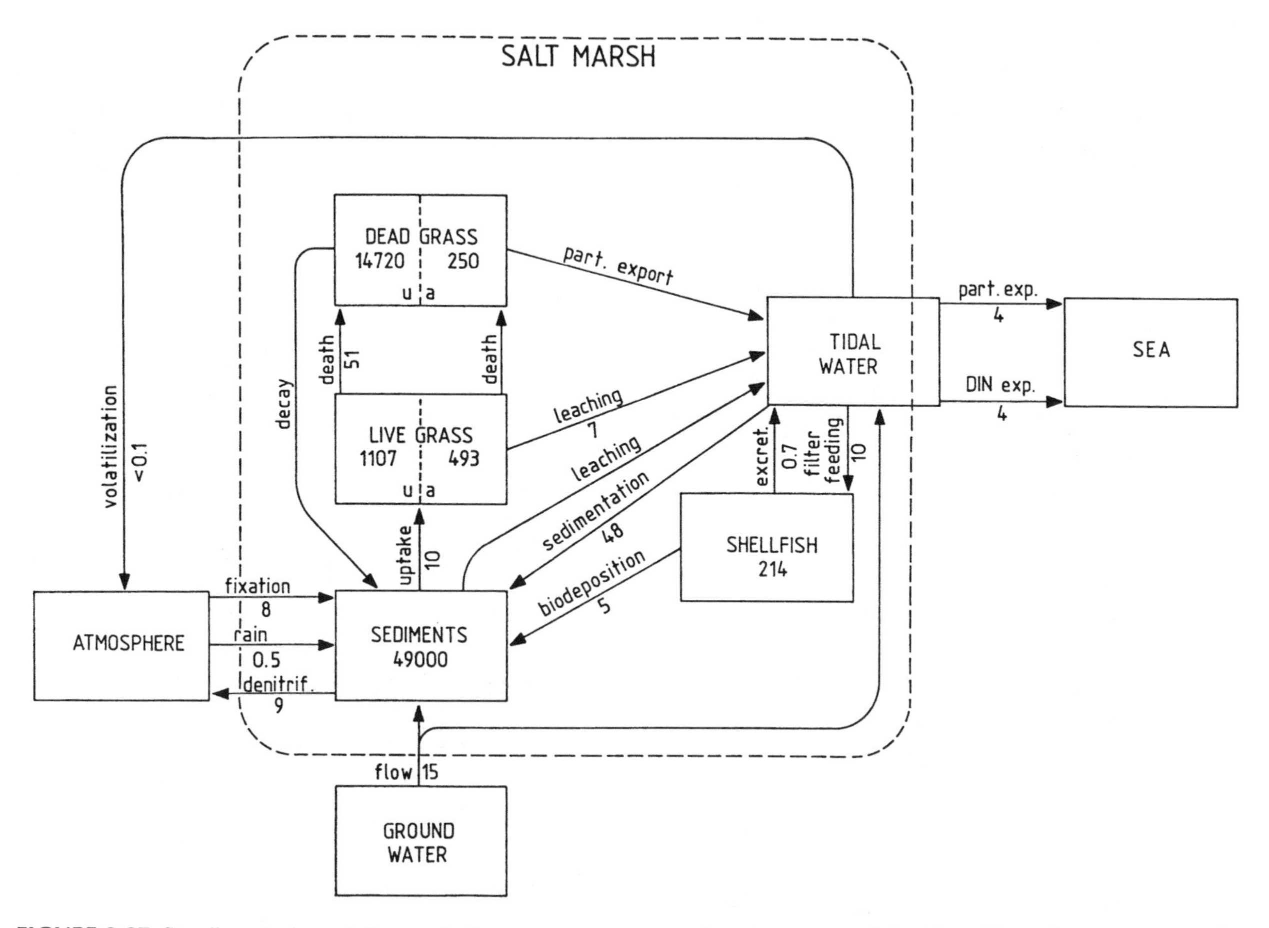

FIGURE 3.87 Standing stocks and fluxes of nitrogen among some major components of the Great Sippewissett ecosystem. The values are in kg N for the entire marsh (226,000 m^2), while exchanges are in kg day^{-1}. The values are for one day in early August. Letters "u" and "a" refer to under- and aboveground, respectively, and "part. export" means particulate matter exported. Sedimentary N is also resuspended, but this is not shown. (Redrawn from Valiela, I. and Teal, J.M., in *Ecological Processes in Coastal Environments*, Jeffries, R.L. and Davy, A.J., Eds., Blackwell Scientific, Oxford, 1979, 409. With permission.)

of nitrogen between the rivers, salt marshes, and the adjacent coastal waters. In addition to tidally mediated exchanges, rainfall, biological transformations within the water column, and exchange with the sediment can affect the various forms of nitrogen in the water.

A New England *Spartina* Salt Marsh — White and Howes (1994) using ^{15}N to trace N turnover in *Spartina alterniflora* roots and rhizomes found a close coupling between plant growth and microbial remineralization. The nitrogen budget for short *Spartina* (Figure 3.88) shows that translocation of nitrogen from the above- to belowground biomass during senescence equalled nearly 38% of the aboveground production. The largest flux of nitrogen was via microbial remineralization, equating to ~80% of the belowground plant production, with the remainder being buried, denitrified, and exported. Input from nitrogen fixation in the rhizosphere was substantial, roughly equal to half of the belowground production. This is in agreement with other studies of salt marsh ecosystems that indicate that, over a period of 1 to 3 year, decaying *Spartina* loses the bulk of its original nitrogen through microbial decomposition, and to a lesser extent, by leaching and burial. Inputs account for only about 60% of the annual plant nitrogen demand, while losses via denitrification, burial, and export remove 40 to 60% of the plant nitrogen demand. The losses are compensated for by retaining nitrogen through remineralization and translocation, which supply roughly 70 to 80% of the annual plant nitrogen demand. Thus nitrogen is conserved within the ecosystem by a variety of recycling mechanisms.

3.9.8 Nutrient Cycling in Sea Grass Ecosystems

Most of the research on nutrient cycling in sea grass communities has been carried out either on the temperate species *Zostera marina* or on the tropical/subtropical species *Thalassia testudinum,* which generally grows in nutrient-poor waters. In a study in Barbados, Patriquin (1972) measured the nitrogen and phosphorus content of *Thalassia* leaves and rhizomes and compared these with leaf tissue production. He found a very close correlation between

TABLE 3.39
Annual Mean Standing Stocks of Phosphorus in the Duplin River Salt Marsh And Estuary[a]

	mg P m^{-3}
Estuarine water	30
Sediment	5×10^5
Spartina	
Shoots	523
Roots	744
Detritus	323
Phytoplankton	24
Benthic microalgae	12
Bacteria in sediments	940
River water	15
Oceanic water	15

[a] 1 cm of sediment is assumed to be active in interchange with living compartments; mean water depth is taken at 1 m.

Source: From Whitney, D.M., Chambers, A.G., Haines, E.B., Hanson, R.B., Pomeroy, L.R., and Sherr, B., in *The Ecology of a Salt Marsh,* Pomeroy, L.R. and Miegert, R.G., Eds., Springer-Verlag, Berlin, 1981, 166. With permission.

production and the water-soluble N and P content of the rhizomes, and concluded that the plants obtained their nutrients from the sediments and did not absorb them from the water. When the nutrient requirements of the plants were compared with the quantities of N and P available in the root layer of the sediment, he found that there was enough phosphate for 300 to 1,000 days growth, but only enough inorganic N for 5 to 15 days growth. He also found a good correlation between leaf production and NH_4^+-N in the interstitial root layer leading to the conclusion that nitrogen was the nutrient limiting *Thalassia* production. Later Patriquin and Knowles (1972) showed that substantial nitrogen fixation occurred in the root zones of *Z. marina* in Canada, and *Diplanth*era, *Thalassia*, and *Syringodium* in the West Indies. They concluded that the fixed N was sufficient to supply the nitrogen needs of the plants.

McRoy et al. (1973) attempted to confirm Patriquin and Knowles' results by examining the leaves, roots, rhizomes, and sediments of a *Z. marina* bed for evidence of nitrogen fixation, but they found very low or undetectable rates of fixation. On the other hand, subsequent studies on *T. testudinum* (Capone et al., 1979; Capone and Taylor, 1980a,b) have provided substantial evidence of the occurrence of nitrogenase in the rhizosphere of this species. They recorded N_2 fixation rates of 5.1 to 37.6 mg N m^{-2} day^{-1} in the rhizosphere of the plants. These rates were an order of magnitude higher than recorded in nearby unvegetated areas. Capone (1982) also found that nitrogen fixation was consistently detectable in the rhizoshpere sediment of *Z. marina* collected from several stations in Great South Bay, Long Island, at various times of the year.

Figure 3.89 presents a conceptual model of nitrogen cycling in a sea grass ecosystem. The major nitrogen source for the sediment is sedimented organic matter, a substantial proportion of which is derived from the detritus formed by the death of sea grass leaves. A proportion of this is mineralized in the surface oxic sediments, while the remainder is eventually buried to become organic nitrogen in the anoxic sediments. Nitrogen leaves the biotic compartments as ammonia in both the oxic and anoxic layers. The ammonia then either diffuses out into the water, where it is oxidized to nitrite or nitrate, or it is

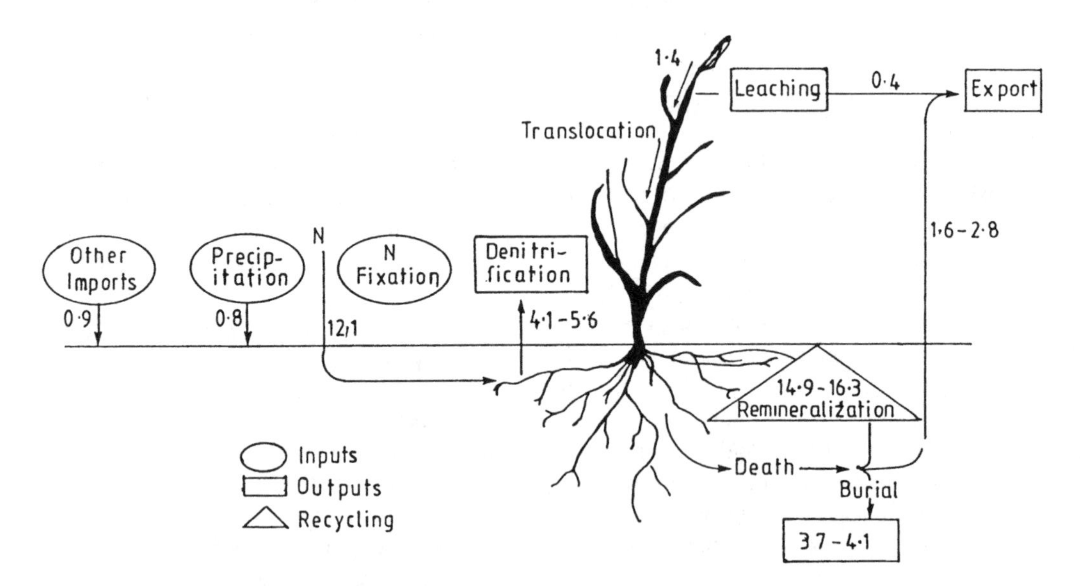

FIGURE 3.88 The annual nitrogen budget for short-form *Spartina alterniflora* in a New England salt marsh. Units are mol N m^{-2} yr^{-1}. (Data from White, D.S. and Howes, B.L., *Limnol. Oceanogr.*, 39, 1978, 1994. With permission.)

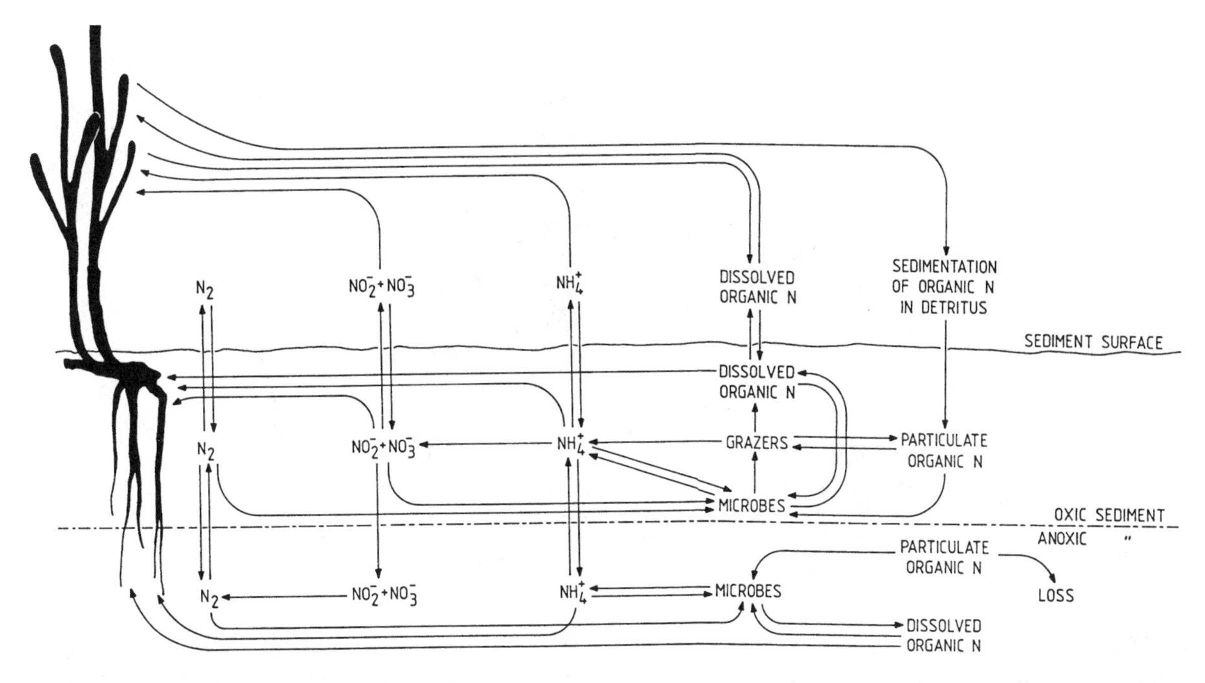

FIGURE 3.89 Circulation of nitrogen in sea grass ecosystems. (Redrawn from McRoy, C.P., Sea grass Ecosystems, Research Recommendations of the Int. Sea grass Workshop, Int. Decade of Ocean Exploration, October 22–26, 1973, Submitted to the National Science Foundation Workshop, Washington, D.C., 1973, 27. With permission.)

retained in the interstitial water by sorption onto sediment particles or perhaps bound to metals. The oxidized forms (nitrate and nitrite) may be denitrified to N_2, which may be returned to ammonia by nitrogen fixation. Ammonia, nitrate, nitrite, and dissolved organic nitrogen can all be used for plant growth.

McRoy and his co-workers (McRoy et al., 1972; McRoy and Goering, 1974) carried out an extensive study of phosphorus cycling in *Z. marina* beds in Alaska. They found that the eelgrass was capable of taking up phosphorus both through the leaves and the root-rhizomes, and estimated that 66% of the phosphorus taken up by the root-rhizomes was translocated to the leaves and released to the surrounding water. Penhale and Thayer (1980) found considerably lower release rates (30 ± 2.3%). This lower release rate was confirmed by Brix and Lyngby (1993) who found a foliar release rate of only 1 to 4%. It is thus clear that the role of eelgrass as a "P pump" may have been considerably overestimated.

3.9.9 Nutrient Cycling in Mangrove Ecosystems

In recent years there have been a number of studies of nutrient cycles in mangrove ecosystems, especially in tropical areas, including Boto and Wellington (1988) at Hitchinbrook Island, Queensland, Australia; Boto and Robertson (1990) in tropical mangrove forests in northern Australia; Alongi (1996) at Coral Creek, Hitchinbrook Island, Queensland, Australia; Balasubramian and Venugopalan (1984) in a tropical mangrove system in India; and Rivera-Monroy et al. (1995) in a fringe mangrove forest in Terminos Lagoon, Mexico. Alongi et al. (1992) have recently reviewed nitrogen and phosphorus cycles in tropical mangrove ecosystems.

Boto (1982) has reviewed nutrient and organic fluxes in mangrove ecosystems and identified the routes whereby organic nutrients can enter mangrove systems as follows:

1. Rainfall
2. Freshwater runoff from the surrounding land, including both dissolved and particulate nutrients.
3. Nitrogen fixation.
4. Mineralization (decomposition — heterotrophic conversion of organic N and P to inorganic forms).
5. Tidal-borne dissolved or particulate nutrients.
6. Chemical release from fixed states in the sediment, mediated by changes in the sediment Eh and pH conditions.
7. Man-made influences, e.g., agricultural drainage, sewage, and clearing of mangrove areas.

Nutrients may leave mangrove ecosystems via a number of possible routes:

1. Tidal transport of dissolved and particulate nutrients and plant litter.
2. Leaching of nutrients from plants (followed by tidal removal).

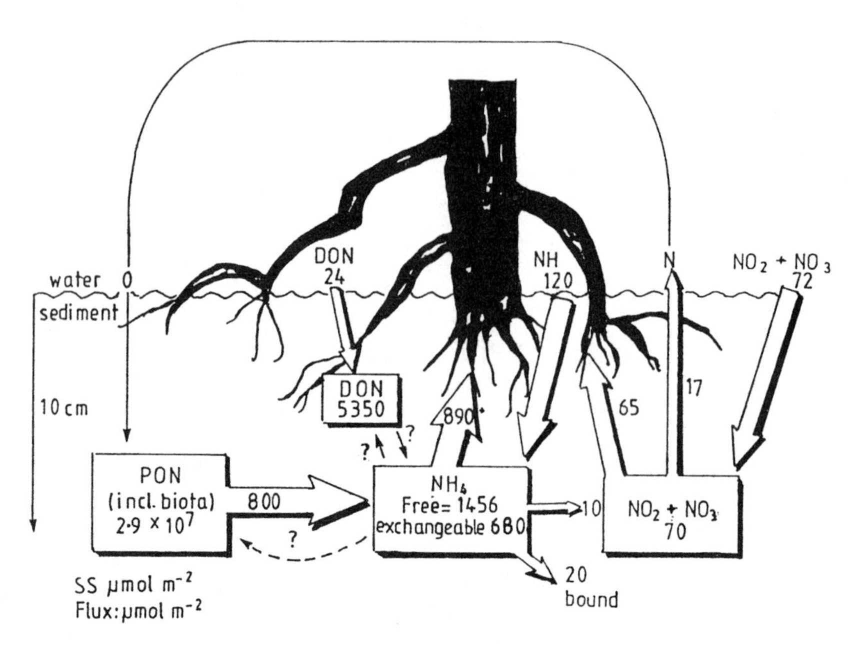

FIGURE 3.90 Nitrogen cycling in the mid-intertidal mangroves of Missionary Bay, Hinchinbrook Island, Queensland. Values are for a depth of 10 cm on a m^2 basis. Standing amounts = mol N m^{-2}; fluxes = mol N m^{-2} day^{-1}. Data from Boto and Wellington (1984), Izumi (1986), Alongi (1985), Boto and Wellington (1989), and Alongi (unpublished data). (Redrawn from Alongi, D.M., Boto, K.G., and Robertson, A.I., in *Tropical Mangrove Ecosystems,* Robertson, A.I and Alongi, D.M., Eds., American Geophysical Union, Washington, D.C., 1992, 4. With permission.)

3. Denitrification and volatization.
4. Immobilization of inorganic nitrogen in the sediment.
5. Leaching of the sediments by freshwater runoff.

Most mangrove forests are very productive and their associated food chains and nutrient cycles are often closely linked with those in adjacent waters (Robertson and Alongi, 1992). The extent to which mangrove forests exchange dissolved and particulate nutrients with adjacent waters depends upon several factors, including geomorphology, tidal regime, climate, and groundwater inputs. Most forests appear to exchange substantial amounts of nutrients (e.g., northern Australia: Alongi, 1996; Kenya: Hemminga et al., 1994; Papua New Guinea: Robertson et al., 1991; Mexico: Rivera-Monroy et al., 1995, but some do not (e.g., southwest Florida: Twilley, 1985). Accounts of two contrasting mangrove systems are discussed below.

Alongi et al. (1992) have summarized data from the long-term study by the Australian Institute of Marine Science at Missionary Bay, Hinchinbrook Island, northern Queensland (Figure 3.90). Nitrogen enters the mangroves of Missionary Bay through the fixation of atmospheric nitrogen by cyanobacteria growing on sediments, on prop roots and on timber lying on the forest floor, by tidal flushing and, to a much lesser extent, by monsoonal rainfall. Tidal flushing brings in nitrogen mainly as DON, with lesser amounts of DIN. Boto and Bunt (1981) indicate that no particulate matter is imported into the system on flood tides. The major losses of N from the system occur via tidal flushing and, to a lesser extent, by denitrification.

It is clear that significant outwelling occurs from Missionary Bay, mainly in the form of dissolved organic nitrogen and particulate N (as mangrove litter). Tidal exchange of these materials has been well documented (Boto and Bunt, 1981; 1982; Boto and Wellington, 1988; Robertson et al., 1988; Alongi et al., 1989; Alongi, 1990b). Additional losses may occur with the emigration of fish and prawns, but these are likely to be relatively small.

Rivera-Monroy et al. (1995) measured the flux of nitrogen and sediment in a fringe mangrove forest in Terminos Lagoon, Mexico (Figure 3.91). They found a net import of dissolved inorganic nitrogen (NH_4^+ and NO_2^- + NO_3^-) from the creek and basin forest, while particulate (PN) and dissolved organic N (DON) were exported to the creek and basin forest. The tidal creek was the principal source of NH_4^+ (0.53 g m^{-2} $year^{-1}$) and NO_2^- + NO_3^- (0.08 g m^{-2} $year^{-1}$) to the fringe forest, while the basin forest was the main source of total suspended sediments (TSS; 210 g m^{-2} $year^{-1}$). Net export of PN occurred from the fringe forest to the tidal creek (0.52 g m^{-2} $year^{-1}$) while less PN was exported from the basin forest (0.06 mg m^{-2} $year^{-1}$). There was a net import of TSS to the fringe forest from both the creek and basin forests, but the net input was 3.5 times higher at the fringe/basin interface. Particulate material exported from the forest during ebb tides generally has a higher C/N ratio than particulate matter imported into the forest on the flooding tide. This suggested that there was a greater nitrogen demand during

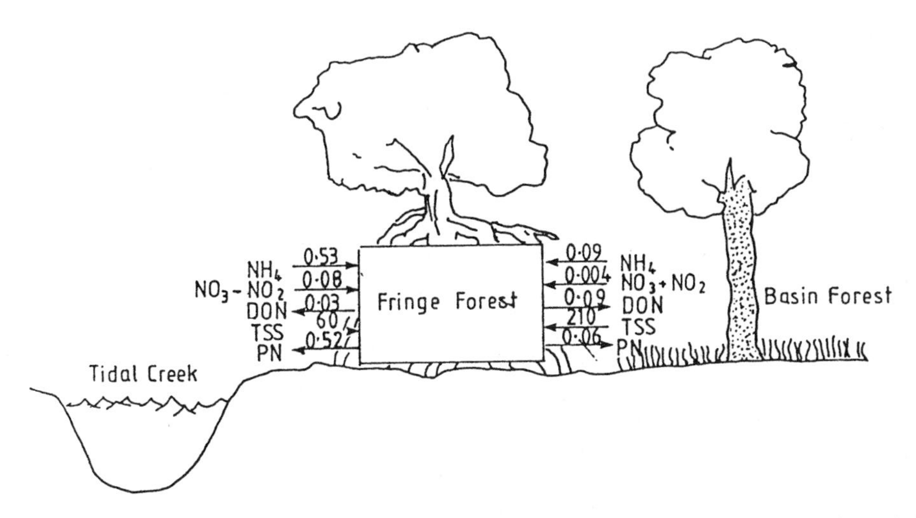

FIGURE 3.91 Net annual fluxes (g m^{-2} $year^{-1}$) in the fringe mangrove at Estero Pargo. Arrows into the forest is import; arrow out of the forest is export. (Based on data from Rivera Monroy, V.H., Twilley, R.R., Bowstang, R.C., Day, J.W., Jr., Vera-Herra, F., and del Carmen Ramirez, M., *Mar. Ecol. Progr. Ser.*, 126, 92, 1995. With permission.)

the ebb tide caused by the export of nitrogen-deficient detritus from the fringe and basin mangroves. The net annual import of inorganic nitrogen and the export of DON and PN suggest, in contrast to other mangrove systems, that the fringe mangrove forest acts as a sink of inorganic nitrogen and as a source of dissolved and particulate nitrogen.

3.9.10 Models of Mangrove-Nutrient Interactions

Knox (1983b) developed a comprehensive model for the major inputs, storages, and pathways for the mangrove-nitrogen system of the Upper Waitemata Harbour, New Zealand (Figure 3.92). In the model, eight major storage compartments are shown: the aboveground biomass (Q_1), the detritus microbial complex (Q_2), the N in the estuarine water (Q_3), the N in the sediment (Q_4), ammonia (NH_4^+) in the sediment (Q_5), nitrate (NO_3^-) in the sediments (Q_6), organic-N in the sediment (Q_7), and the detritus-microbial complex in the tidal water (Q_8). Also included are three atmospheric interactions: rain input (J7), nitrogen fixation (22), and denitrification (20). There are two major terrestrial inputs: river flow and sewage input. These contain both inorganic (N_1 and N_3) and organic (N_2 and N_4) nitrogen. There are two forms of sediment N: organic and inorganic. The organic fraction is a large stable component, while the inorganic fraction is rapidly turned over. Nitrogen processes in the sediment are mineralization (14 and 15), immobilization (storage in deep sediment 9 and 11), nitrification (22), and denitrification (20).

A significant pathway is the incorporation of inorganic N into bacterial protoplasm (16). This occurs during the decomposition of the mangrove detritus (Odum and Heald, 1972). According to Hopkinson and Day (1977), a large proportion of this intake is in the form of NH_4^+ from the sediment. Nitrification and subsequent denitrification take place in the sediment surface.

The model in Figure 3.92 is somewhat complex, and a simpler model (Figure 3.93) was developed for simulation. In this model the radiant energy of sunlight (I) interacts with the available nitrogen (Q_2) and mangrove biomass (Q_1) resulting in the production of organic matter. Some of this production is respired, some is stored as a net increase in mangrove biomass, and some is converted into detritus through litterfall. The nitrogen in this detritus is either exported to adjacent waters, buried in the deep sediment, or incorporated in the sediment as organic nitrogen (Q_3), which is mineralized by bacterial action. Some of the nitrogen thus released is added to the pool of available nitrogen, which can then be utilized by the plants. Other nutrient sources are terrestrial runoff (via river flow) represented by the flows J_1 and J_2 and nitrogen in the tidal waters.

Impact of storm events on nutrient input via river flow: Rutherford (1982), Van Roon (1983), and Knox (1983a,b) have drawn attention to the impact of river flow, as influenced by high rainfall and storm events on nutrient inputs to the Upper Waitemata Harbour. Van Roon reached the following conclusions after analyzing a large volume of data related to such inputs: (1) when there is high rainfall, the catchment concentrations of most parameters in the estuary increase to several times the average fine-weather concentrations, with such increases being particularly noticeable for nitrate, ammonia, total phosphate, and turbidity; and (2) tidal stare is a minor determinant of estuary nutrient concentrations compared with precedent weather conditions.

In view of the significance of storm water inputs of nutrients, a facility for the insertion of such events into the computer simulation program was developed that enabled the incorporation of a specified number of events per year. For each storm the intensity (x normal river flows) and duration of the flow could be specified. Inten-

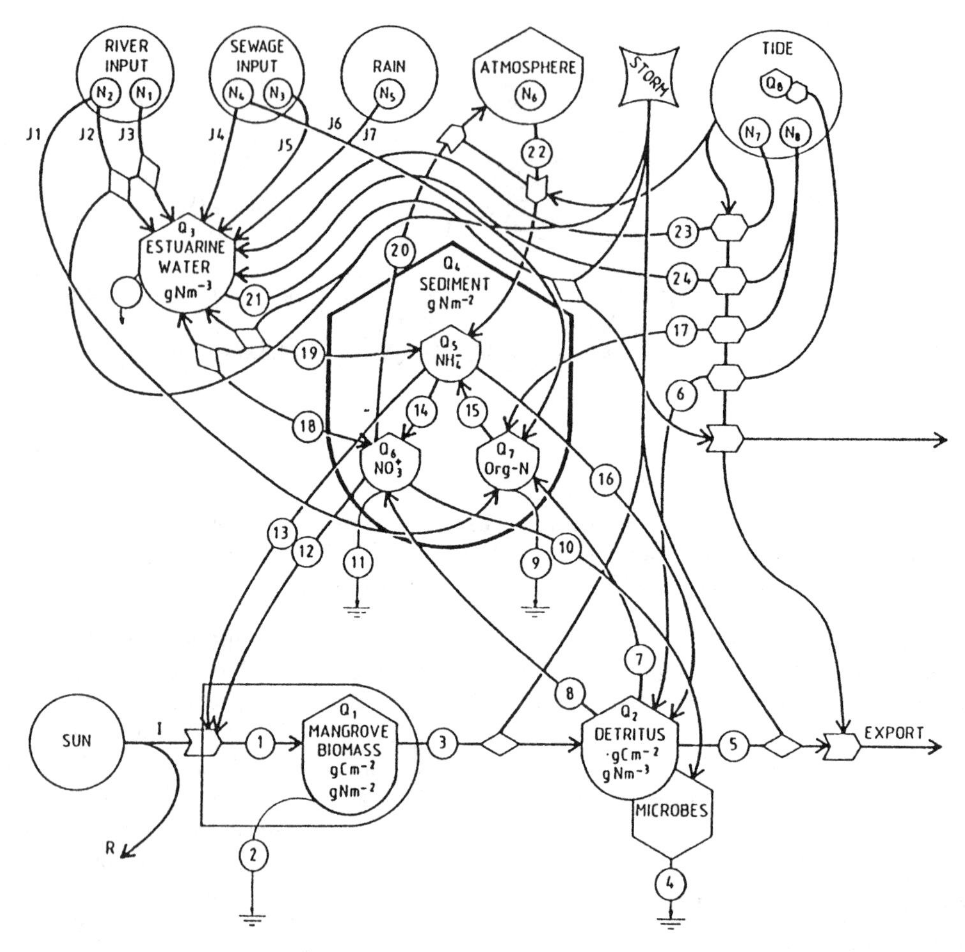

FIGURE 3.92 Mangrove nitrogen model. From Knox, G.A., *Energy Analysis: Upper Waitemata Harbour Catchment Study, Specialist Report*, Auckland Regional Authority, Auckland, New Zealand, 1983b, 23. With permission.)

sities and duration were based on real data from stream flows in the catchment. Figure 3.94 depicts the results of a 10-year simulation of the model under the then-current nutrient input conditions. The model was tested on Lucas Creek, a mangrove-fringed subestuary opening into the Upper Waitemata Harbour. The model was simulated under the following conditions:

1. Current river and tidal input with storm events.
2. Current river and tidal nutrient input without storm events.
3. No river input.
4. Double current river nutrient input with storm events.
5. Treble current river nutrient input with storm events.
6. Year 2000 estimated nutrient input with storm events.

Figure 3.95 depicts the results of the simulated changes in mangrove biomass over a 10-year period. The impact of nutrient input on mangrove biomass is clearly seen, e.g., a doubling of the river flow resulted in a 36% increase in mangrove biomass, whereas the elimination of the river flow resulted in an 18% decrease. The impact of storm events is also striking. When the model was simulated without storm events there was an 11.5% decrease in mangrove biomass over the 10-year period. In the simulation under condition 6, the model run was continued for 16 years (to the year 2000) with the nutrient-input levels increasing annually to reach the year 2000 projected level (50% higher than at present; largely due to residential and industrial developments in the catchment resulting in increased sewage input coupled with intensification of agriculture). Over the 16-year period the mangrove biomass was projected to increase by 13.4%. There were also corresponding changes in the other storage compartments such as sediment organic N and the burial of N in the deep sediment.

3.10 ESTUARINE-SHELF INTERACTIONS

3.10.1 Introduction

It is clear from the preceding section that estuaries are highly productive ecosystems, especially those fringed with salt marsh or mangroves. The question that has moti-

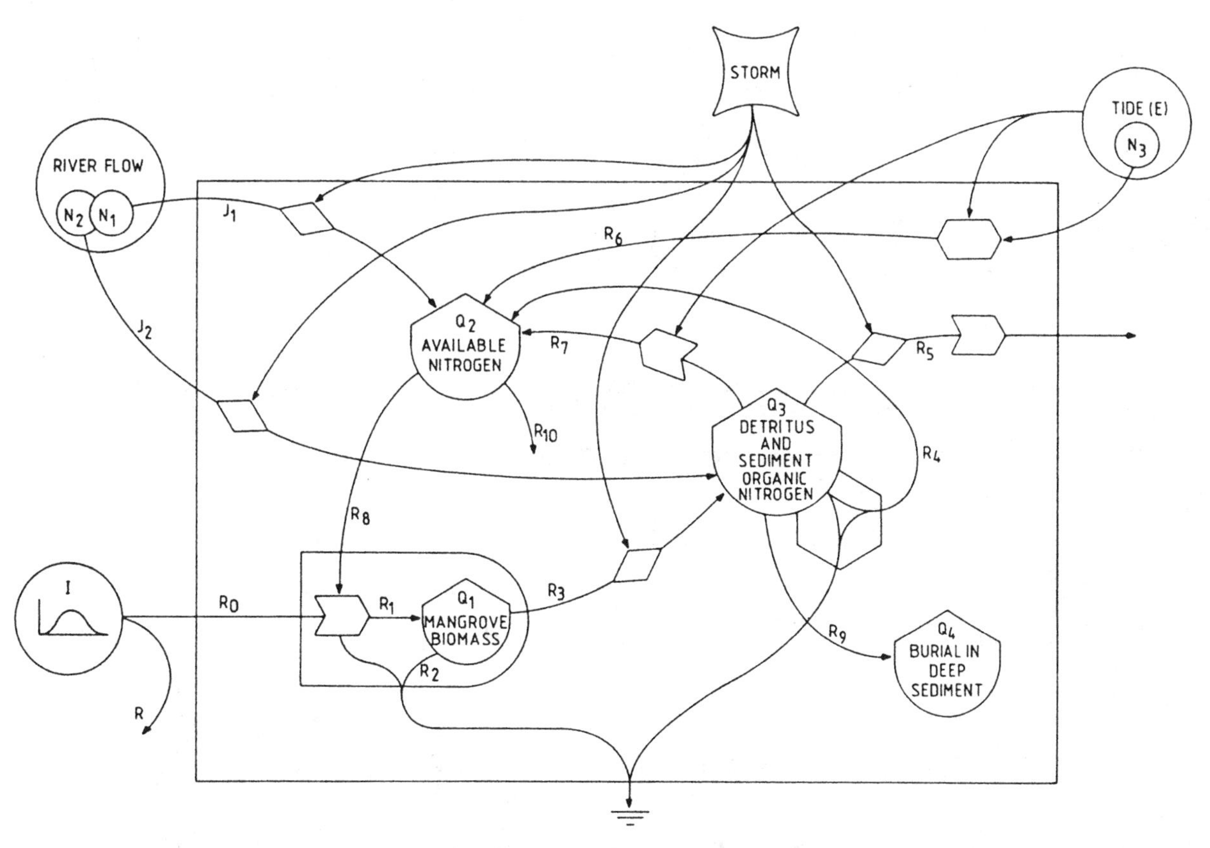

FIGURE 3.93 Simplified mangrove nitrogen model used for the simulations in Figure 3.92. From Knox, G.A., *Energy Analysis: Upper Waitemata Harbour Catchment Study Specialist Report*, Auckland Regional Authority, Auckland, New Zealand, 1983b, 26. With permission.)

vated much of the recent research on these ecosystems has concerned the fate of this productivity. To put it in simpler terms, where does all the carbon go? Estuaries also receive large quantities of nutrients from the surrounding catchment. The bulk of these nutrients is trapped within the estuary, especially in the salt marshes and mangroves and a substantial proportion is incorporated into plant biomass. The marsh plants and the leaves of the mangroves eventually decay to form detritus, which to a major extent is mineralized within the estuary to release its nutrients. This leads us to question the extent to which these nutrients are exported from the estuary to the adjacent coastal waters. A corollary is the extent to which the export of detritus and nutrients sustains the productivity of the shelf waters.

Early classical studies on the coastal salt marshes of Georgia form the basis for the hypothesis that tidal marshes function as net exporters of particulate carbon (Scheleske and Odum, 1961; Teal, 1962; Odum and de la Cruz, 1967; Odum and Fanning, 1973; Head, 1976). This concept was later extended to mangrove systems (Heald, 1969; Odum, 1970a,b; Odum and Heald, 1972). The concept of enrichment of coastal waters from "outwelling" from estuaries has been elaborated by Odum (1971; 1975). He stated the concept as follows: "Most fertile zones in coastal waters capable of supporting expanded fisheries result either from the upwelling of nutrients from deep water, or from outwelling of nutrients or organic detritus from shallow water nutrient traps such as reefs, banks, seaweed of sea grass beds, algal mats, and salt marshes" (Odum, 1968). The evidence upon which this hypothesis was originally based was a study by Thomas (1966) who found that primary production of offshore waters in the Georgia Bight was much higher just offshore from the openings between barrier islands than was the case further out to sea. The highest rate of phytoplankton production in Georgia coastal waters, 547 g C m^{-2} yr^{-1}, occurred in the nearshore zone within 10 to 15 km of the coast. The turbidity and shallow depths of adjacent estuaries limits phytoplankton growth to around 300 g C m^{-2} yr^{-1} (Pomeroy et al., 1981). Further offshore nutrients become limiting, so that phytoplankton photosynthesis on the inner shelf (up to 20 m depth) averaged about 285 g C m^{-2} yr^{-1}, and on the outer shelf (20 to 200 m depth) the photosynthesis was about 130 g C m^{-2} yr^{-1} (Haines and Dunstan, 1975). The figure also depicts the benthic remineralization (ammonium regeneration) rates (Hopkinson and Wetzel, 1982).

However, investigations from other areas and reinterpretations of the Georgia shelf data have provided contra-

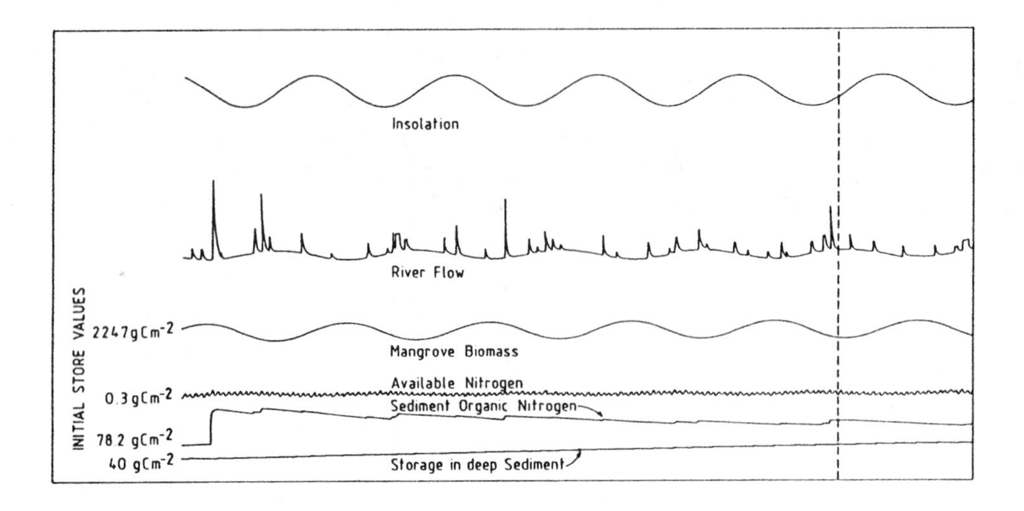

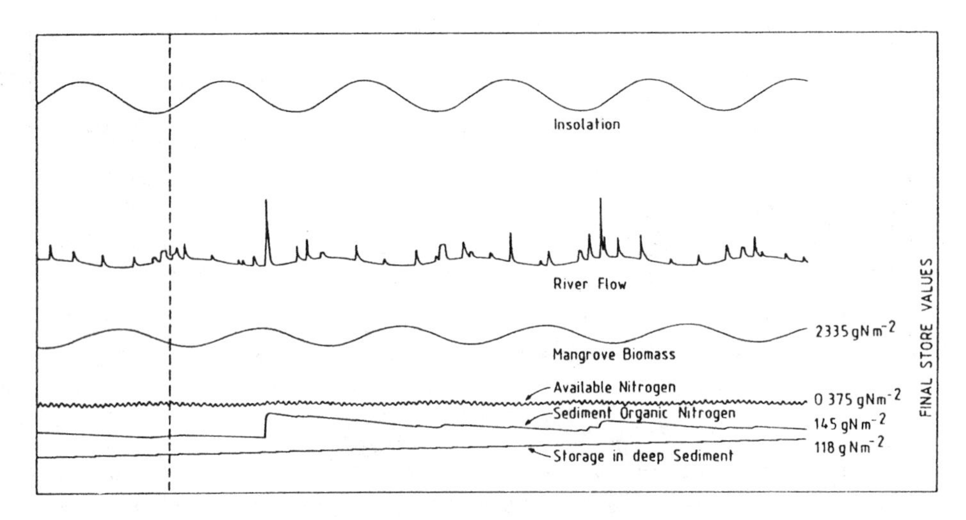

FIGURE 3.94 Graphic printout of a 10-year simulation of the mangrove nitrogen model shown in Figure 3.93. (From Knox, G.A., *Energy Analysis: Upper Waitemata Harbour Catchment Study Specialist Report,* Auckland Regional Authority, Auckland, New Zealand, 1983b, 26. With permission.)

dictory information. As a result, controversy has arisen concerning the correctness of the original "outwelling" hypothesis. For example, Nadeau (1972), Heinle and Flemer (1976), Hackney (1977), Woodwell et al. (1977), and Haines (1979) have been unable to measure net export from tidal wetlands, and in some cases found an apparent net import. On the other hand, Day et al. (1973), Axelrod (1974), Axelrod et al., (1976), Nixon and Oviatt (1977), Valiela et al. (1978), and Knox (1983a) have found a net export of particulate organic carbon from a range of estuaries. A number of case history studies of export-import studies will be considered.

3.10.2 Some Case History Investigations

Historically there have been two basic approaches to the testing of the outwelling hypothesis. One approach is indirect and involves the development of production and consumption budgets for either the marsh-estuarine system (Day et al., 1973; Pomeroy and Wiegert, 1981), or the nearshore system (Hopkinson, 1988). The direct approach involves measurement of the fluxes of water and materials across a cross-section of a creek or inlet (Woodwell et al., 1977; Valiela and Teal, 1979b).

3.10.2.1 North Inlet, South Carolina

Extensive measurements of material concentrations and water velocities along a transect across North Inlet, South Carolina, U.S.A., enabled the estimation of material and water fluxes. North Inlet is the largest marsh-estuarine system from which direct net-flux measurements between a marsh-estuarine system and the coastal zone have been made. Statistical and hydrodynamic models were used to develop flux estimates for specific tidal cycles. The investigation revealed that there was a net discharge of water from the marsh-estuarine system to the Atlantic Ocean coastal waters that is attributed to rainfall runoff and fresh-

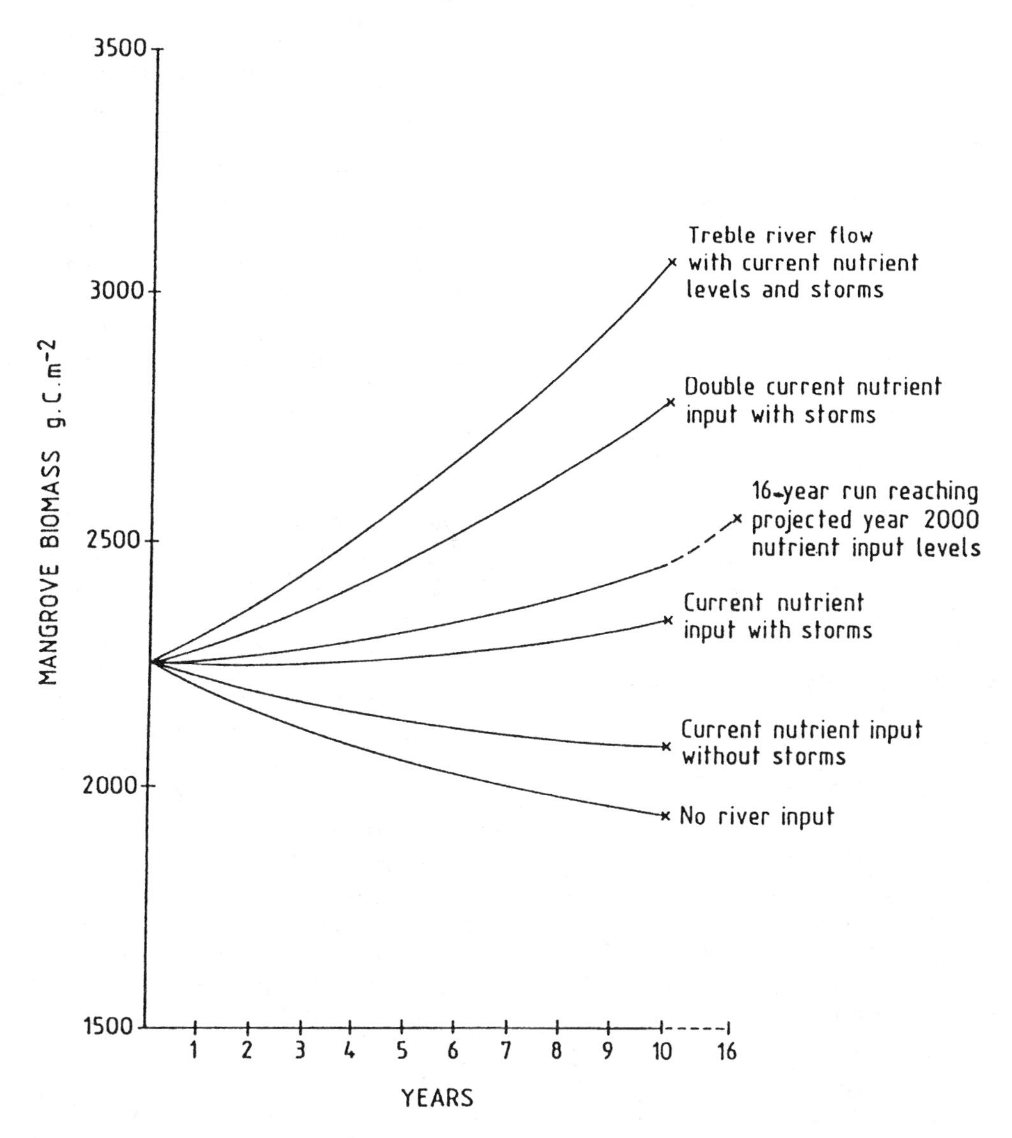

FIGURE 3.95 Changes in mangrove biomass resulting from a 10-year simulation of the mangrove nitrogen model shown in Figure 3.94. Fom Knox, G.A., *Energy Analysis: Upper Waitemata Harbour Catchment Study Specialist Report,* Auckland Regional Authority, Auckland, New Zealand, 1983b, 27. With permission.)

water input from an adjacent estuary. All constituents were exported seasonally and annually from the system, except total sediments, imported during the autumn and winter, and chlorophyll *a* and zooplankton, imported in the summer and autumn (Table 3.40). ATP, bird biomass, and macrodetritus were exported throughout the year. Particulate organic carbon was exported to the Atlantic Ocean on 30 of the 32 tidal cycles sampled during the study. There was an annual export of 128 g C m^{-2} yr^{-1} of POC, which is in the range of 62 g C m^{-2} yr^{-1} import to 303 g C m^{-2} yr^{-1} export reported from 8 studies reviewed by Nixon (1980). The high rates of export of ammonium and orthophosphate along with detritus and microorganisms suggest that major decomposition processes are taking place within the system. Export of ammonium and orthophosphate to the coastal ocean also suggest a feedback loop with phytoplankton utilizing these materials, then in turn phytoplankton are imported into the estuary where they are consumed and remineralized.

Total carbon exported from North Inlet was 435 g C m^{-2} yr^{-1}, or about 42% of the total net primary production within the system. It can be seen that North Inlet is a highly productive system compared to the others. The study found that North Inlet exported carbon, nitrogen, and phosphorus at higher rates than most previously studied estuarine systems.

3.10.2.2 Beaufort, North Carolina

Eelgrass, *Zostera marina*, and shoalgrass, *Halodule wrightii*, occupy about 93 km^2 within the 552-km^2 estuarine system at Beaufort, North Carolina. The area includes

TABLE 3.40
Yearly Carbon, Nitrogen, and Phosphorus Flux in North Inlet on an Area of Marsh Basis

	Carbon	Nitrogen	Phosphorus
POC	128		
DOC	328		
NH_4		6.3	
$NO_2 + NO_3$		0.9	
TN		42.7	
PO_4			1.7
TP			3.1
ATP	22	3.3	0.2
Chlorophyll *a*	1.911	0.27	0.016
Zooplankton	–1.2	–0.26	–0.010
Macrodetritus	0.9	0.01	0.001
Birds	0.05	0.01	0.001

Note: C, N, and P fluxes in g m^{-2} $year^{-1}$; Drainage area = 2,380 ha; Import = –; Export = + (sign not shown).

Source: From Dame, R., Chrzanowski, T., Bildstein, K., Kjerfve, B., McKellar, H., Nelson, D., Spurrier, J., Stancyk, S., Stevenson, H., Vernberg, J., and Zingmark, R., *Mar. Ecol. Prog. Ser.*, 33, 223, 1986. With permission.

the Newport and North River estuaries. *Z. marina* has been estimated to account for approximately 64% of the total primary production in this estuarine system (Thayer et al. 1975; Penhale and Smith, 1977). Approximately 55% of the total production by phytoplankton, benthic algae, and *Zostera* in the Newport River estuary eelgrass bed was estimated to be consumed by the macrofauna (Thayer et al. 1975). The remaining portion was then available for consumption by meiofauna, microfauna, and microbes, for sedimentation or for export to surrounding areas.

Bach et al. (1986) measured the export of detritus (including live and dead sea grass blades, root and rhizomes, macroalgal fronds, *Spartina alterniflora* stems, and particulate organic carbon) from two sheltered embayment-type grass beds and an open water bed. They also estimated export from a much larger part of the estuary (Bogue Sound). Rates of export of *Z. marina* from an embayment-type bed ranged from 4.9 to 6.0 g ash-free dry wgt $month^{-1}$ m^{-2} during the summer of 1972, equivalent to a total of 195 to 240 kg $month^{-1}$ for the bed, or 456 to 558 tons $month^{-1}$ for the 552 km^2 estuary, assuming that sea grasses cover 93 km^2. Rates of export from Harkers Island (exposed) and Middle Marsh (embayment) beds were 13 and 1 g ash-free dry wgt $month^{-1}$ m^{-2}, respectively, equivalent to 1,200 and 112 tons $month^{-1}$ for the whole estuary. The differences of the export between embayment and open water types of sea grass beds are due to entrapment of floating detritus by the embayment beds. The amount of detrital material exported from Bogue Sound (47.4 kg $month^{-1}$) was an extremely small portion of the total estimated biomass (3.6×10^5 kg).

The volume of the export was greatly influenced by meteorological conditions. On days when the wind blew from the south at sufficient strength, export of drifting sea grass detritus was greatly reduced or even negative (i.e., imported) in Bogue Sound. Export of sea grass material from Bogue Sound was much higher on calm days when wind speed was 0 to 5 knots. The export of other material was more variable. Bach et al. (1986) have compared the estimates they made in their study with those reported from the Virgin Islands (Zieman et al. 1979). Rates for open beds in the two studies were similar, with values ranging from 0.23 to 0.57 ash-free dry wgt m^{-2} day^{-1} for *Zostera marina* and *Halodule wrightii*, respectively at Harkers Island, and from 0.13 to 0.25 g ash-free dry wgt m^{-2} day^{-1} for *Syringodium filiforme* in the Virgin islands (Zieman et al., 1979; Thayer et al., 1984).

3.10.2.3 Mangrove Forest Systems

Along parts of the inner shelf of the Central Great Barrier Reef (the GBR), Australia, mangrove litter is exported from the mangrove forests to the adjacent subtidal seabed (Boto and Bunt, 1981; 1982; Robertson et al., 1988; Alongi et al., 1989). Extensive investigations of the ecology of their mangroves and their links with nearshore areas have entered on the tidal forests of Hinchinbrook Island and adjacent continental rivers. Net annual exchange of most dissolved materials between mangrove creeks and coastal waters is negligible (Boto and Wellington, 1988), but there is a significant net export of particulate matter, mostly mangrove litter, on the order of 25,000 tons C yr^{-1}, spread across adjacent bays and the channels in the central GBR lagoon (260 km^2) giving an annual flux of 140 tons C m^{-2} yr^{-1} or 52 g C m^{-2} yr^{-1} (Robertson, 1988).

Alongi (1990b) investigated the effect of tidal outwelling of mangrove detritus on sediment nutrient chemistry, nutrient regeneration and oxygen fluxes in a coastal area of the central Great Barrier Reef Lagoon. Organic carbon and total nitrogen concentrations ranged from 0.2 to 3.9%, and 0.01 to 0.18% by sediment dry wgt, respectively, and were highest at stations receiving the greatest quantities of mangrove litter. Total phosphorus concentrations ranged from 0.013 to 0.048%, by dry wgt but did not relate to outwelling. C:N:P ratios ranged from 29:6:1 at the site receiving the least amount of detritus to a high of 397:17:1 at the station receiving the most litter. Oxygen consumption rates ranged from 8.7 to 60.2 mmol O_2 m^{-2} day^{-1} and were highest at most sites closest to mangrove forests, an agreement with previous measurements of bacterial productivity. Rates of net community primary production were either undetectable or low (as were chlorophyll *a* and phaeopigments), ranging from 12 to 77 mg C m^{-2} day^{-1}.

From the above studies it was concluded that much of the mangrove litter is highly refractory and of poor nutritional quality, with enhancement of bacterial activity and DOC fluxes apparent only in semienclosed areas of highest litter deposition. Rates of litter deposition appear to be low, but appears to be great enough to significantly enrich bulk concentrations of particulate carbon and nitrogen, and rates of oxygen consumption. However, subsurface burial and accumulation of litter may, via long-term degradation, support the highly abundant and productive bacterial communities and efficient bacteria-DOC recycling that has been observed in the region.

Hemminga et al. (1994) investigated carbon outwelling from a mangrove forest with adjacent sea grass beds and coral reefs in Gazi Bay, Kenya. The key factors that emerged from this study may be summarized as follows: (1) spatially restricted outwelling of mangrove carbon; (2) reversed fluxes of POM from the sea grass zone to the mangroves; (3) transported POM characterized by low C:N ratios. Hemminga et al. consider that these features may well be typical for mangrove forests with adjacent sea grass and fringing coral reef systems. The presence of the reef creates a low-energy environment on the landward side, restricting the hydrodynamic transport of particles. In addition, the sea grass vegetation will reduce dispersal of outwelling mangrove-derived particles; the presence of sea grasses enhances the sedimentation of particulate material from the overlying eater column, as the resistance of the canopy causes a reduction in the current velocity of the water (Fonseca et al., 1982; Harlin et al., 1982; Ward et al., 1984). Furthermore, as coral reefs can only exist above salinities of 27, with optimal growth at salinities between 34 and 36 (Fagerstrom 1987, and references therein), the abundant occurrence of coral reefs in the vicinity of mangroves is limited to tide-dominated (non-riverine) environments. According to Woodroffe (1992), tide-dominated mangrove habitats are characterized by bidirectional fluxes of suspended material, whereas river-dominated habitats are characterized by strong outwelling from the mangrove forest.

The concept of outwelling from mangroves has been based on less data than for salt marshes, but most studies conclude that there is a net flux of detritus from these forested wetlands. One reason for this may be related to the greater tidal amplitude and runoff at the sites used to study mangrove export compared to temperate intertidal wetlands. For instance, tidal amplitude for mangroves in Australia is greater than 3 m, and rainfall in south Florida is more than 1,500 mm. Also, litterfall in mangroves is continuous throughout the year and for many species this material is very buoyant. These factors lead to greater potential for exchange in mangrove ecosystems.

3.10.3 Conclusions

Attempts to generalize organic carbon and nutrient transport between shallow estuarine basins and embayment and coastal waters have resulted in data that do not give a clear picture of the magnitude and direction of the fluxes (Woodwell et al., 1977; Haines, 1979; W.E. Odum et al., 1979; Nixon, 1980; Twilley, 1986; Hopkinson, 1988). From the studies discussed above it appears that each system must be individually assessed since basin morphology and episodic meteorological events may dramatically impact on net transport within the estuarine system.

Organic carbon: Table 3.41 lists the net exchange of particulate organic carbon (POC) for different salt marshes, and for those in which the movement of dissolved organic carbon (DOC) was also measured. The values for POC flux range from an import of +120 g C m^{-2} yr^{-1}, to an export of –140 g C m^{-2} yr^{-1}, with a mean flux for all marshes, excluding North Inlet of –70 to 95 g C m^{-2} yr^{-1}. DOC was exported in all the studies in which it was measured with fluxes ranging from +5 to +32 g C m^{-2} yr^{-1}. These DOC fluxes for the marshes as a whole are smaller than might be expected on the basis of the sum of DOC flux measurement from *Spartina* leaves (Turner, 1978) and from the marsh surfaces (Pomeroy et al., 1977). This is due to the rapid utilization of available DOC by the microbial communities.

On the basis of the then reported measurement, Nixon (1980) concluded that the total flux of carbon from salt marshes is likely to be between –100 and –200 g C m^{-2} yr^{-1}. He pointed out that there are a large number of uncertainties in the database and that a number of different approaches have been used by the different investigators. For example, none of the studies has adequately accounted for the export of large rafts of *Spartina* leaves during storms, or for the bed load detrital export (Pickral and Odum, 1977), or for the (probably relatively small) export of carbon in the migration of fish, birds, and other animals.

Global export/import values do not reflect the true export picture. In most systems there is a varied pattern over the year of both import and export with pronounced seasonal peaks. The timing of such peaks may be of crucial importance to the production cycle of coastal waters. In addition, the potential significance of organic carbon exports for the secondary production of adjacent waters is a function of the relative sizes of the marsh and the open water systems. As Nixon (1980) points out, lack of appreciation of this simple relationship has helped to confuse the discussion on the significance of export phenomena.

According to Odum et al. (1979), specific estuarine wetlands may either export or import particulate organic carbon on an annual basis depending on several geophysical factors. These include:

TABLE 3.41
Summary of Organic Carbon Fluxes from Salt Marshes, and Comparisons of TOC Fluxes with Net Aerial Primary Production

System	Organic Carbon Flux			Net Aerial Primary Production ($g\ C\ m^{-2}\ yr^{-1}$)	TOC Export as % of Net Aerial Primary Production	Form of Marsh	Source
	DOC	POC	TOC				
Ware Creek, Virginia, USA	+80	+35	+115	599	19	*Spartina alterniflora*	Axelrod et al. (1976)
Carter Creek, Virginia, USA	+25	+116	+142	599	24	*Spartina alterniflora*	Axelrod et al. (1976)
Flax Pond, New York, USA	+8	−61	−53	372	na	*Spartina alterniflora*	Woodwell et al. (1979)
Canary Creek, Delaware, USA	+104	+55	+159	252	63	*Spartina alterniflora*	Roman and Daiber (1989)
Coon Creek, Texas, USA	+21	+4	+25	550–900	3–5	*Spartina patens* *Distichlis spicata*	Borey et al. (1983)
High Marsh, Rhode River, Maryland, USA	+43	+14	+57	nd	nd	*Spartina patens*	Jordon et al. (1983)
			+49			*Distichlis spicata*	Jordon and Correll (1991)
Ems-Dollard Marsh, The Netherlands	+15	−140	−125	500	na	*Puccinellietum maritima*, *Spartina anglica*	Dankers et al. (1984)
North Inlet, South Carolina, USA	+328	+128	+456	1059	na	*Spartina alterniflora*	Dame et al. (1986)
Bly Creek, South Carolina, USA	+250	−31	+219	1028[a]	21	*Spartina alterniflora*	Williams et al. (1992)
	+272	−30	+242	1028[a]	24		Dame et al. (1991b)
Kariega Marsh, South Africa	+13[a]	+3[a]	+16[a]	200–300	5–8	*Spartina perennis*, *Chenolea diffusa*	Taylor and Allanson (1993)

Note: Because these patterns of fluxes of DOC and POC can be different, and certain studies measured only one, only the data from those studies that measured both components are included. nd: not detectable; na = not applicable; – values: net imports; + values: net exports.

[a] Area-weighted aerial production values for tall, medium, and short forms of *Spartina;* areas from Dame et al. (1991).

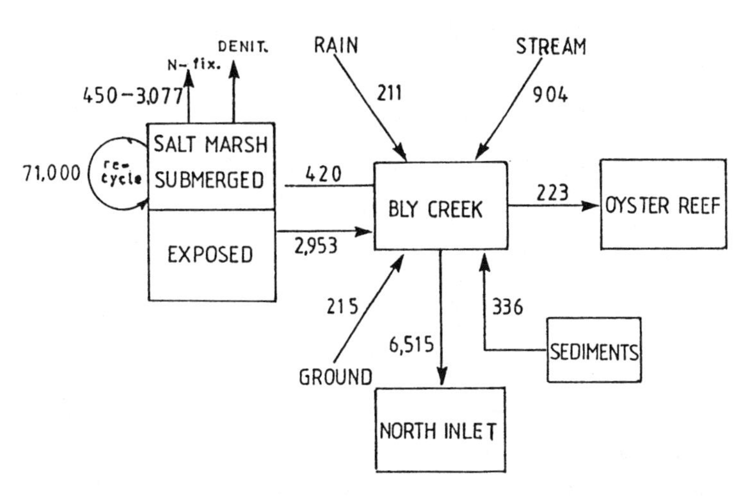

FIGURE 3.96 Nitrogen budget for the Bly Creek Basin, U.S.A. (Redrawn from Dame, R.F., Spurrier, J.D., Williams, T.M., Kjerfve, B., Zingmack, R.G., Wolaver, T.G., Chyzanowski, T.H., McKellar, H.N., and Vernberg, H.N., *Mar. Recol. Prog. Ser.*, 72, 161, 1991. With permission.)

1. The geomorphology of the estuarine basin.
2. The tidal amplitude.
3. The magnitude of the freshwater input to the drainage basin.

They further point out that most published export/import studies have failed to account for two important factors: (1) the pulsed (or intermittent) nature of particulate transport activity in response to irregular storm events; and (2) the particulate loads on or near the bed of estuarine creeks, rivers, and embayment.

Where the channels between offshore and inshore water are narrow and blocked by a sill, or where tidal action is weak, extensive export would not be anticipated. This is apparently the case in the well-studied Flax Pond system (Woodwell et al., 1979), which is often cited as evidence against outwelling. In contrast, where estuaries are more open with extensive exchanges between the estuarine and coastal waters, export would be expected if the estuaries were more productive than the coastal waters.

Nitrogen: The quantification of nitrogen fluxes is more complicated than that of carbon, as this nutrient may exist as organic and inorganic particulates, as well as dissolved organic ammonium, nitrite, and nitrate. Its export is of special interest because of the role it plays as a phytoplankton nutrient. Dame et al. (1991) also investigated the fluxes of nutrients in the Bly Creek ecosystem. Figure 3.96 depicts the nitrogen fluxes between the components of the ecosystem. As with POC, particulate nitrogen (PN) appears to be imported by the tidal waters and most of this may be taken up by the oyster reef and salt marsh. A large proportion of the PN taken up by the salt marsh appears to exit during low tide exposure. Thus, it seems that enough PN is imported via tidal water and recycled via runoff from the exposed marsh to account for the material taken up by the submerged salt marsh and the oyster reef. Of the dissolved inorganic nitrogen, NH_4 appears to be exported from the tidal basin and imported to the salt marsh at a high rate. Studies by Jordon and Correll (1991) indicate that seepage from the creekbanks may contribute significant amounts of dissolved nutrients to tidal creeks. Unlike NH_4, $NO_3 + NO_2$ is imported into the Bly Creek basin at a rate 0.23 g N m^{-2} $year^{-1}$ lower than most of the observed imports of this constituent into other marsh-estuarine systems (Nixon, 1980). Within the Bly Creek basin, the salt marsh appears to be capable of removing all of the $NO_3 + NO_2$ potentially imported into the basin via tidal exchange, streamwater, groundwater, and rain. Most of the nitrogen leaving the Bly Creek basin is in the form of DON, the majority of which appears to be exported by salt marsh runoff.

Figure 3.96 shows a net loss of nitrogen from the Bly Creek basin. However, inputs to the tidal creek explain only 65% of the nitrogen leaving the system. Table 3.42 lists some estimates of the annual flux of nitrogen between salt marshes and coastal waters. There appears to be a total loss of nitrogen from all the marshes listed, with the exception of the Great Sippewissett Marsh, where the net losses amount to only a few grams of N m^{-2} yr^{-1}. On the basis of the information then available, Nixon (1980) concluded that most of the primary production in coastal waters is supported by recycled nitrogen rather than by new inputs. He also makes the point that the path of nutrients from the salt marsh to the open ocean is not through a pipe, but through a complex chain of subsystems, marshes, intertidal flats, estuarine water bodies, estuarine bottom sediments, etc., each of which is char-

TABLE 3.42
Annual Flux of Nitrogen between Salt Marshes and Coastal Waters

Salt marsh	Nitrogen Flux, g N m^{-2} of Marsh $Year^{-1}$						Reference
	PN	DON	NH_4	NO_2	NO_3	N	
Great Sippewissett, Massachusetts	–6.7	–9.8	–4.2	–0.1	–3.8	–24.6	Valiela et al. (1978)
Flax Pond, New York	—	—	–2.0	0	+1.0	—	Woodwell et al. (1979)
Canary Creek, Delaware	–2.9	–0.9	+0.7	+1.9	–1.2	—	Lotrich et al. (1979)
Gott's Marsh, Maryland	–0.3	–2.1	–0.4	0	–0.9	–3.7	Heinle and Flemer (1976)
Ware Creek, Virginia	0	–2.3	–2.9	–0.1	+2.3	–2.8	Moore (1974)
Carter Creek	+4.6	–9.2	–0.3	0	+0.3	–4.0	Moore (1974)

Source: Nixon, S.W., in *Estuarine and Wetland Processes,* Hamilton, P. and MacDonald, K., Eds., Plenum Press, New York, 1980, 491. With permission.

acterized by its own internal cycling as well as by its own inputs, outputs, transformations, and storages.

Phosphorus: Table 3.43 lists nutrient-exchange studies that have included one or more of the forms of phosphorus in their measurements. The results of these studies show a pattern of behavior which is quite different from that of nitrogen. While the oxidized inorganic forms of nitrogen (NO_2 and NO_3) were largely taken up by the marshes, the flux measurements show an export of PO_4 to the offshore waters.

Hobbie et al. (1975) studied the movement of phosphate through the Palmico River estuary in North Carolina. They found that about 42% of the phosphate brought down by the rivers and a local phosphate mine was discharged to the sea (459 tons yr^{-1}). In the Izembek Lagoon, Alaska, McRoy et al. (1972) found a substantial transfer of phosphorus from the bottom sediments into the water from *Zostera* beds. This transfer in turn led to a significant export of 1.47×10^3 kg tidal $cycle^{-1}$ during the growing season to the Bering Sea.

The seasonal pattern of PO_4 exchange is highly variable and Nixon (1980) found no consistent trend in the data from the various studies. He pointed out that nearshore marine and estuarine waters are usually characterized by a summer phosphate maximum and this is the period when some of the studies showed strong PO_4 export.

Conclusions: In an extensive review of organic carbon exchange between coastal ecosystems, Hopkinson (1988) concluded that each of the estuaries he examined was unique. For each estuary there appeared to be a factor that could be used to explain the export characteristics. From his review he drew the following conclusions:

1. Salt marshes are net autotrophic: more carbon is produced than consumed and stored within them.
2. Tidal creeks adjacent to marshes appear to be net heterotrophic: more carbon is degraded than is locally produced.
3. Larger bays and sounds (e.g., Barataria and Narragansett) appear to be net autotrophic.
4. Although marsh-dominated estuaries appear to export organic carbon to coastal waters, such export is usually only of local importance.
5. Like coral reefs, rates of gross primary production and internal recycling are extremely high.

TABLE 3.43
Annual Flux of Phosphorus between Salt Marshes and Coastal Waters

Salt Marsh	PP	DOP	PO_4	P	Reference
Great Sippewissett, Massachusetts	—	—	–0.6	—	Valiela et al. (1978)
Flax Pond, New York	+1.1	—	–1.4	–0.3	Woodwell and Whitney (1979)
Canary Creek, Delaware	—	—	<–0.1	—	Lotrich et al. (1979)
Gotts Marsh, Maryland	–0.1	–0.2	—	–0.3	Heinle and Flemer (1976)
Ware Creek, Virginia	+0.1	–0.2	–0.1	+0.7	Moore (1974)
Carter Creek, Virginia	+0.8	–0.2	–0.6	0	Moore (1974)
Dill Creek, South Carolina	—	—	–6.4	—	Settlemeyer and Gardner (1975)

Source: From Nixon, S.W., in *Estuarine and Wetland Processes,* Hamilton, P. and MacDonald, K., Eds., Plenum Press, New York, 1980, 479. With permission.

4 Adaptations to Shore Life

CONTENTS

4.1 INTRODUCTION

In previous chapters, it was shown that the plants and animals on the shore occupy distinct zones or habitats in which they can survive and obtain the resources they require for growth and reproduction. They thus occupy a specific ecological niche. The ecological niche concept will be explored in the succeeding section. While a limited number of animal species exhibit direct development in which the juveniles hatch directly from the egg, other species have pelagic larvae that need to settle at an appropriate level on the shore in order to maintain viable populations. For sessile species, the choice of a settlement site is irreversible. Hence, such species have evolved behaviors that will ensure that they will settle at the right level on the shore. Other species such as mussels and limpets settle low on the shore and subsequently migrate to occupy the zone in which the adults are found. Most algae reproduce by forming microscopic life cycle stages that are released into the water column and later settle on rocky substrates. If they settle at the appropriate level, they will grow to give rise to the adult plant.

Many motile shore animals on both hard and soft shores have evolved behavioral strategies that enable them to both evade extreme environmental conditions and to undertake feeding migrations in order to utilize available food resources. These behavioral strategies will be discussed in detail later in this chapter.

4.2 ECOLOGICAL NICHES ON THE SHORE

4.2.1 Introduction

It is obvious from the preceding chapters that plants and animals on the shore are not randomly distributed, but occupy distinct vertical zones, and are often restricted to microhabitats within these zones. Basic to an understanding of shore ecology is a knowledge of the ways in which organisms are adapted to the environmental conditions they are subjected to, and the particular functional role or *ecological niche* that they occupy in the ecosystem of which they are an integral part. Here we will consider the twin concepts of "environment" and "ecological niche" in some detail.

The history of the niche concept is well known and documented (see reviews by Whittaker and Levin, 1977; Diamond and Case, 1986). More recently, Price (1980) has reviewed niche and community concepts in the inshore benthos with particular reference to macroalgae. The niche concept includes the ideas of *ecopotential, fundamental niche,* and *realized niche. Ecopotential* can be considered as the unexpressed individual, breeding group, or local population potentiality to occupy a particular role in a community. The *fundamental niche* is the unconstrained expression of that ecopotential in the presence of only those limitations that derive from interactions between the ambient physical environment and the population of the species under consideration. The *realized niche* is the totally constrained living relationships of the population of a species within its delimited community.

4.2.2 The Environment

The term "environment" is not an easy one to define since organisms, populations, and communities form interacting systems within their environments. For the individual organisms, substrate, physical and chemical conditions, its disease organisms, parasites, symbionts and commensals, its associated organisms, competitors and predators, its food resources and other phenomena, all form part of its environment. The environment of a population is more difficult to define, since the individuals within a population do not all respond in the same way to a particular environmental factor. However, it is useful to consider the environment of a population as the sum of all those phenomena to which the population as a whole and its individuals respond. Communities, on the other hand, modify and control the physical and chemical conditions and resources of the areas in which they are found to such an extent that separate consideration of the environment and community is of little value. It is best to view the community and the sum total of the environmental conditions of an area in which it is found as the components of an *ecosystem.*

It is useful to break down the environment of an organism into its component factors, all of which must remain within tolerable limits if the organism is to survive. Any one of these factors may become limiting in the sense that, if it exceeds the tolerable limits for the individual, it will die, although the other factors remain suitable. We will consider this concept of limiting factors in some detail later. In addition to limiting environmental factors we also need to consider regulatory factors that control the size of the population, e.g., disease, competition, or predation may prevent the population from expanding, but it does not threaten its continuous existence.

The environment, then, is a term used to describe in an unspecific way, the sum total of all the factors of an

TABLE 4.1
Classification of Environmental Factors

A. Weather
  1. Immersion
  2. Emersion and water loss
  3. Temperature — heat and cold
  4. Wave action
  5. Salinity
  6. Gases — oxygen and carbon dioxide
  7. Light
  8. Water currents
  9. Nutrients and organic constituents
B. Resources
  1. Food
  2. A place to live
C. Other Organisms
  1. Intraspecific interactions
  2. Interspecific interactions
    (a) Competition
    (b) Parasitism
    (c) Predation
    (d) Commensalism
    (e) Mutualism

area that influence the lives of the individuals present. There have been numerous attempts to classify the important environmental variables, ranging from very general ones (e.g., biotic vs. abiotic), to habitat-specific schemes (e.g., for a rocky shore mussel community, these include tidal emersion and immersion, wave action, water movement, water and air temperatures, salinity, substrate, aspect, etc.). However, a better classification reflecting causal relationships is needed. Perhaps the most useful one is that of Andrewartha and Birch (1954), who separated the environment into four major divisions: weather, food, organism of the same and different kinds, and a place to live. A modified subdivision of these categories as they apply to the shore is given in Table 4.1.

4.2.3 Environmental Stress

4.2.3.1 Desiccation

There is a considerable body of literature on the responses of intertidal communities and individual species to gradients of emersion/submersion (tidal height) and wave exposure. Nevertheless, many of the conclusions reached on the effects of emersion during low tide do not provide completely satisfactory explanations for littoral zonation, species distributions, and abundance patterns (Chapman, 1973; Underwood, 1978a,b; 1985; Underwood and Denley, 1984).

During periods of emersion (exposure to air), desiccation (water loss and temperature stress) may affect the photosynthetic capacities of plants (Schonbeck and Norton, 1979; Dring and Brown, 1982; Smith and Berry, 1986), the nutritional performance of algae (Schonbeck and Norton, 1979), and the ability of animals to grow and carry out the normal functions of feeding and reproduction.

The amount of water lost by algae depends on the duration of exposure to air, the atmospheric conditions (solar insolation, temperature, cloud cover, humidity, etc.), and the surface-to-volume evaporation ratio of the plant (Dromgoole, 1980). While a brief exposure of an alga would have little impact, prolonged exposure could be severe. The higher up the shore that a species grows, the longer it is exposed to desiccation effects. However, desiccation can be minimized by growing in favorable habitats, e.g., under overhangs, in shade, in rock pools, or beneath the canopy of larger algae. Some algae (e.g., fucoids) tolerate desiccation rather than having the ability to avoid stress (i.e., by maintaining a high water potential). Moreover some algae have the ability to harden to drought conditions (Schonbeck and Norton, 1979).

Emersion from the marine environment exposes macroalgae to increased osmotic stress because of tissue water loss (desiccation), increased irradiances, and elevated thallus temperatures as tissues dry. Desiccation stress reduces photosynthetic capacity (Dring and Brown, 1982), as well as altering respiration rates. Increased thallus temperatures are typically associated with emersion stress increases, photosynthesis, and dark respiration rates with a Q_{10} of ca. 2.0. Photosynthetic rates reach temperates above which they rapidly decline.

Many experiments have tested the recovery of algae from emersion, usually by measuring the rates of photosynthesis or respiration (see review of Gesner and Schramm, 1971). Some representative results shown in Figure 4.1 illustrate the recovery of *Fucus vesciculosus* (mid-intertidal) and *Pelvetia canaliculata* (high intertidal). The latter, as expected, was able to withstand longer periods of desiccation. If relative humidity is experimentally maintained at a level high enough to prevent desiccation, the photosynthetic rate may be maintained for long periods, as found in *Fucus serratus* by Dring and Brown (1982). These authors assessed three hypotheses that might explain the effects of desiccation on intertidal plants and zonation: (1) species from the upper shore are able to maintain active photosynthesis at lower tissue water content than are species lower on the shore (this was refuted by the experimental data); (2) the rate of recovery of photosynthesis after a period of emersion is more rapid in species on the upper shore (this was also refuted by the available data); and (3) the recovery of photosynthesis after a period of emersion is more complete in species from the upper shore (this hypothesis was supported by Dring and Brown's data).

Beach and Smith (1997) have studied the ecophysiology of the Hawaiian high-tidal, turf-forming red alga, *Ahnfeltiopsis concinna*. They found that the capacity to recover

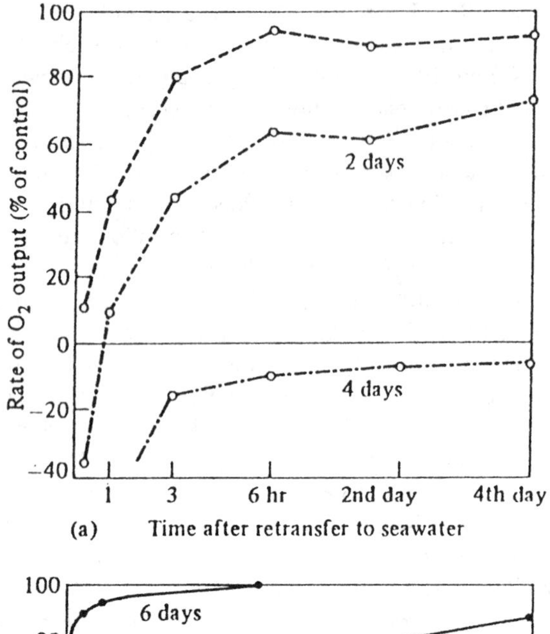

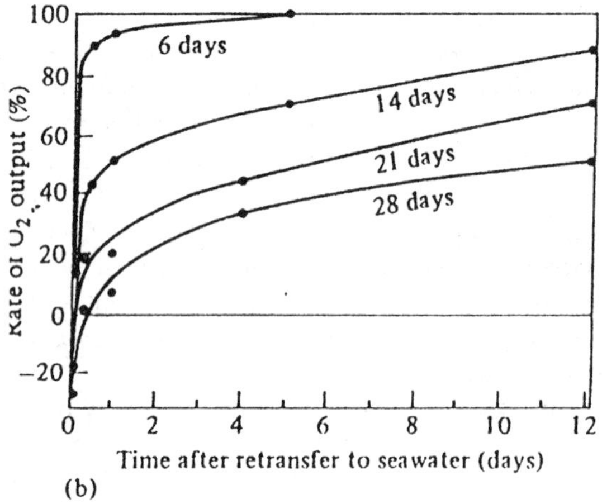

FIGURE 4.1 Recovery of photosynthesis in two intertidal fucoids, *Fucus vesciculosus* (a) and *Pelvetia canaliculata* (b), following desiccation for several days. Upper curve in (a) is rate in a thallus resubmerged immediately after reaching 10 to 12% of the original water content. Photosynthetic rate, as O_2 output, is expressed as percentage of the rate in undehydrated control plants. (From Lobban, C.S., Harrison, P.T., and Duncan, M.J., *The Physiological Ecology of Seaweeds,* Cambridge University Press, Cambridge, 1985, 170. Based on Gessner, F. and Schramm, W., 1971.)

photosynthetic activity from emersion stresses varied between algae from microsites separated by <10 cm. Algae from canopy microsites that were regularly exposed to a greater range of irradiance, temperature, and osmotic stress than algae from understory microsites had greater capacity to recover from these stresses alone or in combination compared to tissues from understory microsites. Net photosynthesis was enhanced by 20% water loss or exposure to 2,150 MosM. kg^{-1} media compared to values for algae that were in a fully immersed state. The temperature optima for net photosynthesis was 33°C, while the upper performance threshold was 40°C. Highly responsive stress acclimation capacity, coupled with microclimate benefits of a turf form, substantially contribute to the ecological success of *A. concinna* as an ecological dominant at high tidal elevations in the Hawaiian archipelago.

The situation concerning the consequences of algae drying out is further complicated in that there is evidence it may be accompanied by an increase in the rate of exudation of organic matter (Siebruth, 1960). The amount of carbon released in 10 minutes by *Fucus vesciculosus* after resubmergence increased in relation to the duration of exposure (and hence the amount of water lost). Algae from higher on the shore lost more water and released less carbon.

Unless rocky-shore animals have special mechanisms to combat water loss, they lose water to the air. If this occurs for extended periods, they eventually die from desiccation.

Death of animals on the shore due to desiccation may be due to disturbances in the metabolism resulting from an increasing concentration of the internal body fluids or more usually from asphyxia. For those organisms that respire by means of gills, a constant water film must be maintained over the respiratory surfaces.

In addition to water loss by evaporation, animals also lose water by excretion. Most marine animals excrete ammonia as their principal nitrogenous waste product, but it is highly toxic, requiring a very dilute urine and the passage out of the body of a large volume of water. Some littoral species of gastropods have been able to reduce their excretory water loss by excreting appreciable amounts of uric acid, which is a soluble and less toxic product requiring less water for its excretion. In British gastropods, those living highest on the shore have the greatest uric acid concentration in their nephridia.

Desiccation stress, of course, varies with position on the shore in relation to the amount of exposure to air over a tidal cycle, as well as to the periods of continuous emersion. Animals on hard shores are much more vulnerable to drying out than those on soft shores, and for the latter the problem is more acute on sandy than on muddy shores. Mudflats rarely dry out, but the upper regions of sandy beaches can become quite dry. However, on sandy beaches, the inhabitants avoid desiccation by burrowing. On muddy shores, surface dwellers such as some mudflat snails burrow into the surface sediments when the tide is out.

On hard shores the animals found in the eulittoral can resist desiccation inside an impervious shell or tube that can be tightly closed up (barnacles and mussels), sealed off by a horny membrane (many gastropods), a calcareous operculum (serpulid tubeworms), or closely pressed to the rock surface (limpets). In many of these species there is a correlation between shell thickness and position on the shore, animals living higher on the shore having, in general, thicker shells than those lower down. Some attached soft-bodied forms, such as anemones, produce a copious secretion of mucus that assists resistance to drying out. In

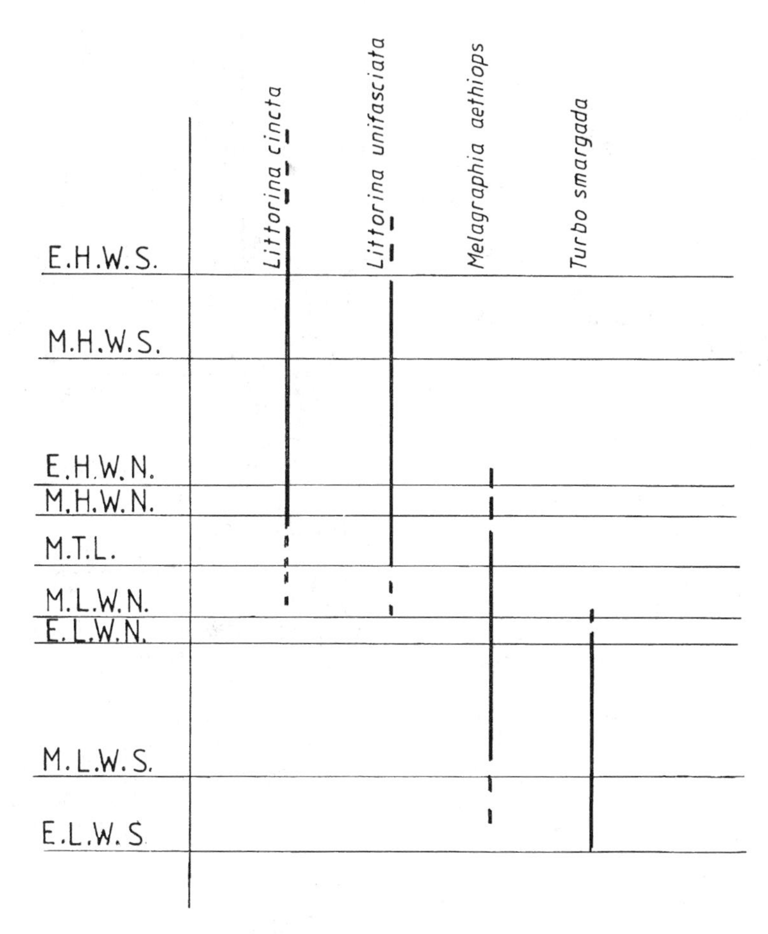

FIGURE 4.2 The vertical distribution of littorinid and trochid gastropods on New Zealand shores.

addition, a number of physiological adaptations have developed to enable animals to withstand the risk of desiccation. Since water loss results in an increasing concentration of the body fluids, an efficient osmoregulatory system is required. Also, since desiccation is usually accompanied by an increase in body temperature, tolerance of high temperatures is also required. It must be pointed out, however, that the latent heat of evaporation helps to cool an animal that is losing water, and this may be a significant factor in reducing body temperature.

On New Zealand shores, littorinids and trochids form a useful series (Figure 4.2), with overlapping vertical ranges and distinctive midpoints, for studying the effects of desiccation on intertidal animals. Rasmussen (1965) has tested the relative amount of desiccation these four species can tolerate by determining the 50% mortality point when they were exposed in sunlight at 35°C. The results are given below:

Littorina cincta	>120 hours
Littorina unifasciata	>120 hours
Turbo smaragda	60–65 hours
Melaraphia aethiops	40–60 hours

He also carried out a series of experiments to test whether there was a differential susceptibility to desiccation with increasing age. The distribution curve for *Melagraphia aethiops* is shown in Figure 4.3. It can be seen that there are four definite size classes and possibly a fifth. *Melagraphia* spat settle over the entire intertidal range and then migrate as they grow toward a central vertical zone. Those that do not reach this zone perish. The first-year class remains well sheltered from desiccation in runnels, pools, and under rocks. Older individuals are found on the open rock surface. Desiccation experiments (Figure 4.3) indicate that there is a definite increase in desiccation tolerance with size and age.

Broekhuysen (1940) studied a series of gastropods ranging, from the upper shore *Littorina africana knysnaensis* through *Oxystele variagata, Thais dubia, O. trgina,* and *Burnupena cincta* to the low shore species *O. sinensis* (Figure 4.4). Broekhuysen (1940) compared the relative tolerance of the six gastropods to desiccation by measuring both the percentage water loss and mortality in the gastropod over a range of temperatures. However, as pointed out by Brown (1960), the water loss was expressed as percentage of total wet weight including the shell, whereas most investigators express the rate of des-

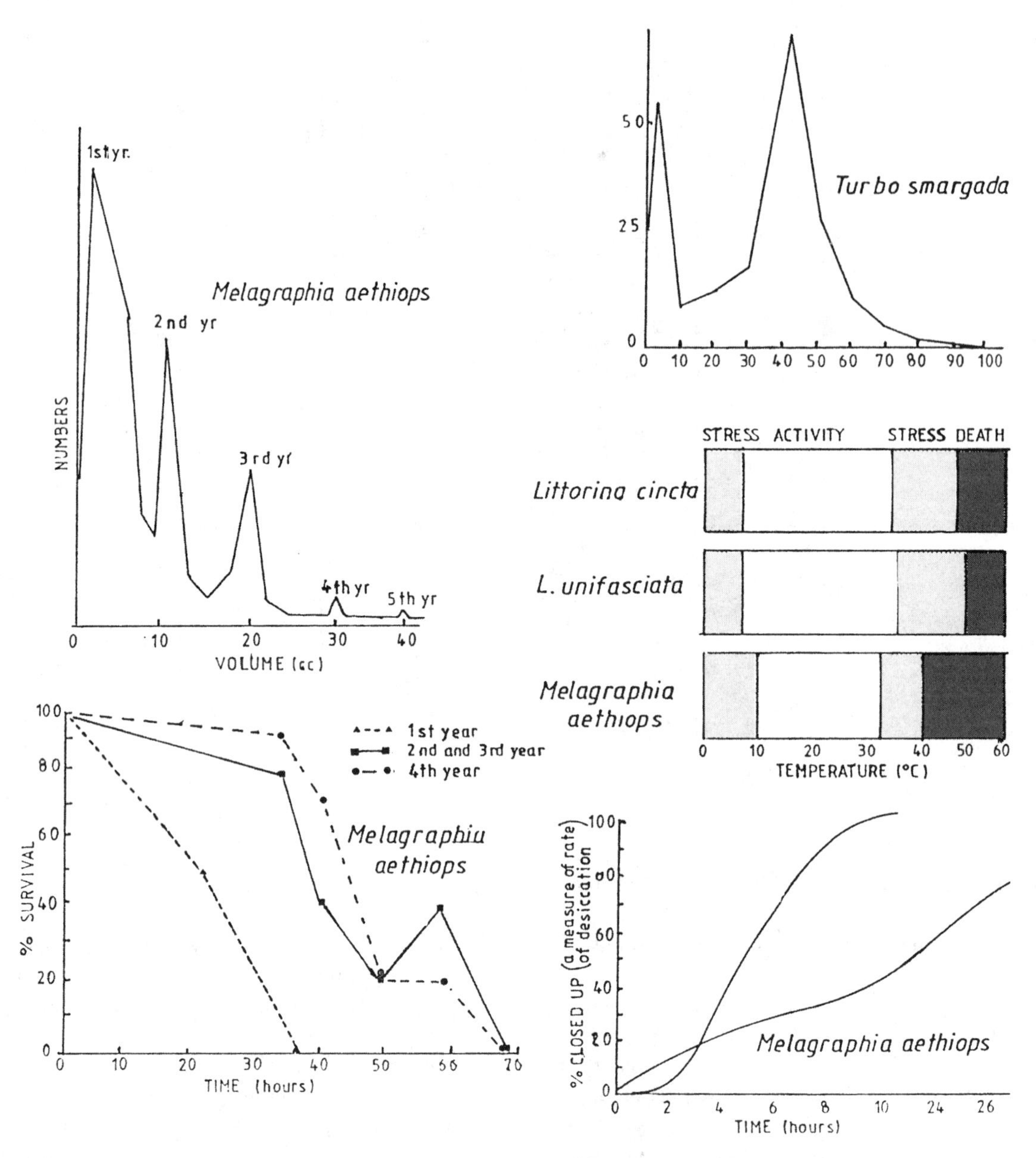

FIGURE 4.3 A. Size class numbers of the trochid, *Melagraphia aethiops*. B. Size classes of the catseye, *Turbo smaragda*. C. Percent survival of the various year classes of *Melagraphia aethiops* exposed in sunlight at 35°C. D. Response to temperature stress of *Littorina cincta*, *L. unifasciata*, and *Melagraphia aethiops*. E. The percentage of individuals of *Melagraphia aethiops* with the operculum closed (a measure of desiccation stress) on a hot windy day and a cloudy day. (After Rasmussen, N., *The Ecology of the Kaikoura Peninsula*, Ph.D. thesis, University of Canterbury, Christchurch, New Zealand, 1965.)

iccation as water loss per unit dry weight including the shell. Brown repeated Broekhuysen's experiments to give the results shown in Table 4.2, where desiccation is expressed as water loss per unit dry weight including shell. From the table, two general conclusions can be drawn. First, there is a correspondence between zonational level and the percentage water loss causing 50% mortality. However, some species, such as *L. africana knysnaensis*, are less tolerant of water loss than would be assumed from their level on the shore, while others such as *Burnupena cincta* are apparently more tolerant than would be indicated by their position on the shore. Second, a tolerance of between 15 and 37% water loss is characteristic of the species series before 50% mortality occurs.

The reasons for the exceptions in the zonational sequence are twofold. *Burnupena cincta* lives in a drier situation on the open rock surface compared to *Oxystele tigrina*, which is restricted to damp situations and pools. Second, *L. africana knysnaensis*, in common with other high tidal species, can cement the rim of the shell to the substratum with mucus and thus limit water loss. Other species including the gastropod *Nerita* and limpets can retain extra-

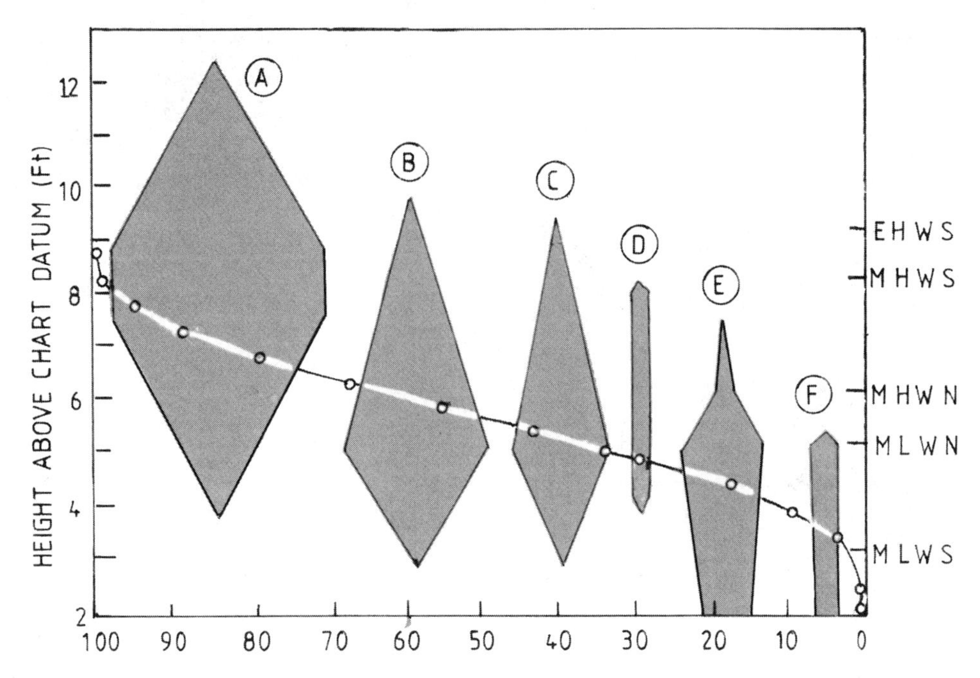

FIGURE 4.4 Graph showing the relation between the distribution of gastropods on the shore at False Bay, South Africa, and tidal level. The curve shows the percentage exposure at the various tidal levels. A. *Littorina africana knysnaensis* B. *Oxystele variegata*. C. *Thaisdubia*. D. *Oxystele tigrina*. E. *Burnupena (=Cominella) cincta*. F. *Oxystele sinensis*. (Redrawn from Newell, R.C., *The Biology of Intertidal Animals*, 3rd ed., Marine Ecological Surveys, Faversham, Kent, 1979, 125. After Broekhuysen, 1940. With permission.)

TABLE 4.2
The Range of Distribution (Height in Feet above Datum) and Water Loss Required to Induce 50% Mortality in a Series of Gastropods from Cape Peninsula, South Africa

Species	Mean Zonational Level	% Water Loss for 50% Mortality
Littorina africana knysnaensis	12.3	33.17
Oxystele variegata	9.7	37.61
Thaia dubia	9.2	34.88
Oxystele tigrina	8.0	24.40
Burnupena cincta	7.5	32.03
Oxystele sinensis	5.2	15.87

Source: After Brown, A.C., *Porta Acta Zool.*, 7, 1960.

corporeal water under the shell for much of the intertidal emersion period and thus reduce desiccation effects.

4.2.3.2 Thermal Tolerance

The temperature tolerance of an intertidal organism is an important factor in determining the upper level at which a particular species can survive when the tide is out. However, the situation is complicated by the interaction of a large number of variables. The magnitude of the temperature stress is dependent on season, the time of day when emersion occurs, the duration of the exposure to air, and other factors. Its effect may be modified by factors such as shape and color of the organism, body size, and the magnitude of the water loss. Desiccation may modify the effects of temperature stress in a variety of ways. For example, each gram of water evaporated from the tissues at 33°C removes 544 calories of heat, and this value increases with temperature so that it represents an important potential method of facilitating heat loss. Many intertidal organisms have evolved structural and physiological adaptations that minimize the impact of thermal stress such as shell shape and the retention of extracorporeal water in the mantle cavities of molluscs.

Much of the extensive literature on the thermal tolerance of intertidal and subtidal organisms has been reviewed by Kinne (1971), Somero and Hochachka (1976), and Newell (1979). The most detailed of these early studies relating the temperature tolerances of intertidal animals to their zonational position on the shore was that of Broekhuysen (1940). He demonstrated that the sequence of thermal death points of a series of South African gastropods showed a general correspondence with their zonational position on the shore much as described for their desiccation tolerance (see Section 5.1.4.2 above) (Figure 4.5). The highest species on the shore, *L. africana knysnaensis*, had the highest upper lethal temperature (48.6°C), while the lowest species, *Oxtstele sinensis*, had the lowest (39.6°C). Since then, numerous studies have confirmed and amplified such sequences in the thermal tolerances for a variety of taxa.

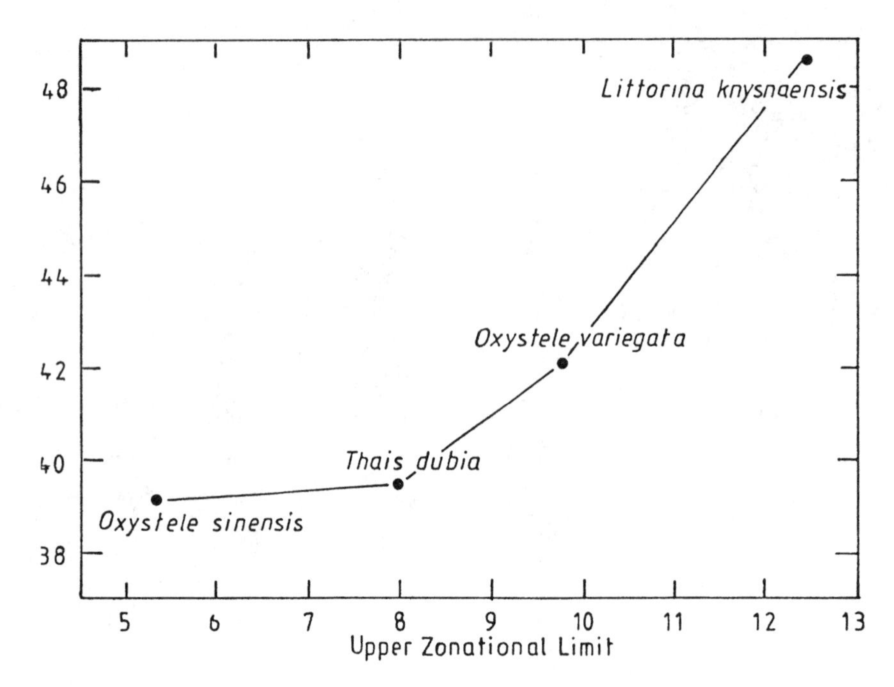

FIGURE 4.5 Graph showing the relationship between upper zonational limit (height in feet above chart datum) and upper limit of thermal tolerance of the series of gastropods illustrated in Figure 4.4. (Redrawn from Newell, R.C., *The Biology of Intertidal Animals*, 3rd ed., Marine Ecological Surveys, Faversham, Kent, 1979. 146. Data from Broekhuysen, 1940. With permission.)

4.2.4 Ecological Niches

4.2.4.1 Introduction

As Warren (1971) points out in referring to animal species, "Each species has evolved as part of an ecosystem: an ecosystem in which it occupies certain spaces during certain times; and ecosystem in which it can tolerate the ranges of physical and chemical conditions; an ecosystem in which it utilizes some of the species for energy and material resources and in which it is utilized by other organisms; an ecosystem in which it has many kinds of relations with different species, and in which it can satisfy its shelter and other needs."

In considering the ecological niche of a species, we concern ourselves with what a species does, what activities characterize its life, how and where it carries out these activities, why or for what purpose they are carried out, and when they occur. The concept of the niche is thus a functional one.

The ecological niche of a species can be described by considering:

1. The interaction of the species populations with the environmental factors listed in Table 4.1
2. The structural, physiological, and behavioral adaptations that enable the species to survive and reproduce in the environment it inhabits
3. The times at which the interactions occur
4. The effects of the species' activities on the ecosystem of which it is a part

While a complete description of the ecological niche of a species is not usually possible, the concept is nevertheless a useful one in that it enables us to gain an understanding of the role of a particular species in the ecosystem in which it is found.

Species can be categorized (Vermeij, 1978) as:

1. *Opportunists*: Such species show high reproductive output, a short life history, high dispersability, reduced long-term competitive abilities, and generally occupy ephemeral or disturbed habitats.
2. *Stress-tolerant forms*: These can tolerate chronic physiological stress, exhibit low rates of recolonization, tend to be long-lived with slow growth rates and, consequently are generally poor competitors.
3. *Biotically competent forms*: These generally live in physiologically favorable environments, have long life spans, are good competitors, and have evolved mechanisms to reduce predation.

In the rocky intertidal zone, stress-tolerant forms are characteristic of the upper intertidal habitat, whereas biotically competent forms are prevalent in the lower intertidal. Opportunistic forms appear ephemerally on disturbed or newly available substrates.

Andrewartha and Birch (1984) make three propositions concerning the way in which the environment works. The first is that the environment can be considered as a

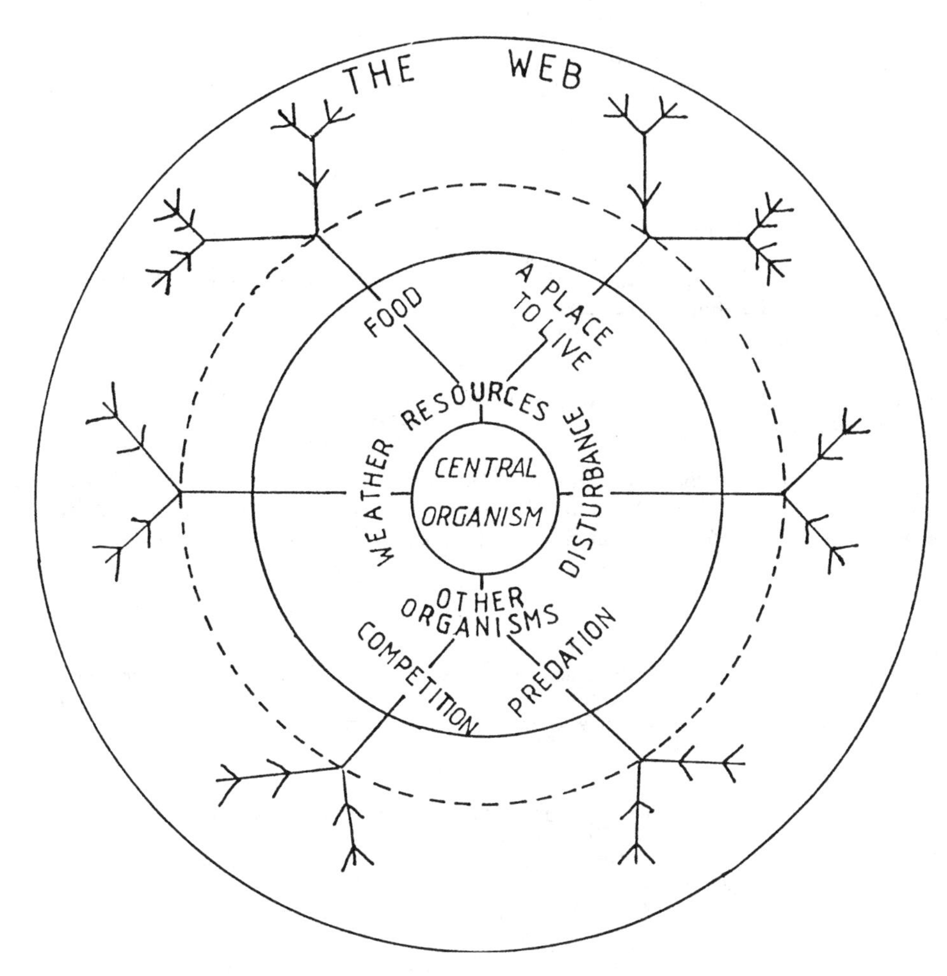

FIGURE 4.6 The environment comprises everything that might influence an animal's chance to survive and reproduce. Only those "things" that are the proximate causes of changes in the physiology or behavior of an animal are placed in the centrum and recognized as "directly acting" components of the environment. Everything else acts indirectly, that is, through an intermediary of chain of intermediaries that ultimately influences the activity of one or other of the components of the centrum. All these indirectly acting components are placed in the web. (Modified from Andrewartha, H.G. and Birch, L.C., *The Ecological Web: More on the Distribution and Abundance of Animals,* University of Chicago Press, Chicago, 1984, 7. With permission.)

centrum of components that act directly on a species together with a *web* of indirectly acting components that affect those in the centrum (Figure 4.6). The second is that the centrum consists of four divisions here modified to include (1) *resources,* with two components (*food* and *a place to live*), (2) *other organisms* (competitors and predators), (3) *weather*, and (4) *disturbances* (accidental events that eliminate an organism or population). The third proposition is that the web is a number of systems of branching chains; a link in the chain may be a living organism (or its artifact or residue), or inorganic matter or energy.

4.2.4.2 The "Envirogram" Concept

According to Andrewartha and Birch (1984), activity in the directly acting components is the proximate cause of the condition of an individual of a species that affects its chance to survive and reproduce. But the distal cause of an individual's condition is to be found in the web, among the indirectly acting components that modify the centrum. A modifier may be one or several steps removed from the centrum, and the pathway from a particular modifier to its target in the centrum may be joined by incoming pathways from other modifiers that may be behind or alongside the first one (n steps away from its target in Figure 4.8). The envirogram is a graphic representation of these pathways.

An example of an envirogram for the food resource of a limpet, *Patelloida latistrigata*, is given in Figure 4.7. The food of this limpet on the rocky shores of southern Australia comprises the spores and young stages of algae that it scrapes from the rock. The envirogram depicts the web of effects determining the supply of food and, hence, indirectly affecting the limpet. Nearby mature algae are the source of the spores and the water currents are required to carry them onto the shores. Another limpet, *Cellana*

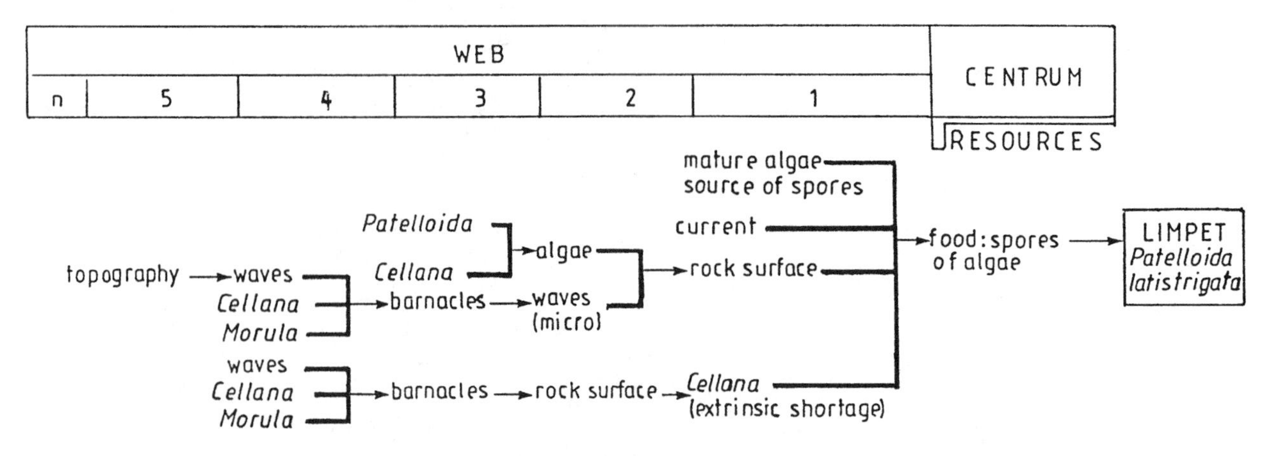

FIGURE 4.7 Part of the envirogram of the limpet *Patelloida latistrigata* on the coast of New South Wales, showing only the interactions that lead to food for the limpet. (Redrawn from Butler, A.J., in *Marine Biology*, Hammond, L.S. and Synnot, R.R., Eds., Longman Chesire, Melbourne, 1994, 156. Adapted from Andrewartha and Birch, 1984. With permission.)

THE BASIC LIFE-CYCLE (TYPE A):

The sporophyte and gametophyte are separate, free-living plants. They may look exactly the same or be very different. In some cases, two stages of one life-cycle have been so different that they have been classified members of different species or even of different families.

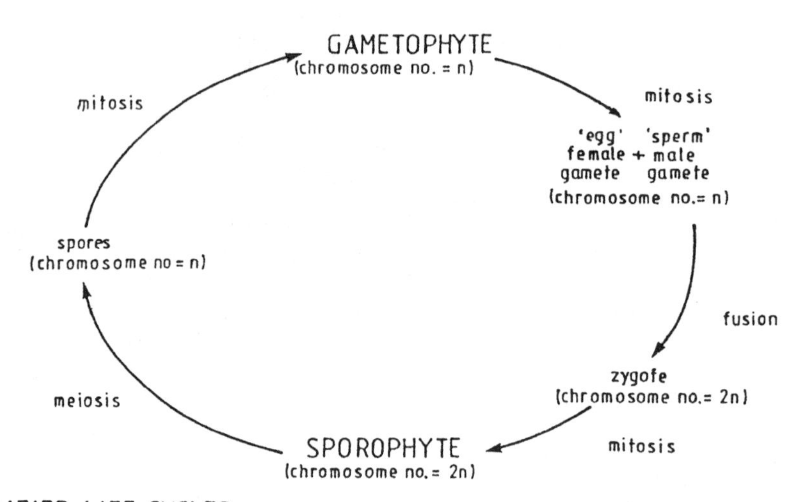

SIMPLIFIED LIFE-CYCLES:

Some seaweeds have simplified life-cycles. There are two types, B and C.

Type B

PLANT	meiosis	Gametes	fusion	Zygote	mitosis	NEW PLANT
2n		n		2n		2n

Type C

PLANT	meiosis	Gametes	fusion	Zygote	meiosis	Spores	mitosis	NEW PLANT
n		n		2n		n		n

These life-cycles are found in the green and brown seaweeds. Red seaweeds have an additional third generation, which is always attached to the gametophyte plant and is not a separate ecological phase.

FIGURE 4.8 Basic algal life cycles. (Redrawn from Hinde, R., in *Coastal Marine Ecology of Temperate Australia*, Underwood, A.J. and Chapman, M.G., Eds., University of New South Wales Press, Sydney, 1995, 127. With permission.)

tramoserica, which eats the same food as *Patelloida*, is an important factor determining the abundance of food available to *Patelloida*. Other variables that determine the action of the first-order interactions (1*n)* lie further out in the web of the environment of *Patelloida*. Barnacles may preempt space needed by *Cellana* (*Patelloida* can graze over the tops of the barnacles) so barnacles indirectly influence the food supply of *Patelloida*. The density of the barnacles in turn is determined by the suitability of the substrate for settling, predation by the whelk *Morula marginata*, and "bulldozing" of newly settled barnacle by *Cellana*. Other 2*n* to 5*n* factors affecting the supply of food for *Patelloida* are also shown.

4.2.4.3 Weather

The various weather factors listed in Table 4.1 will be considered in turn with reference to the ways in which selected organisms are adapted to cope with the problems encountered.

Immersion and emersion: The general effects of submersion and emersion have already been discussed in Sections 1.3.4 and 2.3.3. Some shore organisms require periods of emersion in order to function normally and die if subjected to periods of continuous submergence. On the other hand, as discussed in the sections listed above, emersion poses considerable problems for aquatic organisms and a variety of mechanisms have evolved (morphological, physiological and behavioral) to minimize water loss.

Temperature: The impact of temperature on the zonation patterns of rocky shore intertidal organisms has been considered in Section 2.3.6.2. When the shore is uncovered by the tide, wide and rapid temperature changes are encountered by the species living there. However, it needs to be borne in mind that the body temperatures of intertidal organisms rarely correspond exactly to the ambient air temperatures. In some cases, the body temperatures may exceed those of the air as has been recorded for barnacles; in other cases, the body temperature may be reduced below that of the ambient temperature due to the cooling effect of evaporative water loss. Edney (1951) has demonstrated the importance of transpiration in the control of body temperature in the littoral fringe isopod, *Ligia oceanic*. The thermal sensitivity of most species populations correlates with both latitudinal distribution and level occupied on the shore, as well as the topography and aspect of the shore from which samples have been taken for experimental studies. In such experiments, as Newell (1979) points out, the duration of exposure at each temperature is important; a long exposure at a lower temperature may cause the same percentage mortality as a brief exposure at a higher temperature. In considering the temperature tolerances of intertidal organisms, care in their interpretation needs to be taken since the stresses of temperature and desiccation are interdependent, and other environmental factors, such as wind velocity, are involved.

In considering thermal tolerances, the phenomenon of "acclimation" must also be taken into account. Thus, animals have the ability to adjust their metabolic rates over a period of time to the prevailing temperature conditions. The results of numerous studies have led to general agreement that in many species, cold acclimation involves a compensatory rise in the level of activity in such a way that the rate remains comparably with that of warm-acclimated animals. The result of this is that comparably sized individuals of a species with a wide latitudinal range show similar levels of activity at the cold temperate and warm temperate limits of their ranges. As a measure of metabolic activity, the rates of various functions, such as cirral beat in barnacles, heartbeat, and the rate of water filtration in bivalves, are measured. Since these rate functions all vary with body weight, it is important that comparisons be made with comparable size ranges of the species being compared. In considering the temperature tolerances of intertidal organisms, the stresses of temperature and desiccation are interdependent, and other environmental factors, such as wind velocity and humidity, are also involved.

Wave action: The general effects of wave action on the vertical zonation patterns on hard shores has already been considered in Section 2.3.1. Here we shall examine in more detail some of the adaptations to combat wave shock found in intertidal plants and animals.

Wave action has a considerable effect on the size range of many species of intertidal animals. This is especially notable in the case of littorinid snails. On New Zealand shores, both *Littorina cincta* and *L. unifasciata* show a decrease in average size with increasing wave action. Smith (1958) found that the degree of wave action has a moderating effect on numbers, vertical distribution and range, and size and shape of both species on the shores of Lyttelton Harbour. An increase in wave action was accompanied by an upward shift in the zone of vertical distribution and an increase in the width of the zone. Shading permitted an upward extension of the range of *L. unifasciata*. Maximum and mean densities showed a general increase corresponding with an increase in exposure (see Section 4.2.3.1). *L. cincta* was less tolerant of extreme exposure than *L. unifasciata*. Mean weights were greatest when wave action was least. Larger individuals occupied a wider range of sites than smaller individuals. There were also differences in shell shape that could be correlated with exposure to wave action; shells were significantly more elongated in the more sheltered areas and the rate of increase in the diameter of the whorl, and hence aperture size relative to the rate of growth of the shell, also differed significantly. The elongated shape would presumably be more vulnerable to wave action and more liable to dislodge.

Other species of gastropods appear to be able to modify the form and thickness of the shell in order to withstand the stresses imposed by strong wave action. For British species of limpets, it has been shown that those living high on the shore in exposed situations have shells that are flatter in shape and better adapted to withstand wave action than those in sheltered situations.

On soft shores, especially on sand beaches, wave action influences the distribution of the animals by the effect of the waves on the type of sediment, strong wave action being associated with coarser sediments. Beach profiles and beach stability are also affected, especially after storm events when the levels may change by a meter or more. Such changes can prevent some species from establishing themselves.

Salinity: Organisms on hard shores do not normally have to cope with fluctuations in the salinity of the seawater that covers them during the tidal cycles. Exceptions are those living in high shore tide pools, rock substrates in estuaries, and near the mouths of rivers. A limited number of species, such as the barnacle *Elminius modestus*, are able to adjust to salinity changes and consequently substrates in areas of low or fluctuating salinities are characterized by an impoverished flora and fauna. Often their vertical ranges are extended due to the absence of competing species. Organisms high on the shore may be affected by freshwater seepage and runoff from the land and in such situations, blue-green algae and filamentous greens such as *Enteromorpha spp.* are common. On mudflats, runoff from the land may dilute the surface salinities of the sediment.

Kinne (1967) has analyzed the responses of intertidal organisms to environmental stresses such as salinity and groups them under the following headings: escape, reduction of contact, regulation, and acclimation. Escape may be affected by vertical or horizontal migration into an area where the salinity range is more tolerable. Motile animals, especially rapid movers such as fishes, can respond in this way as can many organisms in soft deposits by simply burrowing into the sediments where conditions are more suitable. Many organisms are able to escape temporary short-term salinity changes by reduction of contact. This may involve the closing of valves (barnacles and bivalves), operculum (gastropods and tubeworms), or by the secretion of a mucus coating.

The most important class of response is that of regulation. Space does not permit a detailed discussion of the ways in which regulation occurs. In general, species react in one of three ways. First, there are those species in which the blood concentration is almost isotonic (i.e., similar in concentration) throughout the salinity range that is tolerated by the organism. Such species are poikilosmotic organisms or conformers. Second, there are the species whose blood is hyperosmotic (higher in concentration) to that of the medium at reduced salinities and isosmotic at higher salinities. Third, there are those whose blood is hyperosmotic in reduced salinities and hyposmotic (lower in concentration) at higher salinities. Reduced and higher salinities are relative to the salinity in which the animal normally lives and to which it is acclimatized. The latter two classes are the regulators or homiosmotic organisms.

Some open shore organisms are conformers; others can regulate to a varying extent. Most are stenohaline (tolerating a narrow range of salinity variation); a few are euryhaline (tolerating wide variations in salinity). These are often the open shore organisms that extend into estuaries.

Gases — oxygen and carbon dioxide: On hard shores, apart from the special case of rock pools, oxygen concentration is rarely a significant factor. In tide pools under bright light, photosynthesis by dense algal vegetation can sometimes raise the oxygen content appreciably and at the same time the withdrawal of carbon dioxide from the water raises the pH. Bacterial degradation of stranded debris, on the other hand, can lead to decreased oxygen, increased carbon dioxide, and reduced pH.

Since the amount of dissolved oxygen in water is a function of temperature, the heating of shallow water over mudflats may result in lowered oxygen levels. This may be accentuated by the presence of decaying organic matter. The infauna of mudflats and estuarine intertidal flats often have to cope with reduced oxygen levels at low tide. Burrowing forms have a variety of mechanisms for circulating water through their burrows but are unable to do so when the tide is out. This situation is tolerated by reducing the metabolic rate to low levels and many burrowing species can withstand surprisingly long anaerobic periods such as the nine days for the lugworm *Arenicola marina* and 21 days for the tubeworm, *Owenia fusiformis*, reported by Dales (1958). If the burrows drain completely, some species, provided the body surface remains wet, can breathe atmospheric air. The meio- and microfauna of soft shores, on the other hand, must cope with anaerobic conditions if they live below the RPD layer.

Light: The amount of light on the shore varies widely with the rise and fall of the tide. Excessive illumination can be damaging due to the ultraviolet and infrared rays, which can be lethal to some organisms. Such rays are, however, rapidly absorbed by seawater. It is difficult, however, to dissociate the effects of radiation from those of heat and desiccation stress. For many organisms, as discussed below, light is of great significance in controlling the movements of animals in maintaining themselves within optimal environmental conditions or microclimates.

The role of light in controlling the photosynthetic rates of algae and the attenuation of light and its spectral composition with water depth have already been covered in Section 2.3.5.1. On sand and mud shores, light is not an important factor in the lives of burrowing species. However, the degree of illumination is an important factor in the photosynthesis of benthic microalgae.

Water currents: Perhaps the most important role of currents in the life of shore organisms is in the distribution of larval stages of animals and the sporelings of algae. Much of the patchy distribution patterns that occur, e.g., bivalves on sandy beaches, can be explained by the vagaries of current systems during the critical settling period. Special conditions occur where tidal currents, which can often be termed rapids, occur in narrows where tidal movement is compressed in a funneling effect between islands or the narrow entrances of some inlets.

4.2.4.4 Resources

Resources are those "things" provided by the environment that enhance the ability of an individual or population to establish itself, survive, and grow to reproduce. The principal resources for an animal are "food" and "a place to live." For plants, "foods" are nutrients and carbon dioxide. Food and feeding have been considered in Section 2.7 and will be dealt with in greater detail in Chapter 6. A place to live is the equivalent of the "habitat" of an organism. For littoral species, this includes the vertical range over which a species can live and includes all those physical factors that limit their distribution.

Food: In general, for most shore animals, food does not become a limiting factor. It is, however, a limiting factor for those filter feeders that grow high on the shore near the upper limits of the species vertical range. Here the limiting effect is the time coverage by water which may be insufficient to enable an adequate supply of food to be obtained.

A place to live: A place to live is synonymous with the *habitat*, or the place you would go to find the species in question. For shore organisms, the types of substratum (rock, shingle, sand, mud, and various intermixtures) are of prime importance in determining whether the appropriate habitat is available for a particular organism.

On hard shores, rock surfaces at the appropriate level are necessary for sessile plants and animals, as well as for relatively sedentary forms such as limpets and some other herbivores. For rock borers, rock of the type that they can bore into is essential and for those species that live in the shade and humidity under boulders, the distribution of such habitats will determine their presence or absence. Other species, such as the epiflora and epifauna that live in association with seaweeds, depend on the presence of other living organisms to provide a suitable habitat.

On soft shores, many animals construct permanent burrows in substrates of the appropriate grain size. Deposit feeders are restricted to sediments of a particular grain size composition. For the interstitial animals on soft shores, a place to live is on or in the spaces between sand and silt particles of the appropriate size. Throughout this book many examples have been given of the importance of habitat in determining the presence or absence of a particular species. As a population regulator, this factor acts in a negative sense since it becomes significant by its absence.

TABLE 4.3
Analysis of Intraspecific Population Interactions

Type of Interaction	Species 1	Species 2	General Nature of the Interaction
1. Neutralism	0	0	Neither population affects the other
2. Competition	–	–	Inhibition of each species by the other
3. Amensalism	–	0	Population 1 inhibited, 2 not affected
4. Parasitism	+	–	Population 1 the parasite, generally smaller than 2, the host
5. Predation	+	–	Population 1, the predator, generally larger than 2, the prey
6. Commensalism	+	–	Population 1, the commensal, benefits, while 2, the host, is not affected
7. Mutualism	+	+	Interaction favorable to both populations

Note: 0 indicates no significant interactions; + indicates growth, survival, or other population attribute benefited; and – indicates population growth, survival, or other attribute inhibited.

4.2.4.5 Other Organisms

The other animals and plants that make up the living component of an organism's environment include members of the population to which it belongs as well members of the populations of other species. On this basis we can divide the interactions that occur into *intraspecific* (between members of the same species) and *interspecific* (between members of different species).

Intraspecific interactions: Such interactions are most obvious when the members of the same species compete for some resource that is in short supply, e.g., food or space. In general food is not as limiting as is space. Intraspecific interactions will be dealt with in Sections 5.3.2.2, 5.3.2.3, and 5.3.4.2.

Interspecific interactions: Populations of two species may interact in basic ways that correspond to combinations of 0 (no significant interaction), + (positive interaction), and – (negative interaction). The different kinds of possible interactions are shown in Table 4.3. All of these interactions are likely to occur in littoral communities. For a given pair of species, the interactions may change under different environmental conditions or during successive stages in their life histories.

1. *Competition.* Competition occurs when two species strive and compete for the same environmental resource, and is best exemplified when it is for living space, especially between sessile species such as barnacles, mussels, and serpulids. The various kinds of competition will be discussed in Sections 6.6.2. and 6.3.4.

2. *Parasitism.* It would be difficult to find a shore organism that does not have its quota of parasites. The most widely studied of the parasitic animals are the flukes or trematodes that occur as larval stages in invertebrates, and the shore mollusc in particular. The final hosts for these parasites are vertebrates, especially fishes and birds. Estuarine molluscs, which form the principal food resource of many wading birds, often carry a great variety of the larval trematodes. Other parasites include protozoans, bacteria, and viruses. The impact of parasites on the population density of sea urchins on the Nova Scotia coast is discussed in Section 5.2.
3. *Predation.* The dense populations of animals on both hard and soft shores offer ideal opportunities to predators and on all types of shore, predation is a significant factor in controlling the densities of many species. Filter-feeding animals can be regarded as indiscriminate predators on small zooplankton as well as the larval stages of many species of shore animals. Other predators are more specific and in some cases the prey may be restricted to a single group of animals and even to a single species, e.g., bivalve molluscs as is the case of some predatory gastropods.

   Shore animals are exposed to a double set of predators. During submergence they are preyed upon by other marine animals, but when uncovered they are subject to terrestrial predators, especially marine birds. Most of the marine invertebrate predators are either anemones, gastropod molluscs, cards, lobsters, and echinoderms, while fishes are important marine predators. The role of predators will not be discussed here but will be dealt with in detail in Sections 6.2.3 and 6.3.5.
4. *Commensalism and Mutualism.* There are a great variety of associations between marine animals in addition to those of competitor, predator and prey, and host and parasite. Broadly speaking, these other relationships can be grouped into those of commensalism and mutualism. In the former association, the two species live together in some degree of harmony with one species generally benefiting to a greater or lesser degree from the association. In mutualism there is a close physiological association between the two species, usually for mutual benefit. It must be remembered, however, that there are many transitional examples that do not fit neatly into either category. *Commensalism:* Based on the type of commensal relationship, we can distinguish three subgroups: *epizoitism, endoecism,* and *iniquilism.*
   - *Epizoitism.* Epizoites are animals that live attached to the surfaces of other animals, such as barnacles and tubeworms and other sessile species found growing on the shells of molluscs. Epizoites sometimes display definite preferences for particular species such as the New Zealand species, the small mudflat limpet, *Notoacmea helmsi*, on the shells of the mudflat snail, *Zediloma subrostrata*, and the hydroid, *Amphisbetia fasciculata,* which attaches itself to the shell of the bivalve, *Paphies donacina*, on sand beaches.
   - *Endoecism.* There are large numbers of commensals that lurk in the burrows, tubes, or dwellings of various animals. Many polynoid worms are such commensals. They include the polynoid *Lepidasthenia aecolus*, found in the burrows of the lugworm, *Abarenicola assimilis*; the short, rather broad and flat *Lepidastheniella comma* living in the tubes of terebellids, especially *Thelepus* species; and the larger, more slender *Lepidasthenia* sp. found in the tubes of the sand beach maldanid, *Axiothella quadrimaculata.* Experimental analyses of host-commensal relationships have shown that some species are attracted to specific chemicals given off by their copartners.

     A number of molluscs, especially bivalves, are commensal with other animals. New Zealand examples include the small bivalves of the genus *Arthritica*, *A. hulmei* which lives under the elytra of the scale worm *Aphtodita australis*, *A. crassiformis* living with the large rock borer, *Anchomasa similis*, and *A. bifurcata* attached to the outer surface of the head end of the perctinarid polychaete, *Pectinaria australis.*
   - *Iniquilism.* This term is applied to the cases where the commensals live in the body cavities or internal cavities of their hosts. Such species benefit by obtaining access to food supplies, by sharing the host's refuge, or by taking advantage of the repellent properties of the host. This is probably one of the routes to parasitism and many inquiline commensals are close to being parasites. One such example are the pinnitherid (pea) crabs, which are found in the mantle cavities of mussels and a range of other bivalves.

4.2.4.6 Disturbance and Patchiness

Two of the salient features of littoral systems are that they are spatially patchy and that important processes such as disturbance and recruitment are spatially and temporarily variable. On rocky shores in particular, spatial patchiness is obvious as a result of differences in substrate, angle of slope, degree of wave action, amount of shade, degree of sand coverage and sand scouring, and in the frequency and intensity of disturbances that result in new substrates being made available for colonization. Aspects of all of these have already been covered in Chapter 2 and they will be further considered in Chapter 5. Patchiness can also be a consequence of random settlement and recruitment in intertidal invertebrates, e.g., barnacles and mussels on rocky shores, and bivalves and polychaetes on soft shores.

4.2.4.7 The Importance of Recruitment

Many recent field studies have shown that, once planktonic larvae are transported to a substrate suitable for settlement, variation in recruitment (the proportion of the settlers that have survived over a time period with the potential to contribute to the adult population) on a relatvely small spatial scale (sites of meters to ten meters apart) can be determined by variation in the supply of planktonic larvae (Grosberg, 1982; Minchinton and Scheibling, 1991; Bertness et al., 1992). Further, it has been shown that variation in recruitment may be directly proportional to the amount of space on the substratum available for colonization (Gaines and Roughgarden, 1985; Minchinton and Scheibling, 1993; Chabot and Bourget, 1988). However, Raimondi (1990) has shown that under certain conditions this relationship does not apply. In addition, a wealth of studies (mostly mechanism-orientated and carried out in the laboratory) have demonstrated that physical (Butman, 1987; Raimondi, 1988a) and biological (Raimondi, 1988b; Andre et al., 1993) interactions between the incoming larvae and established residents and the behavioral responses of the larvae when selecting a settlement site can influence the distribution of the larvae at settlement (Crisp, 1984; Pawlik et al., 1991). A prevalent indicator of the suitability of a habitat for settlement is the presence of conspecific adults (i.e., gregarious behavior: Gabbott and Larman, 1987; Raimondi, 1988b). Resident individuals may exude chemical attractants that stimulate settlement of conspecific larvae, or physical contact between conspecific individuals may be required.

The role of settlement and recruitment in the establishment of intertidal communities will be considered later in this chapter and will be explored in detail in Chapter 5.

4.3 THE ESTABLISHMENT OF ZONATION PATTERNS

In order to maintain a population at a particular zone on the shore, a species must reproduce, disperse its larvae, and the larvae must settle at the appropriate level on the shore and survive to reproduce. In this section we will consider these events in the life histories of intertidal organisms and the various factors that influence them.

4.3.1 Reproduction

4.3.1.1 Developmental Types in Marine Benthic Invertebrates

Numerous classifications have been proposed for the spectrum of developmental types found in marine invertebrates (e.g.. Mileikovsky, 1974; Jablouski and Lutz, 1983). Here we adopt a modified version of that proposed by Thorson (1946; 1950).

1. *Pelagic (long-life) planktotrophic*. These larvae spend a significant proportion of the development time swimming freely in the surface waters and feeding on other planktonic organisms, usually phytoplankton. Duration of larval life varies from a few weeks to two or three months. Eggs are released with little yolk but are produced in great numbers, e.g., up to 85,000 eggs per spawning in the gastropod, *Littorina littorea*, (Bingham, 1972), and up to 70 million eggs per individual in a single spawning of the oyster, *Crassostrea virginica*. Predation, starvation, and other factors take a tremendous toll on planktotrophic larvae, with an estimated mortality exceeding 99% (Thorson, 1950). However, the enormous numbers of larvae that are produced counterbalance this extremely high larval mortality. Scheltema (1967) divided the planktonic stage of planktotrophic species into two phases: (1) growth and development (the pre-competent period of Jackson and Strathmann, 1981), followed by (2) a "delay period" ("competent period") in which development is essentially completed but larval adaptations for a planktonic existence are retained until a suitable substrate for settlement is found.
2. *Short-life planktotrophic*. These larvae are discriminated chiefly on the basis that their size and organization change, hardly, if at all, in the course of the week or less spent in the plankton.
3. *Pelagic lecithotrophic larvae*. These larvae are large and provided with much yolk, hatching from a large yolky egg. The yolk provides all

the energy needed by the larva until metamorphosis into a settled juvenile (thus they are non-planktotrophic). Reproductive effort per offspring is thus much higher than in the planktotrophic species and larval mortality is much lower. Accordingly, far fewer eggs per parent are produced (4,100 eggs per parent in the bivalve, *Nucula proxima*, and 1,200 in the related species, *N. annulata*).

4. *Non-pelagic lecithotrophic*. To the above pelagic species must be added those that show (1) the so-called "mixed development," (2) "direct development," and (3) some form of brooding. Mixed development occurs when early developmental stages are encapsulated, but later stages emerge as free-swimming, pre-metamorphic larvae (Pechenik, 1979; see also Caswell, 1981). Mixed development is prominent in several benthic marine groups, especially polychaetes and gastropods. Direct development takes place within a encapsulated egg from which a benthic juvenile eventually hatches. Among the higher prosobranch gastropods (Neogastropoda), oviparous species may deposit, along with viable eggs, a supplementary food source in the form of nurse eggs. Brooded larvae, which are characteristic of ovoviviparous species, are retained (sometimes encapsulated) within the parent throughout development, emerging as metamorphosed juveniles.

4.3.1.2 Developmental Types in Marine Algae

The basic life cycle of an algal macrophyte is depicted in Figure 4.8. Type A is what is known as a sporic life cycle, which has a spore-producing phase, the sporophyte, which is usually diploid (2n); and a usually haploid (n), gamete-producing phase, the gametophyte. The sporophyte and gametophyte are separate free-living plants, which may look exactly the same, or be very different. The sporophyte produces spores that are liberated into the water column (they may be motile, and as such are called zoospores). The spores give rise to the gametophytes (male and female), which produce eggs and sperm. The fertilized egg results in the production of the zygote (2n) which develops into the mature plant (sporophyte), which by meiosis produced the spores that are the agent for dispersal. Figure 4.9 illustrates the life cycle of a kelp. The adult sporophyte releases spores into the water. These are washed around by waves and currents, eventually to settle on the bottom where they develop into gametophytes.

In addition, some algae can reproduce vegetatively through fragmentation. It is especially common in filamentous species and appears to play an important role in maintaining populations in habitats such as estuaries or salt marshes (Norton and Mathieson, 1983). Other species can regenerate from basal structures after the foliose fronds have been abraded or consumed by herbivores.

4.3.1.3 Reproductive Strategies

Life history patterns are often referred to as "strategies," a viewpoint that has often been criticized. However, as pointed out by Grahame and Branch (1985), if the view is taken that survival and progeny leaving are the outcomes of a series of features (morphological, physiological, and behavioral), then these adaptations can be seen as a "strategy" assembled by natural selection and ensuring survival. Todd (1985) has reviewed reproductive strategies of rocky shore invertebrates with special reference to northern temperate regions. He differentiates the terms life history strategy, life cycle strategy, and larval strategy, and illustrates their interrelations as shown in Figure 4.10. *Reproductive strategy* is a general term encompassing all three strategies listed above. *Life history strategy* has a dichotomous base and refers to organisms as being either semiparpous (reproducing once and then dying) or interparous (undergoing repeated breeding periods) (Cole, 1954). *Larval strategy* applies to the three fundamental larval types — planktotrophic, pelagic lecithotrophic, and non-pelagic lecithotrophic.

Energetic considerations also enter into life history strategies. Since the total energy budget of an individual organism is finite, the proportion allocated to reproduction will vary depending on age, availability of food resources, and environmental variables. Montague et al. (1981) considered that the reproductive strategy of an individual organism is the set of physiological, morphological, and behavioral traits that dictate the "where," "when," "how often," "how many" (and for marine invertebrates, "what kind of") tactics of propagule production. In this context there is the concept of reproductive effort (RE, i.e., a measure of the proportion of somatic effort (in energy terms) allocated specifically to reproduction) (Hirshfield and Tinkle, 1974).

Life cycle and life history strategies: Most temperate rocky shore invertebrates, almost without exception, display extended or perennial life cycles. This contrasts to soft shore invertebrates in which a range of life cycles, especially those of bivalves and gastropods, which are long lived, to some polychaetes which are annuals or less. A distinction needs to be made between *potential* and *realized* longevity in that survivorship of (potentially) long-lived animals, such as barnacles, may be markedly controlled by the activities of predators (e.g., Connell, 1970; 1972; 1975), or abiotic factors such as freezing (e.g., Wethey, 1985). In addition, local habitat patches, separated by perhaps only a few meters, may confer very different survivorship and growth probabilities for conspecific individuals (e.g., *Patella vulgata*, see Lewis and Bowman, 1975).

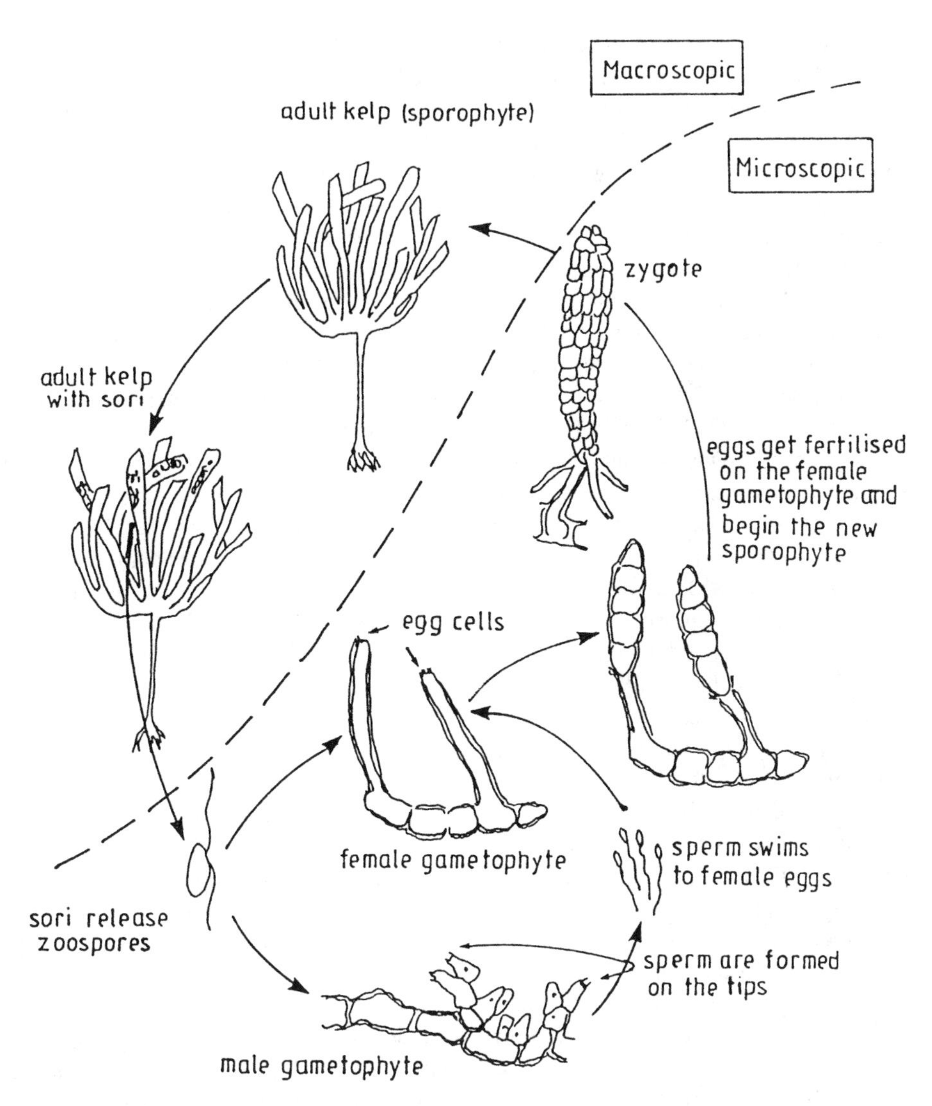

FIGURE 4.9 The life cycle of a kelp. (Redrawn from Kennelly, S.J., in *Coastal Marine Ecology of Temperate Australia,* Underwood, A.J. and Chapman, M.G., Eds., University of New South Wales Press, Sydney, 1995, 108. With permission.)

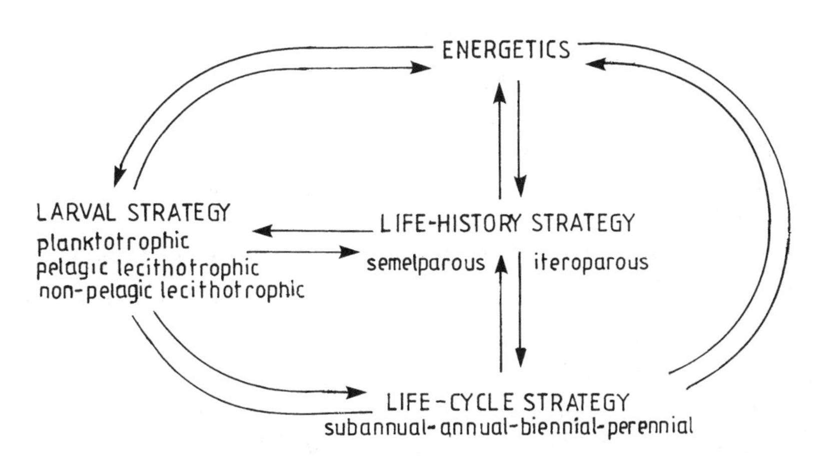

FIGURE 4.10 The interrelationships between reproductive energy traits. (Redrawn from Todd, C.D., in *The Ecology of Rocky Shores,* Moore, P.C. and Seed, R., Eds., Hodder & Straughton, London, 1985, 204. With permission.)

Barnacles, characteristic of rocky shores worldwide, are all potentially long lived and interoparopus. Most species (e.g., *Chthalamus stellatus* and *Elminius modestus*) complete several reproductive cycles at variable intervals depending on a complex of factors such as natality, mortality, food availability, and ambient temperatures (Barnes and Barnes, 1968; Crisp and Patel, 1969). Life spans are highly variable. *Tetraclita squamosa,* which requires ten years to reach maturity, may attain 14 years of age (Hines, 1979), while *Semibalanus balanoides* has a maximum (realized) longevity of only two years on New England shores (Wethey, 1984). Nevertheless, on British shores *S. balanoides* matures in its first year and together with *Chthalamus* spp. has a considerably greater life span than on New England shores. On the North American west coast, the goosenecked barnacle, *Pollicipes polymerus*, has a maximum life span of a least 6 (Paine, 1974) and perhaps up to 20 years.

Temperate shore bivalves are also long lived. The cosmopolitan *Mytilus edulis* may attain sexual maturity within only 1 to 2 months of settling and may live as long as 18 to 24 years (Seed, 1969; Suchanek, 1981). In contrast, the much larger, faster-growing *M. californianus* may require anywhere from 4 months to 3 years to mature (Suchanek, 1981) but certainly lives for 7 to 30, and possibly as many as 50 to 100 years (Suchanek, 1981).

On North American Pacific shores, the large (up to 1,200 g) chiton, *Cryptochiton stelleri*, lives for 16 to 25 years (Branch, 1981). Other chitons are similarly long lived. Branch (1981) in his review of limpet biology found that they invariably displayed extended perennial life cycles. Of the 20 species for which he had data, all except one had life cycles ranging from in excess of one year to approximately 30 years.

Among the archaeogastropoda trochid ("topshell") prosobranchs, Williamson and Kendell (1981) estimated a maximum adult life span of more than 10 years for *Monodonta lineata* and concluded that other major British topshells, *Gibbula cineraria* and *G. umbilicalis*, were similarly long lived. For British littorinids, Hughes and Roberts' (1980) estimated age at first maturity ranged from 8.5 months (*Littorina littorea*) to 18 months (*L. nigrolineata* and *L. saxatilus*) to 3 years (*L. neritoides*), while longevity ranged from 8 years (*L. littorea* and *L. nigrolineata*), to 8 to 11 years (*L. saxtilus*), to 16 years (*L. neritoides*). The predatory dogwhelk, *Nucella lapillus*, matures after 2.5 to 3 years (Hughes, 1972) and individual survivorship does not exceed 6 years (Hughes and Roberts, 1980). The three *Nucella* species on Pacific Northwest shores show variable life spans from 2 to 4 years (Todd, 1985).

Pisaster ochraceus on the Pacific Northwest coast of North America is estimated to have a life span of perhaps 34 years (Menge, 1972), while the smaller *Leptasterias hexactis* has an estimated longevity of from 4 to 18 years.

Larval dispersal and larval strategies: Pelagic larval forms generally display "delay" of metamorphosis in the absence of a specific cue or cues, and as the pause passes the larvae become less and less discriminating with regard to the choice of a settlement site (e.g., Crisp, 1974; Strathmann, 1978; Pechenik, 1984; and others in Chia and Rice, 1978). In this context there is a distinction between the "pre-competent" and "competent" phases of development (e.g., Crisp, 1974). During the pre-competent phase, the larva is morphologically and/or physiologically incapable of settlement and metamorphosis; the competent (= delay) phase commences from the point at which metamorphosis can take place upon the reception of the appropriate stimulatory cues. The competent phase is not, however, of indefinite duration (see review by Jackson and Stratham, 1981) and varies for a range of taxa from a few days to a few months of duration.

All pelagic larvae are subject to dispersal away from the potential micro-habitat, and for rocky shore species in particular this poses considerable risks in subsequently finding a suitable substratum for settlement. Larval transport is unpredictable and suitable habitats are patchy in space and time. However, there are advantages commensurate with a pelagic phase; these include the potential to increase the species' geographical range, an increase in gene flow, the reduction of local extinctions resulting from density-independent perturbations, and (as a result of the necessarily high fecundity) the increase in potential juvenile offspring per unit RE.

Timing of reproduction: In many species the release of eggs and sperm or larvae coincides with particular phases of the tidal cycle; e.g., in the pulmonate snail, *Melampus bidentatus*, which inhabits the higher levels of salt marshes, both hatching of the eggs and settlement of larvae are synchronized with spring tides (Russel-Hunter et al., 1972). Both *Littorina littorea* and *L. melanostoma* have a lunar-tidal rhythm in which the release of eggs coincides with spring tides.

Many species of intertidal organisms reproduce annually, or concentrate their reproduction over a certain period of the year. In many cases, temperature may be the cue for reproduction. However, food availability may, in many instances, be more important than temperature. Many invertebrates spawn so as to coincide with the spring phytoplankton bloom and thus allow the larvae to capitalize on a rich but transient food resource. In the tropics, however, many invertebrates may breed continuously.

4.3.1.4 A Model of Non-pelagic Development Co-adaptive with Iteroparity

Todd (1985) has developed a flowchart illustrating the possible reproductive strategy responses of rocky shore animals (especially prosobranchs), based largely on the extensive body of research on British littorinids (e.g., Hughes and Roberts, 1980; Raffaelli, 1982; Atkinson and Newberry, 1984) (Figure 4.11). In the model, increases and decreases in life span, current fecundity, and juvenile

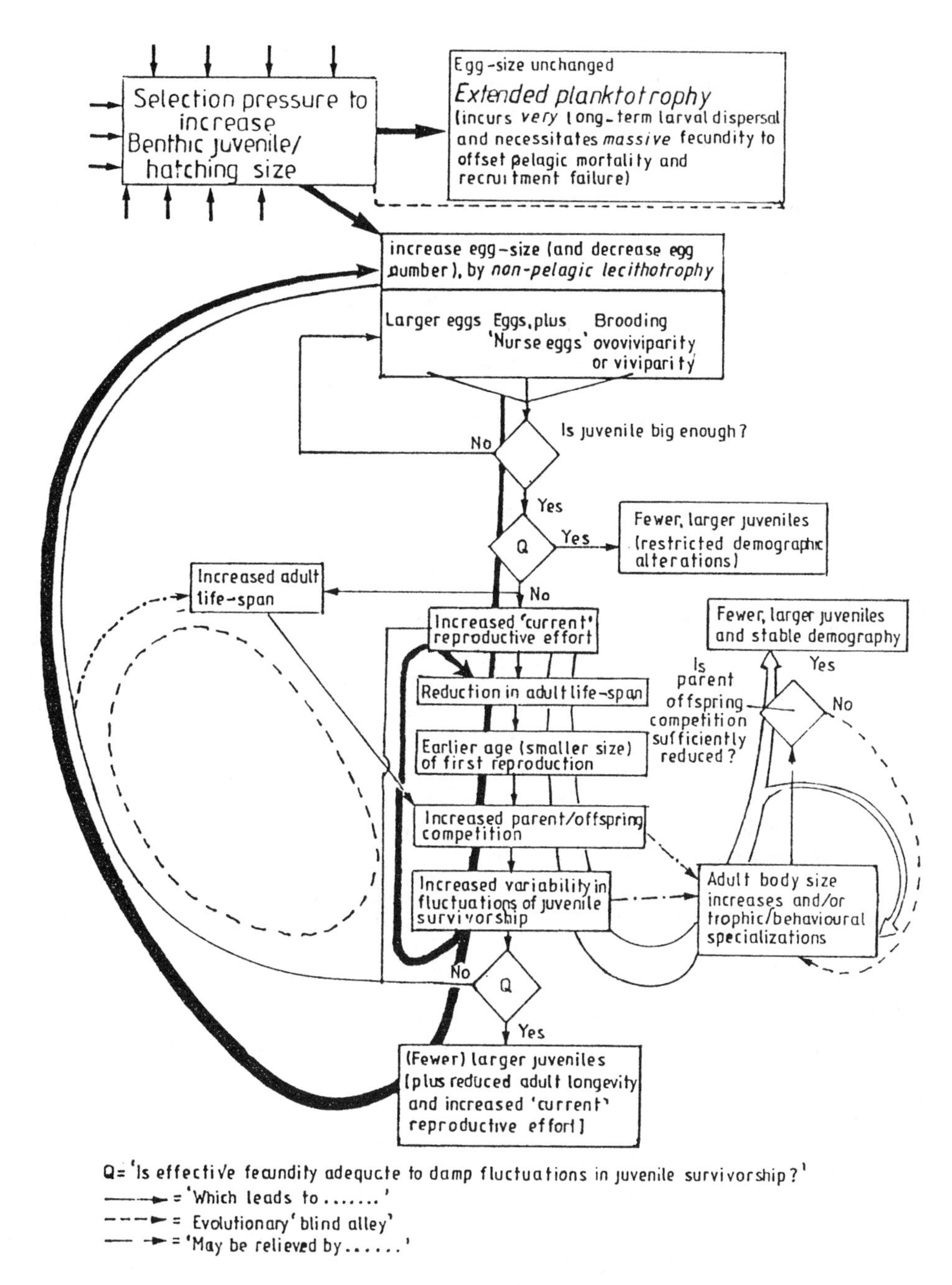

FIGURE 4.11 Flowchart illustrating possible reproductive strategy responses of rocky shore animals (especially prosobranch gastropods). The responses are to selection pressure favoring an increase in hatching size. The diagram should be followed according to the key. (Redrawn from Todd, C.D., in *The Ecology of Rocky Shores*, Moore, P.C. and Seed, T., Eds., Hodder & Straughton, London, 1985, 211. With permission.)

survivorship are balanced according to the prevailing selective regime. Selection (predatory) pressure favors an increase in hatching size (fewer larger individuals, plus reduced adult longevity and increased reproductive effort).

4.3.2 Settlement and Recruitment

4.3.2.1 Introduction

The initial settlement of planktonic propagules of benthic organisms usually varies considerably both in time and space (e.g., Connell, 1961a; 1985; Hawkins and Hartnoll, 1982; Caffey, 1985; Wethey, 1985). As a consequence, over the last few years there has been much interest and research on local variation of settlement and/or recruitment, and its consequences upon the distribution and abundance of intertidal species (e.g., Grosberg, 1982; Keough and Downes, 1982; Keough, 1983; Underwood and Denley, 1984; Gaines and Roughgarden, 1985; Bushek, 1988; Fairweather, 1988a; Bertness et al., 1992; Rodriguez et al., 1993). This emphasis has been termed "supply-side" ecology (see Lewin, 1986; Young, 1987; Underwood and Fairweather, 1989). It is therefore essential that demographic models of intertidal populations incorporate settlement and recruitment as important variables (Caffey, 1985; Underwood and Fairweather, 1989).

Where settlement is dense, the predominant features of the community structure have been attributed to factors such as competition, predation, and disturbance (Underwood and Denley, 1984; Gaines and Roughgarden, 1985; Roughgarden et al., 1985; Lewin, 1986; Sutherland and Ortega, 1986; Roughgarden et al., 1991), and have been the basis of a group of models that have attempted to explain the most important ecological interactions and processes that determine the structure and dynamics of different marine assemblages (e.g., Dayton, 1971; Menge, 1976; Menge and Sutherland, 1976; Paine, 1984; Connell, 1985). Settlement and/or recruitment, in addition to the physical, chemical, and biological factors and processes of the water column that affect them, would determine many patterns and would play an important role in communities with sparse settlement (e.g., Paine, 1974; Keough, 1983; Underwood and Denley, 1984; Gaines and Roughgarden, 1985; Roughgarden et al., 1985; 1991; Menge and Sutherland, 1987).

Population dynamics and community structure may also be affected by variations in the intensity of settlement of competing species or of species interacting through a predator-prey relationship (Connell, 1985; see also Fairweather, 1988a). Thus, changes in the larval source or recruitment levels of relevant species can define a scenario of ecological interactions (i.e., vary the intensity and/or the importance of one type of interaction or the other, both spatially and temporarily; see Keough, 1983; Menge and Sutherland, 1987), by determining *a priori* the participant's population sizes (Caffey, 1985; see also Fairweather, 1988a).

4.3.2.2 Distinction Between Settlement and Recruitment

It is important to distinguish between settlement and recruitment. However, in the literature there are conflicting views with respect to definitions of both processes, particularly those referring to settlement (see Keough and Downes, 1982; Coon et al., 1990a,b; Keough, 1986; Bonar et al., 1990). For benthic marine invertebrates with pelagic larvae, the term "settlement" should be restricted to the events involved in the passage from a pelagic way of life to a benthic way of life. This includes the descent of the larva from the water column and the adoption of a permanent residence on the substratum, and the metamorphic changes which permit this to happen.

Rodrigues et al. (1993) define settlement as "a process beginning with the onset of a behavioral search for a suitable substratum and ending with metamorphosis. Two phases may be distinguished in the process: (1) a behavioral phase of searching for a suitable substratum, and (2) a phase of permanent residence or attachment to the substratum, which triggers metamorphosis and in which metamorphic events take place. Recruitment, on the other hand, implies the lapse of some period of time after settlement. The latter is variable, depending on the definition of the particular researcher. Hence, recruits are newly settled individuals that have survived to a specified size after their settlement (see also Keough and Downes, 1982; Connell, 1985; Hurlbut, 1991). Many studies have used the terms "settlement" and "recruitment" interchangeably, or have defined settlement in a way that includes post-settlement mortality.

4.3.3 Settlement

4.3.3.1 Introduction

The planktonic larval stage can be divided into a period of pre-competence in which larval growth and development occurs, and one of competence in which development has been completed (Jablouski and Lutz, 1983). When the latter state is achieved, the larvae have the ability to respond to the appropriate stimuli that lead to settlement (Coon et al., 1990a; Pechenik et al., 1998). The stimuli necessary for settlement involve a combination of biological and physical factors and the presence of chemical cues (Table 4.4, Figure 4.12). They include the speed of fluids (especially close to the sediment surface), the contours of the substratum surface (e.g., Sebens, 1983; Wethey, 1986; Pawlik et al., 1991), and luminosity and chemical cues (e.g., Hadfield and Pennington, 1970; Morse, 1991). Increases in luminosity trigger the deposition of barnacle cyprid larvae while water currents induce their active swimming and attachment to the substratum

TABLE 4.4
Examples of the Main Factors Acting During the Settlement Response of Benthic Marine Invertebrate Larvae

Factors	Examples
Biological	Larval behavior
Physical	Water flow velocity
	Contour and chemistry of the attachment substrate surface
	Luminous intensity
Chemical	
Natural inducers	Associated with conspecific individuals
	Associated with microbial films
	Associated with prey species
Artificial inducers	Neurotransmitters (e.g., GABA, catecholamines)
	Neurotransmitter precursors (e.g., choline)
	Ions (e.g., potassium)

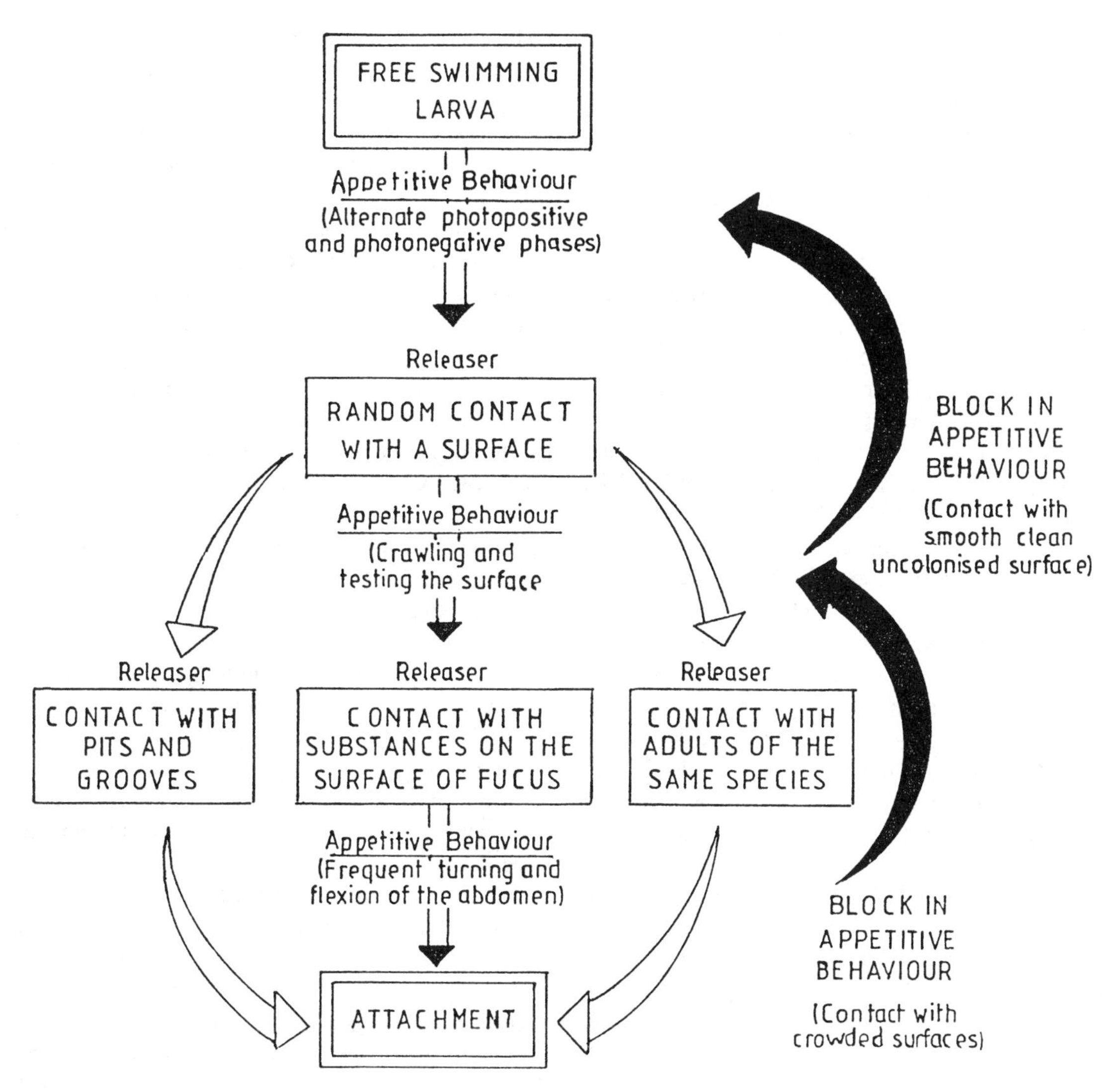

FIGURE 4.12 Diagrammatic analysis of the settlement behavior of the serpulid *Spirorbis borealis*. (Redrawn from Newell, R.C., *The Biologyof Intertidal Animals*, Marine Ecological Surveys, Faversham, Kent, 1979, 210. With permission.)

(Crisp, 1955; Crisp and Ritz, 1973). In the absence of an appropriate stimulus, many larvae delay their settlement (Jensen and Morse, 1990; Highsmith and Emlet, 1986; Coon et al., 1990a).

4.3.3.2 Settlement Inducers

Stimuli that trigger larval settlement are known as settlement inducers and the larvae of different species have different degrees of dependence and specificity with respect to these inducers (Morse, 1990). These chemosensory-type inducers activate the generically scheduled sequence of behavioral, anatomical, and physiological processes that determine settlement (Yool et al., 1986; Morse, 1990). While these chemical inducers are important, physical factors or processes and larval behavior (e.g., geotaxis or phototaxis) are of importance in that they bring the larva close to the substrate at the time of its settlement (Coon et al., 1985; Jackson, 1986). Physical factors such as contouring of the substratum (Crisp and Barnes, 1954; Wethey, 1986) are used by barnacle larvae to select discontinuities in the substratum.

Natural inducers: Many settlement-inducing chemical cues have been identified from studies of larval settlement on different natural substrates (see Rodriguez et al., 1992). The sources of such inducers are of three major types: (1) conspecific individuals (Highsmith, 1982: Burke, 1984; Pawlik, 1986; Jensen and Morse, 1990); (2) microbial films (Kirchman et al., 1982; Morse et al., 1984; Bonar et al., 1990; Pearse and Scheibling, 1991; Unabia and Hadfield, 1999); and (3) prey species (Morse et al., 1979; Morse and Morse, 1984; Todd, 1985; Barlow, 1990; Hadfield and Pennington, 1970).

Inducers associated with conspecifics: Settlement induced by the presence of conspecific adults has been described in a wide range of benthic invertebrates including polychaetes (Jensen and Morse, 1984; Pawlik, 1986), sipunculids (Hadfield, 1986), echinoderms (Highsmith, 1982), molluscs (Seki and Kan-no, 1981), barnacles (Knight-Jones, 1953; Raimondi, 1988b; 1991), and oysters (Hidu et al., 1978). Induction of settlement by conspecifics to a large extent accounts for the aggregated distribution of many benthic invertebrates, especially sessile species such as barnacles.

Settlement of abalone larvae is stimulated by the mucus produced and secreted by adult and juvenile individuals (Seki and Kan-no, 1981). The best example of a conspecific settlement inducer is that of the sabellid polychaete, *Phragmatopoma lapidosa californica*. Larvae settle as a response to free fatty acids (e.g., palmitoleic and linoleic acids) found in the sand and cement matrix of the tubes of adults (Pawlik, 1986). The response is highly species specific and restricted to the genus *Phragmatopoma*. Highsmith (1982) demonstrated that a small peptide of low molecular mass (<10,000), produced by the adults of an irregular echinoid, *Dendraster excentricus*, and sequestered by some compound in the sand in which they are buried, was the settlement-inducing chemical cue for this species. This induces aggregated settlement in areas where the major predators of this species are absent (Highsmith, 1982). Barnacle settlement induced by the presence of conspecifics has previously been discussed in Section 2.8.3.3.

Inducers associated with microbial films: Microbial films are important for the settlement of many marine invertebrate pelagic larvae. A surface submerged in the sea passes through an orderly progression of biofilm formation, beginning with deposition of organic material, then colonization by bacteria, diatoms, and other eucaryote organisms. Settlement is induced by films of diatoms and cyanobacteria (e.g., Morse et al., 1984), and by bacterial films (e.g., Kirchman et al., 1982 for polychaetes; Cameron and Hinegarden, 1974 for sea urchins; Maki et al., 1988 for barnacles; Maki et al., 1989 for bryozoans; and Bonar et al., 1990 for oysters). In the latter case, the response is quite specific and is generated by the presence of extracellular polysaccharides or glycoproteins attached to the bacterial wall (Kirchman et al., 1982; Hadfield, 1986), or soluble compounds released from these films (Bonar et al., 1990). The former induces metamorphosis of oyster larvae of the genus *Crassostrea* (Fitt et al., 1990; Coon et al., 1990b).

Inducers associated with prey species: In some larvae, settlement is induced by potential prey species of juveniles or adults. Many herbivore species are induced to settle by the crustose algae upon which they feed. These include the abalone (Morse and Morse, 1984; Morse, 1990), limpets (Steneck, 1982), and sea urchins (Cameron and Hinegarden, 1974; Rowley, 1989; Pearse and Scheibling, 1990; 1991). Inducing chemicals would possibly be released when the algal epithelial cells are grazed (see Morse and Morse, 1984). Crustose algae are also substrates for the settlement of coral larvae (Sebens, 1983).

Foliose algae also have inducing properties. Morse (1991) described the settlement process of the larvae of the red abalone. After a week of swimming near the surface, the ciliated larva arrests its development and starts swimming near the bottom, in a bouncing trajectory touching the bottom, then moving back up into the water column. In order to break the developmental arrest, the larva must contact an exogenous trigger, a small peptide that is found uniquely on the surface of red algae (including species of *Laurencia, Gigartina,* and *Porphyra*). Even a single contact with this chemical causes the larva to stop swimming. The releaser then triggers a cascade of chemical reactions in the target sensory cells that stimulate the central nervous system, turning on behavioral and cellular processes previously arrested. Within 29 hours, metamorphosis will be completed, beginning an irreversible commitment to a benthic life. Larvae of the chiton, *Ischnochiton heathiana*, metamorphose on foliose algae of the genera *Ulva*. Some carnivorous species, particularly nudibranchs, also settle in response to their prey. For example, *Onchidoris bilamellata* only metamorphoses in the presence of its prey, newly settled barnacles (Todd, 1985).

The above discussion has emphasized the role of positive cues in determining larval settlement. Woodin (1991) has drawn attention to the role of negative cues in determining settlement, causing the larvae to reject some substrates. Thus habitat selection may be more a matter of the absence of strong negative, rather than the presence of a strong positive cue. In addition emigration, either by crawling or by behavior that encourages advection, is a mechanism by which metamorphosed or still metamorphosing juveniles can escape unsuitable habitats.

4.3.3.3 Settlement on Rock Surfaces and Algae

The factors involved in the settlement behavior of barnacles have already been discussed in Section 2.8.3.3, and they will not be considered further here. In addition to the extensive studies on barnacles, the settlement of the larvae of intertidal animals on solid substrates has been investigated in detail in serpulid tubeworms, bryozoans, hydroids, and the colonial polychaete, *Sabellaria alveolata*. As we have seen, there is a hierarchy of factors involved in the settlement process, ranging from generalized responses to light to some more specific orientation to the settlement substratum, which serve to bring the larvae into the vicinity of appropriate settlement sites. Temporary attachment followed by settlement may then be made in response to a variety of physical and chemical properties of the substratum, as well as to some more precise stimuli such as the presence of adults of the same species. This hierarchy, with a gradation of physical and chemical features that act as "releasers" for behavioral

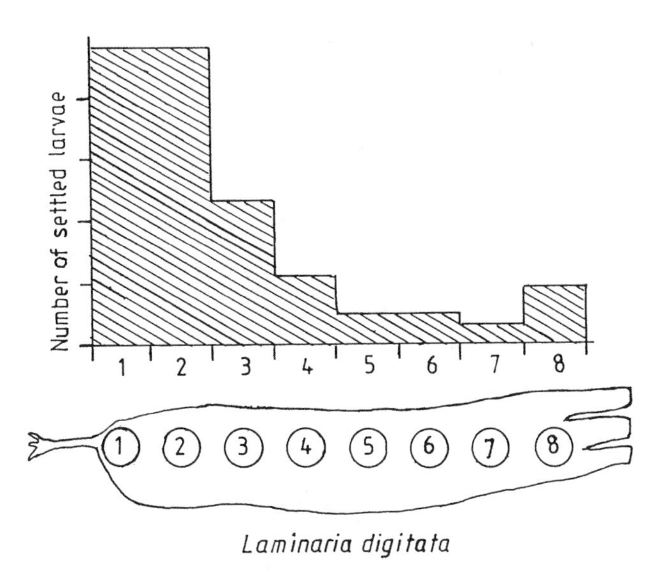

FIGURE 4.13 Settlement of the serpulid *Spirorbis* (two species represented in these data) on discs cut from a frond of the brown alga, *Laminaria digitata*. The youngest part of the frond is that part closest to the stalk. (Modified from Stebbing, A.R.D., *J. Mar. Biol. Assoc. U.K.*, 1962, 52, 765. With permission.)

responses that culminate in settlement (Williams, 1965). The sequence of events together with the relevant releasers in the settlement behavior of the tubeworm, *Spirorbis borealis*, is shown in Figure 4.12. Here we will discuss the settlement behavior of spirorbids in some detail with brief mention of that in some other species.

The settlement responses of tubeworms of the genus *Spirorbis* has remarkable similarities to that shown by barnacles with both physical and chemical properties of the substratum, as well as the presence of adults of the same species being involved in the settlement. Spirorbid tubeworms settle on rock surfaces, the shells of molluscs (e.g., *Mytilus edulis*), and macroalgae, especially laminarians and fucoids. Spirorbid settlement has been extensively studied by Knight-Jones, (1951; 1954), Wisely (1960), Stebbing (1962), and Williams (1965). *Spirorbis* has a short larval life of 6 to 12 hours. Initially it is photopositive, then progressively photonegative, during which it seeks a place to settle. Knight-Jones (1951) showed that on hard substrates, a bacterial film was necessary to induce settlement, while Williams (1965) showed that the larvae of *S. borealis* are attracted by extracts of the brown alga, *Fucus serratus*, on which it prefers to settle. Goss and Knight-Jones (1957) reported that the larvae of *S. borealis* readily attached to *Fucus*, but that few settled on *Laminaria, Himanthalia, Ascophyllum, Rhodomenia*, or stones. De Silva (1962) subsequently carried out an investigation of the substrate preferences of the larvae of *S. borealis, S. corallinae*, and *S. tridentatus*. *S. borealis*, as we have seen, occurs primarily on *Fucus*, *S. corrallinae* on the coralline alga *Corallina officinalis*, while *S. tridentatus* is found on rock and stones. De Silva showed that the occurrence of the three species on their typical substratum was due to the larvae selecting the substratum upon which they normally grow. Some species of *Spirorbis* that settle on the broad fronds of the brown alga, *Laminaria*, preferentially settle on the youngest part of the fronds (Stebbing, 1962). Stebbing cut discs from the length of a blade of *Laminaria*, arranged them evenly in a circular vessel containing *Spirorbis* larvae, and noted that the number of larvae settling was greatest on discs cut from the growing end of the frond (1 to 3 in Figure 4.13). The adaptive value of this is that the youngest part of the frond has the least dense growth of attached organisms, such as algae and sessile invertebrates; hence, the competition for space is less. In addition, since the fronds wear from the free end, the youngest part will provide the longest period of a stable homesite. Similar site selection behavior has been shown to occur in bryozoans and hydroid.

4.3.3.4 Avoidance of Crowding

As we have seen, gregariousness is of widespread occurrence in intertidal organisms. Such behavior, however, can lead to intense intraspecific competition for space and food. Thus, while the presence of adult conspecifics is a good indicator of the suitability of a site for settlement, it would be expected that mechanisms to prevent overcrowding would have evolved. This aspect of settlement has been studied in *Spirorbis borealis* by Wisely (1960) and Knight-Jones and Moyse (1961), and in barnacle cyprids by Knight-Jones and Moyse (1961).

Wisely showed that the distribution of *Spirorbis borealis* on the fronds of *Fucus serratus* was remarkably even and that settlement did not occur closer than 0.5 mm. In natural populations, the distance between the centers of adults was mainly about 1 mm, even under crowded conditions. This settlement behavior ensured that there was

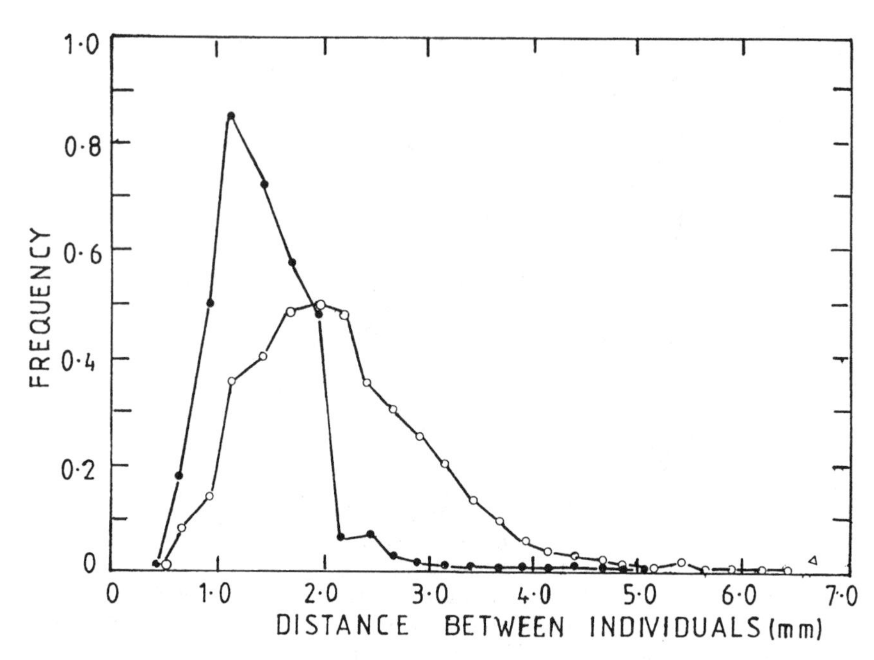

FIGURE 4.14 Graphs showing the frequencies of occurrence in relation to the distance apart of adjacent *Spirorbis borealis* arranged in a linear series of grooves on each side of the midrib of fronds of *Fucus serratus*. The open circles are for a sparse population and the solid circles for a crowded population. (Redrawn from Newell, R.C., *The Biology of Intertidal Animals*, 3rd ed., Marine Ecological Surveys Ltd., Faversham, Kent, 781 pp, 1979. With permission.)

room for the settled individuals to grow to their full size of some 2 mm diameter (Figure 4.14). Crisp (1961) has shown that a similar behavioral pattern occurs in the cyprids of *Balanoides balanoides*. However, extreme high densities of barnacles may occur on the shore. Crisp found that territorial separation becomes reduced as the population density increases. In *Elminius modestus* at low population densities of 3.5 cm^{-2}, the territorial separation was 1.9 mm; at 12.5 cm^{-2} it was 1.42 mm; at 15 cm^{-2} it was 1.25 mm; at 85 cm^{-2} it was only 0.86 mm. This behavior appears to be the result of physical contact between the cyprids and neighboring settled barnacles, the larger cyprids of *Balanus balanoides* and *B. crenatus* having a greater territorial separation than the smaller *Elminius modestus*.

The tendency of barnacle cyprids to space out from older individuals of their own species, but not to other species, may account for the observed sharp boundaries between barnacle species (Newell, 1979). The tendency to settle and space out in zones where there are adults of the same species, and to settle on alien survivors in that zone tends not only to make the barnacle zones distinct, but to make interspecific competition between overlapping species of barnacles more severe than intraspecific competition.

4.3.3.5 Settlement on Particulate Substrates

A number of studies have shown that the larvae of many soft bottom invertebrates settle preferentially on the type of substrate in which the adults occur. Such substrates are usually characterized by particular grain size composition and organic content. It has also been found that some species can delay metamorphosis until a suitable substratum is found. For example, Day and Wilson (1934) and Wilson (1937) found that the larvae of the polychaete, *Scolecolepis fuliginosa*, could postpone metamorphosis for as much as several weeks until a suitable substratum was found. This was followed by the classic work of Wilson (1948; 1952) on the factors influencing metamorphosis in the polychaete, *Ophelia bicornis*.

Wilson (1948) found that *Ophelis bicornis* larvae metamorphosed most readily in response to physical contact with natural sand from the particular site where adults lived. This sediment had a characteristic modal grain size and consisted of well-rounded smooth grains with little organic detritus. Sands with smaller natural grain sizes and angular grains were less favorable to settlement. In subsequent work, Wilson (1955) showed that one of the attractive features of the sand was the presence of a film of microorganisms on the sand grains, and that acid-cleared sand lost its attractiveness. Similar studies have shown that other benthic species such as polychaetes *Mellina cristata* and *Pygiospio elegans*, the phoronid *Phoronis mulleri*, and the gastropod *Nassarius obsoletus* show similar behaviors.

Gray (1966; 1967a,b) studied the responses of the archianellid, *Protodrilus symbioticus*, to sand grains, the surface of which had been modified by various experimental procedures. Treatment with acids, drying, and autoclaving reduced their attractiveness to *Protodrilus*. Figure 4.15 shows how an increase in the bacterial num-

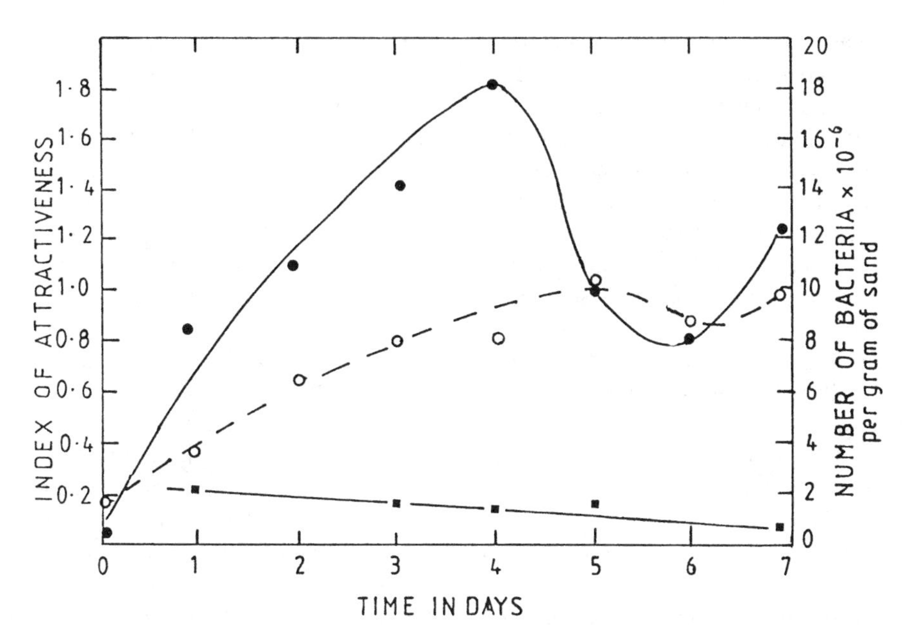

FIGURE 4.15 Graphs showing the increase in bacterial numbers (solid circles) and index of attractiveness for *Protodrilus symbioticus* (open circles) of autoclaved sand inoculated with natural sand bacteria. Triangles denote the attractiveness of sterile controls. Based on the mean of two experiments. (Redrawn from Newell, R.C.,*The Biology of Intertidal Animals,* Marine Ecological Surveys, Faversham, Kent, 1979, 222. Data from Gray, J.S., 1966. With permission.)

bers in autoclaved sand that had been inoculated with sand bacteria increased their attractiveness to *Protodrilus*. It is clear that the bacteria on the sand grains was the source of attractiveness. Furthermore, Gray (1966) showed that the degree of attractiveness was not due so much to the numbers as to the kinds of bacteria. The presence of particular species of bacteria, rather than bacterial numbers, has been shown to be of importance in the distribution of the sand-dwelling harpacticoid copepod, *Leptastacus constrictus* (Gray, 1968), as well as in habitat selection by the interstitial gastrotrich, *Turbanella hyalina* (Gray and Johnson, 1970).

4.3.3.6 Variation in Settlement

The initial settlement of planktonic propagules of marine benthic organisms usually varies considerably both in space and time (e.g., Connell, 1961a; Kendell et al., 1982; Caffey, 1985; Wethey, 1985). Density of newly settled individuals may differ between sites for three general reasons: (1) competent planktonic larvae or spores are brought into the immediate vicinity of the substratum at some sites in greater quantities than at other sites; (2) characteristics of the water in the immediate vicinity of some sites allowed a higher proportion of the propagules to attach than at others; and (3) the substratum was more attractive (more favorable surface microtopography, greater concentrations of settlement inducers) to the settling propagules at some sites.

Gaines et al. (1985) sampled the plankton at a range of sites in close vicinity to the natural rock substrate on which the recruitment of the barnacle, *Balanus glandula*, was measured weekly or biweekly. Variation in planktonic cyprids explained >80% of the variation in its weekly recruitment on the shore. They concluded that variation in settlement density of *B. glandula* among different habitats was more likely to be a function of planktonic larval supply than of the characteristics of either the local water column or substratum.

4.3.4 Recruitment

4.3.4.1 Introduction

Considerable evidence suggests that a large variation in recruitment is common in marine species with planktonic larvae (Caffey, 1985; Connell, 1985; Hunt and Scheibling, 1997). Yet it is unclear what proportion of this variation is attributable to larval dynamics. As many investigators have noted (e.g., Keough and Downes, 1982; Underwood and Denley, 1984; Connell, 1985), larval supply and early settler mortality are often confounded since recruitment is generally measured by censuring juveniles long after they have settled out of the water column. Thus, recruitment combines settlement with early mortality that has occurred on the substratum up to the time of the first census.

4.3.4.2 Components of Recruitment

For benthic organisms with planktonic larvae, recruitment has three components: (1) water column larval supply; (2) the settlement patterns of competent larvae: and (3) the survivorship of the settlers to the time of the initial census.

These three components have received unequal attention. Empirical studies of the connection between larval supply and recruitment are rare (Roughgarden et al., 1988). In contrast, the dynamics of larval settlement preferences have received a great deal of attention (Meadows and Campbell, 1972; Crisp, 1974; Raimondi, 1988a,b; Jensen and Morse, 1990). Active habitat selection by larvae and passive settlement mechanisms driven by fluid dynamics have been shown to have major impacts on settlement patterns (Eckman, 1985).

Factors determining the density of settlers and recruitment: The density of settled individuals may differ between sites for three general reasons: (1) competent planktonic larvae or spores were brought into the immediate vicinity of the substratum at a higher rate at some sites than others; (2) characteristics of the water in the immediate vicinity of the sites allowed a higher proportion of the propagules to attach at some sites; and (3) the substratum was more attractive (or less repellent or hazardous) to these settling propagules at some sites than others (Connell, 1985).

Connell (1985) has analyzed the available data on the range of variation in the densities of recruits and found that in general at a single site they varied by an order of magnitude. However, among different sites and species, the range of variation was much higher, spanning two or three orders of magnitude. For barnacles, they range from 0.9 to 139 [c] for *Semibalanus balanoides* at Millport, Scotland (Pyfinch, 1948), 0 to 9.3 [c] for *Chthalamus montagui* at Millport (Barnes & Powell, 1950), 1 to 25 [c] for high shore *Balanus glandula* at Friday Harbor, Washington, and 0 to 33 [c] for low shore *Semibalanus cariosus* at Friday Harbor (Connell, 1985). For recruitment of *Tesseropora rosea* at seven stations along the New South Wales coast, the ranges were 40 to 330 per 400 m^2 in 1980, 28 to 165 in 1981, and 3 to 274 in 1982 (Caffey, 1985). Figure 4.16 depicts the variation in the recruitment of the first-year class of the clam, *Tivela stultorum*, at three sand beaches at Pismo Beach, California (Fitch, 1965). It can be seen that there was considerable variation from year to year at all three beaches and that years of high recruitment (e.g., 1931, 1935, 1937, and 1948) coincided at all three sites, although the number of recruits differed at the three sites.

Factors affecting the density of juveniles and adults: Connell (1985) examined the relationship between the density of larval settlement, the density of recruits, mortality up to the end of the period of settlement, the mortality between settlement and the age of 1 year, and the density of adults for *Semibalanus balanoides* in Scotland England and Massachusetts, and for *Tesseropora rosea* in Australia. He concluded that mortality during the first few weeks after settlement acted independently of the initial density of settlers; thus, the density of recruits was a direct reflection of that of the settlers. However, the

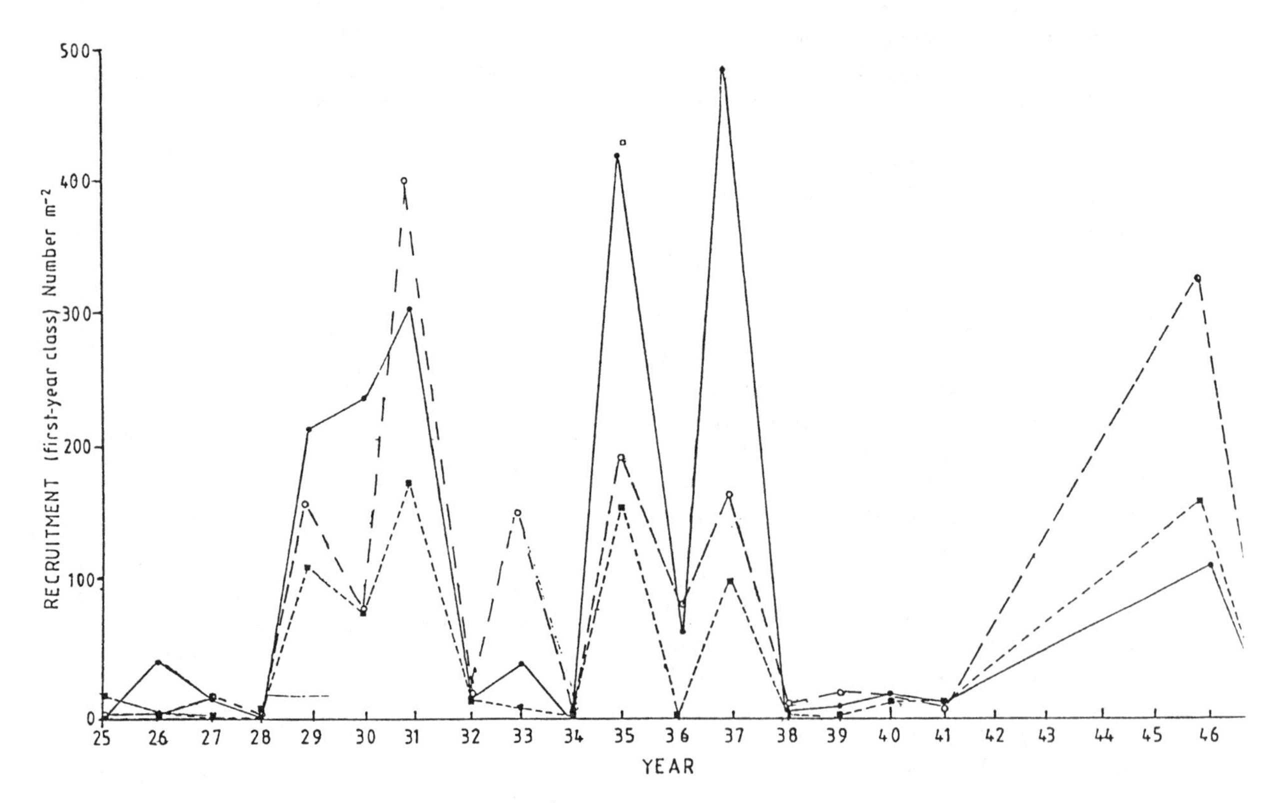

FIGURE 4.16 Variation in recruitment of the first year classes of the clam, *Trivela stultorum*, on three sand beaches at Pismo Beach, California; all years are included in which some clams in the first year were found in at least one site per standard sample. (Based on data from Fitch, T.E., *Proc. Malacol. Soc. Lond.*, 86, 1965. With permission.)

TABLE 4.5
Measures of Persistence (Column 2) and Precision (Columns 3 and 4) of the Free-Running Locomotor Rhythms of Two Burrowing Supralittoral Amphipods (*Talitrus saltator* and *Talorchestia deshaysi*) and Two Surface-Living Forms, the Amphipod *Orchestia gammarella* and the Isopod *Ligia oceanica*, which are Cryptozoic but Nonburrowing

Species	Persistence of Distinct Circadian Rhythm (Days) in DD	Period Length of Circadian Rhythm		Circadian Periodgram Statistic (Arbitrary Units)
		hrs	SD	
Talitrus saltator	35	24.56	0.43	53
Talorchestia deshaysi	20	24.64	0.54	32
Orchestia gammarella	12	24.51	1.0	22
Ligia oceanica	12	24.42	0.88	20

Source: From Naylor, E., in *NATO Advanced Workshop on Behavioral Adaptation to Intertidal Life*, Chelazzi, G. and Vannini, M., Eds., Plenum Press, New York, 1988, 5.

relationship between the densities of settlers and adults was not clear-cut. Where recruitment was sparse, it was highly correlated with the abundance of those adults that survived. In contrast, where recruitment was dense it showed no relationship to subsequent adult numbers.

4.4 THE MAINTENANCE OF ZONATION PATTERNS

4.4.1 Introduction

We have seen that the vertical zones occupied by the plants and animals on the shore are determined by a combination of factors involving the following:

1. The ability of the larvae of animals and the settling stages of algae to select an environment that is suitable for adult life
2. The presence of a suitable substratum or sediments of a suitable grain size in the particular zone occupied
3. The interaction of the plants and animals with the various environmental factors (Table 4.5)
4. The availability of the right energy source (food)
5. Interactions with other organisms, especially predator-prey relationships and competition for space and food

Some species of plants and animals settle over a much wider vertical range than suitable for adult life with the eventual elimination of those outside the range of tolerance. Others, such as barnacles and serpulids, due to their gregarious settling behavior, settle over a vertical range that more closely approximates the adult range. For these sessile species, the choice is irreversible. Yet others, such as mussels and limpets, settle low on the shore, or in the shallow sublittoral among growths of algae, especially corallines. For the British populations of the mussel, *Mytilus edulis*, it has been found that initial settlement takes place in shallow water, where after a period of growth they detach and are carried inshore by waves and currents. Periodic flotation and reattachment eventually establishes the adults within their typical range.

Most of the shore gastropods, even littorinids which are found highest on the shore, settle at lower levels and migrate up the shore as they grow. In general, the size range increases upshore until optimum levels are reached, above which they decrease in size due to unfavorable environmental conditions. These migrations up the shore can be considered to be an extension of the settlement phase. Once the zone of adult distribution is reached, motile species may be faced with the problem of maintaining themselves in the optimal zone in the face of periodic unfavorable environmental conditions such as strong wave action which may dislodge them and transport them to a less optimal level. Investigations have shown that in such species there has evolved a series of behavioral responses to the environmental climate that helps to maintain them within their optimum vertical range. In the succeeding sections we shall examine some examples of the ways in which this zone-maintaining behavior operates.

4.4.2 Elements of Behavior in Littoral Marine Invertebrates

The behavior of marine invertebrates is divisible into two distinct phases of the life cycle: larval behavior and the subsequent behavior of the adults (Branch and Barkai, 1988) (Figure 4.17). Both types of behavior are constrained by both intrinsic factors (e.g., morphology and

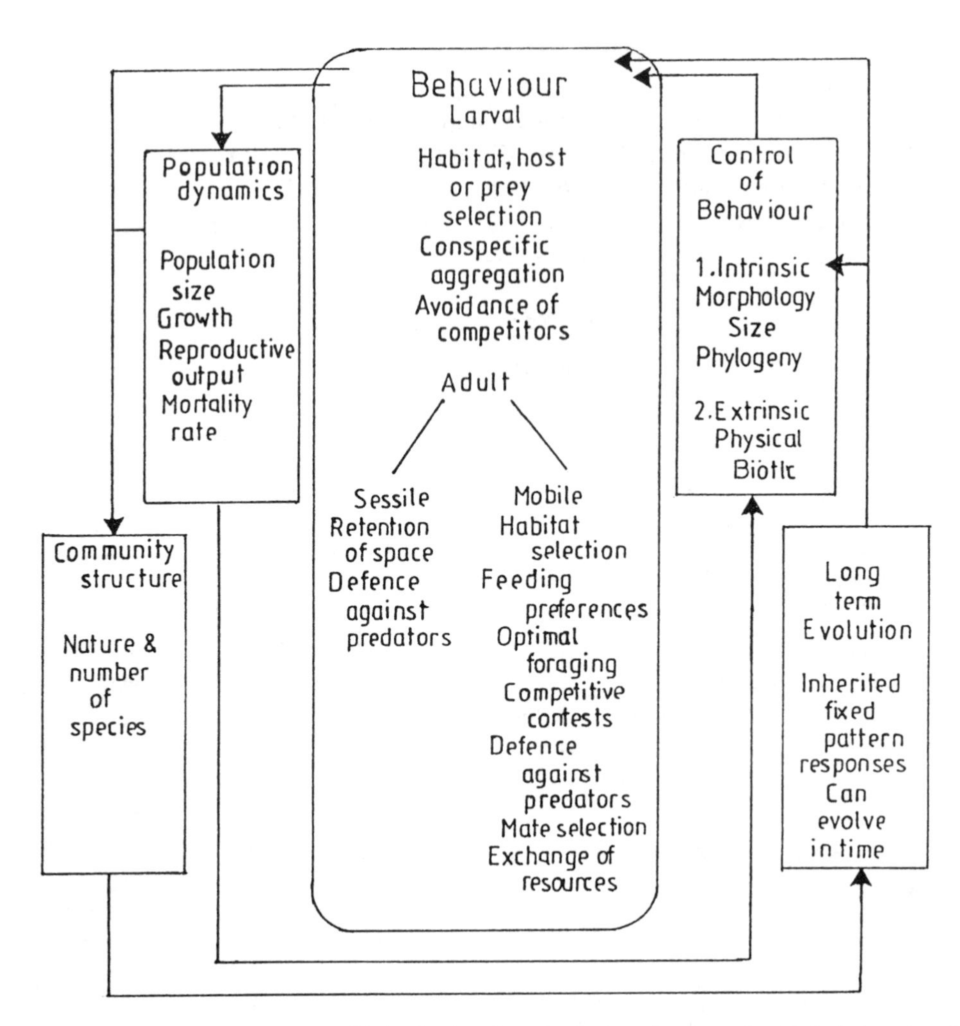

FIGURE 4.17 Flow diagram depicting elements of behavior in a hypothetical marine animal with both larval and adult phases in the life cycle. Although not exhaustive, the examples of behavior are intended to indicate that the adults of sessile species are likely to have a more restricted behavioral repertoire than those of mobile species. Both evolved characteristics and extrinsic physical and biotic factors will influence behavior. Behavior, in turn, affects population dynamics, both of the species concerned and of other species it interacts with, thus potentially modifying community structure. Both population dynamics and community structure may have an input into the factors affecting short-term behavioral responses and the long-term evolution of behavior. (Redrawn from Branch, G.M. and Barki, A., in *NATO Advanced Workshop on Behavioral Adaptations to Intertidal Life,* Chelazzi, G. and Vannini, M., Eds., Plenum Press, New York and London, 1988, 229. With permission.)

phytogenetic history) and extrinsic factors (e.g., the physical environment and other organisms).

Larval behavior. Excellent reviews of such behavior have been provided by Scheltema (1974), Pechenik (1987), Morgan (1995), and Pechinek et al. (1998). Larval settlement behavior was discussed in the account of barnacle settling behavior in Section 2.8.3 and in Section 4.3.3 in this chapter. There it was noted that the settling larvae of many species react to specific chemicals secreted by conspecifics or by other species. Larvae of species that are epibiotic on algae are highly specific in their selection of their host plant, postponing settlement and metamorphosis until the correct algal species is found (Williams, 1965; Nishihira, 1967). In a number of instances, the larvae of slow-moving opisthobranch predatory molluscs have been shown to settle selectively on species that later form their prey (Hadfield, 1986).

Recent investigations have shown that settling larvae respond to the presence of other competitors (Branch and Barkai, 1988). Grosberg (1981) demonstrated that the larvae of many species avoided settling on plates on which the space-dominating colonial ascidian *Botryllus schlosseri* was present. The species that showed such avoidance behavior were all species susceptible to overgrowth by *B. schlosseri*. An almost equal number of other species, all of which were not subject to overgrowth by *B. schlosseri,* settled randomly on plates irrespective of the density of *B. schlosseri*. Bernstein and Jung (1979) have shown that the behavior patterns of larvae that occur on the kelp, *Macrocystis pyrifera,* as adults have evolved behaviors that minimize the threat of overgrowth by the bryozoan, *Membranipora menbranacea.* In addition, Buss (1986) has shown that the larvae of the arborescent bryozoan, *Bugula turrita*, settle preferentially among clumps of its own species (although it avoids large groups). While there is a cost in this behavior in that growth declines with density, the advantage lies in that while large clumps of *B. turrita* are seldom overgrown by the encrusting bryozoan, *Schizoporella errata*, solitary colonies are regularly destroyed in this manner.

Adult behavior. From Table 4.4 it can be seen that to a certain extent innate characteristics, including morphology, size, and inherited characteristics, constrain not only the behavior of a species, but the future evolutionary pathway it is likely to follow. It will be noted that the adults of sessile species have a more restricted behavioral repertoire than those of mobile species.

Behavior is also strongly influenced by extrinsic factors, including the physical environment and interactions with other species. In particular, the tidal cycles with alternate exposure and submergence subject intertidal organisms to stresses in the form of desiccation, extremes of temperature, changes in salinity, and periodic wave action. A wide range of behavioral responses have evolved in response to these stresses. Behavior is further shaped by interaction with other species, such as escape responses by prey, feeding behavior of predators and herbivores, including optimal foraging, and interference competition. Some aspects of such behavior will be discussed below and will be considered further in Chapter 6.

4.4.3 Behavior Patterns in Representative Species

4.4.3.1 Movement Patterns and Orientation Mechanisms in Intertidal Chitons and Gastropods

Chelazzi et al. (1980) have recently reviewed behavior patterns in intertidal chitons and gastropods. These two groups of animals share a large number of adaptations to intertidal life, including morpho-functional behavioral traits, communication, clustering, and aggressiveness. Partial reviews of their behavior may be found in Boyle (1977), Newell (1979), Underwood (1979), Branch (1981), and Hawkins and Hartnoll (1983a). In considering their activity patterns it is important to distinguish between occasional, continuous, and rhythmic phenomena. Occasional activity includes movements in response to such unpredictable ecological events as those caused by storms, the sudden encounter with predators, or the ability to regain their optimal zone after displacement. Another class of occasional behaviors includes the avoidance responses when in contact with, or the approach of, potential predators such as seastars (Branch, 1979), or other gastropods (Hoffman et al., 1978).

Rhythmic activity includes movements related to seasonal, synodic, tidal, or diel fluctuations in the shore environment. Seasonal and synodic movements usually consist of migrations up and down the shore in order to minimize the exposure to stress factors and to optimize the access to resources, particularly food.

The short-term activity of intertidal chitons and gastropods is organized into behaviors determined by tidal and diel variations in physical and biological factors in the shore environment. In general, intertidal animals may adopt one of two alternate strategies (Chelazzi and Vannini, 1985): (1) the "isophasic" pattern, consisting of rhythmic zonal migrations, synchronous and coincident with the tides; or (2) the "isospatial" strategy in which they remain within a meter or less narrow belt in the intertidal zone with a specific amount of alternate submersion and emersion. The activity of the animals is limited to the time when conditions in the zone they occupy are suitable for moving and feeding. Most intertidal chitons and gastropods show combined diel-tidal rhythms of high complexity.

Feeding excursions of rocky shore chitons and gastropods are generally classified into three distinct modes of increasing complexity: ranging patterns, zonal shuttling, and central place foraging. In the *ranging pattern,* the feeding excursions are not orientated toward a constant direction and the animals do not return to the same place or shelter, or to the same shore level. This behavior has been reported by Underwood (1977) for some Australian gastropods and by McQuaid (1981) and Petraitis (1982) for littorinids. *Zonal shuttling*, sometimes referred to as "tidal migration," has been described in neritids, trochids, planaxids, and some limpets. Feeding excursions are usually made upshore during low tide, with a downshore movement before high tide (Chelazzi, 1982). *Central place foraging* involves the establishment of a definite "home" site on the shore and returning to it after feeding excursions. This behavior has been described in limpets and chitons. Such "homing" behavior is discussed in detail in Section 5.2.2.1.

4.4.3.2 Interaction Between the Siphonarian Limpet *Siphonaria Thersites* and Its Host Plant *Iridaea Corriucopiae*

Siphonaria thersites occurs high on the shore on the northern Pacific coast of North America, ranging from 1.0 to 2.3 m above low water during spring tides (when the tidal amplitude is about 3.8 m). Its zonal distribution is closely associated with that of its host plant *Iridaea corriucopiae* (Branch, 1988) (Figure 4.18). *S. thersites* is active only at low tide. Activity commences shortly after the limpets are fully exposed on the ebb tide. During emersion they

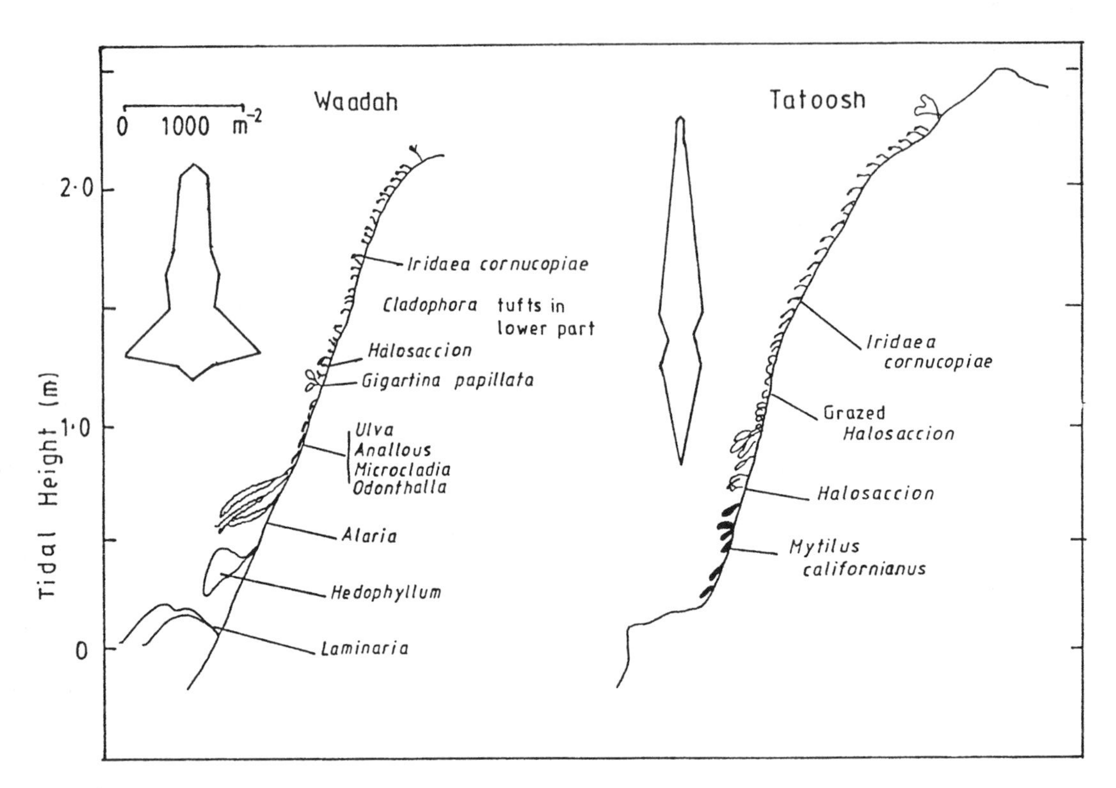

FIGURE 4.18 The density and zonation of *Siphonaria thersites* at Waadah and Tatoosh Islands, in relation to the vertical distribution of dominant bed-forming algae. Tidal height is expressed as height above low water. (Redrawn from Branch, G.M., in *NATO Advanced Workshop on Behavioral Adaptations to Intertidal Life,* Chelazzi, G. and Vannani, M., Eds., Plenum Press, New York and London, 1988, 30. With permission.)

actively feed on the tips of the algal blades. Activity ceases well in advance of immersion when the limpets return to their particular shelters (holes, crevices, and mussel beds), returning faithfully to the same shelter after each period of activity. Activity is inversely correlated with the height of the tide (Figure 4.19).

Figure 4.20 summarizes the interactions of *S. thersites* with its food plant. Its behavior stems from its characteristics, namely: (1) low tenacity, high shore distribution, and activity greatest at the lowest tides, which makes it vulnerable to wave action; (2) the characteristics listed in (1) above make it vulnerable to desiccation effects; (3) its flexible body, large foot, and small shell, while making it vulnerable to desiccation, allows free movement on its food plant; and (4) while it may be exposed to predators (e.g., birds), its tissues contain propionates that make it unpalatable and consequently it is not eaten by most predators. In many respects the behavior and morphology of *S. thersites* closely parallels that of the opisthobranch molluscs of the genera *Onchidium* and *Onchidella*. Species of both these genera occur at similar levels on the shore. Lacking shells, they are vulnerable to desiccation, and also home to holes and crevices after bouts of feeding (McFarlane, 1980). Their activity is also rhythmic and occurs only at low tide (Pepe and Pepe, 1985). Like *S. thersites,* both genera possess repugnatorial glands and are known to repel a number of predators, including starfish, crabs, and fish (Ireland et al., 1984). One species, *Onchidella binneyi,* which is weakly attached and easily dislodged by waves, becomes active shortly after being exposed by the ebbing tide; it also ceases activity well in advance of the time of the rising tide. Its activity rhythms are endogenous and circalunar (Pepe and Pepe, 1985), and the intensity of its activity is inversely related to the height of the tide in a manner similar to that of *S. thersites*. While *O. binneyi* is active only at night, four other species of the genera *Onchidella* and *Onchidium* have been described as being diurnally active.

4.4.3.3 Maintenance of Shore-Level Size Gradients

Many mobile intertidal invertebrates exhibit size gradients with respect to tidal height. Two general patterns are evident: high shore species tend to increase in size with increasing tidal height, while lower shore species show the opposite pattern of increasing size with decreasing tidal height (Vermeij, 1972, 1973). Size gradients may arise from differential growth or mortality at various tidal heights, and also from active segregation through differential migration of size classes in the intertidal zone (Vermeij, 1972; Bertness, 1977). Such active migration occurs in a number of gastropods (Bertness, 1977; McQuaid, 1981; Doering and Phillips, 1983). Static gradients, aris-

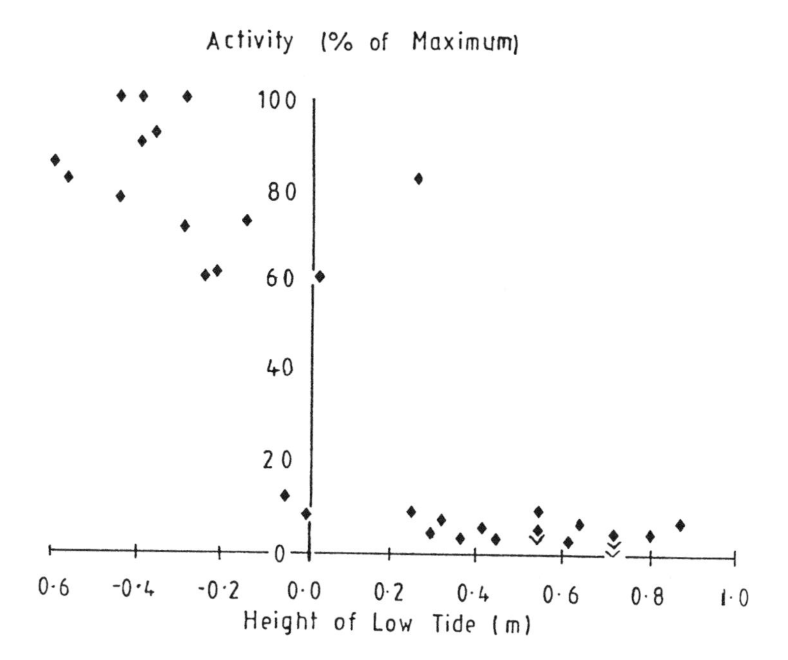

FIGURE 4.19 Activity of *Siphonaria thersites* (shown as the percentage of maximum activity at each site) in relation to the height of the low low tide (m above chart datum). (Redrawn from Branch, G.M., in *NATO Advanced Workshop on Behavioral Adaptations to Intertidal Life,* Chelazzi, G. and Vannini, M., Eds., Plenum Press, New York and London, 1988, 41. With permission.)

ing from differential growth, have been shown to occur in limpets (Sutherland, 1970), mussels (Seed, 1968), and probably barnacles (Bertness, 1977).

The shell size of the herbivorous gastropod, *Tegula funebralis,* on the Californian coast increases in a downshore direction (Paine, 1969; Markowitz, 1980) throughout its intertidal range of +1.8 m to 0.0 m. Paine (1969) has shown that small *Tegula* (1.6 to 1.7 cm shell width) grow better higher up than lower down in the intertidal zone. With increasing size, however, it is increasingly advantageous (in terms of growth and reproduction) to exist lower on the shore even though the snail comes into contact with the predatory seastar, *Pisaster ochraceus.* The density of snails in the lower regions is kept at moderate levels through losses by *Pisaster,* and reduced density apparently increases food availability to surviving individuals. According to Paine (1969), the size gradient results from the settlement of juveniles in the higher intertidal, followed by migration of the larger individuals to lower regions over a period of years.

Doering and Phillips (1983) experimentally investigated the behavior involved in the establishment of the size gradient in *Tegula funebralis*, both in the field and the laboratory. Snails transferred to the zones where they do not usually occur migrated back toward their original zone, thus reestablishing the size gradient and implying differential movement among the size classes. Both large (>2.1 cm shell width) and small (<1.77 cm) snails were photonegative on a horizontal surface and geonegative in the laboratory; there were no statistical differences between the size classes. Light, however, inhibited upward, or caused downward movement of large snails on vertical surfaces. Small snails were unaffected, ranging higher on illuminated vertical surfaces than large snails. Both sizes exhibited similar distributions in the dark. In an experimental chamber providing both emersed and immersed surfaces, *T. funebralis* established vertical size gradients when the chamber was illuminated from above. Doering and Phillips suggested that light is an important factor in the formation and maintenance of *Tegula*'s shore-level size gradient.

In response to water-borne chemicals derived from the seastar, *Pacesetter ochraceus,* large snails moved up vertical surfaces in greater proportion than small ones. In response to contact with the predator, large snails moved away faster than small ones, and individuals collected from crevices in the field moved away slower than those collected from the open rock faces. Although predation may select for a size gradient in *Tegula funebralis,* it is unlikely that responses to predatory seastars directly and proximally cause or maintain them over the short term.

In elucidating the factors involved in the determination of size gradients in gastropods, other factors in addition to those discussed above, such as food and reaction to other physical factors such as emergence time and exposure to high temperatures, may contribute.

4.4.3.4 Behavior Patterns in Sandy Beach Invertebrates

McLachlan (1988) has recently reviewed the behavioral adaptations of sandy beach organisms. He points out that community and population patterns on exposed sandy

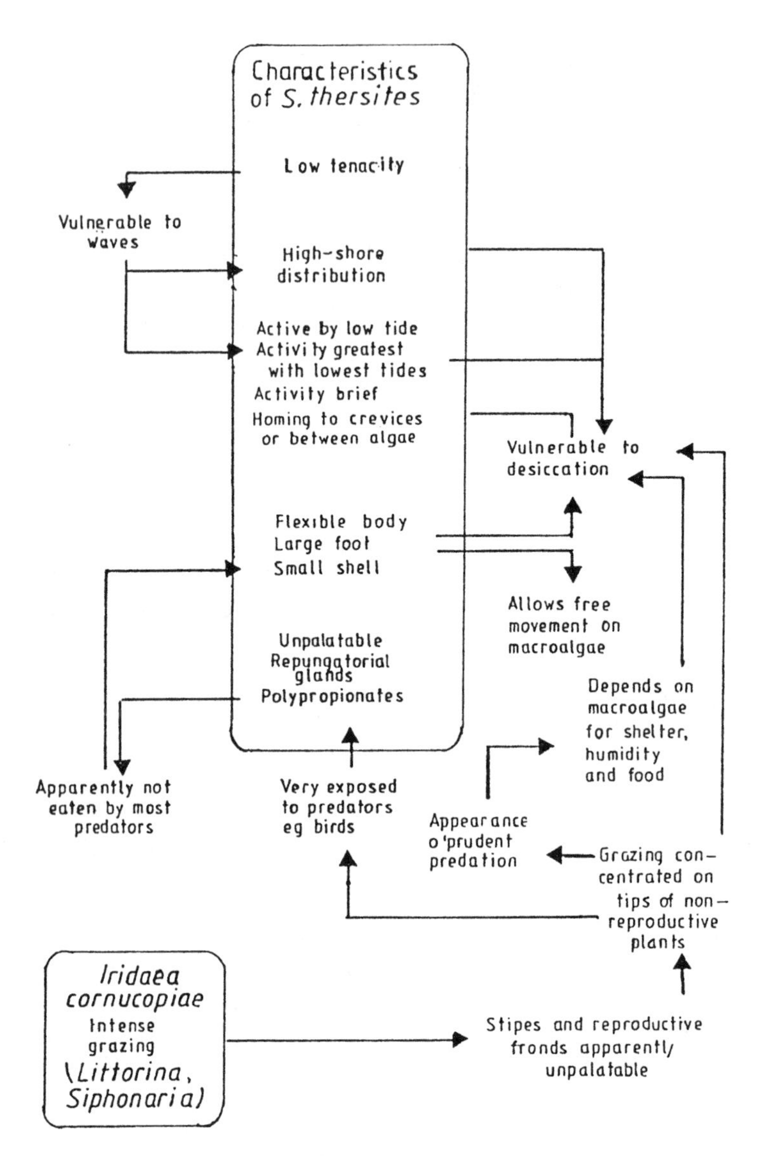

FIGURE 4.20 Summary of the interactions between *Siphonaria Thersites* and its food plant *Iridea cornucopiae,* and the interdependent adaptations of behavior and morphology in relation to physical conditions. (Redrawn from Branch, G.M., in *NATO Advanced Workshop on Behavioural Adaptation to Intertidal Life,* Chelazzi, G. and Vannini, M., Eds., Plenum Press, New York and London, 1988, 43. With permission.)

beaches are the consequences of the behavioral responses of the individual species to the swash climate, and sand movement and liquefaction on the beach face, and that they are not greatly controlled by biological interactions. An important factor is the burrowing rate of the individuals. Burrowing rate is determined by size, temperature, and substrate properties. Besides motility and the ability to burrow rapidly, other adaptations to these dynamic environments include brood protection (Woolridge, 1981; Dexter, 1984), sensitivity to water flow, current direction (Brown, 1982), the degree of thixotrophy, as well as the acoustic shock of breaking waves (Tiffany, 1972; Trueman, 1975). Most crustaceans show positive geotaxis and may be sensitive to changes in hydrostatic pressure (Brown, 1973). In the presence of turbulence, some species burrow while others emerge and swim (Brown, 1983; Jones and Hobbins, 1985). On steep reflective beaches the shock of breaking waves may be a signal for emergence (Tiffany, 1972; Trueman, 1975), whereas on the more dissipative beaches this does not occur. The role of circatidal and circadian rhythms in swimming activity in sandy beach crustaceans is discussed below.

Tidal migration in bivalves: Tides shift the swash zone up and down the shore, and most molluscs in particular follow this oscillation by tidal migratory behavior. In most cases this involves movement up and down the beach following the swash zone. This can take the form of distinct, almost synchronized movements of the whole population, or more gradual shifts of populations as a consequence of scattered individual movements (see Fig-

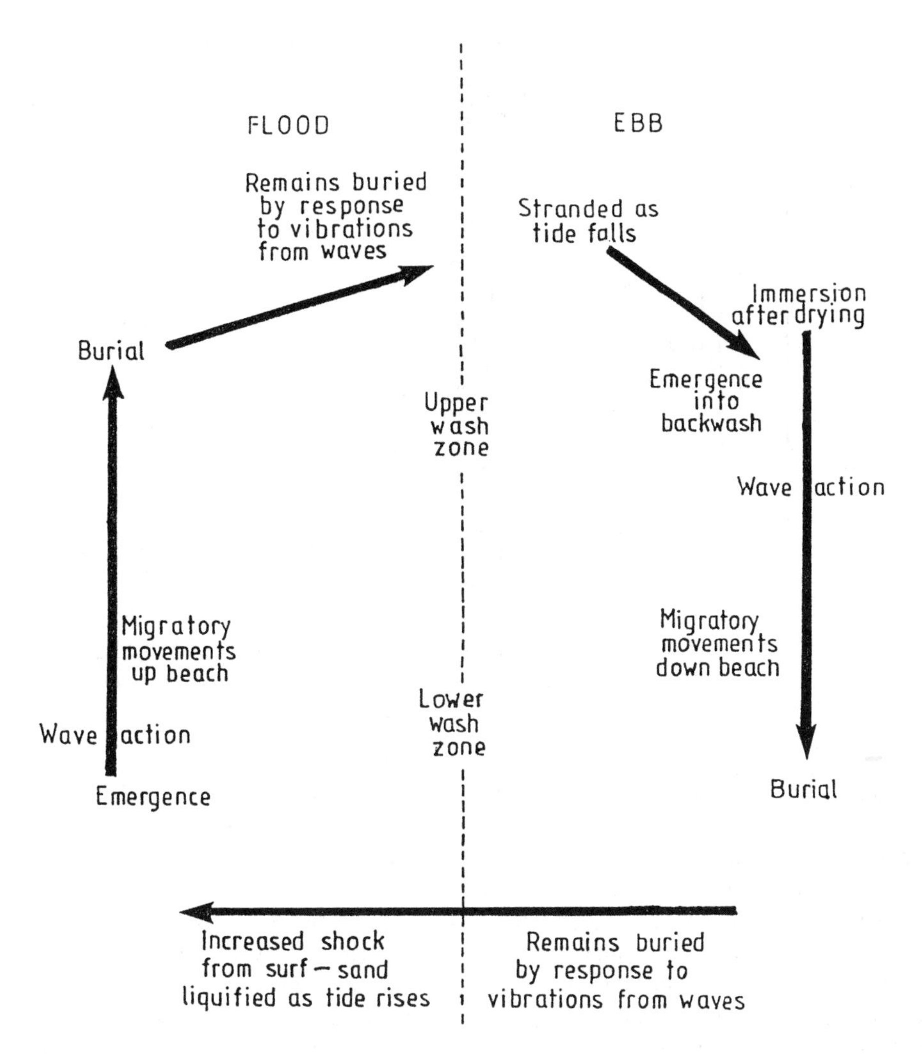

FIGURE 4.21 Main features of the migration cycle of the bivalve *Donax* on sand beaches. (Redrawn from McLachlan, A., in *NATO Advanced Workshop on Adaptations to Intertidal Life,* Chelazzi, G. and Vannini, M., Eds., Plenum Press, New York and London, 1988, 459. After Trueman, 1975. With permission.)

ure 4.21 for the tidal migrations of the bivalve *Donax sordus)*. This migration involves no endogenous rhythm, but only a series of individual responses to changing physical conditions such as the thixothrophy or dilatancy of the sand, and the breaking of waves on the beach face (Figure 4.22) (Ansell and Trevallion, 1969; Tiffany, 1972). The movement of *Donax* and the gastropod *Bullia,* which also undertake similar migrations, is greatly facilitated by the extension of the foot which, by presenting a broad surface area, acts as an underwater sail. They also ensure that the animal is able to dig immediately at the end of a migration up or down the beach (Brown, 1983).

Not all species migrate tidally and there are species that remain at characteristic levels. However, tidal migration has a number of advantages for sandy shore animals. It maintains the animals in the optimal feeding zone, i.e., the swash zone, where water coverage is almost continuous, but wave action is not too severe. The animals are too shallow to be reached by fishes and they pose hazards for bird predators. The migratory animals are less likely to be stranded as the beach changes with the tides, storms, and other environmental events.

Upper beach crustaceans: The supralittoral fauna of sandy beaches consists of crustaceans and insects. The crustaceans (ocypodid crabs, oniscid isopods, and talitrid amphipods) all exhibit circardian rhythms while some also follow circatidal rhythms (see Section 4.4.4.2). In most cases maximum activity coincides with the nocturnal low tide; this reduces predation, desiccation, and the chances of tidal inundation (Marsh and Branch, 1979).

In supralittoral forms, where migrations often cover large distances and represent a transition from terrestrial to intertidal conditions, accurate orientation is necessary. Talitrid amphipods are common dwellers of the upper levels of sandy and pebble beaches in both tropical and temperate shores. Their activity and zonation are determined by two opposing constraints: that of being swept away by waves or tides, or being desiccated when the

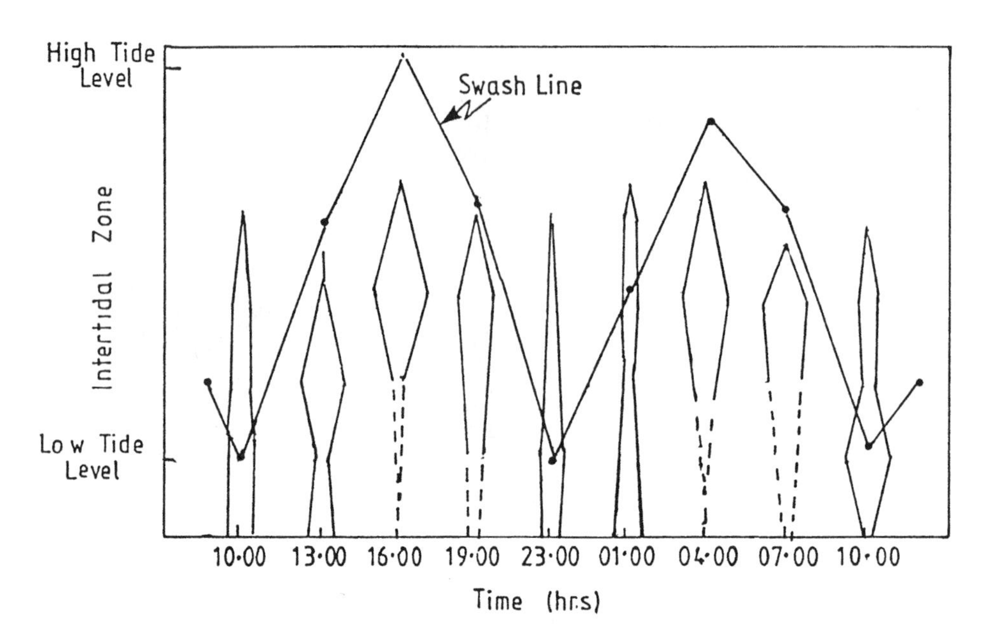

FIGURE 4.22 Tidal migrations of the bivalve, *Donax sordidus*, over 24 hours. Movement of individuals causes a gradual shift of the population up and downshore. (After McLachlan, A., in *NATO Advanced Workshop on Adaptations to Intertidal Life,* Chelazzi, G. and Vannini, M., Eds., Plenum Press, New York and London, 1988, 458. With permission.)

upper layers of the beach dry up. They show a cycle of surface activity at night when the tide is out, wandering over much of the intertidal zone, as well as above the high water mark; followed by burrowing and quiescence by day in optimal humidity conditions. The mechanisms involved in their navigation has attracted considerable attention, especially by British and Italian workers (Williams, 1980; Ugolini et al., 1986; Scapini et al., 1988; 1993). Sandhoppers are capable of zonal orientation by means of universal factors (such as the sun, moon, magnetic compass, polarized light, etc.) and local cues (landscape, wave action, substrate chemical and physical gradients, slope, etc.) (reviewed in Pardi and Ercolini, 1986 and Scapini et al., 1988; 1993) (Figure 4.23).

In experiments with the European species, *Talitrus saltator,* it was found that when released on dry sand above high tide they invariably moved seaward and this was postulated to be a non-compass orientation. Movement also occurred when the sun was obscured and experiments suggested that they orientated toward polarized light. On the other hand, specimens placed on moist intertidal sands migrated landward. Such a landward migration was believed to involve a visual response to coastal topography. It was found that when the eyes of animals were painted black, the animals were unable to navigate and movement was random. In experiments with Mediterranean populations of *T. salator,* it was found that they were capable of orientating geotactically positive on a dry substrate inclined at more than 3° and geotactically negative on a wet substrate of at least 5°, the response being enhanced with increasing slope. Visual stimuli (i.e., an artificial light, view of the sun and sky or of a natural landscape), when tested in a conflict situation, prevail over gravity information (Scapini et al., 1988). Orientation in response to a black strip above the horizon has been shown to be an important mechanism for British sandhoppers in their landward migration at dawn (Edwards and Naylor, 1987). In contrast, Mediterranean sandhoppers avoid a semicircular black strip above the horizon when tested under severe dehydrating conditions (Ugolini et al., 1986). This reaction was compared with the response to natural landscapes, supposing that the black strip simulates a dune/sky boundary. Both responses (to natural landscapes and to a black strip) turned out to be highly variable among different populations (Pardi and Scapini, 1979; Ugolini et al., 1986).

For Mediterranean populations of *Talitrus saltator,* the seaward directional tendency in sun orientation was shown to be genetically determined and adapted to the sea/land axis of the beach (Pardi, 1960; Pardi and Scapini, 1979). This implies genetic differences among natural populations that have a certain degree of local fidelity. Scapini et al. (1993) tested populations of *T. salator* by rearing them under different conditions in order to determine the degree of innateness and modifiability of orientation to substrate slope and landscape features. The results were as follows: the sandhoppers responded to slope only when orientating themselves downslope on a dry substrate and landward on a wet one. These responses appeared to be innate but modifiable by experience. Visual disharmony (half horizon covered with a black strip), with a horizontal substrate, elicited an innate orientation toward the black shape, again modifiable by experience. When a black strip was positioned around the upper half of the inclined arena, the sandhoppers responded differently, depending on their

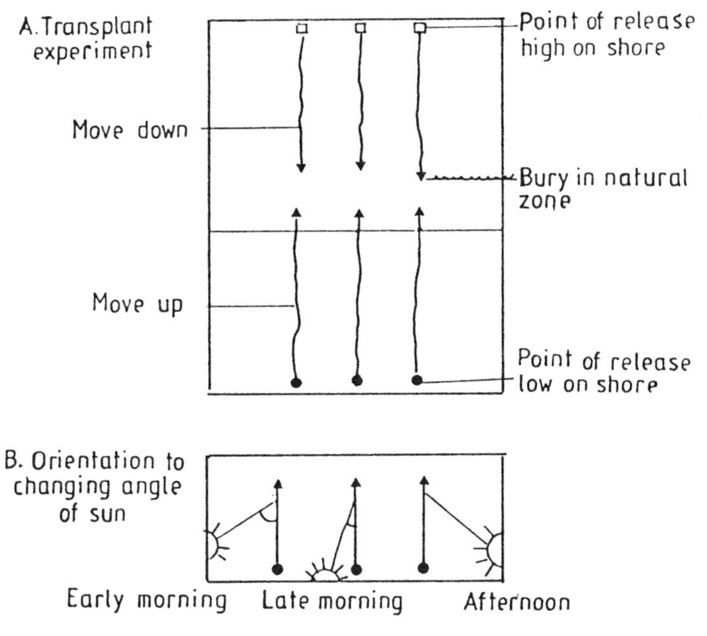

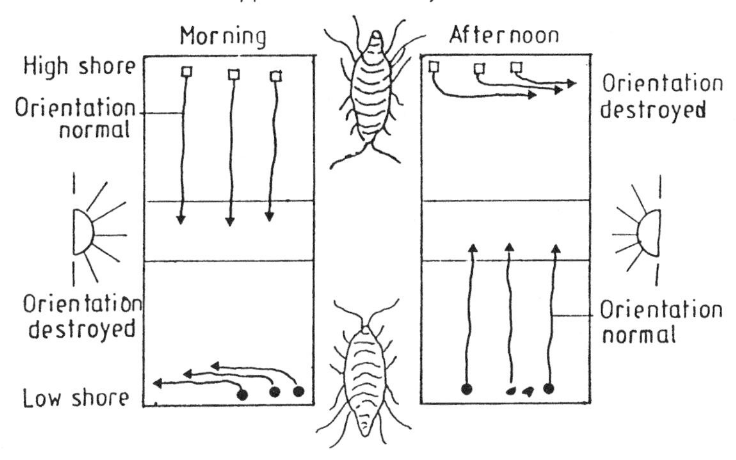

FIGURE 4.23 The orientation of talitrid amphipods on a sandy beach. A. Animals transplanted up or down the shore can find their way back to their original zone. B. They do so by orientating to the sun or moon and must thus allow for its changing position during the night and day, respectively. C. If the eye facing the sun is painted over, the animals are unable to orient in an appropriate manner. (Redrawn from Branch, G.M. and Branch, M.L., *The Living Shores of Southern Africa*, G. Struik, Ed., Capetown, 1981, 60. With permission.)

prior experience; wild individuals responded to the visual stimulus, amphipods reared in optical homogeneity did not make a choice, and those reared in conditions identical to those in the test arena responded to slope by orientating downward and away from the black strip.

All of the migrations, responses, and orientation mechanisms result in sandy beach animals maintaining characteristic levels on the shore when the tide recedes. Their behavior consists of a series of responses to interactions between swash-backwash processes on the one hand and beach face slope, water content, and accretion/erosion on the other. Most of the patterns of animal movements and distribution are the accumulated results of these individual responses.

4.4.4 Clock-Controlled Behavior in Intertidal Animals

4.4.4.1 Introduction

The behavioral patterns of sessile and motile intertidal animals often consist of rhythmic sequences of movements which are often correlated with environmental variables of daily or tidal periodicity (Naylor, 1988). For example, many tubiculous or burrowing polychaete worms such as sabellids and the lugworm, *Arenicola*, irrigate their burrows by actively circulating water through their burrows by a pumping action. Such pumping action occurs in periods of activity cycles with regularly repeating rhythms at

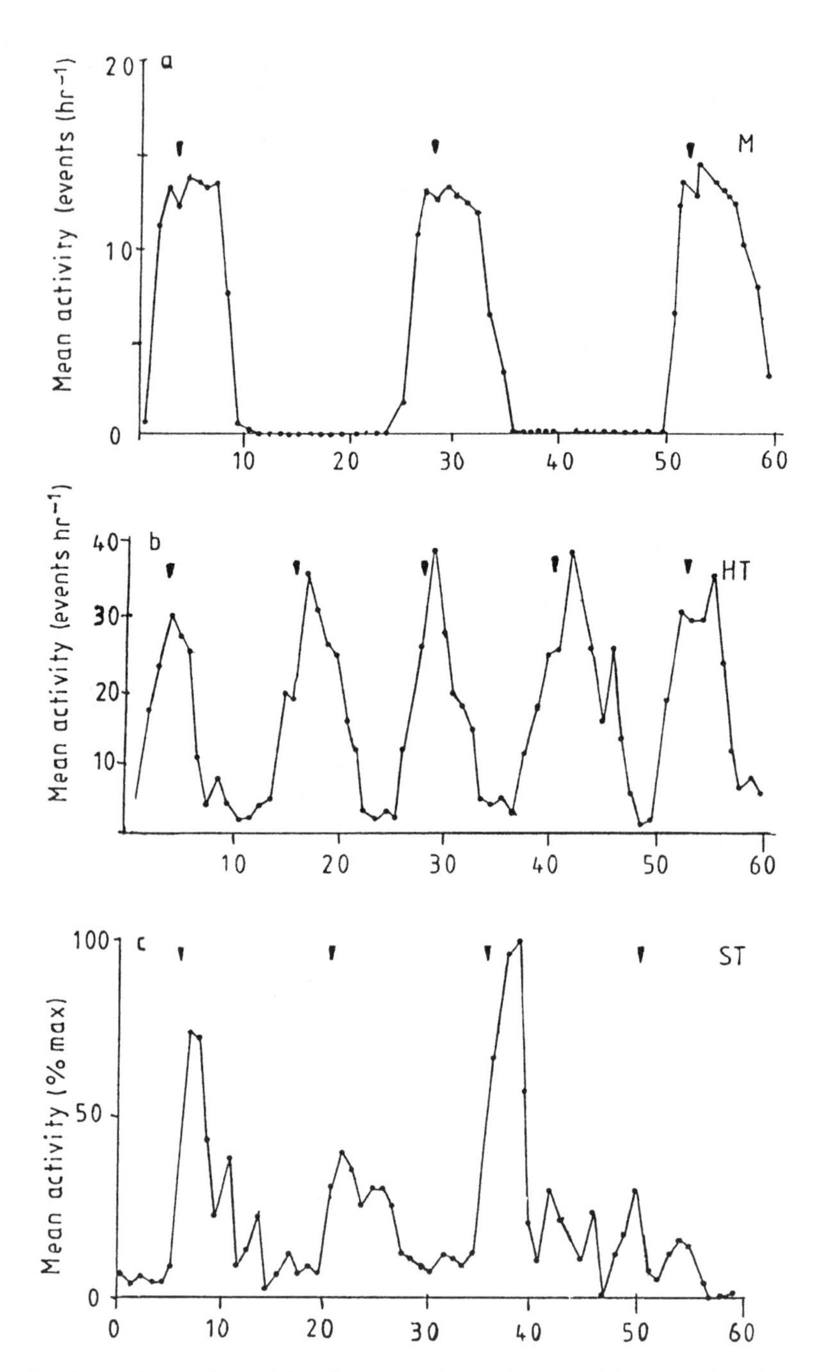

FIGURE 4.24 Free-running locomotor rhythms of circadian, circatidal, and circasemilinar periodicities in coastal crustaceans kept in constant conditions in the laboratory: a — circardian walking rhythm of 32 *Talitrus saltator* recorded for 60 hours (arrows are times of "expected" midnight — M) (after Bregazzi and Naylor, 1972); b — circatidal swimming rhythm of 5 *Eurydice pulchra* recorded for 60 hours (arrows are times of "expected" high tide — HT); c — circa semilunar swimming rhythm of 20 *Eurydicepulchra* recorded for 60 days (arrows are times of "expected" spring tides — ST) (After Reid and Naylor, 1985). (Redrawn from Naylor, E. in *NATO Advanced Workshop on Behavioral Adaptation to Intertidal Life*, Chelazzi, G. and Vannini, M., Eds., Plenum Press, New York, 1988, 4. With permission.)

definite intervals. Many of these rhythms have been found to be tidal, or tidal superimposed with diurnal effects; the rhythms corresponding to the two major events of the tidal environment, change from day to night and the rise and fall of the tide. While many rhythms are driven solely by environmental variables, others occur in constant conditions under the control of internal physiological pacemakers (clock controlled). These latter rhythms corresponding to tidal rise and fall, night and day, and lunar cycles are termed circatidal, circadian, and circasemilunar (Figure 4.24). Evidence suggests that such rhythms are innate and therefore a true genetic adaptation (Naylor, 1987).

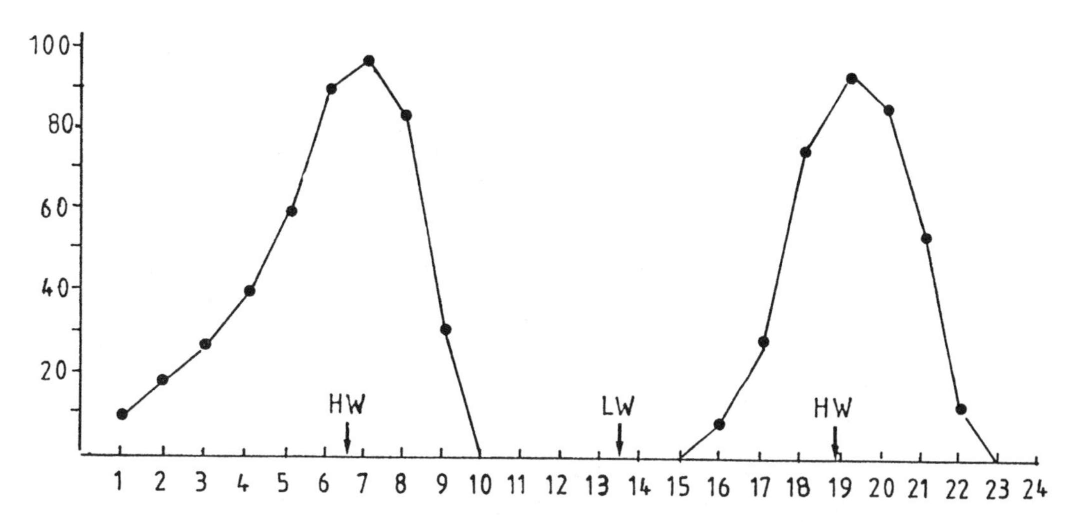

FIGURE 4.25 The percentage of 48 *Austrivenus stutchburyi* cockles with shells agape while kept in constant illumination and a relatively constant temperature. (Redrawn from Bentjes, P.M. and Williams, B.G., *Mar. Behav. Physiol.*, 12, 178, 1986. With permission.)

4.4.4.2 Behavior Rhythms, Tidal Oscillations, and Lunar Cycles

Numerous observations have confirmed that many intertidal animals are active at one phase of the tidal cycle and inactive at the antiphase. Laboratory studies have confirmed that such rhythms persist when the animals are kept continuously under water. Examples from the literature include the New Zealand mud crab, *Helice crassa* (Williams et al., 1985), the mussel, *Mytilus edulis*, (Ameyaw-Akumfi and Naylor, 1987), and the sandy beach isopod, *Eurydice pulchra*, (Jones and Naylor, 1970) (see Figure 4.23b).

Figure 4.25 illustrates shell-gaping behavior in the New Zealand cockle, *Austrovenus stutchburyi*, kept in constant illumination and at relatively constant temperature (Beentjes and Williams, 1986). It will be noted that peak opening with siphons out corresponded with high tide in the environmnt from which the cockles were taken. The value of this circalunidian rhythm to the organism is evident as filter feeding is attuned to the coverage of the beach flats by the tidal waters that transport food to the area occupied by the bivalves.

In many cases these rhythmic activity patterns are often highly persistent without reinforcement in constant conditions in the laboratory. An ecological advantage of diel or tidal rhythms is that they provide a mechanism whereby nocturnal or high tidal active intertidal animals such as the nocturnally active amphipod *Talitrus salator* (Bregazzi and Naylor, 1972; Williams, 1980), the high tide active crab *Carcinus maenus* (Naylor, 1958), and the sand beach isopod *Eurydice pulchra* (Jones and Naylor, 1970), can seek shelter beneath stones or burrow in the sand in anticipation of dawn or tidal ebb. Among mobile intertidal species, burrowing forms exhibit more precise and persistent endogenous rhythmicity than epifaunal forms, e.g., in the burrowing amphipods, *Talitrus saltator* and *Talorchestia deshaysi*, the rhythms are more persistent and more precise than those in the epifaunal amphipod, *Orchestia gammarella*, and the isopod, *Ligna oceana*, which are cryptozoic, but non-burrowing (Table 4.5).

4.4.4.3 Locomotor Rhythms and Maintenance of Zonation

A number of studies have suggested that several coastal animals use their internal clock-controlled locomotor rhythms to maintain their preferred distribution patterns on the shore, notably sand beach crustaceans, which burrow at low tide and swim at high tide. The isopod, *Eurydice pulchra*, which has a "preferred" distribution above MTL, shows cessation of swimming after high tide. A behavioral characteristic ensures that the animals reburrow in the sand and thus avoid being carried too far offshore on ebbing tides (Jones and Naylor, 1970). This semilunar swimming and burrowing in *E. pulchra* has been shown to free run for 60 days in control conditions (Read and Naylor, 1985). A similar pattern of behavior has been recorded for the isopod, *Exocirolina chiltoni*, on Californian beaches (Klapow, 1972).

In a recent review, Jones and Hobbins (1985) (Figure 4.25) concluded that rhythmicity in cirolanid amphipods was essential for the maintenance of zonation in the eulittoral and shallow sublittoral; offshore species showed only circadian and not circatidal rhythms. Endogenous circasemilunar rhythmicity ensures maximum swimming activity on spring tides. In the intertidal, however, zonation is unlikely to be controlled only by rhythmicity, because, should the isopods be swept away from their normal habitat, the timing of their activity to the high tide period alone would not return them to their preferred zone (Jones

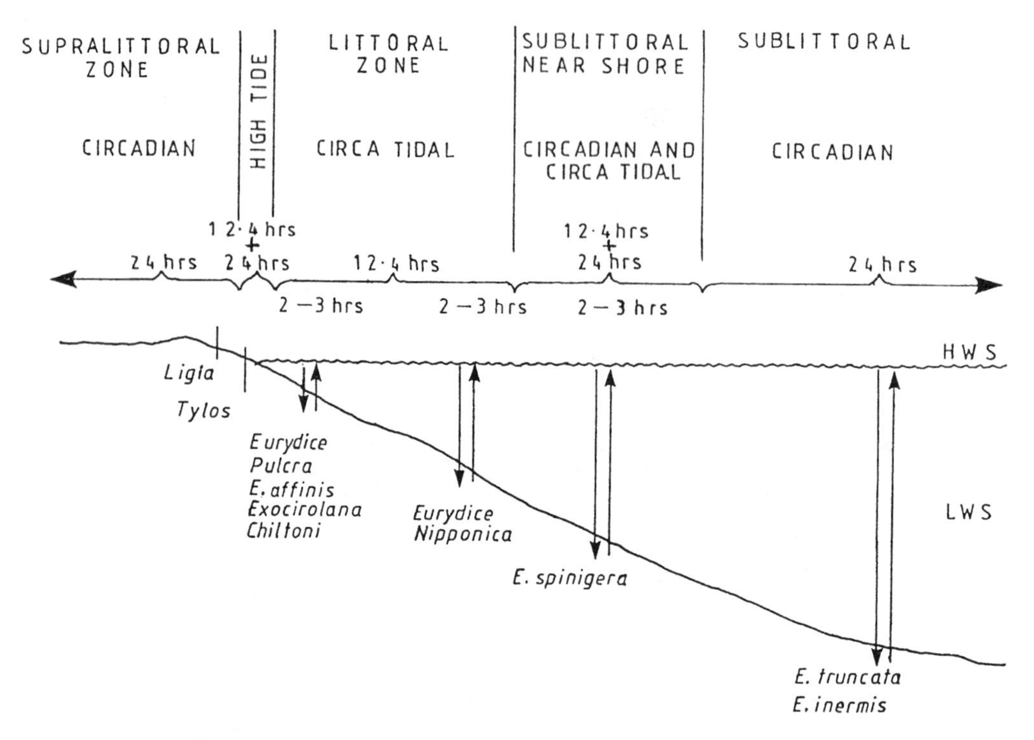

FIGURE 4.26 Zonation and activity rhythms of some sand beach isopods with distributions ranging from aupralittoral to offshore. (After Jones, D.A. and Hobbins, C.St.C., *J. Exp. Mar. Biol. Ecol.*, 93, 57, 1985. With permission.)

and Hobbins, 1985). Photic reactions, substrate selection and responses to turbulence, pressure and beach slope may all maintain zonation.

Naylor (1988) in his review of clock-controlled behavior in intertidal animals concluded that clock-controlled behavior is widespread in eukaryotes; among intertidal animals it varies in persistence between sessile and motile species. It is the mobile, and particularly burrowing forms that exhibit the most pronounced endogenous rhythms of locomotor activity, which may be circatidal, circadian, or circasemilunar periodicity. Such rhythms appear to be correlated with: (1) the avoidance of unfavorable conditions at particular states of the tides; (2) maintenance of preferred patterns of zonation; and (3) anticipatory activity in relation to tidal rise and fall. Circatidal and circadian rhythms of responsiveness to environmental factors have also been demonstrated in migratory intertidal animals. The differences noted in rhythmic behavior patterns between conspecifics and congenerics, which have been discussed above, can be functionally related to the difference in tidal amplitude on the two types of shore.

5 Control of Community Structure

CONTENTS

5.1 INTRODUCTION

In previous sections in this book, the emphasis has been on the impact of physical factors on the distribution of plants and animals on the shore. While these factors set the framework and define the limits over which the various life cycle stages of a particular species can exist, the patterns of distribution are subject to modification by a complex of interacting biological factors. Early studies (for example, reviews by Lewis, 1964; Stephenson and Stephenson, 1972) were oversimplifications relating distributions to tidal rise and fall and wave exposure, and much subsequent research has shown that interactions among species can profoundly modify distribution patterns, and often determine these patterns (see reviews by Dayton, 1971; 1984; Connell, 1972; 1975; 1983; 1985; 1986; Underwood, 1979; 1985; 1991; 1992; 1994; Underwood and Denley, 1990).

In this chapter we will be concerned with the processes that interact to ensure the persistence of species as members of shore communities and the maintenance of community structure. Concepts discussed include the ways in which plants and animals interact on the shore (including grazing, predation, and competition), the role of disturbance, and the interrelated ideas of succession and stability. While some of these processes operate on both hard and soft shores, there are considerable differences between the two shore types that require their separate consideration.

5.2 HARD SHORES

5.2.1 Interactions between Plants and Animals

The interactions between plants and animals on rocky shores are complex. Of all the possible interactions, grazing has received the most attention. Research has focused on topics such as grazer diets, energy flow, foraging behavior, grazing and control of community structure, impacts of grazing on succession after disturbance events, grazing as a factor controlling the vertical distributions of macroalgae, macroalgae as habitats for invertebrate macrofauna and meiofauna, the impact of grazers on sessile animals, especially barnacles, competitive interactions, grazing, and algal functional morphology and chemistry. Space, however, will not allow the consideration of all these topics.

5.2.1.1 Grazing

5.2.1.1.1 Ecological categories of algae

Three ecological groups of algae are recognized, each of which presents grazing animals with different problems.

1. *Microflora.* The film of diatoms, blue-green algae, and the spores and sporelings of macroalgae that carpet rock surfaces and the hard part of many animals.
2. *Encrusting algae.* Many kinds of calcareous (e.g., *Lithothamnion*) and noncalcareous "tar-crusts" such as *Hildenbrandia* form patches or continuous sheets on rocky shores. Generally, the calcareous algae are restricted to low on the shore, except in rock pools or beneath macroalgae.
3. *Erect, foliose algae.* These can be either turf forming or occur as discrete individuals. They are subdivided into ephemeral, short-lived opportunistic species; longer-lived, larger perennial species; and epiphytes (species growing on other algae). Many are opportunistic species.

5.2.1.1.2 Principal types of herbivorous grazers

Grazers eat both micro- and macroalgae; only parts of the latter may be consumed and the plants may survive the grazing. A great variety of grazers are found on rocky shores depending on tidal level, exposure to wave action, and geographic location. They can be categorized as follows:

1. *Littoral fringe species.* The principal species worldwide are small snails, principally species in the Family Littorinidae. Other grazers high on the shore include omnivorous amphipods and isopods.
2. *Eulittoral species.* Molluscs are the dominant grazers. Prosobranch limpets are particularly

important in temperate regions (Branch, 1981). Trochids and mesogastropods, including some littorinids, are also major grazers (see Menge, 1976; Lubchenco, 1978; Underwood, 1979). At lower latitudes and in the Southern Hemisphere, pulmonate limpets such as *Siphonaria* spp. are important (Underwood, 1980; Branch, 1981; Underwood and Jernakoff, 1981). In some regions, e.g., the west coast of South America and Australasia, chitons are significant grazers (Boyle, 1977; Paine, 1980). Various crustaceans (classified as mesoherbivores) can be important grazers. They include amphipods (Pomeroy and Levings, 1980) and isopods (Nicotri, 1977, 1980), which can attain densities of thousands of animals m^{-2}, but due to their nocturnal activity their abundance and importance have not been recognized. Decapods and fishes can also be important grazers, especially in tropical regions (Menge and Lubchenco, 1981).

3. *Mid- and low-shore tide pools, sublittoral fringe, and sublittoral proper.* Here the predominant grazers are regular sea urchins (Mann, 1977; Paine and Vadas, 1969; Dayton et al., 1984; 1992). Some grazing opisthobranchs, such as aplysiomorphs, bullomorphs, and saccoglossans are also found here. On some shores, especially along the temperate West coast of the Americas and Australasia, abalones (gastropods of the genus *Haliotis*) are prominent herbivores feeding on drift algae.

5.2.1.1.3 Diets of grazers

Mollusca: Most chitons, prosobranch limpets, and snails are generalist grazers, feeding on any microalgae or detritus available on the rock surface (Newell, 1979; Underwood, 1979; Branch, 1981; Steneck and Watling, 1982). Much of the microalgal film consists of algal sporelings. Other molluscs, particularly mesogastropods, feed on large erect algae, rather than, or as well as, microflora or encrusting forms, and usually do exhibit choice in diet. Winkles and topshells in general prefer green algae to the tougher, unpalatable reds and browns (Lubchenco, 1978; Underwood, 1979). Siphonarian limpets feed mainly on greens such as *Enteromorpha* and *Ulva* (Underwood and Jernakoff, 1981). Some limpets and many snails such as *Littorina obtusata* and *Lacuna* spp. live and feed epiphytically on a particular host alga (e.g., Branch, 1981; Underwood, 1979). Most molluscan grazers are, however, extremely catholic in their feeding behavior eating the algae that are available if they can manage to eat them. Hawkins and Hartnoll (1983a) following Branch (1981) list four basic feeding patterns: (1) generalists feeding mainly on microalgae, detritus, and encrusting algae on the rock surface; (2) species feeding on macroalgae; (3) terrestrial species closely linked to a food plant; and (4) epiphytic stenotypic species that feed on their host plant.

Echinoderms: Regular sea urchins are the major group of grazing echinoderms. Most species are subtidal and they are generalist browsers feeding on a wide variety of algae and also sessile encrusting animals (although most species have diets dominated by macroalgae). When food is abundant, they exhibit marked food preferences (Menge, 1976; Lubchenco, 1978; Sousa et al., 1981). They can also capture and feed on drift algae.

Crustacea: There are two basic types of feeder: those that scrape the rock surface and those which feed on macroalgae. None of the latter seem to feed on a single species, but they often exhibit clear preferences both in the field and the laboratory.

Fish: Few species of fish feed on intertidal algae in temperate regions, but fish grazing is very important in subtropical and tropical regions (e.g., Montgomery, 1980; Menge and Lubchenco, 1981), especially on coral reefs (Sale, 1980; John et al., 1992). In terms of their feeding methods, herbivorous fishes can be classified either as browsers or grazers (Russ, 1984; 1987; Horn, 1992). Browsers bite or tear off pieces from upright macroalgae and rarely ingest any inorganic material, whereas grazers feed on leafy, filamentous, or finely branched red or green algae (Klump and Polunin, 1990) and may ingest quantities of inorganic material. Herbivorous fishes are more diverse in tropical than in temperate waters and few strictly herbivorous species are found beyond 40°N and S latitudes (Horn, 1989). Browsing species belong to the tropical Families Acanthuridae (sturgeon fishes), Pomatocentritidae (damsel fishes), and Signidae, the tropical/warm temperate Families Grillidae, Odacidae, and Stichaeidae, among others. Grazing species belong to tropical Families Acanthuridae, Pomatocentridae, and Scaridae, the tropical/warm temperate Families Bleniidae and Mugilidae, and the temperate-zone Families Aplodactylidae and Grillidae, among others.

On the West coasts of tropical and subtropical Africa, algivorous or omnivorous fishes often congregate in considerable numbers on broken rocky shores (John et al., 1992) where the physical relief affords them protection from carnivores. In the Caribbean they are associated with coral reefs rather than rocky shores. The majority of the species of nonterritorial fishes belong to two Families, the Acantharidae (sturgeon fishes) and Scaridae (parrot fishes). In addition there are the generalist herbivores belonging to the Family Pomacentridae (damsel fishes). These are territorial "gardening" species (see Section 5.2.1.1.11).

5.2.1.1.4 Foraging behavior

Chapman and Underwood (1992) recently reviewed foraging behavior in marine benthic grazers. Two main aspects of foraging behavior relate to its temporal and spatial components. The temporal component covers variations in the timing or rate of foraging, particularly diurnal or tidal rhythmicity. The spatial component concerns the distances and directions moved while foraging, and the occurrence or otherwise of homing to a fixed site.

The timing of foraging: Temporal patterns of foraging activities have recently been reviewed by Hawkins and Hartnoll (1983a) and Little (1989). A great variety of patterns have been reported. Nevertheless, despite the complexity, certain trends can be observed. Nearly all grazers show a tidal correlation in their grazing pattern. The exception to this are species living high on the shore that respond quickly to external stimuli like wave wash, the weather, or periods of submergence. Low-shore animals are often more predictable in their temporal patterns, frequently showing circatidal rhythms on the rising and falling tide, remaining inactive at low and high tide. Other low-shore species, e.g., the limpet *Patella miniata,* feed independently of either time of day or the tide (Branch, 1971). Some subtidal grazers also show circadian patterns of foraging. The New Zealand sea urchins *Evechinus chloroticus* and *Centrostephanus coronatus* are nocturnal grazers, remaining immobile during the day.

Foraging patterns: Nearly all benthic grazers move in order to feed and their movement patterns during feeding depend on their distribution and the distribution and abundance of their food. Foraging patterns can be classified as follows:

1. *Free range foraging*: Where the animals (e.g., many gastropods) are surrounded by food, they do not need to search for it, and any movements made are essentially foraging excursions. While the movements are random in orientation and extent (Underwood and Chapman, 1985; 1989), over longer periods they do tend to remain within a general area of the shore, usually within a vertical belt.
2. *Foraging in response to patchy distribution of food*: If the food is patchily distributed, some grazers make directional foraging movements. Chemical cues are thought to be important in the location and choice of preferred species of algae (Imrie et al., 1989; Norton et al., 1990). Many species of sea urchins will cover large distances in search of patchily distributed algal food and will form feeding aggregations around individual food items when these are located (Schiel, 1982; Choat and Andrew, 1986; Vadas et al., 1986).
3. *Foraging in response to distribution of a shelter:* Many benthic grazers are limited in the range over which they can forage because of the need to return to shelter, or a different microhabitat after foraging. Many species, particularly limpets, "home" to specific sites (Branch, 1975b), whereas others return to particular habitats different than those in which they feed, e.g., crevices, erect algal shelter (Chelazzi et al., 1985). Homing also occurs in siphonarian limpets, chitons, and some gastropods. Figure 5.1 illustrates the homing of a group of *Patella vulgata* on one high tide.

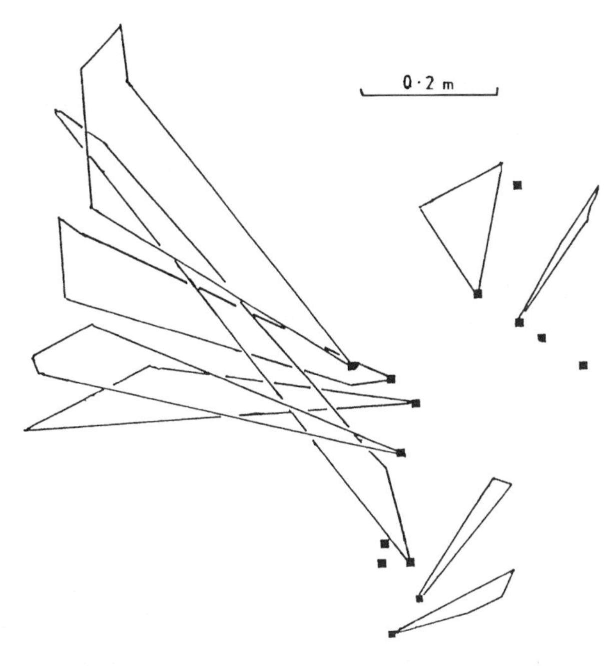

FIGURE 5.1 The homing excursions of a group of the limpet, *Patella vulgata*, recorded by hourly observations during a daytime submersion; the squares indicate the home sites. (Redrawn from Hartnoll, R.G. and Wright, J.R., *Anim. Behav.,* 25, 808, 1977. With permission.)

The widespread occurrence of homing indicates that it must be of considerable advantage to the species concerned. The benefits that have been suggested fall into two basic categories. The first relates to the exploitation and availability of resources. A fixed location provides a starting point for better relocation of preferred feeding areas (McFarlane, 1980). It could also maintain a beneficial even spacing of the individuals that would optimize the use of resources for grazing (Underwood, 1979). The second category of benefits relates to the reduction of physical stress. Continuously returning to a fixed base ensures that the animal remains at a level on the shore where it can survive environmental stress. For limpets on an irregular substrate, homing permits a good fit between the margin of the shell and the rock surface. This good fit should improve protection against predation, dislodgement by wave action, and desiccation (Branch, 1981).

Some limpets establish territories around their home scars (Branch, 1971), usually on patches of their preferred algal food. They vigorously defend both territories and food against invading animals, pushing and occasionally dislodging invaders. The large limpet *Lottia gigantia* maintains a territory on a patch of the encrusting alga *Ralfsia* around its home scar (Stimson, 1970; 1973), defending this against encroachment by pushing out invading *Lottia*, smaller limpets such as *Acmaea digitalis*, and sessile animals such as mussels.

5.2.1.1.5 Sit-and-wait grazers

These species, rather than searching, wait for food in the form of drift algae to come to them. Prominent sit-and-wait grazers are sea urchins and abalones.

5.2.1.1.6 Variability of foraging

Foraging activity in many species appears to be extremely labile, varying from place to place and from time to time (Hawkins and Hartnoll, 1983a; Little, 1989). Some species change feeding patterns at different seasons because of either changes in activity or changes in the availability of food. For example, the sea urchins *Strongylocentrotus droebachiensis* and *S. franciscanus* feed on a range of attached algae, but have a preference for *Nereocystis* (Vadas, 1977). This alga is an annual and it dies back each winter. In summer the urchins mainly eat *Nereocystis* and a variety of other algae. In winter *Nereocystis* forms a smaller proportion of their diet. Many species also show changes in diet foraging behavior as they grow larger.

5.2.1.1.7 Grazing and benthic microalgal distribution, biomass, and diversity

Hunter and Russell-Hunter (1983) found that grazing by the gastropod *Littorina littorea* on benthic microalgae (aufwuchs) grown on glass surfaces resulted in substantial changes in the bioenergetics and community structure of the microalgae. At all grazing densities (13 to 504 snails m^{-2}), the standing crop of the benthic microalgae was markedly reduced compared to that on control (ungrazed) substrata. Both *in situ* dry mass and organic carbon per square decimeter decreased as snail density increased; however, nutritional quality was improved with an increase in the density of grazers. Carbon per unit dry mass of the benthic microalgae was higher at all grazer densities than for controls, and nitrogen per unit dry mass increased as grazer density increased. The C:N ratio of the benthic microalgae decreased from 10:1 at 13 snails m^{-2} to 2:1 at 504 snails m^{-2}.

Both microalgal abundance and richness decreased as *Littorina* density increased. Grazing reduced the number of taxa to 50% of that of control substrates at low snail densities and to <30% of that of controls at high densities. Five taxa (including the genera *Achanthes*, *Nitzschia*, *Amphora*, and *Cocconeis*) appeared to resist grazing and in terms of number of cells cm^{-2}, comprised 73% of the cells on highly grazed substrata, compared to 7.9% on control substrata.

Other species that feed in a similar manner to the littorinids (by using their radulae to scrape the surface of the rock) consume benthic microalgae and the microscopic of sporelings include limpets, siphonariids, and chitons. Most such species have a major impact on macroalgal colonization, depending on the degree of grazing pressure they exert. Grazing pressure is a function of grazer density and behavior. Density can be influenced by the density of predators (generally carnivorous gastropods) as well as the life histories and the recruitment success of the grazers. It is also influenced by the availability of suitable microhabitats, such as crevices, barnacle cover, or algal canopy. Other factors include: (1) height on the shore (many species only graze when submerged); (2) weather (including season); and (3) degree of wave exposure, as heavy wave action can inhibit grazing.

5.2.1.1.8 Algal defenses against grazing

There are three basic ways in which macroalgae can survive herbivore grazing: they can escape grazing by using refuges in time and space; they are able to tolerate grazing; or they can deter herbivores (Hay and Fenical, 1988; 1992). Spatial escape is achieved by growing at a level on the shore inaccessible to grazers, or growing in inaccessible crevices. Temporal escape can be achieved by growing and reproducing at times when grazing pressure is low. Tolerance of grazing may involve growing at a rate that compensates for tissue loss. Other algae that may be intensively grazed maintain a holdfast or encrusting base that can regenerate when grazing pressure is lessened.

Some algae have a growth form that deters grazers. For example, grazers have difficulty in penetrating the encrusting calcified thalli of some calcareous red algae. In such algae the growth layer or meristem and the conceptacles (bearing the reproductive organs) are in pits and are thus protected from grazing. Many coralline algae such as *Corallina* have a large proportion of structural tissue, but also low growth rates and low reproductive output.

Hay and Fenical (1992) have recently reviewed chemical defenses developed by algae to combat grazing. In the field, seaweeds are attacked by many species of herbivores; this is especially true on species-rich coral reefs where rates of herbivory are higher than for any other known habitat (Carpenter, 1986; Hay, 1991). Thus, to be advantageous in such a habitat, a defensive compound would have to function against a broad range of herbivores. Several experimental field studies demonstrate that common seaweed metabolites are often able to do this (Hay and Fenical, 1988; Hay, 1991). As examples, when the following secondary metabolites were coated on the palatable sea grass *Thalassia* and placed on shallow portions of Caribbean coral reefs, all significantly decreased the amount of plant material lost to herbivorous fishes: pachydictyol-A from brown algae in the family Dictyotaceae, cymopol from the green alga *Cymopolia*, stypotriol from the brown alga *Stypopodium*, and elatrol and isolaurinterol from red algae in the genus *Laurencia* (Hay et al., 1987).

To date, over 40 pure compounds from a variety of seaweeds have been tested in the field on Caribbean and Pacific coral reefs, or in the laboratory against fishes, sea urchins, crabs, amphipods, or polychaetes (Hay, 1991; 1992; Steinberg, 1992). Although many seaweed secondary metabolites are broad-spectrum deterrents, several have no known effects against herbivores, and few, if any are deterrents to all herbivores.

Herbivore size and mobility are often correlated with resistance to seaweed chemical defenses. Small, secondary herbivores (mesograzers) that live on the plants they consume often preferentially consume, or specialize on seaweeds that are chemically defended from fishes. These mesograzers avoid or deter predators by associating with chemically noxious host plants and may use compounds that deter fishes as specific feeding or host identification cues.

5.2.1.1.9 Grazing and algal distribution

It has been generally accepted that the upper limits of intertidal algae are set by physical factors, while the lower limits are set by biological interactions such as grazing, predation, and competition (see Section 2.7). However, while physical factors are of major importance in controlling upper limits, it has become evident that grazing also plays a major role in determining algal upper limits (Underwood, 1980; Underwood and Jernakoff, 1981; 1984; Hawkins and Hartnoll, 1983a; Jernakoff, 1983).

Upper limits and grazers — One of the most notable features of the rocky shores of New South Wales, Australia, is the fairly abrupt upper limit of foliose macroalgae at the top of the extensive algal beds on the lower shore (Underwood, 1981b). Above this limit on sheltered shores, grazing gastropods are abundant, and the only algae are crustose species (Underwood, 1980). When cages and fences prevented grazers from entering some areas of the shore, there was rapid development of foliose algae (Underwood, 1980; Jernakoff, 1983). These plants could survive at levels much higher than they were normally found, provided grazers were absent. Intensive grazing by the gastropods remove virtually all the sporelings and microscopic stages of the algae.

Southward and Southward (1978) document the raising of the upper limits of various low shore red algae and the brown algae *Himanthalia* and *Laminaria* following massive mortality of limpets, topshells, and *Littorina* spp. following the Torrey Canyon oil spill (Figure 5.2). Other algae such as *Fucus serratus, Palmaria palmata,* and *Dumontia* have all grown more abundantly higher up the shore than normal after limpet removal experiments (Hawkins, 1981a; Lubchenco, 1980). On the Pacific coast of the U.S., ephemeral algae such as *Enteromorpha* and *Porphyra* survive the summer in areas from which grazers, such as littorinids and limpets and also crabs (*Pachygrapsus*) and dipteran larvae, have been excluded (Robales and Cubit, 1981).

Lower limits and grazers — Many species of algae will grow profusely at lower levels on the shore than normally occupied if grazing lessens or stops due to natural causes or experimental removal of grazers. This is particularly true of North Atlantic ephemeral algae, which are the initial colonizers in grazer exclusion experiments (e.g., *Ulothrix* spp., *Blidingia minima*, *Enteromorpha* spp., *Porphyra* spp.) (Menge, 1976; Lubchenco, 1978; 1980: Little and Smith, 1980; Hawkins and Hartnoll, 1983a). Similar results in molluscan removal experiments have been found in other areas around the world in Australasia (Underwood and Jernakoff, 1981), South Africa (McQuaid, 1980; Branch, 1981), and on the Pacific coast of North America (Dayton, 1971; Paine, 1980).

In conclusion, grazing is in many instances as important as competition or physical factors in determining the vertical distribution patterns of algae, though grazing can act in concert with competition or modify competitive ability (Lubchenco, 1980). Figure 5.3 plots models of variation in grazer importance at various levels on the shore and their relationship to food availability. Food availability increases downshore (Figure 5.3A). On the Pacific coast of North America, grazing declines into the mid-sublittoral zone and then declines again into the deep sublittoral (Foster, 1992). On northeastern Atlantic shores (Hawkins and Hartnoll, 1983a), grazing is most important in the high- and mid-eulittoral and in the mid- to deep-sublittoral, and the algal assemblage in the splash zone is limited directly by physiological stress.

Duggins and Dethier (1985), by manipulations of the density of the chiton *Katharina tunicata,* studied its impact on the species composition and abundance of algal assemblages in the San Juan Islands, Washington, U.S.A. over a period of ten years. *K. tunicata* ranges from about 0.5 m above, to 1.0 m below mean lower low water (MLLW). Prominent algae here are the perennial intertidal kelps *Hedophyllum sessile, Alaria, Laminaria,* and *Nereosyctis*. Over the ten-year period, algal abundance and diversity increased in the areas where *Katharina* was removed; algae of most functional groups proliferated and a multistoried intertidal kelp bed eventually developed. In areas where *Katharina* were added, the abundance of all plants except crusts, diatoms, and surfgrasses decreased and overall diversity declined. Control sites underwent year-to-year fluctuations in the abundance of the most conspicuous algae, *H. sessile,* but otherwise remained

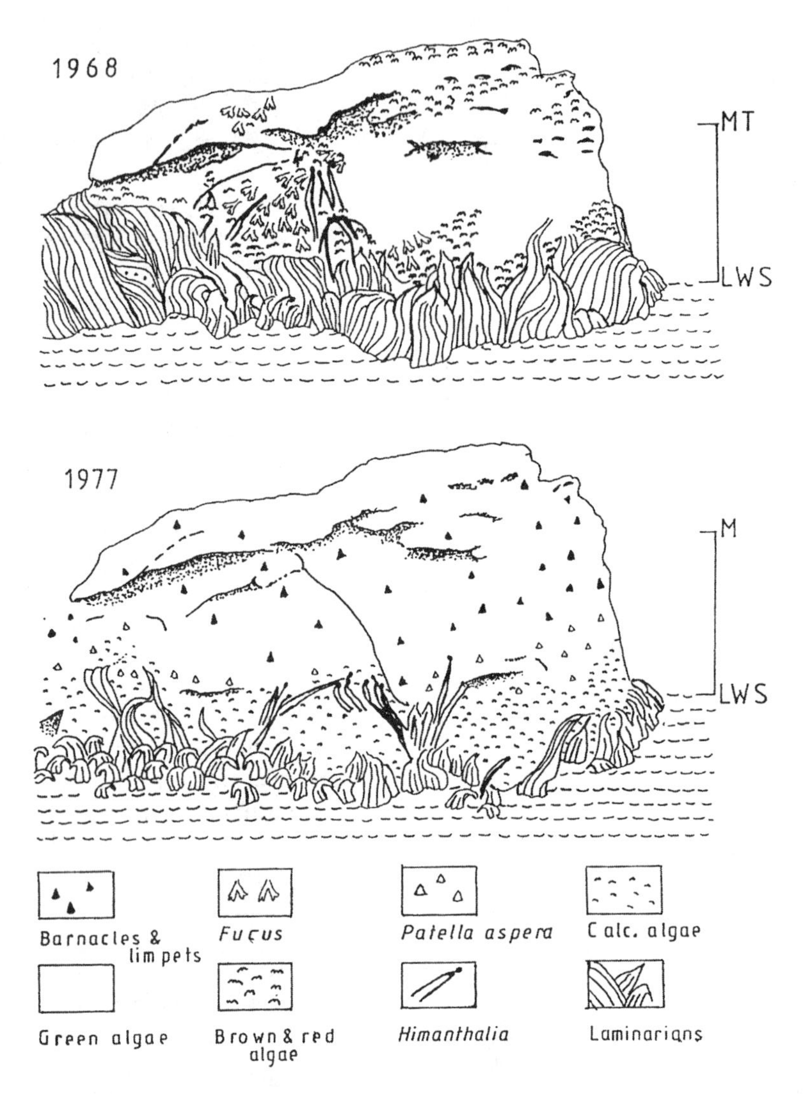

FIGURE 5.2 Grazing and zonation of low shore algae, Cape Cornwall, wave-beaten rocks: sketches showing changes in zonation after the Torrey Canyon disaster: upper — the situation in May 1968, 13 months after all herbivores (limpets, topshells, littorinids) had been killed by dispersants; lower — nine years later in May 1977, showing more normal conditions after full return of herbivorous populations; upper limits of *Laminaria digitata* and *Himanthalia* were 1.5 to 2 m higher in Spring 1968 than in Spring 1977; MT, mean tide level; LWS, mean low water springs level. (Redrawn from Southward, A.J. and Southward, E.C., *J. Fish. Res. Bd. Canada*, 35, 698, 1978. With permission.)

unchanged. It is clear that this grazer has a considerable impact on the diversity and abundance of the algae in the areas in which it grazes.

Levings and Garrity (1984) investigated the impact of the pulmonate *Siphonaria gigas* in the mid-intertidal on rocky, wave-exposed shores on the tropical Pacific coast of Panama. On these coasts, erect macroalgae and sessile invertebrates are rare; crustose algae covers 90% of the rock surface. The relative abundance of a common blue-green algal crust (*Schizothrix calcicola*?) is negatively correlated with *Siphonaria*'s abundance. Large-scale removals of the limpets caused rapid increases in the percent cover of *Schizothrix* and concomitant decreases in other crusts, but no changes in the abundance of erect algae or sessile invertebrates. Removal of *Siphonaria* also (1) increases recruitment of crustose algae and barnacles into new rock and plexiglass surfaces, and (2) decreases the abundance of a calcified form of *Schizothrix*.

5.2.1.1.10 Contrasts in grazing on temperate and tropical shores

Brosnan (1992) has drawn attention to the contrast in zonation patterns and the nature of herbivorous grazers on tropical and temperate shores. Overall, herbivorous fish are more common in the tropics, while invertebrate grazers dominate temperate shores; this difference is responsible for much of the variation in algal composition and abundance between tropical and temperate shores. Brosnan (1992) presented a generalized model to predict the dis-

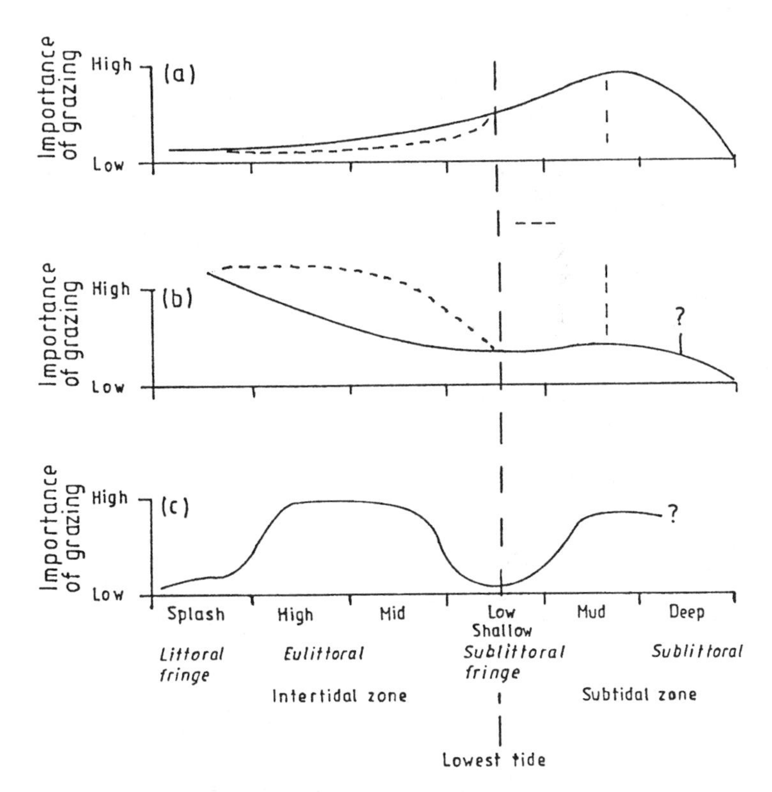

FIGURE 5.3 A graphical summary or model of variation in grazer importance and its relationship to food availability on semiprotected shores in the northeastern Pacific, and a comparison with the model of Hawkins and Hartnoll (1983). (a) Changes in food availability (solid line: maximum availability with bars representing variation; dashed line: effect of increased desiccation). (b) Changes in grazing importance at the spatial scale of assemblage (solid line: minimum importance with bars representing variation; dashed line: effect of reduction in algal availability due to desiccation). (c) Hawkins and Hartnoll's (1983) model of grazer importance. ? = lack of information. (Redrawn from Foster, M.S., in *Plant-Animal Interactions in the Marine Benthos*, John, D.M., Hawkins, S.J., and Price, J.H., Eds., Clarendon Press, Oxford, 1992, 72. With permission.)

tribution of algal types in relation to herbivore size and efficiency (Figure 5.4).

When herbivorous fish are abundant, algal growths will not compensate for herbivory — such shores appear barren or dominated by grazer-resistant crustose forms. If slower-moving invertebrate grazers predominate, the greatest effects occur where algal growth is too slow to compensate for herbivory (e.g., the upper shore, especially during the tropical dry season). Grazer impact increases as algal growth rates increase (e.g., lower down the shore and during the wet season (Lubchenco, 1980; Underwood, 1980; Cubit, 1984).

As fish forage frequently on tropical shores, they have a less patchy appearance than on temperature shores (e.g., Gaines and Lubchenco, 1982; Menge et al., 1985; 1986a,b). Seasonality on tropical shores will be less pronounced as any increase in algal biomass will be quickly eaten, and herbivores will limit the upper distributions of algae more frequently than on temperate shores. In addition, competition among foliose algae should be less important. On temperate shores where molluscs dominate, seasonal effects in algal abundance are more pronounced as algal productivity often exceeds grazing (Underwood, 1980; 1981; Underwood and Jernakoff, 1981; 1984: Branch, 1986), and other factors (e.g., competitive and physical) are relatively more important in setting distributional limits. This is because at low primary productivity, spatial and temporal plant escapes will be important and will tend to result in patchy algal distributions. At high growth rates, algae will be more evenly distributed as productivity will exceed herbivory and algae will dominate spatially. Tropical shores will be located in the right rectangle in Figure 5.4, but their mid- to upper zones could oscillate back and forth between the upper left and lower left rectangles during the wet and dry seasons.

A second prediction of the model relates to herbivore size and the rate of resource renewal. Fast-growing algae will be able to support larger-sized grazers than will slow-growing algae. Larger grazers should be more prevalent lower on the shore, and smaller grazers higher up. Large limpets such as *Patella cochlear* are found low in the intertidal zone where their persistence depends on high algal productivity, while smaller littorinids are found high on the shore. On most shores this pattern applies, although other factors such as desiccation are involved.

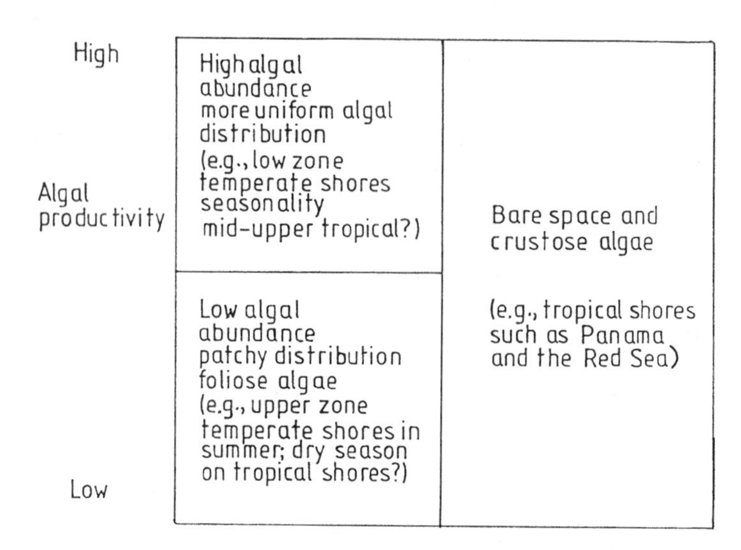

FIGURE 5.4 A model of algal composition and distribution in relation to herbivore type (mollusc or fish, and algal productivity). This model predicts outcomes of interactions between herbivore type and algal productivity. When fish are more important (right rectangle), shores have a uniform appearance (bare space or crustose algae), competition between foliose algae is predicted to be low, and herbivory is important at all tidal heights. When slow-moving molluscan grazers predominate, their effectiveness depends on algal growth rates, and this in turn affects the relative importance of competition and physical stress. (Redrawn from Brosnan, D.M., in *Plant-Animal Interactions in the Marine Benthos*, John, D.M., Hawkins, S.J., and Price, J.H., Eds., Clarendon Press, Oxford, 1992, 111. With permission.)

5.2.1.1.11 Gardening

Grazers that influence the composition or growth of algae are often referred to as "gardeners." Branch et al. (1992) define this behavior as: "modification of plant assemblages, caused by the activities of an individual grazer within a fixed center, which selectively enhances particular plant species and increases the food value of the plants for the grazer." Three major groups of animals have been recorded as gardeners. Tropical reef fish (mainly pomacentrids) are well-documented examples (see Branch et al., 1992 for a list of references). Individual fish defend patches of algae, aggressively repelling other fish and sea urchins. The result is the development of small algal assemblages different than those in the surrounding area.

Some limpets also garden. In California the giant limpet *Lottia gigantea* maintains a territory with a fine algal film and excludes other grazers from this territory. It also hinders invasion of its gardens by sessile species (Stimson, 1973). The South African limpets *Patella longicostata* and *P. tabularis* have specific associations with the encrusting alga *Ralfsia verrucosa* and also defend their gardens against other grazers (Branch, 1975b,c; 1976; 1981). The South African limpet *Patella cochlear*, which lives in dense aggregations, has a narrow fringing garden around each limpet of fine red algae, usually *Herposiphonia heringii* or *Gelidium micropterum*, the latter apparently only occurring in these gardens (Branch, 1975c).

The third example comprises the nereid polychaetes *Platynereis bicanaliculata* and *Nereis vexillosa*, which catch drifting fragments of green algae and attach them to their tubes, which are embedded in soft sediments. The attached algae provide a predictable food supply. As the nereids interact aggressively, their tubes are spaced out, helping to restrict the use of the algae to the individuals gardening them.

Most gardens consist of opportunistic, fast-growing species — often filamentous, delicate red or green algae, but sometimes encrusting forms. Gardens increase local productivity relative to that in adjacent areas (Figure 5.5) (Montgomery, 1980; Klump et al., 1987; Russ, 1987). This shift is brought about by the nature of the algae involved, and because they are maintained in an early rapid phase of growth by continual grazing. Apart from their higher productivity, algae in the gardens of fish species have a higher proportion of protein, and lower ash content and C:N ratios than algae outside the territories (Montgomery, 1980; Klump et al., 1987; Russ, 1987).

Outside damselfish territories, algal biomass is low (Figure 5.5) and is dominated by encrusting corallines. Inside territories, biomass rise and become dominated by filamentous forms. Limpets present a different picture and appear to intensify grazing pressure within their gardens by concentrating on a small area. Algal biomass is invariably lower in their gardens than in adjacent areas (Figure 5.5). They do, however, enhance productivity.

5.2.1.1.12 Grazing and community structure

Patchiness is a fundamental feature of rocky shore communities. Such patchiness can be due to a variety of causes, including grazing. One of the best documented examples of the role of grazing in creating patchiness is on moderately exposed shores in the British Isles where *Patella* spp.

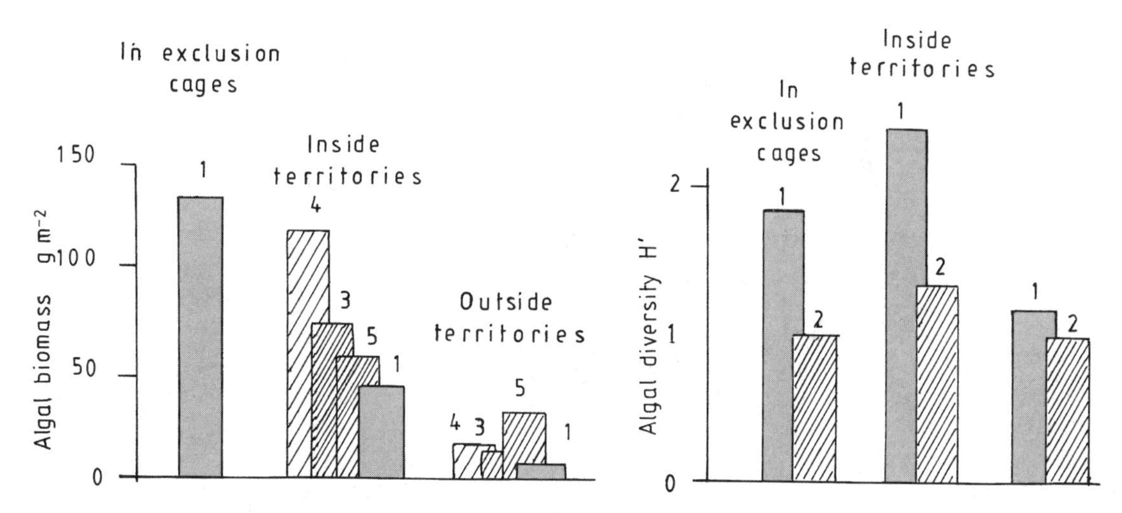

FIGURE 5.5 Algal biomass and diversity inside and outside fish territories and in fish exclusion cages: 1. *Hemiglyphododon plagiometopon* (Sommarco, 1983); 2. *Stegastes fasciolatus* (Hixon and Brostoff, 1983); 3 and 4. *Eupomacentrus lividus* and *H. polagiometopon* (Lassuy, 1980); 5. *S. fasciolatus* (Russ, 1987; biomass estimated by doubling his C values). (Redrawn from Branch, G.M., Harris, J.M., Parkins, C., Bustamante, R.H., and Eckhunt, S., in *Plant-Animal Interactions in the Marine Benthos*, John, D.M., Hawkins, S.J., and Price, J.H., Eds., Clarendon Press, Oxford, 46, 1992, 409. With permission.)

are the dominant grazers among a mosaic of the brown alga *Fucus*, barnacles, and bare rock. Patches of ephemeral algae such as *Enteromorpha* and *Fucus* arise by "escapes" of sporelings from *Patella* grazing (Southward and Southward, 1978; Hawkins, 1981a,b; Hawkins and Hartnoll, 1983a; Hartnoll and Hawkins, 1985). Once the *Fucus* sporelings have reached a length of 3 to 4 cm, they are little affected by limpet grazing at normal densities.

Patches of ephemeral algae and *Fucus* can arise from localized reduction in *Patella* grazing intensity on various scales. Southward and Southward (1978) summarized a seven- to eight-year cycle following large-scale experimental limpet removal from barnacle-dominated western British shores. Ephemeral algae are followed by fucoids, which form a dense canopy in two to three years. Settlement, immigration, and rapid growth of *Patella* follows, so further increases in *Fucus* are prevented by grazing. Canopy and whiplash effects of dense *Fucus* cover can also be as important as grazing in preventing further algal and barnacle recruitment (see also Hawkins, 1981b; Schonbeck and Norton, 1980). Then the fucoid canopy thins, as does the number of limpets as their food supply diminishes. This allows settlement by barnacles and a return to the normal barnacle-limpet community. The massive kills of limpets after the Torrey Canyon disaster prompted a similar cycle, but of a greater scale and a duration of 10 to 12 years (Southward and Southward, 1978) (see Figure 5.2). These cycles can occur naturally on a smaller scale (patch size 5 to 50 m diameter) due to storms or beach movements killing limpets, or fluctuations in recruitment leading to reduced numbers of limpets (Thompson, 1980). Variable recruitment of *Patella* is well documented (Lewis, 1977).

"Escapes" of fucoids are more likely to occur at seasons of peak recruitment, and in years of enhanced reproduction of *Fucus*. Dense cover of barnacle and crevices can increase the chances of *Fucus* escapes (Hawkins, 1981a,b). Figure 5.6 outlines the sequence of events following an "escape" from limpets on a moderately exposed shore. It also shows how recruitment of other components of the shore community can tilt the balance between limpets and fucoids. The balance between limpet grazing and fucoid recruitment and growth is delicately poised and will be tilted readily by fluctuations in recruitment (of limpets, fucoids, and barnacles), climate change, and unusual physical and biological disturbances.

Petraitis (1983) investigated the grazing patterns of the periwinkle *Littorina littorea* and its impact on two abundant sessile organisms, the green alga *Enteromorpha* and the barnacle *Balanus balanoides* on New England coasts. At very low densities, *Littorina* can initially maintain areas clear of *Enteromorpha,* while higher densities are required to eliminate established patches. At low periwinkle densities, *Enteromorpha* interferes with *Balanus* settlement, while at high densities, *Littorina* appears to dislodge newly settled barnacle cyprids. *Balanus* abundance is greatest at intermediate *Littorina* densities. Experimental manipulation of periwinkle behavior showed that *Enteromorpha* can persist due to an interaction between snail behavior and surface irregularities, which are not grazed by *Littorina*. There is no such refuge for *Balanus*. Refuges (regions in which a predator or disturbance does not affect a species or in which mortality rate is lowered) permit the persistence of a species that otherwise might be eliminated (Connell, 1961a,b; Paine, 1969).

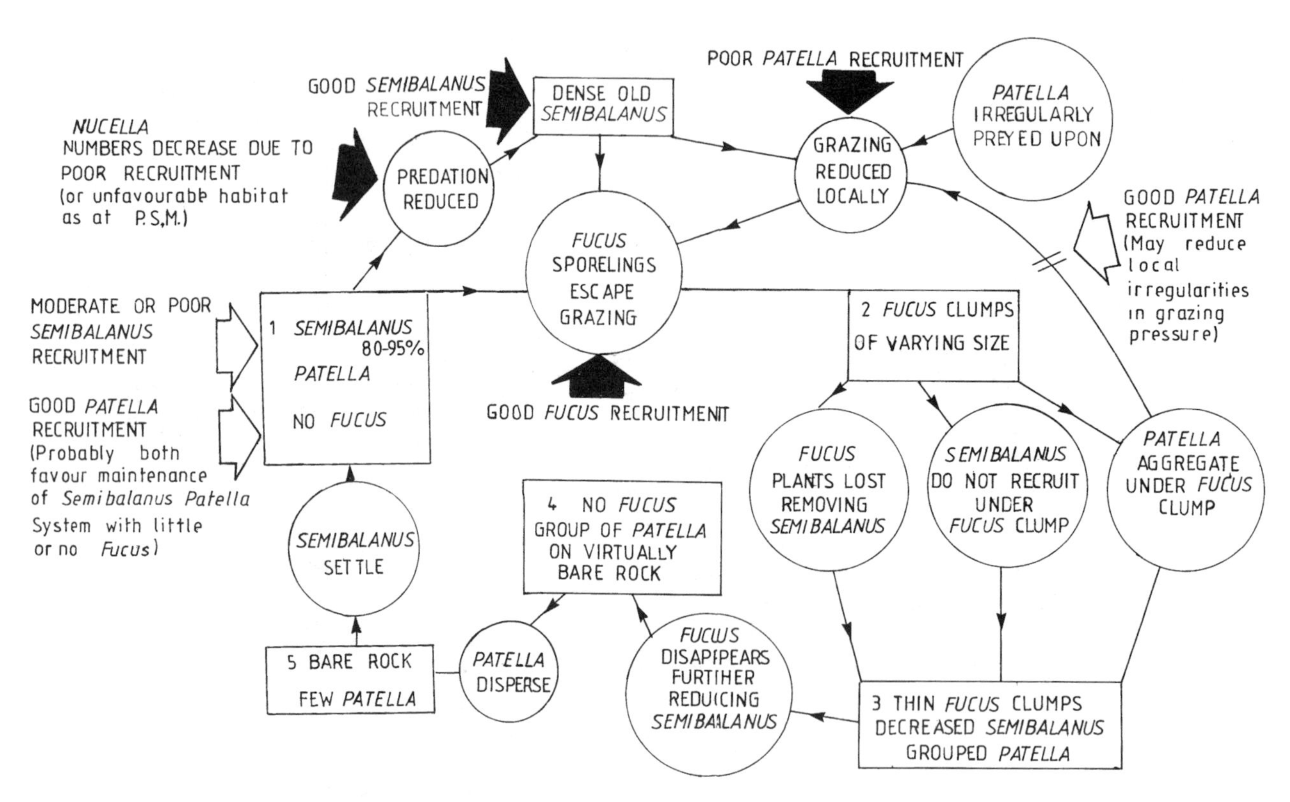

FIGURE 5.6 Flow diagram outlining events following an escape of *Fucus* from *Patella* grazing: processes generating patchiness and governing the observed cycle are shown; the role of variable recruitment is considered; circles denote processes; rectangles denote stages in the cycle; broad arrows indicate the influence of variation in recruitment in generating patchiness (open, suppresses; solid enhances). (Redrawn from Hawkins, S.J. and Hartnoll, R.G., *Oceanogr. Mar. Biol. Annu. Rev.*, 21, 252, 1983a. With permission.)

Jara and Moreno (1984) have investigated the abundance of the red alga *Iridaea boryana* and the role of herbivores in structuring the mid-intertidal community at Mehuin, Chile, by manipulating herbivore density. The mid-intertidal community has two layers of sessile organisms: (1) the basal stratum consists of crustose algae and barnacles (*Chthalamus scabrosus* (70 to 75%), *Ch. carats* (20 to 25%), and *Balanus flocculus* (5%); dark patches of crustose algae, mainly the prostrate phase of *I. boryana* and *Ralfsia* sp., grow both on the barnacles and the rock surface; and (2) the canopy space is dominated by the rhodophyte *I. boryana*. The principal herbivores are *Fisurella picta, Siphonaria lessoni, Chiton granosus,* and *Colisella* spp., while the main predators are *Concholepas concholepas, Nucella calcar, Acanthocyclus gayi,* and *Larus dominicanus*.

Herbivore removal allowed high cover (>80%) of *I. boryana* to develop. Herbivore addition and natural increases in herbivore density caused a drastic decline in this algal species, and the substrate became dominated by barnacles and crustose algae. Ephemeral algae (e.g., *Petalonia* and *Scytosiphon*) were frequent in the presence of herbivores, while in the absence of herbivores, invasion by ephemerals is precluded by the abundance of *I. boryana*. In winter, herbivores and high wave action caused a significant reduction in algal abundance.

Jara and Moreno (1984) have summarized community development on this shore, depending on grazing or time of physical disturbance, in the conceptual model shown in Figure 5.7. Two alternative states are achieved that coexist, but are separated on a local spatial scale. In the absence of the *Iridaea* canopy and associated understory algae, barnacles and crustose algae dominate. These crustose algae are apparently either resistant or unpalatable to grazers (Lubchenco and Cubit, 1980; Dethier, 1981), so that herbivores migrate into surrounding areas where erect algal forms are present (Lewis and Bowman, 1975). Grazing in areas of high algal cover tends to produce the dominance by barnacles and crustose forms. At the same time, however, areas from which herbivores have recently emigrated are recovering their algal cover.

The time of the year in which free space is made available is also important in maintaining the community mosaic (Paine and Levin, 1981). In spring, free space is colonized by ephemeral algae; barnacle recruitment is either inhibited by the algal cover, or less intense during this period, or both. With time this space attains the typical configuration dominated by *Iridaea*, and herbivores immigrate to feed on the algal fronds. However, when free space is made available in fall, it is readily dominated by barnacles. Filtration by barnacles and/or reduced fall settlement would diminish the abundance of ephemeral

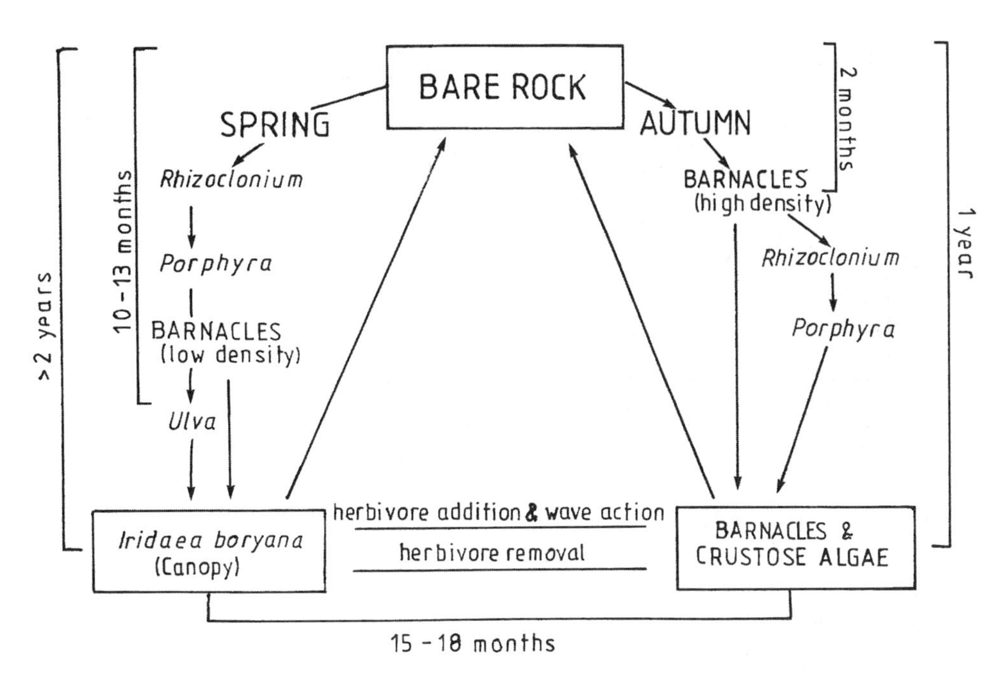

FIGURE 5.7 Diagrammatic representation of a functional model accounting for the temporal dynamics and mechanisms of the mid-intertidal community studied by Jara and Moreno at Mehuin, Chile. (Redrawn from Jara, H.F. and Moreno, C.A., *Ecology*, 65, 36, 1984. With pernission.)

algae, which are present at such low densities that herbivores would not be attracted. Vegetative growth by crustose phases is an effective colonization strategy for algae. Community development leading to *Iridaea* dominance appears to proceed by either of two pathways: a community started in spring is not dominated by *Iridaea* (erect form) for 2 years, while a community initiated in the fall takes 9 to 12 months for *Iridaea* (crustose form) to dominate. Both *Ulva* and *Porphyra* maintained populations of low abundance. On the herbivore exclusion area they were not grazed, but their growth was inhibited by the canopy of *Iridaea*. On the control and herbivore addition areas with abundant light and no *Iridaea* canopy interference, the more rapid growth of these species was probably counteracted by grazing.

A final example of the impact of grazers on algal communities is a sublittoral one. At exposed sites in the Gulf of Maine, U.S.A., subtidal mussels (*Modiolus modiolus*) dominates space on the upper rock surfaces at intermediate depths (11 to 18 m), but at shallow depths (4 to 8 m), the dominants are the kelps *Laminaria digitata* and *L. saccharina*. Results of observations and experiments by Whitman (1987) indicated that storm-generated dislodgement of mussels overgrown by kelps was the mechanism reducing the ability of *Modiolus* to maintain and hold space in the shallow kelp zone. Removal of sea urchins *Strongylocentrotus droebachiensis* from the lower edge of the kelp zone resulted in the downward shift of kelp to a 12.5-m depth, demonstrating that the lower depth limit of the kelp is set by sea urchin grazing. Sea urchin densities were significantly higher inside mussel beds than outside. Removal of the urchins from the mussel beds led to rapid kelp recruitment, resulting in a 30-fold increase of mussel mortality (via kelp-induced dislodgement). This *Modiolus-Strongylocentrotus* interaction can be considered as facultative mutualism that appears to facilitate coexistence of kelps and mussels at shallow depths. The factors maintaining subtidal zonation at exposed sites off the New England coasts are summarized in the conceptual model depicted in Figure 5.8. The three endpoints of the interactions — *Modiolus* dominance, kelp dominance, and coexistence of mussels and kelp — are influenced by the intensity of storm disturbance, sea urchin grazing, and the rate of recovery from grazing.

Branch et al. (1992) have developed a generalized model for algal diversity, biomass, and production per unit biomass in relation to grazing intensity. At high intensity of grazing, both diversity and biomass of algae are low (Figure 5.9). Both increase as grazing declines, although diversity peaks at a relatively low level of grazing and then drops off as competition eliminates some species at very low levels of grazing. Production per unit biomass peaks at an intermediate level of grazing. The relative proportions of the three major functional types of algae can be related to grazing intensity. At very low levels, grazing dense beds of foliose algae can become established. At high grazing intensity, grazer-resistant coralline and other crust-forming species dominate, a condition often termed "the barrens" or "overgrazed." At intermediate levels of grazing, low turfs of fine or filamentous algae are frequent and brown or red algal crusts may occur. The former tend to comprise species that are small, short-lived, opportunis-

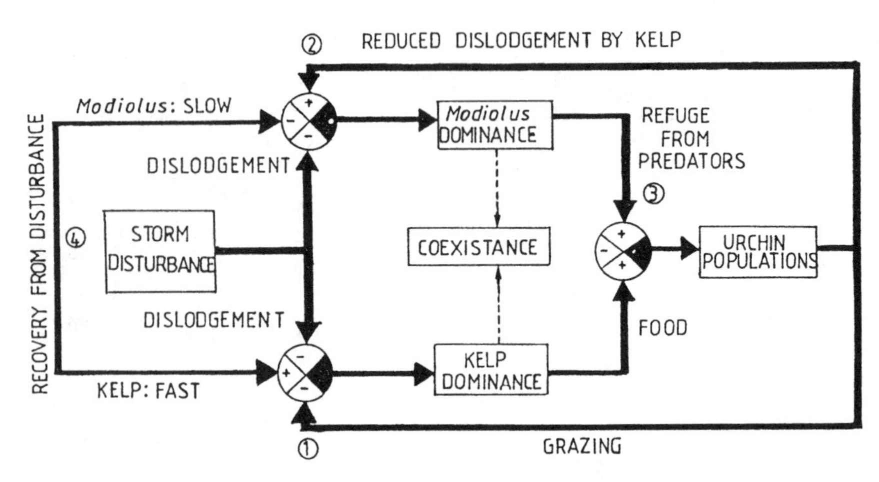

FIGURE 5.8 Conceptual model of interactions between kelps, sea urchins, and *Modiolus*. The three integrator wheels classify the interaction as positive (+) or negative (–). Begin at step 1 and follow arrows through the diagram. The three endpoints of the interactions: *Modiolus* dominance, kelp dominance, and coexistence are influenced by the intensity of storm disturbance, sea urchin grazing, and the rate of recovery from disturbance. As demonstrated by urchin removal experiments, sea urchin grazing has a negative effect on kelp (step 1) by regulating the lower depth limit of the kelp zone and restricting the local distribution of kelp (browse zone observations). Urchin grazing has a positive effect on *Modiolus* (step 2) because it reduces dislodgement mortality (urchin removal and tagging experiments). There is a positive feedback between *Modiolus* and urchins (step 3) as *Modiolus* beds provide a refuge from predation for small urchins (predation experiments in Whitman, 1985), and may decrease the susceptibility of large urchins to predation and dislodgement mortality (urchin dislodgement force measurements). The urchin-*Modiolus* relationship is mutualistic as both species benefit from the association. As demonstrated in the patch recolonization experiments (step 4), fast recovery from disturbance enables kelp to monopolize patch space. The failure of *Modiolus* to fill patches created by dislodgement indicates that, depending on the strength of grazing pressure, *Modiolus* will lose space to rapidly colonizing kelp. The coexistence of kelp and mussels is facilitated by mutualism. (Redrawn from Whitman, J.D., *Ecol. Monogr.*, 57, 182, 1987. With permission.)

tic, fast-growing, and highly productive. They also have minimal structural tissue, high energy and nitrogen content, and few antiherbivore defenses; thus, their value to grazers is high. On the other hand, encrusting and large foliose algae have diminished food value to herbivores.

5.2.1.2 Algae and the Lower Limits of the Distribution of Limpets

From the experiments discussed above it is clear that on some shores, the dense growths of algae at low shore levels restrict the limpets to levels above the algal bed upper limits. Underwood (1994) poses the question: "Why do abundant mid-shore grazing gastropods not venture to lower levels on the shore, but instead are found above a clear boundary with the macroalgae?" Mid-shore limpets on New South Wales coasts can survive for long periods at levels on the shore lower than those at which they are usually found (Underwood and Jernakoff, 1981; Fletcher, 1984a,b). Thus, prolonged periods of submergence are not deleterious to these limpets.

In a series of experiments, Underwood and Jernakoff (1981) demonstrated that the limpet *Cellana tramoserica*, which feeds on microalgae and does not eat macroalgae, tended to move away from areas with well-established algal beds more than from areas with little or no macroalgae. Limpets confined in cages where macroalgae were abundant, lost weight and rapidly died. In contrast, limpets confined in cages without macroalgae retained their tissue weights and survived. Finally, limpets introduced into experimentally cleared areas within the low shore algal zone were unable to keep these cleared areas free of macroalgae. Similar results were obtained by Creese (1982) with the small acmaeid limpet *Patelloida latistrigata*.

5.2.2 Competitive Interactions

5.2.2.1 Introduction

Competitive interactions on rocky shores and the shallow sublittoral have been widely reviewed in recent years (e.g., Connell, 1972; 1983; Underwood, 1979; 1986; 1992; Schoener, 1983; Branch, 1984). In particular, Branch (1984) extensively reviewed competition between marine organisms, while Underwood (1992) reviewed competition with special references to plant-animal interactions. Competitive interactions on rocky shores generally involve competition for limiting resources, and of these the most important are space and food. Competition for space is critical on rocky shores, since space, as a two-dimensional resource, is often in short supply (see reviews by Connell, 1972; Underwood, 1979; 1986; Branch, 1984). Competition for food is especially critical for herbivorous molluscs.

Following Underwood (1992), competitive interactions are considered to be those as defined by Birch (1957). Birch considered that competition would be found

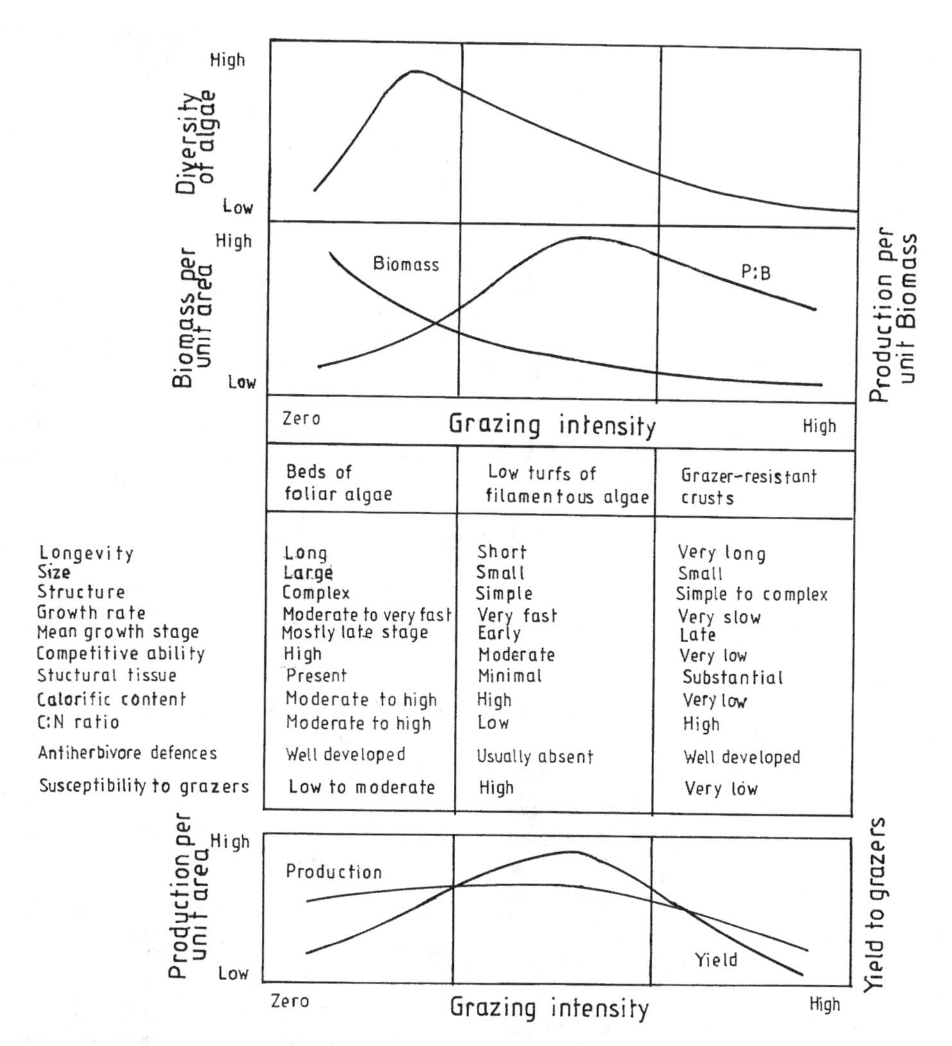

FIGURE 5.9 A generalized model for algal diversity, biomass, and production per unit biomass in relation to grazing intensity. Characteristics of the major functional groups of algae typically associated with each grazing regime are summarized in relation to their productivity per unit area. Yield reflects how edibility of different algal types increases or decreases the gains that grazers may obtain from production. (Redrawn from Branch, G.M., Harris, J.M., Parkins, C., Bustamante, R.H., and Eckhunt, S., in *Plant Animal Interactions in the Marine Benthos*, John, D.M., Hawkins, S.J., and Price, J.H., Eds., Clarendon Press, Oxford, 1992, 415. With permission.)

when any of two or more organisms, which required the same resources occurred together, and the resources were in short supply such that the organisms in some way harmed or impaired each other while trying to use them. Thus, in studying competition, one needs to identify the resources, determine the reasons why they are in short supply, and the harm that the organisms might do to each other under the circumstances being investigated.

5.2.2.2 Intraspecific Competition

When competition involves the exploitation of a limited resource, it will normally be most intense if conspecifics (individuals of the same species) are involved, since they will probably have nearly identical requirements. There is ample circumstantial evidence for the impact of intraspecific competition. For instance, the limpet *Patella cochlear* on South African shores reaches high densities of up to 1,700 m^{-2} and these high densities impact on the individuals in a number of ways (Branch, 1975a,b). As density increases, maximum size declines, thus decreasing mean body weight (Figure 5.10A). Growth rates also decrease, while mortality rises. Densities are usually so high that newly settled juveniles fail to survive unless they settle on the shells of adults where they cannot be grazed. Density also affects the standing stock of *P. cochlear* (Figure 5.10A). Increased density means an increase in total biomass, but once the density reaches 400 m^{-2}, carrying capacity is reached and intraspecific competition becomes intense if the density rises higher. Another important consequence at high densities is that reproductive output per individual decreases, and, in addition, the total reproduc-

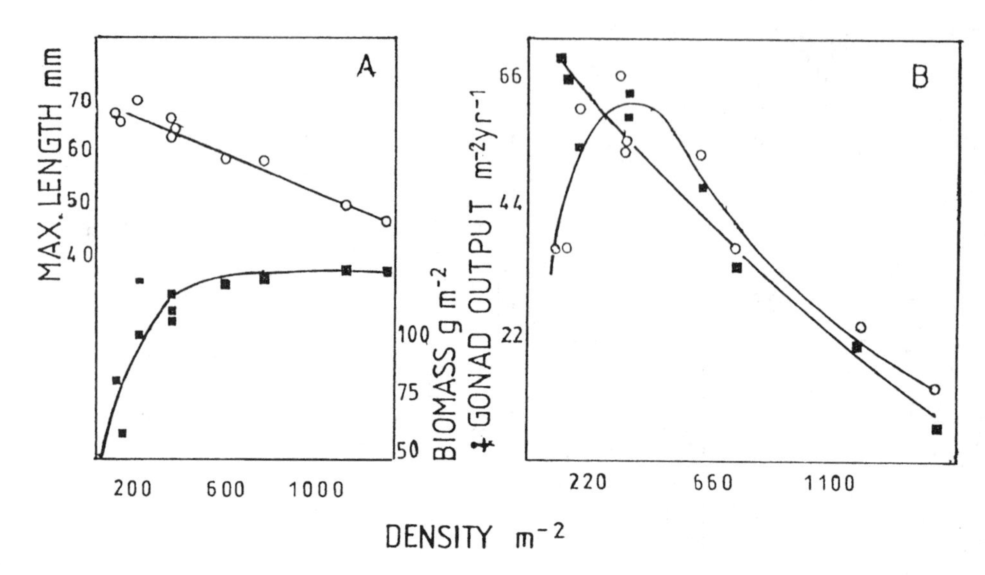

FIGURE 5.10 A. Maximum length and biomass of *Patella cochlear* relative to density; B. *P. cochlear*, effect of density on the total gonad output per year (○) and on the output per unit biomass (■). (Redrawn from Branch, G.M., *J. Anim. Ecol.*, 44, 264, 266, 1975b. With permission.)

tive output of the population per unit of biomass declines. Reproductive output per m^{-2} initially rises with density, but peaks at a density of about 400 m^{-2} and then declines sharply (Figure 5.10B). There are many other examples where limpets exhibit intraspecific competition (e.g., Sutherland, 1970; Black, 1979; Choat, 1977; Creese, 1980; and see review by Branch, 1981). One limpet that has been studied extensively is the New South Wales species *Cellana tramoserica*. Manipulative experiments have shown that an increase in density brings about a decrease in growth and an increase in mortality, and that there is a negative correlation between the density of recruiting juveniles and that of adults (Underwood, 1979; Branch and Branch, 1981; Underwood et al., 1983). Experimental alteration of the population density of the gastropod *Nerita atramentosa* on New South Wales shores has clearly shown that increased densities sharply reduce the growth rate of juveniles, increase the mortality of adults, and decrease the mean tissue rate production of both adults and juveniles (Underwood, 1976).

Often where there is intense competition for space or shelter, intensive and aggressive interference competition may occur between individuals. Grunbaum et al. (1978) have shown that tropical sea urchins aggressively defend the holes in which they live against conspecifics. Other animals that aggressively defend their shelters are hermit crabs, mantis shrimps, and alphaeid shrimps. Instances in which competition over a source of food involves aggressive defense include the amphipod *Erichthonius braziliensis* (Connell, 1963) and the territorial limpets *Lottia gigantea* (Stimpson, 1970; 1973), *Patella longicosta*, and *P. tabularis* (Branch, 1975a,b). Damsel fish are also renowned for their belligerent defense of their territories (see Sale, 1977, for a review).

5.2.2.3 Mechanisms for Reducing Intraspecific Competition

Branch (1984) has listed a variety of mechanisms that bring about a reduction in intraspecific competition. These are discussed below.

5.2.2.3.1 Larval settling patterns

The larvae of many rocky-shore sessile species are gregarious, settling in the vicinity of adults of the same species. In sessile species it is essential that the larvae space out so that there is room for subsequent growth (see Knight-Jones and Moyse, 1961, for a review). For example, larvae of the spirorbid polychaete *Spirorbis borealis* crawl in irregular circles before metamorphosing, swimming away if they contact an irregular projection or one of their own species (Wisely, 1960). This behavior ensures that there is sufficient space for the settled larvae to grow to adult size. Cyprids of the barnacle *Balanus balanoides* space themselves away from adults of their own species when settling, but react quite differently to other species crowding against them (Knight-Jones and Moyse, 1961).

5.2.2.3.2 Dispersal of adults

Motile species can reduce competition by moving away from each other. Dispersal at high densities has been reported for a number of sea urchin species. The crabs *Carcinus maenus* and *Pachygrapsus crassipes* disperse at high densities.

5.2.2.3.3 Dispersal along an environmental gradient

If juveniles settle in a particular habitat and progressive migration occurs away from this site as the animals age, there is the possibility that intraspecific competition can be reduced by lessening overlap between different age groups. In a number of species of molluscs, e.g., *Patella granularis* (Branch, 1975a), *Littorina littorea*, *Littorina unifasciata* (Branch and Branch, 1980), and *Bembicium auratum* (Branch and Branch, 1980), the juveniles occur above and below the adult population.

5.2.2.3.4 Avoidance or ritualization of combat

For species that are highly aggressive, particularly those with the ability to inflict damage on their opponents, ritualization is an important means of avoiding the escalation of intraspecific contacts to the point where one of the contestants is damaged or killed. Ritualized displays have been described for many species, e.g., mantis shrimps, the crab *Carcinus maenus* (an aggressive species in which dispersal is related to density) reduces its level of combat when kept at high densities, fiddler crabs, and hermit crabs.

5.2.2.3.5 Difference in diet

Intraspecific competition may be reduced if different sized animals feed on different sized prey. Correlations between body size and prey size are well known for a number of species including the rock lobster *Jasus lalandii* (Griffiths and Seiderer, 1980), the starfish *Pisaster ochraeus*, the crab *Carcinus maenus*, the whelk *Thais emarginata*, and the gastropod *Conus ebraeus*.

5.2.2.4 Interspecific Competition

The outcome of interspecific competition will vary with the nature of the species involved, the means of competition, the kind of resources competed for, and the amount of overlap of the niches of the species (Branch, 1984). Where competition is extreme, it can result in the exclusion of the weaker competitor from the area occupied by the dominant. The weaker species can be forced to withdraw from part of its potential niche (the "fundamental niche") to a smaller part of its range (its "realized niche") and may be confined to suboptimal areas in the process. In this section we discuss a range of examples of interspecific competition in which different kinds of resources are competed for in different ways.

5.2.2.4.1 Competition between plants

Competition for space occurs among algal species growing at the same level on the shore (interference competition). Exploitative competition for nutrients and light can also take place among algal species. Competitive interactions among macroalgae have been reviewed by Denley and Dayton (1985). In general, the consensus is that plants compete for space.

Turf-forming algae are faster growing than crustose species and they spread over the substratum occupying large amounts of space. In two cases, the competitive interactions of these algae with other species have been studied. In central Chile (Santelices et al., 1981), *Codium dimorphum* forms a thick spongy crust and is able to overgrow and exclude most other lower intertidal algae. In southern California, a red algal turf comprising *Gigartina canaliculata, Laurencia pacifica,* and *Gastroclonium coulteri* outcompetes the brown alga *Egregaria laevigata*. The kelp recruits only from spores and only at certain times of the year, whereas the red algae expand vegetatively at all times of the year, encroaching on any spare space that becomes available. A 100 cm^2 clearing in the middle of a *G. canaliculata* bed was completely filled in 2 years.

Preemption of space is a common phenomenon in interactions between algae (Foster, 1982). For example, Lubchenco (1980) removed the encrusting alga *Chondrus crispis* from areas low on the shore, below the lower boundary of species of *Fucus*. Provided the *Chondrus* area remained clear, *Fucus* could become established at lower levels than normal. Thus, competition for space at low levels on the shore eliminated the higher shore species and prevented them from moving down the shore. However, in other studies, different results occurred in which the effect of the lower shore species was insufficient to account for the lower boundary of the upper shore species (see Underwood, 1991).

5.2.2.4.2 Competition between species of grazers

Because the densities of relatively immobile herbivores (e.g., sea urchins, gastropods) are easy to manipulate and their food supply can be altered experimentally, there is an extensive literature on experiments testing interspecific competition among such species. Limpets in particular, being dominant organisms on most shores, have received much attention.

Interspecific competition among intertidal snails is widespread (reviewed by Underwood, 1979; 1986; 1992; Branch, 1981; 1984). Underwood (1978c; 1984c) demonstrated that the snail *Nerita atramentosa* clearly outcompetes the limpet *Cellana atramoserica* when the two coexist in areas with inadequate food resources. Limpets lost weight and died much faster when *Nerita* were present than when on their own. Increased densities of *Cellana* did not affect *Nerita*. In Underwood's experiments, rates of mortality at different seasons and different heights on the shore conformed to those predicted from a knowledge of the amounts of food available at different levels and seasons (Underwood, 1984b,c), and rates of mortality were entirely consistent with the amount of food in experimental enclosures with different densities of limpets and snails. *C. tramoserica* was also found to be competitively superior to the gastropod *Bembicium nanum* (Underwood, 1978c), and the pulmonate limpets *Siphonaria* spp. (Underwood and Jernakoff, 1981; Creese and Underwood, 1982).

One clear-cut demonstration of the effects of competition has been described by Choat (1977) who showed that the experimental removal of a high-shore limpet *Collisella digitalis* allowed the species occupying the zone immediately below (*C. strigatella*) to expand its vertical range and move upward to occupy part of the zone previously dominated by *C. digitalis*.

Creese (1980) has investigated how the limpets *Cellana atramoserica* and *Patelloida latistrigata* can coexist when both feed indiscriminately on microalgae. Experiments showed that *C. tramoserica* outcompetes *P. latistrigata* unless barnacles are present. *Cellana* is presumably hindered during feeding when barnacles are present, while *Patelloida* is small enough to be unaffected by barnacles and even feeds on the surface of the barnacles. In addition, *Patelloida* finds a refuge from *Cellana* among the barnacles (Figure 5.11). Figure 5.11A shows that the mortality of *P. latistrigata* increased if all or half of the barnacle cover is removed, due to *C. tramoserica* invading the area. This interspecific interaction is further complicated by the presence of the common predatory gastropod *Morula*, which attacks the barnacles. Following an increase in the numbers of *Morula*, Creese (1982) recorded a decrease in barnacle cover, which led to a dramatic decline in *P. latistrigata* as *C. tramoserica* migrated into the area. Subsequent recruitment of *P. latistrigata* also failed due to heavy mortality of juveniles in the area where barnacles had been depleted (Figure 5.11B). Since the recruitment of *C. tramoserica* is decreased by the presence of barnacles, and the immigration of adults increased, and since *P. latistrigata* predominates among barnacles (Underwood et al., 1983), differences in the preferred microhabitats of these two species prevent exclusion of the inferior competitor.

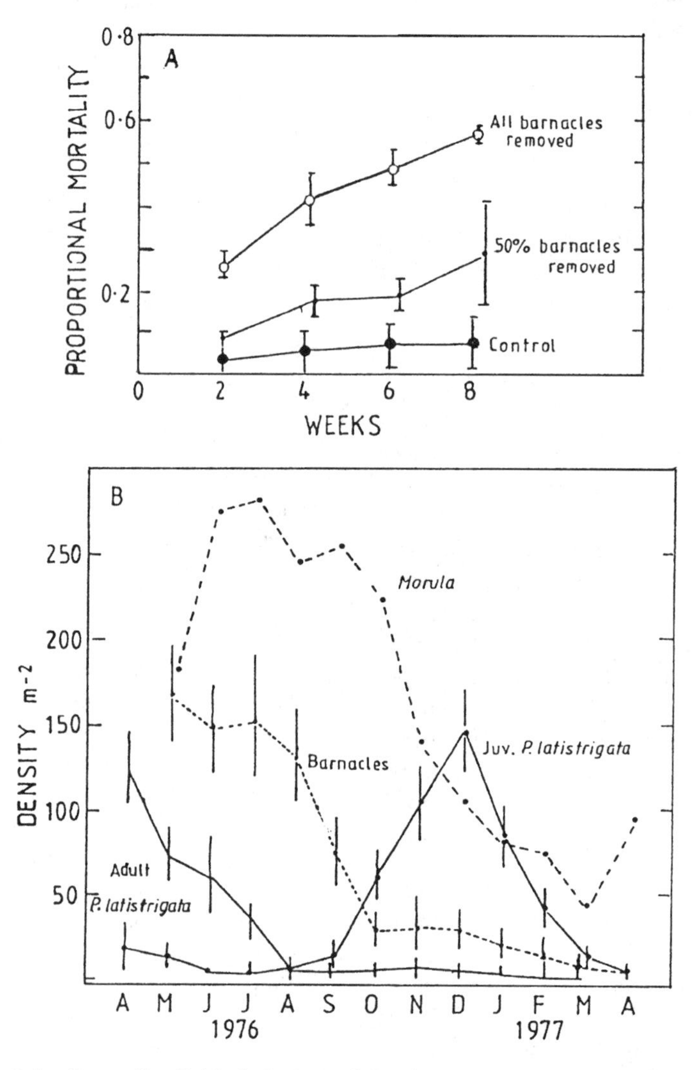

FIGURE 5.11 A. Mortality of the limpet *Patelloida latistrigata* following experimental removal of barnacles. B. Decline in the density of adult *Patelloida latistrigata* and failure of juvenile recruitment following elimination of barnacles by the predatory gastropod *Morula*. (Redrawn after Creese, R.G., Ecology and reproductive biology of limpets, Ph.D. thesis, University of Sydney, 1982. With permission.)

TABLE 5.1
Comparison of the Patterns of Symmetry Found in Five Studies of Competition among Grazing Molluscs

	Source:			
Comparison	**Underwood (1978c; 1984b)**	**Creese and Underwood (1982)**	**Ortega (1984)**	**Fletcher and Creese (1985)**
(1) Simple interspecific competition	*Nerita atramentosa* dominant over *Cellana tramoserica* (asymmetric)	*Cellana tramoserica* dominant over *Siphonaria denticulata* (asymmetric)	*Siphonaria gigas* dominant over *Fissurella virescens* (asymmetric)	*Cellana tramoserica* dominant over *Patelloida alticostata* (asymmetric)
(2) Relative effect of dominant species	*N. atramentosa* affects *C. tramoserica* more than it affects itself (asymmetric)	*C. tramoserica* has a greater effect on itself than on *S. denticulata* (symmetric)	*S. gigas* has a weak effect on both itself and *F. virescens* (asymmetric)	*C. tramoserica* has a great effect on itself and on *P. alticostata* (symmetric)
(3) Relative effect of inferior competitor	*C. tramoserica* has great effect on itself but little or no effect on *N. atramentosa* (asymmetric)	*S. denticulata* has little effect either on itself or *C. tramoserica* (symmetric)	*F. virescens* has strong effect on itself but no effect on *S. gigas* (asymmetric)	*P. alticostata* has moderate effect on both itself and on *C. tramoserica* (asymmetric)

Source: From Fletcher, W.J. and Creese, R.G., *Mar. Biol.*, 86, 189, 1985. With permission.

The examples discussed above indicate that microhabitat differences between herbivores, rather than differences in diet, were the important factors in their competitive interactions. However, some herbivores are very specialized in their diets and thus avoid competition for food.

5.2.2.4.3 Competition for space

In addition to the studies demonstrating direct competitive interactions for food, there are others that demonstrate interference competition between grazing species for space. Organisms involved in competition need to keep competitors away from areas of space that have their preferred food. A good example is provided by the owl limpet *Lottia gigantea* on Californian shores (Stimpson, 1970; 1973). *Lottia* occupy a space surrounded by a thin film of algae that are only found in such spaces. Each territory (space) is approximately 900 cm^2 in area, with the size increasing with the size of the limpet. Stimpson (1973) demonstrated that the densities of other grazing limpets, especially small species of the genus *Acmaea*, were low in the *Lottia* territories compared with other parts of the shore. Stimpson experimentally removed *Lottia* from some areas and introduced them to areas lacking *Lottia*. Acmaeds increased in density in areas from which *Lottia* were removed and conversely decreased in areas where *Lottia* were introduced. Stimson (1973) also noted that when acmaeds occupied areas previously occupied by *Lottia*, the algal film disappeared. He also demonstrated that *Lottia* actively defended their territories against invaders, including members of their own species. These studies demonstrate a process of direct interference competition where one species actively and aggressively pushes another out of part of the habitat. Similar processes have been described for South African limpets (*Patella longicostata* and *P. tabularis*) by Branch (1971; 1984).

Fletcher and Creese (1985) investigated interspecific competition between two species of grazing limpets, *Patelloida alticostata* and *Cellana tramoserica*, in northeastern Australia. Table 5.1 compares the patterns of interaction of these two species with those of four other studies of two competing limpet species. In studies that have specifically investigated interactions among co-occurring limpets using experimental manipulations, simple asymmetry (i.e., where one species adversely affects another but the reciprocal effect is less intense) has been found in every case (Haven, 1971; Underwood, 1978c, 1984c; Black, 1979; Creese, 1982; Creese and Underwood, 1982; Ortega, 1985). However, Fletcher and Creese (1985) found that *P. alticostata* has a relatively smaller effect on other *P. alticostata* than on *C. tramoserica*, while *C. tramoserica* has a greater impact on members of its own species than on *P. alticostata*; both of these interactions are symmetrical.

5.2.2.4.4 Competition between plants and grazers

Since grazers require space over which to feed and since plants need space to grow, there is a potential for competitive interactions between plants and grazers. At low levels on New South Wales shores, it has been demonstrated that plants can grow quickly even in the presence of grazing limpets (Underwood and Jernakoff, 1981). Limpets from higher levels on the shore were able to survive in areas down low where they did not naturally occur if such areas were cleared of foliose plants. The plants, however, were able to colonize the experimentally cleared areas rapidly. Unless the densities of the grazers was very large, the algae occupied all the space, making it impossible for microalgal grazing species such as *Cellana* to continue to survive there. Underwood and Jernakoff's (1981) experiments also demonstrated that *Cellana* would migrate away from areas that were occupied by mature stands of algae. Thus, the low areas of the shore are dominated by foliose plants that exclude microalgal grazing gastropods. At higher levels on the shore, the grazing gastropods are able to eliminate all of the propagules of the lower shore algal species (Underwood, 1985; Underwood and Jernakoff, 1981; 1984).

5.2.2.4.5 Competition between grazers and other organisms

Other organisms, especially sessile species, may compete with, or be out-competed for space by grazers. Connell (1961a,b) demonstrated that the limpet *Patella vulgata,* while grazing, can have an adverse impact on barnacles, by "bulldozing" them off the rock surface. This has been observed in the interactions between many species of limpets and barnacles (e.g., Dayton, 1971). However, there are situations where the presence of barnacles prevents grazing by large molluscan grazers (Choat, 1977; Lubchenco, 1983; Underwood et al., 1983; Jernakoff, 1985b).

5.2.2.4.6 Competition between plants and other organisms

Competition may occur between algae and sessile invertebrates. Colonial invertebrates may overgrow slow-growing perennial algae while the algae may smother barnacles (Underwood et al., 1983). Some foliose algae are washed around by wave action and brush over the surface of the rocks. This "whiplash" effect removes newly settled sessile invertebrates and prevents the survival of the adults (Menge, 1976; Dayton, 1971). Thus, in areas where there are large upright algae, there will be fewer barnacles or mussels and the space between the holdfasts may be available for grazing mollusca (Southward, 1964).

In addition, barnacles can provide direct shelter for the algae from grazers. Thus, the existence of sessile invertebrates, particularly barnacles, will alter the structure of the local community. This is an indirect result of competition for space between sessile invertebrates and grazers. Where sessile species such as barnacles are very abundant, they reduce grazing (they compete for space with the grazers) and thereby enhance the cover, diversity, and abundance of the algae. On some shores, barnacles provide the only refuge from grazing the propagules of algal species, so that, for example, *Fucus* grows only in areas where propagules are safe from grazing by snails (Lubchenco, 1983).

5.2.2.4.7 Competition between sessile filter feeders

Intense competition for space can occur between suspension feeders, often with the consequence that one or two species dominate the filter-feeding communities. However, many of these communities are characterized by instability, with variations in the recruitment of one or a few species leading to changes in dominance.

The classic and often quoted example of competition between sessile filter feeders is the work of Connell (1961a,b; 1970) on barnacles (see also Section 2.8.3). Connell (1961a) showed that *Balanus balanoides* consistently outcompetes *Chthalamus stellatus* in the mid-shore region, smothering, undercutting, or pushing *Chthalamus* off the rocks. As a result, *Chthalamus* is limited to a high-shore band above the tolerance limits of *Balanus*. Experimental removal of *B. balanoides* increases the survival of *C. stellatus* in the mid-shore region, where it achieves a faster growth rate than in its high-shore refuge. On New Zealand shores, the barnacle *Chaemosipho brunnea* is similarly restricted to the high shore by a combination of competition from other barnacles and predation (Luckens, 1975b; see also Section 2.8.3.5).

Since mussels are renowned for their competitive ability, interactions between pairs of mussel species have received special attention. Suchanek (1985) has described how *Mytilus edulis* is confined to a band above *Mytilus californianus*, its lower limit of zonation being set by competition. In another similar situation, Hoshiai (1964) has shown that the upper limits of *M. edulis* may be set by competition with the high-shore mussel *Septifer virgatus.* On some New Zealand shores, the lower limit of *M. galloprovincialis* is set by competition with the larger mussel *Perna canaliculus,* while its upper limit is set by competition with the smaller mussel *Xenostrobus neozelanicus.*

5.2.2.5 Processes Affecting the Outcome of Competition

5.2.2.5.1 Disturbance

As discussed below, physical disturbances of the habitat will affect the outcome of competition between plants, between animals, or between plants and animals. Sousa (1979b) demonstrated that disturbance of intertidal boulders can clearly influence the outcome of competitive processes. Where boulders are frequently overturned, successional processes among algae were disrupted and early colonizing species (especially *Ulva*) tended to predominate since they were able to preempt space and exclude other colonists. At the other end of the successional series on the boulders, there were "grab-and-hold" algae such as *Gigartina canaliculata*, which grows vegetatively from the remnants of plants that had been abraded.

Connell (1975) postulated that when dominant organisms and any associated species were removed, early colonization and subsequent events would depend largely on the physical harshness of the environment. In physically harsh environments, organisms that settled would often be killed by extreme environmental conditions; only when the environmental harshness was ameliorated would the colonists thrive and survive. At the opposite end of the physical spectrum, in physically benign environments, the organisms that settled would be removed by predators and grazers. Thus, the outcome of the successional process would depend on the interaction between environmental conditions and changes in the population densities of predators and grazers. Competition between colonizing species would result in the competitively dominating species occupying the space.

5.2.2.5.2 Grazing and preference for different types of food

Preferences for different types of plants by grazing animals may influence the outcome of competition among other plant species. For example, Paine and Vadas (1969) investigated grazing by sea urchins on sublittoral species of algae. Where sea urchins were relatively rare, the alga *Chondrus* out-competed other species of algae, with the result that algal diversity gradually decreased. On the other hand, where sea urchins were abundant, they consumed *Chondrus* preferentially and thereby made space available for other inferiorly competitive algae, thus increasing algal diversity (see also Lubchenco and Menge, 1978).

Lubchenco (1978) and Lubchenco and Gaines (1981) have addressed the issue of whether grazing or predation can increase or decrease the diversity of algae. Lubchenco (1978) found that algae were relatively diverse in rock pools with grazing snails (*Littorina littorea*), as compared to pools without grazers. It was suggested that the outcome of grazing on competitive interactions among species of food algae would be dependent on whether the grazers were preferentially feeding on a competitively superior species or on a competitively inferior one. In the former case, grazing would have the effect of reducing the intensity of competitive interactions, thereby freeing resources for other species and increasing algal diversity. Alternatively, where grazers fed on competitively inferior species, there would be a reduction in species diversity.

5.2.3 Predation

5.2.3.1 Introduction

Predation (including both herbivores and carnivores) can have a considerable impact on the behavior, ecology and evolution of organisms, and in organizing the structure and diversity of marine communities. Connell (1975) has suggested that it is the single most important factor affecting natural communities. Predators assume an important role in a developing body of ecological theory that attempts to explain local, regional, and global patterns of community organization (e.g., Dayton, 1971; 1975; 1984; Paine, 1974; 1977; 1980; Connell, 1975; Lubchenco, 1978; Lubchenco and Menge, 1978; Menge, 1978a,b; Huston, 1979; Menge and Lubchenco, 1981; Gaines and Lubchenco, 1982; Sih et al., 1985; Menge and Farrell, 1989). This body of research has led to the development of both mathematical (e.g., Cramer and May, 1972; Holt, 1977) and conceptual (Paine, 1969; Connell, 1975; Glassner, 1979) models. Hughes (1985) has presented a behavioral classification of predators (Figure 5.12) that includes grazers, filter feeders, and deposit feeders. Hunters are predators that feed on discrete macroscopic food items; there are three categories of hunters: ambushers, searchers, and pursuers.

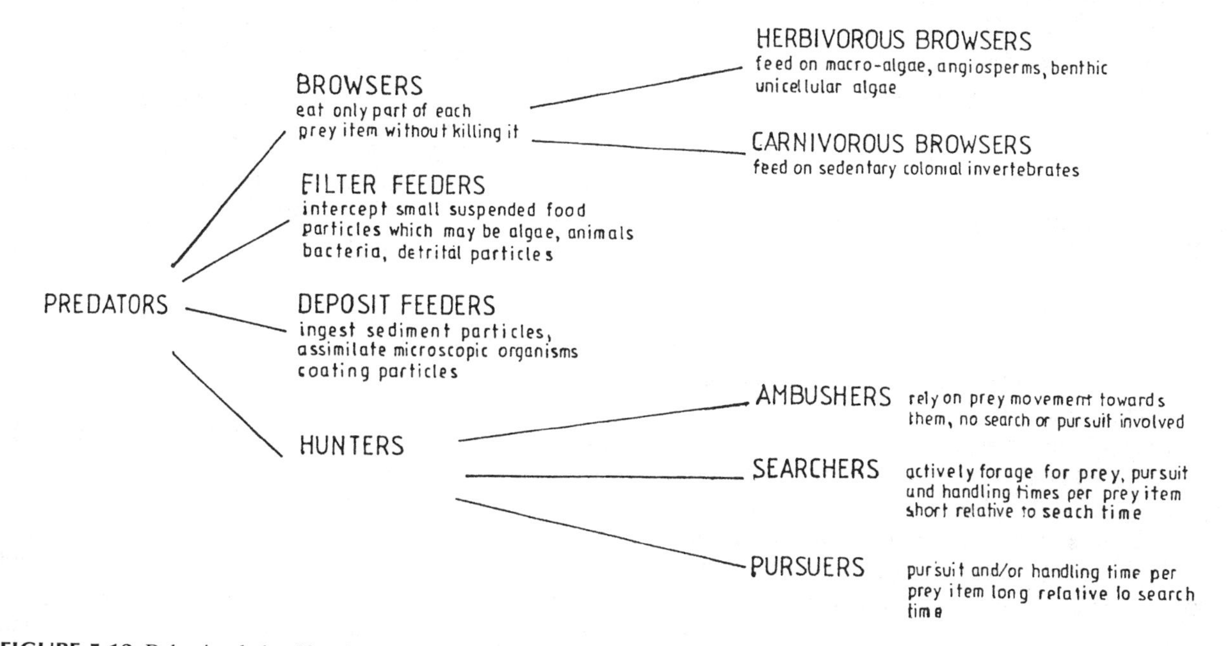

FIGURE 5.12 Behavioral classification of predators. (Modified from Hughes, R.N., *Oceanogr. Mar. Biol. Annu. Rev.*, 18, 423, 1980a. With permission.)

From extensive studies of predators on Northern Hemisphere shores and elsewhere, they have been assigned two main, but nonexclusive (see Hixon and Brostoff, 1983), roles in community organization, especially with regard to species diversity (see Connell, 1978). The first role is one of several "compensatory mechanisms" (Connell, 1978); certain species of predators mediate the outcome of predation in a community as it exists at or near equilibrium. By preferentially consuming the species that is a dominant competitor, predators indirectly favor the coexistence of subordinate species. Such species have been categorized as "keystone predators" (sensu Paine, 1969), and their effects are discussed in detail in Section 5.2.3.5.

The second well-documented effect on community structure has been called biological disturbance (Dayton, 1971; Connell, 1972). Predators that act patchily in space or time, whether they are selective or not, can locally disrupt the pattern of ecological succession (Connell, 1978). In patches subject to predation, species occurring earlier in the succession are favored, and this results in a mosaic of differently aged patches (Connell, 1961a; Dayton, 1971).

Typically, investigators determine the effect of consumers by examining the responses of the system in the absence of the predators. However, predation intensity varies from the extremes of no effect vs. a normal effect. In order to gain a clear understanding of the patterns of community structure as a result of the impact of predators, we need information on the components of predation intensity (Menge, 1983). There are at least four such components, including variation in (1) predator effectiveness (individual level), (2) predator density (species level), (3) number of species of similar morphologies (e.g., snails), and (4) number of consumer types of different morphologies (e.g., snails, seastars, and crabs represent three different types). Two groups of predators, whelks and birds, have been selected to illustrate predation impacts on rocky shores.

5.2.3.2 Predation by Whelks

Intertidal muricid whelks (Neogastropoda) are common predators on rocky shores and their behavior has been studied by many workers (e.g., Connell, 1970; Black, 1978; Menge, 1978a,b; Fairweather et al., 1984; McQuaid, 1985; Fairweather, 1985; 1987; 1988a,b; Fairweather and Underwood, 1983; 1991; Moran, 1985a,b,c). Most species (but not all; see Menge, 1972, and Garrity and Levings, 1981) forage during high tide and respond to either chemical or tactile stimuli. An understanding of the mechanics of foraging has provided information on the feeding biology of predators and their impact on prey species. Aspects of the foraging behavior of whelks in the context of optimal foraging theory will be dealt with in Section 5.4. Here we will be concerned with the impact of whelk predation on prey populations. Figure 2.23 depicts the distribution of barnacles and whelks on three shores. From the figure it can be seen that the barnacles settle over a greater intertidal vertical range than that occupied by the adults. The whelks are largely responsible for the restriction of the barnacles to their adult distributions.

The bulk of the early work on whelk predation has been carried out on mid-shore communities dominated by barnacles and mussels in temperate parts of the Northern Hemisphere. However in recent years, a group of workers in New South Wales, Australia (see Fairweather and Underwood, 1983; Moran, 1985a,b,c; Fairweather, 1988a,b), have carried out extensive investigations on the predatory whelk, *Morula marginalba*. On the New South Wales shores, the mid-shore assemblages consist primarily of barnacles, tubeworms, limpets, and algae (Denley and Underwood, 1979; Underwood et al., 1983; Moran, 1985a,b; reviewed by Underwood, 1991). Many different types of prey are eaten (barnacles, gastropod and bivalve molluscs, and serpulid polychaetes), but these are not chosen randomly (Fairweather and Underwood, 1983; Fairweather et al., 1984; Moran, 1985c; Fairweather, 1987). The impact of *M. marginalba* on the assemblages of prey depends indirectly on the availability of crevices, prevailing conditions of desiccation and wave shock, and the densities and preference ratings of the available prey (see Fairweather et al., 1984; Fairweather, 1985; 1987; 1988a,b; Moran, 1985a,b,c).

In experimental removal of *Morula* (Fairweather, 1986), it was found that certain prey that were readily eaten were relatively rare in a particular habitat with whelks present (e.g., barnacles in the mid-shore, tubeworms in gastropod pools and the mid-shore). When young individuals occurred in these areas, the whelks normally ate them all. In another series of experiments (Fairweather, 1987), either barnacles, limpets, or tubeworms were added to sites where they were quickly consumed by the whelks that were present.

McQuaid (1985) investigated the differential effects of predation by the intertidal whelk *Nucella dubia*, which occurs in the middle and upper balanoid zone of the rocky shores of the Cape of Good Hope, South Africa. Two areas were compared, the mid-balanoid zone and the upper *Littorina* zone. Ten prey species were recorded (Figure 5.13). The principal prey species in the mid-balanoid zone was the barnacle *Tetraclita serrata*, while on the upper shore it was the winkle *Littorina africana knysnaensis*. Predation in the *Littorina* zone was cyclical. *Nucella* migrated upshore from the upper balanoid zone as spring tides approached and downshore during neap tides. Both the number of whelks present and the percentage feeding peaked 1 to 2 days before spring tide. Despite high rates of consumption (mean 0.47 *Littorina* whelk^{-1} day^{-1}), only 12% of the *Littorina* population are lost to predation per year. Caging experiments in the balanoid zone revealed consistently low predation rates (0.02 barnacle whelk^{-1} day^{-1}) for both large and small whelks. However, due to

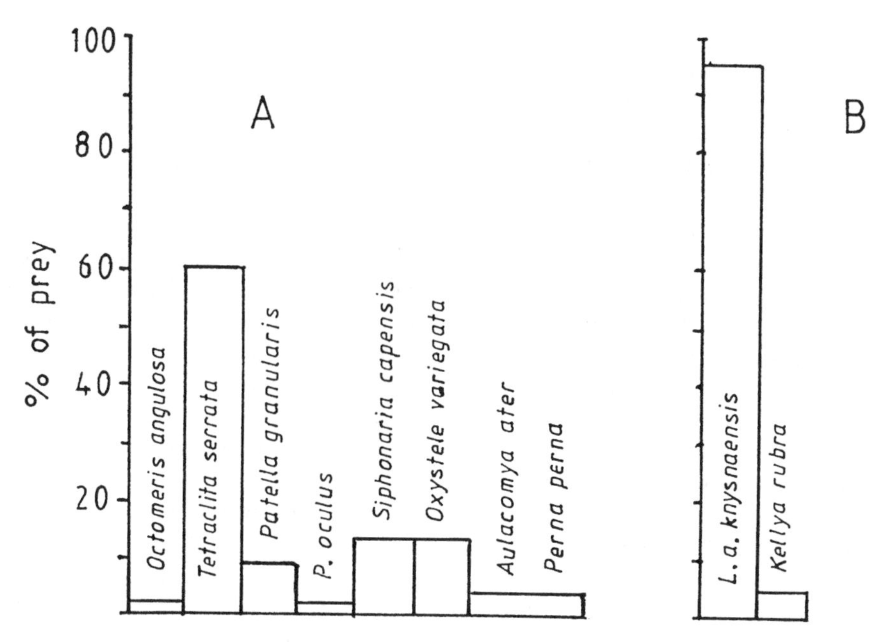

FIGURE 5.13 Percentage frequency with which prey species are taken by *Nucella dubia*. A. Mid-balanoid zone; B. *Littorina* zone; horizontal shading indicates barnacle species; light shading, gastropods; dark stipple, bivalves. (Redrawn from McQuaid, C.D., *J. Exp. Mar. Biol. Ecol.*, 89, 102, 1985. With permission.)

high whelk density (86 m^{-2}) potential predation was estimated at 43.66% of the barnacle standing crop per year. This is probably adequate to control barnacle density and to maintain free space in the barnacle beds. The results of this study support the view that predators have a greater influence on sessile than on mobile prey.

In most habitats and ecosystems, predation varies significantly over space and time. Navarrete (1996) investigated such variation on a mid-tidal successional community at a wave-exposed environment on the Oregon coast. At this site two predatory whelk species, *Nucella canaliculata* and *N. emarginata,* were highly correlated over time and varied greatly within and among years. A suite of direct and indirect effects was observed following the permanent exclusion of the predators, notably a rapid increase in the cover of the bay mussel *Mytilus trossulus,* and a slow, small increase in the cover of the gooseneck barnacles and the Californian mussel *Mytlus californianus.* Temporarily variable predation (medium and low frequencies) produced community composition different than those observed under a constant predation regime or predation exclusions. Both the ability of some prey to escape predators by reaching a large size and the temporal pattern of prey recruitment seemed important in determining the effect of variable predation on individual prey, but there were also many indirect effects. Thus, the temporal variability in predation by whelks can have distinctive effects on prey, create distinctive community composition, and affect successional paths in the intertidal community. It is clear that temporal variability in predation is probably an important, yet poorly understood factor, cause of spatial heterogeneity in shore ecosystems.

5.2.3.3 Predation by Highly Mobile Predators

Predators such as starfish and whelks are slow moving, spend a long time handling their prey, and are influenced by physical factors such as desiccation at the upper limits of their range. In contrast, highly mobile predators, such as crabs and fishes, can forage throughout the entire tidal zone during one high tide (e.g., Robales et al., 1989), and often require only a short time to handle their prey; e.g., Menge (1983) found that the crab *Carcinus maenas* had a per capita consumption that was 25 times that of the drilling whelk *Nacella lapillus.*

On the northeastern Pacific shore, the intertidal distributions of the gastropods *Littorina sitkana* and *L. scutulata* overlap broadly with those of two species of mid-shore crabs *Hemigrapsus nudus* and *H. oregonensis*, but do not extend below the upper limit of two species of shore crabs *Lophopanopeus bellus* and *Cancer productus,* or into the subtidal zone where adult *C. productus* are abundant. Yamada and Boulding (1996) found that adult *L. sitkana* transferred to lower intertidal levels suffered considerable mortality, principally from predation by *C. productus.* Their results indicated that the risk of predation decreased upshore with decreasing immersion time. Thus, mobile predators play a role in controlling the populations of intertidal molluscs and in determining their lower limits on the shore.

5.2.3.4 Predation by Birds

A wide variety of birds feed on rocky shores. For example, during a 3-year period at Robin Hood's Bay, Yorkshire, 47 species of birds were seen feeding in the rocky intertidal along a 1.5 km stretch of coast (Feare and Summers, 1986). However, of these species only nine were seen to feed regularly on the shore and only four of these occurred in appreciable numbers. The principal species groups that feed when the tide is out are sandpipers, turnstones, gulls, and oystercatchers. When the tide is in, the littoral zone becomes available to aquatic birds such as cormorants, ducks, gulls, and terns. The Northern Hemisphere eider duck, *Samateria mollissima*, includes a wide variety of invertebrates in its diet, but mussels generally predominate. Very few birds graze on seaweeds of rocky shores. One notable species is the kelp goose *Chloephaga hybrida* from southern South America, which feeds on *Ulva* spp. and *Porphyra* spp.

Rocky shores support high densities of potential prey for birds. For example, Seed (1969) recorded mussel (*Mytilus edulis*) densities of over 400,000 m^{-2}, while Gibb (1956) found the average density of the periwinkle *Littorina neritoides* to be 25,000 m^{-2}. Densities of limpets can exceed 1,000 m^{-2}. Investigations have shown that large numbers of prey species are eaten by the birds. Gibb (1956) calculated that on the shore he studied, the rock pipits, *Anthus spinoletta petrosus*, ate about 14,300 periwinkles and 3,500 chironomid larvae per day, and Feare (1966) estimated that a flock of 40 purple sandpipers, *Calidrus maritima*, ate about 11.5 million littoral snails during a 2-month winter residence. The question arises as to the impact of such predation on the populations of the prey species.

A large number of studies (e.g., Blankley, 1981; Frank, 1982; Lindberg and Chu, 1983; Hockey and Branch, 1984; Hockey and Underhill, 1984; Luchenbach, 1984; Feare and Summers, 1986; Marsh, 1986a,b; Piersma, 1987; Evans, 1988; Hockey and Bosman, 1988) have shown that bird predation may significantly affect the distribution, abundance, behavior, and habitat selection of rocky shore intertidal invertebrates. In Britain, Feare (1969) estimated that purple sandpipers were responsible for 93% of the 89% winter mortality in first-year dogwhelks. The birds had access to the whelks for only 33 days between January and April. At a nearby locality, Lewis and Bowman (1975) recorded an 81% reduction in populations of the limpet *Patella vulgata* population in 2 months following the winter influx of oystercatchers. A number of studies (e.g., Frank, 1982; Marsh, 1986a) have suggested that birds may be able to reduce prey abundances to levels below which continuous foraging on a patch is unprofitable, thus resulting in changes in foraging patterns over time scales as brief as a few days. Meese (1993), in a study of predation by surfbirds and gulls on the Californian coast, found that surfbirds consumed large numbers of gooseneck barnacles *Pollicipes polymerus*. Bird exclusion experiments showed that winter feeding by these species significantly reduced gooseneck barnacle abundance at middle and low intertidal levels.

On the Pacific Northwest coast of the United States, mussels (*Mytilus edulis* and *M. californianus*) are major prey items of surfbirds (*Aphriza virgata*), gulls (*Larus glaucescens* and *L. occidentalis*), and black oystercatchers (*Haematopus bachmani*). Marsh (1986a) tested the effect of these predators on mussel recolonization of cleared areas, using bird exclusion cages. Three of the five enclosure experiments showed that the birds significantly reduced recruitment of juvenile mussels into the clearings. A sixth experiment showed that birds were responsible for the absence of mussels from an area of smooth substrate lacking mussel beds. In this experiment, clumps of *M. edulis*, 10 to 20 mm in length, became established in all enclosures, but not in any controls. The results of the experiments indicated that the long-term impact of avian predators was greatest in patches where invertebrate predators were uncommon, and where larval settlement, rather than adult encroachment, is the major form of recruitment. In a subsequent study, Marsh (1986b) found that the black turnstones (*Arenaria melanocephalata*) were the major predators on the Oregon coast of small limpets (*Collisella* spp.), while black oystercatchers were the major predators of large limpets (>2 mm). Intermediate-sized limpets were eaten by both species as well as by gulls. Marsh found that the predators were capable of drastically reducing limpet densities in unvegetated high intertidal areas.

On some shores, ducks can have a considerable impact on invertebrate prey, in particular on mussel populations (Feare and Summers, 1986). On the northwestern Atlantic coast, seaducks are major predators on intertidal invertebrates (Gouldie and Ankney, 1986). Seaducks (oldsquaw, common eider, and black scooter) are major predators on mussels (Figure 5.17). The black scooter feeds almost exclusively on mussels, while the common eider also consumes considerable amounts of the sea urchin, *Strongylocentrotus droebachiensis*, as well as a wide range of other invertebrates. The herbivorous gastropod, *Lacuna vincta*, is prominent in the diets of the oldsquaw and harlequin ducks.

Wootin (1992) experimentally manipulated avian (black oystercatchers and gulls) predation pressure at the Tatoosh Island, Washington, U.S.A., intertidal community to determine its direct and indirect influence on the abundance and distribution of three limpet species (*Lottia (Colisella) digitalis, L. pelta,* and *L. strigatella*) and their algal food supply. Bird predation reduced the overall abundance of *L. digitalis* only; *L. strigatella* abundance increased and the abundance of *L. pelta*, the species most commonly consumed, did not change. By consuming limpet grazers and reducing space competition, birds indirectly enhanced the abundance of algae (Figure 5.14). The gooseneck barnacle *Pollicipes polymerus* and the mussel *Mytilus californianus* indirectly affected the abundance

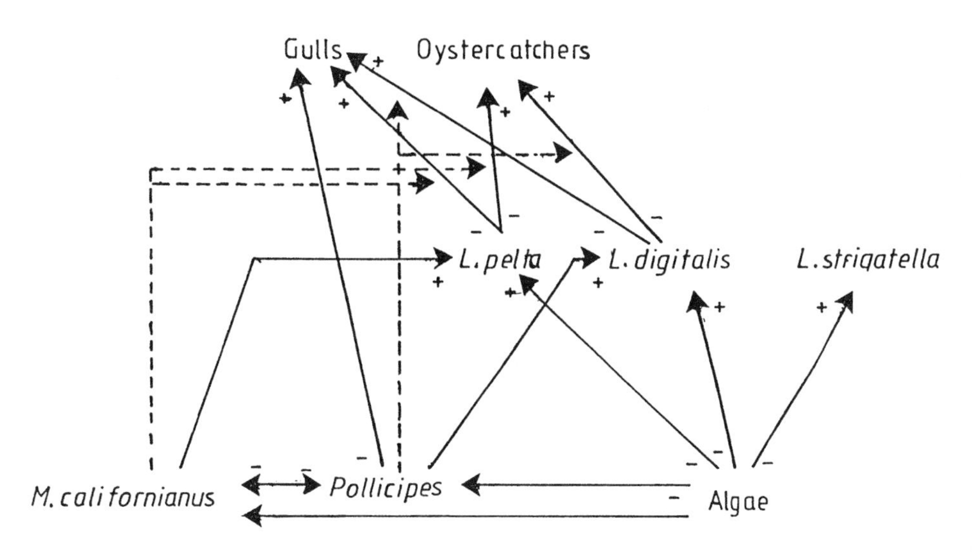

FIGURE 5.14 Diagram of interactions affecting limpet abundance in the middle intertidal zone of Tatoosh Island, Washington. Solid arrows indicate direct interactions among species (e.g., predator-prey, interference competition, habitat preference). Stippled arrows indicate indirect interactions that act by modifying a direct interaction between two other species. Other indirect effects can be visualized by following chains of direct interactions. (Redrawn from Wootin, J.T., *Ecology*, 73, 989, 1992. With permission.)

of *L. digitalis* and *L. pelta* in opposite ways by changing their risk of succumbing to bird predation. These limpets also exhibited strong habitat selection for their cryptic substrates. By altering the amount of preferred habitat, birds indirectly influenced limpet abundance; gull predation reduced the area covered by *P. polymerus*, thus releasing *M. californianus* from space competition. *L. strigatella* sizes generally fell below the range of limpet sizes consumed by birds. Consequently, birds indirectly increased *L. strigatella* density by reducing the intensity of exploitative competition with other limpet species; *L. strigatella* biomass declined significantly with increasing biomass of the other two limpets, and with decreasing algal cover. The results of this study demonstrate that indirect effects and apparently adaptive behaviors can counteract (or reinforce) direct interactions between species pairs, suggesting that conclusions from short-term experiments emphasizing species pairs need to be viewed with caution (Figure 5.14).

The rocky shore avifauna of South African shores may be divided into three categories, based on foraging behavior and dispersion patterns; two categories comprise individual resident species while the other is a group of migrants. Of the dominant resident species, African black oystercatchers *Haematopus moquini* are territorial and specialized in their foraging (on limpets and mussels) in contrast to kelp gulls *Larus dominicanus*, which are itinerant and feed opportunistically. Several smaller migratory species are present during the austral summer, notably turnstones *Arenaris interpres*, sanderlings *Calidris alba*, and curlew sandpipers *C. ferruginea*. These migrant species prey predominantly on the infauna of mussel and algal beds.

Seabirds in the southwestern Cape interact as predators with at least three of the spatially dominant intertidal assemblages of the region, namely mussel beds, areas dominated by limpets, and the infauna of algal mats. Hockey and Bosman (1988) have investigated the role of avian predators in structuring the intertidal communities. African black oystercatchers were assumed to derive 50% of their annual energy needs from mussels (*Chloromytilus meridionalis*) (Hockey and Underhill, 1984), while kelp gulls were estimated to obtain 25% of their energy requirements from small mussels. Apart from gulls and oystercatchers, the only important predators of intertidal mussels are whelks *Nucella* spp. There was a marked decrease in losses due to competition as mussels approach their terminal size. The relative importance of the predators also varied over time. Kelp gulls and *Nucella* both feed on small mussels and account for almost all the predatory losses in the first year of a mussel cohort's existence. In subsequent years, mussels achieve a size refuge from these predators, but enter the preferred size range of oystercatchers, which become the major predator of mussels from year 2 onward. When mussels are small, losses due to predation are less than losses due to competition, but from year 4 onward, oystercatchers reduce the numbers so as to achieve a stable packing density in the mussel population.

Not all oystercatchers that feed on limpets are territorial throughout the year. Frank (1982) investigated the impact of seasonal (winter) foraging by flocks of nonterritorial oystercatchers (*Haematopus bachmani*) on limpets. He found that numbers of *Notoacmea persona* >20 mm in length fell by two orders of magnitude, and *N. scutum* by one order of magnitude during the winter. Thus, oystercatchers have been shown to have two types of effect on limpet prey populations. The predation intensity of a resident, territorial species is attuned to the life-history characteristics of its prey in a manner that leads to long-term, large-scale stability, whereas seasonality flocking species can have a potentially destabilizing, but spatially restricted impact on their prey populations.

The predation of algal infauna by small migrant shorebirds can have a major depletion effect on their prey populations, but only over a short period of the year; the prey are thus afforded a temporal refuge from predation. The high P:B ratios of the algal infauna enable invertebrate numbers to build up prior to the next period of intense predation pressure.

It is clear that birds are an integral and important component of rocky shore communities and that they interact with their prey populations in a number of stabilizing and potentially destabilizing ways. The process involved may vary according to life-history parameters and morphologies of the prey, variations in primary and secondary production rates, and the spatiotemporal incidence of predation (Hockey and Bosman, 1988).

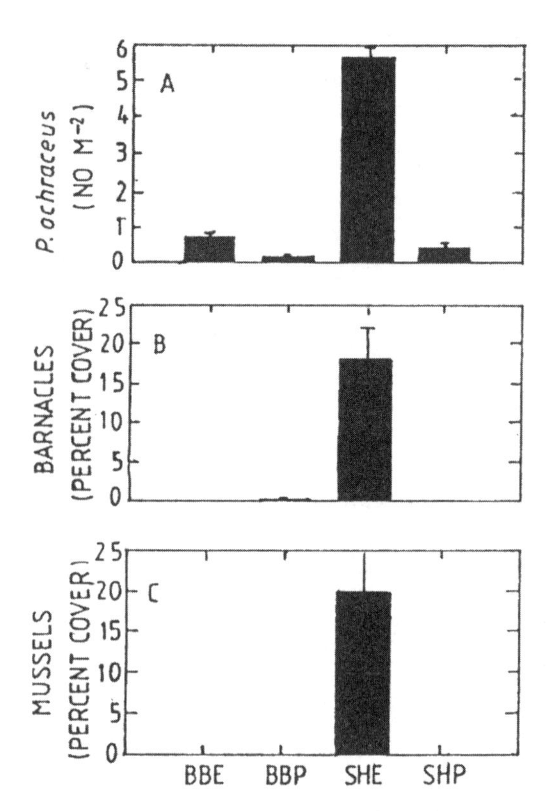

FIGURE 5.15 Abundance of major categories of invertebrates in the low zone (+1.2 to –0.6 m) at Broiler Bay exposed (BBE) and protected (BBP) and Strawberry Hill exposed (SHE) and protected (SHP) sites. (A) Density of the seastar *Pisaster ochraceus*. (B) Percent cover of barnacles (*Balanus glandula, Chthalamus dalli, Pollicipes polymerus, Semibalanus cariosus,* and *Balanus nubilus*). (C) Percent cover of mussels (*Mytilus trossulus*). Percent cover was estimated using the transect-quadrant method (Lubchenco et al., 1984); Ten 0.25-m^2 clear vinyl sheets were placed at random points along a 30-m transect placed parallel to the shore within the low zone. Cover was estimated as the number or randomly plotted dots on the sheet that overlaid each species or type of organism. Error bars are 1 SE. (Redrawn from Menge, B.A., *Ecology*, 73, 758, 1992. With permission.)

5.2.3.5. The "Keystone Species" and "Diffuse Predation" Concepts

Since it was first introduced by Paine (1969), the "keystone species" concept has been a central organizing principle in community ecology (Mills et al., 1993). This concept has been loosely applied to species at all trophic levels that have disproportionately large effects on their communities (Walker, 1991; Mills et al., 1993; Lawton and Jones, 1995). Recently, however, both the usefulness and generality of this concept has been questioned (Strong, 1992; Mills et al., 1993). Menge et al. (1994) recently reviewed the "keystone species" concept with particular reference to variation in interaction strength between the original keystone predator, the seastar *Pisaster ochraeus*, and its primary prey, the mussels (*Mytilus californianus* and *M. trossulus*).

A "keystone predator" is defined as only one of several predators in a community that alone determines most patterns of prey community structure, including distribution, abundance, composition, size, and diversity (Menge et al., 1994). Thus, a predator is not a keystone if: (1) total predation is moderate to strong, but each of the predators alone has little measurable effect (termed "diffuse predation"; see Menge and Lubchenco, 1981; Robales and Robb, 1993), or (2) total predation is slight (termed "weak predation"). Paine (1969) suggested that two important properties of the keystone predator concept were: (1) the predator preferentially consumed and controlled the abundance of the prey species, and (2) this prey species could competitively exclude other species from the community.

Menge et al.'s 1994 study of interaction strength between the keystone predator and its prey was prompted by the differences in community structure at two sites along the central Oregon coast, Broiler Bay (BB) and Strawberry Hill (SH) (Figure 5.15). Predators, especially seastars, were larger and more abundant at SH than at BB. Further, sessile animals were more abundant and macrophytes (macroalgae and sea grasses) were less abundant at SH. Predators were more abundant at wave-exposed sites at both BB and SH, sessile invertebrates were more abundant at wave-exposed locations, and sand cover was high at wave-exposed locations. To test the hypothesis that variation in predation strength explained some of the differences, seastar-mussel interactions were studied at locations with high and low wave exposure at both sites. A series of transplant experiments and density manipulations were carried out to quantify the density and growth of individually transplanted mussels.

Predation intensity varied greatly at all spatial scales. At the two largest spatial scales (10s of kilometers, 100s of meters), differences in both survival of transplanted mussels and prey recolonization depended on the extent of sand burial. Variation at the two smallest scales (meters, 10s of meters) was high when seastars were scarce and low when seastars were abundant. Transplanted mussels suffered 100% mortality in 2 weeks at wave-exposed SH, but took 52 weeks at wave-protected BB. Sister effects on prey recolonization were detected only at the SH wave-exposed site. Here, where prey recruitment and growth were unusually high, the mussel *Mytilus trossulus* invaded and dominated space within 9 months. After 14 months, whelks, which increased both in size and abundance in the absence of *Pisaster,* arrested this increase in mussel abundance. Similar changes did not occur at other site–exposure combinations, evidently because prey recruitment was low and possibly also due to whelk predation on juveniles. Longer term results indicated that, as in Washington State, seastars prevent large adult *M. californianus* from invading lower intertidal regions, but only at wave-exposed, not wave-protected sites. Thus, three distinct predation regimes were observed: (1) strong keystone predation by seastars at wave-exposed headlands; (2) less-strong diffuse predation by seastars, whelks, and possibly other predators at a wave-protected cove; and (3) weak predation at a wave-protected site buried regularly by sand. Figure 5.16 depicts Menge et al.'s interpretation of the interaction webs for Broiler Bay and Strawberry Hill.

An alternative viewpoint to the keystone model suggests that all species might play a similarly small but significant role in the functioning of their communities (Erlich and Wilson,1991; McNaughton, 1993), as a sort of "diffuse" or more equally shared impact on the rest of the species (Lubchenco et al., 1984; Robales and Robb, 1993; Menge et al.,1994).

Examples from a range of habitats suggest that both keystone and non-keystone or diffuse predation are widespread, and that prey production rates may be the primary factor underlying variation in keystone predation. At intermediate to sheltered sites in New England, mid-intertidal community structure appeared to be controlled by the whelk *Nucella lapillus*, despite the presence of other predators (Menge, 1976; 1978a,b; 1983; 1991b; but see Edwards et al., 1982 and Petraitis, 1990, for alternative explanations). In contrast, low intertidal community structure was evidently determined by a guild of predators (whelks, seastars, and crabs), no one of which was consistently dominant (Lubchenco and Menge, 1978, Menge, 1983). Hence, mid and low zones were characterized by keystone and diffuse predation, respectively. Similar variation was observed in southern California (Robles, 1987;

POSTULATED INTERACTION WEBS

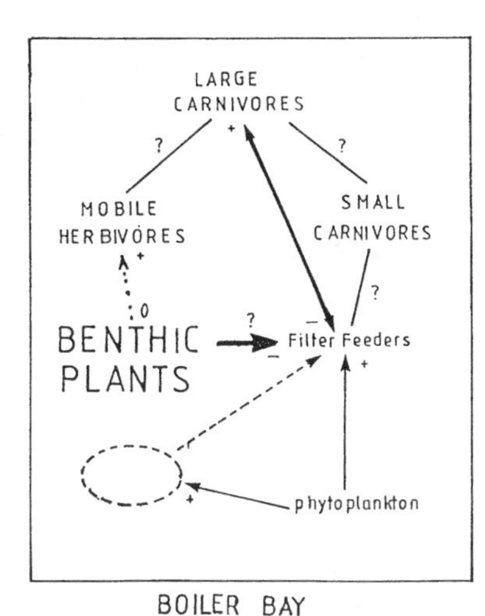

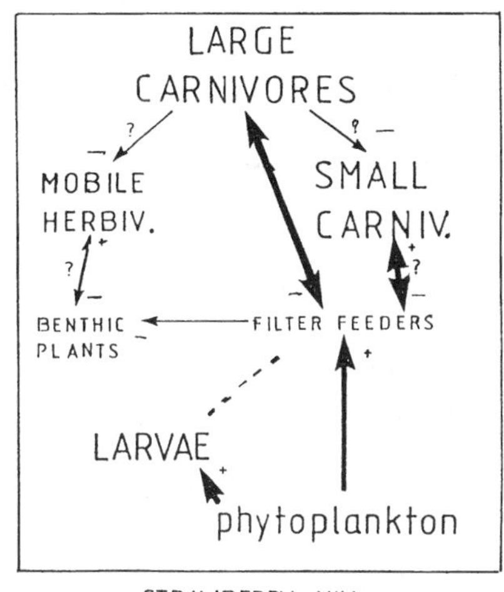

FIGURE 5.16 Hypothetical interaction webs for Broiler Bay and Strawberry Hill. Size of lettering qualitatively indicates relative abundance of each category; thickness of lines qualitatively suggests the strength of the link. +, –, and 0 indicate effects that are positive, negative, and weak or indictable. Question marks indicate that the magnitude of the link is unknown. Links with no question marks are suggested by the data obtained in the study. Arrowheads point to group affected; upward arrowheads indicate energy gains, downward or lateral arrowheads indicate interaction effects. Absence of an arrowhead suggests insignificant effects. Large carnivores are mostly *Pisaster ochraceus*; small carnivores are mostly whelks (*Nucella* spp.); mobile herbivores are primarily limpets (*Lottia* spp.) and chitons (*Katharina tunicata, Tonicella laneata, Mopalia* spp.); filter feeders are largely mussels (*Mytilus* spp.) and barnacles (*Balanus* spp., *Semibalanus cariosus, Chthalamus dalli*, and *Pollicipes polymerus*); benthic plants are seaweeds and surfgrasses. "Larvae" refer to planktrophic larvae of mussels and barnacles; the dashed arrow signifies larval settlement. (Redrawn from Menge, B.A., *Ecology*, 73, 762, 1991b. With permission.)

Robles and Robb, 1993). High sheltered and mid-exposed community structure depended on keystone predation by lobsters, and mid-sheltered community structures depended on diffuse predation (mostly by lobsters and whelks).

Keystone predation was also observed in New Zealand (Paine, 1971), where a seastar-controlled mussel zonation with little apparent impact on other predators, and in Chile (Duran and Castilla, 1989), where a large predatory gastropod evidently controlled mussel abundance and zonation in the mid-intertidal zone. Diffuse predation has been suggested in two other studies. In Panama, with low prey recruitment, total predator exclusion (fishes, molluscan grazers, crabs, and whelks) produced large increases in prey, while single predator exclusions produced little or no change (Menge and Lubchenco, 1981; Lubchenco et al., 1984; Menge et al., 1986b; Menge, 1991a). In Ireland, crabs, seastars, and whelks apparently combined to limit mussels to wave-exposed shores (Kitching et al., 1959).

In some subtidal habitats (Alaska), keystone predation by sea otters evidently controlled sea urchins and thereby indirectly maintained dense kelp beds (Estes et al., 1978, Van Blaricom and Estes, 1988; Estes and Duggins, 1995). In the tropics, the seastar *Acanthaster planci* can alter community structure of coral reefs dramatically during population outbreaks, and in such cases appears to be a keystone predator. Finally, in South Africa, rock lobsters control mussels on one island (Malgas Island), but are eliminated on nearby Marcus Island by whelks, evidently resulting in high mussel abundance (Barkai and Branch, 1988; Barkai and McQuaid, 1988). Both lobsters and whelks appear to be keystone predators.

5.2.3.6 Impact of Predation on Community Structure

Experimental studies of predation (taken in its widest ecological sense to include herbivory; Lubchenco, 1980; Hughes, 1980b) have demonstrated grazing can prevent the establishment of macroalgae. Following the pioneer study of Connell (1961a), which showed that the removal of dogwhelk predation ameliorated the intensity of competition among barnacles, a series of investigations (some of which have been discussed above) have demonstrated that predation, acting synergistically with disturbance, can shape the nature of intertidal communities. Examination of some examples follows.

5.2.3.6.1 New England rocky intertidal

Processes in the intertidal zone of New England rocky coasts have been intensively studied by Menge and his co-workers (Menge, 1976; 1978a,b; 1983; 1991a,b; Menge and Sutherland, 1976; 1987) by carrying out replicated field experiments at different sites along an environmental gradient. On relatively protected low rocky intertidal regions in northern New England, community structure depends on the foraging activities of up to six species (three general types) of predators (Lubchenco and Menge, 1978). These include three species of crabs (*Carcinus maenus, Cancer borealis,* and *C. irroratus*), two species of seastars (*Asterias forbesi* and *A. vulgaris*), and one thaid gastropod (*Nucella lapillus*). This predator guild prevented mussel and barnacle populations from outcompeting the red algae *Chondrus crispis,* which dominates space when predators are present, but is out-competed by mussels when predators are absent.

Predation consumption rates in field experiments indicated that the rank order from the most to the least effective predation type was crabs, seastars, and the gastropod. Estimates of the relative contribution of each species to the total predation intensity indicated that each species was a major predator at one or more sites. Thus, if one predator species in the guild becomes scarce, the other predators may increase their efforts and thus reduce variation in the total predation intensity exerted by the guild. Predator intensity seems to be dependent on variation of at least four predator characteristics, which Menge (1983) terms the components of predation intensity. These range over four levels of complexity: individuals (e.g., among phenotypes in different habitats), populations (e.g., density, size, or age structure), species, and predator types.

Community theory holds that species diversity (s = richness or H') is related to predation and disturbance intensity in at least one or maybe two ways (see Lubchenco and Gaines, 1981, for a summary). Most commonly observed is a "hump-shaped" or quadratic curve (i.e., high diversity at intermediate densities, low diversity at low and high intensities), although an inverse relationship is sometimes found. Predator manipulations can reveal the relationship between diversity and predator intensity. When predators are excluded, one of four diversity responses is observed:

1. No change indicates that predators are ineffective (e.g., Menge, 1976; Paine, 1980);
2. A decline in diversity (e.g., Paine, 1966; 1971; 1974; McCauley and Briand, 1979; Russ, 1980; Peterson, 1979a,b) indicates that predators normally maintain high diversity; such declines are generally due to the expression of competitive dominance by one or two species, which are normally held in check by the predator;
3. An increase in diversity (e.g., Lubchenco, 1978; Reise, 1977; Virnstein, 1977a,b; Peterson, 1979b) suggests that predation is very intense and maintains low diversity; prey coexisting with predators are predator-resistant, occur in refuges, or are highly opportunistic; and
4. Diversity may first increase due to successful invasions and increase in abundance of prey, and then decrease due to competitive exclusion (e.g., Paine and Vadas, 1969). Thus, predation

intensity varies widely in communities differing greatly in complexity.

Similar patterns occur in other communities. For example, Menge and Menge (1974) showed that prey ingestion rates g^{-1} for the seastars *Pisaster ochraceus* and *Lepasterias hexactis* on the West Coast of North America differ by nearly an order of magnitude. In the same region, *Thais* spp. feeding rates vary along a gradient of wave exposure. Further, the *Pisaster*-determined lower limit of mussels on these shores (Paine, 1974) is lower on the more exposed shores than on more protected ones in both Washington (e.g., Dayton, 1971) and Oregon (Menge, 1983). This suggests that the effectiveness of *Pisaster* predation may decline with increased wave exposure. In Oregon, Gaines (1983) found that quantitative differences in foraging among a guild of taxonomically diverse lower intertidal herbivores led to important differences in algal species composition and abundance. In England, Seed (1969) found variations in feeding rates among predators similar to those reported by Menge (1983) among *Nucella lapillus, Asterias rubens,* and two species of crabs.

5.2.3.6.2 North American West Coast

Paine (1966; 1980), in a series of experiments on the exposed rocky shores of Washington, U.S.A., demonstrated that predation might be the key to coexistence of the diverse community of sedentary organisms found there. His predator exclusion experiments (Paine, 1966) showed that the consumption of mussels (*Mytilus californianus*) by the starfish *Pisaster ochraceus* can regenerate primary space for colonization by sedentary organisms. This work led to the development of the "keystone species" concept discussed above, which has had considerable influence on thinking on the role of predation on rocky shores. Figure 5.17A depicts the *Pisaster*-dominated subweb on the Washington coast. Light predation may be ineffectual in preventing competitive exclusion among prey, heavy pre-

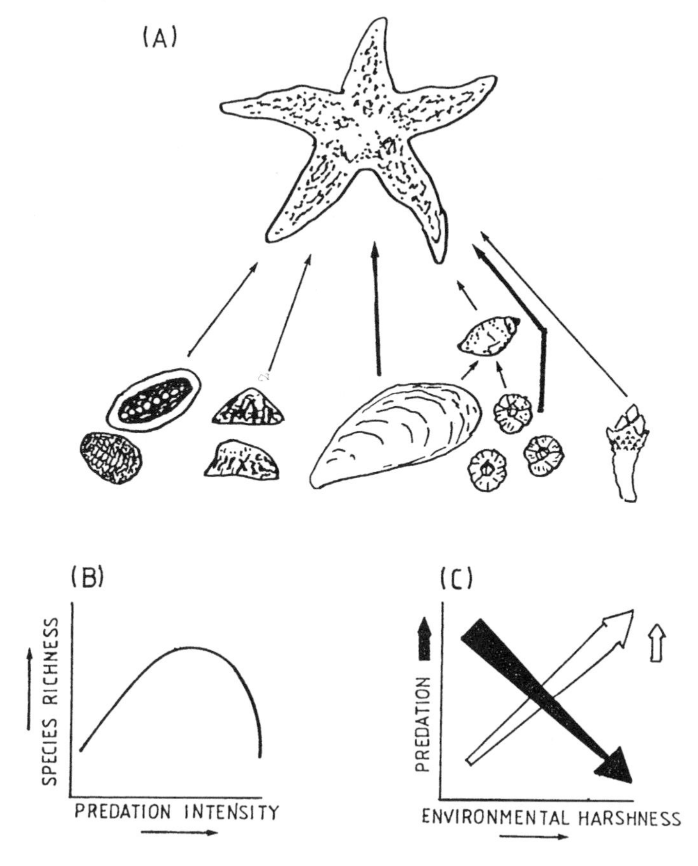

FIGURE 5.17 A. The *Pisaster*-dominated subweb in the exposed rocky shore community studied by Paine (1966) on the West Coast of North America. Arrows point from prey to predator. From left to right, the prey are two species of chiton, two acmaeid limpets, *Mytilus californianus, Nucella emarginata*, three species of sessile barnacle, and *Pollicipes polymerus*. B. The effects of predation intensity on species richness; light predation may be ineffectual in preventing competitive exclusion among prey and heavy predation may eradicate prey, whereas intermediate intensities of predation will promote coexistence of prey (from Lubchenco, 1978). C. Predation diminishes in importance as a structural force in rocky shore communities, while competition among prey increases in importance as the physical conditions become harsher (from Menge and Sutherland, 1976). (Redrawn from Hughes, R.N., in *The Shore Environment*, Vol. 2, *Ecosystems*, Price, J.H., Irvine, D.E.G., and Farnham, W.F., Eds., Academic Press, London, 1980b, 228. With permission.)

dation may eradicate the prey, and intermediate intensities of predation will promote the coexistence of prey and hence species richness (Figure 5.17B) (Lubchenco, 1978). As the physical environment becomes harsher (e.g., increasing wave action), predation diminishes in importance as a structural factor in rocky shore communities, while competition among prey species increases in importance (Figure 5.17C) (Menge and Sutherland, 1976).

In contrast to the single "keystone" predator discussed above in the barnacle-mussel zone, a greater number of "strongly interacting" (Paine, 1980) "foundation" species determined the nature of the richer, algal-dominated community lower on the shore (Dayton, 1975). The competitively dominant alga *Hedophyllum sessile,* and (in areas exposed to greater wave action) *Lessoniopsis littoralis purpuratus*, were prevented from monopolizing the substratum. Sea urchins, however, are intensive grazers and have the capacity to eradicate all the macroalgae and reduce the flora to encrusting corallines. They are prevented from doing this by the starfish *Pyconopodia helianthoides,* which feeds on sea urchins and causes them to vacate local areas.

Increased harshness of the environment seems to be associated with the diminished importance of predation as a factor in promoting species richness in the community. From Figure 5.18 it can be seen that there is a marked contrast between the species-rich, trophically complex rocky shore communities of the Pacific coast of North America in which predation plays a key role in facilitating

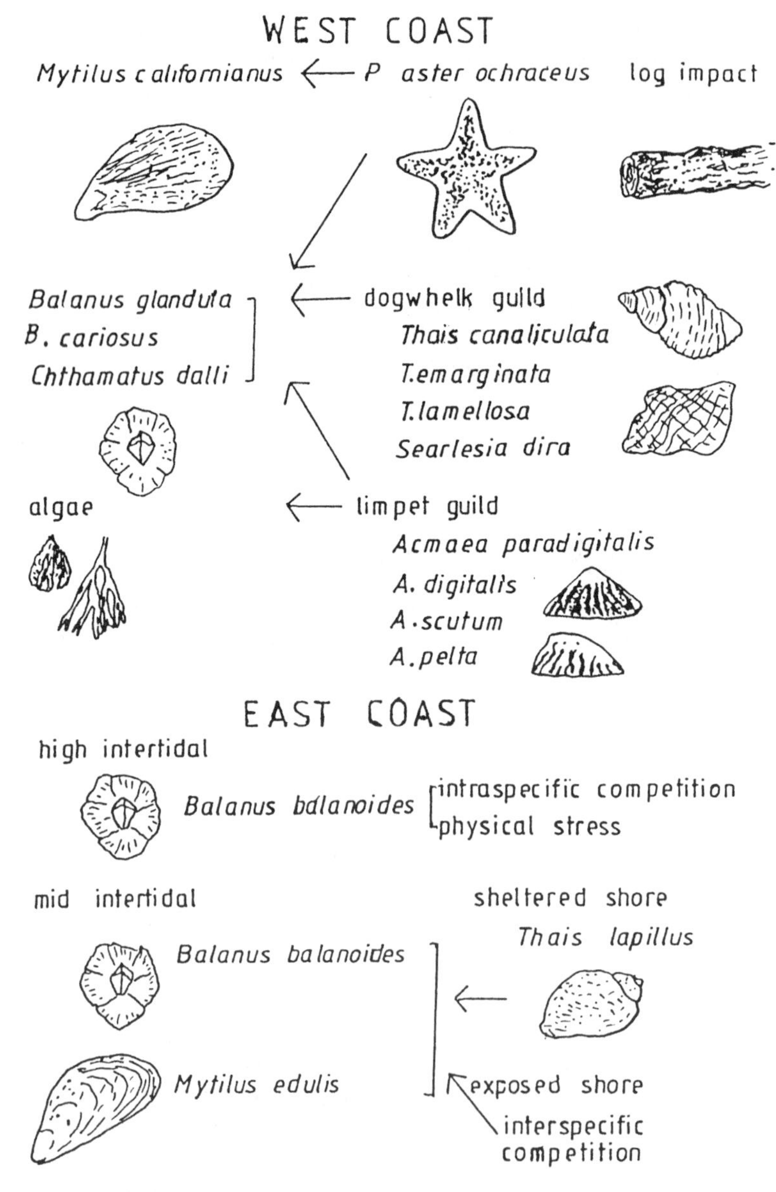

FIGURE 5.18 Determinants of community structure on the Washington coast of North America (Dayton, 1971) and on the New England coast of North America (Menge, 1976). (Redrawn from Hughes, R.N., in *The Shore Environment*, Vol. 2, *Ecosystems,* Price, J.H., Irvine, D.E.G., and Farnham, W.F., Eds., Academic Press, London, 1980b, 724. With permission.)

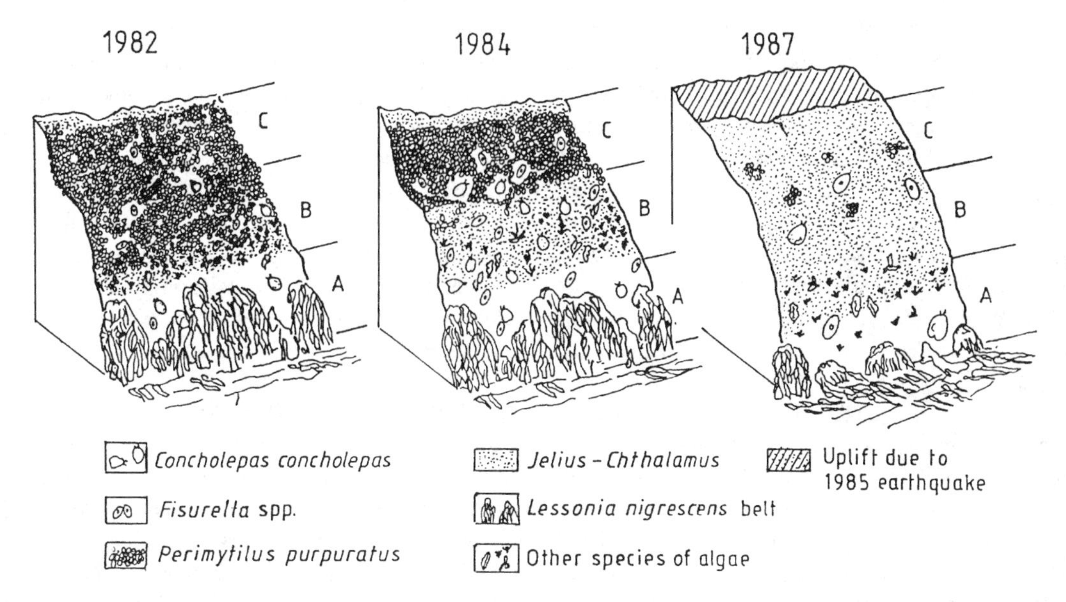

FIGURE 5.19 Schematic view of the unharvested rocky intertidal at Las Cruces showing the three zones described in the text. 1982: Initial state before enclosing; 1984: state ca. 2 yr later; 1987: state ca. 5 yr later, December. (From Duran, L.R. and Castilla, J.C., *Mar. Biol.*, 103, 560, 1989. With permission.)

coexistence among competing prey species, and the relatively species-poor, trophically simple rocky shore communities of the Atlantic coast, in which competition rather than predation is the major structural force (Menge, 1976; Hughes, 1980b). On the northeastern seaboard of North America, the intertidal climate is more extreme than on the West Coast and the more stressful environment has resulted in a more limited number of species with few predator species and a small number of physiologically robust sessile species whose populations are influenced more by competition than by predation (Menge and Sutherland, 1976).

5.2.3.6.3 Panamanian rocky shores

Lubchenco et al. (1984) have described the role of consumers in maintaining the structure of a rocky intertidal community on Taboguilla Island, Bay of Panama. The substratum is basaltic and heterogeneous in texture and topography. During the wet season (May to mid-December), rain is frequently heavy, daily air temperatures range from 24 to 31.5°C, and surface seawater temperatures range from 18 to 27°C. During the dry season (mid-December through April), rain is infrequent, daily air temperatures range from 27 to 32°C, and surface seawater temperatures range from 18 to 27°C.

Taboguilla shore appears barren throughout the year because benthic animals and erect macroalgae are rare on exposed rock surfaces (Figure 5.19). In the high zone, most surfaces are bare (91.5 to 98.1%), with small barnacles, *Chthalamus fissus* and *Euraphia imperatrix*, as the dominant space occupiers. In the mid and low zones, encrusting algae dominate space (25.0 to 92.5% cover) and sessile animals are scarce (<1 to 9.8% cover). Maximum cover (7.0%) of erect algae occurs in the low zone. The plants are short, <5 cm tall. Consumers (e.g., limpets, predacious gastropods, crabs, chitons, fishes) are diverse and abundant in all zones.

Both prey and consumer species composition and abundance change from high to low zones. Densities tend to vary more in the high, and less in the lower zones. The large herbivorous crab *Grapsus grapsus* is usually most abundant at higher intertidal levels. In contrast, fishes are probably effectively denser in lower than in higher zones, at least at high tide, although they may range throughout the intertidal as well as the subtidal zones. Both species richness (S) and diversity (H') increase with decreasing tidal levels. The major differences between the Panama community and its temperate counterparts are twofold: (1) the paucity of macroalgae; and (2) the abundance of herbivorous fishes and crabs and predacious fishes in Panama. Despite seasonal changes in the physical environment, seasonal changes in community structure are small or lacking. Annual changes are sometimes larger, but still small in comparison to temperate regions. The fact that the crustose algae dominate this otherwise relatively barren shoreline seems due to: (1) intense, consistent grazing and predation by a diverse assemblage of vertebrate and invertebrate consumers on the upright algae and sessile animals; (2) desiccation, possibly ultraviolet and heat stress, especially at higher tidal levels; and (3) possible inhibition of settlement by crustose algae.

Menge and Lubchenco (1981) and Menge et al. (1985; 1986a) have studied the effects of predation by the diverse assemblage of consumers on the community structure of sessile prey in the low rocky intertidal zone at Taboguilla Bay. The food web is particularly complex (Lubchenco et al., 1984; Menge et al., 1986a). Animal (but not plant) species richness is high (>100 species occur in the low zone). Trophic links are numerous, occurring both between trophic levels and among species of similar trophic status. Consumers include both slow-moving and fast-moving benthic invertebrates and fishes (Menge and Lubchenco, 1981), with up to 29 species of predators, up to five species of two types of carnivore, and up to 13 species of six types of herbivore. Four functional groups of consumers were defined: (1) large fishes, (2) small fishes and crabs, (3) herbivorous molluscs, and (4) predacious gastropods. Groups (1) and (2) included fast-moving consumers, and groups (3) and (4) included slow-moving consumers. Experimental treatments included: no consumers deleted (all groups present), pairs of groups deleted (two absent, two present), trios of groups deleted (three absent, one present), and the entire consumer assemblage deleted (all groups absent). Changes in abundance (percent cover) of crustose algae, solitary sessile invertebrates, foliose algae, and colonial sessile invertebrates were quantified periodically in 2 to 4 plots of each treatment for a period of three years, February 1977 to January 1980.

After deletion of all consumers, ephemeral green algae increased from 0 to nearly 70% cover. Thereafter, a succession of spatial dominants occurred, with peak abundances as follows: the foliose coralline alga *Jania* spp. by July 1977, the barnacle *Balanus inexpectatus* by April 1978, and the rock oyster *Chama echinata* by January 1980. Although no longer occupying primary rock space, *Jania* persisted as a dominant or co-dominant turf species (with the brown alga *Giffordia mitchelliae* and/or the hydrozoan *Abietinaria* sp.) by colonizing the shells of sessile animals as they became abundant instead of the rock surface. Multivariate analysis variance (MANOVA) indicated that the effect of each group was as follows. Molluscan herbivores grazed foliose algae down to grazer-resistant, but competitively inferior algal crusts, altered the relative abundances of the crusts, and inhibited recruitment of sessile invertebrates. Predacious gastropods reduced the abundance of solitary sessile animals. Small fishes and crabs, and large fishes reduced the cover of solitary and colonial sessile animals, and foliose algae, although they were incapable of grazing the foliose algae down to the rock surface. Many of the effects of each consumer group on prey groups or species were indirect; some effects were positive and some negative. The variety of these indirect effects was due to both consumer-prey interactions among the consumers, and competitive or commensalistic interactions among the sessile prey. Comparison of the sum of the effects of each of the single consumer groups (i.e., the sum of the effect observed in treatments with one group absent, three present) with the total effects of all consumers (i.e., the effect observed in the treatment with all groups present) indicates that a "keystone" consumer was not present in the community. Rather, the impacts of the consumer groups were similar but, due to dietary overlap and compensatory changes among the consumers, not readily detected in deletions of single consumer groups. The normally observed dominance of space by crustose algae is thus maintained by persistent, intense predation by a diverse assemblage of consumers on potentially dominant sessile animals and foliose algae. The large difference in structure between this and temperate intertidal communities appears to be due to differences in degree, not kind of ecological processes that produce the structure.

5.2.3.6.4 Temperate shore at Catalina Island

Robales and Robb (1993) investigated varied predator impacts on a temperate shore at Catalina Island, California, with special reference to mussels and intertidal algal turfs. A red algal turf covers the mid-shore from sheltered to all but the most wave-exposed conditions (Stewart, 1982). Only 1 to 3 cm thick, the turf consists of the branching thalli of the anchor species, principally *Corallina officinalis*, entwined with epiphytes, predominantly *Gigartina canaliculata, Laurencia pacifica,* and *Gelidium coulteri*. Invertebrates are present underneath, or in the gaps in the turf. They include barnacles *Tetraclita squamosa* var. *rubescens,* jewel box oysters *Pseudochama exogyra*, sessile tubiculous gastropods *Aletes squamigerus*, and three species of mussel, *Mytilus californianus, M. edulis,* and *Brachidontes adamsianus*. Predators include the spiny lobster *Palulirus interruptus*, labrid fishes *Halichaeres semisibctus, Oxyjulis californica,* and *Semicossyphus pulcher*, and whelks, *Cerastostroma nuttalli* and *Maxwellia gemma*.

In their experiments, Robales and Robb (1993) demonstrated that predation by spiny lobsters maintained the distinctive red algal turf by killing juvenile mussels (*M. californianus* and *M. galloprovincialis*) that otherwise overgrow and replace the algae. Further investigations (Robales, 1997) revealed that the high recruitment of the predominant mussel (*M. californianus*) occurred only on the most wave-exposed sites in certain years; mussel recruitment was slight to nil on relatively protected sites in most years. A predator exclosure experiment demonstrated that the effects of the predator depended on the spatial differences in recruitment rates. Lobsters on wave-exposed sites functioned as keystone predators; on more sheltered sites, little or no predation, whether by lobsters, fishs, or whelks (also foraging on the sheltered sites), was necessary to maintain the algal assemblage. Thus, similar assemblages can be maintained by markedly different relative levels of crucial ecological rates. In the mid-tidal zone of Catalina Island, the intense species interactions characteristic of the keystone predator hypothesis occurred only at productive, high wave exposure locations; low recruitment of mussels elsewhere preempted both predation and competition between mussel and algal assemblages. Depending on the levels of mussel recruitment, keystone predation, diffuse predation, or no predation may be required to maintain the distinctive assemblages of algae at different times and different locations along the shore of Catalina Island. Thus, red algae dominate the rocky shores through different mechanisms over a range of physical conditions depending on the level of mussel recruitment.

5.2.4 Human Predation on Rocky Shores

Humans have been important predators on rocky shores since prehistoric times, taking shellfish, both gastropods and bivalves (especially mussels and oysters), stalked barnacles (*Pollicipes*), large acorn barnacles, sea cucumbers, sea urchins, octopus, a variety of crustaceans (especially crabs, lobsters, and crayfish), and algae. The impact of humans as top predators on rocky intertidal communities has been the focus of recent studies, particularly in South Africa (Siegfried et al., 1985; Hockey and Bosman, 1986), Chile (Castilla and Duran, 1985; Castilla, 1986; Oliva and Castilla, 1986; Moreno, 1986; Moreno et al., 1986; Castilla and Bustamante, 1989; Duran and Castilla, 1989), Costa Rica (Ortega, 1987), and Australia (Catterall and Poinier, 1987). The intense human predation on the rocky intertidal community of central Chile leads to a complete elimination of some species (a "press-perturbation" effect, sensu Bender et al., 1984). Predation by humans in this ecosystem is mainly directed at the carnivorous muricid gastropod *Concholepas concholepas*, and the herbivorous keyhole limpets, e.g., *Fisurella crassa* and *F. limbata* (Duran and Castilla, 1989).

Collection of seaweed for food is widely practiced in many parts of the world, especially in the Far East, where there is extensive seaweed cultivation. Seaweeds are also eaten in Europe and South America. However, most seaweeds are collected or cultivated for industrial purposes; red algae, such as species belonging to the genera *Gelidium, Gracilaria,* and *Pterocladia,* are used for the manufacture of agar, and *Chondrus* and *Euchema* are harvested for the extraction of carrageenan. Kelps and other brown algae such as *Macrocystis, Laminaria, Ascophyllum,* and *Durvillaea* are harvested for the production of alginates.

In central Chile there is a continuous small-scale harvesting of the bull kelp *Durvillaea antarctica* for human consumption. Castilla and Bustamante (1989) studied the impact of this harvesting by monitoring fenced and unfenced areas over a period of two and a half years. Comparison between harvested and unharvested areas revealed significant differences in density, biomass, and size structure, with a marked reduction of these parameters in the harvested areas.

A five-year study of the impact of human predation on Chilan shores and the effect of the exclusion of humans from areas of the shore has been carried out (Castilla and Duran, 1985; Oliva and Castilla, 1986; Duran and Castilla, 1989). The middle intertidal of harvested and unharvested areas diverged in species diversity and composition during the experiment. In harvested areas the mid-intertidal areas of the rocky shore were dominated throughout by a monoculture of mussels *Peramytilus purpuratus* (Figure 5.19, 1982). When humans were excluded, the mid-intertidal community switched to one dominated by barnacles (predominantly *Jehlius cirratus* and *Chthalamus scabrosus*) (Figure 5.19, 1987). This community persisted for at least three years, despite the presence of forces (e.g., mussel larvae) that had the potential to alter the community structure. Such changes were mediated by the muricid gastropod *Concholepas concholepas*, a keystone predator. As a consequence of the changes outlined above, the species diversity, H', (primary space occupiers) in the unharvested areas increased from $H' = 0$ at the beginning of the study in 1983, when the intertidal community was dominated by mussels, to values of ca. $H' = 2$ toward the middle of the study in 1984 (which coincided with the maximum predatory impact of *C. concholepas*), and subsequently decreased to ca. $H' = 0.5$ at the end of the study in 1987, when the mid-intertidal community was dominated by barnacles.

Fissurelid (keyhole) limpets are among the most important food items harvested by shellfishermen in central Chile. Of the five species that occur, only *Fisurella crassa* and *F. limbata* are truely intertidal species. Oliva and Castilla (1986) studied the impact of human predation on these species for two and a half years by monitoring harvested and exclusion (unharvested) areas. As a consequence of human exclusion, an increase in densities and mean sizes of the two species was seen in the unharvested areas as compared to the harvested areas.

The studies discussed above illustrate a clear case of "cascade effects " (Paine, 1980) due to press perturbation. Hence, the exclusion of a top predator (humans) in the rocky intertidal resulted in an increase of a keystone predator, *Concholepas concholepas*, and the two species of herbivorous keyhole limpets. In turn, the increase of *C. concholepas* reduced the cover of the mussel *Peramytilus purpuratus*, a domimant competitor, favoring the settlement of macroalgae in newly available primary space. This state, however, is transient since the macroalgae were subsequently eliminated from the system (most probably due to the grazing of the keyhole limpets) leading to domination of the community by barnacles.

Thus, harvesting of particular species, e.g., mussels or other molluscs for food, creates free space for succession sequences as described above. Removal of the dominant macroalgae also initiates successional events. The rate of recovery for the impact of harvesting will largely depend on the sizes of the cleared patches and the type of harvesting regime (i.e., whether removal is total or partial) and the the capacity of the algae for vegetative growth.

5.2.5 Environmental Heterogeneity, Community Structure, and Diversity

Both theoretical and empirical studies have demonstrated that system (i.e., community, guild, or species pair) stability, or maintenance of species diversity is strongly influenced by environmental heterogeneity. For example, competition, predator-prey, and community models demonstrate that although species associations in homogeneous environments sometime exhibit limited stability (e.g., species coexistence), stability is higher in heterogeneous model environments (e.g., Hastings, 1980). Empirical studies in general reinforce this theoretical conclusion by demonstrating that persistence of prey or competitors depends on refuges in time, space, size, or behavior (e.g., Menge, 1972; 1976; Lubchenco, 1978; 1980; Lubchenco and Menge, 1978; Menge and Lubchenco, 1981; Hixon, 1980; Russ, 1980; Williams, 1981; Schulman, 1984).

Menge et al. (1985) have investigated the role of environmental heterogeneity on the rocky shores at Taboguilla Island, Gulf of Panama. An earlier study (Menge and Lubchenco, 1981) indicated that the persistence of many intertidal organisms depended on holes and crevices in the rock as refuges from both vertebrate (fishes) and invertebrate (crabs, gastropods, chitons) consumers. Local substratum topography was highly variable, ranging from smooth to irregular surfaces. Number (S) and diversity (H') of sessile species was lower on homogeneous surfaces than on heterogeneous surfaces. Rate of increase of S in areas sampled was positively correlated with substratum heterogeneity. The number of species sampled per transect at a homogeneous site was about 10 vs. 30 to 60 on a heterogeneous site. Large fishes and crabs foraged intensively over both substratum sites, but could not enter holes and crevices to eat prey. Gastropods, chitons, limpets, and small crabs fed on both substrata but varied in abundance from hole to hole. Prey mortality was thus intense and constant on open surfaces, but variable in space and time in holes and crevices. When consumers were excluded from the general rock surface, algal crusts were settled on and overgrown by foliose algae, hydrozoans, and sessile invertebrates, particularly bivalves. Both S and H' first increased, as sessile species invaded and became more abundant, and then decreased as the rock oyster *Chama echinata* began to outcompete other species and dominate the primary space. Hence, consumers normally kept diversity low by removing most sessile prey from open surfaces.

Thus, the diversity patterns in the Taboguilla Island community are maintained by consumers interacting with the spatial heterogeneity of the substratum. Key features of this interaction are the relatively uniform grazing pressure on homogeneous surfaces and resultant restriction of sessile prey to surface irregularities by large, fast-moving consumers. Variability among microhabitats depends both on intrinsic and extrinsic factors. For example, interactions between the physical environment and variability among holes and crevices in orientation, water retention at low tide, depth, width, position of the sun through the day, cloud cover, rainfall, time of low tide, and degree of wave action produces intrinsic variation. Spatial and temporal variation in presence and species composition of herbivores and predators, and patchy recruitment impose extrinsic variation (e.g., Menge et al., 1983). In combination, these intrinsic and extrinsic sources can produce endless variety in microhabitat. In contrast, the physical and biotic environment on homogeneous substrate are more uniform both in space and time.

Another example of the influence of environmental heterogeneity on community composition is the study carried out by Dean and Connell (1987a,b) on marine invertebrate succession in algal mats in southern California. They found that the diversity and abundance of the animals increased between the early and middle stages of algal succession, then remained similar into the later stages. An increase in the complexity of the physical aspects of algal structure (biomass, surface area) caused an increase in invertebrate richness and abundance. This increased complexity influenced the associated invertebrate community through several mechanisms: (1) it decreased mortality caused by predation from fish and crabs; (2) it reduced the severity of physical stresses, primarily wave shock; (3) it increased the accumulation of those species transported passively by wave action; and (4) in mobile species, selection of algal substrates was largely based on physical aspects of algal structure.

5.2.6 Persistence and Stability

Community structure and its dynamics can be described in terms of the successional processes and stability properties that prevail in a community. Successional processes were discussed in the section above. Here, we are concerned with the nature of stability (Holling, 1973; Sutherland, 1981; Pimm, 1984; Connell and Sousa, 1983; Dayton et al., 1984; Johnson and Mann, 1986). These two properties share a common underlying basis in that they relate to the way in which species respond to biological and physical stresses and disturbances in the environment. An allied concept is that of persistence (Connell, 1986), i.e., the degree to which a species or population maintains itself as a member of a community over time.

Of the multiplicity of definitions that have been applied to aspects of stability (e.g., Connell and Sousa, 1983), the terminology of Dayton et al. (1984) is generally accepted. They suggested that community stability may be separated into three parts: (1) *persistence*, referring to a community having approximately constant composition through more than one turnover of the dominant species (Margalef, 1969; Lubchenco and Menge, 1978); (2) *resis-*

tance (preemption), referring to the resistance of the community to disturbance (invasion) or replacement by other species (Underwood et al., 1983): and (3) *resilience (recovery)*, which refers to the ability of a community to return to its original composition following perturbation and invasion by new species. Resistance is a specific form of competition that can be demonstrated experimentally. Sutherland (1981), for example, used fouling plates to show that although larvae were available to settle, a previously established tunicate, *Styela*, normally prevented them from invading. Recovery is the return to a predisturbed state. It describes the response of the community to perturbations; it may be rapid or slow. Johnson and Mann (1986) have expanded the meaning of "resilience" to embody situations in which a community returns to its original composition following a perturbation that changes markedly the relative abundance of species (which may include local extinctions) but where new species do not colonize. Gray (1977; 1981) has discussed the question of stability in benthic communities. He recognizes two types of stability: neighborhood stability and global stability. The neighborhood model represents many cases of local temporal changes in benthic communities. At one point in time, species A (or a particular combination of species) dominates, but is replaced by species B (or a particular combination of species) which may then revert back to A or on to C depending on which factors are operating. Global stability is rare and in general there is a range of possible stable positions

Johnson and Mann (1986) have examined community stability with particular reference to Nova Scotian kelp beds. The rocky subtidal of the Atlantic coast of Nova Scotia has two contrasting community configurations, either of which may span decades in time and extend along hundreds of kilometers of coastline. In the absence of high densities of sea urchins, the hard substratum supports dense and highly productive seaweed beds that grow in a more or less continuous band around the coast and are dominated by *Laminaria longicirrus* to a depth of 15 to 20 m (Mann, 1972a; Novaczek and McLachlan, 1986). The biological and physical structure of these seaweed communities is three-tiered and relatively simple; *Laminaria* forms a closed canopy over smaller perennial and ephemeral algae that may grow to 0.01 to 0.5 m above the bottom, and there is a basal layer of encrusting coralline algae (Johnson and Mann, 1986; Novaczek and McLachlan, 1986). However, grazing by sea urchins (*Strongylocentrotus droebachiensis*) can convert the seaweed beds into unproductive sea urchin/coralline alga communities that are largely devoid of noncalcareous algae (Mann, 1977; Chapman, 1981; Wharton and Mann, 1981). The unproductive state persisted from the late 1960s, but mass mortalities of the sea urchins between 1980 and 1983 (Miller and Colodey, 1983; Scheibling and Stephenson, 1984) attributed to an amoeboid pathogen triggered a switch from the first state to the second and provided a unique opportunity to study: (1) the ability of *L. longicirrus* to recover its former dominant status, and (2) its stability when competing with other seaweeds and when perturbed by storms and grazers other than sea urchins.

Rates of recolonization of *L. longicirrus* depended on the proximity of a refugial source of spores. When reproductive plants were nearby, a closed canopy developed within 18 months of sea urchin mortality. When a reproductive population was several kilometers away, there was sparse recolonization for 3 years, and then a massive recruitment occurred with the closure of the canopy in the fourth year.

Laminaria is clearly the competitive dominant in the seaweed community. Manipulative experiments showed that kelp limits the abundance of several understory species, but there was no evidence that the abundant annual seaweeds limited kelp recruitment. When sea urchins were rare, the density and growth rates of *Laminaria* were influenced mostly by intraspecific competition. When the canopy of adult plants were removed, there was a dramatic increase in kelp recruitment, but the recruits that grew in dense patches in the clearings were significantly smaller than those of a similar age that grew more sparsely beneath the canopy. Once the kelp recovered from destructive grazing and formed a mature forest, it was able to maintain its dominance, even in habitats subject to severe nutrient stress for 8 months of the year. For most of the year, mortality and erosion of laminae outweighed the effects of recruitment and growth, and the canopy declined, especially during winter when storms were frequent. Erosion was exacerbated by grazing of the gastropod *Lacuna vincta*. However, in late winter and early spring, recruitment and rapid growth restored the canopy. When severe storm damage was stimulated by completely removing *Laminaria* in patches, the kelp rapidly recolonized and soon outgrew other seaweeds.

Unlike the competitive dominants in kelp bed systems in the northeast Pacific, *L. longicirrus* in Nova Scotia manifests multiple patterns of adaptation that enable it to dominate early and late stages of succession in a range of habitats of different nutrient stress and of disturbance from storms and grazers. The principal threat to the stability of the kelp beds is destructive grazing by sea urchins. Johnson and Mann (1986) suggested that the considerable differences between the dynamics of kelp beds in Nova Scotia and those of the northeastern Pacific, and the high degree of stability of *L. longicirrus* stands in Nova Scotia, is attributable to the low diversity of kelps and therefore low levels of competition in Nova Scotia, and to the multiple adaptations of *L. longicirrus* that enable it to tolerate several stresses and disturbances.

Johnson and Mann (1986) argue that the dynamics of community organization, and therefore the stability properties of the Nova Scotia system are determined primarily

TABLE 5.2
Comparison in Terms of Stability of the Dynamics of Kelp Beds in Nova Scotia with Those in the Northeastern Pacific

Nova Scotia	California	Northern Northeastern Pacific
	Persistence	
Laminaria longicruris grows in extensive homogeneous stands that may persist for several turnovers	Kelps appear as dynamic mosaic of patches	Kelps appear as dynamic mosaic of patches
Coralline algae abundant beneath seaweed cover and in areas of intensive grazing by urchins	Coralline algae abundant beneath seaweed cover and in areas of intensive grazing by urchins	Coralline algae abundant beneath seaweed cover and in areas of intensive grazing by urchins
	Resistance	
L. longicruris recruitment unaffected by dominant annuals (but limited by canopy of conspecific adults)	Some canopy-forming kelps can be displaced by invasion of other species (several limit their own recruitment)	Some canopy-forming kelps can be displaced by invasion of other species (several limit their own recruitment)
L. longicruris relatively well adapted to wave action	*Macrocystis pyrifera* vulnerable to wave action	*M. integrifolia, Pterygophora californica,* and *Laminaria* spp.; vulnerable to wave action
L. longicruris able to store nitrogen and is tolerant of nutrient stress	*M. pyrifera* has poor capacity for nitrogen storage and is vulnerable to nutrient stress	
L. longicruris and other fleshy seaweeds vulnerable to destructive grazing by sea urchins	*M. pyrifera* and other fleshy seaweeds vulnerable to destructive grazing by sea urchins	Destructive grazing appears to be induced by an increase in urchin density
Change in urchins' behavior from passive detritivore to active destruction of seaweed related to increase in density of urchins and not to change in density of kelp	Change in urchins' behavior in some areas induced by change in density of kelp and unrelated to density of urchins. In other areas destructive grazing may be caused by increase in urchin density	Kelps and other fleshy macroalgae vulnerable to destructive grazing by sea urchins, but some species may attain refuge in size
L. longicruris resistant to grazing by gastropods at high densities. Many understory species increase in abundance following canopy removal	Many understory species increase in abundance following canopy removal	Many understory species increase in abundance following canopy removal, but some decrease their cover
	Resilience	
L. longicruris usually quick to respond to small- and large-scale disturbances and dominates early and late stages of succession	*M. pyrifera* colonizes soon after disturbance only if reproductive conspecifics are nearby; otherwise, other kelps can establish which may preempt superior competitors and other species	Following disturbance, abundant early colonizers are not usually competitive dominants
Sequence of succession relatively invariant; factors affecting sequence in other systems affect only rate of succession	Sequence of succession affected by timing and scale of disturbance, proximity to spore source, dispersal ability, and season of spore production	
Endpoint of succession nearly always *L. longicruris* over wide range of wave exposure and nutrient stress	Endpoint of succession modified by disturbance	Endpoint of succession modified by disturbance

Source: From Johnson, C.R. and Mann, K.H., *Ecol. Monogr.*, 58, 146, 1986. With permission.

by biological interactions and not by physical variables. This differs from the kelp communities in the northeastern Pacific, in which both biological and physical factors influence dynamics significantly at a primary level (Table 5.2). Figure 5.20 is a qualitative model that summarizes the gross behavior and dynamics of the system. In this model, the highly productive *Laminaria* kelp beds and poorly productive sea urchin/coralline community are regarded as two alternative and stable configurations, since either community state can persist for much longer than the life span of the dominant species. A biological mechanism causes switching from one state to the other, and biological interactions determine the stability properties and community structure of each configuration. Destructive grazing by sea urchins causes a transition from the kelp beds to the unproductive state, and an amoeboid

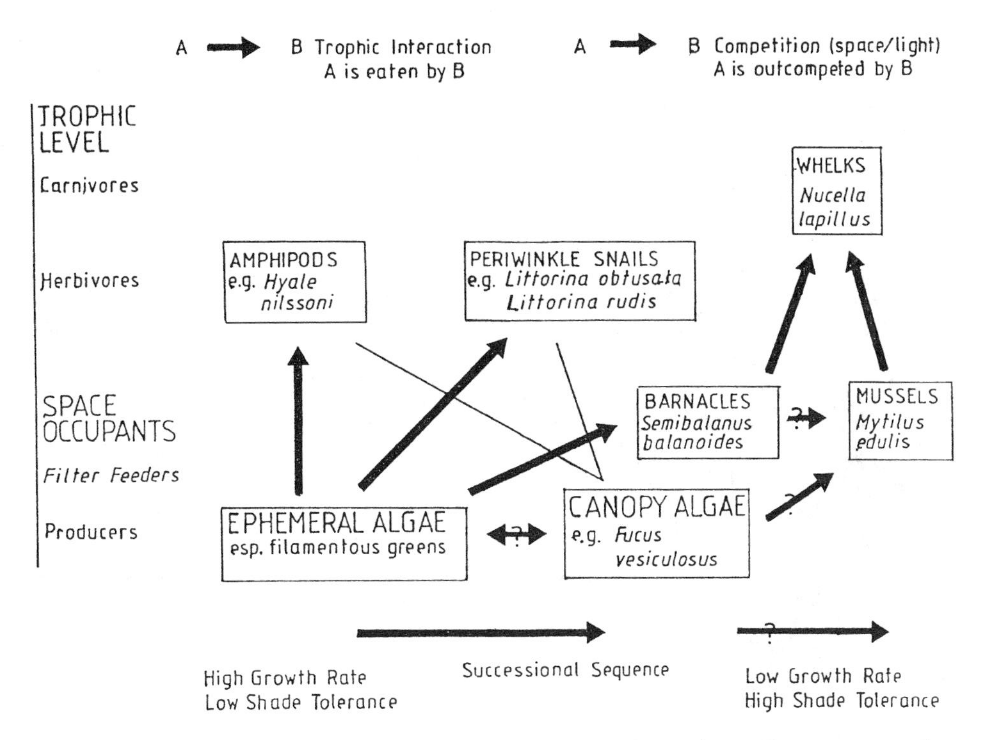

FIGURE 5.20 Simplified structure of the species assemblages on exposed rocky seashores, for emergent rock at midshore level, near Halifax, Nova Scotia, Canada. The competitive hierarchy is in the absence of predation, and is derived from Menge (1976), Lubchenco (1978; 1983; 1986), and Lubchenco and Menge (1978) (see also Menge and Sutherland, 1987; Chapman and Johnson, 1990). Question marks indicate variable or ambiguous outcomes. The successional sequence proposed is based on direct observation and on Menge (1976), Lubchenco (1983), others. It is postulated that mussels may not always succeed *Fucus*. (Redrawn from McCook, L.J. and Chapman, A.R.O., *J. Exp. Mar. Biol. Ecol.*, 154, 138, 1991. With permission.)

disease appears to be the mechanism that destroys the sea urchin population and facilitates recovery of the kelp beds.

5.2.7 Disturbance and Succession

5.2.7.1 Introduction

Species composition and abundance in many communities is strongly influenced by disturbance (reviewed by Sousa, 1984a,b; McCook and Chapman, 1991). Disturbances that create open space (patches) maintain the diversity of many natural communities for which space is a limiting resource, including those of rocky shores (Dayton, 1971; Levin and Paine, 1974; Connell, 1978; 1979; Sousa, 1979a; Paine and Levin, 1981; and many others). Disturbance is the result of any process that deceases the amount of biomass within a given area.

Extensive investigations of disturbance in the rocky intertidal have been carried out along the Pacific Northwest coast of North America (see Paine and Levin, 1981; Dayton, 1984; Sousa, 1984b). Much of the space in the mid-tidal zone of exposed coasts is occupied by beds of the dominant competitor, the mussel *Mytilus californianus* (Paine, 1966; 1974). Complete monopolization of this space is prevented by localized disturbances. The impact of drifting logs, and more commonly, the shearing forces of large winter waves (Dayton, 1971; Paine, 1994) clear patches of open spaces of different sizes within the mussel bed matrix. These patches serve as the foci for the recruitment, growth, and reproduction of many competitively inferior, fugitive species, including algae and sessile invertebrates. These species are doomed to elimination as *M. californianus* gradually invades and closes the patches.

The damage caused by most natural disturbances is localized, so that the open space created is generally in the form of more or less discrete patches, or gaps, within a preexisting background assemblage of organisms. If physical conditions are not too harsh, a patch begins to be colonized soon after it is formed. A series of species replacements (succession) follows, which usually leads inevitably to the local extinction of all but a few dominant species. These species dominate by virtue of being especially vigorous at interference competition or at preempting the open space made available by the deaths of the previous occupants and holding it against invasion (Sousa, 1985). Species whose populations become extinct in the patches during this process are able to persist by dispersing their propagules to other patches where conditions are more favorable for growth and reproduction. Localized disturbances are asynchronous both in space and time, thus maintaining a mosaic of patches varying in characteristics such as size and age (time since the last disturbance). The sequence of colonization of the disturbed areas (patches) is influenced by a complex of factors, including seasonality (Hawkins, 1981a; Sousa, 1984b; Lubchenco, 1986), temporal distribution (Abugov, 1982), intensity (Sousa, 1980), location (Palumbri and Jackson, 1982; Sousa, 1984b; Connell and Keough, 1985), patch size (Sousa, 1984b: Farrell, 1989), and propagule availability (Sousa, 1984b).

In order to be able to predict what will happen in the colonizing sequence of events within the habitats subject to disturbance, knowledge is required of: (1) the disturbance regime, including the size distribution of the patches and the timing of their occurrence; (2) the role of patch size and position within a patch (Farrell, 1989): and (3) the patterns of colonization and succession within the patches. Selected studies that have investigated these phenomena will be discussed below. Investigators have studied such processes both in naturally occurring and artificially created patches. Other experiments have involved the manipulation of the density of grazers (e.g., limpets) and predators (e.g., whelks), or their exclusion from the patches.

5.2.7.2 Size and Location of a Patch

Sousa (1984b) studied the dynamics of algal succession in experimental patches cleared in mussel beds. In particular, the study investigated two potentially important sources of variation in successional dynamics: (1) the size of the patch when first created; and (2) the location of the patch with respect to the potential source of propagules.

Patch size per se, under experimental conditions of reduced grazing (exclusion of limpets), had little influence on the pattern of macroalgal colonization and species replacement. For all but one of the common macroalgal species (Table 5.3), the mean percent cover in small and large patches with copper paint barriers were not significantly different. The sequence of colonization in the 16 patches without reduced grazing is illustrated in Figure 5.22. Soon after clearing, the patches became coated with a mat of diatoms and the filamentous green alga *Urospora*. This mat was rapidly replaced by a turf of *Ulva* 1 cm high and some thalli of *Enteromorpha* mixed in (because of the difficulty of distinguishing these two species in a mixed mat, they are grouped as *Ulva* in Figure 5.21). The *Ulva* turf dominated the patches by October 1979 but declined precipitously in the winter of 1979–80. In the spring of 1980, the *Ulva* turf temporality increased slightly but then continued to decline as other species of red and brown algae invaded the patches. Of the latter, the brown fucoid alga *Pelvitiopsis* and the red alga *Mastocarpus* were the first to appear. The brown alga *Fucus* and the red alga *Iridaea* were not observed in the patches until the spring. *Pelvetiopsis* attained a mean cover of c. 40% by the autumn of 1980. Both *Mastocarpus* and *Pelvetiopsis* declined toward the end of the study while *Iridaea* and *Fucus* increased. The rare species in Table 5.3 were either ephemerals or perennials that were common lower in the intertidal zone.

TABLE 5.3
Macroalgal Species that Colonized the Experimental Patches in the Experiments Carried out by Sousa (1984). Common Species are Those that Attained an Average of >5% Cover in Any of the Experimental Treatments During the Study

Common Species	Rare Species
Chlorophyta	
Urospora penicilliformis	
Ulva californica	
Enteromorpha intestinalis	
Cladophora columbiana	
Phaeophyta	
Pelvetiopsis limitata	*Colpomenia bullosa*
Fucus gardneri (= *distichus*)	*Leathesia difformis*
Analipus japonicus	*Scytosiphon dotyi*
Rhodophyta	
Mastocarpus (= *Gigartina*) *papillata*	*Polysihonia hendryi*
Iridaea flaccida	*Callithamnion pikeanum*
Endocladia muricata	*Microcladia borealis*
	Cryptosiphonia woodii
	Odonthalia floccosa
	Neorhodomela (= *Odonthalia*) *oregona*
	Gelidium coulteri
	Porphyra perforata

Source: From Sousa, W.P., *Ecology*, 65, 1923, 1984b. With permission.

The size of the cleared patch was found to strongly influence the course of algal succession. This effect was largely indirect, resulting from an interaction between patch size and grazing intensity. Small patches, as reported in previous investigations (Suchanek, 1978; 1979; Paine and Levin, 1981), support higher densities of grazers, especially limpets, than do larger patches. As a consequence, the assemblages of algae that develop within small and large patches differ markedly. The assemblage in small patches included grazer-resistant but apparently competitively inferior species (*Analipus, Endocladia,* and *Cladophora*), whereas in large patches it is composed of grazer-vulnerable but competitively superior species. Small patches appear to serve as refuges from competition for grazer-resistant species.

The positioning within a disturbed patch may also influence succession. Areas near the edge of a gap may be affected by their proximity to an intact community. The undisturbed community may provide shading (Rumble, 1985), propagules (Sousa, 1984b), increased competition for nutrients, or act as a refuge for animals that forage in

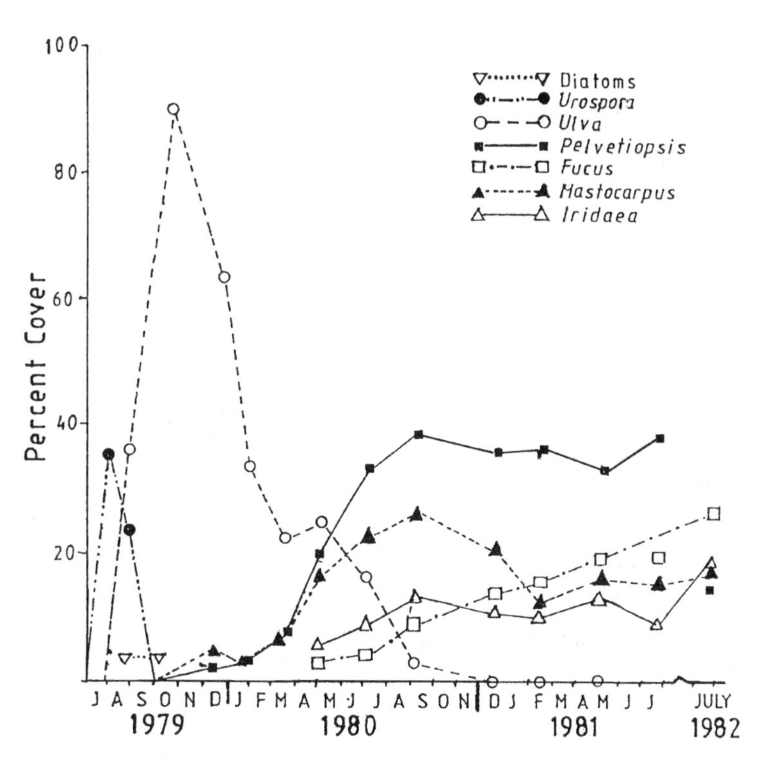

FIGURE 5.21 Seasonal changes in the composition of the algal canopy in the 16 experimental patches from which limpets were excluded. Data are means for species that attained at least 10% cover on some sampling date. (Redrawn from Sousa, W.P., *Ecology*, 65, 1923, 1984b. With permission.)

the gap (Suchanek, 1978; Sousa, 1984b; Rice, 1987). Lateral growth of undisturbed organisms will also have a greater effect at the perimeter than in the center of a gap (Miller, 1982; Connell and Keough, 1985). Edge effects have been suggested by a number of investigators (e.g., Suchanek, 1978; Paine and Levin, 1984; Sousa, 1984b). If proximity to the intact community affects succession, then both gap shape and size will also influence community development. Elongate disturbances will have more of their area near a gap boundary than circular disturbances of equal size (Sousa, 1985). Other investigations (Palumbri and Jackson, 1980; Keough, 1984; Sousa, 1984b) have also indicated that gap size can have large effects on the abundance and species composition of the colonists.

Farrell (1989) investigated the effects on succession of the size of a disturbed patch and position within a disturbed patch in a high intertidal community on the central Oregon coast, U.S.A. Algae and barnacles were scraped off the rocks to create gaps of three sizes (4 × 4, 8 × 8, and 16 × 16 cm). Previous experimental data indicated that gap size can have large effects on the abundance and species composition of the colonists (Osman, 1977; Sousa, 1979a,b; 1984b; Palumbri and Jackson, 1980; Keough, 1984). Farrell (1989) found that limpet densities were higher both in smaller gaps and in the perimeter of larger gaps. Late in the succession, *Pelvetopsis limitata* was the dominant algal cover in the gap perimeters, while *Endocladia muricata* was the dominant algal cover in the gap centers. Neither gap size nor position within a gap influenced the species composition of the recruiting algae or the total amount of barnacle cover. *Balanus glandulus* was relatively more abundant than *Chthalamus dalli* in the larger gaps. Several studies have shown that *Balanus* is more susceptible to limpet caused mortality (Dayton, 1971; Paine, 1981; Farrell, 1988). Table 5.4 summarizes the effects of patch size and position within a patch in Farrell's (1989) experiments. In contrast to Farrell's findings, dense growths of ephemeral algae occur in the larger gaps that occur naturally in mussel beds. These algae are surrounded by a barren, 10 to 20 cm wide zone that is maintained by limpet grazing (Suchanek, 1978; Paine and Levin, 1981; Sousa, 1984b).

5.2.7.3 Succession

Succession has long been a topic of study, especially for plant ecologists, and more recently the question of successional processes on rocky shore has attracted considerable attention (e.g., Sousa, 1984a,b; Turner, 1983a,b; Van Tamlen, 1987; Farrell, 1989; 1991; McCook and Chapman, 1991). Connell and Slayter (1977) proposed three general mechanisms by which species replacement can occur during succession: (1) the resident species enhance the establishment of invading species (*facilitation);* (2) the resident species do not affect the establishment of the invading species (*tolerance*); and (3) the res-

TABLE 5.4
A Summary of Patch Size and Position Within a Patch

Community Attributes	Position in Patch	Size of Patch
Limpet density	Edge > Center*	Small > Large*
Total barnacle cover	No effect	No effect
Relative abundance of *Balanus*	Center > Edge^{NS}	Large > Small*
Total algal cover	Edge > Center*	Small > Large^{NS}
Relative cover of *Pelvetiopsis*	Edge > Center*	Small > Large^{NS}
Algal recruit density	No effect	Large > Small*
Species composition of	No effect	No effect

Note: * = statistically significant effect; NS = trend but not statistically significant.

Source: From Sousa, W.P., *Ecology*, 65, 1928, 1984b. With permission.

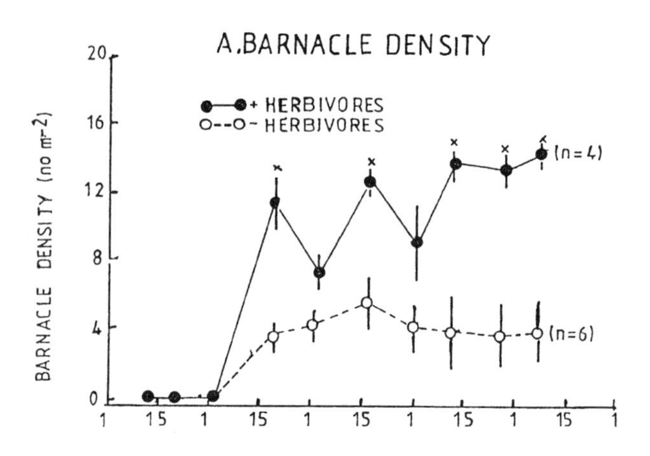

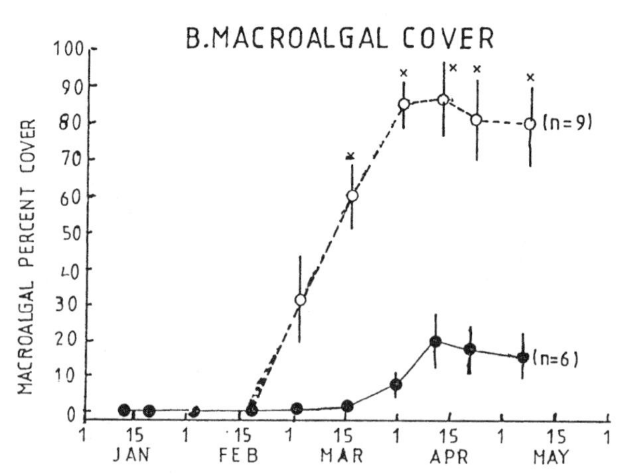

FIGURE 5.22 The effect of herbivores on (A) barnacle density and (B) the percent coverage of macroalgae over the course of the experiment. Solid circles and lines signify the presence of herbivores while herbivore removals are denoted by the open circles and dashed lines. Stars indicate statistically significant differences between the herbivore treatments at the $P < 0.05$ level. Barnacle abundance was tested using Student's t tests and algal cover was arcsine-transformed following Sokal and Rohlf (1981) and tested using a 2-way ANOVA (analyzed using SYSTAT version 3.0). Each date was tested separately due to non-independence of sampling dates. Error bars represent 1 SE of the x and sample sizes are given for each treatment at the end of the curve. (Redrawn from Van Tamelen, P.G., *J. Exp. Mar. Biol. Ecol.*, 112, 43, 1987. With permission.)

ident species depress the recruitment and growth of the invading species (*inhibition*). However, the application of these models are complicated by factors such as seasonality of recruitment, growth, and mortality (Turner, 1983a,b). For example, Palumbri (1985) described an interaction in which two species were competitors in areas of low desiccation stress, but the interaction was commensalistic under more stressful conditions. So depending on the environmental conditions, the interaction would be called either inhibition or facilitation.

Interaction between barnacles, algae, and herbivores have been studied in many intertidal communities in widespread geographic areas (Dayton, 1971; Choat, 1977; Underwood, 1980; Hawkins and Hartnoll, 1983a; Jernakoff, 1983; 1985a,b; Lubchenco, 1983; 1986; Dungan, 1986). It has often been found that barnacles have a facilitative effect on algae (Hawkins, 1981a,b), especially in the presence of herbivores (Creese, 1982). However, if the barnacles are of inappropriate size (either too small or too large), they can have no effect (Jernakoff, 1983), or a negative effect on late successional species, either accelerating succession (Lubchenco, 1978; 1983; Lubchenco and Menge, 1978; Bertness et al., 1992), or delaying it (Lubchenco and Gaines, 1981), depending on the mode of succession.

Van Tamlen (1987) studied direct and indirect interactions among barnacles, algae, and herbivores in a rocky intertidal boulder field in southern California by manipulating barnacle and limpet densities in cleared plots. The removal of a fine algal mat by the herbivores led to higher recruitment of barnacles (Figure 5.22A). The herbivores also prevented the establishment of algae in the absence of barnacles (Figure 5.22). The barnacles provided a refuge from the herbivores for the algae by inhibiting the grazing activities of the herbivores. Thus, herbivores inhibited both the colonization of the rock surface by microalgae and the recruitment and growth of the macroalgae. Both types of algae inhibited recruitment of barnacles, probably by interfering with their settlement. When algae are present in herbivore research plots, barnacles have been found to be less abundant than in the presence of herbivores (Hawkins, 1983; Petraitis, 1983; Jernakoff, 1985b). Because of the inhibition of barnacles by algae, herbivores facilitated barnacle recruitment by removing the algae. Small barnacles had no effect on the algae but when they grew to larger sizes (2 mm basal diameter), they facilitated the recruitment and growth of the macroalgae.

McCook and Chapman (1991) investigated community succession following massive ice scour on the exposed rocky shores near Halifax, Nova Scotia, Canada. On these shores the midshore area is dominated by the brown alga *Fucus vesiculosus*, often with a prolific understory of mussels *Mytilus edulis*, along with barnacles *Semibalanus balanoides*, and patches of the crustose alga *Hildenbrandia rubra*. Other common species include grazing periwinkle snails, particularly *Littorina obtusata* and *L. rudis*, grazing gammaridean amphipods, predominantly *Hyale nilssoni*, and carnivorous dogwhelks *Nucella lapillus*, which eat the filter-feeding barnacles and mussels (Figure 5.23).

Previous work has suggested that in this community, in the absence of predation or disturbance, the mussels would increase and out-compete *Fucus* as the dominant space occupant (Menge, 1976; 1978b; Menge and Sutherland, 1976; Petraitis, 1987; Chapman and Johnson, 1990). The massive ice scour, which denuded >50 km of shoreline, leaving it virtually bare of plants and animals, offered an opportunity to test the above hypothesis. Experimental studies involved the removal of the *Fucus* canopy and mussels that had been reestablished during a later stage of succession 2 to 3 years after the ice scour had occurred.

5.2.7.4 Role of Recruitment in Succession and Its Impact on Community Structure

As we saw in Chapter 4, organisms with planktonic larvae must survive several critical stages before reaching adulthood (i.e., larvae, settlement, recruitment, juvenile). A current issue in rocky shore community ecology is the relative contribution to "adult" community structure of prejuvenile (such as larval production, settlement, and recruitment) vs. postjuvenile factors (such as predation, competition, and physical factors) (Denley and Underwood, 1979; Watanabe, 1984; Caffey, 1985; Connell, 1985; Gaines and Roughgarden, 1985; Sutherland and Ortega, 1986; Menge and Farrell, 1989; Sutherland 1990; Underwood and Denley, 1990). While variation in recruitment on a relatively small spatial scale can be determined by variation in the supply of planktonic larvae (Grosberg, 1982; Bertness et al., 1992), it has been demonstrated that under certain conditions, variation in recruitment may be proportional to the amount of substratum available for colonization (Gaines and Roughgarden, 1985; Chabot and Bourget, 1988). However, in some cases this relationship does not apply. Minchinton (1997) investigated the recruitment of the tubeworm, *Galeolaris caespitosa*, to different sized patches on the New South Wales coast. He found that the density of recruits per unit area was, on average, almost three times greater in small than in large patches, indicating that the recruitment of *G. caespitosa* was not directly related to area. In contrast, the density of recruits per unit perimeter was not significantly different between small and large patches, indicating that recruitment of *G. caespitosa* was related to the presence of conspecific adults in the patch.

A discussion of the relationships between adult abundance of barnacles and their settlement and recruitment densities can be found in Section 4.3.4. Here we will consider Menge's (1991a) evaluation of the relative importance of recruitment in relation to other factors known to cause variation on New England and Panamian shores (see Section 5.2.3.6.3 for a description of these shores).

In Menge's studies, rates of increase in prey abundance in predation exclusion cages in New England were at least an order of magnitude greater than in Panama (e.g., it took 4 to 6 months vs. 60 to 70 months to reach 100% cover, respectively). Recruitment densities of sessile invertebrates and algae were more variable in space and time at both sites, but were lower by at least an order of magnitude and less synchronous in Panama than in New England. Analyses of data from the two shores indicated that, although predation, competition, recruitment, and the level on the shore explained significant amounts of variation in community structure at both sites, the proportionate contribution of these factors differed. In New England, recruitment explained at most 11%, while predation and competition explained 50% to 70% of variation in sessile invertebrate abundance. In Panama, recruitment explained 39 to 87%, while predation and competition explained only 8 to 10% of variation in sessile invertebrate abundance. Thus, sessile dominants in New England tend to have high recruitment while sessile dominants in Panama tend to have low recruitment. What then is the cause of these considerable differences in the rates of increase in the abundance of sessile species? Potential causes for species with planktonic larvae are: (1) low production of larvae per unit area of shore by adults; (2) low rates of survival of larvae in the plankton to settlement stages; (3) low rates of return to the shore due to unfavorable currents; (4) low rates of survival from settlement to recruitment stages (i.e., low recruitment); (5) slow growth rates after recruitment; (6) high mortality of juveniles; (7) high mortality of adults; or (8) some combination of the above.

Larval production is likely to be low in Panama since the abundance of adult barnacles is extremely low compared to temperate sites. For example, maximum barnacle densities in New England were 1,850 m^{-2}, whereas in Panama they were 11.9 to 15.6 m^{-2} (Lubchenco and Menge, 1978; Lubchenco et al., 1984). Comparable bivalve densities were 86,900 m^{-2} and 13.8 m^{-2}, respectively. Recruitment success in Panama was much less than that in New England. The highest density of *Balanus inexpectus* recruits recorded at Panama was 400 m^{-2}, while the highest density of *Semibalanus balanoides* recruits observed in New England was 123,200 m^{-2}, a 308-fold difference. Two biotic factors that could influence survival between settlement and recruitment were predation and

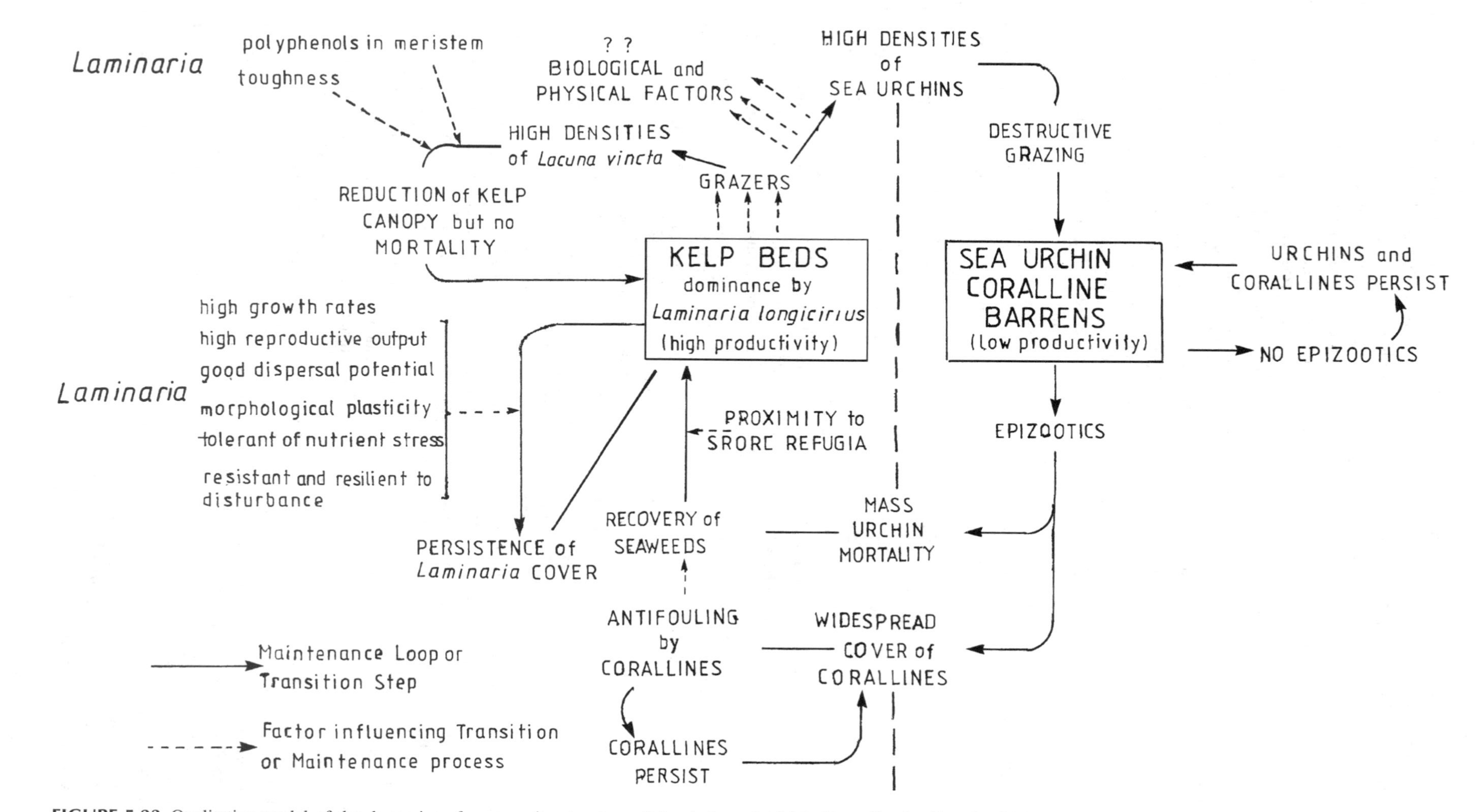

FIGURE 5.23 Qualitative model of the dynamics of community structure of the rocky subtidal in Nova Scotia, Canada. Biological interactions define community structure within the two alternate stable configurations of the community (kelp beds and sea urchin/coralline alga community) and mediate transitions from one state to the other. Notice that the principal void in the model concerns the factors that lead to increased densities of sea urchins and their behavioral changes that result in destructive grazing of seaweeds. (Redrawn from Johnson, C.R. and Mann, K.H., *Ecol. Monogr.*, 58, 149, 1986. With permission.)

biotic disturbance. In New England, limpets, chitons, urchins, and all other grazers except *Littorina* spp. are scarce or absent, predator diversity is low, and consumer activity is highly seasonal. In Panama, on the other hand, grazers are abundant, predator diversity is high, including groups uncommon in New England (e.g., omnivorous fishes), and consumer activity is seasonal and incessant (Menge et al., 1986a,b). Menge concluded that low rates of prey colonization in Panama were probably due to several factors, including low recruitment rates, low survival of juveniles and adult prey, low production of larvae and possibly low numbers of competent larvae. Predation, biological disturbance, lack of settlement, and possibly seasonal transport of larvae away from the shore appear to be the major mechanisms influencing these rates.

5.3 SOFT SHORES

5.3.1 INTRODUCTION

In contrast to rocky shores where the nature and abundance of the biota are considerably influenced by tidal variation in environmental conditions and wave action, the animals living in soft sediments have, to a large extent, evaded the external environmental stresses of temperature, desiccation, and extremes of salinity. On the other hand, the distribution and abundance of the sediment infauna is, to a large extent, controlled by complex interactions between the physicochemical and biological properties of the sediment.

The physicochemical properties include sediment grain size, water content, flushing of water through the sediment, oxidation-reduction state, dissolved oxygen, temperature, organic content, and light. The biological properties include the following:

1. Food availability and feeding activity
2. Reproductive effects on dispersal and settlement
3. Behavior effects that induce movement and aggregation
4. Intraspecific competition
5. Interspecific competition — competitive exclusion effects
6. Predation effects — population reduction or elimination of certain species

As noted previously, soft-bottom communities are characterized by highly variable species composition. The local abundance of a given species may vary by orders of magnitude in time and space, while at the same time the abundance of other species changes little, if at all. The underlying causes of such changes have been one of the central questions in soft-bottom ecology. Year-to-year fluctuations in species composition may reflect catastrophic adult mortality brought about by climatic factors or competitive interactions between species. More often, however, such fluctuations are believed to be the result of factors affecting the success of spawning, larval dispersion patterns, and recruitment success.

Benthic samples taken in close proximity are often found to be less alike in species composition, numbers of individuals, and representation of year classes, than widely separated samples (Eagle, 1975; Stephenson et al., 1971). Such patchiness has been explained as the result of chance dispersal or concentration of planktonic larvae by water movement and favorable currents during settlement, small-scale habitat heterogeneities, intra- and interspecific competition, and amensal interactions (Rhoads and Young, 1971). Many authors have ascribed these spatially heterogeneous distribution patterns to the differential response of species to disturbance. They contend that the early and late colonizers of a new environment created by disturbance have different adaptive strategies: r-strategies (opportunistic), or k-strategies (equilibrium) (McArthur and Wilson, 1967; Grassle and Sanders, 1973).

5.3.2 EXPERIMENTS ON SOFT SHORES AND CAGING METHODOLOGY

In recent years, experimental approaches have been used with varying degrees of success in efforts to understand the mechanisms involved in the determination of the structure and dynamics of intertidal and shallow-water benthic communities. Four excellent reviews of such experimental approaches are those of Dayton and Oliver (1979), Peterson (1979b; 1980), and Reise (1985). Ecologists have used three types of experiments to study the processes underlying the natural pattern of species abundance and distribution: laboratory experiments, field experiments, and so-called natural experiments. In the laboratory, all the independent variables may be controlled and the behavior of the dependent variable can be explored. However, it is not clear whether the results of such experiments are applicable to natural communities. In field experiments, one variable is controlled (a species, or functional group, or an environmental factor), while the rest of the natural community remains intact. Here, the major problem is that such experiments are subject to the natural variability and unpredictably of events, as well as to the full complexity of a usually unknown number of variables (Reise, 1985). Natural experiments rest on the assumption that there are sites that naturally differ in just one variable, while all the others remain the same (Diamond, 1983). Ideally, this can never be true.

Experiments in which one or more species densities are altered, and changes in density of other species in the community are observed in the community, are termed pertubation experiments. Two different kinds of pertubation experiments can be distinguished. In one kind of experiment, termed "pulse" by Bender et al. (1984), one or more species densities are instantaneously changed, and then the system is observed as it returns to equilibrium (or some other attractor). In the second type of experiment, termed "press" by Bender et al. (1984), one or more species densities are continuously altered to a higher or lower level by means of continual addition or removal of members of those species.

On tidal flats, fences or cages are frequently used as an experimental tool to enclose or exclude certain species or functional groups, especially predators. Interpretation of the data obtained by the use of cages poses considerable problems due to the unnatural effects of the cages themselves (Virnstein, 1978). Also, in the case of predator exclusion experiments, the community composition inside the fenced or caged area and in the ambient area may be different for several reasons (see Arnst, 1977; Virnstein, 1977b). Cages can also change the pattern of current flow and thus influence the sedimentation pattern within the cages, increasing the rate of sedimentation, and changing the nature of the sediment by increasing the proportion of fine sediments. While such effects have been documented, no cage artifacts on the enclosed sediment (grain size, water content, organic content) were measured in experiments in the sheltered subtidal of Chesapeake Bay, U.S.A. (Holland et al., 1980; Dauer et al., 1982). Similarly, Mahoney and Livingstone (1982) found no cage effects on sedimentation, granulometry, silt content, and current velocity in a shallow estuary. Fouling of the cage surface can increase local food concentrations. On the other hand, food supplies for suspension feeders may be decreased, and shading may decrease the benthic microalgal production. Drifting algae, especially at subtidal sites may be trapped by the cages (Arnst, 1977; Hulberg and Oliver, 1980). Pelagic and emigrating stages of benthic animals may become attracted to the cages and larval settlement may be affected. Furthermore, although the cages exclude large predators, they can serve as refuges into which predators can recruit as planktonic larvae, or as small individuals (Arnst, 1977). Cages not only exclude predators but may keep out large herbivorous fish and birds, as well as possible disturbers of the sediments such as rays (Van Blaricom, 1982). Thus, the results of caging experiments in soft sediment systems must be interpreted with caution. Nevertheless, such experiments have proved useful.

5.3.3 Grazing on Soft Shores

5.3.3.1 Introduction

Apart from the marginal marsh vegetation on the upper shore and patches of sea grasses lower down, macrophytes are relatively rare on tidal flats. On mudflats, especially in estuaries, green algae, particularly species of *Ulva* and *Enteromorpha*, may be abundant, their density usually being a function of nutrient availability. Red algae belonging to the genus *Gracilaria* may also be common, growing attached to stones and bivalve shells, such as cockles, which grow in the surface sediments. As we have seen in Chapter 3, the dominant producers on soft shores are the benthic microalgae. Diatoms are the dominant taxa, but other important primary producers are flagellates and cyanobacteria. Grazers on soft shores can be categorized as either epibenthic grazers or infaunal grazers.

5.3.3.2 Epibenthic Grazers

The predominant epibenthic grazers are mudflat snails belonging to genera such as *Littorina, Hydrobia, Ilyanassa, Potamopyrgus,* and *Amphibola* on temperate shores, and cerithoid and potamanidid snails on tropical shores. Other epibenthic grazers include amphipods, especially species of the genus *Corophium,* and various burrowing crabs, such as fiddler crabs (*Uca* spp.) and species of *Helice* and *Macrophthalamus* on temperate shores, and oxypodid crabs (*Oxypode, Uca, Scopimera, Mictylis,* and *Macrophthalamus*) on tropical shores. While these crabs are, in general, deposit feeders, gaining their energy requirements from the consumption of a variety of organic matter, they all depend to a varied extent on microalgal food. A number of species of fish are also microalgal feeders; they include mullets (*Mugul* spp.) on both temperate and tropical shores, and the mudskipper *Boteophthalmus* on tropical shores, especially on mangrove flats.

The mudflat snail *Hydrobia ulvae* is known to ingest individual motile diatoms (Lopez and Kofoed, 1980), to scrape microalgae from the surface of sand grains (Lopez and Kofoed, 1980: "epipsammic browsing"), and to swallow entire sediment particles from which the associated microbiota is digested (Lopez and Levinton, 1987). In addition to the microalgae, bacteria can account for up to 10% of consumption (Jensen and Siegismund, 1980). The impact of these epibenthic grazers on the sediment microflora and microfauna was considered in Section 3.6.3.5.

5.3.3.3 Epiphytic Grazers

As discussed in Section 3.1.5.4, epiphytes, especially diatoms, are a prominent feature of sea grass and salt marsh systems. Within these systems grazing is an important process for three main reasons: (1) it provides a major link for the cycling of nutrients; (2) it affects the production of sea grass communities; and (3) it can change the species composition of the sea grass epiphytes (Jernakoff et al., 1996). Sea grass epiphytic grazers include some fish species, molluscs (prosobranch snails), crustaceans (harpacticoid copepods, amphipods, and isopods), and free-living polychaetes. These epiphytic grazers can (1) significantly reduce the epiphytic biomass and production (between 20 and 60% of the total primary production); (2) by selective grazing, change the species composition of the epiphytic community; and (3) increase sea grass production by reducing the shading effect of the epiphytes. Amphipod grazing on sea grass epiphytic diatoms also removes detrital material from the sea grass leaves. Robertson and Mann (1982) found that the reduction in the amount of detritus on the sea grass leaves had a positive effect on sea grass growth.

5.3.3.4 Infaunal Grazers

There are three categories of infaunal grazers: (1) micro- and meiofaunal grazers, (2) infaunal polychaetes and crustaceans that feed within the sediment, and (3) infaunal polychaetes and bivalves that graze on the surface sediments. The micro- and meiofauna move through the interstitial system and capture mobile flagellates and diatoms, or browse on the epigrowth on the sand grains. Important selective grazers are the ciliates that can occur in concentration of up to 4,000 individuals cm^{-2}. According to Fenchel (1968; 1969), several species feed exclusively on flagellates or diatoms, or on both. Diatoms are selected according to size. All major groups of meiofauna (e.g., nematodes, copepods, and platyhelminths) include algivores (Hicks and Coull, 1983; Jensen, 1987; Reise, 1988). Among the 177 species of platyhelminths on sediment shores around the Island of Sylt, Germany, 27% were found to be diatom feeders (Reise, 1988). Admiraal et al. (1983) demonstrated that a nematode species discriminated between congeneric diatoms and preyed more on one than the other.

A number of polychaete species, especially members of the Families Spionidae, Cirratulidae, and Terebellidae, are surface deposit feeders that include sediment microalgae in their diets. Most of these are tube-dwelling species that use a variety of tentacular appendages for obtaining their food. The other major group of surface deposit feeders are the bivalves. Most live at depth in the sediment and use their inhalant siphon to suck up surface sediment with its organic matter and microbial community to be sorted by their gill mechanisms.

Blanchard (1991) measured nematode and harpacticoid grazing rates on benthic microalgae in an oyster pond on the French Atlantic coast. Total meiofauna grazing pressure (33.94 μg C 10 cm^{-2} hr^{-1}) slightly exceed primary production (29.53 μg C 10 cm^{-2} hr^{-1}). Despite the lower biomass (64% of that of the copepods), the grazing rate of nematodes (4.6×10^3 hr^{-1}) was almost twice as high as that of the copepods (2.43×10^3 hr^{-1}). Other food sources are probably involved in C fluxes to meiofauna (e.g., bacteria and protozoa), particularly in copepod nutrition.

Epstein et al. (1992) studied the grazing of ciliates on bacteria, flagellates, and microalgae in a temperate-zone sandy tidal flat. Most ciliate species were herbivorous, preferentially consuming several dinoflagellate, heterotrophic flagellate, and diatom species. Experimental studies confirmed that the ciliates specialized on benthic microalgae with some bactivory. The entire ciliate community consumed 10% of the total daily primary production and 93% of the estimated daily dinoflagellate production. The diatom daily production was consumed at much lower rates (6%), and only four of the 42 extant diatom species were extensively grazed. Thus, ciliate grazing on diatoms, while likely to influence diatom species frequencies, will have little impact on diatom abundance. Heterotrophic flagellates constituted 17% of the entire ciliate assemblage food spectrum.

5.3.4 Competition

5.3.4.1 Introduction

There are only a few demonstrations of competition in the marine soft-bottom benthic literature. Connell (1983) and Schoener (1983) cite only seven field experiments that document competition in soft-bottom benthic communities: Woodin (1974), Fenchel (1975b), Bertness (1981), Kastendiek (1982), Peterson (1979a; 1982a,b), Peterson and Andre (1980). To these can be added the studies of Wilson (1983) and Reise (1985), including references to his previous studies. Reise (1985) concluded: "There is no evidence for reciprocal competition ... an increase in either harms the other, ... from tidal flats." Perhaps the most famous demonstration of competition in the soft-bottom benthos is Fenchel's (1975b) study of character displacement in hydrobiid gastropods, which were competing for benthic diatoms (Fenchel and Kofoed, 1976). However, this classic study has been criticized (Levinton, 1982; Hylleberg, 1986; Cherrill and James, 1987), and it provides only indirect evidence for past competition.

Competition among benthic infaunal species generally involves competition for resources that may be in short supply. The two resources likely to be limiting are food and space. A number of approaches have been used in the study of competition in soft-sediment communities. Field studies have demonstrated intraspecific spacing and aggression among soft-bottom species (Connell, 1983; Peterson, 1979a,b, 1980; Peterson and Andre, 1980; Wilson, 1983). Laboratory and field experiments have been used to test hypotheses concerning competition. In such experiments, the density of a presumed competitor, or the abundance of a potentially limiting resource are manipulated. In order to evaluate the importance of competition, it is necessary to examine the effect of such manipulations on (1) growth, (2) mortality, (3) recruitment, (4) migration, and (5) reproductive effort (Peterson, 1980). Most studies have involved two, or at the most three competing species. However, we need to have information on the impact of competition on all the competing species in a community.

Peterson (1980) recognized four types of competitive interaction in soft-bottom sediments: (1) direct interference competition for space; (2) exploitative competition for food; (3) adult interference with larval settlement; and (4) direct interference through alteration of the physical environment.

5.3.4.2 Intraspecific Competition

Intraspecific competition between members of the same species in soft-bottom communities is not as intense as that which occurs on rocky shores. This is in part due to the three-dimensional nature of the sedimentary environment where different sized individuals of the same species may be separated by vertical stratification, with the newly settled juvenile stages occupying the surface sediments, and burrowing deeper as they grow. The various sizes of deposit feeders may also be separated from competing with each other by feeding on different sized particles.

Peterson (1982b) has demonstrated intraspecific competition in two bivalves, *Chione undulata* and *Prototheca staminea*, which exhibited decreased growth rate, gonadial mass, and (in the case of *Prototheca*) emigration at higher densities.

5.3.4.3 Interspecific Space Competition

In soft-sediment communities, space is not competed for with the same intensity as on hard substrates. However, there are a number of studies that have documented space competition. Peterson (1977), in comparing the vertical stratification of the macrobenthos in a California lagoon, found that only species that overlap vertically had any adverse effect on each other. The numbers of sand prawns *Callianassa californiensis,* and the bivalve *Sanguinolaria nuttalli* were negatively correlated, suggesting that one normally displaces the other. After experimental elimination of *Callianassa*, recruitment of *Sanguinolaria* and the seastar *Dendraster* increased markedly, although other species were not affected.

Manipulative laboratory experiments have shown direct adult interference among several infaunal bivalves. Levinton (1979) observed vertical and horizontal interspecific avoidance between three species of bivalve, *Yoldia limatula, Solemya velum,* and *Nucula proxima*. In addition to direct interference, exploitation of food and sediment reworking also played a role. As a consequence, the distribution of two of the species, *Nucula* and *Yoldia*, is largely mutually exclusive. Direct competition for space has been invoked to explain complementary and largely nonoverlapping distributions of various pairs of infaunal bivalves, e.g., *Mya arenaria* and *Gemma gemma* (Sanders et al., 1962).

Peterson and Andre (1980) performed an elegant series of experiments that demonstrated competitive interference between two bivalve species, *Salquinolaria nuttalli* and *Tresus nuttalli*. These species have widely overlapping vertical stratification in the sediment, but disjunct horizontal distributions. Peterson and Andre transplanted known numbers of the two suspension-feeding bivalves into the same sediment and observed a significantly higher emigration of *S. nuttalli* in the presence of *T. nuttalli* than in its absence. *Sanquinolaria* growth was also reduced by the presence of *Tresus*. In a second experiment that substituted dead shells secured to wooden poles simulating clams with extended siphons for living *Tresus* found similar results. *Sanquinolaria* had a lower growth rate in the presence of these artificial clams and siphons than they did in their absence. This artificial clam experiment suggests that the interaction is mediated by spatial interference and not by competition for food.

The clearest examples of competition in soft sediments involve direct interference. The spionid polychaete *Pseudopolydora paucibranchiata* is territorial and by means of palp fighting and biting, it prevents other spionids from living near it (Levin, 1984). Other spionids are known to pull different species from their tubes if they build their tubes within reach (Whitlach, 1976). Another spionid, *Streblospio benedicti,* shifts its tube to avoid interaction with another spionid *Paraprionospio pinnata* (Dauer et al., 1981). Other species that act aggressively toward other like species include nereid polychaetes (Woodin, 1974), grass shrimps (*Palaemonetes* spp.), mantis shrimps, and grapsid crabs. The crab *Pachygrapsus crassipes* not only displaces another crab, *Hemigrapsus oregonensis,* from the high shore, but takes over burrows dug by *Hemigrapsus* and does not dig its own burrows (Willason, 1981).

In addition to horizontal habitat partitioning, vertical partitioning within the sediment has been frequently found to occur. Croker (1967) described the ecology of many species of haustoriid amphipods that occur together on Georgia beaches. Of the five most abundant species, *Neohaustorius schmitzi* and *Haustorius canadensis* occur together high on the shore, *Parahaustorius longimanus* and *Acanthohaustorius millsi* lower down, while *Lepidactylus dystis* is sandwiched between these and shifts its zone up or down depending on the abundance of the upper or lower group, and only becomes abundant if either of these groups is absent. The pairs of species that share zones occupy different depths within the substratum. Stratification of species in the sediment is more common in deposit feeders and suspension feeders than in tube builders.

5.3.4.4 Exploitive Competition for Food

Levinton (1972) has examined the abundance and composition of the food resource available to soft-bottom benthic communities and its role as a limiting resource. He developed a set of hypotheses related to these food resources. He considers that suspension feeders need to cope with unpredictable and fluctuating food supplies, whereas deposit feeders have relatively constant food supplies. The food of suspension feeders can be phytoplankton, dissolved organic matter, organic aggregates (detritus with its associated microbial community), bacteria from the water column and resuspended benthic microalgae, sedimented phytoplankton, and organics. The primary food source is sedimented phytoplankton. However, phytoplankton can be extremely variable, both in space and time, and also in terms of quantity and species composition, while suspension feeders remain fixed to the substratum, apart from some bivalves that can move limited distances. Thus, suspension feeders must be typically opportunistic, rapidly exploiting favorable conditions by building up large populations. However, these populations can fluctuate strongly over time and dramatic population crashes can occur (Levington, 1970).

Deposit feeders, on the other hand, feed on phytoplankton (dead or alive) deposited on the bottom, benthic microalgae, dissolved organic matter, detritus with its attached microbial community, bacteria, and micro- and meiofauna. There is not only a greater range of potential food items but a major food source, organic detritus, is being constantly supplied to the sediment. There is a large "sink" of organic matter and it is constantly being recycled. On the face of it there would appear to be an abundance of food so that competition for food would be lessened. However, a number of studies have shown that in spite of this, competition does indeed occur.

Deposit feeders produce large quantities of fecal pellets. The mudflat snail *Hydrobia minuta* will not ingest its own fecal pellets until they have broken down (Levinton and Lopez, 1977). Thus, rates of pellet formation and breakdown can be a very important population regulation factor, since if the sediment becomes almost all pellets, then the animals stop feeding, although there is potentially a large organic matter resource present. The breakdown of fecal pellets depends on bacterial activity. According to Levinton, bacteria provide the rate-controlling factor for the generation of food, and the organic matter resides in the sediment as a "sink," buffering deposit feeders against fluctuations in the food supply. Thus, the food resource, though limited, is predictable. Levinton suggests that deposit feeders have competed for this resource over evolutionary time and as a result, feeding specializations have evolved. Also, the availability of food within the sediment has permitted different species to interact and out-compete other species, due to their greater efficiency of exploitation of the food resource. The mobility of the deposit feeders also permits competitive interference to result in the exclusion of one species by another. Levinton argues that species competitive interactions should be important in deposit-feeding communities and relatively unimportant in suspension-feeding communities.

The tube-building polychaete, *Axiothella rubrocincta*, reduces the number of the spionid polychaetes that coexist with it (Weinberg, 1984). *Axiothella* feeds on organic-mineral aggregates as a food source and competes with the spionids for these. In areas where the aggregates are abundant, it has no effect on the spionids; but where they are limiting, *Axiothella* has an adverse effect on the spionids, decreasing their density, survival, and larval recruitment. When *Axiothella* reduces the organic-mineral aggregates, the spionid *Pseudopolydora paucibranchiata* is forced to reach far out of its tube, and in doing so it becomes ready prey for flatfish. Thus, competition is supplemented by predation in bringing about the decline of *Pseudopolydora*.

It is possible that deposit feeders that have similar food requirements may feed selectively on particles of different sizes, thus reducing interspecific competition. Particle selection has been demonstrated in a number of deposit feeders such as the polychaetes *Pectinaria gouldii* (Whitlach, 1974), *Abarenicola pacifica*, and *A. vagabunda* (Hylleberg, 1975). Holothurians also ingest particles of different sizes and organic content.

A detailed comparison between the amphipod *Corophium volutator* and the mud snail *Hydrobia ulvae* (Fenchel et al., 1975) showed that the former selects fine particles and also feeds on smaller diatoms. Because of this, the diet of *H. ulvae* contains far more diatoms and that of *Corophium* has proportionately more bacteria. Using radioactively labeled bacteria, Fenchel et al. (1975) showed that *Corophium* is almost incapable of feeding on bacteria unless they adhere to fine particles. Similar labeling of several strains of diatoms showed that *Hydrobia* is a more efficient feeder on diatoms than *Corophium*.

A comparison of the deposit-feeding gastropods *Ilyanassa obsoleta* and *Hydrobia totteni* (Bianchi and Levinton, 1981) showed that both species depress blue-green algae and diatoms in the sediments they feed upon, but that only *Ilyanassa* has any effect on coccoid blue-green bacteria. When the two species are kept together, *Ilyanassa* depresses the growth of *Hydrobia*, while the latter has no effect on *Ilyanassa*. Bianchi and Levinton propose that the high-shore distribution of *H. totteni* is due to it being confined there by an "interference response" to competition, with *Ilyanassa* and other deposit feeders found lower than the intertidal.

Whitlach (1980) has provided a more penetrating analysis of niche differences between intertidal deposit-feeding polychaetes. He found that the surface feeders divided their food resources more finely than burrowers; their use of feeding tentacles allowing finer selectivity of ingested particles. Burrowers were more diverse and partitioned the habitat both by particle size of the food and by depth — another example of the use of more than one environmental dimension to segregate niches of species in diverse communities (cf. Schoener, 1971). In areas where the organic content of the sediment was high, species with similar trophic needs coexisted, but only dis-

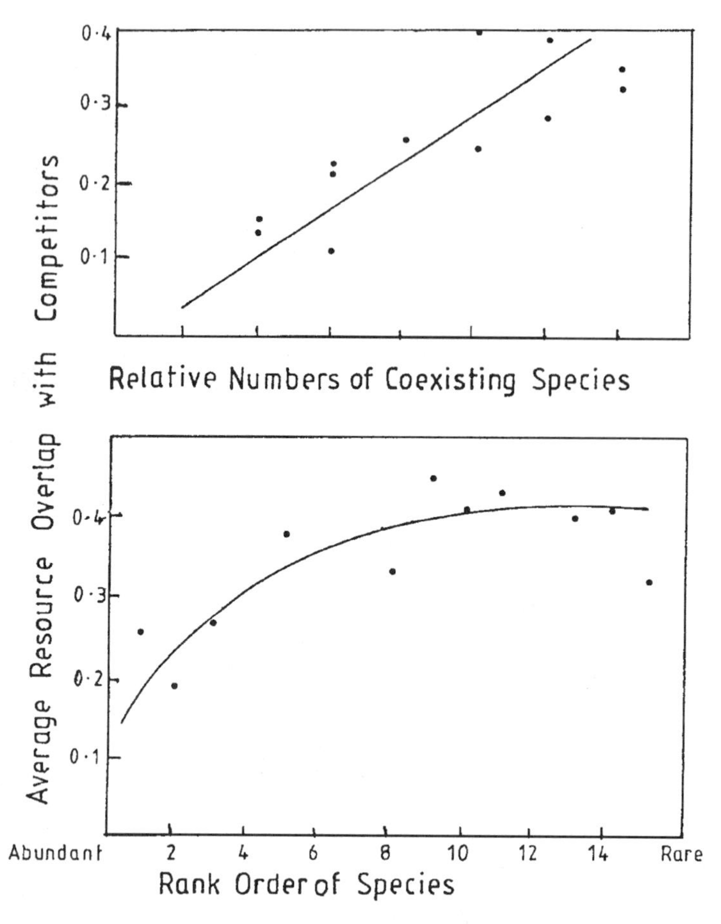

FIGURE 5.24 Relationships between the average resource overlap between deposit-feeding, burrowing polychaetes that coexist and the number of coexisting species (A); B, the abundance rank of the species. (After Whitlach, R.B., *J. Mar. Res.*, 38, 750 and 751, 1980. With permission.)

similar species lived together in areas of low organic content. The implications behind this are, first, that food is limiting in these soft-bottom communities, and secondly, that competition is greatest between species with similar needs. Comparing different areas, Whitlach showed that as diversity increases, so does the overlap between species (Figure 5.24A). This stands in direct contradiction to the generally accepted rule that as communities become more diverse, the mean overlap between the specie niches will reduce, and appears to be related to the fact that higher food levels sustain more diverse communities in these soft-bottom polychaete communities, permitting greater overlap. When these polychaete species are ranked in order of abundance (Figure 5.24B), it is the rarer species that have the highest niche overlaps with other species, and it is these species that are eliminated first when food levels decline and the community becomes less diverse. Since no particular species seems responsible for excluding any of these rarer species, Whitlach suggests that is an example of "diffuse competition," the cumulative effect of several species resulting in the exclusion of the rare species.

5.3.4.5 Interaction Between Deposit and Suspension Feeders and Tube-Builders

Traditionally, soft-bottom benthic communities have been subdivided into those dominated by suspension feeders and those dominated by deposit feeders (Woodin, 1976). There is, however, a third dominant infaunal group, the tube-building species such as many polychaetes, amphipods, tanaids, and phoronids. In contrast to dense assemblages of burrowing deposit feeders that destabilize the sediments, the dense assemblages of tube-builders stabilize the sediments regardless of their trophic type (Mills, 1969; Young and Rhoads, 1971).

Thus, three types of dense infaunal assemblages have been described in soft-bottom sediment (Table 5.5):

1. Infaunal deposit-feeding assemblages of burrowing bivalves, decapods, holothurians, polychaetes, and irregular echinoids
2. Infaunal assemblages of epifaunal suspension-feeding bivalves

TABLE 5.5
Functional Groups of Infauna and the Types of Interaction That Occur

Functional Group	Ingests or Disturbs by Its Feeding Activities, Surface or Near Surface Larvae	Filters Larvae out of Water Prior to Settlement	Changes Sediment	Larval Type at Settlement	Predicted Dense Co-occurring Forms
Deposit-feeding bivalve	Yes	No	Makes more easily resuspended	May be surface or may burrow below surface	Burrowing polychaetes
Suspension-feeding bivalve	No	Yes	No, or much more slowly	Surface	None
Tube-building forms	Yes	Depends on trophic type	Stabilizes; reduces space below surface; increases epifaunal space on surface due to tubes	Surface	Epifaunal bivalves and tube epifauna

Source: From Woodin, S.A., *J. Mar. Res.*, 34, 35, 1976. With permission.

3. Infaunal tube-building assemblages of various trophic types, comprising tube-building and relatively sedentary amphipods, tanaids, phoronids, and polychaetes

Where these assemblages co-occur, they often have sharp boundaries that are not associated with obvious changes in temperature, depth, or sediment grain size. They can vary from assemblages that occur over large areas to those that occupy discrete areas within other assemblage types.

Woodin (1976) considers that the overlapping distributions of the three assemblage types result from the interactions between established individuals and settling or newly settled larvae, rather than between the established individuals. Deposit feeders frequently feed on surface sediments (Levinton, 1972) and during this they would engulf newly settled larvae (Muus, 1973). Thorson (1957) has also suggested that much of the mortality of settling larvae is due to filtration by suspension feeders. Fitch (1965) described two populations of the Pismo clam, *Tivela stultonum,* in which settlement occurred in 1938, 1941, 1944, 1945, 1946, and 1947 in both populations (see Figure 4.16). After 1947, settlement occurred in one population and not in the other. Fitch suggests that the density of the adults in the latter population was sufficient after 1947 to prevent settlement due to the adults straining the settling spat out of the water column. Wilson (1980) has demonstrated that a terebellid polychaete has a major impact on the survivorship of nereid polychaete larvae, while Oliver et al. (1982) have demonstrated that phoxocephalid amphipod crustaceans are major predators on the larvae and juveniles of other species in marine soft-bottom communities.

Table 5.5 lists the three functional groups described above and the types of interactions that occur. Deposit feeders make the sediment more easily resuspended and ingest or disturb surface or near-surface layers, such as those of suspension-feeding bivalves, as well as those of many deposit feeders. Suspension-feeding bivalves filter larvae as well as resuspended materials out of the water column. Tube-building forms, because of their tubes, reduce the ability of larvae to penetrate the sediment surface and, depending on the trophic type of the tube-builder, ingest and/or defecate on the larvae. Woodin predicts on the basis of these adult-larval interactions that:

1. Suspension-feeding epifaunal bivalves, which brood their young and thus release them at sizes larger than those usually manageable by tube-building forms, should reach their highest densities among tube-builders.
2. Small burrowing polychaetes should reach their highest densities among deposit feeders since they are susceptible to larval filtration by suspension feeders, and that they are substrate limited because of the tube density among tube-building forms. Woodin (1974) has demonstrated spatial competition between the burrowing polychaete *Armandla brevis* and several tube-building polychaetes. The density of the burrowing polychaetes increased when the tube-builders were reduced.
3. No infaunal forms should consistently attain their highest densities among densely packed suspension-feeding bivalves.
4. As bivalves are perennials and can thus destroy their own larvae when filtering water or reworking sediments, their assemblages should be persistent but strongly age-class dominated.

There are two general components of the recruitment process: (1) habitat selection, and (2) mortality during and after settlement. Dayton and Oliver (1979) point out that field experimental manipulations of sediment grade, water, and motion result in consistent species-specific recruitment patterns into artificial sedimentary habitats (plastic cups of defaunated sediment). These experimental recruitment patterns correspond remarkably well to the patterns of abundance found in established assemblages, and with larval recruitment into large (400 m^2) defaunated bottom patches (Oliver, 1979). Generally, larvae settled into the particular artificial cup habitat that was the closest mimic to the natural sedimentary environment inhabited by the same species.

Settling larvae and young juveniles also interact with the meiofauna. Most macrofauna start their life in the meiofaunal size category and they may suffer mortality due to possible interaction with the permanent meiofauna. Watzin (1983) showed that increasing the density of turbularians and other meiofauna (nematodes, copepods, and less abundant taxa) significantly altered the density of macrofaunal settlers. In a subsequent study, Watzin (1986) carried out a series of experimental manipulations, which he examined at monthly intervals over a period of a year, in order to determine the impact of increasing densities of (1) predatory turbularians alone, and (2) other meiofauna. Densities of spionid, cirratulid, terebellid, capitellid and maldanid polychaetes, oligochaetes, bivalves, and amphipods were all reduced by high densities of turbellarians, and sometimes by other meiofauna. Turbellarian predation was responsible for significant postsettlement mortality, but spionids, cirratulids, terebellids, and maldanids also appeared to avoid dense turbellarian treatments. Bivalve densities were reduced by high densities of other meiofauna, probably due to their sedimentary disturbance. Syllid polychaete densities were higher in areas with high densities of their prey (small turbellarians). The effects of meiofauna on settlement and survival were greatest in spring and summer. It is therefore clear that meiofauna play an important role in the larval settlement process in soft-bottom communities.

5.3.4.6 Interference by Alteration of the Physical Environment

Rhoads and Young (1970) noticed that high abundances of deposit feeders in Long Island Sound were associated with fine sediments, high organic content, and relatively low densities of suspension feeders. The boundaries between suspension- and deposit-feeding communities were often very sharp. Rhoads and Young further noted that where deposit feeders were abundant, the sediments had a surface layer of semifluid consistency. This surface layer was made up principally of fecal pellets readily suspended by small water movements. The impact of sediment reworking by the deposit feeders is shown in Figure 5.25. From the figure it can be seen that the water content of sediments dominated by suspension feeders is approximately 25% and shows very little variation with depth. On the other hand, sediments dominated by deposit feeders have a high water content that reaches 70% in the semifliud muds at the surface. From their observations, Rhoads and Young developed the trophic group amensalism hypothesis which states that the interactions between two trophic groups (deposit feeders and suspension feeders) causes inhibition of one group, the suspension feeders, due to resuspended sediments clogging their filtering apparatus and preventing larval settlement.

Building on the observations of Rhoads and Young (1970), Wildish (1977) proposed the trophic group mutual exclusion hypothesis. It states that: (1) sublittoral macrofaunal community composition, biomass, and productivity are food limited; and (2) the basic exclusion mechanism is a physical one: the current speed through its control on the supply of suspended food and its inhibitory effect on the development of the later stages of deposit feeding by control of oxygen exchange between sediments and bottom water and by removing biogenically resuspended sediments. According to this hypothesis, the proportion of suspension to deposit feeders is decided primarily by the current speed spectra of the water immediately above the sediment, with predominantly deposit-feeding communities developing in low current speeds, and predominantly filter-feeding communities developing in areas with higher current speeds.

5.3.5 Predation

5.3.5.1 Introduction

The importance of predator/disturbance as a structuring component of marine soft-bottom communities is now well established. Experimental exclusion of predators often results in higher abundances of sediment-dwelling fauna, underscoring the role of predation in maintaining population levels below the carrying capacity of the habitat (Virnstein, 1977a,b; Peterson, 1979a,b). Large numbers of benthic meiofauna and macrofauna in the guts of predators, especially larval and juvenile demersal fishes, is evidence of substantial epifaunal predation on the benthos.

Habitat complexity is also important in benthic communities. Numerous investigations have documented higher abundances and diversity of both macrofauna and meiofauna is associated with more heterogeneous environments. Manipulative experiments have confirmed that habitat complexity affects the abundance, distribution, and diversity of the benthic fauna by providing: (1) an additional niche axis (Gilinsky, 1984; Marinelli and Coull, 1987), or (2) a partial refuge from predation/disturbance (Woodin, 1978, 1981; Coull and Wells, 1983; Minello and Zimmerman, 1983; Bell and Woodin, 1984). The spatial heterogeneity found in soft-bottom benthic environments can be primarily biogenic, consisting of structures such as worm tubes, sea grasses, feeding pits, and fecal mounds.

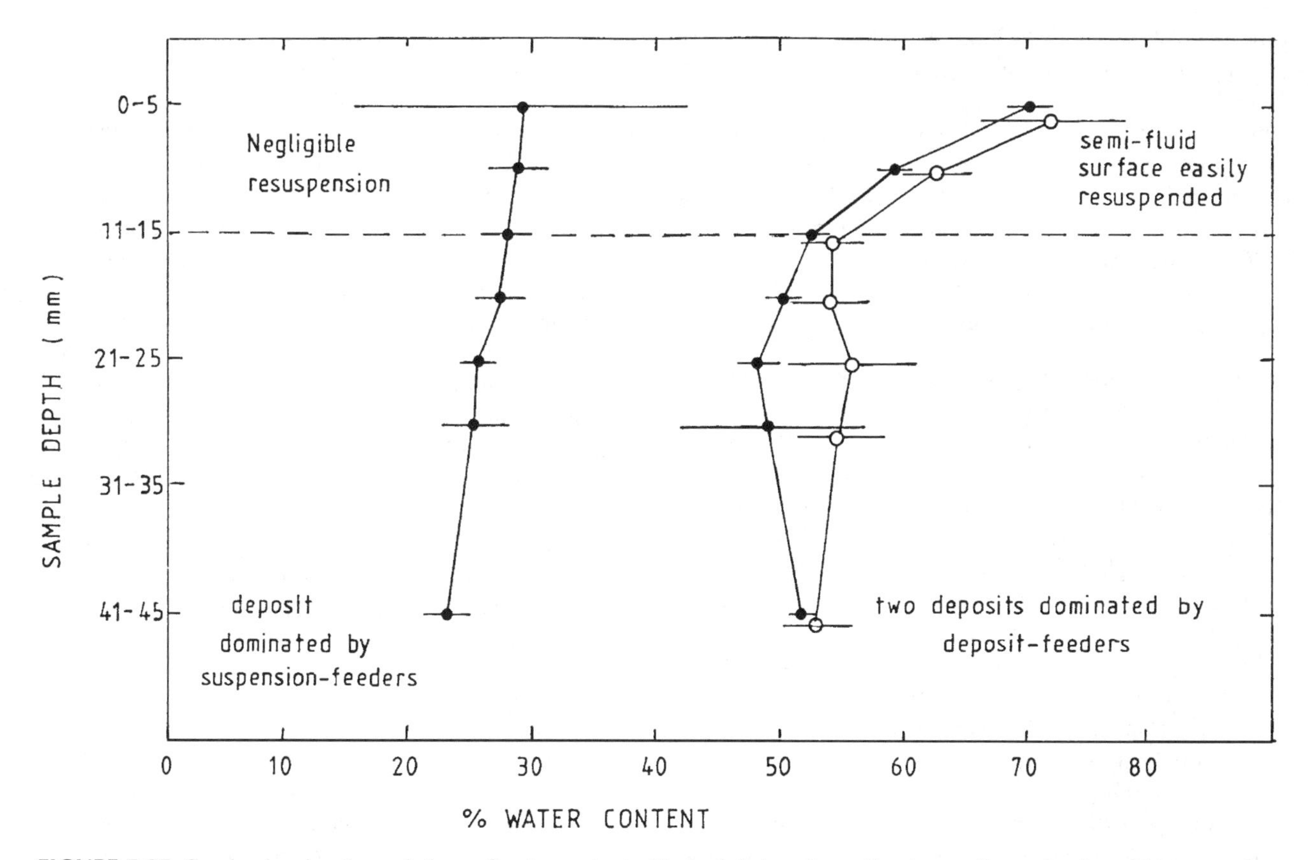

FIGURE 5.25 Graphs showing the variations of water content with depth below the surface in a sediment dominated by suspension feeders and in two sediments dominated by deposit feeders. (Redrawn from Newell, R.C., *The Biology of Intertidal Animals*, 3rd ed., Marine Ecological Surveys, Fasherman, Kent, 1979, 42. With permission.)

5.3.5.2 Meiofaunal Predation

Among the permanent meiofauna, the turbellarians in particular are voracious predators and they may prey on macrofaunal larvae and young juveniles. Watzin (1983) found that increased densities of both turbellarians and other meiofauna (tested separately) significantly reduced densities of juvenile spionids and deposit feeders. Syllid abundances increased at high-density turbellarian treatments. Nereid polychaetes, other predatory polychaetes, and bivalve densities showed no significant differences among the treatments. By both altering densities and acting selectively on various groups of macrofaunal juveniles, the meiofauna can significantly affect the structure of macrofaunal communities.

While it is well known that meiofauna are an important food source for higher trophic levels (see Section 3.6.2), the role of predation in determining the species composition and abundance of meiobenthic communities is relatively unknown (see Hicks and Coull, 1983, for a review; Sogard, 1984; Fitzhugh and Fleeger, 1985; Gee et al., 1985). Several studies have shown significant reductions in meiofaunal abundance in response to predation (Feller and Kaczynski, 1975; Fitzhugh and Fleeger, 1985; Smith and Coull, 1987), while others have concluded that predation has little or no direct effect on meiofaunal densities (Alheit and Scheibel, 1982; Hicks, 1984; Sogard, 1984).

Marinelli and Coull (1987) and Ellis and Coull (1989) have investigated the impact of predation by juveniles of the fish species *Leiostomus xanthumus* (spot) on meiofauna in the North Inlet, South Carolina. Gut content analyses of juvenile spot revealed significant consumption of meiofaunal taxa, particularly harpacticoid copepods (Chestnut, 1983). Since meiobenthic taxa are known to be vertically stratified in the sediment (Joint et al., 1982; Palmer and Molloy, 1986), and since an average spot pit is 2.2 mm deep (Billheimer and Coull, 1988), only those taxa that occur in the surface sediment will be vulnerable to predation. On North Inlet mud sites, 64% of the total copepods and 62% of the nauplii were found in the top 4 mm of the sediment, while only 42% of the nematodes and 38% of the total meiofauna were found in this layer.

Predation by spot in the experiments conducted by Ellis and Coull (1989) affected the abundance of copepods and nauplii but not that of nematodes, foraminifera, and total meiofauna. The greatest predation effect was found on the two most abundant epibenthic copepod species, *Halicyclops coulii* and *Pseudobradya pulchella*. In their laboratory experiments, Marinelli and Coull (1987) tested the effectiveness of polychaete tube mimics as refuges from predation and disturbance activity by juvenile spot. Four of the six meiofaunal categories (copepods, copepod nauplii, juvenile macrofauna, and ostracods) experienced significant reductions due to fish predation in the samples with tube mimics, while there was no statistically significant predation on any meiofaunal taxa in the habitats containing no tube mimics. This result (structure facilitating predation) was a surprising one.

5.3.5.3 Predation by Infauna

Although the effect of epibenthic predators on infauna is well known, the impact of infaunal predators on benthic community composition and abundance is poorly understood (Commito and Ambrose, 1985a,b). Feller et al. (1979) provided immunological evidence that infaunal organisms eat a variety of benthic recruits. Laboratory studies have shown that terebellid polychaetes ingest nereid polychaete larvae (Wilson, 1980) and that nereid polychaetes consume a wide variety of infaunal organisms (Witte and de Wilde, 1979). Rae (1976) demonstrated that the nemertine *Paranemertes peregrina* consumes one-third of the nereid polychaetes and virtually all of the spionid polychaetes annually at Garrison Bay, Washington, U.S.A.

Reise (1979), Commito (1982), Commito and Schraeder (1985), and Ambrose (1984a,b) have all manipulated predatory nereid polychaete abundances in the field. They have demonstrated that *Nereis diversicolor* in the Wadden Sea (Reise, 1979) and *N. virens* in New England (Commito, 1982; Commito and Schraeder, 1985; Ambrose, 1984a,b) play significant roles in regulating the densities of infaunal amphipods, bivalves, and annelids. At Bok's Cove, Maine, Commito (1982) found that the removal of *N. virens* led to an increase in the abundance of the amphipod *Corophium volutator* (which normally comprises 63% of the individuals in the community). The addition of *N. virens* resulted in a decline in *C. volutator* populations. Commito and Schraeder (1985) added *N. virens* to a community in which *C. volutator* was absent or extremely rare. Three taxa, tubificid oligochaetes and the polychaetes *Streblospio benedicti* and *Capitella capitata,* comprised 95% of the individuals of this community. The addition of *N. virens* did not result in any measurable change in the relative proportions of the three taxa. It had been anticipated that, in the absence of *C. volutator,* the addition of *N. virens* would reduce the density of other infaunal species. This did not occur. The reason for this was probably that *N. virens* reduced the population of an intermediate polychaete, probably *Nephtys incisa.* Ambrose (1984b) found reduced densities of *N. incisa* when *N. virens* was added to enclosures, and Reise (1979) observed lowered abundance of macrofaunal annelids in *Nephtys hombergi* enclosures.

Ambrose (1984b) carried out an elaborate series of field manipulations designed to determine the influences of two predatory infaunal polychaetes, *Glycera dibranchiata* and *Nereis virens,* and epifaunal predators on the structure of a soft-bottom community in a Maine estuary. Densities of the polychaetes *Nephtys incisa, Polydora ligni, Streblosoma benedicti,* and *Scoloplos robustus,* and phyllodicids and bivalves were highest in cages containing elevated *G. dibranchiata* densities and lowest in cages containing elevated *N. virens* densities. *N. virens* was the only species whose abundance was reduced in the presence of *G. dibranchiata* and this may account for high infaunal abundances in the *G. dibranchiata* treatments. Laboratory experiments demonstrated that *G. dibranchiata* is capable of feeding on *N. virens.* Other experiments demonstrated that the presence of *N. virens* reduced the abundances of some taxa within 10 days. These reductions may have been due to predators and/or disturbance at the sediment surface. It is clear that the interactions of infaunal predators and other infauna are complex.

5.3.5.4 Predation by Epifauna

Epifaunal predators include prawns, crabs, a variety of adult and juvenile fishes, and birds. The impact of epibenthic predators has been investigated by using cages to exclude them or to increase their density. The experiments that have been carried out can be grouped into two categories: those carried out on intertidal or subtidal soft-bottoms (e.g., Reise, 1977; 1978; Virnstein, 1977a,b; Schneider, 1978; Peterson, 1979a,b; Gilinsky, 1984; Wiltse et al., 1984; Gee et al., 1985; Marinelli and Coull, 1987; Sanchez-Salazar et al., 1987; Frid and James, 1988; Raffaelli et al., 1989), and those carried out on sea grass beds (e.g., Reise, 1977; Virnstein, 1977a,b; 1978; Nelson, 1979; Heck and Orth, 1980; Bell and Westby, 1986). As an example of the former, that of Woodin (1974), who used cages to exclude predators on a muddy intertidal shore will be discussed. Table 5.6 shows the results of one such experiment over the course of which only *Platynereis bicaniculata* and *Armandia brevis* reproduced. From the table it can be seen that there was a clear change in the dominance pattern within the cages where burrowing species increased in abundance. Woodin suggested that the tube-builders settled on the cage surface while the burrowers passed through the cage onto the sediment surface. Normally, tube-builders out-compete burrowers for space by means of interference competition. In her experiments, Woodin (1974) demonstrated two other important interactions: (1) that when a predator (the crab, *Cancer magister*) was placed inside a cage, the abundance of tube-builders increased dramatically, whereas burrowers maintained the same densities; and (2) that variations in larval recruitment can strongly influence the community structure.

TABLE 5.6
Abundance of Three Tube-Building and One Burrowing Species within 0.05 m^{-2} of Caged and Uncaged Sediments

	Tube-Building Species				Burrowing Species
	Lumbrinerus inflata	*Axiothella rubrocinctata*	*Platynereis bicaniculata*	Total	*Armandia brevis*
No cage	168	123	358	649	47
	92	158	313	563	52
Cage	168	136	47	351	143
	132	153	25	310	160
	113	141	19	273	129
	64	104	54	222	139

Note: Samples taken from August to November–December.

Source: From Woodin, S.A., *Ecol. Monogr.*, 44, 143, 1974. With permission.

In nine of the eleven studies on soft-bottom caging experiments cited by Peterson (1979b), the density of the macroinvertebrates was significantly higher inside the predator exclusion cages than in control areas after some period of time. In the majority of the experiments, species richness also increased after predators were excluded, and remained higher than in control areas after substantial periods of time at the elevated densities. Apparently, adult-adult interactions, even at the relatively high densities achieved within the exclusion cages, are incapable of simplifying the soft-bottom communities. This is in direct contrast to the generally accepted model of community organization developed from experimental work on intertidal rocky shore communities, which predicts that significant simplification of the community should occur as a consequence of intense competition where density had increased substantially following the exclusion of predators.

Gee et al. (1985) have studied the role of epibenthic predators in determining prey densities in the Lynher Estuary, England. The benthic macrofauna of the mudflats of this estuary is dominated numerically by small annelids and there is evidence that adult crabs (*Carcinus maenus*) can cause significant increases in the oligochaete component of the assemblage, probably as a result of its burrowing activity. Juvenile *C. maenus*, on the other hand, significantly reduced the abundance of small annelids, particularly the dominant polychaete *Mayanukia aestuarina* and could be responsible for the year-to-year variations in the abundance of this species. Fish predators (principally *Pomatoschistus microps*) had little effect on the abundance of prey species. In the same estuary, Gee (1987) investigated the impact of epibenthic predation on harpacticoid copepod populations. Harpacticoid copepods are the only component of the meiofauna that form a significant part of the diet of the early juvenile stages of the predators. Flatfish, gobies, and shrimp were estimated to consume daily 0.01 to 0.1% of the standing stock of the copepod *Asellopsis intermedia* (the dominant species and the sole species on sand flats) and they accounted for between 12 and 22% of the observed reduction in the populations of this species between July and October.

Laboratory studies have shown that predation rates in vegetated aquatic systems are lower in structurally complex habitats (Nelson, 1979; Heck and Thoman 1981; Stoner, 1982; Coull and Wells, 1983). This has led to the hypothesis that predation is the proximate cause of correlations between prey abundance and structural complexity of sea grasses (Heck and Orth, 1980; Stoner, 1982). However, habitat preferences could be responsible. Stoner (1980) showed that amphipods could discriminate between sea grasses of different complexity. Moreover, if survival is increased among sea grass beds with better cover, then preference for that habitat would be selected (Stoner, 1980; 1982). When predators have been excluded, or included in field experiments on sea grass beds, prey abundances have often failed to alter as predicted by laboratory experiments (Young and Young, 1978; Nelson, 1979; Summerson and Peterson, 1984). Bell and Westby (1986), in field experiments, reduced the density of sea grasses in the presence and in the absence of predators. All of the six main prey species decreased simultaneously in the presence and absence of predators. Thus, it would appear that habitat choice of the prey, rather than increased predation when sea grass density is reduced, is responsible for reduced densities of prey.

Peterson (1979b) points out that the reasons why competitive exclusion is not shown in caging experiments when predators are excluded might include the following: (1) the experiments have not been carried out for a sufficient length of time for interference mechanisms to show up; (2) soft sediments provide much reduced opportunities for interference mechanisms to be employed (in contrast to the essentially two-dimensional rocky shore situation, soft sediments are three-dimensional and thus there is greater opportunity for spatial separation); (3) the extreme importance of adult-larval interactions in soft sediments maintains densities below the carrying capacity, even in the absence of predators; and (4) under food limitation, marine invertebrates exhibit extreme plasticity in growth and reproduction and possess low-energy needs, both of which help to ensure high survivorship.

5.3.5.5 Impact of Predation on Bivalves

Many studies have focused on the role of predation in structuring bivalve communities and they have demonstrated that predators can influence the spatial distribution and abundance of their bivalve prey (Holland et al., 1980; Reise, 1978; 1981; 1985; Jensen and Jensen, 1985; Phil, 1985; Sanchez-Salazar et al., 1987). Many of these bivalve predators are also known to alter the bivalve population structure by foraging selectively on particular size classes (Griffith and Seiderer, 1980; Seed, 1982; Moller and Rosenberg, 1983).

Cerastoderma (= Cardium) edule is a widely distributed bivalve found throughout northwestern Europe. In the Wadden Sea, it is one of the major benthic producers (Beukema, 1976; 1981). Sanchez-Salazar et al. (1987) found that the size and age structure of this species varied dramatically. In areas low on the shore, up to 96% of cockle spat fail to survive their first summer, but the mortality rate subsequently declines and remains at a low level. By contrast, cockles high on the shore suffer moderate mortality during their first year (47%), but increasing rates thereafter. High-shore populations consequently consist mainly of smaller (younger) individuals and low-shore ones of a transient spatfall, plus a few larger and older individuals. Shore crabs (*Carcinus maenas)* move up into the intertidal to feed with each flood tide from about April to December. They selectively consume cockles,15 mm in length, taking an estimated 256 × 10^3 cockles, or 2,432 g dry flesh $year^{-1}$ per linear meter of shoreline, mostly from lower shore levels. Oystercatchers (*Haemotopus ostralegus*), the other major predator, are present only during the winter and they preferentially select large cockles of at least 20 mm in length. They are estimated to remove 9 × 10^3 cockles, or 1,204 g dry flesh year [1] per linear meter of shoreline, most of this from mid- and high-shore levels. The results of this study indicated that shore crabs are the major predators, taking 25 times the numbers and 2 times the biomass consumed by the oystercatchers. Predation appears the be a key factor in controlling the structure of the *C. edule* population (Figure 5.26). Crabs consume almost all the cockles settling low on the shore during their first summer, but avoid older individuals, which subsequently survive and grow well under low levels of oystercatcher predation. On the high shore, crabs are unimportant and the cockles survive well as they slowly grow into the size range attractive to oystercatchers. Thereafter, they suffer increasingly severe winter mortality and are soon eliminated.

5.3.5.6 Multiple Predation on Tidal Flats

As Reise (1985) points out, with special reference to the Wadden Sea, tidal flats are rich in predators, both within the sediment and the water above. In addition to preying on nonpredatory infauna, many of the predators prey on each other (Figure 5.27). Birds prey on small epibenthic predators, and both prey on infaunal predators. In addition, there is predation within the various predator groups, e.g., gulls prey on eggs and chicks of other birds. Among the high-tide carnivores, cannibalism and size-dependent reciprocal predation are particularly common.

Prey is primarily selected on size. Thus, large benthic invertebrates will have survived the entire spectrum of predator groups until they attain adult body mass. An example is the lugworm *Arenicola marina* (Figure 5.28). The platyhelminth *Monocelis fusca* preys on the early juvenile stages of this polychaete, and probably other Platyhelminthes do the same. The errant polychaete *Nereis diversicolor* preys on both juvenile lugworms and Platyhelminthes (Witte and de Wilde, 1979; Reise 1979). Crabs attack juvenile lugworms if they live close to the sediment surface. They also prey on the platyhelminths and nereids (Reise, 1978; 1979). Flatfish and birds with long bills prey on the tail ends of the lugworms (Pienkowski, 1982) and also on crabs and nereids.

The predator-prey relationships outlined above demonstrate the high complexity of the predatory interactions on tidal flats. Specialized feeding is rare. This multiple predation between various predators and the extensive prey overlap among them, may well serve to dampen both prey and predator fluctuations.

5.3.5.7 Predation by Birds

An important group of top predators on sediment shores are the birds. Many studies indicate that the predator pressure exerted by the birds may be considerable (e.g., Milne and Dunnet, 1972; Wolff et al., 1976; Wolff, 1983; Goss-Custard et al., 1988; Raffaelli et al., 1990). They can be divided into two groups, the permanent residents such as gulls, cormorants, some ducks, and marsh rails, or migratory species such a wading birds, geese, and many ducks. Consequently, the numbers and species of birds on a shore vary throughout the year. Many coastal mudflats and estuaries are stopover points for many species on their annual migration pathways. Other species may overwinter before migrating to breeding areas in the spring and summer.

Bird predators are highly mobile and typically exhibit clear tidal rhythms of activity that are associated with water movements and the activity of their prey in relation to the tides. On the basis of their food and feeding meth-

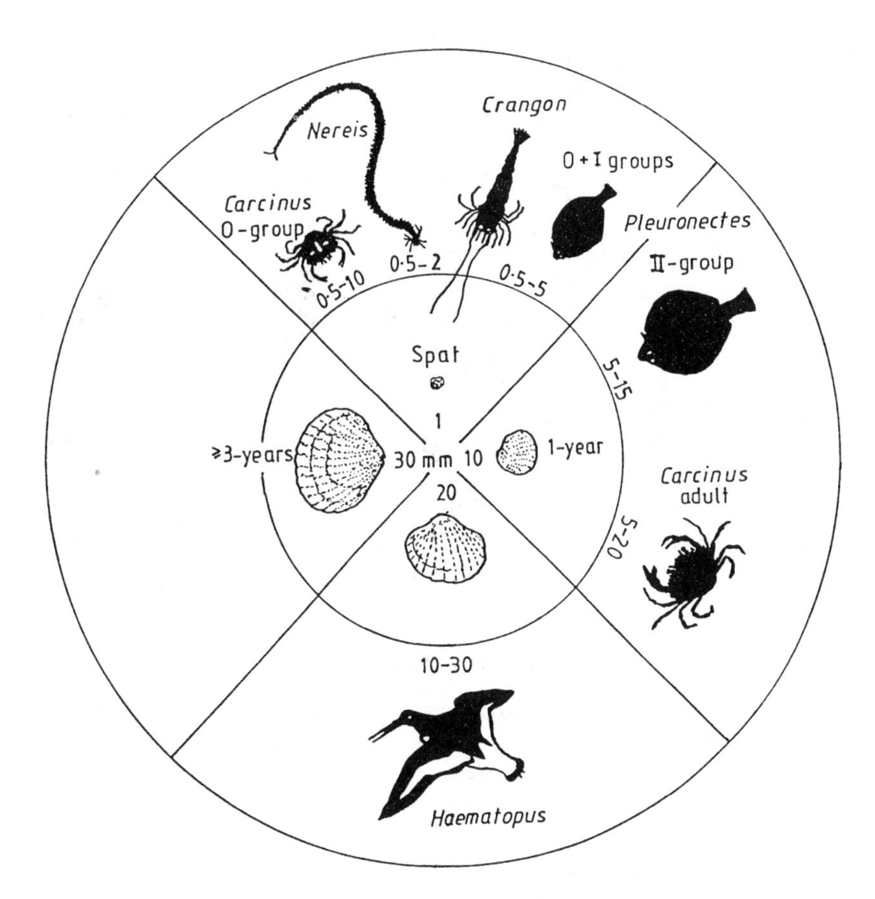

FIGURE 5.26 Exemplified predator spectrum of *Cerastoderma edule* growing from spat to adult size on tidal flats in Konigshafen, Island of Sylt. Prey spectra are given in shell length (mm) at the *inner circle*. (Redrawn from Reise, R., *Tidal Flat Ecology: An Experimental Approach to Species Interactions*, Springer-Verlag, Berlin, 1985, 113. With permission.)

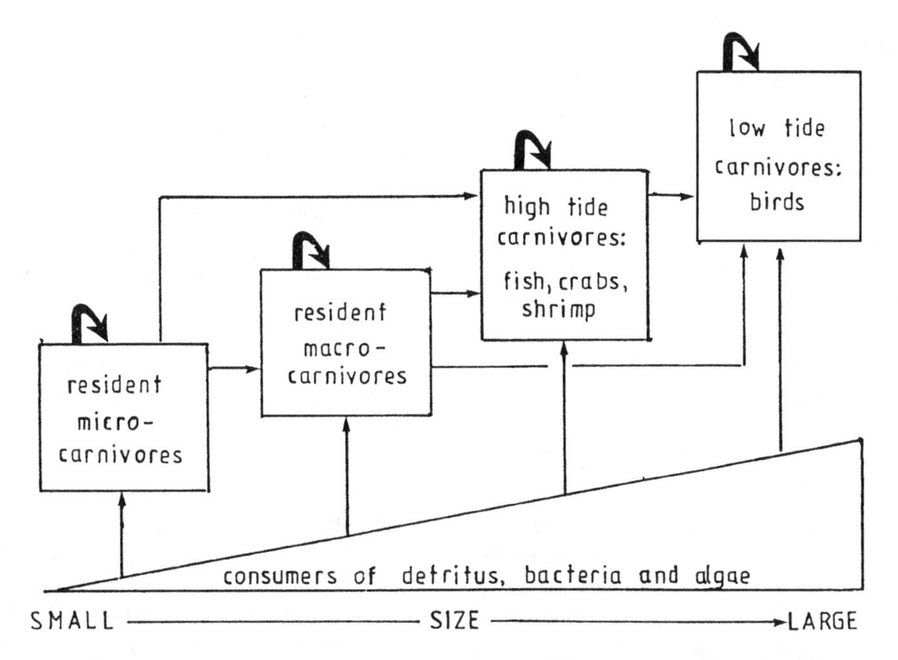

FIGURE 5.27 Predator interactions on tidal flats. *Arrows* indicate the transfer of prey biomass. Predation within compartments is primarily on each other's juveniles with size-dependent reversals. Some prey species pass through the entire size range until they attain adult body size. (Redrawn from Reise, R., *Tidal Flat Ecology: An Experimental Approach to Species Interactions,* Springer-Verlag, Berlin, 1985, 106. With permission.)

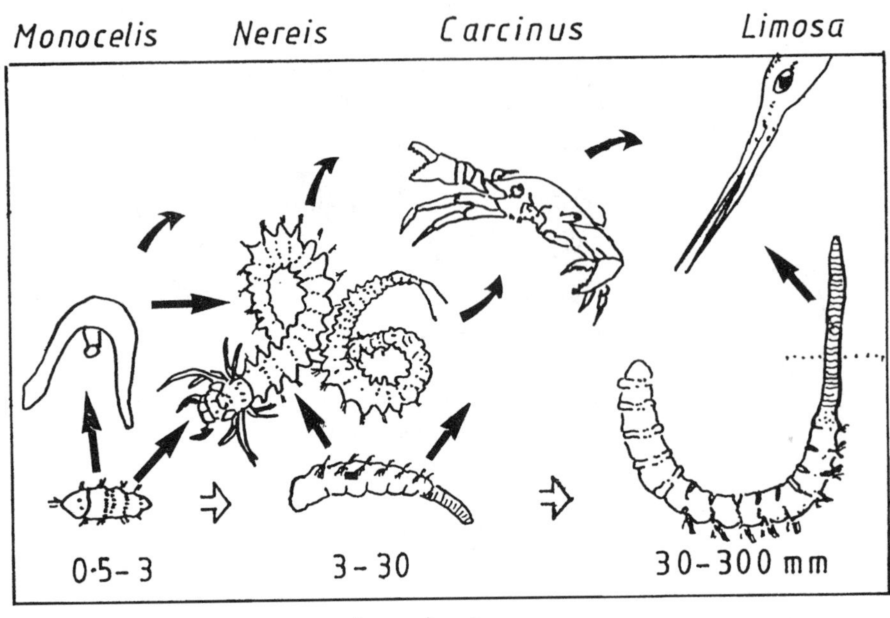

FIGURE 5.28 Exemplified predator spectrum of the lugworm, *Arenicola marina*, developing from benthic larvae in tidal flat sediments. In reality there are more predators and the lugworm is one more prey species among many. Based on observations from tidal flats at Konigshaften, Island of Sylt. (Redrawn from Reise, R., *Tidal Flat Ecology: An Experimental Approach to Species Interactions*, Springer-Verlag, Berlin, 1985, 107. With permission.)

ods, the birds can be divided into wading birds, water fowl, fishing birds, rails, and other marsh species. Wading birds can be divided into three groups: (1) the herons, egrets, and ibis which feed principally on small fishes, crabs, and snails; (2) the shore birds, comprising resident stilts and migratory species such as plovers, curlews, turnstones, godwits, red shanks, dunlin, knot, sandpipers, whimbrels, stints, etc., which feed principally on small crustaceans and polychaete worms; and (3) the oystercatchers, which feed principally on bivalves. The fishing birds such as cormorants and terns consume principally fishes, but also take larger pelagic crustacea. Ducks and geese consume plant material (up to 75%) and take smaller amounts of benthic animals, mainly snails. However, some species, e.g., the eider duck, feed almost exclusively on mussels. Some swamp hens feed on the roots and shoots of marsh plants, while marsh rails feed on snails, crabs, other crustaceans, and insects.

Goss-Custard (1977a,b) and Goss-Custard et al. (1977) have investigated the *feeding ecology* of wading birds in the Wash, Great Britain. Table 5.7 lists the bird predators with their principal prey species. From this table it can be seen that there is only limited overlap in the species taken by the predators. Within the Morecombe Bay estuarine area in northern England, there is a large (70,000+) overwintering population of the knot (*Calidris camatus*). These birds feed principally on the bivalve *Macoma balthica*, supplemented by the mussel *Mytilus*

TABLE 5.7
Principal Prey Species of the Main Wading Birds in the Wash (England)

Bird Species	Prey Species	
Oystercatcher (*Haematopus ostralegus*)	Bivlaves	*Cardium edule*
		Mytilus edulis
Knot (*Calidris canutus*)	Bivalves	*Macoma balthica*
		Cerastoderma edule
Dunlin (*Calidris alpina*)	Snail	*Hydrobia ulvae*
	Polychaetes	*Nereis diversicolor*
Redshank (*Tringa totanus*)	Crustacea	*Carcinus maenus*
		Crangon sp.
	Snail	*Hydrobia ulvae*
	Polychaetes	*Nereis* spp.
Bar-tailed Godwit (*Limosa lapponica*)	Polychaetes	*Lanice conchilega*
		Nereis spp.
	Bivalves	*Macoma balthica*
Turnstone (*Arenaris interpres*)	Bivalves	*Cerastoderma edule*
		Among mussel beds
Grey Plover	Polychaetes	*Lanice conchilega*
		Various spp.
Curlew (*Numenius arquata*)	Crustacea	*Carcinus maenus*
	Polychaeta	*Arenicola marina*

Source: Data from Goss-Custard, T.D., Jones, R.E., and Newberry, D.E., *J. Appl. Ecol.*, 14, 681, 1977. With permission.

edulis and the mudflat snail *Hydrobia ulvae*. The choice of *Macoma* appears to be dictated by its availability. On the Morecombe Bay flats, the knot feed predominantly on the lower half of the intertidal zone where they have access to *Macoma*, which are generally lower down than *Hydrobia*. Prater (1981) suggests that while the knot have a clear preference for *Hydrobia* in Morecombe Bay, *Macoma* is the dominant food item because of its greater accessibility. The preferred diet of the red shank, *Tringa totanus*, at 30 estuarine sites in the British Isles is the amphipod *Corophium volutator*, with the feeding rate depending mainly on the density of the prey (Goss-Custard et al., 1977a). Where the density of *Corophium* is low, red shanks take polychaetes worms, *Nereis diversicolor,* and *Nephtys hombergi* instead. They also sporadically take the bivalves *M. balthica* and *Scrobicularia plana*.

One of the most intensively studied species of estuarine birds is the oystercatcher (*Haematomus ostralegus*) in northern European estuaries (Heppleston, 1971; Zwarts and Drent, 1981; Hulscher, 1982; Goss-Custard and Dit Durell, 1988). In these estuaries, oystercatchers feed on mussel beds when they are exposed at low tide. In a study of oystercatcher feeding in The Netherlands, Zwarts and Drent (1981) found that when mussels taken by oystercatchers were compared with the size distribution of mussels present, there was a clear preference for larger mussels; when compared to the "modal mussel" of 44 mm, the relative risk of a mussel of 50, 54, or 59 mm being taken by oystercatchers was 3.6, 6.7, and 10.5 times as high, respectively. They also found that the feeding rate decreased significantly as the density of the oystercatchers increased.

A detailed study of the feeding ecology of the South Island pied oystercatcher, *Haematopus finschii*, has been carried out in the Avon-Heathcote Estuary, South Island, New Zealand (Baker, 1966; 1969; 1974). A maximum of about 4,000 birds overwinter on this 8 km^2 estuary. Their feeding behavior follows a regular tidal cycle. During high water, the birds congregate on high water roosts, while the periods of ebb and flood tides are spent feeding in a narrow band averaging about 3 m from the water's edge down to the maximum depth at which the birds can stand (about 6 to 7 cm). While the range of food items taken includes cockles (*Austrovenus stutchburyi*), pipis (*Amphidesma australe*), wedge shells (*Tellina liliana*), mudflat snails (*Amhibola crenata*), a range of polychaete worms, amphipods, shrimps, mudflat crabs, and juvenile flounders, the predominant food item is the cockle. Baker found that when feeding, the birds were not uniformly distributed, but were concentrated in areas of high prey density. The effect of prey concentrations on bird distributions and feeding rates is shown in Table 5.8.

A detailed study of feeding rates and the factors affecting the rates was carried out. Feeding rates were highest when the tide was retreating or advancing over the feeding grounds and lowest when the cockle beds were completely exposed. Feeding rates also varied with season and thus with ambient temperatures. Mean monthly feeding rates were highest in the midwinter months of June, July, and August (40+ cockle hr^{-1} in July), and lowest in the midsummer months of December, January, and February (27 cockles hr^{-1} in February). Similar variations in feeding rates have been noted for the European oystercatcher (Davidson, 1967; Hulscher, 1975). Energy requirements are greatest in the colder months and this is met by higher food intake. In addition to prey availability, prey size also has a direct effect on intake. Feeding rates of 108 cockles hr^{-1} were recorded when the cockle modal size was 3.1 cm (1.9 ml flesh volume) as compared to a rate of 24 cockles hr^{-1} when the modal size was 5.3 cm (6.0 ml flesh volume).

During the winter months, the oystercatchers on the Avon-Heathcote Estuary ingested an average of 40.9 cockles hr^{-1}. Predated cockles had a mean valve length of 3.1

TABLE 5.8
The Effect of Prey Concentration (*Chione stutchburyi*) in the Avon-Heathcote Estuary, New Zealand, on the Distribution and Feeding Rate of the South Island Pied Oystercatcher (*Haematopus finschii*)

Mean Number of Birds/Study Site	Mean Cockle Concentration (no. m^{-2})	Mean Feeding Rate (Cockles hr^{-1})	Number of Observations
22	78	96	10
16	54	60	14
10	42	48	12
5	33	24	14
1	20	24	11

Source: From Baker, A.J., The comparative biology of New Zealand oystercatchers, M.Sc. thesis, University of Canterbury, Christchurch, New Zealand, 1969. With permission.

TABLE 5.9
Oystercatcher Predation on Cockles in the Avon-Heathcote Estuary, New Zealand

	Winter	Summer
No. of cockles per hour	40.9	29.0
No. of cockles per day	368	261
Grams of cockle flesh per day	174.9	124.0
Percentage of body weight eaten per day	35.2	25.0
Mean daily winter cockle predation for 4,000 oystercatchers	1,472,000	
Mean yearly food intake per oystercatcher	190,000	
Annual cockle predation by 4,000 oystercatchers	438,876,000	

Source: From Baker, A.J., The comparative biology of New Zealand oystercatchers, M.Sc. thesis, University of Canterbury, Christchurch, New Zealand, 1969. With permission.

cm and a mean flesh volume of 1.9 ml. The average length of the daily feeding period was 12 hr, but half of this occurred in darkness when the feeding rate was assumed to be half that during daylight hours (Heppleston, 1971). Thus, in the winter, the birds ingested an average of 368 cockles day^{-1}, or an estimated intake of 174 g dry wgt cockle flesh, corresponding to 35.2% of their dry body weight (Table 5.9). In summer, the daily cockle intake was reduced to 261 cockles day^{-1}, or 124 g dry wgt cockle flesh. The mean daily winter cockle predation by the oystercatchers in the small estuary was 1,472,000.

Workers in Great Britain (Davidson, 1967; Norton-Griffiths, 1967; Goss-Custard et al., 1977) who have investigated oystercatcher feeding rates have found rates comparable to those discussed above. In Morecombe Bay, it was found that the percentage annual cockle mortality due to oystercatcher predation was 21.9% of the total population.

Bird predation on the mussel bed community of the Ythan Estuary in Scotland has been studied by Milne and Dunnet (1972) who found that the mussel (*Mytilus edulis*), the gammarid amphipod (*Marinogammarus marinus*), the shore crab (*Carcinus maenus*), the tube-dwelling amphipod (*Corophium volutator),* and fish (butterfish, blenny) were the principal prey of the birds (oystercatchers, eider ducks, gulls, and turnstones). The partitioning of the *Mytilus* among the various predators is shown in Figure 5.29. Annual gross production was 268 g dry wgt m^{-2}, or 1340.8 kcal m^{-2}. Of this, 600 kcal represented winter metabolism, leaving a net production of 740 kcal m^{-2}. Of this, 400 kcal m^{-2}, or 65% of the net production, was consumed by bird predators. The feeding of these bird predators was separated both in space and time as the different species fed on different sized mussels. Eider ducks fed when the mussels were submerged and consumed mussels averaging 18 mm in length (2-year-old mussels). Oystercatchers and gulls fed when the mussel beds were exposed, the oystercatchers taking mussels averaging 33 mm (3- to 10-year-old mussels), while gulls took mainly very small mussels 2 to 10 mm long (1-year-old mussels). The oystercatchers were estimated to consume 93 kcal m^{-2}, eider ducks 275 kcal m^{-2}, and gulls 112 kcal m^{-2}, representing 12.5, 37.2, and 14.5% of mussel net production, respectively.

5.3.5.8 Role of Predation in Structuring Soft-Bottom Communities

Ambrose (1984a,b; 1986) has discussed in detail the role of predatory infauna and epifauna in structuring soft-bottom communities. Early studies of epifaunal predator impacts classified crabs, fishes, and birds as predators and the infaunal species as prey. Many infaunal species, however, are themselves predators and, as prey for epibenthic predators and predators on other infauna, may function as intermediate predators. The predatory infauna are capable of influencing the abundance of other infaunal species (see Section 5.3.5.3 above) (Reise, 1979; Ambrose, 1982; Commito, 1982; Oliver et al., 1982). Most predatory infauna do not form protective tubes (exceptions include the polychaete Families Nereidae, Phyllodocidae, Ouphinidae, and Polyodontidae). Those that do have protective burrows frequently extend large portions of their bodies onto the sediment surface, or leave their burrows altogether (e.g., nereids, phyllodicids), making themselves more susceptible to predation from epibenthic predators than other nonpredatory infauna, many of which have protective tubes or burrows and rarely venture from them (e.g., capitellids, chaetopterids, maldanids, *Corophium*).

Ambrose (1984b) analyzed data from seven studies that employed cages to exclude epibenthic predators. This revealed that predatory infauna became proportionately more abundant following the exclusion of epibenthic predators from muddy sand and sea grass habitats. Increases in the proportion of predatory infauna could be a consequence of: (1) preferential predation on these predatory infauna by epibenthic predators; (2) preferential predation

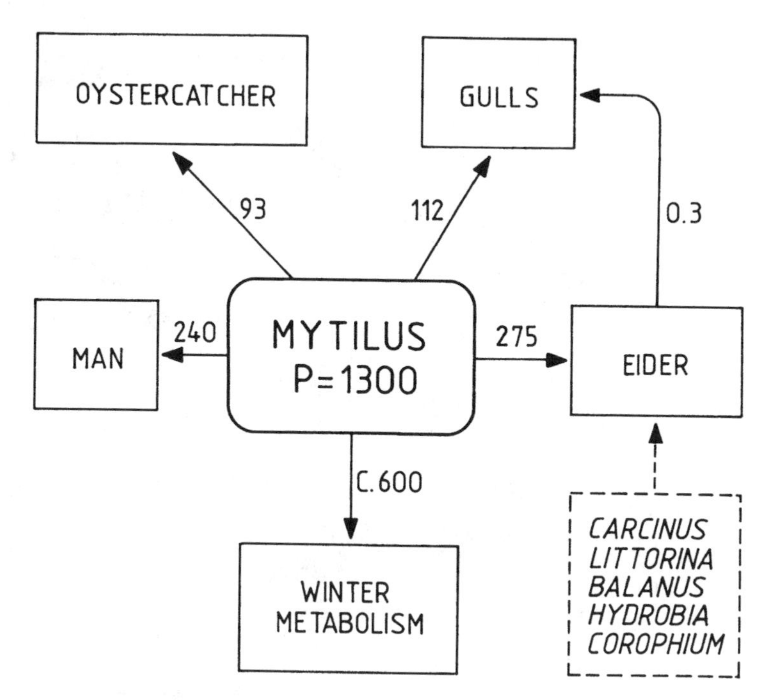

FIGURE 5.29 The partitioning of production of the mussel (*Mytilus edulis*) among the various predators of the mussel bed community in the Ythan Estuary, Scotland. (Redrawn from Milne, H. and Dunnet, G., in *The Estuarine Environment*, Barnes, R.S.K. and Green, J., Eds., Applied Science Publication, London, 1972, 90. With permission.)

on the predatory infauna and predation by predatory infauna on other infaunal species; (3) equal predation on predatory and nonpredatory infauna with additional predation by the predatory infauna on other infaunal species; or (4) competition between predatory and nonpredatory infauna, with the predatory infauna outcompeting nonpredatory infauna. While the degree to which these mechanisms operate is unclear, the increase in the proportion of predatory infauna following the exclusion of epibenthic predators did occur in most studies despite several factors favoring an increase in the abundances of nonpredatory infauna under exclusion cages. First, because the cages reduce the rate of water flow, they will accumulate fine sediment and organics (Virnstein, 1978), which would favor deposit-feeding species. Second, life history characteristics of many of the nonpredatory infauna make them more likely than predatory infauna to settle and reach high densities under cages. Opportunistic species sometimes account for the largest absolute increases under cages (Holland et al., 1980), a result of these species' high reproductive rates.

Ambrose (1984b) proposes a 3-level interactive model for soft-sediment marine systems (Figure 5.30) comprising epibenthic predators, predatory infauna, and nonpredatory infauna. However, the impact of the predators on community structure may not be confined to predation itself. Certain epibenthic predators, such as blue crabs, *Callinectes sapidus*, and the horseshoe crab *Limulus polyphemus* on the U.S. East Coast may be responsible for as much mortality through sediment disturbance as by predation. Many fish species, e.g., eagle rays also disturb the surface sediments. This disturbance makes the infauna, both predatory and nonpredatory, more accessible to both epifaunal and infaunal predators.

While "keystone" predators are a feature of most hard shores, they are generally lacking in soft sediments (Peterson, 1979b). In soft sediments, high predation rates commonly result in decreases in abundance and changes in the size structure of prey populations rather than changes in species diversity (Peterson, 1979a,b; Kvitek et al., 1992; Micheli, 1997). Indirect effects of predation on soft-sediment communities have also been shown to occur through sediment disturbance (Woodin, 1978; Oliver and Slattery,1985; Kvitek et al., 1992) because of trophic interaction chains. In the salt marshes of the southeastern U.S., for example, the kill fish *Fundulus heteroclitus* has been shown to decrease predation rates on benthic macrofauna by consuming, shrimp the intermediate predators in the system (Kneib, 1987). In these examples, direct and indirect effects of predation on soft-sediment benthic communities did not produce species replacement or competitive dominance.

5.3.6 Influence of Resident Fauna on the Development of Soft-Bottom Communities

Interactions between established adults and settling larvae have long been considered to be important in structuring soft-bottom communities (Woodin, 1976; Peterson,

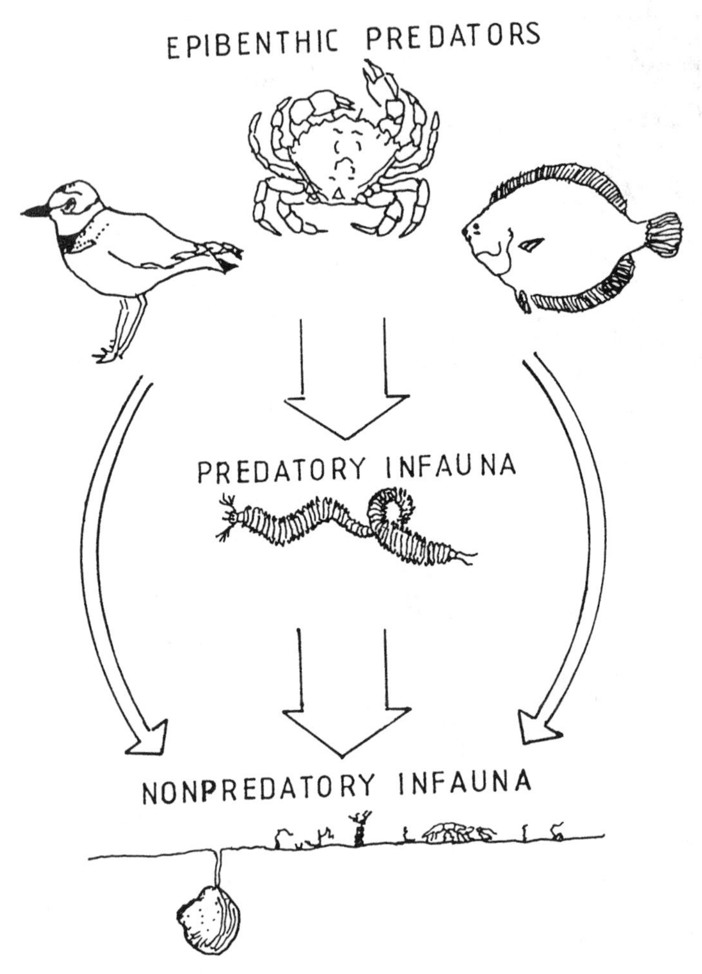

FIGURE 5.30 Diagram of a 3-level interactive model of predators in soft sediments. Curved arrows: interactions between epibenthic predators and nonpredatory infauna. Broad arrows: relations that may be particularly important in predicting the outcome of interactions in soft-bottom communities. (Redrawn from Ambrose, W.G., Jr., *Mar. Ecol. Progr. Ser.*, 17, 113, 1984b. With permission.)

1979a; Brenchley, 1981; Reise, 1985). Manipulative experiments have demonstrated that many adult invertebrates can depress settlement and recruitment directly by ingesting larvae and recent settlers and indirectly by burying larvae and juveniles with sediment displaced by feeding, burrowing, and defecating (Williams, 1980; Wilson, 1981; Brenchley, 1981; Commito, 1982; Peterson, 1982a,b; Ambrose, 1984a,b).

Hunt et al. (1987) manipulated the densities of a deposit-feeding gastropod, *Ilyanassa obsoleta*, and a suspension-feeding bivalve, *Mercenaria mercenaria*, to test the importance of interactions between adults and settling larvae in affecting the abundance patterns of infaunal invertebrates. The abundance of postlarvae and juveniles was significantly lower in the presence of *Ilyanassa* than in snail-free controls, with the reduction in abundance being as much as 45%. Surface-dwelling tubiculous polychaetes, gastropods, harpacticoid copepods, and mobile surface-dwelling infauna all declined in abundance with increasing *I. obsoleta* density. *Mercenaria,* on the other hand, had little or no effect on larval settlement. Disturbance and ingestion of newly settled larvae by *Ilyanassa*, larval settlement selectivity, and disturbance-induced emigration are probable mechanisms causing the differences in abundance between *Ilyanassa* treatments.

Deposit feeders and suspension feeders affect the recruitment of the infauna on very different spatial scales. Suspension feeders, which feed on larvae in the water column, probably have their impact on recruitment averaged over large areas. Deposit feeders, on the other hand, have very local effects on recruitment by causing mortality of recruits within the area of feeding activity (Peterson, 1982a,b). Woodin's (1976) predictions concerning adult-larval interactions were discussed previously. The results of Hunt et al.'s (1987) study did not support the prediction (Woodin, 1976) that no infaunal group should consistently attain their highest densities among densely packed suspension-feeding bivalves. In one enclosure where virtually all the clams were touching, the abundance of only one group, the Syllidae, was negatively affected. Hunt et al.'s (1987) results showed that burrowing polychaetes (Cirratulidae, Syllidae), sedentary tube-builders (Capitellidae,

Spionidae, Terebellidae), gastropods, and epibenthic crustaceans (harpacticoid crustaceans) were all negatively affected by *Ilyanassa*. Regardless of functional or taxonomic group, juveniles of all these taxa spend at least some time near the sediment surface and are therefore subjected to predation and disturbance by *Ilyanassa*. In light of their findings, Hunt et al., (1987) modified Woodin's predictions to include consideration of individual species characteristics, such as: (1) mobility of resident infauna, not just their trophic group (Brenchley, 1981); (2) relative and absolute sizes of the interacting species (Wilson, 1981); (3) species feeding behavior (Weinberg, 1984); (4) the spatial scale of the interactions (Wilson, 1981; Peterson, 1982a,b); and (5) possible nonadditive interaction between residents that might modify their separately measured effect on recolonization.

Ambrose (1984a) carried out a series of field experiments designed to determine if resident infauna affected the settlement and survivorship of colonizing infauna. This involved controls of defaunated sediment and similar sediment to which the polychaetes *Nereis virens* and *Glycera dibranchiata* were added in different densities. Large numbers of infauna colonized the defaunated sediment while over the same period, infaunal density usually declined in the sediment containing infauna. Sediment with low initial densities, however, always had a greater net change in density than sediment with high initial densities, suggesting that residents influence density changes. Highest densities of most infauna were recorded in the *Glycera* addition treatments and the lowest in the *Nereis* addition treatments. Nereis may have reduced infaunal abundances by direct predation, disturbance of the sediment, fecal deposition, and by influencing larval site selection. *Glycera*, on the other hand, had little effect on the physical environment and did not deposit large quantities of fecal material on the surface since it maintains a semipermanent burrow. Persistence of some colonizing polychaete species such as *Hobsonia florida*, *Polydora ligni*, and *Streblospio benedicti* was probably due to their rapid reproduction and direct development. From the experiments it was evident that residents were important determinants of successional events. However, the impact of the residents on recruitment and colonization is very much dependent on their species composition and mode of life (e.g., the extent to which they displace the physical structure of the sediment, deposit large quantities of feces on the surface, and their method of feeding (e.g., predator, filter feeder, deposit feeder, omnivore).

5.3.7 Role of Recruitment Limitation in Soft-Sediment Communities

Olafsson et al. (1994) have recently reviewed the role of recruitment limitation in soft-bottom benthic communities. Thorson (1957) was the first to advance the thesis that variable success during a risky planktonic life stage was the primary determinant of variability in the abundance of adult soft-bottom invertbrates with planktonic larval reproduction. In their review of the literature, Olafsson et al. (1994) found that the empirical evidence did not provide compelling support for the argument that the impact of settlement variations persists to shape adult soft-bottom populations and communities. The authors identified several processes of post-settlement mortality that are significant contributors to density regulation and pattern generation in soft sediments:

1. The impact of large mobile epibenthic predators
2. The presence of adult invertebrates, especially deposit feeders and infaunal predators inhibits the recruitment of potential colonists
3. Physical disturbance of the surface sediments subjects some smaller infaunal invertebrates, including many newly settled individuals to erosion and mortality
4. There is some evidence that juvenile invertebrates experience density-dependent mortality as a consequence of food limitation

Consequently, there exist multiple postsettlement processes that cause density-dependent regulation of soft-sediment invertbrates after settlement.

5.3.8 Disturbance and Succession

5.3.8.1 Introduction

The distribution of soft-bottom communities tends to be patchy. Johnson (1971) envisaged the benthos as a temporal mosaic, with different parts of the habitat disturbed at different times containing a different species assemblage. "The community," he says, "is conceived of as continually varying in response to a history of disturbance." In his view, the community is a collection of relics (and recoveries) from former disasters. Sutherland (1974) has likewise emphasized the role of history in determining the structure of a variety of natural communities. For a recent review of disturbance, sediment stability, and trophic structure in soft-bottom communities, see Probert (1984).

Studies of soft-bottom communities have shown: (1) that natural disturbances (especially biologically mediated disturbances that are usually localized and reoccur reasonably frequently) help maintain spatiotemporal heterogeneity of communities; and (2) that biogenic modification of the sediment can affect sediment stability with respect to fluid forces and geotechnical properties, and that this is an important factor in determining community organization (Probert, 1984).

5.3.8.2 Disturbance

A disturbance is defined as an event initiating species population change from mortality or removal, and/or a change in the resource base of the community (Zajac and Whitlach, 1982a,b). Soft-bottom communities can be affected by a wide variety of disturbances. Typically of large scale, with the area affected often being km^2, are natural disturbances caused by physical factors such as storms (Eagle, 1975; Rees et al., 1977; McCall, 1978), tidally induced sand movement (Grant, 1983), internal waves associated with pycnoclines, unusually low winter temperatures, salinity reduction sometimes associated with hypoxia, and deposition of fluvial sediment.

In contrast, natural disturbances that are biologically caused are typically small in scale, with the affected area usually m^2 or cm^2. Sediment processing by enteropneusts can result in disturbed patches of about 20 cm^2 (Thistle, 1981) to 80 cm^2 (Grant, 1983), and that of large holothurians 80 to 700 cm^2 (Rhoads and Young, 1971). Mounds resulting from burrow excavations by thalassinidans can be of similar size (e.g., Aller and Dodge, 1974), or larger (up to 0.3 m^2). Bottom-feeding fish, such as rays, can produce disturbed areas on the order of 500 to 700 cm^2 (Gregory et al., 1979). Of larger magnitude (up to several thousand m^2) are disturbances caused by schools of bottom-feeding fish (Orth, 1975) and biogenic alterations of sediment grain-size leading to increased susceptibility to erosion. Occasionally, biologically induced disturbances can be of large scale, such as the defaunation caused by red tides (Simon and Dauer, 1977), or low dissolved oxygen levels (Santos and Simon, 1980).

Disturbances of soft-bottom communities are also caused by a variety of human activities such as dredging (Connor and Simon, 1979), spoil and mining waste disposal (Probert, 1975; Rhoads et al., 1978), organic pollution (Pearson and Rosenberg, 1978), oil spills (Sanders et al., 1980; Dauvin, 1982), and bait digging (McLusky et al., 1983).

5.3.8.3 Levels of Faunal Disturbance

Assessing the effects of disturbance depends on the level of biological organization that is the focus: individual, population, or community. Clearly, not all effects of disturbance are negative; e.g., it can make food resources available to predators that were not previously available. Table 5.10 lists the potential classes of disturbance effects from the individual to the community level. A dominant paradigm for the effects of disturbance shows that, because disturbance events are unevenly distributed in space and time, a mosaic of patches is generated at different stages in a successional sequence (e.g., Johnson, 1973; Grassle and Sanders, 1973). The characteristics of the mosaic will depend on the spatial scale and level (kind, frequency, and intensity) of disturbance and the subsequent ways in which the community responds in the recovery process. The responses are to a large extent governed by the availability of colonizers and this in turn is a function of the time of the year in which the disturbance occurs.

5.3.8.4 Impact of Disturbance on Productivity

Few studies have examined the impact of disturbance on ecosystem productivity. A notable exception to this, however, is the work of Emerson (1989) who examined the relationship between annual coastal benthic secondary production and local wind field data. He found a significant

TABLE 5.10
The Possible Effects of Disturbance at Various Levels of Biological Organization

Level of Organization	Possible Effects
Individual	Increased possibility of death or injury
	Energetic cost of reestablishment
	Effect of reproductive development
	Effects on food availability
	Exposure to predation or displacement
	Provision of colonizable space
	Competitive release
Population	Changes in density
	Changes in recruitment intensity and/or variability
	Changes in dispersion patterns
Community	Changes in species diversity
	Changes in overall abundance
	Changes in productivity
	Changes in the patterns of energy flow or nutrient recycling

negative correlation between total, macro-, and meiobenthic production and wind stress. Further multiple regression analyses showed that approximately 90% of the variance in secondary production estimates could be accounted for by wind stress, exposure indices, tidal height, and mean annual water temperature. These findings support the hypothesis that secondary production in shallow-water benthic communities may be controlled indirectly by wind stress, which regulates environmental factors such as water temperature, mixing depth, food supply, and sediment transport. This supports the Trophic Group Mutual Exclusion hypothesis advanced by Wildish (1977), which proposed that benthic productivity is food limited, the supply of food being controlled by hydrodynamic factors.

5.3.8.5 Succession and Sediment Stability

Probert (1984) has listed a number of factors that influence sediment stability. They are:

1. *Microorganisms and algae.* Webb (1969) drew attention to the significant effects that bacterial films can have on sediment properties, particularly through increasing the adhesion between particles and altering granulometry. Extracellular and autolytic products of microorganisms living on the grains and within the interstices can also foster sediment stability through the accumulation of mucilaginous materials. The seasonal accretion of estuarine sediment can also be mediated by filamentous and unicellular algae that slow the near-bed flow and bind sediment through the accretion of mucilage. Benthic microalgae are known to secrete copious amounts of mucilage (Coles, 1979).
2. *Meiofauna.* Nematodes are usually the dominant meiofaunal taxon in marine sediments, especially if they are muddly. Cullen (1973) observed that meiofaunal nematodes can rapidly establish "an intricate, closely spaced network of thread-like intergranular burrows" within the surface layer of the sediment. Many meiofaunal nematodes are known to produce copious secretions of mucus that agglutinate the sediment. Other meiofaunal taxa also secrete mucoid materials for locomotion, adhesion, and protection (and in some cases for feeding) that may assist in the binding of sediments, e.g., ciliates, turbellarians, nemertineans, gastrotrichs, kinorhynchs, and harpacticoid copepods.
3. *Macrofauna.* It has frequently been reported that tube-dwelling infauna increase sediment stability, and tubiculous polychaetes have been considered particularly effective in this respect (Rhoads and Young, 1971; Probert, 1975), although Eckman et al. (1981) suggest that this may be the exception rather than the rule. In addition, biogenic reworking can decrease the critical erosion velocity of muds and sands.

5.3.8.6 Postdisturbance Responses of Microorganisms

A small-scale biogenic disturbance, such as the creation of an opened patch by bottom-feeding fish, could result in substantial bacterial growth as nutrients are made available through the mixing of the sediments. Few data are available on the response of benthic diatoms to disturbance of soft-bottom communities. Lee and Lee (unpublished, cited by Davis and Lee, 1983) observed that defaunated sediment developed a diatom layer within a few days. From experiments on an estuarine intertidal flat of well-sorted fine sand, Davis and Lee (1983) found that defaunated sediment was rapidly colonized by microalgae; chlorophyll *a* and gross primary production returned to control levels within 10 days.

5.2.8.7 Postdisturbance Response of Meiofauna

In an experimental study of nematode distribution in a beach, Gerlach (1977) used 50-ml samples of sand in which the fauna had been killed, and found that after one day nematode densities of 11% in unbaited samples and 34% in samples that had been baited with a piece of fish. Normal population densities were attained after 7 days and 14 days, respectively. Sherman and Coull (1980) dug over a 9 m^2 area of intertidal mud to a depth of 15 cm, raked it smooth, and observed a rapid (within 12 hr) recolonization by the meiofauna (91% nematodes) following this disturbance. The methods whereby meiofauna colonize available substrates were discussed in Section 3.6.2.4. Meiofauna taxa that are most active near the sediment-water interface, such as harpacticoid copepods, tend to be the most susceptible to resuspension and are the most rapid recolonizers of disturbed patches (Bell and Sherman, 1980; Alongi et al., 1983; Chandler and Fleeger, 1983).

5.3.8.8 Postdisturbance Response of Macrofauna

There have been many studies of the colonization sequence of disturbed substrates in benthic communities (e.g., Rhoads, 1976; McCall, 1977; Simon and Dauer, 1977). In this colonization sequence, there are progressive changes in trophic structure and life habits of the macrofauna, accompanied by diagenic changes in sediment properties (Yingst and Rhoads, 1980). Pioneering species tend to be tube dwelling or otherwise sedentary species that live near the sediment surface and feed on the surface layer or from the overlying water. Their influence on the

TABLE 5.11
Colonization of Defaunated Sediments Placed on the Bottom of Long Island Sound, U.S.A. (No. m^{-2})

Species	Peak Abundance	Sampling Interval (days)	Final Abundance
	Group I		
Streblospio benedictii	418,315	10	335
Capitella capitata	80,385	29–50	995
Ampelisca abdita	9,990	29–50	0
	Group II		
Nucula proxima	3,375	50	50
Tellina agilis	1,400	86	0
	Group III		
Nephtys incisa	220	175	129
Ensis directus	30	50–223	0

Source: From McCall, P.L., *J. Mar. Res.*, 35, 231, 1977. With permission.

sediments through biogenic reworking may be intense, but is restricted to the top few centimeters. In contrast, high order or equilibrium successional stages tend to be dominant by errant infaunal deposit feeders that feed well below the sediment surface and intensely rework the sediment to a depth of 10 cm or more (Pearson and Rosenberg, 1976; Rhoads et al., 1978).

Much of the theoretical focus on the colonizing sequence in soft sediments has been on species-adaptive strategies, especially in relation to their life history characteristics (Grassle and Grassle, 1974; 1977; McCall, 1977; Pearson and Rosenberg, 1978; Santos and Simon, 1980; Thistle, 1981). A central concept is that species with r-selected life history traits have been adapted to respond in an opportunistic fashion following a disturbance and will dominate the early stage of succession (Grassle and Grassle, 1974; McCall, 1977). Grassle and Grassle (1974) define the characteristic of these "opportunistic" species as: initial response to disturbed conditions, ability to increase rapidly, large population size, early maturation (short generation time), and high mortality. The major opportunists are deposit-feeding polychaetes, and using the criteria listed above, Grassle and Grassle (1974) rank the opportunistic species that colonized the sediments of Falmouth Harbour following an oil spill as follows: (1) *Capitella capitata*, (2) *Polydora ligni*, (3) *Syllides verrilli*, (4) *Microphthalamus abberans*, (5) *Streblospio benedicti*, and (6) *Mediomastus ambiseta*.

McCall (1977) placed 0.1 m^2 samples of defaunated mud on the bottom of Long Island Sound to simulate a local disaster and sampled after 10 days and periodically for up to 384 days. Some representative results are given in Tables 5.11 and 5.12. After 10 days, a total of 14 species were present with a total abundance of 4.7×10^5 individuals m^{-2}. McCall (1978) recognizes three species groups: Group I, opportunistic species; Group III, equilibrium species; and Group II with characteristics intermediate between the two other groups. The attributes of Group I and Group III are listed in Table 5.13. McCall found that the differences in the distribution and abundance of the benthic communities of Long Island Sound could be explained in terms of two different adaptive strategies: opportunistic or equilibrium.

Santos and Simon (1980) studied the macrofaunal recolonization of a large (.3 km^2) soft-bottom area of Florida bay (water depth of 4 to 5 m) following summer defaunation attributable to anoxic conditions. They initiated a recolonization experiment immediately after the annual die-off using cups containing azoic sediment. After 7 days when the first cups were recovered, 15 species were present with a total density of about 8×10^4 individuals m^{-2}. In such experiments, appreciable mortality is generally observed after the initial colonization. In Santos and Simon's (1980) study, total density after 14 days was about 2.5×10^4 individuals m^{-2}. In McCall's (1977) study, the next samples were not taken until 86 days after the start of the experiment when total density had fallen to about 3×10^4 m^{-2}. McCall (1977) and Santos and Simon (1980) found that more than 90% of the individuals were recruited as larvae, but in both cases the containers used were mounted above the bottom, which presumably impeded the colonization of adults. However, in the case of large areas of open habitat, initial recolonization by motile adults invading from the edges will be less important than that effected through the water column (Santos and Simon, 1980). More recent results of Bell and Derlin (1983) show that certain adult macrofauna can recolonize small patches (100 cm^2) of defaunated sediment within 7.5 hours.

TABLE 5.12
A Comparison of the Colonization of Empty Sediment and of Sediment with Animals Present

	Colonization of Empty Sediment		Colonization of Sediment with Animals Present	
	Colonization in August 1972	Colonization in August 1973 (partially empty)	Natural Bottom in August 1973	1-year-old Previously Empty Sediment in August 1973
Capitella capitata	36,120	170,585	5,868	8,940
Streblospio benedicti	418,315	12,575	13,310	4,015
Owenia fusiformis	0	86,650	240	0
Ampelisca abdita	5,130	45	15	472
Tellina agilis	275	60	206	55
Nucula proxima	205	35	205	75
Yoldia limatula	0	10	24	45
Nepthys incisa	0	95	266	15
Ensis directus	10	0	36	0
Totals	459,565	269,855	19,433	13,427

Note: Abundances are no. m^{-2}.

Source: From McCall, P.L., *J. Mar. Res.*, 35, 233, 1977. With permission.

TABLE 5.13
Summary of the Attributes of Opportunistic (Group I) and Equilibrium (Group III) Adaptive Macrobenthic Species Types

Group I	Group III
Opportunistic species	Equilibrium species
Many reproductions per year	Few reproductions per year
High recruitment	Low recruitment
Rapid development	Slow development
Early colonizers	Late colonizers
High death rate	Low death rate
Small	Large
Sedentary	Mobile
Deposit feeders (mostly surface feeders)	Deposit and suspension feeders
Brood protection; lecithophic larvae	No brood protection; plankotrophic larvae

Zajac and Whitlach (1982a) have studied the responses of estuarine benthic infauna to disturbance in a small estuary in Connecticut. They concluded that the physical and biological processes that were important determinants of estuarine succession could include: (1) the timing of the disturbance; (2) the habitat in which the disturbance takes place; (3) the reproductive periodicity of the infauna; (4) the ambient population dynamics that generate the pool of recolonizes; and (5) abiotic and biotic factors (e.g., food and space resources affecting the preceding four factors).

5.3.8.9 Distribution along a Gradient of Organic Enrichment

The general distribution of sediments and benthic macrofauna along a gradient of organic enrichment has been the subject of a number of studies over the last two decades in widely distributed geographic areas including Sweden and Scotland (Bagge, 1969; Pearson, 1971a,b; 1980; Rosenberg, 1972; 1976; Pearson and Rosenberg, 1976; 1978), France (Bellan and Bellan-Santini, 1972), Canada (Otte and Levings, 1975), the United States (Soule et al. 1978; Grizzle and Penniman, 1979; Reisch, 1979; Soule and Soule, 1981; Grizzle, 1984), Japan (Kitamouri, 1975; Tsutsumi and Kukichi, 1983; Tsutsumi, 1987; 1990), Australia (Poore and Kudenov, 1978; Ritz et al., 1989), and New Zealand (Knox, 1972; Knox and Bolton, 1979; Knox and Fenwick, 1978; 1982; Knox, 1988a,b).

Bottom sediments and waters in heavily organically polluted areas tend to become anaerobic as temperatures rise due to the accelerated decomposition of deposited organic matter. The typical response of benthic communities to organic pollution is the reduction in density or

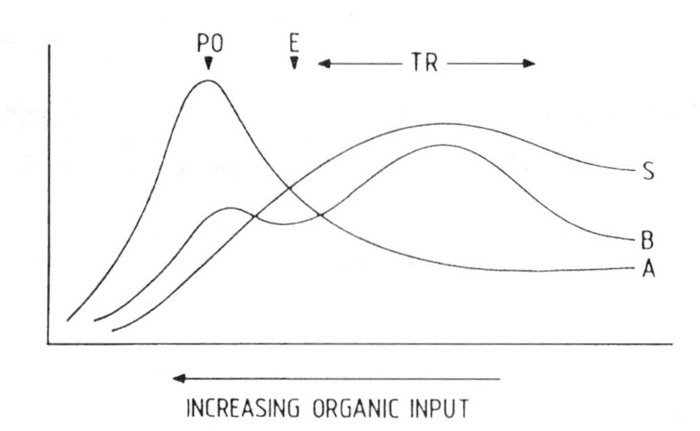

FIGURE 5.31 Generalized SAB (species number, abundance, and biomass) of changes along a gradient of organic enrichment. S = species numbers; A = total abundnce; PO = peak of opportunists; E = ecotone point; and TR = transition zone. (Redrawn from Pearson, H.T. and Rosenberg, R., *Oceanogr. Mar. Biol. Annu. Rev.*, 16, 234, 1978. With permission.)

the elimination of species characteristic of oxygenated conditions, and consequent changes in the taxonomic composition, including initial exploitation of the polluted areas by so-called "opportunistic species" (Grassle and Grassle, 1974; Tsutsimi and Kukuchi, 1983; Tsutsumi, 1987). If the pollutant source is removed, the opportunists are subsequently replaced in a successional process that eventually leads to the return of the original species (McCall, 1977; 1978).

The studies listed above have led to a number of generalizations concerning the impact of increased organic loading, both spatially and temporarily, on macrobenthic communities; namely, they will exhibit: (1) a decrease in species richness and an increase in total individuals as the result of high densities of a few opportunistic species; (2) a general reduction in biomass, although there may be an increase in biomass corresponding to a dense assemblage of opportunists; (3) a decrease in the body size of the average species or individual; (4) a shallowing of that portion of the sediment column occupied by infauna; and (5) shifts in the relative dominance of a trophic guild, e.g., a decrease in suspension feeders and an increase in deposit feeders. Pearson and Rosenberg (1978) predicted dominance of subsurface feeders in enriched areas, while Rhoads and Boyer (1982) predicted dominance of species feeding at, or near (<2 cm) the sediment-water interface. Many of the studies listed above have shown that most of the above predictions have been realized, i.e., a decrease in species richness, high densities of a few opportunistic species, a predominance of small-sized species in the most heavily impacted areas, the elimination of suspension feeders and their replacement by deposit feeders, and often a shallowing of that portion of the sediment column occupied by the infauna.

However, caution needs to be exercised in the application of these general principles to all organically enriched environments. Firstly, as Weston (1990) points out, inadequate sampling does not detect large, rare, or deep-burrowing species. Secondly, it has been shown that some species may be significantly larger in organically enriched areas. This has been shown for *Capitella capitata* (Young and Young, 1978; Tsutsumi, 1990).

Pearson and Rosenberg (1978) have reviewed the literature on macrobenthic succession in relation to organic enrichment. They identified a steady faunal transition along a gradient of organic enrichment from "abiotic" at the enrichment source to "normal" some distance away. Depending on the level of organic enrichment, the abiotic zone may be absent or may vary from a few meters to several kilometers (Poore and Kudenov, 1978). Within the extremes of the transition from abiotic to normal, three zones can be recognized: (1) an "opportunistic zone" with high numbers of a few species; (2) an "ecotone point" of overlap between opportunists and normal faunal species, characterized by low abundance of a few equally dense species; and (3) a "transition zone" of "normal" species at abnormal densities (Knox and Fenwick, 1982) (Figure 5.31). Figures 5.32 and 5.33 depict this pattern of faunal transition from selected studies. The pattern of faunal transition along a gradient of organic enrichment may also occur along similar gradients in time (Pearson and Rosenberg, 1978). With increased organic enrichment, the faunal zones succeed each other from "normal" through "transition," and opportunistic zones and eventually to abiotic zones when the rate of organic input is sufficiently high. As this occurs, the extent and distance from the source of each zone will change becoming further away with increased organic inputs (Rosenberg, 1976). The reverse sequence will occur following the reduction or cessation of the input. Once the source of the organic input has abated or has been removed, recovery to a more normal diverse community can be relatively rapid (Figure 5.34).

Thus, benthic communities in organically polluted areas are characterized by low diversity, but often high numbers of a few species of small polychaetes (the capitellids *Capitella capitata* and *Heteromastus filiformis*, the spionids *Paraprionospio pinnata*, and species of the gen-

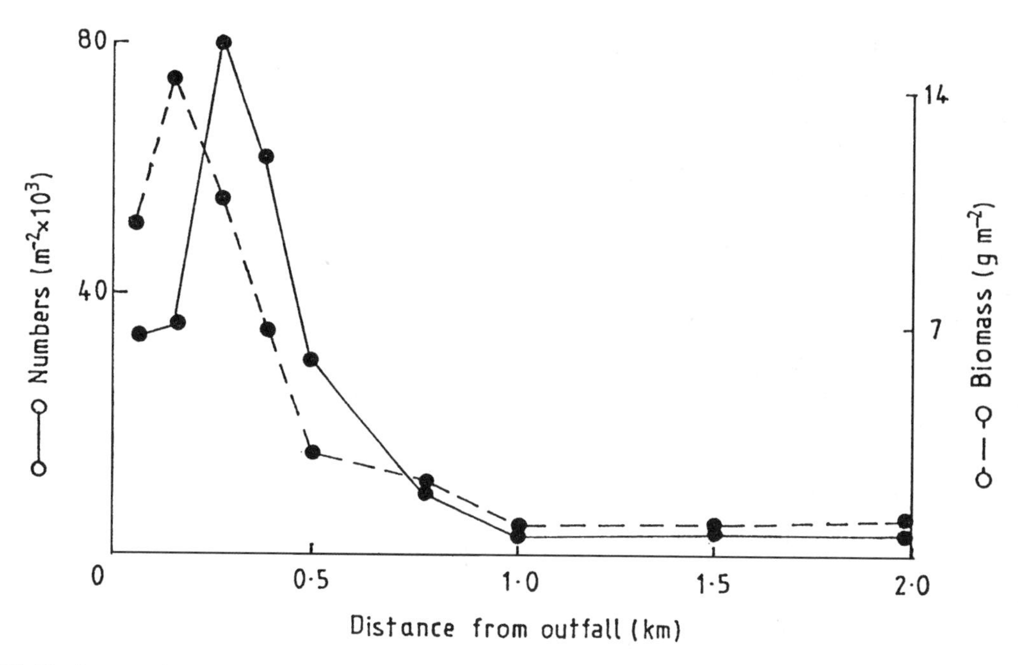

FIGURE 5.32 The impact of sewage discharge on the numbers and biomass of the benthic fauna in Keil Bay. (Redrawn from Gray, J.S., *The Ecology of Marine Sediments*, Cambridge University Press, Cambridge, England, 1981, 88. With permission.)

era *Scolelepis, Polydora, Streblospio, Prionospio,* and *Aquilospio*), and sometimes small bivalves of the genera *Gemma, Ampelisa, Maorimactra,* and *Theora*. In general, these species are deposit feeders ingesting sediment at the surface or deeper in the sediment.

In particular, *Capitella capitata*, a subsurface deposit feeder, has been reported to occur in high densities in heavily polluted areas throughout the intertidal and shallow subtidal areas of the world's oceans (Reisch, 1971, 1979; Rosenberg, 1972; 1976; Grassle and Grassle, 1974; Knox and Bolton, 1978; Pearson and Rosenberg, 1978; Yamamoto, 1980; Tsutsumi and Kukuchi, 1983; Tsutsumi, 1987; Knox, 1988a,b). Studies of the life history and population dynamics of *C. capitata* (Reisch, 1974; Grassle and Grassle, 1974; 1977; Kikuchi, 1979;Yamamoto, 1980; Tsutsumi and Kikuchi, 1984; 1987; Tsutsumi, 1987), electrophoretic analysis (Grassle and Grassle, 1974), and chromosomal studies (Grassle et al., 1987) have shown that this cosmopolitan species actually consists of a number of sibling species with similar adult morphologies, but distinct reproductive modes, generally, but not always occurring sympatrically (Grassle and Grassle, 1974; 1977). While *C. capitata* is widespread, especially in estuaries, occurring both intertidally and in shallow waters in a range of sediment types, only a few sibling species occur in organically polluted habitats and exhibit hugely opportunistic characteristics. Such sibling species in general are tube-builders, producing lecithtrophic larvae with rather short planktonic stages, ranging from a few hours to one day, and have a short generation time of 30 to 40 days (Grassle and Grassle, 1976; Tsutsumi and Kikuchi, 1983). The exclusive domination of *C. capitata* in organically polluted areas is primarily a reflection of these life-history characteristics rather than its tolerance of hypoxia (Warren, 1984; Tsutsumi, 1987).

Other opportunistic polychaete species exhibit similar life-history characteristics. Thus a life-history strategy that combines multiple generations each year, often brood protection, ease of dispersal (pelagic or benthic larvae that can be transported by tide-, wind-, or density-drive circulation), enables near-continuous dispersal and recolonization of disturbed areas. However, a number of studies have shown that the populations of these species are subject to wide fluctuations (Kikuchi 1979; Santos and Simon, 1980; Tsutsumi and Kikuchi, 1983; Chesney and Tenore, 1985a,b; Tsutsumi, 1987). Tenore (1977a,b; 1981; 1983), in experiments with *Capitella* sp. I, which apparently has the most opportunistic life history (Grassle and Grassle, 1976) has shown nutritional control of growth rates, population sizes, and annual production. In experimental studies, the size of laboratory populations consistently reflected the nutritional levels available to the populations (Chesney and Tenore, 1985a,b; Tenore and Chesney, 1985; Gremore et al., 1989; Marsh et al., 1989; Chareonpanich et al., 1993). Tsutsumi (1987) concluded, "Therefore, it appears that the absence of *Capitella* sp. in less organically polluted areas with stable environmental conditions, and the rapid decline of their populations in the benthic recovery process that follows the regression of pollution, can be ascribed to the shortage of organic intake and to their physiological demand for organic matter rather than their poor ability to compete with other benthic animals considered to be less opportunistic." Thus, a requirement for large quantities of organic matter of high nutritional qual-

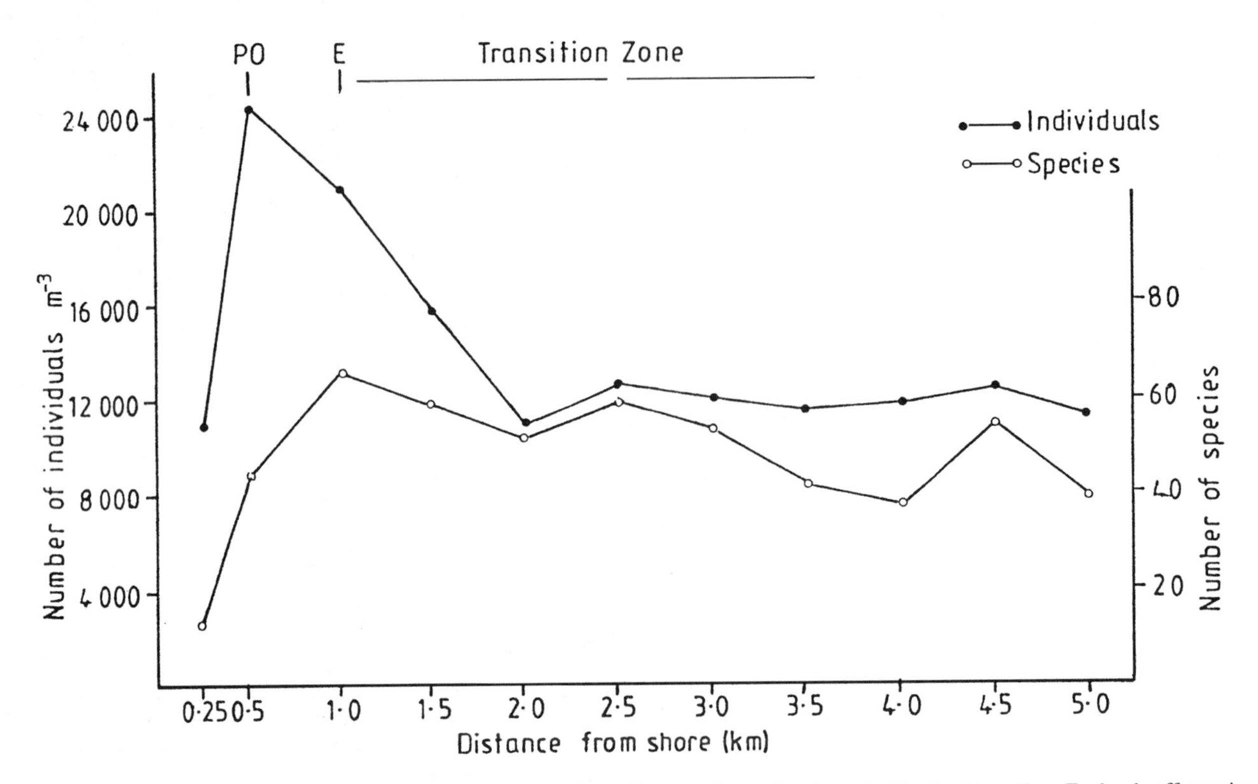

FIGURE 5.33 Faunal density and number of species at various distances from the shore in Hawke Bay, New Zealand, off a point where raw sewage was being discharged at mean low water. (Redrawn from Knox, G.A. and Fenwick, G.D., *N.Z. J. Mar. Freshw. Res.*, 15, 429, 1982. With permission.)

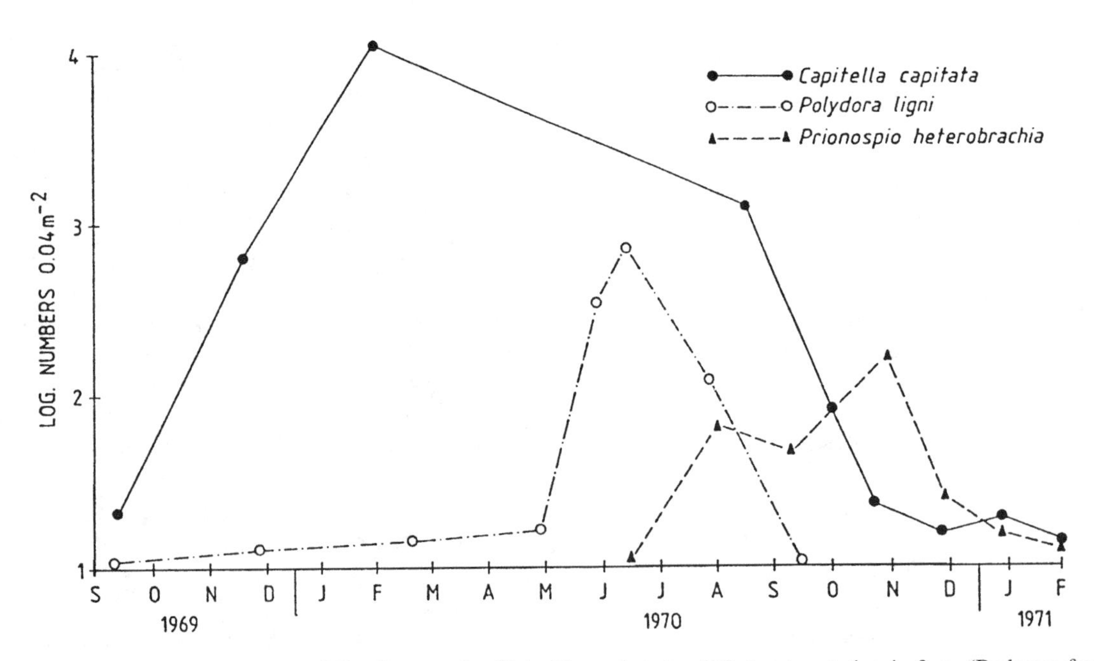

FIGURE 5.34 Recolonization sequence following an oil spill in Massachusetts, U.S.A., at a station in 3 m. (Redrawn from Gray, J.S., *The Ecology of Marine Sediments*, Canbridge University Press, Cambridge, England, 1981, 83. With permission.)

ity may be an important factor controlling the distribution of the so-called "opportunists."

5.3.8.10 Models of Postdisturbance Succession

The probability of recolonization being affected chiefly by adult macrofauna will be inversely related to the size of the disturbed area. The smaller the area, the greater the likelihood that the process of recolonization will be dominated by adult macrofauna invading from the edge of the disturbed patch; and the smaller the area, the greater will be the eventual resemblance in community structure between the ambient fauna and that of the previously disturbed area. The responses of macrofaunal larvae, meio-, and microbenthos are less dependent on the patch area, so that for a given habitat there will be a critical patch size above which larval macrofauna, meio-, and microbenthos may mainly determine eventual community structure.

Trueblood et al. (1994) investigated successional processes in a benthic infaunal mudflat community in Boston Harbor, U.S.A. They found that there were three groups of species that produced a three-stage or triangular succession pattern. Stage 1 comprised a March bloom of harpacticoid copepods that closely followed a benthic diatom bloom. Harpactocoid copepod abundance rapidly declined in the late spring and was followed by the recruitment of four opportunistic polychaete annelids (*S. benedicti*, *Pygiospio elegans*, *Capitella*, and an oligochaete). Stage 2 was a dense assemblage of four surface deposit-feeding and shallow subsurface deposit-feeding annelids that reached a peak abundance in June and declined in late summer, marking the break between stages 2 and 3. Stage 3 populations (comprising *P. ligni* (17%), ostracods (8%), *M. aestuarina* (3%), *L. robustus* 2%), and *C. setosa* (2%)), are more diverse than stage 2, reach peak abundance in the autumn, and decline in the late autumn. The infaunal structure of December resembled that of the community in the previous January. Succession on this mudflat is a fast-paced and dynamic process affected by benthic microalgal production, the timing and duration of juvenile recruitment, and the ability of the infauna to survive in dense assemblages of tube builders.

Probert (1984) has advanced a conceptual scheme for the influence of sediment stability on the timing of micro-, meio-, and macrobenthos responses following a disturbance (Figure 5.35). If there is a paucity of suitable larvae of the macrofauna available for immediate colonization, as may often be the case, initial postdisturbance events will be characterized mainly by a period of increased microbenthic production and rapid repopulation by a nematode-dominated meiofauna. Activity of the microbenthos and nematodes will tend to increase sediment stability through mucus binding. The increased sediment stability, combined with stimuli provided by bacteria, diatoms, and/or their mucus secretions, may make the substrate attractive to the larvae of filter feeders. On the other hand,

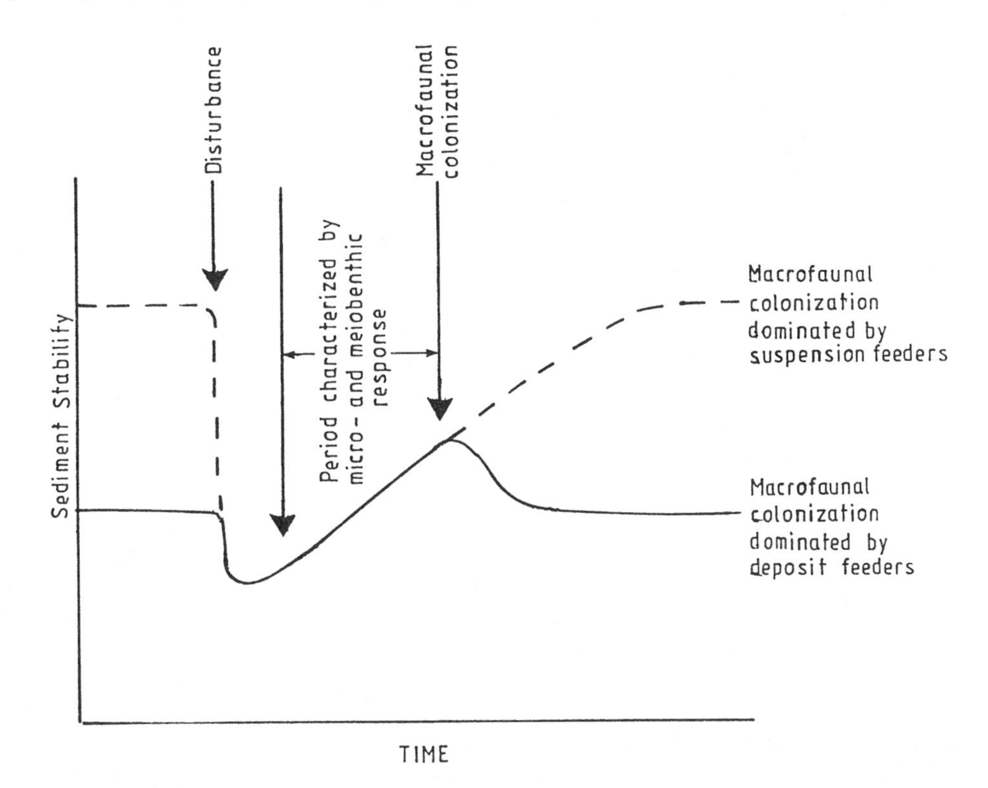

FIGURE 5.35 Conceptual scheme of the influence of sediment stability on the timing of micro-, meio-, and macrobenthic responses following a disturbance. (Redrawn from Probert, P.K., *J. Mar. Res.*, 42, 907, 1984. With permission.)

if the initial settlers are the larvae of deposit feeders, the physical nature of the sediment-water interface and the degree of mucus binding may provide appropriate physicochemical stimuli to the larvae of suspension feeders as to the suitability of the substrate for settlement, and they may delay metamorphosis until a suitable substratum is found.

5.4 SYNTHESIS OF FACTORS INVOLVED IN CONTROLLING COMMUNITY STRUCTURE

5.4.1 Hard Shores

5.4.1.1 Introduction

Community structure on hard substrates varies at several scales in space and time (De Angelis and Waterhouse, 1987). Spatial scales can range from a few centimeters to hundreds of kilometers. Temporal scales of variation are also broad, ranging from hours to centuries. Community structure can also vary due to physical stress (wave exposure, nutrient availability, tidal height, disturbance) and biological factors (recruitment, predation, competition, mutualism, and the indirect effects of these factors). Investigations of the roles of these factors in the control of community structure were discussed previously.

There is considerable literature on the various factors that determine the spatial and temporal variation in community structure on rocky shores. Such shores viewed both as a total ecosystem or as separate communities at various levels on the shore vary in species diversity, relative abundance, trophic structure, size structure, and spatial structure. Community structure can vary due to physical and physiological stress (e.g., sand scour and coverage, wave exposure, tidal height, nutrient availability, and temperature stress), a suite of biological factors (e.g., larval settlement, recruitment, grazing, predation, competition, and mutualism), and the indirect effects of these factors. Another important factor is the availability of space on the two-dimensional shore, coupled with disturbances that create bare space available for colonization. The successional processes that occur and the various factors that determine them (such as season, grazing, predation, and competition) are important in determining the kind of community that develops.

The situation then is one of great complexity, but much progress has been made in our understanding the relative roles of the different factors that are involved in the regulation of rocky shore communities (see reviews by Dayton, 1984; Menge and Sutherland, 1987; Underwood and Denley, 1984; Menge and Farrell, 1989). Studies concerning the elucidation of these factors have involved two main approaches: (1) the investigation of a particular region or level on the shore, or a species group over time; and (2) experimental manipulation of species densities, i.e., excluding or adding selected species to an area on the shore. Most of the experiments have investigated the effects of intense competition or predation on cool temperate shores with seasonally high production (reviews in Connell, 1972; Paine, 1994; Menge, 1995). Much of the theory underlying such experiments addresses mechanisms that could disrupt the process of competitive exclusion (discussion in Connell, 1978; Paine, 1994), including "keystone predators" (Paine, 1966; 1974; Menge and Lubchenco, 1981), "diffuse predation" (Menge and Lubchenco, 1981; Robales and Robb, 1993), "intermediate disturbance" (Connell, 1978; Sousa, 1979; 1984a,b), and "inhibition" hypotheses (Connell and Slayter, 1977; Sousa, 1979a). All of these theories contend that some agent, whether physical disturbance, a single keystone predator, or a more diverse guild of predators, removes local concentrations of a competitively dominant species thus allowing the populations of subordinate species to grow.

Early experiments supported the view that the control of community structure on rocky shores resulted from the interplay between competition and predation or disturbance (discussion in Connell, 1975; 1978; 1983; Paine, 1994). However, recent studies have emphasized the role of recruitment in the determination of community structure (see discussion in Raimondi, 1990; Menge, 1992; Robles, 1997). Studies have shown that dominant competitors recruit in large numbers and either preempt the colonization of less prolific species, or quickly overgrow and displace previously established, but less robust species. High rates of recruitment can also swamp the impact of a predator by ensuring that significant numbers of juveniles will survive to grow to sizes that resist subsequent attacks by predators (Paine, 1976; Robles et al., 1989).

In subsequent sections we will consider two models of the factors controlling rocky shore community structure, and finally consider the comparative roles of direct and indirect actions in rocky shore interaction webs.

5.4.1.2 The Menge-Sutherland Model

Menge and Sutherland (1987) and Menge and Farrell (1989) have advanced a model of the relative importance of several ecological factors (disturbance, predation, competition, recruitment) and environmental factors (environmental stress, productivity, habitat complexity) in influencing community structure (Figure 5.36). This model proposes that variation in community structure depends directly on variation in the effects of abiotic disturbance, competition, and predation; and indirectly on variation in recruitment density (or for clonal organisms, individual growth rates) and environmental stress. The strengths of competition and predation are predicted to be a function of the complexity of interaction webs (the subset of link-

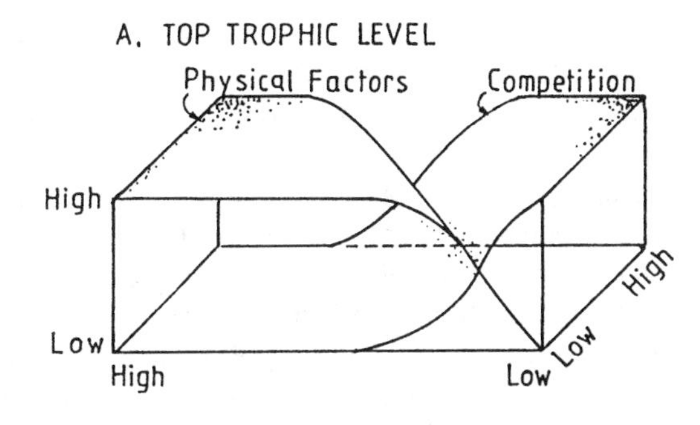

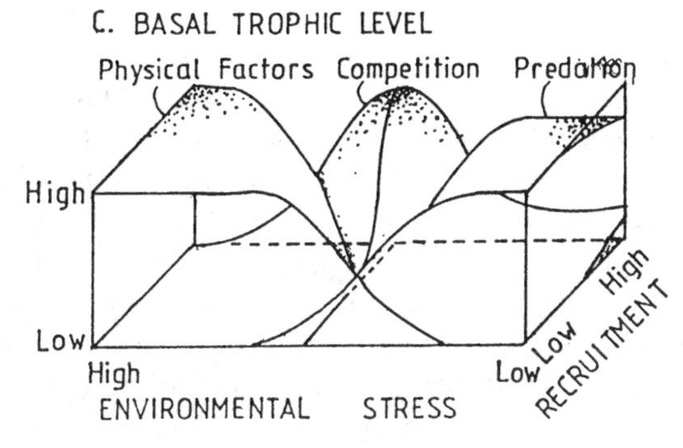

FIGURE 5.36 Model of community regulation. The model predicts the relative importance of physical factors, competition, and predation in controlling community structure as a function of environmental stress and recruitment density (or growth rates of recruits) by trophic level. (Redrawn from Menge, B.A. and Sutherland, J.P., *Am. Nat.*, 130, 739, 1987. With permission.)

ages in food webs that have strong effects on the interacting species). Mobile organisms are assumed to be more sensitive to environmental stress than are sessile organisms, leading to the prediction that a given level of stress will have stronger effects on mobile consumers than on sessile prey. Interaction web complexity is thus assumed to increase with a decrease in environmental stress, and therefore, community structure is indirectly dependent on environmental stress.

The relative importance of physical factors, competition, and predation is also predicted to vary with trophic level. At higher trophic levels, competition and physical factors are predominant under conditions of low and high stress, respectively. Predation has no influence at high trophic levels but increases in importance at lower trophic levels under low, but not high environmental stress.

Additional assumptions are (1) important consumers tend to be generalized in diet, (2) consumers are mobile while species in the lowest trophic level (= basal species) are sessile, (3) recruitment densities of major species are positively correlated, and (4) basal species may include both autotrophs and heterotrophs.

Specific predictions include: (1) under the severest conditions with high recruitment, food web structure was postulated to be under the direct control of the physical environment via physical or physiological disturbance; (2) consumer presence and/or activity should be prevented while sessile organisms are kept scarce or even eliminated; and (3) under less harsh conditions, consumers were predicted to still be inhibited while sessile organisms, more tolerant of physical or physiological stress, could become dense, leading to competition. Thus, under severe physical conditions, the highest trophic level would be sessile organisms.

At a more moderate position on the environmental gradient, consumers would no longer be completely suppressed, so that their effects on prey would increase. Increased consumer effectiveness should reduce the competition among sessile organisms, enhancing coexistence of prey (Paine, 1966). Finally, near the favorable end of the environmental gradient, consumer effectiveness would peak. This would be a consequence of both relaxed constraints on consumer activity and increases in the abundance and diversity of consumers. Strong predation, would, with the exception of organisms capable of escaping consumers, greatly reduce competition among species at lower trophic levels.

With low recruitment (or slow growth in clonal organisms), some of these predictions are modified (Menge and Sutherland, 1987). The primary effect of low recruitment was predicted to be the reduction of the importance of competition at all levels of environmental stress. The primary controllers of food web structure were thus predicted to be physical and/or physiological disturbance and predation. These models thus suggest that (1) as environmental harshness lessens, physical conditions shift from exerting control on food web structure to exerting indirect control through influences on biotic interactions, to exerting little effect; (2) at any point along the environmental gradient, several factors will contribute to control of food web structure; and (3) the relative contribution of each factor varies in a predictable manner.

The model reconciles seemingly contradictory predictions that plants are controlled by competition (Hairston et al.,1960) and sessile organisms are controlled by predation (Menge and Sutherland, 1976). The model argues that each prediction is correct, depending on the environmental conditions. Plants are predicted to compete when omnivory by top consumers is weak or absent, or when moderate environmental stress allows a less complex three-trophic-level interaction (i.e., with no omnivory). Plants will not compete when omnivory by top consumers is strong, or when intermediate environmental stress allows a simple two-trophic-level interaction web (Menge and Sutherland, 1987).

Space on the rock surface ("primary space") is an important resource for the sessile organisms whose distribution and abundance taken together produce the observed patterns of spatial community structure. Processes producing increases in space utilized by sessile species include recruitment and individual growth. Processes producing decreases in space utilized by sessile organisms include competition, predation, and disturbance.

The main process producing increases in sessile species abundance is *recruitment* of new individuals. Recruitment can influence adult community structure in three ways. Firstly, if the recruiting organism is generally free of interspecific sources of mortality (predation, interspecific competition) it will be favored. This can be seen in studies of barnacles occurring in high intertidal zone that are free of competitors or predators (e.g., Roughgarden et al., 1985), or mussels or surfgrasses occurring in mid- or low intertidal zones that are either naturally or artificially free of predators and dominant competitors for space (e.g., Dayton, 1971; Lubchenco and Menge, 1978; Turner, 1983a,b).

Secondly, recruitment can influence adult community structure if the recruiting organism either swamps its predators with high recruitment or keeps pace with the rate of predation (e.g., Lubechenco and Menge, 1978; Sutherland, 1978; 1981; Dayton et al., 1984). Thirdly, recruitment can influence adult community structure if recruitment rates of all species in the community are low and those organisms that do settle are prevented from exerting competitive dominance (e.g., Connell, 1985; Gaines and Roughgarden, 1985; Menge et al., 1985; 1986a,b; Menge, 1988, where intertidal settlement was low and competition did not occur).

Connolly and Roughgarden (1999) have modeled the relationship between physical oceanographic processes that transport larvae to the shore and the strength of species interactions on the shore. From studies on the Pacific coast of North America, they predict that a latitudinal gradient in upwelling intensity produces a gradient in the intensity of species interactions in rocky intertidal communities. Recent studies along the Oregon coast suggest that among-site variation species composition and abundance at intermediate, or "meso-scales" (i.e., 10s to 100s of kilometers) may be driven by consistent among-site differences in nearshore oceanographic processes (Menge, 1991a; Menge et al., 1994; 1996; 1997a,b). Inverse between-site differences in relative abundance of sessile invertebrates and macrophytes appear to vary with upwelling intensity, currents, and phytoplankton concentrations. Specifically, high abundances of sessile invertebrates occurred on shores adjacent to a region characterized by gyres and eddies, which both may concentrate zooplankton and phytoplankton and transport them to shore during upwelling relaxations. High phytoplankton abundance supports increased growth of filter feeders, potentially reducing macrophyte abundance through competition for space. High sessile invertebrate abundance also attracts high concentrations of predators, leading to high rates of predation. Menge et al. (1999) concluded that the studies suggested that pelagic and benthic ecosystems are coupled via the linkage: plankton concentrating gyres → phytoplankton and invertebrate larvae → filter-feeding invertebrates predators.

Menge et al. (1999) tested the above prediction in a study of the effects and rates of predation, grazing, and recruitment on rocky intertidal community dynamics at upwelling and non-upwelling sites on the West and East coasts of the South Island of New Zealand. Experiments and observations indicated that predation, grazing, prey recruitment, and mussel growth were greater on the West coast than on the East. The differences are probably due to the fact that summer upwelling is relatively frequent on the West coast but rare on the East.

Evidence discussed in Section 2.2.5 suggests that community structure can be strongly influenced by *physical disturbance* (defined as sources of mortality imposed by the abiotic environment, e.g., wave action, sand scour or burial, cobble scour, but excluding direct and indirect sources of mortality, such as predation and limpet bulldozing, respectively). Keough and Connell (1984) and Sousa (1984a,b) note that disturbance regimes can be quantified with respect to the size of the clearance, intensity or degree of damage, and frequency and seasonality of disturbance. Patterns of colonization are affected by patch characteristics (patch type [type 1 = those surrounded by occupied areas, type 2 = those isolated from occupied areas]; location; surface characteristics; size and shape; and time of creation), life history characteristics of potential colonists, and mobile consumers.

In order to test the validity of the model, four phenomena need to be examined in relation to trophic complexity and environmental stress: (1) the direct impact of physical and/or physiological disturbance; (2) the effectiveness of consumers as controlling agents; (3) the strength of intraspecific and interspecific competition; and (4) the role of structuring agents at both juvenile and adult stages of basal species. In order to test their model, Menge and Farrell (1989) analyzed 25 experimental investigations of marine intertidal rocky food webs and 11 hard-bottom subtidal food webs. The results of this analysis are depicted in Figure 5.37.

The data were subjected to both linear and quadratic regression (Figure 5.37a). Linear regressions between web complexity and the percent importance of physical factors,

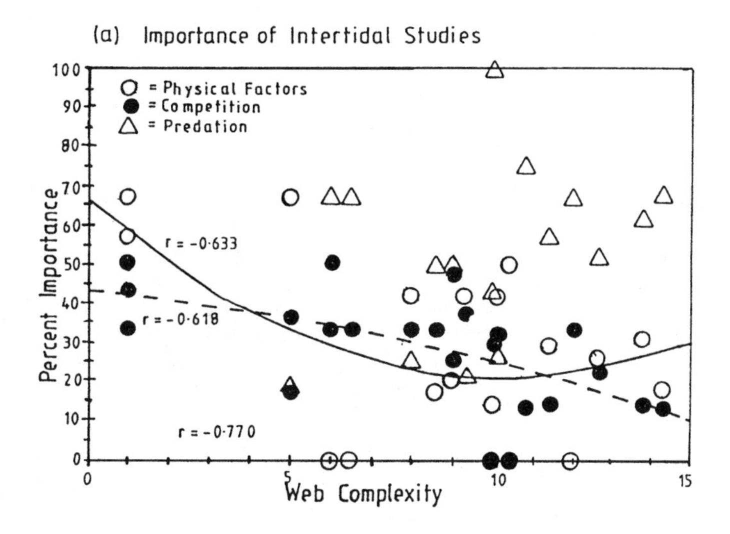

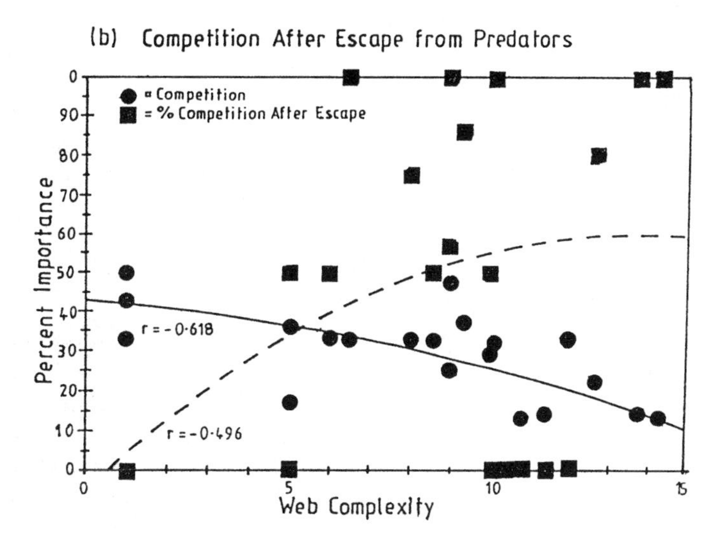

FIGURE 5.37 Test of the Menge and Sutherland model (1987) of community organization. The plots represent the food web listed in Table 3 in Menge and Sutherland (1987). Correlation coefficients (*r*) are shown next to each quadratic regression line. Solid line = physical factors; dashed line = competition; dotted line = predation. (a) Percent importance of physical factors, competition, and predation. (b) Percent importance of competition and "escape" competition (i.e., competition among adults after passing through a predation "bottleneck" as juveniles. Competition = solid line; escape competition = dashed line. (Redrawn from Menge, B.A. and Farrell, T.M., *Adv. Ecol. Res.*, 19, 248, 1989. With permission.)

competition, and predation, were all highly significant. Web complexity was strongly correlated to physical factors, competition, and predation.

The model predicts that as recruitment density decreases, the relative importance of competition declines (Menge and Sutherland, 1987). Menge and Farrell (1989) compared sites with moderate wave exposure and low recruitment to sites with intermediate wave exposure and high recruitment. As predicted, the average percent of competition with high recruitment was higher (32.1%) than with low recruitment (9%).

The prediction that the proportion of space occupiers that compete as adults after passing through a predator-vulnerable stage (i.e., have achieved a coexistence escape) increases with increasing trophic complexity, was also supported by the analysis. However, only 19% of the variance in the proportion escaping is explained by the index.

The prediction that trophic complexity increases with a decrease in environmental stress was also supported by the analysis. Interaction web complexity generally increased with decreases in environmental stress. The analysis suggested that in rocky intertidal habitats, effective trophic complexity tends to be greatest in low zones at intermediate wave exposures and decreases with tidal level and with both increased and decreased wave exposure.

Two other predictions of the model — that top consumers should be most strongly influenced by competition and that trophically intermediate consumers should be

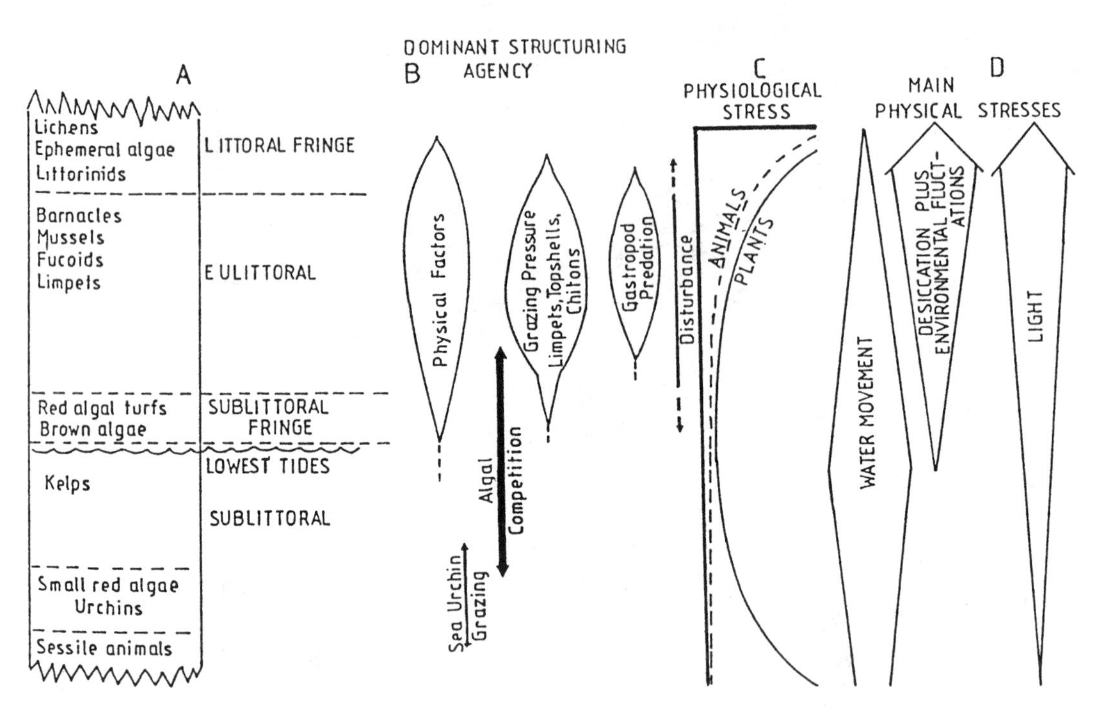

FIGURE 5.38 The relative importance of structural factors on the vertical distribution of rocky shore communities. (Modified from Russell, G., in *Intertidal and Littoral Systems, Ecosystems of the World,* 24, Mathieson, A.C. and Nienhuis, P.H., Eds., Elsevier, Amsterdam, 1991, 61. With permission.)

influenced by both competition and predation — could not be tested owing to a lack of information. However, on the whole, the analysis indicated that the model was capable of describing some major organizational features of rocky intertidal communities.

5.4.1.3 The Relative Importance of Various Structural Agencies on the Vertical Distribution of Rocky Shore Communities

Hawkins and Hartnoll (1983a) and Hawkins et al. (1992) have developed a model of the different forces organizing communities on the vertical gradient on exposed shores. This model was developed from studies on European temperate shores. Figure 5.38 summarizes the model concepts with some additions.

Physical factors such as wave shock are greatest in the mid- and upper eulittoral zones and attenuate throughout the littoral fringe above and into the sublittoral below. Grazing by limpets, chitons, and gastropods is most intense in the mid-eulittoral and tapers off both upshore and downshore. In the sublittoral zone, sea urchin grazing is an important structural force. Algal competition is most intense in the sublittoral zone and the sublittoral fringe. Gastropod predation on barnacles and bivalves is most intense in the mid-eulittoral and is less important both upshore and downshore. Disturbance is a major structural force at all levels, but is most pronounced in the mid-eulittoral.

Physiological stress for both plants and animals increases upshore and is most pronounced in the littoral fringe. The main physical stresses (water movement, desiccation, environmental fluctuations, and light intensities) are shown on the right of the figure. Desiccation and environmental fluctuations increase markedly at the top of the eulittoral and into the littoral fringe. Light intensity increases upshore, while deeper in the sublittoral it may be limiting to algal growth.

5.4.2 Soft Shores

5.4.2.1 Introduction

Over the last few decades, many studies have correlated infaunal benthic distributions with sediment grain size, leading to the generalizations of distinct associations between animals and specific sediment types. Snelgrove and Butman (1994) recently reviewed the data on animal-sediment relationships and concluded that they are much more variable than traditionally postulated. In contrast to the conclusions of previous reviews (Gray, 1974; 1981; Rhoads, 1974; Peres, 1982b; Probert, 1984), Snelgrove and Butman (1994) found little evidence that animal soft-bottom distributions are determined by any of the classical parameters of grain size, organic content, microorganisms, or sediment "stability" alone. They point out that in investigations of animal-sediment relationships, sediment grain size has usually been determined on completely disaggregated samples, which may have little relevance to what an organism actually encounters in nature. Likewise, distribution patterns have been documented using primarily sediment and biological samples that were not integrated over the same vertical scales within the substrate, or on samples that were integrated over much larger vertical scales than those relevant to the organisms. Thus, the grain-size distributions described for a given habitat may be very different than those actually experienced by the organism.

5.4.2.2 Evidence of Distinct Correlation Between Infauna and Sediments

In addition to grain size, other factors that have been proposed as determinants of benthic infaunal distributions include organic content, microbial content, food supply, and trophic interactions.

Organic content: A number of studies have correlated distribution patterns with sediment organic content (e.g., Sanders et al., 1962; Buchanan and Warwick, 1974). However, bulk carbon measurements may not accurately reflect the amount of carbon that may be actually utilized by an organism (Tenore et al., 1982; Cammen, 1989; Mayer, 1989). Nevertheless, many studies of animal-sediment relationships have documented a strong relationship between animal and sediment organic-carbon distributions. In addition to the role of organic matter as a food resource, organic matter may limit the distribution of organisms through differential settlement of larvae (or postlarvae) or differential postlarval survival. Controlled laboratory experiments of larval settlement suggest that preference for a particular grain size may be related to organic content; fine sediments tend to have larger amounts of organic matter (Butman et al., 1980).

Perhaps the most compelling evidence that the supply of organic matter influences infaunal distribution patterns comes from community-level pollution studies (see Section 5.3.8.9). Such studies have identified a suite of species that often show dramatic responses to organic loading (either by decreasing or increasing in abundance). It is thus clear that the supply of organic matter is important to these species.

The role of organic matter as a food resource for benthic invertebrates was discussed in Section 3.8.7. The ways in which organisms are able to utilize different types of organic matter is a complex issue (Lopez and Levinton, 1987), and organic matter may take many different forms (Whitlach and Johnston, 1974; Mayer, 1989). A good example of well-controlled experiments on the role of organic matter as a food resource is a recent study by Taghorn and Greene (1992). They tested the hypothesis that switching from deposit feeding to suspension feeding for two infaunal polychaete species was energetically profitable because suspended particles have a greater food value in terms of mass-specific concentrations of total organic matter, organic carbon, labile protein, nitrogen, and chlorophyll *a*. In these laboratory flume experiments, both species fed at significantly lower volumetric rates when suspension feeding (evidently because of the increased food gain per unit time when suspension feeding) than when deposit feeding. A number of deposit feeders living in muddy sediments will suspension feed in response to suspended sediment flux (Levinton, 1991), and some species once thought to be suspension feeders actually utilize deposited sediment as well (Mills, 1967; Hughes, 1970).

Microorganisms: The role of microorganisms (including bacteria and microalgae) in soft-bottom food webs was discussed in Section 3.8 and will be discussed further in Section 6.9.3. In fine sediments with high organic content, deposit feeders may obtain the bulk of their nutrition from the microbial community (Fenchel, 1972; Cammen, 1980; 1989; Taghorn, 1982). Elevated bacterial levels and growth in the feces of benthic detritovores can quickly replenish a depleted food resource (Fenchel, 1970; 1972; Hargrave, 1970b; Juniper, 1981).

While bacteria can be an important food resource for deposit feeders, those living in sandy environments, they can probably satisfy only a minor proportion of their nutritional requirements with sedimentary bacteria alone (Plante et al., 1989). Low microbial biomass is well documented in sandy sediments. However, especially in fine sand beaches, benthic microalgae may be important. Lopez and Levinton (1987) conclude that only in intertidal mudflats where benthic microalgae are extremely abundant (Cammen, 1980; 1989), would microbial food alone satisfy nutritional requirements.

Biological structures in sediments, such as tubes and sea grass shoots, enhance the local boundary shear stress and fluid flux near the bed (Eckman et al., 1981). Enhanced nutrient flux leads to increased bacterial biomass near the structures (Eckman, 1985). Examples have been given in Section 3.7.5 of the enhanced levels of bacteria, protozoa, and meiofauna in the vicinity of tubes and burrows.

Functional species groups: In their pioneering study, Rhoads and Young (1970) interpreted their results in terms of "amensalistic interactions" and classified the benthic fauna in terms of "functional groups." Snelgrove et al. (1994) criticize these concepts on a number of points:

1. There is little evidence for the generalization that deposit feeders are restricted largely to muddy sediments and suspension feeders to sandy sediments. Furthermore, in view of the

observations on switching behavior discussed above, categorizing infaunal organisms into simple functional groups such as deposit or suspension feeders, irrespective of the hydrodynamic and sediment transport regime, is no longer meaningful.
2. There is little evidence for the generalization that muddy sediments are detrimental to larval and adult suspension feeders. Many species are not always associated with a simple sediment type.
3. The concept of "trophic-group amensalism" in which tube-building suspension feeders are considered to exclude deposit feeders has had to be extensively modified and qualified to account for their coexistence in numerous instances.

5.4.2.3 Processes that Determine the Sedimentary Environment

The boundary layer flow and sediment-transport regime play a critical role in a variety of benthic ecological processes. Snelgrove and Butman (1994) have discussed these processes in detail, and the following account is based on their discussion of these processes.

Boundary layer flow: As a fluid moves across a fixed surface such as the sea floor, frictional drag retards the motion of the fluid such that velocity is zero at the sediment surface. As a result, horizontal velocities very close to the sediment are much lower than at distances further up in the water column. Increase in velocity with increasing height above the bottom is referred to as a shear, and the shear region adjacent to the bottom is referred to as the bottom boundary layer. In "depth-limited" boundary layers in shallow water, the effect of the bottom drag (and thus enhanced mixing) extends all the way to the water surface. In deeper water, however, boundary drag affects only a portion of the water column. The boundary-layer thickness (i.e., the region of shear) is a function of the turbulent mixing in the flow and the periodicity of the force driving the flow. Thus, currents, tides, and waves all contribute to boundary-layer formation, flow characteristics, and thickness.

The boundary layer can be divided into three regions based on the shape of the velocity profile (Figure 5.39). At the sediment-water interface is a region called the "viscous sublayer"; this region is of paramount relevance to the benthos because it is the interface between the water column and the sediment. Thus, the settling organic material, sediments, and larvae are transported along and must pass through this layer en route to the bottom. This is also a region where vertical shear in velocity is highest; this has important ramifications for the horizontal transport of organic matter (Muschenheim, 1987a,b) and larvae (Johnson et al., 1991). Directly above the viscous layer is a region referred to as the "log layer." Because the viscous sublayer is very thin (on the order of a few millimeters), the log layer is also extremely important for the transport of material to and from the benthos. It is often convenient to characterize the boundary-layer flow by the boundary shear velocity (U_v), which is a measure of the magnitude of the turbulent mixing in the flow and is directly related to the shear. The shear velocity is proportional to the square root of the boundary shear stress (T), the tangential force per unit of bottom area. Because u roughly characterizes vertical mixing within the boundary layer, vertical transport of materials increases with increasing u. Nowell

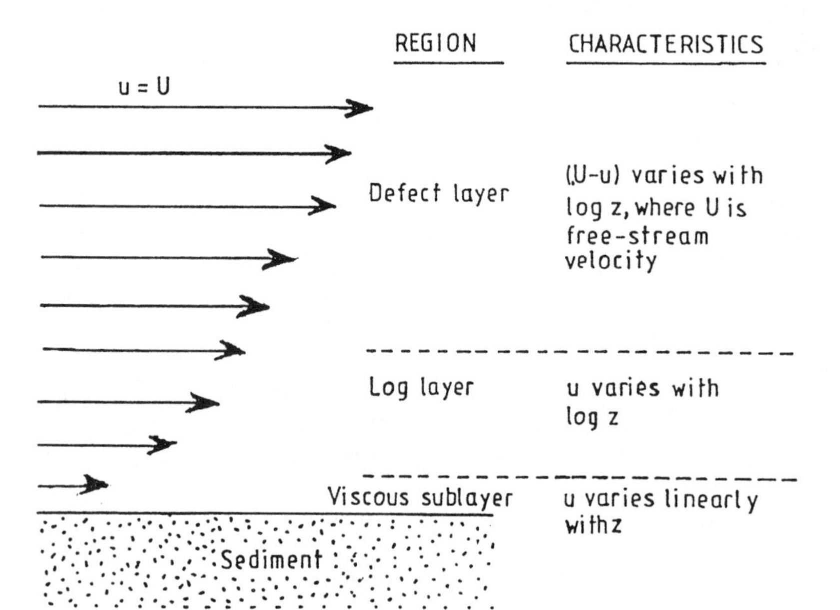

FIGURE 5.39 Boundary layer flow. (Redrawn from Snelgrove, P.V.R. and Butman, C.A., *Oceanogr. Mar. Biol. Annu. Rev.*, 32, 147, 1994. With permission.)

et al. (1984), Muschenheim (1987a,b), and Butman (1987) give general descriptions of the basic features of hydrodynamic and sediment processes within the bottom boundary layer, particularly with respect to potential effects on benthic organisms.

Distribution of sediments: Superficial sediment distributions are determined by (a) the sediment source (relic or modern), (b) interactions between sediment particles (including adsorption of chemicals), (c) the hydrodynamic regime, and (d) biological effects. All four of these factors can potentially determine whether sediment remains in the bed or is transported by the flow. Regardless of sediment type, the sediment mixture within a given locale generally is not static, but is in dynamic equilibrium with flow conditions at that site. Surface particles, ranging from sand to clay, are constantly being removed (through resuspension and burial) and added (through deposition or regeneration from depth in the bed). Relatively coarse beds generally occur in regions that regularly experience high U_v (where fine sediments are prevented from settling on the bed) and, likewise, relatively fine sediments occur in regions that rarely experience high U_v (where deposition can occur).

Laboratory flume experiments on the effects of individual benthic organisms (Nowell et al., 1981), groups of individuals of the same species (Rhoads and Boyer, 1982), and cores of sediments containing natural macrofaunal (Grant et al., 1982; Luckenbach, 1986) and meiofaunal (Palmer and Gust, 1985; Palmer, 1988b) communities demonstrated substantial biological effects on sediment transport. Similarly, diatom films, mats of purple sulfur bacteria (Grant and Gust, 1987), and exopolymer adhesion can significantly affect the entrainment of sand. These studies indicate that benthic biological processes can significantly increase or decrease u_{vcrit} of marine sediment.

An important parameter is the distribution of suspended material as a function of height above the bottom (Muschenbeim, 1987b; Frechette et al., 1989). Benthic organisms can directly affect suspended sediment concentration profiles, for example, by pelletizing the bed and changing the transport characteristics of the particles, by directly ejecting particles into suspension (Rhoads, 1963), and by affecting the vertical distribution of grain size within the sediment through their feeding activities such that superficial sediments differ from the sediment mixture deposited originally (Rhoads, 1967).

5.4.2.4 The Hydrodynamic Regime and Benthic Infaunal Species

As discussed above, there is mounting evidence for plasticity in feeding mode of many species as a function of the flow and sediment-transport regime. Many species of surface deposit feeders, for example, are now known to be facultative suspension feeders (Hughes, 1969; Fauchild and Jumars, 1979; Dauer et al., 1981), evidently in response to flow and elevated fluxes of suspended particles (Olafsson, 1986; Thompson and Nichols, 1988; Levinton, 1991; Taghorn and Greene, 1992). Switching between deposit and suspension feeding can occur over a tidal cycle, as observed for the bivalve *Macoma balthica* (Brafield and Newell, 1961). Epifaunal gastropods also show changes in motility with increasing oscillatory flow and sediment transport (Snelgrove and Butman, 1994). Thus, the conventional concepts of "feeding guilds" (*sensu* Fauchild and Jumars, 1979) and "functional groups" (*sensu* Rhoads and Young, 1970) should be revised to account for behavior as a function of the flow and sediment-transport regime (Jumars and Nowell, 1981; Nowell et al., 1984; Okamura, 1990; Miller et al., 1992). The boundary-layer flow regime may also directly affect animal distributions through drag and lift forces on aboveground structures, such as tubes, or on the animals themselves. Passive, suspension-feeding tube dwellers are also known to utilize the flow regime to enhance food capture.

The exchange of pore waters, and thus the depth within the sediment that is oxygenated, are related to the near-bed flow regime and the geometry of the pore spaces, both of which correlate with sediment particle size. Microbial population growth is another important variable that is strongly correlated with near-bed flow (reviewed by Nowell and Jumars, 1984). Small, near-surface-dwelling organisms that are susceptible to erosion, such as some meiofaunal species, may be transported directly by the near-bed flow regime. Palmer and Gust (1985) have shown, for example, that meiofauna can be resuspended and transported by everyday tidal flows on intertidal mudflats.

5.4.2.4.1 *Larval supply*

Larval settlement and recruitment were discussed in Section 4.3. There it was demonstrated that there is active habitat selection by the settling larvae and that this process contributes to adult distribution patterns. Results of efforts to conduct field sampling on a fine scale also suggest that hydrodynamics may impose considerable constraints on distribution and eventual settlement site. Fine-scale sampling of larvae in estuaries suggests that distributions are constrained by physical processes. Several studies have observed comparable settlement over adjacent areas with different sediment types (Muus, 1973), or biological structures (Luckenbach, 1984), suggesting that the larvae of at least some species may be nonselective. Given that flow may determine where larvae are transported, the potential importance of hydrodynamics in distributions of infaunal species is clear. Larvae may be sorted like passive particles, and thus may be associated with a given sediment type for this reason alone. Flow may also redistribute newly settled individuals and this may be an important means of dispersal, particularly for direct developers (Sigurdsson et al., 1976). Postlarval transport has been noted for a number of species, including *Cerastoderma edule* (Baggerman, 1953), *Mya arenaria* (Emerson and Grant, 1991), and *Macoma balthica* (Gunther, 1990).

5.4.2.4.2 Food supply

Food supply to benthic organisms is heavily dependent on local flow conditions, which have been shown to be primary determents of sediment distributions. Muddy sediments generally have a higher organic content than sandy sediments because organic matter tends to be more closely associated with lighter, depositional sediment fractions that accumulate in low-flow areas. Benthic microalgal biomass and production is also correlated with fine sediments. In many environments, fine particles may still deposit, but they tend to resuspend easily and are transported both vertically (upward mixing) and horizontally, resulting in little accumulation of fine sediments and organics.

The availability of food in suspension may also be limiting to the distribution of many organisms. Because many suspension feeders depend on the horizontal transport of organic matter, their distributions may be confined to areas of relatively high fluid velocity (Wildish, 1977); such high-flow areas also tend to be dominated by relatively coarse, low-organic sediments. Rates of suspension feeding and growth are a function of food supply in a variety of taxa (Grizzle and Lutz, 1989; Grizzle and Morin, 1989; Peterson and Black, 1991), clearly influencing the distribution of the benthos. Resuspension of bottom material may augment phytoplankton as a food for some suspension-feeding bivalves, though other species have shown decreased growth in relatively high turbidity.

The process of filter feeding is complex, however, because there are a variety of different particle-trapping mechanisms, and the type of suspension feeder may range from flow-dependent, facultative suspension feeders, to organisms that resuspend depositional material for feeding, to active and passive suspension feeders (Jorgensen, 1990). Thus, the relationship between suspension feeder distribution and the flow regime may be extremely complex.

5.4.2.5 The Importance of Recruitment

The current paradigm for recruitment to soft-bottom communities invokes passive transport of the larvae in the water column followed by larval choice of habitat after deposition of the larvae on or near the bottom. Larval choice is typically considered to be based on positive cues (Butman, 1987). However, Woodin (1991) considers negative cues to be as important as positive ones. In addition, metamorphosed or still metamorphosing juveniles (and even adults) are able to escape from unsuitable habitats. Biogenic events that induce such movements fall into two basic categories: those mediated by chemical changes and those mediated by the physical disruption to stability of the sediments. Physical disruptions include the disturbances discussed in Section 5.3.8, and movements and changes in sediment associated with movement, feeding, and/or defacation of the infauna (Thayer, 1983). Increased sediment stability is typically associated with tube-builders and may be associated with increased microbial growth (Eckman, 1985). Chemical changes can occur through the activities of the infauna such as irrigation effects on pore water constituents (Woodin and Marinelli, 1991), or through the release of chemical constituents into the sediments by the infauna (gregarious settlement compounds or halogenated aromatics).

It is clear that small-scale movements of newly settled larvae and juveniles can be an effective means to escape locally unsuitable sites (Woodin, 1991). Surface disruptions, whether biotic or abiotic, appear to make sites unsuitable in that the larvae and juveniles emigrate away from such disrupted areas, and biogenic chemical contamination can cause rejection. Therefore, habitat acceptance may be more a matter of the absence of a strong negative, rather than the presence of a strong positive cue, with a considerable amount of secondary movement possible even after the initiation of metamorphosis.

Olafsson et al. (1994) have recently reviewed the role of recruitment limitation in soft-bottom benthic ecosystems. In their review of the literature, they found that the empirical evidence did not provide compelling support for the argument that the imprint of settlement variations persist to shape adult soft-bottom populations and communities. The authors identify several processes of postsettlement mortality that are significant contributions to density regulation and pattern generation in soft sediments:

1. The impact of large mobile epibenthic predators
2. The presence of adult invertebrates, especially deposit feeders and infaunal predators, inhibits the recruitment of potential colonists
3. Physical disturbance of surface sediments subjects some smaller infaunal invertebrates, including many newly settled individuals, to erosion and mortality
4. There is some evidence that juvenile invertebrates may experience density-dependent mortality as a consequence of food limitation

Consequently, multiple postsettlement processes exist that cause density-dependent regulation of soft-sediment invertebrates after settlement.

5.4.2.6 Conclusions

Snelgrove and Butman (1994) consider that the complexity of soft-sediment communities may defy any simple paradigm relating to any single factor and they propose a shift in focus toward understanding relationships between organism distribution and the dynamic sedimentary and

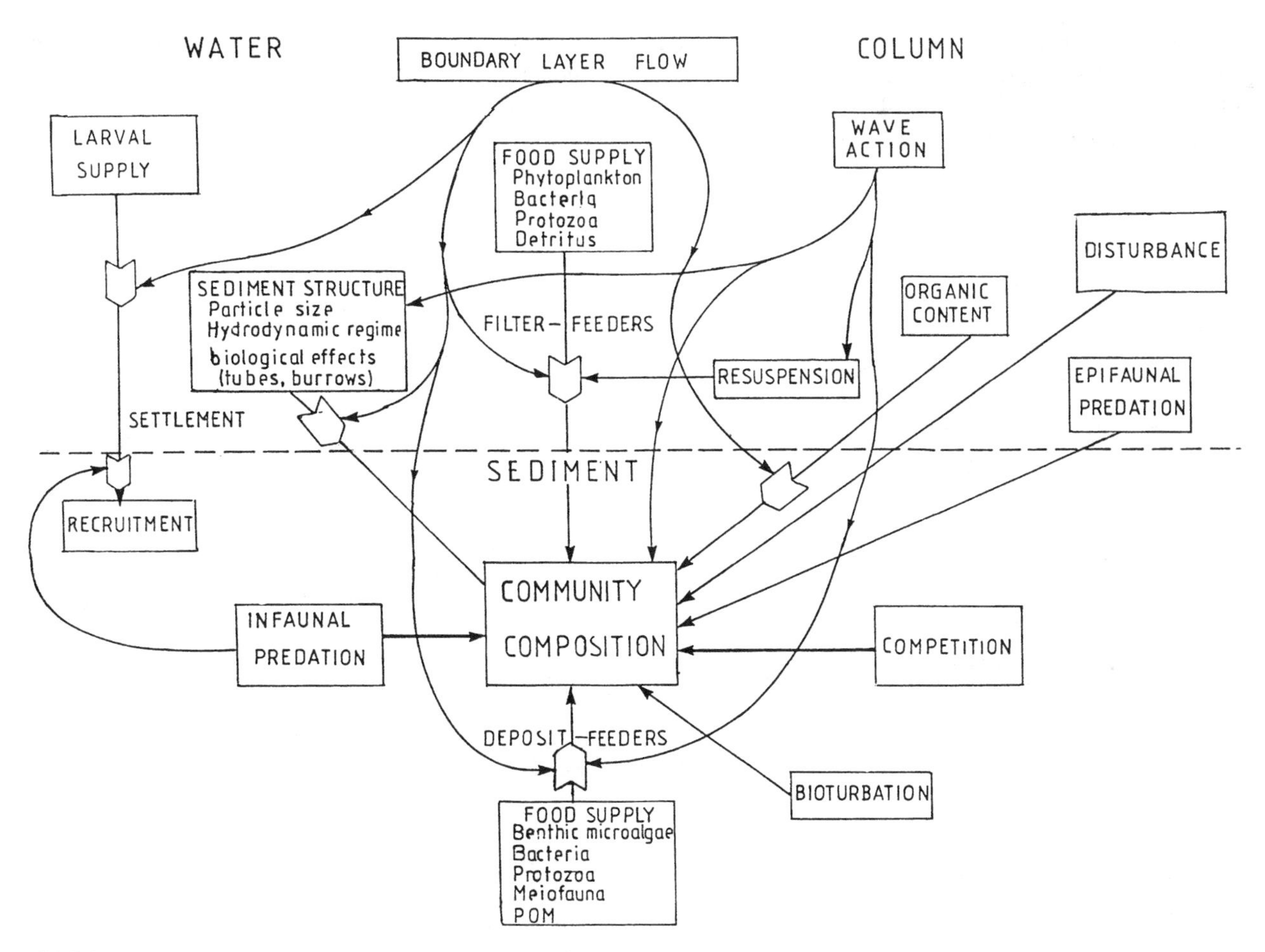

FIGURE 5.40 Model of the factors determining macrofaunal community composition on soft shores.

hydrodynamic environment. Grain size co-varies with sedimentary organic matter content, pore water chemistry, and microbial abundance and composition, all of which are influenced by the near-bed flow regime. These variables could directly or indirectly influence faunal distributions via several compelling mechanisms. Moreover, because the sedimentary environment in a given area is, to a large extent, the direct result of near-bed flow conditions, factors such as larval supply and particle flux that are similarly determined by the boundary-layer flow may be particularly important determinants of species distributions. It is unlikely that any one of these factors alone can explain the patterns of distribution across all sedimentary habitats; however, meaningful and predictive relationships are more likely to emerge once the influence of these dynamic variables is examined systematically in controlled experiments.

5.4.2.7 Synthesis of Factors Determining Macrofaunal Community Composition on Soft Bottoms

Figure 5.40 is an attempt to model the principal factors that interact to determine macrofaunal community composition in coastal soft sediments. On the left, larval supply interacting with the complex of factors discussed above determines settlement, which interacts with predation to determine recruitment. The principal sedimentary factors influencing community composition, sediment composition and structure, and organic content are influenced by wave action. Sediment structure features include particle size distribution, the hydrodynamic regime, and biological effects such as the presence of tubes and burrows (increasing habitat diversity) and bioturbation. The organic content of the sediment is dependent on the complex of factors discussed earlier in this chapter.

The available food resources also influence the growth rates and densities of the benthic fauna. Food supply comprises the resources available both in the water column and in the sediment. The former, which includes bacteria, phytoplankton, protozoa, zooplankton, and detritus (particulate organic matter), is consumed by the benthic filter feeders, while the latter, which includes bacteria, benthic microalgae, protozoa, meiofauna, and particulate organic matter, is consumed by the benthic deposit feeders. The water column food resource available to the filter feeders is augmented by resuspension of the surface sediments. Boundary layer flow impacts on settlement, sediment structure, organic content, and water column food supply. Three other biological factors are important in determining the composition of the benthic community: predation (both infaunal and epifaunal), competition, and bioturbation. Finally, disturbances with subsequent successional events are most important structural influences.

6 Energy Flow, Food Webs, and Material Cycling

CONTENTS

6.1 INTRODUCTION

In this chapter we shall first look at food sources in the intertidal zone. Then we shall examine energy budgets for intertidal animals, leading to a discussion of the trophic structure and food webs of intertidal ecosystems, on both soft and hard shores.

Early models of these processes in the coastal zone assumed linear food chains of the Lindeman (1942) type, consisting of phytoplankton, zooplankton, benthos, and fish (Clarke, 1946; Riley, 1963). The compartments of such models were equated with trophic levels and ecological efficiency transfers were used to evaluate energy flux. Ryther (1969) attempted to show how fish production was limited by the number of transfers of energy from one trophic level to another. Steele (1974) developed a compartmental bifurcated model with one pathway involving phytoplankton, zooplankton herbivores, zooplankton carnivores, and pelagic fish; and the other pathway involving fecal pellets, bacteria, benthic meiofauna, benthic macrofauna, epibenthos, and demersal fishes. This model pointed out two unknown factors: first the efficiency of the bacteria in breaking down organic matter, and second the trophic link between the meiofauna and the macrofauna.

In a landmark paper, Pomeroy (1979) presented a compartmental model of energy flow through a continental shelf ecosystem postulating the potential for substantial energy flow through dissolved organic matter (DOM), detritus (POM), and microorganisms to terminal consumers. This model was further developed by Pace et al. (1984). Both of these models have previously been discussed in Chapter 3. These models involved the abandonment of the classical idea of trophic levels and replaced it with the concept of food webs as anastomosing structures that defy classification into trophic levels. Pomeroy demonstrated that it was possible for energy to flow either through the grazer, or alternate pathways, to support all major trophic groups at a reasonable level and to maintain fish production at about the levels commonly seen. As discussed earlier, benthic microalgal production, organic detritus, dissolved organic matter, the microbial community, and the meiofauna play more important roles than had been previously thought.

6.2 FOOD SOURCES

6.2.1 Hard Shores

The food resources available on hard shores can be subdivided into the categories listed in Table 6.1. The principal *in situ* primary producers are the benthic microalgae growing on the rock surfaces, barnacle tests, molluscan shells and other hard surfaces, and on the attached macroalgae. The contribution of the benthic microalgae depends on the availability of suitable surfaces for their growth and the level on the shore. At certain times the sporelings of the attached macroalgae are an important component of the microalgal films. The production of the benthic microalgae is highly variable, depending on the species composition (which varies geographically), the intertidal level on the shore, and competition. Microalgal films are grazed by gastropod molluscs (especially limpets, top shells, and chitons), and some fish species. The attached macroalgae are consumed directly by a variety of molluscs (limpets, top shells, chitons, abalones), sea urchins, crustaceans (especially isopods and amphipods), and fishes. Many algal species are highly productive, e.g., on an exposed rocky shore on the West coast of South Africa, Gibbons and Griffiths (1986) recorded a maximum algal standing crop of 403 g m^{-2}. Many epifaunal crustaceans are adapted to feed on particular parts or tissues of algal species. However, some macroalgae have developed chemical defense mechanisms to limit grazing.

Bustamante et al. (1995) have documented *in situ* production of coastal phytoplankton, epithithic microflora (chlorophyll *a* production cm^{-2} $month^{-1}$) (Figure 6.1) and the standing stock of the different functional groups of macroalgae around the South African coast (Figure 6.2).

A well-documented productivity gradient exists in the pelagic ecosystem around southern Africa, due to the existence of strong upwelling on the west coast and its virtual absence on the east coast (e.g., Shanon, 1985; Branch and Branch, 1981; Moloney, 1992) (see Figure 2.12). In a review of the published productivity data for the Benguela and Agulhas ecosystems, Branch and Branch (1981) demonstrated the existence of this gradient for water lying inshore of the 200 m isobath. The northwestern coast is highly productive, supporting chlorophyll biomass up to 16.43 mg chlorophyll *a* m^{-3}, whereas intermediate concentrations (about 5.0 mg chlorophyll *a* m^{-3} occur off the southwestern and southern coasts. Off the southeastern coast, chlorophyll concentrations are an order of magnitude lower than the northwestern coast (<2.0 mg chlorophyll *a* m^{-3}).

Primary production of the intertidal epilithic microalgae showed a similar pattern to that of the phytoplankton (Figure 6.1) and was correlated with nutrient availability. The dominance patterns of the different functional groups of macroalgae changed around the coast (Figure 6.2), with

TABLE 6.1
Food Sources on Hard Shores

Food Type	Source	Consumers	Remarks
Benthic microalgae	Microalgal films on rocks, mollusc shells, etc.	Microfauna Meiofauna Gastropod molluscs, fish	Principally limpets
Phytoplankton	The sea	Filter feeders	Principally bivalves and barnacles
Water column bacteria	The sea	Filter feeders	
Water column particulate organic matter	The sea	Filter feeders and detritovores	Bivalves, amphipods, and other crustaceans
Water column dissolved organic matter	The sea		
Detritus	Intertidal rocks, sea grass beds	Detrital consumers	
Water column micro-zooplankton	The sea	Filter feeders	
Attached macroalgae	Intertidal rocks	Crustaceans, and molluscs	Principally amphipods, gastropods, and sea urchins
Intertidal sea grasses	Intertidal rocks		
Meiofauna	Intertidal rocks, macroalgae	Meiofauna, invertebrate consumers, fish	
Macrofaunal invertebrates	Intertidal rocks	Invertebrate and vertebrate consumers	

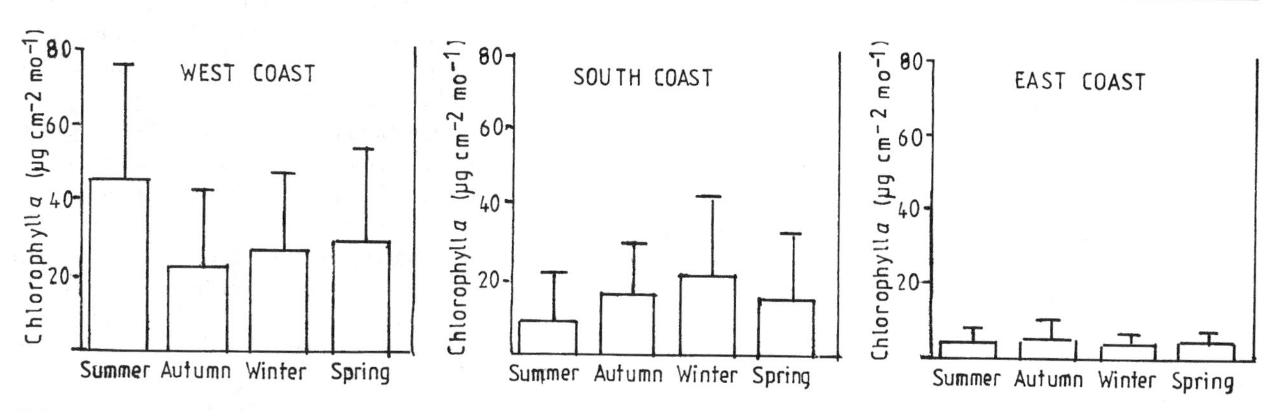

FIGURE 6.1 Seasonal epilithic chlorophyll *a* production month^{-1} in the three biogeographic provinces around South Africa. (Redrawn from Bustamante, R.H., Branch, G.M., Eckhout, S., Robertson, B., Zoutendyk, R., Schleyer, M., Dye, A., Hanekon, N., Keats, D., Jurd, M., and McQuaid, C., *Oecologia (Berlin)*, 102, 193, 1995. With permission.)

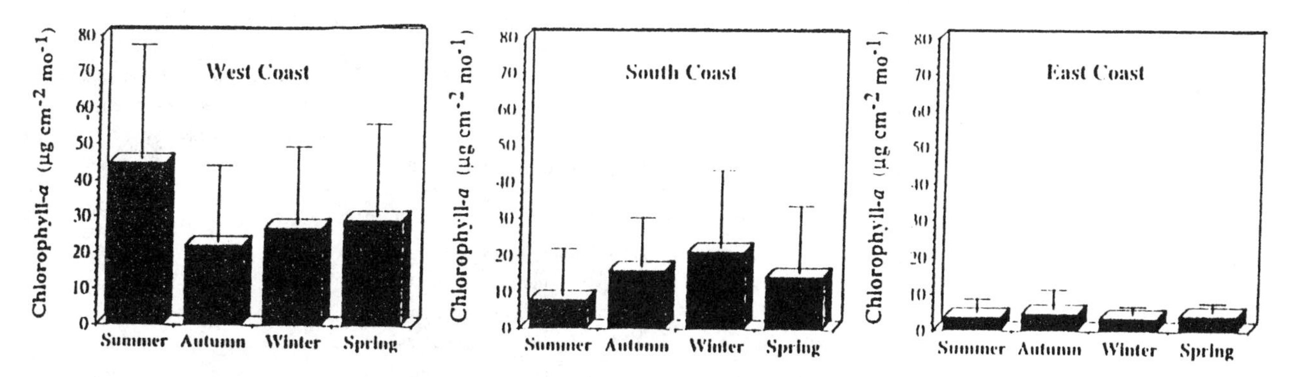

FIGURE 6.2 Macroalgal standing stocks around South African shores. *Bars* represent the mean (SD) dry biomass for each of the three functional groups of algae. (Redrawn from Bustamante, R.H., Branch, G.M., Eckhout, S., Robertson, B., Zoutendyk, R., Dye, A., Hanekon, N., Keats, D., Jurd, M., and McQuaid, C., *Oecologia (Berlin)*, 102, 194, 1995. With permission.)

foliose algae prevalent on the West coast and coralline algae on the East coast. However, overall macroalgal standing stocks did not reflect the productivity gradient, which was equally high on the East and West coasts, and low in the South.

A specific algal food resource that is of importance to some herbivores are the small epiphytic algae growing on other algae and on hard substrates such as the shells of molluscs such as limpets. Filter feeders feed on the water column bacteria, phytoplankton, detritus, and the microzooplankton, especially the protozoans.

Particulate organic matter (POM), or detritus, is derived from the water column, or the *in situ* breakdown of algae and the dead bodies of animals, as well as the feces of the invertebrate secondary consumers. POM in the water column is derived from a variety of sources (see Section 3.8.2.1), especially macroalgae, submerged macrophytes, and zooplankton fecal production. Dissolved organic matter, again, is derived from a variety of sources (see Section 3.8.2.1) and is utilized primarily by bacteria both within the water column and in the microbial film on the rock surface, molluscan shells, barnacle tests, and the macroalgae.

There is a wide range of predators on rocky shores including gastropod molluscs, seastars, many crustaceans, fishes, and shore birds.

6.2.2 Soft Shores

Since sand beaches lack macrophytes (seed plants and macroalgae) below the drift line, the basis of the food web is *in situ* microalgal production or food inputs from the sea or the land. The food inputs can be divided into the categories shown Table 6.2. The principal primary producers on sand beaches are the epipsammic diatoms attached to the sand grains. Their contribution is greatest on sheltered and fine sand flats. Recorded values range from 0 to 50 g C m^{-2} (Steele and Baird, 1968). On very exposed beaches, production is practically zero while on those exposed to wave action, values are less than 10 g C m^{-2} (Brown and McLachlan, 1990). This food resource is consumed by the meiofauna and deposit feeders, e.g., polychaetes and callianassid shrimps.

Surf-zone primary production is highly variable. Where surf-zone diatom accumulations occur, production rates may be very high, on the order of 200 to 500 g C m^{-3} yr^{-1}, with instantaneous rates of between 5 and 10 g C m^{-3} hr^{-1} (Lewin and Schaefer, 1983; Campbell, 1987; Brown and McLachlan, 1990) (see Section 3.5.5.5). Where such accumulations (patches) are absent, primary production rates in the surf zone are much lower in the range of 20 to 200 g C m^{-3} yr^{-1} (Brown and McLachlan, 1990). On the Eastern Cape, South Africa, the surf diatoms produced 120 kg C m^{-1} yr^{-1} within the surf zone (250 m), while mixed phytoplankton in the water column, mainly autotrophic flagellates, produced 110 kg C m^{-1} yr^{-1} in the rip-head zone (250 m) (McLachlan, 1983; Campbell, 1987). This surf phytoplankton is an important food resource for benthic and planktonic filter feeders and some fishes, especially where it is concentrated into foam (Romer and McLachlan, 1986).

Particulate organic matter, or detritus, (POM), generally has a higher biomass than the microalgae and it constitutes a relatively constant food resource. It is derived from the breakdown of plants and animals, "sloppy" feeding, and the aggregation of DOM. In Eastern Cape waters, South Africa, values of between 1 and 5 g

TABLE 6.2
Soft Shore Food Sources

Food Type	Source	Consumers	Remarks
Benthic microflora	Sediments	Microfauna, meiofauna, and deposit feeders	More abundant on sheltered beaches
Phytoplankton	Coastal water	Filter feeders	
Surf diatoms	Surf water	Filter feeders	In well-developed surf zones
Stranded macrophytes (sea grasses, macroalgae)	The sea	Detrital feeders	Near kelp beds, rocky coasts, and sea grass beds
Detritus (particulate organic matter)	The sea	Detrital feeders filter feeders	
Dissolved organic matter	The sea		
Insects	The land	Particularly during offshore winds	
Meiofauna	Sediments	Other meiofauna, macrofauna	
Macrofaunal invertebrates	Sediments	Invertebrate and higher consumers	
Bacteria	The sea and the sediments	Microfauna, meiofauna, and macrofauna	
Microzooplankton	The sea	Filter feeders	
Carrion	The sea	Fish, crabs, and birds	

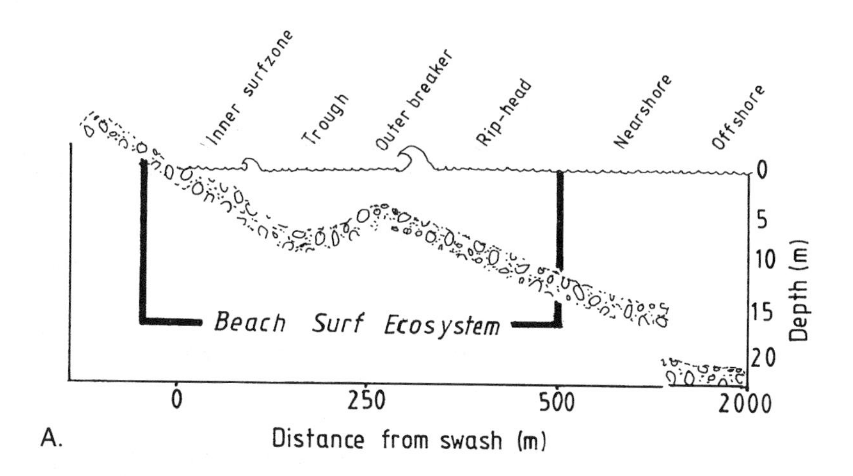

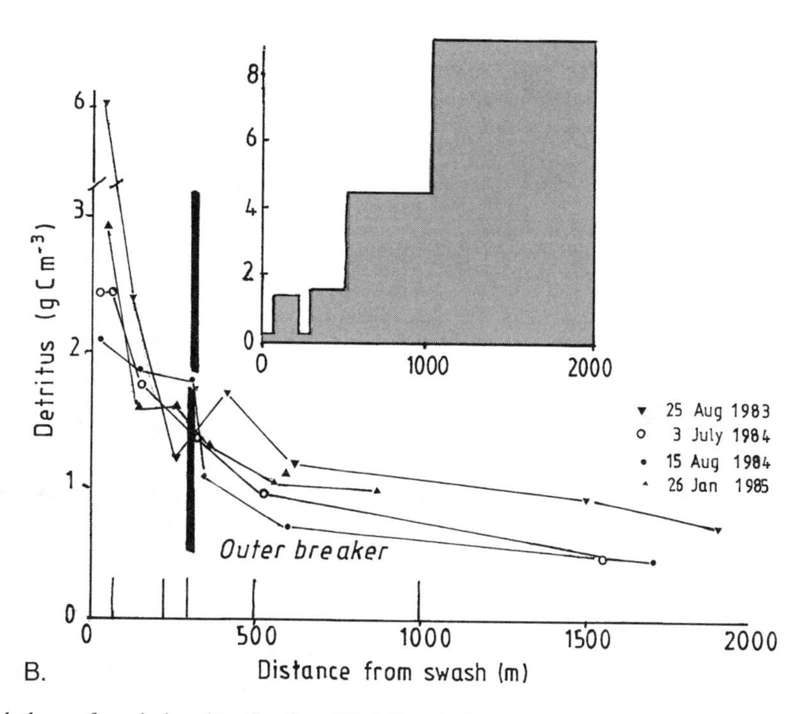

FIGURE 6.3 A. Morphology of sandy beach at Sundays River Beach, South Africa, showing the various zones sampled for detrital C concentrations. B. Shore normal distribution of detrital C concentrations on four sampling occasions. Each value is the mean of at least 20 replicates. Inset ordinate axis represents detrital standing mass m^{-1} of each zone (thereby taking dimensions of the various zones into account). Zones are: (1) inner surf; (2) trough; (3) outer breaker; (4) rip-head; (5) nearshore; and 6) offshore. (Redrawn from Talbot, M.M.B. and Bate, G.C., *J. Exp. Mar. Biol. Ecol.*, 121, 257 and 259, 1988d. With permission.)

C m^{-3} have been recorded (McLachlan and Bate 1984; Talbot and Bate, 1988d). Talbot and Bate (1988d) measured detrital standing mass along Sundays River Beach, South Africa, and found that it was consistently high in the surf zone (averaging 3.5 kg per running meter of beach, m^{-1}), exceeding values recorded in the immediate offshore zone by a factor of four and comprising 91% of the total POC (Figure 6.3). In contrast, in the inner surf zone, nearly 50% of the POC was composed of the surf diatom *Anaulus birostratus*.

In most parts of the world, sand beaches receive large inputs of drift algae from offshore kelp beds or nearby rocky coasts, (Brown et al., 1989). Studies of the input of drift algae to sand beaches have been carried out in a diverse range of localities, e.g., in California (Zobell, 1959), South Africa (Koop and Field, 1981; Griffiths and Stenton-Dozey, 1981; Stenton-Dozey and Griffiths, 1983; Griffiths et al., 1983), New England (Behbehani and Croker, 1982), Australia (Lenanton et al., 1982: Robertson and Hansen, 1981), and New Zealand (Inglis, 1989; Marsden, 1991a,b). Sten-

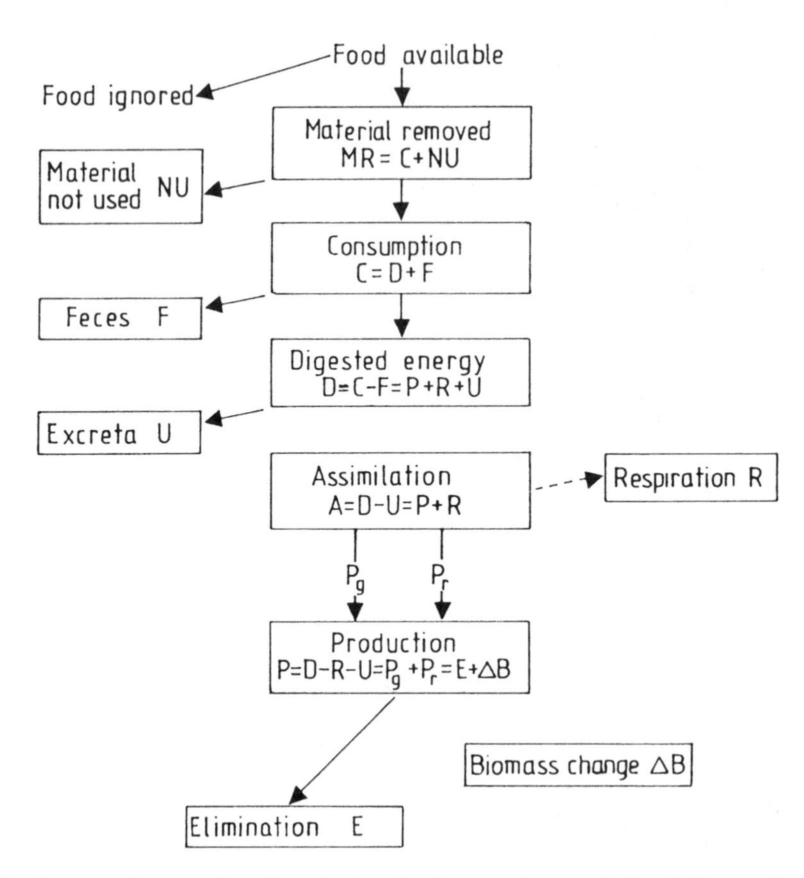

FIGURE 6.4 Schematic diagram of energy flow through an animal or species population. *MR* = total material removed by the population; *NU* material removed but not used (not consumed); *C* = consumption; *FU* = rejecta; *U* = excreta; *A* = assimilation; *D* = digested energy (materials); *P* = production; P_g = production due to body growth; P_r = production due to respiration; *R* = respiration (cost of maintenance); *B* = changes in biomass (standing crop) of the individual or population; and *E* = elimination. After Petrusewicz (1967) and Petrusewicz and Macfadyen (1970).

ton-Dozey and Griffiths (1983) estimated the seasonal and annual biomass of macroalgae deposited on a 300 m sandy beach at Kommetjie, South Africa. Highest deposition values occurred in autumn and winter and lowest values in summer. The mean standing stock of kelps on the beach was 25.07 kg m^{-1}. Using a residence time of 14 days, Griffiths and Stenton-Dozey (1981) estimated a total deposition rate of 2,179 kg wet mass m^{-1} yr^{-1}, equivalent to 4.07×10^6 kJ deposited per running meter of beach each year. The food webs associated with such algal drifts will be discussed in Section 6.9.3.3.

Carrion, usually of marine origin, is a highly erratic food supply. Jellyfish, siphonophores, bivalve molluscs, seabirds, cetaceans, and other animals are cast up on beaches at various times. Sometimes after storms that disturb sublittoral sediments, burrowing species such as polychaetes, holothurians, and echiuroids can be deposited in large quantities (Knox, 1957). In the absence of other major inputs, or on beaches adjacent to seal or seabird colonies, carrion inputs may be seasonally significant, but generally are of minor importance. McGwynne (1980), for beaches on the Eastern Cape, South Africa, estimated an annual input of carrion of about 120 g C m^{-1} yr^{-1}.

Dissolved organic matter in the water column may be concentrated by wave action into a rich yellow foam, which accumulates in the surf or on the beach. It has been shown that such foam is utilized by the bivalve *Donax serra*. However, it is principally used by the water column and sediment bacteria.

Two organic land sources, insects and plant litter, though usually not found in significant concentrations, are often found on beaches and in the surf waters.

6.3 ENERGY BUDGETS FOR INDIVIDUAL SPECIES

6.3.1 Introduction

The sequence of food transformations by an individual or species population can be represented by a schematic flow diagram as depicted in Figure 6.4 (Petrusewicz, 1967; Petrusewicz and Macfadyen, 1970). Ingested food may be assimilated, egested, excreted, respired, and ultimately forms new biomass.

Energy budgets of an individual organism or population relate the intake of food energy and its subsequent

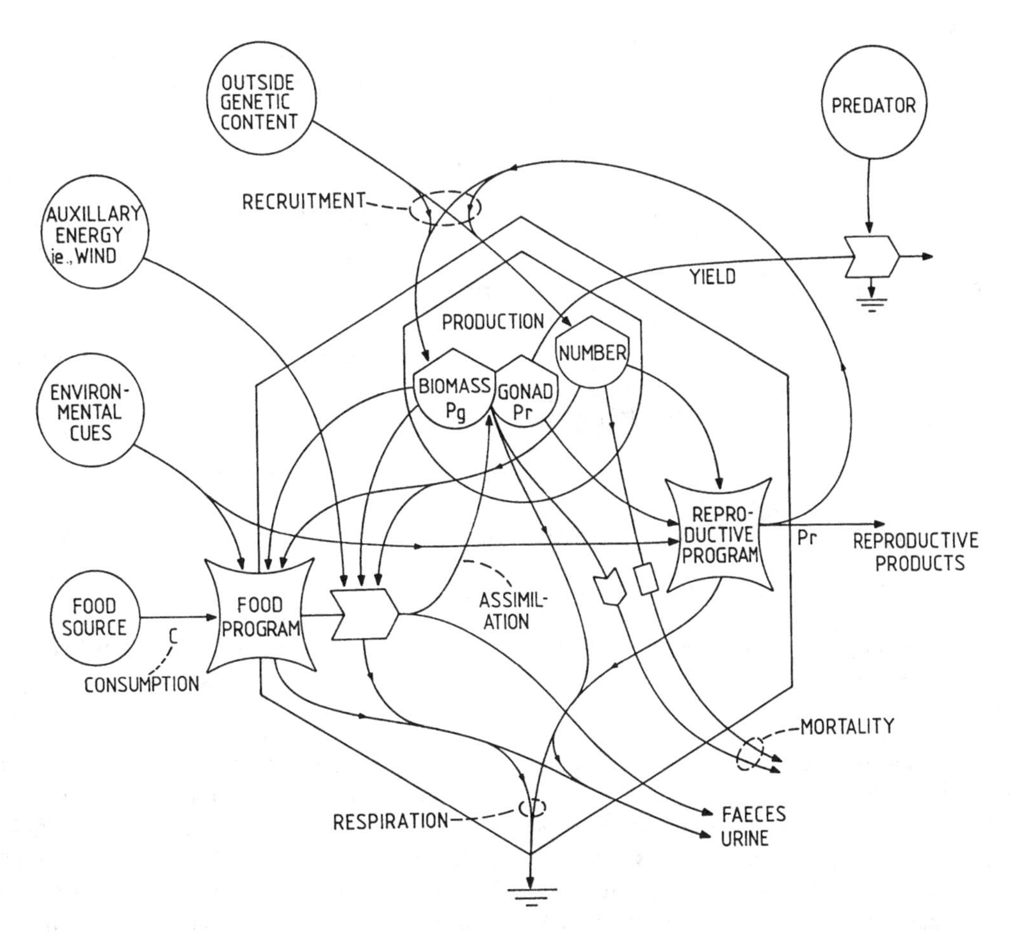

FIGURE 6.5 Diagram of the major sources and flows for a typical population of consumer units. After E.P. Odum (1983).

utilization according to the well-known balanced energy equation of Winberg (1956) (see also Ricker, 1968 and Grodzinsky et al., 1975).

$$C = P + (R + F + U)$$

where C = consumption (or intake) (energy content of the food absorbed), P = production (energy utilization in growth or gamete production), R = respiration (energy loss through metabolism), F = feces (energy loss through feces egested), U = urine (energy loss through dissolved organic matter, including urine), and expressed in units of energy (calories or joules).

Odum (1983) presented a diagram of the main energy sources and flows for a typical population of consumer units in which the influence of additional energy sources (such as recruitment and environmental parameters) were included. Stephenson (1981) adapted this diagram to conform with the terminology of the International Biological Programme (Petrusewicz and Macfadyen, 1970) (Figure 6.5).

6.3.2 Suspension-Feeding Bivalves

Stephenson (1981) developed an energy budget for a filter-feeding bivalve, the cockle *Austrovenus stutchburyi* in the Avon-Heathcote Estuary, New Zealand (Figure 6.6). This estuary is a small (8 km^2), bar-built estuary with a drainage basin of approximately 200 km^2), drained by two rivers entering the estuary. The cockle is the dominant macrobenthic species in the estuary. Densities range up to over 3,000 m^{-2} with a biomass (total ash free) dry weight of up to 1,200 g m^{-2}. The flow diagram for the energy budget of an individual *A. stutchbury*i depicted in Figure 6.7. For an *Austrovenus* population, inputs from food intake and recruitment result in standing crop through growth, reproduction, egestion, respiration, and mortality. This is the net organic production of the population.

The concept of production, as usually understood, refers to the amount of biomass produced over a given time period and is assumed to be a measure of the food energy potentially contributed to the succeeding stages of the food chain (Macfadyen, 1963). However, the methods of specific measurement and expression of "net production" in the literature are numerous. Petrusewicz and Macfadyen (1970) list five different definitions "each of them characterizing different ecological views of the concept in question." In its most general sense, "net production" may be considered to be organic matter available to be utilized by the next stage in the food chain, divided by the time taken for the organic matter to be produced.

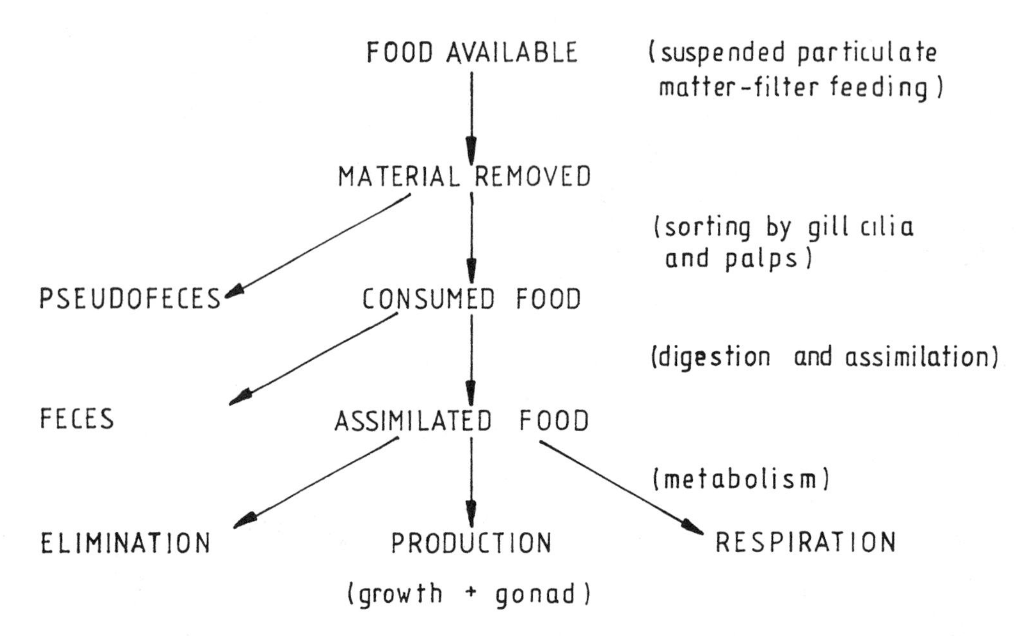

FIGURE 6.6 Schematic diagram of the functional components of an energy budget for the New Zealand cockle *Austrovenus stutchburyi*. (After Stephenson, R.L., Ph.D. thesis, Zoology Department, University of Cantaerbury, Christchurch, New Zealand, 1981. With permission.)

Figure 6.8 summarizes the major compartments and paths of energy flow in the Avon-Heathcote Estuary cockle population (Stephenson, 1981). Stephenson estimated the spatial distribution of "net production" of *A. stutchburyi* by applying a previously established length-age relationship to the mean shell length to estimate age at 200 sample sites. Net production (g ash-free dry wgt m^{-2} yr^{-1}) was estimated for each site as:

$$\frac{\text{Accumulated organic biomass}}{\text{Mean age of the population}}$$

The maximum net production value was about 15 g ash-free dry wgt m^{-2} yr^{-1}.

Net production estimated in this manner represents only accumulated organic matter and omits the part of production that has gone into mortality, elimination, and reproduction. This is similar to the concept of "yield" of Petrusewicz and Macfadyen (1970). On this basis the total winter organic biomass of cockles in the Avon-Heathcote Estuary (8 km^{-2}) was estimated at being between 8.2×10^4 and 1.7×10^6 kg (ash-free dry wgt), or 1.62×10^6 to 3.4×10^{10} kJ yr^{-1}.

Stephenson (1981) has also modeled the yearly flow of energy through the *A. stutchburyi* population. *Austrovenus,* which have very short siphons, filter organic matter from the layer of water immediately above the sediment surface. This water layer will contain suspended organic matter (microalgae and detritus) of terrestrial, marine, and estuarine origin. Major inputs are the input from the City of Christchurch sewage treatment oxidation ponds, the two rivers entering the estuary, *in situ* microalgal production (phytoplankton and suspended sediment microalgae), and sea phytoplankton production brought into the estuary on incoming tides (Stephenson and Lyon, 1982). This organic matter is filtered from the overlying water and processed by *Austrovenus,* which passes sediment, nutrients, organic matter, and mucus to the surface sediments as feces and pseudofeces. The assimilated organic matter is passed on to the predators (especially oystercatchers, fish, and whelks), or upon death to scavengers and decomposers.

For a second example of a filter-feeding bivalve, we shall consider *Macoma balthica*, a lamellibranch mollusc (Tellinidae) that colonizes intertidal and subtidal zones in different climatic regions of the Northern Hemisphere. Its distribution extends from San Francisco Bay (Nichols and Thompson, 1982) to Hudson Bay (Green, 1973) in North America and from the Gironde estuary, France, (Bachelet, 1980) to the White Sea and other parts of northern Russia (Beukema and Meehan, 1985). Individuals of this species are interoparous. Reproduction is indirect and longevity ranges from 5 to 50 years, depending on geographic location. Food is acquired by suspension and/or detrital feeding (Hummel, 1985a,b; Olafsson, 1986).

Hummel (1985b) calculated seasonal and annual budgets for a tidal flat population of *Macoma balthica* in the western part of the Dutch Wadden Sea. The budget was calculated as $C = P + R + G + F$, where C = consumption, P = somatic production, R = respiration, G = gonad output, and F = feces (including excreta, U). Values for the energy budget were obtained by summing monthly values (Table 6.3). In Table 6.3 these values are compared to those obtained for three other tellinid bivalves. The energy bud-

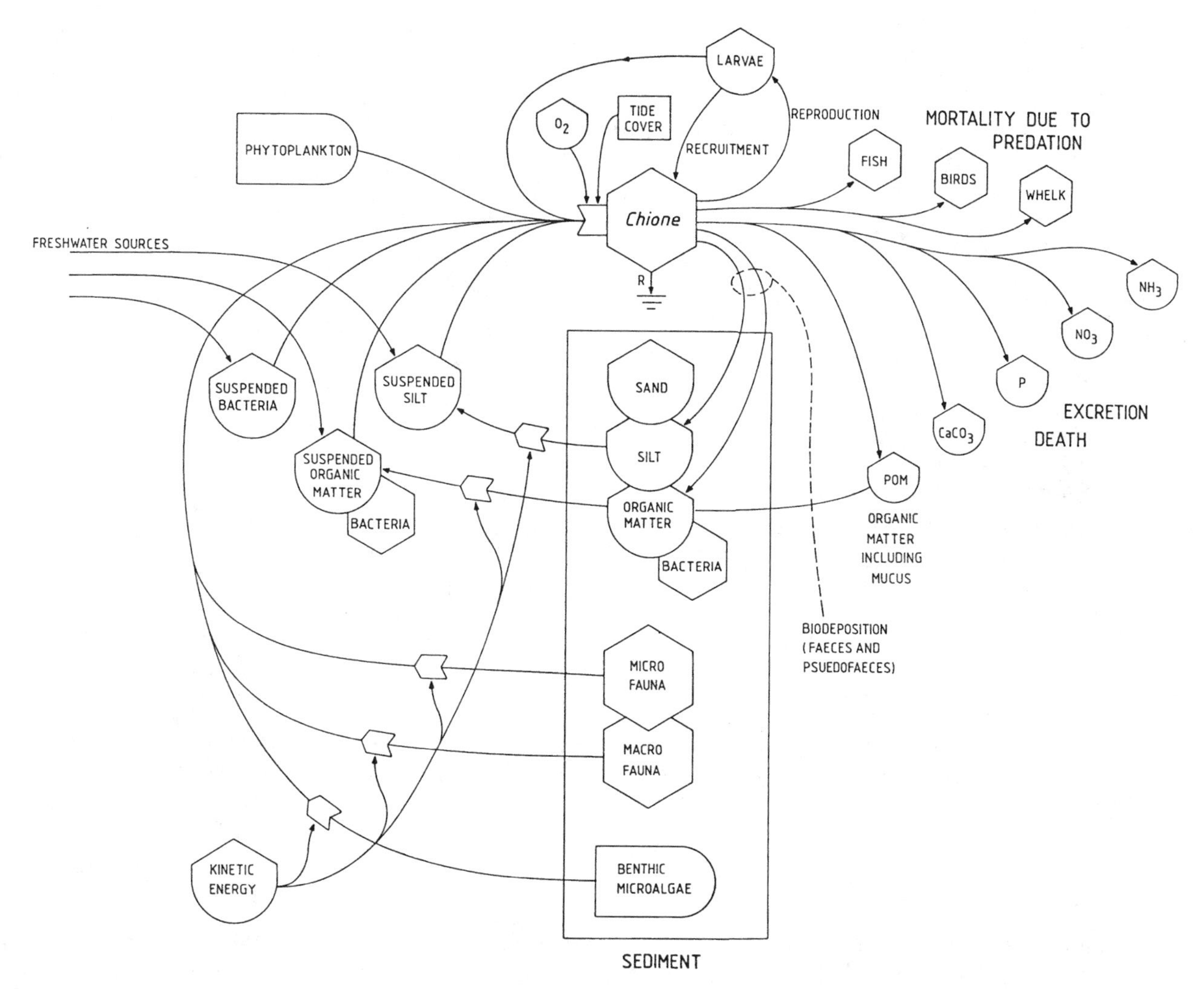

FIGURE 6.7 Major components and paths of energy flow relating to the cockle *Austrovenus stutchburyi* in the Avon-Heathcote Estuary, New Zealand. (After Stephenson, R.L., Ph.D. thesis. Zoology Department, University of Canterbury, Christchurch, New Zealand, 1981. With permission.)

get for *Macoma* as shown in Table 6.3 can be calculated in two ways: (1) when absorption (assimilation) is calculated from $A = P + G + R$, a value for absorption (A) of 71.7 kJ m^{-2} is obtained; and (2) when the absorption value is calculated independently from total consumption multiplied by the absorption efficiency, it amounts to 107.4 kJ m^{-2}. The former value is close to the values for absorbed chlorophyll *a* related food (71.4 kJ m^{-2}). This close fit suggests that the energy utilized by *Macoma* is primarily chlorophyll *a* related food. Thus, the main food of *Macoma* on the Wadden Sea tidal flat consists of microalgae, "fresh" algal detritus, and closely associated microorganisms. Various investigations have established that the preferred food items of *Macoma* appear to be diatoms, bacteria, and protozoa (Fenchel, 1972). Stable carbon isotope measurements at Pecks Cove, Bay of Funday, indicated that *Macoma* were feeding either on diatoms or fresh *Spartina* detritus and its associated microorganisms (Schwinghamer et al., 1983). Excretion (U) is thought to be quantitatively of little importance. Based on data for other bivalves, e.g., *Mytilus edulis* (12%) (Bayne and Widdows, 1978) and *Mytilus chilensis* (3%) (Navarro and Winter, 1982), it was estimated to be maximally 8 kJ m^{-2}.

Based on the annual value of 71.7 kJ m^{-2} for the absorbed food, only 28% of the consumed (ingested) food (257.7 kJ m^{-2}) was assimilated. This low assimilation may have been due to a large inert component ($C - Cc$) with little or no nutritional value in contrast to the chlorophyll *a* related food (Cc) (Figure 6.9). The production ($P + G$) amounted to 28% of the assimilated chlorophyll *a* related food. The values for consumption (C), absorption (A), production ($P + G$), and respiration (R) for *Macoma* are all within the range of those found for other tellinid bivalves listed in Table 6.3.

Table 6.4 compares average biomass, production, and P:B ratios for *Macoma balthica* for entire estuaries of

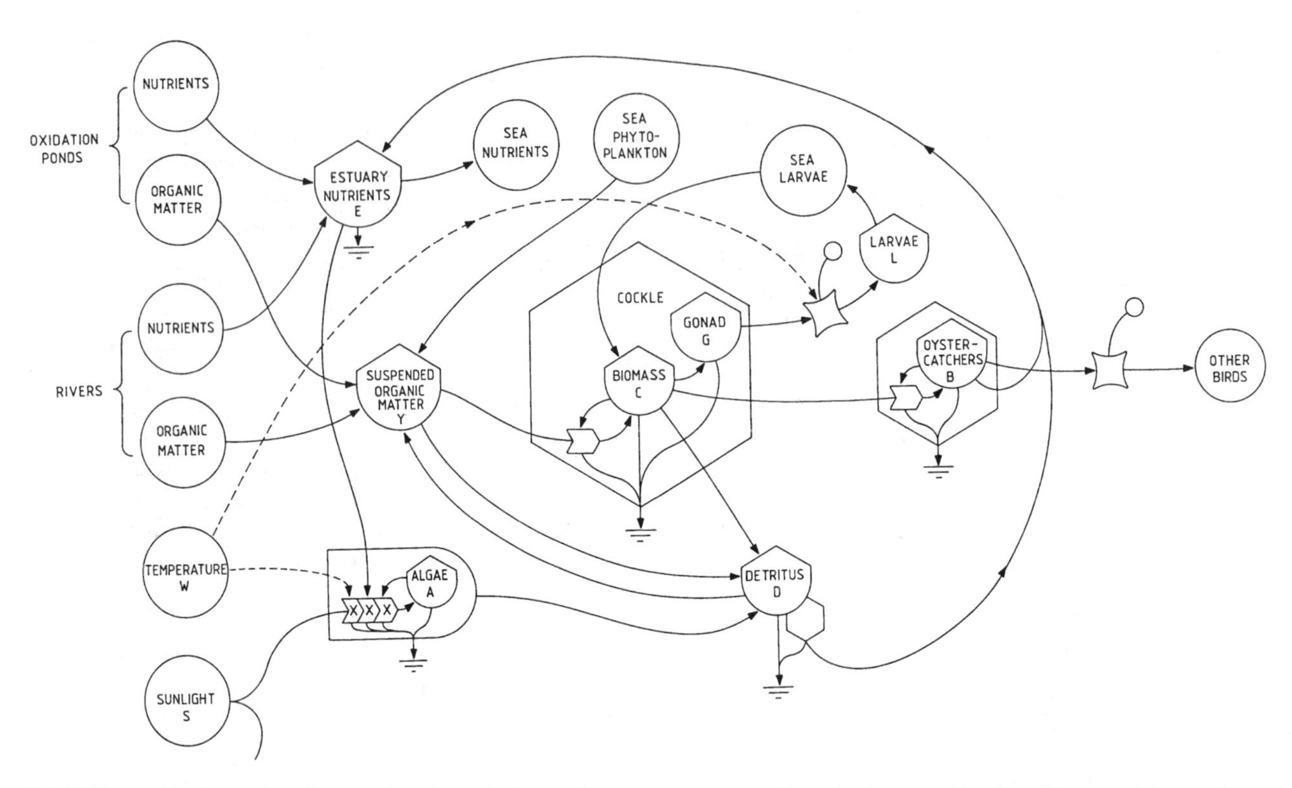

FIGURE 6.8 Yearly model of energy flow through the cockle *Austrovenus stutchburyi* in the Avon-Heathcote Estuary, New Zealand. (After Stephenson, R.L., Ph.D. thesis, Zoology Department, University of Canterbury, Christchurch, New Zealand, 1981. With permission.)

TABLE 6.3
Annual Values for Populations of Tellinid Bivalve Consumption (C: kJ m^{-2}), Absorption Efficiency (A/C), Absorption (A; kJ m^{-2}), Gonad Output (G; kJ m^{-2}), Respiration (R; kJ m^{-2}), Growth (K_1) and Net Growth Efficiency (K_2), Number (n; m^{-2}), and Biomass (B; kJ m^{-2})

		C	A/C	A	P	G	R	K_1	K_2	n	B
Macoma balthica	a	257.7	0.35–0.56	107.4	12.6	7.8	(87.0)	0.80	0.19	60	48.1
Western Wadden Sea	b		(0.28)	(71.7)	12.6	7.8	51.3	0.08	0.28	60	48.1
Hummel (1985b)											
Scrobicularia plana	c	3,565.7	(0.71)	(2,514.0)	251.4	268.2	1994.4	0.15	0.21	150	855.6
North Wales	d	436.6	(0.69)	(302.1)	55.7	17.6	228.8	0.17	0.24	55	81.3
Hughes (1970)											
Tellina fabula	e				(148.8)	63.2	4.8	80.8	0.46	987	39.2
German Bight	f				(216.5)	51.1	25.9	139.5	0.36	980	83.4
Salzweldel (1980)											
Tellina tenius	1966				(186.9)	27.6	2.7	156.6	0.16	111	74.9
N.W. Scotland	1967				(84.4)	4.3	7.1	73.0	0.14	59	59.7
Trevallion (1971)	1968				(111.9)	14.7	17.4	79.8	0.29	47	56.6

Note: Values in parentheses are not directly estimated but calculated from, e.g., $A = P + G + R$ or $R = A - P - G$. Data for *Macoma* are calculated for a population of an average 60 1+ year-old individuals. Explanation: a. A is calculated from results given by Hummel (1985a) (Table 6.1) and $R = A - P - G$. b. R is calculated from results given by De Wilde (1975), and $A = R + P + G$. c. Low level. d. High level. e. *Tellina* ground. f. Fine sand center.

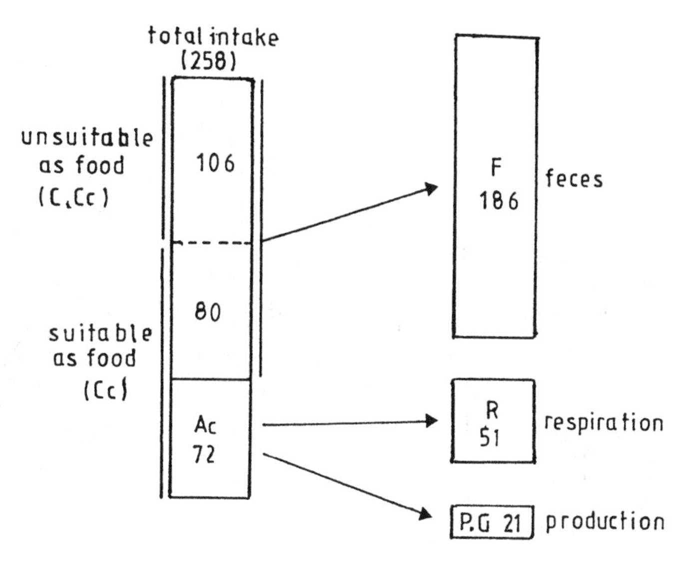

FIGURE 6.9 Schematic model of the main pathways in the energy budget for a mean population of 60 1+ year-old *Macoma balthica* in the Dutch Wadden Sea. All values are in kJ m^{-2} $year^{-1}$. (Redrawn from Hummel, H., *Neth. J. Sea. Res.*, 19, 89, 1985b. With permission.)

TABLE 6.4
***Macoma* Mean Biomass, Annual Production Estimates, and P:B Ratios Averaged for the Entire Mudflat or Estuaries from High to Low Water Level**

Location	Biomass ($g\ m^{-2}$)	Production ($g\ m^{-2}\ yr^{-1}$)	P:B
Grevellingen estuary, The Netherlands (Wolfe and De Wolf, 1977)	1.1	0.7	0.7
Wadden Sea, The Netherlands (Beukema, 1981)	2.4	1.7	0.7
Ythan Estuary, Scotland (Chambers and Milne, 1975)	2.8	5.7	2.1
Lynher River estuary, England (Warwick and Price, 1975)	0.4	0.3	0.9
Petpeswock Inlet, Canada (Burke and Mann, 1974)	1.3	1.9	1.5
Cumberland Basin, Bay of Funday, Canada (Cranford et al., 1985)	0.6	0.5	0.8
Shepody Bay, Bay of Funday, Canada (Cranford et al., 1985)	4.7	3.9	0.8

Note: Weights are in units of flesh dry wgt (including ash). Data originally presented in units of carbon and ash-free dry wgts were multiplied by 2.44 and 1.1, respectively.

mudflats from a range of locations in North America and Europe. As expected, high production correlated with high biomass.

6.3.3 Scope for Growth

Newell (1979) points out that the growth of invertebrates is dependent upon an interaction between net energy gained from the environment minus losses from metabolism and excretion, and that the net energy gain itself, as well as energy losses, is curtailed by a variety of factors including body size, food acquisition activity (e.g., filtration rate), ration, and assimilation efficiency. The product of these variables then yields an index of energy gain or "scope for growth" and reproduction. He reformulated the energy equation as follows:

$$A = C - (F + U + R) = P_p + P_r$$

A	$= C$	$- (F + U + R)$	$= P_p + P_r$
= Net assimilated energy	= Food consumed	= Losses through feces, dissolved organic matter, respiration	= Energy available for growth and reproduction

Experiments with suspension-feeding invertebrates have shown that for such species it is relatively easy to determine the fate of uptake of algal suspensions of known concentrations and energy content, the energetic losses from feces, dissolved organic matter, and respiration. Consequently, they have been widely used in studies on factors contributing energy gain and loss by marine invertebrates. Dame (1972) has shown that the growth of the estuarine oyster *Crassostrea virginica* from South

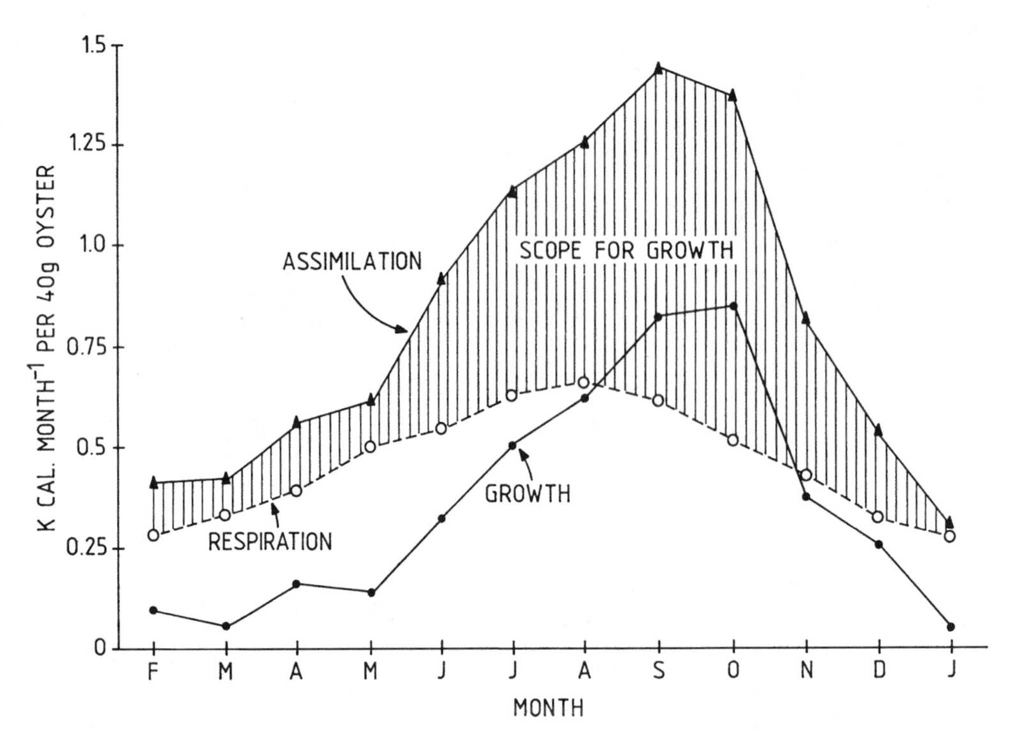

FIGURE 6.10 Graphs showing the seasonal variation in assimilation, energy losses through respiration, and resultant scope for growth (shaded) in the oyster *Crassostrea virginica*. (Redrawn from Newell, R.C., *The Biology of Intertidal Animals*, 3rd ed., Marine Ecological Surveys, Faversham, Kent, 1979, 494. With permission.)

Carolina, U.S.A., varies seasonally, reaching a maximum in September to October. The difference between net energy gain for the assimilated ration (A) and energy losses from respiration (R) can be considered as "scope for growth" (Warren and Davies, 1967; Bayne et al., 1973; Newell, 1979) (Figure 6.10).

For suspension feeders, the energy gain may be expressed as the clearance rate (V_w liters of water per hour), and oxygen consumption, or energy loss by respiration may be expressed as V_o ml hr^{-1}. The ratio V_w/V_o gives the "convection ratio," which can then be used to compare the energy gain between animals under different conditions. The influence of body size on filtration rate in a range of species has been well established (Winter, 1978), with the V_w/V_o greater in small animals than in large ones. Thus, small individuals can channel more energy into growth than larger ones feeding on the same ration.

Humel (1985b) has discussed the "scope for growth" in *Macoma balthica* in the Dutch Wadden Sea. Production observed in the field ($P + G$) was high only during the months of April, May, and June (Figure 6.11A). From July to February, *Macoma* steadily loses weight (Beukema and de Bruin, 1977), i.e., the energy balance is negative. When "scope for growth" is estimated from the total ingested food minus respiration ($A - R$ in Figure 6.11B), a large discrepancy emerges. This poor fit may be due to a relatively high percentage of poorly digestible detritus present during these months in the food resource. During this time the ratio of chlorophyll *a* to organic carbon was low in the overlying water (Hummel, 1985a). On the other hand, "scope for growth" as calculated from the net difference between the observed chlorophyll *a* related food (Ac) and respiration ($Ac - R$) in Figure 6.11B indicates that rapid growth is restricted to the months of April, May, and June, which coincides with the observed production.

6.3.4 Carnivorous Molluscs

As we have seen, predatory whelks are prominent predators on intertidal rocky shores. Here we shall consider the energetics of two common European species, the dogwhelk *Nucella lapillus* and the common whelk *Buccinum undulatum*.

Dogwhelks are well suited to intertidal life from the viewpoint of their physiological energetics. When uncovered by the tide, feeding dogwhelks do not become inactive, nor do they actively seek shelter, but remain in the open feeding on mussels and barnacles. Stickle (1985) found that the rate of consumption of mussel prey was higher during a simulated aerial exposure experiment than when continuously covered by seawater. Dogwhelks retain fluid in their mantle cavity, which reduces the effects of desiccation. Stickle and Bayne (1987) studied the energetics and "scope for growth" in this species. Scope for growth was estimated from the following relationship:

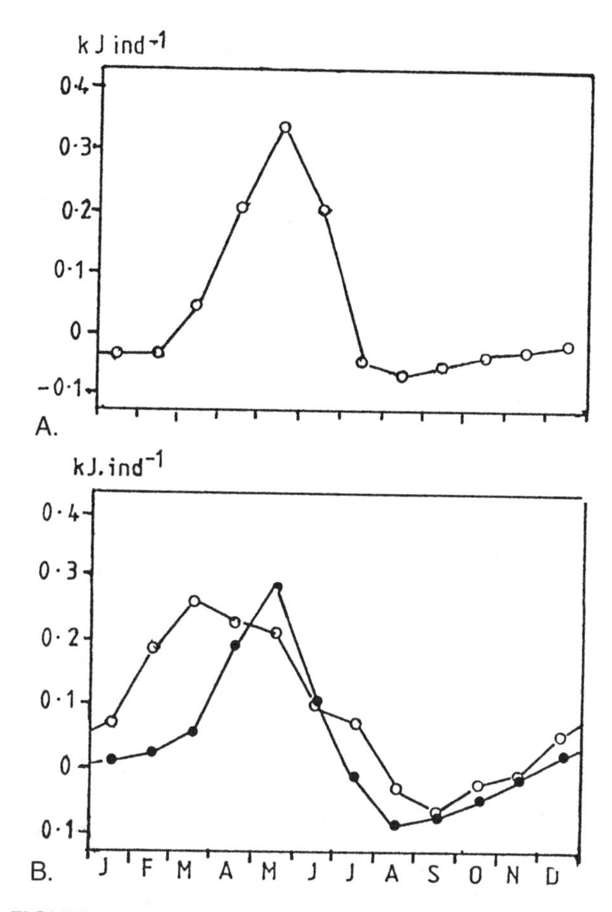

FIGURE 6.11 A field population of *Macoma balthica*. A. Production ($P + G$; kJ individ^{-1}; □). B. The "scope for growth" (kJ individ^{-1}) calculated from all absorbed food ($A - R$; ○) and from the absorbed chlorophyll *a* related food ($A_c - R$; ■). (Redrawn from Hummel, H., *Neth. J. Sea. Res.*, 19, 87, 1985b. With permission.)

$$SFG = Ab - (R + U)$$

where Ab is the absorbed energy ($C - F$, consumption or total intake – unabsorbed energy lost in feces), R is the energy lost as heat through respiration, and U is energy lost in excretory products.

Most of the energetic expenditure by *N. lapillus* was due to aerobic respiration, which varied between 41 and 83% of the total metabolic loss (Table 6.5). Energy losses due to ammonium excretion ranged from 9 to 17% of total energy expenditure. Stickle and Bayne (1987) found good agreement between the ability of *N. lapillus* to maintain a positive energy balance and its field distribution with respect to salinity. Dogwhelks are capable of maintaining a positive energy balance at a salinity of 25 but not at 20. Survival at a salinity of 25 is predicated in their ability to tolerate 5 months of a negative energy balance over the winter months. Stickle (1971) found that *N. lapillus* can withstand 150 to 180 days starvation before 50% mortality occurs. Growth is suppressed at minimum winter (5°C) and maximum summer temperatures (20°C), largely due to suppression of feeding.

In their study of the common whelk *Buccinum undulatum*, Kideys and Hartnoll (1991) have emphasized the important role of mucus production in this species. Energy consumption in this species is partitioned according to the following equation:

$$C = P_g + P_r + R + F + U + M$$

where C = consumption, P_P = somatic growth, P_r = reproductive investment, F = feces (including fecal mucus which is used in compacting feces), U = excretion, and M = hypobranchial and pedal mucus (Branch, 1981). Mucus production has largely been ignored in energy studies (Hughes, 1971a; Huebner and Edwards, 1981). Investigations have shown that molluscan mucus, which is a protein-carbohydrate complex, serves many purposes including locomotion, adhesion, location of home position, a medium for microbial growth, reduction of desiccation when animals are exposed to air, removal of sediment and feces from the pallial space, reproduction, reduction of exposure to environmental stress, as offensive and defensive agents, and in home scar formation (Edwards and Welsh, 1982; Horn, 1986; Peck et al., 1987; Davies et al., 1990; 1991; Kideys and Hartnoll, 1991). Thus it is clear that mucus production is a major energy drain in molluscan species. The existence of mucopolysaccharides in the shell composition indicates that mucus might also play a part in shell formation.

Kideys and Hartnoll (1991) estimated the importance of mucus in individual energy budgets (see Table 6.6). The energy allocated to mucus (pedal and hypobranchial) was about 30% of the total energy uptake. Energy loss via hypobranchial mucus secretion was appreciably higher than that of pedal mucus secretion in all size groups. Mean pedal mucus production ranged from 0.8 to 1.3 mg dry mucus g^{-1} dry tissue hr^{-1}. These are higher values than those found by Culley and Sherman (1985), 0.01 to 0.58 mg dry mucus g^{-1} dry tissue hr^{-1}, but comparable to those of Peck (1983): 0.04 to 2.00 mg dry mucus g^{-1} dry tissue hr^{-1}.

6.3.5 Grazing Molluscs

Marine grazing gastropods show great variations in energy intake and growth rates, both within a species and between species (Vermeij, 1980; Chow, 1987). Growth rates can be related to food resources, which vary in response to the population density of the consumers (Underwood, 1976, 1978c; Creese, 1980), or seasonal changes in environmental conditions (Stickle, 1985). Several studies suggest that physical stress may restrict gastropod foraging (Menge, J.L., 1974, 1975; Menge, B.A., 1978a,b) and thus reduce the growth rate of individual snails (Lewis and

TABLE 6.5
A Summary of the Ecological Energetics of *Nucella lapillus* Exposed to Temperature and Salinity Gradients

Temperature (°C)	Salinity	Ingestion Rate[a] (C) x	Ingestion Rate[a] (C) DMR	Absorption Efficiency (%)	Absorption Rate[a] (Ab) x	Absorption Rate[a] (Ab) DMR	Respiration Rate[b] (R) x	Respiration Rate[b] (R) DMR	Excretion Rate[b] NH_4^+ x	Excretion Rate[b] NH_4^+ DMR	Excretion Rate[b] PA x	Excretion Rate[b] PA DMR	Total Energy Loss (R + U) x	Total Energy Loss (R + U) DMR	Scope for Growth [P = Ab – (R + U)] x	Scope for Growth [P = Ab – (R + U)] DMR
5	17.5	0.0	F	—	0.0	F	3.6	HI	1.0	F	4.4	ABCD	9.0	FG	–9.0	DEFG
	25	3.9	F	—	0.8	F	4.6	HI	1.1	F	1.5	EF	7.2	G	–6.5	CDEF
	35	12.8	F	19	2.5		4.4		1.0		4.2		9.5		–7.1	
10	17.5	0.0	F	—	0.0	H	8.5	GHI	1.0	F	2.5	DEF	12.0	FG	–12.0	EFG
	25.0	48.3	DE	41	19.9	CD	16.8	GCDE	2.9	BC	3.8	BCDEF	23.4	CD	–3.5	CDE
	35.0	76.2		40	30.7		12.1		2.1		0.8		15.0		15.7	
15	17.5	17.4	EF	3	6.6	F	16.7	CDE	1.8	E	3.3	CDEF	21.8	CD	–21.3	GH
	25.9	116.7	B	40	46.1	B	21.9	ABC	3.4	AB	7.2	A	32.5	A	13.6	B
	30.0	189.7	A	42	78.9	A	22.5	AB	2.4	CDE	5.7	ABC	30.6	AB	46.2	B
20	17.5	7.7	F	2	0.1	F	9.3	FGH	2.1	DE	6.9	A	18.3	DE	–18.2	FGH
	25.0	48.3	DE	11	13.4	DE	18.4	BCD	3.2	AB	4.6	ABCD	26.3	BC	–12.9	
	30.0	92.7	BC	42	38.6	B	26.5	A	3.7	A	3.0	CDEF	33.3	A	5.3	

Note: All energy budget values are given for a uniform size animal (23–28 cm long) as J $snail^{-1}$ day^{-1}. Sample size was usually of 10 snails per treatment. Values sharing the same letter in each column under DMR, excluding the 35 salinity treatment at 5 and 10°C, are not significantly different from each other, — indicates no data; PA = primary amine; DMR = Duncan's Multiple Range Test.

[a] From Stickle et al., 1985.
[b] From Stickle and Bayne, 1982.

Source: From Stickle, W.B. and Bayne, B.L., *J. Exp. Mar. Biol. Ecol.*, 107, 257, 1987. With permission.

TABLE 6.6
***Buccinum undulatum*: Estimated Percentage of Total Energy Intake at 10.5°C Invested by Various Sizes (Shell Length 30 to 90 mm) of Whelk**

Animal Dry Wgt (g)	Mucus Type		
	Hypobranchial	Pedal	Total
1	18.1	8.9	27.0
2	14.0	8.0	22.0
3	15.7	10.4	26.1
4	19.7	15.1	34.8
5	16.9	10.6	17.5

Note: Based on equations derived for mucus production and food consumption.

Source: From Kideys, A.E. and Hartnoll, R.G., *J. Exp. Mar. Biol. Ecol.*, 150, 102, 1991. With permission.

Bowman, 1975: Roberts and Hughes, 1980). In this section the energetics of three grazing molluscs will be considered, the gastropods *Littorina* spp. and *Haliotis tuberculata*, and the chiton *Chiton pelliserpentis*,

Chow (1987) studied the seasonal patterns of growth and energy allocation in three species of western North American rocky shore littorinids: *Littorina keenae*, *L. scutulata*, and *L. plena*. Relative to male snails, female *L. keenae* appeared to sacrifice shell growth to spend more energy on the production of large numbers of gametes during spring; growth rates of females and large males were highest during the autumn when microalgal food supplies were seasonally abundant. In contrast, spring tended to be the best season for shell growth of *L. scululata* and *L. plana*. However, reproductive output was poor and growth rates in the field for all seasons were low compared to growth rates of snails held under experimental conditions with abundant food. High population densities of snails at the shore levels occupied by *L. scutulata* and *L. plana* may greatly reduce food levels upon which the snails rely for growth and reproduction.

The acquisition of usable energy by gastropods depends directly upon the quantity and quality of food resources as well as on foraging abilities. Where resources vary spatially or temporarily, gastropod growth rates vary correspondingly. Habitats that produce large, rapidly growing snails have ample food resources relative to habitats in which food is scarce and patchily distributed. Many herbivorous gastropods from temperate environments show seasonal variations in growth related to seasonal productivity of food plants (e.g., Sutherland, 1970; McQuaid, 1980). Moreover, food supplies may be affected by the gastropod consumers themselves. High densities of snail consumers can collectively limit the availability of food resources left for an individual (Sutherland, 1970; Branch, 1975b); density-dependent population effects on individual growth rates are widespread and have been established experimentally for a wide range of gastropod species (e.g., Choat, 1977; Underwood, 1978b; Creese, 1980).

Haliotis tuberculata, the European ormer or abalone, is a large gastropod mollusc (up to 120 mm in length) living in shallow marine habitats. It feeds on a wide range of macroalgae but shows distinct preferences for certain food species (Culley and Peck, 1981). Peck et al. (1987) constructed a laboratory energy budget for *H. tuberculata* based on the energy equation:

$$I = E + P_g + P_r + R + U + M$$

The above components were assessed for the whole size range of individuals held at 15°C in a 13-hr light to 12-hr dark regime with the green seaweed *Ulva lactuca* as the food source. Energy budget parameters are given in Table 6.7. Ingestion rates ranged from 1.94 to 9,972 cal animal^{-1} day^{-1} in 0.01 to 50 g dry wgt (3.71×10^3 to 17.3×10^3 g dry flesh wgt) animals, respectively. The major component of the energy budget was somatic growth (37.5% of I) in a 0.01 g dry wgt animals, while it was respiration (31% of I) in a 50 g dry wgt animal. Mucus production formed a large part of the budget (from 23.3% of I in a 0.01 g dry wgt animals to 29.1% of I in a 50 g dry wgt animal). Scope for growth, $I - (E + R + M)$ was calculated as ranging from 24.5% of ingestion in a 50 g dry wgt animal to 36.8% in a 0.01 g dry wgt animal.

The caloric value of the food was 3,419 cal g dry wgt^{-1} and for feces was 2,817 cal g dry wgt^{-1}. Absorp-

TABLE 6.7
***Haliotis tuberculata*: Energy Budget Parameters (cal day^{-1}) and Animal Size (dry wgt)**

Parameter	Animal Dry Wgt (g)				
	0.01	0.1	1	10	50
I	1.994	10.51	56.78	306.68	997.22
E	0.314	1.78	10.02	56.65	189.99
P_g	0.736	2.91	11.50	45.47	118.85
P_r	0	0	0.77	8.90	49.07
R	0.424	2.47	14.38	83.71	286.73
U	0.032	0.15	0.67	3.04	8.77
M	0.458	2.56	14.35	80.33	267.72
Total energy used	1.964	9.94	51.69	287.10	921.33
% of I not accounted	–1.0	5.4	9.0	9.3	7.6
SG (% of I)	36.8	33.8	30.6	27.0	24.5

Note: SG = scope for growth = $I - (E + R + U = M)$; see text for details.

Source: From Peck, L.S., Culley, M.B., Helm, M.M., *J. Exp. Mar. Biol. Ecol.*, 101, 117, 1987. With permission.

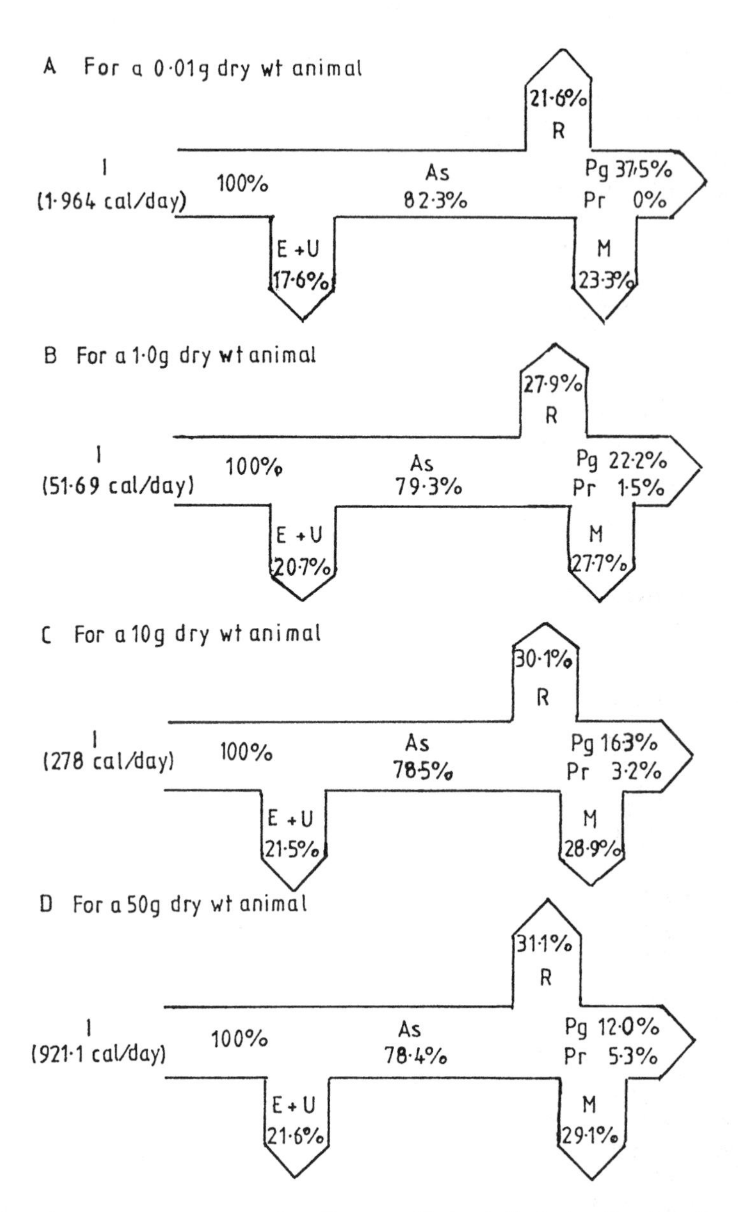

FIGURE 6.12 *Haliotis tuberculata*: percentage energy flow diagrams (A_s (assimilation) = $R + P_g + P_r + M$); ingestion rates shown (I) are sum totals of budget parameters, not measured ingestion rates. (Redrawn from Peck, L.S., Culley, M.B., and Helm, M.M., *J. Exp. Mar. Biol. Ecol.*, 106, 119, 1987. With permission.)

tion as a percentage of ingestion in terms of dry wgt ranged from 78% for 95 mm length animals (50 g dry wgt) to 81% in a 6 mm (0.01 g dry wgt). Peck et al. (1987) developed energy flow diagrams to illustrate the relative importance of the various components of the energy budget throughout the whole size range (Figure 6.12). They show the following trends with increasing size:

1. P_g decreases with size from 37.5 to 12.9% (a reduction of 66%);
2. P_r increases from 0 to 5.3%;
3. R increases by 44%, from 23.3 to 31.1%;
4. M increases less by 25%, from 23.3 to 29.1%;
5. $E + U$ losses increase by 23%, from 17.6 to 21.6%.

Thus, all the components of the budget increase with animal size at the expense of somatic growth. It can also be seen that the major components of the energy balance changes with age. Mucus production also plays a more prominent part in the energy budget of older animals.

Horn (1986) measured energy budgets for high and low shore populations of the chiton *Chiton (Sypharochiton) pelliserpentis* on the Kaikoura Peninsula, New Zealand. *C. pelliserpentis* is a dominant species on New Zealand intertidal rocky shores. Components of the energy budget were calculated using the following equation:

TABLE 6.8
Components of the Annual Energy Budget for Groups of the Chiton *Chiton pelliserpentis* from the High and Low Shore Zones

	C_m	C_s	P_g	P_r	P_m	R	F	U
High shore	1258	721	32.9	7.5	493	135	51	1.2
	471	532	18.6	5.3	370	102	34	0.9
	47	392	8.5	3.6	279	80	20	0.7
Low shore	3516	1510	60.9	11.8	828	397	210	2.0
	1521	1131	36.2	8.7	658	294	132	1.6
	456	840	18.8	6.1	525	215	74	1.3

Note: The upper and lower bounds identify the 95% confidence intervals around each component; C_m = measured consumption; other components are identified in the text; all units are kJ m^{-2} yr^{-1}.

Source: From Horn, P.L., *J. Exp. Mar. biol. Ecol.*, 101, 137, 1986. With permission.

$$C = P_g + P_r + P_m + R + F + U$$

The calculated values for all the components of the energy budget for both high and low shore chiton populations are shown in Table 6.8. Annual energy flow through the high shore population (532 kJ m^{-2}) was about half that of the low shore population (1,131 kJ m^{-2}). The largest single component of the energy budgets of both chiton populations was the production of mucus, which accounted for 74 to 66% of the assimilated energy in the high and low shore populations, respectively, whereas respiration (generally assumed to be the largest component of energy expenditure) accounted for only 21 to 19% of assimilation, respectively.

Estimates of energy investment in mucus by a range of molluscs are given in Table 6.9. Most of these are <10% of the ingested energy. Exceptions are pedal mucus production by *H. tuberculata* (23 to 29% of energy intake) (Peck et al., 1987) and *Patella vulgata* (23% of energy intake) (Davies et al., 1990). The high values for pedal mucus production (22.7% of the ingested energy) for the deposit-feeding gastropod *Ilyanassa obsoleta* (Edwards and Welsh, 1982) are perhaps compensated for by regaining the lost energy from feeding (probably after enrichment of the mucus by microorganisms).

6.3.6 Deposit-Feeding Molluscs

Deposit-feeding snails are dominant epifaunal species on mudflats, especially in estuaries. On the North American East Coast, *Ilyanassa obsoleta* and *Hydrobia totteni* are two of the most abundant species. Edwards and Welsh (1982) investigated carbon, nitrogen, and phosphorous cycling and energy transfer by a field population of *I.*

TABLE 6.9
Percentage Total Energy Intake Invested in Mucus in Various Gastropods

Species	Type of Mucus	Percentage of Ingested Energy	Reference
Ilyanassa obsoleta	Pedal	22.7	Edwards and Welsh (1982)
Ilyanassa obsoleta	Fecal	7.9	Edwards and Welsh (1982)
Hydrobia ventrosa	Pedal + fecal	9.0	Kofoed (1975)
Navanax inermis	Pedal	7.0	Paine (1965)
Ancylus fluviatilis	Pedal	9.0	Calow (1975)
Ancylus fluviatilis	Fecal	4.0–6.0	Calow (1975)
Haliotis tuberculata	Pedal	23.3–29.1	Peck et al. (1987)
Patella vulgata	Pedal	23.0	Davies et al. (1990)
Buccinum undulatum	Pedal	10.6	Kideys and Hartnoll (1991)
Buccinum undulatum	Hypobranchia	16.9	Kideys and Hartnoll (1991)
Buccinum undulatum	Total	27.5	Kideys and Hartnoll (1991)
Chiton pellisserpentis	Total	66–74	Horn (1986)

TABLE 6.10
Production Estimates and Losses for the Mud Snail *Ilyanassa obsoleta* at Branford Harbour, Connecticut, U.S.A.

Yearly Production Estimates	Units (mean g dry wgt m^{-2})
Tissue production ($P_g + P_r$)	20.0
Reproduction (P_r)	13.2
Somatic tissue (P_g)	6.8
Losses to emigration and mortality	23.2

Source: From Edwards, S.F. and Welsh, B.L., *Est. Coastal Shelf Sci.*, 14, 680, 1982. With permission.

TABLE 6.11
Annual Carbon, Nitrogen, and Phosphorus Budgets for a Field Population of the Deposit-Feeding Snail *Ilyanassa obsoleta*

	Element		
Components of the Budget	C	N	P
Assimilated components, A	306	32.3	0.868
Production, P	306	24.8	0.018
Mucus, P_m	298	23.9	—
Mucus trail	237	19.0	—
Fecal envelope	61	4.9	—
Reproductive tissue, P_r	5.2	0.55	0.012
Somatic tissue, P_g	2.5	0.26	0.006
Shell periostracum, P_p	0.2	0.07	—
Respiration, R	57	—	—
Excretion, Ex	2.8	7.48	0.85
DOM	2.8	1.05	0.27
DIM	—	6.43	0.58
Unassimilated components, UF	618	44.7	8.8
Fecal POM	596	42.4	7.5
Fecal DOM	22	0.6	—
Gross ingestion, I	984	77.0	9.67

Note: Values are g m^{-2} $year^{-1}$.

Source: From Edwards, S.F., Welsh, B.L., *Est. Coastal Shelf Sci.*, 14, 683, 1982. With permission.

obsoleta throughout the year at Branfield Harbour, Connecticut, U.S.A. Snail densities varied seasonally, reaching a maximum of 2,500 m^{-2}. Yearly production estimates are given in Table 6.10. The Petruzewicz (1967) energy budget equation was modified to fully represent the energy investment in mucus production:

$$P = P_g + P_r + P_p + P_m$$

where P_p = shell periostricum production and P_m = mucus production. Carlow and Fletcher (1972) showed that the F term (fecal production) should be modified to exclude fecal mucus, which contributes a significant component of the pellets and represents a product of metabolism. It was therefore included by Edwards and Welsh in the P_m component.

As in other mud snails, *Ilyanassa* forms continuous trails of mucus as it moves across the mud surface. In Branford Harbour, Edwards and Welsh (1982) found that in August at 21°C, the production of the mucus trail was 21.8 g cm^{-2} ash-free dry wgt (range 21.0 to 22.6). The average trail width was 0.5 cm, and the crawling velocities were 3.0 ± 1.9 cm min^{-1}. On an annual basis, the snail tissue production was 20.0 g m^{-2} (see Table 6.10). Slightly more tissue (23.2 g m^{-2}) was lost to mortality and/or emigration. Weight loss of adults during spawning (13.2 g m^{-2}) indicated that a major share (66%) of the annual production was allocated to reproduction.

Defecation rates for the snails ranged seasonally from 29.8 to 81.1 mg dry wgt feces g^{-1} dry wgt snail hr^{-1}, which is similar to values reported for other deposit feeders. The organic content of the feces ranged from 10.6 to 16.6%, while the mucus envelope of the fecal pellets weighed 0.016 0.005 g dry wgt mucus g^{-1} dry wgt feces, representing 10 to 17% of their organic content. Calow (1975) reported mucus contributions of up to 20% for fecal organics of the prosobranch snails *Ancylus fluviatilis* and *Planorbis contortus*.

Table 6.11 lists the annual carbon, nitrogen, and phosphorus budgets for a field population of *Ilyanassa*. In atomic ratios, the C:N:P ratio was 106:95.5:10 for snail tissue, and 106:6.5:0.54 for fecal organics; thus, the N content of the tissue was enriched relative to the fecal organics. By far the greatest proportion of all three chemical constituents was unassimilated (*UF* in Table 6.11). Within the assimilated category, mucus production was the largest single category. Mean annual P:B ratios were 0.63 for *C*, 0.67 for *N*, and 0.58 for *P*.

Edwards and Welsh (1982) developed a population energy budget (Table 6.12) from their monthly estimates. Fecal POM averaged 4,001 ± 107 cal g^{-1}, which is similar to that reported by Hargrave (1971) (3,600 ± 1,500 cal g^{-1}) for the deposit-feeding amphipod *Hyallela azteca*. Mucus was a major component of the energy budget at 3,521 kcal m^{-2} yr^{-1}, or 80% of the assimilated energy.

The elements and energy returned by *Ilyanassa* to the mudflat ecosystem were 95% particulate organic matter (894 g C m^{-2} yr^{-1}, plus 66 g N m^{-2} yr^{-1} in mucus plus feces) (9,541 kcal m^{-2} yr^{-1}), 4% dissolved organic matter (26 g DOC m^{-2} yr^{-1} + 1.7 g DON m^{-2} yr^{-1}), and 1% dissolved inorganic matter (2.8 g C(CO_2) m^{-2} yr^{-1}, plus 8.13 g N (NH_4^+ + NO_2^-) m^{-2} yr^{-1}, plus 1.9 g P (PO_4^{3-}) m^{-2} yr^{-1}. These would then be available for use by the microbial community and secondary consumers. Ammonia released by *Ilyanassa* increased over the summer months and peaked in September. The large allocation of resources

TABLE 6.12
Annual Energy Budget and Trophic Efficiencies for a Field Population of the Deposit Feeding Mud Snail, *Ilyanassa obsoleta*

Components of the Budget	kcal m^{-2} yr^{-1}	Trophic Efficiencies: Fraction of Consumption	Trophic Efficiencies: Fraction of Consumption
Assimilation components, A	4,049	0.381	—
Production, P	3,350	0.315	0.827
Mucus, P_m	3,251	0.306	0.803
Mucus trail	2,409	0.227	0.595
Fecal envelope	842	0.079	0.208
Somatic tissue, P_s	31	0.003	0.008
Shell periostracum, P_p	3	0.0003	0.0007
Respiration, R	617	0.060	0.157
Excretion, Ex	62	0.006	0.15
DOM	36	0.003	0.009
NH_3	26	0.002	0.006
Unassimilated components, UF	6,567	0.619	—
Fecal POM	6,290	0.592	—
Fecal DOM	271	0.026	—
Gross ingestion, I	10,616	—	—

Source: From Edwards, S.F. and Welsh, B.I., *Est. Coastal Shelf Sci.*, 14, 673, 1982. With permission.

to mucus trail production by *Ilyanassa* raises the question of the value of such energy expenditure over and above its role in locomotion. Carlow (1975) has shown that mucus trails left by gastropods enhance bacterial growth in freshwater systems. Juniper (1982) has also shown that the passage of sediment through the gut stimulates bacterial growth in the fecal strings of the New Zealand mudflat snail *Amphibola crenata*. The mucus in the fecal strings and trails of *Ilyanassa* and other deposit-feeding snails may therefore play an important role in the mudflat ecosystem as substrates for bacterial growth.

6.3.7 Fishes

Du Preez et al. (1990) have developed energy budgets for five teleost and two elasmobranch species, as well as for the main ichthyofaunal groups of the surf-zone ecosystem of the Eastern Cape beaches in South Africa. These species were a benthos and plankton feeder *Lithognathus mormurus*, the benthic feeders *L. lithognathus, Pomadasys commersonni, Rhinobatos annulatus*, and *Myliobatus aquila*, a pisicivorous fish *Lichia cemia*, and a plankton feeder *Liza richardsonii*. Using the equation, $C = F + U + R_d + R_g + AB$, where C = food consumption, F = feces, R_d = apparent specific dynamic action, R_g = routine standard metabolism, and AB = growth, the following general energy budgets were derived for the fishes:

$$C = F + U + R_d + R_g + AB$$

Teleosts: $100 = 10 + 4 + 21 + 23 + 42$

Elasmobranchs: $100 = 11 + 2 + 16 + 24 + 48$

These budgets show that most of the energy consumed is used in metabolism ($R_d + R_g$) and growth (AB), whereas excretion accounts for only a small proportion.

The main feeding groups of the surf-zone ichthyofauna are the southern mullet *Liza richardsonii*, the sandshark *Rhinobatus annulatus*, benthic feeders, zooplankton feeders, omnivores, and piscivores, with biomass values of 1,000, 1,000, 3,000, 2,400, and 400 kJ m^{-1}, respectively, and annual consumption budgets of 22,107, 13,725, 65,710, 65,476, and 8,517 kJ m^{-1} yr^{-1}, respectively. *L. richardsonii* feeds mainly on surf diatoms, consuming 0.5% of the total diatom production. Zooplankton production (mainly shoals of mysids *Mesopodoria slabberi* and swimming prawns *Macropetasma africanus*) supplies 91% and macrobenthic production (mainly bivalves, *Donax serra* and *D. sordidus*) and 9% of the energy needs of the other non-piscivorous fishes. Piscivorous fishes consume 30% of the available fish production. Fecal energy production (30,3412 kJ m^{-1} yr^{-1}) is utilized in the microbial loop. Nonfecal production (U = 8,229 kJ m^{-1} yr^{-1}) of inorganic nitrogen as ammonium is utilized by the surf-zone diatoms. Within the beach/surf-zone system, fishes are therefore (1) the major predators on the macrofauna; (2) important transformers of carbon and nitrogen; and (3) agents for the transfer of materials across the nearshore boundary.

6.4 OPTIMAL FORAGING

6.4.1 Introduction

Consumers, especially predators, are often exposed to fluctuating resource availability, where food may be spatially unpredictable, scarce, and even absent (Menge 1972; Connell, 1975; Murdock and Oaten, 1975). The feeding activities and strategies of consumers is of great theoretical interest generating questions such as: (1) the extent to which herbivores or carnivores depress the populations of their food organisms; (2) the functional relationships between feeding rate and the density of food (Steele, 1974); (3) to what extent are different components of the predation process stabilizing or destabilizing; (4) what are the mechanisms that prevent extinction of the prey population; and (5) to what extent do predators switch diets according to the relative abundances of the prey? Diverse aspects of the foraging activities of animals, ranging from choice of foraging location, through choice of searching methods, to choice of diets have been modeled using the premise that all foraging activities adopted by an animal are those that maximize the net energy intake (Hughes, 1980a). This is the Energy Maximization Premise (Townsend and Hughes, 1981), and the theories are collectively known as Optimal Foraging Theory (Schoener, 1971; Pulliman, 1974; Krebs, 1978; Pyke, 1984). The optimal foraging theory has profoundly influenced experimental investigations of feeding behavior and its predictions have frequently been confirmed by laboratory data. Nevertheless, it has been deemed worthless by some and its basic premise, that maximizing net rate of energy intake promotes fitness, has seldom been verified (Hughes and Burrows, 1990).

Schleiper (1981) outlined what he considered to be the three most important predictions of Optimal Foraging Models: (1) at high food densities, a forager should concentrate solely on the energetically most valuable type of prey; (2) the inclusion of a food type in the optimal diet does not depend on its abundance, but only on its value and the absolute abundance of foods of higher value; and (3) as food abundance declines, the diversity of foods in the diet should increase. Horn (1983) tested these predictions in a study of the seasonal diets of *Cebidichthys violaceus* and *Xiphister mucosus*, two herbivorous fishes from the central California rocky intertidal zone. The prediction that at high food densities a forager should concentrate solely on the energetically most valuable items was incompletely met by these two fish species. *C. violaceus* and *X. mucosus* increased their consumption of energy-rich annual macrophytes during periods (summer and autumn) of high food abundance, but nevertheless continued to take a mixed diet. The prediction that abundance of lower-valued foods does not determine their inclusion in the diet was largely upheld by the feeding habits of these two intertidal fishes. The probability of an item being consumed apparently depended upon its abundance as well as its chemical composition. The prediction that foragers will generalize as food abundance declines was largely met by the two fishes since their diets broadened considerably during periods of reduced food supply (e.g., winter). Furthermore, the diets of the two species converged during periods of high food abundance and diverged during months of low food abundance.

Studies on foraging have followed two routes, one leading to the study of prey choice (optimal diets), and the other leading to the study of feeding locations (optimal foraging). Hughes (1980a,b) has reviewed these concepts in the marine context.

6.4.2 Optimal Diets

For a particular predator, each type of prey will have a characteristic dietary "value" defined as the ratio of energy yield to handling time. Handling time includes all the events from perception of the prey item to the resumption of searching for more prey. According to the type of predator, handling time could include events such as pursuit, attack, subduing, ingestion, and digestive pause. The theory assumes that the predator can rank all prey items in terms of their prey value, and it can use this ranking to decide whether the prey should be eaten or not. Simple models predict the (1) the highest ranking prey should always be eaten when encountered, and (2) lower ranking prey should only be eaten when better prey are too scarce to meet the energy requirements of the predator, irrespective of their own relative abundance.

Hughes (1980b) considers that the Optimal Foraging Theory is applicable, with suitable modifications, to all free-living animals whether they are herbivores, filter feeders, deposit feeders, or hunters. He has provided the behavioral classification shown in Figure 5.14. Predators are classified into browsers, filter feeders, and hunters. The browsers are subdivided into herbivorous browsers and carnivorous browsers, while the hunters are subdivided into ambushers, searchers, and pursuers.

Browsers (or grazers) are either herbivores feeding on microalgae (planktonic, benthic, or epiphytic) or macroalgae, or carnivores feeding on sedentary invertebrates, e.g., coelenterates, sponges, and bryozoans. Herbivorous browsers feeding on microalgae or easily digestible macroalgae generally have wide diets, whereas those feeding on the more intractable macroalgae are generally specialists. Since sessile invertebrates often have well-developed defense mechanisms such as nematocysts, skeletal encrustations and spines, and the production of toxic chemicals, carnivorous browsers may have very restricted diets, often feeding on a single species. Hunters that attack and kill

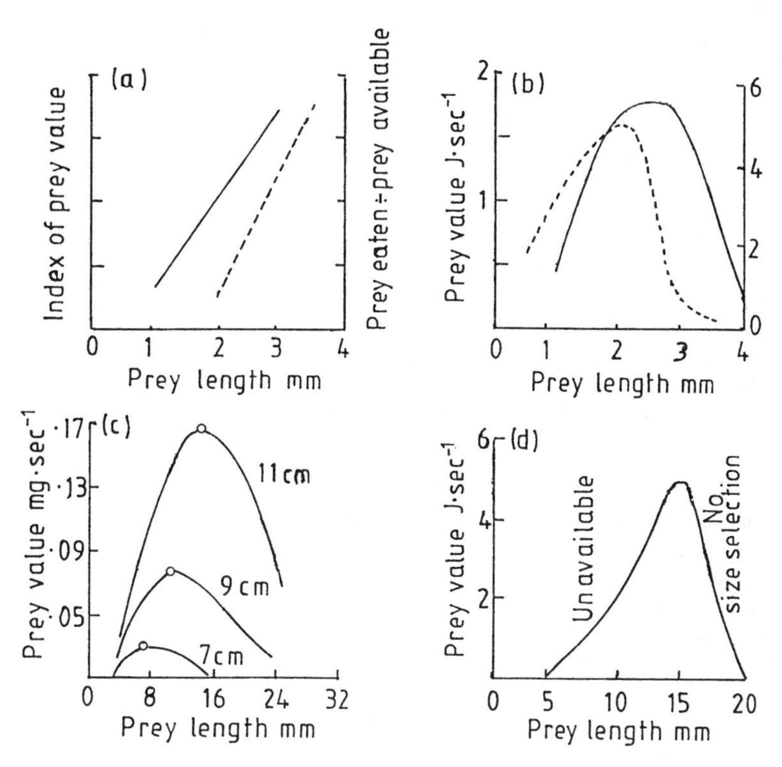

FIGURE 6.13 Prey values and preferences of some predators. A. *Balanus glandula* eaten by the whelk *Thais emarginata* (Emlen, 1966). Solid line = prey value; dashed line = prey preference. B. *Mytilus edulis* eaten by the crab *Carcinus maenus* (Elner and Hughes, 1978). Solid line = prey value; dashed line = prey preference. C. *Praunus flexuosus* eaten by the stickleback *Spinachia spinachia* (Kialalioglu and Gibson, 1976). Prey value curves for sticklebacks of 7, 9, and 11 cm in length, respectively. Open circles = mean preferred prey sizes. D. The whelk *Thais lapillus* eaten by the crab *Carcinus maenus* (Hughes and Elner, 1979). Solid line = prey value. (Redrawn from Hughes, R.N., in *The Shore Environment*, Vol. 2, *Ecosystems*, Price, J.H., Irvine, D.E.G., and Farnham, W.F., Eds., Academic Press, London, 1980b, 709. With permission.)

their prey fall into three categories: (1) *ambushers,* which often are concealed by cryptic coloring and/or morphology, rely on the prey to move toward them so that no search or pursuit is involved; (2) *searchers,* which include a wide variety of animals that actively search for food items but which normally do not spend much time pursuing or handling food items; and (3) *pursuers,* which generally spend a considerable amount of time pursuing and handling each prey item, and on this basis would be expected to have specialized, narrow diets.

Some of the most successful applications of the optimal diet theory have been those dealing with intertidal carnivores feeding on discrete prey items. Preference for prey items with the highest dietary values, defined as the ratio of energy yield to handling time, have been demonstrated for a number of littoral predators (Figure 6.13). These include the starfish *Asteria rubens* feeding on the bivalve *Macoma balthica* (Anger et al., 1977), the starfish *Lepasterias hexactis* feeding on a variety of gastropods and barnacles (Menge, 1972), the shore crab *Carcinus maenus* feeding on the mussel *Mytilus edulis* (Elner and Hughes, 1978), the dogwhelk *Nucella lapillus* feeding on *Mytilus edulis* (Dunkin and Hughes, 1984; Hughes and Dunkin, 1984; Hughes and Burrows, 1990), the dogwhelk *Acanthina punctulata* feeding on winkles *Littorina* spp. (Menge, 1974), the dogwhelk *Thais emarginata* feeding on barnacles, *Balanus glandula* (Emlen, 1966), the fifteen-spined stickleback *Spinachia spinachia* feeding on mysids *Neomysis integer,* the redshank *Tringa totanus* feeding on the amphipod *Corophium volutator* and polychaetes (Goss-Custard, 1977a,b), the oystercatcher *Haemotopus ostralegus* feeding on cockles *Cerastoderma edule* (O'Connor and Brown, 1977), and the black oystercatcher *H. bachmani* feeding on mussels *Mytilus californianus* and limpets *Acmaea* sp. Dogwhelks (*Nucella* spp.) are known to prefer larger barnacles and mussels when offered a range of prey sizes (Dunkin and Hughes, 1984; Hughes and Dunkin, 1984). Larger prey yield more food per handling time. In field experiments, dogwhelks grow faster when allowed to feed on preferred prey items (Palmer, 1983). The evidence, therefore, suggests that dogwhelks feed selectively in a way that maximizes the net rate of energy intake.

Hughes and Burrows (1990), in a field study of dogwhelks and their prey mussels (*Mytilus edulis*) and barnacles at two intertidal sites (sheltered and exposed) in North Wales, demonstrated a direct relationship between cumulative food intake and individual growth rate. It was

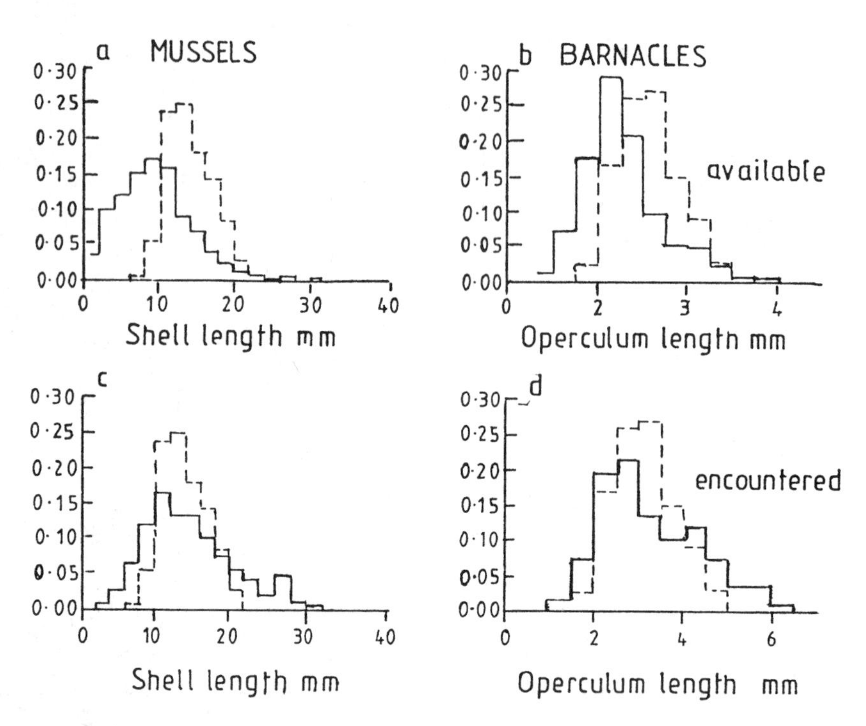

FIGURE 6.14 Size frequency distributions of mussels on exposed shores (A., C.) and barnacles (B., D.), estimated encountered (solid line) and included in diets of dogwhelks (broken line). Proportions taken in the diet differed significantly from both available and encountered distributions; mussels in diet (n = 418) vs. available, G (goodness of fit) = 597.3, vs. encountered, G = 297.5; barnacles in diet (N = 425) vs. available, G = 343.4; vs. encountered, G = 194.0; all $P < 0.0001$. (Redrawn from Hughes, R.N. and Burrows, M.T., in *Trophic Relationships in the Marine Environment*, Barnes, M. and Gibson, R.V., Eds., Aberdeen University Press, Aberdeen, 1990, 519. With permission.)

found that dogwhelks prefer larger, more profitable prey (see Figure 6.14), but are often forced by unfavorable weather to feed close to the shelter of crevices where only small prey are available. Two factors lower food intake during adverse conditions: the reduced amount of time spent on foraging and the enforced consumption of smaller prey. When foraging activity is reduced to 20% of maximum, the mean size of mussels eaten falls from 12 to 1.5 mg and the mean size of barnacle prey from 2.5 to 1.5 mg, representing a 33% and 40% reduction in gain per prey item, respectively. Energy maximization behavior evidently plays a major role in the foraging behavior of dogwhelks, but it is modified by the need to avoid desiccation, elevated metabolism costs, or dislodgement by waves.

Perry (1987) investigated the behavior of the predatory snail *Acantina spirata* when feeding on the barnacles *Balanus glandula* and *Chthalamus fissus* at Lunada Bay in Southern California, under conditions of satiation and starvation. It was found that the foraging behavior of starved *A. spirata* differs from that of satiated individuals in prey selectivity, attack frequency, and handling times. When given a choice, starved snails attack both barnacle species equally, whereas satiated individuals preferentially attack *B. glandula*, the more profitable prey (ash-free dry wgt of barnacles ingested per unit of handling time). It appears that as the snails pass from satiation to hunger, behavioral changes occur that translate into an energetic disadvantage during feeding for hungry snails for two reasons. First, higher prey handling times result in a reduced energy intake. Second, alteration in the relative attack frequency between barnacle species, combined with a decrease in attack success on the more profitable prey, leads to more frequent ingestion of the less profitable prey. Optimal diet models do not have an explicit term for hunger; in fact, the effects of hunger are excluded from the models (Hughes, 1980a). The experiments of Perry highlighted the need to include starvation in optimal diet models and consider it as part of a hunger-satiation continuum where the feeding behavior of predators fluctuates with position along this continuum.

6.4.3 Optimal Patch Use

The energy maximization principle can also be applied to a predator's allocation time to foraging activities among different environmental patches and containing specific prey mixtures and prey densities. When choosing foraging sites, energy intake is maximized by the predator's concentrating on the *most profitable patches,* staying in each patch until the net rate of energy intake falls to the average intake for the habitat. As better patches become depleted to the average productivity of the habitat, more patches should be visited. The greater cost of travel between the

TABLE 6.13
Mean Biomass and P:B Ratios for the Most Important Macrobenthic Invertebrates in a Mudflat of the Lynher Estuary, England

Species	Feeding Type	Mean Biomass ($g\ m^{-2}$)	Production ($g\ m^{-2}\ yr^{-1}$)	P:B Ratio
Nephtys hombergii	Carnivore?	3.95	7.34	1.9
Mya arenaria	Suspension feeder	5.54	2.66	0.5
Amphitrite acutifrons	Deposit feeder	0.43	2.32	5.5
Scrobicularia plana	Deposit feeder	2.15	0.48	0.2
Macoma balthica	Deposit feeder	0.32	0.31	0.9
Cerastoderma edule	Suspension feeder	0.85	0.21	0.2

Source: From Warwick, R.M. and Price, R., *J. Mar. Biol. Assoc. U.K.*, 55, 16, 1975. With permission.

patches, or the greater the search costs within the patches, the longer the predator should stay in each patch.

In selecting searching methods, the predator should use the most efficient, but most costly methods only at the highest prey densities, switching to the less costly, but less efficient searching methods as the prey density decreases. Two studies on the application of optimal patch use theory are those of birds feeding in estuaries. O'Connor and Brown (1977) demonstrated that oystercatchers, *Haematopus ostralegus*, feeding on cockles, *Cerastoderma edule*, in Stanford Lough, Northern Ireland, concentrated on patches where two-year-old cockles were abundant. Two-year-old cockles were preferred because they gave the highest average net energy return. The oystercatchers left these profitable patches once the average yield for the area had been taken. They apparently assessed the quality of the patches as they flew over the area by perceiving the densities of the anvils where the cockles were smashed open. A similar concentration on the most profitable patches was observed in redshanks, *Tringa totanus*, feeding on the amphipod *Corophium volutator*, (Goss-Custard, 1977a) and polychaetes (Goss-Custard, 1977a,b) in the Ythan Estuary, Scotland. The redshanks spent more time foraging on the most productive patches where their ingestion rates were highest. A small proportion of the foraging time was also spent in poor patches throughout the period studied (autumn to spring), a behavior consistent with the hypothesis that the average productivity of the habitat was being measured by direct sampling rather than by visual cues as in the oystercatcher example.

6.5 SECONDARY PRODUCTION

6.5.1 MACROFAUNA

The biomass of macrobenthic invertebrates varies widely according to the type of substrate and exposure to wave action. However, within any given ecosystem it remains relatively constant from year to year (Beukema et al., 1978). Wolff (1983) lists biomass values for various soft-bottom macrobenthic assemblages averaged over large areas ranging from values of 0.3 to 5.0 g ash-free dry wgt m^{-2} for deposit feeders and 3.1 to 15.2 for suspension feeders. It appears that the highest values are found among the suspension-feeding species, especially those that live in dense aggregations. Dare (1976) gave a series of biomass values for the mussel (*Mytilus edulis*) beds in Great Britain with maxima ranging from 400 to 1,420 g (dry flesh) m^{-2}. Walne (1972) found 970 g (ash-free dry wgt) m^{-2} for the bivalves *Ostrea edulis* and *Mercenaria mercenaria* in England, while Bahr (1976) found 970 g (ash-free dry wgt) m^{-2} for the oyster *Crassostrea virginica* in Georgia, and Dame et al. (1977) recorded 165 g m^{-2} for the same species. Higher biomasses generally occur within estuaries and on mudflats.

Systems with high values are those with a high biomass of suspension-feeding bivalves. Ecosystems with high values have high levels of available food (phytoplankton, benthic microalgae and detritus) and often a preponderance of opportunistic species characterized by high growth rates and a rapid turnover.

Warwick and Price (1975) measured the biomass and production of the macrofauna on a mudflat in the Lynher estuary, Cornwall, England. They found a mean biomass of 13.24 g m^{-2}. Table 6.13 shows that six species, a polychaete carnivore/deposit feeder, two suspension-feeding bivalves, two deposit-feeding bivalves, and one deposit-feeding polychaete, contributed the bulk of both the biomass and production. Biomass estimates for estuarine invertebrates range from 0.2 g (ash-free) dry wgt m^{-2} for *Hydrobia ulvae* to 40.8 for the bivalve *Scrobicularia plana*. In general, the species with high biomass values are filter-feeding bivalves such as *S. plana, Mytilus edulis*, and *Cerastoderma edule*.

6.5.2 MICRO- AND MEIOFAUNA

Estimates of the production of micro- and meiofauna are usually made by multiplying the standing crop by turnover times, which have been determined from studies of the life

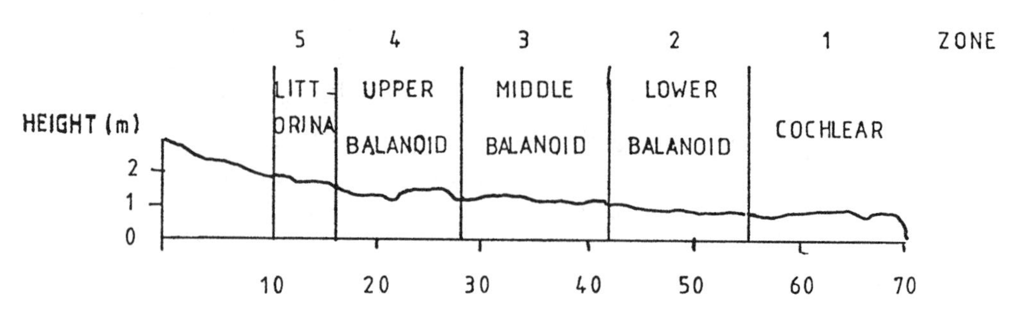

FIGURE 6.15 Transect of the rocky shore site studied by Gibbons and Griffiths at Dalebrook, South Africa, showing the zonation patterns of the biota. Zones are numbered 1 to 5. (Redrawn from Gibbons, M.J. and Griffiths, C.L., *Mar. Biol.*, 93, 182, 1986. With permission.)

cycles of the organisms (Gerlach, 1971). However, there is great variation in the duration of the life cycles, with some meiofauna (e.g., nematodes) reproducing in as little as 5 days (Tietjen, 1980), while others require up to 3 months (Gerlach, 1971). Assuming carbon to be 40% of dry wgt, McIntyre (1964) found that the production of subtidal meiobenthos was between 2.5 and 3.8 g C m^{-2} yr^{-1} with a P:B ratio of 10. On the other hand, Arlt (1973) estimated a P:B ratio of 6 for meiofauna, while Dye and Fustenberg (1981) considered that, on average, life cycles of 44 days, or a turnover of 8 per year is a reasonable estimate for the meiofauna. However, there are many species in which reproduction occurs continuously and asynchronously, and where cohorts (distinct size classes) cannot be separated. In such species it is very difficult to estimate production using the traditional methods.

There are a number of published studies that have estimated meiofaunal harpacticoid copepod production. Using equations generated from P:B ratios, Heip (1976; 1980) derived production values for four benthic copepods of 0.20 to 0.11 g C m^{-2} yr^{-1}. For the harpacticoid *Huntemannia judensis,* Feller (1982) calculated its production in a Washington estuary at 0.7 to 1.7 g C m^{-2} yr^{-1}; while for the smaller *Microarthridion littorale* in South Carolina, Fleeger and Palmer (1982) reported a production of 0.06 g C m^{-2} yr^{-1}. In the Lynher estuary (Warwick and Price, 1979), seven harpacticoid copepod species contributed 5.70 g C m^{-2} yr^{-1} to the total meiofaunal production of 13.5 g C m^{-2} yr^{-1}. Ankar and Elmgren (1976) estimated total meiofaunal production in the shallow water benthos of the Baltic Sea at 2.7 g C m^{-2} yr^{-1}, 20% of which was contributed by harpacticoid copepods. In the Lynher estuary, Warwick and Price (1979) found that the total nematode production was significantly (about 12 times) greater than that of the entire macrofaunal production.

6.5.3 A Comparison of the Macrofaunal and Meiofaunal Standing Stock and Production Across a Rocky Shore

While the relative densities, distribution patterns, and functional roles of the macrofauna and meiofauna on soft shores is reasonably well known, the same cannot be said for rocky shores. In contrast to soft shores, rocky shores are characterized by large amounts of *in situ* primary production. While the physical and biological factors determining the distribution of algae and macrofauna are well documented, there is a paucity of information on the roles of macro- and meiofauna. As a contribution to remedying this situation, Gibbons and Griffiths (1986) determined algal, macrofaunal, and meiofaunal standing stocks and production on an exposed rocky shore in False Bay, South Africa. The shore was subdivided into five major zones (McQuaid, 1980) as shown in Figure 6.15.

The shore supported a rich growth of algae, particularly in the summer, when a maximum standing crop of 403 g m^{-2} was recorded on the low shore (Figure 6.16). Biomass levels were consistently higher in the summer than in the winter. For a 1 m^2 transect across the shore, the overall standing crop increased from 7,754 g to 11,264 g, mainly as a result of large increases in zones 1 and 2. There was a general trend for algal biomass to increase in a downshore direction in both summer and winter, except in the cochlear zone due the the grazing pressure of the limpets.

The distribution patterns of the macrofaunal groups across the shore is depicted in Figure 6.17. In winter, the largest component of the macrofaunal biomass comprised the filter-feeding barnacle *Tetraclita serrata*, which attained 75 g m^{-2} in the middle balanoid zone; but as a result of late recruitment and high mortality of this species, the shore was dominated by herbivorous gastropods, particularly *Patella cochlear,* which attained a maximum biomass of 66 g m^{-2} on the low shore. However, despite a 30% increase in the mean numbers between winter and summer (1,150 to 2,100 m^{-2}), mean macrofaunal biomass remained virtually constant at 40 g m^{-2} throughout the year.

The distribution patterns and composition of the meiofauna across the shore are depicted in Figure 6.18. Numbers and biomass were closely correlated with the distribution of algal turfs and associated trapped sediments. Numerically, the most important components of the meiofauna were nematodes and copepods, while the biomass was more evenly shared among foraminifera, minute gas-

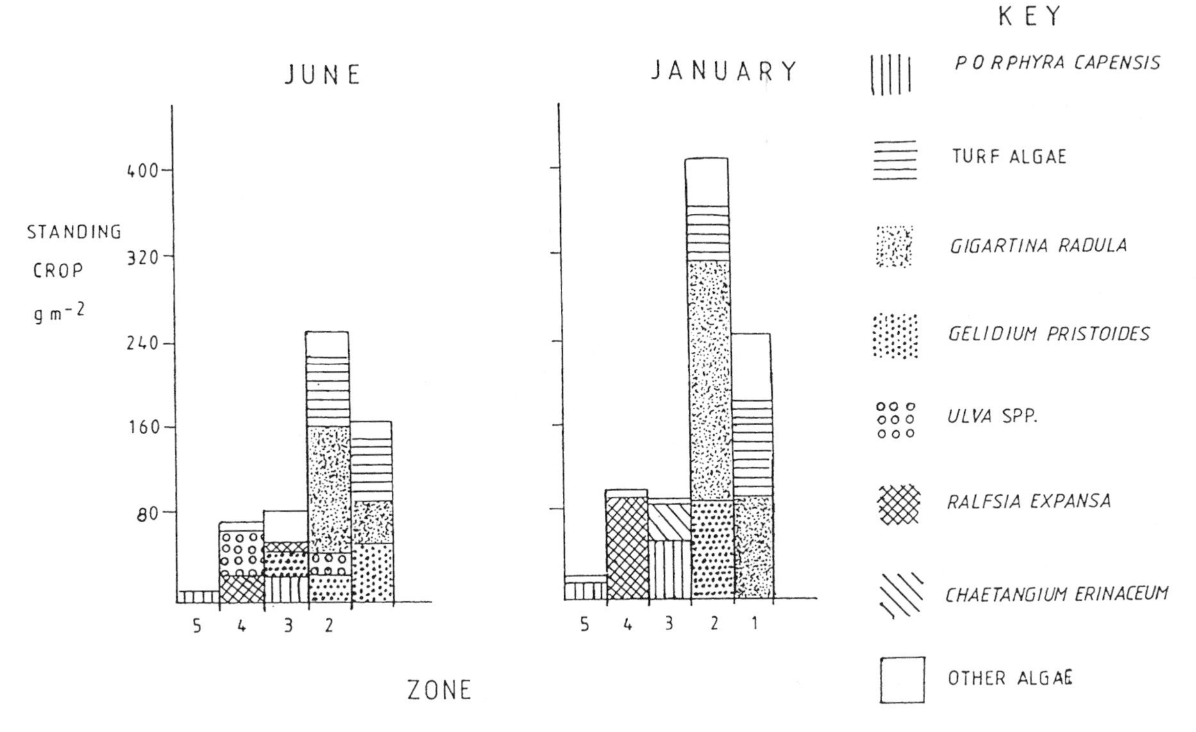

FIGURE 6.16 Distribution of the algal biomass m^{-2} on the rocky shore at Dalebrook, South Africa, during June 1984 and January 1985. Zones are numbered 1 to 5, as in Figure 6.15. Only algae representing a minimum of 5% of the zonal biomass have been included. (Redrawn from Gibbons, M.J. and Griffiths, C.L., *Mar. Biol.*, 93, 183, 1986. With permission.)

tropods, copepods, and insect larvae. Numbers and biomass in the lower balanid zone was greatest during winter (1.9×10^6 individuals or 8.5 g m^{-2}). The meiofaunal densities and biomass figures recorded are comparable to those recorded by McLachlan (1977a,b) on sandy shores. It is thus clear that rocky shore meiofauna are of considerable importance in rocky shore secondary production.

Numerically, meiofauna exceed macrofauna by an overall ratio of 1:391, with values ranging from 1:556 in the lower balanid zone to 1:18 in the *Littorina* zone. Macrofaunal biomass exceeded that of meiofauna by an overall ratio of 10:1, but values range from 2.1:1 in the upper balanoid zone to 48.1:1 in the middle balanoid zone. By incorporating turnover ratios extrapolated from the literature, mean annual productivity values were calculated. These indicate that macrofauna account for 75% of the total secondary production and meiofauna for 25%.

6.5.4 Relative Contributions of the Benthic Macroinfauna, Permanent and Temporary Meiofauna, and Mobile Epifauna to Soft Shore Secondary Production

Moller et al. (1985) investigated the relative contributions of the infauna, including macrofauna and temporary and permanent meiofauna, and the mobile epifauna to secondary production in shallow (0 to 1.5 m) bottom areas on the Swedish West Coast between 1977 and 1982. The areas investigated were grouped into three types of habitats having little or no vegetation: (1) exposed, (2) semiexposed, (3) sheltered, and one habitat where vegetation (*Zostera marina*) dominated.

Infaunal annual production varied between and within habitats depending on temperature, recruitment strength, available space, and predation pressure. Production varied considerably from year to year (12.1 to 69.4 g AFDW m^{-2} yr^{-1} over a six-year period). Epibenthic faunal production was similar within the habitats; highest in vegetated areas (about 6 g AFDW m^{-2} yr^{-1}), followed by semiexposed areas (4 to 5 g AFDW m^{-2} yr^{-1}). In most years (1977, 1978, 1980, and 1981), 51 to 75% of the production of the dominant infaunal prey invertebrates was consumed by epibenthic carnivores. However, in 1979 and 1982, when infaunal production was high, the corresponding figures were 3 and 10%, respectively.

Figure 6.19 is a schematic representation of the annual consumption of the infauna (including temporary meiofauna), permanent meiofauna, "small" epifauna (including mysids), and detritus by mobile epibenthic carnivores. It can be seen that infauna is quantitatively the most important food category for the epibenthic carnivores in exposed, semiexposed, and sheltered areas, whereas the small epifauna dominates as prey in vegetated areas. The most important larger predators are cod *Gadus morhua* and flounder *Platichthys flesus*. In 1978 it was found that

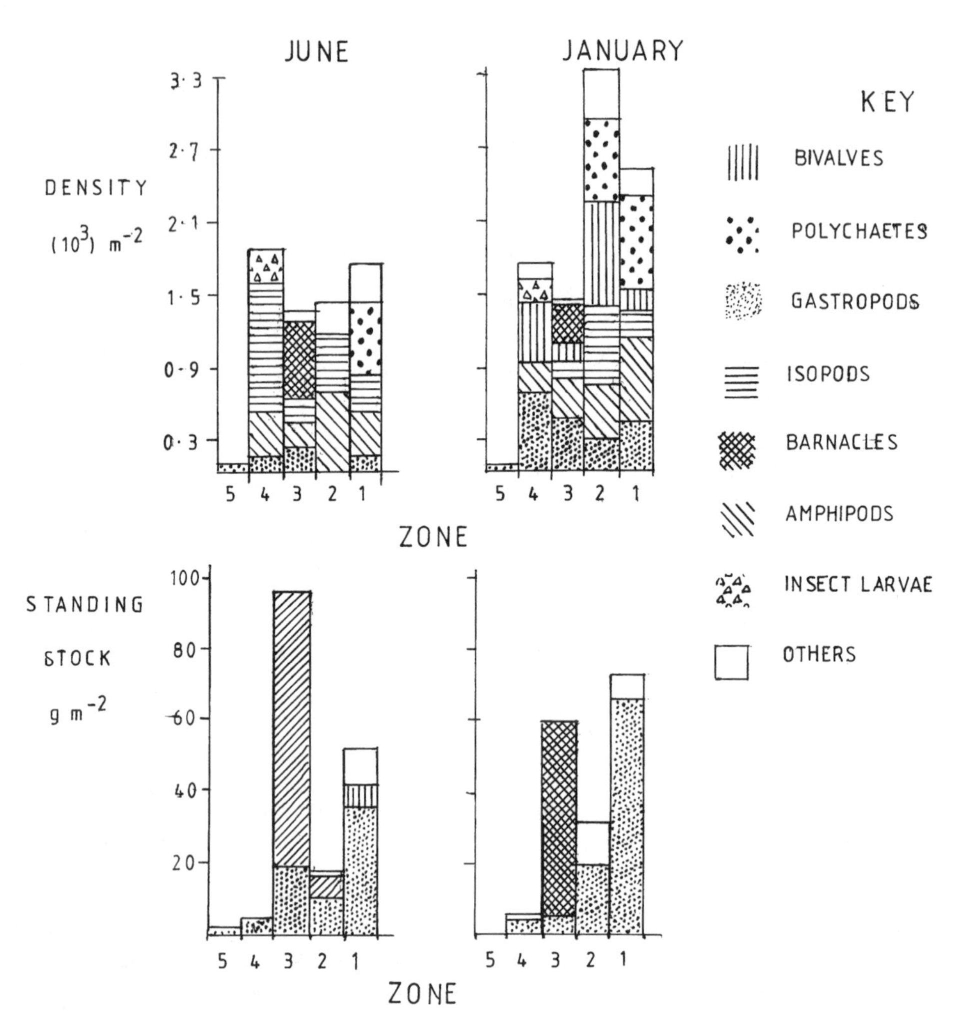

FIGURE 6.17 Mean distribution of macrofaunal numbers and biomass on the rocky shore at Dalebrook, South Africa, in June 1984 and January 1985. Zones are numbered 1 to 5, as in Figure 6.15. Only animals representing a minimum of 5% of the zonal total (density and biomass) have been included. (Redrawn from Gibbons, M.J. and Griffiths, C.L., *Mar. Biol.*, 93, 184, 1986. With permission.)

the predators consumed 98% of the annual production of the amphipod *Corophium volutator*, 92% of that of the bivalve *Cerastoderma edule*, and 12% of that of the bivlave *Mya arenaria* (Phil and Rosenberg, 1984; Phil, 1985).

6.6 P:B RATIOS AND PRODUCTION EFFICIENCY

Production to biomass (P:B) ratios are a measure of the turnover time of the biomass of a species or a community. The higher the ratio, the shorter the turnover time. The ratios are higher for short-lived, short-generation time species and lower for long-lived, long-generation time species. As long-lived species (e.g., some bivalves) grow older, a larger proportion of their energy intake is used in maintenance (respiration) and reproduction, and only a very small proportion is utilized for adding new tissue (biomass).

The P:B ratios for the Lynher River estuary study by Warwick and Price (1975) are given in Table 6.13. P:B ratios for the six most abundant species range from 0.2 to 5.5. The bivalves *Mya arenaria* (0.5) and *Cerastoderma edule* (0.2) had the lowest ratios, while the polychaetes *Nephtys hombergi* (1.9) and *Ampharete acutifrons* (5.5) had the highest. The polychaetes are short-lived species, whereas the bivalves are long-lived. In the same study, the P:B ratio for the total community was approximately 1.0.

Table 6.14 lists the P:B ratios for a number of estuarine benthic communities. These range from 0.6 to 1.6. From these studies and other data it appears that a macrofaunal P:B ratio of approximately 1.0 is a general rule. Warwick et al. (1979) gave additional data on the numbers, biomass, and rate of respiration of the meiofauna in the Lynher River estuary. Using various assumptions and literature values, they calculated that the meiofauna produced 14.7 g C m^{-2} yr^{-1} and that the P:B ratio for the meiofauna was 11.1; with the small annelids included, it was 7.7. There is agreement from other studies (McIntyre, 1969; Gerlach, 1971) that the P:B ratio for the meiofauna is about 10.0.

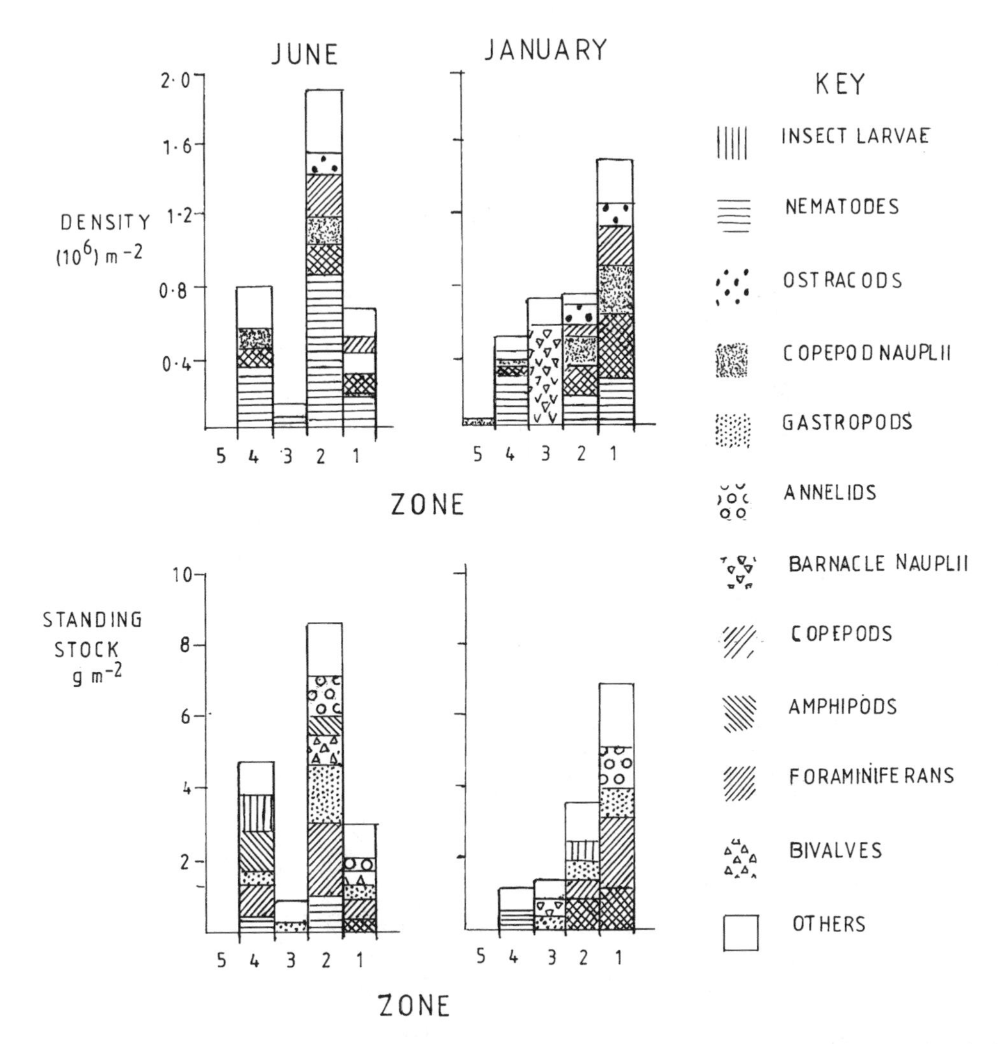

FIGURE 6.18 Mean distribution of the meiofaunal numbers and biomass on the rocky shore at Dalebrook, South Africa, in June 1984 and January 1985. Zones are numbered 1 to 5 as in Figure 6.15. Only animals representing a minimum of 5% of the zonal totals (density and biomass) have been included. (Redrawn from Gibbons, M.J. and Griffiths, C.L., *Mar. Biol.*, 93, 185, 1986. With permission.)

The measurement of secondary production is tedious and time-consuming, and consequently various investigators have tried to develop shortcuts by investigating the rules governing P:B ratios. Robertson (1979) analyzed about 80 estimates of P:B ratios for marine macroinvertebrates. For 49 of these, information was available on the length of life and it was found that the P:B ratio was significantly (r = 0.835) correlated with the life span according to the equation:

$$\log_{10} \text{P:B} = 0.660 - 0.726 \log_{10}$$

From species energy budgets, estimates can be made of production efficiency. The ratio of production/consumption (P/C) gives production as a fraction of the total food consumed. However, it is well known that fecal material is utilized by detritivores, so the ratio of production to energy assimilation (A), which is given by $P/(P + R)$, where $A = P + R$, is probably a more useful and meaningful measure of the ecological efficiency of a population. Table 6.15 lists some values for the production efficiency of marine macrobenthic species. They range from 12 to 63%; for bivalves the range is 21 to 43%, while for gastropods it is 12 to 27%. The efficiency ratios can vary with the type of food eaten, the season, and other variables such as temperature. In general, efficiencies on the order of 20% would appear to be the rule.

6.7 RELATIVE CONTRIBUTIONS OF SOFT SHORE BENTHIC INFAUNA TO SECONDARY PRODUCTION

There are very few studies in which all the components of the benthic community have been considered together when arriving at estimates of benthic secondary production. This would involve the simultaneous measurement

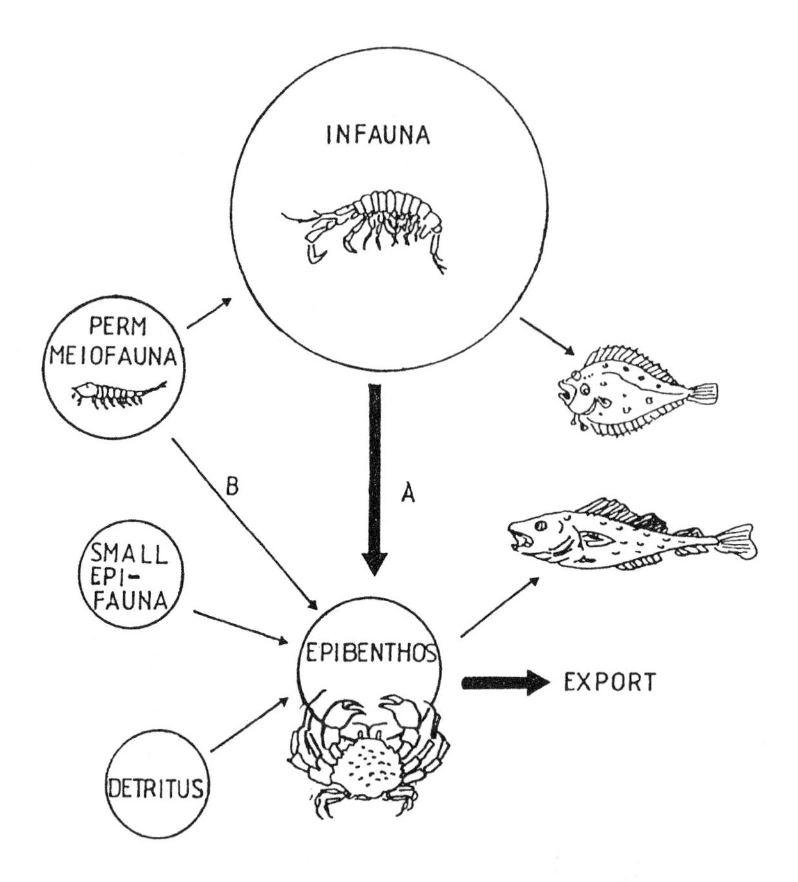

Habitat	Epibenthic consumption				
	Total g AFDW m^{-2} yr^{-1}	Different loops A	B	C	D
Exposed	9	55	10	35	0
Semi-exposed	28	70	10	10	10
Sheltered	7	60	30	0	10
Vegetated	30	15	10	60	15

FIGURE 6.19 Schematic representation of the annual consumption by mobile epibenthic carnivores of the infauna (including temporary meiofauna), permanent meiofauna, "small" epifauna (including mysids), and detritus. Total consumption and relative importance of the different consumption loops are also indicated. The exposed, semiexposed, and sheltered habitats have little or no vegetation. (Redrawn from Moller, P., Phil, L., and Rosenberg, R., *Mar. Ecol. Prog. Ser.*, 27, 115, 1985. With permission.)

TABLE 6.14
Production, Biomass, and Production:Biomass (P:B) Ratios for Total Benthic Communities in Various Localities

Depth (m)	Location	P (g C m^{-2} yr^{-1})	B (g C m^{-2})	*P:B*	Reference
0	Grevelingen estuary, The Netherlands	16.5	10.0	1.6	Wolff and De Wolff (1977)
0	Lynher Estuary, England	5.3	5.3	1.0	Warwick and Price (1975)
17	Severn Estuary, England	10.0	17.0	0.6	Warwick et al. (1974)
0	Upper Waitemata Harbour, New Zealand	10.92	7.15	1.5	Knox (1983a)

Note: Converted from biomass data by assuming that carbon is 40% of ash-free dry wgt.

TABLE 6.15
Some Values for Population Production Efficiency of Macrobenthic Invertebrates

Species	Type	Habitat	Efficiency (%)	Reference
Littorina irrotata	Gastropod	Salt marsh	14	Odum and Smalley (1959)
Geukensia demissa	Bivalve	Salt marsh	30	Kuenzler (1961a)
Scrobicularia plana	Bivalve	Mudflat	21	Hughes (1969)
Neanthes virens[a]	Polychaete	Mudflat	63	Kay and Bradfield (1977)
Mytilus edulis[b]	Bivalve	Estuarine	10.01	Tenore et al. (1973)
Crassostrea virginica[b]	Bivalve	Estuarine	18.38	Tenore et al. (1973)
Mercenaria mercenaria[b]	Bivalve	Estuarine	26.69	Tenore et al. (1973)

Note: Calculated as $P/(P + R) \times 100\%$.

[a] The only carnivore in the list.
[b] Determined in an aquaculture food chain.

TABLE 6.16
Ratios Between Macrofauna, Meiofauna, and Ciliates on the Basis of Numbers, Biomass, and Metabolism at Three Sites in Denmark

Location	Proportions		
	Numbers	Biomass	Metabolism
Alsgarde Beach (fine sand 175 μm)	1:40:1,500	190:1.5:1	2.7:1.1:4
Helsinger Beach (fine sand 200 μm)	1:28:3,000	3.9:1.6:1	1:3:8
Niva Bay (medium sand 350 μm)	1:10:50	170:10:1	4:2:1

Source: From Fenchel, T., *Ophelia*, 6, 1, 1969. With permission.

of the production of bacteria, microalgae, protozoa, meiofauna, and macrofauna. There have been differing opinions as to the importance of meiofauna in energy flow in benthic systems as compared to the macrofauna. McIntyre (1969) concluded that their role was probably an unimportant one, accounting for a rather small proportion of the energy flow in the benthos. He considered that the regeneration of nutrients for use by autotrophs was probably their main function. This view would appear to be supported by data from Dye et al. (1977) for the Swarthops Estuary, South Africa, where of the mean total secondary production of 38.7 g C m^{-2} yr^{-1} on the sand flats, the meiofauna accounted for only 1.2% or 0.46 g C m^{-2} yr^{-1}. In muddy areas, their contribution was smaller, 0.3% of the total production of 88 g C m^{-2} yr^{-1}, or 0.24 g C m^{-2} yr^{-1}.

Fenchel (1969) separately considered the numbers of ciliates (microfauna), meiofauna, and macrofauna in three different soft-bottom habitats. In terms of numbers, the meiofauna outnumbered the macrofauna by factors of 10 to 100 (Table 6.16). However, in terms of biomass, the situation was reversed, with the macrobenthos having 10 to 100 times or more biomass than the meiofauna. Comparisons of the metabolism of the three groups revealed that in spite of the considerable difference in biomass, the macrofauna, meiofauna, and microfauna (mainly ciliates) had roughly equal shares of the metabolism. In reviewing production data from sublittoral silty sands, Gerlach (1971) estimated that the meiobenthos share in the consumption and production of food was 15% that of the macrobenthos which had more than 97% of the total benthic biomass. In a later review, Gerlach (1978) revised these estimates on the basis of more recent data and concluded that the meiofaunal biomass was about 10% of the macrofaunal biomass, but that the production of the meiofauna and the deposit-feeding macrofauna were roughly equal (Table 6.17). Fenchel (1978), however, gives a much higher estimate of the importance of the micro- and meiofauna, 41% of the total community metabolism in the sandy sublittoral and from 69 to 97.5% in estuarine sediments.

Beukema (1976) estimated that seven dominant macrobenthic species on the tidal flats of the Dutch Wadden Sea had a mean density of 242 individuals m^{-2}, with an average biomass of 24.4.g ash-free dry wgt m^{-2}, with the average biomass of one individual being 1 g ash-free dry wgt. The

TABLE 6.17
Estimated Biomass, Turnover, and Production of Various Components of the Infaunal Benthos in a Generalized Silty-Sand Station

Component	Mean Biomass ($g\ m^{-2}$)	Turnover P:B	Annual Production ($g\ m^{-2}$)
Microfauna and bacteria	5.5	19	105
Meiofauna	1.0	10	10
Macrofauna	10	—	—
Subsurface deposit feeders	6	2	12

Source: Mann, K.H., *The Ecology of Coastal Waters: A Systems Approach*, Blackwell Scientific, Oxford, 1982, 188. With permission.

meiofauna, on the other hand, had an average biomass of 1 g ash-free dry wgt m^{-2}, with the nematodes dominating (2×10^6 individuals m^{-2}). The average weight of a nematode is 0.5×10^{-6} g ash-free dry wgt (Kuipers et al., 1981). While the body weights of a macrobenthic invertebrate and a nematode have a ratio of 200,000:1, their individual metabolisms (according to $R = {}_aW^{0.75}$; R = respiration and W = body weight) have a ratio of 9,475:1. Kuipers et al. (1981) state that a weight exponent of 0.75 is universal for a wide range of animal groups. Thus, a nematode has a metabolic rate 21 times that of a macrobenthic invertebrate. According to Kuipers et al. (1981), if the factor *a* on which the ratio depends is left out, it is likely that 1 g of nematodes consumes an amount of organic matter of the same order as that consumed by the total macrofauna.

Warwick et al. (1979) developed an energy budget for the benthic community of a mudflat in the Lynher River Estuary in England. Respiration rates were measured for a range of meiobenthic species and production was inferred from the respiration data following the methods in McNeil and Lawton (1970). The total annual production of the meiofauna, dominated by small polychaetes, nematodes, and copepods, was 20.17 g C m^{-2} yr^{-1}, nearly four times as much as the macrofaunal production of 5.46 g C m^{-2} yr^{-1}. Warwick et al. calculated that 16% of the meiofaunal production was consumed by the sediment macrofauna, leaving 16.83 g C m^{-2} yr^{-1}, or three times the production of the macrofauna, available to mobile carnivores such as fish and birds.

The partitioning of respiration and production among benthic bacteria, micro- and meiofauna, and macrofauna has only been described for localized areas in a few studies (Fenchel, 1969; Gerlach, 1971; 1978; Ankar, 1979; Warwick et al., 1979; Dye, 1981; Asmus, 1982). Data from most of these has been discussed above. The most comprehensive recent study of the partitioning of production and respiration among size groups of organisms in soft sediments is that carried out by Schwinghamer et al. (1986) in Pecks Cove, Bay of Funday, Canada. Published data on the production of natural populations of benthic organisms was used to derive allometric equations relating annual production per unit biomass (P:B ratios) to mean individual body mass (time and body mass weighted) in the population on which production was measured. Separate equations were derived for meiofauna and macrofauna. P:B ratios for bacteria were calculated by extrapolation from an all-inclusive regression. *In situ* respiration was then calculated from production assuming the two to be approximately equal over an annual cycle for bacteria and benthic microalgae, and using an empirical relation between annual respiration and production in benthic meiofauna and macrofauna. Data for mean biomass, production, and P:B ratios, based on monthly observations of benthic biomass spectra and the relationships discussed above, are given in Table 6.18. Calculated values for production by the bacteria were of an expected order of magnitude, if between 1 and 10% of their total biomass was assumed to be active. The contribution of the meiofauna and macrofauna to total community production (8 to 19%) was disproportionately small compared to their relative biomass (17 to 52%). Microalgal production was also high in relation to their biomass, ranging 64 to 25.8%. Estimates of respiration were higher than measured sediment oxygen consumption (2.5 to 5.5 times). These discrepancies may be due to the activities of the microalgae.

Table 6.18 compares estimates of production for Pecks Cove with those of two other northern temperate benthic studies. Apart from the estimates for bacteria, which are dependent on estimates of the percentage of active bacteria, the annual production calculated for Pecks Cove is similar in magnitude to that determined for the other two sites. There is a particularly striking correspondence between production values estimated by Warwick et al. (1979) for the Lynher Estuary and the Pecks Cove estimates. The relatively high P:B ratio for Pecks Cove macrofauna is due to the small average size of the macrofauna at the study site where the amphipod *Corophium volutator* and young bivalves *Macoma balthica* were dominant.

TABLE 6.18
Average Biomass (*B*, kcal m^{-2}) and Annual Production (*P*, kcal m^{-2} yr^{-1}) for Groups of Benthic Organisms in Three Coastal Locations

Organisms		Baltic Sea[a] Asko	English Channel[a] Lynher Estuary	Bay of Funday[c] Pecks Cove
	P	256 (76.2%)	34	147–1,474 (16–66%)[d]
Bacteria	*B*	10 (16.1%)	1	0.05–5 (1–9%)
	P:B	26	34	292
	P	—	1,716	573
Microalgae	*B*	—	99	24
	P:B	—	17	24
	P	26 (7.7%)	245	147 (7–16%)
Meiofauna	*B*	6 (9.7%)	26	12 (22–24%)
	P:B	4.3 (9.7%)	8.4	12
	P	54 (16.1%)	66	14 (1–3%)
Macrofauna	*B*	46 (13.7%)	54	29 (25–28%)
	P:B	1.2	1.2	2.1
	P	336	1,961	896–1 223
Total	*B*	62	180	50–55
	P:B	5.4	9.12	17.9–44.5

[a] Anker (1977).
[b] Warwick et al. (1979).
[c] Schwinghamer et al. (1986).
[d] Metabolically active biomass assumed to be 1 to 10% of total biomass.

Source: From Schwinghamer, P., Hargrave, B., Peer, D., and Hawkins, C.M., *Mar. Ecol. Prog. Ser.*, 31, 138, 1986. With permission.

Asmus (1987) calculated secondary production for each size class of the benthic macrofauna of an intertidal community of an intertidal mussel bed in the northern Wadden Sea. The major food and energy source of predators in a bed of the blue mussel *Mytilus edulis* is composed of juvenile mussels and associated species. This part of the community is characterized by high P:B ratios and a relatively high turnover. In his model of energy flow, these were combined in a "turnover compartment" (see Figure 6.20). The production of the other part of the community consisting of larger mussels is subject to little predator pressure. In addition, larger mussels have low P:B ratios. Consequently, this part of the total production will be stored as biomass and is included in the "storage compartment."

The mean biomass of the community was about 25 times higher (1,243 g ash-free dry wgt m^{-2}) than in other communities in the Wadden Sea. Of this biomass, 97% was represented by the mussel *Mytilus edulis* (Table 6.19), a biomass comparable to that reported by Dare (1976), which are the highest values reported for natural mussel beds. Although the associated fauna had a share of only 3%, their absolute biomass (35 g ash-free dry wgt m^{-2}) was very similar to that reported for other communities in the Dutch Wadden Sea (Asmus, H., 1982; Asmus, H., and Asmus, 1985; Beukema, 1974; 1976; 1981; Beukema et al., 1978). Annual production of the macrofaunal community (468 g ash-free dry wgt m^{-2} yr^{-1}) was dominated by mussels (436 g ash-free dry wgt m^{-2} yr^{-1}). The P:B ratio was very low for the total community (0.36). Annual production of the associated macrofauna was much lower (31 g ash-free dry wgt m^{-2} yr^{-1}), but the P:B ratio (0.89) indicates a much higher turnover rate. The storage compartment in the model (Figure 6.20) utilizes a large part of its energy intake for metabolic and reproductive requirements. This compartment also governs the import and export of nutrients as well as oxygen.

6.8 COMMUNITY METABOLISM

Here we will be concerned with methods of estimation of the metabolism of benthic communities. Because of the methodological problems involved in the estimation of community metabolism on rocky shores, the bulk of the studies to date have been carried out on sandy beaches, estuarine mudflats, and shallow sublittoral benthic areas. Benthic soft-bottom population energetics have been extensively studied, but the number of groups of organisms involved — bacteria, benthic microalgae, fungi, protozoa and other microorganisms, meiofauna,

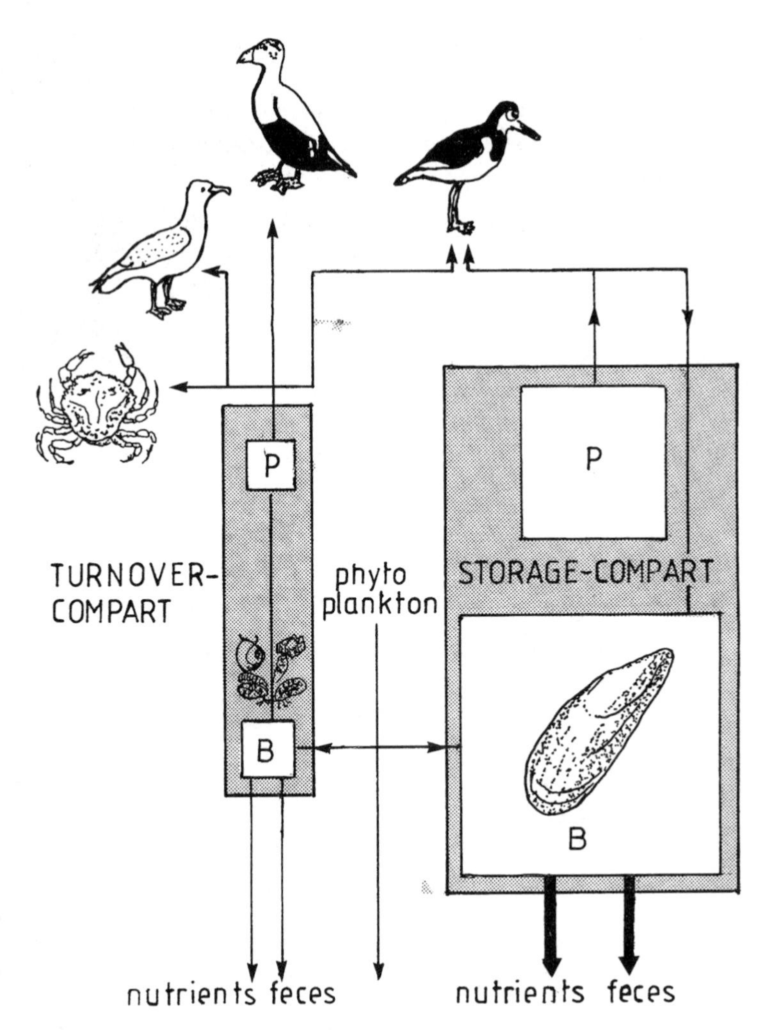

FIGURE 6.20 Simplified model of the partitioning of production of an intertidal mussel bed, showing storage and turnover compartments and their availability to predators. The boundary between these two compartments is drawn at an individual weight of 0.1 g AFDW. B is the mean biomass (turnover compartment: 49.75 g AFDW m^{-2}; storage compartment 1,193.40 g AFDW m^{-2}); P is annual production (turnover compartment: 42.30 g m^{-2} yr^{-1}; storage compartment 425.80 g AFDW m^{-2} yr^{-1}). P:B ratio is 0.85 for the turnover compartment and 0.36 for the storage compartment. (Redrawn from Asmus, H., *Mar. Ecol. Prog. Ser.*, 39, 265, 1987. With permission.)

macroalgae, and the various macrofaunal groups — make an overall synthesis of community energetics difficult. Therefore, holistic approaches to the measurement of community metabolism as a whole have increasingly been used. Benthic community metabolism has been measured primarily in terms of oxygen consumption, *in situ* or using intact cores in the laboratory. Carbon dioxide liberation has also been used.

Oxygen exchange in undisturbed sediments is generally accepted as a measure of autotrophic and heterotrophic (aerobic as well as anaerobic) activity in the sediments (Jorgensen, 1983; Andersen and Hargrave, 1984; Kristensen and Blackburn, 1987). It may also provide the basis for an estimate of primary production and decomposition by the benthic community. The results can be converted to carbon flow by using a conversion factor of 1 (Andersen and Kristensen, 1988). However, as Hargrave (1980) points out, such estimates may be in error for two reasons: (1) if metabolic processes occur, reduced substances are formed that may not immediately be oxidized; and (2) if energy released is not converted into heat but stored as high-energy compounds, actual heat production is less than conversion of the oxygen uptake by usual oxycalorific equivalents.

Table 6.20 shows the results of Pamatmat and Banse's (1969) investigation of oxygen consumption by the seabed in Puget Sound. From the table it can be seen that there is no relationship between oxygen consumption and mean grain size or silt-clay fraction, nor with organic content or organic nitrogen. However, temperature did have an influence with higher temperatures giving higher oxygen consumption. The lack of correlation between oxygen consumption and organic matter seems to be a common phenomenon. In sediments with low organic content but

TABLE 6.19
Mean Biomass and Total Annual Production of the Mussel Bed Compared to Other Intertidal Communities in the Konigshafen Area and Their Trophic Components

Community	Total Annual Mean Biomass (g AFDW m^{-2})	Annual Mean Biomass of Trophic Groups (g AFDW m^{-2}): Grazers	Suspension Feeders	Detritus Feeders	Predators
Nereis-Corophium belt	16.48	10.59	0.16	4.12	1.61
Sea grass bed	30.17	21.20	0.74	7.92	0.31
Arenicola flat	27.55	18.85	0.79	7.38	0.53
Mussel bed	1,243.15 (34.68)[a]	21.22	1,218.49 (9.95)[a]	1.07	2.44
Community	**Total Annual Production (g AFDW m^{-2})**	**Secondary Production of Trophic Groups (g AFDW m^{-2})**			
Nereis-Corophium belt	17.48	13.16	0.01	2.43	1.68
Sea grass bed	48.22	38.85	1.29	7.63	0.45
Arenicola flat	50.21	33.13	1.80	12.74	1.17
Mussel bed	468.09	11.43	429.27 (12.43)[a]	2.25	5.12

[a] Biomass and production values without *Mytilus edulis*.
Source: Data from Asmus (1982; 1987) and Asmus and Asmus (1985).

TABLE 6.20
Relationship Between Oxygen Consumption and Various Environmental Factors on the Seabed of Puget Sound, Washington, U.S.A.

Station	Mean Grain Size (phi)	Percentage Silt-Clay	Organic Matter: Total (% dry wgt)	Organic Matter: Nitrogen (% dry wgt)	Macrofauna (AFDW)	Mean Oxygen Consumption (ml m^{-2} hr^{-1})
4	4.0	36.9	2.78	0.059	10.6	33
5	4.3	42.8	3.92	0.083	31.4	25
6	3.4	14.6	2.45	0.046	4.6	19
7	1.8	70.0	6.87	0.162	19.6	17
10	5.6	72.3	6.45	0.148	9.3	35
14	1.2	6.3	1.43	0.023	6.6	35

Source: From Pamatmat, M.M. and Banse, K., *Limnol. Oceanogr.*, 14, 1204, 1969. With permission.

high rates of oxygen consumption, the organic input must be rapidly consumed so as to prevent accumulation.

Schwinghamer et al. (1991) measured the impact of detritus deposition on soft-bottom oxygen flux and community biomass in laboratory microcosms in order to determine the response of a salt-marsh sediment community to individual and combined effects of *Spartina* detritus burial and constant, artificial illumination. Detritus was added to stimulate the development of a decomposition community within the sediment and the illumination was designed to promote the development of a microalgal photosynthetic community near the sediment surface. Detrital enrichment resulted in significantly increased O_2 flux in both illuminated and dark cores. In contrast, ATP and community biomass were significantly increased only by the combined effects of illumination and detritus. The addition of detritus or illumination acting alone did not result in a significant increase in ATP biomass over the long term. While diatom biomass responded positively to illumination and bacteria and nematode biomass responded positively to detritus addition, microflagellates took advantage of increased food supply in both situations. Multivariate ANOVA indicated that the response of the benthic community was dominated by the growth of non-pigmented, flagellated protozoans in the size range of 1 to 8 μm. Heterotrophic microflagellates are well documented as bacteriovores in the water column (Sherr and Sherr, 1984; Caron, 1987; Caron et al., 1988; Sibbald and Albright, 1988). It would appear that they play a similar important role in sediment systems. Anderson and Har-

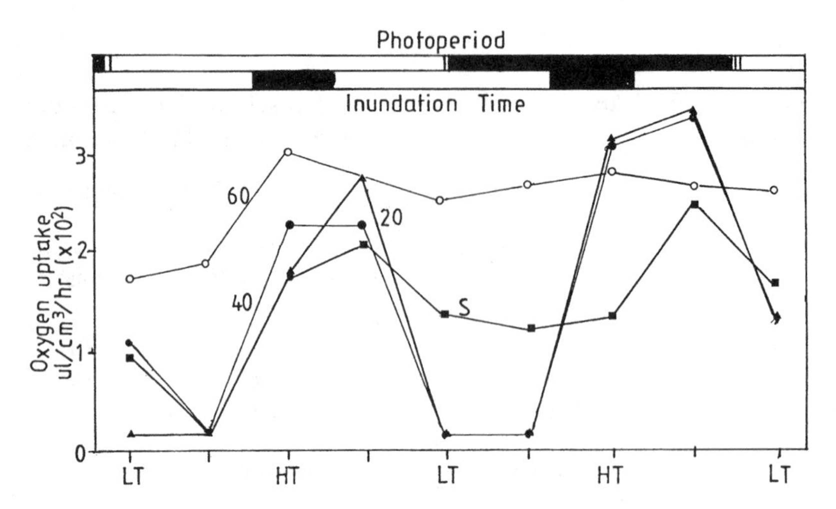

FIGURE 6.21 Fluctuations in intertidal oxygen demand as a function of tidal inundation on an exposed sandy beach. S = surface; 20 cm, 40 cm, and 60 cm are depths below the sand surface. (Redrawn from Dye, A.H., in *Sandy Beaches as Ecosystems,* McLachlan, A. and Erasmus, T., Eds., Dr. W. Junk Publishers, The Hague, 1983b, 696, With permission.)

grave (1984), Kepkay and Anderson (1985), and Schwinghamer and Kepkay (1987) have all found that detrital burial increases the flux of O_2 and CO_2 at the rate of 7% of the original material day^{-1} over periods of up to 105 days. Given that extensive microalgal blooms can form on the surface of the sediment (Hargrave et al., 1983), it is important to combine studies of *Spartina* litter decomposition and microalgal production.

Dye (1983b) reviewed studies of sandy beach community metabolism prior to 1983. He cautions on applying the results from experimental studies in the laboratory directly to sandy beaches in the field, as many of the laboratory studies have not adequately simulated the natural conditions, particularly with respect to water flow characteristics, tidal rise and fall, and wave action. Figure 6.21 depicts the fluctuations occurring at various depths on an exposed sandy beach as a function of tidal inundation. Dye (1981) found a positive correlation between oxygen consumption and water saturation above 30%. The greatest effects were found near the surface of the sand at the upper tidal levels. Maximum oxygen consumption at all levels within the sediment coincided with periods of inundation.

Table 6.21 summarizes studies carried out by Dye (1979; 1980; 1981; 1983a,b) on South African exposed sandy beaches with those obtained by Smith et al. (1972). Total O_2 uptake was highest on the South African sheltered beach. In addition, there were significant differences in the proportion of the total uptake contributed by the various taxonomic groups. Macrofaunal uptake on the sheltered beach was only 3.9% of the total, compared to 59.2% on the exposed beach. Bacterial uptake percentage on the sheltered beach (60.2%) was nearly three times that of the exposed beach (22,2%). Meiofaunal uptake percentage on the sheltered beach (37.7%) was about double that of the exposed beach (18.5%).

TABLE 6.21
Comparison of Community Respiration Studies from Two South African Beaches (Core Measurements) with Those from Similar (Sublittoral) Sediments in Bermuda and Georgia (Bell-Jars)

Area	Temperature (°C)	Total O_2 Uptake (ml m^{-2} hr^{-1})	Macrofaunal Uptake (ml m^{-2} hr^{-1})	Bacterial Uptake (ml m^{-2} hr^{-1})	Meiofaunal Uptake (ml m^{-2} hr^{-1})	Reference
Sheltered beach	20	17.6	0.7 (3.9%)	10.6 (60.2%)	6.3 (35.7%)	Dye (1981)
Exposed beach	20	13.5	8.0 (59.2%)	3.0 (22.2%)	2.5 (18.6%)	Dye (1981)
Sublittoral[a]	23	25.8	0.6 (2.3%)	7.8 (30.2%)	17.4 (67.4%)	Smith et al. (1973)
Sublittoral[b]	15–20	81.41[c]	6.2 (7.6%)	49.1 (60.3%)	26.1 (32.1%)	Smith et al. (1972)

[a] Bermuda.
[b] Georgia.
[c] 4% silt/clay.

Source: Dye, A.H., in *Sandy Beaches as Ecosystems,* McLachlan, A. and Erasmus, T., Eds., Dr. W. Junk Publishers, The Hague, 1983b, 695. With permission.

The percentage of total oxygen uptake that can be attributed to the macrofauna is generally surprisingly small and is usually less than 29% (Banse et al., 1971; Smith et al., 1972). In an *in situ* study of oxygen consumption on an intertidal and flat in Washington, Pamatmat (1968) found that the microflora contributed 47%, the bacteria/microflora plus meiofauna 33%, and the macrofauna 20%. This contrasts with a study of oxygen uptake in a coral sand community (Smith et al., 1972) where the microflora plus microfauna contributed over 60%, the bacteria about 37%, and the macrofauna only about 2.5%.

As discussed in Section 3.8.6.1, a large proportion of the energy flow in anaerobic sediments is involved in sulfate reduction. Thus, because of sulfate reduction and other anaerobic metabolic processes, carbon cycling is not proportional to energy flow, and this means that the "conventional use of benthic O_2 uptake and ideal RQ values (e.g., 0.85) sweeps an important category of energy under the conceptual rug" (Rich, 1979). There will thus be considerable error in the use of oxygen uptake measurements as an energy flow in marsh, estuarine, and mudflat sediments. Only in sandy sediments with strong water flow through the sediment, where the oxygenated layer may extend to 30 cm or more, does the measurement of oxygen uptake yield a measure of benthic community metabolism.

6.9 TROPHIC STRUCTURE AND FOOD WEBS

6.9.1 Introduction

The relative densities, distribution patterns, and functional roles of sediment microalgae, bacteria, microfauna (principally protozoans), meiofauna, and macrofauna of soft sediments are now relatively well known. Bacteria, depending on the exposure to wave action, the sediment grain size distribution, and the input of organic detritus (POM), often occur at very high densities (Dale, 1974; Meyer-Reil et al., 1974; Rublee and Dornseif, 1978; Schwinghamer et al., 1986). They are now considered to be responsible for the bulk of the secondary production (Koop and Griffiths, 1982), essentially fueling the interstitial food chain (McIntyre and Murison, 1973; Gerlach, 1978). The meiofauna is largely responsible for maintaining the bacteria in a continued state of growth by grazing and nutrient cycling (McIntyre, 1969). The meiofaunal commumnity is typically diverse, dense, and relatively stable (Swedmark, 1964; McLachlan, 1983). The macrofaunal community characteristically shows poor species diversity on sandy beaches but greater diversity on mudflats. Population densities are highly variable. In terms of the numbers of individuals, the meiofauna are always more abundant than the macrofauna, while the macrofauna generally dominate the meiofauna in terms of biomass. Nevertheless, as a result of the faster turnover rates of the meiofauna (Gerlach, 1978), they can be as important in terms of secondary production as the macrofauna and in many instances be considerably more productive (Koop and Griffiths, 1982; McLachlan, 1983).

Unlike beaches and mudflats, rocky shore ecosystems are characterized by large amounts of *in situ* primary production. The biological and physical factors determining algal and macrofaunal distribution and community structure are reasonably well known (e.g., Menge and Sutherland, 1976; Newell, 1979; Norton, 1985). In spite of this large body of information, our understanding of the dynamics of rocky shore ecosystems, especially in relation to energy flow is inadequate. In particular, data on the meiofaunal and bacterial components and their role in energy flow are severely lacking (Gibbons and Griffiths, 1986). Rocky shore meiofauna occupy a wide range of habitats, including algae, rock crevices, sessile animals, and entrapped sediments. They can attain considerable densities in the phytal (algal) component, and it is now known that algal morphology, age, condition, distribution, and entrapped sediment determine their abundance, distribution, and composition (Mukai, 1971; Hicks, 1980; 1986; Gunhill, 1982; 1983; Edgar, 1983a,b,c). Estimates of diversity vary from alga to alga. The patterns of distribution and densities of the macrofauna on rocky shores has already been considered in Chapter 2.

6.9.2 Hard Shores

6.9.2.1 Coastal Water–Rocky Shore Interactions

Rocky shores are the interface between the coastal marine and the terrestrial environments. The offshore waters provide nutrients and food (bacteria, phytoplankton, zooplankton, and detritus) for the shore flora and fauna, and a habitat for the larval stages, which are the source of potential settlers. In turn, death and decay of the shore plants and animals provide organic matter and nutrients to the nearshore waters. The major contributors are the large kelps that are a feature of the lower eulittoral, sublittoral fringe, and the upper sublittoral zones on temperate shores.

After settlement on the shore, the early stages of the plants and animals are subject to a range of physical and biotic influences that determine recruitment, i.e., the portion of the settlers that survive to be recruited into the population. Tidal rise and fall acts in conjunction with temperature, wind, and wave action to subject the settlers to desiccation stress, which results in the elimination of those that settle high on the shore above the zone occupied by adults.

Water movements, such as waves and currents, are important physical factors that play roles in determining the distribution of the plants and animals on the shore. As we have seen, wave action also plays a role in deteremiming the size to which individual species can grow in a a given locality. Wave shock can modify the feeding activities of predators, while swash and spray up the shore enables some species to extend their ranges higher up the shore. As detailed in Section 2.3.1.3, wave action determines the degree to which plants and animals on the lower shore are buried by sand, as well as the amount of sand abrasion that impacts on algae and invertebrates.

Water currents transport larval stages to coastal waters from the shore and, in conjunction with wave action and tidal rise and fall, bring the settlement stages to suitable settlement sites. They thus play an important role in settlement.

A final interaction is that of mobile predators, especially fishes, that feed on the shore when the tide rises. Algal-consuming fishes also play a dominant role on many shores, especially tropical shores. Apart from settlement and recruitment studies, the coastal water–rocky shore interaction discussed above has received little emphasis in models of rocky shore communities.

6.9.2.2 Hard Shore Ecosystem Models

There are a limited number of investigations that have studied food webs and energy flow on rocky shores. Exceptions are those of Taylor (1984), Gibbons and Griffiths (1986), Field and Griffiths (1991), and Hawkins et al. (1992).

6.9.2.2.1 Northeastern Atlantic shores

Hawkins et al. (1992) have recently summarized data on plant-animal interactions on northeastern Atlantic shores with special reference to the British Isles. The best-studied communities are those where there is a relatively simple set of plant-animal interactions. The primary production consists of the water column phytoplankton (consumed by filter-feeding barnacles) and two *in situ* components, the microbial film on the rock surface and barnacle tests, and a macroalgal population dominated by fucoids. On British shores, these are grazed respectively by the limpet *Patella* and the winkle *Littorina*, other grazers playing very minor roles. Hartnoll (1983) proposed a tentative energy budget for this community, which has been upgraded by Hawkins et al. (1992).

The budget for *Patella vulgata* (Wright and Hartnoll, 1981) was modified in the light of recent measurements of mucus production (Davies et al., 1990), which turned out to be much greater than previously thought. The revised budget is:

$$C = P_g + P_r + R + F + U + M$$

$$>3{,}082 = 68 + 96 + 498 + 1{,}698 + 2 + >720$$

This and the budgets for *Littorina obtusata* and *Nucella lapillus* are listed in Table 6.22. The energy budget for

TABLE 6.22
Estimated Values for Biomass, Consumption, and Production on Exposed and Very Sheltered Shores on U.K. Coasts

	Exposed			Very Sheltered		
	B	C	P	B	C	P
Algal film	700		4,900	430		3,000
Fucoids	600		820	45,000		62,000
Patella	430	3,800	200	170	1,500	80
Semibalanus	3,450	23,000	5,250	215	1,440	330
Littorina	0	0	0	640	6,240	1,470
Nucella	24	120	25	70	370	75

Note: All energy values in kJ m^{-2} $year^{-1}$; biomass in kJ m^{-1}.

Source: From Hawkins, S.J., Hartnoll, R.G., Kain, J.M., and Norton, T.A., in *Plant-Animal Interactions in the Marine Benthos,* John, D.M., Hawkins, S.J., and Price, J.H., Eds., Systematics Assoc. Spec. Publ., Vol. No. 46, Clarendon Press, Oxford, 1992, 6. With permission.

Semibalanus balanoides is based on a number of gross assumptions. There is a dearth of relevant information on microalgal production. Values in the literature range from 6,250 to 17,588 kj m^{-2} yr^{-1} . The energy budget in Table 6.22 is for a semiexposed shore on the Isle of Man. The table lists derived estimates for the energy flow on exposed and very sheltered shores. These estimates suggest that the exposed shores are net consumers of energy by more than 20,000 kJ m^{-2} yr^{-1}, and that the sheltered shore has, by contrast, a very high net production >55,000 kJ m^{-2} yr^{-1}. Thus, on English rocky shores, the balance between algal production and consumption by primary consumers varies within wide limits. At the mid-tide level, exposed shores are net consumers and sheltered shores net producers; shores of intermediate wave action are near trophic balance.

6.9.2.2.2 South African littoral and sublittoral ecosystems

Field and Griffiths (1991) have recently reviewed information on food webs and energy flow on the rocky shores of South Africa. Biomass data collected by McQuaid and Branch (1985) at 12 sites around Cape Peninsula and by McLachlan et al. (1981) at two sites on the south coast show a relationship between biomass and trophic level at both exposed and sheltered shores (Figure 6.22). At sheltered sites there is a semi-logarithmic relationship (r_i^2 = 0.65, n = 28) between biomass and trophic level (with filter feeders nominally at level 3). At exposed sites the biomass of algae, herbivores, and carnivores is much the same, but there is a marked increase in the filter-feeding biomass. The South Coast sheltered site has biomass values in the range of Cape Peninsula sites, but the exposed site has the lowest biomass values for all trophic groups except grazers. This is confirmation of the conclusion of

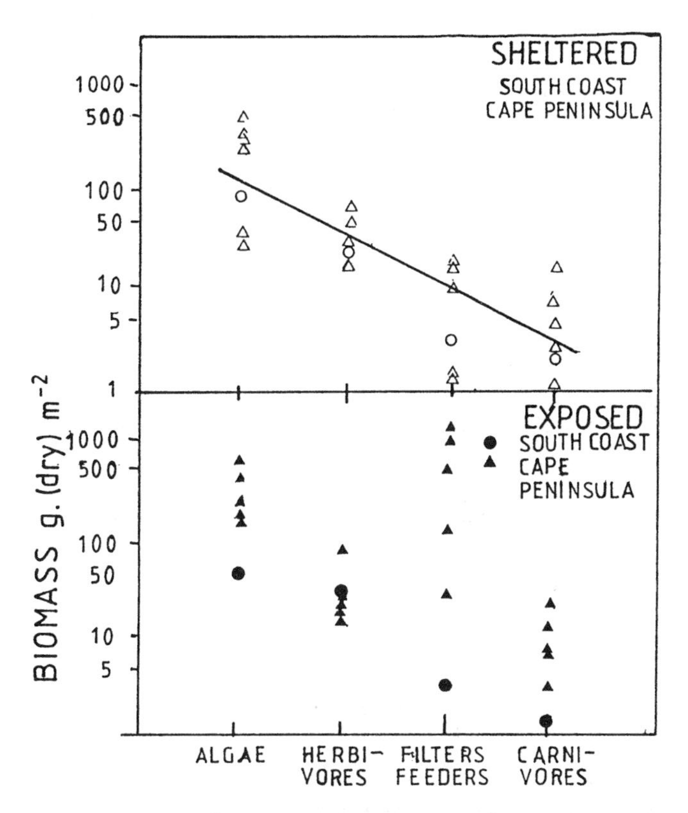

FIGURE 6.22 Biomass of trophic components in the rocky eulittoral, at sites exposed to and sheltered from wave action along the South Coast and the Cape Peninsula, South Africa. At sheltered sites there is a semilogarithmic relationship between dry biomass and trophic level, with filter feeders at nominal trophic level 3 (r^2 = 0.65, n = 28). At exposed sites, filter-feeder biomass is much increased, whereas other trophic categories have similar biomasses to sheltered sites. Data from McLachlan et al. (1981) and McQuaid and Branch (1985). (Redrawn from Field, J.G. and Griffiths, C.L., in *Intertidal and Littoral Systems, Ecosystems of the World* 24, Mathieson, A.E. and Nienhuis, P.H., Eds., Elsevier, Amsterdam, 1991, 336. With permission.)

McQuaid and Branch (1985) that the degree of wave exposure determines the trophic structure of a rocky-shore community, but biogeographic factors determine the species composition of the community.

Figure 6.23 depicts the flows of energy through a hypothetical generalized South African rocky shore littoral system exposed to wave action. The primary producers are the phytoplankton, macroalgae, sporelings, and benthic microalgae. Branch (1981) has shown that the latter provide an important food resource for grazing molluscs. While the biomass of sporelings and benthic microalgae is negligible, their turnover rate is fast and is maintained by the continued grazing and "gardening" activities of the limpets. Their productivity is estimated to be about equivalent to that of the macroalgae (Field, 1983). The utilization of the coastal water phytoplankton by the rocky-shore filter feeders is dependent on the rate at which it is transported to them by tide, waves, and currents. Thus, the "work-gates" in the flow diagram depict the importance of water movements in transporting phytoplankton and detritus, which is augmented by the resuspension of animal species. This rate of transport has not yet been measured.

The rate of nutrient cycling can roughly be estimated if it is assumed that all filter feeders excrete ammonium at the same rate as mussels (5 to 40 μg NH_4-N g^{-1} hr^{-1}; Bayne and Widdows, 1978; Kautsky and Wallentinus, 1980). Based on this assumption, the mean value for nitrogen released by all animals should support primary production of about 2,220 g C m^{-2} yr^{-1}, the estimated production of small littoral algae in the Cape Peninsula (Field, 1983).

On the South African East Coast, energy flow has been studied on a shallow reef ecosystem near Durban. Figure 6.24 shows the estimated biomasses of the principal components and some of the main energy flows (Berry et al., 1979; Berry, 1982; Schleyer, 1986). Accumulations of detached macroalgae and terrestrial plant material carried by currents and flooding rivers are characteristic of the area. High populations of bacteria and high indices of bacterial activity, based on the uptake of radio-labeled glucose and labeled algal exudate lead to the conclusion that the animals depend heavily on detrital input, since phytoplankton primary productivity in the area is low. Macroalgal cover and growth are intermittent. The main grazers of the macroalgae are fish, including the omnivorous blenny *Blennius cristatus*. The fish fauna is rich and diverse, with 62 species

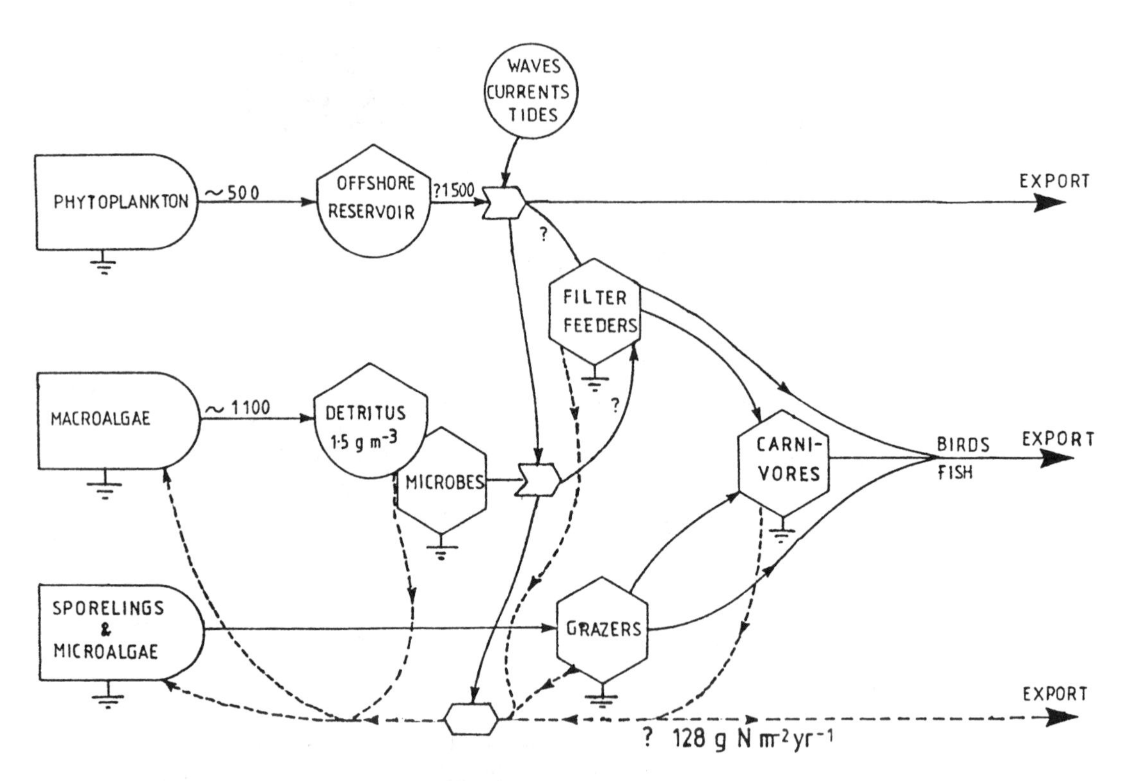

FIGURE 6.23 Carbon and nitrogen flow through a hypothetical South African rocky shore ecosystem exposed to wave action. Carbon flows are g C m^{-2} yr^{-1}. The influence of external energy sources (tides, waves, and currents) is depicted by uni- or bidirectional workgate symbols, indicating their control in advecting and suspending phytoplankton, detritus, and nutrients to filter feeders or out of the system. Excretions of heterotrophs are indicated by dashed arrows leading down to the macro- and microalgae, or export from the system (g N m^{-2} yr^{-1}). (Redrawn after Field, J.G., in *Marine Ecology*, Vol. 5, Kinne, O., Ed., Wiley Interscience, Chichester, 1983, 781. With permission.)

having been recorded. Herbivores are poorly represented, and omnivores and carnivores dominate.

Suspension feeders dominate the reef biomass. These include the brown mussel *Perna perna*, which has a large and variable biomass, fast growth, and annual P:B ratio of 4.8 (Berry, 1982). Other major suspension feeders are the ascidian *Pyura stolonifera* and the oyster *Saccostrea margaretacea*, both of which have a P:B ratio of 3.54 and production of some 2.4 g dry mass m^{-2} yr^{-1}. Carnivores include the spiny lobster *Panulinus homarus* and fish such as the small blenny *Blennius cornutus* and the rock cod *Epinephelus andersoni*. None of the top carnivores are permanent residents, but move offshore as they grow larger, or migrate periodically (e.g., the spiny lobster) or seasonally (e.g., the Cape coromant *Phalacrocorax capensis*). Thus, there is an export of energy at the highest trophic level, and import of detrital material and plankton to support the filter feeders. The system is thus a very open one, with physical forces such as waves tides, currents, and flooding rivers playing an important role.

Figure 6.25 summarizes the factors that are thought to control the structure of the reef community (Berry, 1982). Although the ascidian *Pyura stolonifera* and the mussel *Perna perna* may coexist for short periods, heavy settlement of *P. perna* occurs occasionally, smothering *P. stolonifera* with the result that mussels dominate the community. Normally, however, mussel spat settles on established mussel beds and does not replace *P. stolonifera*. After disturbance by wave or sand, the primary colonizers are algae, principally *Hypnea sicifera, Ulva* sp., *Enteromorpha* sp., and corallines such as *Cheilosporum cultratum* (Jackson, 1976).

An energy flow model has been developed for the kelp beds that dominate the rocky subtidal areas on the west coast of South Africa, which are bathed by the cold Benguela Current. The structure and biomass of these kelp beds have been documented by Field et al. (1980). Cluster analysis shows that there are two principal types of biotic association within the kelp beds: areas dominated by macroalgae (including kelp) and areas dominated by animals. These tend to form a mosaic, with inshore areas dominated by kelps and an understorey of algae to 5 to 10 m depth, and dense patches of animals populate the rock surfaces between the kelp in deeper water. There are three species of kelps: *Ecklonia maxima*, reaching a length of 10 m; *Laminaria pallida*, forming a subcanopy 1 to 2 m off the bottom; and, in calmer protected waters, the rarer *Macrocystis angustifolia*, which can grow in deep water and

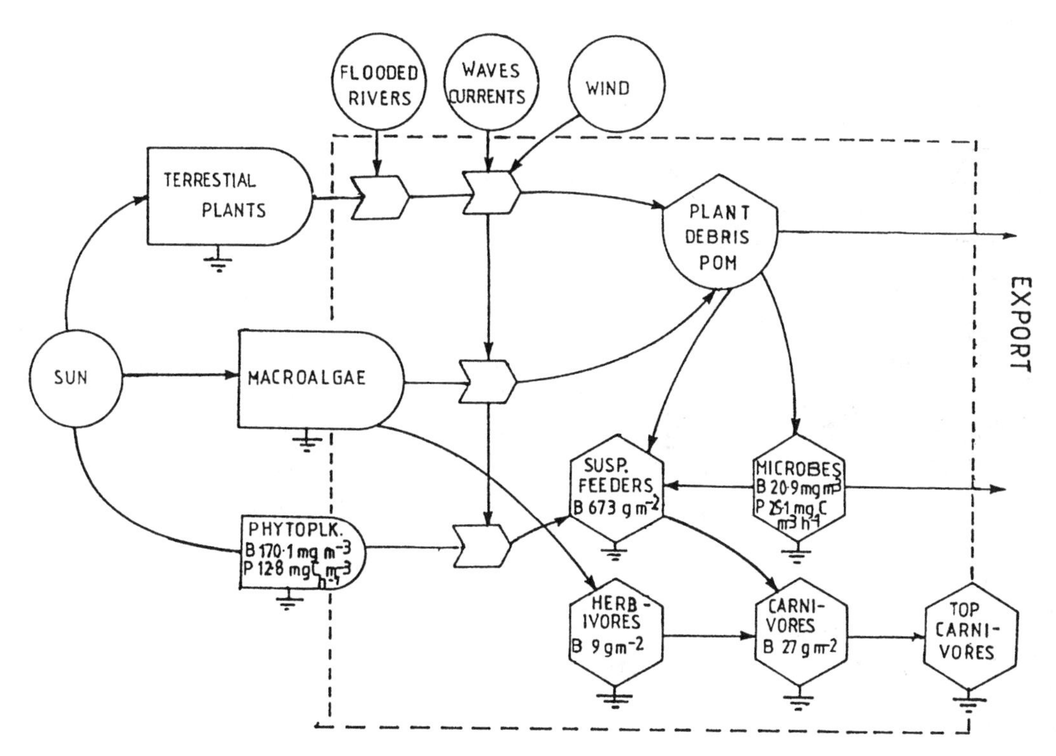

FIGURE 6.24 Carbon flows through a sublittoral rocky reef community at Durban on the South African East Coast. B = biomass; P = production (annual unless indicated otherwise). Microbial production estimates are based on radio-labeled algal exudate (Schleyer, 1980). (Redrawn from Field, J.G. and Griffiths, C.L., in *Intertidal and Littoral Systems, Ecosystems of the World* 24, Mathieson, A.E. and Nienhuis, P.H., Eds., Elsevier, Amsterdam, 1991, 353. With permission.)

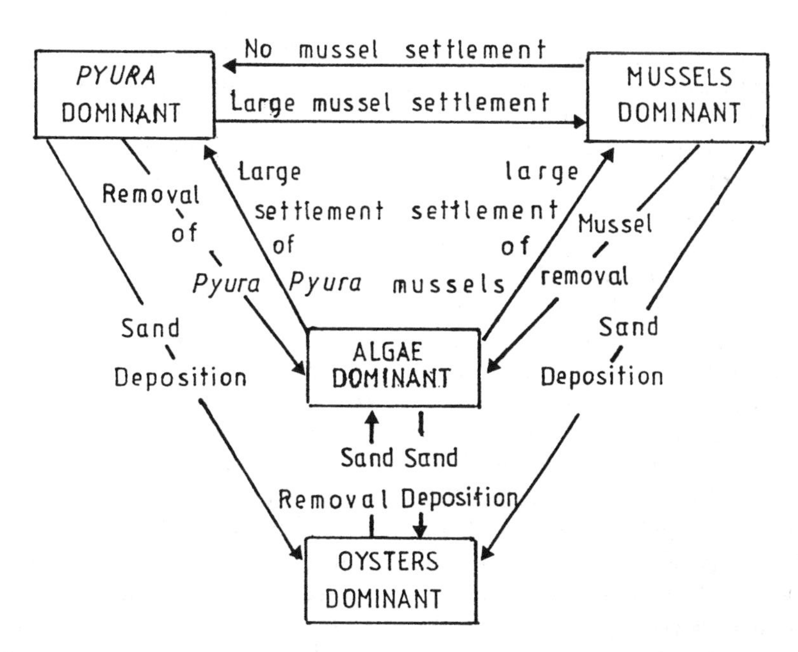

FIGURE 6.25 Schematic representation of the factors influencing community structure of a subtropical rocky sublittoral reef at Durban on the South African East Coast. The four species that may dominate the community are represented in the boxes, and the factors influencing the transition from one dominant to another are represented on the arrows. (Redrawn from Berry, P.F., *Invest. Rep. Oceanogr. Res. Instit.* Durban, S. Af., 53, 1, 1982. With permission.)

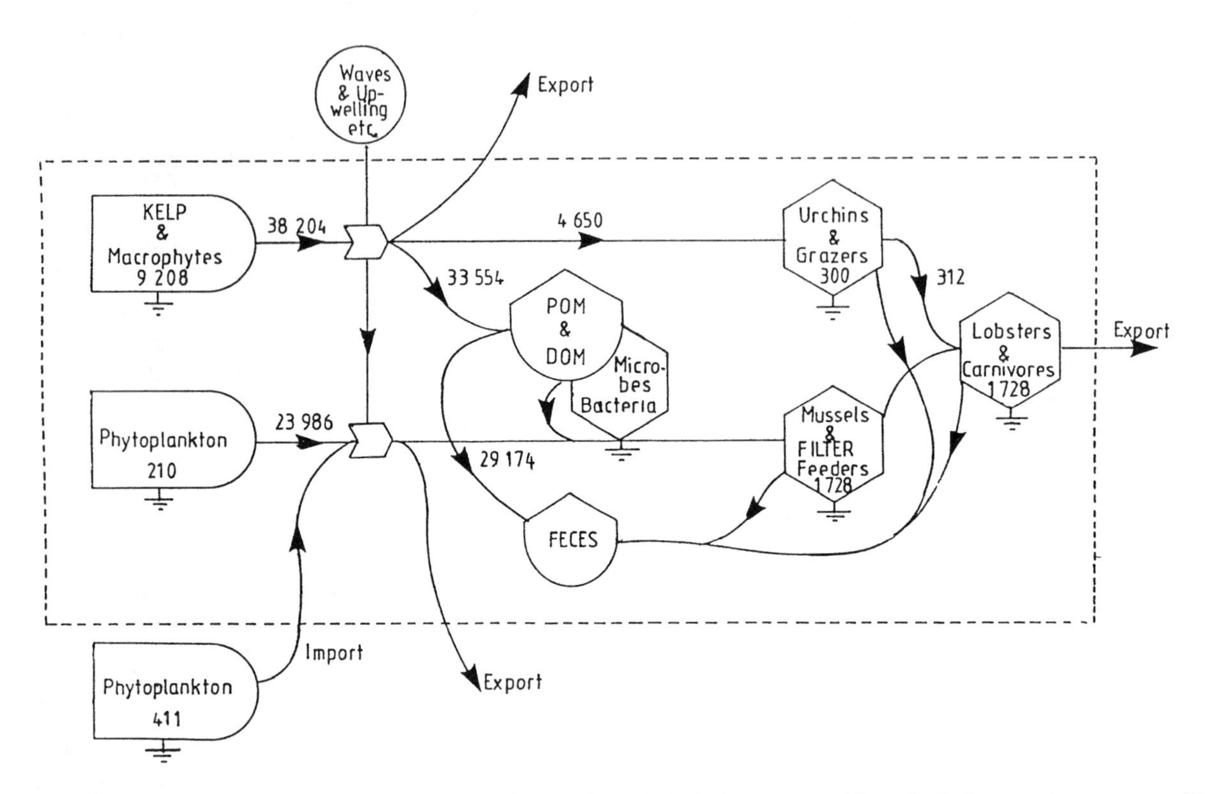

FIGURE 6.26 Energy flow in a Bengula (West Coast South Africa) kelp bed ecosystem. Note the influence of waves, upwelling water movements, and other physical factors in controlling fragmentation of macrophytes, export of suspended matter, and import of phytoplankton. Standing stocks are in kJ m^{-2}, and average energy flows in kJ m^{-2} yr^{-1}. The kelp bed is taken as having a mean depth of 10 m, and the boundaries of the system are represented as dashed lines. The "feces loop" of animal feces is depicted as contributing substantially to the particulate and dissolved organic matter pools (POM and DOM, respectively). The values are from Newell et al. (1982). (Modified from Field, J.G. and Griffiths, C.L., in *Intertidal and Littoral Systems, Ecosystems of the World*, 24, Mathieson, A.E. and Nienhuis, P.H., Eds., Elsevier, Amsterdam, 1991, 335. With permission.)

extend to the surface. Under the kelps there is an understory of red and brown algae such as *Botryocarpa prolifera, Epymenia obtusa, Gigartina radula,* and *Hymenena venosa* (Field et al., 1980). The animals are dominated by suspension feeders, especially the ribbed mussel *Aulacomya ater,* sponges such as *Polymastia mamillaris* and *Tethya* spp., holothurians *Pentacta doliolum* and *Thyone aurea*, and the ascidian *Pyura stolonifera.* The barnacle *Notomegabalans algicola* may be common, as well as grazers, including the limpet *Patella compressa,* the gastropod *Turbo cidaris,* the abalone *Haliotis midae,* and the sea urchin *Parechinus angulosus.*

Energy flow in the West-Coast kelp beds has been studied by Newell and Field (1983) and Newell et al. (1983). Figure 6.26 incorporates their data for energy balance, in which the kelp bed system is depicted within a broken line border. Kelps, understory algae, and epiphytes with a standing stock of 9,208 kJ m^{-2} are estimated to produce an average of some 38,000 kJ m^{-2} yr^{-1}. The kelp fronds behave as moving belts of tissue, growing at the bases and eroding at the tips due to strong wave action. The eroded particles form detritus, depicted as particulate organic matter (POM), dissolved organic matter (DOM), and the community of bacteria and other microorganisms that colonize the organic matter. Wave action also breaks free whole plants and large pieces of drift algae, which are fed on by herbivores such as sea urchins and abalones; but their biomass is relatively small (300 kJ m^{-2}) and their food consumption is a modest 4,650 kJ m^{-2} yr^{-1}. Tidal and upwelling currents transport quantities of drift algae out of the system. The remaining macrophyte production (33,554 kJ m^{-2} yr^{-1}) forms detritus that may be kept in suspension by wave action and consumed by suspension feeders such as mussels, barnacles, and holothurians, totaling 5,150 kJ m^{-2} yr^{-1}, or 72% of the animal biomass. Another component of the diet of the suspension feeders is the phytoplankton, which has an annual estimated productivity of 23,986 kJ m^{-2} yr^{-1} inside the kelp bed. This is available mainly under conditions of calm winds or downwelling conditions, when offshore phytoplankton is carried into the kelp beds. Under upwelling conditions, both detritus and phytoplankton are transported out of the kelp beds. Grazers and filter feeders are preyed upon by carnivores of different sizes, pooled in the diagram into one compartment with a biomass of 1,728 kJ m^{-2}. The production of lobsters and other carnivores is consumed by various top predators, including humans, seals, and seabirds, and is depicted by the arrow out of the system. Feces from the animal population adds to the detrital pool, depicted by arrows leading back to the POM/DOM compartments. This organic matter is colonized by bacteria and microfauna and thus enriched in terms of nitrogen and protein (Newell et al., 1983; Newell and Field, 1983).

Simulation models were developed in which a realistic but simplified kelp bed food web was modeled with ranges of electivity and assimilation efficiencies to test the effect of water transport under closed-system, upwelling, and downwelling conditions (Wulff and Field, 1983). A striking result was that the transport of phytoplankton, detritus, and feces by water movements had a greater effect on simulated filter-feeder feeding and growth rates than the full range of biological electivity and assimilation efficiencies. Surprisingly, it was also found that there was a shortage of food under conditions of continuous upwelling, when detritus is exported from the kelp beds, whereas under downwelling conditions, the vast reservoir of phytoplankton from offshore provided a rich source of carbon and nitrogen for the filter feeders. When realistic weekly and seasonal rates of upwelling and downwelling were included in the models, filter feeders fluctuated in scope for growth and reproduction, increasing in the winter downwelling season and decreasing during the upwelling season. These results were confirmed by analyses of the lipid content of the mussels and the fact that their main spawning activity is at the end of winter, after the downwelling season. The study thus emphasized the importance of physical factors such as the upwelling/downwelling patterns in influencing the ecosystems in the sublittoral zone of the Benguela region.

6.9.3 Soft Shores

6.9.3.1 Exposed Beaches

Brown and McLachlan (1990) have recently summarized a large body of information on food webs and energy flow on exposed sandy beaches with special reference to South Africa where the bulk of such research has been carried out. The general patterns of distribution of the primary producers, macrofauna and meiofauna, have already been discussed in Chapter 3.

Macroscopic food webs: The components of the macroscopic food web comprise benthos, zooplankton, fish, and birds. Among the benthos, filter feeders, herbivore scavengers, and carnivorous scavengers/predators are the main groups, although deposit feeders occur on the more sheltered shores. If the sediment is sufficiently stable, the sediment microalgal populations are high, organic detritus abundant, and deposit feeders such as the lugworm *Arenicola*, the burrowing shrimp *Callianassa*, and other polychaetes such as spionids and orbiniids can occur in high densities.

Where the beach slope is not steep, filter feeding molluscs are usually dominant and abundant. Feeding on surf phytoplankton and/or detritus, they represent the main pathways of energy flow through the benthos, and can account for up to 90% of this flow (Ansell et al., 1978; McLachlan et al., 1981). Scavengers and predators seldom attain great abundance. Notable exceptions are beaches with large inputs of drift macroalgae where scavenging herbivores may dominate the fauna.

Where surf zones are present, the benthic macrofauna and zooplankton occupy key positions in sandy beach food chains. The zooplankton of the surf zone is dominated by relatively large species, for example, mysid shrimps, prawns, and bentho-pelagic species. The aggregation of these zooplankters into swarms make them the chief prey items of a variety of fishes (Lasiak and McLachlan, 1987).

The turnover rates of the various fauna components differ geographically. Scottish workers have highlighted great differences in the turnover rates of benthos between temperate and tropical beaches (Ansell et al., 1978). They showed that the P:B ratios on tropical Indian beaches were about ten times greater than on temperate Scottish beaches. Temperate bivalves (*Tellina* and *Donax*) have P:B ratios of less than 1, whereas tropical *Donax* have ratios of about 10. On intermediate beaches in warm temperate regions, P:B ratios of 1 to 2 are typical.

Sandy beach macroscopic food chains end in fishes, carnivorous invertebrates, and birds. Birds are significant predators on tallitrid amphipods and oniscid isopods (Griffiths et al., 1983) in the supralittoral and on bivalve molluscs lower down (McLachlan, 1983; Hockey et al., 1983), although marine predators such as fish, crabs, and gastropods are usually more important. Slow-moving predatory gastropods generally attack clams such as *Donax* and *Mesodesma*. Ocypodid crabs are both scavengers and predators preying on bivalves and gastropods.

Fishes are the only higher predators on zooplankton and on most beaches are the main predators on intertidal and subtidal benthos (McLachlan, 1983; McDermott, 1983). Juvenile flatfishes crop the siphons of bivalves and the palpi of polychaetes, and take whole smaller prey such as harpacticoid copepods. Adult flatfishes and elasmobranch feed extensively on bivalves. Planktivores include both phytoplankton and zooplankton feeders. However, most grazing on zooplankton is by opportunists that mostly feed on the benthos. Predatory fishes, cetaceans, seals, and diving birds such as cormorants constitute the top of the beach surf-zone food chains.

6.9.3.1.1 *Examples of macroscopic food webs (Figure 6.27)*

Tropical Indian beaches, studied by Steele (1976) and by Ansell et al. (1978), are low energy, reflective to intermediate beaches during calm weather, but they may erode extensively during monsoon storms. The surf zone is neg-

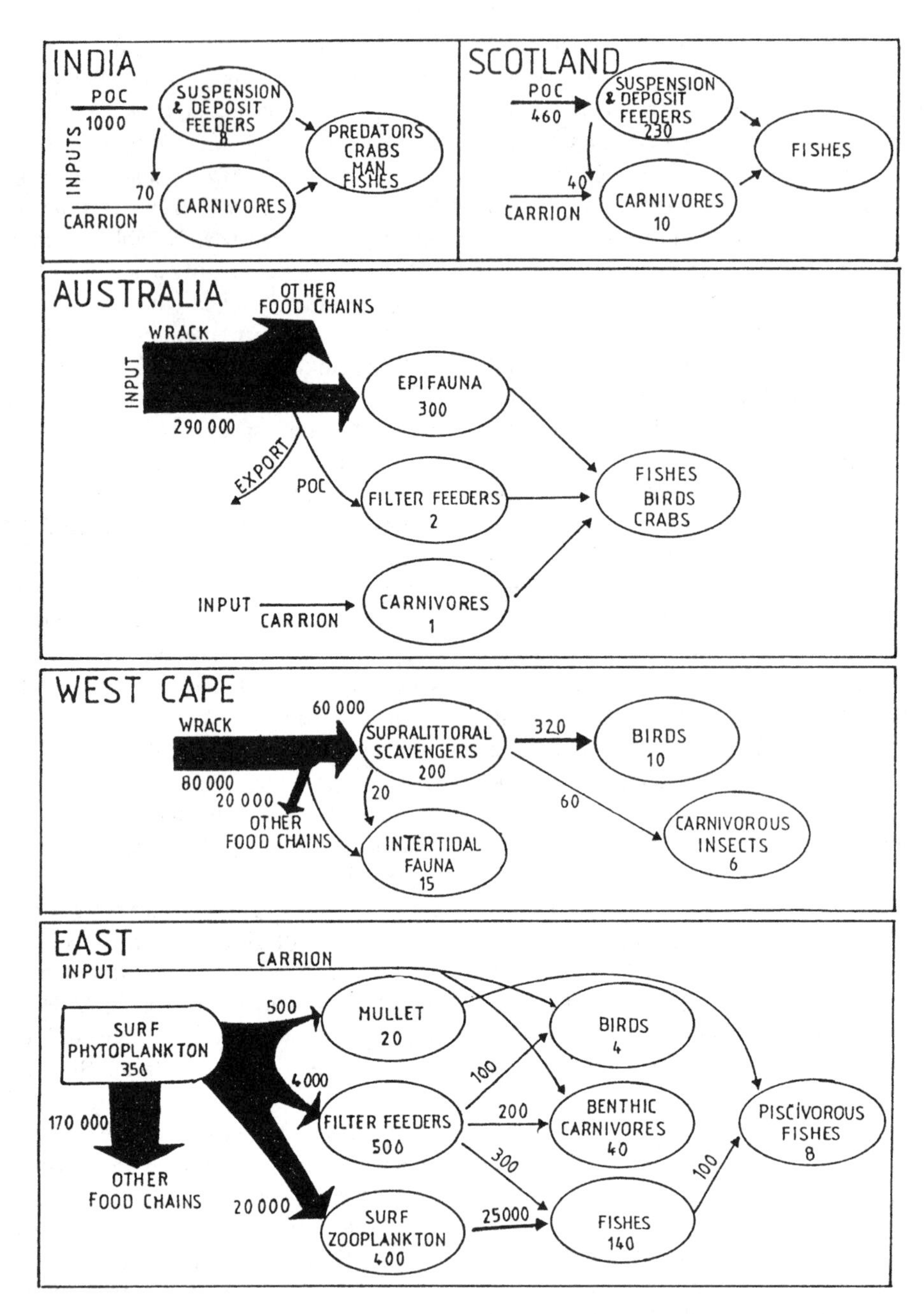

FIGURE 6.27 Macroscopic food chains and carbon flows through some sandy beach ecosystems. All values are given in g C m^{-1} yr^{-1}. (Redrawn from Brown, A.C. and McLachlan, A., *Ecology of Sandy Shores,* Elsevier, Amsterdam, 1990, 203. With permission.)

ligible and there is no pronounced driftline. These investigators divided the benthos into particulate feeders and carnivores, the former consuming on average about 1000 g C m^{-2} yr^{-1}, with more than 90% of this due to the activities of filter feeders. Crabs were the chief predators both in the supralittoral and sublittoral, followed by fishes, with birds having a negligible impact. Man was also a major predator. Despite a low biomass (5 to 10 g C m^{-1}), these beaches nevertheless process large amounts of organic matter derived from the sea due to the high turnover rates of the invertebrates, the overall P:B ratio being 20. This production is cropped and exported by marine predators.

In low energy dissipative Scottish beaches studied by the same workers, biomass was higher than on the tropical beaches at about 240 g C m^{-1}; but in contrast, the turnover was lower, the average P:B ratio being about 2. Total consumption was about 500 g C m^{-1} yr^{-1}, approximately half that of the Indian beaches. Most of this consumption was again due to particulate feeders, especially filter-feeding bivalves, although deposit feeders and carnivores were present in smaller numbers. Another difference was in the composition of the predators in that fishes, particularly juvenile flatfishes, were more important on Scottish than on Indian beaches. The demersal fishes prey on *Tellina* siphons, polychaete palps, and whole, small crustaceans.

The coast near Perth, in western Australia, is characterized by microtidal, low energy, reflective beaches without surf zones, with many of them receiving high macrophyte, especially sea grass, inputs. The intertidal fauna is impoverished, consisting of small populations of donacid bivalves and hippid crabs, while sparse populations of ocypodid crabs occupy the drift line. Total biomass is only about 1 g C m^{-1}. While the energetics of these populations have not been quantified, they are known to consume small quantities of particulate organic matter and stranded organisms cast ashore. These beaches are therefore dependent upon inputs of food from the sea. However, the stranded wrack, which totals about 290,000 g C m^{-1} yr^{-1}, accumulates on the beaches, with an average standing biomass of 10,000 to 50,000 g C m^{-1} (Robertson and Hansen, 1981; McLachlan, 1985). This wrack supports large populations of epifauna, especially the amphipod *Allorchestes compressa*, which comprises more than 90% of the fauna and averages about two individuals per gram of wrack, reaching densities of up to 100 per gram of carbon. The wrack is broken down by amphipod grazing, physical processes, and microbial degradation.

Since there are no surf zone and sparse intertidal populations, predators are not numerous on these western Australian beaches; they include wading birds, portunid and ocypodid crabs, and fishes. The great abundance of amphipods constitute a valuable food resource for fishes, which are the main predators. These Australian beaches thus serve as interfaces processing organic material, predominately macrophytes, derived from the sea (Robertson and Lenenton, 1984).

In contrast to the western Australian beaches, where much of the macrophyte remains in the shallows, on the South African sandy beaches studied by Griffiths et al. (1983), the macrophytes are cast up on the driftline. This shore also differs from the Australian ones in that it has a mesotidal regime, in displaying strong wave action, and in its close proximity to rocky shore and reefs, thereby ensuring large inputs of kelps *Ecklonia*, and *Laminaria*. Input of kelp averaged about 80,000 g C m^{-1} yr^{-1}. While the intertidal fauna is sparse, an abundance of insects, tallitrid amphipods, and oniscid isopods occur along the driftline, with an average biomass of 200 g C m^{-1}. The supralittoral macrofauna are estimated to consume up to 70% of the kelp, the remainder being processed by the interstitial meio- and microfauna. Much of the consumption is returned to the detritus pool as feces. The production of the driftline fauna (400 g C m^{-1} yr^{-1}) is consumed by birds (80%), by carnivorous coleopterans (15%), and by carnivorous intertidal isopods (5%). Of the large input of macrophytes from the sea, only 0.5% of the input goes to terrestrial predators with the remaining 99.5% being either respired or returned to the sea.

In a similar situation in California, Hayes (1974) estimated that kelp inputs amounted to 30,000 g C m^{-1} yr^{-1}, of which only 4 to 5% was consumed by the supralittoral isopod *Tylos*.

On the Eastern Cape, South Africa, the exposed sandy shores are of the high-energy intermediate type tending toward the dissipative extreme and becoming fully dissipative during rough conditions. They are characterized by surf diatoms and support rich macrofaunas both on the beach and in the surf zone. They are, however, not as rich as the Brazilian or Washington beaches studied by Gianuca (1983) and Lewin et al. (1979). These beaches are dissipative and have continuous, heavy diatom patches.

The ecology and energetics of the Eastern Cape beaches have been reviewed by McLachlan (1983; 1987) and by McLachlan and Bate (1984). Food inputs include surf diatom production, stranded carrion, as well as a variety of minor sources. The surf diatoms, although not as rich as on the sandy shores in Washington, Brazil, and New Zealand, are consistent and highly productive. Campbell (1987) estimated diatom production at 120,000 g C m^{-1} yr^{-1} within a surf zone 250 m wide and 0 to 3 m deep, equivalent to a production of 500 g C m^{-2} yr^{-1}. Production, however, was patchy and highly variable. Phytoplankton primary production outside the breaker zone, due to both diatoms and flagellates, totaled 11,000 g C m^{-1} yr^{-1} in a zone 250 m wide and 3 to 10 m deep. Carrion input was estimated to be in the range of 100 to 200 g C m^{-1} yr^{-1}.

Benthos, zooplankton, and fishes occupy dominant positions in the food web. The benthos is dominated by the filter-feeding bivalve *Donax*, which makes up over 90% of the biomass, with scavengers and predators making up the balance. Supralittoral populations are sparse due to the absence of a driftline. Total faunal biomass is estimated to be in the region of 540 g C m^{-1}, of which some 90% is intertidal with the remainder occurring just beyond the surf zone, at a depth of about 10 m, 500 m from the shore. Zooplankton biomass, totaling some 400 g C m^{-1}, is dominated by omnivorous prawns (45%) in the surf zone and mysids (32%) in the rip-head zone, both of these grazing surf diatoms. One fish species, the mullet *Liza richardsoni*, with an average biomass of 20 g C m^{-1}, also grazes diatom accumulations. It has been estimated that about 25% of the surf-zone primary production (or 15% of the surf and rip-head zone total primary production) is consumed by these groups. In total, zooplankton are the most important consumers (16%), followed by benthic filter feeders (16%), and mullet (4%). Fishes, with a biomass of about 130 g C m^{-1}, birds (4 g C m^{-1}), and crabs (6 g C m^{-1}) are the predators cropping, respectively, about 89%, 5%, and 6% of the production of the zooplankton and benthos. The latter in turn are consumed by predatory fishes, which have a biomass of some 10 g C m^{-1}.

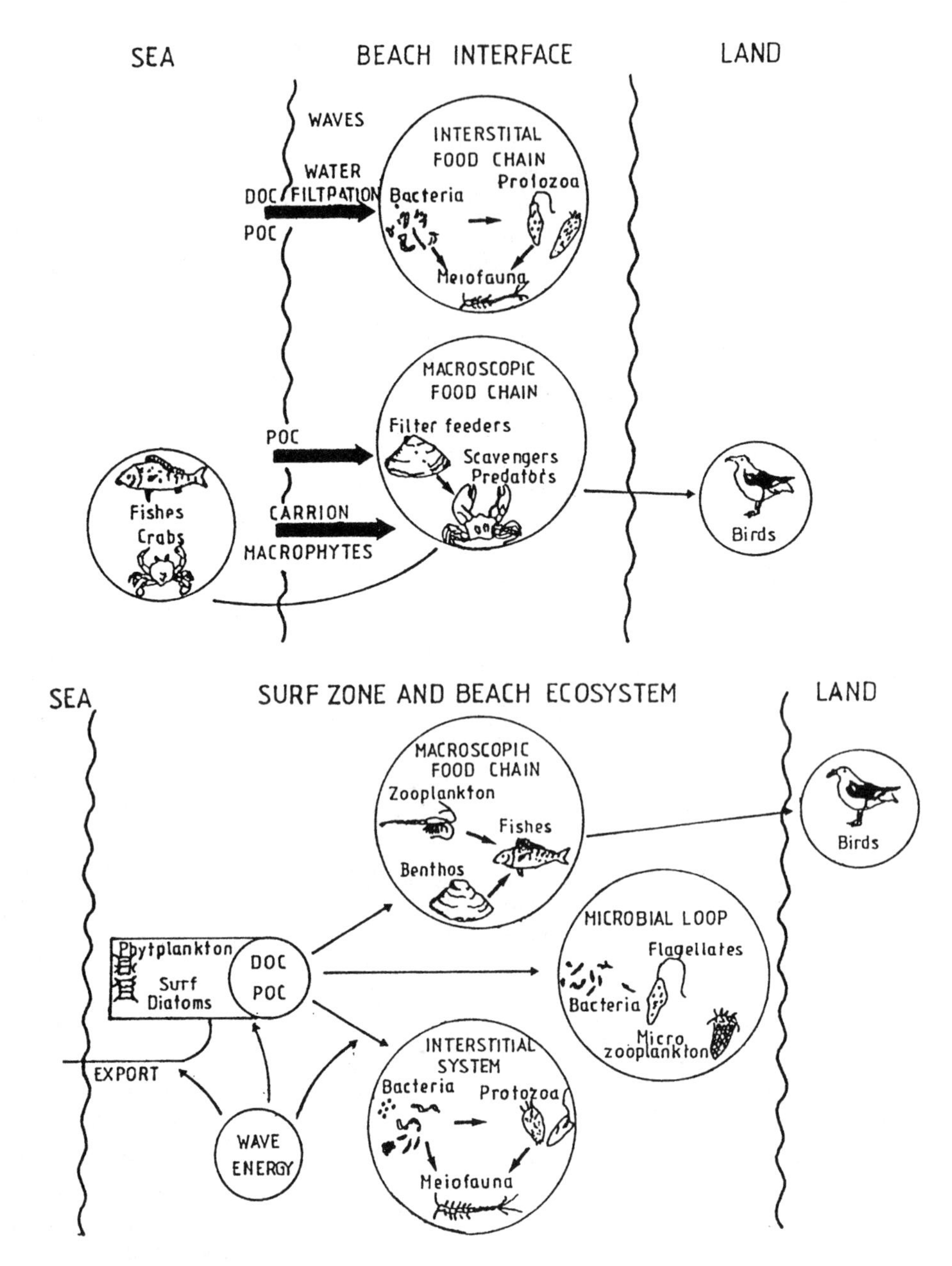

FIGURE 6.28 Carbon flow through the two main types of sandy beach ecosystems. (Redrawn from Brown, A.C. and McLachlan, A., *Ecology of Sandy Shores*, Elsevier, Amsterdam, 1990, 215. With permission.)

6.9.3.1.2 Energy flow in beach and surf-zone ecosystems

Brown and McLachlan (1990) have summarized the available data of energy flow in beach and surf zone ecosystems. The partitioning of energy flow between the diatoms, water-column microbes, the interstitial biota, and the macrofauna differ with beach type and particularly with the degree of coupling between the beaches and the surf zones. Two major types of sandy shore ecosystems can be recognized: beaches with little or no surf zone, which are dependent on food imports from the seas, and beaches with extensive surf zone sustaining sufficient primary production to be self-supporting. The fundamental difference between the open and closed beach systems lies in the incident wave energy and surf circulation patterns.

Figure 6.28 summarizes the key components and key processes of these beach systems. Characteristics of the open beach type are:

1. Dependency on marine inputs
2. The presence of only two food webs: interstitial and macrofaunal
3. The major importance of the interstitial system
4. The importance of wave energy in driving the interstitial system via water filtration
5. Large macrophyte inputs in some cases

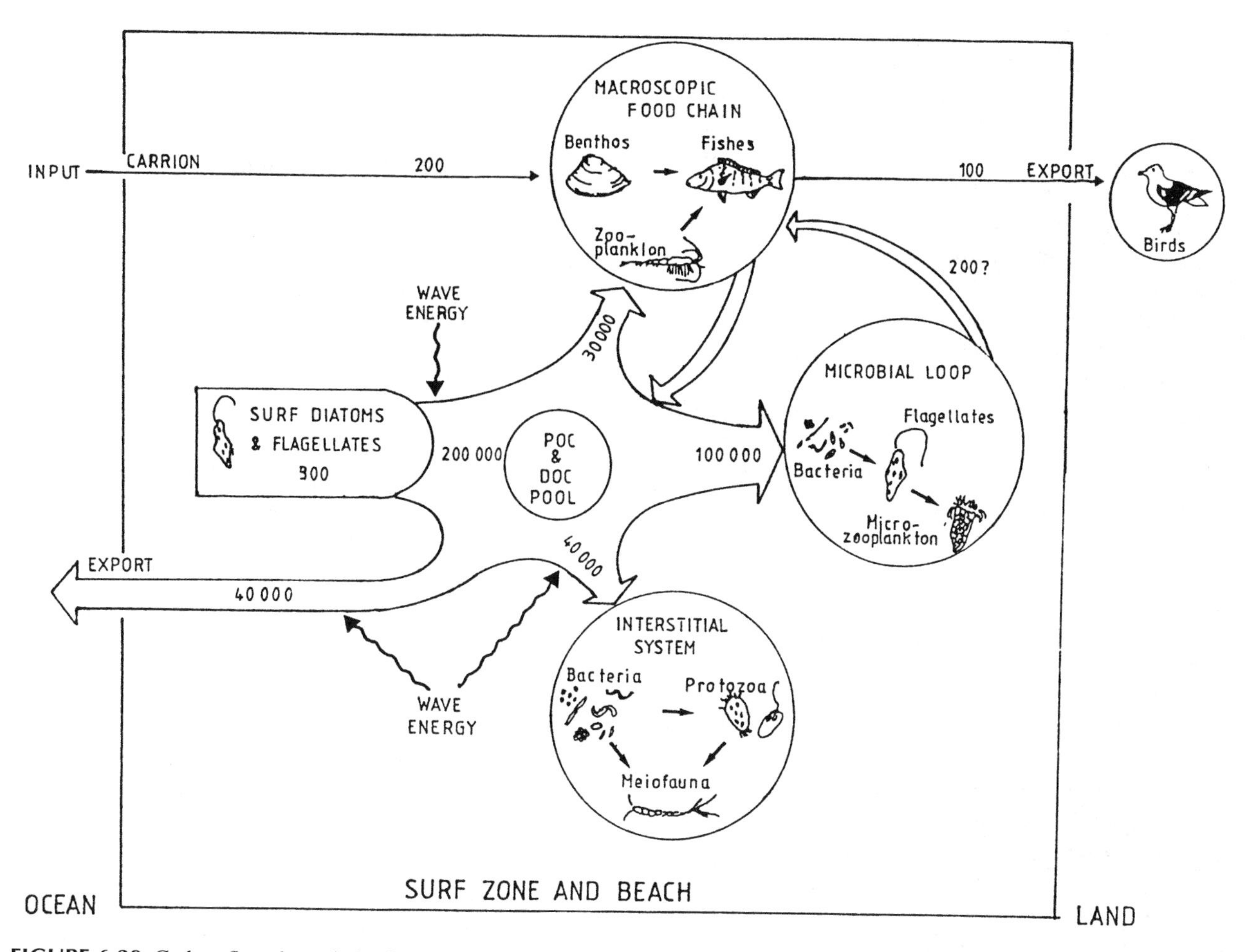

FIGURE 6.29 Carbon flow through the food chains of the Eastern Cape beaches, South Africa. All values in g C m^{-2} yr^{-1}. (Redrawn from Brown, A.C. and McLachlan, A., *Ecology of Sandy Shores*, Elsevier, Amsterdam, 1990, 218. With permission.

6. The export of macrofaunal production to marine and terrestrial predators
7. The generally low faunal biomass

These features contrast with those of the self-sustaining beach/surf-zone ecosystems, which are:

1. A wave-driven surf circulation that retains material
2. The presence of no less that four biotic systems: phytoplankton, microbial loop, interstitial system, and macrofauna
3. High primary production
4. The major importance of the microbial loop
5. The significance of wave energy in controlling primary production, the interstitial system, the export of surplus production, etc.
6. The generally high biomass

Intermediate beaches with moderately developed surf zones and negligible primary production are transitional and include elements of both beach types.

6.9.3.1.3 The sandy beaches of the Eastern Cape, South Africa (Figure 6.29)

The only sandy-beach/surf-zone ecosystem for which all components have been studied is in the Eastern Cape, South Africa. Brown and McLachlan (1990) have summarized the structure and dynamics of this system based on the reviews of McLachlan et al. (1981), McLachlan (1983), McLachlan and Bate (1984), and Cockroft (1988).

The Sundays River beach is a high-energy, intermediate to dissipative beach with a well-developed surf zone averaging 250 m in width. The rip-head zone, which lies beyond this to the outer limit of the surf circulation cells, extends an additional 250 m. Water turnover in the surf zone takes only a few hours, but turnover of water between the combined surf and rip-head zones and the nearshore zone takes days. The ecosystem then comprises the sand beach and the water envelope of the surf and rip-head zones, its boundaries being along the dune/beach interface and the outer limit of the surf circulation cells.

The chief primary producer in the system is the surf diatom *Anaulus*, while flagellates are also important. Total production in the surf zone is 120,000 g C m^{-1} yr^{-1} (Campbell, 1987), 90% of this due to *Anaulus*. Of this production, 40% is in the form of mucus and 5% as exudates, the remaining 55% being accounted for by cell division. In the outer surf zone or rip-head zone, primary production is estimated at up to 110,000 g C m^{-1} yr^{-1}. If mucus sloughed off the diatoms to produce both particulate and dissolved organic carbon are included, a conservative estimate of the available primary production in the ecosystem as a whole is 50,000 g C m^{-1} yr^{-1} as DOC and 15,000 g C m^{-1} yr^{-1} as POC, including dead particulates, living *Anaulus*, and other phytoplankton species.

The interstitial system filters 3 to 4 × 10^4 m^3 m^{-1} yr^{-1}. Studies of oxygen consumption (Dye, 1981) indicate that the total amount of carbon metabolized is about 40,000 g C m^{-1} yr^{-1}. No significant interaction occurs between the interstitial meiofauna and the macrofauna.

The macroscopic food web has been discussed previously. Estimates indicate an input of about 30,000 g C m^{-1} yr^{-1} into this food web, most originating from the phytoplankton. With an average assimilation efficiency of about 70%, some 10,000 g C m^{-1} yr^{-1} are thought to be recycled as feces into the water column, becoming available to the microbial loop and other food chains. The microbial loop is well developed. Total bacterial utilization of POC and DOC is estimated at 100,000 g C m^{-1} yr^{-1}.

Thus, in this system, primary production exceeds faunal requirements and the surplus is probably exported as living and dead POC across the boundary of the ecosystem. The microbial loop is the most significant of the food webs, accounting for about 50% of carbon production, followed by the interstitial system (20%) and the macroscopic food web (15%).

6.9.3.2 Tidal Flats

Tidal flats are characterized by high biomass and high productivity of the macrofauna reflecting high food availability. They are subsidized systems, depending largely on the import of nutrients and food (Hickel, 1984: Reise, 1985). Tidal flats receive detritus and plankton from the adjacent sea, rivers, and salt marshes. Reise (1985) depicts this as the tidal flat turntable for organic matter between the land and the sea. Considerable work has been carried out on the biomass and energy flow on the tidal flats of the Wadden Sea, especially in the Dutch sector. Estimates of the budgets for secondary and primary production have been reported for the Western Wadden Sea (Cadee, 1980; 1984; De Wilde and Beukema, 1984). In the Northern Wadden Sea, biomass as well as primary and secondary production of the tidal flats were measured in the mid-1980s (Asmus and Asmus, 1985; Asmus, 1987; Asmus and Asmus, 1990). Intensive studies of seston and plankton dynamics led to the conclusion that large quantities of organic material are transported from the North Sea through the tidal inlet onto the tidal flats (Hickel, 1984).

TABLE 6.23
Benthic Food Sources in the Grevelingen Estuary, The Netherlands, Revised after Wolff (1977), Expressed as g (organic carbon) m^{-2} yr^{-1}

Food Source	Amount
Net import of detritus from salt marshes	0
Production of detritus by sea grass beds	5–30
Net import of detritus from the North Sea	155–225
Detritus from land runoff	2
Primary production, benthic microalgae	25–57
Primary production, phytoplankton	130
Total	317–444

Source: From Wolff, W.J., in *Estuaries and Enclosed Seas. Ecosystems of the World 26*, Ketchum, B.H., Ed., Elsevier, Amsterdam, 1983, 166. With permission.

Wolff (1977) published a survey of food sources for the benthic macrofauna of the Grevelingen Estuary, The Netherlands (Table 6.23). From this table it can be seen that the net import of detritus from the North Sea exceeded that of the phytoplankton and benthic microalgae combined. Van Es (1977) developed a carbon budget for a brackish embayment of the Ems Estuary, The Netherlands. In this estuary, detritus is the major organic input. Benthic animals are consumed by benthic predators (crabs, starfish), fishes, birds, and by man. Klein-Bueteler (1976) estimated the consumption by the crab *Carcinus maenus* at a minimum of 1.5 g ash-free dry wgt m^{-2} yr^{-1}, constituting about 5% of the annual production of the benthic macrofauna. Van der Schoot (1974) found that the starfish *Asterias rubens* in the Grevelingen Estuary consumed a maximum of 0.3 to 1.8 g ash-free dry wgt m^{-2} yr^{-1} of the mussel *Mytilus epulis* and other molluscs. The consumption of benthic animals on the Balgraud tidal flats in the Dutch Wadden Sea by plaice (*Pleuronectes platessa*) varied between 1.7 and 5.0 g ash-free dry wgt m^{-2} yr^{-1} in four consecutive years (Kuipers, 1977). Gobiid fishes (*Pomatoschistus minutus* and *P. microps*) may consume some 1.4 g m^{-2} yr^{-1} in the same area (Van Beek, 1976).

Nilsson (1969) recorded that diving ducks in the coastal zone (less than 3 m depth) of the western Baltic, Sweden, consumed about 3% of the benthic biomass that had been present in November during the following winter months. In the same period, 30 to 70% of the biomass disappeared due to other unknown causes. Hibbert (1976) estimated that from an intertidal area in Southhampton Water, England, birds take about 37% of the bivalve production, principally *Cerastoderma edule*. Hulscher (1982) estimated the consumption by carnivorous birds in the Dutch Wadden Sea at about 3.3 to 4.3 g ash-free dry wgt m^{-2} yr^{-1} of benthic animals (an average value for the whole Dutch Wadden Sea, 2,600 km^2). Wolff et al. (1976) made a similar calculation for the Grevelingen Estuary, estimating that the birds consumed about 3.4 g ash-free dry wgt m^{-2} yr^{-1}, constituting about 6% of the total production of the area's biomass.

Asmus and Asmus (1990) measured the biomass and production of the macrobenthos in four intertidal areas of "Koningshagen" (Island of Sylt) at monthly intervals in 1980 and 1984. The areas were characterized by different macrofaunal assemblages: *Nereis-Corophium* belt, sea grass bed, *Arenicola* flat, and mussel bed. Figure 6.30 depicts data for the annual average biomass and secondary production of the four assemblages, and the annual production of the suspension feeders and the other trophic groups. In the mussel beds, macrofaunal biomass was dominated by *Mytilus edulis*. In the *Nereis-Corophium* belt, the *Arenicola* flat, and the sea grass bed, only 0.97% (0.16 g ash-free dry wgt m^{-2}), 2.45% (0.74 g ash-free dry wgt m^{-2}), and 2.86% (0.79 g ash-free dry wgt m^{-2}), respectively, of the macrofaunal biomass was represented by suspension feeders; whereas in the mussel beds 98.02% (1,218 g ash-free dry wgt m^{-2}) of the total macrofaunal biomass was accounted for by this trophic group.

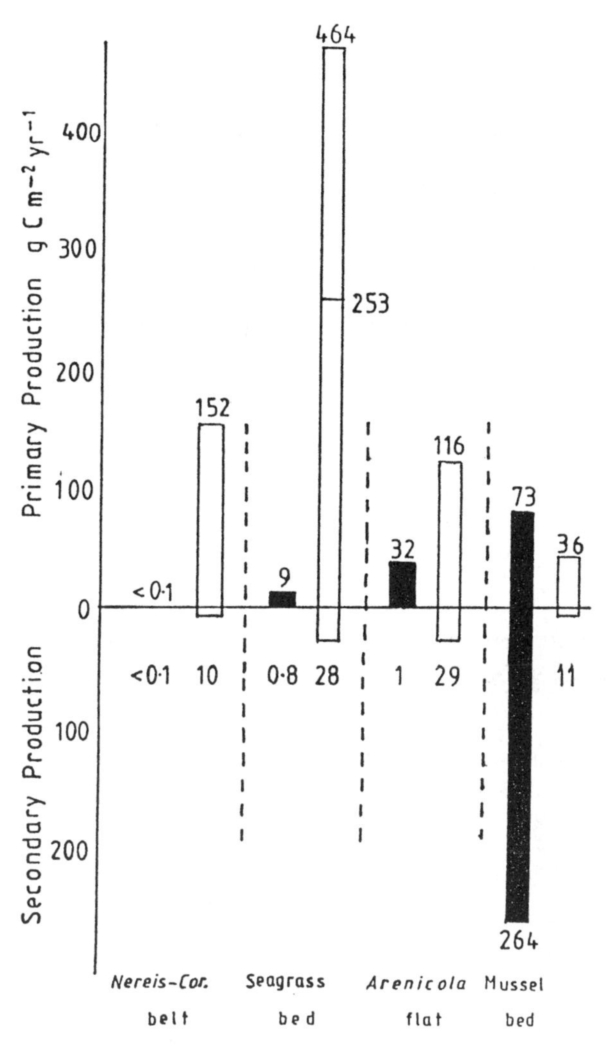

FIGURE 6.30 Annual autochthonous primary production (g C m^{-2} yr^{-1}) and annual secondary production (g C m^{-2} yr^{-1}) (lower bars) of dominant macrobenthic assemblages in the northern Wadden Sea. Light bars indicate production of benthic primary producers as well as secondary production of animals feeding at the bottom. Primary production of *Zostera* is shown by a checkered bar. Black bars mark autochthonous primary production in the water directly above the investigated assemblage as well as the secondary production of the suspension feeders of this macrobenthic assemblage. (Redrawn from Asmus, H. and Asmus, R.M., *Neth J. Sea Res.*, 39, 98, 1985. With permission.)

Secondary production of the macrobenthic fauna in Koningshagen ranged from 17.48 to 468.09 g ash-free dry wgt m^{-2} yr^{-1} (Asmus, 1987). The secondary production followed the same pattern as the mean annual biomass. Low secondary production characterized the *Nereis-Corophium* belt near the high water mark, whereas the mussel beds near the low water mark had a secondary production more than ten times higher.

6.9.3.3 Beach Wrack Communities

As discussed above in Section 6.9.3.1, stranded kelp can provide a considerable input of organic carbon to high-energy beaches. Marsden (1991a) estimated an annual input to a South Island, New Zealand, East Coast beach of >400 kg m^{-1}, while Hansen (1985) recorded an input of 1,900 kg m^{-1} for West Australian beaches. Along the West Coast of South Africa, over 2 metric tons (wet mass), or 4×10^6 kJ, may be deposited on each meter of beach per annum. Reviews of energy flow on these beaches in South Africa are to be found in Griffiths and Stenton-Dozey (1981), Koop and Griffiths (1982), Griffiths et al. (1983), and Stenton-Dozey and Griffiths (1983). Behbehani and Croker (1982) investigated the ecology of beach wrack communities in northern New England, while Inglis (1989) has studied the colonization and degradation of stranded kelp on a New Zealand sandy beach.

Inglis (1989) used litter-bag techniques to assess the faunal role in the breakdown of the large brown kelp *Macrocystis pyrifera*. The wrack was colonized by the supralittoral fauna in two distinct phases. The macrofauna, including the talitrid amphipod *Talorchestia quoyana*, the dipteran *Leptocera (Limosina) aucklandica*, the centipede *Nesogeophilus xylophagus*, and the beetles *Lagrioida brouni*, *Sitonia humeralis*, and *Bledius* sp., colonized the kelp within one day, with the highest numbers being recorded after three days. Following this, their presence in the samples declined and meiofauna, consisting of nematodes, enchtraeids, dipteran larvae, and mites, became increasingly abundant. After 18 days, the meiofauna dominated the kelp surface. Algal material was lost from the bags at a rapid rate, with only 36 to 59% of the original mass remaining after 18 days. Exclusion of the macrofauna from the litter-bags had no appreciable effect on the rate of dry matter loss.

The wrack fauna on Kommetjie Beach, South Africa, had a greater diversity of species than that found on the New Zealand beach (Stenton-Dozey and Griffiths, 1983); 35 macrofaunal species were recorded, comprising 4 species of amphipods, with *Talorchestia capensis* accounting for over 90% of the macrofaunal numbers, 2 isopods, 7 molluscs, 4 dipterans, and 19 coleopteran beetles. The meiofauna was dominated by nematodes and oligochaetes, with their joint contribution never being less than 90%, followed by harpacticoid copepods and turbellarians (both with less than 3%). Other forms such as gastrotrichs, acarines, archiannelids, and polychaetes were found in extremely low numbers. Bacterial densities ranged from 15×10^{12} m^{-1} to 366×10^{12} m^{-1}. The productivity of the principal components of the sandy beach biota at Kommetjie are listed in Table 6.24.

Griffiths et al. (1983) have provided an energy diagram for the energy flow in the high kelp-input beach at Kommetjie (Figure 6.31). Imports provide most of the energy requirements of the beach community. The most

TABLE 6.24
Mean Annual Standing Stocks of Macrofaunal Organisms on Kommetjie Beach

Macrofaunal Group	Mean Annual Biomass (g dry mass m^{-1})
Herbivores	
Amphipoda	2,083
Dipteran larvae	63
Herbivorous Coleoptera	28
Molluscs	11
Filter feeders	
Bivalve molluscs	8
Carnivores	
Isopoda	42
Carnivorous Coleoptera	21
Total	2,256

Source: From Stenton-Dozy, J.M.E. and Griffiths, C.L., in *Sandy Beaches as Ecosystems,* McLachlan, A. and Erasmus, T., Eds., Dr. W. Junk Publishers, The Hague, 1983, 566. With permission.

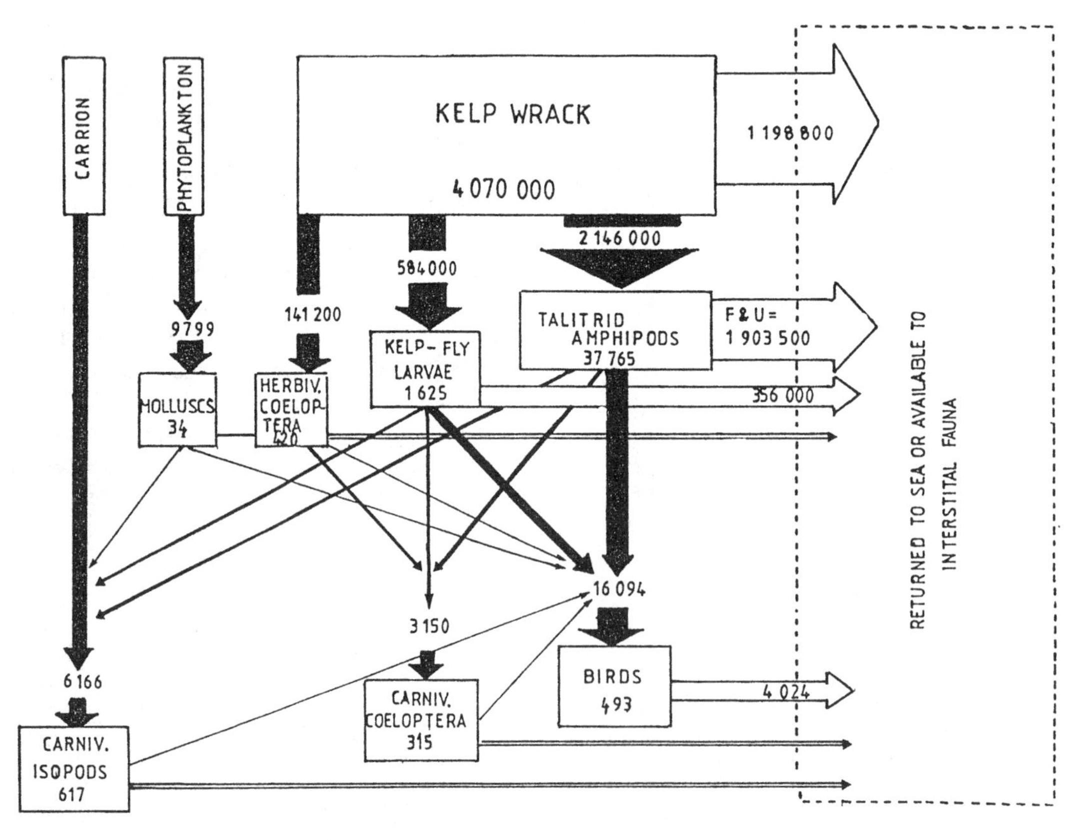

FIGURE 6.31 Energy flow diagram for Kommetjie Beach, South Africa. Figures in boxes are mean annual standing stocks in kJ m^{-1}, or in the case of kelp wrack, annual input in kJ m^{-1} yr^{-1}. (Redrawn from Griffiths, C.L., Stenton-Dozey, T.M.E., and Koop, K., in *Sandy Beaches as Ecosystems*, McLachlan, A. and Erasmus, T., Eds., Dr. W. Junk Publishers, The Hague, 1983, 553. With permission.)

important component was the kelp wrack, which had an energy equivalent of over 4×10^6 kJ m^{-1} yr^{-1}. This is more than 30 times the combined input of detritus, phytoplankton, and carrion reported for a beach near Port Elizabeth by McLachlan et al. (1981). Since the major food resource, the kelp wrack, is deposited high on the beach, the macrofauna is concentrated around the driftline. Calculated kelp consumption by the herbivores amounts to 71.7% of the deposition. Amphipods of the genus *Talorchestia* are the major herbivores (95% of the energy) and they consume 52.7% of the kelp deposited on the beach, as opposed to 14.5% for kelp fly larvae and 3.5% for the herbivorous Coleoptera. The remaining 29.3% is thought to be degraded by the bacteria and is either washed back into the sea by the tide or enters the sand column in dissolved and particulate form. The assimilation of the major herbivores, however, is low so that much of the material they eat is also returned to the beach in the form of feces or excretory products.

The macrofaunal herbivores are consumed by predatory birds, isopods, and Coleoptera. These are thought to consume 40%, 15% (maximum), and 8%, respectively, of the overall standing stock per year, or 63% of the total biomass, but since the P:B ratio of the macrofauna is approximately 2.5, this is only 25% of the macrofaunal production.

The total amount of organic material entering the sand is estimated at 3.5×10^6 kJ m^{-1} yr^{-1}, or 85% of the total input to the beach. Dissolved and particulate kelp debris that enters the sand column is rapidly utilized by the bacteria and the rich meiofauna. Bacteria can convert kelp detritus with up to 28% efficiency and potentially they have a P:B ratio of 30 to 70. Meiofaunal food requirements total approximately 32 times the bacterial standing stock of 91 g m^{-1}, indicating that they consume virtually all of the bacteria.

6.9.3.4 Estuarine and Coastal Soft Shore Food Webs

From investigations of estuarine primary production, it is clear that they can range from ones in which phytoplankton dominates as the principal primary producer, to ones in which macrophytes (marsh grasses, sea grasses, and mangroves) are domimant producers, or to others in which the benthic microalgae provide the bulk of the primary production. However, even in phytoplankton-dominated systems, a considerable proportion of the primary production is not consumed by the herbivores, but becomes available to the detrital pathway. Sorokin (1978; 1981) has estimated that, in general, the maximum direct utilization by grazing on phytoplankton production is about 50%.

Organic carbon may enter estuarine waters in a variety of ways:

1. Emergent macrophytes (marsh grasses, rushes, sedges, etc.), submerged grasses (*Zostera*, etc.), macroalgae (green, especially species of *Ulva* and *Enteromorpha*, reds, and browns), epiphytic microalgae, benthic microalgae, and phytoplankton fix inorganic carbon which is replenished by respiration and decomposition and some diffusion of CO_2 from the air to the water.
2. Phytoplankton, epiphytic and benthic microalgae, and macroalgae release some fraction of their fixed carbon as DOC.
3. Living shoots of *Spartina* and submerged grasses such as *Zostera* secrete some DOC into the water during high tide (Gallagher et al., 1976; Turner, 1978).
4. Dead marsh macrophytes, submerged macrophytes, and mangrove litter are degraded by bacteria and fungi with the loss of both DOC and POC.
5. Bacteria form large quantities of extracellular particulate material.
6. Feces and excretory products of both the meiofauna and macrofauna add to both the DOC and POC pools.
7. Exuvia of crustaceans and the dead bodies of animals add to the POC pool.
8. In contrast to the above autochthonous inputs, organic matter may enter from outside the system via the input of POC and DOC from rivers flowing into the estuaries (allochthonous processes). As we have seen, such inputs dominate in many systems.

As an end result of the processes of primary production, detrital formation, and decomposition, a particle pool is produced (Figure 6.32) consisting principally of bacteria, detrital particles (POM), and microalgae, which become available to the next trophic level, the particle consumers (Correll, 1978). In estuaries the traditional grazing and detrital energy flow pathways are not clearly defined. As Haines (1979) points out, there has been confusion in that (1) the role of bacteria has been misunderstood, and (2) estuarine food webs have been oversimplified.

Odum (1980) has reviewed concepts concerning detritus food chains in estuarine ecosystems. He emphasizes that: "While few will challenge the concept that organic detritus is the link between autotrophic and heterotrophic parts of the ecosystem in most estuaries, there is considerable disagreement about how the producer detritus energy is actually used by the heterotrophs themselves." E.P. Odum presents a graphic model (Figure 6.33) showing the way in which estuarine food chains are coupled. The unique feature of this model is the integration of three distinct groups of primary producer, macrophytes, phy-

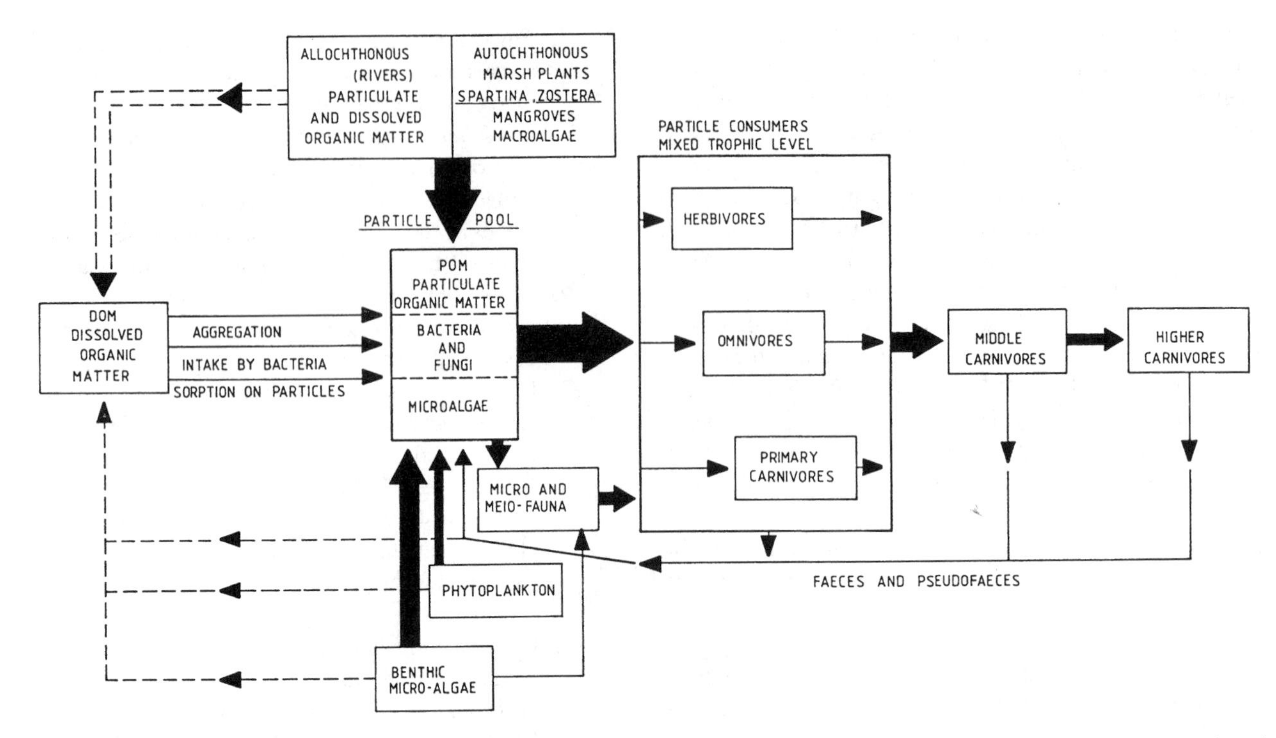

FIGURE 6.32 A conceptual model of a salt marsh or mangrove-fringed estuary showing the most important energy flows as broad arrows, less important energy flows as narrow arrows, and the pathways of dissolved organic matter as dotted lines. Based on Odum and Heald, 1975. (From Knox, G.A., *Estuarine Ecosystems: A Systems Approach,* Vol. I, CRC Press, Boca Raton, Florida, 1986, 97. With permission.)

toplankton, and benthic microalgae linked together by detritus and tidal action. The figure shows how food chains originating from each of the three autotrophic groups are linked together to form the estuarine food web. Since the benthic microalgal component is tightly coupled to the heterotrophic microorganisms, the "detritus complex" is depicted as a single subsystem that provides the primary energy source for the macrofauna in the detritus food web. This is the dominant energy-flow pathway in most estuaries.

In the conceptual model in Figure 6.32, the central role of bacteria in estuarine food webs is emphasized. Bacteria provide a substantial food resource for consumer species and consequently their proper trophic level is

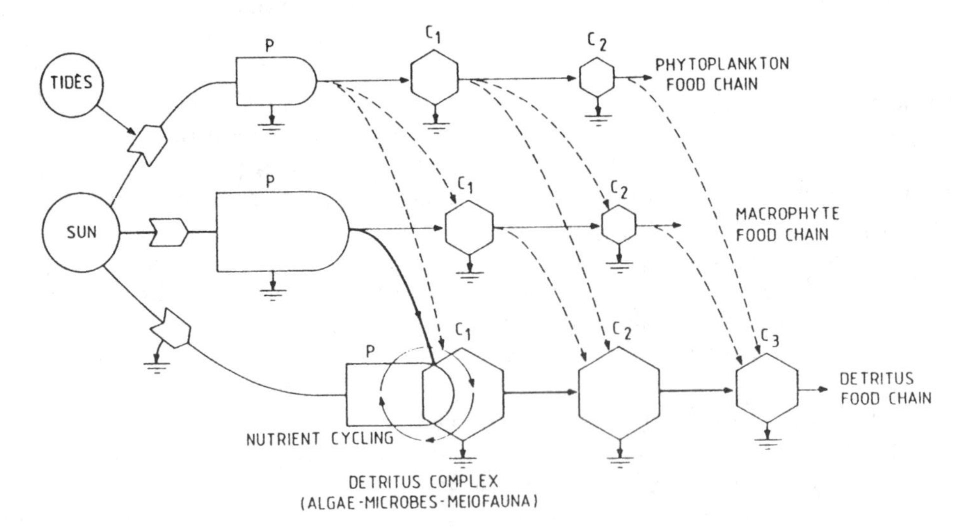

FIGURE 6.33 Flow diagram of an estuarine food web showing the three coupled food chains, namely two grazing food chains (phytoplankton, macrophyte) and a detritus food chain. In many estuaries, the detritus pathway via macrophytes is the dominant one (as shown). (Redrawn from Odum, E.P., in *Estuarine Perspectives,* Kennedy, V.S., Ed., Academic Press, New York, 1980, 493. With permission.)

equivalent to that of the "primary producers." Input is organic matter from all trophic levels and the output is particulate food for consumers in addition to their role as "mineralizers." As we have seen, bacteria serve several roles other than that of decomposers. They break plant material and scavenge dissolved and particulate organic matter. In the process, bacteria produce high cell populations (particulates). Bacteria also break down large detrital particles into fine particulates suitable for utilization by a wide range of filter feeders and deposit feeders. Thus, bacteria also pay an important role as particle producers.

Correll (1978) envisages the estuarine food web as being composed of three levels: particle pool, particle consumers, and predators (Figure 6.32). The particle pool comprises microalgae, protozoans, bacteria, and particulate organic matter. The particle consumers include herbivores (mainly filter feeders), a wide range of detritovores including deposit feeders and many omnivores, and some primary carnivores that feed on microfauna (protozoans) and meiofauna. This is the consumer group that W.E. Odum and Heald (1972) termed the "mixed trophic level," which included herbivores, omnivores, and primary carnivores. Very few of these organisms feed selectively on a single component of the particle pool, e.g., phytoplankton or benthic microalgae. The majority of the benthic epifaunal and infaunal primary consumers have a mixed diet of detrital particles, benthic microalgae, bacteria, fungi, protozoa and other microfauna, and meiofaunal species (nematode, annelids, copepods, etc.). Thus, such consumers are perhaps best termed "opportunistic omnivores" (Haines, 1979).

Within estuarine ecosystems, DOM plays a very important role. As it becomes available, the DOM pool is rapidly metabolized by bacteria and hence at any one instant, the concentrations are never very large. However, there are a variety of mechanisms, as outlined above, whereby the pool is replenished.

Kuipers et al. (1981) have drawn attention to the importance of the temporary meiofauna, the newly settled stages of macrobenthic invertebrates, in estuarine ecosystems. In 1979 in a sheltered area of the western Wadden Sea, the total number of newly settled juveniles of macrobenthic species retained by a 400 μm sieve exceeded 150,000 m^{-2}. Comparable numbers have been recorded for other localities (Perkins, 1974). Kuipers et al. (1981) advance the view that bacteria, microfauna, meiofauna, temporary meiofauna, and small macrofauna should be regarded as one complex, characterized by the small size of the individuals, a high turnover rate (high P:B ratios), relatively short life spans, and a complicated trophic structure. They call this "the small food web": a web in which no significant biomass is accumulated. Figure 6.34 depicts their view of energy flow through a mudflat ecosystem.

In their model, the macrobenthos consumes only 25% of the organic material that is available and it actually forms a side track of the main carbon flow through the

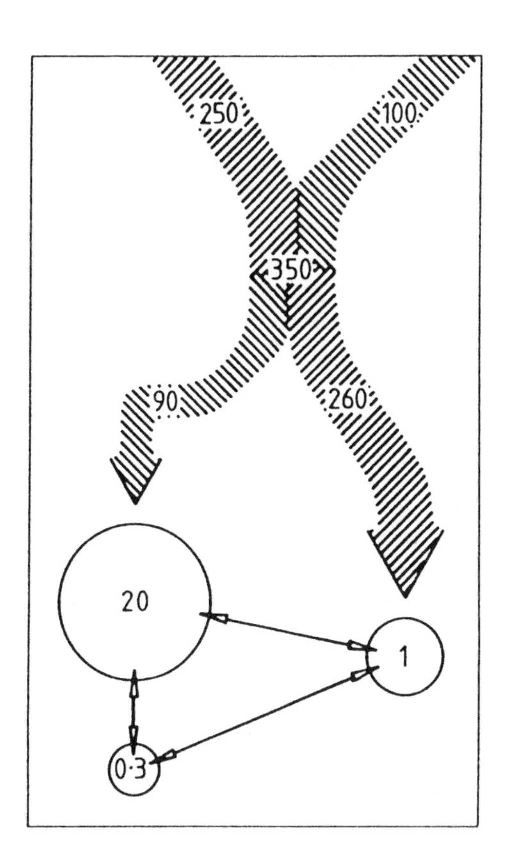

FIGURE 6.34 Carbon flow through an estuarine tidal flat ecosystem. Shaded arrow is the flow of detritus input (250) plus autochthonous production (100), toward macrobenthos (90), and into the "small food web" (260). Circles are the relative biomass of the macrobenthos (20), small food web (1), and macrobenthic predators (0.3). (Redrawn from Kuipers, B.R., de Wilde, P.A.J.W., and Creutzberg, F., *Mar. Ecol. Prog. Ser.*, 3, 219, 1981. With permission.)

system. This accounts for the relatively low macrobenthic productivity found an some mudflat ecosystems. Most of the organic matter is consumed by "the small food web," consisting of bacteria, microfauna, and permanent and temporary meiofauna. The components of the small food web live mainly in the aerobic surface layer of the sediments. Anaerobic bacteria living deeper in the sediment form the last link in the food web, breaking down the organic material not consumed by the aerobic part of the small food web.

Thus, the traditional trophic level concept has little relevance in estuarine food webs where even zooplankton can obtain a considerable proportion of their energy requirements through the consumption of bacteria and detritus with its associated microbial community. Macrobenthic animals (both filter and deposit feeders) feed on bacteria, organic particles, benthic microalgae, micro- and meiofauna, and even on their own eggs, larvae, and newly settled juveniles. The availability of this mixed diet is aided by bioturbation process mixing and reworking the surface layers (Rhoads, 1974). The fluid surface layer of these muddy sediments is readily suspended by turbulent water movement at the sediment-water interface and the microalgae, bacteria, detritus, etc. is then made readily available to filter feeders.

Haines (1979) has drawn attention to the important role of estuaries, and in particular estuarine marshes, as nursery areas for the larval and juvenile stage of many species, especially prawns and fish. In addition at high tide, many species including large and small fish, shrimps, and other crustaceans invade the marshes to feed and avoid predators. Mention has previously been made of the important role of the meiofauna of estuarine mudflats as a major food source for larval fishes such as juvenile flatfishes, e.g., meiofauna, especially harpacticoid copepods, are an important food for juvenile chum salmon during the first critical weeks of their life (Healey, 1979; Naimen and Sibert, 1979). Thus, a major export of estuarine plant production occurs as living organisms. Seasonal waterfowl and wading birds fill a similar role.

Odum (1980) has advanced the hypothesis that tides in estuaries provide an energy subsidy that enhances estuarine productivity. He contends that adapted, coevolved populations in estuarine ecosystems utilize tidal power as an ancillary energy source coupled with solar power, the principal driving force. A number of studies have demonstrated that at least within the moderate range of tidal amplitudes, the greater the tidal range the higher the primary production is likely to be. Steever et al. (1976) found that within a range of 0.5 to 2.0 m, an increase in tidal amplitude of 1 m approximately doubled the standing crop.

6.10 CARBON FLOW MODELS

6.10.1 A Salt Marsh Ecosystem in Georgia

A pioneer study of the energy flow through an estuarine ecosystem was that of Teal (1962). His studies have been expanded by Montague et al. (1981) in their studies of the Sapelo Island salt marsh (Figure 6.35). In the intertidal zone, the biomass of aquatic macroconsumers may exceed 15 g C m^{-2}. Eighty to 200 mud fiddler crabs (*Uca pugnax),* 400 to 700 marsh snails (*Littorina irrorata*) or mud snails (*Ilyanassa obsoleta),* and seven to eight ribbed mussels (*Geukensia demissa*) per m^{-2} of the marsh are not unusual (Teal, 1962; Kuenzler, 1961a). In addition quantities of the pulmonate snail *Melampus bidentatus*, the marsh clam *Polynesoda caroliniana*, and several polychaete worms are abundant. On mudflats, densities of 500 to 1,600 mud snails (*Ilyanassa obsoleta*) per m^{-2} are common, along with other consumers (Pace et al., 1979). Many commercial species such as the white shrimp (*Penaeus setiferus*), brown shrimp (*P. aztecus*), blue crabs (*Callinectes sapidua*), and oysters (*Crassostrea virginica*) are abundant. Many fur-bearing mammals — racoon, muskrat, mink, and otter — inhabit the salt marshes. Shorebirds,

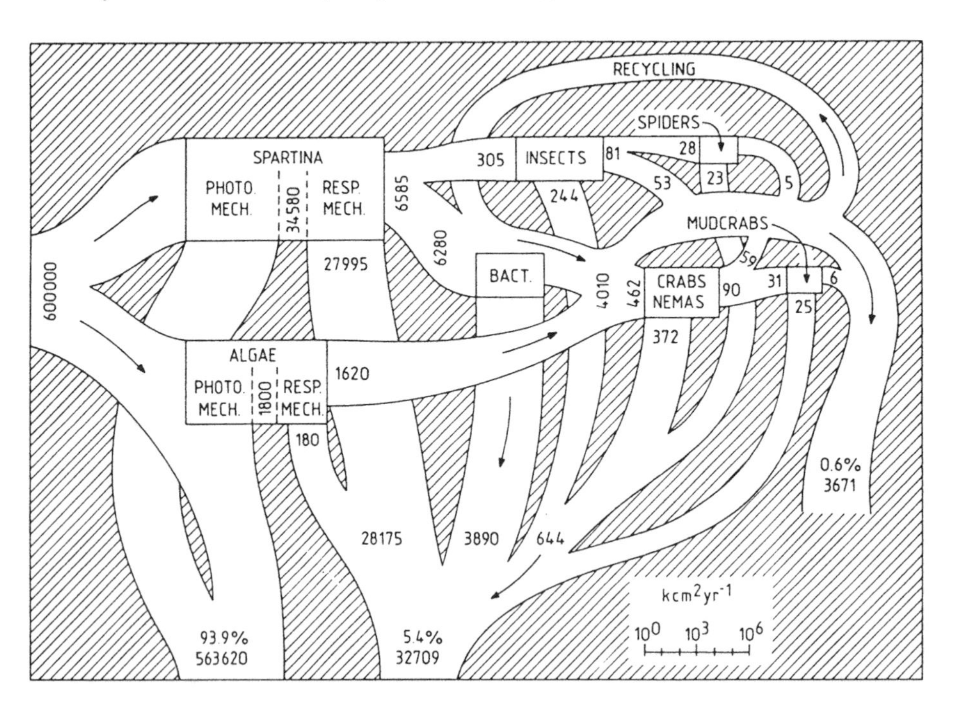

FIGURE 6.35 Box model of energy flow in the Georgia salt marshes. The two outward flows at the bottom are losses as heat, while the outward flow at the right is net secondary production exported from the marsh. (Redrawn from Montague, C.L., Bunker, S.M., Haines, E.B., Pace, M.L., and Wetzel, R.L., in *The Ecology of a Salt Marsh*, Pomeroy, L.R. and Wiegert, R.G., Eds., Springer-Verlag, Berlin, 1981, 72. With permission.)

waterfowl, reptiles, and mammals depend on the salt marshes for food and shelter. Alligators, porpoises, and manatees inhabit the tidal creeks.

Spartina productivity amounts to 6,585 kc m^{-2} yr^{-1}; of this, herbivores (insects) consume only 305 kc m^{-2} yr^{-1} leaving 6,584 kc m^{-2} yr^{-1} to ultimately enter the detritus food web as dead plant material. The microalgae contribute 1,620 kc m^{-2} yr^{-1}. Of the combined microalgal and *Spartina* detritus production, the mud crabs consume 3,476 kc m^{-2} yr^{-1} while 3,671 kc m^{-2} yr^{-1} is exported from the marsh. Macroinvertebrate species degrade much less energy than the microbes (Teal, 1962; Pomeroy et al., 1977), but the behavior of the macroconsumers may influence ecosystem energy flow far more than this would imply. For example, the ribbed mussel is more important as a builder of phosphorus-rich surface marsh sediments than as a consumer of energy (Kuenzler, 1961b). This filter feeder daily removes one-third of the phosphorus in the overlying water and deposits it on the marsh surface as pseudofeces. Macroconsumers prey on, and thus control, standing stocks of certain resources. Standing stocks of benthic microalgae, as chlorophyll *a* m^{-2}, and their production as carbon fixed m^{-2} hr^{-1}, increased when mud snails were removed from experimental plots on the mudflat (Pace et al., 1979). Conversely, feeding on overcrowded prey populations can increase the production rate of the remaining organisms by temporarily reducing intraspecific competition for available space and food (Wiegert et al., 1975).

Many salt marsh macroconsumers concentrate organic matter in their feces and pseudofeces. This provides decomposers particle rich in organic carbon. Oysters, clams, and ribbed mussels remove organic matter from the water column and concentrate it on the sediment surface (Bahr, 1976; Haven and Morales-Alamo, 1966; Krauter, 1976). Grass shrimp, mullet, fiddler crabs, and marsh snails all deposit feces on the marsh and mudflats, enhancing the carbon available to decomposers and providing a renewable food resource for the deposit feeders. In addition to accelerating decomposition, the conditioning and packaging of detritus in the macroconsumer feces facilitates the regeneration of nutrients by microbes (Hargrave, 1980). Sediment reworking by macroconsumers increases the turnover rate of the benthic microalgae and nutrients by bringing limiting nutrients into contact with the microalgae. Marsh macroconsumers also significantly influence the marsh sediments, especially at the sediment-water interface. Bioturbation also provides habitat for other organisms, alters water turbidity, affects primary production, detrital decomposition, and nutrient recycling.

6.10.2 Barataria Bay Marsh-Estuarine Ecosystem

Hopkinson and Day (1977) have developed a carbon flow model for the marsh-estuarine ecosystem of Barataria Bay, Louisiana. This model, depicted in Figure 6.36, is a simplification of earlier conceptual and diagrammatic models based on extensive field work (Day et al., 1973). The marsh subsystem is subdivided into four compartments: (1) primary producers; (2) dead standing grass; (3) a detrital-miocrobial complex; and (4) the macrofuana. The detritus-microbial complex includes detritus (particulate organic matter of plant or animal origin) and the organisms associated with it (bacteria, yeasts, fungi, protozoa, and meiofauna). The macrofaunal compartment includes insects, snails, crabs, mussels, polychaetes, birds, racoons, and muskrats. The aquatic subsystem is subdivided into three compartments: (1) aquatic microalgae (phytoplankton); (2) a detrital-microbial complex; and (3) the aquatic fauna (zooplankton, shrimps, and fish). There are two main sources of organic matter in the aquatic subsystem: that exported from the marsh and *in situ* production (mainly phytoplankton).

As Hopkinson and Day (1977) point out, the grouping of the fauna into two compartments affects the realism of the model, depending on the degree of functional homogeneity of each. The marsh fauna is a fairly homogeneous group with practically all of the species being either herbivores or detritovores. The aquatic fauna, on the other hand, is a much more diverse group, with all trophic levels being represented.

Of the net production of the marsh plants (1546 g dry wgt m^{-2}), mainly *Spartina alterniflora*, 77 g (5%) is consumed by hervivores and the rest eventually passes via the dead standing grass to the detrital-microbial complex. Respiration by the microbes is 611 g m^{-2}, while 312 g m^{-2} of detritus is consumed by the marsh fauna and 749.4 g m^{-2} is flushed out of the estuarine waters. The net production of the aquatic flora is 879.5 g dry wgt m^{-2}, and of this 88.3 g m^{-2} (10%) is consumed directly by the aquatic fauna, 45.6 g m^{-2} (5.2%) is exported to the Gulf, and 745.6 g m^{-2} (84.8%) goes to the aqautic detrital-microbial complex. The water fauna consumes 446.2 g dry wgt m^{-2} of detritus, while 576.3 g m^{-2} are exported to the Gulf.

6.10.3 Upper Waitemata Harbour Carbon Flow Model

The Upper Waitemata Harbour is a mangrove-fringed estuarine systam (7.4 km^2) in northern New Zealand (Knox, 1983a,b). Various aspects of this system have previously been discussed.

The Upper Waitemata Harbour carbon-flow model (Figure 6.37) is subdivided into five compartments: (1) primary producers; (2) mangrove litter; (3) dissolved organic matter (DOM); (4) detrital-microbial complex (particulate organic matter (POM), plus its associated microbial community); and (5) the macrofaunal consumers. The horizontal line in the figure separates sediment processes from those occurring in the water column. Three

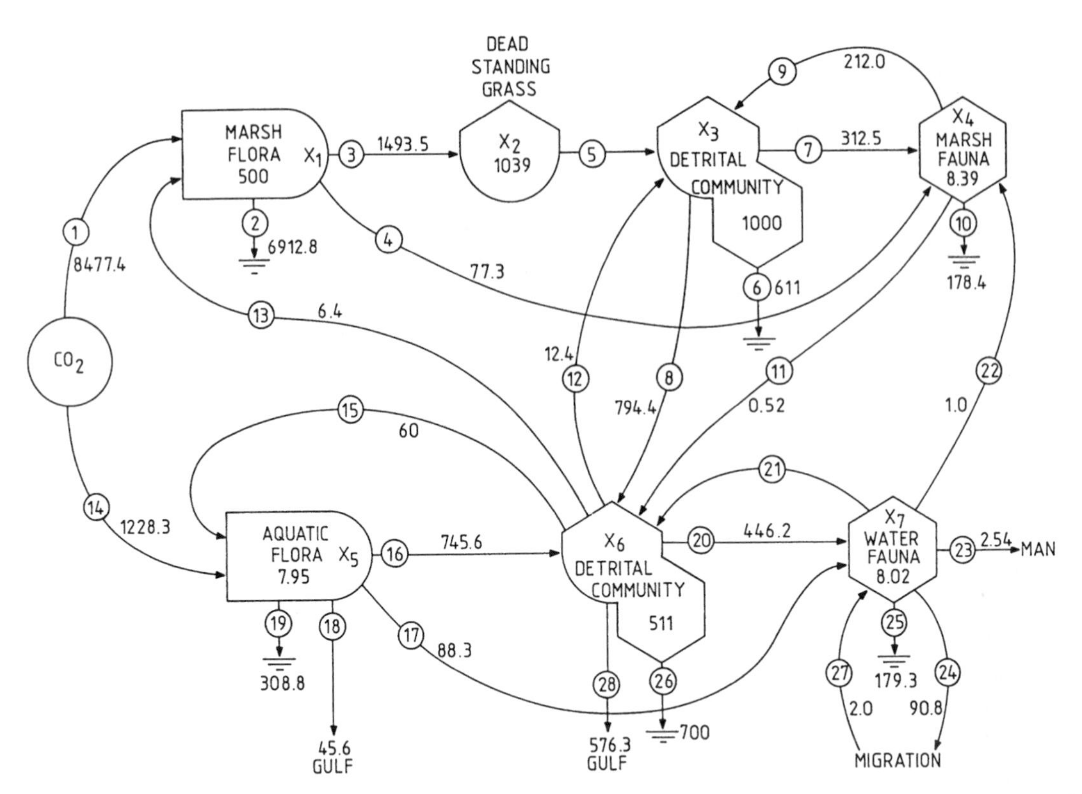

FIGURE 6.36 Carbon flow model for Barataria Bay, Louisiana. Steady-state values for carbon flows as g dry wgt m^{-2} yr^{-1} and storages as g dry wgt m^{-2}. (Modified from Hopkinson, C.S. and Day, J.W., Jr., in *Ecosystem Modelling in Theory and Practice,* Hall, C.A.S. and Day, J.W., Jr., Eds., John Wiley & Sons, New York, 1977, 242. With permission.)

major producer units are shown: the phytoplankton, the mangrove (*Avicennia marina*), and the benthic microalgae. The detrital-microbial complex includes detritus (particulate organic matter) predominantly of plant origin, but also includes material of animal origin and the organisms associated with it (bacteria, yeasts, fungi, protozoa, and meiofauna). The sediment detrital-microbial complex is similar. The macrofaunal consumer compartment includes the benthic meiofauna, macrofauna (principally polychaetes and bivalves), epifauna, zooplankton, fish, and birds.

A number of minor pathways have not been included in order to simplify the model. Some of the values for storages and flows were obtained during a 3-year study of the ecology of the Upper Waitemata Harbour (Knox, 1983a). Where data was not available, estimates were based on data from studies carried out elsewhere in New Zealand and other countries.

Of the mangrove net production of 310.25 g C m^{-2} yr^{-1}, 13.15% (40.8 g C m^{-2} yr^{-1}) becomes available to the pool of detrital organic matter via litterfall. Of the 40.8 g C m^{-2} yr^{-1} litterfall, 6.11 g C m^{-2} (15%) is released to the water DOM pool, and the balance of 26.53 g C m^{-2} (65%) becomes incorporated into the sediment POM. Of the benthic microalgal net production of 170.6 g C m^{-2} yr^{-1}, 76.86 g C m^{-2} (44.9%) is estimated to be consumed by the benthic epifauna (especially the mudflat snail *Amphibola crenata*), 32.86 g C m^{-2} (19.26%) is consumed by the benthic infauna, 24.4 g C m^{-2} (14.3%) is released as DOM, and 36.8 g C m^{-2} (22.64%) becomes part of the POM upon death. Of the net phytoplankton production of 140 g C m^{-2} yr^{-1}, 13.2 g C m^{-2} (11%) is exported from the harbor, 24.9 g C m^{-2} (20.5%) is estimated to be consumed by the zooplankton, 16.6 g C m^{-2} (15.5%) is consumed by the benthic infaunal suspension feeders, 20.0 g C m^{-2} (14.3%) is released to the water column as DOM, and 63.3 g C m^{-2} (53%) is not consumed but becomes part of the POM and eventually sinks to the bottom. Totals of 19.9 g C m^{-2} yr^{-1} of DOM and 33.25 g C m^{-2} yr^{-1} of POM are estimated to be exported from the harbor. Of the annual input to the water DOM pool of 142.81 g C m^{-2} yr^{-1}, approximately 14% is exported from the harbor, 2.1% is taken up by the zooplankton, 39.4% is taken up by the water microbial community, 13.4% is exported, and 45.5% is estimated to be aggregated by physical processes to form fine POM.

Before the construction of the model and the estimation of the carbon flows, it had been assumed that because of their large net production and litterfall, the mangroves were the major contributors to the organic matter pool within the harbor. However, from the data discussed above, it can be seen that the order in terms of their contribution is benthic microalgae, phytoplankton, and mangroves. This highlights the need to assess the contribution of all the primary producers when constructing carbon budgets for estuarine ecosystems. In the Upper

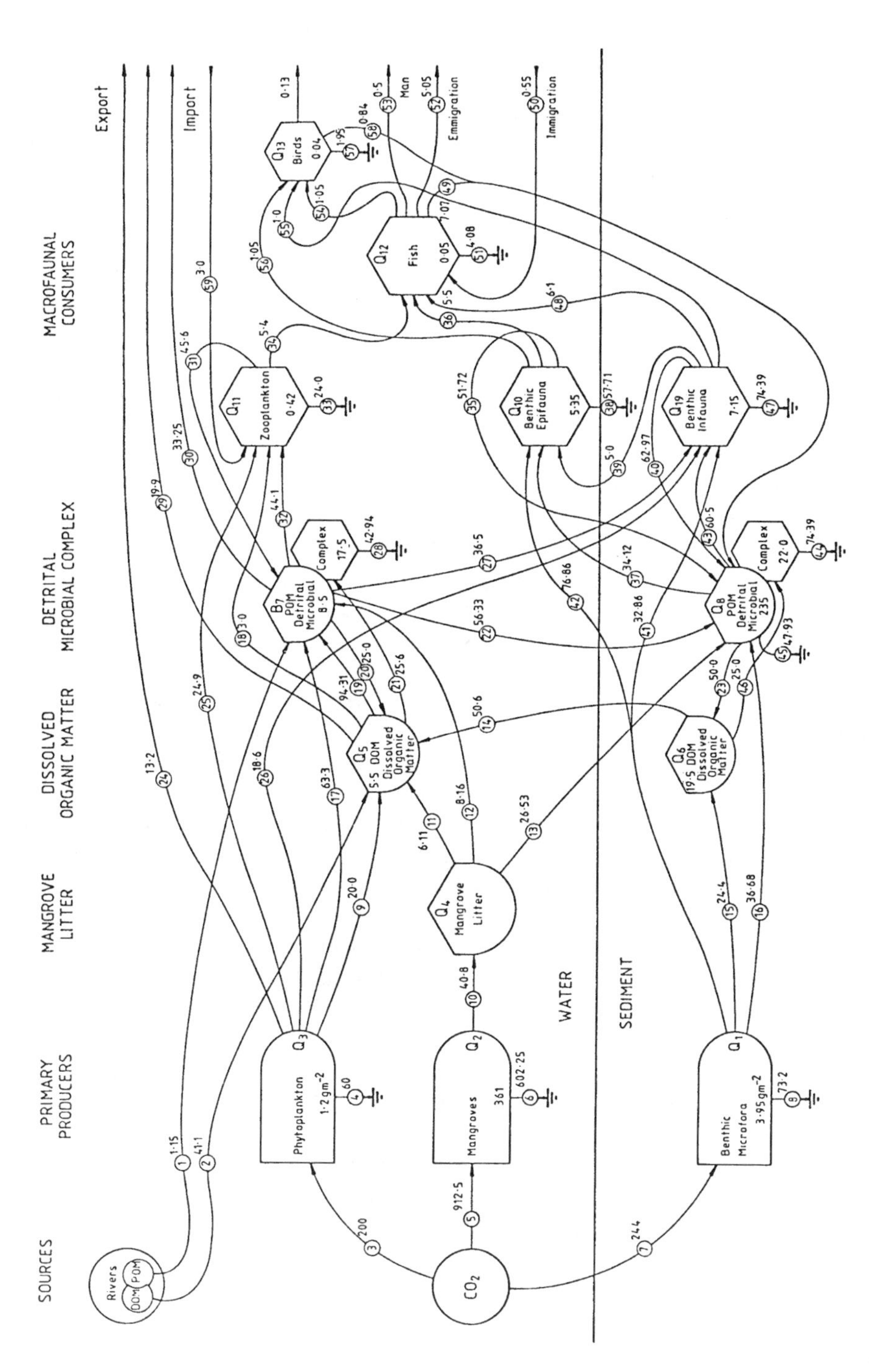

FIGURE 6.37 Carbon flow model for the Upper Waitemata Harbour, New Zealand. Steady-state values for carbon flows a g C m^{-2} yr^{-1}, and storages as g C m^{-2}. (Redrawn from Knox, G.A., *Energy Analysis: Upper Waitemata Harbour Catchment Study Specialist Report*, Auckland Regional Authority, Auckland, 1983b, 481. With permission.)

Waitemata Harbour, the mean freshwater input in relation to the tidal prism is low, although inputs from flood events can be high (see Section 3.9.9). In systems where freshwater input is high, inputs of DOM and POM from the rivers constitute the major carbon input (e.g., the Nainamo River estuary, see Section 3.8.3).

6.10.4 Interstitial Communities

The interstitial food web comprises benthic microalgae, bacteria, protozoa, and meiofauna. On exposed high-energy beaches, the benthic microfauna are of minor importance and the food web is fueled by dissolved organic matter and particulate organic matter flushed into the sand by wave and tidal action. On fine sand beaches and mudflats, the benthic microalgae are a major food resource. The importance of the interstitial system in processing organic materials is determined by four factors: (1) the volume of water filtered by the sediment; (2) its residence time; (3) the organic content of the sediment; and (4) the temperature that controls the metabolic rates of the bacteria and the interstitial fauna.

Water filtration by beach and surf zone sand has been described in Section 3.2. Essentially, the volume of water filtered by a beach increases from dissipative to reflective states, from fine to coarse sand, with decreasing tidal height and increasing wave energy (Brown and McLachlan, 1990). Residence times of this water in the sand show the reverse trends. Coarse-grained, exposed reflective beaches filter most water, generally exceeding 100 m^3 m^{-1} day^{-1}, and fine-grained, sheltered dissipative beaches the least, often less than 1 m^3 m^{-1} day^{-1}. The volumes of water passed through the bed of the surf zone are greater where the sand is coarse, where wave action is heavy, and where the water is shallow. The water input introduces large quantities of organic matter into the sediment. This organic matter is entirely or partially mineralized by the interstitial community. As shelter increases and as the sediments become finer, benthic microalgal production increases.

Trophic interactions among the benthic microbes and the meiofauna are poorly understood. Most benthic interstitial research has focused on the roles of bacteria and benthic microalgae as food for meiofauna (see reviews by Hicks and Coull, 1983, and Heip et al., 1985). Meiofauna feed primarily on bacteria, microalgae, and small detrital particles (Alongi and Tietjen, 1980; Montagna, 1984), and perhaps protozoans (Hicks and Coull, 1983), but the interactions among the organisms involved are complex and incorporate a variety of mechanisms in addition to predation (Lee, 1980; Alongi, 1988c). The meiofauna may be broadly divided into herbivores, detritovores, bacterivores, and predators (Swedmark, 1964). Nematodes, which often dominate, can, on the basis of their feeding methods, be divided into selective deposit feeders, nonselective deposit feeders, and omnivore-predators (Platt and Warwick, 1980). In terms of carbon consumption, the proportions of these groups differ in mud and sand sites. At the mud site, nonselective deposit feeders accounted for 55.9% of the carbon consumption, whereas at the sand site, where omnivore-predators dominated (57.4%), they accounted for only 16.1%. In the Exe Estuary, U.K., Kennedy (1994) found that the omnivore-predators derived a significant proportion of their diet from Metazoa and consequently they can play a more important role in the interstitial community than had previously been acknowledged. However, the harpacticoid copepods are not preyed upon by members of the interstitial community but are subject to predation by the epifauna, especialy larval flatfish.

In its simplest form, the interstitial food web consists of bacteria utilizing POC and DOC, protozoans preying on bacteria, as well as consuming benthic macroalgae, POC and DOC, meiofauna feeding on all of these, and carnivorous meiofauna, especially turbellarians, taking other meiofaunal forms. However, the situation is more complicated than this, as can be seen in Figure 6.38, which is an attempt to diagram this complexity. Bacteria take up DOC not only from the interstitial water, but also that absorbed onto sand grains. The rate at which they utilize this DOC and POC is limited by oxygen and nutrient availability (Fenchel, 1972). Interstitial protozoans utilize both benthic microalgae and bacteria, and in some cases may even prey on small members of the meiofauna.

Because of the difficulties in working on the small scale of the interstitial environment and the minute size of the biota, studies on interstitial energetics have tended to adopt a "black box" approach measuring overall interstitial activity rather than attmpting to quantitfy individual feeding relationships. This has been carried out by measuring benthic metabolism or oxygen demand in the sediment *in situ*, in cores, or in sand columns in the laboratory, and partitioning this demand between the various components of the interstitial biota. These studies have indicated that bacteria generally account for 90 to 95% of the total interstitial demand and the meiofauna some 5% (McIntyre, 1969; McLachlan et al., 1981). Detailed studies of interstitial energetics have been undertaken by only four groups of workers: on Scottish and Indian beaches (Monro et al., 1978); on the macrophyte-loaded beaches of the Western Cape, South Africa (Koop and Griffiths, 1982; Koop et al., 1982); on the exposed, diatom-fueled beaches of the Eastern Cape, South Africa (Dye, 1979; 1980; 1981; McLachlan et al., 1981); and tropical mangrove habitats in northeastern Australia (Alongi, 1988a,c).

The comparative study of Scottish and Indian beaches has revealed a number of important differences. The temperate beach had abundant benthic microalgae, which were absent on the tropical beach. DOC comprised 80% of the organic input to the Scottish beach, but only 39% on the Indian beach. Community respiration on the Scottish

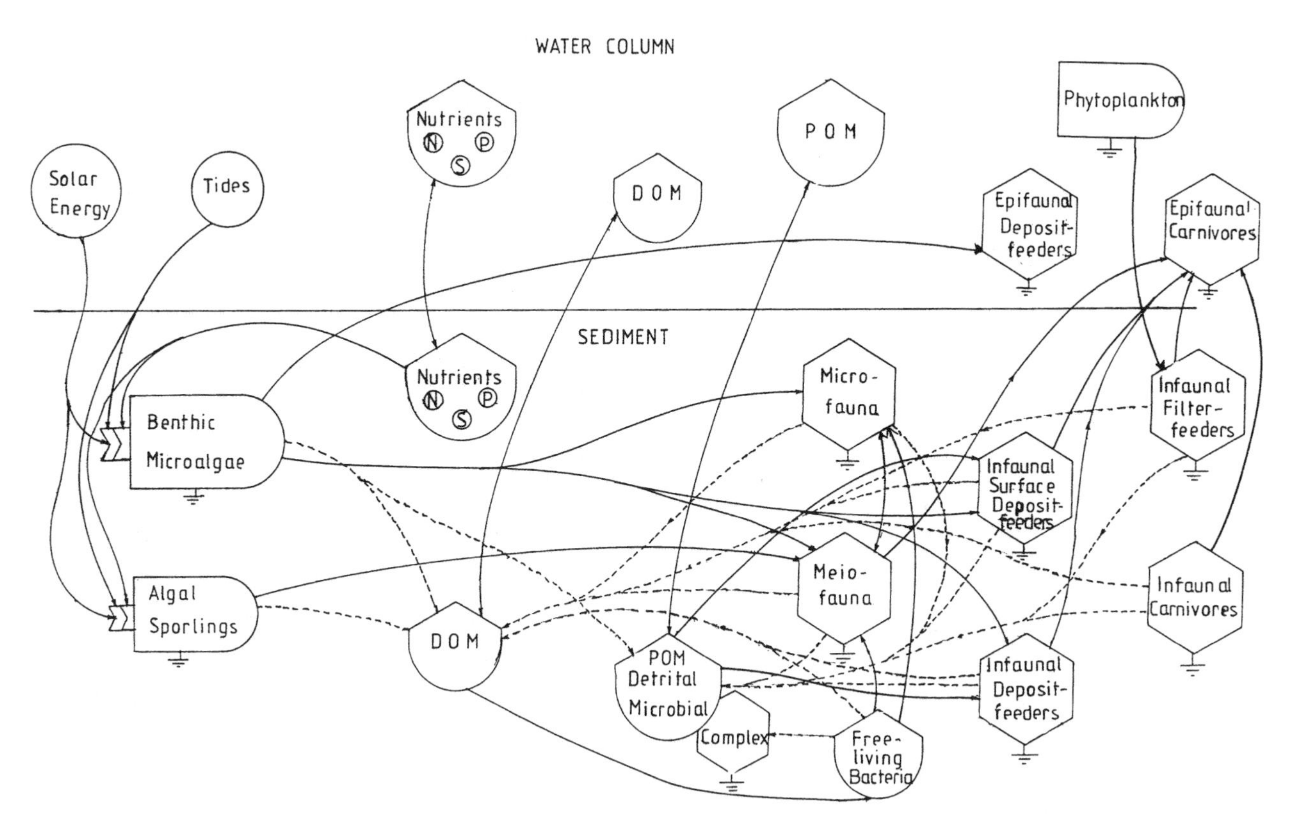

FIGURE 6.38 Model of energy flow through the interstitial soft shore ecosystem.

beach (163 g C m^{-2} yr^{-1}) was four times that on the temperate beach; however, respiration was similar if calculated for the total width of the beach, as the temperate beach was 200 m wide compared with only 40 m for the tropical beach. On this basis, community respiration on the Indian beach was 6,500 g C m^{-1} yr^{-1}, compared with 8,000 g C m^{-1} yr^{-1} on the Scottish beach. Other differences included microbial production, estimated at 72 and 15 g C m^{-2} yr^{-1} on the tropical and temperate beaches, respectively, and meiofaunal biomass, which was an order of magnitude higher in Scotland (0.2 g C m^{-2}) than in India (0.02 g m^{-2}). It was suggested that the more vigorous wave action on the Indian beach enhanced interstitial metabolism and stripped bacteria from the sediment. It can be concluded that the tropical beach, experiencing higher temperatures, high rates of water flushing, and higher surf organic levels (0.9 to 3.7 as opposed to between 0.5 to 1.0 mg C l^{-1}), promoted much greater microbial intertstitial activity.

In the Western Cape, there were rich interstitial populations that developed in association with the kelp inputs to the beach. On a beach 60 m wide, meiofaunal biomass was about 200 g C m^{-1} (or 3 to 4 g C m^{-2}) and bacterial biomass close to 300 g C m^{-1} (or 5 g C m^{-2}). Kelp leachates and breakdown products draining through the beach concentrated bacteria at lower tidal levels (Koop and Griffiths, 1982; Griffiths et al., 1983). Meiofaunal oligochaetes and nematodes were concentrated at the top of the shore, suggesting that their grazing pressure had depressed bacterial levels there. Koop et al. (1982) and Koop and Lucas (1983) showed that bacteria not only were involved in kelp breakdown on the drift line but also rapidly assimilated breakdown products (up to 5 g C l^{-1}) draining into the sand. Bacteria accounted for more than 90% of the carbon input into the sand, converting it into bacterial biomass with an efficiency of about 30%. Griffiths et al. (1983) estimated total carbon input to the interstitial system from leachates and macrofauna feces at 70,000 g C m^{-1} yr^{-1}, or 85% of the total input to the beach. They calculated from this input and the bacterial biomass present that the bacteria might have a turnover (P:B ratio) of 70 yr^{-1}, whereas meiofaunal food requirements could be met by 30 bacterial turnovers a year. It is clear that this beach processes a vast amount of kelp material via its interstitial system.

On the high-energy beaches of the Eastern Cape, where there is a well-developed surf zone, with diatom "blooms," interstitial energetics have been studied both in the intertidal and subtidal. The biomass estimates for bacteria, protozoans, and meiofauna in the 60-m wide intertidal beach averaged approximately 10, 3, and 15 g C m^{-1}, respectively. Water filtration flushes some 10 m^3 m^{-1} day^{-1} through the system (McLachlan, 1989), which introduces large quantities of POC and DOC. Surf diatoms also migrate into the sand at night and they may exude significant amounts of mucus there. Studies of oxygen consumption (Dye, 1981) showed that the interstitial system consumes, on average, 3,200 g C m^{-1} yr^{-1}, partitioned between bacteria (60%),

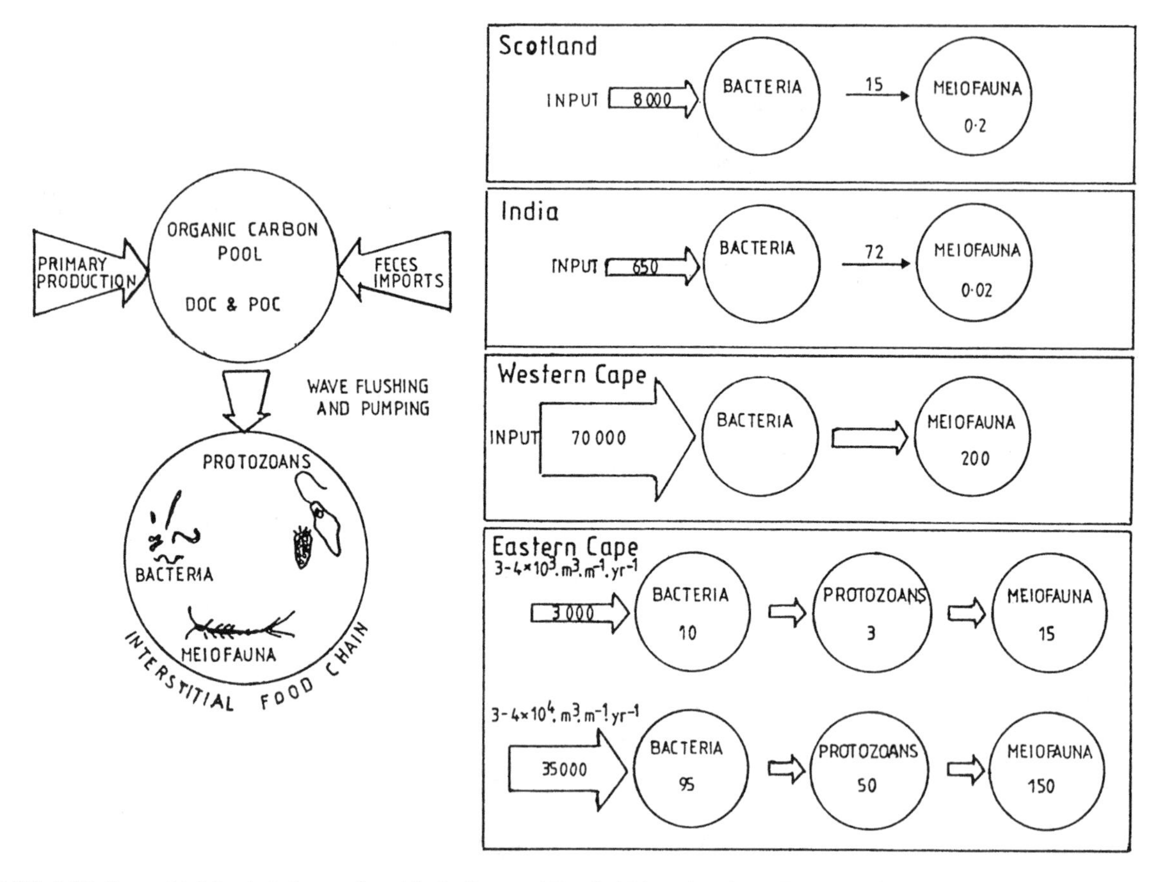

FIGURE 6.39 Interstitial food chains on Scottish, Indian, and South African beaches. (Redrawn from Brown, A.C. and McLachlan, A., *Ecology of Sandy Shores*, Elsevier, Amsterdam, 1990, 209. With permission.)

protozoans (20%), and meiofauna (20%). The interstitial fauna is abundant in the upper 1 to 2 cm in the intertidal sediments, but becomes concentrated in the top 20 to 50 cm in the subtidal (Malan and McLachlan, 1985).

Figure 6.39 summarizes some of these interstitial processes. On the vast majority of beaches, the interstitial food web consumes more carbon than does the macroscopic food web. Approximate proportions are 77% interstitial web to 23% macrofaunal web on the Scottish beach, 62% to 38% on the Indian beach, 97% to 3% on the South African Western Cape beach studied by Griffiths et al. (1983), and 63% to 37% on Eastern Cape beaches excluding the surf zone.

Alongi (1988c) has reviewed microbial-meiofaunal interrelationships in some tropical intertidal sediments, and considers that abundances of meiofaunal consumers relative to microbial prey, rates of nutrient supply, and species composition of consumers groups are only a few of the factors determining the extent of microbe-meiofaunal interactions. On the basis of his findings he concluded that Gerlach's (1978) hypothesis that meiofauna stimulated bacterial productivity is not readily applicable to microbial-meiofaunal food chains on tropical mangrove beaches and sand flats in northern Australia. Many of the attributes that he described in these tropical systems (Alongi, 1989) have little parallel in temperate communities, which were used as the basis of Gerlach's calculations due to: (1) very fast (>0.5 day^{-1}) rates of bacterial growth and very high rates of bacterial productivity (15 to 1,725 mg C m^{-2} day^{-1}); (2) low densities of hard-bodied meiofauna (10^2 to 10^5 m^{-2}); (3) very low (generally 5 μg C chl *a* g^{-1} dry wgt sediment) microbial standing stocks; (4) lack of tidal migration; (5) numerical dominance of turbellarians; and (6) lack of seasonality in most populations. Thus, generalizations based on temperate work may not necessarily be relevant to tropical shores (Alongi, 1988c).

6.11 STABLE ISOTOPES AND FOOD WEB ANALYSIS

Identification of food sources for many consumers, especially in estuaries, poses a major problem for ecologists. Early studies (Odum, 1968; Tagatz, 1968; Odum and Heald, 1972) were based on gut analyses, which provided only rough estimates of the percentages of the various materials ingested by the species under investigation, and were complicated by the presence of material in the guts that was not potentially assimilated. While the proportion of detrital material present in the gut contents can be determined, the biomass of the associated microorganisms

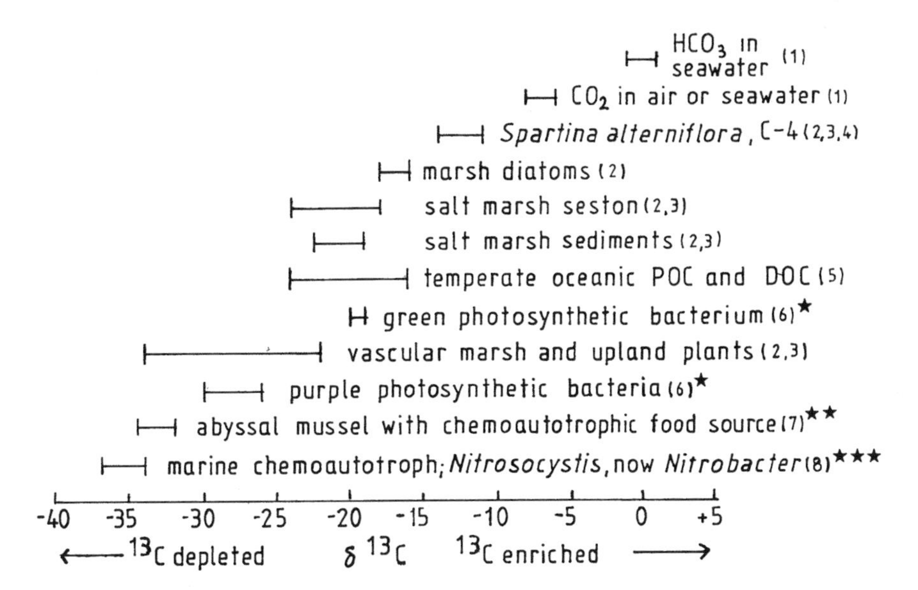

FIGURE 6.40 Carbon isotope ratios for selected components of marine ecosystems, with emphasis on salt marshes. Results are expressed a parts per thousand deviation from the PDB carbonate standard:

$$\delta^{13}C \frac{^{13}C/^{12}C \text{ sample}}{^{13}C/^{12}C \text{ standard}} - 1 \times 1,000$$

(Redrawn from Peterson, B.T., Howarth, R.W., Lipschultz, F., and Ashendorf, D., *Oikos*, 34, 175, 1980. With permission.)

(bacteria, fungi, microalgae, and protozoans) are not estimated. A further complication for those species that consume microalgae, macroalgae, and detritus is that ingestion studies do not usually provide adequate information on the original plant sources.

A central tenet of estuartine ecology over the past three decades has been that organic detritus, derived mainly from vascular plants, is a major food source for estuarine consumers. Reviews of the role of organic detritus in estuarine ecosystems can be found in Mann (1972a,b; 1982), Nixon (1980), Knox (1986b,); Mitsch and Gosselink (1986), Day et al. (1987), and Valiela (1995). The concept of "outwelling" of organic matter from marshes to estuarine and nearshore waters (see Sections 3.5.5.4 and 3.9) has been the subject of many studies and reviews (e.g., Odum and de la Cruz, 1967; Nixon, 1980; Odum 1980). However, this concept, as discussed previously, has frequently been challenged.

Determining carbon flow in estuaries in particular is difficult because of the problems encountered in quantifying the total net primary production by producers, in measuring the dispersal of the carbon thus produced by tides, and in determining the consumption of this carbon by bacteria, fungi, and other consumers both in the water column and in the sediments (Peterson and Howarth, 1987). The development of the use of stable carbon isotopes as an analytical tool in the mid- and late-1970s (Haines, 1976a,b; 1977; 1979; Haines and Dunstan, 1975; Haines and Montague, 1979) showed considerable promise in interpreting the sources of organic carbon in marine ecosystems and their contributions to the diet of consumers. This approach involved the measurement of the ratios of the naturally occurring isotopes of carbon (^{12}C and ^{13}C) in different producers and consumers. Different groups of plants have different photosynthetic pathways (C_3, C_4, etc.), different sources of CO_2 (air vs. seawater), and consequently have different ratios of the stable isotopes ^{12}C and ^{13}C. The isotopic ratio is thought to be conserved in the organisms that consume a given plant, so that the ratio can be used as a quantitative tracer of food pathways. The conventional way of representing this ratio is the quantity of ^{13}C defined as:

$$\delta^{13}C = \frac{^{13}C/^{12}C \text{ sample}}{^{13}C/^{12}C \text{ standard}} \times 1{,}000$$

where the standard is a marine belemnite limestone referred to as the "Chicago PDB standard"; C_3 vascular plants have low $\delta^{13}C$ values of –24 to –34‰, while C_4 plants have higher values of –6 to –19‰. Algae tend to have intermediate values of –12 to –23% (see Figure 6.40 and Table 6.25).

On the Sapelo Island marshes (Haines, 1976a,b; Haines and Montague, 1979), it was found that the C_4 marsh plants (*Spartina alterniflora, S. cynosuroides, Distichlis spicata,* and *Sporobolus virginicus*) had $\delta^{13}C$ values of –12.3 to –13.6‰, while C_3 marsh plants (*Juncus roemerianus, Salicornia virginicus, Batis maritima,* and *Borrochia frutescens*) had values of –22 to –25‰ (see Table

TABLE 6.25
Carbon Isotope Composition of Organisms in the Marshes and Coastal Waters of Georgia

	$\delta^{13}C$ (‰)
Marsh grasses	
Spartina alterniflora (live)	–12.7 to –13.6
Spartina alterniflora (dead)	–12.3
Distichlis spicata	–13.3
Juncus roemerianus	–22.8
Salicornia virgincus	–26.0
Sediment	
Sediment surface (0–5 cm)	–16.0 to –22.6
Organic matter in sediment (0–5 cm)	–13.2 to –13.4
Benthic microflora	–16.2 to –17.9
Marsh fauna	
Orchelium fudicinum	–13.2
Palaemonetes pugio	–13.6
Littorina irrorata	–14.7
Ischnodemus badius	–15.0
Gammonota trivittata	–15.2
Nassarius obsoletus	–15.9
Uca pugilator	–16.2
Uca pugilator	–16.8
Uca pugnax	–16.8
Geukensia demissa	–17.0
Uca pugnax	–17.8
Sesarma sp.	–19.1
Mercinaria mercinaria	–18.5
Crassostrea virginica	–21.0
Phytoplankton	
Diatom bloom (*Skeletonema costatum*)	–22.1 to –22.7
Dinoflagellate bloom (*Kryptoperidium* sp.)	–20.0
Green flagellate bloom	–26.3
Suspended particulate matter	
Estuary tidal creeks	–19.8 to –22.8
Shelf (0 to 10 km offshore)	–18.0 to –22.3
Shelf (20 km offshore)	–21.0 to –23.9

Source: After Haines (1976b) and Haines and Dunstan (1975).

6.25 and Figure 6.41A). Benthic microalgae, phytoplankton, and organic seston in the estuary had $\delta^{13}C$ value of –16 to –26‰, with the $\delta^{13}C$ of the benthic microalgae averaging –17‰ and that of the phytoplankton –212‰. Grazers that consume live *Spartina*, such as the grasshopper *Orchelimum fudicinum*, have $\delta^{13}C$ values (–13.2‰) of the vascular plants nearby. Detritovores, on the other hand, have values related to the detrital source, e.g., they reflect the higher proportion of *Juncus* detritus in a *Juncus* stand and of *Spartina* detritus in a *Spartina* stand. The $\delta^{13}C$ values of fiddler crabs (*Uca* spp.) are skewed in the direction of the benthic microalgae. However, it is possible that the crab's diet consisted of a mixture of about 50% each of *Spartina* and *Juncus* detritus, or 50% each of *Spartina* detritus and phytoplankton that had sedimented to the marsh surface. In contrast to the deposit feeders, filter-feeding mussels, clams, and oysters have $\delta^{13}C$ values in the lower range characteristic of algae and chemoautrophic microorganisms, which suggests an intermediate to no influence of *Spartina* carbon. Ribbed mussels (*Geukensia demissa*), which occur on the salt marsh itself, have the highest values, averaging –17‰. Edible clams (*Mercenaria mercenaria*) that are embedded in the tidal creek bottoms have $\delta^{13}C$ values of –18 to –19‰. The oyster (*Crassostrea virginica*) has an isotopic composition of –21‰.

Hughes and Sherr (1983) have analyzed the stable carbon isotope composition of the subtidal food web of the Sapelo Island marshes. The results for consumers (invertebrates and fishes) were –17.8 and –17.5‰, respectively, in a *Spartina* creek, and –20.1 and 20.0‰, respectively, in two *Juncus* creeks. These intermediate $\delta^{13}C$ values (higher than that of *Spartina* and lower than that of *Juncus*) in the fauna implied an alternate carbon input to the food web, most probably benthic microalgae ($\delta^{13}C$ of –16.0 to –18.0‰), and phytoplankton ($\delta^{13}C$ of –20.0 to –22.0‰). In the *Spartina* marsh creek, animals with $\delta^{13}C$ values less negative than –17.0‰, e.g., mummichag (*Fundulus heteroclitus*), striped mullet (*Mugil cephalus*), and mud crabs (*Panopeus herbsti*) were most closely linked to the *Spartina* detritus/benthic microalgal food web, while species with $\delta^{13}C$ values more negative than –19.0‰, e.g., brown shrimp (*Penaeus aztecus*), menhaden (*Brevoortia tyrannus*), and oysters (*Crassostrea virginicus*) were apparently more dependent on the phytoplankton food web. Hughes and Sherr (1983) concluded that: (1) marsh plant carbon is a component of the subtidal food web; (2) phytoplankton is a more important food source for subtidal animals than for intertidal animals; and (3) the subtidal food web is structured such that individual invertebrate and fish species show varying degrees of dependence on the detritus/benthic macroalgal-based food web of the marsh vs. the phytoplankton-based food web of the tidal creek.

McConnaughey and McRoy (1978) examined the stable carbon isotope composition of the food web of Izemkek Lagoon, Alaska. They found that the $\delta^{13}C$ for eelgrass (*Zostera marina*) was –10.3 to 0.04‰, while phytoplankton had values from 22.1 to 17.3‰ depending on tide, net mesh, and turbulence. Ratios in a range of estuarine animals were substantially enriched when compared with those of the adjacent Bering Sea. The values obtained were consistent with the known diets of the species concerned, and it was concluded that the animal community of the lagoon was sustained largely by eelgrass detritus. Stephenson and Lyon (1982) used stable carbon isotope analysis to define food sources of the cockle *Austrovenus stutchburyi* in the Avon-Heathcote Estuary, New Zealand. At five locations, separated by less than 4 km, but subject to different hydrological regimes and different

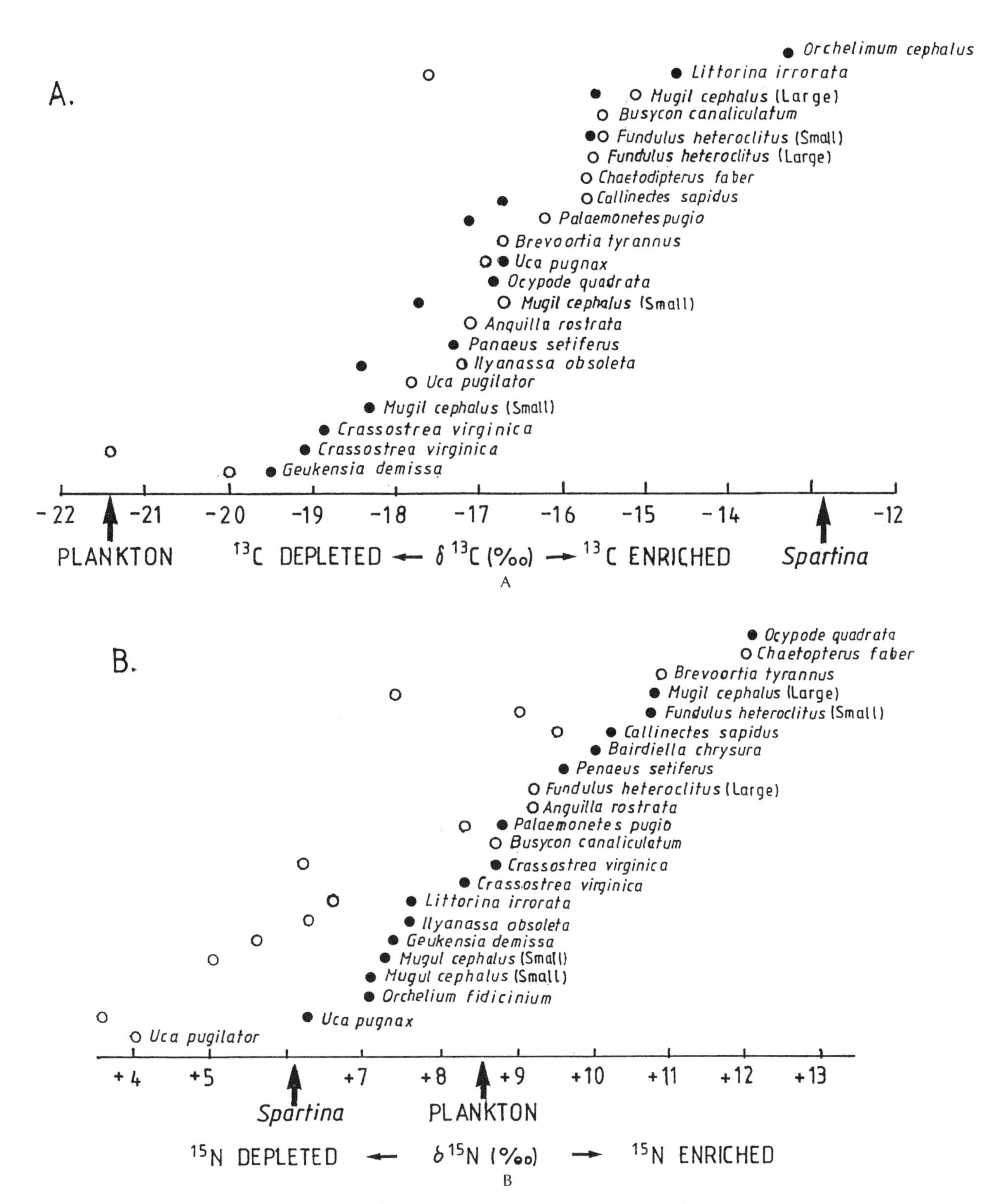

FIGURE 6.41 A: The ^{13}C values for consumers in the salt marshes at Sapelo Island in May (●) and September (○). B: As for A, but for ^{15}N values. C: As for A, but for ^{34}S values. (Redrawn from Peterson, B.J. and Howarth, R.W., *Limnol. Oceanogr.*, 32, 1205 and 1206, 1987. With permission.)

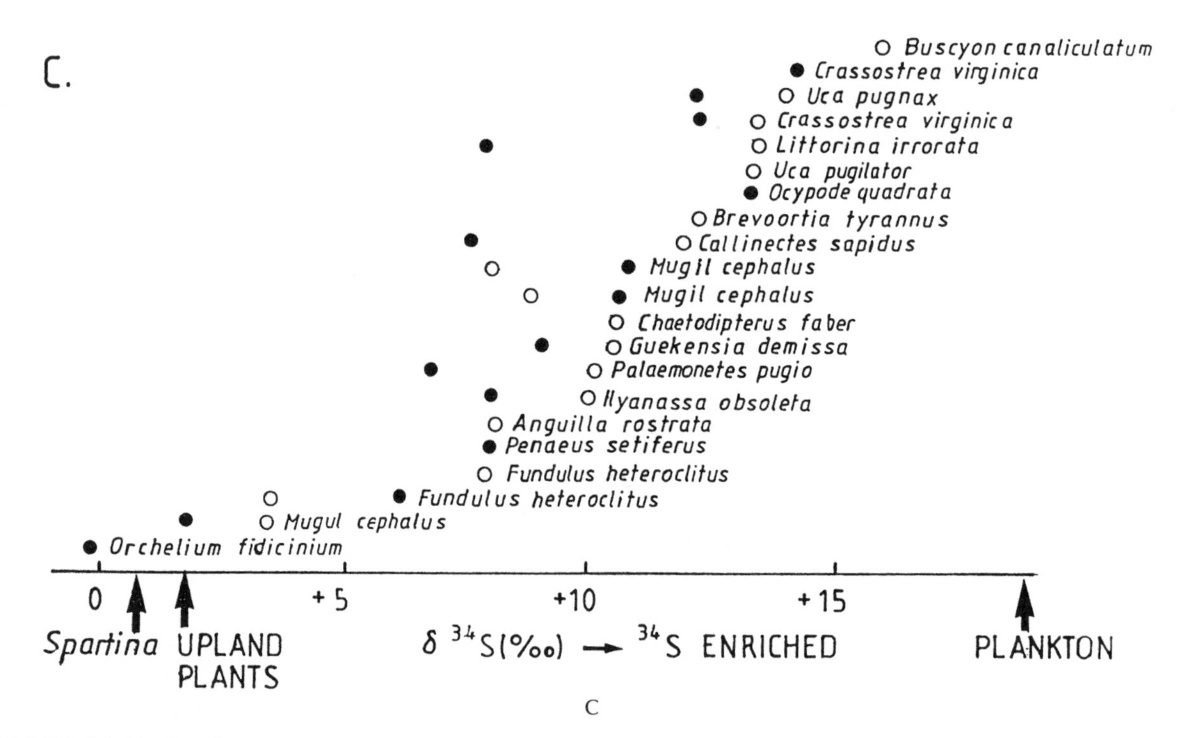

FIGURE 6.41 *Continued.*

isotopic ratios in the particulate matter suspended in the inflowing water, the $\delta^{13}C$ values of *A. stutchburyi* were significantly different (from –16.7 to –23.5‰). *A. stutchburyi* was thus shown to utilize carbon of riverine, oxidation pond effluent ouflow, and marine origin in varying proportions depending on its location in the estuary.

Incze et al. (1982) have compared trends in $\delta^{13}C$ values for particulate organic carbon and the tissues of several filter-feeding bivalves (*Mytilus edulis, Mya arenaria, Geukensia demissa, Modiolus modiolus*, and *Mercenaria mercenaria*) in two estuarine systems in Maine; one, the Sheepscot estuary (with a large volume of freshwater input) and the other, the Damariscotta River Estuary (with little freshwater input). The POC samples showed little isotopic changes over the length of the Damariscotta estuary, but in the Sheepscot estuary, a systematic landward decrease was found from –23.8‰ at the mouth to –25.0‰, 32.8 km upstream. This is consistent with an increase in the relative contribution of isotopically light, terrigenous detritus to the POC pool. These differences were reflected in the tissues of the bivalves. In the Damariscotta, $\delta^{13}C$ values ranged from 16.5 to 18.8‰ with no upstream trend, whereas in the Sheepscot, they showed progressively lighter isotopic composition upstream, ranging from –18.7 to –22.5‰ for *Mytilus edulis* and from –18.4 to 21.8‰ for *Mya arenaria*. Variations in the above pattern could be attributed to: (1) the influence of locally abundant, isotopically distinct aquatic macrophytes such as *Zostera*; and (2) surface deposit feeding on benthic microalgae with high $\delta^{13}C$ values.

The early work on the Sapelo Island marshes has been updated by Peterson and Howarth (1987), and their results will be discussed below. Once fixed during photosynthesis, the carbon isotopes show little further fractionation during the death and decay of the plant material, or during the assimilation of organic matter by the animals. Work on stable carbon isotopes up to 1980 has been reviewed by Stephenson (1980). While the technique has proved valuable in cases where there are few food sources and where these food sources are isotopically distinct, results have been difficult to interpret in cases of multiple carbon sources or overlap in the isotopic composition of the food (e.g., Schwinghamer et al., 1983). One reason for this is that a mixture of, say, carbon from *Spartina*, and that derived from upland plants would very likely have a carbon isotope ratio similar to that of phytoplankton (Peterson et al., 1985). As we have seen in Chapter 3 there are often large inputs of allochthonous carbon from rivers into estuaries and coastal waters.

The chemoautotrophic bacteria that use sulfide as an energy source to fix CO_2 may also complicate the carbon isotope story. These bacteria use energy provided by the anaerobic belowground decomposition of *Spartina* to fix CO_2 in marsh sediments, or at the sediment surface, and their production in marshes may be high (Howarth, 1984). The energy is derived indirectly from *Spartina*, but the carbon isotopic ratio of the bacteria may be similar to that of plankton (Peterson et al., 1980; 1986).

However, these early studies using just the stable carbon isotope were somewhat limited in their ability to

TABLE 6.26
$\delta^{13}C$ Values (‰) of Common Macrophytes of St. Margarets Bay, Nova Scotia, Canada

Macrophyte	$\delta^{13}C$
Algae	
Desmarestia virides	–29.1
Agarum crobosum	–25.3
Fucus distichus	–23.7
Fucus vesiculosus	–14.3
Ascophullum nodosum	–21.1
Palmaria palmata	–19.0
Alaria esculenta	–17.8
Laminaria longicruris	
Blade tip	–15.4 to –17.5
Blade base	–16.8 to –17.3
Stipe	–16.7 to –19.4
Holdfast	–17.5 to –18.4
Sea Grass	
Zostera marina	
Green leaves	–7.5 to –8.0
Blackened leaves	–9.7 to –12.4
Dead (buried) leaves	–10.2 to –12.1
	–6.0 to –10.0

Source: After Stephenson, R.L., Tan, F.C., and Mann, K.H., *Mar. Biol.*, 81, 228, 1984. With permission.

differentiate organic matter sources because a mixture of terrestrial ($\delta^{13}C$ –27‰) and salt marsh ($\delta^{13}C$ –13‰) organic matter yielded organic matter with a $\delta^{13}C$ similar to marine phytoplankton ($\delta^{13}C$ –21‰) (Peterson et al., 1985; Peterson and Fry, 1987). Sea grass $\delta^{13}C$ values reported in the literature show variation over a range of approximately 20‰ (Hemminga and Mateo, 1996). A frequency histogram on the basis of the reported data shows a unimodal dstribution, with values between –10 and –11‰ being found with the highest frequency. Fry et al. (1982), who studied Caribbean sea grass meadows, found that the isotopic values of a large proportion of the algal species were in the same range as the dominant sea grass species.

A further complication is due to some plants showing considerable variation in stable carbon isotopes, both in space and time. Stephenson et al. (1984) investigated stable carbon isotope variability in the marine macrophytes, the kelp *Laminaria longicruris*, and the eelgrass *Zostera marina* in St. Margarets Bay, Nova Scotia (Table 6.26). The isotopic composition of *Z. marina* varied seasonally from a mean of –6‰ for leaves formed in June to a mean of –10‰ for leaves formed in February. The maximum range for individual leaves was from 5 to –11.4‰.Once a leaf is formed, its isotopic composition did not appear to change. This variation is as large as (and includes most of) the geographical isotopic range of the species (–7.8 to –12.4‰) reported by McMillan et al. (1980). In *L. longicruris,* there was no clear seasonal pattern of variation, but in any given blade there was a spatial pattern of variation, with the central thickened band tending to be the least negative and the margins most negative. In one blade, the range was from –12 to –20‰. Since the variation in *L. longicruris* is large and unpredictable, and overlaps the literature values for phytoplankton (Tan and Strain, 1979), as well as the values obtained in St. Margarets Bay for seston (–19‰) and other common macrophytes, the isotope technique appears to be of limited usefulness for tracing this species through the food web. *Z. marina*, on the other hand, has a unique isotopic ratio that is seasonally predictable, so that the isotope ratio might be used as a tracer if interpreted with care.

Peterson and Howarth (1987) consider that one measure for resolving ambiguities such as those discussed above is to use additional stable isotopic ratios. Multiple stable isotope comparisons provide significantly more power to resolve food web structure than a single isotope approach (Peterson et al., 1985). Sulfur is a second isotopic tracer of the important organic matter sources in estuaries and on mudflats. The sulfur isotopic approach is based on the fact that the sulfur available to and taken up by upland plants, estuarine phytoplankton, and *Spartina*, have different isotopic ratios. Upland plants use sulfate from precipitation ($\delta^{34}S$ = +2 to 8‰), estuarine phytoplankton use sulfate in seawater ($\delta^{14}S$ = +21‰), and *Spartina* takes up sulfate from the sediments for a fraction of its total sulfur uptake. Sulfide produced via sulfate reduction in sediments averages 50% higher than the sulfate in pore waters in a Massachusetts marsh (Peterson et al., 1986). The dual isotope approach using C and S isotopes can be extended to include the isotopes of nitrogen, ^{15}N and ^{14}N. Atmospheric nitrogen has a $\delta^{15}N$ value of zero, and upland vegetation tends to have $\delta^{15}N$ close to 0‰, whereas *Spartina* has $\delta^{15}N$ values around +6‰, and nearshore phytoplankton have values around +6 to +10‰ (Macko et al., 1984; Altabet and McCarthy, 1985).

Table 6.27 lists the $\delta^{13}C$, $\delta^{15}N$, and $\delta^{34}S$ for producers and consumers in the Sapelo Island marshes. The general impression gained from Table 6.27 is that with the exception of the leafhopper, all of the consumers fall between the $\delta^{13}C$ and $\delta^{34}S$ values for the potential foods *Spartina* and phytoplankton. The $\delta^{15}N$ values tend also to fall within the range found for *Spartina* and phytoplankton, but about one third of the consumers have a $\delta^{15}N$ value of +10‰ higher than those expected for *Spartina* and for phytoplankton. One reason for this is that consumers fractionate nitrogen by from +1 to +5‰ per trophic transfer. A second potential reason for the $\delta^{15}N$ shifts is that during decomposition, the $\delta^{15}N$ values of detrital organic matter may change due to microbial immobilization of nitrogen (Macko and Zieman, 1983).

TABLE 6.27
$\delta^{13}C$, $\delta^{15}N$, and $\delta^{34}S$ for Producers and Consumers Collected at Sapelo Island, Georgia, in May 1982 and September 1983; The Feeding Mode is Indicated for the Fauna

	$\delta^{13}C$ (%)		$\delta^{14}N$ (%)		$\delta^{34}S$ (%)	
	May	Sept	May	Sept	May	Sept
Producers						
Spartina alterniflora (cordgrass)	−13.1	−12.8	+4.3	+7.2	−2.3	+2.6
Juncus roemerianus	−23.6	−25.0	+2.8	+5.1	—	+6.1
Pinus taeda (yellow pine)	−27.1	−29.7	−1.4	+0.6	—	+2.5
Quercus virginiana (live oak)	−30.1	−29.9	+0.22	+0.9	—	+1.1
Creekbank algae	−16.7	—	+3.5	—	—	—
Consumers						
Herbivore						
Orchelimum fidcinium (grasshopper)	—	−13.2	—	+7.1	—	−0.2
Suspension feeders						
Geukensia demissa (ribbed mussel)	−20.0	−19.5	+5.6	+7.4	+10.5	+9.0
Crassostrea virginica (oyster)	−21.4	−19.1	+6.2	+8.7	+12.2	+13.4
Brevootia tyrannus (menhaden)	−16.7	—	+10.8	—	+12.2	—
Deposit-suspension feeders						
Palaemonetes pugio (grass shrimp)	−16.2	−17.1	+8.3	+8.8	+10.1	+6.8
Penaeus setiferus (white shrimp)	-	−17.1	—	+9.6	—	+8.0
Mugil cephalus (large mullet)	−15.1	−15.6	+7.4	+10.8	+8.0	+10.8
M. cephalus (small mullet)	−16.7	−17.7	+4.8	+7.3	+3.4	+1.8
Deposit feeders						
Uca pugnax (mud fiddler crab)	−16.9	−16.7	+3.6	+6.3	+12.1	+13.9
Uca pugilator (sand fiddler crab)	−17.8	—	+4.0	—	+13.3	—
Ilyanassa obsoleta (mud snail)	−17.2	−18.4	+6.3	+7.6	+10.0	+8.0
Littorina littorea (marsh periwinkle)	−17.6	−14.6	+6.4	+7.6	+13.4	+7.9
Omnivores						
Fundulus heteroclitus (small killifish)	−15.5	−15.6	+9.0	+11.5	+3.5	+6.2
F. heteroclitus (larger killifish)	−15.6	-	+9.2	—	+7.9	—
Callinectes sapidus (blue crab)	−15.7	−16.7	+9.5	+10.5	+11.9	+7.6
Ocypode quadrata (ghost crab)	—	−16.8	—	+12.3	—	+13.3
Predators						
Anguilla rostrata (American eel)	−17.1	—	—	+9.2	—	+8.1
Bardiella chrysura (silver perch)	—	−16.6	—	+10.8	—	+8.0
Chaetodipterus faber (spade fish)	−15.5	—	+12.0	—	+15.9	—

Source: From Peterson, B.J. and Howarth, R.W., *Limnol. Oceanogr.*, 32, 1204, 1987. With permission.

Figure 6.41A plots the May and September $\delta^{13}C$ values for the marsh consumers, with values spanning the range from −22 to −12‰. The marsh fauna fall in between the $\delta^{13}C$ values characteristic of phytoplankton (−21‰) and *Spartina* (−13‰). The seasonal differences are small, apart from values for *Littorina irrorata* and *Crassostrea virginica*. These shifts may indicate dietary changes throughout the year, changes in the isotopic composition of the foods, or perhaps metabolic shifts with the production of gametes. By and large, the $\delta^{13}C$ values fall in the range of −18 to −14‰, which appears to indicate that a mixture of *Spartina* and phytoplankton is used, or that other foods such as benthic microalgae ($\delta^{13}C = -16.7‰$) are important. Benthic deposit feeders (see Table 6.27) have $\delta^{13}C$ values that range from −16.9 to −17.8‰, which are close to the benthic microalgae value. Suspension feeders, on the other hand, such as the ribbed mussel *Geukensia demissa* and the oyster *Crassostrea virginica*, have values (−18.8 to −21.8‰) that would indicate that phytoplankton was a major component of their diet.

The $\delta^{15}N$ values are centered around the range expected for phytoplankton (+7 to +11‰) (Figure 6.41B). Very few of the fauna have $\delta^{15}N$ values as low as the mean value for *Spartina* (+6‰). However, animal metabolism tends to shift consumer tissue values in the positive direction (Minagawa and Wada, 1984). A surprising and unexplained result was that the $\delta^{15}N$ values for animals collected in May were considerably lower than values for animals

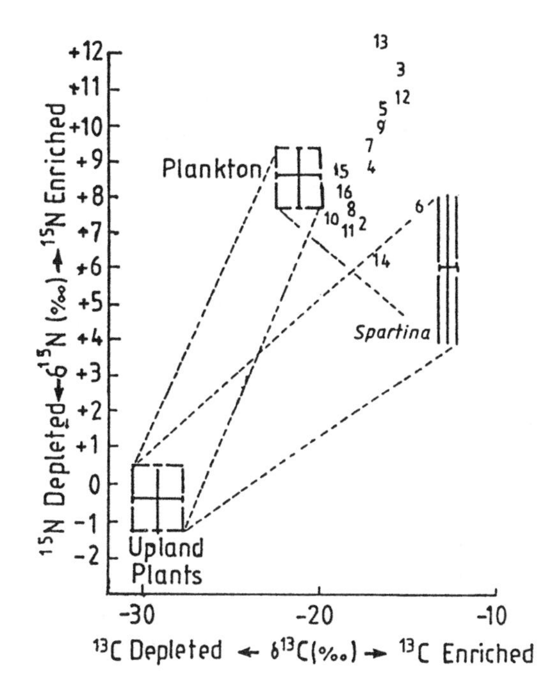

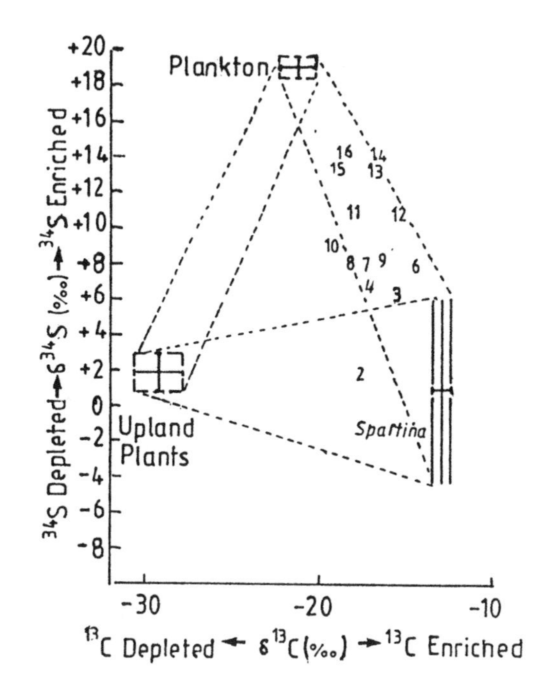

FIGURE 6.42 A. Carbon and nitrogen isotopic ratios in relation to the ratios in potential organic matter sources in September 1983. B. Carbon and sulfur isotopic ratios in relation to the ratios in potential organic matter sources. Key to consumers: 1. *Orchelimum fidicinium*; 2. *Mugil cephalus*; 3. *Fundulus heteroclitus*; 4. *Palaeminetes pugio*; 5. *Callinectes sapidus*; 6. *Littorina irrorata*; 7. *Penaeus setiferus*; 8. *Ilyanassa obsoleta*; 9. *Bairdiella chrysura*; 10. *Geukensia demissa*; 11. *Mugil cephalus* (small); 12. *Mugil cephalus* (large); 13. *Ocypode quadrata*; 14. *Uca pugnax*; 15. *Crassostrea virginica*; 16. *C. virginica*. (Redrawn from Peterson, B.J. and Howarth, R.W., *Limnol. Oceanogr.*, 32, 1208, 1987. With permission.)

collected in September. This difference may indicate an important seasonal change in the marsh nitrogen cycle.

From Figure 6.41C it can be seen that the upland plants and *Spartina* have low $\delta^{34}S$ values while the phytoplankton has a high value of +18 to +20‰. The consumer values range from +6 to +14‰, with the majority concentrated in the range from +6 to +14‰, which are between the values for *Spartina* and phytoplankton. This pattern is very similar to the situation for the consumer $\delta^{15}N$ values shown in Figure 6.41B. The herbivore, *Orchelimum fidicinium*, is isotopically similar to *Spartina* and is known to feed directly on *Spartina*. The organisms with relatively high $\delta^{34}S$ values include the whelk *Busycon canaliculatum*, fiddler crabs *Uca* spp., and the marsh periwinkle *L. irrorata*. The crabs and the periwinkle are deposit feeders and the benthic microalgae they feed on obtain most of their sulfur from seawater.

Peterson and Howarth (1987) consider that when there are three or more potentially important sources of food for animals, it is not possible to discriminate between different sources of organic matter through the use of a single tracer, unless the isotopic ratios in consumers cluster about one end member. However, a combination of two or more isotopes should make it possible to identify the important organic matter sources more clearly (Rau et al., 1981; Fry, 1984; Peterson et al., 1985). The relationships between marsh consumers and their potential foods on diagrams of $\delta^{13}C$ vs. $\delta^{34}S$, and $\delta^{13}C$ vs. $\delta^{15}N$ for September 1983, are shown in Figure 6.42 In Figure 6.42A the consumer organisms are numbered in order of lowest to highest $\delta^{34}S$ values. The locations of the potential foods are identified as rectangles indicating the mean ±1 SD for potentially important foods. The macroconsumers are distributed systematically within or close to the band of carbon and sulfur values expected if these organisms derive most of their nutrition from *Spartina*, from phytoplankton, or most often a mixture of the two. In the plot of C vs. N (Figure 6.42B), the distribution of consumers describes a vertical band of values intermediate between phytoplankton and *Spartina* on the $\delta^{13}C$ axis and ranging from *Spartina* to greater than phytoplankton the $\delta^{15}N$ axis. The distributions support the conclusion that marsh consumers derive most of their organic matter from a mixture of *Spartina* and phytoplankton.

Sullivan and Moncreiff (1990) measured carbon, sulfur, and nitrogen stable isotope ratios in an irregularily flooded Mississippi salt marsh to evaluate the importance of vascular plants and algae in the food web. Primary producers had distinct $\delta^{13}C$ values: *Spartina alterniflora* (–13‰), *Juncus roemerianus* (–26‰), and benthic microalgae (–21‰). A pure zooplankton sample, which should closely approximate the phytoplankton, had a $\delta^{13}C$ value of –23‰. Values for $\delta^{34}S$ ranged from 0 to +2 for vascular plants while those for benthic microalgae and zooplankton were +14 to

+11‰, respectively; $\delta^{15}N$ for all primary producers ranged from +5 to +6‰. Forty-nine of the 57 (88‰) consumers sampled had $\delta^{13}C$ values within a range of –22 to –18‰; this range centered around benthic microalgae and zooplankton, but were distinct from the $\delta^{13}C$ of *Spartina* and *Juncus*. Values of $\delta^{34}S$ for 48 of the 56 (86%) estuarine consumers ranged from +9 to +16‰, a range that included that of the phytoplankton and zooplankton, but was 8 to 15% more enriched in $\delta^{34}S$ than *Spartina* or *Juncus*. Dual isotope plots showed the fauna to be tightly clustered around values for benthic microalgae and zooplankton, with no consumer even moderately resembling *Spartina* or *Juncus*. Sullivan and Moncreiff concluded that the ultimate food sources for the marsh invertebrate and fish fauna are the benthic and plankton microalgae. However, although detrital feeders may consume *Spartina* or *Juncus* detritus, the assimilated food largely consists of bacteria, microalgae, fungi, and protozoans.

The food web structure of a mangrove forest and adjacent sea grass beds on Gazi Bay, Kenya, was examined by Marguillier et al. (1997) with stable carbon and nitrogen isotope ratio techniques. A carbon isotope gradient was found from mangroves with mean (±SD) $\delta^{13}C$ value of –26.75 ± 1.64‰ to sea grass beds with –16.23 ± 4.35‰. Macroinvertebrates collected along this mangrove/sea grass bed transect showed a simiar $\delta^{13}C$ gradient. Analysis of the fish community using both isotope techniques showed that the sea grass stands were the main feeding grounds providing food for all the fish species studied. $\delta^{15}N$ signatures enabled the identification of the following trophic levels: (1) fish species feeding on sea grasses and macroalgae (herbivores) ($\delta^{15}N$ +5.88 to +6.62‰); (2) fish feeding on zoo/benthos = plankton (zoobenthic planktivores) ($\delta^{15}N$ +8.55 to +10.39‰); and (3) other fish and/or macrocrustacea (piscivores/benhtivores) ($\delta^{15}N$ +11.05 to +11.36‰).

Deegan and Garritt (1997) studied the estuarine food web in Plum Island Sound, Massachusetts, U.S.A., using the stable isotopes of carbon, nitrogen, and sulfur. Three distinct regions of the estuary were examined: the oligohaline upper estuary with high freshwater inputs with fringing fresh marsh; the middle estuary with extensive salt marsh; and the lower estuary with a greater proportion of open bay and a direct connection to the open ocean. Consumers in all regions relied most heavily on locally produced organic matter. In the upper estuary, most consumers had $\delta^{13}C$ values of –29 to –21‰ and $\delta^{34}S$ values of about 8‰, which indicated a dependence on a mixture of fresh marsh emergent vegetation amd phytoplankton from the oligohaline region. In the middle and lower estuary, consumers had $\delta^{13}C$ values of –23 to 15‰ and $\delta^{34}S$ values of 5 to 25‰, resembling a mixture of *Spartina* spp., benthic microalgae, and marine phytoplankton. Terrestrial organic matter was of minimal importance in the upper estuary and was not evident in the food web of the middle and lower estuary. The isotopic values of the pelagic consumers reflected a greater dependence on phytoplankton than those of the benthic consumers, which were closer to fresh or salt marsh emergent vegetation and benthic microalgae. It was conclued that: (1) consumers tend to utilize sources of organic matter produced in the same region of the estuary in which they reside; (2) consumer dependence on terrestrially derived riverine organic matter is mimimal; and (3) benthic and pelagic organisms rely on different mixes of organic matter sources.

Stable isotope studies including meiofauna are few (Spies and Des Maraîs, 1983; Gearing et al., 1984; Schwinghamer et al., 1983; Simenstad and Wissmar, 1985). Couch (1989) used carbon and nitrogen stable isotope ratios to measure the *in situ* assimilation of *Spartina alterniflora* and benthic microalgae by meiofauna harpacticoid copepods and nematodes in North Inlet Estuary, South Carolina, U.S.A. (Figure 6.43). Meiofauna, detrital *Spartina*, benthic microalgae and sediment exhibited no differences between mean seasonal $\delta^{13}C$ values (meiofauna –13.8 to 15.83‰, benthic microalgae –12.11 to 13.46‰, *Spartina* detritus –15.94 to 16.68‰, sediment –18.89 to 20.53‰) (see Figure 6.43). The $\delta^{13}C$ of populations of harpacticoids and nematodes were similar, indicating their reliance on similar foods throughout the year. The $\delta^{15}N$ of detrital *Spartina* and benthic microalgae were not distinct and thus it was not possible to identify the source of the nitrogen assimilated by the meiofauna. Populations of meiofauna assimilated a mixture of food resources. However, it is thought that detrital *Spartina* may be the predominant source of carbon with microalgae contributing some portion of the assimilated carbon and nitrogen. This is supported by studies showing preferential ingestion or assimilation of bacterial over algal carbon by meiofauna.

The results of the studies discussed above indicate that a number of uncertainties surround the use of isotopic analysis. Fry and Sherr (1984) reviewed the use of $\delta^{13}C$ measurements and came to the following conclusions:

1. When an organism consumes detritus, it may consume both the dead organic substrate (such as a fragment of *Spartina* or other macrophyte) as well as the associated microbial community. It has been established that this community is the major food resource of the organic detritus. The microbial community may contain microalgae, which have been established to have a different $\delta^{13}C$ value than that of the detritus. Bacteria and fungi may derive some of their food from a source other than the substrate (such as the algae) and thus have still a different $\delta^{13}C$ value. The substrate material may serve mainly as a carrier and decompose very slowly.
2. There may be considerable variability of isotopic ratio values within a functional food source. For example, phytoplankton from Georgia aver-

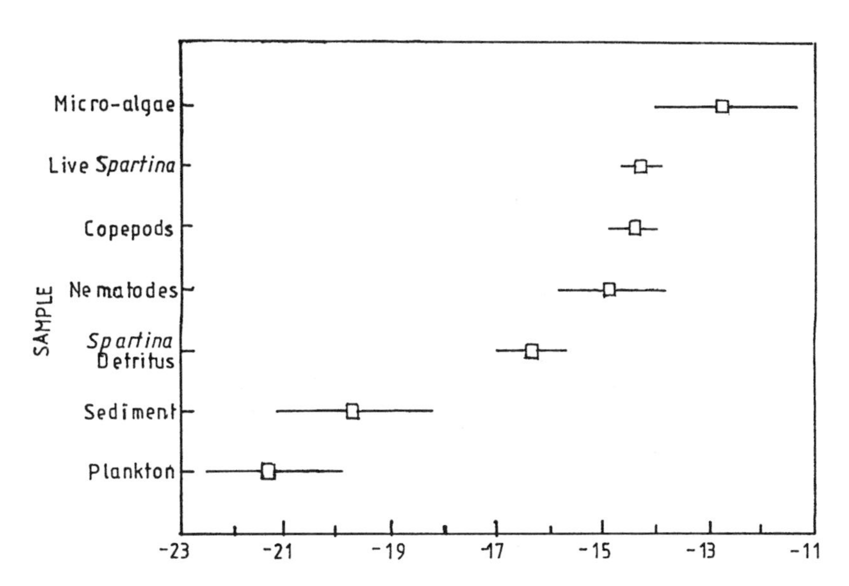

FIGURE 6.43 Mean $\delta^{13}C$ ($\pm$SD) of meiofauna and potential food resources. All data except for plankton (data from Peterson et al., 1985) were measured in the study by Couch (1989). (Redrawn from Couch, C.A., *Est. Coastal Shelf Sci.*, 28, 438, 1989. With permission.)

aged –20 to –22‰, while those from the Gulf of Mexico ranged from –15 to –18‰ (Fry and Parker, 1979).

3. There are indications that $\delta^{13}C$ values may not be conserved and that there is further fractionization along a food chain (Fry, 1977; DeNiro and Epstein, 1978).
4. The isotopoic ratio may not be constant for different tissues of a single organism. For example, Fry and Parker (1979) found that the $\delta^{13}C$ for lipid-free muscle of red snappers (*Lutjanus campechanus*) was –15.3‰, while that of total lipids was –20.9‰.
5. In addition, the isotopic ratios are an average value for all food consumed. For example, an organism with a $\delta^{13}C$ of –15.0‰ could have eaten a single food source of –15.0‰, or equal amounts of two food sources with values of –10.0 and –20.0‰, respectively.

It is thus clear that the use of stable carbon isotopes, if used and interpreted with caution, can provide a powerful tool for tracing the origin of particulate carbon in estuarine ecosystems in particular and its incorporation in food chains. The incorporation of multiple isotopes such as $\delta^{34}S$ and $\delta^{14}N$ can further enhance the reliability of the technique.

6.12 TOP-DOWN AND BOTTOM-UP CONTROL OF TROPHIC STRUCTURE

6.12.1 Introduction

Ecologists have long debated the importance of trophic interactions in determining the distribution and abundance of organisms. The question is whether the primary control is by resources (e.g., nutrients, bottom-up forces), or by trophic interactions (e.g., predators, top-down forces). The top-down view holds that organisms at the top of the food chains are food limited and that at successively lower levels, they are alternately predator, then food limited (Browley and Roff, 1986). The "top-down" view, first introduced by Hairston et al. (1960), predicts that whether or not organisms are predator or resource limited depends on their position in the food chain. Hairston et al. argued that "green biomass" accumulated (in mature communities) because predators kept herbivores in check. While this theory has been highly productive, it has had its critics. Those ecologists who contended that trophic levels are nonoperational with no useful correspondence to reality (e.g., Murdock, 1966; Peters, 1977; Polis, 1991) have asked why, despite omnivory and the complex linkages of real food webs, manipulations of top predators in communities sometimes trigger chain-like trophic cascades (discussed below). The "top-down" views, along with the later trophic cascade models (Paine, 1980; Carpenter et al., 1985) hold that plant standing crops are largely regulated by top-down forces. By their presence or absence, higher trophic levels will determine whether or not substantial plant biomass accumulates in communities.

However, Hunter and Price (1992) have presented a compelling argument for the primacy of bottom-up forces in food webs: " ... the removal of higher trophic levels leaves the lower levels intact (if perhaps greatly modified) whereas the removal of primary producers leaves no system at all." Fretwell (1987) proposed that food chains can have fewer or more than three trophic levels, and that top trophic levels and those an even number of steps below them are resource limited; trophic levels an odd number of steps below the top are predator limited. He also predicted that trophic levels would be added sequentially as primary productivity increased. Thus, plant productivity constrains the number of trophic levels. This model, then, predicts that top-down forces will dominate trophic dynamics, but that food web structure will be set by the fundamental bottom-up attributes of ecosystem plant productivity. A number of mathematical models of trophic stacks (e.g., Okansen et al., 1981; Gertz, 1984) incorporate top-down/bottom-up dualities. The literature on the ideas discussed above have recently been reviewed by Power (1992). The relative efficacy of top-down vs. bottom-up forces in food webs will depend in part on the efficiency with which consumers exploit their prey. Interactions among consumers, between consumers and resources, and between nonadjacent trophic levels can affect consumer efficiency, and thereby modify top-down forces in food webs. Figure 6.44 illustrates the mechanisms modulating top-down and bottom-up forces in food chains.

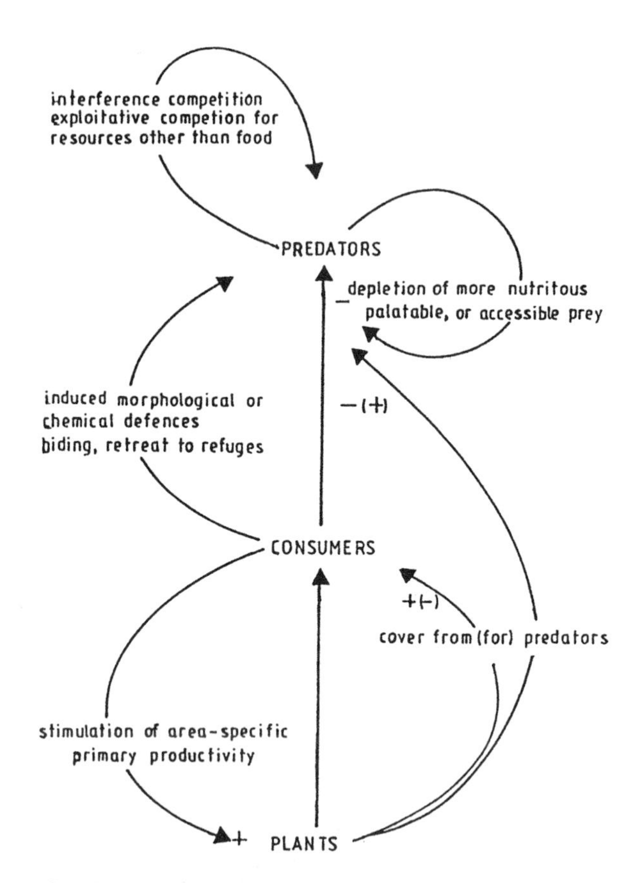

FIGURE 6.44 Mechanisms (curved arrows) modulating top-down and bottom-up forces (straight arrows) in food chains. (Redrawn from Power, M.E., *Ecology*, 73, 735, 1992. With permission.)

6.12.2 Top-Down and Bottom-Up Community Regulation on Rocky Shores

Menge (1992) has recently reviewed data on top-down and bottom-up control of community structure on rocky shores. He defines "control" as having a major quantitative and/or qualitative effect on a community. Quantitative effects are changes in abundance (numbers, cover, biomass), while qualitative effects are changes in community composition and/or the nature of interactions among community components. Menge defines "top-down control" as referring to "… situations where the structure (abundance, distribution, and/or diversity) of lower trophic levels depends directly or indirectly on trophic activity of higher trophic levels." For instance, Paine's (1966; 1974) demonstration of the effects of a keystone predator on sessile organisms on rocky shores is a classic example of top-down control. Bottom-up control, on the other hand, refers to "… direct or indirect dependence of community structure on factors producing variation at lower trophic levels or in their resources." For instance, increased abundance or diversity of hervivores and/or predators with high levels of nutrients demonstrates bottom-up control (e.g., Crowder et al., 1988). Interactions at lower trophic levels can "cascade up" food webs to regulate higher levels (Hunter and Price, 1992). Physical or physiological stresses on plants can also exert bottom-up control (Menge and Olsen, 1990).

Menge (1992) proposed that bottom-up and top-down factors may be tightly linked in rocky intertidal communities, and that this could account for much of the unexplained variation in communities. He illustrates this from his studies on rocky shores along the Oregon coast (see Figure 5.15) where the upper shore is dominated by barnacles and seaweeds, the mid-shore by mussels, and the low shore by seaweeds, sea grasses, and invertebrates (Farrell, 1991). Mobile invertebrates (gastropods, seastars, sea urchins, chitons, crabs, isopods, and amphipods) are most diverse in the low zone. The experiments of Paine (1966; 1974) and Dayton (1971; 1975) in Washington State and Farrell (1991) provided an explanation for some of these patterns. At the low shore levels, predators, primarily the seastar *Pisaster ochraceus*, maintained community structure by restricting mussels to the mid-shore and allowing the invasion and persistence of other invertebrates and seaweeds. At high shore levels, grazing, recruitment, and competition interacted to produce the patterns of community structure.

In Oregon, Menge's studies at Broiler Bay (44° 50'N, 124° 03'W) and Strawberry Hill (44° 15'N, 124° 07'W) revealed an interesting difference in low shore communities. At Broiler Bay, benthic plants were dominant, while at Strawberry Hill benthic plants were scarce and sessile invertebrates were abundant. These differences were not easily explained by physical factors; for example, differences in wave action at the two sites were small. The most puzzling differences between sites was the high abundance of barnacles and mussels at exposed locations at Strawberry

Hill compared to protected locations at Strawberry Hill, and both exposed and protected locations at Broiler Bay.

Field observations and experimental data support the hypothesis that variation in prey production underlies differences in seastar density and thus in predator rates. Predator density, prey abundance, mussel mortality rate, mussel recruitment, and mussel and barnacle growth and abundance are all highest at Strawberry Hill exposed locations (see Figure 5.15). Thus, differences in prey densities may depend on bottom-up processes, including recruitment and growth. Higher prey abundance presumably supports higher density of predators. Dense predator populations exert high top-down effects on prey. It is concluded that observations and experiments suggest that bottom-up factors may underlie both quantitative (abundances of filter feeders and consumers) and qualitative (strengths of interactions, species composition) differences in the rocky intertidal communities of the Oregon coasts. However, although recruitment differences may in part reflect differences in abundance of phytoplankton food, different rates of larval delivery to each site (e.g., Roughgarden et al., 1988) are an equally viable alternative.

In the low intertidal at Strawberry Hill exposed locations, prey abundance fluctuates inversely with predator intensity each year. From late autumn through early spring, when seastars are less active and mussel recruitment is high, prey increases. From spring through autumn, when seastars are active and mussel recruitment is low, prey decreases. At this time, massive numbers of *Pisaster* feed on mussels and barnacles at low tide and the lower edge of the *Mytilus trossulus* zone retreats upward until all are consumed. Thus, mussels are restricted to a mudshore refuge in space by the depredations of *Pisaster*.

Menge hypothesizes that between-site dissimilarities in community structure such as the example discussed above, after correcting for differences in wave forces, reflect variations in top-down and bottom-up forces. The simplest possibility is that higher recruitment, abundance, and growth of prey at Strawberry Hill all result from higher food supplies for filter-feeding prey. Under this scenario, the resulting higher secondary production could fuel higher predator abundances and thus higher predation intensities.

Field observations are consistent with the idea that food supply for filter feeders is higher at Strawberry Hill. Densities of large barnacles, including *Pollicipes polymerus, Balanus nubilus*, and *Semibalanus cariosus*, are much greater on the low shore at Strawberry Hill than at Broiler Bay. Lower on the shore in the sublittoral fringe, abundances of sponges and colonial tunicates are also much greater at Strawberry Hill than Broiler Bay. Organic foam, which usually reflects higher abundances of phytoplankton, is both more abundant and occurs more frequently at Strawberry Hill.

The data discussed above supports the arguments of Hunter and Price (1992) and others that bottom-up and top-down forces are joint, rather than alternate determinants of community structure and energy flow. Thus, marine community regulation may depend not only on top-down forces such as predation, competition, recruitment, and abiotic conditions (e.g., Paine, 1980; Menge and Sutherland, 1987), but also on bottom-up processes.

6.12.3 Trophic Cascades

Trophic cascades mean runaway comsumption, downward dominance through the food chain. Especially vulnerable are the producers (autotrophs). Standing crop and coverage of the plants are reduced wholesale when one or two species of potent herbivores are not suppressed. In architypical trophic cascades, overwhelming effects propagate down through three trophic levels. Primary carnivores, by supressing herbivores, switch the substrate from a sparse plant cover to one with dense cover. In some instances, it may encompass four trophic levels (Carpenter et al., 1985, Power, 1990); secondary carnivores by suppressing primary carnivores, unleash herbivores that clear the substrate and greatly decrease the standing crop of the plants. In true trophic cascades, pervasive top-down influence combines with always strong bottom-up influence to produce acute intertwining of population, community, and ecosystem processes (Carpenter and Kitchell, 1987; 1988; Okansen, 1990; Power, 1992). True trophic cascades also imply keystone species (Paine, 1980), taxa with such top-down dominance that their removal causes precipitous change in the system. It should, however, be pointed out that all trophic interactions do not cascade, and simple top-down dominance is not the norm of communities or ecosystems.

Strong (1992) has recently reviewed trophic cascades and points out that the majority of examples of true trophic cascades have algae at the base and are aquatic. Strong lists a number of potential and factual trophic cascades. Among the former are rocky seashores with epilithic algae and gastropod molluscs, and fish herbivores (Lubchenco and Gaines, 1981; Carney, 1990). Macroalgae and epilithic forms are more vulnerable to herbivory than macroalgae. A second example concerns coastal kelp beds with sea urchins as the herbivore and sea otters as the first carnivores (Dayton, 1985b; Elner and Vadas, 1990). The interactions of sea urchins and otters with kelp forests are discussed in Section 2.9.4.5.

7 Ecosystem Models

CONTENTS

7.1 INTRODUCTION

Throughout the previous chapters in this volume, models have been used to illustrate aspects of the dynamics of ecosystems, ranging from energy or nutrient flows through individual species populations, through abstracted food chains to partial models of whole systems. A model is any abstraction or simplification of a system, i.e., it is a simplified version of the real world. Building models of total ecosystems, in contrast to models of component populations or trophic levels, focuses on the ways in which the individual components of the ecosystem are linked to each other, i.e., the ways in which the state variables are linked together and how they interact with the forcing functions.

Reichle et al. (1975) have listed the characteristics of ecosystem function that ensure their persistence. They are: (1) the establishment of an energy base, (2) development of an energy reservoir, (3) recycling of elements, and (4) rate regulation. All ecosystems have evolved regulative mechanisms whereby they establish a secure energy base (autotrophic production) in the presence of a fluctuating environment. The autotrophs display one or two basic properties (or combinations thereof): (1) small individuals with rapid turnover (phytoplankton, benthic microalgae, and short-lived epiphytes), and (2) large individuals with slow turnover (macroalgae, sea grasses, marsh macrophytes, mangroves). A combination of these properties serve to provide homeostatic mechanisms that facilitate both rapid responses and long-term stability in the photosynthetic base. Estuarine ecosystems with their combination in varying degrees of phytoplankton, benthic microalgae, epiphytic algae, emergent macrophytes (marsh plants, mangroves), submerged macrophytes (sea grasses and macroalgae) display these characteristics better than most ecosystems.

In addition, ecosystems have developed mechanisms for energy storage that provide for homeostasis. Such energy storage reservoirs are characteristically large with slow response times (Reichle et al., 1975). In most ecosystems, especially in estuaries, this is the large particulate organic matter pool (detritus) resulting from the death and decay of primary producers, especially the macrophytes which, by and large, are not grazed by herbivores. In estuaries, biodeposits from filter feeders, especially bivalves, contribute substantial amounts to this pool. In addition, many systems receive large allochthonous inputs of organic matter via freshwater input. This detrital pool has the capability of storing large quantities of nutrients. Mudflat, salt marsh, mangrove, and estuarine sediments, with their large biomass of decomposers, both aerobic and anaerobic, coupled with processes of meiofaunal activity and rapid turnover, macrofaunal biodeposition, and resuspension of surface sediments, ensure that organic matter and nutrients are continuously recycled.

The development of a comprehensive model for a coastal marine ecosystem will be illustrated with reference to a model of ecosystem processes in the mangrove-fringed Upper Waitemata Harbour, New Zealand (Figure 7.1) (see Knox, 1983a,b, for full details). Table 7.1 lists the energy sources, forcing functions, storages, and main interactive processes within this ecosystem. The energy sources as depicted clockwise from the bottom left are:

1. "Natural energies" — sun, wind, river input, sewage input, rain, geological uplift.
2. "Human applied energies" — government, commercial fishing, aesthetic use, and recreation.

The dotted line separates the sediment processes from those in the overlying estuarine waters. On the left, the principal primary producers shown are epiphytic algae (microalgae and small macroalgae), *Zostera* (eelgrass), marsh plants (*Juncus, Leptocarpus, Salicornia*), mangroves (*Avicennia*), and the benthic microflora (diatoms, flagellates, and blue-green algae). Inputs to these are solar energy (the amount to a large extent controlled by turbidity), and nutrients (mainly nitrogen and phosphorous) required for plant growth. In the top center, the estuarine water storage compartment is shown with component sub-storages of salt, silt, sand, nitrogen, phosphorous, dissolved organic matter (DOM), particulate organic matter (POM),

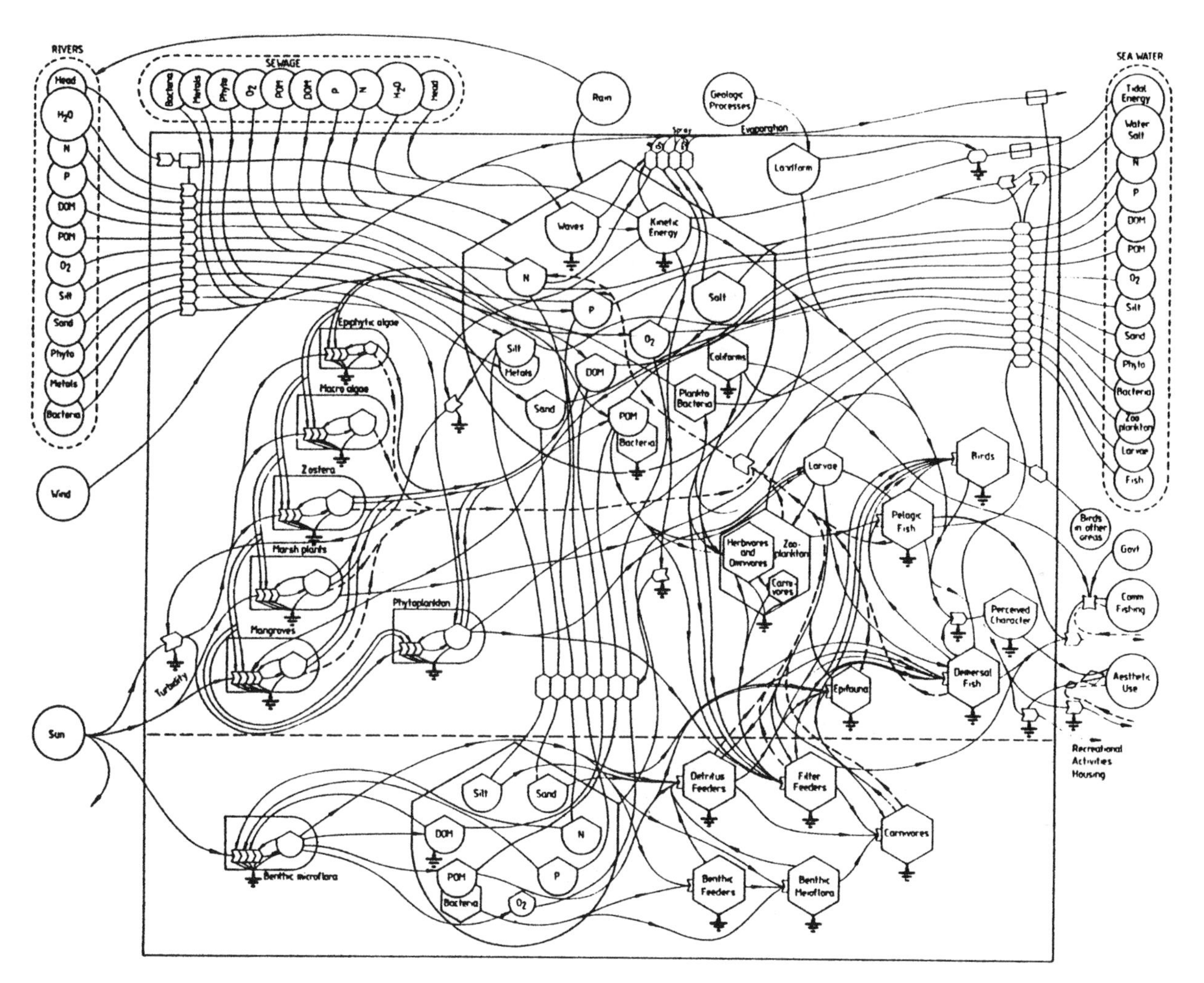

FIGURE 7.1 Energy flow model of the Upper Waitemata Harbour, a mangrove fringed estuary in northern New Zealand. (From Knox, G.A., *Estuarine Ecosystems: A Systems Approach,* Vol. II, Frontispiece, 1986b. With permission.)

plankto-bacteria, and oxygen. Consumer food chains are shown on the right side of the water compartment. In the sediment compartment, the storages include silt, sand, nitrogen, phosphorous, DOM, POM, and oxygen. Food chains of the benthic sediment fauna are shown to the right.

While flow diagrams, and the measurement of the values of the storage compartments and the flows to and from them, can identify important processes that occur in the ecosystem, reveal data deficiencies, and suggest laboratory and field experiments that need to be performed (Walsh, 1975), mathematical relationships (generally differential equations) need to be assigned to the various processes if the model is to be used for predictive purposes. The model can then be computer simulated, permitting the identification of probable outcomes as the parameters in the model are changed, as new parameters are added, or as old ones are removed.

The model shown in Figure 7.1 is a very complex one and it would need to be simplified for the purpose of simulation. Such simplification can involve the elimination of storages and flows that are relatively insignificant and the aggregation of processes and components that are similar into functional groups such as trophic (feeding) levels, particle size, functional guilds, and so on. However, if the model is too simple, it may not adequately illustrate the key system processes. Examples of model simulation will be discussed in the succeeding sections of this chapter.

7.2 HARD SHORES

Partial models of hard shores have been discussed in Chapter 6 (Section 6.9.2). However, to date there has not been a comprehensive model of a rocky shore ecosystem. Figure 7.2 is an attempt to devise a model of a rocky shore ecosystem that includes most of the important forcing functions, storages, and interactive processes. Table 7.2 lists the forcing functions, storages, and main interactive processes.

As can be seen from Figure 7.2, there is a significant exchange of organic material both living and non-living,

TABLE 7.1
List of Energy Sources, Storage, and Main Interactive Processes for the Upper Waitemata Harbour

Energy Sources (Forcing Functions)	
Sun	Wind
Rain	Nutrients in rain
Rivers (water, nitrogen, phosphorus, DOM, POM, oxygen, silt, sand, phytoplankton, bacteria, metals)	
Sewage (water, nitrogen, phosphorus, DOM, POM, oxygen, phytoplankton, bacteria, metals)	
Seawater (tidal energy, water, salts, nitrogen, phosphorous, DOM, POM, oxygen, silt, sand, phytoplankton, zooplankton, fish)	
Storage	
Mangroves	Marsh plants
Eelgrass (*Zostera*)	Macroalgae
Epiphytic algae	Benthic microalgae
Phytoplankton	
Water (salt, silt, sand, nitrogen, phosphorus, DOM, POM, oxygen, planktobacteria)	
Zooplankton	Epifauna
Demersal fish	Pelagic fish
Birds	
Sediment (silt, sand, DOM, POM, nitrogen, phosphorus, oxygen, bacteria)	
Benthic microfauna	Benthic meiofauna
Benthic macrofauna	
Perceived character (image)	
Main Intermediate Processes (connecting energy sources and storage with the system)	
Plant production	Water flow
Production of DOM and POM	Interchange between the sediment and the overlying water
Estuarine food web	
Recreational use	Interchange between the Upper and Middle Waitemata Harbour

Source: From Knox, G.A., *Energy Analysis: Upper Waitemata Harbour Catchment Study Specialist Report*, Auckland Regional Authority, Auckland, New Zealand, 1986b, 135. With permission.

between the rocky shore and the nearshore waters. With the inundation of the shore, the water carries with it bacterioplankton, phytoplankton, and zooplankton that are consumed by the filter-feeding barnacles, bivalves, bryozoans, and polychaetes. It also carries the nutrients necessary for the growth of the rocky shore plants. Consumers from the water column, in particular herbivorous and carnivorous fishes, graze on micro- and macroalgae and consume shore invertebrates. In addition, where there are sandy deposits in the lower intertidal and subtidal, sand burial and sand scour can impact in the ways described in Section 2.3.4. Wave action and the ebb and flow of the tide carries particulate organic matter and sediment that may be deposited in crevices, algal holdfasts and fronds, and the matrix of mussel beds where it may provide a food source for deposit feeders, or may be remineralized to provide nutrients to the water column.

To the flux of organic matter described above there must be added the large number of larvae and algal propagules that are exported by the rocky shore invertebrates and fishes and the shore macroalgae. These larvae utilize food resources in the water column and in turn provide a food source for offshore pelagic and benthic systems. A small percentage of the larvae will return to the shore where they settle and develop into adult organisms.

7.3 SAND BEACHES

Various aspects of energy flow in beach ecosystems with special reference to South African beaches are discussed

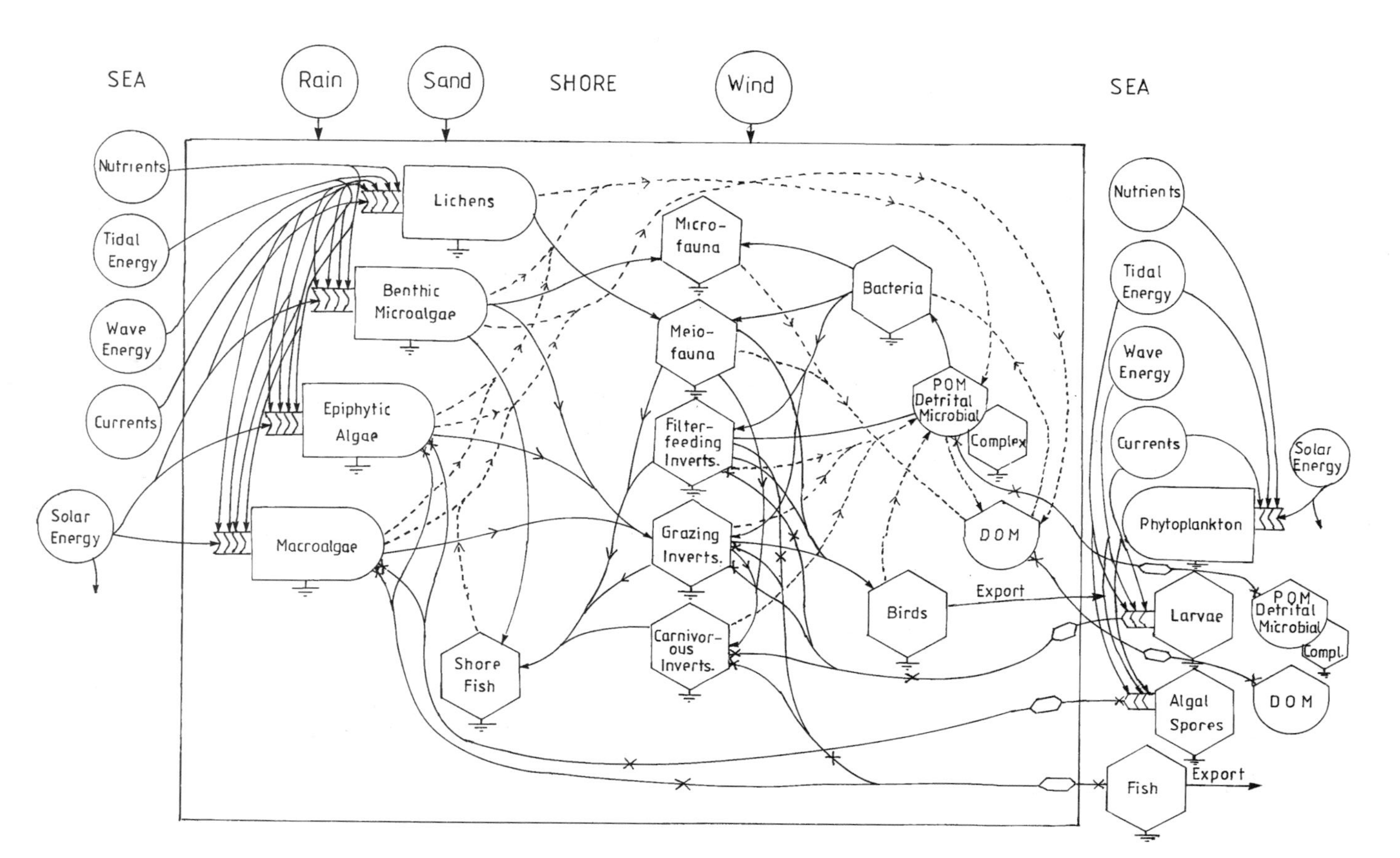

FIGURE 7.2 Energy flow model of a typical rocky shore ecosystem.

TABLE 7.2
List of Energy Sources, Storages, and Main Interactive Processes for a Rocky Shore Ecosystem

Energy Sources (Forcing Functions)	
Sun	Wind
Rain	Nutrients in rain
Currents	Waves
Seawater (tidal energy, water salts, nutrients (nitrogen, phosphorus and silicon), DOM, POM, oxygen, silt, sand, bacteria, phytoplankton, zooplankton, and fish)	
Storages	
Benthic microalgae	Macroalgae
Epiphytic algae	Lichens
Bacteria	Microfauna
Meiofauna	Macrofauna
Fish	Birds
Sediment (silt, sand, DOM, POM, nutrients, oxygen)	
Main Intermediate Processes (connecting energy sources and storages within the system)	
Microalgal production	Macroalgal production
Production of DOM and POM	Tidal energy
Consumption	

in Section 7.9.3.1. Here we discuss a recent model of a South African high-energy beach/surf-zone ecosystem.

Heymans and McLachlan (1996) have recently constructed a food web model and carbon budget for a *high-energy beach surf-zone ecosystem at Sundays Beach, South Africa* (Figure 7.3). This system has been extensively studied and the carbon budget and food web has been covered in many reviews (McLachlan et al., 1981; McLachlan, 1983; 1990a; McLachlan and Bate, 1984; Malan and McLachlan, 1985; 1991; Lasiak, 1986; Bate et al., 1990; Du Preez et al. 1990; McLachlan and Romer, 1990; Romer, 1990). Aspects of the dynamics of this beach were discussed in Chapter 3 (Section 3.7.2.1).

The beach environment: The Sundays Beach/surf-zone ecosystem is a high-energy environment with a surf zone of an intermediate type characterized by bars, channels, and rip currents (McLachlan and Bate, 1984). The sand of Sundays Beach is fine to medium, well sorted, and primarily quartz. The maximum tidal range is 2.1 m and the wave height mainly ranges from 1 m to 3 m. The climate is temperate and the seawater temperature ranges between 10 and 26°C (monthly means: 16 to 21°C).

The ecosystem extends to the outer limit of the surf circulation cells 500 m seaward of the high-water mark, at a depth of 10 m. Water turnover in the surf zone is on the order of hours, but turnover of the water between the combined surf and rip-head zones and the nearshore zone can take days. Groundwater from the dune field enters the surf zone at a mean rate of 1,000 l m^{-1} day^{-1} (McLachlan and Illenberger, 1986), resulting in a net flow of nitrogen to the surf zone of 1,050 g N m^{-1} $year^{-1}$ (Campbell and Bate, 1991).

Primary production, inputs, dissolved organic carbon and particulate organic carbon (Table 7.3): Phytoplankton biomass and production (mainly from the surf diatom *Anaulus*) are 696 mg C m^{-2} day^{-1} and 1,205 mg C m^{-2} day^{-1}, respectively (McLachlan and Romer, 1990), and fecal input into the suspended particulate organic carbon (POC) pool totals 112 mg C m^{-2} day^{-1}. As there are no benthic producers, the total available input to the carbon pool is thus about 1,317 mg C m^{-2} day^{-1}, of which only approximately 68 mg C m^{-2} day^{-1} is consumed by the macrofauna. The interstitial fauna utilize 263 mg C m^{-2} day^{-1} (Malan, 1991), and 523 mg C m^{-2} day^{-1} (Malan, 1991) is absorbed by the microbial loop (McGwynne, 1991), leaving a surplus of approximately 332 mg C m^{-2} day^{-1} available for export, probably as detritus outwelled to nearshore waters. Phytoplankton is also exported from the system, at a rate of 63 mg C m^{-2} day^{-1}.

Detritus composition and quantity amounts to approximately 3,500 g C m^{-1}, with the highest concentrations near the beach, but greatest quantities in the rip-head zone, which encompasses the largest water volume (Talbot and Bate, 1988a). This average detrital load makes up to 91% of the total POC pool, which amounts to 7,692 mg C m^{-2} (Talbot and Bate, 1988b). Carbon removed from this pool by the interstitial system totals 219 mg C m^{-2} day^{-1} (153 mg C m^{-2} day^{-1} from the dissolved organic carbon [DOC] pool, and 66 mg C m^{-2} day^{-1} from the POC pool). The

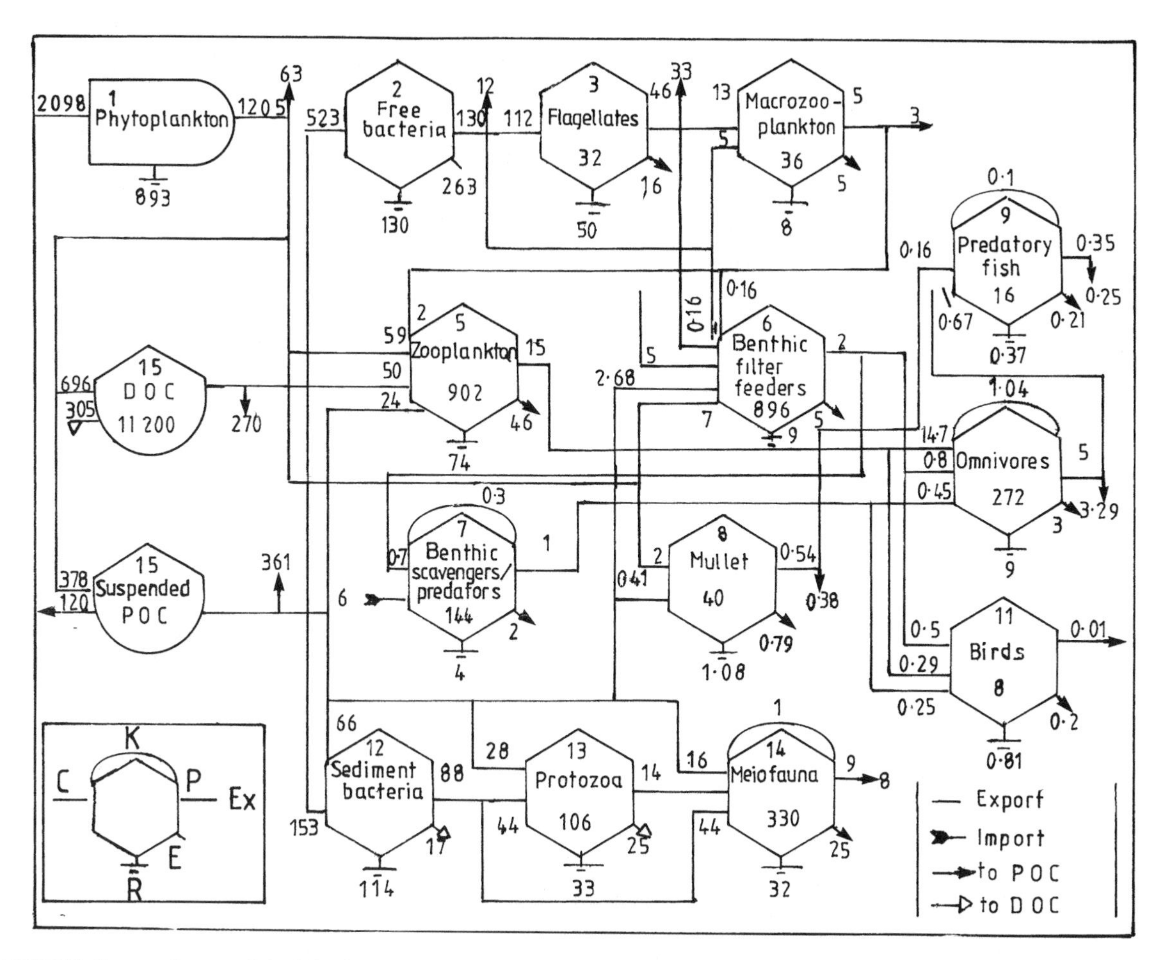

FIGURE 7.3 Energy flow model of the Sundays Beach ecosystem, with biomass (B) in mg C m^{-2}, consumption (C), production (P), respiration (R), egestion (E), and flows between compartments in mg C m^{-2} day^{-1}. K, cannibalism; POC, particulate organic carbon; DOC, dissolved organic carbon. (Redrawn from Heymans, J.J. and McLachlan, A., *Est. Coastal Shelf Sci.*, 43, 486, 1996. With permission.)

total amount of carbon exported via the POC pool is 361 mg C m^{-2} day^{-1}.

Dissolved organic carbon averages 2.24 mg l^{-1}, which converts to 11,200 mg C m^{-2}. Primary production exceeds faunal requirements and the surplus is exported as living and dead POC across the boundary of the ecosystem (McLachlan, 1990b). Of the 45% of phytoplankton production, or 696 mg C m^{-2} day^{-1} that is converted into phytoplanktonic dissolved organic carbon (PDOC), 523 mg C m^{-2} day^{-1} is utilized by the free bacteria, and the remainder is consumed by the macroscopic assemblage (McLachlan, 1990b). The DOC pool is utilized by the interstitial system (153 mg C m^{-2} day^{-1}), zooplankton (50 mg C m^{-2} day^{-1}), and benthic filter feeders (5 mg C m^{-2} day^{-1}).

Fauna: The fauna of the beach and surf zone includes three distinct trophic assemblages: macroscopic fauna, interstitial fauna in the sediment, and the microbial loop in the water. The microbial loop, comprising bacteria, flagellates, ciliates, and other microzooplankton in the water column (McLachlan, 1990b), subsists largely on phytoplankton exudates and other DOC, which are utilized by water-column bacteria (McLachlan, 1990b).

Microbial loop: The free bacterial biomass in the surf zone is 0.086 mg C l^{-1} and in the rip-head zone is 0.01 mg C m^{-1} (McLachlan and Romer, 1990). Thus, the total free bacterial biomass is 146 mg C m^{-2}. Free bacteria produce 0.007 mg C l^{-1} day^{-1} in the surf zone and 0.009 mg C l^{-1} day^{-1} in the rip-head zone, giving a total production of 130 mg C m^{-2} day^{-1} (McLachlan and Romer, 1990). Free bacteria are consumed by the flagellates (112 mg C m^{-1} day^{-1}) and microzooplankton (5 mg C m^{-2} day^{-1}) as well a benthic filter feeders (1 mg C m^{-2} day^{-1}), leaving 12 mg C m^{-2} day^{-1} to be exported from this system in the form of free bacteria.

Heterotrophic nanoplankton and microplankton are dominated by flagellates, tintinnids, and micrometazooans (McLachlan and Romer, 1990). Flagellates have biomass and production values of 0.008 mg C l^{-1} and 0.011 mg C l^{-1}, respectively, in the surf zone; and 0.006 mg C l^{-1} and 0.008 mg C l^{-1} in the rip-head zone (McLachlan and Romer, 1990), which translates to 32 mg C m^{-2} and 45

TABLE 7.3
Carbon Throughput (in mg C m^{-2} day^{-1}) for Each Compartment in the Sundays Beach Ecosystem

Compartment	Throughput
1. Phytoplankton	2,100.00
2. Free bacteria	523.00
3. Flagellates	112.00
4. Microzoopankton	18.00
5. Zooplankton	115.00
6. Benthic filter feeders	16.00
7. Benthic scavengers and predators	7.00
8. Mullet	2.41
9. Predatory fish	0.93
10. Omnivorous fish	17.00
11. Birds	1.04
12. Sediment bacteria	19.00
13. Protozoa	72.00
14. Meiofauna	75.00
15. DOC	1,000.00
16. Suspended POC	498.00

Source: From Heymans, J.J., McLachlan, A., *Est. Coastal Shelf Sci.*, 43, 500, 1996. With permission.

mg C m^{-2} day^{-1} over the whole system. Flagellates derive all their carbon requirements from free bacteria and are consumed by microzooplankton (134 mg C m^{-2} day^{-1}), with the remaining 32.84 mg C m^{-2} day^{-1} being exported.

Microzooplankton biomass amounts to 36 mg C m^{-2} (McLachlan and Romer, 1990), and production and respiration are approximately 5 mg C m^{-2} day^{-1} and 8 mg C m^{-2} day^{-1}, respectively. Microzooplankton derive all their carbon requirements from free bacteria (5 mg C m^{-2} day^{-1}) and flagellates (13 mg C m^{-2} day^{-1}), and are consumed by zooplankton (2 mg C m^{-2} day^{-1}) and benthic filter feeders (0.16 mg C m^{-2} day^{-1}), with the remaining 2.84 mg C m^{-2} day^{-1} being exported from the system.

Macrofauna: Zooplankton biomass is dominated by omnivorous prawns (45%) in the surf zone, and mysids (32%) in the rip-head zone, both species grazing surf diatoms (McLachlan, 1990b). Zooplankton biomass is 902 mg C m^{-2} (McLachlan and Romer, 1990) and production amounts to 15 mg C m^{-2} day^{-1}. Carbon consumed by zooplankton is derived from phytoplankton (59 mg C m^{-2} day^{-1}), microzooplankton (2 mg C m^{-2} day^{-1}), and suspended POC (4 mg C m^{-2} day^{-1}) (McLachlan and Romer, 1990). Zooplankton is consumed by birds (0.29 mg C m^{-2} day^{-1}) and omnivorous fish (14.7 mg C m^{-2} day^{-1}) (McLachlan and Romer, 1990).

The macrofauna consists of filter feeders and predator-scavengers; filter feeders include the clams *Donax serra* and *D. sordidus*, and *G. psammodytes* that together make up 97.8% of the macrofauna biomass (McLachlan et al., 1981). The biomass, consumption, and production of the benthic filter feeders are given by McLachlan and Bate (1984), and McLachlan and Romer (1990) as 866 mg C m^{-2}, 16 mg C m^{-2} day^{-1}, 2 mg C m^{-2} day^{-1}, and 9 mg C m^{-2} day^{-1}, respectively. Benthic production is consumed by birds, benthic scavengers and predators, and omnivorous fishes in the order 25%, 35%, and 40%, respectively (McLachlan and Romer, 1990). The benthic filter feeders, which produce 2 mg C m^{-2} day^{-1}, consume suspended particulates; free bacteria (1 mg C m^{-2} day^{-1}); heterotrophic flagellates (0.16 mg C m^{-2} day^{-1}); microzooplankton (0.16 mg C m^{-2} day^{-1}); DOC (4 mg C m^{-2} day^{-1}); and POC (2.68 mg C m^{-2} day^{-1}).

The scavenger component of this system includes the gastropods *Bullia rhodostoma*, *B. digitalis*, and *B. pura*, and the swimming crab *Ovalipes punctatus*. Input of carrion to the system totals 6 mg C m^{-2} day^{-1}. Benthic scavenger biomass, consumption, production, and respiration are 144 mg C m^{-2}, 7 mg C m^{-2} day^{-1}, 1 mg C m^{-2} day^{-1}, and 4 mg C m^{-2} day^{-1}, respectively (McLachlan and Romer, 1990; McLachlan and Bate, 1984). Of this, 0.3 mg C m^{-2} day^{-1} (or 35%) is consumed by benthic scavengers, 0.45 mg C m^{-2} day^{-1} (or 40%) by omnivorous fish, and 0.25% by birds (McLachlan and Romer, 1990).

The main feeding groups of surf-zone ichthyofauna are the southern mullet *Liza richardson*, the sandshark *Rhinobatos annulatus*, benthic feeders, zooplankton feeders, and omnivorous and piscivorous fish. They have biomass values of 1,000, 1,000, 3,000, 2,400, 400, and 400 kJ m^{-1}, respectively and annual consumption values of 22,107, 13,725, 65,710, 65,476, 9,758, and 8,517 kJ m^{-1} $year^{-1}$, respectively (Du Preez et al., 1990). The sandshark (*R. annulatus*), benthic feeders, zooplankton feeders, and omnivorous feeders are grouped together in a compartment called omnivorous fish, with a biomass of 272 mg C m^{-2} (McLachlan and Romer, 1990). They consume 17 mg C m^{-2} day^{-1}, produce 5 mg C m^{-2} day^{-1}, and their respiration amounts to 9 mg C m^{-2} day^{-1} (Du Preez et al., 1990). The mullet and predatory fishes have biomass values of 40 mg C m^{-2} and 16 mg C m^{-2}, consumption values of 2.41 mg C m^{-2} day^{-1} and 0.93 mg C m^{-2} day^{-1}, production values of 0.54 mg C m^{-2} day^{-1} and 0.35 mg C m^{-2} day^{-1}, and respiration values of 1.08 mg C m^{-2} day^{-1} and 0.37 mg C m^{-2} day^{-1}, respectively (Du Preez et al., 1990).

Omnivorous fish also include zooplanktivorous and benthic feeding fish, which consume zooplankton (14.71 mg C m^{-2} day^{-1}), benthic filter feeders (0.8 mg C m^{-2} day^{-1}), benthic predators and scavengers (0.45 mg C m^{-2} day^{-1}). The remaining 1.04 mg C m^{-2} day^{-1} was assumed to be derived from cannibalism. About 30% of the fish production is consumed by predatory fish (Du Preez et al., 1990). Thus, of the 0.54 mg C m^{-2} day^{-1} and 0.35 mg C m^{-2} day^{-1} produced by mullet and predatory fish, respectively, 0.16 mg C m^{-2} day^{-1} and 0.1 mg C m^{-2} day^{-1} is consumed by predatory fish. The remaining 0.67 mg C

m^{-2} day^{-1} consumed by predatory fish was derived from omnivorous fish. The 0.38 mg C m^{-2} day^{-1}, 0.25 mg C m^{-2} day^{-1}, and 3.29 mg C m^{-2} day^{-1} that remains of the mullet, predatory fish, and omnivore production, respectively, was assumed to be exported from the system.

The bird population of this ecosystem consists mostly (95%) of four species: the Southern Blackbacked Gull (*Larus dominicanus*), Black Oystercatcher (*Haematopus moquini*), Whitefronted Sandplover (*Charadrius marginatus*), and the Sanderling (*Calidris alba*) (McLachlan et al., 1981). Bird biomass totals 8 mg C m^{-2} (McLachlan and Romer, 1990). Birds consume 1.04 mg C m^{-2} day^{-1} (McLachlan and Bate, 1984), of which 0.5 mg C m^{-2} day^{-1} and 0.25 mg C m^{-2} day^{-1} is derived from benthic filter feeders and benthic scavengers and predators, respectively, and the remaining 0.29 mg C m^{-2} day^{-1} is derived from zooplankton.

Interstitial fauna: Interstitial faunal carbon requirements amount to 219 mg C m^{-2} day^{-1} as measured by sediment oxygen consumption (Malan, 1991), and this input into the system is partitioned among bacteria (60%), protozoans (20%), and meiofauna (20%) (McLachlan, 1990b). The biomass figures for bacteria, protozoans, and meiofauna in the 60 m wide intertidal beach are 10, 3, and 15 g C m^{-1}, respectively, and in the subtidal zone 95, 50, and 150 g C m^{-1}, respectively (Malan and McLachlan, 1985; McLachlan, 1990b). Thus, overall interstitial biomass amounts to 210, 106, and 330 mg C m^{-2} as bacteria, protozoans, and meiofauna, respectively. Intertidal respiration for bacteria, protozoa, and meiofauna amounts to 1,500 g C m^{-1} $year^{-1}$, 444 g C m^{-1} $year^{-1}$, and 445 g C m^{-1} $year^{-1}$, respectively (Dye, 1979). Subtidal respiration for bacteria, protozoa, and meiofauna amounts to 19,448 g C m^{-1} $year^{-1}$, 5 469 g C m^{-1} $year^{-1}$, and 5,469 g C m^{-1} $year^{-1}$, respectively (Dye, 1979), and the overall respiration for the inter- and subtidal bacteria, protozoa, and bacteria was calculated at 114, 33, and 32 mg C m^{-2} day^{-1}, respectively.

Sediment bacterial production was estimated at 88 mg C m^{-2} day^{-1}. Of this, 44 mg C m^{-2} day^{-1} is consumed by both protozoa and meiofauna (McLachlan, 1990b). The meiofauna community is dominated by nematodes (56% of total numbers) (Malan and McLachlan, 1985). Meiofaunal production was estimated at 9 mg C m^{-2} day^{-1} and meiofaunal consumption was calculated at 75 mg C m^{-2} day^{-1}, of which 44 mg C m^{-2} day^{-1} is derived from sediment bacteria, 14 mg C m^{-2} day^{-1} from protozoa, 1 mg C m^{-2} day^{-1} (4%) from meiofauna, and the remaining 16 mg C m^{-2} day^{-1} from suspended POC.

7.4 INTERTIDAL MUDFLAT IN THE LYNHER ESTUARY, CORNWALL, U.K.

Warwick et al. (1979) and Warwick and Radford (1989) have developed a comprehensive model of carbon flow through the benthic community of a mudflat in the Lynher Estuary, Cornwall, U.K. (Figure 7.4). This mudflat has been the subject of intensive investigation over the past 20 years. Joint (1978) quantified autotrophic and heterotrophic microbial production both in the water column and in the sediments. Several detailed studies have been carried out on energy flow within the macrobenthos and meiobenthos (Teare and Price, 1969; Warwick and Price, 1975; 1979; Price and Warwick, 1980a,b; Warwick, 1981). Field caging experiments have been conducted to investigate the relationships between the floral and faunal components of the ecosystem (Warwick et al., 1982; Warwick and Price, 1975; 1979).

From the model described above, Warwick and Radford (1989) developed the model shown in Figure 7.5 in which some of the components were aggregated after testing the effects of various types of aggregation.

Joint (1978) estimated that there was a production of 143 g C m^{-2} $year^{-1}$ by the benthic microalgae and 82 g C m^{-2} $year^{-1}$ in the water column when the tide was in. Table 7.4 lists the standing stock, consumption, and production of the various ecosystem components. Total annual input of C to the consumers was 163 g C m^{-2} $year^{-1}$. The total annual production of the macrofaunal species was 5.49 g C m^{-2}, all of which was available to mobile predators such as birds, fish, and crabs. The total annual production of the meiofauna that feed on primary carbon sources was 20.17 g C m^{-2} but 3.34 g of this was utilized within the system by predators (*Nepthys, Protohydra*), so that the output was 16.8 g, three times greater than that of the macrofauna. The macrofauna utilized 55.74 g of primary C, 35% from the water column and 65% from the sediment. Of this, 29.49 g were recycled as feces and 12.12 g were respired. The meiofauna utilized 107.09 g of primary C, 19% from the water column and 80% from the sediment. Of this, 45.10 g were recycled as feces and 41.82 g were respired. Thus, the net production of the meiofauna was 3.7 times that of the macrofauna and the total carbon turnover of the meiofauna was 1.9 times that of the macrofauna, whereas the meiofauna biomass was only 0.49 times that of the macrofauna.

7.5 SALT MARSHES

7.5.1 INTRODUCTION

Salt marshes, as we have seen, produce large quantities of detritus, much of which is incorporated into the sediments and decomposed within the marsh and adjacent mudflats. Thus, a large proportion of the energy flow is regulated by the rate of increase of the microbial populations (Sikora and Sikora, 1982). This in turn is controlled by the grazing of microbial populations by micro-, meio-, and macrofauna. Bacterial numbers and detritus decomposition have been shown to increase in the presence of

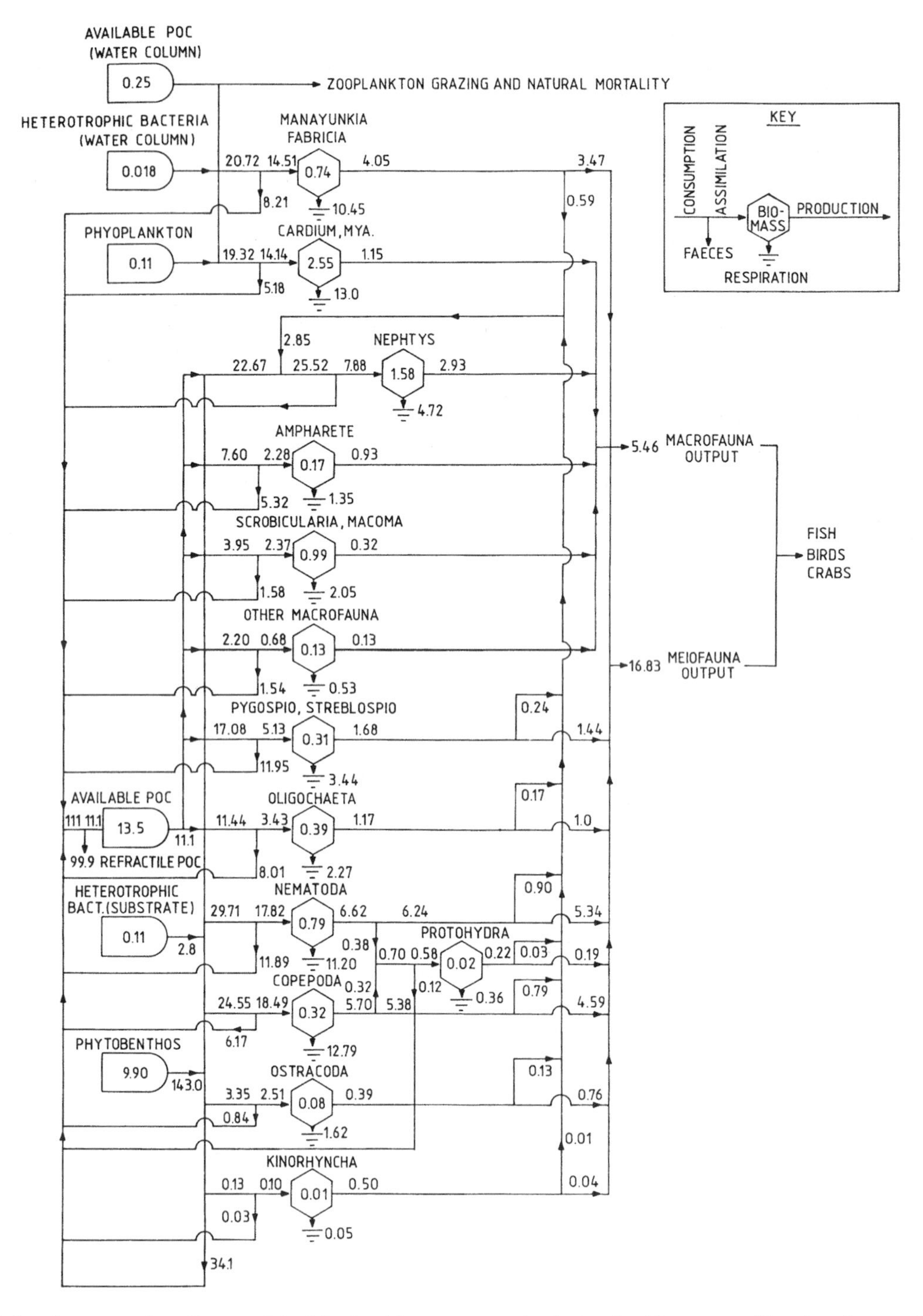

FIGURE 7.4 Steady-state carbon flow model of the benthic community in the Lynher Estuary mudflat. Standing stock biomass in g C m^{-2}, flow rates in g C m^{-2} year^{-1}. (Redrawn from Warwick, R.M., Joint, I.R., and Radford, P.J., in *Ecological Processes in Coastal Enviroments*,Jeffries, R.L. and Davy, A.J., Eds., Blackwell Scientific, Oxford, 1979, 443. With permisssion.)

grazers (Fenchel, 1970; 1972; Hargrave, 1970a; Fenchel and Harrison, 1976), as does nutrient regeneration (Pomeroy, 1970). As noted in Chapter 3 (Section 3.6.3.2), nematodes reach their highest densities in salt marshes (Platt, 1981). Thus, the grazing of the microbial populations by the nematodes regulates the rate of energy transfer from marsh-plant detritus to the food webs in the marsh as well as the rates of nutrient regeneration. In addition, the activities of the deposit and suspension feeders also play key roles in the energy transfer processes.

Howarth and Teal (1980) have drawn attention to the important role of reduced inorganic sulfur compounds in salt marsh and estuarine systems. Under aerobic conditions, energy flow is proportional to, and largely mediated

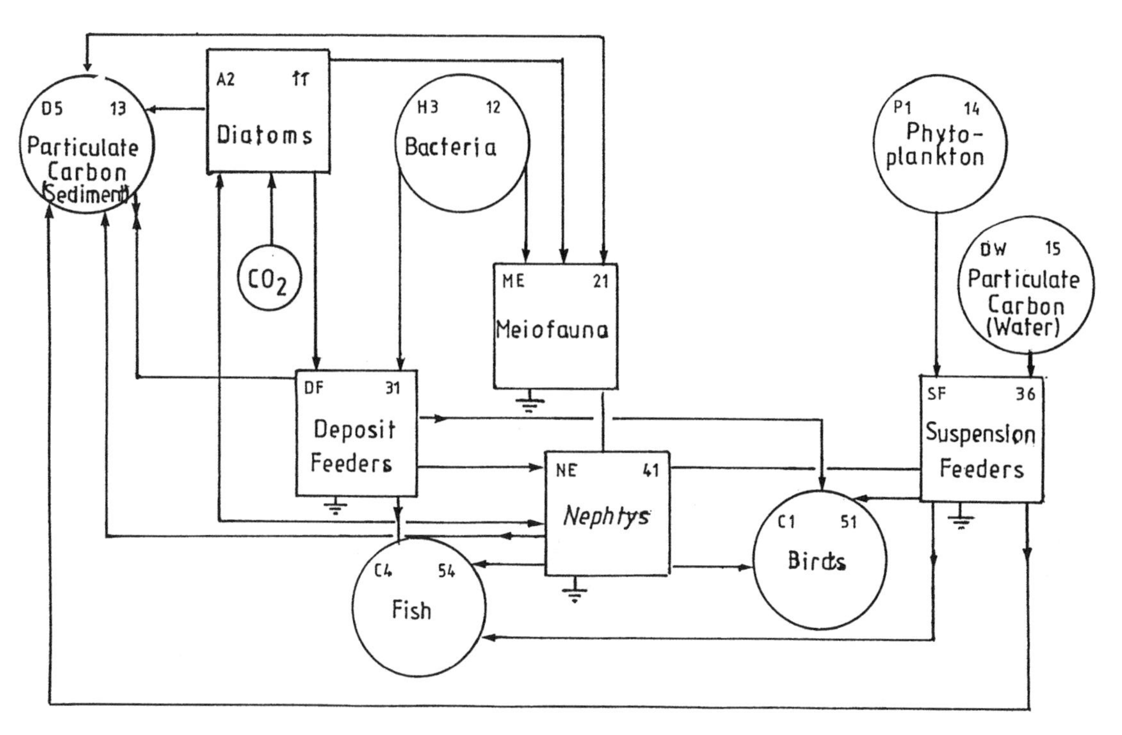

FIGURE 7.5 Process flow diagram for the dynamic simulation model of the Lynher Estuary. (Redrawn from Warwick, R.M. and Radford, P.J., in *Network Analysis in Marine Ecology,* Wulff, F., Field, J.G., and Mann, K.H., Eds., Springer-Verlag, Boston, 1989, 224. With permission.)

TABLE 7.4
Components of the Lynher Estuary Mudflat. Output is the Energy Available to the Next Trophic Level

Component	Standing Stock (g C m^{-2})	Consumption (g C m^{-2} year^{-1})	Output (Production) (g C m^{-2} year^{-1})
1. Heterotrophic bacteria (water column)	0.018	14.51	4.05
2. Phytoplankton	0.11	—	19.32
3. Heterotrophic bacteria (sediment)	0.11	—	2.8
4. Benthic microalgae	14.30	9.90	143.0
5. *Manayunkia* and *Fabricia* (small sabellid polychaetes: meiobenthos)	0.74	14.51	4.05
6. *Cardium* and *Mya* (suspension-feeding bivalves: macrobenthos)	2.55	14.14	1.15
7. *Nephtys* (omnivorous polychaete: macrobenthos)	1.58	7.88	2.93
8. *Ampharete acutifrons* (deposit-feeding polychaete: macrobenthos)	0.17	2.28	0.93
9. *Scrobicularia* and *Macoma* (deposit-feeding bivalves: macrobenthos)	0.99	2.37	0.32
10. Other macrobenthos (six species)	0.13	0.68	0.13
11. *Pygiospio* and *Strebospio* (deposit-feeding polychaetes: meiobenthos)	0.31	5.13	1.68
12. Oligochaeta (two species: meiobenthos)	0.39	3.43	1.17
13. Nematoda (40 species meiobenthos)	0.79	17.82	6.62
14. *Protohydra leuckarti* (predatory hydroid: meiobenthos)	0.02	0.58	0.22
15. Copepoda (8 species: meiobenthos)	0.32	18.49	5.70
16. Ostracoda (2 species: meiobenthos)	0.08	2.51	0.39
17. Kinorhyncha (2 species: meiobenthos)	0.01	0.10	0.50
18. Available Particulate Organic Carbon (water column)	0.25	—	—
19. Available Particulate Organic Carbon (sediment)	13.5	—	11.10

Source: From Warwick, R.R. and Radford, P.J., in *Network Analysis in Marine Ecology: Models and Applications,* Wulff, F., Field, J.G., and Mann, K.H., Eds., Springer-Verlag, Berlin, 1989, 224. With permission.

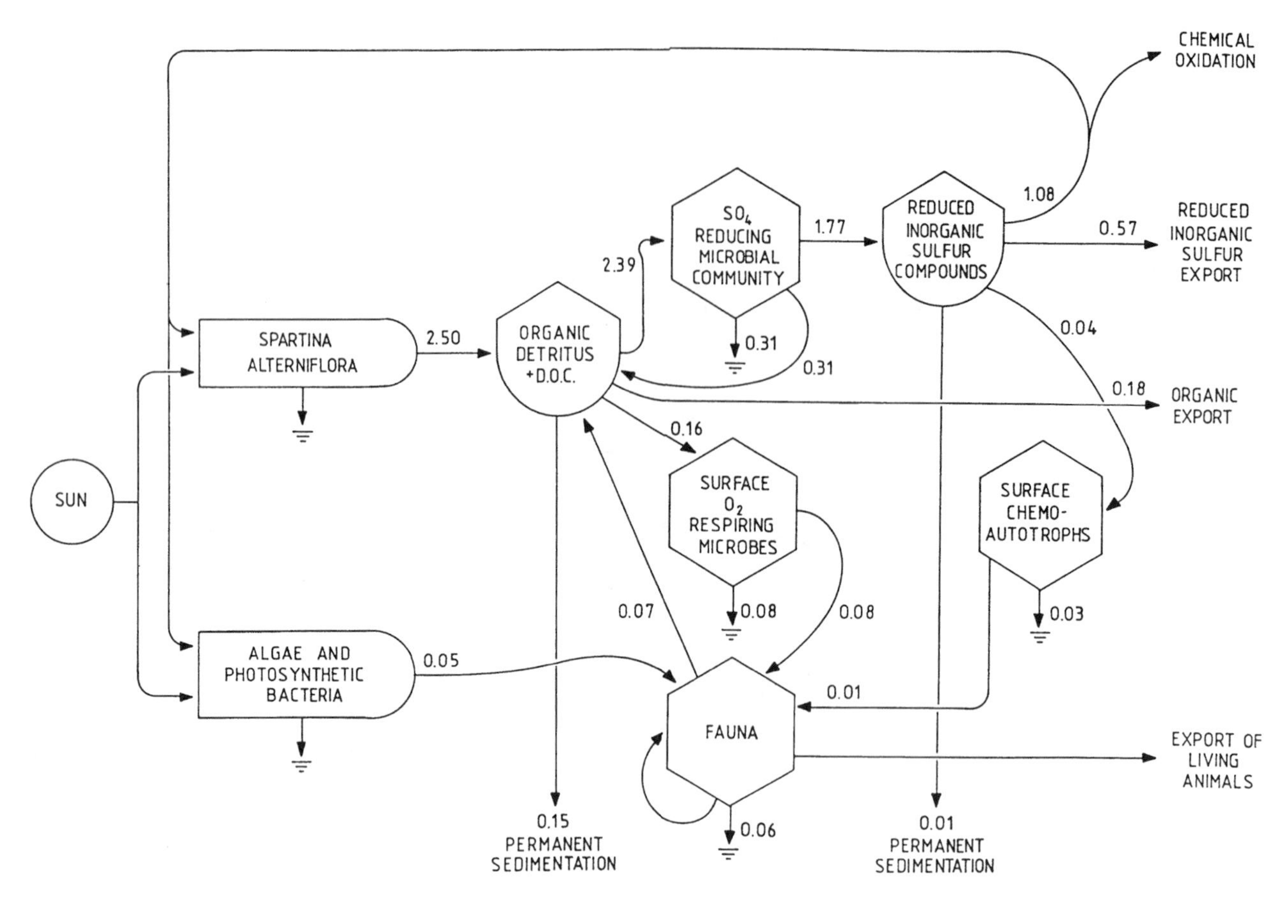

FIGURE 7.6 Energy flow model of a salt marsh illustrating the role of reduced inorganic sulfur compounds. (After Howarth, B.L. and Teal, J.M., *Am. Nat.*, 116, 865, 1980. With permission.)

by, organic carbon cycling. Under such conditions, the respiration of 1 g of organic carbon-C will yield approximately 42 kJ (10 kcal). Other studies have shown that anaerobic processes (see Section 3.8.6.2) are of even greater importance in energy flow within marsh and estuarine systems (Jorgensen, 1977a,b; Howarth and Teal, 1979; Skyring et al., 1979; Wiebe et al., 1981). Sulfide reduction probably degrades 12 times more organic matter than do respiration and denitrification combined (Howarth, 1979). Howarth and Teal (1980) in a study on the Great Sippeswissett salt marsh, Massachusetts, found that respiration of 1 g of organic carbon-C by the sulfate-reducing community yielded 11 kJ (2.6 kcal), only one-fifth of the energy released in oxygen-mediated respiration. Energy is stored as reduced inorganic sulfur compounds such as H_2S. When the reduced inorganic compounds are subsequently oxidized, energy is released. This energy can be trapped by a variety of organisms that fix CO_2 as organic biomass (Howarth and Teal, 1980).

Figure 7.6 is a simple model of energy flow demonstrating the role of reduced inorganic compounds in a marsh ecosystem. The model is limited to energy flows involving reduced sulfur and carbon compounds, and omits a number of minor elements. From the energy flows, a number of points emerge. First, most of the energy export from the marsh appears to be in the form of reduced inorganic sulfur compounds. Second, as Howarth and Teal (1980) point out, if the energy available from the oxidation of the reduced inorganic compounds was used with 25% efficiency by chemoautrophs in nearby creeks and estuaries, it would support a bacterial production equivalent to more than half of the aboveground production of *Spartina alterniflora*.

7.5.2 Sapelo Island Salt Marsh Ecosystem Carbon Model

Wiegert (1979a,b), Wiegert and Wetzel (1979), and Wiegert et al. (1981) discuss the series of models of the Sapelo Island marsh system, which were developed and used by the large interdisciplinary study of this ecosystem. Figure 7.7 depicts version six of these models. It reflects the three distinct regions or divisions of the salt marsh, divisions based on processes occurring in the sediment, water, and air. The 14 components of the model are listed in Table 7.5. Transfers of material or energy between the compart-

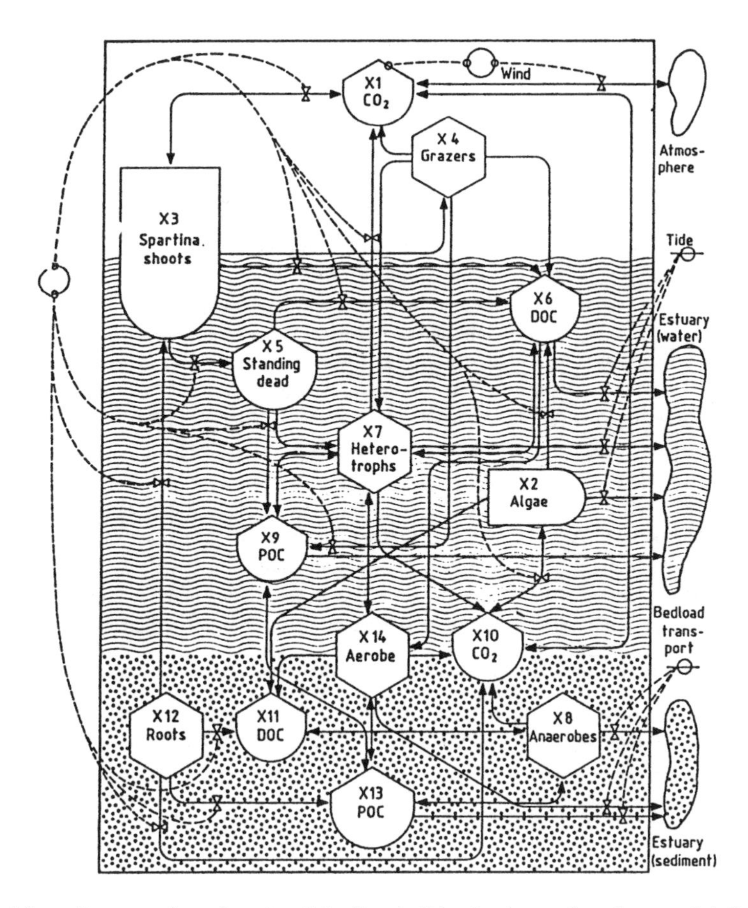

FIGURE 7.7 Influence and flow diagram of version six of the Sapelo Island salt marsh carbon model. Solid lines are carbon flows; dashed lines are flows of information. Bullets are autotrophs; hexagons are heterotrophs; tanks are abiotic storages. Circles are auxiliary variables supplying information but not simulated. Sources of information are indicated by small circles; material source sinks as small clouds. Wave and sediment boundaries indicate the demarcation between the three primary divisions of the marsh. (Redrawn from Wiegert, R.G., Christian, R.R., and Werzel, R.L., in *The Ecology of a Salt Marsh*, Pomeroy, L.R. and Wiegert, R.G., Eds., Springer-Verlag, Boston, 1981, 189. With permission.)

ments include active feeding, passive absorption, egestion, and mortality, as well as purely physical transfers such as diffusion or sedimentation. Wherever necessary, factors such as temperature, light, and nutrients have been included as implicit controls operating through seasonal changes in the maximum permitted rates of carbon transfer or transformation.

Six different versions of the model (MSHR) were tested with the same basic structural plan, but with minor structural changes to the pathways, and some major changes in certain functional parameter values and control functions. Details of these versions are given in Wiegert et al. (1981). Of the 81 parameters in the model, only 21 produced a change of more than 5% in the standing stock of any compartment within the fifth year. Table 7.6 shows the predictions of the annual mean standing crop (g C m^{-2}) of the 14 state variables for the first and third versions of the model. It can be seen that the agreement with the field data was best in version 3.

In version 6 of the model, the results of a hydrological study (Pomeroy and Imberger, 1981) were incorporated in the model to more accurately reflect the patterns of water movement to and from the marsh. In addition, storm events, which were shown to be significant, were incorporated into the model. Simulations suggested and observations confirmed that the difference between carbon-fixed and carbon-degraded *in situ* left a net several hundred g C m^{-2} $year^{-1}$ (out of the total net fixation from all sources of about 1,500 g C m^{-2} $year^{-1}$) (Wiegert, 1979a). Model simulations of the effects of heavy rainstorms showed that the normal complement of storms (up to 6 per year), with accompanying freshwater flushing the Duplin River, could account for the export of much of the surplus carbon. Reasonable assumptions about bacterial degradation and assimilation of bacteria into food chains leading to large, mobile, aquatic macrofauna could account for the remaining 200 to 300 g C m^{-2} $year^{-1}$.

7.5.3 Kaituna Marsh Estuarine Model

Knox (1983c) has developed a comprehensive model of the Kaituna marsh estuarine ecosystem in Pelorus Sound,

TABLE 7.5
The 14 Compartments of the Sapelo Island 1st Generation Salt Marsh Ecosystem

Symbol	Abbreviation	Name
X1	CO_2-air	Atmospheric carbon dioxide
X2	Algae	Benthic algae + phytoplankton
X3	*Spartina* shoots	Living shoots of *Spartina alterniflora*
X4	Grazers	Consumers of living *Spartina* + predators
X5	*Spartina* dead	*Spartina* dead + wrack[a]
X6	DOC-H_2O	Dissolved organic carbon
X7	Heterotrophs-H_2O	All aerobic heterotrophs — bacteria to fish
X8	Aerobes-sediment	All aerobic organisms in the sediment
X9	POC-H_2O	Abiotic particulate organic carbon
X10	CO_2-H_2O	Carbon dioxide in creek, marsh, and interstitial water
X11	DOC-soil	Dissolved organic carbon in the interstitial water of the sediment
X12	*Spartina* roots	Roots and rhizomes of *Spartina alterniflora*
X13	POC-sediment	Abiotic particulate organic carbon in the sediment
X14	Aerobes-sediment	All aerobic heterotrophs in or on the sediment

[a] Wrack comprises the large rafts and windrows of uncomminuted *Spartina* stems.

Source: From Wiegert, R.G., Christian, R.R., and Wetzel, R.L., in *The Ecology of a Salt Marsh,* Pomeroy, L.R. and Wiegert, R.G., Eds., Springer-Verlag, Berlin, 1981, 191. With permission.

TABLE 7.6
Annual Mean Standing Crop (g C m^{-2}) of the 14 State Variables of the Sapelo Island Salt Marsh Ecosystem

		MRSHIVI		MRSHIV3		Field Data	
Compartment	Abbreviation	Annual Average	Annual Change	Annual Average	Annual Change	Annual Average	Annual Change
X1	CO_2-air	782	—	875	—	975	—
X2	Algae	15	—	0.5	—	1	—
X3	*Spartina* shoots	125	—	118	—	135	—
X4	Grazers	0.5	—	0.4	—	1	—
X5	Dead *Spartina*	98	—	115	—	130	—
X6	DOC-H_2O	995	+225	4	—	5.6	—
X7	Heterotrophs-H_2O	10	—	12	—	7.5–30	—
X8	Aerobes-sediment	25	+1	44	—	30–40	—
X9	POC-H_2O	240	—	8	—	9	—
X10	CO_2-H_2O	46	—	8	—	?	—
X11	DOC-sediment	0.8	—	36	—	26	—
X12	*Spartina*-roots	137	—	475	—	450	—
X13	POC-sediment	20,291	—	17,733	—	18,200	+50
X14	Aerobes-sediment	10	—	9	—	3–6	—

Note: Predictions of the Sapelo Island simulation models MRSHIV1 and MRSHIV3, compared with field data.

Source: From Wiegert, R.G., Christian, R.R., and Wetzel, R.L., in *The Ecology of a Salt Marsh,* Pomeroy, L.R. and Weiget, R.G., Eds., Springer-Verlag, Berlin, 1981, 201. With permission.

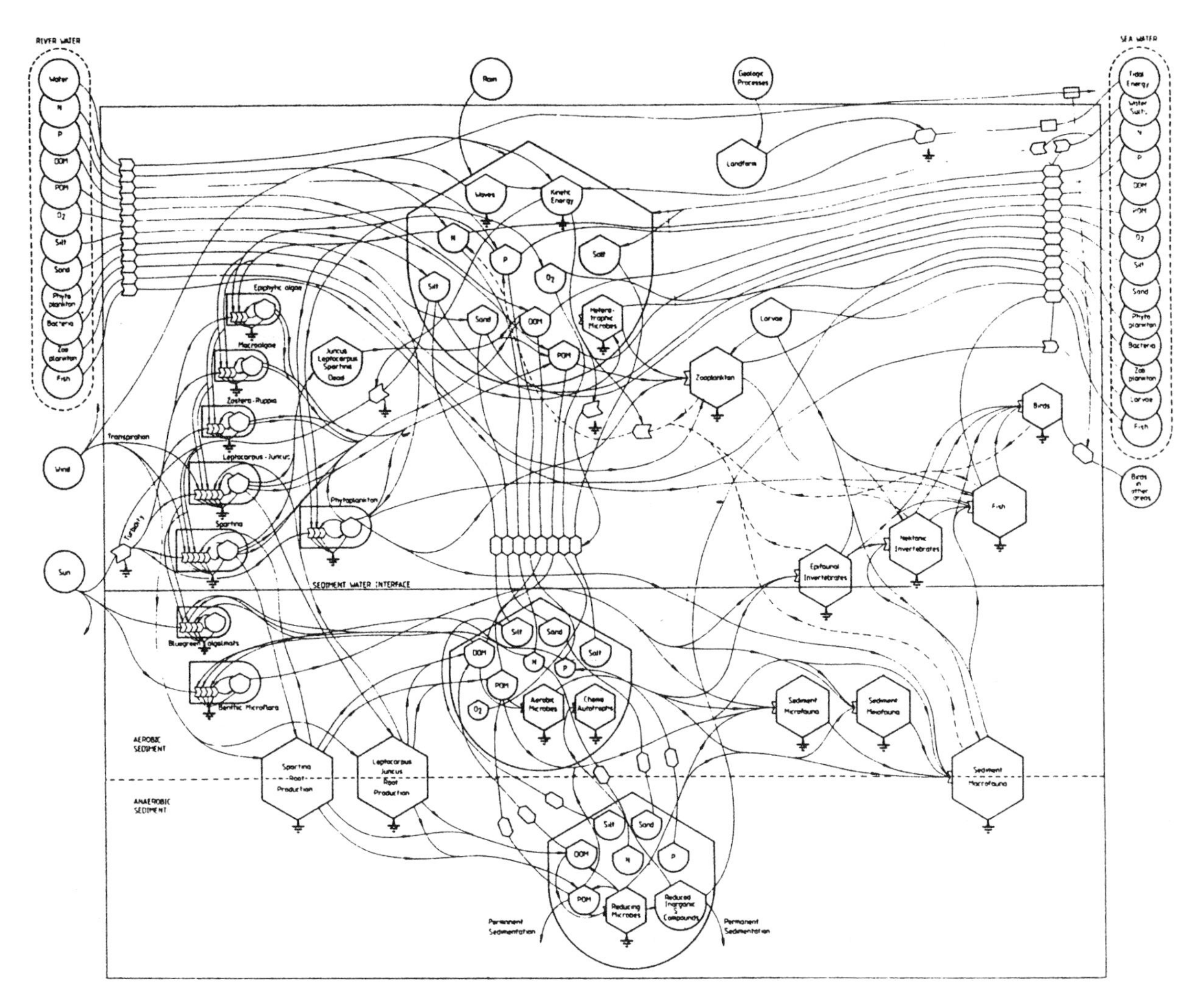

FIGURE 7.8 Energy flow model of the Kaituna marsh estuarine ecosystem in Pelorus Sound, New Zealand. (From Knox, G.A., *Estuarine Ecosystems: A Systems Approach*, Vol. II, 140, 1986. With permission.)

New Zealand (Figure 7.8). The forcing functions in this model are: (1) on the left — sun, wind, and river input; (2) on the top — rain and geological processes; and (3) on the right — seawater. Within the ecosystem boundary, the model is subdivided into a water component and a sediment component, with the latter being subdivided into an upper aerobic sediment component and a lower anaerobic sediment component. Producers shown on the left are epiphytic algae, macroalgae, submerged macrophytes (*Zostera* and *Ruppia*), emergent macrophytes (*Juncus, Leptocarpus,* and the introduced cordgrass *Spartina anglica*), and the sediment benthic microalgae and blue-green algal mats. Exchanges of silt, clay, salt, nitrogen, phosphorus, dissolved organic matter (DOC), and particulate organic matter (POC) between the water and the sediment compartments are also depicted. An important aspect of the model is the incorporation of the energy flow via reduced inorganic sulfur compounds as discussed by Howarth and Teal (1979). The various consumers are shown on the right of the figure.

7.5.4 North Inlet Salt Marsh Ecosystem

The salt marshes near North Inlet, South Carolina, are relatively pristine and have been the focus of much ecological research in recent years. The marshes are bounded partly by scrub oak and long-leaf pine forests on the landward edges and by barrier beaches on the eastern border separating the marsh from the Atlantic Ocean. The 3,200-ha marsh is dominated by intertidal stands of the cordgrass (*Spartina alterniflora*) interlaced with numerous tidal creeks with subtidal benthic communities. Substantial oyster reefs also occur in patches along the intertidal areas between the vegetated marsh and the creek low tide levels.

Early studies, which attempted to describe the holistic structure and functioning of the North Inlet salt marsh, resulted in a set of deterministic simulation models examining the role of subsystem interactions in controlling carbon exchange with the sea (Summers and McKellar, 1979; 1981a,b; Summers et al., 1980). Simulation results

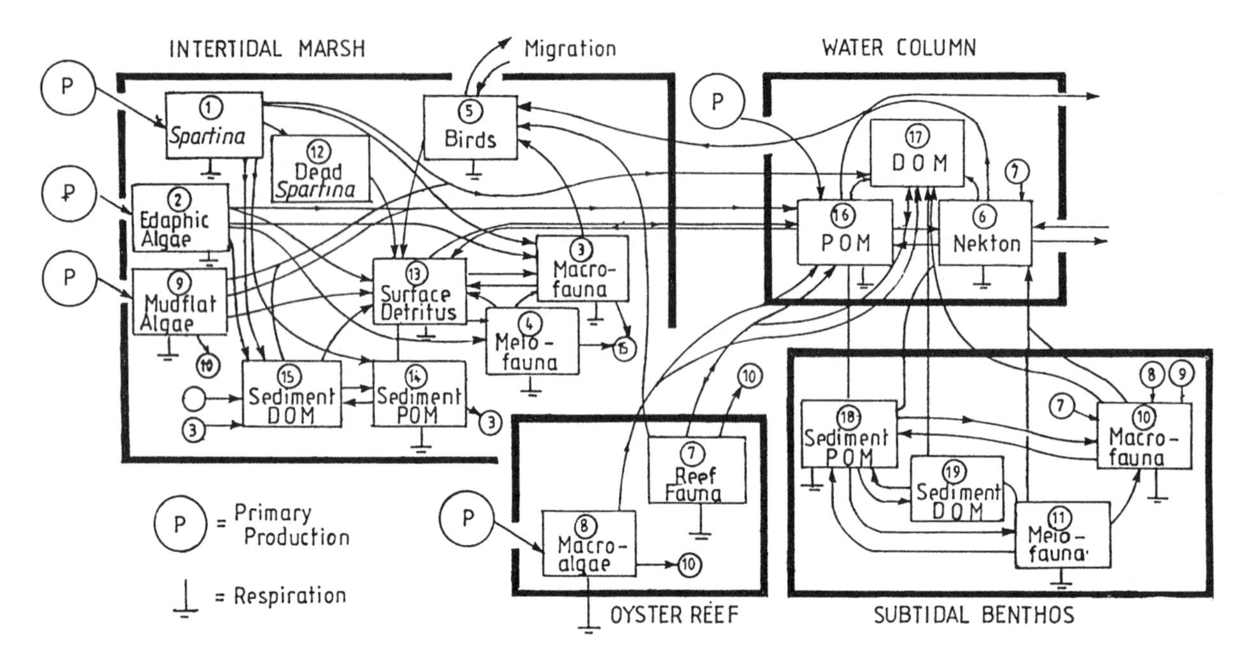

FIGURE 7.9 Model of energy flow in the North Inlet salt marsh ecosystem. (Redrawn from Asmus, H. and McKellar, H.N., Jr., in *Network Analysis in Marine Ecology, Methods and Applications,* Wulff, F., Field, J.G., and Mann, K.H., Eds., Spring-Verlag, Berlin, 1989, 209. With permission.)

indicated the importance of tidal driving forces, coupled with seasonal changes in sea level and the areal distribution of subsystem in controlling carbon exchange. The variables may be critical in explaining the large variability in system "outwelling" that is observed for different marsh-estuarine systems (see review by Nixon, 1980). More recent simulation analyses have focused on more mechanistic interactions within the tidal creek water columns (Childers and McKellar, 1987) emphasizing the coupling of carbon and nitrogen exchanges.

The simulation model that was developed by Summers et al. (1980) is reproduced in Figure 7.9. The model has four interconnected systems: the intertidal marsh, the tidal creek water column, the oyster reef, and the subtidal benthos. The model includes a spectrum of intersubsystem exchanges among the 19 compartments of energy storages. The result of the model simulation will be discussed further in this chapter.

7.6 SEA GRASS ECOSYSTEMS

Sea grass production and aspects of decomposition in sea grass ecosystems have previously been considered in Section 4.7.5. Harrison and Mann (1975b) found that very little of the live plant material was consumed, and that after death, the bulk of the leaves were buried in the sediment so that a considerable reservoir of slowly decaying organic matter accumulates in the sediment. In a *Posidonia australis* sea grass bed, Kirkman and Reid (1979) found that 37% of the year's production sank to the bottom where it was utilized by benthic detritovores, 12% was exported from the bed as floating leaves, and only 3% of the production was eaten by herbivores.

Comprehensive reviews of the kinds and quantities of consumer organisms in temperate sea grass beds have been given by Kikuchi and Peres (1977), Kikuchi (1980),Virnstein et al. (1983; 1984), Larkum et al. (1989), Howard and Edgar (1994), and Keough and Jenkins (1995), and in tropical beds in Ogden (1980).

Kikuchi and Peres (1977) subdivided the sea grass community into a number of subunits. The first category is the biota attached to the leaves. This group consists of: (1) epiphytic felt flora (small macroalgae, diatoms, and blue-green algae) and the micro- and meiofauna living in it; (2) sessile fauna attached to the leaves (hydrozoans, actinians, bryozoans, tube-building polychaetes, and ascidians); (3) mobile epifauna crawling on the leaves (gastropods, crustaceans, polychaetes, and a variety of other invertebrates); and (4) a group of swimming animals that often rest on the leaves (mysids, hydromedusae, small squid, and some fishes). The second category is the biota attached to stems and rhizomes such as nest-building polychaetes and amphipods, decapod crustaceans, gastropods, and bivalves. In addition, three other categories can be recognized: (1) motile epifauna — motile animals such as amphipods, errant polychaetes, and gastropods associated with the surface of the sediment, often living among sea grass debris; (2) epibenthic fauna — larger animals such as crabs, cephalopods, shrimps, and fishes, which are highly mobile and not closely associated with individual sea grass plants; and (3) infauna — including the smaller micro- and meiofauna and the macrofauna (bivalves, sedentary polychaetes,

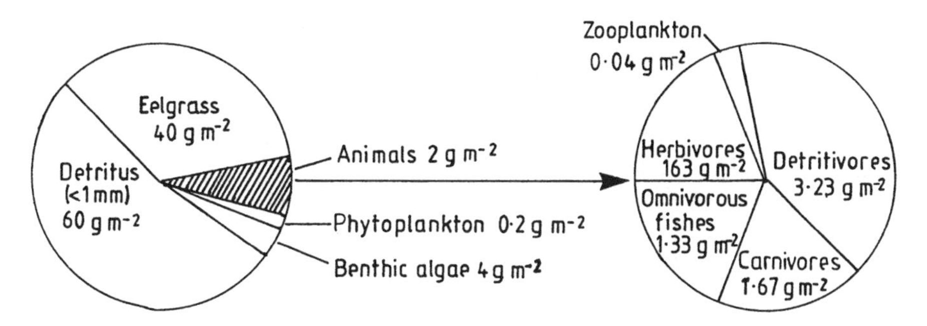

FIGURE 7.10 Composition of the biomass in an eelgrass bed in California. (Based on data in Thayer, G.W., Adams, S.M., and La Croix, M.W., in *Estuarine Research*, Vol. 1, Cronin, L.E., Ed., Academic Press, New York, 1975, 518. With permission.)

holothurians, and other species) living in the sediment among the rhizomes of the sea grasses (Howard and Edgar, 1994). In a *Posidonia* bed in the Gulf of Naples, the total density of decapod crustacea in the rhizome layer reached 500 to 700 m^{-2}, while the total density of medium-sized crustacea (excluding harpacticoidea) reached more than 8,000 individuals m^{-2}. Generally speaking, total density, biomass, and community diversity of the infauna of a sea grass bed are higher than those of the surrounding unvegetated areas (Thayer et al., 1975; Kullenberg, 1983). Thayer et al. (1975) made quantitative estimates of the fauna of an eelgrass bed in California. Biomass of the functional groups of organisms are shown in Figure 7.10. Polychaetes and bivalves dominated the biomass of detritovores; gastropods were the most abundant herbivores; while decapods were the dominant carnivores. Respiration of each consumer species was measured and used to calculate consumption and assimilation rates. The total net production of phytoplankton, benthic microalgae, and eelgrass was estimated at 1,545 kcal m^{-2} $year^{-1}$, and of this, eelgrass contributed about two-thirds. The standing crop of detritus both on and in the sediment was estimated at about 21,000 kcal m^{-2}. From a measured primary production of 1,545 kcal m^{-2} yr^{-1} it was estimated that the macrofauna consumed 841 kcal, or approximately 55%. Microbial metabolism accounted for 467 kcal m^{-2} $year^{-1}$, or 30%, leaving a small surplus for export. However, it should be noted that the primary production of the epiphytic flora and the benthic microalgae were not measured in this study.

Ogden (1980) has reviewed faunal relationships in tropical Caribbean sea grass beds and has pointed out some fundamental differences between them and those of temperate areas. The dominant species is the turtle grass *Thalassia testudinium*, with lesser amounts of manatee grass *Syringodium filiforme*, and shoal grass *Halodule wrightii*. Tropical sea grass beds differ from temperate ones in that a large percentage of the production is consumed by herbivores, including conch gastropods and herbivorous vertebrates, mainly parrotfishes, turtles, dugong, and manatees (Laynon et al., 1989). The green turtle, *Chelonia mydas*, is the most abundant large vertebrate consumer of sea grasses in the world, with over 90% of its diet comprising such grasses. When feeding, it produces conspicuous scars in the grass beds. Over 30 species of tropical fishes, of which parrotfishes (Scaridae) are the most important, have been recorded as feeding on sea grasses. Among the invertebrates, sea urchins are significant grazers on sea grass beds (Ogden, 1980; Nojima et al., 1987). Greenaway (1976), working on *Thalassia* beds in Kingston Harbour, Jamaica, found a blade production of 53.8 g dry wt m^{-1} $week^{-1}$, of which 24.0 g m^{-2} $week^{-1}$ was eaten by the sea urchin *Lytechinus variegatus*, 0.06 g m^{-1} $week^{-1}$ was eaten by the parrotfish *Sparisoma radians*, and 27.7 g m^{-2} $week^{-1}$ went to detritus. Like its Caribbean relative, the manatee *Trichetus manatus*, the dugong, *Dugong dugong* feeds almost exclusively on sea grass.

Short (1980) has developed a numerical production model of a sea grass ecosystem to investigate the mechanism of sea grass production in a quantitative manner. Carbon flow, initiated by photosynthetic uptake, is depicted in Figure 7.11, together with its subsequent modification by auxiliary energy sources. The structure of the photosynthetic unit (eelgrass blades) is subdivided into five leaf-size groups with production rates for each determined, taking into account the limitations of temperature, light, and current speed. The subsurface root and rhizome storage is fed by transfer of carbon from the leaves. McRoy (1970) found a nearly constant rate of translocation of about 17% for experimental plants. The remaining 83% of the total production determines the average rate of leaf growth. The biomass of a new leaf in the model increases until it reaches a weight equal to the weight of a leaf of the smallest measurable size. When the storage is large enough, a new leaf is added to the smallest compartment and existing leaves move up one size class. Leaf loss to the benthic detrital storage is assumed to be the result of physical damage caused by wind-driven circulation and turbulence. Root and rhizome losses result from uprooting and senescent death.

Two simulations of the model were run on data collected in Charlestown Pond, Rhode Island, during 1974 and 1975 at two stations with different environmental conditions. Calculated leaf biomass reached a maximum

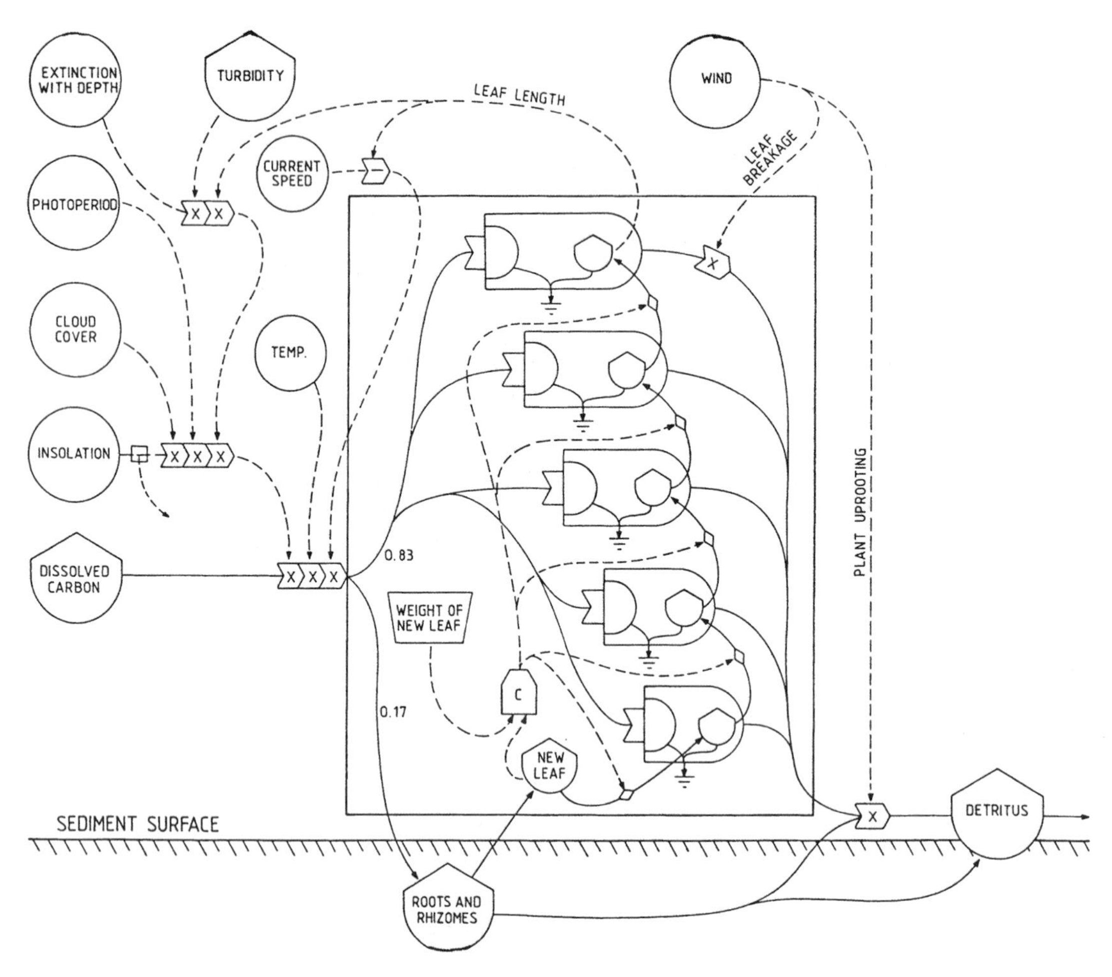

FIGURE 7.11 A simplified carbon flow of a sea grass ecosystem. (Modified from Short, F.T., in *Handbook of Seagrass Biology: An Ecosystems Approach,* Phillips, R.C. and McRoy, C.P., Eds., Garland STMP Press, New York, 1980, 280. With permission.)

of 400 g dry wgt m^{-2} at Station F and 250 g dry wgt m^{-2} at Station C in late August. These are comparable to results obtained by Brown (1962) in Charlestown Pond showing a summer standing stock of 380 g dry wgt m^{-2}. The simulation model calculated growth rates for *Zostera marina* in Charlestown Pond that reached a maximum production rate of 20 g dry wgt m^{-2} day^{-1} (8 g C m^{-2} day^{-1}) near the end of July. However, the average production rate calculated over a 12-month period was 2.7 g C m^{-2} day^{-1} (with a range of 0.05 to 8.0 g C m^{-2} day^{-1}). McRoy (1970) found 4.0 g C m^{-2} day^{-1} as an average rate for a number of sea grasses reported in the literature. Total annual production was estimated at 2,200 g dry wgt m^{-2} for high-density beds. A similar estimate of 1,525 g dry wgt m^{-2} was made for an average grass bed in nearby Long Island, New York (Burkholder and Doheny, 1968).

Figure 7.12 is a conceptual model of energy flow in a sea grass ecosystem. The major energy flow (a), solar radiation (which drives the photosynthetic production), is influenced by auxiliary energy sources such as photoperiod, cloud cover, light attenuation (c), temperature, salinity, and current speed. Nutrients also act to reduce production when one or more is low. The primary producers, represented as leaves, epiphytic algae, and benthic microalgae, incorporate energy into biomass and lose energy via respiration. Sea grass production is amplified by reward loops that change through seasonal development. Some part of the energy received by the leaves is transferred to the roots and rhizomes (b). Energy stored as biomass in the compartment labeled "LEAVES" is lost as a result of mechanical damage to benthic detritus (d) and the associated flora and fauna, to water as DOM and inorganic nutrients (e), and to vertebrate grazers (f) that eat the grass. Outside sources and internal recycling provide nutrients to the surrounding water where they become available to the sea grass leaves. Sediment nutrients, replenished by detrital decomposition, are absorbed directly by the roots. As Short (1980) points out, the actual dynamics of the ecosystem are much more complex than shown in Figure 7.12.

7.7 MANGROVE ECOSYSTEMS

Mangrove production models as developed by Lugo et al. (1974) and Knox (1983a,b) have previously been dis-

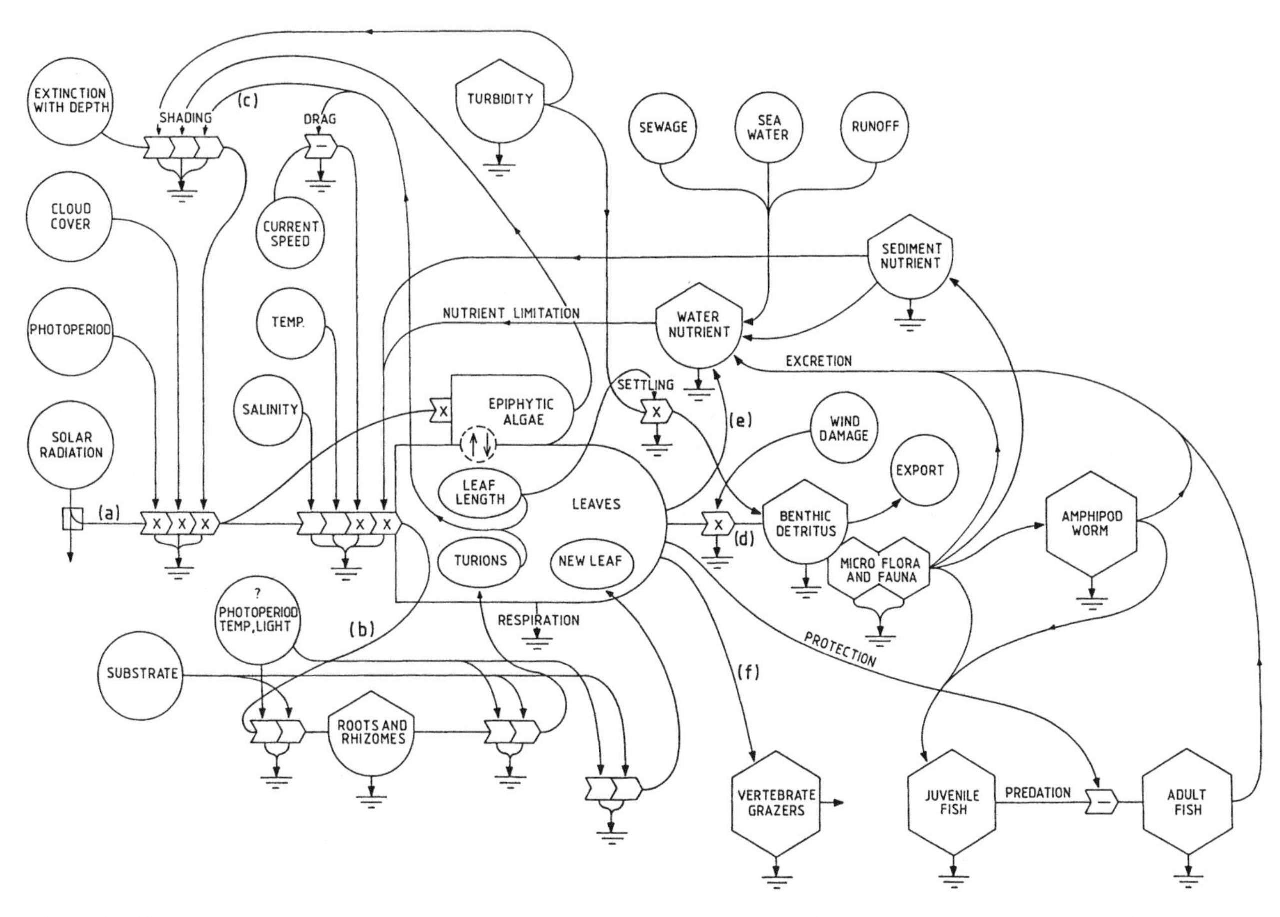

FIGURE 7.12 A conceptual energy flow model of a sea grass ecosystem. (Modified from Short, F.T., in *Handbook of Seagrass Biology: An Ecosystem Approach*, Phillips, R.C. and McRoy, C.P., Eds., Garland STMP Press, New York, 1980, 280. With permission.)

cussed in Section 4.7.7, and the impact of storm events on litterfall and nutrient supply have also been outlined.

Lugo et al. (1975), by varying tidal level in their simulations, found that tides had less effect on gross photosynthesis as compared to the impact of tides on detritus export. High tidal levels slightly increased the rate of gross photosynthesis, but significantly increased the rate of detritus export. As the tide levels increased, the amount of detritus stored in the mangrove forest decreased. Maximum detritus accumulation in the forest was observed with no tidal rise when detritus export was zero. Simulation of the model also demonstrated that the mangroves exert significant controls over the amount of nutrients in adjacent waters, but when terrestrial runoff was reduced, the mangroves did not have the capacity to maintain themselves at the same level of production. This was due to the loss of nutrients in detritus export.

In simulation runs with mean rates of metabolism, the time taken to reach a steady state was ten years from the stage that the mangroves had reached at the beginning of the simulation and 23 years from an early stage of succession. Literature reports on the age of the mangroves in Florida indicated that most mangroves reach maturity in 20 to 25 years (Craighead, 1971), which agrees well with the simulation prediction. Wadsworth (1972) reported that mangrove stands were generally 25 years old with usually a few large individuals exceeding 50 years. These older mangroves represent survivors from previous hurricanes. A moderate hurricane (with wind speeds of up to 40 m sec^{-1}) reduced mangrove biomass to 9,000 g C m^{-2}, and the recovery time to attain maximum biomass levels was about 16 years. A major hurricane (wind speeds of up to 90 m sec^{-1}) reduced the mangrove biomass to 3,000 g C m^{-2}, and the recovery time to attain maximum biomass levels was 20 years. The level after the major hurricane was slightly greater than the maximum level reached without any hurricane occurring. According to the model simulations, mangroves may attain their greatest biomass when there is a 25-year cycle of hurricanes.

7.8 ESTUARIES

7.8.1 INTRODUCTION

Some examples of estuarine whole ecosystem models such as the Upper Waitemata Harbour mangrove-estuarine model have been discussed previously. Here we will consider in detail a number of examples, including four large estuarine systems (Barataria Bay, Narragansett Bay, and Chesapeake Bay in the United States, and the Baltic Sea), and four smaller estuarine systems (the Ythan Estuary in Scotland, the Ems Estuary in The Netherlands, and the Swartkops and Kromme Estuaries in South Africa).

The functioning of tidal estuaries involves the complex interactions of physical, biogeochemical, and biological processes. The situation is complicated by freshwater input, which not only varies between estuaries but also seasonally, and the amount of interchange between estuaries and the sea including the percentage of export from the estuarine system. Primary production within estuaries may include phytoplankton, benthic microalgae, epiphytic micro- and macroalgae, benthic macroalgae, marsh macrophytes, mangroves, and sea grasses. Not all of these have been measured in many of the studies. Additional problems include: (1) difficulties in estimating belowground production and decomposition of marsh plants and sea grasses; and (2) adequate estimates of exchanges across the sediment-water interface.

7.8.2 Barataria Bay

Barataria Bay is located at the seaward end of a typical Louisiana interdistributary coastal basin that exhibits well-developed physical gradients and vegetation zones. The shallow bay (average depth 1.0 to 1.5 m) is characterized by turbid waters with mean salinities ranging from 15 to 25. The margins of the bay are characterized by extensive salt marshes. Of the primary producers within the system, the marsh grasses (predominantly *Spartina alterniflora*) are the most productive. Kirby (1972) calculated net production to be as high as 2,800 g dry wgt m^{-2} $year^{-1}$, with a mean of 2,180 g m^{-2} $year^{-1}$. The net production of the phytoplankton and benthic microalgae, mainly diatoms, are 418 and 488 g dry wgt m^{-2} $year^{-1}$, respectively (Day et al., 1973). Most of the marsh grass net production enters the detrital food chain.

In the simulation model developed by Hopkinson and Day (1977), the carbon flow and nitrogen flow models developed earlier were combined (Figure 7.13). Forcing functions included in the model are: insolation, water level, temperature, migration of the water fauna, nitrogen in the rain, nitrogen in the river, and a log function for marsh grass decomposition. A switching function (16) is used to simulate the seasonal entrance of larval and juvenile fishes, shrimp, crabs, etc. into the estuary. Whenever the slope of the temperature curve exceeds a certain value, the function becomes operable. Practically, this occurs in the spring. A lag of 6 months based on field measurements is used to control the timing of the flow of live marsh grass to the dead-standing compartment. Hopkinson and Day (1977) considered that the simulations of the model indicated that it was a fairly realistic representation of many, but not all, of the important components of the natural system. Simulations were run for 10 years, and the model stabilized after 4 years. There was good agreement in the simulations with field data on net marsh grass production (x_1), production of dead standing grass (x_2), and flow from dead standing marsh grass to detritus. One interesting feature of the simulation was the sequential lag that carbon exhibited as it was transferred from one compartment to the next. There was a 14-month lag in the simulations between peak carbon uptake by marsh plants and subsequent flushing into adjacent bays.

The effects of varying the three forcing functions of temperature, insolation, and nutrients on primary production and biomass of the marsh grasses were investigated (Figure 7.13). Temperature as a single forcing function gave the best fit to field measurements of standing crop, supporting Turner's (1976) findings that temperature alone can explain 95% of the variation of peak standing crop in a variety of North American Atlantic and Gulf Coast marshes.

Hopkinson and Day (1977) asked what was gained from the modeling exercise and was it worthwhile? In the formation of the carbon and nitrogen budgets in the combined model, they were forced into careful consideration of the functioning of the estuarine ecosystem. They became aware of additional data needs and these were pursued. The simulation results tested their understanding of the system, raised several interesting questions, and suggested new areas for further research.

7.8.3 Narragansett Bay

Narragansett Bay, Rhode Island, is an irregular basin with a length of 45 km, a maximum width of 18 km, a mean depth of 9 m, and semidiurnal tides with a mean range of 1.1 m at the mouth and 1.4 m at the head. A drainage basin of some 4,660 km^{-2} provides freshwater input varying seasonally around an average of 37 m^3 sec^{-1}. The mean tidal prism is about 13% of the mean volume and over 250 times the mean total river flow during a tidal cycle. As a result, the waters are well mixed from top to bottom, and salinities range from about 24 at the head of the bay to 32 at the mouth.

It is a phytoplankton-based ecosystem with a seasonal cycle characterized by much greater standing crops than are found in adjacent waters, and by a marked winter-early spring bloom that precedes a series of intense summer blooms. The winter population is usually dominated by diatoms, while the summer flora is composed predominantly of flagellates and microflagellates. The major consumers of the phytoplankton are zooplankton, dominated alternately by the ubiquitous copepods, *Acartia clausi* and *A. tonsa*, which may make up to 95% of the population. The zooplankton are preyed upon by the ctenophore *Mnemiopsis leidyi* (see Section 4.8.5), and by various fish species, especially the menhaden *Brevoortia tyrannus*.

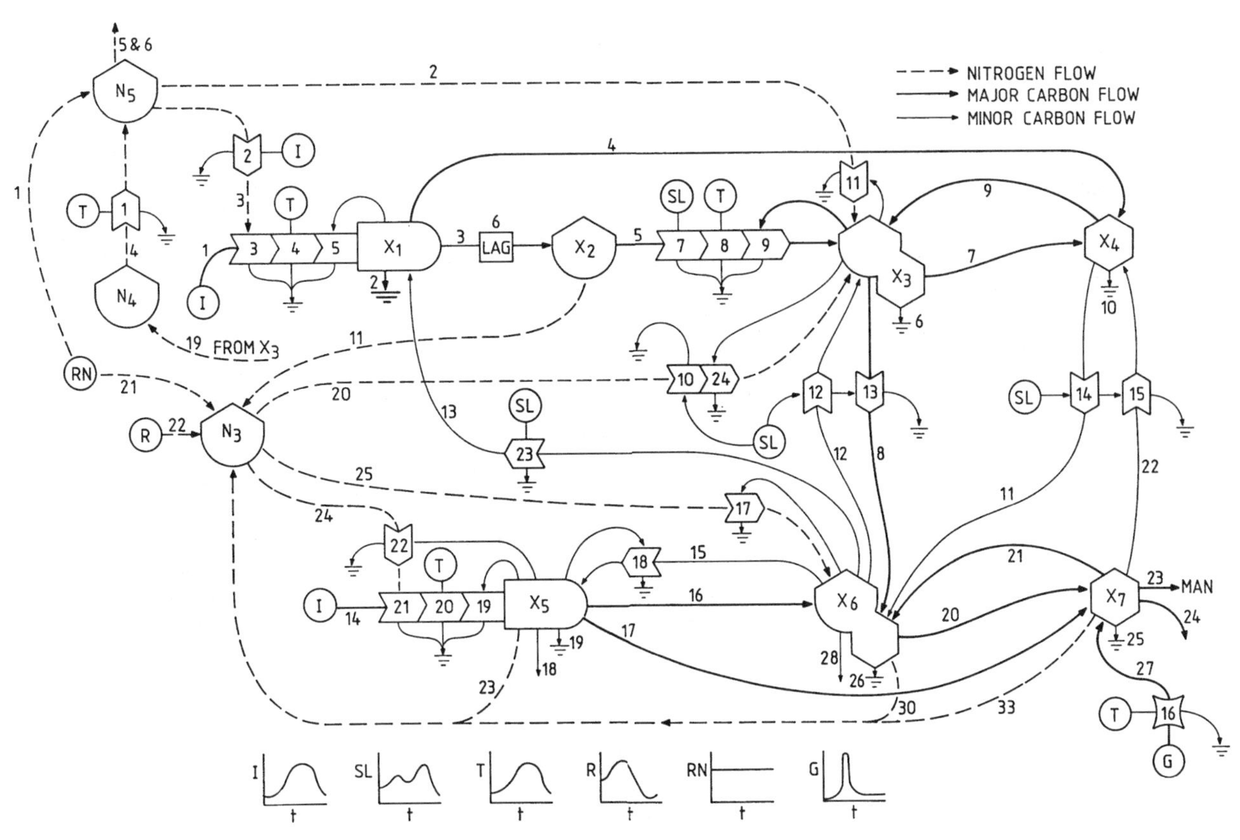

FIGURE 7.13 The Barataria Bay simulation model. 1-year patterns of forcing functions are shown at the bottom; I = insolation; SL = sea level; T = temperature; R = river; RN = rain; G = migration form the Gulf of Mexico; t = 1 year; N = inorganic nitrogen in water; N_4 = total N in marsh sediment; N_3 = extractable ammonia in marsh sediment; X_1 = live marsh plants (mainly *Spartina alterniflora*); X_2 = dead standing *S. alterniflora*; X_3 = marsh microbial-detritus complex; X_4 = marsh macrofauna; X_5 = water flora; X_6 = water microbial-detritus complex; X_7 = water fauna. Numbers refer to work gates and flow functions; 6 = 0.4-year log controlling input of grass to dead standing compartment; 10 = sea level controlling N-rich water cover on marsh; 12 = effect of sea level on washing organics onto marsh; 13 = sea level controlling detritus flushing from marsh; 14 = sea level affecting washout of marsh fauna into water; 15 = sea level affecting the stranding or washing of aquatic fauna onto marsh; 16 = switch based on slope of temperature curve controlling immigration of fishes to estuary from Gulf; 18 = nitrogen associated with detritus washed onto the marsh from water; 19 = immobilization in subsurface sediment; 22 = aquatic fauna stranded or trapped on marsh; 24 = emigration of fish to Gulf; 27 = immigration of fish from Gulf; 28 = loss of detritus to Gulf via tidal exchange; 5, 9, 11, and 19 = feedback loops controlling process rates. (Modified from Hopkinson, C.S., Jr. and Day, J.W., Jr., in *Ecosystem Modelling in Theory and Practice*, Hall, C.A.S. and Day, J.W., Jr., Eds., John Wiley & Sons, New York, 1977, 248. With permission.)

Predators on the ctenophores include the butterfish (*Preptilus triacanthus*), and those on menhaden include striped bass (*Morone saxatilis*) and blue fish (*Pomatomus saltatrix*). The macrobenthos is dominated by dense populations of the hard clam *Mercenaria mercanaria.* The bay bottom consists largely of empty clam, mussel, scallop, and oyster shells, with an epifauna of starfishes, lobsters, conches, scattered sponge beds, crabs, and swarms of shrimps. There is an abundant infauna dominated by the bivalves *Nucula proxima* and *Yoldia limatula*, and the polychaete *Nephtys incisa*, with a biomass of about 8.0 g dry wgt m^{-2}.

On the basis of standing crop estimates, it was concluded that much of the high phytoplankton production of the bay was directed to benthic food chains and the support of an abundant infauna and large populations of bivalves and flounders. A preliminary energy-flow model for the Narragansett Bay ecosystem (Figure 7.14) indicates that the interrelations between storages and flows within the system. The major forcing functions, apart from sunlight, are upstream inputs, tidal inputs, tidal mixing, and offshore exchange. Inputs of energy from marsh detritus or sewage are also important in the upper part of the bay. Studies of marsh production in the bay indicate a potential total input of detritus of 1.9×10^6 kg $year^{-1}$. The secondary production of bacteria is considered to be substantial and it may serve as an important food source for some of the smaller species.

The Narragansett Bay computer modeling program consisted of three separate efforts, dealing with economic activity and effluent loads, the dynamics of water movements, and the chemistry and biology of the bay ecosystem. Of the conceptual model in Figure 7.14, only the

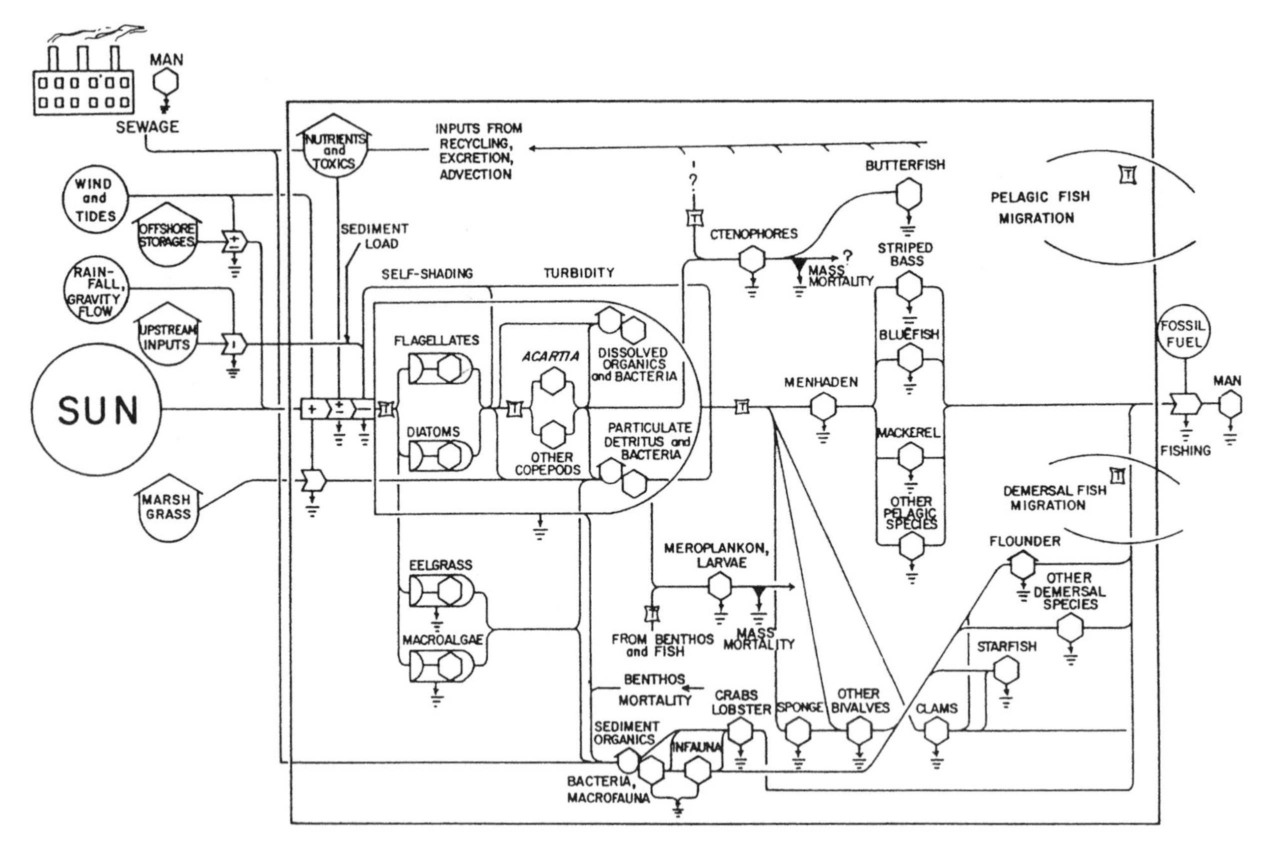

FIGURE 7.14 A complex but still greatly simplified energy flow model of the Narragansett Bay ecosystem on a summer day. (Redrawn from Kremer, J.N. and Nixon, S.W., *A Coastal Marine Ecosystem: Simulation and Analysis,* Springer-Verlag, Berlin, 1978, 12. With permission.)

phytoplankton, zooplankton, and nutrient compartments were simulated in full detail. The bay was divided into eight spatial elements (or geographic areas) and appropriate values and forcing functions were inserted simultaneously into each of the spatial elements. All were assumed to be vertically and horizontally homogeneous and were integrated into the hydrodynamic model driven by variations in tide height at the mouth. River flow was programmed to follow a seasonal cycle.

A simplified conceptual model (Figure 7.15) (Kremer and Nixon, 1980; Kremer, 1989) was developed to schematically represent the material, energy flow, and controlling interactions among the compartments of the ecological simulation model. The model reflects the plankton nature of the dominant parts of the bay system. The phytoplankton compartment incorporates a seasonal shift in population dominance from cold water primary diatom, to warmwater flagellate species groups. Choices of growth parameters were made so as to reflect the respective advantages of these characteristic populations. Herbivores consist of smaller zooplankton, mainly copepods, along with the clam component of the benthos. Carnivorous organisms include seasonal pulses of meroplankton, primarily larval fish and ctenophores. Apart from menhaden, pelagic and demersal fish were not included in the model.

The zooplankton compartment, representing primarily herbivorous copepods is subdivided into adults, eggs, and juveniles. Adults consume available food — phytoplankton, detritus, their own eggs, and juveniles — in a density-dependent manner and as a function of temperature. Eggs hatch after a delay determined by ambient temperature and this is followed by juvenile development. Development is controlled by temperature. Food-limitation and respiration were incorporated in the calculation of the realized daily growth rate. Predation by carnivores depletes adults, juveniles, and eggs, while the additional pressure of adult cannibalism further reduces eggs and juveniles. The unassimilated fraction of ingestion is transferred to the benthos.

The benthos in the model affects the plankton system in two ways. First, the impact of bivalve filtering on the standing crop of phytoplankton is included, and second, the benthos plays a role in the regeneration of nutrients. Fluxes of ammonia, phosphate, and silicate pass into the water column as a function of temperature. Particulate organic matter is not fully represented in the model; it is included with no specification of sources.

The seasonally important predation pressure of three carnivores, the ctenophore *Mnemiopsis leidyi,* the menhaden, and fish larvae on the zooplankton, is represented in

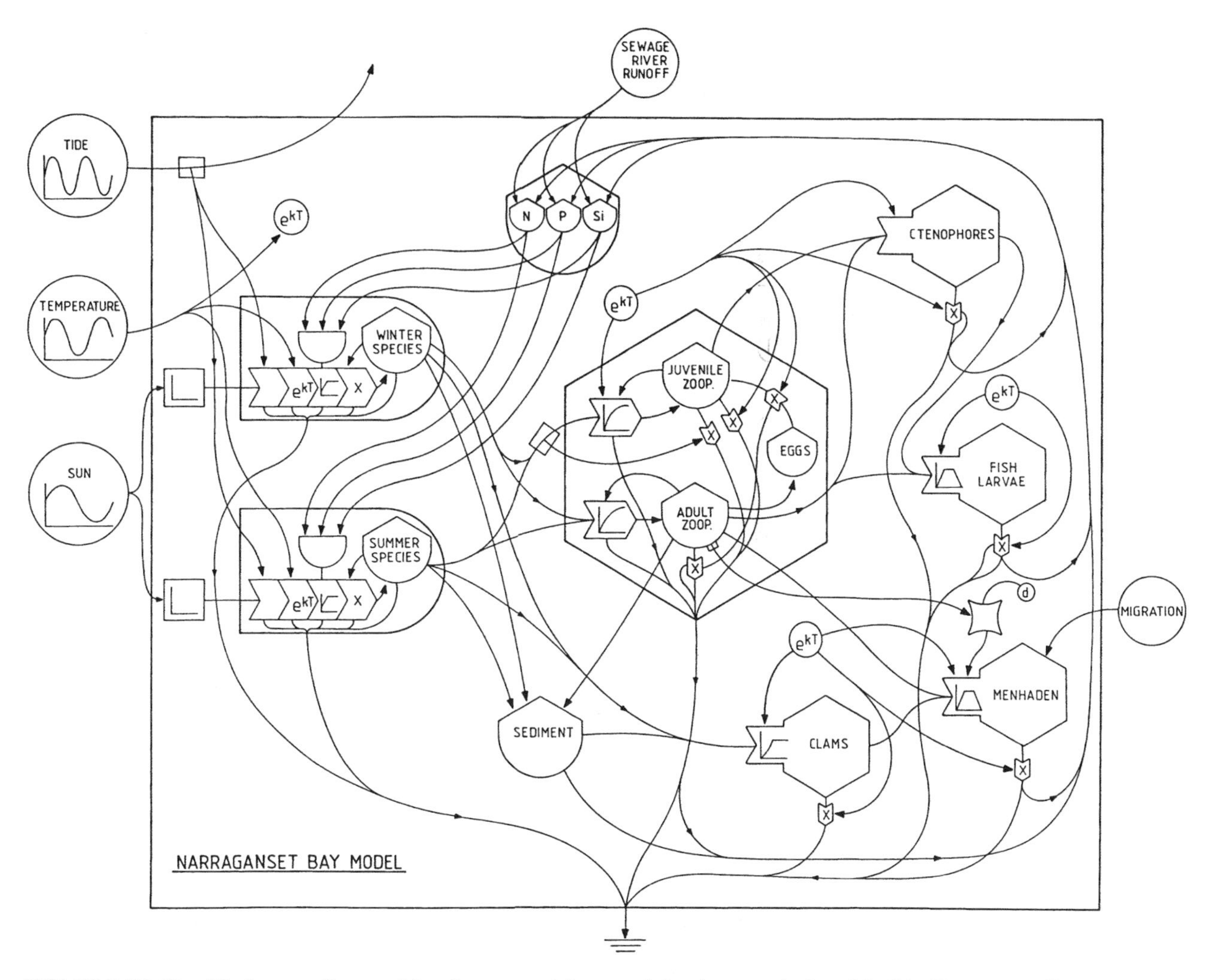

FIGURE 7.15 Simplified energy flow model and conceptual framework for the numerical model of the Narragansett Bay ecosystem. (Modified from Kremer, J.N. and Nixon, S.W., *A Coastal Marine Ecosystem: Simulation and Analysis*, Springer-Verlag, 1978, 16. With permission,)

the model. The final model compartment and the one providing a critical feedback loop interconnecting all compartments in the bay ecosystem is that of chemical nutrients. Excretion by animals, uptake by algae, and fluxes from the benthos are all included as contributions to the nutrient balances, which are determined by the dynamics of the respective compartments. In addition, the input of nutrients in domestic and industrial sewage entering throughout the bay is included. No organic nutrients were included in the model.

In all simulation runs, it was found that the timing and magnitude of the winter to spring phytoplankton bloom was well-represented in the model. All runs consistently indicated that the prebloom winter phytoplankton biomass was insufficient to maintain even the observed low zooplankton stock. It was suggested that alternative food sources, probably detritus, were being utilized at that time. Simulations also provided information on the relative importance of controlling factors on phytoplankton production. Figure 7.13 depicts the relative importance of the three nutrients, nitrogen, phosphorus, and silica, and demonstrates that of these, nitrogen remained the single most-limiting nutrient throughout the simulation.

When the model was run with all zooplankton grazing removed, the phytoplankton biomass still followed a "reasonable pattern," and this led to the conclusion that nutrient limitation was the prime factor terminating the bloom; the actual removal of phytoplankton biomass being caused by flushing plus some grazing by other filter feeders such as clams. In contrast, when all carnivores were removed from the system, dramatic changes occurred in which phytoplankton and zooplankton populations became unstable and went into a series of oscillations that did not occur in the natural system. It was therefore concluded that zooplankton predators, mainly ctenophores and menhaden, exerted a strong stabilizing influence on the system.

Field (1983) has made estimates of the phytoplankton carbon production in Narragansett Bay based on the mean

TABLE 7.7
Estimated Annual Carbon and Nitrogen Fluxes in Narragansett Bay

Category	Carbon ($g\ C\ m^{-2}\ year^{-1}$)	Nitrogen ($g\ N\ m^{-2}\ year^{-1}$)
Primary production		
Phytoplankton	98.1	16.3
Secondary production		
Consumption		
Zooplankton on phytoplankton	58.1	(9.7)
Clams, benthos on phytoplankton	24.9	4.1
Benthos on detritus	(24)	(>9.4)
Production		
Zooplankton	7.1	(1.7)
Clams, benthos	(5)	(1.3)
Respiration		
Zooplankton	(34.9)	—
Clams, benthos	(24.5)	—
Feces and urine		
Zooplankton	(11.6)	6.60
Clams, benthos	(24.5)	8.10
Detritus formation		
(Primary production – consumption + feces) 98.1 – 83.0 + 21.6 = 36.7 g C)	—	?
Detritus respiration and other losses		
(Formation – consumption by benthos) = 36.7 – 24 = 12.7 g C	—	—
Carnivores		
Ctenophores, larvae	—	0.053
Fish	—	?

Note: Averaged over eight compartments.

Source: From Field, T.G., in *Mar. Ecol. Vol. 5,* Kinne, O., Ed., 1983, 780. With permission.

values of the energy flows in the eight compartments (Table 7.7): 59% is grazed by the zooplankton, 25% by benthic filter feeders, and the remaining 16% becomes detritus directly, while some 22% of primary production is voided as feces and added to detritus. Thus, although the system is based on phytoplankton production and the important planktonic food chains that are traditionally regarded as being centered around herbivores, at least one third of the carbon available to consumers as food is detritus. It should be pointed out, however, that bacteria and heterotrophic protozoans were not included in the carbon-flow budget.

Flows of nitrogen are also detailed in Table 7.7. The zooplankton excrete some 40% of the nitrogen required by the phytoplankton annually (16.34 $g\ N\ m^{-1}\ year^{-1}$), 50% is released by the benthic community, and the remaining 10% is not accounted for, but may come from planktonic decomposition or carnivore excretion.

7.8.4 Chesapeake Bay

Chesapeake Bay is the largest drowned river valley estuary in the United States (12,500 km^2). The bay is shallow with an average depth of about 9 m and exceeding 60 m only in a few places along the channel. The tidal range is small, averaging 0.6 m. Thus, the estuary is tidally mixed with a characteristic flushing time of 22 days. During the summer months there is a moderately strong stratification. The model analysis discussed below concentrates on the mesohaline region of the bay, which encompasses about 48% of the tidal area, and in which the salinity ranges from 6 to 18. The water temperature varies from 2.3 to 29.0°C (Baird and Ulanowicz, 1989). The flushing time of the mesohaline regions is about 42 days. The Chesapeake Bay is rich in natural resources such as finfish, blue crabs, oysters, and clams, which have been heavily exploited. The bay also receives large amounts of pollutants and nutrients in the runoff from the catchment. Over the years it has become increasingly eutrophic.

The ecosystem model: The ecosystem model depicted in Figure 7.16 is based on those of Wulff and Ulanowicz (1989) and Baird et al. (1991). The model comprises 15 compartments. All arrows in the figure represent carbon flows and are expressed in $mg\ C\ m^{-2}\ day^{-1}$, as averaged over an annual cycle. For the heterotrophic compartments (the hexagon symbols), the inputs (feeding) are balanced

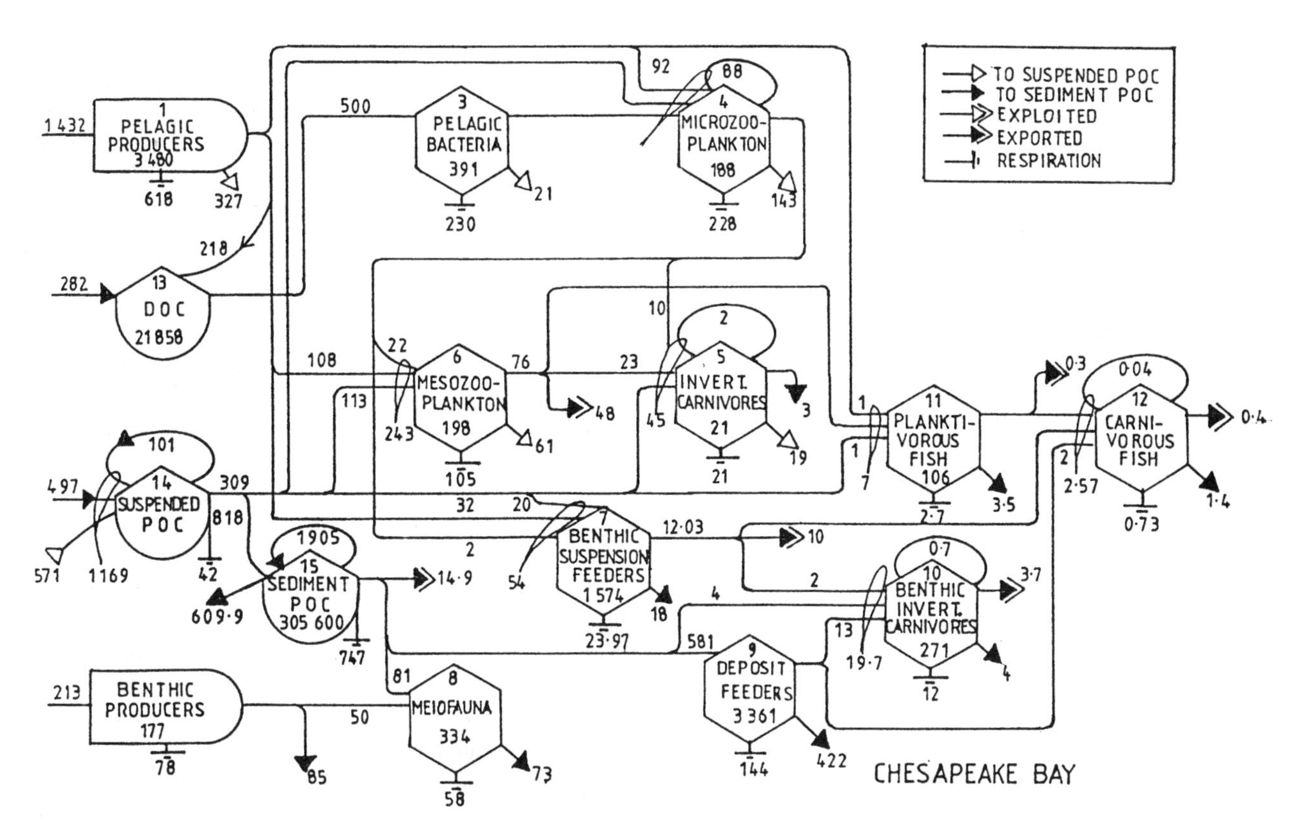

FIGURE 7.16 Energy flow network of the Chesapeake Bay ecosystem (mg C m^{-2} day^{-1} for flows). (Redrawn from Baird, D., McGlade, T.M., and Ulanowicz, R.E., *Philos. Trans. R. Soc. London Ser. B*, 333, 21, 1991. With permission.)

with exports (respiratory losses, feces production, and feeding by organisms in other compartments).

Phytoplankton: Phytoplankton biomass is dominated by flagellates and small centric diatoms during the warm season and by larger diatoms during the winter. On a seasonal basis the greatest productivity occurs throughout the summer (39% of annual production), followed in order of magnitude by spring (27%), autumn (20%), and winter (14%).

Microbial loop: Free-living bacterial biomass in the Chesapeake Bay area is generally higher than that observed in any comparable estuary. At times, the bacterial production exceeds primary production (Malone et al., 1986).

Carnivores: This compartment is dominated by the ctenophore *Mnemiopsis leidyi*, which becomes very abundant during the summer and autumn. It feeds mostly on mesozooplankton and to a lesser extent on microzooplankton and suspended detritus. Its high stocks during the summer extend heavy grazing pressure on the mesozooplankton, and this depresses the copepod populations during the summer (Baird and Ulanowicz, 1989).

The benthic subsystem: Rooted vegetation (*Potomogeton, Ruppia, Myriophyllum*) has declined drastically over the past 10 to 12 years. Benthic microalgal biomass and production are highest in the summer. About 35% of the benthic microfloral production is consumed by the meiofauna, and the remainder becomes sediment POM. There are three main categories of benthic invertebrates: the suspension feeders (including the oyster *Crassostrea virginica*, and the clam *Mya arenaria*), the deposit feeders (polychaetes such as *Nereis succinea*, bivalves like *Macoma balthica*, and various amphipod crustaceans, and the important predator/scavenging crab *Callinectes sapidus*. The benthic feeding carnivores are dominated by the fish *Leiostomus xanthurus*.

The nektonic subsystem: Filter-feeders (11) such as the bay anchovey (*Anchoea mitchelli*, a major forage fish) and the Atlantic menhaden (*Brevoortia tyrannus*) comprise over half the finfish biomass, and they constitute a significant forage base for the nektonic carnivores (12). The carnivorous fishes are of two types: (1) the benthic predators Atlantic croaker (*Micropogonias undulatus*), hogcroaker (*Trinectes maculata*), spot (*Morone americana*), white perch (*Ictalurus catus*), and white catfish; and (2) the pelagic feeders, bluefish (*Pomatomus saltatrix*), weakfish (*Cynoscion regalis*), summer flounder (*Paralichthys dentatus*), and striped bass (*Morone saxtilis*). The latter group reaches its maximum biomass and its members emigrate from the mesohaline region during the winter.

Table 7.8 lists the total carbon throughput (mg C m^{-2} day^{-1}) for each model compartment while Figure 7.17 depicts the percentage of primary production that leaves the system at each compartment. The significance of the data in Table 7.8 and Figure 7.17 will be discussed later in comparison with the Baltic ecosystem.

TABLE 7.8
Total Carbon Throughput (mg C m^{-2} day^{-1}) for Each Compartment in the Baltic and Chesapeake Networks

Compartment	Baltic	Chesapeake
Pelagic producers (1)	570	1,430
Benthic producers (2)	178	213
Pelagic bacteria (3)	144	499
Microzooplankton (4)	166	495
Invertebrate carnivores (5)	11.1	44
Mesozooplankton (6)	330	253
Benthic suspension feeders (7)	69.3	53.9
Meiofauna (8)	28	131
Deposit feeders (9)	73	581
Benthic invertebrate carnivores (10)	2.5	19.8
Planktivorous fish (11)	10.5	7.3
Carnivorous fish (12)	3.8	2.6
Dissolved organic matter, DOC (13)	76.8	499
Suspended particulate organic matter, POC	300	1,100
Sediment POC	170	3,380

Source: From Wulff, F. and Ulanowicz, R.E., in *Network Analysis in Marine Ecology: Methods and Applications*, Wulff, F., Field, J.G., and Mann, K.H., Eds., 1989, 242. With permission.

7.8.5 The Baltic Sea Ecosystem

The Baltic Sea is one of the most intensively studied and most often reviewed marine areas in the world. For recent reviews, see Jansson (1972; 1976; 1978), Margard and Rheinheimer (1974), Jansson and Wulff (1977), Voipio (1981), Kullenberg (1983), Elmgren (1984), Jansson et al. (1984), and Knox (1986b). It is the largest brackish-water area in the world, covering 365,000 km^2 with a volume of roughly 200,000 km^3, and an average depth of 60 m with a maximum of 459 m. The average surface salinity is 6 to 7, while that of the deeper water is 8 to 12, maintained through large freshwater outflows from the rivers and saltwater inflows over the sill in the Orensund off Copenhagen. As there are no tides in the Baltic, the salinity regime is stable and the water stratified; this results in a residence time for the water of 25 to 40 years. The seasonal pulse is pronounced with ice cover in the winter, a spring circulation, a summer thermocline, and an autumn circulation. Maximum summer temperatures are about 20°C above the thermocline and 2.5 to 6.0°C below.

The Baltic ecosystem is comprised of low-diversity assemblages of euryhaline marine and brackish estuarine species, freshwater species, and glacial marine or freshwater relics (Sergerstraale, 1957). Many species tend to occupy broader ecological niches than in their original habitats.

Owing to increasing inputs of nutrients over the past 50 years, the originally oligotrophic Baltic proper has become slightly eutrophic, and anoxic conditions now occur with greater frequency and duration in the deep water basins (Stigebrandt and Wulff, 1987). There has been an increase in benthic biomass in the littoral areas above the halocline, but massive mortalities in benthic

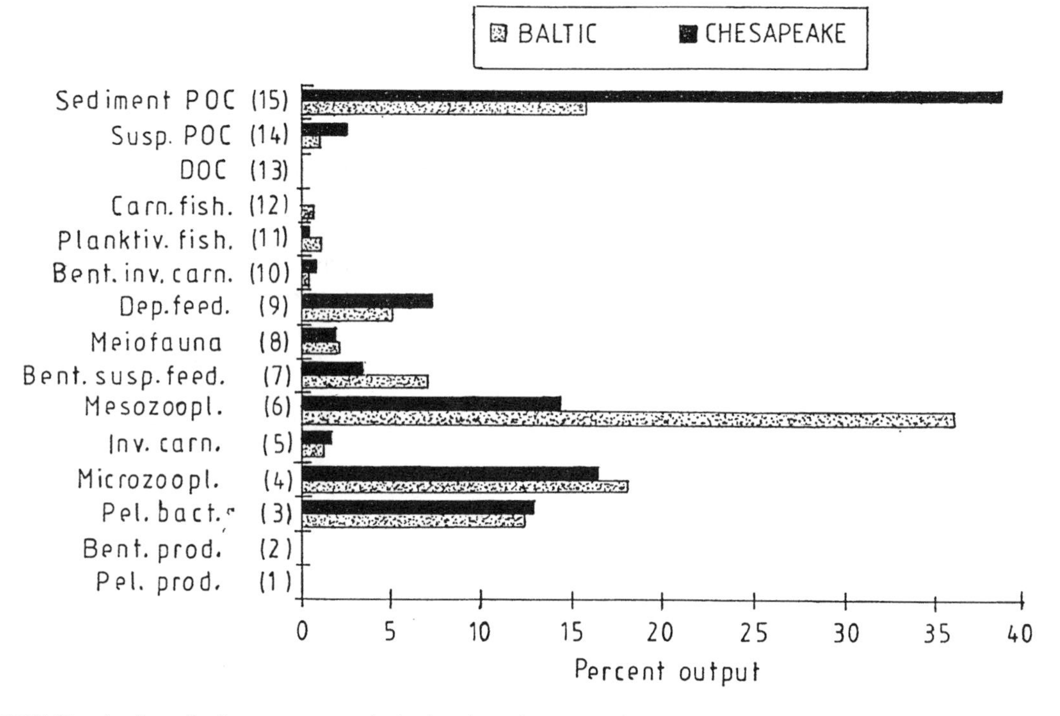

FIGURE 7.17 Results from the input-output analysis showing where net primary production leaves the Baltic and Chesapeake Bay ecosystems. (Redrawn from Wulff, F. and Ulanowicz, R.E., in *Network Analysis in Marine Ecology: Models and Applications*, Wulff, F., Field, J.C., and Mann, K.H., Eds., Springer-Verlag, Berlin, 1989, 243. With permission.)

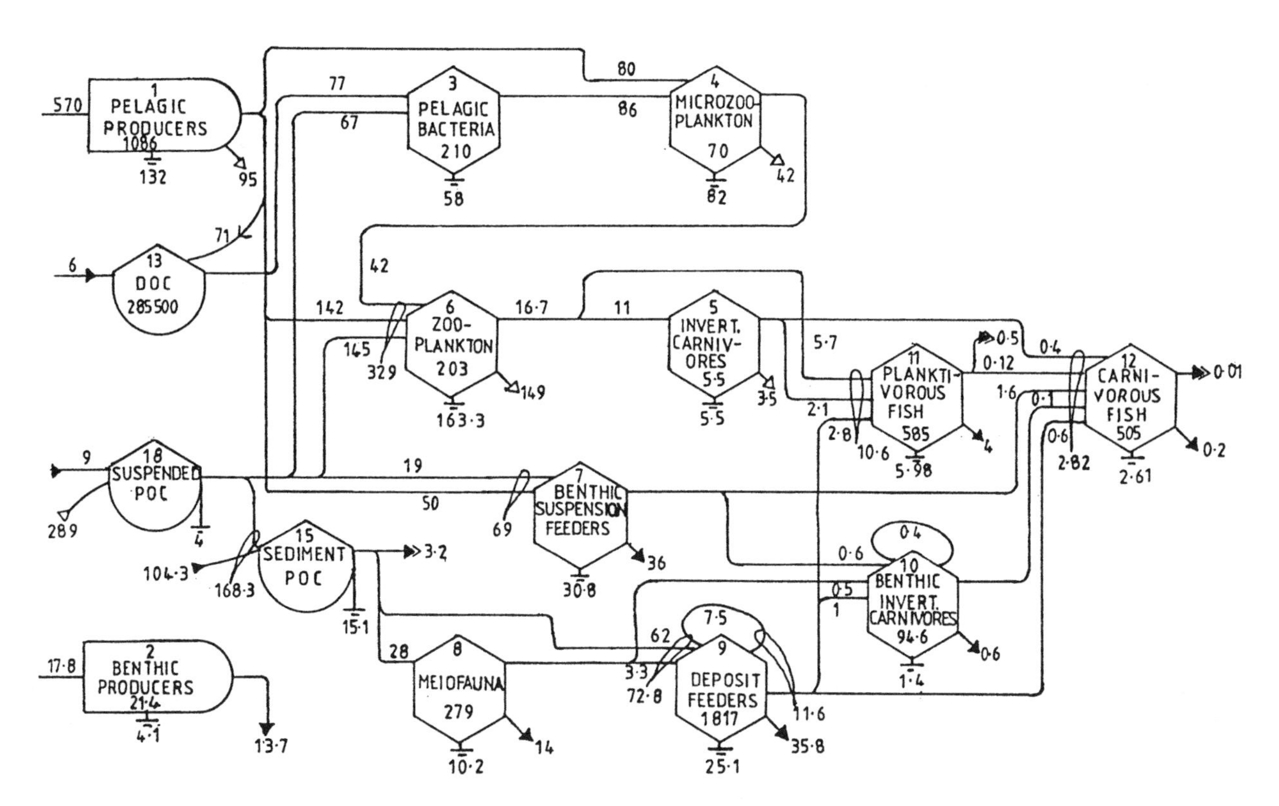

FIGURE 7.18 Energy flow network of the Baltic Sea (mg C m^{-2} for biomass, mg C m^{-2} day^{-1} for flows). (Redrawn from Baird, D., McGlade, T.M., and Ulanowicz, R.E., *Philos. Trans. R. Soc. London Ser. B*, 333, 237, 1991. With permission.)

populations have occurred in those areas subject to anoxia (Stigebrandt and Wulff, 1987; Wulff and Ulanowicz, 1989). Fish catches have increased since the 1930s to about 999,000 tons in 1984, but subsequently decreased to 730,000 tons by 1987 (Sparholdt, 1989).

The ecosystem model: The model depicted in Figure 7.18 has the same basic structure as that of Chesapeake Bay discussed above. See Table 7.10 for the total carbon throughput (mg C m^{-2} day^{-1}) for each model compartment and Figure 7.21 for the percentage of primary production that leaves the system at each compartment.

Phytoplankton: The Baltic pelagic system (see Hallfors and Niemi, 1981, and Wulff, 1986 for reviews) is subject to large seasonal production pulses, starting with the spring diatom bloom in March–April, then followed by a minimum in June and a second pulse during July–August that is dominated by small flagellates. Sometimes, very conspicuous blooms of nitrogen-fixing blue-green algae occur during the late summer. Autumn blooms are sporadic, except in the southern parts, where they occur with regularity. The amount of net primary production has been estimated at 160 g C m^{-2} $year^{-1}$ by Elmgren (1984).

Microbial loop: The importance of the microbial food chain, in which phytoplankton exudate is utilized by bacteria and then channeled into the flagellate-ciliate-mesozooplankton food chain (Azam et al., 1983) has been well established in the Baltic. In the model, all bacteria are considered to be eaten by heterotrophic flagellates and ciliates, which are lumped together into the microzooplankton compartment (4). An annual biomass production of 15 g m C m^{-2} and an assimilation efficiency of 50% was assigned to the microzooplankton.

Mesozooplankton: The overwintering mesozooplankton population (6) is very small in the Baltic, with the maximum biomass occurring in the late summer. This compartment is dominated by copepods (*Acartia, Temora, Pseudocalanus*), but rapidly reproducing (pathenogenetic) populations of rotifers (*Synchaeta*) and cladocerans (*Bosmia, Podora, Evadne*), sometimes important during the late spring and summer. The carbon flows relating to the mesozooplankton are calculated by applying a C:P ratio of 0.5 to an estimated annual biomass production of 15 g C m^{-2}.

Carnivores: Planktonic invertebrate carnivores (5) in the Baltic consist primarily of medusae (*Aurelia aurita*) in the southern parts and mysids in coastal areas. A C:P ratio of 0.22 and a P:B ratio of 1 were used in conjunction with an annual biomass production of 2.4 g C m^{-2} to estimate the carbon flows through this compartment. The only food source for the medusae is the mesozooplankton and all their net production is consumed by fish.

Allochthonous inputs: Of the allochthonous organic carbon input to the Baltic, one third is channeled into the pool of dissolved organic matter (DOC) (13), and the rest to the pool of suspended organic matter, POC (14). The POC pool receives additional inputs from the feces of pelagic organisms and from the phytoplankton. Small

TABLE 7.9
Properties of the Ythan, Swartkops, Kromme, and Ems Estuaries

Attribute	Ythan	Swartkops	Kromme	Ems
Area (km^{-2})	2–4	4.0	3.0	500
Temperature range (°C)	5–15	13–26	13–28	4–20
Salinity (%)	15–35	10–35	33–35	14–34
No. of compartments	14	15	16	15
Net primary production (mg C m^{-2} day^{-1})	1,729	1,823	2,312	203
Total standing stock (g C m^{-2})	131.1	398.1	213.9	6.7
Total production (primary and secondary, mg C m^{-2} day^{-1})	1,926.6	3,981.0	327.9	268.0
NPP efficiency (%)	10.5	38.0	9.2	98.0
Detrivory:herbivory ratio	10:1	1.5:1	6.7:1	0.5:1
P:B ratio				

Source: From Baird, D. and Ulanowicz, R.E., *Mar. Ecol. Progr. Ser.*, 99, 228, 1993. With permission.

fractions form these two pools are lost in the flow of water out of the Baltic.

The benthic subsystem: Shallow sandy littoral areas are typical in the eastern and southern parts of the Baltic. The shallow soft bottom vegetation is dominated by freshwater plants (*Phragmites, Potamogeton, Chara, Ruppia*), although dense beds of the eelgrass (*Zostra*) may occur in some of the southern parts. Dense algal vegetation covers the shallow rocky shores of the barren archipelagos found in the western and northern parts of the sea. The typical zonation starts with a filamentous green algal belt (*Cladophora*) occupying the first meter, followed by a brown alga (*Fucus*) down to about 8 m. Below that depth there is a scattering of red algae. Although this vegetation plays an important role as a spawning and nursery ground, few herbivores utilize these plants directly, and the net production (5 g C m^{-2} $year^{-1}$ as an average for the whole Baltic proper) is channeled into the sediment detrital pool (15).

Two bivalves (*Mytilus edulis* and *Macoma balthica*) overwhelmingly dominate the benthic filter-feeding community (7) and two amphipods (*Pontoporeia affinis* and *P. femorata*) comprise the deposit-feeding community. The polychaete *Harmothoe sarsi*, the isopod *Mesidotea antomon*, and the priapuloid *Halicryptus spinulosus* are the important benthic carnivores (10) (Anker, 1977; Elmgren, 1978). Echinoderms and decapod crustaceans are totally absent, except in the southernmost part. The meiofaunal compartment (8) is a heterogeneous mixture of predominantly deposit feeders but includes some carnivores. Their food sources were assumed to be primarily bacteria and sediment POC. They are consumed by deposit feeders and benthic invertebrate carnivores.

The nektonic subsystem: Like the rest of the fauna, the fishes found in the Baltic are a mixture of fresh- and saltwater species, especially in the littoral communities. The commercial fishery, however, is totally dominated by three marine species: the sprat (*Sprattus sprattus*), herring (*Clupea harengus*), and cod (*Gadius morhua*). The two former species constitute the filter-feeding fish compartment (11), while cod dominate the carnivorous fish compartment (12). Other exploited fish species such as flounder, turbot, salmon, pike, and perch have been lumped with the carnivorous fishes.

7.8.6 Four Tidal Estuaries

The four estuaries considered here have several features in common. They are tidally dominated, have extensive intertidal mudflats and salt marshes, and exhibit salinity gradients (with the exception of the Kromme Estuary) from the tidal inlet to the upper reaches. They differ, however, in some respects. Two of the systems, the Swartkops and Ythan estuaries, are mildly polluted by inputs of agricultural, industrial, and sewage effluents (Baird and Milne, 1981; Lord and Thompson, 1988). The Kromme Estuary is a fairly pristine system, in that the river flowing into it drains a forested catchment area with little or no agricultural or industrial activity taking place. It is, however, subject to reduced freshwater inflow due to two large impoundments in the catchment, resulting in a homogeneous estuarine water body (salinity range 32 to 35). The Ems Estuary is rich in pelagic and benthic organisms and there is little evidence of pollution or eutrophication despite it receiving a quantity of nutrients from the catchment. The carbon flow models of these four estuaries are depicted in Figures 7.19 through 7.22, while Table 7.9 lists the properties of the four ecosystems. Discussions of these will be found in Sections 7.8.6.1 to 7.8.6.4.

7.8.6.1 The Swartkops Estuary (Figure 7.19)

This estuary is located on the southeast coast of southern Africa near Port Elizabeth, and discharges into the Indian Ocean through a constricted but permanently open inlet.

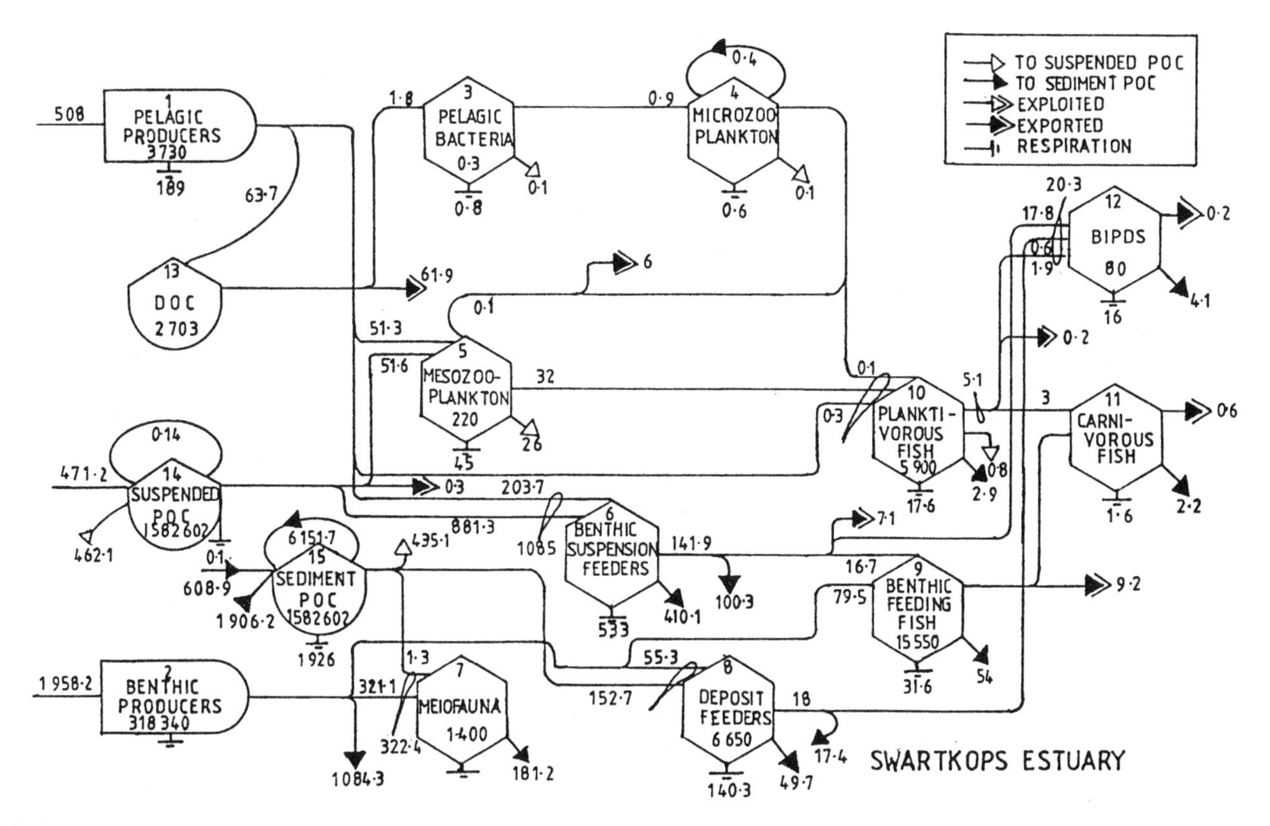

FIGURE 7.19 Carbon flow network of the Swartkops Estuary (mg C m^{-2} for biomass, mg C m^{-2} day^{-1} for flows). Biomass given in mg C m^{-2} and flows in mg C m^{-2} day^{-1}. Note feedback loops in compartments 4, 14, and 15. (Redrawn from Baird, D. and Ulanowicz, R.E., *Mar. Ecol. Progr. Ser.*, 99, 231, 1993. With permission.)

The sediments of the lower reaches and mouth regions consist mainly of coarse sand of marine origin, whereas the sediments of the middle and upper regions are composed of fine-grained sand and mud of fluvial origin. The tidal range at the inlet is 1.6 m but decreases up-estuary. The average tidal prism is ca. 2.88 × 10^{6} m^{3} in volume and the flushing time is ca. 22 hours on a spring tide. It is a small, shallow, temperate, turbid, well-mixed estuary (Baird and Winter, 1989) that attracts many migrating Paleartic birds (Martin and Baird, 1987), and is a popular recreational area for boating, angling, and swimming. The river and its estuary flow through heavily populated urban areas, and are subject to agricultural and industrial pollution (Lord and Thompson, 1988). The salinity ranges from 35 at the mouth to about 10 at the head, and the water temperature varies from 28°C in summer to 13.5°C in winter. The estuary is about 15 km long and has a total surface area of about 4 km^{2}, including an intertidal area of 3.63 km^{2} composed of sand and mudflats (1.81 km^{2}) and *Spartina maritima*-dominated marshes (1.82 km^{2}).

7.8.6.2 The Kromme Estuary (Figure 7.20)

The estuary discharges into St. Francis Bay in the Indian Ocean, through a constricted permanently open inlet. The sediments in the lower and mouth reaches consist mainly of sand grains of marine origin, while organic-rich muddy substrates of fluvial origin are distinctive of the middle and upper reaches of the estuary. The temperature fluctuates between 11.7°C in the winter and 28.0°C in the summer, while the salinity rarely drops below 32.

The Kromme Estuary is shallow (average depth at low water spring tide of 2.5 m), well mixed, and about 14 km long with a total surface area of about 3 km^{2}. The tidal prism and flushing time during spring tides are on average 1.87 × 10^{6} m^{3} in volume and 26 hours, respectively. The tidal amplitude is 1.6 m at the tidal inlet but decreases upstream. The intertidal area of ca.1.4 km^{2} consists of a salt marsh of about 0.8 km^{2}, which is dominated by the halophyte *Spartina maritima*, and 0.6 km^{2} of mud and sand flats. Dense beds of eelgrass *Zostera capensis* occur inter- and subtidally all along the estuary.

7.8.6.3 The Ythan Estuary (Figure 7.21)

This estuary, which is located about 20 km north of Aberdeen, Scotland, is a drowned river valley (Raffaelli, 1992). Aspects of energy flow through the estuarine mudflats of this estuary were dealt with previously in Section 6.5.1. The surface area of the estuary is about 256 ha, with an

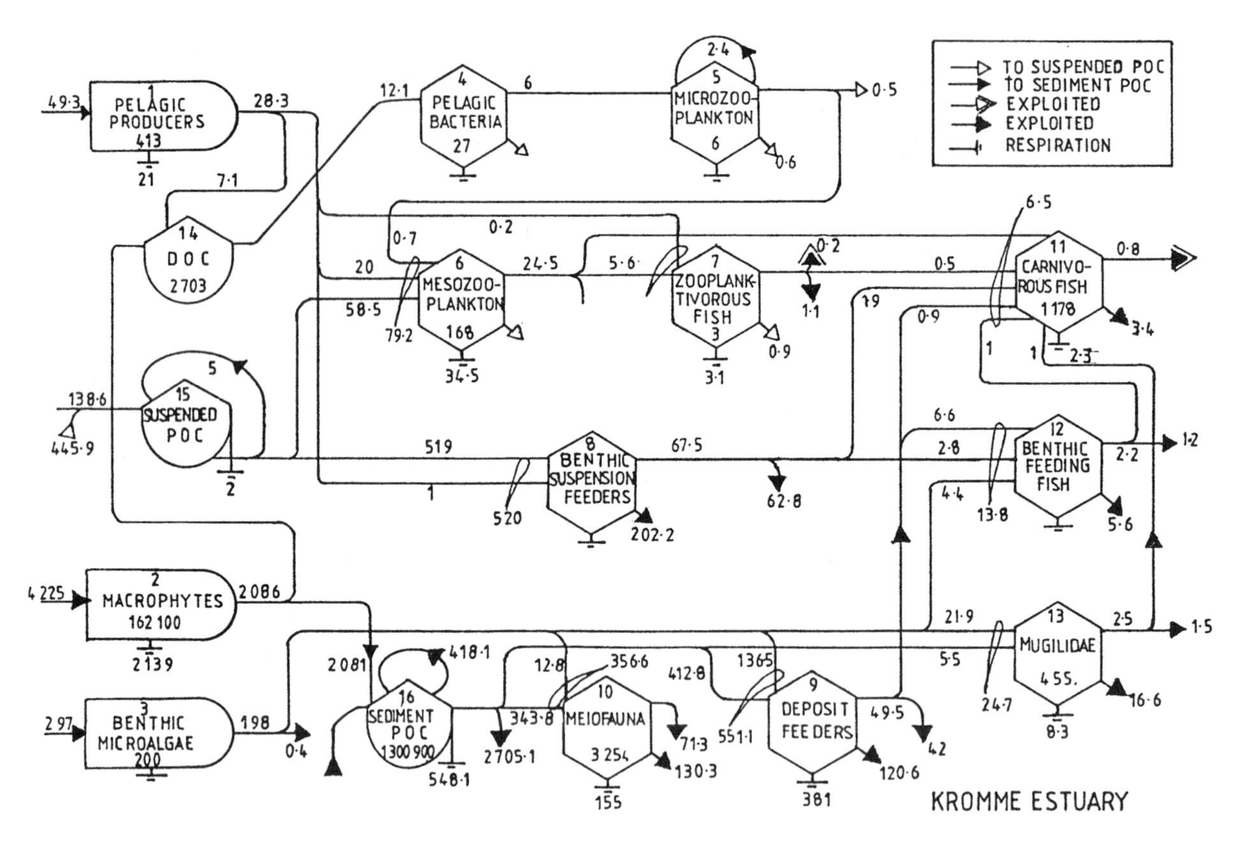

FIGURE 7.20 Carbon flow network of the Kromme Estuary (mg C m^{-2} for biomass, mg C m^{-2} day^{-1} for flows). (Redrawn from Baird, D. and Ulanowicz, R.E., *Mar. Ecol. Progr. Ser.*, 99, 221, 1993. With permission.)

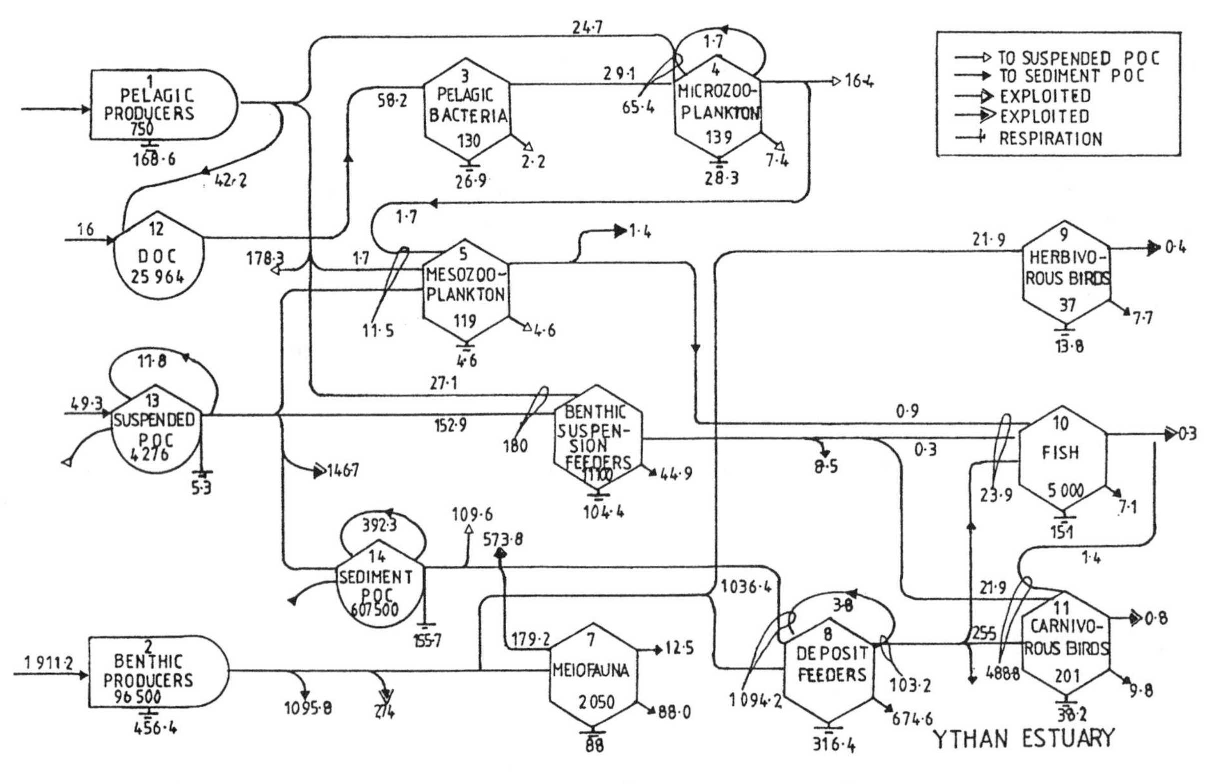

FIGURE 7.21 Carbon flow network of the Ythan Estuary (mg C m^{-2} for biomass, mg C m^{-2} day^{-1} for flows). (Redrawn from Baird, D. and Ulanowicz, R.E., *Mar. Ecol. Progr. Ser.*, 99, 230, 1993. With permission.)

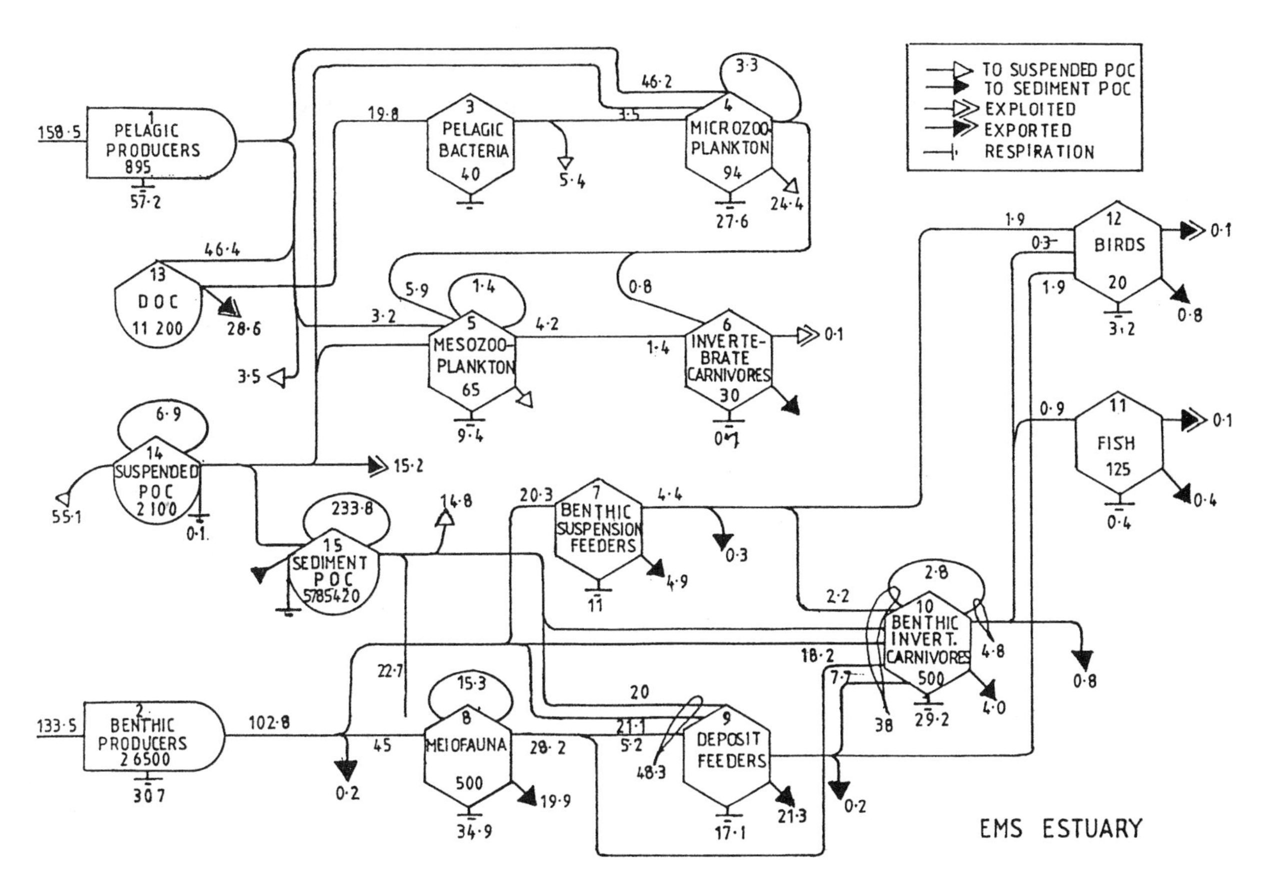

FIGURE 7.22 Carbon flow network of the Ems Estuary (mg C m^{-2} for biomass, mg C m^{-2} day^{-1} for flows). (Redrawn from Baird, D. and Ulanowicz, R.E., *Mar. Ecol. Progr. Ser.*, 99, 230, 1993. With permission.)

intertidal area of about 185 ha. The substrate in the mouth and lower reaches consists mainly of sand, with the rest of the estuarine sediments composed mainly of mud and a mud-sand mixture. The mean flushing time of ca. 15 hours is relatively brief because of strong tidal currents and the small volume of freshwater drainage. The water volume of the estuary at LWST is 3.8×10^5 m^3, and 3.8×10^6 m^3 at high water spring tide (HWST) (Baird and Milne, 1981).

The estuary is tidal for about 8 km from the mouth and it receives about 1,000 m^3 of primary treated sewage per day from upstream towns. The salinity ranges from 5.0 at the head to 35.0 at the mouth, while the temperature fluctuates from about 3°C during the winter to 20°C in the summer. Baird and Milne (1981) give detailed information on the tidal characteristics, and on the standing stocks and energy flow between the various biotic components.

7.8.6.4 The Ems Estuary (Figure 7.22)

This is a shallow semidiurnal tidal estuary draining into the Wadden Sea. The sediments in the mouth region consist of sand, whereas the clay content increases in the middle and upper reaches of the estuary. Freshwater runoff to the estuary is variable, ranging from 1.2×10^6 m^3 to 7.9×10^8 m^3 annually. The tidal amplitude is 2.20 m near the mouth but increases upstream to 3.03 m about 35 km from the inlet. Due to the asymmetry in the tidal currents, i.e., currents of higher velocities during flood than ebb tides, and drift currents, deposition, and resuspension of sediment particles are common in the Ems Estuary. Low density particles in suspension are exported to the sea, while for heavier particles, a net import can occur (De Jonge, 1988). The tidal prism at the inlet is ca. 900×10^6 m^3, and the flushing time of the water in the estuary varies between 12.1 and 72.1 days at low and high freshwater discharge rates (De Jonge, 1988).

The Ems Estuary contains large tidal flats of about 245.7 km^2 and has a total area of about 500 km^2. It is rich in benthic and pelagic life and although it receives large amounts of nutrients annually from both the sea and the catchment, there is little evidence of eutrophication or pollution (De Jonge, 1988). Water temperatures range from 4 to 20°C, and the salinity from 34 in the Wadden Sea to 14 in the upper reaches.

7.8.6.5 Comparison of the Ecosystem Flow Networks

The Ems and Swartkops Estuaries consist of 15 compartments, the Ytham Estuary of 14 compartments, and the

Kromme Estuary of 16 compartments. The reason for the 16 compartments in the latter ecosystem is that the benthic producers are separated into macrophytes (Compartment 2) and benthic microalgae (mainly diatoms, Compartment 3). The Ythan network has one less compartments than the Ems and Swartkops Estuaries, because the fish compartment (Compartment 10) consists mainly of benthic feeding fish. The fish compartment (Compartment 11) in the Ems Estuary similarly consists mainly of benthic feeding fish.

7.9 NETWORK ANALYSIS

7.9.1 INTRODUCTION

As Baird et al. (1991) point out, biological throughput and the physical scales over which individuals and populations interact are key factors in determining the spatial and temporal variability of ecosystems. Ecologists have attempted to simulate such changes in whole ecosystems by using sets of coupled differential equations of physical, behavioral, and physiological processes. Although this approach has had some success in describing the interactions between ecosystem levels, it has increasingly been recognized that deeper insights into the dynamics of whole ecosystems will require new approaches. This had been recognized by the Scientific Committee on Oceanic Research (SCOR) which, in 1977, established its Working Group 59 on Mathematical Models in Biological Oceanography under the co-chairmanship of Kennith Mann and Trevor Platt. The work of this group resulted in the publication of a book in 1981 under the same title (Platt et al., 1981). The group concluded that while simulation modeling was a useful activity, simulation modeling of ecosystems had strict limitations and was not predictive because qualitative changes of state variables often occurred in ecosystems and these could not usually be predicted or modeled.

In addition, many of the biological parameters in such models are highly variable because they are derived from systems that are spatially extended, hierarchial (O'Neill et al., 1986), and self-organizing (Mann et al., 1989). When a large number of the ecosystem processes are coupled together in a simulation model, the confidence limits on the output are very wide indeed. So far, most ecosystem comparisons have generally examined only temporal changes in a single system, e.g., days (Field et al., 1989) and seasons (Baird and Ulanowicz, 1989; Kremer, 1989; Warwick and Radford, 1989). However, as Baird et al. (1991) emphasize, the finding of a suitable methodology for inter-ecosystem comparisons still remains a major challenge.

In 1982, SCOR disbanded Working Group 59 and established a new group to explore ecosystem theory in relation to biological oceanography and search for alternatives to simulation modeling as an approach to understanding ecosystem processes. Five topics were identified with potential to reveal ecosystem-level properties: *thermodynamics, statistical mechanics, input-output analysis, information theory,* and *ataxonomic aggregations*. A colloquium was held in 1984 to discuss these topics and the proceedings were later published (Ulanowicz and Platt, 1985). There emerged from the discussions at the colloquium a recognition that input-output analysis, flow analysis, information networks, and dynamic network analysis were a family of related techniques with some potential to provide insights into the properties of whole ecosystems. Work on these topics culminated in the publication of *Network Analysis in Marine Ecology. Methods and Applications* (Wulff et al., 1989).

Baird et al. (1991) emphasize that: "Recent work on the measurement of material and energy flows between the various ecosystem components has shown that the efficiency with which material is transferred, assimilated and dissipated conveys significant information about the fundamental structure of the whole system (Longhurst, 1984; Ulanowicz and Platt, 1985; Ulanowicz, 1986; Wulff et al., 1989; Mann et al., 1989)." The algorithms used, known collectively as network analysis, are derived from input-output, trophic and cycle analysis, and the calculation of whole system properties such as ascendency, throughput, and development capacity (Kay et al., 1989). Although some parts of network analysis are strictly linear, weighted flow networks have been useful in exploring the spatial and temporal patterns of the interconnections and food webs of ecosystems, either at the level of individual elements, intra-level interactions, or the whole ecosystem (Mann et al., 1989; Baird and Ulanowicz, 1989). Detailed information on the underlying theoretical concepts and methodology of network analysis can be found in Ulanowicz (1986; 1989; 1997), Kay et al. (1989), and Mann et al. (1989).

7.9.2 NETWORK ANALYSIS — SYSTEM PROPERTIES AND INDICES

The following is a brief summary of the properties and indices measured by network analysis.

7.9.2.1 Comparing Trophic Position in a Network

1. *Effective Trophic Position*: In a simple food chain, the species can often be easily assigned to a trophic position, e.g., herbivore, carnivore, detritovore, etc. However, in most ecosystems, the situation is much more complicated because many organisms have mixed diets. In spite of

this, it is of interest to know the trophic position of the living components of the ecosystem. To this end the *Effective Trophic Position*, or weighted average number of trophic steps that separate a living component from the primary producers, or non-living components of the system, e.g., a fish may get 90% of its diet from zooplankton and 10% from diatoms, placing it as 90% at position 3 (carnivores) and 10% at position 2 (herbivores); thus the average trophic position of the fish would be 2.9. This index allows one to compare the trophic positions of similar species in different ecosystems.

2. *The "Lindeman Spine"*: The *Lindeman Spine* results from the mapping of the complex network of trophic levels into a linear food chain with discrete trophic levels (Figure 7.23). Each level receives a net amount of energy from the previous trophic level and creates exports, respiration, detritus from recycling, and net production for transfer to the next level. By convention, autotrophs, nonliving organic matter, and networks are allocated to trophic level one.
3. *The Trophic Efficiency*: The *Trophic Efficiency* of a food web can be calculated from a Lindeman Spine by comparing the inflow to a trophic category with its outflow to the next category. Thus index, coupled with the structure of the Lindeman Spine (e.g., number of trophic levels), enables meaningful comparisons to be made between ecosystems.

7.9.2.2 The Finn Cycling Index (FCI)

A cycling index shows the proportion of the flow that is recycled compared with the total flow. The *FCI* gives the proportion of the total system activity (*T*) that is devoted to recycling of media, e.g., nutrients, carbon, energy (Finn, 1976). Thus $FCI = T_c/T$, where T_c is the amount of system activity involved in recycling and T is the total system throughput (see 7.9.2.4.1 below). The cycling index measures the retentiveness of a system; the higher the index, the greater the proportion recycled and possibly the more mature or less stressed the system (Odum, 1969). An individual cycle is a unique pathway that ends and begins in the same compartment.

7.9.2.3 The Average Path Length (APL)

This index measures the average number of steps or transfers a unit of flux will experience from its entry into a system until it leaves the system. This index is derived from $APL = ((T - Z)/Z)$ where Z is the sum of all exogenous inputs. The APL includes the effects of recycling within the system.

7.9.2.4 Global Measures of Ecosystem Organization or Indices of Growth and Development

Margalef (1968) and Odum (1969) have documented that many changes occur as the ecosystem food web develops with succession. As these successional changes occur, the network of flows changes to a more complex network with richer connections. Ulanowicz (1980; 1986) proposed that the maturity of an ecosystem can be assessed by using measures based on information theory. Of these, the first three are the basic measures, with the others being derived from them.

1. *Total System Throughput* (*T*)*:* This index measures the size or growth of the system in terms of the flows through all its components; the greater, the flows the bigger the *T*.
2. *Network Ascendency* (*A*): The index represents both the size and organization of the flows in a single measure, the product of *T* and the diversity of flows within a system. Ascendency is the central concept of network analysis. It has the form of a thermodynamic work function, which means that it can be expressed as the sum of products of existing flows times their conjugate "forces" (propensities). Wherever the fluxes are expressed as energy, their logrithmic cofactors represent the increase in the quality of energy as it enters a higher trophic compartment. That is, the product of the energy exchange times its increase in quality is equivalent to the rate at which energy is being stored in the receiving compartment. This is the same as H.T. Odum's (1983b; 1996) concept of embodied energy or "emergy." The network or system ascendency, being the aggregation of all such products, therefore represents the overall rate at which the system is doing work (i.e., storing energy). According to Ulanowicz (1980; 1986), as an ecosystem matures and goes through a series of successional stages, the ascendency of the system should increase. On the other hand, if the system is disturbed, e.g., by organic pollution, this should be reflected by a change in the flows of energy and materials through the system, and thus the ascendency should decrease.
3. *Development Capacity* (*C*): This index measures the potential of a particular network to develop, given its particular set of connections and total throughflow. *C* gives the highest possible value of *A*.
4. *Relative Ascendency* (*A:C*). This index measures the fraction of the possible organization

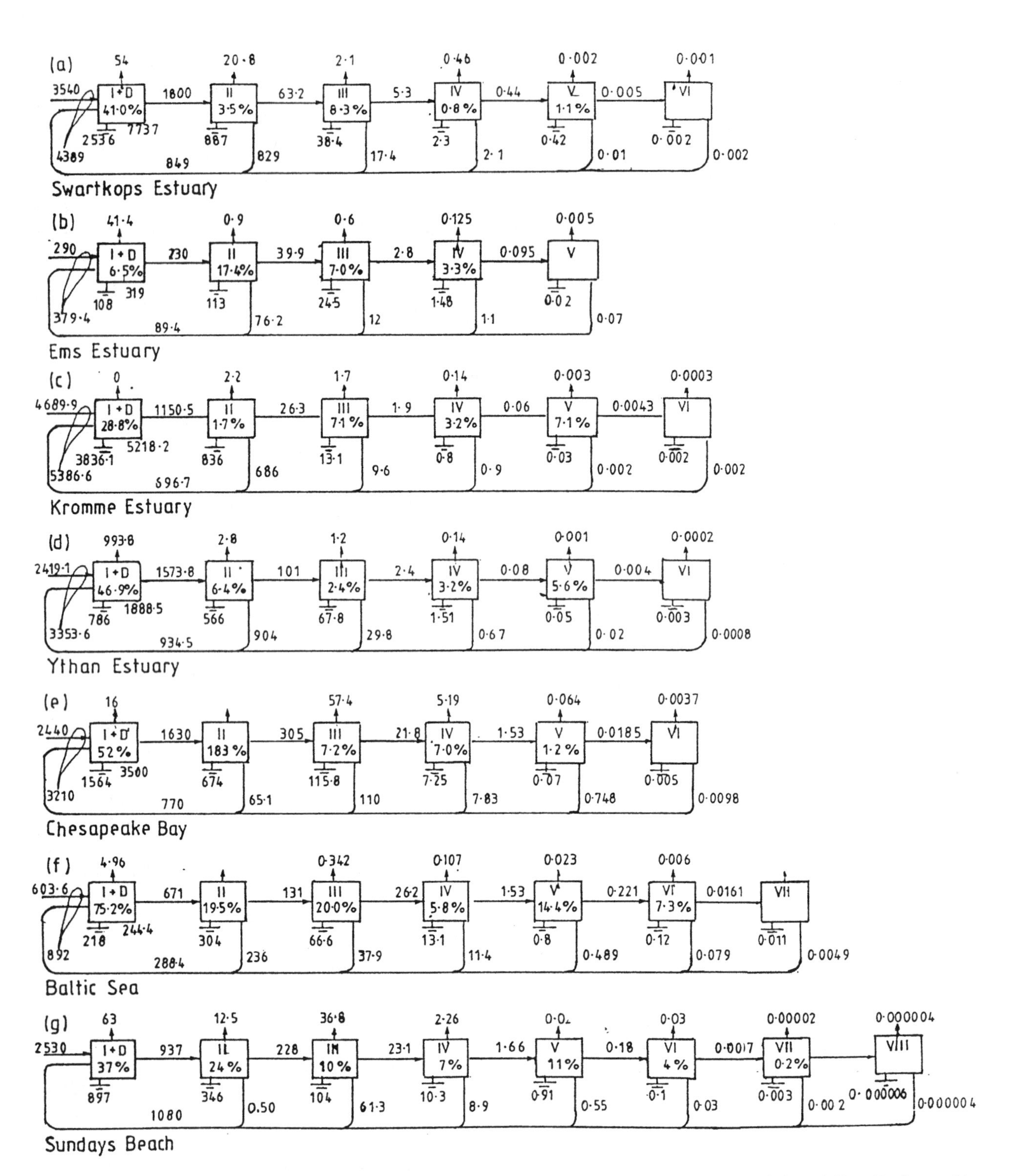

FIGURE 7.23 Trophic aggregation into Lindeman Spines, with the autotrophs merged with the detrital pool (1 + 1); the trophic levels are designated by Roman numerals, and flows given in mg C m^{-2} day^{-1}. The percentages in the boxes represent the daily trophic efficiencies. POC, particulate organic carbon; DOC, dissolved organic carbon. (Redrawn: (a) to (e) from Baird, D. and Ulanowicz, R.E., *Mar. Ecol. Progr. Ser.*, 99, 229, 1993; and (g) from Heymans, J.J. and McLachlan, A., *Est. Coastal Shelf. Sci.*, 43, 497, 1996. With permisssion.)

that is realized (Ulanowicz and Mann, 1987; Ulanowicz, 1986; 1987).

5. *Internal Ascendency* (A_i): This is a measure of those components of ascendency that are composed solely of internal exchanges, and thus excludes exogenous transfers into and out of the system (Baird and Ulanowicz, 1993).
6. *Internal Capacity* (C_i): This is similar to 5 in that it is a measure of those components of capacity that are composed solely of internal exchanges.
7. *Internal Relative Ascendency* (A_i:C_i): Field et al. (1989) consider that this ratio is the quantity most representative of the system's developmental status. Systems with a high value of A_i:C_i are considered by Baird et al. (1991) to be well organized and therefore unlikely to disintegrate spontaneously, i.e., they possess significant internal stability and resistance to perturbation.

   Kay (1984) has proposed two additional indices.
8. *Flow Diversity* (D): This index measures the diversity with which components of the system are used as food resources by other living components. In other words, it measures the evenness of inputs to living components and reaches a maximum when all flows are of equal importance.
9. *Flow Specialization* (S): This index measures the diversity of flows, averaged over all living components in the system; it is the average resource niche breadth. When S is large, most species or components are generalists. As a system matures, D should increase and S decrease (Kay, 1984).

7.9.3 Some Examples of Coastal Ecosystem Network Analysis

7 9.3.1 Introduction

Network analysis has been carried out on seven coastal marine ecosystems: Sundays Beach, South Africa (Heymans and McLachlan, 1996); two large estuarine systems, Chesapeake Bay, U.S.A. and the Baltic Sea (Wulff and Ulanowicz, 1989; Baird et al., 1991), and four smaller estuarine systems (Baird and Ulanowicz, 1993), the Ythan Estuary, Scotland, the Kromme and Swartkops Estuaries in South Africa, and the Ems Estuary in The Netherlands. Details of the carbon flow models for these ecosystems were discussed earlier in this chapter. Here we will discuss the results of the network analyses carried out on these systems and compare the various ecosystems. Table 7.10 lists the system properties and the indices of system organization derived from the network analyses. For comparative purposes two open ocean upwelling systems have been added to this table, the Southern Benguela and the Peruvian upwelling systems (Baird et al., 1991).

7.9.3.2 The Four Small Tidal Estuaries

7.9.3.2.1 Ecosystem flow networks and trophic structure

The carbon flow networks for the four ecosystems are illustrated in Figures 7.19 through 7.22 and their properties in Table 7.10. The Ems and Swartkops Estuaries consist of 15 compartments each, the Ythan Estuary has 14 compartments, while the Kromme Estuary has 16 compartments. The reason for the extra compartment in the Kromme network is due to the separation of the benthic producers into macrophytes (Compartment 2) and benthic microalgae (mainly diatoms, Compartment 3). The Ythan network has one less than the Ems and Swartkops Estuaries, because the fish component (Compartment 10) consists mainly of benthic feeding fish. The fish compartment (Compartment 11) in the Ems Estuary similarly consists mainly of benthic feeders. The ubiquitous fish family Mugilidae was separated from the other benthic-feeding fish in the Kromme Estuary (Compartment 11) due to their smaller size and mode of feeding; they feed exclusively on sediment bacteria and microalgae.

It is of interest to examine the variability in the standing crop and rate of primary production between the four ecosystems (see Table 7.10). The Kromme Estuary has the highest rate of primary production (2,312 mg C m^{-2} day^{-1}) with the Ythan and Swartkops Estuaries being very similar (1,729 and 1,823 mg C m^{-2} day^{-1}, respectively), while the Ems Estuary has the lowest (203 mg C m^{-1} day^{-1}). However, phytoplankton production in the Kromme Estuary is exceptionally low (28.3 mg C m^{-2} day^{-1}) in comparison with the other estuaries, primarily due to the low nutrient levels in this estuary. On the other hand, macrophyte production (2,086 mg C m^{-2} day^{-1}), in particular that of the eelgrass *Zostera capensis* is higher than in any of the other estuaries (Hanekon and Baird, 1988). The net primary production in all four systems is dominated by the benthic microalgal production, with the exception of the Ems Estuary where the ratio of pelagic phytoplankton production to benthic primary production is 1:1.04. This ratio is particularly high in the Kromme Estuary (1:80). The total standing crop (i.e., the sum of the biomass of all living components) also differs markedly between the four systems. The standing stocks of the Swartkops (398.1 mg C m^{-2}) and the Kromme Estuaries (213.9 mg C m^{-2}) are much higher than those of the Ythan (131.9 mg C m^{-2}) and the Ems Estuaries (6.7 mg C m^{-2}). These differences may be due to the large temperature range experienced by the Ythan and Ems estuaries compared to the more temperate southern estuaries. Also of interest is the range of net primary production (NPP)

TABLE 7.10
System Properties and Indices of System Organization Derived from Network Analysis of the Various Ecosystems

	Sundays	Chesapeake	Baltic	Ythan	Kromme	Swartkops	Ems
1. Area (km^2)	20	5,980	257,000	2.4	3.0	4.0	500
2. Temperature range (°C)	10–26	2–29	2–20	5–15	11.7–28	13–26	4–20
3. Salinity (%)	35	6–18	7.15	15–35	33–35	10–35	14–35
4. Components	16	15	15	14	16	15	15
5. Total standing stock (*TTS*) (g m^2)	22.7	10.1	7.8	131.1	213.9	398.1	6.7
6. Net primary production (*NPP*) (mg C m^{-2} day^{-1})	1,521.20	950	425	1,729	2,312	1,859	202
7. *NPP* efficiency (%)	40.0	42.0	87.0	10.5	9.2	38.0	98.0
8. Average Path Length (*APL*)	2.30	3.61	3.27	2.86	2.38	3.95	3.42
9. Finn Cycling Index (*FCI*)	12.80	29.70	22.80	25.50	25.9	43.8	28.0
10. Production:Biomass (*P:B*) (day^{-1})	0.04	0.14	0.08	0.014	0.0015	0.01	0.04
11. Number of cycles	18.7	20.0	22.0	15.1	19.0	14.0	26.0
12. Trophic efficiency (log mean) (%)	2.60	9.30	16.20	6.64	6.02	4.00	12.30
13. Total System Throughout (*T*)	6,899.6	11,224	2,577	9,350	16,879	17,541	1,283
14. Development Capacity (*C*)	26,076	35,000	8,007	39,264	59,422	62,252	5,971
15. Ascendency (*A*)	11,660	16,335	4,452	13,511	20,022	17,565	2,307
16. *A:I*	1.60	1.46	1.73	—	—	1.0	1.8
17. Relative Ascendency (*A:C*)	45.0	49.5	55.6	34.4	33.7	26.0	38.6
18. Overhead on imports (%)	0.23	2.6	0.9	—	—	14.1	4.8
19. Overhead on exports (%)	9.0	0.4	0.6	—	—	0.4	2.5
20. Overhead on respiration (%)	—	19.4	21.4	—	—	21.6	19.2
21. Redundancy	25.57	28.1	22.0	—	—	36.0	35.0
22. Internal Capacity (C_i)	13,025	28,005	7,007	19,950	29,880	32,359	3,400
23. Internal Ascendency (A_i)	5,313.2	9,969	2,810	6,774	8,597	9,842	1,353
24. Internal Relative Ascendency (A_i:C_i) (%)	40.8	35.0	39.7	33.8	29.4	30.4	39.4
25. Internal Redundancy (%)	59.2	63.0	60.3	—	—	69.6	60.1
26. Flow Diversity (*C:T*)	3.8	2.9	3.1	4.1	3.5	3.6	4.6
27. Gross primary production (*GPP:T*) (%)	30.4	14.7	30.4	—	—	14.0	22.6
28. Detritovory:*T* (%)	12.6	12.3	15.5	—	—	6.2	6.1
29. Detritovory:Herbivore ratio	13:1	4.8:1	1.5:1	10:1	6.7:1	1.5:1	0.5:1
30. Food web connectance	1.36	1.87	1.85	—	—	1.46	1.85

consumption efficiencies, i.e., the ratio between NPP and herbivore consumption. This ratio is relatively high for the Ems and Swartkops Estuaries, but relatively low in the Ythan and Kromme Estuaries.

7.9.3.2.2 Effective trophic position (ETP)

In all estuaries, consumer species occupy the second and third trophic levels. The top carnivores in the Swartkops and Ems Estuaries appear to feed at slightly higher trophic levels than their counterparts in the other two estuaries. This is due to the fact that carnivorous fish in the Ythan and Kromme Estuaries (Compartment 11) obtain more of food indirectly through suspended and sedimentary POC and benthic primary producers at level one, than the top predators in the other two estuaries.

7.9.3.2.3 The Lindeman Spine

The energy flow networks for the seven estuaries with their associated rates of recycling have been mapped into simplified Lindeman Spines as shown in Figure 7.23 (Ulanowicz, 1986; Baird and Ulanowicz, 1989; 1993; Baird et al.,1991; Heymans and McLachlan, 1996). The detrital pool (D) in the spines is merged with the primary producers (I) to represent the first trophic levels. Detrital returns from all other levels are shown, as well as the loop of detritus contribution from plants and the utilization of organic material by microorganisms. Respiration and exports from each level are also shown. The trophic efficiencies generally decrease with ascending levels, although the efficiency at the fifth trophic level in the Swartkops, Kromme, and Ythan Estuaries is unexpectedly higher than that of the preceding one. This is especially pronounced in the Kromme and Ythan Estuaries, indicating that these estuaries are quite effective in delivering resources to higher trophic levels despite their relatively poor performance over the first few transfers. The efficiencies at the highest trophic levels can be ascribed to the feeding activities of birds on planktivorous fish in the Swartkops Estuary (Martin, 1991), the predation of carnivorous birds on fish in the Ythan (Baird and Milne,

1981), and the predation of carnivorous fish on benthic feeding fish and on species of Mugilidae in the Kromme Estuary. Examination of the Lindeman Spines indicates that the efficiency at the first trophic level is highest in the Ems Estuary (60%) and the lowest in the Kromme Estuary (28.8%), while the efficiencies for the Ythan and Swartkops Estuaries are 46.9% and 41.0%, respectively. This indicates that there is a more effective direct utilization of the primary production in the Ems Estuary. This is also reflected in the low detritovory:herbivory ratio, i.e., the ratio between detrital consumption (uptake from suspended and sediment POC) and direct grazing on primary producers (Table 7.12). The geometric-mean trophic efficiencies range from 4.02% in the Swartkops to 12.49% in the Ems Estuary (see Table 7.12). The Ythan and Kromme Estuaries have very similar efficiencies of 6.64 and 6.02%, respectively. It is thus clear that the estuaries differ in their trophic efficiencies.

7.9.3.2.4 Structure and magnitude of cycling

The cycling of material and energy in natural ecosystems is generally considered to be an important process in ecosystem function (Odum, 1969; Ulanowicz, 1986). This cycling occurs through a number of cycles of different path lengths. Short cycles are usually indicative of fast cycling rates and longer path lengths of slower rates (Baird and Ulanowicz, 1989). The number of cycles per nexus and the total number of cycles in each system are given in Table 7.11A and the amount of cycled flow over various path lengths in Table 7.11B. The number of cycles ranges from 26 in the Ems Estuary to 14 in the Kromme Estuary. Of importance is the distribution of cycle per nexus and the amount of cycled flow over various path lengths. The results show that all of the cycling in the Swartkops Estuary occurs through single-cycle nexuses and that 80% of the cycled flow is via short, fast loops involving mainly bacteria and sediment particulate organic matter (Table 7.11). The Ems estuary, on the other hand, shows a more even distribution of the cycles per nexus (Table 7.11) and cycles about 38% of its throughput via longer and slower loops. In the Kromme and Ythan Estuaries, about 40 and 83% of their activity, respectively, is devoted to the recycling of material via longer paths (Table 7.14B).

TABLE 7.11
Cycles Distributions in the Four Estuaries

A. Distribution (%) of Cycle per Nexus

Cycles Nexus^{-1}	Ythan	Swartkops	Kromme	Ems
1	60	100	89	34
2	40	0	11	23
3	0	0	0	12
4	0	0	0	31

B. Percent of Cycled Flow Through Loops of Various Path Lengths

Path Length	Ythan	Swartkops	Kromme	Ems
1	16.9	80.0	61.9	61.8
2	74.0	2.2	17.8	30.0
3	9.3	16.1	19.7	7.4
4	0.7	1.0	0.5	0.6
5	0.01	0.3	0.1	0.2
Total cycled flow (mg C m^{-2} day^{-1})	2,389	7,679	4,378	390

Source: From Baird, D., Ulanowicz, R.E., *Mar. Ecol. Progr. Ser.*, 99, 232, 1993. With permission.

7.9.3.2.5 The Finn Cycling Index (FCI) and Average Path Length (APL)

*FCI*s ranged from 25.5% in the Ythan, with those of the Kromme and Ems Estuaries slightly higher at 25.9 and 30.0%, respectively, to 43.8% in the Swartkops Estuary (Table 7.10 and Figure 7.24. The APL for two of the estuaries, the Ythan and Kromme, are 2.86 and 2.38, respectively, while those of the Swartkops and Ems Estuaries are longer at 3.95 and 3.42, respectively (Table 7.10 and Figure 7.25). This indicates that a unit of carbon will be transferred at least one step more in the latter two estuaries where it thus has a slightly longer residence time. In terms of cycle distributions (the *FCI* and the *APL*), it appears that the mildly polluted Swartkops Estuary shows cycle characteristics slightly different from the other three systems. It recycles a relatively large proportion (43.8%) of its total flows in short cycles and it has a high system *APL* value of 3.95. A large proportion of the material is thus retained within the system, and this retention takes place over a few short recycling circuits within the system. The high *FCI* thus suggests a relatively simple cycling structure. The cycle distributions, *FCI*s, and *APL*s in the other estuaries are very similar. The Kromme Estuary, which is the most pristine of the four estuaries, shows a more even distribution of cycled flows through different path lengths, although its *FCI* and *APL* are on the same order as the other estuaries. The composite cycling structures of the Ythan, Swartkops, and Kromme Estuaries are very similar. In these systems, virtually all living compartments, with the exception of pelagic and microzooplankton, participate in recycling of material through the suspended and sediment POC pools. A common feature of estuaries is the resuspension of sedimentary particulate material, which is then cycled through the suspended POC compartment to other components such as mesozooplankton and benthic suspension feeders. This process appears to be characteristic of tidal estuaries with constricted inlets and relatively fast tidal currents in excess of 1 m sec^{-1}, features common to the Ythan, Kromme, and Swartkops Estuaries. The Ems Estuary, on the other hand, exhibits two separate cycling structures, one in the water column and one benthic-pelagic cycle. Benthic suspension feeders (Compartment 7) do not participate in any cycling and

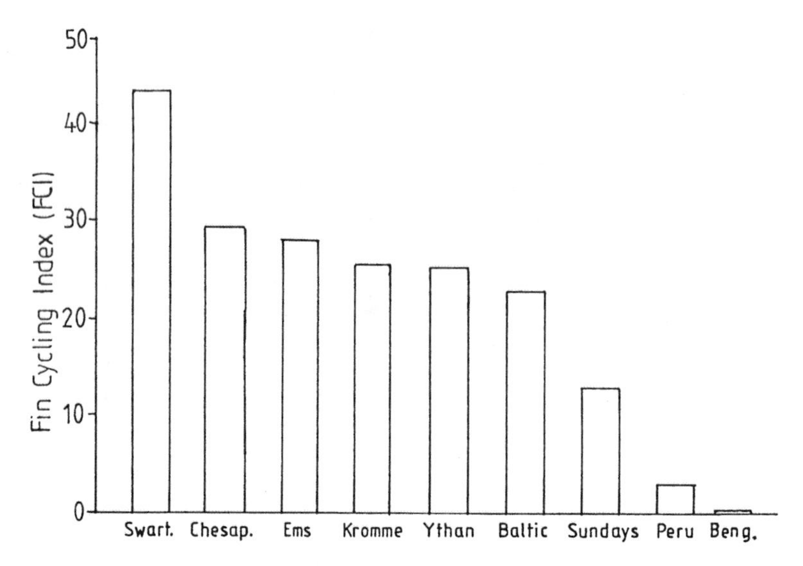

FIGURE 7.24 The Finn Cycling Indices (FCI) for the various ecosystems.

serve as a link between the pelagic and benthic systems. There is, however, a link between the sediment and suspended POC compartments, due to the resuspension and deposition of organic material.

7.9.3.2.6 System properties

The global measures of systems organization of the four estuaries are listed in Table 7.10. The development capacity (*C*), total throughput (*T*), ascendency (*A*), and redundancy (*R*) of the Swartkops and Kromme Estuaries are of the same order but much higher than those of the Ythan and Ems Estuaries (Figure 7.26). The Relative Ascendency (*A*:*C*), a dimensionless ratio that excludes the influence of *T* and is a suitable index for comparing different ecosystems (Mann et al., 1989; Baird et al., 1991), is highest for the Ems Estuary (38.2%), 34.3% and 33.7% for the Ythan and Kromme Estuaries, respectively, and lowest for the Swartkops Estuary (28%) (see Table 7.10). Table 7.10 also lists values for internal capacity (C_i), the internal ascendency (A_i), and the normalized internal ascendency ratio (C_i/A_i). The latter ratio remains approximately the same as the relative ascendency ratio (*A*/*C*) for all the systems. It increases slightly by 2.4% in the Swartkops and 0.7% in the Ems Estuaries, but decreases by 0.6 and 4.3% in the Ythan and Kromme Estuaries, respectively.

7.9.3.3 Chesapeake Bay and the Baltic Sea

Wulff and Ulanowicz (1989) and Baird et al. (1991) have compared the system properties of these two systems and the following discussion is based on their analyses. The results from the network analyses are listed in Tables 7.8 and 7.10.

7.9.3.3.1 Trophic structure

The rate of net primary production in the Chesapeake (950 mg C m^{-2} day^{-1}) is double that of the Baltic Sea (452 mg C m^{-2} day^{-1}). On the other hand, the Net Primary Productivity (*NPP*) Efficiency in the Baltic (87%) is double that of the Chesapeake (42%), reflecting an effective utiliza-

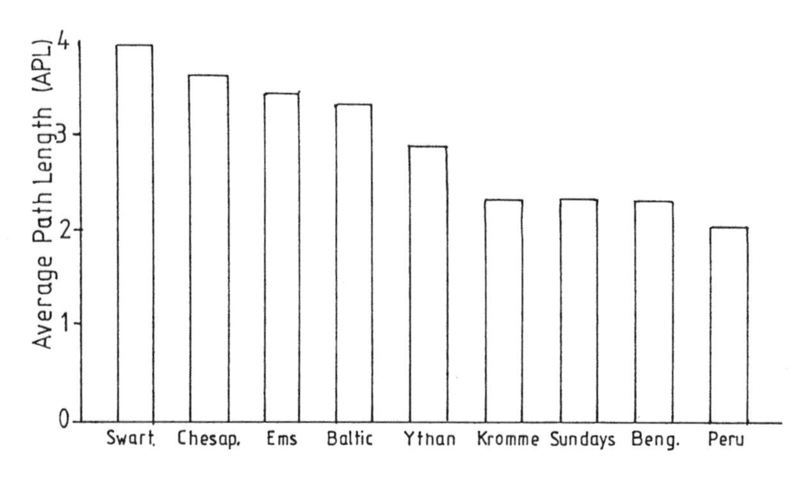

FIGURE 7.25 The Average Path Lengths (APL) for the various ecosystems.

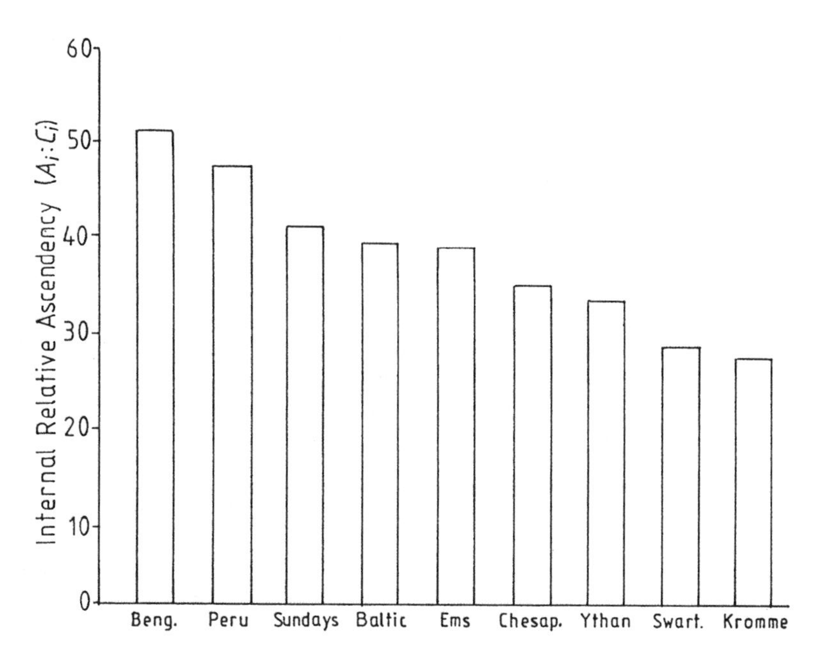

FIGURE 7.26 The Internal Relative Ascendency (A_i:C_i) values for the various ecosystems.

tion of primary production (Figure 7.27). The low utilization of primary production in the Chesapeake Bay is offset by its much higher detritovory to herbivore ratios, 4.8:1 compared to 1.5:1 in the Baltic Sea. The Baltic system, in addition to its high detritovory to herbivore ratio, also has high detritovory to *T*, gross primary production (*GPP*) to *T* ratio, and *NPP* consumption efficiency. If detritovory *T* and *GPP*:*T* are added together, it appears that about 46% of the total throughput is utilized in the Baltic system, which is higher than that of the Chesapeake. Furthermore, allochtonous inflows to the Chesapeake are almost 50% of the magnitude of its primary production, while such inputs to the Baltic are less than 3% of its primary production. Natural factors contributing to the contrasts discussed above are the higher mean temperatures in the Chesapeake and the ratio of the drainage basin to the water area, 20:1 in the Chesapeake and 4:1 in the Baltic.

Figure 7.27 depicts the percentage of the primary production that leaves the system at each compartment. From the figure, it is obvious that most of the important "sinks" for net primary production differ in the two systems. The

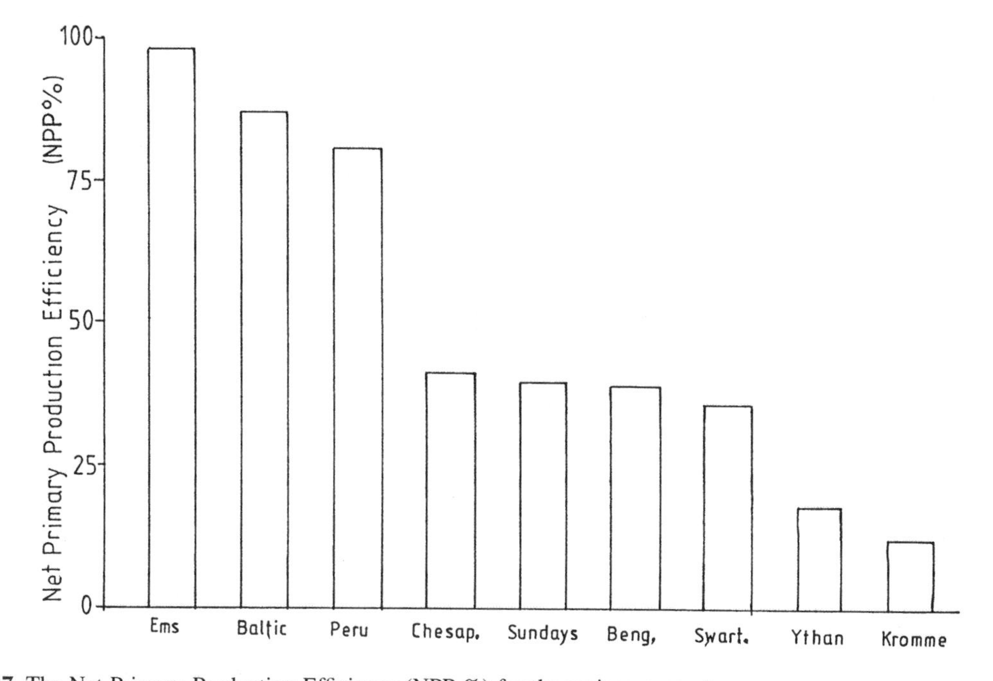

FIGURE 7.27 The Net Primary Production Efficiency (NPP %) for the various ecosystems.

TABLE 7.12
The Average Path Lengths (APL) in the Baltic and Chesapeake Networks

Compartment	Baltic	Chesapeake
Pelagic producers (1)	3.28	3.14
Benthic producers (2)	2.80	3.52
Pelagic bacteria (3)	2.55	2.78
Microzooplankton (4)	2.59	3.19
Invertebrate carnivores (5)	2.45	3.42
Mesozooplankton (6)	2.73	2.59
Benthic suspension feeders	2.28	2.50
Meiofauna (8)	2.50	3.36
Deposit feeders (9)	2.53	4.14
Benthic invertebrate carnivores (10)	1.93	1.91
Planktivorous fish (11)	2.09	3.40
Carnivorous fish (12)	1.72	3.26
Dissolved organic matter, DOC (13)	3.35	3.78
Suspended organic matter, POC (14)	3.54	4.48
Sediment POC	2.34	4.24

Source: From Wulff, F. and Ulanowicz, R.E., in *Network Analysis in Marine Ecology: Methods and Applications*, Wulff, F., Field, J.G., and Mann, K.H., Springer-Verlag, Berlin, 1989, 245. With permission.

respiration of the mesozooplankton (6) dominates the Baltic sinks, whereas bacterial respiration in the sediments (15) eventually removes most of the autochthonous production from Chesapeake Bay. This illustrates the predominance of the pelagic system in the Baltic versus the key role of the benthos in the more shallow Chesapeake Bay. The picture for the eventual fates of the allochtonous inputs is similar for that for the *in situ* production.

7.9.3.3.2 Average Path Length (APL) (Figure 7.25)

The APLs in both the Chesapeake Bay (3.61) and the Baltic Sea (3.27) are similar to those of Swartkops (3.95) and Ems (3.42) but higher than those of Kromme (2.38) and Ythan (2.86). Table 7.12 lists the APLs for the two systems. The APL for higher members of the Baltic food chain tend to be low. In both systems the APL of the benthic invertebrate carnivores is low (1.93 for the Baltic and 1.91 for the Chesapeake); this reflects both the large metabolic losses and the large exports (i.e., the harvesting of the blue crab in the Chesapeake and predation by Baltic carnivores [cod], which in turn are heavily exploited). The large APL for Chesapeake fishes (11, 12) can be attributed to their high production of feces and natural mortality, both of which contribute to intensive benthic recycling. Exploitation of and respiratory losses by the Baltic fishes give them a reduced APL.

7.9.3.3.3 The Lindeman Spine

The number of trophic levels in the Chesapeake Bay (VI) is similar to that of the Swartkops, Kromme, and Ythan Estuaries but one less than that of the Baltic Sea (VII) (Baird et al., 1991) (Figure 7.23) The larger chain in the Baltic is probably due to the generally higher trophic efficiencies inherent in that system.

7.9.3.3.4 Status and magnitude of cycling

The two systems possess a comparable number of cycles (22 in the Baltic versus 20 in the Chesapeake); however, most of the recirculation in the Chesapeake occurs over cycles of very short length, whereas in the Baltic the cycles are generally longer (and hence distributed among more trophic levels). The primary cycling route in the Baltic is in the water column between the zooplankton and the suspended POC (6-14-6). In Chesapeake Bay the dominant loop lies in the benthos between the deposit feeders and the sediment bacteria-POC (9-15-9). The pelagic bacteria in the Chesapeake Bay do not participate in any recycling. They function as a sink for excess productivity (Ducklow et al., 1986). Filter-feeding fish and benthos are likewise absent from Chesapeake cycling; their apparent function is to shunt material and energy from the planktonic system into the benthic-nektonic system.

7.9.3.3.5 Finn Cycling Index (FCI)

The *FCI* in the Chesapeake Bay (29.7) is similar to that of the Ems Estuary and slightly higher than that of the Baltic (22.8) (Figure 7.24).

7.9.3.3.6 System properties

The development capacity (C) represents the maximum potential evenness and diversity of flows and the total system throughput (T) measures the size (or activity) of the system in terms of all its flows (Ulanowicz, 1980; 1996; Kay et al., 1989). The greater the amount of material flowing through the system, the greater the value of T and thus the activity of the system. The T value of the Chesapeake is five times that of the Baltic, while the value of C is 4.73 times that of the Baltic Sea. The Chesapeake is thus a much more productive system.

7.9.3.4 Sundays Beach

7.9.3.4.1 Trophic structure

The attributes of the Sundays Beach/surf-zone ecosystem are summarized in Table 7.13. The total standing stock of this ecosystem (22.7 g m^{-2}) is similar to that of the Baltic Sea and Chesapeake bay, but much lower than that of the Swartkops and Kromme Estuaries (Baird et al., 1991). The net primary production (*NPP*) is 1,521 mg C m^{-2} day^{-1}, with *NPP* efficiency of 40%, and a P:B ratio of 0.40, similar to that of the Kromme Estuary (Heymans, 1992). The *NPP* efficiency of the Sundays ecosystem (40%) is similar to the Swartkops Estuary (38%), Chesapeake Bay (38%), and the Southern Benguela ecosystem (39%), but much lower than the Ems Estuary (98%), the Baltic Sea (87%), and the Peruvian ecosystem (82%) (Baird et al., 1991).

TABLE 7.13
Attributes of the Sundays Beach Ecosystem

Attributes		Units
Area	km^2	20.00
Temperature	°C	11–25
Salinity	%	35.00
Number of compartments		16.00
Total standing stock	g C m^{-2}	22.70
NPP efficiency	%	40.00

Source: From Heymans, J.J. and McLachlan, A., *Est. Coastal Shelf Sci.*, 43, 500, 1996. With permission.

7.9.3.4.2 Average Path Length (APL)

The average path length in this system is 2.3 (Figure 7.25). A unit of carbon will be transferred 2.3 times between entering and leaving the system, thus having a residence time longer than in the Kromme Estuary (Heymans, 1992), but shorter than in the Swartkops or Ems Estuaries, Chesapeake Bay, or the Baltic Sea (Wulff and Ulanowicz, 1989; Baird et al., 1991).

7.9.3.4.3 Total system throughput (T)

Total system throughput (T) (Table 7.14) is the total amount of carbon flowing through each compartment and is a measure of the importance of each compartment. On this basis, the components of the system can be arranged in a natural ranking of their ecological importance (Szyrmer and Ulanowicz, 1987). Thus, phytoplankton ($T = 2{,}100$ mg C m^{-2} day^{-1}) and DOC (1,000 mg C m^{-2} day^{-1}) are the two most important components in the ecosystem, followed by free bacteria (523 mg C m^{-2} day^{-1}), and suspended POC (498 mg C m^{-2} day^{-1}). This is different from the Baltic Sea, where the pelagic producers, mesozooplankton, and suspended POC are the most important components, and the Chesapeake Bay where pelagic producers, sediment POC, and suspended POC are the most important (Wulff and Ulanowicz, 1989). It is also evident that the microbial loop in the Sundays Beach/surf-zone ecosystem is of greater importance than the interstitial fauna or the macrofauna; it dominates the trophic assemblages of this system.

In contrast to the Kromme Estuary, where the benthic food web is dominant (Heymans and Baird, 1995), the water column food web is more important in this system, with even the interstitial fauna utilizing suspended food sources. The total system throughput (T) per unit area (6,900 mg C m^{-2} day^{-1}, see Table 7.10) exceeds that of the Ems Estuary, Baltic Sea, and Southern Benguela system, but is less than that of the Swartkops and Kromme Estuaries, Chesapeake Bay, and the Peruvian system (Baird et al., 1991; Heymans, 1992).

TABLE 7.14
Carbon Throughput (in mg C m^{-2} day^{-1}) for Each Compartment in the Sundays Beach Ecosystem

Compartment	Throughput
1. Phytoplankton	2,100.00
2. Free bacteria	523.00
3. Flagellates	112.00
4. Microzooplankton	18.00
5. Zooplankton	135.00
6. Benthic filter feeders	16.00
7. Benthic scavengers and predators	7.00
8. Mullet	2.41
9. Predatory fish	0.93
10. Omnivorous fish	17.00
11. Birds	1.04
12. Sediment bacteria	219.00
13. Protozoa	72.00
14. Meiofauna	75.00
15. DOC	1,000.00
16. Suspended POC	498.00

Note: DOC, dissolved organic carbon; POC, particulate organic carbon.

Source: From Heymans, L.L. and McLachlan, A., *Est. Coastal Shelf Sci.*, 43, 494, 1996. With permission.

7.9.3.4.4 Dependency coefficients

Dependency coefficients identify the fraction of total carbon leaving one compartment that eventually enters another compartment, through both direct food chains and cycling. As such, this defines the "extended diet" of a compartment and yields information on the degree to which the diet of any particular component depends indirectly on the production of any other compartment of the ecosystem (Field et al., 1989). Thus, although free bacteria consume only DOC directly, they indirectly derive 27% of their carbon requirements from free bacteria. Flagellates and microzooplankton derive all their carbon requirements directly from free bacteria, but also feed indirectly on phytoplankton and DOC. Microzooplankton similarly fed indirectly on free bacteria. Of the interstitial fauna, sediment bacteria indirectly derive 99.9% of their carbon diet from phytoplankton, 21% from free bacteria, 74% from DOC, and 32% from suspended DOC. Sediment protozoa feed indirectly on phytoplankton (99.8%), free bacteria (16%), sediment bacteria (64%), DOC (50%), and suspended POC (52%).

7.9.3.4.5 Effective trophic level

Table 7.15 gives the effective trophic level for each compartment (Ulanowicz, 1987). These trophic positions are measures of the average trophic level at which each com-

TABLE 7.15
The Effective Trophic Level of Each Living Compartment in the Sundays Beach Ecosystem

Compartment	Effective Trophic Level
1. Phytoplankton	1.00
2. Free bacteria	2.00
3. Flagellates	3.00
4. Microzooplankton	3.72
5. Zooplankton	2.01
6. Benthic filter feeders	2.11
7. Benthic scavengers and predators	1.22
8. Mullet	2.00
9. Predatory fish	3.82
10. Omnivorous fish	3.02
11. Birds	2.88
12. Sediment bacteria	2.00
13. Protozoa	2.61
14. Meiofauna	2.90
15. DOC	1.00
16. Suspended POC	1.00

Note: DOC, dissolved organic carbon; POC, particulate organic carbon.

Source: From Heymans, L.L. and McLachlan, A., *Est. Coastal Shelf Sci.*, 43, 496, 1996. With permission.

partment is receiving energy (Levine, 1980). No compartment has a higher effective trophic position than level four. Predatory fish (3.8) and microzooplankton (3.7) are the highest since both are top predators in their respective subsystems. However, interstitial meiofauna has a lower effective trophic level than flagellates or omnivorous fish, despite being the top trophic level in the interstitial assemblage. Carnivorous fish have a higher effective trophic level in the Sundays Beach/surf-zone ecosystem than in the Ythan, Swartkops, Kromme, or Ems Estuaries. This is because carnivorous fish in the Sundays ecosystem derive less sustenance indirectly through suspended POC at trophic level one than do carnivorous fish in the other systems; further, there are no benthic primary producers to fuel them via indirect pathways.

7.9.3.4.6 The Lindeman Spine

The Lindeman Spine (Figure 7.23) for Sundays Beach has eight compartments, more than the Baltic and Chesapeake Bay, but less than the Kromme Estuary (Baird et al., 1991; Heymans, 1992). Thus, of the 228 mg C m^{-2} day^{-1} consumed at the third level, only 23.1 mg C m^{-2} day^{-1} is transmitted to level four, giving a trophic efficiency of 10%. The efficiency at the first trophic level (37%) in this system is lower than that of the Ems (60%), the Ythan (46.9%), or the Swartkops (41%) Estuaries, indicating a less effective utilization of primary production (Baird and Ulanowicz, 1993). This is also reflected by the high detritivory to herbivory ratio of this ecosystem.

The productivity of higher trophic levels is overwhelmingly dependent on the recycling of carbon through the nonliving compartments of the ecosystem (Heymans and McLachlan, 1996). Detritivory, defined as the flow from the detrital pool to the second trophic level (Ulanowicz, 1987), amounts to 868 mg C m^{-2} day^{-1}. The detritivory to herbivory ratio is 13.1, and 92% of the carbon entering the second trophic level is recycled. It is, therefore, clear that detritivory is more important in the Sundays ecosystem than herbivory, with most of the net primary production being broken down into detritus.

7.9.3.4.7 Cycle analysis

Heymans and McLachlan (1996) have analyzed the recycling process in the Sundays ecosystem. They identified 187 distinct cycles, considerably more than those recorded in other ecosystems (99 found in the Kromme Estuary, Heymans, 1992; 61 in Chesapeake Bay, Baird and Ulanowicz, 1989; and less than 25 in other systems, Baird et al., 1991). Heymans and McLachlan (1996) consider that the high number of cycles in this system could be due to the considerable amount of information and knowledge of links that might not be known in other, less well-studied systems.

7.9.3.4.8 The Finn Cycling Index (FCI)

A Finn Cycling Index (*FCI*) of 13% has been calculated for the Sundays ecosystem (Figure 7.24). This index, indicating the percentage of all flows in the system that are recycled (Finn, 1976), is below that of the Ythan (25.5%), Swartkops (43.8%) and Ems (30%) Estuaries (Baird and Ulanowicz, 1993), Chesapeake Bay (29.7%), and the Baltic Sea (22.8%), but greater than the Southern Benguela (0.01%) and Peruvian (3.2%) systems.

7.9.3.4.9 Other global attributes

A number of other global attributes that were calculated for this system are listed in Table 7.10 (Heymans and McLachlan, 1996). They include:

1. *Development Capacity* (*C*): This is derived by multiplying the total systems throughput by the entropy of the individual flows. A development capacity of 26,076 mg C m^{-2} day^{-1} was calculated on this basis; it is a measure of the network to develop, given its particular set of connections and total through-flow, and an upper boundary for ascendency (see below) (Ulanowicz, 1980).
2. *Ascendency* (*A*): The ascendency (*A*) is a measure of the degree of system development (Ulanowicz, 1986; Baird and Ulanowicz, 1989). The ascendency of the Sundays ecosystem (11 660 mg C m^{-2} day^{-1}) is lower than that of the Swartkops Estuary, Chesapeake Bay, or the Peruvian system, but greater than that of the

Ems Estuary, the Baltic Sea, or the Southern Benguela system (Baird et al., 1991).

3. *Relative Ascendency* (A:C): An important index for comparing different ecosystems is the relative ascendency (A:C), a dimensional ratio that excludes the influence of the total system throughput on ascendency and development capacity (Field et al., 1989). This ratio is 45% for the Sundays ecosystem (Figure 7.26), higher than the Swartkops or Ems Estuaries, but lower than for the Southern Benguela or Peruvian systems (Baird et al., 1991), or the Lynher Estuary (Warwick and Radford, 1989).
4. *Internal Capacity* (C_i) and *Internal Ascendency* (A_i): One aspect of highly organized systems is the tendency to internalize most of the activity, thereby becoming relatively indifferent to outside supplies and demands (Baird et al., 1991; Heymans and McLachlan, 1996). Thus, internal capacity and internal ascendency are recasts of ascendency and development capacity as internal indices, i.e., they are functions of internal exchanges alone (Ulanowicz and Norden, 1990). For the Sundays ecosystem, the internal capacity is 13,025 mg C m^{-2} day^{-1} and the internal ascendency is 5,313.20 mg C m^{-2} day^{-1}.
5. *Normalized Internal Ascendency Ratio* (A_i:C_i): This ratio for the Sundays ecosystem is 40.8%, which again is higher than that of the Kromme or the Swartkops Estuaries, but less than that of the upwelling systems. The significance of this ratio for ecosystem maturity and stability will be discussed below (Figure 7.26).

7.9.4 Conclusions

Mann et al. (1989) made an assessment of the accomplishments and potential of network analysis in marine ecology while Baird et al. (1991) in their comparative study of six marine ecosystems discussed in detail the application of the network analysis of food web fluxes as an important tool in the analysis and comparison of different ecosystems. The following account is based on their conclusions. Ecosystem analysis was originally concerned with questions of stability, resistance-resilience, nutrient cycling as key constraints, functions determining structure, energy and available water as explanatory controls on system function, food chain, and food web structure. More recently, ecosystem ecologists have developed ideas concerning succession and maturity and the evolution of ecosystems through the work of E.P. Odum and others (Odum, E.P., 1969; Odum, H.T., 1983b).

E.P. Odum (1969) presented a summary of 24 attributes believed to characterize mature ecosystems. They can be grouped under four headings as the tendencies: (1) to internalize flows; (2) to increase cybernetic feedback; (3) to increase specialization of the components; and (4) to increase the number of components in the system. Such concepts were developed primarily from the study of freshwater and terrestrial ecosystems. In such systems, the P:B ratio appears to decrease over the process of succession leading to ecosystem maturity. However, in marine ecosystems, the opposite appears to be the case, i.e., the P:B ratio increases with the measure of ecosystem maturity as defined by the ratio A_i:C_i (internal ascendency: internal capacity).

Baird et al. (1991) consider that the maturity of marine ecosystems in particular can be assessed from their sets of topological indices. A number of these indices such as total system throughput (T), ascendency (A). Development capacity (C), and diversity of flows (D) have been discussed above. However, all these indices are sensitive to the model structure and the degree of aggregation adopted by the investigator. Qualitative differences among systems or groups of systems usually appear as significant discontinuities in system properties. For example, there are radical contrasts in recycling between upwelling systems on the one hand, and estuaries on the other. The Finn Cycling Index (*FCI*) displays a significant change that can give a clue to the physical differences that discriminate between them. For example, the two upwelling systems have *FCI* values of less than 5%, whereas the remaining coastal systems, apart from Sundays beach (13.0%), all have indices in excess of 20%.

Quantitative indices of ecosystem status often appear as correlations among measured system attributes. The most significant is the inverse rank-order correlation between the *FCI* and the normalized internal ascendency (A_i:C_i). Baird et al. (1991) suggest that the normalized internal ascendency can be taken as the measure of system efficiency or maturity (Mann et al., 1989), and so we should expect to see an inverse relation between *FCI* and A_i:C_i. This can be seen when comparing Figures 7.24 and 7.26; the order in Figure 7.24 is the reverse of that in Figure 7.26.

Next we turn to the *NPP* Efficiencies and the P:B ratios in the systems discussed above. Figure 7.27 graphs the Net Primary Production Efficiency (*NPP*) for the various ecosystems. The Ems Estuary, the Baltic, the Peruvian Upwelling System, and Chesapeake Bay have the highest *NPP* efficiencies, while the Swartkops, Ythan, and Kromme Estuaries have low values. In the latter three systems, macrophyte and benthic microalgae dominate the primary production, whereas in the others the phytoplankton dominates. The bulk of the phytoplankton is consumed by the herbivores whereas the macrophyte production is not consumed, but enters the detrital food web. In going from an open ocean system to an estuarine marsh, the P:B ratio goes down dramatically (Figure 7.28), and from then on into terrestrial systems the P:B ratio will be lower.

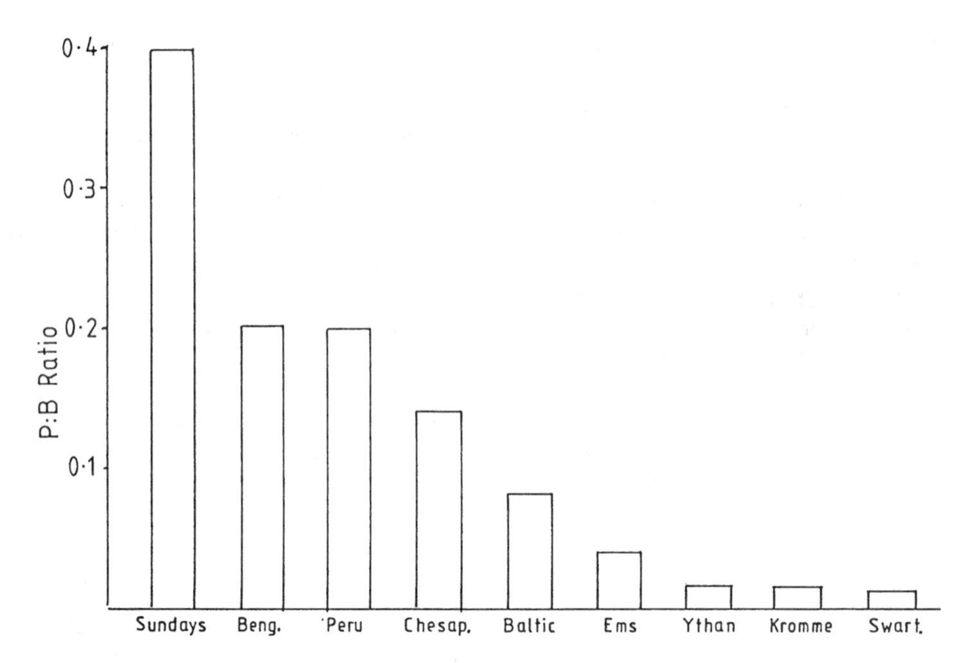

FIGURE 7.28 The P:B ratios for the various ecosystems.

Baird et al. (1991) consider that this is not a reflection of system maturity, but rather an adaptation to the physical environment and the degree to which the biomass is retained within the system. They suggest that an ecological equivalence potentially exists across terrestrial and aquatic biotas, when biomass is retained for approximately the same characteristic time.

Along the spectrum of biomass retention, there will be a range of ecosystems evolving under a variety of constraints. Evolutionary advancement or maturity is a measure of where a system lies along this continuum. The studies of the wide range of marine ecosystems discussed above have shown that evolutionary advancement (as measured by increasing A_i:C_i) has to do with the ability of an ecosystem to develop a topological network that can mobilize resources and put then into production. On the other hand, the development of a complex ecosystem refers to efficiency and refinements of the underlying topology. Thus, some of the qualitative changes that would be expected to be seen as a system matures are: (1) increasing number of cycles; (2) longer cycles; (3) greater specificity in predator-prey interactions; and (4) a topological structure that can withstand large perturbations in available resources through time. On the other hand, the development of a climax ecosystem will result in intensification of existing parts of the network.

From their analysis of marine network analyses, Baird et al. (1991) have made some key observations. First, the development status of a system can be assessed only from a suite of indices. It is also clear that only dimensionless indices, such as A_i:C_i and *FCI*, can be used for comparative purposes. Real system values, such as the development capacity, total system throughput, etc., are usually not useful for intersystem comparisons, although they may be applicable when the same system is compared through time.

Secondly, an assessment combining the magnitude of cycling structure and the normalized relative ascendency values can give a clear yardstick for comparison, but the most revealing correlation lies between *FCI* and A_i:C_i.

Thirdly, a definite distinction must be made between the assessment of system maturity (A_i:C_i) and the current functional status of an ecosystem (its P:B ratio). It has been demonstrated that there is a direct relation between these two indices, which contradicts conventional wisdom about system evolution.

Finally, as Ulanowicz (1984) has argued, contrary to what E.P. Odum (1969) has suggested, that the aggregate amount of cycling is not an indication of system maturity, but rather of stress within an ecosystem. The analyses, as discussed above, support this idea by showing that stressed systems are associated with a higher *FCI*. How this comes about is unclear, but one possibility is that perturbations often impact higher-level species to a greater extent, thereby releasing resources bound up in standing biomass. The homeostatic response of the less-disturbed lower trophic components is to retain the resources as best they can, usually by recycling the material among themselves in short, intense loops. Thus, should *FCI* be inversely correlated with system maturity, as noted above, it would imply that the Swartkops is the most severely impacted system, and the Benguela, the least impacted.

The above points can be illustrated with reference to the Sundays Beach ecosystem (Heymans and McLachlan, 1996). This ecosystem is a mature system, having nexuses with large numbers of cycles, many higher-order cycles, and low *FCI*. It is not under anthropogenic stress but is under perpetual physical stress. This is a system that has evolved in response to physical stress, whereas anthropogenic influences on other systems (e.g., the Baltic, Chesapeake Bay, and some of the smaller estuaries), such as pollution and organic enrichment, are recent. The relative and normalized internal ascendancy values for the Sundays Beach ecosystem are higher than for most estuaries, but lower than for upwelling systems. Mature systems possess high speciation, specialization, internalization, and cycling, and are thus dramatically fragile, since extreme disturbances can destroy them. Thus, any severe change in wave climate, which drives the beach system, could break down cycling, causing it to become stressed and switch to a different ecosystem type.

In their review, Mann et al. (1989) come to the following conclusions:

1. Network analysis of ecosystems facilitates study of one or more of the many feedbacks and controls that operate between the ecosystem level of organization on the one hand, and community, population, and even physiological levels on the other. However, it should be realized that flows of energy and materials are not the only controls that operate in ecosystems, e.g., keystone predators may influence community structure, visual cues may affect animal behavior, and flows of energy and matter may have an effect vastly greater than their volumes.
2. Network analysis offers formal methods for analyzing ecosystem structure. The networks drawn by ecologists can be viewed as hypotheses of ecosystem structure that can be analyzed formally to see which hypotheses are most consistent with observations and measurements. Even more important, however, is the use of network analyses to suggest key variables (such as particular flows of energy or matter, and internal variables such as assimilation efficiency) that need to be measured accurately to test the hypothesis.
3. Component-level and bilateral-level measures show clear patterns that can be used to compare different ecosystems if they are carefully studied in the same way. General trends become apparent when the succession of ecosystems is analyzed by compressing the food webs into Lindeman Spines and their associated measures of average path length and effective trophic position. These patterns include a tendency to decreased flows and decreased trophic efficiency with greater distance along the Lindeman Spine. As ecosystems mature in succession, there is a trend toward lengthening of the Lindeman Spine, increased effective trophic positions, and more recycling along longer loops.
4. At the level of the whole ecosystem, there are also a number of information theory measures that have been found to be useful in assessing the status of an ecosystem. Flow diversity (D) and resource niche breadth (S) are useful indices for following the development of ecosystem structure in the succession of a single system after disturbance, whereas relative ascendency, a dimensionless ratio, can be used for comparing the growth and development status of different ecosystems, provided they have been carefully studied and modeled in the same way and at the same level of aggregation.

7.10 POTENTIAL APPLICATIONS OF THE ASCENDENCY CONCEPT

Ulanowicz (1997) has listed a numbr of potential applications of the ascendency concept. According to Ulanowicz, ascendency "derives from a very concrete mathematical formula that can be applied to any ecosystem where connections can be identified and quantified. Once we know which system components affect which others and by how much, we can *measure* ascendency." A list of the potential applications follows:

1. *Quantifying ecosystem status:* Possibly the most pertinent use of the ascendency measures is to quantitfy the effects of perturbations on ecosystems. As mentioned above it is expected that ascendency would increase under relatively unimpeded growth and development, and that a significant perturbation should have the opposite effect on information indices, i.e., the total activity (total system throughput, TST) and development capacity of a stressed system would both be negatively affected. Comparisons of perturbated vs. unperturbed systems will enable the identification of what changes in information variables correspond to characteristic qualitative shifts in ecosystem behavior.
2. *Assessing eutrophication*: Ulanowicz (1986) defines eutrophication in terms of the ascendency theory as "any increase in systems ascendency due to a rise in total system throughput that more than compensates for a concomitant fall in the mutual information in the flow network." Thus, eutrophication would be detected if ascendency in the systen increased while at the same time the mutual information of the flows fell.

3. *Ecosystem health and integrity:* A healthy ecosystem is vigorous with a potential to grow. A healthy ecosystem is also one that resists perturbation, or, if perturbed, recovers toward its unperturbed configuration. Costanza et al. (1989) have suggested that the several aspects of healthy or well-functioning ecosystems can be aggregated into three fundamental system attributes: (1) vigor, (2) organization, and (3) resilience. Vigor is well represented by total system throughput (*TST*). System organization is represented by the mutual information of flow networks. The conditional indeterminancy of the network (indeterminancy minus mutual information) is related to the ability of the system to persist and be resilient.
4. *Comparisons between ecosystems*: If the flow networks in two ecosystems have been analyzed in nearly identical fashion, then meaningful comparisons of ecosystem indices can be made such as total system throughput (*TST*), flow diversity, and development capacity.
5. *Trophic status and the identification of perturbations*: Systems with greater resistance and resilience from perturbations are more complex, in the sense that they contain more and larger loops of connections that cycle at lower frequencies. Perturbed ecosytems possess fewer such cycles, and, on the average, each cycle will involve fewer transfers.

   Ulanowicz and Kemp (1979) have described a mathematical mapping that apportions any particular trophic exchange according to how much of it is flowing after two, three, or more trophic transfers; this algorithm converts a complicated web of trophic exchanges into an equivalent trophic chain or pyramid (the Lindeman Spine) (Ulanowicz, 1989). The trophic pyramids of perturbed systems (carbon or energy) will reach upper levels.
6. *Identification of the sensiitivity of the system to limiting flows (e.g., a nutrient)*: Conventional wisdom suggests that the largest inflow of a limiting element (e.g., nitrogen) should be the transfer that most affects system activity and organization. However, when the sensitivity of the overall ascendency to the change in any particular flow is calculated, the result is unexpected; it turns out that the controlling inflow of the limiting nutrient is that which depleted its donor pool at the fastest relative rate. Control, therefore, is exercised by this donor, even though the flows of other sources may be larger in absolute magnitude. Ulanowicz (1997) gives an example from mesoplankton dynamics from the Chesapeake Bay (Baird et al., 1991).

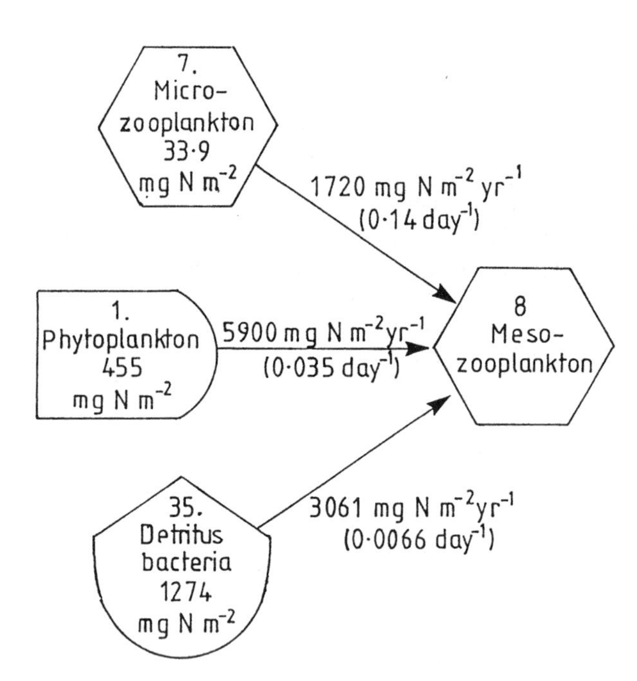

FIGURE 7.29 The three sources of nitrogen that sustain the mesozooplankton in the Chesapeake Bay ecosystem. The numbers in parentheses associated with each input are the rates (per diem) at which that flow is depleting the stock of that particular source (indicated inside the donor box). (Redrawn from Ulanowicz, R.E., *Ecology the Antecedent Perspective*, Columbia University Press, New York, 1997, 136. With permission.)

   In this system, nitrogen is retained the longest; hence, the mesozooplankton is limited more by the availability of nitrogen than by either carbon or phosphorus. The mesozooplankton obtain their nitrogen from three sources (Figure 7.29): (1) phytoplankton, (2) the dettitus-bacteria assemblage, and (3) the microzooplankton. The largest input of nitrogen into the mesoplankton flows from the phytoplankton at a rate of ca. 5,900 mg N m^{-2} yr^{-1}. The donor pool of phytoplankton N averages 455 mg N m^{-2} over the year. The smallest inflow of nitrogen to the mesozooplankton comes from the microzooplankton at the rate of 455 mg N m^{-2} yr^{-1}; however, this flow issues from a nitrogen pool that averages only 33.9 mg N m^{-2} over the year. Thus, the store of nitrogen in the phytoplankton is being depleted by the mesozooplankton at a rate of about 3.5% per day, whereas the microzooplankton yields its stock of nitrogen to the same predators at the much faster rate of 14% per day. The microzooplankton are thus identified as the key donor in the system.
7. *Ecosystem values*: In the developing discipline of ecological economics (Costanza, 1991), a

TABLE 7.16
Designs and Mechanisms that Contribute to Maximum Power

Depreciation pathways are always required according to the second law of thermodynamics.
Systems are selected to operate at an optimum efficiency that generates maximum power.
That optimum efficiency will be less during growth phases than at maturity.
There is selection for autocatalytic processes that reinforce production.
These reinforcements may operate at scales above, at, or below the processes.
Energy transformation webs are hierarchical.
Storages increase with scale.
Transformities increase with scale.
Different kinds of sources interact rather than add.
Items of different transformity interact multiplicatively.
Energy transformations converge spatially.
Production pathways generate their own storages.
No consumption occurs without reinforcement to production.
Small-scale pulses are filtered at larger scales.
Territory of input and influence increases with scale.
All systems pulse; thus, sustainability must incorporate pulses.

Source: From Hall, C.A.S., in *Maximum Power. The Ideas and Applications of H. T. Odum,* Hall, C.A.S., Ed., University of Colorado Press, Niwat, Colorado, 1995, xiii and xiv. With permission.

primary goal is to place an economic value on those biotic resources that normally lie outside traditional markets. Ascendency provides a way to estimate the relative value of stocks as members of a functioning ecosystem. Ascendency, being a work function, serves as a clear analog to the "capital generation" function in economics. An alternative approach is that of the H.T. Odum "emergy" analysis, as discussed in the next section.

7.11 EMERGY ANALYSIS

Throughout this book the Odum energy analysis approach has been used to analyze food webs and coastal ecosystems (see Appendix). The ecosystem models discussed in this chapter all utilized this approach and the network analysis concepts that were developed had their origins in the concepts developed by H.T. Odum. The basic unifying principle in the work of Odum is that there has been, and is, strong selection for systems that can obtain and efficiently use the most energy. This is the "maximum power concept," which Odum has used and expanded as a general systems hypothesis as to the way in which systems are organized in both space and time (Odum, 1975; 1982; 1983a,b; 1986; 1987; 1991; 1994; 1996; Hall, 1995). Table 7.16 lists the designs and mechanisms that contribute to maximum power.

The central concept is that of *emergy.* Emergy includes the energy of one type directly and indirectly used in the production of a resource, product, or service. Solar emergy in a resource, product, or service is the sum of the solar energies required. Emergy includes both fossil fuel energies and environmental energies (such as sunlight, rain, and tides) that are necessary to most processes of energy transformation. Thus, the energy of a fish sold at the market includes the prorated share of the emergy spent for goods and services consumed by the fisherman, and all of the emergy from direct sunlight and tidal action that is necessary to provide essential ecological support for the fish prior to its being caught (Brown et al., 1995).

Two of the most important properties of systems are (1) the total function and (2) the complexity of the systems' parts and connections. As discussed above, Ulanowicz (1986) combined these atributes in a single measure, *ascendency,* in which the bits of information measuring complexity are multiplied by the energy flow in the pathways. However, according to Odum (1996), the importance of a flow is not measured by energy but by its emergy. The Odum approach multiplies the bits of complexity by emergy to weight complexity measures. It will be of interest to reanalyze the network analyses using emergy as the multiplier.

Odum contends that decision-making in resource management should be based on the maximizing of emergy production. This is particularly pertinent to decision-making in the coastal zone. In coastal ecosystems, the evidence of environmental overuse and degredation typically includes (Olsen, 1995):

- Declining water quality in rivers, estuaries, and nearshore waters
- Destruction of habitats important to the production of food, fiber, and fuel, and for maintaining ecosystem integrity and of habitats with important roles in maintaining the physical stability of the coastline

- Declines in nearshore fisheries resources
- Mounting user conflicts
- Increasing inability of governmental organizations to cope by either mitigating adverse effects or mediating among conflicting groups

In his recent book, H.T. Odum (1996) outlined the procedures for emergy accounting for evaluation of environmental and economic use. In this approach, an energy diagram of a region, subsystem, ecosystem, or process is constructed and energy flow data is obtained for the various flows. Next, an emergy analysis is drawn up with four columns: (1) the flows by name; (2) the energy of the flow; (3) transformities (which when multiplied by the energy values, convert them into solar emergy equivalents); and (4) the product of the multiplication listed in (3). Differential equations are used for the simulation of the emergy flow in the system diagram. The transformities are derived from the collective work of numerous studies (Odum and Odum, 1983; Odum et al., 1987a,b).

Brown et al. (1995) list several emergy-based indices that have been found useful for the evaluation of resource questions; these are *net emergy yield ratio*, the *emergy investment ratio*, and the *ratio of emergy to money. Net emergy yield ratio* is the ratio of yield to inputs supplied by the monied economy. The *emergy investment ratio* is the ratio of inputs from the economy to the free inputs from environmental resources. By summing the main energy inputs of a country and dividing by the gross national product (GNP), an emergy/dollar ratio is derived for that country (expressed as sej/GNP$). The ratio of *emergy per capita* is a useful measure of total emergy contributions to a population's existence. Rural populations characteristic of developing countries receive more energy directly from the environment without money payment than do those of industrialized urban populations. Thus, money does not measure their relative standards of living. Therefore, the total emergy flow supporting humans and their local economies may be a more comprehensive measure of the standard of living than dollar income alone.

Emergy analyses of the coastal zone carried out to date include:

1. Shrimp mariculture in Ecuador (Olsen, 1994)
2. Productivity of the Upper Sea of Cortex, Mexico (Brown et al., 1995)
3. Impact of a proposed hydroelectric dam on the Mekong River (Brown et al., 1995)
4. Evaluation of coastal alternatives (Brown et al., 1992)
5. Environmental emergy signature of the Texas coastal zone (Odum et al., 1987)
6. Emergy evaluation of the *Valdiz* oil spill (Brown et al., 1993)
7. Emergy analysis properties, public perceptions, and development guidelines for the coastal zone of Nayant, Mexico (Brown et al., 1992)

From these studies of other development projects and national economies based on emergy analysis, Brown et al. (1995) concluded:

1. A resources contribution to an economy is often inverse to its dollar value.
2. GNP per capita as a measure of living does not include important contributions from the unmonied sectors of the economy and provides a false impression of the relative well-being of a community's inhabitants. A better measure is emergy per capita.
3. Currency exchange rates do not reflect the real buying power of a country's currency relative to another's. A better measure is emergy per dollar. When a country with a high emergy/dollar ratio exports resources and then purchases goods with the money obtained from a country with a lower ratio, more total value leaves the first country than is received.
4. Raw resources should not be exported in exchange for finished products. The resulting net trade deficit drains the resource-exporting country in favor of the importing economy.
5. For a development project that involves export to have a positive benefit to a local economy and not drain it of resources for export, the investment of economically derived energy should be equal to that characteristic of the economy as a whole.

From the examples discussed in the concluding sections of this chapter, it is clear that both network analysis and energy analysis (based on emergy flows) can both contribute to questions concerning development projects and the sustainable development of natural resources. They provide a means for making informed public policy decisions.

Appendix: Energy Concepts (Table A.1)

Over his lifetime, H.T. Odum (1971a,b; 1983; 1996) has influenced ecosystem research to a greater extent than any other scientist. He developed the diagramming of ecosystems using energy systems symbols (Figure A.1). The pathways in such energy system diagrams may indicate causal relationships, interactions, show material cycles, or carry information, but always with some energy. These diagrams also define the equations that are used for systems simulation.

Table A.2 gives definitions of energy and related concepts, while Table A.3 states the principles of energy networks that may eventually be recognized as laws of energy systems. Figure A.1 illustrates the symbols that are used in energy systems diagrams

TABLE A.1
Energy Definitions and Concepts

Heat: The collective motions of molecules, whose average intensity is the temperature that may be measured by expansion of matter in a thermometer.

Energy: May be defined as anything that can be 100% converted into heat.

First law: Energy is neither created or destroyed in circulation and transformation in systems. That energy is conserved is a consequence of its definition above.

Available energy (exergy): Potential energy capable of doing work and being degraded in the process (units: joules, kilocalories, etc.),

Work: An energy transformation process that results in a change in concentration or form of energy.

Power: Flow of energy per unit of time.

Second law (times's arrow): Concentrations (storages) of available energy are continuously degraded (depreciating). Available energy is degraded in any energy transformation process.

Third law: As the heat content approaches zero, the temperature on the Kelvin scale approaches absolute zero (–273°C); molecules are in simple crystalline states, and the entropy of the state is defined as zero.

Molecular entropy: A measure of the complexity of a molecular state due to heat energy. Starting at zero degrees Kelvin (0°K), molecular entropy is the integrated sum of heat added, divided by the Kelvin temperature at which it is added (units: kilocalories per degree).

Source: From Odum, H.T., *Environmental Accounting. Energy and Environmental Decision Making*, John Wiley & Sons, New York, 1996, 16.

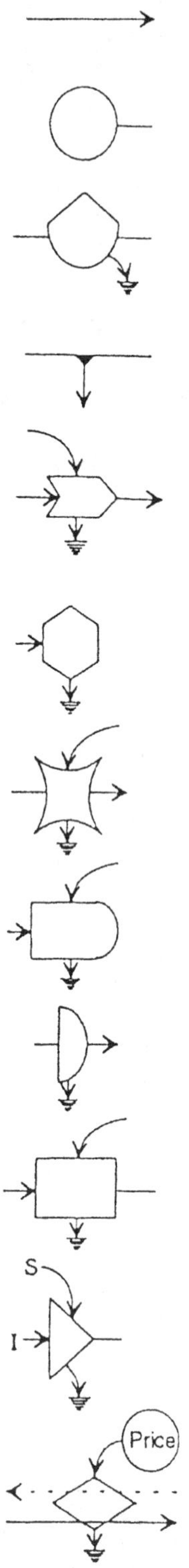

Energy circuit: A pathway whose flow is proportional to the quantity in the storage or source upstream.

Source: Outside source of energy delivering forces according to a program controlled from outside; a forcing function.

Tank: A compartment of energy storage within the system storing a quantity as the balance of inflows and outflows; a state variable.

Heat sink: Dispersion of potential energy into heat that accompanies all real transformation processes and storages; loss of potential energy from further use by the system.

Interaction: Interactive intersection of two pathways coupled to produce an outflow in proportion to a function of both; control action of one flow on another; limiting factor action; work gate.

Consumer: Unit that transforms energy quality, stores it, and feeds it back autocatalytically to improve inflow.

Switching action: A symbol that indicates one or more switching actions.

Producer: Unit that collects and transforms low-quality energy under control interactions of high-quality flows.

Self-limiting energy receiver: A unit that has a self-limiting output when input drives are high because there is a limiting constant quality of material reacting on a circular pathway within.

Box: Miscellaneous symbol to use for whatever unit or function is labeled.

Constant-gain amplifier: A unit that delivers an output in proportion to the input *I* but is changed by a constant factor as long as the energy source *S* is sufficient.

Transaction: A unit that indicates a sale of goods or services (solid line) in exchange for payment of money (dashed line). Price is shown as an external source.

FIGURE A.1 Symbols of the energy systems language (H.T. Odum, 1971a,b, 1983b).

TABLE A.2
Summary of Energy Systems Definitions[a]

Available energy: Potential energy capable of doing work and being degraded in the process (exergy) (units: kilocalories, joules, etc.).

Useful energy: Available energy used to increases system production and efficiency.

Power: Useful energy flow per unit of time.

EMERGY: Available energy of one kind previously required directly or indirectly to make a product or service.

Empower: Emergy flow per unit of time (units: emjoules per unit time).

Transformity: Emergy per unit of available energy (units emjoule per joule).

Solar emergy: Solar energy required directly and indirectly to make a product or service (units: solar emjoules).

Solar empower: Solar energy flow per unit of time (units: solar emjoule per unit time).

Solar transformity: Solar emergy per unit available energy (units: solar emjoules per joule).

[a] Another commonly used unit of energy is the kilocalorie (4,186 Joules). If the kilocalorie, abbreviated kcal, is used, then the following are the appropriate units:

Power: kcal/time.
EMERGY: emkcal
Empower: emkcal/time
Transformity: emkcal/kcal (numerically equal to emjoule/Joule).
Solar emergy: semkcal.
Solar empower: semkcal/time.
Solar transformity: semkcal/kcal (numerically equal to sej/J).

Source: From Odum, H.T., *Environmental Accounting. Energy and Environmental Decision Making*, John Wiley & Sons, New York, 1996, 13.

TABLE A.3
Network Energy Concepts in Emergy Accounting

Time's speed regulator: Power in an energy transformation depends on the workload. Maximum output occurs with an optimum intermediate efficiency.[a]

Maximum empower principle (fourth law?): In the competition among self-organizing processes, network designs that maximize empower will prevail.

Energy transformation hierarchy (fifth law?): Energy flows of the universe are organized in an energy transformation hierarchy. The position in the energy hierarchy is measured with transformities.

[a] For a quantitative treatment of the optimum efficiency for maximum power, see Chapter 7 in Odum, H.T., *Systems Ecology. An Introduction*, John Wiley & Sons, New York, 1983.

Source: From Odum, H.T., *Environmental Accounting. Energy and Environmental Decision Making*, John Wiley & Sons, New York, 1996, 16.

References

Abd Aziz, S.A., Nedwell, D.B., The nitrogen cycle of an east coast U.K. saltmarsh. II. Nitrogen fixation, nitrification, denitrification, and tidal exchange, *Est. Coastal Shelf Sci.*, 22, 689, 1986.

Abugov, R., Species diversity and phasing of disturbance, *Ecology*, 63, 289, 1982.

Adam, P., *Saltmarsh Ecology*, Cambridge University Press, Cambridge, 1990.

Adams, S.M., Feeding ecology of eelgrass fish communities, *Trans. Am. Fish. Soc.*, 105, 514, 1976a.

Adams, S.M., The ecology of eelgrass, *Zostera marina* (L.), fish communities. 2. Functional analysis, *J. Exp. Mar. Biol. Ecol.*, 22, 293, 1976b.

Admiraal, W., The ecology of estuarine sediment-inhabiting diatoms, in *Progress in Phycological Research Volume 3*, Raind, F., Chapman, G., Eds., Biopress Ltd., Bristol, 1984, 269.

Admiraal, W., Bouwman, L.A., Hockstra, L., Romcyn, K., Quantitative and qualitative interactions between microphytobenthos and herbivorous meiofauna on a brackish intertidal beach, *Int. Rev. Ges. Hydrobiol.*, 68, 175, 1983.

Alheit, J., Scheibel, W., Benthic harpacticoids as a food source for fish, *Mar. Biol.*, 70, 141, 1982.

Allen, G.W., Estuarine destruction — a monument to progress, *Proc. 26th North Am. Wildlife Conf.*, 29, 324, 1964.

Aller, R.C., The influence of macrobenthos on the chemical diagenesis of marine sediments, Ph.D. thesis, Yale University, New Haven, CT, 1977.

Aller, R.C., Relationships of tube-dwelling benthos with sediment and overlying water chemistry, in *Marine Benthic Dynamics*, Tenore, K.R., Coull, B.C., Eds., University of South Carolina Press, Columbia, SC, 1980, 385.

Aller, R.C., The effects of macrobenthos on chemical properties of marine sediment and overlying water, in *Animal-Sediment Relations. The Biogenic Alteration of Sediments*, McCall, P.L., Teresz, M.J.S., Eds., Plenum Press, New York, 1982, 53.

Aller, R.C., The importance of diffusive permeability of animal burrow linings in determining marine sediment chemistry, *J. Mar. Res.*, 41, 299, 1983.

Aller, R.C., Aller, J.Y., Microfauna and solute transport in marine muds, *Limnol. Oceanogr.*, 17, 1018, 1992.

Aller, R.C., Dodge, R, E., Animal-sediment relations in a tropical lagoon, Discovery Bay, Jamaica, *J. Mar. Res.*, 32, 209, 1974.

Aller, R.C., Yingst, J.Y., Biogeochemistry of tube-dwellings: A study of the sedentary polychaete *Amphitrite ornata* (Leidy), *J. Mar. Res.*, 35, 201, 1978.

Aller, R.C., Yingst, J.Y., Effects of the marine deposit-feeders *Heteromastus filiformis* (Polychaeta), *Macoma balthica* (Bavalvia) and *Tellina texana* (Bivalvia) on averaged sedimentary solute transport, reaction rates and microbial distribution, *J. Mar. Res.*, 43, 615, 1985.

Aller, R.C., Yingst, J.Y., Ullman, W.J., Comparative geochemistry of water in intertidal *Onuphis* (Polychaeta) and *Upogebia* (Crustacea) burrows: temporal patterns and causes, *J. Mar. Res.*, 41, 571, 1983.

Alongi, D.M., Microbes, meiofauna and bacterial productivity on tubes constructed by the polychaete *Capitella capitata*, *Mar. Ecol. Prog. Ser.*, 23, 207, 1985a.

Alongi, D.M., Effect of pysical disturbance on population dynamics and trophic interactions among microbes and meiofauna, *J. Mar. Res.*, 43, 351, 1985b.

Alongi, D. M., *Coastal Ecosystem Processes,* CRC Press, Boca Raton, FL, 1988.

Alongi, D.M., Bacterial productivity and microbial biomass in tropical mangrove sediments, *Microbial Ecol.*, 15, 59, 1988a.

Alongi, D.M., The role of soft-bottom benthic communities in tropical mangrove and coral reef ecosystems, *Crit. Rev. Mar. Sci.*, 1, 20, 1988b.

Alongi, D.M., Microbial-meiofaunal interrelationships in some tropical intertidal sediments, *J. Mar. Res.*, 46, 349, 1988c.

Alongi, D.M., The fate of bacterial biomass and production in marine benthic food chains, in *Recent Advances in Microbial Ecology*, Hattori, T., Ishida, Y., Momyama, Y., Morita, R.Y., Uchida, A., Eds., Japan Sciences Society Press, Tokyo, 1989, 355.

Alongi, D.M., The ecology of tropical soft-bottom benthic ecosystems, *Oceanogr. Mar. Biol. Annu. Rev.*, 28, 381, 1990a.

Alongi, D.M., Effect of mangrove detrital outwelling on nutrient regeneration and oxygen fluxes in coastal sediments of the Central Barrier Reef Lagoons, *Est. Coastal Shelf Sci.*, 31, 58, 1990b.

Alongi, D.M., The dynamics of benthic nutrient pools and fluxes in tropical mangrove forests, *J. Mar. Res.*, 54, 123, 1996.

Alongi, D.M., *Coastal Ecosystem Processes*, CRC Press, Boca Raton, FL, 1998.

Alongi, D.M., Boesch, D.F., Diaz, R.J., Colonization of meiobenthos in oil-contaminated subtidal sands in lower Chesapeake Bay, *Mar. Biol.*, 72, 325, 1983.

Alongi, D.M., Boto, K.G., Robertson, A.I., Nitrogen and phosphorus cycles, in *Tropical Mangrove Ecosystems,* Robertson, A.I., Alongi, D.M., Eds., American Geophysical Union, Washington, DC, 1992, 251.

Alongi, D.M., Boto, K.G., Tirendi, F., Effect of exported mangrove litter on bacterial productivity and dissolved organic carbon fluxes in adjacent tropical nearshore sediments, *Mar. Ecol. Prog. Ser.*, 56, 133, 1989.

Alongi, D.M., Tietjen, J.H., Population growth and trophic interactions among free- living nematodes, in *Marine Benthic Dynamics,* Tenore, K.R., Coull, B.C., Eds., University of South Carolina, Columbia, SC, 1980, 167.

Altabet, M.A., McCarthy, J.J., Temporal and spatial variations in the natural abundance of ^{15}N in PON from a warm-core ring, *Deep-Sea Res.*, 32, 755, 1985.

Ambrose, W.G., Jr., The influnce of the predatory polychaetes *Glycera dibranchiata* and *Nereis virens* on the structure of a soft-bottom community, Ph.D. thesis, University of South Carolina, Chapel Hill, SC, 1982.

Ambrose, W.G., Jr., Influence of residents on the development of a marine soft-bottom community, *J. Mar. Res.*, 42, 633, 1984a.

Ambrose, W.G., Jr., Role of predatory infauna in structuring marine soft-bottom communities, *Mar. Ecol. Prog. Ser.*, 17, 109, 1984b.

Ambrose, W.G., Importance of predatory infauna in more soft-bottom communities: reply to Wilson, *Mar. Ecol. Prog. Ser.*, 32, 41, 1986.

Ameyaw-Akumfi, C.A., Naylor, E., Temporal patterns in shell gape in *Mytilus edulis*, *Mar. Biol.*, 95, 237, 1987.

Andersen, F.Ø., Hargrave, B.T., Effects of *Spartina* detritus enrichment on exchange of nutrients between sediment and water in an intertidal area of the Bay of Fundy, *Mar. Ecol. Prog. Ser.*, 16, 161, 1984.

Andersen, F.Ø., Kristensen, E., The influence of macrofauna on estuarine benthic community metabolism — a microcosm study, *Mar. Biol.*, 99, 596, 1988.

André, C., Jonsson, P.P., Lindegarth, M., Predation on settling bivalve larvae by benthic suspension feeders: the role of hydrodynamics and larval behaviour, *Mar. Ecol. Prog. Ser.*, 97, 183, 1993.

Andrew, N.L., Choat, J.H., The influence of predation and conspecific adults on the abundance of juvenile *Evechinus chloroticus* (Echinoidea: Echinometridae), *Oecologia (Berlin)*, 54, 80, 1982.

Andrew, N.L., Choat, J.H., Habitat related differences in the survivorship and growth of juvenile sea urchins, *Mar. Ecol. Prog. Ser.*, 27, 151, 1985.

Andrew, N.L., Jones, G.P., Patch formation by herbivorous fish in a temperate Australian kelp forest, *Oecologia (Berlin)*, 85, 57, 1990.

Andrewartha, H.G., Birch, L.C., *The Distribution and Abundance of Animals*, University of Chicago Press, Chicago, IL, 1954.

Andrewartha, H.G., Birch, L.C., *The Ecological Web. More on the Distribution and Abundance of Animals*, University of Chicago Press, Chicago, IL, 1984.

Anger, K., Rogal, H., Schreiver, G., Valentin, C., *In situ* investigations on the echinoderm *Asterias rubens* as a predator on soft bottom communities in the western Baltic Sea, *Helg. Wiss. Meeresunters*, 29, 439, 1977.

Ankar, S., The soft bottom ecosystem of the Northern Baltic Proper with special reference to the macrofauna, *Contr. Askö Lab, Univ. of Stockholm*, 19, 1, 1977.

Ankar, S., Annual dynamics of a Northern Baltic bottom, in *Cyclic Phenomena in Marine Plants and Animals*, Naylor, E., Ed., Pergamon Press, Oxford, 1979, 29.

Ankar, S., Elmgren, R., The benthic macro- and meiofauna of the Askö-Landsort area (Northern Baltic proper). A stratified random sampling survey, *Contr. Askö Lab, Univ. of Stockholm*, 1, 1976.

Ansell, A.D., McLusky, D.S., Stirling, A., Trevallion, A., Production and energy flow in the macrobenthos of two sandy beaches in South West India, *Proc. Roy. Soc. Edinb.*, 76B, 269, 1978.

Ansell, A.D., Tevallion, A., Behavioural adaptations of intertidal molluscs from a tropical sandy beach, *J. Exp. Mar. Biol. Ecol.*, 4, 9, 1969.

Arlt, G., Zur produktions biologishen Bedentung der meiofauna in Kustengewassen, *Wiss. Z. Univ. Rostok*, 22, 1141, 1973.

Arnaud, P.M., Contribution à la bionomie marine venthique des régions antarctiques et subantarctiques., *Tehtys*, 6, 465, 1974.

Arnold, K.E., Manley, S.L., Carbon allocation in *Macrocystisc pyrifera* (Phaeophyta): intrinsic variability in photosynthesis and respiration, *J. Phycol.*, 21, 154, 1985.

Arnst, W.E., Results and problems of an "unsuccessful" benthos cage experiment (Western Baltic), in *Biology of Benthic Organisms*, Keegam, B.F., Ceidigh, P.O., Boaden, P.J.S., Eds., Pergamon Press, New York, 1977, 31.

Asmus, H., Field measurements on respiration and secondary production of a benthic community in the northern Wadden Sea, *Neth. J. Sea Res.*, 15, 403, 1982.

Asmus, H., Secondary production in an intertidal mussel bed community related to its storage and turnover compartments, *Mar. Ecol. Prog. Ser.*, 39, 251, 1987.

Asmus, H., Asmus, R.M., The importance of grazing food chain for energy flow and production in three intertidal sand communities, *Helg. Wiss. Meeresunters*, 39, 273, 1985.

Asmus, H., Asmus, R.M., Trophic relationships in tidal flat areas: to what extent are tidal flats dependent in imported foods?, *Neth. J. Sea Res.*, 27, 93, 1990.

Asmus, M.L., McKellar, H.N., Jr., Network analysis of the North Inlet salt marsh ecosystem, in *Network Analysis in Marine Ecology*, Wulff, F., Feild, J.G., Mann, K.H., Eds., Springer-Verlag, Berlin, 1989, 206.

Asmus, R.M., Asmus, H., Are mussel beds limiting or promoting phytoplankton?, *J. Exp. Mar. Biol. Ecol.*, 148, 215, 1991.

Astles, K.L., Patterns of abundance and distribution of species in intertidal rock pools, *J. Mar. Biol. Ass. U.K.*, 73, 555. 1993.

Atkinson, W.D., Newbury, S.F., The adaptations of the rough winkle, *Littorina rudis*, to desiccation and dislodgement by wind and waves, *J. Anim. Ecol.*, 53, 93, 1984.

Ayling, A.M., The role of biological disturbance in temperate subtidal encrusting communities, *Ecology*, 62, 830, 1981.

Axelrod, D.M., Nutrient flux through the salt marsh ecosystem. Ph.D. thesis, College of William and Mary, Williamsburg, VA, 1974.

Axelrod, D.M., Moore, K.A., Bender, M.E., Nitrogen, phosphorus and carbon flux in Chesapeake Bay marshes, *Virginia Wat. Bd. Center Bull.*, 79, 1976.

Azam, F., Fenchel, T., Field, J.G., Gray, J.S., Meyer-Reil, A., Thingstad, F., The ecological role of water-column microbes in the sea, *Mar. Ecol. Prog. Ser.*, 10, 257, 1983.

Bach, S.D., Thayer, G.W., La Croix, M.W., Export of detritus from eelgrass (*Zostera marina*) beds near Beaufort, North Carolina, *Mar. Ecol. Prog. Ser.*, 28, 265, 1986.

Bachelet, G., Growth and recruitment in the tellinid bivalve *Macoma balthica* at the southern limit of its geograhical distribution, the Gironde estuary (SW France), *Mar. Biol.*, 59, 105, 1980.

Bada, J.L., Lee, C., Decomposition and alteration of organic compounds dissolved in seawater, *Mar. Chem.*, 5, 523, 1977.

Bagge, P., Effects of pollution on estuarine ecosystems. 1. Effect of effluents from wood processing industries on the bottom fauna of Salthallefjord (W. Sweden). 2. The succession of the bottom fauna communities in polluted estuarine habitats in the Baltic-Skagerrak region, *Merentnkimskitokson Julkaiso Harsforskningstitutets Skrift*, 228, 1–130, 1969.

Baggerman, B., Spatfall and transport of *Cardium edule* L., *Arch. Neerland Zool.*, 10, 315, 1953.

Bahr, K.-J., Eine Massensiedlung von *Lanice conchlegia* (Polychaeta: Terebellidae) im Weser-Astuar, *Veroff Inst. Meeres. Bremerhaven*, 17, 101, 1979.

Bahr, L.M., Energetic aspects of the intertidal oyster reef community at Sapelo Island, Georgia (U.S.A.), *Ecology*, 57, 121, 1976.

Bahr, L.M., Lanier, W., The intertidal oyster community of the South Atlantic: A community profile, U.S. Fish and Wildlife Service, Office of Biological Sciences, Washington, DC, FWS/OBS-81/15, 195 pp., 1976.

Baillie, P.W., Welsh, B.L., The effect of tidal resuspension on the distribution of intertidal epipelic algae in an estuary, *Est. Coastal Mar. Sci.*, 10, 165, 1980.

Baines, S.B., Pace, M.L., The production of dissolved organic matter and by phytoplankton and its importance to bacteria: Patterns across marine and freshwater systems, *Limnol. Oceanogr.*, 36, 1078, 1991.

Baird, D., McGlade, J.M., Ulanowicz, R.E., The comparative ecology of six marine ecosystems, *Phil. Trans. Roy. Soc. London, B*, 333, 15, 1991.

Baird, D., Milne, H., Energy flow in the Ythan estuary, Aberdeenshire, Scotland, *Est. Coastal Shelf Sci.*, 13, 455, 1981.

Baird, D., Ulanowicz, R.E., A network analysis of the Chesapeake Bay ecosystem, Ref. No. UMCEES. 97-77, Centre for Environmental and Estuarine Studies, Slomons, MD, 1986.

Baird, D., Ulanowicz, R.E., The seasonal dynamics of the Chesapeake Bay ecosystem, *Ecol. Monogr.*, 59, 329, 1989.

Baird, D., Ulanowicz, R.E., Comparative study on the trophic structure, cycling and ecosystem properties of four tidal inlets, *Mar. Ecol. Prog. Ser.*, 99, 221, 1993.

Baird, D., Winter, P.E.D., Annual flux of dissolved inorganic nutrients through a well-mixed estuary, in *Estuarine Water Quality Measurement: Monitoring, Modelling and Research*, Michaelis, W., Ed., Springer-Verlag, Heidelburg, 1989, 335.

Baker, A.J., Observations on the winter feeding on the South Island Pied Oystercatcher (*Haemotopus ostralegus* finschi), B.Sc. Hons project in Zoology, University of Canterbury, Christchurch, NZ, 1966.

Baker, A.J., The comparative biology of New Zealand oystercatchers, M.Sc. thesis, University of Canterbury, Christchurch, NZ, 134 pp, 1969.

Baker, A.J., Prey-specific feeding methods of New Zealand oystercatchers, *Notornis*, 21, 219, 1974.

Balasubramian, T., Venugopalan, V.K., Dissolved organic matter in Pitchawaram mangrove environment, Tamil Nadu, South India, in *Proceedings of the Asian Symposium on Mangrove Environment: Research and Management*, Soepadme, E., Rao, A.N., Macintosh, P.J., Eds., University of Malaya, Kuala Lumpur, 1984, 596.

Ballantine, W.J., A biologically-defined scale for the comparative description of rocky shores, *Field Stud.*, 1, 1, 1961.

Bally, R., The ecology of sandy beaches on the west coast of South Africa, Ph.D. thesis, University of Cape Town, 404 pp., 1981.

Bally, R., Intertidal zonation on sandy beaches of the west coast of South Africa, *Cashiers Biol. Mar.*, 24, 85, 1983.

Bally, R., McQuaid, C.P., Pierce, S.M., Primary production in the Bot River Estuary, South Africa, *Trans. Roy. Soc. S. Af.*, 45, 33, 1985.

Banse, K., Nichols, F.H., May, D.R., Oxygen consumption by the seabed. III. On the role of the macrofauna at three stations, *Vie Milieu (Suppl.)*, 22, 31, 1971.

Barkai, A., Branch, G.M., The influence of predation and substrate complexity on recruitment to settlement plates: a test of the theory of alternative states., *J. Exp. Mar. Biol. Ecol.*, 124, 215, 1988.

Barkai, A., McQuaid, C., Predator-prey role reversal in a marine benthic ecosystem, *Science*, 242, 62, 1988.

Barlow, J.P., Physical and biological factors determining the distribution of zooplankton in a tidal estuary, *Biol. Bull.*, 109, 211, 1955.

Barlow, L.A., Electrophysiological and behavioural responses of larvae of the red abalone (*Haliotis rufescens*) to settlement inducing substances, *Bull. Mar. Sci.*, 46, 527, 1990.

Barnes, H., Surface roughness and the settlement of *Balanus balanoides* L., *Arch. Soc. Zool. Bot. Fenn.Vanamo*, 10, 2, 1956.

Barnes, H., Note on variations in the release of nauplii of *Balanus balanoides* with special reference to the spring diatom outburst, *Crustaceana*, 4, 118. 1962.

Barnes, H., Barnes, M., Recent spread and distribution of the barnacle *Elminius modestus* Darwin in northwest Europe, *Proc. Zool. Soc. Lond.*, 135, 137, 1960.

Barnes, H., Barnes, M., The growth rate of *Elminius modestus* (Crust., Cirripedia) in Scotland, *Int. Rev. Ges. Hydrobiol. Hydrogr.*, 47, 481, 1962.

Barnes, H., Barnes, M., Egg numbers, metabolic efficiency of egg production and fecundity: local and regional variation in, a number of common cirripedes, *J. Exp. Mar. Biol. Ecol.*, 2, 135, 1968.

Barnes, H., Crisp, D.J., Powell, H.T., Observations on the orientation of some species of barnacles, *J. Anim. Ecol.*, 20, 227, 1951.

Barnes, H., Powell, H.T., The development, general morphology and subsequent elimination of barnacle populations, *Balanus crenatus* and *Balanus balanoides*, after an initial heavy settlement, *J. Anim. Ecol.*, 19, 175, 1950.

Barnes, M., Gibson, R.N., Eds., *Trophic Relationships in the Marine Environment*, Aberdeen University Press, Aberdeen, 1990.

Barnett, B.E., Crisp, D.J., Laboratory studies of gregarious settlement in *Balanus balanoides* and *Elminius modestus* in relation to competition between these species, *J. Mar. Biol. Assoc., U.K.*, 59, 581, 1979.

Bate, G.C., Campbell, E.E., Talbot, M.M.B., Primary productivity of the sandy beach surf zones of southern Africa, in *Trophic Relationships in the Marine Environment*, Barnes, M., Gibson, R, Eds., Proceedings of the 24th European Marine Biology Symposium, 1990, 41.

Batham, E.J., Ecology of a southern New Zealand sheltered rocky shore, *Trans. Roy. Soc. N.Z.*, 84, 447, 1956.

Batham, E.J., Ecology of a southern New Zealand rocky shore at Little Papanui, Otago Peninsula, *Trans. Roy. Soc. N.Z.*, 85, 647, 1958.

Bayne, B.L., Hawkins, A.J.S., Herbivory in suspension-feeding molluscs, in *Plant-Animal Interactions in the Marine Benthos* (Systematics Association Special Publication No. 46), John, D.M, Hawkins, A.J.S., Price, J.H., Eds., Clarendon Press, Oxford, 1992, 266.

Bayne, B.L., Scotland, C., Rates of feeding by *Thais* (*Nucella*) *lapillus* (L.), *J. Exp. Mar. Biol. Ecol.*, 32, 113, 1975.

Bayne, B.L., Thompson, R.J., Widdows, J., Some effects of temperature and food on the rate of oxygen consumption by *Mytilus edulis* L., in *Effects of Temperature on Ectothermic Organisms*, Wieser, W., Ed., Springer-Verlag, Berlin, 1973, 181.

Bayne, B.L., Widdows, J., The physiological ecology of two populations of *Mytilus edulis* L., *Oecologia (Berlin)*, 37, 137, 1978.

Beach, K.S., Smith, C.M., Ecophysiology of a tropical rhodophyte. III. Recovery from emersion stress in *Ahnfeltiopsis concinna* (J.Ag.) Silva et De Crew, *J. Exp. Mar. Biol. Ecol.*, 211, 151, 1997.

Beckley, L.E., McLachlan, A., Studies on the littoral seaweed epifauna of St. Croix Island. 2. Composition of the summer standing stock, *S. Af. J. Zool.*, 15, 170. 1980.

Beentjes, M.P., Williams, B.G., Endogenous circatidal rhythmicity in the New Zealand cockle *Chione stutchburyi* (Bivalvia: Veneridae), *Mar. Behav. Physiol.*, 12, 171, 1986.

Beers, J.R., Stewart, G.L., Microzooplankters in the plankton communities of the upper waters of the tropical Pacific, *Deep-Sea Res.*, 18, 861, 1971.

Behbehani, M.I., Croker, R.A., Ecology of the beach wrack in northern New England with special reference to *Orchestia platensis*, *Est. Coastal Shelf Sci.*, 15, 611, 1982.

Bell, S.S., Meiofaunal-macrofaunal interactions in a high marsh habitat, *Ecol. Monogr.*, 50, 487, 1980.

Bell, S.S., Derlin, D.J., Short-term mcrofaunal recolonization of sediment and epibenthic habitats in Tampa Bay, Florida, *Bull. Mar. Sci.*, 33, 102, 1983.

Bell, S.S., Sherman, K.M., A field investigation of meiofaunal dispersal: tidal resuspension and implications, *Mar. Ecol. Prog. Ser.*, 3, 245, 1980.

Bell, S.S., Westby, M., Abundance of macrofauna in dense seagrass is due to habitat preference, not predation, *Oecologia (Berlin)*, 68, 205, 1986.

Bell, S.S., Woodin, S.A., Community unity: experimental evidence of meiofauna and macrofauna, *J. Mar. Res.*, 42, 605, 1984.

Bellan, G., Bellan-Santini, D., Influence de la pollution sur les peuplements marins de la region Marseille, *Marine Pollution and Sea Life,* Ruvio, M., Ed., Fishing News (Books) Ltd., London, 1972.

Bellan, G., Reisch, D. J., Foret, J. P., The sublethal effect of a detergent on reproduction and settlement of the polychaetous annelid *Capitella capitita*, *Mar. Biol.*, 14, 183, 1972.

Bender, E.A., Case, T.J., Gilpin, M.E., Perturbation experiments in community ecology: theory and practice., *Ecology*, 65, 1, 1984.

Bennett, B.A., Griffiths, C.L., Factors affecting the distribution, abundance and diversity of rock-pool fishes on the Cape Peninsula, *S. Af. J. Zool.*, 19, 97. 1984.

Berman, T., Holm-Hansen, O., Release of photoassimilated carbon as dissolved organic matter by marine phytoplankton, *Mar. Biol.*, 28, 305, 1974.

Berner, R.A., Calvert, S.E., Chesselet, R., Cooke, R.C., Groot, A.J. de, Duinker, J.C., Lermann, A., Martin, J.M., Price, N.B., Sales, F.L., Suess, E., Wollast, R., Solution-sediment chemical interactions, in *The Benthic Boundary Layer*, McCone, I.N., Ed., Plenum Press, New York, 1976, 261.

Bernstein, B.B., Jung, N., Eds., Selective processes and coevolution in a kelp canopy community in southern California, *Ecol. Monogr.*, 19, 335, 1979.

Bernstein, B. B., Jung, N., Selective processes and coevolution in a kelp community in southern California, *Ecol. Monogr.*, 49, 335, 1979.

Bernstein, B.B., Wiliams, B.E., Mann, K.H., The role of behavioural responses to predators in modifying urchin's (*Strongylocentrotus droebachiensis*) destructive grazing and seasonal foraging patterns, *Mar. Biol.*, 63, 39, 1981.

Berry, P.F., Biomass and density of detritivores on a littoral rocky reef on the Natal coast, with an estimate of population production for the ascidian *Pyura stolonifera*, *Invest. Rep. Oceanogr. Res. Inst. (Durban, South Africa)*, 53, 1, 1982.

Berry, P.F., Hanekom, P., Joubert, C., Joubert, M., Schleyer, M., Smale, M., Van der Elst, R., Preliminary account of the biomass and major energy pathways through a Natal nearshore reef community, *S. Af. J. Sci.*, 75, 565, 1979.

Bertness, M.D., Behavioural and ecological aspects of size gradients in *Thais lamellosa* and *Thais marginata*, *Ecology*, 58, 86. 1977.

Bertness, M.D., Competitive dynamics of a tropical hermit crab assemblage, *Ecology*, 62, 751, 1981.

Bertness, M.D., Fiddler crab regulation of *Spartina alterniflora* production in a New England salt marsh, *Ecology*, 66, 1042, 1985.

Bertness, M.D., Gaines, S.D., Stephens, E.G., Components of recruitment in populations of the acorn barnacle *Semibalanus balanoides* (Linnaeus), *J. Exp. Mar. Biol. Ecol.*, 156, 199, 1992.

Beukema, J.J., Seasonal changes in the biomass of the macrobenthos of a tidal flat in the Dutch Wadden Sea, *Neth. J. Sea Res.*, 19, 236, 1974.

Beukema, J.J., Biomass and species richness of the macrobenthic animals living on the tidal flats of the Dutch Wadden Sea, *Neth. J. Sea Res.*, 10, 236, 1976.

Beukema, J.J., The role of the large invertebrates in the Wadden Sea ecosystem, in *Invertebrates of the Wadden Sea*, Dunkers, N., Wolff, W.J., Eds., Balkema, Rotterdam, 1981, 211.

Beukema, J.J., de Bruin, W., Seasonal changes in dry weight and chemical composition of the soft parts of the tellinid bivalve *Macoma balthica* in the Dutch Wadden Sea, *Neth. J. Sea Res.*, 11, 42, 1977.

Beukema, J.J., de Bruin, W., Jansen, J.J.M., Biomass and species richness of the macrobenthic animals living on the tidal flats of the Dutch Wadden Sea: long-term changes during a period of mild winters, *Neth. J. Sea Res.*, 12, 58, 1978.

Beukema, J.J., Meehan, B.W., Effect of temperature on the length of the annual growing season in the tellinid bivalve, *J. Exp. Mar. Biol. Ecol.*, 90, 27, 1985.

Beverage, A.E., Chapman, V.J., The zonation of marine algae at Piha, New Zealand, in relaton to the tidal level factor. (Studies in intertidal zonation 2), *Pacific Sci.*, 4, 188, 1950.

Bianchi, T.S., Levinton, J.S., Nutrition and food limitation of deposit feeders. II. Differential effects of *Hydrobia totteni* and *Ilyanassa obsoleta* on microbial communities, *J. Mar. Res.*, 39, 547, 1981.

Biddanda, B.A., Microbial synthesis of macroparticulate matter, *Mar. Ecol. Prog. Ser.*, 20, 241, 1985.

Biddanda, B.A., Pomeroy, L.R., Microbial aggregation and dedgradation of phytoplankton derived detritus in seawater. I. Microbial succession, *Mar. Ecol. Prog. Ser.*, 42, 79, 1988.

Biebel, R., Vergleichende Untersuchungen zur Temperaturrsistenz von Meeresslagenentlang der pazifischen Kuste Nordamerikas, *Protoplasma*, 69, 61, 1970.

Billen, G., Nitrification in the Scheldt estuary (Belgium and The Netherlands), *Est. Coastal Mar. Sci.*, 3, 79, 1975.

Billen, G., A budget of nitrogen recycling in North Sea sediments off the Belgian coast, *Est. Coastal Mar. Sci.*, 7, 127, 1978.

Billheimer, L.E., Coull, B.C., Bioturbation and recolonization of meiobenthos in juvenile spot (Pisces) feeding pits, *Est. Coastal Shelf Sci.*, 27, 335. 1988.

Bingham, F.O., The mucus holdfast of *Littorina irrotata* and its relationship to relative humidity and salinity, *Veliger*, 15, 48, 1972.

Birch, L.C., The meaning of competition, *Am. Nat.*, 91, 5, 1957.

Blaber, S.J.M., The food and feeding ecology of mullet in the St. Lucia Lake system, in *St. Lucia Scientific Advisory Workshop — Charters Creek, February, 1976*, Heydom, A.E.F., Ed., Natal Parks Advisory Board, 1976.

Blaber, S.J.M., Blaber, R.G., Factors affecting the distribution of juvenile estuarine and inshore fishes, *J. Fish Biol.*, 17, 143, 1980.

Blaber, S.J.M., Brewer, D.T., Salini, J.P., Species composition and biomass of fishes in different habitats of a tropical northern Australian estuary: their occurrence in the adjourning sea and estuarine dependence, *Est. Coastal Shelf Sci.*, 29, 509. 1989.

Black, R., Tactics of whelks preying on limpets, *Mar. Biol.*, 46, 147, 1978.

Black, R., Competition between intertidal limpets an intrusive niche on a steep marine gradient, *J. Animal Ecol.*, 3, 200, 1979.

Blackburn, T.H., Henriksen, K., Nitrogen cycling in different types of sediments from Danish waters, *Limnol. Oceanogr.*, 28, 477, 1983.

Blanchard, G.F., Measurement of meiofauna grazing rates on microphytobenthos: is primary production a limiting factor?, *J. Exp. Mar. Biol. Ecol.*, 147, 37, 1991.

Blanchard, G.F., Sauriau, P.-G., Cariou-Le Gall, V., Gauleau, D., Garet, M.-J., Oliver, F., Kinetics of tidal resuspension of microbiota testing the effect of sediment cohesiveness and bioturbation using flume experiments, *Mar. Ecol. Prog. Ser.*, 151, 17, 1997.

Blanchette, C.A., Size and survival of intertidal plants in response to wave action: a case study with *Fucus gardneri*, *Ecology*, 78, 1563, 1997.

Blankley, W.O., Marine food of kelp gulls, lesser sheathbills and imperial cormorants at Marion Island (Subantarctic), *Cormorant*, 9, 77, 1981.

Blankley, W.O. and Branch, G.M., Ecology of the limpet *Nacella delesserti* (Philippi) at Marion Island in the sub-Antarctic Southern Ocean, *J. Exp. Mar. Biol. Ecol.*, 92, 259, 1985.

Blinks, L.R., Photosynthesis and productivity of littoral marine algae, *J. Mar. Res.*, 14, 363, 1955.

Bloom, S.A., Simon, J.L., Hunter, V.D., Animal-sediment relations and community analysis of a Florida estuary, *Mar. Biol.*, 13, 43, 1972.

Bodkin, J.L., Fish assemblages in *Macrocystis* and *Nereocystis* forests off central California, *Fish. Bull., U.S.*, 84, 799, 1986.

Bodkin, J.L., Effects of kelp forest removal on associated fish fauna in central California, *J. Exp. Mar. Biol. Ecol.*, 117, 227, 1988.

Boesch, D.F., Species diversity of the macrobenthos in the Virginia area, *Chesapeake Sci.*, 13, 206, 1972.

Boesch, D.F., Classification and community structure of the macrobenthos in the Hampton Roads area, Virginia, *Mar. Biol.*, 21, 226, 1973.

Boesch, D.F., A new look at the zonation of benthos along an estuarine gradient, in *Ecology of the Marine Benthos*, Coull, B.C., Ed., University of South Carolina Press, Columbia, SC, 1977, 245.

Boesch, D.F., Turner, R.E., Dependence of fishery species on salt meadows: the role of food and refuges, *Estuaries*, 7, 460, 1984.

Bonar, D.B., Coon, S.L., Walch, M., Weiner, R.M., Fitt, W., Control of oyster settlement and metamorphosis by endogenous and exogenous chemical cues, *Bull. Mar. Sci.*, 46, 484, 1990.

Boonruang, P., The degradation rates of mangrove leaves of *Avicennia marina* (Forsk) Vierh. at Puket Island, *Puket Mar. Biol. Center Res. Bull.*, 26, 1, 1978.

Booth, D.J., Brosnan, D.M., The role of recruitment dynamics in rocky shore and coral reef communities, *Adv. Ecol. Res.*, 25, 309, 1995.

Borey, R.B., Harcombe, P.A., Fisher, F.M., Water and organic carbon fluxes from an irregularily flooded brackish marsh on the upper Texas coast, U.S.A., *Est. Coastal Shelf Sci.*, 16, 379, 1983.

Boto, K.G., Nutrient and organic fluxes in mangroves, in *Mangrove Ecosystems in Australia: Structure, Function and Management*, Clough, B.F., Ed., Australian University Press, Canberra, 1982, 239.

Boto, K.G., Alongi, D.M., Nott, A.L.J., Dissolved organic carbon-bacteria interactions at sediment-water interface in a tropical mangrove system, *Mar. Ecol. Prog. Ser.*, 57, 243, 1989.

Boto, K.G., Bunt, J.S., Tidal export of particulate organic matter from a northern Australian mangrove system, *Est. Coastal Shelf Sci.*, 13, 247, 1981.

Boto, K.G., Bunt, J.S., Carbon export from mangroves, in *Cycling of Carbon, Nitrogen and Sulfur and Phosphorus in Terrestrial and Aquatic Ecosystems*, Galbally, I.E., Freney, J.R., Eds., Australian Academy of Sciences, Canberra, 1982, 105.

Boto, K.G., Robertson, A.I., The relationship between nitrogen fixation and tidal exports of nitrogen in a tropical mangrove ecosystem, *Est. Coastal Shelf Sci.*, 31, 351, 1990.

Boto, K.G., Wellington, J.T., Seasonal variations in concentrations and fluxes of dissolved organic and inorganic materials in a tropical, tidally-dominated, mangrove waterway, *Mar. Ecol. Prog. Ser.*, 50, 151, 1988.

Boucot, A.J., *Principles of Benthic Marine Paleoecology*, Academic Press, New York, 463 pp., 1981.

Bousefield, E.L., Ecological control and occurrence of barnacles on the Miramichi estuary, *Bull. Nat. Mus. Canada,* 132, 112, 1954.

Bouvy, M., Contribution of the bacterial and microphytobenthic microflora in the energetic demand of the meiobenthos in an intertidal muddy sediment (Kerguelen Archipelago), *Mar. Biol.*, 9, 109, 1988.

Boyle, P.R., The physiology and behaviour of chitons (Mollusca:Polyplacophora), *Oceanogr. Mar. Biol. Annu. Rev.,* 15, 461, 1977.

Boynton, W.R., Kemp.W.M., Keefe, C.W., A comparative analysis of nutrients and other factors influencing estuarine phytoplankton production, in *Estuarine Comparisons*, Kennedy, V.S., Ed., Academic Press, New York, 1982, 305.

Boynton, W.R., Kemp.W.M., Osborne, C.G., Nutrient fluxes across the sediment water interface of a coastal plain estuary, in *Estuarine Perspectives*, Kennedy, V.S., Ed., Academic Press, New York, 1980, 93.

Bradfield, A.E., The oxygen content of interstitial water in sandy shores, *J. Anim. Ecol.,* 37, 97, 1964.

Bradshaw, J.S., Environmental parameters and marsh formation, *Limnol. Oceanogr.*, 13, 26, 1968.

Brafield, A.E., Newell, G.E., The behaviour of *Macoma balthica* (L.), *J. Mar. Biol. Assoc., U.K.*, 42, 81, 1961.

Branch, G.M., The ecology of *Patella* Linnaeus from Cape Peninsula, South Africa. I. Zonation, movements and feeding, *Zool. Af.*, 6, 1, 1971.

Branch, G.M., Notes on the ecology of *Patella concolor* and *Cellana capensis*, and the effect of human consumption on limpet populations, *Zool. Af.*, 10, 75, 1975a.

Branch, G.M., Intraspecific competition in *Patella cochlear* Born, *J. Anim. Ecol.*, 44, 263, 1975b.

Branch, G.M., Mechanisms reducing intraspecific competition in *Patella* spp.: migration, differentiation and territorial behaviour, *J. Anim. Ecol.*, 44, 575, 1975c.

Branch, G.M., Interspecific competition experienced by South African *Patella* spp., *J. Anim. Ecol.,* 45, 507, 1976.

Branch, G.M., The responses of South African patellid limpets to invertebrate predators, *Zool. Af.*, 13, 221, 1978.

Branch, G.M., Aggression by limpets against invertebrate predators, *Anim. Behav.*, 27, 408, 1979.

Branch, G.M., The biology of limpets: physical factors, energy flow and ecological interactions, *Oceanogr. Mar. Biol. Annu. Rev.,* 19, 235, 1981.

Branch, G.M., Competition between marine organisms: ecological and evolutionary implications, *Oceanogr. Mar. Biol. Annu. Rev.*, 22, 429, 1984.

Branch, G.M., Limpets: Their role in littoral and subtidal community dynamics, in *The Ecology of Rocky Shores*, Moore, P.G., Seed, R., Eds., Hodder & Straughton, London, 1986, 97.

Branch, G.M., Activity rhythms in *Siphonaria theristes*, in *NATO Advanced Workshop on Behavioural Adaptation to Intertidal Life*, Chelazzi, G., Vannini, M., Eds., Plenum Press, New York and London, 1988, 27.

Branch, G.M., Barkai, A., Interspecific behaviour and its reciprocal interaction with evolution, population dynamics and community structure, in *NATO Advanced Workshop. Behavioural Adaptation to Intertidal Life,* Chelazzi, G., Vannini, M., Eds., Plenum Press, New York and London, 1988, 225.

Branch, G.M., Branch, M.L., Competition in *Bembicium auratum* (Gastropoda) and its effects on microalgal standing stock in margrove mud flats, *Oecologia (Berlin)*, 46, 106. 1980.

Branch, G.M., Branch, M.L., *The Living Shores of Southern Africa,* C. Struik, Ed., Capetown, 272 pp, 1981.

Branch, G.M., Harris, J.M., Perkins, C., Bustamante, R.H., Eckhout, S., Algal 'gardening' by grazers: a comparison of the ecological effects of territorial fish and limpets, in *Plant-Animal Interactions in the Marine Benthos* (Systematics Accociation Special Publication Vol. 46), John, D.M., Hawkins, S.J., Price, J.H., Eds., Clarendon Press, Oxford, 1992, 406.

Branch, G.M., Marsh, A.C., Tenacity of shell shape in six *Patella* species: adaptive features, *J. Exp. Mar. Biol. Ecol.,* 34, 111, 1978.

Branch, G.M., Pringle, A., The impact of the sand prawn *Callianassa kraussi* Stebbing on sediment turnover and on bacteria, meiofauna amd benthic macrofauna, *J. Exp. Mar. Biol. Ecol.*, 107, 219, 1987.

Brazeiro, A., Defeo, O., Macroinfauna zonation in microtidal sandy beaches: is it possible to identify patterns in such variable environments?, *Est. Coastal Shelf Sci.,* 42, 523, 1996.

Breen, P.A., Seasonal migration and population regulation in the limpet *Acmaea (collisella) digitalis*, *Veliger,* 15, 133, 1972.

Breen, P.A., Mann, K.H., Destructive grazing of kelp by sea urchins in eastern Canada, *J. Fish. Res. Bd. Canada*, 33, 1298, 1976.

Bregazzi, P.K., Naylor, E., The locomotor activity rhythm of *Talitrus saltator* (Montagu) (Crustacea, Amphipoda), *J. Exp. Mar. Biol. Ecol.*, 57, 375, 1972.

Bregnballe, F., Plaice and flounders as consumers of the microscope bottom fauna, *Medd. Dan. Fisk. Havunders,* 3, 133, 1961.

Brenchley, G., Disturbance and community structure: an experimental study of bioturbation in marine soft-bottom environments, *J. Mar. Res.*, 39, 767, 1981.

Briggs, I., *Assessment of the Current Trophic Status of the Upper Waitemata Harbour,* Upper Waitemata Harbour Catchment Study, Working Report No. 33, Auckland Regional Authority, Auckland, NZ, 65 pp, 1982.

Briggs, S.V., Estimates of biomass in a temperate mangrove community, *Aust. J. Ecol.,* 2, 369, 1977.

Brix, H., Lyngby, J.E., Uptake and translocation of phosphorus in eelgrass (*Zostera marina*), *Mar. Biol.*, 90, 111, 1985.

Brock, V., *Crassostrea gigas* (Thunberg) hepatopancreas: cellulase kinetics and cellulolysis of living monocellular algae with cellulose walls, *J. Exp. Mar. Biol. Ecol.*, 55, 47, 1989.

Broekhuysen, C.J., A preliminary investigation of the importance of desiccation, temperature and salinity as factors controlling the vertical distribution of certain gastropods in False Bay, South Africa, *Trans. Roy. Soc. S. Af.*, 28, 255, 1940.

Brooks, R., Brezonik, H., Putman, H., Keirn, M., Nitrogen fixation in an estuarine environment: the Wassasassa on the Florida coast, *Limnol. Oceanogr.*, 16, 701, 1971.

Broome, S.W., Woodhouse, W.W., Jr., Seneca, E.D., The relationship of mineral nutrients to growth of *Spartina alterniflora* in North Carolina. II. The effects of N, P and Fe fertilizers, *Proc. Am. Soil Sci. Soc.*, 39, 301, 1975.

Brosnan, D.M., Ecology of tropical rocky shores, plant-animal interactions in tropical and temperate latitudes, in *Plant-Animal Interactions in the Marine Benthos*, John, D., Hawkins, S.J., Price, J., Eds., Clarendon Press, Oxford, 1992, 101.

Browley, J.N., Roff, J.C., Trophic structure in southern Ontario streams, *Ecology*, 67, 1670, 1986.

Brown, A.C., Desiccation as a factor influencing the distribution of some South African Gastropoda from intertidal rocky shores, *Port. Acta Biol.*, 7, 11, 1960.

Brown, A.C., Food relationships on the intertidal sandy beaches of the Cape Peninsula, *S. Af. J. Sci.*, 60, 35, 1964.

Brown, A.C., The ecology of sandy beaches on the Cape Peninsula, South Africa. Part I. Introduction, *Trans. Roy. Soc. S. Af.*, 39, 247, 1971.

Brown, A.C., The ecology of sandy beaches of the Cape Peninsula, South Africa. Part 4. Observations on two intertidal isopoda, *Eurydice longicornis* (Studer) and *Exosphaeroma truncatitelson* Barnard, *Trans. Roy. Soc. S. Af.*, 40, 309, 1973.

Brown, A.C. The biology of the sand-beach whelk *Bullia digitalis* (Nassariidae), *Oceanogr. Mar. Biol. Annu. Rev.*, 20, 309, 1982.

Brown, A.C., The ecophysiology of sandy beaches — A partial review, in *Sandy Beaches as Ecosystems*, McLachlan, A., Erasmus, T., Eds., Dr. W. Junk Publishers, The Hague, 1983, 575.

Brown, A.C., Jarman, N., Coastal marine habitats, in *Biogeography and Ecology of South Africa*, Werger, M.J.A., Ed., Dr. W. Junk Publishers, The Hague, 1978, 1239.

Brown, A.C., McLachlan, A., *Ecology of Sandy Shores*, Elsevier, Amsterdam, 328 pp, 1990.

Brown, A.C., Stenton-Dozey, J.M.M.E., Trueman, E.R., Sandy beach bivalves and gastropods: a comparison between *Donax serra* and *Bullia digitalis*, *Adv. Mar. Biol.*, 25, 179, 1989.

Brown, C.L., The ecology of aufwuchs on *Zostera marina* in Charlstown Pond, Rhode Island, M.S. thesis, University of Rhode Island, Kingston, RI, 1962.

Brown, D.H., Gibby, C.E., Hickman, M., Photosynthetic rhythms in epipelic algal populations, *Brit. J. Phycol.*, 7, 37, 1972.

Brown, M.T., Green, P., Gonzales, A., Veregas, J., *Emergy Analysis Perspectives, Public Policy Options, and Development Guidelines for the Coastal Zone of Nayarit, Mexico,* Center for Wetlands and Water Resources, University of Florida, Gainesville, FL, Vol. 1, 215 pp., Vol. 2, 145 pp., 1992.

Brown, M.T., Nyman, M.AS., Keogh, J.A., Chin, N.K.M., Seasonal growth of the giant kelp *Macrocystic pyrifera* in New Zealand, *Mar. Biol.*, 129, 417, 1997.

Brown, M.T., Odum, H.T., Murphy, R.C., Christiansen, R.A., Doherty, S.T., McClannahan, T.R., Terrenbaum, S.E., Rediscovering the world: developing an interface of ecology and economics, in *Maximum Power. The Ideas and Applications of H.T. Odum,* Hall, C.A.S., Ed., University Press of Colorado, Niwat, CO, 1995.

Brown, M.T., Woithem, R.D., Montague, C.L., Odum, H.T., Odum, E.C., EMERGY Analysis Perspectives of the EXXON Valdez Oil Spill in Prince William Sound, Alaska, Report to the Cousteau Society, 1993.

Brown, R.A., Seed, R., *Modiolus modiolus* (L.) — an autecological study, in *Biology of Benthic Organisms,* Keegan, B.F., O'Ceidigh, P., Boaden, P.J.S., Eds., Pergamon Press, Oxford, 1977, 93.

Buchanan, J.B., Warwick, R.M., An estimate of benthic macrofaunal production in the offshore sand of the Northumberland coast, *J. Mar. Biol. Assoc. U.K.*, 54, 197, 1974.

Buesa, R.J., Population biomass and metabolic rates of marine angiosperms on the northwest Cuban shelf, *Aquatic Bot.*, 1, 11, 1975.

Bulthius, D.A., Woelkerling, W.J., Biomass accumulation and shading effects of epiphytes on the leaves of the seagrass *Heterozostera tasmanica*, in Western Port and Port Phillip Bay, Victoria, Australia, *Aquatic Bot.*, 16, 116, 1983.

Bulthuis, D.A., Woelkerling, W.J., Seasonal variation in standing crop and leaf growth rate of the seagrass, *Heterozostera tasmanica*, in Port Phillip Bay, Victoria, Australia, *Aquatic Bot.*, 16, 111, 1983.

Bunt, J.S., Studies of mangrove litter fall in tropical Australia, in *Mangrove Ecosystems in Australia: Structure, Function and Management*, Clough, B.F., Ed., Australian National University Press, Canberra, 1982, 223.

Bunt, J.S., Boto, K.G., Boto, G., A survey method for estimating potential level of mangrove forest primary production, *Mar. Biol.*, 52, 123, 1979.

Buresh, R.J., De Laune, R.D., Patrick, W.H., Jr., Nitrogen and phosphorus distribution and utilization by *Spartina alterniflora* in a Louisiana Gulf coast marsh, *Estuaries*, 3, 111, 1980.

Burke, R.D., Pheromonal control of metamorphosis in the Pacific sand dollar *Dendraster excentricus*, *Sci.*, 225, 442, 1984.

Burkill, P.H., Ciliates and other mocroplankton components in a nearshore food web: standing stocks and production processes, *Ann. Instit. Oceanogr.*, 58, 335.

Burkhill, P.H., Mantoura, R.F.C., Llewellyn, C.A., Owens, N.J.P., Microzooplankton grazing and selectivity of phytoplankton in coastal waters, *Mar. Biol.*, 93, 581, 1987.

Burkholder, P.R., Doheny, T.E., *The Biology of Eelgrass, with Special Reference to Hempstead and South Oyster Bays, Nassau County Long Island, New York,* Contribution No. 3, Department of Conservation and Waterways, Town Hempstead, L.I., N.Y., 120 pp, 1968.

Burrows, E., Conway, E., Lodge, S.M., Powell, H.T., The raising of intertidal algal zones on Fair Isle, *J. Ecol.*, 42, 283, 1954.

Bushek, D., Settlement as a major determinant of intertidal oyster and barnacle distributions along a horizontal gradient, *J. Exp. Mar. Biol. Ecol.*, 122, 1, 1988.

Buss, L.W., Competition and comunity organization on hard surfaces in the sea, in *Community Ecology*, Diamond, L.C., Ed., Harper & Row, New York, 1986, 577.

Bustamante, R.H., Branch, G.M., The dependence of intertidal consumers on kelp-derived organic matter on the west coast of South Africa, *J. Exp. Mar. Biol. Ecol.*, 196, 1, 1996.

Bustamante, R.H., Branch, G.M., Eckhout, S., Robertson, B., Zoutendyk, P., Schleyer, M., Dye, A., Hanekon, N., Keats, D., Jurd, M., McQuaid, C., Gradients of intertidal primary production around the coast of South Africa and their relationships with consumer biomass, *Oecologia (Berlin)*, 102, 189, 1995.

Butler, A.J., Populations and communities, in *Marine Biology*, Hammond, L.S., Synot, R.R., Eds., Longman Cheshire, Australia, 1994, 152.

Butman, C.A., Larval settlement of soft-sediment invertebrates: the spatial scales of pattern explained by active habitat selection and the emerging role of hydrodynamic processes, *Oceanogr. Mar. Biol. Annu. Rev.*, 25, 113, 1987.

Butman, C.A., Grassle, J.P., Webb, C.M., Substrate choices made by marine larvae in still water and in a flume flow, *Nature (London)*, 333, 771, 1980.

Cadée, G.C., Reappraisal of the production and import of organic carbon in the western Wadden Sea, *Neth. J. Sea Res.*, 14, 305, 1980.

Cadée, G.C., Has import of organic matter into the western part of the Dutch Wadden Sea increased during the last decades, *Neth. Instit. Sea Res. Publ. Series*, 10, 71, 1984.

Cadée, G.C., Recurrent and changing seasonal patterns in phytoplankton of the westernmost inlet of the Dutch Wadden Sea from 1959 to 1985, *Mar. Biol.*, 93, 281, 1986.

Cadée, G.C., Feeding traces and bioturbation by birds on a tidal flat, Dutch Wadden Sea, *Ichnos*, 1, 22, 1990.

Cadée, G.C., Hegeman, J., Primary production on the benthic microflora living on the tidal flats in the Dutch Wadden Sea, *Neth. J. Sea Res.*, 8, 260, 1974.

Cadée, G.C., Hegeman, J., Distribution of primary production of the benthic microflora and accumulation of organic matter on a tidal flat area, Bilzand, Dutch Wadden Sea, *Neth. J. Sea Res.*, 11, 24, 1977.

Caffey, H.M., No effect of naturally occurring rock types on settlement or survival in the intertidal barnacle *Tessorospora rosea* (Krauss), *J. Exp. Mar. Biol. Ecol.*, 63, 119, 1982.

Caffey, H.M., Spatial and temporal variation in settlement and recruitment of intertidal barnacles, *Ecol. Monogr.*, 55, 313, 1985.

Calow, P., The feeding strategies of two freshwater gastropods, *Ancylus fluviatilis* Mull. and *Planorbis contortus* Linn. (Pulmonata) in terms of ingestion rates and absorption efficiency, *Oecologia (Berlin)*, 20, 33, 1975.

Carlow, P., Fletcher, C.R., A new radiotracer involving ^{14}C and ^{51}Cr, for estimating the assimilation efficiencies of aquatic primary producers, *Oecologia (Berlin)*, 9, 155, 1972.

Cameron, R.A., Hinegarden, R., Inhibition of metamorphosis in laboratory cultured sea urchins, *Biol. Bull.*, 146, 335, 1974.

Camilleri, J.C., Leaf-litter processing by invertebrates in a mangrove forest in Queensland, *Mar. Biol.*, 114, 139, 1992.

Cammen, L.M., The significance of microbial carbon in the nutrition of the deposit feeding polychaete *Nereis succinea*, *Mar. Biol.*, 61, 9, 1980.

Cammen, L.M., The relationship between ingestion rate of deposit feeders and sediment nutritional value, in *Ecology of Marine Deposit Feeders*, Lopez, G., Taghorn, G., Levinton, J., Eds., Springer-Verlag, New York, 1989, 201.

Campbell, E.E., The estimation of phytomass and primary production of a surf zone, Ph.D. thesis, University of Port Elizabeth, South Africa, 429 pp, 1987.

Campbell, E.E., Bate, G.C., Factors influencing the magnitude of primary production of the surf zone phytoplankton, *Est. Coastal Shelf Sci.*, 24, 741, 1987.

Campbell, E.E., Bate, G.C., The estimation of annual primary production in a high energy surf-zone, *Bot. Marina*, 31, 337, 1988.

Campbell, E.E., Bate, C.C., Groundwater in the Alexandria tide-induced resuspension of sediment and its potential influence on the adjacent surf zone, *Water S. Af.*, 17, 155, 1991.

Capone.D.G., Nitrogen fixation (acetylene reduction) by rhizosphere sediments of the eelgrass *Zostera marina*, *Mar. Ecol. Prog. Ser.*, 10, 17, 1982.

Capone, D.G., N_2 fixation in seagrass communities, *Mar. Tech. Soc. J.*, 17, 32, 1983.

Capone, D.G., Penhale, P.A., Overland, R.S., Taylor, B.F., Relationships between productivity and $N_2(C\ _2H_2)$ fixation in *Thalassia testudinum* community, *Limnol. Oceanogr.*, 24, 117, 1979.

Capone, D.G., Taylor, B.F., N_2 fixation in the rhizosphere of *Thalassia testidinium*, *Can. J. Microbiol.*, 26, 998, 1980a.

Capone, D.G., Taylor, B.F., Microbial nitrogen cycling in a seagrass community, in *Estuarine Perspectives*, Kennedy, V.S., Ed., Academic Press, New York, 1980b, 153.

Capruilo, G., Carpenter, E., Grazing by 35–102 m microzooplanton in Long Island Sound, *Mar. Biol.*, 56, 319, 1980.

Caraco, N., Tamse, A., Boutris, O., Valiela, I., Nutrient limitation of phytoplankton growth in brackish coastal ponds, *Can. J. Fish. Aquatic Sci.*, 44, 473, 1987.

Carefoot, T., *Pacific Seashores: A Guide to Intertidal Life*, University of Washington Press, Vancouver, 208 pp, 1977.

Carlow, P., The feeding strategies of two freshwater gastropods, *Ancylus fluviatilis* Mull. and *Planorbis contortus* Linn, (Pulmonata) in terms of ingestion rates and absorption efficiency, *Oecologia*, 20, 33, 1975.

Carney, H.J., A general hypothesis for the strengthening of food web interactions in relation to trophic state, *Internationale Veringung fur theoretische und angewande Limnologie, Verhandlungen*, 24, 487, 1990.

Caron, D.A., Grazing of attached bacteria by heterotrophic microflagellates, *Microb. Ecol.*, 13, 203, 1987.

Caron, D.A., Goldman, J.C., Dennett, M.R., Experimental demonstration of the roles of bacteria and bacterivorous Protozoa in plankton nutrient cycles, *Hydrobiologia*, 159, 27, 1988.

Carpenter, E.J., Nitrogen fixation by *Oscillatoria (Trichdesmonium) theibautii* in the southwestern Sargasso Sea, *Deep-Sea Res.*, 20, 285, 1986.

Carpenter, E.J., Price, I.V., Nitrogen fixation, distribution and production of *Oscillatoria* (*Trichdesmium*) spp. in the Western Sargasso and Caribbean Sea, *Limnol. Oceanogr.*, 20, 389, 1977.

Carpenter, E.J., Van Raalte, C.D., Valiela, I., Nitrogen fixation by algae in a Massachusetts salt marsh, *Limnol. Oceanogr.*, 23, 318, 1978.

Carpenter, R.C., Partitioning herbivory and its effects on coral reef algal communities, *Ecol. Monogr.*, 56, 345, 1986.

Carpenter, S.R., Kitchell, J.K., The temporal scale of variance in limnetic primary production, *Am. Nat.*, 129, 417, 1987.

Carpenter, S.R., Kitchell, J.F., Consumer control of lake productivity, *Bioscience*, 38, 764, 1988.

Carpenter, S.R., Kitchell, J.F., Hodgson, J.R., Cascading trophic interactions and lake productivity, *Bioscience*, 35, 634, 1985.

Carricker, M.R., Ecology of estuarine benthic invertebrates: a perspective, in *Estuaries*, Lauff, G.H., Ed., American Association for the Advancement of Science, Washington, D.C., Publication No. 83, 1967, 442.

Cassie, R.M., Cassie, V., Primary productivity in a New Zealand west coast phytoplankton bloom, *N.Z. J. Sci.*, 3 , 173, 1968.

Castenholz, R.W., The effect of grazing on marine littoral diatom populations, *Ecology*, 42, 783. 1961.

Castenholz, R.W., An experimental study of the vertical distribution of littoral marine diatoms, *Limnol. Oceanogr.*, 8, 450, 1963.

Castilla, J.C., Sigue existiendo la necesidad de establecer porques y resenas maritimas en Chile, *Am. y Des.*, Vol. II, 53, 1986.

Castilla, J.C., Bustamante, R.H., Human exclusion from rocky intertidal of Las Cruces, central Chile: effects on *Durvillaea antarctica* (Phaeophyta: Durvilleales), *Mar. Ecol. Prog. Ser.*, 50, 203, 1989.

Castilla, J., Duran, R., Human exclusions from the intertidal zone of central Chile: the effects on *C. concholepus concholepus* (Gastropoda), *Oikos*, 45, 391, 1985.

Caswell, H., The evaluation of 'mixed' life histories in marine invertebrates and elsewhere, *Am. Nat.*, 117, 528, 1981.

Catterall, C.P., Poiner, I.R., The potential impact of human gathering on shellfish populations, with referance to some NE Australian intertidal flats, *Oikos*, 50, 114, 1987.

Cecchi, L.B., Cinelli, F., Canopy removal experiments in *Cystoseira*-dominated rockpools from the western coast in the Mediterranian (Ligurian Sea), *J. Exp. Mar. Biol. Ecol.*, 155, 69, 1982.

Chabot, R., Bourget, E., The influence of substratum heterogeneity and settled barnacle density on the settlement of cypris larvae, *Mar. Biol.*, 97, 45, 1988.

Chaloupka, M.Y., Hall, D.N., An examination of species dispersion patterns along the intertidal gradient on Macquarie Island (Sub-antarctic) using a restricted occupancy model, *J. Exp. Mar. Biol. Ecol.*, 84, 33, 1984.

Chambers, A.G., Soil dynamics and the productivity of *Spartina alterniflora*, in *Estuarine Comparisons*, Kennedy, V.S., Ed., Academic Press, New York, 1982, 231.

Chambers, A.G., Wiegert, R.G., Wolf, P.L., Carbon balance in a salt marsh: interactions of diffusive export, tidal deposition and rainfall-caused erosion, *Est. Coastal Shelf Sci.*, 21, 757, 1985.

Chambers, D.G., An analysis of the nekton communities in the upper Barataria Basin, Louisiana, M.S. thesis, Louisiana State University, Baton Rouge, LA, 286 pp, 1980.

Chandler, G.T., Fleeger, J.W., Meiofaunal colonization of azoic estuarine sediments in Louisiana: mechanisms of dispersal, *J. Exp. Mar. Biol. Ecol.*, 82, 15, 1983.

Chaplin, D., Some observations of predation on *Acmea* species by the crab *Pachygrapsus crassipes*, *Veliger*, 11, 67, 1968.

Chapman, A.R.O., A critique of prevailing attitudes towards the control of seaweed zonation, *Mar. Biol.*, 16, 80, 1973.

Chapman, A.R.O., Stability of sea urchin dominated barren grounds following destructive grazing of kelp in St. Margarets Bay, eastern Canada, *Mar. Biol.*, 62, 307, 1981.

Chapman, A.R.O., Reproduction, recruitment and mortality in two species of *Laminaria* in southwest Nova Scotia, *J. Exp. Mar. Biol. Ecol.*, 78, 307, 1984.

Chapman, A.R.O., Effects of grazing, canopy cover and substratum type on the abundance of seaweeds inhabiting littoral fringe tide pools, *Bot. Mar.*, 33, 319, 1990.

Chapman, A.R.O., Johnson, C.R., Disturbance and organization of macroalgal assemblages in the Northwest Atlantic, *Hydrobiologia*, 192, 77, 1990.

Chapman, G., The thixotrophy and dilatancy of a marine soil, *J. Mar. Biol. Assoc. U.K.*, 28, 123, 1949.

Chapman, M.G., Underwood, A.J., Foraging behavior of marine benthic grazers, in *Plant- Animal Interactions in the Marine Benthos* (Systematics Association Special Publication Vol. 46), John, D.M., Hawkins, S.J., Price, J.H., Eds., Clarendon Press, Oxford, 1992, 289.

Chapman, V. J., Mangrove phytosociology, *Trop. Ecol.*, 11, 1, 1970.

Chapman, V.J., *Salt Marshes and Salt Deserts of the World*, 2nd edition, J. Cramer, Vadiz, 392 pp, 1974.

Chapman, V.J., *Coastal Vegetation*, 2nd edition, Pergamon Press, Oxford, 292 pp, 1976a.

Chapman, V.J., *Mangrove Vegetation*, J. Cramer, Vadiz, 447 pp, 1976b.

Chapman, V.J., Ed., *Wet Coastal Ecosystems. Ecosystems of the World 1*, Elsevier, Amsterdam, 428 pp, 1977a.

Chareonpanich, C., Montani, S., Tsutsumi, H., Matsuko, S., Modification of chemical characteristics of organically enriched sediments by *Capitella* sp. I, *Mar. Poll. Bull.*, 26, 375, 1993.

Chavance, P., Yañez-Aranciba, A., Flores, D., Lora-Dominguez, A., Anezena, F., Ecology, biology and population dynamics of *Archosargus rhomboldalis (Pisces: Sparidae)* in a tropical lagoon system, southern Gulf of Mexico, *An. Inst. Ciene de Mer. y Limnol. Univ. Nat. Auten, Mexico*, 13-11-30, 1986.

Chelazzi, G., Behavioural adaptation of the gastropod *Nerita polita* L. on diferent shores at Aldabra Attol, *Proc. Roy. Soc., London, B*, 215, 451, 1982.

Chelazzi, G., Della Santina, P., Vannini, M., Long-lasting substrate marking in the collective homing of the gastropod *Nerita textilis*, *Biol. Bull. Mar. Biol. Lab., Woods Hole*, 168, 214, 1985.

Chelazzi, G., Focardi, S., Deneubourg, J.-L., Analysis of movement patterns and orientation mechanisms in intertidal chitons and gastropods, in *NATO Advanced Research Workshop on Behavioural Adaptation to Intertidal Life*, Chelazzi, G., Vannini, M., Eds., Plenum Press, New York and London, 1980, 175.

Chelazzi, G., Vannini, M., Space and time in the behavioural ecology of intertidal animals. *SITE Atti*, 5, 689, 1985.

Cherrill, A.J., James, R., Character displacement in *Hydrobia*, *Oecologia (Berlin)*, 71, 618, 1987.

Chesney, E.J., Tenore, K.R., Oscillations of laboratory populations of the polychaete *Capitella capitata* (type I): their cause and implications for natural populations, *Mar. Ecol. Prog. Ser.*, 20, 289, 1985a.

Chesney, E.J., Tenore, K.R., Effects of predation and disturbance on the population growth and population dynamics of the polychaete *Capitella capitata* (type I), *Mar. Biol.*, 85, 77, 1985b.

Chestnut, D., Feeding habits of juvenile spot *Leistomus xantharus* (Lacelede) in the North Inlet estuary, Georgetown, South Carolina, M.S. thesis, University of South Carolina, Columbia, 68 pp, 1983.

Chia, F.-S., Rice, M.E., *Settlement and Metamorphosis of Marine Invertebrate Larvae*, Elsevier North Holland Biomedical Press, New York, 290 pp, 1978.

Childers, D.L., McKellar, H.N., Jr., A simulation of saltmarsh water column dynamics, *Ecol. Modelling,* 36, 211, 1987.

Childers, D.l., McKellar, H.N., Dame, R.F., Sklar, F.H., Blood, E.R., A dynamic nutrient budget of subsysten interactions in a salt marsh estuary, *Est. Coastal Shelf Sci.*, 36, 105, 1993.

Choat, J.H., The influence of sessile organisms on the population biology of three species of acmaeid limpets, *J. Exp. Mar. Biol. Ecol.*, 26, 1, 1977.

Choat, J.H., Andrew, N.L., Interactions amongst species in a guild of subtidal benthic herbivores, *Oecologia (Berlin)*, 68, 387, 1986.

Choat, J.H., Ayling, A.M., The relationship between habitat structure and fish on New Zealand reefs, *J. Exp. Mar. Biol. Ecol.*, 110, 257, 1987.

Choat, J.H., Schiel, D.R., Patterns of distribution and abundance of large brown algae and invertebrate herbivores in subtidal regions of northern New Zealand, *J. Exp. Mar. Biol. Ecol.*, 60, 129, 1982.

Chow, V., Patterns of growth and energy allocation in northern California populations of *Littorina* (Gastropoda: Prosobranchia), *J. Exp. Mar. Biol. Ecol.*, 11, 69, 1987.

Christian, R.R., Bancroft, K., Wiebe, W.J., Distribution of microbial adenosine triphosphate in a salt marsh sediment at Sapelo Island, *Soil Science,* 119, 89-97, 1978.

Christian, R.R., Bancroft, K., Wiebe, W.J., Resistance of the microbial community within salt marsh soils to selected perturbations, *Ecology*, 59, 1200, 1978.

Christian, R.R., Wiebe, W.J., Aerobic microbial community metabolism in *Spartina alterniflora* soils, *Limnol. Oceanogr.*, 23, 238, 1978.

Christian, R.R., Wiebe, W.J., Three experimental regimes in the study of sediment microbial ecology, in *American Society for Testing Materials, STP 673*, 148, 1979.

Christensen, B., Biomass and primary production of *Rhizophore apiculata* Bl. in a mangrove in southern Thailand, *Aquatic Bot.*, 4, 43, 1978.

Christie, N.D., A numerical analysis of the distribution of a shallow sublittoral sand macrofauna along a transect at Lamberts Bay, South Africa, *Trans. Roy. Soc. S. Af.*, 42, 149, 1976.

Christie, N.D., Primary production in Langebaan Lagoon, in *Estuarine Ecology: with Particular Reference to South Africa,* Day, J.H., Ed., A.A. Balkema, Rotterdam, 1981, 101.

Claridge, P.N., Potter, I.C., Hardisty, M.W., Seasonal changes in movements, abundance, size composition and diversity of the fish fauna of the Severn Estuary, *J. Mar. Biol. Assoc. U.K.*, 66, 229, 1986.

Clarke, G.L., Dynamics of production in a marine area, *Ecol. Monogr.*, 16, 323, 1946.

Climberg, R.L., Mann, S., Straughan, D., A reinvestigation of southern California rocky intertidal beaches three and one-half years after the 1969 Santa Barbara oil spill: a preliminary report, in *Proc. Joint Conf. on the Prevention and Control of Oil Spills,* American Petroleum Institute, Washington, DC, 1973, 697.

Clokie, J.J.P., Boney, A.D., The assessment of changes in intertidal ecosystems following major reclamation work: framework for the interpretation of algal-dominated biota and the use and misuse of data, in *The Shore Environment, Vol. 2, Ecosystems,* Price, J.H., Irvine, D.E.G., Farnham, W.F., Eds., Academic Press, London, 1980, 609.

Clough, B.F., Ed., *Mangrove Ecosystems in Australia,* Australian National University Press, Canberra, 302 pp, 1982.

Clough, B.F., Andrews, T.J., Cowan, I.R., Physiological processes in mangroves, in *Mangrove Ecosystems in Australia,* Clough, B.F., Ed., Australian National University Press, Canberra, 1982, 193.

Clough, B.F., Attiwell, P.N., Nutrient cycling in a community of *Avicennia marina* in a temperate region of Australia, in *Proceedings of the First International Symposium on the Biology and Management of Mangroves*, 1975, 137.

Cockcroft, A.C., The role of macrofauna in nitrogen cycling in a high energy surf zone, Ph.D. thesis, University of Port Elizabeth, South Africa, 1988.

Cockroft, A.C., Mclachlan, A., Nitrogen budget for a high-energy ecosystem, *Mar. Ecol. Prog. Ser.*, 100, 287, 1993.

Cole, J.J., Cloern, J.E., Significance of biomass and light availability to phytoplankton productivity in San Fransisco Bay, *Mar. Ecol. Prog. Ser.,* 17, 15, 1988.

Cole, L.C., The population consequences of life history phenomena, *Quart. Rev. Biol.*, 29, 103, 1954.

Coles, S.M., Benthic microalgal populations on intertidal sediments and their role as precursors to salt marsh development, in *Ecological Processes in Coastal Environments*, Jeffries, R.L., Davey, A.J., Eds., Blackwell Scientific Publications, Oxford, 1979, 25.

Coljin, F., van Buurt, G., Influence of light and temperature on the photosynthesis of marine benthic diatoms, *Mar. Biol.*, 31, 209, 1975.

Colijn, F., Dijkema, R.S., Species composition of benthic diatoms and distribution of chlorophyll *a* on an intertidal flat in the Dutch Wadden Sea, *Mar. Ecol. Prog. Ser.,* 4, 9, 1981.

Coljin, F., Jonge, N. N. de, Primary production of microphytobenthos in the Ems-Dollard estuary, *Mar. Ecol. Progr. Ser.,* 14, 185, 1984.

Colman, J., The nature of the intertidal zonation of plants and animals, *J. Mar. Biol. Assoc. U.K.*, 24, 129, 1933.

Colman, J., On the faunas inhabiting intertidal seaweeds, *J. Mar. Biol. Assoc., U.K.*, 24, 129, 1940.

Commito, J.A., Importance of predation by infaunal polychaetes in controlling the structure of a soft-bottom community, *Mar. Biol.*, 68, 77, 1982.

Commito, J.A., Ambrose, W.G., Jr., Multiple trophic levels in soft-bottom communities, *Mar. Ecol. Prog. Ser.*, 26, 289, 1985a.

Commito, J.A., Ambrose, W.G., Jr., Predatory infauna and trophic complexity in soft bottom communities, in *Proceedings of the Nineteenth European Marine Biology Symposium*, Gibbs, P.E., Ed., Cambridge University Press, Cambridge, 1985b, 323.

Commito, J.A., Schrader, P.B., Benthic comunity response to experimental additions of the polychaete *Nereis virens*, *Mar. Biol.*, 86, 101, 1985.

Connell, J.H., Effects of competition, predation by *Thais lapillus*, and other factors on natural populations of *Balanus balanoides*, *Ecol. Monogr.*, 31, 61, 1961a.

Connell, J.H., The influence of interspecific competition and other factors on the distribution of the barnacle *Balanus balanoides*, *Ecol. Monogr.*, 31, 710, 1961b.

Connell, J.H., Territorial behaviour and dispersion in some marine invertebrates, *Res. Pop. Ecol.*, 28, 87, 1963.

Connell, J.H., A predator-prey system in the marine intertidal regions. I. *Balanus glandula* and several predatory species of *Thais*, *Ecol. Monogr.*, 40, 49, 1970.

Connell, J.H., Community interactions on marine rocky intertidal shores, *Annu. Rev. Ecol. Systematics*, 3, 169, 1972.

Connell, J.H., Some mechanisms producing structure in natural communities: a model and evidence from a field experiment, in *Ecology and Evolution of Communities*, Cody, M.C., Diamond, J.M., Eds., Harvard University Press, Cambridge, MA, 1975, 460.

Connell, J.H., Diversity in tropical rain forests and coral reefs, *Science*, 199, 1302, 1978.

Connell, J.H., Tropical rain forests and coral reefs as open nonequilibrium systems, in *Popoulation Dynamics*, Turner, B.D., Taylor, L.R., Eds., Blackwell Scientific Publications, Oxford, 1979, 141.

Connell, J.H., On the prevalemce and relative importance of interspecific competition: evidence from field experiments, *Am. Nat.*, 122, 661, 1983.

Connell, J.H., The consequences of variation in initial settlement vs. post-settlement mortality in rocky intertidal communities, *J. Exp. Mar. Biol. Ecol.*, 93, 11, 1985.

Connell, J.H., Variation and persistence in rocky shore populations, in *The Ecology of Rocky Coasts*, Moore, P.C., Seed, R., Eds., 1986, 54.

Connell, J.H., Keough, M.J., Disturbance and parch dynamics of subtidal marine animals on hard substrates, in *Natural Disturbance: An Evolutionary Perspective*, Pritchard, S.T.A., White, P.S., Eds., Academic Press, New York, 1985, 125.

Connell, J.H., Slayter, R.O., Mechanisms of succession in natural communities and their role in community stability and organization, *Am. Nat.*, 111, 1119, 1977.

Connell, J.H., Sousa, W.P., On the evidence needed to judge ecological stability or persistence, *Am. Nat.*, 121, 789, 1983.

Connolly, S.R., Roughgarden, J., Theory of marine communities: competition, predation, and recruitment-dependent interaction strength, *Ecol. Monogr.*, 69, 277, 1999.

Connor, M.S., Edgar, R.L., Selective grazing by the mud snail *Illyanassa obseleta*, *Oecologia*, 53, 271, 1982.

Connor, W.G., Simon, J.L., The effects of oyster shell dredging on an estuarine benthic community, *Est. Coastal Mar. Sci.*, 9, 749, 1979.

Conover, J.T., The ecology, seasonal periodicity, and distribution of benthic plants in some Texas lagoons, *Bot. Marina*, 8, 4, 1964.

Coon, S.L., Primary productivity of macroalgae in North Pacific America, in *Handbook of Biosolar Resources, Vol. 1, Part II*, Mitsui, A., Blank, C.C., Eds., CRC Press, Boca Raton, FL, 447, 1982.

Coon, S.L., Bonar, D.B., Weiner, R.M., Induction of settlement and metamorphosis of the Pacific oyster, *Crassostrea gigas* (Thunberg), by L-DOPA catecholamines, *J. Exp. Mar. Biol. Ecol.*, 94, 211, 1985.

Coon, S.L., Fitt, W.K., Bonar, D.B., Competence and delay of metamorphosis in the Pacific oyster *Crassostrea gigas*, *Mar. Biol.*, 196, 379. 1990a.

Coon, S.L., Walch, M., Fitt, W.K., Weiner, R.M., Bonar, D.B., Ammonia induces settlement behaviour in oyster larvae, *Biol. Bull.*, 179, 297, 1990b.

Correll, D.L., Estuarine productivity, *Bioscience*, 28, 646, 1978.

Correll, D.L., Ford, D., Comparison of precipitation and land runoff as sources of estuarine nitrogen, *Est. Coastal Shelf Sci.*, 15, 45, 1982.

Correll, D.L., Faust, M.A., Devera, D.J., Phosphorus flux and cycling in estuaries, in *Estuarine Research, Vol. 1*, Cronin, L.E., Ed., Academic Press, New York, 1975, 108.

Correll, D.L., Pierce, J.W., Faust, M.A., A quantitative study of the nutrient, sediment, and coliform bacterial constituents of water runoff from Rhode River watershed, in *Non Point Sources of Water Pollution*, Water Resources Research Center, Virginia Polytechnic Institute and State University, Blacksberg, 1976, 131.

Costanza, R., *Ecological Economics*, Columbia University Press, New York, 525 pp, 1991.

Costanza, R., Forher, S.C., Maxwell, J., The valuation and management of ecological systems, *Ecol. Econ.*, 1, 335, 1989.

Cotton, A.D., On the growth of *Ulva latissima* in excessive quantity with special reference to Belfast Louch, Botanical Report to the Royal Commission on Sewage Disposal, 7th Report, Appendix IV, 1911.

Couch, C.A., Carbon and nitrogen stable isotopes of meiobenthos and their food resources, *Est. Coastal Shelf Sci.*, 28, 253, 1989.

Coull, B.C., Estuarine meiofauna. A review: trophic relationships amd microbial interactions, in *Estuarine Microbial Ecology*, Stephenson, L.H., Colwell, R.R., Eds., Belle Barauch Coastal Research Institute, University of South Carolina, 1973, 499.

Coull, B.C., Ecology of meiofauna, in *Introduction to the Study of Meiofauna*, Higgens, R.J., Thiel, H., Eds., 1988.

Coull, B.C., Are members of the meiofauna food for higher trophic levels, *Trans. Am. Micros. Soc.*, 109, 233, 1990.

Coull, B.C., Role of meiofauna in estuarine soft-bottom communities, *Aust. J. Ecol.*, 24, 327, 1999.

Coull, B.C., Bell, S.S., Perspectives of marine meiofaunal ecology, in *Ecological Processes in Coastal and Marine Ecosystems*, Livingstone, R.J., Ed., Plenum Press, New York, 1979, 189.

Coull, B.C., Chandler, G.T., Pollution and meiofauna: field, laboratory and mesocosm studies, *Oceanogr. Mar., Biol. Annu. Rev.*, 30, 191, 1992.

Coull, B.C., Creed, E.L., Eskin, R.A., Montagna, P.A., Palmer, M.A., Wells, J.B., Phytal meiofauna from the rocky intertidal at Murella Inlet, South Carolina, *Trans. Am. Micros. Soc.*, 102, 380, 1983.

Coull, B.C., Wells, J.B.J., Refuges from fish predation: experiments with phytal meiofauna from the New Zealand rocky intertidal, *Ecology,* 64, 1599, 1983.

Cowen, R.K., The effect of the sheepshead (*Semicossyplus pulcher)* predation of the sea urchin (*Strongylocentrotus franciscanus*): an experimental analysis, *Oecologia (Berlin)*, 58, 249, 1983.

Craighead.F.C., *The Trees of South Florida*, University of Miami Press, Coral Gables, FL, 1971.

Cramer, N.F., May, R.M., Interspecific competition, predation and species diversity: a comment, *J. Theoret. Biol.*, 34, 289, 1972.

Crawford, C.C., Hobbie, J.E., Webb, K.L., The utilization of dissolved free amino acids by estuarine microorganisms, *Ecol.*, 55, 551, 1974.

Creese, R.G., Ecology and reproductive biology of intertidal limpets, Ph.D. thesis, University of Sydney, Sydney, Australia, 380 pp, 1982.

Creese, R.G., An analysis of distribution and abundance of populations of the high-shore limpet *Notoacmaea petterdi* (Tenison-Wood), *Oecologia (Berlin),* 45, 212, 1980.

Creese, R.G., Distribution and abundance of the acmaeid limpet *Patelloida latistrigata,* and its interaction with barnacles, *Oecologia (Berlin),* 52, 85, 1982.

Creese, R.G., Underwood, A.J., Analysis of inter- and intraspecific competition amongst intertidal limpets with different methods of feeding, *Oecologia (Berlin),* 53, 337, 1982.

Crisp, D.J., The behaviour of barnacle cyprids in relation to water movement over a surface, *J. Exp.Biol.,* 32, 569, 1955.

Crisp.D.J., Territorial behaviour in barnacle settlement, *J. Exp. Mar. Biol.Ecol.,* 38, 429, 1961.

Crisp, D.J., Factors influencing the settlement of marine invertebrate larvae, in *Chaemoreception in Marine Organisms*, Grant, P.T., Machie, A.M., Eds., Academic Press, London, 1974, 177.

Crisp, D.J., Energy flow measurements, in *Methods for the Study of Marine Benthos,* Hore, N.A., McIntyre, A.D., Eds., IBP Handbook 16, Blackwell, Oxford, 1984, 284.

Crisp, D.J., Barnes, H. The orientation and distribution of barnacles at settlement with special reference to surface contour, *J. Anim. Ecol.,* 23, 142, 1954.

Crisp, D.J., Patel, B., Environmental control of breeding by three Boreo-Arctic cirripedes, *Mar. Biol.,* 2, 283, 1969.

Crisp, D.J., Ritz, D.A., Responses of cirripede larvae to light. I. Experiments with white light, *Mar. Biol.,* 23, 327, 1973.

Croker, R.A., Niche diversity in five sympatric species of intertidal amphipods (Crustacea: Haustoriidae), *Ecol. Monogr.,* 37, 193, 1967.

Cronin, L.E., Daiber, J.C., Hulbert, E.M., Quantitative seasonal aspects of zooplankton in the Delaware River estuary, *Chesapeake Sci.,* 3, 63, 1962.

Crowder, L.B., Drenner, R.W., Kerfoot, W.C., McQueen, D.J., Mills, E.L., Sommer, U., Spencer, C.N., Vannini, M.J., Food web interactions in lakes, in *Complex Interactions in Lake Communities*, Carpenter, S.R., Ed., Springer-Verlag, New York, 1988, 141.

Cubit, J.D., Herbivory and the seasonal abundance of algae on a high intertidal rocky shore, *Ecology,* 65, 1984, 1975.

Cullen, D.J., Bioturbation of superficial marine sediments by interstitial marine animals, *Nature (London),* 242, 323, 1973.

Culley, M.R., Peck, L.S., The feeding preference of the ormer, *Haliotis tuberculata* L. *Kieler Meeresforsch,* 5, 570, 1981.

Culley, M.R., Sherman, K., The effect of substrate particle size on the production of mucus in *Haliotis tubereculata* L. and the importance of this in a culture system, *Aquaculture*, 47, 327, 1985.

Cushing, D.H., The seasonal variation in oceanic production as a problem of population dynamics, *J. Conseil Exp. Mer.,* 24, 255, 1959.

Cuzon du Rest, R., Distribution of zooplankton in the salt marshes of southwest Louisiana, *Publ. Inst. Mar. Sci. Univ. Texas*, 9, 132, 1963.

Dahl, E., On the smaller Arthropoda of marine algae, especially in the polyhaline waters off the Swedish west coast, *Undersokninger over Oresund,* 35, 1948.

Dahl, E., Some aspects of the ecology and zonation of the fauna of sandy beaches, *Oikos*, 4, 1, 1952.

Dalby, D.H., Monitoring and exposure scales, in *The Shore Environment, Vol. 1. Methods*, Price, J.H., Irvine, D.E.G., Farnham, W.F., Eds., Academic Press, New York, 1980, 117.

Dale, N.C., Bacteria in intertidal sediments. Factors related to their distribution, *Limnol. Oceanogr.,* 19, 509, 1974.

Dales, R.P., Survival of anaerobic periods by two intertidal polychaetes, *Arenicola marina* (L.) and *Owenia fusiformis* Delle Chiaje, *J. Mar. Biol. Assoc., U.K.*, 37, 521, 1958.

Daly, M.A., Mathieson, A.C., The effects of sand movement on intertidal seaweeds and selected invertebrates at Bound Rock, New Hampshire, U.S.A, *Mar. Biol.,* 43, 45, 1977.

Dame, R.F., The ecological energies of growth, respiration and assimilation in the intertidal American oyster *Crassostrea virginica*, *Mar. Biol.*, 17, 243, 1972.

Dame, R.F., Chyzanowski, T., Bildstein, K., Kjerfve, B., McKellar, H., Nelson, D., Spurrier, T., Slancyk, S., Stevenson, H., Vernberg, J., Zingmark, R., The outwelling hypothesis and North Inlet, South Carolina, *Mar. Ecol. Prog. Ser.*, 33, 217, 1986.

Dame, R.F., Dankers, N., Prins, T., Jongsana, H., Smaal, A., The influence of mussel beds on nutrients in the Western Wadden Sea and Eastern Scheldt Estuary, *Estuaries*, 14, 130, 1991a.

Dame, R.F., Spurrier, J.D., Williams, T.M., Kjerfve, B., Zingmark, R.G., Wolaver, T.G., Chyzanowski, T.H., McKellar, H.N., Vernberg, F.J., Annual material processing by a saltmarsh-estuarine basin in South Carolina, *Mar. Ecol. Prog. Ser.*, 72, 153, 1991.

Dame, R.F., Spurrier, J.D., Wolaver, T.G., Carbon, nitrogen and phosphorus processing by an oyster reef, *Mar. Ecol. Prog. Ser.*, 54, 249, 1986.

Dame, R.F., Vernberg, F., Bonnell, R., and Kitchens, W., The North Inlet marsh-estuarine ecosystem: a conceptual approach, *Helgol. Wiss. Meeresunters,* 30, 343, 1977.

Dame, R. F., Wolaver, T. G., Williams, T. M., Spurrier, J. D., Miller, A. B., The Bly Creek ecosystem study: phosphorous transport within a euryhaline saltmarsh basin, North Inlet, South Carolina, *Neth. J. Sea. Res.,* 27, 73, 1991b.

Dankers, N., Binsbergen, M., Zegers, K., Laane, R., Van der Loeff, M., Transportation of water, POM, DOM, and inorganic matter between a saltmarsh and the Ems-Dollard estuary, *Est. Coastal Shelf Sci.,* 19, 143, 1984.

Dare, P.J., Settlement, growth and production of the mussel *Mytilus edulis* L., in Morecombe Bay, England, *Fish. Invest., London II*, 29, 1, 1976.

Darley, W.M., Dunn, E.L., Holmes, K.S., Larew, H.G., III., A ^{14}C method for measuring primary microalgal productivity in air, *J. Exp. Mar. Biol. Ecol.*, 25, 207, 1976.

Darley, W.M., Montague, C.L., Plumley, F.G., Sage, W.W., Psalidas, A.T., Factors limiting edaphic algal biomass and productivity in a Georgia salt marsh, *J. Phycology*, 17, 122, 1981.

Darnell, R.M., The organic detritus problem, in *Estuaries,* Lauff, G.H., Ed., American Association for the Advancement of Science, Washington, DC, Publication No. 83, 1967a, 374.

Darnell, R.M., Organic detritus in relation to the estuarine ecosystem, in *Estuaries*, Lauff, G.H., American Association for the Advancement of Science, Washington, DC, Publication No. 83, 1967b, 376.

Dauer, D.M, Ewing, R.M., Tourellotte, G.H., Harlan, W.T., Soubeer, J.W., Barker, H.R., Predation, resource limitation and the structure of benthic infaunal communities of the lower Chesapeake Bay, *Int. Rev. Gesamten Hydrobiol.*, 67, 477, 1982.

Dauer, D.M., Maybury, C.A., Ewing, R.M., Feeding behaviour and general ecology of several spionid polychaetes from Chesapeake Bay, *J. Exp. Mar. Biol. Ecol.*, 54, 21, 1981.

Dauvin, J.-C., Impact of Amico Cadiz oil spill on the muddy fine sand *Abra alba* and *Mellina palmata* community, *Est. Coastal Shelf Sci.*, 14, 517, 1982.

Davidson, K., Modelling microbial foodwebs, *Mar. Ecol. Prog. Ser.*, 145, 279, 1996.

Davidson, P.E., A study of the oystercatcher (*H. ostrelagus)* in relation to the fishery for cockles (*Cardium edule*) in the Burrey Inlet, South Wales, *Ministry of Agriculture, Fisheries and Food, Fisheries Investigations,* Ser. II. 25, 1967, 1.

Davies, J.L., 1973, *Geographical Variation in Coastal Development,* Hafner, New York, 1973.

Davies, J.M., Energy flow through the benthos of a Scottish loch, *Mar. Biol.*, 31 353, 1975.

Davies, M.S., Hawkins, S.J., Jones, H.D., Mucus production and physiological energetics in *Patella vulgata* L., *J. Moll. Stud.*, 56, 499, 1990.

Davies, M.S., Hawkins, S.J., Jones, H.D., Seasonal variation in the composition of pedal mucus from *Patella vulgata* L., *J. Exp. Mar. Biol. Ecol.,* 114, 101, 1991.

Davis, M.W., Lee, H., II, Recolnization of sediment-associated microalgae and effects of estuarine infauna on macroalgal production, *Mar. Ecol. Prog. Ser.,* 11, 227, 1983.

Davis, M.W., McIntire, C.D., Effects of physical gradients on the production dynamics of sediment-associated algae, *Mar. Ecol. Prog. Ser.,* 13, 103, 1983.

Dawes, C.J., Moon, R.E., Davis, M.A., The photosynthetic and respiratory rates and tolerances of benthic algae from a mangrove and salt marsh estuary: a comparative study, *Est.Coastal Mar. Sci.*, 26, 175, 1976.

Day, J.H., The ecology of South African estuaries. Part I. General considerations, *Trans. Roy. Soc. S. Af.*, 35, 475, 1951.

Day, J.H., What is an estuary?, *S. Af. J. Sci.*, 76, 198, 1980.

Day, J.H., Ed., *Estuarine Ecology, with Special Reference to South Africa,* A.A. Balkema, Rotterdam, 411 pp, 1981a.

Day, J.H., Summaries of current knowledge of 43 estuaries in southern Africa, in *Estuarine Ecology, with Particular Reference to Southern Africa*, Day, J.H., Ed., A.A. Balkema, Rotterdam, 1981b, 251.

Day, J.H., The nature, origin and classification of estuaries, in *Estuarine Ecology, with Particular Reference to South Africa*, Day, J.H., Ed., A.A. Balkema, Rotterdam, 1981c, 77.

Day, J.H., The estuarine flora, in *Estuarine Ecology, with Special Reference to South Africa*, Day, J.H., Ed., A.A. Balkema, Rotterdam, 1981d, 77.

Day, J.H., Blaber, S.J., Wallace, J.H., Estuarine fishes, in *Estuarine Ecology: with Special Reference to South Africa*, Day, J.H., Ed., Balkema, Cape Town, 1981e, 198.

Day, J.H., Wilson, D.P., On the relation of the substratum to the metamorphosis of *Scolecolepis fuliginosa* (Claparade), *J. Mar. Biol. Assoc. U.K.*, 19, 695, 1934.

Day, J.W., Jr., Hall, C.A.S., Kemp, M.W., Yanez-Aranciba, A., *Estuarine Ecology,* John Wiley & Sons, New York, 558 pp, 1987.

Day, J.W., Smith, P., Wagner, P., Stowe, W., *Community Structure and Carbon Budget of a Salt Marsh and Shallow Bay Estuarine System in Lousiana,* Louisiana State University, Center for Wetland Resources, Sea Grant Publication No. LSU-SG-72-04, 79 pp, 1973.

Dayton, P.K., Competition, disturbance, and community organization: the provision and subsequent utilization of space in a rocky intertidal community, *Ecol. Monogr.*, 41, 351, 1971.

Dayton, P.K., Experimental evolution of ecological dominance in a rocky intertidal algal community, *Ecol. Monogr.*, 45, 137, 1975.

Dayton, P.K., Processes structuring some marine communities: are they general?, in *Ecological Communities: Conceptual Issues and the Evidence*, Strong, D.R., Jr., Simberloff, D., Abele, L.G., Coull, B.C., Eds., Princetown University Press, Princetown, 1984, 181.

Dayton, P.K., The structure and regulation of some South African kelp communities, *Ecol. Monogr.*, 55, 447, 1985a.

Dayton, P.K., Ecology of kelp communities, *Annu. Rev. Ecol. Systematics,* 16, 215, 1985b.

Dayton, P.K., Currie, T., Gerodette, B.D., Keller, R. Rosenthal, R., Van Tresca, D., Patch dynamics and stability of some California kelp communities, *Ecol. Monogr.*, 54, 253, 1984.

Dayton, P.K., Oliver, J.S., An evaluation of experimental analysis of population and community patterns in benthic marine environments, in *Marine Benthic Dynamics*, Tenore, K.R., Coull, B.C., Eds., University of South Carolina Press, Columbia, SC, 1979, 93.

Dayton, P.K., Rosenthal, R.J., Mahon, L.C., Antenzana, T., Population structure and foraging biology of the predaceous Chilean asteroid *Meyenasterias gelatinosus* and the escape biology of its prey, *Mar. Biol.*, 39, 361, 1977.

Dayton, P.K., Tegner, M.J., Parnell, P.E., Edwards, P.B., Temporal and spatial patterns of disturbance and recovery in a kelp forest community, *Ecol. Monogr.*, 62, 421, 1992.

Dean, R.L., Connell, J.H., Marine invertebrates in an algal succession. I. Mechanisms linking habitat complexity with diversity, *J. Exp. Mar. Biol. Ecol.,* 109, 195, 1987a.

Dean, R.L., Connell, J.H., Marine invertebrates in an algal succession. II. Tests of hypotheses to explain changes in diversity with succession, *J. Exp. Mar. Biol. Ecol.*, 109, 217, 1987b.

Dean, R.L., Connell, J.H., Marine invertebrates in algal succession. III. Mechanisms linking habitat complexity with diversity, *J. Exp. Mar. Biol. Ecol.*, 109, 249, 1987c.

Dean, T.A., Schroeter, S.C., Dixon, J.D., Effects of grazing by two species of sea urchin (*Strongylocentrotus franciscanus* and *Lytechinus anamesus*) on recruitement and survival of two species of kelp (*Macrocystis pyrifera* and *Pterygophora californica*), *Mar. Biol.*, 78, 301, 1984.

De Angelis, D.L., Waterhouse, J.C., Experiments on factors influencing settlement, survival and growth of two species of barnacles in New South Wales., *J. Exp. Mar. Biol. Ecol.*, 36, 269, 1987.

Deason, E.E., Smayada, T.J., Ctenophore-zooplankton-phytoplankton interactions in Narrangansett Bay, Rhode Island, USA, during 1972–1977, *J. Plankton Res.*, 4, 203, 1982.

Deegan, L.A., Garritt, R.H., Evidence for spatial variability in estuarine food webs, *Mar. Ecol. Prog. Ser.*, 147, 31, 1997.

De Jonge, D.J., De Jonge, V.N., Dynamics and distribution of microphytobenthic chlorophyll-*a* in the western Scheldt estuary, (SW Netherlands), *Hydrobiologia*, 311, 21.

De Jonge, V.N., The abiotic environment, in *Tidal Flat Estuaries: Simulation and Analysis of the Ems Estuary*, Baretta, J., Rudrdij, P., Eds., Springer-Verlag, Berlin, 1988, 14.

De Jonge, V.N., von Beuseben, T.E.E., Wind- and tide-induced resuspension of sediment and microphytobenthos from tidal flats in the Ems estuary, *Limnol. Oceanogr.*, 40, 766–778, 1992.

de la Cruz, A.A., Banaag, J.F., The ecology of a small mangrove patch in Matabungbay Beach, Balongus Province, *Nat. Appl. Sci. Bull.*, 20, 486, 1967.

de la Cruz, A.A., Hackney, C.T., Energy value, elemental composition, and productivity of belowground biomass of a *Juncus* tidal marsh, *Ecology*, 58, 1165, 1977.

DeLaune, R.D., Patrick, W.H., Jr., Brannon, J.M., *Nutrient Transformations in Louisiana Salt Marsh Soils*, Sea Grant Publ. No. LSU-T-76-009, Center for Wetland Resources. Louisiana State University, U.S.A., 38 pp, 1976.

DeLaune, R.D., Smith, C.J., Patrick, W.H., Jr., Sikora, W.B., Sikora, J.P., Hambrick, G.A., III, The use of chemical dispersants as a method of salt marsh restoration following an oil spill, *Report Axon Research & Engineering*, Center for Wetland Resources, Baton Rouge, LA, 64 pp, 1981.

Delepine, R., Hureau, J.-C., Comparative study of the vertical distribution of marine vegetation of Archipel de Kerguelen and Isle Crozet, in *Symposium on Anarctic Oceanography*, Scott Polar Research Institute, Cambridge, 1968, 164.

den Hartog, C., The epilithic algal communities occurring along the coast of The Netherlands, *Wentia*, 1, 1, 1959.

den Hartog, C., The structural aspect in the ecology of seagrass communities, *Helgol. Wiss. Meeresunters*, 15, 648, 1967.

den Hartog, C., Structure, function and classification in seagrass communities, in *Seagrass Ecosystems: A Scientific Perspective*, McRoy, C.P., Helfferich, C., Eds., Marcel Dekker, New York, 1977, 89.

de Niro, M., Epstein, S., Influence of diet on the distribution of carbon isotopes in animals, *Geochemica et Cosmochemica Acta*, 42, 485, 1978.

Denley, E.J., Dayton, P.K., Competition among macroalgae, in *Handbook of Phycological Methods. 4. Ecological Field Methods: Macroalgae*, Littler, M.M., Littler, D.S., Eds., Cambridge University Press, Cambridge, 1985, 511.

Denley, E.J., Underwood, A.J., Experiments on factors influencing settlement, survival and growth in two species of barnacles in New South Wales, *J. Exp. Mar. Biol. Ecol.*, 36, 269, 1979.

Denny, M.W., *Biology and Mechanics of Wave-Swept Shores*, Princetown University Press, Princetown, 352 pp, 1988.

Denny, M.W., A limpet shell shape that reduces drag. Laboratory demonstration of a hydrodynamic mechanism and an exploration of its effectiveness in nature, *Can. J. Zool.*, 67, 2098, 1989.

Denny, M.W., Biology, natural selection and prediction of maximal wave induced forces, *S. Af. J. Mar. Sci.*, 10, 353, 1991.

Denny, M.W., Disturbance, natural selection and the prediction of maximal wave induced forces, *Cont. Math.*, 141, 65, 1993.

Denny, M.W., Predicting physical disturbance: mechanistic approaches to the study of survivorship on wave-exposed shores, *Ecol. Monogr.*, 65, 371, 1995.

Denny, M.W., Daniel, T.D., Koehl, M.A.R., Mechanical limits to size in wave-swept organisms, *Ecol. Monogr.*, 55, 65, 1985.

Denny, M.W., Gaines, S.D., On the production of maximal wave forces, *Limnol. Oceanogr.*, 35, 1, 1990.

Denny, M.W., Shibata, M.F., Consequences of surf zone turbulence for settlement and external fertilization., *Am. Nat.*, 134, 859, 1989.

De Silva, P.H.D.H., Studies on the biology of Spirorbinae (Polychaeta), *J. Zool. Lond.*, 152, 169, 1967.

de Sylva, D.P., Nektonic food webs in estuaries, in *Estuarine Research, Vol. 1*, Cronin, L.E., Ed., Academic Press, New York, 1975, 420.

Dethier, M.N., Tidepools as refuges: predation and the limits of the harpacticoid copepod *Tigriopus californicus* (Baker), *J. Exp. Mar. Biol. Ecol.*, 42, 99, 1980.

Dethier, M., Heteromorphic algal life histories: the seasonal pattern and response to herbivory of the brown crust, *Ralfsia californica*, *Oecologia (Berlin)*, 49, 333, 1981.

Dethier, M.N., Pattern and process in tidepool algae: factors influencing seasonality and distribution, *Bot. Marina*, 25, 55, 1982.

Dethier, M., Disturbance and recovery in intertidal pools: maintenance of mozaic patterns, *Ecol. Monogr.*, 54, 99, 1984.

De Villiers, A.F., Littoral ecology of Marion and Prince Edward Islands, *S. Af. J. Ant. Res., Suppl.*, 1, 40 pp, 1976.

Devinny, J.S., Volse, L.A., Effects of sediment on the development of *Macrocystis pyrifera* gametophytes, *Mar. Biol.*, 48, 343, 1978.

De Wilde, P.A., Beukema, J.J., The role of zoobenthos in the consumption of organic matter in the Dutch Wadden Sea, *Neth. J. Sea Res.*, 10, 145. 1984.

De Witt, T.R., Levinton, J.S., Disturbance, emigration and refugia: how the mud snail *Iltanassa obsoleta* (Say) affects the habitat distribution of an epifaunal amphipod, *Microdeutopus gryllotabta* (Costa), *J. Exp. Mar. Biol. Ecol.*, 92, 97, 1985.

De Wneede, R.E., The phenology of three species of *Sargassum* (Sargassaceae: Phaeophyta) in Hawaii, *Phycologia*, 15, 175, 1976.

De Wolf, P., Ecological observations on the mechanisms of dispersal of barnacle larvae during planktonic life and settling, *Neth. J. Sea Res.*, 6, 1, 1973.

Dexter, D.M., Temporal and spatial variability in the community structues of four sandy beaches in south-eastern New South Wales, *Aust. J. Mar. Freshw. Res.*, 35, 663, 1984.

Diamond, J.M., Laboratory, field and natural experiments, *Nature (London)*, 304, 586, 1983.

Diamond, J., Case, T.J., *Community Ecology*, Harper and Row, New York, 1986.

Diechmann, G.S., Aspects of the ecology of *Laminaria pallida* (Grev.) off Cape Peninsula (South Africa). I. Seasonal growth, *Bot. Marina*, 23, 579, 1980.

Dijkema, K., Geography of the salt marshes in Europe, *Z. Geomorph. N.F.*, 31, 489, 1987.

Dillon, R.C., A comparative study of the primary productivity of estuarine phytoplankton and macrobenthic plants, Ph.D. dissertation, Department of Botany, University of North Carolina, Chapel Hill, North Carolina, 112 pp, 1971.

Dobbs, F.C., Whitlach, R.B., Aspects of deposit-feeding by the polychaete *Clymenella torquata*, *Ophelia*, 21, 159, 1982.

Dodge, C.W., Ecology and geographic distribution of Antarctic lichens, in *Biologie Antarctique*, Carrick, R., Holdgate, M., Prevost, J., Eds., Herman, Paris, 1962, 165.

Doering, P.H., On the contribution of the benthos to pelagic production, *J. Mar. Res.*, 47, 371, 1989.

Doering, P.H., Phillips, D.W., Maintenance of shore-level size gradient in the marine snail *Tegula funebralis* (A.Adams): importance of behavioural responses to light and seastar predators, *J. Exp. Mar. Biol. Ecol.*, 67, 159, 1983.

Doty, M.S., Critical tide levels that are correlated with the vertical distribution of marine algae and other organisms along the Pacific coast, *Ecology*, 27, 315, 1946.

Doty, M.S., Archer, J., An experimental test of the tide factor hypothesis, *Am. J. Botany*, 37, 458, 1950.

Dring, M.J., Brown, F.A., Photosynthesis of intertidal brown algae during and after periods of emersion: a renewed search for the physiological causes of zonation, *Mar. Ecol. Prog. Ser.*, 6, 301, 1982.

Dring, M.J., Jewson, D.H., What does the ^{14}C-uptake by phytoplankton really measure? A fresh approach using a theoretical modal, *Brit. Phycol. J.*, 14, 122, 1979.

Dromgoole, F.I., Desiccation resistance of intertidal and subtidal algae, *Bot. Marina*, 23, 281, 1980.

Druehl, L.D., The pattern of Laminariales distribution in the northeast Pacific, *Phycologia*, 9, 237, 1970.

Druehl, L.D., The distribution of *Macrocystis integrifolia* in British Columbia as related to environmental parameters, *Can. J. Bot.*, 56, 69, 1978.

Druehl, L.D., Green, J.M., Vertical distribution of seaweeds as related to patterns of submersion and emersion, *Mar. Ecol. Prog. Ser.*, 9, 163, 1982.

Druehl, L.D., Hsiao, S.I.C., Intertidal kelp response to environmental changes in a British Columbia inlet, *J. Fish. Res. Bd. Can.*, 34, 1207, 1977.

Ducklow, H.W., Purdie, D.A., Williams, P.J.L., Davies, J.M., Bacterioplankton: a sink for carbon in a coastal marine plankton comunity, *Science*, 232, 865, 1986.

Duggins, D.O., Kelp beds and sea otters: an experimental approach, *Ecology*, 61, 447, 1980.

Duggens, D.O., Starfish predation and the creation of mozaic patterns in a kelp-dominated community, *Ecology*, 64, 1610, 1983.

Duggins, D.O., Dethier, M.N., Experimental studies of herbivory and algal competition in a low intertidal habitat, *Oecologia (Berlin)*, 67, 183, 1985.

Duggins, D.O., Eckman, J.E., Is kelp detritus a good food for suspension feeders? Effect of kelp species, age and secondary metabolites, *Mar. Biol.*, 128, 489, 1997.

Duggens, D.O., Smenstad, C.A., Estes, J.A., Magnification of secondary production by kelp detritus in coastal marine ecosystems, *Science*, 245, 170, 1989.

Dungan, M.L., Three-way interactions: barnacles, limpets, and algae in a Sonoran desert rocky intertidal, *Am. Nat.*, 127, 292, 1986.

Dunkin, S. de B., Hughes, R.N., Behavioural components of prey selection by dogwhelks, *Nucella lapillus* (L.) feeding on the barnacle, *S. balanoides* (L.) in the laboratory, *J. Exp. Mar. Biol. Ecol.*, 79, 91, 1984.

Du Preez, H.H., McLachlan, A., Marais, J.F.K., Bioenergetics of fishes in a high energy surf-zone, *Mar. Biol.*, 106, 1, 1990.

Duran, L.R., Castilla, J.C., Variation and persistance of the middle rocky intertidal community of central Chile, with and without human harvesting, *Mar. Biol.*, 103, 553, 1989.

Dye, A.H., An ecophysiological study of the benthic meiofauna of the Swarthops estuary, M.Sc. thesis, University of Port Elizabeth, South Africa, 1977.

Dye, A.H., Measurement of biological oxygen demand in sandy beaches, *S. Af. J. Zool.*, 14, 55, 1979.

Dye, A.H., Tidal fluctuations in biological oxygen demand on exposed sandy shores, *Est. Coastal Mar.* Sci., 11, 1, 1980.

Dye, A.H., A study of benthic oxygen consumption on exposed sandy beaches, *Est. Coastal Shelf Sci.*, 13, 671, 1981.

Dye, A.H., Composition and seasonal fluctuations in a Southern African mangrove estuary, *Mar. Biol.*, 73, 165, 1983a.

Dye, A.H., A synopsis of community respiration on exposed sandy beaches, in *Sandy Beaches as Ecosystems*, McLachlan, A., Erasmus, T., Eds., Dr. W. Junk Publishers, The Hague, 1983b, 693.

Dye, A.H., Erasmus, T., Fustenberg, J.P., An ecophysiological study of the benthic meiofauna of the Swartkops Estuary. III. Partition of oxygen consumption and the relative importance of the meiofauna, *Zool. Africana*, 13, 187, 1977.

Dye, A.H., Furstenberg, J.P., Estuarine meiofauna, in *Estuarine Ecology with Particular Referance to Southern Africa*, Day, J.H., Eds., A.A. Balkema, Rotterdam, 1981, 179.

Dyer, K.R., *Estuaries: A Physical Introduction*, John Wiley & Sons, New York and London, 140 pp, 1973.

Eagle, R.A., Natural fluctuation in a soft bottom benthic community, *J. Mar. Biol. Assoc., U.K.*, 55, 865, 1975.

Ebling, F.J., Kitching, J.A., Muntz, L., Taylor, M.C., Ecology of Louch Ine. XIII. Experimental observations of the destruction of *Mytilus edulis* and *Nucella lapillus* by crabs, *J. Anim. Ecol.*, 33, 73, 1964.

Eckman, J.E., Flow disruption by an animal-tube mimic affect sediment bacterial colonization, *J. Mar. Res.*, 43, 419, 1985.

Eckman, J.E., Nowell, A.R., Jumars, P.A., Sediment destabilization by animal tubes, *J. Mar. Res.*, 39, 361, 1981.

Edelstein, T., McLachlan, J., Autecology of *Fucus distichus* spp. *disthicus* (Phaeophyceae: Fucales) in Nova Scotia, Canada, *J. Mar. Biol.*, 30, 305, 1975.

Edgar, G.J., The ecology of south-east Tasmanian phytal communities. I. Spatial organization on a local scale, *J. Exp. Mar. Biol. Ecol.*, 70, 129, 1983a.

Edgar, G.J., The ecology of south-east Tasmanian phytal communities. II. Seasonal changes in plant and animal communities, *J. Exp. Mar. Biol. Ecol.*, 70, 159, 1983b.

Edgar, G.J., The ecology of south-east Tasmanian phytal communities. III. Patterns of species diversity, *J. Exp. Mar. Biol. Ecol.*, 70, 181, 1983c.

Edgar, G.J., The ecology of south-east Tasmanian phytal communities. IV. Factors affecting the distribution of amphipoid amphipods among algae, *J. Exp. Mar. Biol. Ecol.*, 70, 205, 1983d.

Edney, E.B., The body temperature of woodlice, *J. Exp. Biol.*, 28, 271, 1951.

Edwards, D.C., Conover, D.O., Sutter, F., III., Mobile predators and the structure of marine intertidal communities, *Ecology*, 63, 1175, 1982.

Edwards, J., Naylor, E., Endogenous circadian changes and orientational behaviour of *Talitrus saltator*, *J. Mar. Biol. Ass., U.K.*, 67, 17, 1987.

Edwards, P., An investigation of the vertical distribution of selected benthic marine algae with a tide-simulating apparatus, *J. Phycol.*, 13, 62, 1972.

Edwards, S.F., Welsh, B.L., Trophic dynamics of a mud snail [*Illyanassa obsoleta* (Say)] population on an intertidal mudflat, *Est. Coastal Mar. Sci.*, 14, 663, 1982.

Edzwald, J.K., Upchurch, J.B., O'Melia, C.R., Coagulation in estuaries, *Envt. Sci. Tecj.*, 8, 58, 1974.

Eilers, H.P., Production ecology of an Oregon coastal salt marsh, *Est. Coastal Mar. Sci.*, 8, 399, 1979.

Ellis, M.J., Coull, B.C., Fish predation on meiobenthos: field experiments with juvenile spot *Leiostomus xanthuris* Lacepede, *J. Exp. Mar. Biol. Ecol.*, 130, 19, 1989.

Elmgren, R., Time energy and risk in two species of carnivorous gastropods, Ph.D. dissertation, University of Washington, Seattle, 128 pp, 1978.

Elmgren, R., Structure and dynamics of Baltic benthos communities, with particular reference to the relationships between macro- and meiobenthos, *Kieler Meeresforschungen*, 1, 1, 1978.

Elmgren, R., Trophic dynamics in the enclosed Baltic Sea, *Rapport et proces-verbeaux des reunions Conseil Permenent Internationale pour l'exploration de la Mer*, 183, 152, 1984.

Elner, R.W., Hughes, R.N., Energy maximization in the diet of the shore crab, *Carcinus maenas*, *J. Anim. Ecol.*, 47, 103, 1978.

Elner, R.W., Vadas, R.L., Sr., Interference in ecology, the sea urchin phenomenon in north-west Atlantic, *Am. Nat.*, 136, 108, 1990.

Emerson, C.W., Wind stress limitation of benthic secondary production in shallow, soft-sediment communities, *Mar. Ecol. Prog. Ser.*, 53, 65, 1989.

Emerson, C.W., Grant, , The control of soft-shell clam (*Mya arenaria*) recruitment on intertidal sand-flats by bedload sediment transport, *Limnol. Oceangr.*, 36, 1280, 1991.

Emlen, J.M., Time energy and risk in two species of carnivorous gastropods, Ph.D. dissertation, University of Washington, Seattle, 128 pp, 1966.

Emmerson, W.R., McGwynne, L.E., Feeding and assimilation of mangrove leaves by the crabs *Sesarma meinerti* de Man in relation to leaf-litter production in Mgazana, a warm temperate southern Africa mangrove swamp, *J. Exp. Mar. Biol. Ecol.*, 157, 41, 1992.

Eppley, R.W., Temperature and phytoplankton growth in the sea, *Fish. Bull. Nat. Oceanic Atmos. Admin.*, 70, 1063, 1972.

Eppley, R.W., Estimating phytoplankton growth in oligitrophic oceans, in *Primary Productivity in the Sea*, Falowski, P.K., Ed., Plenum Press, New York, 1980, 231.

Epstein, S.S., Burkovsky, I.V., Shiaris, M.P., Ciliate grazing on bacteria, flagellates and microalgae in a temperate zone sandy tidal flat: ingestion rates and food niche partitioning, *J. Exp. Mar. Biol. Ecol.*, 165, 103, 1992.

Epstein, S.S., Shiris, M.P., Size-selective grazing of coastal bacterioplankton by natural assemblages of pigmented flagellates, colourless flagellates, and ciliates, *Microb. Ecol.*, 13, 211, 1992.

Eriksson, S., Sellei, C., Wallstrom, K., Structure of the plankton community of the Oregrundsgrepen (South-west Bosnian Sea), *Helg. Wiss. Meeresunters.*, 30, 582, 1977.

Erlich, P.R., Wilson, E.O., Biodiversity studies: science and policy, *Science*, 253, 758, 1991.

Estes, J.A., Duggins, D.O., Sea otter and kelp forests in Alaska: generality and variation in a community ecological paradigm, *Ecol. Monogr.*, 65, 75, 1995.

Estes, J.A., Smith, N.S., Palmisano, J.S., Sea otter predation and community organization in the western Aleutian Islands, Alaska, *Ecology*, 59, 828, 1978.

Evans, M.S., Granger, E.H., Zooplankton in a Canadian Arctic estuary, in *Estuarine Perspectives*, Kennedy, V., Ed., Academic Press, New York, 1980, 199.

Evans, P.R., Predation of intertidal fauna by shorebirds in relation to time of day, tide and year, in *NATO Advanced Research Workshop on Behavioural Adaptation to Intertidal Life*, Chelazzi, G., Vannini, M., Eds., Plenum Press, New York and London, 1988, 65.

Evans, R.G., The intertidal ecology of selected localities in the Plymouth neighbourhood, *J. Mar. Biol. Assoc. U.K.*, 27, 173, 1947.

Fagerstrom, J.A., *The Evolution of Reef Communities*, John Wiley & Sons, New York, 1987.

Fahey, E.M., The repopulation of intertidal transects, *Rhodora*, 55, 102, 1953.

Fairbridge, A., The estuary: its definition and geodynamic cycle, in *Chemistry and Biochemistry of Estuaries*, Olausson, E., Cato, I., Eds., John Wiley & Sons, New York, 1980, 1.

Fairweather, P.G., Differential predation on alternative prey, and the survival of rocky shore intertidal organisms in New South Wales, *J. Exp. Mar. Biol. Ecol.*, 89. 135, 1985.

Fairweather, P.G., Interactions between predators and prey, and the structure of rocky intertidal communities, *Aust. J. Ecol.*, 11, 321, 1986.

Fairweather, P.G., Experiments on the interaction between predation and the availability of different prey on rocky seashores, *J. Exp. Mar. Biol. Ecol.*, 114, 261, 1987.

Fairweather, P.G., Predation creates haloes of bare space among prey on rocky shores in New South Wales, *Aust. J. Ecol.*, 13, 401, 1988a.

Fairweather, P.G., Movements of intertidal whelks (*Morula marginalba* and *Thais orbita*) in relation to availability of prey and shelter, *Mar. Biol.*, 100, 63, 1988b.

Fairweather, P.G., Underwood, A.J., The apparent diet of predators and biases due to different handling times of their prey, *Oecologia (Berlin)*, 56, 169, 1983.

Fairweather, P.G., Underwood, A.J., Experimental removals of a rocky intertidal predator: variations within two habitats in the effects on prey, *J. Exp. Mar. Biol. Ecol.*, 154, 29. 1991.

Fairweather, P.G., Underwood, A.J., Moran, M.J., Preliminary investigations of predation by the whelk. *Morula marginalba* (Blainville), *Mar. Ecol. Prog. Ser.*, 17, 143, 1984.

Farrell, T.M., Community stability: the effects of limpet removal and reintroduction in a rocky intertidal community, *Oecologia (Berlin)*, 75, 190, 1988.

Farrell, T.M., Succession in a rocky intertidal community: the importance of disturbance size and position within a disturbed patch, *J. Exp. Mar. Biol. Ecol.*, 128, 53, 1989.

Farrell, T.M., Models and mechanisms of succession: an example from a rocky intertidal community, *Ecol. Monogr.*, 61, 95, 1991.

Fauchild, K., Jumars, P.A., The diet of worms: a study of polychaete feeding guilds, *Oceangr. Mar. Biol. Annu. Rev.*, 17, 193, 1979.

Feare, C.J., The winter feeding of the Purple Sandpiper, *Br. Birds*, 59, 165, 1966.

Feare, C.J., The dynamics of an exposed shore population of dogwhelks *Nucella lapillus* (L.), *Oecologia (Berlin)*, 5, 1, 1969.

Feare, C.J., Summers, R.W., Birds as predators on rocky shores, in *Ecology of Rocky Coasts*, Moore, P.G., Seed, R., Eds., Columbia Press, New York, 1986, 249.

Feder, H.M., Turner, C.H., Limbaugh, C., Observations on fishes associated with kelp beds in southern California, *Calif. Depart. Fish Game, Fish. Bull.*, 160, 144, 1974.

Feller, R. J., Empirical estimates of carbon production for a meiobenthic harpacticoid copepod, *Can. J. Fish. Aquatic Sci.*, 39, 1435, 1982.

Feller, R.J., Coull, B.C., Non-selective ingestion of meiobenthos by juvenile spot (*Leiosroma xanthurus*) and their daily ration, *Vie Milieu*, 45, 19, 1995.

Feller, R.J., Kaczynski, V.W., Size selection predation by juvenile chum salmon (*Onchorynchus chaeta*) on epibenthic prey in Puget Sound, *J. Fish. Res. Bd. Can.*, 32, 1419, 1975.

Feller, R.J., Taghorn, G.L., Gallagher, E.D., Kenny, G.E., Jumars, P.A., Immunological methods for food web analysis in a soft-bottom community, *Mar. Biol.*, 54, 61, 1979.

Femino, R.J., Mathieson, A.C., Investigations of New England marine algae. IV. The ecology and seasonal succession of tide pool algae at Bald Head Cliff, York, Maine, U.S.A., *Bot. Mar.*, 23, 319, 1980.

Fenchel, T., The ecology of the marine microbenthos. II. The food of marine benthic ciliates, *Ophelia*, 5, 73, 1968.

Fenchel, T., The ecology of marine meiobenthos. IV. Structure and function of the benthic ecosystem, its physical factors and the microfaunal communities with special reference to the ciliated Protozoa, *Ophelia*, 6, 182, 1969.

Fenchel, T., Studies on the decomposition of organic detritus derived from the turtle grass *Thallassia testudinium*, *Limnol. Oceanogr.*, 15, 14, 1970.

Fenchel, T., Aspects of decomposer food chains in marine benthos, *Deutsh Zool. Gesell. Vehr.*, 14, 14, 1971.

Fenchel, T., Aspects of decomposer food chains in marine benthos, *Sonderhefte Verhandlungen Deutschen Zoologischen Gesellschaft*, 65, 14, 1972.

Fenchel, T., Factors determining the distribution patterns of mud snails (Hydrobiidae), *Oecologia (Berlin)*, 20, 1, 1975a.

Fenchel, T., Characteristic displacement and coexistance in mud snails (Hydrobiidae), *Oecologia (Berlin)*, 20, 19, 1975b.

Fenchel, T., The quantitative importance of benthic microflora of an Arctic tundra pond, *Hydrobiologica*, 45, 445, 1975c.

Fenchel, T., The ecology of micro and meiobenthos, *Annu. Rev. Ecol. Systematics*, 9, 99, 1978.

Fenchel, T., Ecology of heterotrophic microflagellates. I. Some important forms and their functional morphology, *Mar. Ecol. Prog. Ser.*, 8, 211, 1982a.

Fenchel, T., Ecology of microflagellates. III. Adaptations to heterogeneous environments, *Mar. Ecol. Prog. Ser.*, 9, 25, 1982b.

Fenchel, T., Blackburn, T.H., *Bacteria and Mineral Cycling*, Academic Press, London, 1979.

Fenchel, T., Harrison, P., The significance of bactrial grazing and mineral recycling for the decomposition of particulate detritus, in *The Role of Terrstrial and Aquatic Organisms in Decomposition Processes*, Anderson, J.M., McLachlan, A., Eds., Blackwell Scientific, London, 1976, 285.

Fenchel, T., Kofeod, L.H., Evidence for exploitative interspecific competition in mud snails (Hydrobiidae), *Oikos*, 27, 367, 1976.

Fenchel, T., Kofoed, L.H., Lappalainen, A., Particle size-selection of two deposit-feeders: the amphipod *Corophium volulator*, and the prosobranch *Hydrobia ulvae*, *Mar. Biol.*, 30, 119, 1975.

Fenchel, T., Riedl, R.J., The sulphide system: a new biotic community underneath the oxidized layer of marine sand bottoms, *Mar. Biol.*, 7, 255, 1970.

Fenwick, G.D., The effect of wave exposure on the amphipod fauna of the alga *Caulerpa brownii*, *J. Exp. Mar. Biol. Ecol.*, 25, 1, 1976.

Ferguson, R.L., Murdock, M.B., Microbial ATP and organic carbon in sediments of the Newport River estuary mudflats, in *Estuarine Research*, Vol. 1, Cronin, L.E., Ed., Academic Press, New York, 1975, 229.

Ferguson, R.L., Rublee, P., Contribution of bacteria to standing crop of coastal plankton, *Limnol. Oceanogr.*, 21, 141, 1976.

Ferreira, J.G., Ramos, L., A model for the estimation of annual production rates of macrophyte algae, *Aquatic Bot.*, 33, 53, 1989.

Field, J.G., Coastal ecosystems: flow patterns of energy and matter, in *Mar. Ecol., Vol. 5*, Kinne, O., Ed., Wiley Interscience, Chichester, 1983, 758.

Field, J.G., Griffiths, C.L., Littoral and subtidal ecosystems of South Africa, in *Intertidal and Littoral Systems. Ecosystems of the World 24*, Mathieson, A.C., Nienhuis, P.H., Eds., Elsevier, Amsterdam, 1991, 323.

Field, J.G., Griffiths, C.L., Griffiths, R.J., Jarman, N., Zoutendyk, P., Velimirov, B., Bowes, A., Variation in structure and biomass of kelp communities along the south west Cape coast, *Trans. Roy. Soc. S. Af.*, 44, 145, 1980.

Field, J.G., Wulff, F., Mann, K.H., The need to analyse ecological networks, in *Network Analysis in Marine Ecology: Methods and Applications,* Wulff, F., Field, J.G., Mann, K.H., Eds., Springer-Verlag, Berlin, 1989, 3.

Fielding, P.J., Davis, C.L., Carbon and nitrogen resources available to kelp bed filter feeders in an upwelling environment, *Mar. Ecol. Prog. Ser.*, 55, 181, 1989.

Finlay, S., Tenore, K.R., Nitrogen source for a detritovore: detritus substrate versus associated microbes, *Science,* 218, 371, 1982.

Finn, J.T., Measures of ecosystem structure and function derived from analysis of flows, *J. Theoret. Biol.,* 56, 363, 1976.

Fisher, T.R., Carlson, P., The importance of sediments in the nitrogen cycle of estuarine systems, Abstracts, *42nd Annual Meeting Am. Soc. Limnol. Oceanogr.,* Stony Brook, New York, 1979, 19.

Fisher, T.R., Carlson, P.R., Barber, R.T., Sediment nutrient regeneration in three North Carolina estuaries, *Est. Coastal Shelf Sci.,* 14, 101, 1982a.

Fisher, T.R., Carlson, P.R., Barber, R.T., Carbon and nitrogen primary production in three North Carolina estuaries, *Est. Coastal Shelf Sci.,* 15, 621, 1982b.

Fitch, J.E., A relatively unexploited population of Pismo clams, *Tivela stultorum* (Moore, 1823) (Veneridae), *Proc. Malacol. Soc. London,* 36, 309, 1965.

Fitt, K.W., Coon, S.L., Walch, M., Weiner, R.M., Colwell, R.R., Bonar, D.B., Settlement behaviour and metamorphosis of oyster larvae (*Crassostrea gigas*) in response to bacterial supernatants, *Mar. Biol.*, 106, 389, 1990.

Fitzhugh, G.R., Fleeger, J.W., Goby (Pisces: Gobiidae) interactions with meiofauna and small macrofauna, *Bull. Mar. Sci.*, 36, 436, 1985.

Flach, E.C., Disturbance of benthic infauna by sediment reworking activities of the lugworm *Arenicola marina*, *Neth. J. Sea Res.,* 30, 81, 1992.

Fleeger, J.W., Palmer, M.A., Secondary production of the estuarine meiobenthic copepod *Microarthridium littorale*, *Mar. Ecol. Prog. Ser.*, 7, 157, 1982.

Fletcher, A., Marine and maritime lichens of rocky shores: their ecology, physiology and biological interactions, in *The Shore Environment, Volume 2. Ecosystems*, Price, J.H., Irvine, D.E.G., Farnham, W.F., Eds., Academic Press, London and New York, 1980, 789.

Fletcher, W.J., Variability in the reproductive effort of the limpet, *Cellana tramoserica, Oecologia (Berlin),* 61, 259, 1984a.

Fletcher, W.J., Intraspecific variation in the population dynamics and growth of the limpet, *Cellana tramoserica*, *Oecologia (Berlin),* 63, 110, 1984b.

Fletcher, W.J., Creese, R.G., Competitive interactions between co-occurring herbivorous gastropods, *Mar. Biol.*, 86, 183, 1985.

Flint, R.W., Kaike, R.D., Biological enhancement of estuarine benthic community structure, *Mar. Ecol. Prog. Ser.,* 31, 22, 1986.

Fogg, G.E., The ecological significance of extracellular products of phytoplankton photosynthesis, *Bot. Mar.*, 26, 3, 1983.

Fogg, G.E., Nalewajko, C., Watt, W.D., Extracellular products of phytoplankton photosynethsis, *Proc. Roy. Soc., London, Series B*, 162, 517, 1965.

Folk, R.L., A review of grain size parameters, *Sedimentology*, 6. 73, 1966.

Folk, R.L., *Petrology of Sedimentary Rocks*, Hemphills, Texas, 1974.

Fong, W.C., Mann, K.H., Role of gut flora in the transfer of amino acids through a marine food chain, *Can. J. Fish. Aquat. Sci.,* 37, 88, 1980.

Fonseca, M.S., Fisher, J.S., Zieman, J.C., Influence of the seagrass, *Zostera marine* L., on current flow, *Est. Coastal Shelf Sci.,* 15, 351. 1982.

Foolad, M.M., The biology of certain macrofauna of Longstone Harbour, Ph.D. thesis, Portsmouth Polytechnic, 1983.

Foret, J.P., Étude des effects à long terme de quelques detergents sur la sequence du developpement de la polychete sédentaire *Capitella capitata* (Fabricius), *Tethys*, 6, 751, 1974 (in French).

Forster, B.A., Tolerance of high temperature by some intertidal barnacles, *Mar. Biol.,* 4, 326, 1969.

Forster, B.A., Desiccation as a factor in the intertidal zonation of barnacles, *Mar. Biol.,* 8, 12, 1971a.

Forster, B.A., On the determination of the upper limit of intertidal distribution of barnacles, *J. Anim. Ecol.,* 40, 33, 1971b.

Forster, B.A., Barnacle ecology and adaptation, in *Barnacle Biology*, Crustacean Issues 5, Southward, A.J., Ed., A.A. Balkema, Rotterdam, 1987, 113.

Forster, G.L., Indications regarding the source of combined nitrogen for *Ulva lactuca*, *Ann. Missouri Bot. Garden*, 1, 229, 1914.

Foster, M.S., Microscopic algal food of *Littorina planaris* Phillipi and *Littorina scutulata* Gould (Gastropoda: Prosobranchia), *Veliger,* 7, 149, 1975.

Foster, M.S., Factors controlling the intertidal zonation of *Iridaea flaccida* (Rhodophyta), *J. Phycol.*, 18, 285, 1982.

Foster, M.S., How important is grazing to seaweed evolution and assemblage structure in the northeast Pacific, in *Plant-Animal Interactions in the Marine Benthos*, John, D.M., Hawkins, S.J., Price, J.H., Eds., Oxford University Press, Oxford, 1992, 61.

Foster, M.S., De Vogelacre, A.P., Harrold, C., Pearse, J.S., Causes of spatial and temporal patterns in rocky intertidal communities of central and northern California, *Mem. Calif. Acad. Sci.*, 9, 1988.

Foster, M.S., De Vogelacre, A.P., Oliver, J.S., Pearse, J.S., Harrold, C., Open coast, intertidal and shallow subtidal ecosystems of the Northeast Pacific, in *Intertidal and Littoral Ecosystems: Ecosystems of the World 24,* Mathieson, S.A., Nienhuis, D.H., Eds., Elsevier, Amsterdam, 1991, 235.

Foster, M.S., Schiel, D.R., The ecology of giant kelp forests in California, U.S. Fish. Wildlife Service, Biol. Service Program, Washington, D.C., 85, 7.2, 1985.

Foulds, J.B., Mann, K.H., Cellulose digestion in *Mysis stenplepis* and its ecological implications, *Limnol. Oceanogr.,* 23, 760, 1978.

Fralick, R.A., Mathieson, A.C., Physiological ecology of four *Polysiphonia* species (Rhodophyta, Ceramiales), *Mar. Biol.,* 29, 29, 1975.

Frank, W.P., Effects of winter feeding on limpets, by black oystercatchers *Haemotopus bachmani*, *Ecology,* 63, 1352, 1982.

Frechette, M., Bourget, E., Energy flow between pelagic and benthic zones: factors controlling particulate organic matter available to an intertidal mussel bed, *Neth. J. Sea Res.,* 19, 265, 1985.

Frechette, M., Butman, C.A., Geyer, W.R., The importance of boundary-layer flows in supplying phytoplankton to the suspension feeder, *Mytilus edulis* L., *Limnol. Oceanogr.,* 33, 1182, 1989.

Fretwell, S.D., Food chain dynamics: the central theory of ecology, *Oikos*, 50, 291, 1987.

Frid, C.L.J., James, R., The role of epibenthic predators in structuring the marine invertebrate community of a British coastal salt marsh, *Neth. J. Mar. Res.,* 22, 307, 1988.

Friesen, J.A.K.H., Mann, K.H., Novitsky, J.A., *Mysis* digests cellulose in the absence of a gut microflora, *Can. J. Zool.,* 64, 442, 1986.

Fry, B., Stable carbon isotope ratios — A tool for tracing food chains, MS thesis, University of Texas, Austin, 126 pp, 1977.

Fry, B., $^{13}C/^{12}C$ ratios and the trophic importance of algae in Florida *Syringodium filiforme* seagrass meadows, *Mar. Biol.,* 79, 11, 1984.

Fry, B., Lutes, R., Northam, M., Parker, P.L., Ogden, J., A $^{13}C/^{12}C$ comparison of food webs in Caribbean seagrass meadows and coral reefs, *Aquatic Bot.,* 14, 389, 1982.

Fry, B., Parker, P.L., Animal diet in Texas seagrass meadows: ^{13}C evidence for the importance of benthic plants, *Est. Coastal Shelf Sci.,* 8, 499, 1979.

Fry.B., Sherr, E., ^{13}C measurements as indicators of carbon flow in marine and freshwater ecosystems, *Contr. Mar. Sci.,* 27, 15, 1984.

Fulton, R., Distribution and community structure of estuarine copepods, *Estuaries*, 7, 38, 1984.

Furnas, M.J., Hitchcock, G.L., Smayada, T.J., Nutrient-phytoplankton relationships in Narragansett Bay during the 1974 summer bloom, in *Estuarine Processes*, Vol. 1, Wiley, M.L., Ed., Academic Press, New York, 1976, 118.

Gabbott, P.A., Larman, V.N., The chemical basis of gregariousness in cirripedes: a review (1953–1984), in *Barnacle Biology,* Crustacean Issues 5, Southward, A.J., Ed., A.A. Balkema, Rotterdam, 1987, 377.

Gaines, S.D., Diverse consumer guilds in intertidal communities of Oregon and the Republic of Panama and their effects on prey assemblages, Ph.D. thesis, Oregon State University, Corvallis, Oregon, 133 pp, 1983.

Gaines, S., Brown, S., Roughgarden, J., Spatial variation in larval concentrations as a cause of spatial variation for the barnacle, *Balanus glandula*, *Oecologia (Berlin),* 67, 267, 1985.

Gaines, S.D., Lubchenco, J., A unified approach to marine plant-herbivore interactions. II. Biogeography, *Annu. Rev. Ecol. Systematics*, 13, 111, 1982.

Gaines, S., Roughgarden, J., Larval settlement rate: a leading determinant of structure in an ecological community of the marine intertidal zone, *Proc. Nat. Acad. Sci.,* 82, 3707, 1985.

Gallagher, J.L., Sampling macro-organic matter profiles in salt marsh plant root zones, *Proc. Am. Soil Sci. Soc.*, 38, 154, 1974.

Gallagher, J.L., Zonation of wetlands vegetation, in *Coastal Ecosystem Management,* Clark, J.R., Ed., John Wiley & Sons, New York, 1977, 752.

Gallagher, J.L., Daiber, F.C., Primary production of edaphic communities in a Delaware salt marsh, *Ecology,* 54, 1160, 1974.

Gallagher, J.L., Pfeiffer, W.J., Pomeroy, L.R., Leaching and microbial utilization of dissolved organic matter from leaves of *Spartina alterniflora*, *Est.Coastal Shelf Sci.*, 4, 467, 1976.

Gallagher, J.L., Plumley, F.G., Underground biomass profiles and productivity in Atlantic coastal marshes, *Am. J. Bot.,* 66, 156, 1979.

Gallagher, J.L., Plunley, F.G., Wolf, P.L., Underground biomass dynamics and substrate selective properties of Atlantic coastal marsh plants, MS Report TR-D-77-28, U.S. Army Engineers Waterways Experimental Station, Vicksburg, VA, 1978.

Gallagher, J.L., Reimbold, R.J., Linthurst, R.A., Pfeiffer, W.J., Aerial production, mortality and mineral accumulation-export dynamics in *Spartina alterniflora* and *Juncus roemerianus* plant stands, *Ecology,* 61, 303, 1986.

Gallagher, J.L., Reimbold, R.J., Thompson, D.E., A comparison of four remote sensing methods for assessing salt marsh productivity, in *Proc. 8th Int. Symp. Remote Sensing Environment*, Cook, J., Ed., 1973, 1287.

Galstoff, P.S., The American oyster *Crassostrea virginicata* (Gmelin), *U.S. Fish Wildlife Ser. Bull.,* 64, 1–480, 1964.

Ganning, B., Short time fluctuations of the microfauna in a rock pool in the northern Baltic proper, *Veroff. Inst. Meeresforsch Bremerhaven (Sunderbad),* 2, 149, 1966.

Gardner, L.R., Runoff from an intertidal marsh during exposure-recession curves and chemical characteristics, *Limnol. Oceanogr.,* 20, 81, 1975.

Garner, J.L., Lewin, J., Persistent blooms of surf diatoms along the Pacific coast, USA. 1. Physical characteristics of the coastal region in relation to the distribution and abundance of the species, *Est. Coastal Shelf Sci.*, 12, 217, 1981.

Garrity, S.D., Levings, S.C., A predator-prey interaction between two physically and biological constrained tropical rocky shore gastropods: direct and indirect community effects, *Ecol. Monogr.*, 51, 267, 1981.

Gaylord, B.C., Blanchette, C., Denny, M., Mechanical consequence of size in wave-swept algae, *Ecol. Monogr.*, 64, 287, 1994.
Gearing, J.N., Gearing, P.J., Rudnick, D.T., Requejo, A.G., Hutchins, M.J., *Geochemica et Cosmochemica Acta*, 19, 1089–1098, 1984.
Gee, J.M., Impact of epibenthic predation on intertidal harpacticoid copepod populations, *Mar. Biol.*, 96, 497, 1987.
Gee, J.M., An ecological and economic review of meiofauna as food for fish, *Zool. J. Linn. Soc.*, 96, 243, 1989.
Gee, J.M., Warwick, R.M., Davey, J.T., George, C.L., Field experiments on the role of epibenthic predators in determining prey densities in an estuarine mudflat, *Est. Coastal Shelf Sci.*, 21, 429, 1985.
Geraci, S., Romairone, V., Barnacle larvae and their settlement in Genoa Harbour (North Tyrrenian Sea), *Mar. Ecol.*, 3, 225, 1982.
Gerlach, S.A., On the importance of marine meiofauna for benthic communities, *Oecologia (Berlin)*, 6, 176, 1971.
Gerlach, S.A., Attraction to decaying organisms as a possible cause for patchy distribution of nematodes in a Bermuda beach, *Ophelia*, 16, 151, 1977.
Gerlach, S.A., Food chain relationships in subtidal silty sand marine sediments and the role of meiofauna in stimulating bacterial productivity, *Oecologia (Berlin)*, 33, 55, 1978.
Gesner, F., *Hydrobotanik 2*, Stoffhaushalt VEB, Deutscher, 1959.
Gesner, F., Schramm, W., Salinity: plants, in *Marine Ecology, Vol.1, Pt. 2*, Kinne, O., Ed., Wiley, New York, 1971, 705.
Gertz, W.M., Population dynamics: a per capita resource approach, *J. Theoret. Biol.*, 108, 623, 1984.
Ghelardi, R.J., Species structure of the holdfast community, in *The Biology of Giant Kelp Beds (Macrocystic) in California*, North, J.W., Ed., *Nova Hedwiga*, 32, 361, 1971.
Gianuca, N.M., A preliminary account of the ecology of sandy beaches in southern Brazil, in *Sandy Beaches as Ecosystems*, McLachlan, A., Erasmus, T., Eds., Dr. W. Junk Publishers, The Hague, 1983, 413.
Gibb, J., Food, feeding habits and territories of the Rock-Pipit *Anthus spinoletta*, *Ibis*, 98, 506, 1956.
Gibbons, M.J., The impact of sediment accumulations, relative habitat complxity and elevation on rocky shore meiofauna, *J. Exp. Mar. Biol. Ecol.*, 122, 225, 1988.
Gibbons, M.J., The impact of wave exposure on the meiofauna of *Gelidium pristoides* (Turner) Kuetzing (Gelidiales: Rhodophyta), *Est. Coastal Shelf Sci.*, 27, 581, 1989.
Gibbons, M.J., Griffiths, C.L., A comparison of the macrofaunal and meiofaunal distribution and standing stock across a rocky shore with an estimate of their productivities, *Mar. Biol.*, 93, 181, 1986.
Gibson, R.N., Recent studies on the biology of intertidal fishes, *Oceanogr. Mar. Biol. Annu. Rev.*, 20, 363, 1982.
Gieskes, W.W.C., Kraay, G.W., Baars, M.A., Current ^{14}C methods for measuring primary production: gross underestimates in oceanic waters, *Neth. J. Sea Res.*, 13, 58, 1979.
Gifford, D.J., Dagg, M.J., Feeding of the estuarine copepod *Acartia tonsa* Dana: carnivory versus herbivory in natural microplankton assenblages, *Bull. Mar. Sci.*, 43, 458, 1988.
Gilbert, P.M., Regional studies of daily, seasonal and size function variability in ammonium reminerlization, *Mar. Biol.*, 70, 209, 1982.
Gilinsky, E., The role of fish predation and spatial heterogeneity in determining benthic community structure, *Ecology*, 65, 455, 1984.
Gill, A.M., Tomlinson, P.B., Aerial roots: an array of forms and functions, in *The Development and Function of Roots*, Torrey, J.C., Clarkson, D.T., Eds., Academic Press, London, 1969, 1.
Gillespie, M.C., Analysis and treatment of zooplankton of estuarine waters of Lousiana, in *Cooperative Gulf of Mexico Estuarine Inventory and Study, Lousiana. Phase IV. Biology*, Louisana Wildlife Fisheries Commission, 1971.
Gillespie, P.A., MacKenzie, A.L., Autotrophic and heterotrophic processes on an intertidal mud-sand flat, Delaware Inlet, New Zealand, *Bull. Mar. Sci.*, 31, 648, 1981.
Glassner, J.W., The role of predation in shaping and maintaining the structure of communities, *Am. Nat.*, 113, 631, 1979.
Glynn, P.W., Community composition, structure, and interrelationships in the marine intertidal *Endocladia muricata — Balanus glandula* association in Monterey Bay, California, *Beaufortia*, 12, 1, 1965.
Godin, J.-G.J., Daily patterns of feeding behavior, daily rations, and diets of juvenile pink salmon (*Oncorhynchus gorbuscha*) in two marine bays of British Columbia, *Can. J. Fish. Aquatic Sci.*, 38, 10, 1981.
Goldman, J.C., Temperature effects on steady-state growth, phosphate uptake, and the chemical composition of marine phytoplankton, *Microb. Ecol.*, 5, 955, 1979.
Goldman, J.C., Oceanic nutrient cycles, in *Flow of Energy and Materials in Marine Ecosystems: Theory and Practice*, Fasham, J.R., Ed., Plenum Press, New York, 1984a, 130.
Goldman, J.C., Conceptual role for microaggregates in pelagic waters, *Bull. Mar. Sci.*, 38, 462, 1984b.
Golley, F.B., McGinnis, J.T., Clements, R.G., Childs, G.I., Duener, M.J., *Mineral Cycling in a Tropical Moist Forest Ecosystem*, University of Georgia Press, Athens, Georgia, 1975.
Golley, F.B., Odum, H.T., Wilson, R.F., The structure and metabolism of a Puerto Rico red mangrove forest in May, *Ecology*, 43, 9, 1962.
Gong, K.-W., Ong, J.-E., Plant biomass and nutrient flux in a managed mangrove forest in Malaysia, *Est. Coastal Shelf Sci.*, 31, 519, 1990.
Good, R.E., An Experimental Assessment of the Proposed Reconstrction of State Route 152 (Somers Point-Long Point) Atlantic County, New Jersey, Report to E. Lionel Parlo Engineering, New York, 41 pp, 1977.
Good, R.E., Frasco, B.R., Estuarine valuation study: a four year report on production and decomposition. Dynamics of salt marsh communities, Ocean County, New Jersey, report to New Jersey Department of Environmental Protection, Division of Fisheries, Game and Shellfisheries, Trenton, NJ, 195 pp, 1979.
Good, R.E., Walker, R., Relative Contribution of saltwater and freshwater tidal marsh communities to estuarine productivity, final report (071171), submitted to Rutgers University Center for Coastal and Environmental Studies, New Brunswick, NJ, unpublished manuscript, 1977.
Gordon, D.C., The effects of the deposit feeding polychaete *Pectinaria gouldii* on the intertidal sediments of Barnstable Harbour, *Limnol. Oceanogr.*, 11, 327, 1966.

Goreau, T.F., Goreau, N.V., The physiology of skeleton formation in corals. 2. Calcium deposition by hematypic corals under various conditions in the reef, *Biol. Bull. Mar. Biol. Lab. Woods Hole,* 117, 239, 1959.

Goss-Custard, J.D., The energetics of prey selection by redshank, *Tringa totanis* (L.), in relation to prey density, *J. Anim. Ecol.*, 46, 1, 1977a.

Goss-Custard, J.D., Optimal foraging and size selection of worms by redshank, *Tringa totanus*, in the field, *Anim. Behav.*, 25, 10, 1977b.

Goss-Custard, J.D., Dit Durell, S.E.A. LeV., The effect of dominance and feeding method on the intake rates of oystercatchers, *Haematopus ostrelagus,* feeding on mussels, *J. Anim. Ecol.*, 57, 827, 1988.

Goss-Custard, J.D., Jones, J., Kitching, J.A., Norton, T.A., Tide pools of Carrigathorn and Barloge Creek, *Phil. Trans. Roy. Soc., London (B),* 287, 1, 1979.

Goss-Custard, J.D., Jones, R.E., Newberry, P.E., The ecology of the Walsh. I. Distribution of the wading birds (Chariidae), *J. Appl. Ecol.*, 14, 681, 1977.

Goss-Custard, J., Knight-Jones, E. W., The settlement of *Spirorbis borealis* larvae, *Challenger Soc. Rep.*, 3, 18, 1957.

Goss-Custard, J.D., Moser, M.E., Rates of change in the numbers of dunlin *Calidris alpina* wintering in British estuaries in relation to the spread of *Spartina anglica*, *J. Appl. Ecol.*, 25, 95, 1988.

Goulder, R., Blanchard, A.S., Sanderson, P.L., Wright, B., Relationships between heterotrophic bacteria and pollution in an industrial estuary, *Water Res.*, 14, 591, 1980.

Goudie, R.I., Ankney, C.D., Body size, activity budgets, and diets of seaducks wintering in Newfoundland, *Ecology*, 62, 1475, 1986.

Goulter, P.F.E., Alloway, W.G., Litterfall and decomposition in a mangrove stand, *Avicennia marina* (Forsk.) Vierh., in Middle Harbour, Sydney, *Aust. J. Mar. Freshw. Res.* 30, 541, 1979.

Grahame, J., Branch, G., Reproductive patterns of marine invertebrates, *Oceanogr. Mar. Biol. Annu. Rev.*, 23, 373, 1985.

Grant, J., The relative magnitude of biological and physical sediment reworking in an intertidal community, *J. Mar. Res.*, 41, 673, 1983.

Grant, J., Cust, G., Prediction of coastal sediment stability from photopigment content from the mats of purple sulphur bacteria, *Nature (London),* 330, 224, 1987.

Grant, W.D., Boyer, L.F., Sanford, L.P., The effects of bioturbation on the initiation of motion of intertidal sands, *J. Mar. Res.,* 40, 659, 1982.

Grant, W.S., High intertidal community organization on a rocky headland in Maine, USA, *Mar. Biol.,* 44, 15, 1977.

Grassle, J.F., Grassle, J.P., Opportunistic life histories and genetic systems in marine benthic polychaetes, *J. Mar. Res.*, 32, 253, 1974.

Grassle, J.F., Grassle, J.P., Temporal adaptations in sibling species of *Capitella*, in *Ecology of the Marine Benthos*, Coull, B.C., Ed., Columbia University of South Carolina Press, Columbia, SC, 1977, 177.

Grassle J.F., Sanders, H.L. Life histories and the role of disturbances, *Deep-Sea Res.*, 20, 643, 1973.

Grassle, J.P., Polychaete sibling species, in *Aquatic Oligochaete Biology,* Brinkhurst, R.O., Ed., 1980, 25.

Grassle, J.P., Gelfman, C.E., Mills, S, W., Karyotypes of *Capitella* sibling species, and of several species of the related genera *Capitellides* and *Capitomastus* (Capitellidae: Polychaeta), *Proc. Biol. Soc. Wash.,* 1987.

Grassle, J.P., Grassle, J.F., 1976, Sibling species of the marine indicator *Capitella* (Polychaeta), *Science*, 192, 567, 1976.

Gray, A.H., Benham, P.E.M., Raybould, A.R., *Spartina anglica* — the evolutionary and ecological background, in *Spartina anglica — A Research Review*, Gray, A.H., Benham, P.E.M., Eds., H.M.S.O., London, 1990, 5.

Gray, J.S., The attractive factor of intertidal sands to *Protodrilus symbioticus* (Giard). *J. Anim. Ecol.*, 46, 627, 1966.

Gray, J.S., Substrate selection by the archiannelid *Protofrilus rubropharyngeus* Jagersten, *Helgo. Wiss. Meeresunters,* 15, 235, 1967a.

Gray, J.S., Substrate selection by the archiannedid *Protodrilus hypoleucus* Armenante, *J. Exp. Mar Biol. Ecol.*, 1, 47, 1967b.

Gray, J.S., An experimental approach to the ecology of the harpacticoid *Lepastacus constrictus* Lang, *J. Exp. Mar. Biol. Ecol.*, 2, 278, 1968.

Gray, J.S., Animal-sediment relationships, *Oceanogr. Mar. Biol. Annu. Rev.,* 12, 223, 1974.

Gray, J.S., The stability of benthic ecosystems, *Helg. Wiss. Meeresunters.*, 39, 427, 1977.

Gray, J.S., *The Ecology of Marine Sediments*, Cambridge University Press, Cambridge, 185 pp, 1981.

Gray, J.S., Johnson, R.M., Bacteria of a sandy beach as an ecological factor affecting the interstitial gastrotrich *Turbanella hyalina* Schultze, *J. Exp. Mar, Biol. Ecol.*, 4, 119, 1970.

Green, M.J., Local distribution of *Oligottus maculosus* Girard and other tidepool cottids of the west coast of Vancouver Island, *Can. J. Zool.*, 49, 1111, 1971.

Green, R.H., Growth and mortality in an arctic intertidal population of *Macoma balthica* (Pelecypoda, Tellinidae), *J. Fish. Res. Bd. Can.*, 30, 1345, 1973.

Greenaway, M., The grazing of *Thalassia testudinium* in Kingston Harbour, Jamaica, *Aquatic Biol.,* 2, 117, 1976.

Gregory, M.R., Ballance, P.F., Gibson, G.W., Ayling, A.M., On how some rays (Elasmobranchia) excavate feeding depressions by jetting water, *J. Sed. Petrol.*, 49, 112, 1979.

Grémore, A., Marsh, A.G., Tenore, K.R., Secondary production and reproduction of *Capitella capitata* type I (Annelida; Polychaeta) during a population cycle, *Mar. Ecol. Prog. Ser.,* 51, 99, 1989.

Grenon, J.-F., Walker, G., Biochemical and rheological properties of the pedal mucus of the limpet *Patella vulgata* L., *Comp. Biochem. Physiol.*, 66B, 451, 1980.

Griffiths, C.L, Seiderer, J.L., Rock-lobsters and mussels — limitations and preferences in a predator-prey relationship, *J. Exp. Mar. Biol. Ecol.*, 44, 95, 1980.

Griffiths, C.L., Stenton-Dozey, J.M.E., The fauna and rate of degradation of stranded kelp, *Est. Coastal Shelf Sci.*, 12, 645, 1981.

Griffiths, C.L., Stenton-Dosey, J.M.E., Koop, K., Kelp wrack and the flow of energy in a sandy beach ecosystem, in *Sandy Beaches as Ecosystems*, McLachlan, A., Erasmus, T., Eds., Dr. W. Junk Publishers, The Hague, 1983, 547.

Griffiths, R.J., Population dynamics and growth of the bivalve *Chloromytilus meridionalis*, *Est. Coastal Shelf Sci.*, 12, 477, 1981a.

Griffiths, R.J., Production and energy flow in relation to age and shore level in the bivalve *Chloromytilus meridionalis* (Rr.), *Est. Coastal Shelf Sci.*, 13, 477, 1981b.

Grime, J.P., Evidence for the existence of three primary strategies in plants and its relevance to ecological and evolutionary theory, *Am. Nat.*, 111, 1169, 1977.

Grindley, J.R., Effect of low-salinity water on the vertical migration of estuarine plankton, *Nature*, 203, 781, 1964.

Grindley, J.R., Residence time tests, in *Knysna Lagoon Model Investigation 1*, NRIO, Stollenbosch, South Africa, 1977, 28.

Grindley, J.R., Estuarine plankton, in *Estuarine Ecology with Particular Reference to South Africa*, Day, J.H., Ed., A.A. Balkema, Rotterdam, 1981, 117.

Grindley, J.R., Woolridge, T., The plankton of Richards Bay, *Hydrobiol. Bull.*, 8, 201, 1974.

Grizzle, R.E., Pollution indicator species of macrobenthos in a coastal lagoon, *Mar. Ecol. Prog. Ser.*, 18, 191, 1984.

Grizzle, R.E., Lutz, R.A., A statistical model relating horizontal seston fluxes and bottom sediment characteristics to growth of *Mercinaria mercinaria*, *Mar. Biol.*, 102, 95, 1989.

Grizzle, R.E., Morin, P.J., Effects of currents, seston, and bottom sediments on growth of *Mercinaria mercinaria*: results of a field experiment, *Mar. Biol.*, 102, 85, 1989.

Grizzle, R.E., Penniman, C.A., Effects of inorganic enrichment on estuarine macrofaunal benthos: a comparison of sediment profile imaging and traditional mehtods, *Mar. Ecol. Prog. Ser.*, 74, 249, 1991.

Grodzinsky, W., Klekowski, R.Z., Duncan, A., *Methods for Ecological Bioenergetics*, IBP Handbook No. 24, Blackwell Scientific Publications, Oxford, 367 pp, 1975.

Grosberg, R.K., Competitive ability influences habitat choice in marine invertebrates, *Nature (London)*, 290, 700, 1981.

Grosberg, R.K., Intertidal zonation of barnacles: the influence of planktonic zonation of larvae on vertical distribution of adults, *Ecology*, 63, 894, 1982.

Gross, J., Knight-Jones, E.W., Settlement of *Spirorbis borealis* on algae, *Challenger Soc. Annu. Rep.*, 3, IX, 18, 1957.

Gross, M.G., *Oceanography: A View of the Earth*, 3rd edition, Prentice-Hall, Englewood Cliffs, NJ, 498 pp, 1982.

Grossman, G.D., Dynamics and organization of a rocky fish assemblage: the persistence and resilience of taxocene structure, *Am. Nat.*, 119, 611, 1982.

Grossman, G.D., Coffin, R., Moyle, P.B., Feeding ecology of the bay goby (Pisecs: Gobiidae). Effects of behavioural, ontogenetic, and temporal variation on diet, *J. Exp. Mar. Biol. Ecol.*, 44, 47, 1980.

Grunbaum, H., Bergman, G., Abbott, D.P., Ogden, J.C., Intraspecific agonistic behaviour in the rock-boring sea urchin *Echinometra lucunter* (L.) (Echinodermata: Echinoidea), *Bull. Mar. Sci.*, 28, 181, 1978.

Guiler, E.R., 1959, The intertidal ecology of the Montmartre area, Chile, *Pap. Proc. Roy. Soc. Tasmania*, 93, 165, 1959.

Gunhill, F.C., Effect of plant size and distribution on the numbers of invertebrate species and individuals inhabiting the brown alga *Pelvetis vastigiata*, *Mar. Biol.*, 69, 263, 1982.

Gunhill, F.C., Seasonal variations in the invertebrate faunas of *Pelvetia fastigiats* (Fucaceae): effects of plant size and distribution, *Mar. Biol.*, 73, 115, 1983.

Gunther, C.-P., Distribution patterns of juvenile macrofauna on an intertidal sandflat: an approach to the variability of predator-prey interactions, in *Trophic Relationships in the Marine Environment*, Barnes, M., Gibson, R.N., Eds., Aberdeen University Press, Aberdeen, TX, 1990, 77.

Gunther, G., Temperature, in *Treatise on Marine Ecology and Paleoecology (Ecology) I.*, Hedgpeth, J.W., Ed., Memoir Geological Society of America 67, Washington, D.C., 1957, 159–184.

Gunther, G., Some relations of estuaries to the fisheries of the Gulf of Mexico, in *Estuaries*, Lauff, G., Ed., American Association for the Advancement of Science, Washington D.C., 1967, 621.

Hackney, C.T., Energy flux in a tidal creek draining an irregularly flooded *Juncus* marsh, Ph.D. thesis, Mississippi State University, 1977.

Hackney, C.T., de la Cruz, A.A., *In situ* decomposition of roots and rhizomes of two tidal marsh plants, *Ecology*, 61, 226, 1980.

Hadfield, M.G., Settlement and recruitment of marine invertebrates: a perspective and some prospects, *Bull. Mar. Sci.*, 39, 418, 1986.

Hadfield, M.G., Pennington, J.T., Nature of metamorphic signal and its internal transduction in larvae of the nudibranch *Phestilla sibogae*, *Bull. Mar. Sci.*, 40, 455, 1970.

Haedrick, R.L., Estuarine fishes, in *Estuaries and Enclosed Seas. Ecosystems of the World*, Ketchum, B.H., Ed., Vol. 26, Elsevier, Amsterdam, 1983, 183.

Hagerman, L., The macro- and meiofauna associated with *Fucus serratus* L., with some ecological remarks, *Ophelia*, 3, 1, 1966.

Haines, E.B., Relation between the stable carbon isotope composition of fiddler crabs, plants and soils in a salt marsh, *Limnol. Oceanogr.*, 21, 880, 1976a.

Haines, E.B., Stable carbon isotope ratios in the biota, soils and tidal water of a Georgia salt marsh, *Est. Coastal Shelf Sci.*, 4, 609, 1976b.

Haines, E.B., The origins of detritus in Georgia salt marsh estuaries, *Oikos*, 29, 254, 1977.

Haines, E.B., Interactions between Georgia salt marshes and coastal waters: a changing paradigm, in *Ecological Processes in Coastal and Marine Systems*, Livingston, R.J., Ed., Plenum Press, New York, 1979, 35.

Haines, E.B., Dunn, E.L., Growth and resource allocation response of *Spartina alterniflora* to three levels of NH_4-N, Fe and NaCl in solution culture, *Bot. Gazette*, 137, 224, 1976.

Haines, E.B., Dunstan, W.M., The distribution and relation of particulate organic material and primary productivity in the Georgia Bight, *Est. Coastal Shelf Sci.*, 3, 431, 1975.

Haines, E.B., Montague, C.L., Food sources of estuarine invertebrates analysed using $^{13}C/^{12}C$ ratios, *Ecology*, 60, 48, 1979.

Hairston, N.G., Smith, F.E., Slobodkin, L.B., Community structure, population control, and competition, *Am. Nat.*, 94, 421, 1960.

Hall, C.A.S., Introduction, in *Maximum Power. The Ideas and Applications of H.T. Odum,* Hall, C.A.S., Ed., University Press of Colorado, Niwat, Colorado, 1995, 205.

Hall, S.J., Physical disturbance and marine benthic communities: life in unconsolidated sediments, *Oceanogr. Mar. Biol. Annu. Rev.,* 32, 179, 1994.

Halldal, P., Ultraviolet action spectra of photosynthesis and photosynthetic inhibition in a green alga and a red alga, *Physiol. Plant.,* 17, 414, 1964.

Hallifors, G., Niemi, A., Vegetation and primary production, in *The Baltic Sea,* Voipio, A., Ed., Elsevier, Amsterdam, 1981, 220.

Hammond, L. S., Synott, R. N., Eds., *Marine Biology,* Longman Cheshire, Melbourne, 518 pp., 1994.

Hanekon, N.M., Baird, D., Distribution and variations in the seasonal biomass of eelgrass *Zostera capensis* in the Kromme Estuary, St. Francis Bay, South Africa, *S. Af. J. Sci.,* 7, 51, 1988.

Hansen, J.A., Accumulation of macrophyte wrack along sandy beaches in Western Australia: biomass, decomposition rates and significance in supporting near shore production, Ph. D. thesis, University of Western Australia, Perth, 1985.

Hansen, J.I., Henricksen, K.J., Backburn, T.H., Seasonal distribution of nitrifying bacteria and rate of nitrification in coastal marine sediment, *Microb. Ecol.,* 7, 297, 1981.

Hanson, R.B., Comparsion of nitrogen fixation activity in tall and short *Spartina alterniflora, Applied Envt. Microbiol.,* 33, 846, 1977a.

Hanson, R.B., Nitrogen fixation (acetylene reduction) in a salt marsh amended with sewage sludge and organic carbon and nitrogen compounds, *Applied Envt. Microbiol.,* 33, 846, 1977b.

Hanson, R.B., Tenore, K.R., Microbial metabolism and incorporation by the polychaete *Capitella capitata* of aerobically and anaerobically decomposed detritus, *Mar. Ecol. Prog. Ser.,* 6, 299, 1981.

Hardwick-Whitman, M.N., Mathieson, A.C., Intertidal macroalgae and macroinvertebrates: seasonal and spatial abundance patterns along an estuarine gradient, *Est. Coastal Shelf Sci.,* 16, 113, 1983.

Harger, J.R.E., The effect of wave impact on some aspects of the biology of sea mussels, *Veliger,* 12, 401, 1970.

Hargrave, B.T., The utilization of benthic microflora by *Hyatella azteca* (Amphipoda), *J. Anim. Ecol.,* 39, 427, 1970a.

Hargrave, B.T., The effect of a deposit-feeding amphipod on the metabolism of benthic microflora, *Limnol. Oceanogr.,* 15, 21, 1970b.

Hargrave, B.T., An energy budget for a deposit feeding amphipod, *Limnol. Oceanogr.,* 16, 99, 1971.

Hargrave, B.T., Factors affecting the flux of organic matter to sediments in a marine bay, in *Marine Benthic Dynamics,* Coull, B.C., Ed., University of South Carolina Press, Columbia, SC, 1980, 243.

Hargrave, B.T., Prouse, N.J., Phillips, G.A., Neame, P.A., Primary production and respiration in pelagic and benthic communities at two intertidal sites in the Bay of Funday, *Can. J. Fish. Aquatic Sci.,* 40, 229, 1983.

Harlin, M.M., Thorne-Miller, B., Boothroyd, J.C., Seagrass-sediment dynamics of a flood-tidal delta in Rhode Island (U.S.A.), *Aquatic Bot.,* 14, 127, 1982.

Harris, L.G., Ebling, A.W., Lur, D.R., Rowley, R.J., Community recovery after storm damage: a case of facilitation in primary succession, *Science,* 224, 1336, 1984.

Harris, V.A., *Sessile Animals of the Sea Shore,* Chapman and Hall, 379 pp., 1990.

Harrison, P.G., Mann, K.H., Chemical changes during the seasonal cycle of growth and decay in eelgrass (*Zostera marina*) on the Atlantic coast of Canada, *J. Fish. Res. Bd. Can.,* 32, 615, 1975a.

Harrison, P.G., Mann, K.H., Detritus formation from eelgrass (*Zostera marina* L.): the relative effects of fragmentation, leaching and decay, *Limnol. Oceanogr.,* 20, 924, 1975b.

Harrison, W.G., Nutrient regeneration and primary production in the sea, in *Primary Productivity in the Sea,* Falkowshi, P.G., Ed., Plenum Press, New York, 1980, 433.

Harrold, C., Pearse, J.S., The ecological role of echinoderms in kelp forests, in *Echinoderm Studies,* Vol. 2, Lawrence, M., Lawrence, J., Eds., 1982, 137.

Hartnoll, R.G., Bioenergetics of a limpet-grazed intertidal community, *S. Af. J. Sci.,* 79, 166, 1983.

Hartnoll, R.G., Hawkins, S.J., Monitoring rocky-shore communities: a critical look at spatial and temporal variation, *Helg. Wiss. Meeresunters.,* 33, 484, 1980.

Hartnoll, R.G., Hawkins, S.J., Patchiness and fluctuations on moderately exposed rocky shores, *Ophelia,* 24, 53, 1985.

Hartnoll, R.G., Wright, J.R., Foraging movements and homing in the limpet *Patella vulgata* L., *Anim. Behav.,* 25, 806, 1977.

Hartwig, E.O., Factors affecting respiration and photosynthesis by the benthic community of a subtidal siliceous sediment, *Mar. Biol.,* 46, 283, 1978.

Hastings, A., Spatial heterogeneity and the stability of predator-prey systems: predator-mediated coexistence, *Theor. Pop. Biol.,* 14, 380, 1980.

Hatcher, B.G., Mann, K.H., Above-ground production of the marsh cordgrass (*Spartina alterniflora*) near the northern end of its range, *J. Fish. Res. Bd. Can.,* 82, 83, 1975.

Haven, D.S., Morales-Alamo, R., Biodeposition as a factor in sedimentation of suspended solids in estuaries, *Mem. Geol. Soc. Am.,* 133, 121, 1966.

Haven, S.B., Effects of land-level changes on intertidal invertebrates, with discussion of post earthquake ecological succession, in *The Great Alaskan Earthquake of 1964: Biology,* National Academy of Sciences, Washington, D.C., 1971, 82.

Haven, S.B., Competition for food between the intertidal gastropods *Acmaea scabra* and *Acmaea digitalis, Ecology,* 54, 143, 1973.

Hawkins, S.J., The influence of *Patella* grazing in the fucoid-barnacle mozaic on moderately exposed rocky shores, *Kieler Meeresforsch Sonderheim,* 5, 537, 1981a.

Hawkins, S.J., The influence of season and barnacles on algal colonization of *Patella vulgata* exclusion areas, *J. Mar. Biol. Assoc. U.K.,* 61, 1, 1981b.

Hawkins, S.J., Interactions of *Patella* and macroalgae with settling *Semibalanus balanoides* (L.), *J. Exp. Mar. Biol. Ecol.,* 71, 55, 1983.

Hawkins, S.J., Hartnoll, R.G., The influence of barnacle cover on the numbers and behaviour of *Patella vulgata* on a vertical pier, *J. Mar. Biol. Assoc. U.K.,* 62, 855, 1982.

Hawkins, S.J., Hartnoll, R.G., Grazing of intertidal algae by marine invertebrates, *Oceanogr. Mar. Biol. Annu. Rev.*, 21, 195, 1983a.

Hawkins, S.J., Hartnoll, R.G., Settlement patterns of *Semibalanus balanoides* (L.) in the Isle of Man (1977–1981), *J. Exp. Mar. Biol. Ecol.*, 62, 271, 1983b.

Hawkins, S.J., Hartnoll, R.G., Kain, J.M., Norton, T.A., Plant-animal interactions on hard substrata in the north-east Atlantic, in *Plant-Animal Interactions in the Marine Benthos* (Systematics Association Special Volume No. 46), John, D.M., Hawkins, S.J., Price, J.H., Eds., Clarendon Press, Oxford, 1992, 1.

Hay, C.H., Some factors affecting the upper limit of the southern bull kelp *Durvillaea antarctica* (Chamisso) on two New Zealand shores, *J. Roy. Soc. N.Z.*, 9, 279, 1982.

Hay, M.E., Fish-seaweed interactions on coral reefs: effects of herbivorous fishes and adaptations of their prey, in *The Ecology of Coral Reef Fishes*, Sale, P.F., Ed., Academic Press, New York, 1991, 96.

Hay, M.E., Seawed chemical defences: their role in the evolution of feeding specialization and in mediating complex interactions, in *Ecological Roles of Marine Natural Products*, Paul, V.J., Ed., Comstock Publishing, Ithaca, NY, 1992.

Hay, M.E., Fenical, W., Marine plant-herbivore interactions: the ecology of chemical defence, *Annu. Rev. Ecol. Syst.*, 19, 111, 1988.

Hay, M.E., Fenical, W., Chemical mediation of seaweed-herbivore interactions, in *Plant-Animal Interactions in the Marine Benthos*, John, M.D., Hawkins, S.J., Price, J.H., Eds., Systematics Association Special Publication No. 46, Clarendon Press, Oxford, 1992, 319.

Hay, M.E., Fenical, W., Gustafson, K., Chemical defences against diverse coral-reef herbivores, *Ecology*, 68, 1581, 1987.

Hayes, W.B., Sand beach energetics: importance of the isopod *Tylos punctatus*, *Ecology*, 55, 838. 1974.

Hayes, W.B., Factors affecting the distribution of *Tylos punctatus* (Isopoda, Oniscoidea) on the beaches in Southern California and Mexico, *Pac. Sci.*, 31, 165, 1977.

Hayward, P.J., Invertebrate epiphytes of coastal marine algae, in *The Shore Environment, Vol. 2: Ecosystems*, Price, J.H., Irvine, D.E., Farnham, W.F., Eds., Academic Press, London, 1980, 761.

Head, G.C., Organic processes in estuaries, in *Estuarine Chemistry*, Burton, J.D., Liss, P.S., Eds., Academic Press, New York, 1976, 53.

Heald, E.J., The production of organic detritus in a South Florida estuary. Ph.D. thesis, University of Miami, Miami, FL, 110 pp, 1969.

Heald, E.J., The production of organic detritus in a South Florida estuary, *Univ. Miami Sea Grant Tech. Bull.*, 6, 110 pp, 1971.

Healey, M., Detritus and juvenile salmon production in the Nanaima Estuary. I. Production and feeding of juvenile chum salmon, *J. Fish. Res. Bd. Can.*, 36, 488, 1979.

Heck, K.L., Jr., Orth, R.J., Seagrass habitats: the roles of habitat complexity, competition and predation in structuring associated fish and motile macroinvertebrate assemblages, in *Estuarine Perspectives*, Kennedy, V.S., Ed., Academic Press, New York, 1980, 449.

Heck, K.L., Jr., Thoman, T.A., Experiments on predator-prey interactions in vegetated aquatic habitats, *J. Exp. Mar. Biol. Ecol.*, 53, 125, 1981.

Hedgpeth, J.W., Intertidal zonation and related matters, in *Between Pacific Tides* (3rd edition), Ricketts, E.F., Calvin, J., Eds., (appendix), Californian University Press, Stanford, CA, 1962, 345.

Hedgpeth, J.W., Ecological aspects of the Laguna Madre, a hypersaline estuary, in *Estuaries*, Lauff, G., Ed., American Association for the Advancement of Science, Publication No. 85, 1967, 408.

Heinbokel, J.F., Beers, J.R., Studies on the functional role of tintinnids in the Southern California Bight. III. Grazing impact of natural assemblages, *Mar. Biol.*, 52, 23, 1979.

Heinle, D.R., Production of a calanoid copepod *Acartia tonsa* in the Patuxent River Estuary, *Chesapeake Sci.*, 15, 146, 1966.

Heinle, D.R., Flemer, D.A., Carbon requirements of a population of the estuarine copepod *Eurytemora affinis*, *Mar. Biol.*, 31, 235, 1976a.

Heinle, D.R., Flemer, D.A., Flows of materials between poorly flooded tidal marshes and an estuary, *Mar. Biol.*, 35, 359, 1976b.

Heip, C., The life cycle of *Cyripideis torosa* (Crustacea, Ostracoda), *Oecologia (Berlin)*, 24, 229. 1976.

Heip, C., The influence of competition and predation on the production of meiobenthic copepods, in *Marine Benthic Dynamics*, Tenore, K.R., Coull, B.C., University of South Carolina Press, Columbia, SC, 1980, 167.

Heip, C., Smol, N., On the importance of *Protohydra leuckarti* as a predator of meiobenthic populations, *Proc. 10th European Mar. Biol. Symp. (Ostend)*, 2, 285, 1976.

Heip, C., Vine, M., Vraken, G., The ecology of marine nematodes, *Oceanogr. Mar. Biol. Annu. Rev.*, 23, 399, 1985.

Heip, H.R., Gossen, N.K., Herman, P.M.J., Kromkamp, J., Middleburg, J.J., Soetaert, K., Production and conservation of biological particles in temperate tidal estuaries, *Oceanogr. Mar. Biol. Annu. Rev.*, 33, 1, 1995.

Hemminga, M.A., Harrison, P.G., van Leut, F., The balance of nutrient losses and gains in seagrass meadows, *Mar. Ecol. Prog. Ser.*, 71, 85, 1991.

Hemminga, M.A., Mateo, M.A., Stable carbon isotopes in seagrass: variability in ratios and use in ecological studies, *Mar. Ecol. Prog. Ser.*, 140, 285, 1996.

Hemminga, M.A., Slim, F.J., Kazunga, J., Ganssen, G.M., Neuwenhuise, J., Knight, N.M., Carbon outwelling from a mangrove forest with adjecent seagrass beds and coral reefs (Gazi Bay, Kenya), *Mar. Ecol. Prog. Ser.*, 106, 191, 1994.

Henricksen, K., Hansen, J.I., Blackburn, T.H., The influence of benthic infauna on exchange rates of inorganic nitrogen between sediment and water, *Ophelia, Suppl.*, 1, 249, 1980.

Henricksen, K., Rasmussen, M.B., Jensen, A., Effect of bioturbation on microbial nitrogen transformations in the sediment and fluxes of ammonium and nitrate to overlying water, *Envt. Biogeochem.*, 35, 193, 1983.

Henriques, P.R., The Vegetated Tidelands of the Manakau Harbour, Report to the Manakau Harbour Study, May, 1978, 50 pp, 1978.

Heppleston, P.B., The feeding ecology of oystercatchers (*Haematopus ostralegus*) in winter in northern Scotland, *J. Anim. Ecol.*, 40, 561, 1971.

Herman, S.S., Mihursky, J.A., McEvlean, A.J., Zooplankton and environmental characteristics of the Patuxent River estuary, 1963–1965, *Chesapeake Sci.*, 9, 67, 1968.

Heymans, J.J., Energy flow model and network analysis of the Kromme Estuary, St. Francis Bay, South Africa, M.Sc. thesis, University of Port Elizabeth, Port Elizabeth, South Africa, 1992.

Heymans, J.J., Baird, D., Energy flow in the Kromme estuarine ecosystem, *Est. Coastal Shelf Sci.*, 42, 39, 1995.

Heymans, J.J., Mclachlan, A., Carbon budget and network analysis of a high-energy beach/surf zone ecosystem, *Est. Coastal Shelf Sci.*, 43, 485, 1996.

Hibbert, C.J., Biomass and production of a bivalve community in an intertidal inlet, *J. Exp. Mar. Biol. Ecol.*, 25, 249, 1976.

Hickel, W., Seston in the Wadden Sea of Sylt (German Bight, North Sea), *Neth. J. Sea Res. Publ. Ser.*, 10, 113, 1984.

Hicks, G.R.F., Species composition and zoogeography of marine phytal harpacticoid copepods from Cook Strait, and their contribution to total meiofauna, *N.Z. J. Mar. Freshw. Res.*, 11, 645, 1977a.

Hicks, G.R.F., Species associations and seasonal population densities of marine phytal harpacticoid copepods from Cook Strait, *N.Z. J. Mar. Freshw. Res.*, 11, 621, 1977b.

Hicks, G.R.F., Breeding activity of marine phytal harpacticoid copepods from Cook Strait, New Zealand, *N.Z. J. Mar. Freshw. Res.*, 11, 645, 1977c.

Hicks, G.R.F., Structure of phytal harpacticoid copepod assemblages and the influence of habitat complexity and turbidity, *J. Exp. Mar. Biol. Ecol.*, 44, 147, 1980.

Hicks, G.R.F., Spatio-temporal dynamic of a meiobenthic copepod and the impact of predation-disturbance, *J. Exp. Mar. Biol. Ecol.*, 81, 47, 1984.

Hicks, G.R.F., Meiofauna associated with rocky shore algae, in *The Ecology of Rocky Shores*, Moore, P.C., Seed, R., Eds., Hodder & Straughton, London, 1985, 36.

Hicks, G.R.F., Distribution and behaviour of meiofaunal copepods inside and outside seagrass beds, *Mar. Ecol. Prog. Ser.*, 31, 159, 1986.

Hicks, G.R.F., Coull, B.C., The ecology of marine meiobenthic harpacticoid copepods, *Oceanogr. Mar. Biol. Annu. Rev.*, 21, 67, 1983.

Hidu, H., Williams, G.V., Veitch, F.P., Gregarious settling in European and American oysters — response to surface chemistry versus waterbourne pheromones, *Proc. Nat. Shellfish Assoc.*, 68, 11, 1978.

Higgens, R.P., Thiel, H., Eds., *Introduction to the Study of Meiofauna*, Smithsonian Institution Press, Washington, D.C., 1988.

Highsmith, R.C., Induced settlement and metamorphosis of sand dollar (*Dendraster excentricus*) larvae in predator-free sites: adult sand dollar beds, *Ecology*, 63, 329, 1982.

Highsmith, R.C., Emlet, R.B., Delayed metamorphosus: effect on growth and survival of juvenile sand dollars (Echinoides; Clypeasteroida), *Bull. Mar. Sci.*, 39, 347, 1986.

Hill, A.S., Hawkins, S.J., An investigation of methods for sampling macrobial films on rocky shores, *J. Mar. Biol. Assoc. U.K.*, 70, 77, 1990.

Hily, C., Is the activity of benthic suspension feeders a factor controlling water quality in the Bay of Brest?, *Mar. Ecol. Prog. Ser.*, 69, 179, 1991.

Hinde, R., Seaweeds and other algae. I., in *Coastal Marine Ecology of Temperate Australia*, Underwood, A.J., Chapman, M.G., Eds., University of New South Wales Press, Sydney, 1995, 121.

Hines, A.H., The comparative reproductive ecology of three species of intertidal barnacles, in *Reproductive Ecology of Marine Invertebrates*, Staneyk, S.E., Ed., University of South Carolina Press, Columbia, SC, 1979, 213.

Hines, M.E., Jones, G.E., Microbial biogeochemistry and bioturbation in sediments of Great Bay, New Hampshire, *Est. Coastal Shelf Sci.*, 20, 729, 1985.

Hines, M.E., Owen, W.H., Lyons, W.BV., Jones, G.E., Microbial activities and bioturbation-induced oscillations in pore water chemistry of estuarine sediments in spring, *Nature*, 299, 433, 1982.

Hirschfield, M.F., Tinkle, D.W., Natural selection and the evolution of reproductive effort, *Proc. Nat. Acad. Sci.*, 72, 2227, 1974.

Hixon, M.A., Competitive interactions between Californian reef fishes of the genus *Embiotoca*, *Ecology*, 61, 918, 1980.

Hixon, M.A., Brostoff, W.N., Damselfish as a keystone species in reverse: intermediate disturbance and diversity of reef algae, *Science*, 220, 511, 1983.

Hobbie, J.E., Copeland, B.J., Harrison, W.G., Sources and fates of nutrients of the Palmico River estuary, North Carolina, in *Estuarine Rsearch*, Vol. 1, Cronin, L.E., Ed., Academic Press, New York, 1975, 265.

Hockey, P.A.R., Feeding techniques of the African black oystercatcher *Haemotopus moquini* on rocky shores, in *Proceedings of the Symposium Birds, Sea and Shore*, Cooper, J., Ed., African Seabird Group, Cape Town, 1983, 99.

Hockey, P.A.R., Bosman, A.L., Man as an intertidal predator in Transkei: disturbance, community convergence and management of a natural food resource, *Oikos*, 46, 3, 1986.

Hockey, P.A.R., Bosman, A.L., Stabilizing processes in bird-prey interactions on rocky shores, in *NATO Advanced Research Workshop on Behavioural Adaptation to Intertidal Life*, Chelazzi, C., Vannimi, M., Eds., Plenum Press, New York and London, 1988, 297.

Hockey, P.A.R., Branch, G.M., Oyster catchers and limpets: impact and implications, *Ardea*, 72, 199, 1984.

Hockey, P.A.R., Siegfried, W.R., Crowe, A.A., Cooper, J., Ecological structure and energy requirements of the sandy beach avifauna of Southern Africa, in *Sandy Beaches as Ecosystems*, McLachlan, A., Erasmus, T., Eds., Dr. W. Junk Publishers, The Hague, 1983, 507.

Hockey, P.A.R., Underhill, L.G., Diet of African black oystercatcher *Haemotopus moquini* on rocky shores: spatial, temporal and sex-related variation, *S. Af. J. Zool.*, 19, 1, 1984.

Hockin, D.C., Ollason, J.C., The colonization of artificially isolated volumes of intertidal estuarine sand by harpacticoid copepods, *J. Exp. Mar. Biol. Ecol.*, 53, 9, 1981.

Hodgkin, E.P., Patterns of life on rocky shores, *J. Roy. Soc. West. Aust.*, 43, 35, 1960.

Hodgkin, E.P., Rippingdale, R.J., Interspecies conflicts in estuarine copepods, *Limnol. Oceanogr.*, 16, 573, 1971.

Hoffman, D.L., Homan, W.C., Swanson, J., Weldon, P.J., Flight responses of three congeneric species of intertidal gastropods (Prosobranchia: Neritidae) to sympatric predatory gastropods from Barbados, *Veliger*, 21, 293, 1978.

Hoffman, J.A., Katz, J., Bertness, M.D., Fiddler crab deposit-feeding and meiofaunal abundance in salt marsh habitats, *J. Exp. Mar. Biol. Ecol.,* 82, 161, 1984.

Holdren, G.C., Armstrong, D.E., Factors affecting phosphorus release from intact sediment cores, *Envt. Sci. Tech.,* 14, 79, 1982.

Holland, A.F., Mountford, N.K., Hiegal, M.H., Kaumeyer, K.R., Mihursky, J.A., Influence of predation on infaunal abundance in upper Chesapeake Bay, U.S.A, *Mar. Biol.*, 57, 221, 1980.

Holling, C.S., Resilience and stability of ecological systems, *Annu. Rev. Ecol. Systematics*, 4, 1, 1973.

Holm-Hansen, O., Booth, R.C., The measurement of adenosine triphosphate in the ocean and its ecological significance, *Limnol. Oceanogr.*, 11, 510. 1966.

Holt, R.W., Predation, apparent competition and the structure of prey communities, *Theoret. Pop. Biol.*, 11, 197, 1977.

Hopkinson, C.S., Jr., Patterns of organic carbon exchange between ecosystems. The mass balance approach to salt marsh ecosystems, in *Coastal-Offshore Ecosystem Interactions, Lecture Notes on Coastal and Estuarine Studies 22*, Jannson, B.-O., Springer-Verlag, Berlin, 1988, 122.

Hopkinson, C.S., Day, J.W., A model of the Barataria Bay salt marsh ecosystem, in *Ecosystem Modelling in Theory and Practice,* Hall, C.A.S., Day, J.W., Jr., Eds., John Wiley & Sons, New York, 1977, 236.

Hopkinson, C.S., Gosselink, G., Parrondo, R., Above-ground production of seven marsh species in coastal Louisiana, *Ecology*, 61, 1091, 1978.

Hopkinson, C.S., Wetzel, R.L., *In situ* measurements of nutrient and oxygen fluxes in a coastal marine benthic community, *Mar. Ecol. Prog. Ser.*, 19, 29, 1982.

Hoppe, H.G., Relations between active bacteria and heterotrophic potential in the sea, *Neth. J. Sea Res.*, 12, 78, 1978.

Horn, M.H., Optimal diets in complex environments: feeding strategies of the herbivorous fishes from a temperate rocky shore, *Oecologia*, 58, 345, 1983.

Horn, M.H., Biology of marine herbivorous fishes, *Oceanogr. Mar. Biol. Annu. Rev.*, 27, 167, 1989.

Horn, M.H., Herbivorous fishes: feeding and digestive mechanisms, in *Plant-Animal Interactions in the Marine Benthos*, John, D.M., Hawkins, S.J., Price, J.H., Eds., (Systematics Association Special Volume, No. 46), Clarendon Press, Oxford, 1992, 339.

Horn, P.L., Energetics of *Chiton pelliserpentis* (Quoy and Gaimard, 1835) (Mollusca; Polyplacophora) and the importance of mucus in its energy budget, *J. Exp. Mar. Biol. Ecol.*, 101, 119, 1986.

Hoshiai, T., Synecological study of intertidal communities. IV. An ecological investigation of the zonation in Matsuchiman Bay concerning the so-called covering phenomenon, *Bull. Biol. Stn. Asamushi*, 12, 93, 1964.

Houghton, R.A., Woodwell, G.M., The Flax Pond ecosystem study: exchanges of carbon dioxide between a salt marsh and the atmosphere, *Ecology,* 61, 1434. 1980.

Howard, R.K., Edgar, G.J., Seagrass meadows, in *Marine Biology,* Hammond, L.S., Synnot, R.N., Eds., Longman Cheshire, Melbourne, Australia, 1994, 257.

Howarth, R.W., The ecological significance of sulfur in the energy dynamics of salt marsh and coastal marine sediments, *Biogeochem.*, 1, 5, 1984.

Howarth, R.W., Cole, J., Molybdenum availability, nitrogen limitation, and phytoplankton growth in coastal waters, *Science*, 229, 653, 1985.

Howarth, R.W., Hobbie, J.E., The regulation of decomposition and heterotrophic activity in salt marsh soils: a review, in *Estuarine Comparisons*, Kennedy, V.S., Ed., Academic Press, New York, 1982, 183.

Howarth, R.W., Teal, J.M., Energy flow in a salt marsh ecosystem: the role of reduced inorganic sulfur compounds, *Am. Nat.*, 116, 862, 1980.

Hudson, C., Bourget, E., Legendre, P., An integrated study of the factors influencing the choice of settling site of *Balanus crenatus* cyprid larvae, *Can. J. Fish. Aquatic Sci.,* 40, 1186, 1983.

Huebner, J.D., Edwards, D.C., Energy budget of the predatory marine gastropod *Polinices duplicatus*, *Mar. Biol.*, 61, 221, 1981.

Huggett, J., Griffiths, C.L., Some relationships between elevation, physico-chemical variables and biota of intertidal rock pools, *Mar. Ecol. Prog. Ser.*, 29, 189, 1986.

Hughes, E.H., Sherr, E.B., Subtidal food webs in a Georgia estuary: ^{13}C analysis, *J. Exp. Mar. Biol. Ecol.*, 67, 227, 1983.

Hughes, R.G., Thomas, M.L., The classification and ordination of shallow-water benthic samples from Prince Edward Island, Canada, *J. Exp. Mar. Biol. Ecol.*, 7, 1, 1971.

Hughes, R.N., A study of feeding in *Scrobicularia plana*, *J. Mar. Biol. Assoc., U.K.*, 49, 895, 1969.

Hughes, R.N., An energy budget for a tidal-flat population of the bivalve *Scrobularia plana* (DaCosta), *J. Anim. Ecol.*, 39, 357, 1970.

Hughes, R.N., Ecological energetics of the keyhole limpet *Fissurella barbadensis* Gmelin, *J. Exp. Mar. Biol. Ecol.*, 67, 167, 1971a.

Hughes, R.N., Ecological energetics of *Nerita* (Archaeogastropoda: Neritacea) populations on Barbados, West Indies, *Mar. Biol.*, 11, 12, 1971b.

Hughes, R.N., Annual production of two Nova Scotian population of *Nucell lapillus* (L.), *Oecologia (Berlin)*, 8, 356, 1972.

Hughes, R.N., Optimal foraging in the marine context, *Oceanogr. Mar. Biol. Annu. Rev.*, 18, 423, 1980a.

Hughes, R.N., Predation and community structure, in *The Shore Environment,* Vol. 2: *Ecosystems,* Price, J.H., Irvine, D.E.G., Farnham, W.H., Eds., Academic Press, London, 1980b, 699.

Hughes, R.N., Rocky shore communities: catalysts to understanding predation, in *Ecology of Rocky Shores*, Moore, P.G., Seed, R., Eds., Hodder & Straughton, London, 1985, 223.

Hughes, R.N., Burrows, M.T., Energy maximization in the natural foraging behaviour of the dogwhelk, *Nucella lapillus*, in *Trophic Relationships in the Marine Environment,* Barnes, M., Gibson, R.N., Eds., Aberdeen University Press, Aberdeen, 1990, 517.

Hughes, R.N., Dunkin, S.deB., Behavioural components of prey selection by dogwhelks, *Nucella lapillus* (L.), in the laboratory, *J. Exp. Mar. Biol. Ecol.*, 79, 45, 1984.

Hughes, R.N., Elner, R.W., Tactics of a predator, *Carcinus manas,* and morphological responses of the prey, *Nucella lapillus, J. Anim. Ecol.*, 48, 65, 1979.

Hughes, R.N., Roberts, D.J., Reproductive effort of winkles (*Littorina* spp.) with contrasted methods of reproduction, *Oecologia (Berlin),* 47, 130, 1980.

Hui, E., Studies on intertidal Lepadomorpha and Balanomorpha (Cirripedia: Thoracica), Ph.D. thesis, University of Wales.

Hui, E., Moyse, J., Settlement patterns and competition for space, in *Barnacle Biology,* Crustacean Issues 5, Southward, A.J., Ed., A.A. Balkema, Rotterdam, 1987, 363.

Hulberg, L.W., Oliver, J.S., Caging manipulations in marine soft-bottom communities: importance of animal interactions of sedimentary habitat modifications, *Can. J. Fish. Aquatic Sci.*, 37, 1130, 1980.

Hull, S.C., Macroalgal mats and species abundance: a field experiment, *Est. Coastal Shelf Sci.*, 25, 519, 1987.

Hulscher, J.B., Hetswad, een, overbedig of schaars gedekie tafel voorvogels?, in *Symp.Wassenonderzoek, Arnhem,* Swenneb, C., de Wilde, P.A.W.J., Hacek, J., Eds., 1975, 57.

Hulscher, J.B., The Oystercatcher *Haemotopus ostralegus* as a predator of the bivalve *macoma balthica* in the Dutch Wadden Sea, *Ardea*, 70, 89, 1982.

Hummel, H., Food intake of *Macoma balthica* (Mollusca) in relation to seasonal changes in its potential food on a tidal flat in the Dutch Wadden Sea, *Neth. J. Sea Res.*, 19, 52, 1985a.

Hummel, H., An energy budget for a *Macoma balthica* (mollusca) population living on a tidal flat in the Dutch Wadden Sea, *Neth. J. Mar. Res.,* 19, 84, 1985b.

Hunt, J.H., Ambrose, W.G., Jr., Peterson, C.H., Effect of the gastropod, *Ilyanassa obsoleta (Say)*, and the bivalve, *Mercenaria mercenaria* (L.), on larval settlement and juvenile recruitment of infauna, *J. Exp. Mar. Biol. Ecol.*, 108, 229, 1987.

Hunt, H.L., Scheibling, R.E., Role of early post-settlement mortality in the recruitment of benthic marine invertebrates, *Mar. Ecol. Prog. Ser.,* 155, 269, 1997.

Hunter, M.D., Price, P.W., Playing chutes and ladders: heterogeneity and the relative roles of bottom-up and top-down forces in natural communities, *Ecology,* 73, 724, 1992.

Hunter, R.D., Russel-Hunter, W.D., Bioenergetic and community changes in intertidal aufwachs grazed by *Littorina littorea, Ecology,* 64, 761, 1983.

Hurlbut, C.J., Community recruitment: settlement and juvenile survival of seven co-occurring species of sessile marine invertebrates, *Mar. Biol.*, 109, 509, 1991.

Hurley, A.C., The establishment of populations of *Balanus pacificus* Pilsbury (Cirripedia) and their elimination by predatory Turbellaria, *J. Anim. Ecol.,* 44, 521, 1975.

Huston, M., A general hypothesis of species diversity, *Am. Nat.*, 113, 81, 1979.

Hutchings, L., Nelson, G., Hartsmann, D., Tarr, R., Interactions between coastal plankton and sand mussels along the Cape coast, South Africa, in *Sandy Beaches as Ecosystems,* McLachlan, A., Erasmus, T., Eds., Dr. W. Junk Publishers, The Hague, 1983, 481.

Hutchinson, E.G., Concluding remarks, *Cold Spring Harbor Symposium on Quantitative Biology,* 22, 415, 1952.

Hutchins, J.B., Thompson, M., *The Marine and Estuarine Fishes of South-Western Australia,* Western Australian Museum, Perth, Australia, 1983.

Huttel, M., Influence of the lugworm *Arenicola marina* on porewater nutrient profiles of sandflat sediments, *Mar. Ecol. Prog. Ser.,* 62, 241, 1990.

Hylleberg, J., Selective feeding by *Abarenicola pacifica* with notes on *Abarenicola vagubunda* and a concept of gardening in lugworms, *Ophelia,* 14, 113, 1975.

Hylleberg, J., Distribution of hydrobiid snails in relation to salinity, with emphasis on size and coexistance of the species, *Ophelia (Suppl.),* 4, 85, 1986.

Hylleberg, J., Henriksen, K., The central role of bioturbation in sediment remineralization and element recycling, *Ophelia,* (Suppl. 1), 1, 1980.

Imrie, D.W., Hawkins, S.J., McCrohan, C.R., The olfactory-gustatory basis of food preference in the herbivorous gastropod *Littorina littorea* (Linnaeus), *J. Mollusc. Stud.*, 55, 217, 1989.

Incze, L.S., Mayer, L.M., Sherr, E.B., Macko, S.A., Carbon inputs to bivalve molluscs: a comparison of two estuaries, *Can. J. Fish. Aquatic Sci.*, 39, 1348, 1982.

Inglis, G., The colonization and degradation of stranded *Macrocystis pyrifera* (L.) C.Ag. by the macrofauna of a New Zealand sandy beach, *J. Exp. Mar. Biol. Ecol.,* 125, 203, 1989.

Ireland, C., Biskupiak, J.E., Hite, G.J., Rapprosch, M., Scheuer, P.J., Rublee, J.R., Likonapyrone esters, likely defence allomones of the mollusc *Onchidium verruculatum, J. Organic Chem.*, 49, 539, 1984.

Ito, S., Imai, T., Ecology of oyster beds. I. On the decline of productivity due to repeated cultures, *Tokohu J. Agric. Res.,* 5, 251, 1975.

Izumi, H., Hattori, A., Growth and organic production of eelgrass (*Zostera marina L.*) in temperate waters of the Pacific coast of Japan. III. The kinetics of nitrogen uptake, *Aquatic Bot.*, 12, 245, 1982.

Jablouski, D., Lutz, R.A., Larval ecology of marine benthic invertebrates: paleobiological implications, *Biol. Rev.*, 58, 21, 1983.

Jackson, G.A., Interaction of physical and biological processes in the settlement of planktonic larvae, *Bull. Mar. Sci.*, 39, 202, 1986.

Jackson, G.A., Strathmann, R.R., Larval mortality from offshore mixing as a link between precompetent and competent periods of development, *Am. Nat.*, 118, 16, 1981.

Jackson, L.F., Aspects of the international ecology of the east coast of South Africa, *Invest. Rep. Oceanogr. Res. Institut.,* (Durban, South Africa), 45, 1, 1976.

Jansson, A.-M., Community structure, modelling and simulation of the *Cladophora* ecosystem in the Baltic Sea, *Contr. Asko Lab. Univ. Stockholm,* 5, 1, 1974.

Jansson, B.-O., The Baltic — a systems analysis of a semi-enclosed sea, in *Advances in Oceanography,* Charnock, H., Deacon, G., Eds., Plenum Press, New York, 1978, 181.

Jansson, B.-O., Modelling the Baltic ecosystem, *Ambio Spec. Rep.,* 4, 1976.

Jansson, B.-O., Wilmat, W., Wulff, F., Coupling of the subsystems — the Baltic Sea, a case study, in *Flows of Energy and Materials in Marine Ecosystems*, Fasham, M.J.R., Ed., Plenum Publishing, New York, 1984, 549.

Jansson, B.-O., Wulff, F., Baltic ecosystem modelling, in *Ecosystem Modelling in Theory and Practice*, Hall, C.A.S., Day, J.W., Jr., Eds., John Wiley & Sons, New York, 1977, 323.

Jara, H.F., Moreno, C.A., Herbivory and structure in a midlittoral rocky community: a case in southern Chile, *Ecology*, 65, 28, 1984.

Jaramillo, E., McLachlan, A., Community and population responses of the macroinfauna to physical factors over a ranges of exposed sandy beaches in south-central Chile, *Est. Coastal Shelf Sci.*, 37, 615, 1993.

Jaspar Sidik Bujang, Studies on leaf litter decomposition of the mangrove *Rhizophora apiculata* Bl., Ph.D. thesis, Universiti Sains Malaysia, Malaysia, 322 pp, 1989.

Jeffries, H.P., Saturation of estuarine zooplankton by congeneric associates, in *Estuaries*, Lauff, G.H., Ed., American Association for the Advancement of Science, Washington, DC, 1967, 560.

Jennings, J.N., Bird, E.C., Regional geomorphological characteristics of some Australian estuaries, in *Estuaries*, Lauff, G., Ed., Publication No. 85, American Association for the Advancement of Science, Washington, DC, 1967, 121.

Jensen, K.T., Jensen, J.N., The importance of some epibenthic predators on the density of juvenile benthic macrofauna in the Danish Wadden Sea, *J. Exp. Mar. Biol. Ecol.*, 89, 157, 1985.

Jensen, K.T., Siegsmund, H.R., The importance of diatoms and bacteria in the diet of *Hydrobia* species, *Ophelia (Suppl.)*, 17, 1983, 1980.

Jensen, P., Ecology of free-living aquatic nematodes, *Mar. Ecol. Prog. Ser.*, 35, 180, 1987.

Jensen, R.A., Morse, D.E., Intraspecific facilitation of larval recruitment: gregarious settlement of the polychaete *Phragmatopoma californica* (Fewkes), *J. Exp. Mar. Biol. Ecol.*, 83, 107, 1990.

Jephson, N.A., Gray, P.W.G., Aspects of the ecology of *Sargassum muticum* (Yendo) Fensholt in the Solent region of the British Isles. I. The growth cycle and epiphytes, in *Biology of Benthic Organisms*, Keegan, B.F., Ceodigh, P.O., Boaden, P.H.S., Eds., Pergamon Press, Oxford, 1977, 367.

Jernakoff, P., Factors affecting the recruitment of larvae in a midshore region dominated by barnacles, *J. Exp. Mar. Biol. Ecol.*, 67, 17, 1983.

Jernakoff, P., Interactions between the limpet *Patelloida latistrigata* and algae on an intertidal rock-platform, *Mar. Ecol. Prog. Ser.*, 23, 71, 1985a.

Jernakoff, P., An experimental evaluation of the influence of barnacles, crevices and seasonal patterns of grazing on algal diversity and cover in an intertidal zone, *J. Exp. Mar. Biol. Ecol.*, 88, 287, 1985b.

Jernakoff, P., Brearly, A., Nielsen, J., Factors affecting grazer-epiphyte interactions in temperate seagrass meadows, *Oceanogr. Mar. Biol. Annu. Rev*, 34, 109, 1996.

John, D.M., Hawkins, S.J., Price, J.H., Eds., *Plant-Animal Interactions in the Marine Benthos*, Systematics Association Special Volume No. 46, Clarendon Press, Oxford, 570 pp, 1992.

John, D.M., Price, J.H., Lawson, G.W., Tropical East Atlantic and islands: plant-animal interactions on shores free of biotic reefs, in *Plant-Animal Interactions in the Marine Benthos*, John, D.M., Hawkins, S.J., Price, H.P, Eds., (Systematics Association Special Volume, No. 46), Clarendon Press, Oxford, 1992, 87.

Johnson, C.R., Mann, K.H., The crustose coralline alga *Phymatolithen* Foslie inhibits the overgrowth of seaweeds without relying on herbivores, *J. Exp. Mar. Biol. Ecol.*, 96, 127, 1986.

Johnson, C.R., Mann, K.H., Diversity patterns of adaptation, and stability of Nova Scotia kelp beds, *Ecol. Monogr.*, 58, 129, 1986.

Johnson, C.R., Muir, D.G., Reysenbach, A.L., Characteristic bacteria associated with the surfaces of coralline algae: a hypothesis for bacterial induction of marine invertebrate larvae, *Mar. Ecol. Prog. Ser.*, 74, 281, 1991.

Johnson, D.S., Skutch, A.F., Littoral vegetation on a headland of Mt. Desert Island, Maine. II. Tidepools and the environment and classification of submersible plant commmunities, *Ecology*, 9, 307, 1928.

Johnson, L.E., Strathman, R.R., Settling barnacle larvae avoid substrata previously occupied by a mobile predator, *J. Exp. Mar. Biol. Ecol.*, 128, 87, 1989.

Johnson, R, G., Animal-sediment relations in shallow water benthic communities, *Mar. Geol.*, 11, 93, 1971.

Johnson, R.G., Conceptual models in benthic communities, in *Models in Paleobiology*, Schopf, T.J.M., Ed., Freeman Cooper & Co., San Francisco, 1973, 148.

Johnson, S.C., Scheibling, R.E., Structure and dynamics of epifaunal assemblages in intertidal macroalgae *Ascophyllum nodosum* and *Fucus vesciculosus* in Nova Scotia, Canada, *Mar. Ecol. Prog. Ser.*, 37, 209, 1987.

Joint, I.R., Microbial production of an estuarine mudflat, *Est. Coastal Mar. Sci.*, 7, 185, 1978.

Joint, I.R., Growth and survival of estuarine microalgae, in *Feeding and Survival Strategies of Estuarine Organisms*, .Jones, N.V., Wolf, W.J., Eds., Plenum Press, New York, 1981, 17.

Joint, I.R., Gee, I.M., Warwick, R.M., Determination of the fine-scale vertical distribution of microbes and meiofauna in an intertidal sediment, *Mar. Biol.*, 72, 157, 1982.

Joint, I.R., Morris, R.J., The role of bacteria in the turnover of organic matter in the sea, *Oceanogr. Mar. Biol. Annu. Rev.*, 20, 65, 1982.

Joint, I.R., Pomeroy, A.J., Primary production in a turbid estuary, *Est. Coastal Shelf Sci.*, 13, 303, 1981.

Jones, A.R., Short, A.D., Sandy beaches, in *Coastal Marine Ecology of Temperate Australia*, Underwood, A.J., Chapman, M.G., Eds., University of New South Wales Press, Sydney, Australia, 1995, 136.

Jones, D.A., Hobbins, C., St.C., The role of biological rhythms in some sand beach cirolanid Isopoda, *J. Exp. Mar. Biol. Ecol.*, 93, 47, 1985.

Jones, D.A., Naylor, E., The swimming behaviour of the sand beach isopod *Eurydice pulchra*, *J. Exp. Mar. Biol. Ecol.*, 4, 189, 1970.

Jones, J.A., Primary productivity of the tropical marine turtle grass *Thalassia testudinium* Konig, and its epiphytes, Doctoral dissertation, University of Miami, Miami, FL, 196 pp, 1968.

Jones, J.C., *A Guide to Methods for Estimating Microbial Numbers and Biomass in Fresh Water*, Freshwater Biological Association, Scientific Publication No. 39, 112 pp, 1979.

Jones, K., Nitrogen fixation in a salt water marsh, *J. Ecol.*, 62, 553, 1974.

Jones, M.B., Greenwood, L.G., Greenwood, J., Distribution, body size, and brood characteristics of four species of mysids (Crustacea: Peracaridae) in the Avon-Heathcote Estuary, *N.Z. J. Mar. Freshw. Res.*, 23, 195, 1989.

Jones, W.E., Demetropoulos, A., Exposure to wave action: measurements of an important ecological parameter on rocky shores of Anglesey, *J. Exp. Mar. Biol. Ecol.*, 2, 46, 1968.

Jonsson, P.R., Particle size selection, feeding rates and growth dynamics of marine planktonic oligotrichous citiates (Ciliophora: Oligotrichina), *Mar. Ecol. Prog. Ser.*, 33, 265, 1986.

Jordon, J.E., Correll, D.L., Continuous automated sampling of tidal exchanges of nutrients by brackish marshes, *Est. Coastal Shelf Sci.*, 17, 651, 1991.

Jordon, T.E., Correll, D.L., Miklas, J. Weller, D.E., Nutrients and chlorophyll at the interface of a watershed and an estuary, *Limnol. Oceanogr.*, 36, 251, 1991.

Jordon, T.E., Correll, D.L., Whigham, D.F., Nutrient flux in the Rhode River: tidal exchange of nutrients by brackish marshes, *Est. Coastal Shelf Sci.*, 17, 651, 1983.

Jorgensen, B.B., The sulfur cycle of a coastal marine sediment (Limfjorden, Denmark), *Limnol. Oceanogr.*, 22, 814, 1977a.

Jorgensen, B.B., Bacterial sulphate reduction within reduced microniches of oxidized marine sediments, *Mar. Biol.*, 41, 7, 1977b.

Jorgensen, B.B., Processes at the sediment-water interface, in *The Major Biochemical Cycles and Their Interactions*, Bolin, B., Cook, R.B., Eds., John Wiley & Sons, Chichester, 1983, 477.

Jorgensen, C.B., *Bivalve Filter Feeding: Hydrodynamics, Bioenergetics, Physiology and Ecology*, Olsen & Olsen, Frendebsborg, Denmark, 1990.

Joris, C., On the role of heterotrophic bacteria in marine ecosystems: some problems, *Helogl. Wiss. Meeresunters*, 30, 611, 1977.

Joris, C., Billen, G., Lancelot, C., Doro, M.H., Mommaerts, J.P., Bertels, A., Bossicart, M., Nijs, J., Hecq, J.H., A budget of carbon cycling in the Belgian coastal zone: relative roles of zooplankton, bacterioplankton and benthic in the utilization of primary production, *Neth. J. Sea. Res.*, 16, 260, 1982.

Joseph, E.B., Analysis of a nursery ground, in *Proceedings of a Workshop on Eggs, Larval and Juvenile Stages of Fish in Atlantic Coastal Estuaries*, Technical Publication No. 1., NMFS Mid-Atlantic Coast Fisheries Center, Highlands, NJ, 1973, 118.

Josselyn, M., The ecology of the San Francisco Bay tidal marshes: a community profile, U.S. Fish Wildl. Ser. Rep. No. FWS/OBS — 83/23, Washington, D.C., 102 pp, 1983.

Jumars, P.A., Nowell, A.R.M., Fluid and sediment dynamic effects on marine benthic community structure, *Am. Zool.*, 24, 45, 1984.

Juniper, S.K., Stimulation of sediment bacterial grazing in two New Zealand inlets, *Bull. Mar. Sci.*, 31, 691, 1981.

Juniper, S.K., Regulation of microbial production in intertidal mudflats — the role of *Amphibola crenata*, Ph.D. thesis, Department of Zoology, University of Canterbury, Christchurch, NZ, 152 pp, 1982.

Kaczynshi, V.W., Feller, R.J., Clayton, J., Gerke, R.-J., Trophic analysis of juvenile pink and chum salmon (*Oncorhynchus gorbuscha* and *O. keta*) in Puget Sound, *J. Fish. Res. Bd. Can.* 30, 1003, 1973.

Kain, J.M., Algal recolonization of some cleared subtidal areas, *J. Ecol.*, 63, 739, 1975.

Kaplan, W., Valiela, I., Teal, J.M., Denitrification in a salt marsh ecosystem, *Limnol. Oceanogr.*, 24, 726, 1979.

Karentz, D., Smayda, T.J., Temperature and seasonal occurance of 30 dominant phytoplankton species in Narrangansett Bay over a 22-year period, 1959–1980, *Mar. Ecol. Prog. Ser.*, 18, 277, 1984.

Kastendiek, J., Competitor-mediated coexistance: interactions among three species of macroalgae, *J. Exp. Mar. Biol. Ecol.*, 62, 201, 1982.

Katz, L.C., Effects of burrowing by the fiddler crab, *Uca pugnax* (Smith), *Est. Coastal Mar. Sci.*, 11, 233, 1980.

Kautsky, N., Evans, S., Role of biodeposition by *Mytilus edulis* in the circulation of matter and nutrients in a Baltic coastal ecosystem, *Mar. Ecol. Prog. Ser.*, 38, 201, 1987.

Kautsky, N., Wallentinus, I., Nutrient release from a Baltic *Mytilus* — red algae community and its role in benthic and pelagic productivity, *Ophelia (Supplement)*, 1, 17, 1980.

Kay, J.H., Graham, L.A., Ulanowicz, R.E., A detailed guide to network analysis, in *Network Analysis in Marine Ecology: Methods and Applictions*, Wulff, F., Field, J.G., Mann, K.H., Eds., Springer-Verlag, Berlin, 1989, 51.

Kay, J.J., Self organization in living systems, Ph.D. thesis, Univerity of Waterloo, Waterloo, Canada, 1984.

Kay, P.G., Bradfield, A.E., The energy relations of the polychaete species *Neanthes* (= *Nereis*) *virens* (Sars), *J. Anim. Ecol.*, 42, 673, 1977.

Keats, D.W., Steele, D.H., Smith, G.R., Ocean pout (*Macrozoares americanus*) (Bloch & Schneider) (Pisces:Zoarcidae) predation on green urchins *Strongylocentrotus droebachiensis* (O.F. Mull.) (Echinodermata:Echinoidea) in eastern Newfoundland, *Can. J. Zool.*, 65, 1515, 1987.

Keefe, C.W., Boynton, W.R., Kemp, W.M., A review of phytoplankton processes in estuarine environments, unpublished report, UMCEES Ref, No, 81-193- CBL, Chesapaeke Biological Laboratory, Solomon, MD, 1981.

Kemp, P.F., Direct uptake of detrital carbon by the deposit-feeding polychaete *Euzonus mucronata* (Treadwell), *J. Exp. Mar. Biol. Ecol.*, 99, 49, 1986.

Kemp, P.F., Bactivory by benthic ciliates: significance as a carbon source and impact on sediment bacteria, *Mar. Ecol. Prog. Ser.*, 49, 163, 1988.

Kemp, W.M., Wetzel, R., Boynton, W., D'Elia, C., Stevenson, J., Nitrogen cycling and estuarine interfaces: some current research, in *Estuarine Interactions*, Kennedy, V.S., Ed., Academic Press, New York, 1982, 209.

Kench, P.S., Geomorphology of Australian estuaries: review and prospect, *Aust. J. Ecol.*, 24, 367, 1999.

Kendall, M.A., Bowman, R.S., Williamson, P., Lewis, J.R., Settlement patterns, density and stability in the barnacle *Balanus balanoides*, *Neth. J. Sea Res.*, 16, 119, 1982.

Kennedy, A.D., Carbon partitioning within meiobenthic nematode communities in the Exe estuary, U.K., *Mar. Ecol., Prog. Ser.*, 105, 71, 1994.

Kennedy, V.S., Desiccation, higher temperatures and upper intertidal limits of three species of sea mussels (Mollusca:Bivalvia) in New Zealand, *Mar. Biol.*, 35, 127, 1976.

Kennelly, S.J., Kelp beds, in *Coastal Marine Ecology of* Temperate Australia, Underwood, A.J., Chapman, M.G., Eds., University of New South Wales Press, Sydney, 1995, 106.

Kennelly, S.J., Larkum, A.W.D., A preliminary study of temporal variation in the colonization of subtidal algae in an *Ecklonia radiata* community, *Aquatic Bot.*, 17, 257, 1984.

Kennish, M.J., *Ecology of Estuaries*. Vols. I and II, CRC Press, Boca Raton, FL, 254 pp, 1984.

Kentula, M.E., McIntire, C.D., The autoecology and production dynamics of eelgrass (*Zostera marina* L.) in Netarts Bay, Oregon, *Estuaries*, 9, 188, 1986.

Kenworthy, W.J., Thayer, G.W., Production and decomposition of the roots and rhizomes of seagrasses, *Zostera marina* and *Thalassia testudinium*, in temperate and subtropical marine systems, *Bull. Mar. Sci.*, 5, 364, 1984.

Keough, M.J., Patterns of recruitment of sessile invertebrates in two distinct subtidal habitats, *J. Exp. Mar. Biol. Ecol.*, 66, 213, 1983.

Keough, M.J., Effects of patch size on the abundance of sessile marine invertebrates, *Ecology*, 65, 423, 1984.

Keough, M.J., The distribution of a bryozoan on seagrass blades: settlement, growth and mortality, *Ecology*, 68, 846, 1986.

Keough, M.J., Connell, J.H., Disturbance and patch dynamics of subtidal marine animals on hard substrata, in *Natural Disturbance: An Evolutionary Perspective,* Pickett, S.T.A., White, P.S., Eds., Academic Press, New York, 1984, 125.

Keough, M.J., Downes, B.J., Recruitment of marine invertebrates: role of active larval choices and early mortality, *Oecologia (Berlin)*, 51, 348, 1982.

Keough, M.J., Jenkins, G.P., Seagrass meadows and their inhabitants, in *Coastal Marine Ecology,* Underwood, A.J., Chapman, M.G., Eds., New South Wales University Press, Sydney, 1995, 221.

Kepkay, P.E., Anderson, F.Ø., Aerobic and anaerobic metabolism of a sediment enriched with *Spartina* detritus, *Mar. Biol.*, 21, 153, 1985.

Kern, J.C., Bell, S.S., Short-term temporal variation in population structure of two harpacticoid copepods *Zausodes arenicolus* and *Paradctylopodia brevicornis*, *Mar. Biol.,* 84, 53, 1984.

Keser, M., Larson, B.R., Colonization and growth dynamics of three species of *Fucus*, *Mar. Ecol. Prog. Ser.*, 15, 125, 1984a.

Keser, M., Larson, B.R., Colonization and growth of *Ascophyllum nodosum* (Phaeophyta) in Maine, *J. Phycol.*, 20, 83, 1984b.

Khailov, K.M., Burlakova, Z.P., Release of dissolved organic matter by marine seaweeds and distribution of their total organic production to inshore communities, *Limnol. Oceanogr.*, 14, 521, 1969.

Kideys, A.E., Hartnoll, R.G., Energetics of mucus production in the common whelk *Buccinium undulatum* L., *J. Exp. Mar. Biol. Ecol.*, 150, 91, 1991.

Kikuchi, T., Some aspects of the ecology, life cycle and taxonomy of the polychaete *Capitella cpitata,* so-called pollution indicator: a review, *Benthos Res.*, 17/18, 33, 1979.

Kikuchi, T., Faunal relationships in temperate seagrass beds, in *Handbook of Seagrass Biology: An Ecosystem Perspective,* Phillips, R.C., McRoy, C.P., Eds., Garland STMP Press, 1980, 153.

Kikuchi, T. Contribution of the polychaete, *Neanthes japonica* (Izuka), to the oxygen uptake and carbon dioxide production in an intertidal mudflat of Nanakita estuary, Japan, *J. Exp. Mar. Biol. Ecol.*, 97, 81, 1986.

Kikuchi, T., Pérès, J.M., Consumer ecology of seagrass beds, in *Seagrass Ecosystems: A Scientific Perspective,* McRoy, C.P., Helfferich, C., Eds., Marcel Dekker, New York, 1977, 147.

Kindley, M.J., Physiological ecology of surf-zone diatoms, M.Sc. thesis, University of Auckland, NZ, 109 pp, 1983.

King, C.M., *Introduction to Marine Geology and Geomorphology,* Edward Arnold, London, 370 pp, 1975.

King, C.M., Wiebe, W.J., Regulation of sulphate concentrations and methogenesis in salt marsh soils, *Est.Coastal Mar. Sci.*, 10, 215, 1978.

Kinne, O., Physiology of estuarine organisms with special reference to salinity and temperature: general aspects, in *Estuaries*, Lauff, G.H., Ed., Publication 83 American Association for the Advancement of Science, Washington, DC, 1967, 525.

Kinne, O., Temperature — Invertebrates, in *Marine Ecology Vol. 1, Pt. 1, Environmental Factors*, Kinne, O., Ed., Wiley Interscience, London, 1971, 407.

Kinne, O., Salinity — Invertebrates, in *Marine Ecology*, Vol. 1, Pt. 2, Kinne, O. Ed., Wiley Interscience, New York, 1971, 800.

Kiolaliegla, M.G., Gibson, R.N., Prey "handling time" and its importance in food selection by the 15-spined stickleback, *Spinachia spinachia* (L.), *J. Exp. Mar. Biol. Ecol.*, 15, 115, 1976.

Kirby, C.J., The annual net primary production and decomposition of the salt marsh grass *Spartina alterniflora* Loisel in the Barataria Bay estuary of Louisiana, Ph. D. dissertation, Louisiana State University, Baton Rouge, LA, 87 pp, 1972.

Kirby, C.J., Gosselink, J.C., Primary production in a Louisiana Gulf Coast *Spartina alterniflora* marsh, *Ecology*, 47, 1052, 1976.

Kirchman, D., Graham, S., Reisch, D., Mitchell, R., Bacteria induce settlement and metamorphosis of *Jania (Dexiospira) braziliensis* Grube (Polychaeta:Spirorbidae), *J. Exp. Mar. Biol. Ecol.*, 56, 153, 1982.

Kirkman, H., Standing stock and production of *Ecklonia radiata* (C.Ag.) J.Agardh, *J. Exp. Mar. Biol. Ecol.*, 76, 119, 1984.

Kirkman, H., Reid, P.D., A study of the role of a seagrass *Posidonia australis* in the carbon budget of an estuary, *Aquatic Bot.*, 7, 173, 1979.

Kita, T., Harada, E., Studies on the epiphytic communities. I. Abundance and distribution of microalgae and small animals in *Zostera* blades, *Publ. Seto Mar. Biol. Lab.*, 10, 245, 1962.

Kitamori, R., Benthos as an environmental indicator with special reference to indicator species, in *Special Committee of the Japanese Ecological Society on Environmental Problems (Environment and Indicator Species)*, 1975, 265.

Kitching, J.A., Ecological studies at Lough Ine, *Adv. Ecol. Res.*, 17, 115, 1987.

Kitching, J.A., Ebling, F.J., Ecological studies at Louch Ine, *Adv. Ecol. Res.*, 4, 197, 1967.

Kitching, J.A., Macan, T.T., Gibson, H.G., Studies in sublittoral ecology. I. A submarine gully in Wembury Bay, South Devon, *J. Mar. Biol. Ass. U.K.*, 19, 677, 1934.

Kitching, J.A., Sloane, J.F., Ebling, F.S., The ecology of Lough Ine. VII. Mussels and their predators, *J. Anim. Ecol.*, 28, 113, 1959.

Kjeldsen, C.K., Pinney, H.K., Effects of variations in salinity and temperature on some estuarine macroalgae, in *Proc.7th Internat. Seaweed Symp.*, 1972, 301.

Kjeldsen, S.P., Olsen, G.B., Breaking waves, M.Sc. thesis, Technical University of Denmark, 1972.

Klapow, L.A., Natural and artificial rephasing of a tidal rhythm, *J. Comp. Physiol.*, 79, 233, 1972.

Klein-Bueteler, W.C.M., Settlement, growth and production of the shore crab, *Carcinus maenus*, on tidal flats in the Dutch Wadden Sea, *Neth. J. Sea Res.*, 10, 354, 1976.

Klump, D.W., McKinnon, D., Daniel, P., Damselfish territories: zones of high productivity on coral reefs, *Mar. Ecol. Progr. Ser.*, 40, 41, 1087.

Klump, D.W., Polunin, N.V.C., Algal production, grazers, and sand habitat partitioning on a coral reef: positive correlation between grazing rate and food availability, in *Trophic Relationships in the Marine Environment,* Barnes, M., Gibson, R.N., Eds., Aberdeen University Press, Aberdeen, 1990, 372.

Klump, J.V., Martens, C.S., Biogeochemical cycling in an organic rich coastal marine basin. II. Nutrient sediment-water exchange processes, *Geochemica Cosmochemica Acta,* 45, 101, 1981.

Knauer, G.A., Ayers, A.V., Changes in carbon, nitrogen, adenosine trophosphate, and chlorophyll *a* in decomposing *Thalassia testudinium* leaves, *Limnol. Oceanogr.*, 22, 408, 1977.

Kneib, R.T., Predation risk and use of intertidal habitats by young fishes and shrimps, *Ecology*, 68, 379, 1987.

Knight-Jones, E.W., Gregariousness and some other aspects of the settling behaviour of *Spirorbis*, *J. Mar. Biol. Assoc. U.K.*, 30, 201, 1951.

Knight-Jones, E.W., Laboratory experiments on gregariousness during settlement of *Balanus balanoides* and other barnacles, *J. Exp. Biol.*, 30, 584, 1953.

Knight-Jones, E.W., Decreased discrimination during settling after planktonic life in larvae of *Spirorbis borealis* (Serpulidae), *J. Mar. Biol. Assoc., U.K.*, 32, 337. 1954.

Knight-Jones, E.W., Moyse, J., Intraspecific competition in sedentary marine animals, in *Mechanisms of Biological Competition (*Symposium Society of Experimental Biology, No. 15), Milthorpe, F.L., Ed., 1961, 72.

Knox, G.A., Studies on a New Zealand Serpulid *Pomatoceros coeroleus* Schmarda, M. Sc. thesis, University of Canterbury, Christchurch, NZ, 1948.

Knox, G.A., The intertidal ecology of Taylors Mistake, Banks Peninsula, *Trans. Roy. Soc. N.Z.*, 83, 189, 1953.

Knox, G.A., The intertidal flora and fauna of the Chatham Islands, *Nature*, 174, 871, 1954.

Knox, G.A., *Urechis novae-zealandiae:* a New Zealand Echiuroid, *Trans. Roy. Soc. N.Z.*, 85, 141, 1957.

Knox, G.A., Littoral ecology and biogeography of the Southern Ocean., *Proc. Roy. Soc. London, Series B,* 152, 577, 1960.

Knox, G.A., Littoral ecology and biogeography of Australasian shores, *Oceanogr. Mar. Biol. Annu. Rev.*, 1, 341, 1963a.

Knox, G.A., Tides and intertidal zones, in *Proc. SCAR Symp. on Antarctic Oceanogr., 1966,* Cambridge, Scientific Committee on Antarctic Research, Cambridge, 1968, 131.

Knox, G.A., Beaches, in *The Natural History of Canterbury,* Knox, G.A., Ed., A.H. & A.W. Reed, Wellington, NZ, 1969a, 524.

Knox, G.A., Plants and animals of the rocky shores, in *The Natural History of Canterbury,* Knox, G.A., Ed., A.H. & A.W. Reed, Wellington, NZ, 1969b, 536.

Knox, G.A., *Effects of Freezing Works Effluent on Shakespeare Bay, Queen Charlotte Sound,* Estuarine Research Unit, Zoology Department, University of Canterbury, Christchurch, NZ, Report No. 2, 33 pp, 1972.

Knox, G.A., Marine benthic ecology and biogeography, in *Biogeography and Ecology of New Zealand*, Kuschel, G., Ed., Dr. W. Junk Publishers, The Hague, 1975, 353.

Knox, G.A., The role of polychaetes in benthic soft-bottom communities, in *Essays on Polychaetous Annelids in Memory of Dr. Olga Hartman*, Reisch, D.J., Fauchild, K., Eds., Allan Hancock Foundation, University of Los Angeles, Los Angeles, CA, 1977, 547.

Knox, G.A., Distribution patterns of Southern Hemisphere marine biotas: some commentary on their origin and evolution, in *Proc. Int. Symp. Mar. Biogeography Evol. South. Hemisphere, Vol. 1,* DSIR Information Series, 137, Wellington, NZ, 43, 1979a.

Knox, G. A., *Ahuriri Estuary: An Environmental Study,* Estuarine Research Unit, Zoology Department, University of Canterbury, Christchurch, NZ, Report No. 20, 84 pp, 1979b.

Knox, G.A., Plate tectonics and the evolution of intertidal and shallow-water benthic biotic distribution patterns in the South Pacific, *Palaeog. Palaeocl. Palaeocol.*, 31, 267, 1980.

Knox, G.A., The ecology of the Kaituna Marsh, Pelorus Sound, with special reference to the introduced cordgrass *Spartina*, in *Nutrient Processing and Biomass Production in New Zealand Estuaries,* Gillespie, P.A., Ed., Water and Soil Misc. Publ. No. 60, National Water and Soil Conservation Organization, Wellington, NZ, 1983c, 32.

Knox, G.A., The ecology of the Kaituna marsh, Pelorus Sound, with special reference to the introduced cordgrass *Spartina*, in *Nutrient Processing and Biomass Production in New Zealand Estuaries,* Gillespie, P. A., Ed., Water and Soil Miscellaneous Publications, No. 60, National Water and Soil Conservation Organization, Wellington, NZ, 1983d, 32.

Knox, G.A., *Estuarine Ecosystems: A Systems Approach*, Vol. I, CRC Press, Boca Raton, FL, 287 pp, 1986a.

Knox, G.A., *Estuarine Ecosystems: A Systems Approach*, Vol. II, CRC Press., Boca Raton, Florida, 256 pp, 1986b.

Knox, G.A., *A Resurvey of the Benthic Fauna off Clive, Hawke Bay,* Report Commissioned by the Hastings City Council, Hastings, Hawke Bay, NZ, 108 pp, 1988a.

Knox, G.A., *A Survey of the Benthic Fauna in the Vicinity of the Napier Sewage Outfall,* Report Commissioned by the Napier City Council, Hawke Bay, NZ, 69 pp, 1988b.

Knox, G.A., The littoral ecology of the Subantarctic Region: with special reference to the New Zealand Subantarctic Islands, *Colloque sur l'Ecologie des Isles Subantarctique (Paris 25 juin, 1985). CINFRA,* 57, 47, 1988c.

Knox, G.A., *The Ecology of the Avon-Heathcote Estuary,* Report prepared for the Christchurch City Council and the Canterbury Regional Council, Christchurch, NZ, 158 pp, 1992.

Knox, G.A., *The Ecology of New Zealand Seashores*, University of Canterbury Press, Christchurch, NZ, in preparation.

Knox, G.A., Bolton, L.A., *The Ecology of Shakespeare Bay, Queen Charlotte Sound,* Estuarine Research Unit, Zoology Department, University of Canterbury, Christchurch, NZ, Report No. 10, 144 pp, 1979.

Knox, G.A., Duncan, K., A study of species zonation patterns along intertidal gradients at the Snares Islands, New Zealand, *J. Exp. Mar. Biol. Ecol.,* in preparation.

Knox, G.A., Fenwick, G.D., *A Quantitative Study of the Benthic Fauna off Clive, Hawke Bay,* Estuarine Research Unit, Zoology Department, University of Canterbury, Christchurch, NZ, Report No. 14, 91–99, 1978.

Knox, G.A., Fenwick, G.D., Zonation of inshore benthos off a sewage outfall in Hawke Bay, New Zealand, *N.Z. J. Mar. Freshw. Res.,* 15, 417, 1982.

Knox, G.A., Kilner, A.R., *The Ecology of the Avon-Heathcote Estuary,* Estuarine Research Unit, Zoology Department, University of Canterbury, Christchurch, NZ, Report No. 1, 358 pp, 1973.

Knox, G. A., Miyabara, T., *Coastal Zone Resource Development and Conservation in Southeast Asia, with Special Reference to Southeast Asia,* UNESCO Office for Science and Technology for Southeast Asia, Jakarta, Indonesia, and Resource Systems Institute, East-West Center, Honolulu, Hawaii, 182 pp, 1983.

Knox, G.A., Robb, J.A., Eds., *The Ecology of the Avon-Heathcote Estuary, Christchurch, New Zealand,* in preparation.

Kofoed, L.H., The feeding biology of *Hydrobia ventrosa* (Montagu). I. The assimilation of different components of the food, *J. Exp. Mar. Biol. Ecol.,* 19, 233, 1975.

Kokkinn, M.J., Allanson, B.H., On the flux of organic carbon in a tidal salt marsh, Kowie River estuary, Port Alfred, South Africa, *S. Af. J. Sci.,* 84, 613, 1985.

Koop, K., Field, J.G., Energy transformation by the supralittoral isopod *Ligia dillatata* Brandt, *J. Exp. Mar. Biol.Ecol.,* 53, 221, 1981.

Koop, K., Griffiths, C.L., The relative significance of bacteria, meio- and macrofauna on an exposed sandy beach, *Mar. Biol.,* 66, 295, 1982.

Koop, K., Lucas, M.I., Carbon flow and nutrient regeneration from the decomposition of macrophyte debris in a sandy beach microcosm, in *Sandy Beaches as Ecosystems,* McLachlan, A., Erasmus, T., Eds., Dr. W. Junk Publishers, The Hague, 1983, 349.

Koop, K., Newell, R.C., Lucas, M.I., Biodegradation and carbon flow based on kelp debris (*Ecklonia maxima*) in a sandy beach microcosm, *Mar. Ecol. Prog. Ser.,* 7, 315, 1982.

Korringa, P., Estuarine fishes of Europe as affected by man's multiple activities, in *Estuaries,* Lauff, G., Ed., Publication No. 83, American Association for the Advancement of Science, Washington, DC, 1967, 658.

Krauter, J.N., Biodeposition by salt marsh invertebrates, *Mar. Biol.,* 35, 215, 1976.

Krank, K., Sediment deposition from flocculated suspension, *Sedimentology,* 22, 111, 1975.

Krebs, J.R., Optimal foraging: Decision rules for predators, in *Behavioural Ecology,* Krebs, J, R., Davies, N.B., Eds., Sutherland, England, 1978, 23.

Kremer, J.N., Network information indices with an estuarine model, in *Network Analysis in Marine Ecology; Methods and Applications,* Wulff, F., Field, J.G., Mann, K.H., Eds., Springer-Verlag, Berlin, 1989, 119.

Kremer, J, N., Nixon, S, W., *A Coastal Marine Ecosystem: Smulation and Analysis,* Springer-Verlag, Berlin, 217 pp, 1978.

Kristensen, E., Benthic fauna and biogeochemical precesses in marine sediments: microbial activities and fluxes, in *Nitrogen Cycles in Coastal Marine Environments,* Blackburn, T.H., Sorenson, J., Eds., John Wiley & Sons, London, 1988, 275.

Kristensen, E., Blackburn, T.H., The fate of organic carbon and nitrogen in an experimental marine sediment systems: influence of bioturbation and anoxia, *J. Mar. Res.,* 45, 231, 1987.

Kristensen, E.M., Jensen, M.H., Anderson, T.K., The impact of polychaete (*Nereis virens Sars*) burrows on nitrification and nitrate reduction in sediments, *J. Exp. Mar. Biol. Ecol.,* 85, 75, 1985.

Krom, M.D., Berner, R.A., Adsorption of phosphate in anoxic marine sediments, *Limnol. Oceanogr.,* 25, 327, 1980.

Kromberg, I., Structure and adaptation of the fauna in the black zone (littoral fringe) along rocky shores in northern Europe, *Mar. Ecol. Prog. Ser.,* 49, 95, 1988.

Kruczyzsko, W.L., Subrahmanyan, C.B., Drake, S.H., Studies on the plant community of a Florida salt marsh, *Bull. Mar. Sci.,* 28, 316, 1978.

Kuenzler, E.J., Structure and energy flow of a mussel population in a Georgia salt marsh, *Limnol. Oceanogr.,* 6, 191, 1961a.

Kuenzler, E.J., Phosphorus budget of a mussel population, *Limnol. Oceanogr.,* 6, 400, 1961b.

Kuenzler, E.J., Staley, D.W., Koenings, J.P., Nutrient kinetics of phytoplankton in the Pamlico River, North Carolina, Water Resources Research Institute, Raleigh, NC, Report No. UNC-WRRI-79-139, 1979.

Kuipers, B.R., On the ecology of juvenile plaice on a tidal flat in the Wadden Sea, *Neth. J. Sea Res.,* 11, 56, 1977.

Kuipers, B.R., de Wilde, P.A.W.J., Creutzberg, F., Energy flow in a tidal flat ecosystem, *Mar. Ecol. Prog. Ser.,* 5, 215, 1981.

Kullenberg, G., The Baltic Sea, in *Estuaries and Enclosed Seas. Ecosystems of the World 26,* Ketchum, B.H., Ed., Elsevier, Amsterdam, 1983, 309.

Kvitek, R.C., Oliver, J.S., DeGauge, A.R., Anderson, B.S., Changes in Alaskan soft-bottom communities along a gradient in sea otter predation, *Ecology,* 73, 413, 1992.

Lancelot, C., Extracellular release of small and large molecules by phytoplankton in the Southern Bight of the North Sea, *Est. Coastal Shelf Sci.,* 18, 65, 1984.

Lang, C., Mann, K.H., Changes in sea urchin populations after the destruction of kelp beds, *Mar. Biol.*, 36, 321, 1976.

Langdon, G.J., Newell, R.I.E., Utilization of detritus and bacteria as food sources by two bivalve suspension-feeders, the oyster *Crassostrea virginica* and the mussel *Geukensia demissa*, *Mar. Ecol. Prog. Ser.*, 58, 299, 1990.

Langford, R.R., Coastal lagoons of Mexico: their origin and classification, in *Est. Processes*, Vol. II, Wiley, M.L., Ed., Academic Press, New York, 1976, 182.

Lanyon, J., Limpus, C.J., Marsh, H., Dugons and turtles: grazers in the seagrass ecosystem, in *Biology of Seagrasses: A Treatise on the Biology of Seagrasses with Special Reference to the Australian Region*, Larkum, A.W.D., McComb, A.J., Sheppard, S.A., Eds., Elsevier, Amsterdam, 1989, 610.

Larkum, A.W.D., A study of growth and primary production in *Ecklonia radiata* (C, Ag,) (Laminariales) at a sheltered site in Port Jackson, *J. Exp. Mar. Biol. Ecol.*, 96, 177, 1986.

Larkum, A.W.D., McCombs, A.J., Shepard, S.A., Eds., *Biology of Seagrasses: A Treatise on the Biology of Seagrasses with Special Reference to the Australian Region,* Elsevier, Amsterdam, 1989.

Larman, V.N., Gabbott, P.A., East, J., Physico-chemical properties and the settlement factor protein for the barnacle *Balanus balanoides*, *Comp. Biochem. Physiol.*, 72B, 329, 1982.

Lasiak, T.A., Juveniles, food, and surf zone habitat: implications for teleost nursery areas, *S. Af. J. Zool.*, 21, 51, 1986.

Lauff, G., Ed., *Estuaries,* American Association for the Advancement of Science, Publication No. 83, 757 pp, 1967.

Lawrence, J.M., On the relationship between marine plants and sea-urchins, *Oceanogr. Mar. Biol. Annu. Rev.*, 13, 213, 1975.

Lawton, J.H., Jones, C.B., Linking species and ecosystems: organisms as ecosystem engines, in *Linkng Species and Ecosystems*, Jones, C.G., Lawton, J.H., Eds., Chapman & Hall, New York, 1995, 141.

Leach, J.H., Epibenthic algal production in an intertidal mudflat, *Limnol. Oceanogr.*, 15, 514, 1970.

Lee, H., Swartz, R.C., Biological processes affecting the distribution of pollutants in marine sediments. Part II. Biodeposition and bioturbation, in *Contaminants and Sediments*, Baker, R.A., Ed., University of Michigan, Ann Arbor, MI, 1980, 553.

Lee, J.J., A conceptual model of marine detrital decomposition and the organisms associated with the process, in *Advances in Aquatic Microbiology*, Droop, M.R., Jannesch, H., Eds., Academic Press, New York, 1980, 257.

Lee, S.J., Litter production and turnover of the mangrove *Kandelia candel* (L.) Druce in a Hong Kong tidal shrimp pond, *Est. Coastal Shelf Sci.*, 29, 75, 1989.

Lee, S.Y., Net aerial primary productivity, litter production and decomposition of the reed *Phragmites communis* in a nature resrve in Hong Kong: management implications, *Mar. Ecol. Prog. Ser.*, 66, 161.

Lee, S.Y., Mangrove outwelling: a review, *Hydrobiologia*, 259, 203, 1995.

Lee, S.Y., Potential trophic importance of the faecal material of the mangrove sesamine crab *Sesarma messa*, *Mar. Ecol. Prog. Ser*, 159, 274, 1997.

Lee, S.Y., Tropical mangrove ecology: physical and biotic factors influencing ecosystem structure and function, *Aust. J. Ecol.*, 24, 355, 1999.

Lee, W.Y., McAlice, B.J., Sampling variability of marine zooplankton in a tidal estuary, *Est. Coastal Mar. Sci.*, 8, 565, 1979.

Legendre, L., Demers, S., Therriault, J.-C., Boudreau, C.-A., Tidal variation in the photosynthesis of estuarine phytoplankton isolated in a tank, *Mar. Biol.*, 88, 301, 1985.

Leh, C.M.H., Sasekumar, A., The food of sesarmid crabs in Malaysia mangrove forests, *Malayan Nat. J.*, 39, 135, 1985.

Leighton, D.L., Grazing activities of benthic invertebrates in Southern California kelp beds, *Nova Hedwiga (Suppl.)*, 32, 421, 1971.

Lenanton, R.C.J., Robertson, A.I., Hansen, J.A., Nearshore accumulations of detached macrophytes as nursery areas for fishes, *Mar. Ecol. Prog. Ser.*, 9, 51, 1982.

Lerman, A., Migrational processes and chemical reactions in interstitial water, in *The Sea, Vol. 6. Marine Modelling*, Goldberg, E.D., McCave, I.N., O'Brien, J.J., Dteele, J.H., Eds., Wiley Interscience, New York, 1977, 695.

Lessard, E.J., Swift, E., Species-specific grazing rates of heterotrophic dinoflagellates in oceanic waters measured with dual-isotope techniques, *Mar. Biol.*, 87, 289, 1985.

Levin, L.A., Life history and dispersal patterns in a dense infaunal polychaete assemblage: community structure and response to disturbance, *Ecology*, 65, 1185, 1984.

Levin, S.A., Paine, R.T., Disturbance, patch formation and community structure, *Proc. Natl. Acad. Sci. (USA)*, 71, 2744, 1974.

Levine, S., Several measures of trophic structure applicable to complex food webs, *J. Theor. Biol.*, 83, 195, 1980.

Levings, S.C., Garrity, S.D., Grazing patterns in *Siphonaria gigas* (Mollusca, Pulmonata) on the rocky Pacific coast of Panama, *Oecologia (Berlin)*, 64, 152, 1984.

Levinton, J.S., The paleological significance of opportunistic species, *Lethaia*, 3, 69, 1970.

Levinton, J.S., Stability and trophic structure in deposit-feeding and suspension feeding communities, *Am. Nat.*, 106, 472, 1972.

Levinton, J.S., Ecology of shallow water deposit-feeding communities, Quisset Harbor, Massachusetts, in *Ecology of the Marine Benthos*, Coull, B.C., Ed., University of South Carolina Press, Columbia, SC, 1977, 191.

Levinton, J.S., Deposit-feeders, their responses, and the study of resource limitation, in *Ecological Processes in Coastal and Marine Systems*, Livingston, R.J., Ed., Plenum Publishing, New York, 1979, 117.

Levinton, J.S., Particle feeding by deposit feeders: models, data and its prospectus, in *Marine Benthic Dynamics*, Tenore, K.R., Coull, B.C., Eds., University of South Carolina Press, Columbia, SC, 1980, 423.

Levinton, J.S., The body-prey size hypothesis: the adequacy of body size as a vehicle for character displacement, *Ecology*, 63, 869, 1982.

Levinton, J.S., Variable feeding in three species of *Macoma* (Bivalvia: Tellinacea) as a response to water flow and sediment transport, *Mar. Biol.*, 110, 375, 1991.

Levinton, J.S., Bianchi, T.S., Nutrition and food limitation of deposit-feeders. I. The role of microbes in the growth of mud snails (Hydrobiidae), *J. Mar. Res.*, 39, 531, 1981.

Levinton, J.S., DeWitt, T.H., Relation of particle-size spectrum and food abundance to particle selectivity in the mud snail *Hydroibia totteni* (Prosobranchia: Hydrobiidae), *Mar. Biol.*, 100, 449, 1989.

Levinton, J.S., Lopez, G.R., A model of renewable resources and limitation of deposit-feeding benthic populations, *Oecologia (Berlin)*, 31, 177, 1977.

Lewin, J., Eckman, J.E., Ware, G.N., Blooms of surf zone diatoms along the coast of the Olympic Peninsula, Washington. XI. Regeneration of ammonium in the surf environment of the Pacific razor clam *Siliqua patula*, *Mar. Biol.*, 52, 1, 1979.

Lewin, J., Hurby, T., Blooms of surf zone diatoms along the coast of Olympic Peninsula, Washington. III. A diel periodicity in buoyancy shown by the surf zone diatom species *Chaetoceros armatum* T. West., *Est. Coastal Mar. Sci.*, 1, 101, 1973.

Lewin, J., Rau, V.R.W., Blooms of surf-zone diatoms along the coast of Olympic Peninsula, Washington. VI. Daily periodicity phenomena associated with *Chaetoceros armatum* in its natural habitat, *J. Phycol.*, 11, 330, 1975.

Lewin, J., Schaefer C.T., The role of phytoplankton in surf ecosystems, in *Sandy Beaches as Ecosystems*, Mclachlan, A., Erasmus, T., Eds., Dr. W. Junk Publishers, The Hague, 1983, 381.

Lewin, R., Supply-side ecology, *Science*, 234, 25, 1986.

Lewis, J.R., Observations on a high level population of limpets, *J. Ecol.*, 23, 85, 1954a.

Lewis, J.R., The ecology of exposed rocky shores of Caithness, *Trans. Roy. Soc. (Edinburgh)*, 62, 695, 1954b.

Lewis, J.R., The mode of occurrence of the universal intertidal zones in Great Britain, with a comment by T.A. and A. Stephenson, *J. Ecol.*, 43, 270, 1955.

Lewis, J.R., The littoral zone on a rocky shore — a biological or physical entity?, *Oikos*, 12, 280, 1961.

Lewis, J.R. *The Ecology of Rocky Shores*, English University Press, London, 303 pp, 1964.

Lewis, J.R., The role of physical and biological factors in the distribution and structure of rocky shore communities, in *Biology of Benthic Organisms*, Keegan, B.F., Ceidig, P.O., Boaden, P.J.S., Eds., Pergamon Press, Oxford, 1977, 417.

Lewis, J.R., Composition and functioning of benthic ecosystems in relation to the assessment of long-term effects of oil pollution, *Phil. Trans. Roy. Soc., London B*, 287, 257, 1982.

Lewis, J.R., Bowman, R.S., Local habitat-induced variation in the population dynamics of *Patella vulgata* L., *J. Exp. Mar. Biol. Ecol.*, 17, 165, 1975.

Li, W.K.W., Experimental approaches to field measurements and interpretations, *Can. J. Fish. Aquatic Sci.*, 214, 251, 1986.

Lignall, R., Excretion of organic carbon by phytoplankton: its relation to algal biomass, primary productivity and bacterial secondary production in the Baltic Sea, *Mar. Ecol. Prog. Ser.*, 68, 85, 1990.

Lindberg, D.R., Chu, E.W., Western gull predation on owl limpets: different methods at different methods at different localities, *Veliger*, 25, 229, 1983.

Lindeman, R.L., The trophic-dynamic aspect of ecology, *Ecology*, 23, 399, 1942.

Linley, E.A.S., Newell, R.C., Microheterotrophic communities associated with the degradation of kelp debris, *Kieler Meeresforsch*, 5, 345, 1981.

Linthurst, R.A., The effects of aeration on the growth of *Spartina alterniflora* Losiel, *Am. J. Bot.*, 66, 685, 1979.

Linthurst, R.A., A growth comparison of *Spartina alterniflora* Loisel ecophenes under aerobic and anaerobic conditions, *Am. J. Bot.*, 67, 883, 1980a.

Linthurst, R.A., An evaluation of aeration, nitrogen, pH and salinity as factors affecting *Spartina alterniflora* growth: a summary, in *Estuarine Perspectives*, Kennedy, V.S., Ed., Academic Press, New York, 1980b, 235.

Linthurst, R.A., Reimold, R.J., Estimated net aerial productivity for selected marine angiosperms in Maine, Delaware and Georgia, *Ecology*, 59, 945, 1978.

Linthurst, R.A., Seneca, E.D., Aeration, nitrogen and salinity as determinants of *Spartina alterniflora* Loisel growth response, *Estuaries*, 4, 53, 1981.

Lipschultz, E., Cunningham, J.J., Steveneon, J.C., Nitrogen fixation associated with four species of submerged angiosperms in central Chesapeake Bay, *Est. Coastal Shelf Sci.*, 810, 1979.

Litchfield, C.D., Seyfried, P.L., Eds., *Methodology for Biomass Determinations and Microbial Activities in Sediments*, ASTM STP 673, American Society for Testing and Materials, Philadelphia PA, 199 pp, 1979.

Little, C., Factors governing patterns of foraging activity in littoral marine herbivorous molluscs, *J. Mollusc. Stud.*, 55, 273, 1989.

Little, C., Smith, L.P., Vertical zonation on rocky shores in the Severn Estuary (UK), *Est. Coastal Shelf Sci.*, 11, 651, 1980.

Littler, M.M., Southern California rocky intertidal ecosystems: method, community structure and variability, in *The Shore Environment, Vol. 2. Ecosystems*, Price, J.H., Irvine, D.E.G., Farnham, F., Eds., Systematics Association Special Volume No. 17(b), 1980a, 565.

Littler, M.M., Morphological form and photosynthetic performance of marine macroalgae: tests of a functional/form hypothesis, *Bot. Marina*, 22, 161, 1980b.

Littler, M.M., Arnold, K.E., The primary productivity of marine macroalgal functional-form groups: sampling and interpretative problems, *J. Phycol.*, 18, 307, 1982.

Littler, M.M., Littler, D.S., The evolution of thallus form and survival strategies in benthic marine macroalgae: field and laboratory test of a functional-form model, *Am. Nat.*, 116, 25, 1980.

Littler, M.M., Littler, D.S., Intertidal macrophyte communities from Pacific Baja California: relatively constant vs. environmentally fluctuation systems, *Mar. Ecol. Prog. Ser.*, 4, 145, 1981.

Littler, M.M., Littler, D.S., Murray, S.N., Seapy, R.R., Southern California rocky intertidal ecosystems, in *Intertidal and Littoral Systems. Ecosystems of the World 24*, Mathieson, A.C., Nienhuis, P.H., Eds., Elsevier, Amsterdam, 1991, 273.

Littler, M.M., Matz, D.R., Littler, D.S., Effects of recurrent sand deposition on rocky intertidal organisms: importance of substrate heterogeneity in a fluctuating environment, *Mar. Ecol. Prog. Ser.*, 11, 129, 1983.

Littler, M.M., Murray, S.M., The primary productivity of marine macrophytes from a rocky intertidal community, *Mar. Biol.*, 27, 131, 1974.

Littler, L.M., Murray, S.N., Impact of sewage on the distribution, abundance and community structure of two intertidal macro-organisms, *Mar. Biol.*, 30, 277, 1975.

Livingston, R.J., Ed., *Ecological Processes in Coastal and Marine Systems,* Plenum Press, New York, 548 pp, 1979.

Livingston, R.J., Trophic response of fish to habitat variability in coastal seagrass systems, *Ecology*, 65, 1258, 1984.

Livingston, R.J., Organization of fishes in seagrass systems: the response to stress, in *Fish Community Ecology in Estuaries and Coastal Lagoons*, Yanez-Aranciba, A., Ed., Mexico, D.F., Editorial Universitaria, UNMA-PUAL-ICML, 1985, 307.

Lobban, C.S., Harrison, P.T., Duncan, M.J., *The Physiological Ecology of Seaweeds*, Cambridge University Press, Cambridge, 242 pp, 1985.

Long, S.P., Mason, C.F., *Saltmarsh Ecology*, Blackie & Son, Glascow and London, 160 pp, 1983.

Long, S.P., Woolhouse, H.W., Primary production in *Spartina* marshes, in *Ecological Processes in Coastal Environments*, Jeffries, R.J., Davey, A.J., Eds., Blackwell Scientific Publications, Oxford, 1983, 333.

Longhurst, A.R., The importance of measuring rates and fluxes in marine ecosystems, in *Flows of Energy and Materials in Marine Ecosystems, Theory and Practice,* Fasham, M.J.R., Ed., Plenum Press, New York, 1984, 1.

Lopez, G.R., Kofoed, L.H., Epipsammic browsing and deposit-feeding in mud snails (Hydrobiidae), *J. Mar. Res.*, 38, 585, 1980.

Lopez, G.R., Levinton, J.S., Ecology of deposit-feeding animals in marine sediments, *Quart. Rev. Biol.*, 62, 235, 1987.

Lopez, G.R., Levinton, J.S., Slobodkin, L.B., The effect of grazing by the detritivore *Orchestis grillus* on *Spartina* litter and its associated microbial community, *Oecologia (Berlin)*, 30, 111, 1977.

Lopez, G.R., Taghorn, G. Levinton, J., Eds., *Ecology of Marine Deposit Feeders,* Springer- Verlag, Berlin, 1989.

Lopez-Colet, J., Vieitez, J.M., Diaz-Pineda, F., Tipos de comunidades benthonicas de al Playa de Puntal (Bahia se Santander), *Cashiers Biol. Mar.*, 23, 53, 1982.

Lord, C.J. III., The chemistry and cycling of iron, manganese and sulfur in salt marsh sediments, Ph.D dissertation, University of Delaware, 117 pp, 1980.

Lord, D.A., Thompson, G.A., The Swartkops Estuary pollution status, in *The Swartkops Estuary: Proceedings of a Symposium held on 14 and 15 September, 1987 at the University of Port Elizabeth, South Africa*, National Scientific Report, No. 156, 1988, 16.

Lotrich, V.A., Meridith, W.H., Weisberg, S.B., Hurd, L.E., Dauber, F.C., Dissolved and particulate nutrient fluxes via tidal exchange between a salt marsh and lower Delaware Bay, in *The Fifth Biennial International Estuarine Research Conference Abstracts,* 1979, 7.

Lowell, R.B., Desiccation of intertidal limpets: effects of shell size, fit to substratum, and shape, *J. Exp. Mar. Biol. Ecol.*, 77, 197, 1984.

Lubchenco, J., Plant species diversity in a marine intertidal community: importance of herbivore good preference and algal competitive abilities, *Am. Nat.*, 112, 23, 1978.

Lubchenco, J., Algal zonation in the New England rocky intertidal community: an experimental analysis, *Ecology*, 61, 333, 1980.

Lubchenco, J., Effects of grazers and algal competitors on fucoid colonization in tidal pools, *J. Phycology,* 18, 544, 1982.

Lubchenco, J., *Littorina* and *Fucus*: effects of herbivores, substratum heterogeneity and plant escapes during succession, *Ecology*, 64, 1116, 1983.

Lubchenco, J., Relative importance of competition and predation: early colonization by seaweeds in New England, in *Community Ecology*, Diamond, J., Case, T.J., Eds., Harper & Row, New York, 1986, 537.

Lubchenco, J., Cubit, J.D., Heteromorphic life histories of certain marine algae as adaptations to variations in herbivory, *Ecology*, 64, 676, 1980.

Lubchenco, J., Gaines, S.D., A unified approach to plant-herbivore interactions. I. Populations and communities, *Annu. Rev. Ecol. Systematics*, 12, 405, 1981.

Lubchenco, J., Menge, B.A., Community development and persistance in a low rocky intertidal zone, *Ecol. Monogr.*, 59, 67, 1978.

Lubchenco, J., Menge, B.A., Garrity, S.R., Lubchenco, P.T., Ashkenas, L.R., Gaines, S.D., Emlet, R., Lucas, J., Strauss, S., Structure, persistence and role of consumers in a tropical rocky intertidal community (Toboguilla Island), Bay of Panama, *J. Exp. Mar. Biol. Ecol.*, 78, 23, 1984.

Lucas, M.I., Newell, R.C., Shumway, S.E., Seiderer, L.J., Bally, R., Particle clearance and yield in relation to bacterioplankton and suspended particulate availability in estuarine and open coast populations of the mussel *Mytilus edulis*, *Mar. Ecol. Prog. Ser.*, 36, 215, 1987.

Luckenbach, M.W., Biogenic structures and foraging in five species of shorebirds (Charadai), *Est. Coastal Mar. Sci.*, 19, 681, 1984.

Luckenbach, M.W., Sediment stability around animal tubes: the roles of biotic and hydrodynamic processes, *Limnol. Oceanogr.*, 31, 799, 1986.

Luckens, P.A., Breeding, settlement and survival of barnacles at artificially modified shore levels at Leigh, New Zealand, *N.Z. J. Mar. Freshw. Res.*, 4, 497, 1970.

Luckens, P.A., Predation and intertidal zonation of barnacles at Leigh, New Zealand, *N.Z. J. Mar. Freshw. Res.*, 9, 355, 1975a.

Luckens, P.A., Competition and intertidal zonation of barnacles at Leigh, New Zealand, *N.Z. J. Mar. Freshw. Res.*, 9, 379, 1975b.

Luckens, P.A., Settlement and succession on rocky shores at Auckland, North Island, New Zealand, *N. Z. Oceanogr. Instit. Mem.*, 70, 1, 1976.

Lugo, A.E., Sell, M., Snedaker, S.C., Mangrove ecosystem analysis, in *Systems Analysis 3*, Patten, B.C., Ed., Academic Press, New York, 1974, 114.

Lugo, A.E., Snedaker, S.C., The ecology of mangroves, *Annu. Rev. Ecol. Syst.*, 5, 170, 1974.

Lugo, A.E., Snedaker, S.C., Properties of a mangrove forest in southern Florida, in *Proc. Internat. Symp. Biol. Managt. Mangroves, Vol. 1*, Walsh, G.E., Snedaker, S.C., Teas, H.J., Eds., Institute of Food and Agricultural Sciences, University of Florida, Gainesville, FL, 1975, 170.

Lugo, A.E., Evink, G., Brinson, M.M., Brane, A., Snedaker, S.C., Diurnal rates of photosynthesis, respiration and transpiration in mangrove forests of South Florida, in *Tropical Ecological Systems*, Golley, F.B., Mendina, E., Eds., Springer-Verlag, New York, 1975, 535.

Lund, E.J., Effects of bleedwater, "soluble fraction" and crude oil on the oyster, *Publ. Instit. Mar. Sci., Univ. of Texas*, 4, 296, 1957.

MacArthur, R.H., Wilson, E.O., *Theory of Island Biogeography*, Princeton University Press, Princeton, NJ, 203 pp, 1967.

McCaffrey, R.J., Myers, A.C., Davey, E., Morrison, G., Bender, M., Luedke, N., Cullen, D., Froelich, P., Klinkhammer, G., The relationship between porewater chemistry and benthic fluxes of nutrients and manganese in Narragansett Bay, Rhode Island, *Limnol. Oceanogr.*, 25, 31, 1980.

McCall, P.L., Community patterns and adaptive strategies of infaunal benthos of Long Island Sound, *J. Mar. Res.*, 35, 221, 1977.

McCall, P.L., Spatial-temporal distributions of Long Island Sound infauna: the role of bottom disturbance in a near-shore habitat, in *Estuarine Interactions*, Wiley, M.L., Ed., Academic Press, New York, 1978, 191.

McCall, P.L., Fleeger, J.W., Predation by juvenile fish on hyperbenthic meiofauna: a review with data on post-larval *Leiostomus*, *Vie Millieu*, 45, 61, 1995.

McCall, P.L., Tevesz, M.J.S., Eds., *Animal-Sediment Relationships*, Plenum Press, New York, 336 pp, 1982.

McCauley, E., Briand, F., Zooplankton grazing and phytoplankton species richness: field tests of the predation hypothesis, *Limnol. Oceanogr.*, 24, 243, 1979.

McClatchie, S., Juniper, S.K., Knox, G.A., Structure of a mudflat diatom community in the Avon-Heathcote Estuary, New Zealand, *N.Z. J. Mar. Freshw. Res.*, 16, 299, 1982.

McConnaughey, T., McRoy, C.P., ^{13}C label identifies eelgrass (*Zostera marina*) carbon in an Alaskan estuarine food web, *Mar. Biol.*, 53, 263. 1978.

McCook, L.J., Chapman, A.R.O., Community succession following massive ice-scour on an exposed rocky shore: effects of *Fucus* canopy algae and mussels during late succession, *J. Exp. Mar. Biol. Ecol.*, 154, 137, 1991.

McDermott, J.J., Food web in the surf zone of an exposed sandy beach along the mid-Atlantic coast of the United States, in *Sandy Beaches as Ecosystems*, McLachlan, A., Erasmus, T., Eds., Dr. W. Junk Publishers, Boston, 1983, 529.

Macfadyen, A., *Animal Ecology: Aims and Methods*, Pitman & Sons, London, 344 pp, 1963.

McFarlane, I.D., Trail-following and trail-searching behaviour in homing of the intertidal gastropod mollusc, *Onchidium verruculatum*, *Mar. Behav. Physiol.*, 7, 95, 1980.

McGregor, D.D., Physical ecology of some New Zealand rock pools, *Hydrobiologia*, 25, 277, 1965.

McGuiness, K.A., The species-area relations of communities on intertidal boulders: testing the null hypothesis, *J. Biogeogr.*, 11, 439, 1984.

McGuiness, K.A., Disturbance and organisms on boulders. I. Patterns in the environment and the community, *Oecologia (Berlin)*, 71, 409, 1987a.

McGuiness, K.A., Disturbance and organization on boulders. II. Causes of patterns in diversity and abundance, *Oecologia (Berlin)*, 71, 420, 1987b.

McGuiness, K.A., Underwood, A.J., Habitat structure and the nature of communities on intertidal boulders, *J. Exp. Mar. Biol. Ecol.*, 104, 97, 1986.

McGwynne, L.E., A comparative ecophysiological study of three sandy beach gastropods in the Eastern Cape, M.Sc. thesis, University of Port Elizabeth, South Africa, 144 pp, 1980.

McGwynne, L.E., The microbial loop: its role in a diatom-enriched surf zone, Ph.D. thesis, University of Port Elizabeth, South Africa, 1991.

McHugh, J.L., Estuarine nekton, in *Estuaries*, Lauff, G., Ed., Publication No. 83, American Association for the Advancement of Science, Washington, DC, 1967, 581.

McIntire, C.D., Reimer, C.W., Some marine and brackish water *Acanthes* from Yaquina estuary, Oregon (U.S.A.), *Botanica Marina*, 17, 164, 1974.

McIntyre, A.D., Meiobenthos of sublittoral muds, *J. Mar. Biol. Assoc., U.K.*, 44, 625, 1964.

McIntyre, A.D., Ecology of marine meiobenthos, *Biol. Rev.*, 44, 245. 1969.

McIntyre, A.D., The ranges of biomass in intertidal sand with special reference to the bivalve *Tellina tenuis*, *J. Mar. Biol. Assoc., U.K.*, 50, 561, 1970.

McIntyre, A.D., Murison, D.J., The meiofauna of a flatfish nursery ground, *J. Mar. Biol. Assoc., U.K.*, 53, 93, 1973.

McKee, K.L., Patrick, W.H., Jr., The relationship of smooth cordgrass (*Spartina alterniflora*) to tidal datums: a review, *Estuaries*, 11, 143, 1988.

McLachlan, A., Studies of the psammolittoral fauna of Angola Bay, South Africa. II. the distribution, composition and biomass of meiofauna and macrofauna, *Zool. Af.*, 12, 33, 1977a.

McLachlan, A., Composition, distribution, abundance and biomass of macrofauna and meiofauna of four sandy beaches, *Zool. Af.*, 12, 279, 1977b.

McLachlan, A., Sediment particle size and body size in meiofaunal harpacticoid copepods, *S. Af. J. Sci.*, 75, 75, 1978.

McLachlan, A., The definition of sandy beaches in relation to exposure: a simple rating system, *S. Af. J. Sci.*, 76, 137, 1980a.

McLachlan, A., Exposed sandy beaches as semi-enclosed ecosystems, *Mar. Envt. Res.*, 4, 59, 1980b.

McLachlan, A., A model for the estimation of water filtration and nutrient regeneration by exposed sandy beaches, *Mar. Envt. Res.*, 6, 37, 1982.

McLachlan, A., Sandy beach ecology — A review, in *Sandy Beaches as Ecosystems*, McLachlan, A., Erasmus, T., Eds., Dr. W. Junk Publishers, The Hague, 1983, 321.

McLachlan, A., The biomass of macro- and interstitial fauna in clean and wrack-covered beaches in Western Australia, *Est. Coastal Shelf Sci.*, 21, 587, 1985.

McLachlan, A., *Sandy Beach Research at the University of Port Elizabeth: 1975–1986*, University of Port Elizabeth, Institute of Coastal Research, Report No. 14, 111 pp, 1987.

McLachlan, A., Behavioural adaptations of sandy beach organisms: an ecological perspective, in *NATO Advanced Research Workshop on Behavioural Adaptation to Intertidal Life*, Chelazzi, C., Vannini, M., Eds., Plenum Press, New York, 1988, 449.

McLachlan, A., Water filtration by dissipative beaches, *Limnol. Oceanogr.*, 34, 774, 1989.

McLachlan, A., Dissipative beaches and macrofauna communities on exposed intertidal sands, *J. Coastal Res.*, 6, 57, 1990a.

McLachlan, A., Sandy beaches as ecosystems, in *Ecology of Sandy* Shores, Brown, A.C., McLachlan, A., Eds., 1990b, 5.

McLachlan, A., Bate, G.C., Carbon budget for a high energy surf zone, *Vie et Milieu*, 34, 67, 1984.

McLachlan, A., Cockroft, A.C., Malan, D.E., Benthic faunal response to a high energy gradient, *Mar. Ecol. Prog. Ser.*, 16, 57, 1984.

McLachan, A., Erasmus, T., Eds., *Sandy Beaches as Ecosystems*, Dr. W. Junk Publishers, The Hague, 757 pp, 1983.

McLachlan, A., Erasmus, T., Dye, A.H., Wooldridge, T., van der Horst, G., Lasiak, T., McGwynne, L., Sand beach ecosystems: an ecosystem approach towards a high energy interface, *Est. Coastal Shelf Sci.*, 13, 11, 1981.

McLachlan, A., Hesp, P., Faunal response to morphology and water circulation of a sandy beach with cusps, *Mar. Ecol. Prog. Ser.*, 19, 133, 1984.

McLachlan, A., Illenberger, W.K., Significance of groundwater nitrogen input to a beach surf ecosystem, *Stygologia*, 2, 291, 1986.

McLachlan, A., Jaramillo, E., Zonation on sandy beaches, *Oceanogr. Mar. Biol. Annu. Rev.*, 33, 305, 1995.

McLachlan, A., Lewin, J., Observations on surf phytoplankton blooms along the coasts of South Africa, *Bot. Marina*, 24, 291, 1981.

McLachlan, A., Lombard, H.W., Louwrens, S., Trophic structure and biomass distribution on two East Cape rocky shores, *S. Af. J. Sci.*, 16, 85, 1981.

McLachlan, A., McGwynne, L.E., Do sandy beaches acculate nitrogen?, *Mar. Ecol. Prog. Ser.*, 34, 191, 1986.

McLachlan, A., Romer, G., Trophic relationships in a high energy beach and surf zone ecosystem, in *Trophic Relationships in the Marine Environment*, Barnes, M., Gibson, R.N., Eds., Aberdeen University Press, Aberdeen, 1990, 356.

McLachlan, A., Winter, D., Botha, L., Vertical and horizontal distribution of sublittoral meiofauna in Angola Bay, South Africa, *Mar. Biol.*, 40, 355, 1977.

McLachlan, A., Woolridge, T., Dye, A.H., The ecology of sandy beaches in South Africa, *S. Af. J. Z ool.*, 16, 219, 1981.

McLachlan, A., Woolridge, T., Schramm, M.M., Kuhn, M., Seasonal abundance, biomass and feeding of shorebirds on sandy beaches on the Eastern Cape, South Africa, *Ostrich*, 51, 44, 1980.

MacLulich, J.H., Aspects of the ecology of intertidal epilithic microflora at Green Point, New South Wales, M.Sc. thesis, University of Sydney, Sydney, Australia, 1983.

MacLulich, J.H., Variations in the density and variety of intertidal epilithic microflora, *Mar. Ecol. Prog. Ser.*, 40, 285, 1987.

McLusky, D.S. *The Estuarine Ecosystem*, 2nd edition, Blackwell, London, 150 pp, 1981.

McLusky, D.S., Anderson, F.E., Wolfe-Murphy, S., Distribution and population recovery of *Arenicola marina* and other benthic fauna after bait digging, *Mar. Ecol. Prog. Ser.*, 11, 173, 1983.

McLusky, D.S., Nair, S.A., Stirling, A., Bharoava, R., The ecology of a Central West Indian beach with particular reference to *Donax incarnatus*, *Mar. Biol.*, 30, 267, 1975.

McMillan, C., Parker, P.L., Fry, B., $^{13}C/^{12}C$ ratios in seagrasses, *Aquatic Bot.*, 9, 263, 1980.

MacNae, W., A general account of the fauna and flora of mangrove swamps and forests in the Indo-Pacific Region, *Adv. Mar. Biol.*, 6, 73, 1968.

McNaughton, S.J., Biodiversity and function of grazing ecosystems, in *Biodiversity and Ecosystem Function*, Schultze, E.-D, Mooney, A., Eds., Springer-Verlag, Berlin, 1993, 361.

McNeill, S., Lawton, J.H., Annual production and respiration in animal populations, *Nature*, 225, 472, 1970.

McQuaid, C.D., Spatial and temporal variation in rocky intertidal comunities, Ph.D. thesis, University of Cape Town, 1980.

McQuaid.C.D., Population dynamics of *Littorina africana knysaensis* (Philippi) on an exposed rocky shore, *J. Exp. Mar. Biol. Ecol.*, 54, 65, 1981.

McQuaid, C.D., Differential effects of predation by the intertidal whelk *Nucella duba* (Kr.) on *Littorina africana knysnaensis* (Phillipi) and the barnacle *Tetraclita serrata* Darwin, *J. Exp. Mar. Biol. Ecol.*, 89, 97, 1985.

McQuaid, C.D., Branch, G.M., Trophic structure of rocky intertidal communities: response to wave action and implications for energy flow, *Mar. Ecol. Prog. Ser.*, 22, 153, 1985.

McQuaid, C.D., Dower, K.M., Enhancement of habitat heterogeneity and species richness on rocky shore inundated by sand, *Oecologia (Berlin)*, 84, 142, 1990.

McRoy, C.P., Standing stocks and related features of eelgrass populations in Alaska, *J. Fish. Res. Bd. Canada*, 27, 1811, 1970.

McRoy, C.P., Seagrass productivity: carbon uptake experiments in eelgrass, *Zostera marina*, *Aquaculture*, 4, 131, 1974.

McRoy, C.P., Barsdate, R.J., Phosphate absorption in eelgrass, *Limnol. Oceanogr.*, 15, 14, 1970.

McRoy, C.P., Barsdate, R.J., Nebert, M., Phosphorus cycling in an eelgrass (*Zoatera marina* L.) ecosystem, *Limnol. Oceanogr.*, 17, 58, 1972.

McRoy, C. P., Ed., *Seagrass Ecosystems. Research Recommendations of the International Seagrass Workshop*, Report of a workshop held under the auspices of the International Decade of Ocean Exploration 22–26 Oct. 1973, submitted to the National Science Foundation, Washington, DC, 1973.

McRoy, C.P., Goering, J.J., Chaney, B., Nitrogen fixation associated with seagrasses, *Limnol. Oceanogr.*, 18, 998, 1973.

McRoy, C.P., Goering, J.J., Nutrient transfer between the seagrass *Zoatera marina* and its epiphytes. *Nature (London)*, 248, 173, 1974.

McRoy, C.P., Helfferich, C., Eds., *Seagrass Ecosystems: A Scientific Perspective*, Marcel Dekker, New York, 314 pp, 1977.

McRoy, C.P., McMillan, C., Production ecology and physiology in seagrasses, in *Seagrass Ecosystems. A Scientific Perspective*, McRoy, C.P., Helfferich, G., Eds., Marcel Dekker, New York, 1977, 53.

Macko, S.A., Entzeroth, L., Parker, P.L., Regional differences in nitrogen and carbon isotopes on the continental shelf of the Gulf of Maine, *Naturwissenschaften*, 71, 374, 1984.

Macko, S.A., Zieman, J., Stable isotope composition and aminoacid analysis of estuarine plant litter undergoing decomposition (abstract), *Estuaries*, 6, 304, 1983.

Magnum, C.P., Studies on speciation in maldanid polychaetes of the North American Atlantic coast. II. Distribution and competitive interaction of five sympatric species, *Limnol. Oceanogr.*, 9, 12, 1964.

Mahoney, B.M.S., Livingston, R.J., Seasonal fluctuations of benthic macrofauna in the Apalachicola Estuary, Florida, USA: the role of predation, *Mar. Biol.*, 69, 207, 1982.

Main, S.P., McIntyre, C.D., The distribution of epiphytic diatoms in Yaquina estuary, Oregon, *Bot. Marina*, 17, 88, 1974.

Maki, J.S., Rittschof, D., Costlow, J.D., Mitchell, R., Inhibition of attachment of larval barnacles, *Balanus amphitrite*, by bacterial surface films, *Mar. Biol.*, 97, 199, 1988.

Maki, J.S., Rittschof, D., Schmidt, A.R., Snyder, A.G., Mitchell, R., Factors controlling attachment of bryozoan larvae: a comparison of bacterial films and unfilmed surfaces, *Biol. Bull.*, 177, 295, 1989.

Malan, D.E., The role of the subtidal benthic system in the functioning of exposed sandy beaches, Ph.D. thesis, University of Port Elizabeth, Port Elizabeth, South Africa.

Malan, D.E., McLachlan, A., Vertical gradients of meiofauna and bacteria from the sediments of two high energy surf zones on Angola Bay, South Africa, *S. Af. J. Mar. Sci.*, 3, 43, 1985.

Malan, D.E., McLachlan, A., *In situ* benthic oxygen fluxes in a nearshore coastal marine system: a new approach to the quantitative of wave action, *Mar. Ecol. Prog. Ser.,* 73, 69, 1991.

Mallin, M.A., Paerl, H.W., Planktonic trophic transfer in an estuary: seasonal, diel, and community structure effects, *Ecology*, 75, 2168, 1994.

Malone, T.C., Environmental regulation of phytoplankton productivity in the lower Hudson Estuary, *Est. Coastal Shelf Sci.,* 5, 157, 1977.

Malone, T.C., Kemp, W.M., Ducklow, H.W., Boynton, W.R., Tuttle, J.H., Jonas, R.B., Lateral variation in the production and fate of phytoplankton in a partially stratified estuary, *Mar. Ecol. Prog. Ser.*, 31, 149. 1986.

Manas, T.F.K., Feeding ecology of major carnivores from four Eastern Cape estuaries, *S. Af. J. Zool.*, 19, 210, 1984.

Mann, K.H., Macrophyte production and detritus food chains in coastal waters, in *Detritus and Its Role in Aquatic Ecosystems*, Melchorri-Santalini, U. Hopton, J., Eds., Memoir Instituto Italiano Idriobiologie (Supplement), 29, 325, 1972a.

Mann, K.H., Ecological energetics of the seaweed zone in a marine bay on the Atlantic coast of Canada. I. Zonation and biomass of the seaweeds, *Mar. Biol.*, 12, 1, 1972b.

Mann, K.H., Seaweeds, their productivity and strategy for growth, *Science*, 182, 978, 1973.

Mann, K.H., Destruction of kelp-beds by sea-urchins: a cyclical phenomenon or irreversible degradation, *Helg. Wiss. Meeresunters.*, 30, 455, 1977.

Mann, K.H., *The Ecology of Coastal Waters. A Systems Approach,* Blackwell Scientific Publications, Oxford, 322 pp, 1982.

Mann, K.H., Production and use of detritus in various freshwater, estuarine, and coastal marine ecosystems, *Limnol. Oceanogr.,* 33, 910, 1988.

Mann, K.H., Field, J.G., Wulff, F., Network analysis in marine ecology: an assessment, in *Network Analysis in Marine Ecology: Methods and Applications,* Wulff, F., Field, J.G., Mann, K.H., Eds., Springer-Verlag, Berlin, 1989, 259.

Marchant, C.J., Evolution of *Spartina* (Gramminae). I. The history and morphology of the genus in Britain, *J. Linnean Soc. London, Bot.*, 60, 1, 1967.

Mare, M.F., A marine benthic community, with special reference to the microorganisms, *J. Mar. Biol. Assoc. U.K.,* 25, 517, 1942.

Margalef, R., El ecosistema, in *Ecologica Marina*, Foundation la salle de ciencias naturales, Caracas, 1967, 377.

Margalef, R., *Perspectives in Ecological Theory*, University of Chicago Press, Chicago, 111 pp, 1968.

Margalef, R., Diversity and stability: a practical proposal and a model of interdependence, *Brookhaven Symp. Biol.*, 22, 25, 1969.

Marguillier, S., Van der Velde, G., Dehairs, F., Heminga, M.A., Rajagopal, S., Trophic relationships in an interlinked mangrove-seagrass ecosystem as traced by ^{13}C and ^{15}N, *Mar. Ecol. Prog. Ser.*, 151, 115, 1997.

Marinelli, R.L., Effects of polychaetes on silicate dynamics and fluxes in sediments: importance of species, animal activity and polychaete effects on benthic diatoms, *J. Mar. Res.*, 50, 745, 1992.

Marinelli, R.L., Coull, B.C., Structural complexity and juvenile fish predation on meiobenthos: an experimental approach, *J. Exp. Mar. Biol. Ecol.*, 108, 67, 1987.

Markowitz, D.V., Predator influence on shore-level size gradients in *Tegula funebralis* (A.Adams), *J. Exp. Mar. Biol. Ecol.*, 45, 1, 1980.

Marsden, I.D., Kelp-sandhopper interactions on a sand beach in New Zealand. I. Drift composition and distribution, *J. Exp. Mar. Biol. Ecol.*, 152, 61, 1991a.

Marsden, I.D., Kelp-sandhopper interactions on a sand beach in New Zealand. II. Population dynamics of *Talorchestia quoyana* (Milne-Edwards), *J. Exp. Mar. Biol. Ecol.*, 152, 75, 1991b.

Marsh, A.G., Gremare, A., Tenore, K.P., Effect of food type and ration on growth of juvenile *Capitella* sp. I (Annelida: Polychaeta): macro- and micronutrients, *Mar. Biol.*, 102, 519, 1989.

Marsh, B.A., Branch, G.M., Circadian and circatidal rhythms and oxygen consumption in the sandy beach isopod *Tylos granulatus* Krauss, *J. Exp. Mar. Biol. Ecol.*, 37, 77, 1979.

Marsh, C.P., Rocky intertidal community organization: the impact of avian predators on mussel recruitment, *Ecology,* 67, 771, 1986a.

Marsh, C.P., Impact of avian predators on high intertidal limpet populations, *J. Exp. Mar. Biol. Ecol.*, 67, 276. 1986b.

Marshall, B.E., Park, R.B., The ecotone between *Spartina foliosa* Trin. and *Salicornia virginica* L. in salt marshes in northern San Francisco Bay. I. Biomass and production, *J. Ecol.*, 64, 421, 1976.

Marshall, N., Food transfer through the lower trophic levels of the benthic environment, in *Marine Food Chains*, Steele, J.H., Ed., Oliver and Boyd, Edinburgh, 1970, 52.

Marshall, N., Oviatt, C.A., Skauen, D.M., Productivity of the benthic microflora of shoal and estuarine environments in southern New England, *Int. Rev. Ges. Hydrobiol. Hydrogr.*, 56, 947, 1971.

Martin, A.P., Feeding ecology of the birds in the Swartkops Estuary, Ph. D. thesis, University of Port Elizabeth, South Africa, 267 pp, 1991.

Martin, A.P., Baird, D., Seasonal abundance and distribution of birds on the Swartkops Estuary, Port Elizabeth, *Ostrich*, 58, 122, 1987.

Mateo, M.A., Romero, J., Detritus dynamics in the seagrass *Posidonia oceanica*: elements for an ecosytem carbon and nutrient budget, *Mar. Ecol. Prog. Ser.*, 151, 43, 1997.

Mathieson, A.C., Penniman, C.A., Harris, L.G., Northwest Atlantic shores, in *Intertidal and Littoral Systems. Ecosystems of the World 24*, Mathiesen, A.C., Nienhuis, P.H., Eds., Elsevier, Amsterdam, 1991, 347.

Mathieson, A.C., Tveter, E., Daly, M., Howard, J., Marine algal ecology in a New Hampshire tidal rapid, *Bot. Mar.*, 20, 277, 1977.

Matisoff, G., Mathematical models of biotutbation, in *Animal-Sediment Relations: The Biogenic Alteration of Sediments*, McCall, P.L., Teresz, M.J.S., Eds., Plenum Press, New York, 336 pp, 1982.

Mayer, L.M., The nature and determination of non-living sedimentary organic matter as a food source for deposit feeders, in *Ecology of Marine Deposit Feeders*, Lopez, G., Taghorn, G., Levinton, J., Eds., Springer-Verlag, Berlin, 1989, 98.

Meadows, P.N., Campbell, J.I., Habitat selection by aquatic invertebrates, *Adv. Mar. Biol.*, 10, 271, 1972.

Medlin, L.K., Effects of grazers on epiphytic distom communities, in *Proceedings of the 6th International Diatom Symposium, Budapest, 1980*, Ross, R., Ed., Otto Koeltz, Koenigstein, 1981, 319.

Mees, J., Jones, M.B., The hyperbenthos, *Oceanogr. Mar. Biol. Annu. Rev.*, 35, 221, 1997.

Meese, R.J., Effects of predation by birds on the gooseneck barnacle *Pollicipes polymerus* Sowerby distribution and abundance, *J. Exp. Mar. Biol. Ecol.*, 166, 47, 1993.

Mendelssohn, I.A., The influence of nitrogen level, form and application method on the growth response of *Spartina alterniflora* in North Carolina, *Estuaries*, 2, 106, 1979.

Mendelssohn, I.A., Seneca, E.D., The influence of soil drainage on the growth of salt marsh cordgrass *Spartina alterniflora* in North Carolina, *Est. Coastal Mar. Sci.*, 11, 27, 1980.

Menge, B.A., Foraging strategy of a starfish in relation to prey availability and environmental predictability, *Ecol. Monogr.*, 42, 25, 1972.

Menge, B.A., Organization of the New England rocky intertidal community: the role of predation, competition and environmental heterogeneity, *Ecol. Monogr.*, 46, 355, 1976.

Menge, B.A., Predation intensity in a rocky intertidal community. Relation between predator foraging activity and environmental harshness, *Oecologia (Berlin)*, 34, 17, 1978a.

Menge, B.A., Predation intensity in a rocky intertidal community: effect of an algal canopy on predator feeding rates, *Oecologia (Berlin)*, 34, 17, 1978b.

Menge, B.A., Coexistence between the seastars *Asterias vulgaria* and *A. forbesi* in a heterogeneous environment: a non-equilibrium explanation, *Oecologia (Berlin)*, 46, 245, 1979.

Menge, B.A., Components of predation intensity in the low zone of the New England rocky intertidal region, *Oecologia (Berlin)*, 58, 141, 1983.

Menge, B.A., Relative importance of recruitment and other causes of variation in intertidal community structure: a multivariate evaluation, unpublished manuscript, 1988.

Menge, B.A., Relative importance of recruitment and other causes of variation in rocky intertidal community structure, *J. Exp. Mar. Biol. Ecol.*, 146, 69, 1991a.

Menge, B.A., Generalizations from experiments: is predation strong or weak in a New England rocky intertidal, *Oecologia (Berlin)*, 88, 1, 1991b.

Menge, B.A., Community regulation: under what conditions are bottom-up factors important on rocky shores, *Ecology*, 73, 755, 1992.

Menge, B.A., Indirect effects in marine rocky intertidal interaction web: patterns and importance, *Ecol. Monogr.*, 65, 21, 1995.

Menge, B.A., Ashkenas, L.R., Matson, A., Use of artificial holes in studying community development in cryptic marine habitats in a tropical rocky intertidal region, *Mar. Biol.*, 77, 129, 1983.

Menge, B.A., Berlow, E.L., Blanchette, C., Navarrete, S.A., Yamada, S.B., The keystone species concept variation in interaction strength in a rocky intertidal habitat, *Ecol.* Monogr., 64, 249, 1994.

Menge, B.A., Daley, B.A., Lubchenco, J., Sanford, Dahloff, E., Halpin, P.M., Hudson, G., Burnaford, J.L., Top-down and bottom-up regulation of New Zealand rocky shore communities, *Ecol. Monogr.*, 69, 297, 1999.

Menge, B.A., Daley, B.A., Wheeler, P.A., Control of interaction strength in marine benthic communities, in *Food Webs: Integration of Pattern and Dynamics*, Polis, G.A., Winemiller, R., Eds., Chapman & Hall, New York, 1996, 258.

Menge, B.A., Daley, B.A., Wheeler, P.A., Dahlhoff, E., Sanford, E., Strub, P.T., Benthic pelagic links and rocky intertidal communities: evidence for bottom-up effects on top-down control, *Proc. Nat. Acad. Sci. (USA)*, 94, 14530, 1997a.

Menge, B.A., Daley, P.A., Wheeler, P.A., Strub, P.T., Rocky intertidal oceanography: an association between community structure and nearshore phytoplankton concentration, *Limnol. Oceanogr.*, 42, 57, 1997b.

Menge, B.A., Farrell, T.M., Community structure and interaction webs in shallow marine hard-bottom communities: tests of an environmental stress model, *Adv. Ecol. Res.*, 19, 189, 1989.

Menge, B.A., Lubchenco, J., Community organization in temperate and tropical rocky intertidal habitats: prey refuges in relation to consumer pressure gradients, *Ecol. Monogr.*, 51, 429, 1981.

Menge, B.A., Lubchenco, J., Ashkenas, L.R., Diversity, heterogeneity and consumer pressure in a tropical rocky shore intertidal community, *Oecologia (Berlin)*, 65, 394, 1985.

Menge, B.A., Lubchenco, J.A., Ashkensas, L.R., Ramsay, F., Experimental separation of effects of consumers on sessile prey in the low zone of a rocky shore in the Bay of Panama: direct and indirect consequences of food web complexity, *J. Exp. Mar. Biol. Ecol.*, 100, 225, 1986a.

Menge, B.A., Lubchenco, J., Gaines, S.D., Ashkensas, L.R., A test of the Menge-Sutherland model of community organization in a tropical intertidal food web, *Oecologia (Berlin)*, 71, 75, 1986b.

Menge, B.A., Olsen, A.M., Role of scale and environmental factors in regulation of community structure, *Trends Ecol. Evol.*, 5, 52, 1990.

Menge, B.A., Sutherland, J.P., Species diversity gradients: synthesis of the roles of predation, competition, and temporal heterogeneity, *Am. Nat.*, 110, 357, 1976.

Menge, B.A., Sutherland, J.P., Community vegetation: variation in disturbance, competition, and predation in relation to environmental stress, *Am. Nat.*, 130, 730, 1987.

Menge, J.L., Prey selection and foraging of the predaceous rocky intertidal snail, *Acanthina punctulata, Oecologia (Berlin)*, 17, 293, 1974.

Menge, J.L., Effects of herbivores on community structure of the New England rocky intertidal region: distribution, abundance and diversity of algae, Ph.D. thesis, Harvard University, Cambridge, MA, 164 pp, 1975.

Menge, J.L., Menge, B.A., Role of resource allocation, aggression and spatial heterogeneity in coexistence of two competing starfish, *Ecol. Monogr.*, 44, 189, 1974.

Metaxas, A., Hunt, H.L., Scheibling, R.E., Spatial and temporal variability of macrobenthic communities on a rocky shore in Nova Scotia, Canada, *Mar. Ecol. Prog. Ser.*, 115, 89, 1994.

Meyer-Reil, L.A., Dawson, R., Liebezeit, G., Tiedge, H., Fluctuations and interactions of bacterial activities in sandy beach sediments and overlying waters, *Mar. Biol.*, 48, 161, 1974.

Mgaya, Y.D., Density and production of *Clinocottus gobiceps* and *Oligoclottus maculosus* (Cottidae) in tide-pools at Helby Island, British Columbia, *Mar. Ecol. Prog. Ser.*, 88, 219, 1992.

Micheli, F., Feeding ecology of mangrove crabs in Northeastern Australia: mangrove litter consumption by *Sesarma messa* and *Sesama smithii.*, *J. Exp. Mar. Biol. Ecol.*, 171, 161, 1993.

Micheli, F., Effects of predator forging behaviour on patterns of prey mortality in marine soft bottoms, *Ecol. Monogr.*, 67, 203, 1997.

Mileikovshy, S.A., On predation of pelagic larvae and early juveniles of marine bottom invertebrates by adult benthic invertebrates, and their passing alive through their predators, *Mar. Biol.*, 26, 303, 1974.

Miller, C.B., The zooplankton of estuaries, in *Estuaries and Enclosed Seas. Ecosystems of the World 26*, Ketchum, B.H., Ed., Elsevier, Amsterdam, 1983, 103.

Miller, D.C., Bock, M.J., Tuener, E.J., Deposit and suspension feeding in oscillatory flows and sediment fluxes, *J. Mar. Res.*, 50, 489, 1992.

Miller, R.J., Succession in sea urchin and seaweed abundance in Nova Scotia, Canada, *Mar. Biol.*, 84, 275, 1983.

Miller, R.J., Colodey, A.G., Widespread mass mortalities of green sea urchin in Nova Scotia, Canada, *Mar. Biol.*, 73, 263, 1983.

Miller, T.E., Community diversity and interactions between size and frequency of disturbance, *Am. Nat.*, 120, 533, 1982.

Millero, F.J., Sohn, M.L., *Chemical Oceanography*, CRC Press, Boca Raton, FL, 531 pp, 1991.

Milliman, J.D., Summerhayes, C.P., Barrelto, H., Oceanography and suspended matter of the Amazon River, February–March 1973, *J. Sed. Petrol.*, 45, 189, 1975.

Mills, E.L., The biology of an ampeliscid amphipod crustacean subling species pair, *J. Fish. Res. Bd. Canada*, 26, 1415, 1967.

Mills, E.L., The community concept in marine ecology, with comments on continua and instability in some marine communities: a review, *J. Fish. Res. Bd. Canada*, 24, 305, 1969.

Mills, L.S., Soule, M.E., Doak, D.F., The key-stone species concept in ecology and conservation, *Bioscience*, 43, 219, 1993.

Milne, H., Dunnet, G., Standing crop, productivity and trophic relations of the fauna of the Ythan estuary, in *The Estuarine Environment*, Barnes, R., .Green, J., Eds., Applied Science Publications, London, 1972, 86.

Minagawa, M., Wada, E., Stepwise enrichment of ^{15}N along food chains: further evidence of the relationship between ^{15}N and animal age, *Geochemica et Cosmochemica Acta*, 48, 1135, 1984.

Minchinton, T.E., Life on the edge: conspecific attraction and recruitment of populations to disturbed habitats, *Oecologia (Berlin)*, 111, 45, 1997.

Minchinton, T.C., Scheibling, R.E., The influence of larval supply and settlement on the population structure of barnacles, *Ecology*, 72, 1867, 1991.

Minchinton, T.E., Scheibling, R.E., Free space availability and larval substratum selection as determinants of barnacle population structure in a developing rocky intertidal community, *Mar. Ecol. Prog. Ser.*, 95, 233, 1993.

Minello, T.J., Zimmerman, R.J., Fish predation on juvenile brown shrimp, *Penaeus azetecus* Ives: the effect of simulated *Spartina* structure on predation rates, *J. Exp. Mar. Biol. Ecol.*, 72, 211, 1983.

Mitsch, W., Gosselink, J., *Wetlands*, Van Nostrand Reinbold, New York, 539 pp, 1986.

Moe, R.L., Silva, P.C., Antarctic marine flora: uniquely devoid of kelps, *Science*, 196, 1206, 1977.

Moebius, K., Johnson, K.M., Exudation of dissolved organic carbon by brown algae, *Mar. Biol.*, 26, 117, 1974.

Moller, P., Phil, L., Rosenberg, R., Benthic energy flow and biological interactions in some shallow marine soft bottom habitats, *Mar. Ecol. Prog. Ser.*, 27, 109, 1985.

Moller, P., Rosenberg, R., Recruitment, abundance and production of *Mya arenaria* and *Cardium edule* in marine shallow waters, western Sweden, *Ophelia*, 22, 33, 1983.

Moloney, C.L., Simulation studies of trophic fluxes and nutrient cycles in Benguela upwelling foodwebs, *S. Af. J. Mar. Sci.*, 12, 457, 1992.

Monro, A.L.S., Wells, J.B.J., McIntyre, A.D., Energy flow in the flora and meiofauna of sandy beaches, *Proc. Roy. Soc., Edinb.*, 76, 297, 1978.

Montagna, P.A., *In situ* measurements of meiobenthic grazing rates on sediment diatoms and edaphic diatoms, *Mar. Ecol. Prog. Ser.*, 18, 119, 1984.

Montagna, P.A., Rates of metazoan meiofaunal microbivory: a review, *Vie Milieu*, 15, 1, 1995.

Montagna, P.A., Coull, B.C., Herring, T.C., Dudley, B.W., The relationship between abundances of meiofauna and their suspected microbial food (diatoms and bacteria), *Est. Coastal Shelf Sci.*, 13, 381, 1983.

Montague, C.L., A natural history of the temperate western Atlantic fiddler crabs (Genus *Uca*) with reference to their impact on the salt marsh, *Contr. Mar. Sci., Univ. Texas*, 23, 25, 1980.

Montague, C.L., The influence of fiddler crab burrows and burrowing on metabolic processes in salt marsh sediments, in *Estuarine Comparisons*, Kennedy, V.S., Ed., Academic Press, New York, 1982, 283.

Montague, C.L., Bunker, S.M., Haines, E.B., Pace, M.L., Wetzel, R.L., Aquatic macroconsumers, in *The Ecology of a Salt Marsh*, Pomeroy, L.R., Wiegert, R.C., Eds., Springer-Verlag, New York, 1981, 69.

Montgomery, W.L., Comparative feeding ecology of two herbivorous damselfishes (Pomacentridae: Teleostei) from the Gulf of California, Mexico, *J. Exp. Mar. Biol. Ecol.*, 47, 9, 1980.

Moore, C.G., The zonation of psammolittoral harpacticoid copepods around the Isle of Man, *J. Mar. Biol. Assoc. U.K.*, 59, 711, 1979.

Moore, D., Basher, H.A., Trent, L., Relative abundance, seasonal distribution and species composition of demersal fishes off Louisiana and Texas, 1962–1964, *Contr. Mar. Sci. Univ. Texas*, 15, 45, 1970.

Moore, H.B., Kitching, J.A., The biology of *Chthamalu stellatus* (Poli), *J. Mar. Biol. Assoc. U.K.*, 23, 521, 1939.

Moore, K.A., Carbon transport in two York River, Virginia Marshes, M.S. thesis, University of Virginia, 102 pp, 1974.

Moore, P.G., The kelp fauna of Northeast Britain. II. Multivariate classification: turbidity as an ecological factor, *J. Exp. Mar. Biol. Ecol.*, 13, 127, 1973.

Moore, P.G., The nematode fauna associated with the holdfasts of kelp (*Lamanaria hyperborea*) in north-east Britain, *J. Mar. Biol. Assoc. U.K.*, 51, 589, 1978.

Moore, P.G., Seed, R., *The Ecology of Rocky Coasts*, Columbia University Press, New York, 467 pp, 1986.

Mooring, M.T., Cooper, A.W., Seneca, E.D., Seed germination response and evidence for height ecophenes in *Spartina alterniflora* from North Carolina, *Am. J. Bot.*, 58, 48, 1971.

Moran, M.J., Distribution and dispersion of the predatory intertidal gastropod *Morula marginalba* Blainville (Muricidae), *Mar. Ecol. Prog. Ser.*, 22, 41, 1985a.

Moran, M.J., The timing and significance of sheltering and foraging behaviour of the predatory intertidal gastropod *Morula marginalba* Blainville (Muricidae), *J. Exp. Mar. Biol. Ecol.*, 93, 103, 1985b.

Moran, M.J., Effects of prey density, prey size and predator size on rates of feeding by an intertidal gastropod *Morula marginalba* Blainville (Muricidae), *J. Exp. Mar. Biol. Ecol.*, 90, 97, 1985c.

Moreno, C.A., Un resumen de las consecuenca ecológicas de al exclusion de hombre en la zona intermareal de Mehuin-Chile, *Chile Estud. Oceanol.*, 5, 59, 1986.

Moreno, C.A., Luneche, K.N., Lopez, M.I., The response of the intertidal *Concholepas concholepas* (Gastropoda: Muricidae) to protection from man in southern Chile, *Oikos*, 42, 155, 1986.

Morgan, S.G., Life and death in the plankton, in *Ecology of Marine Invertebrate Larvae*, McEdward, L., Ed., CRC Press, Boca Raton, FL, 1995, 279.

Moriarty, D.J.W., Pollard, P.C., Hunt, W.G., Moriarty, C.M., Wassenberg, T.J., Productivity of bacteria and microalgae and the effect of grazers by holothurians in sediments and a coral reef flat, *Mar. Biol.*, 85, 293, 1985.

Morita, R.Y., The role of microorganisms in the environment, in *Oceanic Sand Scattering Predicting*, Anderson, N.R., Zahurance, B.J., Eds., Plenum Press, New York, 1977, 445.

Morris, J.T., The nitrogen uptake kinetics of *Spartina alterniflora* in culture, *Ecology*, 61, 1114, 1980.

Morris, S., Taylor, A.C., Diurnal and seasonal variation in physico-chemical conditions within intertidal rockpools, *Est. Coastal Shelf Sci.*, 17, 339, 1983.

Morrison, S.T., White, D.C., Effects of grazing by Gammaridean amphipods on the microbiota of allochthonous detritus, *Appl. Envt. Microbiol.*, 40, 659, 1980.

Morse, A.N.C., How do planktonic larvae know where to settle?, *Am. Scientist*, 79, 154, 1991.

Morse, A.N.C., Froyd, C.A., Morse, D.E., Molecules from cyanobacteria and red algae that induce larval settlement and metamorphosis in the mollusc *Haliotis rufescens*, *Mar. Biol.*, 81, 293, 1984.

Morse, A.N.C., Morse, D.E., Recruitment and metamorphosis of *Haliotis* larvae induced by molecules uniquely available at the surfaces of crustose algae, *J. Exp. Mar. Biol. Ecol.*, 75, 191, 1984.

Morse, D.E., Recent progress in larval settlement and metamorphosis: closing the gaps between molecular biology and ecology, *Bull. Mar. Sci.*, 46, 465, 1990.

Morse, D.E., Hooker, N., Duncan, H., Jensen, L., γ-Amino butyricacid, a neurotransmitter induces planktonic abalone to settle and begin metamorphosis, *Science*, 304, 407, 1979.

Morton, J., Miller, M., *The New Zealand Seashore*, Collins, London and Auckland, 638 pp, 1968.

Mukai, H., The phytal animals on the thalli of *Sargassum serratifolium* in the *Sargassum* regions, with reference to their seasonal fluctuations, *Mar. Biol.*, 8, 170, 1971.

Mukai, H., Aloi, K., Ishida, Y., Distribution and biomass of the eelgrass (*Zostera marina* L.) and other seagrasses in Odawa Bay, Central China, *Neth. J. Sea Res.*, 14, 102.

Murdock, W.W., Community structure, population control, and competition — a critique, *Am. Nat.*, 100, 219, 1966.

Murdock, W.W., Oaten, A., Predation and population stability, *Adv. Ecol. Res.*, 9, 1, 1975.

Murphy, R.C., Kremer, J.N., Benthic community metabolism and the role of deposit-feeding callianassid shrimp, *J. Mar. Res.*, 50, 321, 1992.

Muschenheim, D.K., The dynamics of near-bed seston flux and suspension-feeding benthos, *J. Mar. Res.*, 45, 473, 1987a.

Muschenheim, D.K., The role of hydrodynamic sorting of seston in the nutrition of a benthic suspension feeder, *Spio setosa* (Polychaeta: Spionidae), *Biol. Oceanogr.*, 4, 265, 1987b.

Muus, K., Settling growth and mortality of young bivalves in the Orensund, *Ophelia*, 12, 79, 1973.

Myers, A.C., Sediment processing in a marine subtidal bottom community. I. Physical aspects, *Mar. Geol.*, 452, 133, 1977.

Nadeau, R.J., Primary production and export of plant material in a salt marsh ecosystem, Ph.D. thesis, Rutgers University, New Brunswick, NJ, 1972.

Naiman, R.J., Sibert, J.R., Transport of nutrients and carbon from the Nanaimo River and its estuary, *Limnol. Oceanogr.*, 23, 102, 1978.

Naiman, R.J., Sibert, J.R., Detritus and juvenile salmon production in the Nanaimo Estuary. III. Importance of detrital carbon in the estuarine ecosystem, *J. Fish. Res. Bd. Canada*, 36, 504, 1979.

Navarrete, S.A., Variable predation: effects of whelks on a mid-intertidal successional community, *Ecol. Monogr.*, 66, 301, 1996.

Navarro, J.M., Winter, J.E., Ingestion rate, assimilatory efficiency and energy balance in *Mytilus chilensis* in relation to body size and different algal concentrations, *Mar. Biol.*, 67, 255, 1982.

Naylor, E., Tidal and diurnal rhythms of locomotory activity in *Carcinus maenas* (L.), *J. Exp. Mar. Biol. Ecol.*, 35, 602, 1958.

Naylor, E., Temporal aspects of adaptation in the behavioural physiology of marine animals, in *21st European Marine Biology Symposium*, Gdansk, 1987.

Naylor, E., Clock-controlled behaviour in intertidal animals, in *NATO Advanced Research Workshop on Behavioural Adaptation to Intertidal Life*, Chelazzi, G., Vannini, M., Eds., Plenum Press, New York and London, 1988, 1.

Neihof, R.A., Loeb, G.I., The surface charge of particulate matter in seawater, *Limnol. Oceanogr.*, 17, 7, 1972.

Nelson, W.G., Experimental studies of decapod and fish predation on seagrass macrobenthos, *Mar. Ecol. Prog. Ser.*, 5, 441, 1979.

Nestler, J., Interstitial salinity as a cause of ecospheric variation in *Spartina alterniflora*, *Est. Coastal Mar. Sci.*, 5, 707, 1977.

Newcombe, C.L., A study of the community relationships of the sea mussel, *Mytilus edulis* L., *Ecology*, 16, 234, 1935.

Newell, R.C., *The Biology of Intertidal Animals*, 3rd ed., Marine Ecological Surveys Ltd., Faversham, Kent, 781 pp, 1979.

Newell, R.C., Field, J.G., Energy balance and significance of microorganisms in a kelp bed community, *Mar. Ecol. Prog. Ser.*, 8, 103, 1983.

Newell, R.C., Linley, E.A.S., Significance of microheterotrophs in the decomposition of phytoplankton estimates of carbon and nitrogen flow based on the biomass of plankton communities, *Mar. Ecol. Prog. Ser.*, 16, 105, 1984.

Newell, R.C., Linley, E.A.S., Lucas, M.I., Bacterial production and carbon conversion based on saltmarsh plant debris, *Est. Coastal Shelf Sci.*, 17, 405, 1983.

Newell, R.C., Lucas, M.I., Linley, E.A.S., Rate of degradation and efficiency of conversion of phytoplankton debris by marine micro-organisms, *Mar. Ecol. Prog. Ser.*, 6, 123, 1983.

Newman, W.A., The paucity of intertidal barnacles in the tropical western Pacific, *Veliger*, 2, 89, 1960.

Nichols, F.H., Thompson, J.K., Seasonal growth in the bivalve *Macoma balthica* near the southern limit of its range, *Estuaries*, 5, 110, 1982.

Nicotri, M.E., Grazing effects of four marine intertidal herbivores on the microflora, *Ecology*, 58, 1020, 1977.

Nicotri, M.E., Factors involved in herbivore food preference, *J. Exp. Mar. Biol. Ecol.*, 42, 13, 1980.

Nienhuis, P.H., Daemen, E.A.M.J., De Jong, S.A., Hofman, P.A.G., *Biomass and Production of Macxrophytobenthos*, Progress Report 1985, Delta Institute for Hydrobiological Research, The Netherlands, 1985.

Nienhuis, P.H., De Bree, H.H., Consumption of the eelgrass, *Zostera marina* by birds and invertebrates during the grazing season in Lake Grevelingen (S.W. Netherlands), *Neth. J. Sea Res.*, 12, 1980.

Nilsson, L., Food consumption of diving ducks wintering on the coast of South Sweden in relation to food resources, *Oikos*, 20, 128, 1969.

Nishihira, M., Observations on the selection of algal substrata by hydrozoan larvae, *Sertularella miurensis* in nature, *Bull. Mar. Biol. Sta. Asamushi*, 13, 34, 1967.

Nishio, T., Koike, I., Hattori, A., Denitrification, nitrate reduction, and O_2 consumption in coastal and estuarine sediments, *Appl. Environ. Microbiol.*, 43, 648, 1982.

Nixon, S.W., Between coastal marshes and coastal waters — a review of twenty years of speculation and research on the role of salt marshes in estuarine productivity and water chemistry, in *Estuarine and Wetland Processes*, Hamilton, P., MacDonald, K., Eds., Plenum Press, New York, 1980, 437.

Nixon, S.W., Remineralization and nutrient cycling in a coastal marine ecosystems, in *Nutrient Enrichment in Estuaries*, Neilson, B., Cronin, L.E., Eds., Humana Press, Clifton, NJ, 1981, 111.

Nixon, S.W., The ecology of New England high salt marshes: a community profile, *U.S. Fish & Wildlife Service, Report No. FWS.OBS-81-55*, Washington, DC, 70 pp, 1982.

Nixon, S.W., Kelly, J.R., Furnas, B.N., Oviatt, C.A., Hale, S.S., Phosphorus regeneration and the metabolism of coastal marine bottom communities, in *Marine Benthic Dynamics*, Tenore, K.R., Coull, B.C., Eds., University of South Carolina Press, Columbia, SC, 1980, 209.

Nixon, S.W., Oviatt, C.A., Ecology of a New England salt marsh, *Ecol. Monogr.*, 43, 463, 1973.

Nixon, S.W., Oviatt, C.A., Hale, S., Nitrogen regeneration and metabolism of coastal marine bottom communities, in *The Role of Terrestrial and Aquatic Organisms in Decomposition Processes*, Anderson, J., Macfadyn, J., Eds., Blackwell Scientific Publications, London, 1976, 269.

Nixon, S.W., Oviatt, J.N., Kremer, J.N., Perez, K., The use of numerical methods and laboratory microcosms in estuarine systems: simulation of a winter phytoplankton bloom, in *Marsh-Estuarine Systems in Marine Eutrophication*, Dame, R.F., Ed., University of South Carolina Press, Columbia, SC, 1979, 165.

Nixon, S.W., Oviatt, C.A., Rogers, C., Taylor, K., Mass and metabolism of a mussel bed, *Oceologia (Berlin)*, 8, 21, 1971.

Nojima, S., Nishihara, M., Makai, H., Outflow of seagrass leaves and effects of sea urchin grazing, in *Studies on the Dynamics of the Biological Community in Tropical Seagrass Ecosystems in Papua New Guinea*, Hattori, A., Ed., Ocean Research Institute, University of Tokyo, 1987.

Norhko, A., Bonsdorff, E., Population responses of coastal zoobenthos to stress induced by drifting algal mats, *Mar. Ecol. Prog. Ser.*, 140, 141, 1996.

North, J.W., Ed., The biology of giant kelp bed (*Macrocystis*) in California, *Nova Hedwigia*, 32, 600 pp., 1971a.

North, J.W., Introduction and background, *Nova Hedwegia*, 32, 1971b.

Norton, T.A., An ecological study of the fauna inhabiting the subtidal marine alga *Saccorhiza polyschides* (Lightf.) Batt., *Hydrobiologica*, 37, 215, 1971.

Norton, T.A., The zonation of seaweeds on rocky shores, in *The Ecology of Rocky Coasts*, Norton-Griffiths, M., Moore, P.C., Seed, R., Eds., Hodder & Straughton, London, 1985, 7.

Norton, T.A., Hawkins, S.J., Manley, N.L., Williams, G.A., Watson, D.G., Scraping a living: a review of littorinid grazing, *Hydrobiologica*, 193, 117, 1990.

Norton, T.A., Mathieson, A.C., The biology of unattached seaweeds, *Progr. Phycol. Res.*, 2, 333, 1983.

Norton, T.A., Mathieson, A.C., Neuschel, M., Some aspects of form and function in seaweeds, *Bot. Marina*, 25, 501, 1981.

Novaczek, I., McLachlan, J., Recolonization by algae of the sublittoral habitat of Halifax County, Nova Scotia, following the demise of sea urchins, *Bot. Mar.*, 29, 69, 1986.

Nowell, A.R.M., Jumars, P.A., Eckman, J.E., Effects of biological activity on the entrainment of marine sediments, *Mar. Geol.*, 42, 133, 1981.

Noye, B.J., The Coorong — past, present, and future, Publication 38, Department of Adult Education, the University of Adelaide, South Australia, 1973.

O'Connor, R.J., Brown, R.A., Prey depletion and foraging strategy in the oystercatcher *Haemotopus ostralegus*, *Oecologia (Berlin)*, 27, 75, 1977.

Odum, E.P., A research challenge: evaluating the productivity of coastal and estuarine water, *Proc. 2nd Sea Grant Conf.*, Graduate School of Oceanography, University of Rhode Island, Kingston, RI, 1968.

Odum, E.P., The strategy of ecosystem development, *Science*, 164, 262, 1969.

Odum, E.P., *Fundamentals of Ecology*, W.B.Saunders, Philadelphia, 547 pp, 1971.

Odum, E.P., Halophytes, energetics and ecosystems, in *Ecology of Halophytes*, Reinold, R.J., Queen, W.H., Eds., Academic Press, New York, 1974, 599.

Odum, E.P., *Ecology*, 2nd ed., Holt, Rinehart & Winston, New York, 243 pp, 1975.

Odum, E.P., The status of three ecosystem-level hypotheses regarding salt marsh estuaries: tidal subsidy, outwelling, and detritus-based food chains, in *Estuarine Perspectives*, Kennedy, V.S., Ed., Academic Press, New York, 1980, 485.

Odum, E.P., *Basic Ecology*, Saunders College Publishing, Philadelphia, 513 pp, 1983.

Odum, E.P., de La Cruz, A.A., Particulate detritus in a Georgia salt marsh, in *Estuaries*, Lauff, G.H., Ed., American Association for the Advancement of Science, Publication No. 83, Washington, DC, 1967, 383.

Odum, E.P., Fanning, M.E., Comparison of the productivity of *Spartina alterniflora* and *Spartina cynosuroides* in Georgia coastal marshes, *Bull. Georgia Acad. Sci.*, 31, 1, 1973.

Odum, E.P., Smalley, A. E., Comparison of the population energy flow of a herbivorous and deposit-feeding invertebrate in a slat-marsh ecosystem, *Proc. Nat. Acad. Sci., USA*, 55, 617, 1959.

Odum, H.T., An energy circuit language for ecological and social systems, in *Systems Analysis and Simulation in Ecology*, Academic Press, New York, 1971a, 139.

Odum, H.T., *Environment, Power and Society*, John Wiley & Sons, New York, 331 pp, 1971b.

Odum, H.T., Combining energy laws and corollaries of the maximum power principle with visual system mathematics, in *Ecosystems: Analysis and Prediction*, Levin, S., Ed., Proceedings of the Conference on Ecosystems at Alta, Utah, SIAM Institute for Mathematics and Society, Philadelphia, 1975, 90.

Odum, H.T., Pulsing, power and hierarchy, in *Energetics and Systems*, Mitsch, J., Ragade, R.K., Bosseman, R.W., Dillon, J.A., Jr., Eds., Ann Arbor Science, Ann Arbor, MI, 1982, 33.

Odum, H.T., Maximum power and efficiency: a rebuttal, *Ecol. Modelling*, 20, 71, 1983a.

Odum, H.T., *Systems Ecology: An Introduction*, John Wiley & Sons, New York, 644 pp, 1983b.

Odum, H.T., Energy analysis evaluation of coastal alternatives, *Water Sci. Tech.*, 16, 717, 1984.

Odum, H.T., *Emergy* in ecosystems, in *Ecosystem Theory and Application*, Polunin, N., Ed., John Wiley & Sons, New York, 1986, 337.

Odum, H.T., Living with complexity, in *Cranford Prize in Biosciences, Combined Lectures*, Royal Swedish Academy of Sciences, Stockholm, 1987, 19.

Odum, H.T., Self organization, transformity and information, *Science*, 242, 1132, 1988.

Odum, H.T., Emergy and biogeochemical cycles, in *Ecological and Physical Chemistry*, Rossi, C., Tiezzi, E., Eds., Proceedings of an International Workshop, November 1990, Sena, Italy, Elsevier, Amsterdam, 1991, 25.

Odum, H.T., *Ecological and General Systems: An Introduction to Systems Ecology*, University of Colorado Press, Niwat, CO, 646 pp., 1994.

Odum, H.T., *Environmental Accounting. Energy and Environmental Decision Making*, John Wiley & Sons, New York, 370 pp, 1996.

Odum, H.T., Copeland, B.J., McMahon, B.W., *Coastal Ecological Systems in the United States*, Publication Department of the Conservation Foundation, Washington, DC, 1974.

Odum, H.T., Knox, G.A., Campbell, D.E., *Organization of a New Ecosystem, Exotic Spartina Marsh in New Zealand*, Report to the National Science Foundation International Exchange Program New Zealand-United States. INT-8122010, NST, Center for Wetlands, Gainesville, FL, 107 pp, 1983.

Odum, H.T., Odum, E.C., *Energy Analysis Overview of Nations*, with sections by G. Bosch, L. Braat, W. Dunn, G. de R. Innes, J.R. Richardson, D.M. Scienceman, J.P. Sendzimer, D.J. Smith, M.V. Thomas, Working Paper, International Institute of Applied Systems Analysis, Laxenberg, Austria (WP-83-82), 469 pp, 1983.

Odum, H.T., Odum, E.C., Blissett, M., *Ecology and Economy: Energy Analysis and Public Policy in Texas*, L.B.J. School of Public Affairs and Texas Department of Agriculture (Policy Research Publication 78), University of Texas, Austin, 178 pp, 1987a.

Odum, H.T., Wang, F.C., Alexander, J.F., Jr., Gilliland, M., Miller, M., Sendzimer, J., *Energy Analysis of Environmental Value*, Center for Wetlands, University of Florida, Gainesville, 97 pp, 1987b.

Odum, W.E., Pathways of energy flow in a South Florida estuary, Doctoral dissertation, University of Miami, Florida, 1970a.

Odum, W.E., Insidious alteration of the estuarine environment, *Trans. Am. Fish. Soc.*, 99, 836, 1970b.

Odum, W.E., Fisher, J.S., Pickral, J.C., Factors controlling the flux of particulate organic carbon from estuarine wetlands, in *Ecological Processes in Coastal and Marine Systems*, Livingstone, R.J., Ed., Plenum Press, New York, 1979, 69.

Odum, W.E., Heald, E.J., Trophic analysis of an estuarine mangrove community, *Bull. Mar. Res.*, 22, 671, 1972.

Odum, W.E., Heald, E.J., The detritus-based food web of an estuarine mangrove community, in *Estuarine Research Volume I*, Cronin, L.E., Ed., Academic Press, New York, 1975, 265.

Officer, C.B., Ryther, J.H., The possible importance of silicon in marine eutrophication, *Mar. Ecol. Prog. Ser.*, 3, 83, 1980.

Ogden, J.C., Faunal relationships in Caribbean seagrass beds, in *Handbook of Seagrass Biology: An Ecosystem Perspective*, Phillips, R.C., McRoy, C.P., Eds., STMP Press, New York, 1980, 173.

Ohgaki, S., Vertical movement of the littoral fringe periwinkle *Nodilittorina exigua* in relation to wave height, *Mar. Biol.*, 100, 443, 1989.

Ojeda, F.P., Santelices, B., Ecological dominance of *Lessonia nigrescens* (Phaeophyta) in central Chile, *Mar. Ecol. Prog. Ser.*, 19, 83, 1984.

Okamura, B., Behavioural plasticity in suspension feeding benthic animals, in *Behavioural Mechanisms of Food Selection*, Hughes, R.N., Ed., Springer-Verlag, Berlin, 1990, 637.

Okansen, T., Exploitation ecosystems in heterogeneous habitat complexes, *Evol. Ecol.*, 4, 220, 1990.

Okansen, T., Fretwell, S.D., Armda, J., Niemela, P., Exploitation ecosystems in gradients of primary productivity, *Am. Nat.*, 118, 220, 1981.

Olafsson, E.B., Density dependence in suspension-feeding and deposit-feeding populations of the bivalve *Macoma balthica*: a field experiment, *J. Anim. Ecol.*, 55, 517, 1986.

Olafsson, E.B., Peterson, C.H., Ambrose, W.G., Jr., Does recruitment limitation structure populations and communities of macro-invertebrates in marine soft sediments: the relative significance of pre- and post-settlement processes, *Oceanogr. Mar. Biol. Annu. Rev.*, 32, 65, 1994.

Oliva, D., Castilla, J.C., The effect of human exclusion on the populaion structure of key-hole limpets *Fissurella crassa* and *F. limbata* on the coast of central Chile, *PSZNI*: *Mar. Ecol.*, 7, 201, 1986.

Oliver, J., The geomorphic and environmental aspects of mangrove communities, in *Mangrove Ecosystems in Australia*, Clough, B.F., Ed., Australian National University, Canberra, 1982, 19.

Oliver, J., The geomorphic and environmental aspects of mangrove communities, in *Mangrove Ecosystems in Australia*, Clough, B.F., Ed., Australian National University Press, Canberra, 1982, 302.

Oliver, J.S., Processes affecting the organization of soft-bottom communities in Monterey Bay, California and McMurdo Sound, Ph.D. thesis, University of California, San Diego, 1979.

Oliver, J.S., Oakden, J.M., Slattery, P.N., Phoxocephalid amphipod crustaceans as predators on larvae and juveniles in Monterey Bay, California and McMurdo Sound, *Mar. Ecol. Prog. Ser.*, 7, 179, 1982.

Oliver, J.S., Slattery, P.N., Effects of crustacean predators on species composition and population structure of soft-bodied infauna from McMurdo Sound, Antarctica, *Ophelia*, 24, 155, 1985.

Olsen, S.B., Struggling with an emergy analysis: shrimp mariculture in Ecuador, in *Maximum Power. The Ideas and Applications of H.T. Odum*, Hall, C.A.S., Ed., University of Colorado Press, Niwat, CO, 1994, 216.

O'Neill, R.V., DeAngelis, D.L., Waide, J.B., Allen, T.F.H., *A Hierarchical Concept of Ecosystems*, Princetown University Press, Princetown, 1986.

Oren, A., Blankburn, T.H., Estimates of sediment denitrification rates at *in situ* nitrate concntrations, *Limnol. Oceanogr.*, 32, 525, 1979.

Ortega, S., Competitive interactions among tropical intertidal limpets, *J. Exp. Mar. Biol. Ecol.*, 90, 11, 1985.

Ortega, S., The effects of human predaton on the size distribution of *Siphonaria gigas* (Mollusca, Pulmonata) on the Pacific coast of Costa Rica, *Veliger*, 29, 251, 1987.

Orth, R.J., Destruction of eelgrass, *Zoatera marina* by the cownose ray, *Rhinoptera bonasus* in Chesapeake Bay, *Chesapeake Sci.*, 16, 205, 1975.

Orth, R.J., The importance of sediment stability in seagrass communities, in *Ecology of the Marine Benthos*, Coull, B.C., Ed., University of South Carolina Press, Columbia, SC, 1977, 281.

Orth, R.J., A perspective on plant-animal interactions in seagrasses: physical and biological determinants influencing plant and animal abundance, in *Plant-Animal Interactions in the Marine Benthos* (Systematics Association Special Publications Vol. 46), John, D.M., Hawkins, S.J., Price, J.H., Eds., Clarendon Press, Oxford, 1992, 147.

Orth, R.J., Heck, H.L., Jr., Van Montfrans, J., Faunal communities in seagrass beds: a review of the influence of plant structure and predator-prey relationships, *Estuaries*, 7, 273, 1984.

Osman, R.S., The establishment and development of a marine epifaunal community, *Ecol. Monogr.*, 47, 7, 1977.

Ott, J.A., Maurer, L., Strategies of energy transfer from marine macrophytes to consumer levels, in *The Biology of Benthic Organisms*, Keegan, B.B. et al., Eds., Pergamon Press, London, 1977, 495.

Otte, G., Levings, C.D., The distribution of macroinvertebrate communities on a mud flat influenced by sewage, Fraser River Estuary, British Columbia, *Fisheries Marine Service, Research Division, Ottawa, Canada, Tech. Rep. No. 476*, 88 pp, 1975.

Ovaitt, C.A., Nixon, S.W., The demersal fish of Narragansett Bay: an analysis of community structure, distribution and abundance, *Est. Coastal Mar. Sci.*, 1, 361, 1973.

Pace, M.L., Glasser, J.E., Pomeroy, L.R., A simulation analysis of continental shelf food webs, *Mar. Biol.*, 82, 47, 1984.

Pace, M.L., Shimmel, S., Darley, W.M., The effect of grazing by the gastropod *Nassarius obsoletus* on the benthic microbial community of a saltmarsh, *Est. Coastal Mar. Sci.*, 9, 121, 1979.

Packham, J.R., Liddle, M.J., The Cefni salt-marsh, Anglesey, and its recent development, *Field Studies*, 3, 311, 1970.

Paine, R.T., Food web complexity and species diversity, *Am. Nat.*, 100, 65, 1966.

Paine, R.T., The *Pisaster-Tegula* interaction: prey patches, predator food preference, and intertidal community structure, *Ecology*, 50, 950, 1969.

Paine, R.T., A short-term experimental investigation of resource partitioning in a New Zealand rocky intertidal habitat, *Ecology*, 52, 1096, 1971.

Paine, R.T., Intertidal community structure. Experimental studies on the relationship between a dominant competitor and its principal predator, *Oecologia (Berlin)*, 15, 93, 1974.

Paine, R.R., Size-limited predation: and observational and experimental approach with the *Mytilus-Pisaster* association, *Ecology*, 57, 858, 1976.

Paine, R.T., Controlled manipulatons in the marine intertidal zone and their contributions to ecological theory, in *Changing Scenes in Natural Sciences, 1776–1976*, Goulden, C.E., Ed., Natl. Acad. Sci., U.S.A., Spec. Publ. No. 12, 1977, 245.

Paine, R.T., Disaster, catastrophe, and local persistence of the sea palm *Postelsia palmiformis*, *Science*, 205, 685, 1979.

Paine, R.T., Food webs: linkage, interaction strength and community infrastructure, Third Tansley Lecture, *J. Anim. Ecol.*, 49, 667, 1980.

Paine, R.T., Barnacle ecology: is competition important? The forgotten roles of predation and disturbance, *Paleobiology*, 7, 553, 1981.

Paine, R.T., Ecological determinism in the competition for space, *Ecology*, 65, 1339, 1984.

Paine, R.T., *Marine Rocky Shores and Community Ecology: An Experimental Perspective*, Ecology Institute, Oldendorf/Luhe, Germany, 1994.

Paine, R.T., Castilla, J.C., Concina, J., Perturbation and recovery patterns of starfish dominated intertidal-assembleages in Chile, New Zealand and Washington State, *Am. Nat.*, 125, 679, 1985.

Paine, R.T., Levin, S.A., Intertidal landscapes: disturbance and the dynamics of pattern, *Ecol. Monogr.*, 51, 145, 1981.

Paine, R.T., Vadas, R.L., The effects of grazing by sea urchins, *Strongylocentrotus* spp. on benthic algal populations, *Limnol. Oceanogr.*, 14, 710, 1969.

Palmer, A.R., Growth rate as a measure of food value in thaidid gastropods: assumption and implications for prey morphology and distribution, *J. Exp. Mar. Biol. Ecol.*, 73, 94, 1983.

Palmer, A.R., Prey selection by thaidid gastropods: some observational and experimental field tests of foraging models, *Oecologia (Berlin)*, 62, 162, 1984.

Palmer, J.D., Round, F.E., Persistent, vertical migration rhythms in benthic microflora. I. The effect of height and temperature on rhythmic behaviour of *Euglena obtusa*, *J. Mar. Biol. Assoc. U.K.*, 45, 567, 1965.

Palmer, M.A., Epibenthic predators and marine meiofauna separating predation disturbance, and hydrodynamic effects, *Ecology*, 69, 1251, 1988a.

Palmer, M.A., Dispersal of marine meiofauna: a review and conceptual model explaining passive transport and active emergence with implications for recruitment, *Mar. Ecol. Prog. Ser.*, 48, 81, 1988b.

Palmer, M.A., Gust, G., Dispersal of meiofauna in a turbulent tidal creek, *J. Mar. Res.*, 43, 179, 1985.

Palmer, M.A., Molloy, R.M., Water flow and the vertical distribution of meiofaunal: a flume experiment, *Estuaries*, 9, 225, 1986.

Palumbri, S.R., Spatial variation in an algal-sponge interaction and the evolution of intertidal organisms, *Am. Nat.*, 126, 267, 1985.

Palumbri, S.R., Jackson, J.B.C., Ecology of cryptic coral reef communities. 2. Recovery from small disturbance events by encrusting bryozoa: the influence of "host" species and lesion size, *J. Exp. Mar. Biol. Ecol.*, 64, 103, 1982.

Pamatmat, M.M., Ecology and metabolism of a benthic community on an intertidal sandflat, *Int. Rev. Gesamt. Hydrob. Hydrog.*, 53, 211, 1968.

Pamatmat, M.M., Banse, K., Oxygen consumption by the seabed. II. *In situ* mesurements to a depth of 180 m, *Limnol. Oceanogr.*, 14, 250, 1969.

Pardi, L., Innate components in solar orientation of littoral amphipods, *Cold Spring Harb. Symp. Quant. Biol.*, 25, 394, 1960.

Pardi, L., Ercolini, A., Zonal recovery mechanisms in talitrid crustaceans, *Boll. Zool.*, 53, 139, 1986.

Pardi, L., Scapini, F., Solar orientation and landscape visibility in *Talitrus saltater*, *Monotore Zool. Ital. (NS)*, 13, 210, 1979.

Parry, G.D., The evolution and life histories of four species of intertidal limpets, *Ecol. Monogr.*, 52, 65, 1982.

Parsons, T.R., Seki, H., Importance and general implications of organic matter in aquatic environments, in *Organic Matter in Natural Waters*, Hood, D.W., Ed., Institute of Marine Science (Alaska), Occasional Publication No. 1, 1970, 1.

Parsons, T.R., Takahashi, M., Hargrave, B.T., *Biological Oceanographic Processes*, 2nd ed., Pergamon Press, Oxford, 322 pp, 1979.

Patriquin, D., The origin of nitrogen and phosphorus for the growth of *Thalassia testudinium*, *Mar. Biol.*, 15, 35, 1972.

Patriquin, D., Estimation of growth rate, production and age of the marine angiosperm *Thalassia testudinium* Konig., *Carribean J. Sci.*, 13, 111, 1973.

Patriquin, D.G., Knowles, R., Nitrogen fixation in the rhizosphere of marine angiosperms, *Mar. Biol.*, 16, 49, 1972.

Patriquin, D.G., Mclung, C.R., Nitrogen accretion, and the nature and possible significance of N_2 fixation (acetylene reduction) in a Nova Scotia *Spartina alterniflora* stand, *Mar. Biol.*, 47, 237, 1978.

Patten, B.C., Mulford, R., Warinner, J.E., An annual phytoplankton cycle in lower Chesapeake Bay, *Chesapeake Sci.*, 4, 1, 1963.

Pauley, D., Theory and management of tropical multispecies stocks: a review with emphasis on the Southeastern Asian demersal fisheries, *Stud. Rev. Int. Cent. Living Aquat. Resour. Mgmt., Manila*, 1, 1, 1979.

Pawlik, J.R., Chemical induction of larval settlement and metamorphosis in the reef-building tube worm *Phragmatopoma californica* (Sabellidae: Polychaeta), *Mar. Biol.*, 91, 59, 1986.

Pawlik, J.R., Butman, Ch.A., Starczak, V.R., Hydrodynamic facilitation of gregarious settlement of a reef-building tubeworm, *Science*, 251, 421, 1991.

Pearse, C.M., Scheibling, R.E., Induction of metamorphosis of larvae of the green sea urchin *Strohgylocentrotus droebachiensis* by corallinr and red algae, *Biol. Bull.*, 179, 304, 1990.

Pearse, C.M., Scheibling, R.E., Effect of macroalgae, microbial films and conspecifics on the induction of metamorphosis of the green sea urchin *Strongylocentrotus droebachiensis* (Muller), *J. Exp. Mar. Biol.Ecol.*, 147, 147, 1991.

Pearse, J.S., Clark, M.E., Leighton, D.L., Mitchell, C.J., North, W.J., Marine waste disposal and sea urchin ecology, A Report W.M. Heck Laboratory of Environmental Health Engineering, Californian Institute of Technology. Kelp Habitat Improvement. Project 179-1970, 89 pp, 1970.

Pearse, J.S., Costa, D.P., Yellum, M.B., Agegian, C.R., Localized mass mortality of red sea urchins, *Strongylocentrotus franciscanus*, near Santa Cruz, California, *Fish. Bull.*, 75, 645, 1977.

Pearson, H.T., Studies on the ecology of the macrobenthic fauna of Lochs Linne and Eil, west coast of Scotland. II. Analysis of the macrobenthic fauna by comparison of feeding groups, *Symposium d'Arcachon, Vie Milieu, Suppl.*, 20, 1, 1971a.

Pearson, H.T., The benthic ecology of the Loch Linne and Loch Eil, a sea-lock system on the west coast of Scotland, III, The effect on the benthic fauna of the introduction of pulp mill effluent, *J. Exp. Mar. Biol. Ecol.*, 6, 211, 1971b.

Pearson, H.T., Marine pollution effects of pulp and paper industry wastes, in *Protection of Life in the Sea*, Kinne, O., Bulnheim, H.-P., Eds., *Helgol. Meerrsunters*, 33, 340, 1980.

Pearson, H.T., Rosenberg, R., A comparative study of the effects on the marine environment of wastes from cellulose industries in Scotland and Sweden, *Ambio*, 5, 77, 1976.

Pearson, H.T., Rosenberg, R., Macrobenthic succession in relation to organic enrichment and pollution of the marine environment, *Oceanogr. Mar. Biol. Annu. Rev.*, 16, 229, 1978.

Pechenik, J.A., Role of encapsulation in invertebrate life histories, *Am. Nat.*, 114, 859, 1979.

Pechenik, J.A., The relationship between temperature, growth rate, and duration of planktonic life for larvae of the gastropod *Crepisula fornicata* (L.), *J. Exp. Mar. Biol. Ecol.*, 74, 241, 1984.

Pechenik, J.A., Environmental influences on larval survival and development, in *Reproduction of Marine Invertebrates, Vol. 9*, Gierse, A.C., Pearse, J.S., Pearse, V.B., Eds., Blackwell Scientific Publications, Palo Alto, CA, 1987, 551.

Pechenik, J.A., Wendt, D.E., Jarrett, J.N., Metamorphosis is not a new beginning, *Bioscience*, 48, 901, 1998.

Peck, L.S., An investigation into the growth and early development of the ormer, *Haliotis tuberculata* L., Ph.D. thesis, CNAA Portsmouth Polytechnic, 386 pp, 1983.

Peck, L.S., Culley, M.B., Helm, M.M., A laboratory energy budget for the ormer *Haliotis tuberculata* L., *J. Exp. Mar. Biol. Ecol.*, 106, 103, 1987.

Penhale, P.A., Macrophyte-epiphyte biomass and productivity in an eelgrass (*Zostera marina* L.) community, *J. Exp. Mar. Biol. Ecol.*, 42, 113, 1977.

Penhale, P.A., Smith, W.O.Jr., Excretion of dissolved organic carbon by eelgrass (*Zoatera marina*) and its epiphytes, *Limnol. Oceanogr.*, 22, 400, 1977.

Penhale, P.A., Thayer, G.W., Uptake and transfer of carbon and phosphorus by eelgrass (*Zostera marina* L.) and its epiphytes, *J. Exp. Mar. Biol. Ecol.*, 42, 113, 1980.

Pepe, P.J., Pepe, S.M., The activity patterns of *Onchidella binneyi* Stearns (Mollusca: Opisthobranchiata), *Veliger*, 27, 375, 1985.

Percival, M., Womersley, J.S., *Floristics and Ecology of the Mangrove Vegetation of Papua New Guinea*, Department of Forestry, New Guinea, Lae, 96 pp, 1975.

Pérès, J.M., Zonations, in *Marine Ecology*, Vol. 5, pt. 1, Kinne, O., Ed., John Wiley & Sons, New York, 1982a, 9.

Pérès, J.M., Major benthic assemblages, in *Marine Ecology*, Vol. 5, pt. 1, Kinne, O., Ed., John Wiley & Sons, New York, 1982b, 373.

Perkins, E.J., *The Biology of Estuaries and Coastal Waters*, Academic Press, London, 678 pp, 1974.

Perry, D.M., Optimal diet theory: behaviour of a starved predatory snail, *Oecologia (Berlin)*, 72, 360, 1987.

Peters, R.H., The unpredictable problems of tropho-dynamics, *Envt. Biol. Fish.*, 2, 97, 1977.

Peterson, B.J., Aquatic primary productivity and the $^{14}C-CO_2$ method: a history of the productivity problem, *Annu. Rev. Ecol. Syst.*, 11, 359, 1980.

Peterson, B.J., Fry, B., Stable isotopes in ecosystem studies, *Annu. Rev. Ecol. Syst.*, 18, 293, 1987.

Peterson, B.J., Howarth, R.W., Sulfur, carbon, and nitrogen isotopes used to trace organic matter in the salt-marsh estuaries of Sapelo Island, Georgia, *Limnol. Oceanogr.*, 32, 1195, 1987.

Peterson, B.J., Howarth, R.W., Garritt, R.H., Multiple stable isotopes used to trace the flow of organic matter in an estuarine food webs, *Science*, 227, 1361, 1985.

Peterson, B.J., Howarth, R.W., Garritt, R.H., Sulfur and carbon isotopes as tracers of salt-marsh organic matter flow, *Ecology*, 67, 865, 1986.

Peterson, B.J., Howarth, R.W., Lipschultz, F., Ashendorf, D., Salt marsh detritus: an alternative interpretation of stable carbon isotope ratios and the fate of *Spartina alterniflora*, *Oikos*, 34, 173, 1980.

Peterson, C.H., Competitive organization of the soft-bottom macrobenthic communities of southern California lagoons, *Mar. Biol.*, 43, 343, 1977.

Peterson, C.H., The importance of predation and competition in organizing the intertidal epifauna of Barnegat Inlet, New Jersey, *Oecologia (Berlin)*, 39, 1, 1979a.

Peterson, C.H., Predation, competitive exclusion and diversity in the soft-sediment benthic communities of estuaries and lagoons, in *Ecological Processes in Coastal and Marine Systems*, Livingston, R.J., Ed., Plenum Press, New York, 1979b, 223.

Peterson, C.H., Approaches to the study of competition in benthic soft sediments, in *Estuarine Perspectives*, Kennedy, V.S., Ed., Academic Press, New York, 1980, 291.

Peterson, C.H., The importance of predation and interspecific competition in the population biology of two infaunal suspension feeders in a soft-sediment environment, *Ecology*, 61, 129, 1982a.

Peterson, C.H., The importance of predation and intra- and interspecific competition in the population biology of two infaunal suspension-feeding bivalves, *Protothaca staminea* and *Chione undatella*, *Ecol. Monogr.*, 52, 437, 1982b.

Peterson, C.H., Andre, S.V., An experimental analysis of interspecific competition among marine filter feeders in a soft-sediment environment, *Ecology*, 61, 129, 1980.

Peterson, C.H., Black, R., Preliminary evidence for progressive sestonic food depletition in incoming tide over a broad tidal sand flat, *Est. Coastal Shelf Sci.*, 32, 405, 1991.

Peterson, D.H., Conomos, T.J., Broenkon, W.W., Scrivani, E.P., Processes controlling the dissolved silica distribution in San Francisco Bay, in *Estuarine Research Vol. 1*, Cronin, L.E., Ed., Academic Press, New York, 1975, 153.

Petraitis, P.S., Occurrence of random and directional movements in the periwinkle *Littorina littorea* (L.), *J. Exp. Mar. Biol. Ecol.*, 59, 207, 1982.

Petraitis, P.S., Grazing patterns of the periwinkle and their effect on sessile intertidal organisms, *Ecology*, 64, 522, 1983.

Petraitis, P.S., Factors organizing rocky intertidal communities of New England: herbivory and predation in sheltered bays, *J. Exp. Mar. Biol. Ecol.*, 109, 117, 1987.

Petraitis, P.S., Direct and indirect effects of predation, herbivory and surface rugosity on mussel recruitment, *Oecologia (Berlin)*, 83, 405, 1990.

Petrusewicz, K., Suggested list of more important concepts in productivity studies (definitions and symbols), in *Secondary Productivity in Terrestrial Ecosystems (Principles and Methods)*, Petrusewicz, K., Ed., Warszawa, Krakow, 1967, 51.

Petrusewicz, K., Macfadyen, A., *Productivity of Terrestrial Animals: Principles and Methods*, IBP Handbook. No. 13, Blackwell Scientific Publications, Oxford, 190 pp, 1970.

Phil, L., Food selection and consumption of mobile epibenthic fauna in shallow marine areas, *Mar. Ecol. Prog. Ser.*, 22, 169, 1985.

Phil, L., Rosenberg, R., Food selection and consumption of the shrimp *Cragnon cragnon* in some shallow marine areas in western Sweden, *Mar. Ecol. Prog. Ser.*, 15, 159, 1984.

Phillips, R.C., Observations on the ecology and distribution of Florida seagrasses, professional paper, *Florida Board of Conservation*, 2, 1, 1960.

Phillips, R.C., McRoy, C.P., *Handbook of Seagrass Biology: An Ecosystem Perspective*, Garland STMP Press, New York, 353 pp, 1980.

Phillips, R.J., Burke, W.D., Keener, E.J., Observations on the trophic significance of jelly fishes in Mississippi Sound with quantitative data on the associative behaviour of small fishes with medusae, *Trans. Am. Fish. Soc.*, 98, 703, 1969.

Pickney, J., Zingmark, R.G., Effect of tidal stage and sun angles on intertidal benthic microalgal productivity, *Mar. Ecol. Prog. Ser.*, 76, 81, 1991.

Pickney, J., Zingmark, R.G., Biomass and production of benthic microalgal communities in estuarine habitats, *Estuaries*, 16, 887, 1993a.

Pickney, J., Zingmark, R.G., Modelling intertidal benthic microalgal production in estuarine ecosystems, *J. Phycol.*, 29, 296, 1993b.

Pickral, J.C., Odum, W.E., Benthic detritus in a salt marsh tidal creek, in *Estuarine Processes, Vol. 2*, Wiley, M., Ed., Academic Press, New York and London, 1977, 280.

Pielou, E.C., *Ecological Diversity*, John Wiley & Sons, New York, 1975.

Pielou, E.C., Routledge, R.D., Salt marsh vegetation: latitudinal gradients in the zonation patterns, *Oecologia (Berlin)*, 24, 311, 1976.

Pienkowski, M.W., Diet and energy intake of Grey and Ringed Plovers, *Pluvialis squatarola* and *Charadrius hiaticula*, in the non-breeding season, *J. Zool. (Lond.)*, 197, 541, 1982.

Piersma, T., Production by intertidal benthic animals and limits to their predation by shorebirds: a heuristic model, *Mar. Ecol. Prog. Ser.*, 38, 187, 1987.

Pimm, S.L., The complexity and stability of ecosystems, *Nature (London)*, 307, 321, 1984.

Pineda, J., Spatial and temporal patterns in barnacle settlement rate along a southern Caifornia rocky shore, *Mar. Ecol. Prog. Ser.*, 107, 125, 1994.

Plante, C.J., Jumars, P.A., Baross, J.A., Rapid growth in the hind gut of a marine deposit feeder, *Microbial Ecol.*, 18, 29, 1989.

Plante, C.J., Jumars, P.A., Baross, J.A., Digestive associations between marine detritovores and bacteria, *Mar. Ecol. Prog. Ser.*, 44, 149, 1990.

Platt, H.M., Meiofauna dynamics and the origin of the metazoa, in *The Evolving Biosphere*, Forey, P.L., Ed., Cambridge University Press, Cambridge, 1981, 207.

Platt, H.M., Warwick, R.M., The significance of free-living nematodes to the littoral ecosystem, in *The Shore Environment. Vol 2: Ecosystems*, Price, J.H., Irvine, D.E.G., Farnham, W.F., Eds., Academic Press, London, 1980, 729.

Platt, T., Mann, K.H., Ulanowicz, R.E., Eds., *Mathematical Models in Biological Oceanography*, The UNESCO Press, Paris, 156 pp, 1981.

Polis, G.A., Complex trophic interactions in deserts: an empirical critique of the food web theory, *Am. Nat.*, 138, 123, 1991.

Pollock, L.W., Hammon, W.D., Cyclic changes in interstitial water content, atmospheric exposure and temperature in a marine beach, *Limnol. Oceanogr.*, 16, 522, 1971.

Pomerat, C.M., Renier, E.R., The influence of surface angle and light on the attachment of barnacles and other sedentary organisms, *Biol. Bull. Mar. Biol. Lab., Woods Hole*, 91, 57, 1942.

Pomeroy, L.R., Algal productivity in salt marshes in Georgia, *Limnol. Oceanogr.*, 4, 386, 1959.

Pomeroy, L.R., The strategy of mineral cycling, in *Annual Reviews*, Palo Alto, California, Johnson, R.F., Frank, P.W., Michner, C.D., Eds., 1970, 171.

Pomeroy, L.R., The ocean's food web, a changing paradigm, *Bioscience*, 24, 499, 1974.

Pomeroy, L.R., Mineral cycling in marine ecosystems, in *Mineral Cycling in Southeastern Systems*, Howell, F.G., Gentry, J.B., Smith, M.H., Eds., ERDA Symposium Series, CONF-740513, 1975, 209.

Pomeroy, L.R., Secondary production mechanisms of continental shelf communities, in *Ecological Processes in Coastal and Marine Systems*, Livingston, P.J., Ed., Plenum Press, New York, 1979, 163.

Pomeroy, L.R., Bancroft, K., Breed, J., Christian, R.R., Frankenberg, D., Hall, J.R., Maurer, L.C., Wiebe, W.J., Wiegert, R.G., Wetzel, R.L., Flux of organic matter through a salt marsh, in *Estuarine Processes, Vol. 2*, Wiley, M., Ed., Academic Press, New York, 1977, 270.

Pomeroy, L.R., Darley, W.M., Dunn, E.L., Gallagher, J.L., Haines, E.B., Whitney, D.M., Primary production, in *The Ecology of a Salt Marsh*, Pomeroy, L.R., Wiegert, R.G., Eds., Springer-Verlag, New York, 1981, 39.

Pomeroy, L.R., Imberger, J., The physical and chemical environment, in *The Ecology of a Salt Marsh*, Pomeroy, L.R., Wiegert, R.G., Eds., Springer-Verlag, New York, 1981, 22.

Pomeroy, W.M., Levings, C.D., Association and feeding relationships between *Eogammarus confervicolus* (Arthropoda: Gammaridae) and benthic algae on Sturgen and Roberts Banks, Fraser River estuary, *Can. J. Fish. Aquatic Sci.*, 37, 1, 1980.

Pomeroy, L.R., Shenton, L.R., Jones, R.D.H., Reimold, R.J., Nutrient fluxes in estuaries, in *Nutrients and Eutrophication, Special Symposium, Vol. 1*, American Society of Limnology and Oceanography, Allen Press, Lawrence, 1972, 272.

Pomeroy, L.R., Wiegert, R.G., *The Ecology of a Salt Marsh*, Springer-Verlag, New York, 271 pp, 1981.

Poorachiranon, S., Chansang, H., Structure of the Ao Yon mangrove forest and its contribution to the coastal ecosystem, *Biotropica*, 17, 101, 1982.

Poore, G.C.B., Kudenov, J.D., Benthos around an outfall of the Werribe sewage-treatment farm, Port Philip Bay, Victoria, *Aust. J. Mar. Freshw. Res.*, 29, 157, 1970.

Porter, K.G., Sherr, E.B., Sherr, B.F., Pace, M., Saunders, P.W., Protozoa in planktonic food webs, *J. Protozool.*, 32, 409, 1985.

Postma, H., Sediment transport and sedimentation in estuaries, in *Estuaries*, Lauff, G., Ed., Publication No. 83, American Association for the Advancement of Science, Washington, DC, 1967, 158.

Potter, I.C., Claridge, P.N., Warwick, R.M., Consistency of seasonal change in an estuarine fish assemblage, *Mar. Ecol. Prog. Ser.*, 32, 217, 1986.

Power, M.E., Effects of fish on river food webs, *Science*, 250, 811, 1990.

Power, M.E., Top-down and bottom-up forces in food webs: so plants have primacy?, *Ecology*, 73, 733, 1992.

Prater, A.J., *Estuary Birds of Britain and Ireland*, Poyser, Carlton, 439 pp, 1981.

Pregnall, A.M., Release of dissolved organic carbon from the estuarine intertidal macroalgae *Enteromorpha prolifera*, *Mar. Biol.*, 73, 37, 1983.

Pregnall, A.M., Rudy, D.P., Contribution of green macroalgal mats (*Enteromorpha* spp.) to seasonal production in an estuary, *Mar. Ecol. Prog. Ser.*, 24, 167, 1985.

Price, J.H., Niche and community in the inshore benthos, with emphasis on the macroalgae, in *The Shore Environment*, Vol. 2: *Ecosystems*, Price, J.H., Irvine, D.E.G., Farnham, W.F., Eds., Academic Press, London, 1980, 487.

Price, R.M., Warwick, R.M., Temporal variations in annual production and biomass in estuarine populations of two polychaetes, *Nephtys hombergi* and *Ampharete acutifrons*, *J. Mar. Biol. Assoc., U.K.*, 60, 481, 1980a.

Price, R.M., Warwick, R.M., The effect of temperature on the respiration rate of meiofauna, *Oecologia (Berlin)*, 44, 145, 1980b.

Primavera, J.H., Mangrove as nurseries: shrimp populations in mangrove and non-mangrove habitats, *Est. Coastal Mar. Sci.*, 46, 457, 1998.

Pringle, J.D., Mathieson, A.C., *Chondrus crispus* Stackhouse, in *Case Studies of Several Commercial Seaweed Resources*, FAO Fisheries Tech. Rep. 281, Food and Agricultural Organization, United Nations, Rome, 1987, 49.

Pritchard, D.W., What is an estuary, physical viewpoint, in *Estuaries*, Lauff, G., Ed., American Association for the Advancement of Science, Publication No. 85, Washington, DC, 1967a, 37.

Pritchard, D.W., Observations of the circulation in estuaries, in *Estuaries*, Lauff, G., Ed., American Association for the Advancement of Science, Publication No. 85, Washington, DC, 1967b, 57.

Probert, P. K., The bottom fauna of china clay waste deposits in Mevagissey Bay, *J. Mar. Biol. Assoc., U.K.*, 55, 19, 1975.

Probert, P.K., Disturbance, sediment stability and trophic structure of soft-bottom communities, *J. Mar. Res.*, 42, 893, 1984.

Propp, M.V., Tarasoff, V.G., Gherbadgi, I.I., Lootzik, N.V., Benthic pelagic oxygen and nutrient exchange in a coastal region of the Sea of Japan, in *Marine Benthic Dynamics*, Tenore, K.R., Coull, B.C., Eds., University of South Carolina Press, Columbia, SC, 1980.

Prosch, R.M., McLachlan, A., The regeneration of surf-zone nutrients by the sand mussel, *Donax serra* Rodiney, *J. Exp. Mar. Biol. Ecol.*, 80, 221, 1984.

Pulliman, H.R., On the theory of optimal diets, *Am. Nat.*, 108, 59, 1974.

Pyefinch, K.A., Notes on the biology of cirripedes, *J. Mar. Biol. Assoc., U.K.*, 27, 464, 1948.

Pyke, G.H., Optimal foraging theory: a critical review, *Annu. Rev. Ecol. Syst.*, 15, 523, 1984.

Quasim, S.Z., Some problems related to the food chain in a tropical estuary, in *Marine Food Chains*, Steele, J.H., Ed., Oliver & Boyd, Edinburgh, 1970, 45.

Quasim, S.Z., Sankaranarayanan, V.N., Organic detritus in a tropical estuary, *Mar. Biol.*, 15, 193, 1972.

Raffaelli, D.G., Recent ecological research on some European species of *Littorina*, *J. Molluscan Stud.*, 48, 342, 1982.

Raffaelli, D.G., Functional feeding groups of some intertidal molluscs defined by gut content analysis, *J. Molluscan Stud.*, 51, 233. 1985.

Raffaelli, D.G., Conservation of Scottish estuaries, *Proc. Roy. Soc., Edinburgh*, 100B, 55, 1992.

Raffaelli, D.G., Conacher, A., McLachlan, H., Emes, C., The role of epibenthic crustacean predators in an estuarine food web, *Est. Coastal Shelf Sci.*, 28, 149, 1989.

Raffaelli, D.G., Faley, V., Galbraith, C., Eider predation and the dynamics of mussel bed communities, in *Trophic Relationships in the Marine Environment*, Barnes, M., Gibson, R.N., Eds., Aberdeen University Press, Aberdeen, 1990, 157.

Raffaelli, D.G., Hawkins, S., *Intertidal Ecology*, Chapman & Hall, London, 356 pp, 1996.

Raffaelli, D.G., Karakassis, I., Galloway, A., Zonation schemes on sandy shores: a multivariate approach, *J. Exp. Mar. Biol. Ecol.*, 148, 241, 1991.

Ragotzkie, R., Plankton productivity in estuarine waters of Georgia, *Pub. Instit. Mar. Sci., Univ. Texas*, 6, 146, 1959.

Raimondi, P.T., Rock type affects settlement, recruitment and zonation of the barnacle *Chthamalus anisopoma* Pilsbury, *J. Exp. Mar. Biol. Ecol.*, 123, 253, 1988a.

Raimondi, P.T., Settlement cues and determination of the vertical limit of an intertidal barnacle, *Ecology*, 69, 400, 1988b.

Raimondi, P.T., Patterns, mechanisms, consequences of variability in settlement and recruitment of an intertidal barnacle, *Ecol. Monogr.*, 60, 283, 1990.

Raimondi, P.T., Settlement behaviour of *Chthalamus anisopoma* larvae largely determines the adult distribution, *Oecologia (Berlin)*, 85, 349, 1991.

Rasmussen, N., The ecology of the Kaikoura Peninsula, Ph.D. thesis, University of Canterbury, Christchurch, NZ, 1965.

Rassoulzadegan, F., Lavat-Peuto, M., Sheldon, R.W., Partition of the food ration of marine ciliates between picoplankton and nanoplankton, *Hydrobiologia*, 159, 75, 1988.

Rau, G.H., Mearns, A.J., Young, D.T., Differences in animal ^{13}C, ^{15}N and abundance between a polluted and an unpolluted coastal site. Likely indications of sewage uptake by a marine food web, *Est. Coastal Shelf Sci.*, 13, 701, 1981.

Redfield, A.C., On the proportions of organic derivations in seawater and their relation to the composition of phytoplankton, in *James Johnston Memorial Volume,* Liverpool, 1934, 171.

Redfield, A.C., Ketchum, B.H., Richards, F.A., The influence of organisms on the composition of sea-water, in *The Sea*, Vol. 2, Hill, M., Ed., Wiley Interscience, New York, 1963, 67.

Rees, C.B., A preliminary study of the ecology of a mudflat, *J. Mar. Biol. Assoc. U.K.*, 21, 185, 1940.

Rees, E.I.S., Nicholaidou, A., Laskaridon, P., The effects of storms on the dynamics of shallow water benthic associations, in *Biology of Benthic Organisms*, Keegan, B.F., Cledigh, P.O., Boaden, P.J.S., Eds., Pergamon Press, Oxford, 1977, 503.

Reeve, M.R., The ecological significance of zooplankton in the shallow subtropical waters of South Florida, in *Estuarine Research Vol. 1*, Cronin, L.E., Ed., Academic Press, New York, 1975, 352.

Reeve, M.R., Cooper, E., The plankton and other seston in Card Sound, South Florida in 1971, *Univ. Miami Tech. Rep. UM-RSM AS-73007*, 24 pp, 1973.

Reichardt, W., Microbial aspects of bioturbation, *Scient. Mar.*, 53, 301, 1984.

Reichart, W., Polychaete tube walls: zonated microhabitats for marine bacteria, *Actes Colloques INFREMER,* 3, 415, 1986.

Reichardt, W., Impact of bioturbation by *Arenicola marina* on microbial parameters in intertidal sediments, *Mar. Ecol. Prog. Ser.*, 44, 149, 1988.

Reichle, D.E., O'Neill, R.V., Harris, W.F., Principles of energy and material exchange in ecosystems, in *Unifying Concepts in Ecology*, van Dobben, W.H., Lowe-McConnell, R.H., Eds., Dr. W. Junk Publishers, The Hague, 1975, 17.

Reid, D.G., Naylor, E., Free-running, endogenous semilunar rhythmicity in a marine isopod crustacean, *J. Mar. Biol. Assoc., U.K.*, 65, 85, 1985.

Reid, D.G., Naylor, E., An entraiment model for semilunar rhythmic swimming behaviour in the marine isopod *Eurydice pulchra* (Leach), *J.Exp. Mar. Biol. Ecol.*, 100, 25, 1986.

Reimold, J., The movement of phosphorus through the marsh cord grass *Spartina alterniflora*, *Limnol. Oceanogr.*, 17, 606, 1972.

Reisch, D.J., Effect of pollution abatement in Los Angeles harbors, *Mar. Poll. Bull.*, 2, 71, 1971.

Reisch, D.J., The establishment of laboratory colonies of polychaetous annelids, *Thalassia Jugosl.*, 10, 181, 1974.

Reisch, J.D., Bristle worms (Anellida: Polychaeta), in *Pollution Ecology of Estuarine Invertebrates,* Hunt, C.W., Fuller, S.L.H., Eds., Academic Press, New York, 1979, 78.

Reise, K., Predation pressure and community structure of an intertidal soft-bottom fauna, in *Biology of Benthic Organisms*, Keegan, B.F., O'Connor, P.O., Boaden, P.J.S., Eds., Pergamon Press, New York, 1977, 573.

Reise, K., Experiments on epibenthic predation in the Wadden Sea, *Helgol. Wiss. Meeresunters,* 32, 55, 1978.

Reise, K., Moderate predation on meiofauna by the macrobenthos of the Wadden Sea, *Helgol. Wiss. Meeresunters*, 32, 435, 1979.

Reise, K., High abundance of small zoobenthos around biogenic structures in tidal sediments of the Wadden Sea, *Helgol. Wiss. Meeresunters,* 34, 413, 1981.

Reise, K., Experimental removal of lugworms from marine sand affects small zoobenthos, *Mar. Biol.,* 74, 327, 1983a.

Reise, K., Biotic enrichment of intertidal sediments by experimental aggregates of the deposit-feeding bivalve *Macoma balthica*, *Mar. Ecol. Prog. Ser.*, 12, 119, 1983b.

Reise, K., Indirect effects of sewage on a sandy flat in the Wadden Sea, *Neth. Instit. Sea Res. Publ. Ser.*, 10, 159, 1984.

Reise, K., *Tidal Flat Ecology: An Experimental Approach to Species Interactions,* Springer-Verlag, Berlin, 1985.

Reise, K., Platyhelminth diversity on littoral sediments around the island of Sylt in the North Sea, *Progr. Zool.*, 36, 469, 1988.

Relevante, N., Gilmartin, M., Characteristics of microplankton and nanoplankton communities of an Australian coastal plain estuary, *Aust. J. Mar. Freshw. Res.*, 29, 9, 1978.

Remane, A., The ecology of brackish water, *Die Binnengewasser,* 25, 1, 1971.

Relexans, J.C., Meybeck, M., Billen, G., Brugeville, M., Etcheber, H., Somerville, M., Algal and microbial processes involved in particulate organic matter dynamics in the Loire Estuary, *Est. Coastal Mar. Sci.,* 27, 625, 1988.

Rhoads, D.C., Rates of sediment reworking by *Yoldia limatula* in Buzzards Bay, Massachusetts, and Long Island Sound, *J. Sedim. Petrol.*, 33, 727, 1963.

Rhoads, D.C., Biogenic working of intertidal and subtidal sediments in Barnstable Harbor and Buzzards Bay, *J. Geol.*, 75, 461, 1967.

Rhoads, D.C., The influence of deposit feeding benthos on water turbidity and nutrient recycling, *Am. J. Sci.,* 273, 1, 1973.

Rhoads, D.C., Organism-sediment relations on the muddy sea floor, *Oceanogr. Mar. Biol. Annu. Rev.,* 12, 263, 1974.

Rhoads, D.C., Allen, R.C., Goldhaber, M.B., The influence of colonizing benthos on the physical properties and chemical diagenesis of the estuarine sea floor, in *Ecology of the Marine Benthos*, Coull, B.C., Ed., University of South Carolina Press, Columbia, SC, 1977, 113.

Rhoads, D.C., Boyer, L.E., The effects of marine benthos on the physical properties of sediments. A successional perspective, in *Animal Sediment Relations. The Biogenic Alteration of Sediments,* McCall, P.L., Teresz, M.J.S., Eds., Plenum Press, New York, 1982, 3.

Rhoads, D.C., McCall, P.L., Yingst, J.Y., Disturbance and production on the estuarine seafloor, *Am. Sci.*, 60, 577, 1978.

Rhoads, D.C., Young, D.K., The influence of deposit-feeding organisms on sediment stability and comunity trophic structure, *J. Mar. Res.*, 28, 150, 1970.

Rhoads, D.C., Young, D.K., Animal sediment relationships in Cape Cod Bay Massachusetts. II. Reworking by *Molpadia oolitica*, *Mar. Biol.*, 11, 225, 1971.

Rice, K.J., Interaction of disturbance patch size and herbivory in *Evadium* colonization, *Ecology*, 68, 1113, 1987.

Rich, P.M., Differential CO_2 and O_2 benthic community metabolism in a soft-water lake, *J. Fish. Res. Bd. Canada*, 36, 1377, 1979.

Ricker, W.E., Ed., *Methods for the Assessment of Fish Production in Freshwaters,* IBP Handbook No. 3, Blackwell Scientific Publications, Oxford, 313 pp, 1968.

Ricketts, F.F., Calvin, J., Hedgpeth, J.W., Phillips, D.W., *Between Pacific Tides*, Stanford University Press, Stanford, CA, 652 pp, 1985.

Riedl, R.J., How much water passes through sandy beaches?, *Int. Rev. Ges. Hydrobiol.*, 56, 923, 1971.

Riedl, R.J., Huang, N., McMahan, R., The subtidal pump: a mechanism of interstitial water exchange by wave action, *Mar. Biol.*, 13, 179, 1972.

Rieman, F., Schrage, M., The mucus-trap hypothesis on feeding of aquatic nematodes and implications for biodegradation and sediment texture, *Oecologia (Berlin)*, 34, 75, 1978.

Riley, G.A., Oceanography of Long Island Sound, 1952–1954. IX. Predation and utilization of organic matter, *Bull. Bingham Oceanogr. Coll.*, 18, 324, 1956.

Riley, G.A., Theory of food chain relations in the ocean, in *The Sea Vol. 2*, Hill, M.N., Ed., Wiley Interscience, New York, 1963, 438.

Riley, G.A., The plankton of estuaries, in *Estuaries*, Lauff, G. Ed., American Association for the Advancement of Science, Publication No. 85, Washington, DC, 1967, 316.

Riley, G.A., Conover, S.M., Phytoplankton of Long Island Sound, 1954-55, *Bull. Bingham Oceanogr. College*, 19, 5, 1967.

Rittschof, D., Branscomb, E.S., Costlow, J.D., Settlement and behaviour in relation to flow and surface in larval barnacles, *Balanus amphitrite* Darwin, *J. Exp. Mar. Biol.*, 82, 131, 1984.

Ritz, D.A., Lewin, M.E., Shen, Ma., Response to organic enrichment of infaunal macrobenthic communities under salmonid cages, *Mar. Biol.*, 103, 211, 1989.

Rivera-Monroy, V.H., Twilley, R.R., Boustany, R.C., Day, J.W., Vers-Herra, F., del Carmen Ramirez, R.C., Direct denitrification in mangrove sediments in Terminos Lagoon, Mexico, *Mar. Ecol. Prog. Ser.*, 126, 97, 1995.

Robales, C., Predator foraging characteristics and prey population structure on a sheltered shore, *Ecology*, 68, 1502, 1987.

Robales, C.D., Changing recruitment in coastal species assemblages: implications for predation theory in intertidal communities, *Ecology*, 78, 1400, 1997.

Robales, C., Cubit, J., Influence of biotic factors in an upper intertidal community: dipteran larvae grazing on algae, *Ecology*, 62, 1536, 1981.

Robales, C., Robb, J., Varied carnivore effects and the prevalence of intertidal algal turfs, *J. Exp. Mar. Biol. Ecol.*, 166, 65, 1993.

Robales, C., Sweethson, D., Dittman, D., Diel variation of intertidal foraging by *Cancer productus* in British Columbia, *J. Nat. Hist.*, 23, 1041, 1989.

Roberts, D.J., Hughes, R.N., Growth and reproductive rates on *Littorina rudis* from three contrasted shores on North Wales, UK, *Mar. Biol.*, 58, 47, 1980.

Robertson, A.I., The relationship between annual production: biomass ratios and life-spans for marine macrobenthos, *Oecologia (Berlin)*, 38, 193, 1979.

Robertson, A.I., Leaf-burying crabs: their influence on energy flow and export from mixed mangrove forests (*Rhizophora* spp.) in northeastern Australia, *J. Exp. Mar. Biol. Ecol.*, 102, 237, 1986.

Robertson, A.I., The determination of trophic relationships in mangrove dominated systems: areas of darkness, in *Mangrove Ecosystems of Asia and the Pacific: Status, Exploitation and Management*, Institute of Marine Science, Townsville, Australia, 1987, 293.

Robertson, A.I., Decomposition of leaf litter on tropical Australia, *J. Exp. Mar. Biol. Ecol.*, 116, 235, 1988.

Robertson, A.I., Alongi, D.M., Eds., *Tropical Mangrove Ecosystems*, American Geophysical Union, New York, 1992.

Robertson, A.I., Alongi, D.M., The role of riverine mangrove forests inncarbon flux to the preliminary mass balance for the Fly River Delta, *Geo-Mar. Lett.*, 6, 1996.

Robertson, A.I., Alongi, D.M., Boto, K.G., Food chains and carbon fluxes, in *Tropical Mangrove Ecosystems*, Robertson, A.I., Alongi, D.M., Eds., American Geophysical Union, New York, 1992, 293.

Robertson, A.I., Alongi, D.M., Daniel, P.A., Boto, K.G., How much mangrove detritus reaches the Great Barrier Reef lagoon, in *Proceedings of the Vth International Coral Reef Conference,* Vol. 2, Choat, J.H., Ed., Sixth International Coral Reef Executive Committee, Sydney, 1988, 601.

Robertson, A.I., Daniel, P.A., The influence of crabs on litter processing in mangrove forests in tropical Australia, *Oecologia (Berlin)*, 78, 191, 1989.

Robertson, A.I., Daniel, P.A., Dixon, P., Mangrove forest structure and productivity in the Fly River Estuary, Papua New Guinea, *Mar. Biol.*, 111, 147, 1991.

Robertson, A.I., Duke, N.C., Mangroves as nursery sites: comparisons of the abundance and species composition of fish and crustaceans in mangroves and other nearshore habitats in tropical Australia, *Mar. Biol.*, 96, 193, 1987.

Robertson, A.I., Hansen, J.A., Decomposing seaweed: a nuisance or vital link in coastal food chains, *CSIRO Marine Laboratory Research Report*, 75, 1981.

Robertson, A.I., Lenenon, R.C.J., Fish community structure and food chain dynamics in the surf-zone of sandy beaches: the role of detatched macrophyte detritus, *J. Exp. Mar. Biol. Ecol.*, 84, 265, 1984.

Robertson, A.I., Mann, K.H., Population dynamics and life history adaptations of *Littorina neglecta* Bean in an eelgrass meadow (*Zostera marina* L.) in Nova Scotia, *J. Exp. Mar. Biol. Ecol.*, 63, 151, 1982.

Rodriguez, S.R., Ojeda, F.D., Ineestrosa, N.C., Inductores quimicos del asentamiento de invertbradia marinos bentonicos: importancia y necesidad de sur estudio en Chile, *Revia Chil. Hist. Nat.*, 65, 297, 1992.

Rodriguez, S.R., Ojeda, F.P., Inestrosa, N.C., Settlement of benthic marine invertebrates, *Mar. Ecol. Prog. Ser.*, 97, 193, 1993.

Roe, P., Life history and predator-prey interactions of the nemertine *Paranemertes peregrina, Coc. Biol. Bull.*, 150, 80, 1976.

Roman, C.T., Able, K.W., Lazzari, M.A., Heck, K.L., Primary productivity of angiosperm and macroalgal dominated habitats in a New England salt marsh: a comparative analysis, *Est. Coastal Shelf Sci.*, 30, 35, 1990.

Roman, C.T., Daiber, F.C., Organic carbon flux through a Delaware Bay salt marsh: tidal exchange, particle size distribution, and storms, *Mar. Ecol. Prog. Ser.*, 54, 149, 1989.

Romer, G.L., Surf zone fish community and species response to a wave energy gradient, *J. Fish Biol.*, 36, 279, 1990.

Romer, G.S., McLachlan, A., Mullet grazing on surf diatom accumulations, *J. Fish Biol.*, 28, 93, 1986.

Rosenberg, R., Benthic faunal recovery in a Swedish fjord following the closure of a sulphate pulp mill, *Oikos*, 23, 92, 1972.

Rosenberg, R., Benthic faunal dynamics during succession following pollution abatement in a Swedish estuary, *Oikos*, 27, 424, 1976.

Rosenthal, R.J., Clarke, W.D., Dayton, P.K., Ecology and natural history of a stand of giant kelp, *Macrocystis pyrifera* off Del Mar, California, *Fish. Bull.*, 72, 670, 1974.

Roughgarden, J., Gaines, S.D., Possingham, H., Recruitment dynamics in a complex life cycle, *Science*, 241, 1460, 1988.

Roughgarden, J., Iwaza, Y., Baxter, C., Demographic theory for an open marine population with space-limited recruitment, *Ecology*, 66, 54, 1985.

Roughgarden, J., Pennington, J.T., Stoner, D., Alexander, S., Miller, K., Collisions of upwelling fronts with the intertidal zone: the cause of recruitment pulses in barnacle populations of central California, *Acta Oecol.*, 12, 35, 1991.

Round, F.E., Benthic marine diatoms, *Oceanogr. Mar. Biol. Annu. Rev.*, 9, 83, 1971.

Rowe, G.T., Clifford, C.H., Smith, K.L., Hamilton, D.L., Benthic nutrient regeneration and its coupling to primary productivity in coastal waters, *Nature*, 255, 215, 1975.

Rowe, G.T., Clifford, C.H., Smith, K.L., Jr., Nutrient regeneration in sediments off Cap Blanc, Spanish Sahara, *Deep-Sea Res.*, 255, 215, 1977.

Rowley, R.J., Settlement and recruitment of a sea urchin (*Strongylocentrotus* spp.) in a sea-urchin barren ground and a kelp bed: are populations regulated by settlement or post-settlement processes, *Mar. Biol.*, 100, 485, 1989.

Rublee, P.A., Bacteria and microbial distribution in estuarine sediments, in *Estuarine Comparisons*, Kennedy, V.C., Ed., Academic Press, New York, 1982a, 159.

Rublee, P.A., Seasonal distribution of bacteria in salt marsh sediments in North Carolina, *Est. Coastal Shelf Sci.*, 15, 67, 1982b.

Rublee, P.A., Dornseif, B.E., Direct counts of bacteria in the sediments of a North Carolina salt marsh, *Estuaries*, 1, 188, 1978.

Rumble, J.R., Disturbance regimes in temperate forests, in *Ecology of Natural Disturbance and Patch Dynamics*, Pickett, S.T.A., White, P.S., Eds., Academic Press, New York, 1985, 17.

Russ, G., Effects of predation by fishes: competition and structural complexity of substratum on the establishment of a marine epifaunal community, *J. Exp. Mar. Bio. Ecol.*, 42, 55, 1980.

Russ, G., Distribution and abundance of herbivorous grazing fishes in the central Great Barrier Reef. I. Levels of variability across the entire continental shelf, *Mar. Ecol. Progr. Ser.*, 20, 23, 1984.

Russ, G., Is the rate of removal of algae by grazers reduced inside territories of the tropical damselfish?, *J. Exp. Mar. Biol. Ecol.*, 110, 1, 1987.

Russell, G., Vertical distribution, in *Intertidal and Littoral Ecosystems of the World 24*, Mathieson, A.C., Nienhuis, P.H., Eds., Elsevier, Amsterdam, 1991, 43.

Russell-Hunter, W.D., Alpey, M.L., Hunter, R.D., Early life-hictory of *Melampus acun* and the significance of semilunar synchrony, *Biol. Bull. Mar. Biol. Lab., Woods Hole*, 143, 623, 1972.

Rutgers van der Loeff, M.M., Anderson, L.G., Hull, P.O.J., Iverfeldt, A., Josefson, A.B., Sundby, B., Wesdterlund, S.F.G., The asphyxiation technique: an approach to distinguishing between molecular diffusion and biologically mediated transport at the sediment-water interface, *Limnol. Oceanogr.*, 29, 675, 1984.

Rutherford, J.C., *Modelling Investigations of Potential Eutrophication in the Upper Waitemata* Harbour, Upper Waitemata Harbour Catchment Study Working Report, No. 22, Auckland Regional Authority, Auckland, New Zealand, 1982.

Ryther, J.H., Geographical variations in productivity, in *The Sea, Vol. II*, Hill, M.N., Ed., Wiley Interscience, New York, 1963, 347.

Ryther, J.H., Photosynthesis and fish production in the sea, *Science*, 166, 72, 1969.

Saenger, P., Morphological, anatomical and reproductive adaptations of Australian mangroves, in *Mangrove Ecosystems in Australia*, Clough, B.F., Ed., Australian National University Press, Canberra, 1982, 153.

Saenger, P., Specht, M.M., Specht, R.L., Chapman, V.J., Australia, New Zealand, in *Wet Coastal Ecosystems, Ecosystems of the World 1*, Clapman, V.J., Ed., Elsevier, Amsterdam, 1977, 293.

Sainsbury, K.J., Supralittoral zone ecology of the Snares Island, New Zealand, B.Sc. (Hons) Project in Zoology, University of Canterbury, Christchurch, NZ, 41 pp, 1972.

Sale, P.F., The ecology of fishes on coral reefs, *Oceanogr. Mar. Biol. Annu. Rev.*, 18, 337, 1977.

Salmon, D.G., Moser, M.E., *Wildfowl and Wader Counts 1984–85*, Slimbridge Wildlife Trust, Slimbridge, U.K., 1985.

Salvat, B., Les conditions hydrodynamiques interstitielles des sediment meubles intertidaux et la repartition verticale de la jeune endogee, *C.R. Acad. Sci., Paris*, 259, 1576, 1964.

Salvat, B., *Eurydice pulchra* (Leach 1815) et *Eurydice affinis* (Hanson 1905) (Isopodes: Cirolanidae). Taxonomie, ethologie, reparition vertical et cycle reproducteur, *Acta. Soc. Linn. Bordeau A*, 195, 1, 1966.

Salvat, B., La macrofaune carcinologique endogee des sediments meubles intertidaux (Tanaidaces, Isopodes et Ampohipodes): ethologie, bionomie et cycle biologique, *Mem. Mus. Natur. Hist. Nat. Ser. A*, 45, 1, 1967.

Samuel, A.C., Verhagen, J.H.G., Cossen, J., Hass, H.A., Interaction between seston quantity and quality and benthic suspension feeders in the Osterschelde, The Netherlands, *Ophelia*, 26, 385, 1986.

Sanchez-Salazar, M.E., Griffiths, C.L., Seed, R., The interactive roles of predation and tidal elevation in structuring populations of the edible cockle, *Cerastoderma edule*, *Est. Coastal Shelf Sci.*, 25, 245, 1987.

Sand Jensen, K., Biomass, net production and growth dynamics in a eel grass (*Zostera marina* L.) population in Vellerup Vig, Denmark, *Ophelia*, 14, 185, 1975.

Sanders, H.L., Grassle, J.F., Hampson, G.R., Morse, L.S., Garner-Price, S., Jones, C.C., Anatomy of an oil spill: long-term effects from the grounding of the barge *Floeida* off West Florida, Massachusetts, *J. Mar. Res.*, 38, 265, 1980.

Sanders, H.L., Gouldsmit, E.M., Mills, E.l., Hampton, G.E., A study of the intertidal fauna of Barnstable Harbor, Massachusetts, *Limnol. Oceanogr.*, 7, 63, 1962.

Santilices, B., Montalva, S., Oliger, P., Competitive algal community organization in exposed intertidal habitats from central Chile, *Mar. Ecol. Prog. Ser.*, 6, 267, 1981.

Santos, S.L., Simon, J.L., Response of soft-bottom benthos to annual catastrophic disturbance in a South Florida estuary, *Mar. Ecol. Prog. Ser.*, 3, 347, 1980.

Sasekumar, A., Loi, J.J., Litter production in three mangrove forest zones in the Malay Peninsula, *Aquatic Bot.*, 17. 283, 1983.

Savidge, W.B., Taghorn, G.L., Passive and active components following two types of disturbance on an intertidal sand flat, *J. Exp. Mar. Biol. Ecol.*, 115, 137, 1988.

Scapini, F., Buiatti, M., Ottaviano, O., Phenotypic plasticity in sun orientation of sandhoppers, *J. Comp. Physiol., A*, 163, 739, 1988.

Scapini, F., Lagar, M.C., Mezzetti, M.C., The use of slope and visual information in sandhoppers: innateness and plasticity, *Mar. Biol.*, 115, 545, 1993.

Scheibling, R.E., Stephenson, R.L., Mass mortality of *Strongylocentrotus draebachensis* (Echinodermata: Echinoidea) of Nova Scotia, *Mar. Biol.*, 78, 153, 1984.

Scheleske, C.L., Odum, E.P., Mechanisms maintaining high productivity in Georgia estuaries, *Proc.Gulf Caribbean Fish. Instit.*, 14, 75, 1961.

Scheltema, R.S., The relationship of temperature to the larval development of *Nassarius obsoletus* (Gastropoda), *Biol. Bull. Mar. Biol. Lab., Woods Hole,* 132, 253, 1967.

Scheltema, R.S., Biological interactions determining larval settlement of marine invertebrates, *Thalass. Jugoslav.*, 10, 263, 1974.

Schiel, D. R., Demographic and experimental evaluation of plant and herbivore interactions in subtidal algal stands, Ph.D. thesis in zoology, University of Auckland, 1981.

Schiel, D.R., Selective feeding by the echinoid *Evechinus chloroticus*, and the removal of plants from subtidal algal stands in northern New Zealand, *Oecologia (Berlin)*, 54, 377, 1982.

Schiel, D.R., Algal interactions on shallow subtidal reefs in northern New Zealand: a review, *N.Z. J. Mar. Freshw. Res.*, 22, 481, 1988.

Schiel, D.R., Macroalgal assemblages in New Zealand, structure, interactions and demography, *Hydrobiologica*, 192, 59, 1990.

Schiel, D.R., Kelp Communities, in *Marine Biology*, Hammond, L.S., Synnot, R.N., Eds., Longman Cheshire, Sydney, 1994, 345.

Schiel, D.R., Foster, M., The structure of subtidal algal stands in temperate waters, *Oceanogr. Mar. Biol. Annu. Rev.*, 24, 265, 1986.

Schleyer, M.H., Decomposition in estuarine ecosystems, *J. Limnol. Soc. S. Af.*, 112, 90, 1986.

Schlieper, C., Physiology of brackish water, *Die Binnengewasser*, 25, 211, 1971.

Schleiper, D., Does the theory of optimal diets apply in complex environments?, *Am. Nat.*, 118, 1981.

Schneider, D.C., Equilization of prey numbers by migrating shorebirds, *Nature (London)*, 271, 371, 1978.

Schoenberg, S.A., MacCubbin, A.E., Hodson, R.E., Cellulose digestion by freshwater microbiota, *Limnol. Oceanogr.*, 29, 1132, 1984.

Schoener, T.W., Theory of feeding strategies, *Annu. Rev. Ecol. Syst.*, 11, 369, 1971.

Schoener, T.W., Field experiments in interspecific competition, *Am. Nat.*, 122, 240, 1983.

Scholander, P.F., Hammel, H.T., Hemmingsen, E.A., Garey, W., Sap concentrations in halophytes and some other plants, *Plant Physiol.*, 41, 529, 1962.

Schonbeck, M.W., Norton, T.A., Factors controlling the upper limits of fucoid algae on the shore, *J. Exp. Mar. Biol. Ecol.*, 31, 303, 1978.

Schonbeck, M.W., Norton, T.A., Drought-hardening in the upper shore seaweeds *Fusus spiralis* and *Pelvetia canaliculana*, *J. Ecol.*, 67, 687, 1979.

Schonbeck, M.W., Norton, T.A., Factors controlling the lower limits of fucoid algae on the shore, *J. Exp. Mar. Biol. Ecol.*, 43, 131, 1980.

Schroeter, S.C., Dixon, J., Kastendeik, J., Effects of the starfish, *Patiria miniata,* on the distribution of the sea urchin *Lytechinus anamesus* in a southern California kelp forest, *Oecologia (Berlin)*, 56, 141, 1983.

Schubauer, J.P., Hopkinson, C.S., Above- and belowground emergent macrophyte production and turnover in a coastal marsh ecosystem, Georgia, *Limnol. Oceanogr.*, 29, 1052, 1984.

Schulman, M.J., Resource limitation and recruitment patterns in a coral reef fish assemblage, *J. Exp. Mar. Biol. Ecol.*, 74, 85, 1984.

Schwinghamer, P., Francis, C.T., Gordon, D.C., Stable carbon isotope studies on the Pecks Cove mudflat ecosystem in Cumberland Basin, Bay of Fundy, *Can. J. Fish. Aquatic Sci.*, 40 (Suppl.), 262, 1983.

Schwinghamer, P., Hargrave, B., Peer, D., Hawkins, C.M., Partitioning of production and respiration among size groups of organisms in an intertidal benthic community, *Mar. Ecol. Prog. Ser.*, 31, 131, 1986.

Schwinghamer, P., Kepkay, P.E., Effects of experimental enrichment with *Spartina* detritus on sediment community biomass and metabolism, *Biol. Oceanogr.*, 4, 289, 1987.

Schwinghamer, P., Kepkay, P.E., Foda, A., Oxygen flux and community biomass associated with benthic photosynthesis and detritus decomposition, *J. Exp. Mar. Biol. Ecol.*, 147, 9, 1991.

Schienceman, D.M., Energy and emergy, in *Environmental Economics*, Pillet, G., Muraton, T., Eds., Roland Leimgruber, Geneva, 308 pp, 1987.

Sebens, K.P., The larval and juvenile ecology of the temperate octocoral *Alcyonium siderium* Verrill. I. Substratum selection by benthic larvae, *J. Exp. Mar. Biol. Ecol.*, 71, 73, 1983.

Sebens, K.P., Lewis, J.R., Rare events and population structure of the barnacle *Semibalanus cariosus* (Pallas, 1788), *J. Exp. Mar. Biol. Ecol.*, 87, 55, 1985.

Seed, R., Factors influencing shell-shape in the mussel *Mytilus edulis*, *J. Mar. Biol. Assoc. U.K.*, 48, 561, 1968.

Seed, R., The ecology of *Mytilus edulis* L. (Lamellibranchiata) on exposed rocky shores. II. Growth and mortality, *Oecologia (Berlin)*, 3, 317, 1969.

Seed, R., Ecology, in *Marine Mussels: Their Ecology and Physiology*, Bayne, B.L., Ed., Cambrige University Press, Cambridge, 1976, 13.

Seed, R., Predation of the ribbed mussel *Geukensia demissa* by the blue crab *Callinectes sapidus*, *Neth. J. Sea Res.*, 16, 103, 1982.

Seitzinger, S.P., Denitrification in freshwater and coastal marine ecosystems: ecological and geochemical significance, *Limnol. Oceanogr.*, 33, 702, 1988.

Seitzinger, S.P., Nixon, S.W., Pilson, M., Burke, A., Denitrification and N_2O production in near-shore marine sediments, *Geochem. Cosmochem. Acta*, 44, 1053, 1980.

Seki, T., Kan-no, H., Induced settlement of the Japanese abalone, *Haliotis discus hannai*, veliger by the mucus of the juvenile and adult abalones, *Bull. Tohoku Reg. Fish. Res. Lab.*, 43, 29, 1981.

Sellner, K.G., Plankton productivity and biomass in a tributary of the upper Chesapeake Bay. I. Importance of size-fractionated phytoplankton productivity, biomass and species composition in carbon export, *Est. Coastal Shelf Sci.*, 17, 197, 1983.

Sellner, K.G., Zingmark, R.G., Interpretations of the ^{14}C method of measurement of the total annual production of phytoplankton in a South Carolina estuary, *Bot. Marina*, 19, 119, 1976.

Sen Gupta, R., Nitrogen and phosphorus budget in the Baltic Sea, *Mar. Chem.*, 1, 267, 1973.

Sergerstraale, S.G., Brackish water classification, a historical summary, *Archivio di Oceanografia e Limnologica*, 11 (Suppl.), 7, 7, 1959.

Settlemyre, J.L., Gardner, L.R., A field study of chemical budgets for a small tidal creek — Charlestown Harbour, S.C., in *Marine Chemistry in the Coastal Environment*, Church, T.M., Ed., ACS Symposium Series, No., 18, 1975, 152.

Shanks, A.L., Surface slicks associated with tidally forced internal waves may transport pelagic larvae of benthic invertebrates and fishes shorewards, *Mar. Ecol. Prog. Ser.*, 13, 311, 1983.

Shanon, I.V., The Benguela ecosystem, I. Evolution of the Benguela physical features and processes. *Oceanogr. Mar. Biol. Annu. Rev.*, 23, 105, 1985.

Sharp, J.H., Cuthbertson, C.H., The physical definition of salinity: a chemical evaluation, *Limnol. Oceanogr.*, 27, 385, 1982.

Sheith, M.-S.J., Nutrients in Narragansett Bay sediments, M.S. thesis, University of Rhode Island, Kingston, RI.

Sheldon, R.W., Prakash, A., Sutcliffe, W.H., Jr., Size distribution of particles in the ocean, *Limnol. Oceanogr.*, 17, 327, 1972.

Shenton, M. A., Nitrifying and denitrifying activity in the sediments of estuarine littoral zones, M.S. thesis, University of Maryland, College Park, 51 pp., 1982.

Sheridan, P.F., Livingstone, R.J., Cyclic trophic relationships of fishes in an unpolluted river-dominated estuary in North Florida, in *Ecological Processes in Coastal and Marine Systems*, Livingstone, R.J., Ed., Plenum Press, New York, 1979, 143.

Sherman, K.S., Coull, B.C., The response of meiofauna to sediment disturbance, *J. Exp. Mar. Biol. Ecol.*, 45, 59, 1980.

Sherr, B.F., Sherr, E.B., The role of heterotrophic protozoa in carbon and energy flow in aquatic systems, in *Current Perspectives in Microbial Ecology*, King, M.J., Reddy, C.A., Eds., American Society for Microbiology, Washington, DC, 1984, 412.

Sherr, B.F., Sherr, E.B., Andrew, T.L., Fallon, R.D., Newell, S.Y., Trophic interactions between heterotrophic Protozoa and bacterioplankton in estuarine water analyzed with selective metabolic inhibitors, *Mar. Ecol. Prog. Ser.*, 32, 169, 1986.

Sherr, B.F., Sherr, E.B., Berman, T., Grazing, growth, and ammonium excretion rates of a heterotrophic microflagellate fed with four species of bacteria, *Applied Envt. Microbiol.*, 45, 1196, 1983.

Sherr, E.B., Sherr, B.F., McDaniel, J., Clearance rates of <6 μm fluorescently labelled algae (FLA) by grazing protozoa: potential grazing impact of flagellates and ciliates, *Mar. Ecol. Prog. Ser.*, 60, 81, 1991.

Shew, D.M., Linthurst, D.A., Seneca, E.D., Above- and below-ground production and turnover in a coastal marsh ecosystem, Georgia, *Linmol. Oceanogr.*, 29, 1052, 1981.

Shinn, E.A., Burrowing in recent lime sediments in Florida and the Bahamas, *J. Paleont.*, 42, 879, 1968.

Short, A.D., Wright, I.D., Physical variability of sandy beaches, in *Sandy Beaches as Ecosystems*, McLachlan, A., Erasmus, T., Eds., Dr. W. Junk Publishers, The Hague, 1983, 133.

Short, A.D., Wright, L.D., Morphodynamics of high energy beaches — an Australian perspective, in *Coastal Geomorphology in Australia*, Thom, B.G., Ed., Academic Press, Sydney, 1984, 43.

Short, F.T., A simulation model of the seagrass production system, in *Handbook of Seagrass Biology: An Ecosystem Perspective*, Phillips, R.C., McRoy, C.P., Eds., Garland STMP Press, New York, 1980, 277.

Short, F.T., McRoy, C., Nitrogen uptake by leaves and roots of the seagrass *Zostera marina* L., *Bot. Marina*, 27, 547, 1984.

Sibbald, M.J., Albright, L., Aggregated and free bacteria as food sources for heterotrophic microflagellates, *Appl. Envt. Microbiol.*, 54, 613, 1988.

Sibert, J., Brown, M.C., Healey, M.C., Kask, B.A., Naimen, R.J., Detritus based food webs: exploitation by juvenile chum salmon (*Oncorhynchus keta*), *Science*, 196, 649, 1977.

Siebruth, J.McN., Soviet aquatic bacteriology: a review of the past decade, *Quart. Rev. Biol.*, 35, 179, 1960.

Siebruth, J.McN., Studies on algal substances in the sea. III. The production of extracellular organic matter by littoral marine algae, *J. Exp. Mar. Biol. Ecol.*, 3, 290, 1969.

Siebruth, J.McN., Bacterial substrates and productivity in marine ecosystems, *Annu. Rev. Ecol. Syst.*, 7, 259, 1976.

Siebruth, J.McN., Smetacek, V., Lenza, J., Pelagic ecosystem structure: heterotrophic compartments and their relationship to plankton size fractions, *Limnol. Oceanogr.*, 23, 1256, 1978.

Siegfried, W.R., Kockey, P.A.R., Crowe, A.A., Exploitation and conservation of brown mussel stocks by coastal people of Transhei, *Envt. Conservation*, 12, 303, 1985.

Sigurdsson, J.B., Titman, C.W., Dawks, P.A., The dispersal of young post-larval bivalve mollusca by byssal threads, *Nature (London)*, 262, 386, 1976.

Sih, A., Crowley, P., McPeck, M., Petranka, J., Strohmeier, K., Predation, competition and prey communities: a review of field experiments, *Annu. Rev. Ecol. Syst.*, 16, 111, 1985.

Sikora, W.B., The ecology of *Palaemonetes pugio* in a southeastern salt marsh ecosystem with particular emphasis on production and trophic relationships, Ph.D. thesis, University of South Carolina, Columbia, SC, 122 pp, 1977.

Sikora, W.B., Sikora, J.P., Ecological implications of the vertical distribution of meiofauna in salt marsh sediments, in *Estuarine Comparisons*, Kennedy, V.S., Ed., Academic Press, New York, 1982, 219.

Silva, P. M. de, Experiments on choice of substrate by *Spirobis* larvae (Serpulidae), *J. Exp. Mar. Biol. Ecol.*, 39, 483, 1962.

Simenstad, C.A., Wissmar, R.C., ^{13}C evidence of origins and fates of organic carbon in estuarine and nearshore food webs, *Mar. Ecol. Prog. Ser.*, 22, 141, 1985.

Simon, J.L., Dauer, D.M., Reestablishment of a benthic community following natural defaunation, in *Ecology of the Marine Benthos,* Coull, B.C., Ed., University of South Carolina Press, Columbia, SC, 1977, 139.

Simpson, H.J., Hammons, D.E., Deck, B.l., Williams, S.C., Nutrient budgets in the Hudson River estuary, in *Marine Chemistry in the Coastal Environment,* Church, T.M., Ed., American Chemical Society, New York, 1975, 618.

Sinsabaugh, R.L., Linhius, A.E., Benfield, E.F., Cellulose digestion and assimilation by three leaf-shredding aquatic insects, *Ecology*, 66, 1464, 1985.

Skottsberg, G.J.F. Communities of marine algae in Subantarctic and Antarctic waters, *Klung. Svenka. Vatenskap. Handlinger. Ser. 3*, 19, 1, 1941.

Skyring, G.W., Oshrun, R.l., Wiebe, W.J., Assessment of sulfate reduction in Georgia marshland soils, *J. Geomicrobiol.*, 1, 389, 1979.

Slim, F.J., Hemminga, M.A., Ochieng, C., Jannink, N.T., Cecheret de la Moriniere, E., van der Velde, G., Leaf litter removal by the snail *Terebrakia palustris* (Linnaeus) and sesarmid crabs in an East African mangrove forest (Gazi Bay, Kenya), *J. Exp. Mar. Biol. Ecol.*, 215, 35, 1997.

Slocum, C.J., Differential susceptability to grazers in two phases of an intertidal alga: advantages of heteromorphic generations, *J. Exp. Mar. Biol. Ecol.*, 46, 99, 1980.

Smalley, A.E., The growth cycle of *Spartina* and its relation to insect populations in the marsh, in *Proceedings of the Salt Marsh Conference*, Ragotzkie, R.A., Teal, J.M., Pomeroy, L.R., Scott, B.C., Eds., University of Georgia, Athens, GA, 1959, 96.

Smart, R.M., Barko, J.W., Nitrogen nutrition and salinity tolerance of *Distichlis spicata* and *Spartine alterniflora*, *Ecology*, 61, 620, 1978.

Smayda, T.J., The phytoplankton of estuaries, in *Estuaries and Enclosed Seas. Ecosystems of the World 26,* Ketchum, B.H., Ed., Elsevier, Amsterdam, 1983, 65.

Smetacek, V.S., The annual cycle of protpzooplankton in the Keil Bight, *Mar. Biol.*, 63, 1, 1981.

Smith, C.M., Berry, J.A., Recovery of photosynthesis after exposure of intertidal algae to osmotic and temperature stress. Comparative studies of species with different distributional limits, *Oecologia (Berlin)*, 70, 6, 1986.

Smith, F.G.W., Effect of water currents upon the attachment and growth of barnacles, *Biol. Bull. Mar. Biol. Lab., Woods Hole,* 90, 51, 1946.

Smith, F.G.W., Surface illumination and barnacle attachment, *Biol. Bull. Mar. Biol. Lab., Woods Hole,* 94, 33, 1948.

Smith, K.K., Good, R.E., Good, N.F., Production dynamics for above- and belowground components of a New Jersey *Spartina alterniflora* tidal marsh, *Est. Coastal Mar. Sci.*, 9, 189, 1979.

Smith, K.L., Jr., Burns, H.A., Teal, J.M., *In situ* respiration of benthic communities in Castle Harbour, Bermuda, *Mar. Biol.*, 12, 196, 1972.

Smith, K.L., Rowe, G.T., Nichols, J.A., Benthic community respiration near Woods Hole sewage outfall, *Est. Coastal Shelf Sci.*, 1, 65, 1973.

Smith, L.D., Coull, B.C., Juvenile spot (Pisces) and grassshrimp predation on meiobenthos in muddy and sandy substrates, *J. Exp. Mar. Biol. Ecol.*, 17, 183, 1987.

Smith, S.V., Phosphorus versus nitrogen limitation in the marine environment, *Limnol. Oceanogr.*, 29, 1149, 1984.

Smith, S.V., Wiebe, W.J., Hollibough, J.T., Dollor, S.J., Hager, S.W., Cole, B.E., Tribble, G.W. Wheeler, P., Stoichiometry of C, N, P, and Si fluxes in a temperate-climate environment, *J. Mar. Res.*, 45, 427, 1987.

Smith, T.J., III, Boto, J., Fraser, S.D., Giddens, R.L., Keystone species and mangrove forest dynamics: the influence of burrowing by crabs on soil nutrient status and forest productivity, *Est. Coastal. Shelf Sci.*, 33, 419, 1991.

Smith, T.J., III, Odum, W.E., The effects of grazing by snow geese on coastal salt marshes, *Ecology,* 62, 98, 1981.

Snelgrove, P.V.R., Butman, C.A., Animal-sediment relationships revisited: cause versus effect, *Oceanogr. Mar. Biol. Annu. Rev.*, 32, 111, 1994.

Snelling, B., The distribution of intertidal crabs in the Brisbane River, *Aust. J. Mar. Freshw. Res.*, 10, 67, 1959.

Sogard, S.M., Feeding ecology, population structure, and community relationships of a grassbed fish, *Callionymus pauciradiatus* in southern Florida, M.Sc. thesis, University of Miami, Coral Gables, FL, 1982.

Sogard, S.M., Utilization of meiofauna as a food source by a grassbed fish, the spotted dragonet *Callionymus pauciradiatus*, *Mar. Ecol. Prog. Ser.*, 17, 183, 1984.

Somero, G.N., Hochachka, P.W., Biochemical adaptation to temperature, in *Adaptations to Environment: Esaays on the Physiology of Marine Mammals,* Newell, R.C., Ed., Butterworth, London, 1976, 125.

Sondergaard, M., Extracellular organic carbon (EOC) in the genus *Carpophyllum* (Rhodiohyceae): diel release patterns and EOC lability, *Mar. Biol.*, 104, 143, 1990.

Sorenson, J., Denitrification rates in a marine sediment by the acetylene inhibition technique, *Appl. Environ. Microbiol.*, 35, 201, 1978.

Sorokin, Yu.I., Decomposition of organic matter and nutrient regeneration, in *Mar. Ecol., Vol. 4. Dynamics*, Kinne, O., Ed., John Wiley & Sons, New York, 1978, 501.

Sorokin, Yu.I., Microheterotrophic organisms in marine environments, in *Analysis of Marine Ecosystems*, Longhurst, A.R., Ed., Academic Press, London, 1981, 293.

Soule, D.F., Oguri, M., Soule, J.D., Urban and fish-processing wastes in the marine environment: bioenhancement studies at Terminal Island, California, *Bull. Calif. Wat. Poll. Cont. Assoc.*, 15, 58, 1978.

Soule, D.F., Soule, J.D., The importance of non-toxix urban wastes in the estuarine detrital food chain, *Bull. Mar. Sci.*, 31, 786, 1981.

Sousa, W.P., Disturbance in marine intertidal boulder fields: the nonequilibrium maintenance of species diversity, *Ecology*, 60, 1225, 1979a.

Sousa, W.P., Experimental investigations of disturbance and ecological succession in a rocky intertidal algal community, *Ecol. Monogr.*, 49, 221, 1979b.

Sousa, W.P., The responses of a community to disturbance: the importance of successional age and species life histories, *Oecologia (Berlin)*, 45, 72, 1980.

Sousa, W.P., The role of disturbance in natural communities, *Annu. Rev. Ecol. Syst.*, 15, 353, 1984a.

Sousa, W.P., Intertidal mozaics: patch size, propagule availability, and spatially variable patterns of succession, *Ecology*, 65, 1918, 1984b.

Sousa, W.P., Disturbance and patch dynamics on rocky intertidal shores, in *Natural Disurbance: The Patch Dynamics Perspective*, Pickett, S.T.A., White, P.S., Eds., Academic Press, New York, 1985, 104.

Sousa, W.P., Schroeder, C.S., Gaines, S.D., Latitudinal variation in intertidal algal community structure: the influence of grazing and vegetative propogation, *Oecologia (Berlin)*, 48, 297, 1981.

Southward, A.J., The zonation of plants and animals on rocky shores, *Biol. Rev.*, 33, 137, 1958.

Southward, A.J., Limpet grazing and the control of vegetation on rocky shores, in *Grazing in Terrestrial and Marine Environments*, Crisp, D.J., Ed., Blackwell, Oxford, 1964, 265.

Southward, A.J., Southward, E.C., Recolonization of rocky shores in Cornwall after use of toxic dispersants to clean up the torry Canyon spill, *J. Fish. Res. Bd. Can.*, 35, 163, 1978.

Sparholdt, H., The ICES multispecies VPA's for the Baltic fish stocks, in *ICES Symp. Multispecies Models Relevant to Management of Living Resources*, No. 6, 1, 1989.

Spencer, A.L.M., On the wisdom of calculating annual material budgets in intertidal wetlands, *Mar. Ecol. Prog. Ser.*, 150, 207, 1997.

Spies, R.B., DesMarais, D.J., Natural isotope study of trophic enrichment of marine benthic communities by petroleum seepage, *Mar. Biol.*, 73. 67, 1983.

Stanley, S.M., Newman, W.A., Competitive exclusion in evolutionary time: the case of acorn barnacles, *Paleobiology*, 6, 173, 1980.

Staples, D.J., Polzin, H.G., Heales, D.S., Habitat requirements of juvenile penaeid prawns and their relationships to offshore fisheries, in *Second Australian National Prawn Seminar*, Rothlisberg, P.C., Hill, B.J., Staples, D.J., Eds., NPS2, Cleveland, OH, 1985, 47.

Stebbing, A.R.D., Preferential settlement of a bryozoan and serpulid larvae on the younger parts of *Laminaria* fronds, *J. Mar. Biol. Assoc., U.K.*, 52, 765, 1962.

Steele, J.H., *The Structure of Marine Ecosystems*, Harvard University Press, Cambridge, MA, 128 pp, 1974.

Steele, J.H., Comparative studies of beaches, *Phil. Trans, Roy. Soc. Edinb.*, 274B, 401, 1976.

Steele, J.H., Baird, I.E., Production ecology of a sandy beach, *Limnol. Oceanogr.*, 13, 14, 1968.

Steeman-Nielsen, E., The use of radioactive carbon for measuring organic production in the sea, *J. Cons. Explor. Mer.*, 18, 117, 1952.

Steeman-Nielsen, E., The balance between phytoplankton and zooplankton in the sea, *J. Cons. Explor. Mer.*, 23, 178, 1958.

Steever, Z.E., Warren, R.S., Niering, W.A., Tidal energy subsidy and standing crop production of *Spartina alterniflora*, *Est. Coastal Mar. Sci.*, 4, 473, 1976.

Steffensen, D.A., The *Euglena* of the Avon-Heathcote Estuary, B.Sc. (Hons) project in Botany, University of Canterbury, Christchurch, NZ, 1969.

Steffensen, D.A., An ecological study of *Ulva lactuca* L. and other benthic algae on the Avon-Heathcote Estuary, Ph.D. thesis, Botany Department, University of Canterbury, Christchurch, NZ, 217 pp, 1974.

Steffensen, D.A., The effect of nutrient enrichment and temperature on the growth in culture of *Ulva lactuca* L., *Aquatic Bot.*, 2, 337, 1976. *Mar. Biol.*, 68, 199, 1982.

Steinberg, P.D., Geographical variation in the interaction between marine herbivores and brown algal secondary metabolites, in *Ecological Roles of Marine Natural Products*, Paul, V.J., Ed., Constock Publishing Associates, Ithaca, NY, 1992.

Steinke, T.D., Rajh, A., Holland, A.J., The feeding behaviour of the red mangrove crab *Sesarma meinerti* de Mann (Crustacea: Decapoda: Grapsidae) and its effect on the degradation of mangrove leaf litter, *S. Af. J. Mar. Sci.*, 13, 151, 1993.

Steneck, R.S., A limpet-coralline algal association: adaptations and defences between selective herbivore and its prey, *Ecology*, 63, 507, 1982.

Steneck, R.S., Escalating herbivory and resulting adaptive trends in calcaneous algal crusts, *Paleobiology*, 9, 41, 1983.

Steneck, R.S., The ecology of coralline algal crusts: convergent patterns and adaptive strategies, *Annu. Rev. Ecol. Syst.*, 17, 273, 1986.

Steneck, R.S., Watling, L., Feeding capabilities and limitation of herbivorous mollusca: a functional group approach, *Mar. Biol.*, 68, 299, 1982.

Stenton-Dozey, J.M.E., Griffiths, C.L., The fauna associated with kelp stranded on a sandy beach, in *Sandy Beaches as Ecosystems*, McLachlan, A., Erasmus, T., Eds., Dr. W. Junk Publishers, The Hague, 1983, 559.

Stephens, K.R., Sheldon, R.H., Parsons, T.R., Seasonal variation in the availability of food for the benthos in a coastal environment, *Ecology*, 48, 852, 1967.

Stephenson, R.L., *A Stable Isotope Study of Chione (Austrovenus) Stutchburyi and Its Food Source in the Avon-Heathcote Estuary*, Estaurine Research Unit, Zoology Department, University of Canterbury, Report No. 22. 48 pp, 1980.

Stephenson, R.L., Aspects of the energetics of the cockle *Chione (Austrovenus) stutchburyi* in the Avon-Heathcote Estuary, Christchurch, New Zealand, Ph.D. thesis, University of Canterbury, Christchurch, NZ, 165 pp, 1981.

Stephenson, R.L., Lyon, G.L., C^{13} depletion in an estuarine bivalve: detection of marine and terrestrial food source, *Oecologia (Berlin)*, 1982, 110, 1982.

Stephenson, R.L., Tan, F.C., Mann, K.H., Stable carbon isotope variability in marine macrophytes and its implications for food web studies, *Mar. Biol.,* 81, 223, 1984.

Stephenson, T.A., The marine ecology of South African coasts, with special reference to the habits of limpets, *Proc. Linn. Soc., Lond.,* 148, 74, 1936.

Stephenson, T.A., The constitution of the intertidal fauna and flora of South Africa. III, *Annu. Natal Mus.,* 11, 207, 1948.

Stephenson, T.A., Stephenson, A., The universal feaures of zonation between the tidemarks on rocky coasts, *J. Ecol.,* 37, 289, 1949.

Stephenson, T.A., Stephenson, A., Life between the tidemarks in North America. IIIA. Nova Scotia and Prince Edward Island: description of the region, *J. Ecol.,* 42, 14, 1954a.

Stephenson, T.A., Stephenson, A., Life between the tidemarks in North America. IIIB. Nova Scotia and Prince Edward Island: the geographic features of the region, *J. Ecol.,* 42, 46, 1954b.

Stephenson, T.A., Stephenson, A., Life between the tidemarks in North America. IVA. Vancouver Island, *J. Ecol.,* 49, 1, 1961.

Stephenson, T.A., Stephenson, A., *Life between the Tidemarks on Rocky Shores,* W.H. Freeman, San Francisco, 425 pp, 1972.

Stephenson, W., Williams, W.T., Cook, S.G., Computer analysis of Petersen's original data on bottom communities, *Ecol. Monogr.,* 42, 387, 1971.

Stevenson, J.C., Comparative ecology of submerged grass beds in freshwater, estuarine and marine environments, *Limnol. Oceanogr.,* 33, 867, 1988.

Steward, C.C., Pickney, J., Piceno, Y., Lovell, C.R., Bacterial numbers and activity, microbial biomass and productivity and meiofaunal distribution in sediments naturally contaminated with biogenic bromophenols, *Mar. Ecol. Prog. Ser.,* 90, 61, 1992.

Stewart, J.G., Anchor species and epiphytes in intertidal algal turf, *Pac. Sci.,* 36, 45, 1982.

Stickle, W.B., The metabolic effects of starving *Thais lamellosa* immediately after spawning, *Comp. Biochem. Physiol.,* 40A, 627, 1971.

Stickle, W.B., Effects of environmental factor gradients on scope for growth in several species of carnivorous marine invertebrates, in *Marine Biology of Polar Regions and Effects of Stress on Marine Organisms,* Gray, J.S., Christiansen, M.E., Eds., John Wiley & Sons, Chichester, 1985, 601.

Stickle, W.B., Bayne, B.L., Energetics of the muricid gastropod *Thais (Nucella) lapillus* (L.), *J. Exp. Mar. Biol. Ecol.,* 107, 263, 1987.

Stickney, R.R., Taylor, G.L., Heard, R.W., III, Food habits of Georgia estuarine fishes. I. Four species of flounders (Pleutonectiformes: Bothidae), *Fish. Bull.,* 72, 515, 1974.

Stigebrandt, A., Wulff, E., A model for the dynamics of nutrients and oxygen in the Baltic proper, *J. Mar. Sci.,* 42, 729, 1987.

Stimson, J.S., Territorial behaviour of the owl limpet *Lottia gigantia, Ecology,* 51, 113, 1970.

Stimson, J. S., The role of territory in the ecology of the intertidal limpet *Lottia gigantea* (Gray), *Ecology,* 54, 1020, 1973.

Stobbs, R.E., Feeding habits of the giant clingfish *Chorisochismus dentex* (Pisces: Gobiescocidae), *S. Af. J. Zool.,* 15, 146, 1980.

Stoner, A.W., The role of seagrass biomass in the organization of benthic macrofaunal assemblages, *Bull. Mar. Sci.,* 30, 537, 1980.

Stoner, A.W., The influence of benthic macrophytes on the foraging behaviour of the pinfish *Lagodon rhomboides* (Linnaeus), *J. Exp. Mar. Biol. Ecol.,* 38, 271, 1982.

Stout, J.P., An analysis of the annual growth and productivity of *Juncus roemerianus* Scheele and *Spartina alterniflora* in coastal Alabama, Ph.D. dissertation, University of Alabama, 95 pp, 1978.

Strathman, R.R., Larval settlement in echinoderms, in *Settlement and Metamorphosis of Marine InvertebrateLarvae,* Chia, F.-S, Rice, M.E., Eds., Elsevier/North Holland Biomedical Press, New York, 1978, 235.

Strathman, R.R., Branscomb, E.S., Adequacy of cues to favourable sites by settling larvae of two intertidal barnacles, in *Reproductive Ecology of Marine Invertebrates,* Stanger, S.E., Ed., University of South Carolina Press, Columbia, SC, 1979, 77.

Strathman, R.R., Branscomb, E.S., Vedder, K., Fatal errors in set as a cost of dispersal and the influence of intertidal flora on set of barnacles, *Oecologia (Berlin),* 48, 13, 1981.

Strickland, J.H., Parsons, T.R., *A Practical Handbook of Seawater Analysis,* 2nd ed., *Fish. Res. Bd. Canada, Bull., No. 17,* 310 pp, 1972.

Strong, D.R., Jr., Are trophic cascades all wet? Differentiation and donor-control in species ecosystems, *Ecology,* 73, 747, 1992.

Stroud, L.M., Net primary production of belowground material and carbohydrate patterns of two height forms of *Spartina alterniflora* in two North Carolina marshes, Ph.D. dissertation, North Carolina State University, Raleigh, NC, 140 pp, 1976.

Stroud, L.M., Cooper, A.W., Colour-infrared aerial photographic interpretation and net primary productivity of a regularly flooded North Carolina salt marsh, *University of North Carolina Water Resource Institute Rep. No. 14,* 81 pp, 1968.

Stuart, V., Field, J.C., Newell, R.C., Evidence for the absorption of kelp detritus by the ribbed mussel *Aulacomya ater* using a new ^{51}Cr-labelled microsphere technique, *Mar. Ecol. Prog. Ser.,* 9, 263, 1982.

Suchanek, T.H., The ecology of *Mytilus edulis* L. in exposed rocky intertidal communities. *J. Exp. Mar. Biol. Ecol.,* 31, 103, 1978.

Suchanek, T.H., The *Mytilus californianus* community: studies on the composition, structure, organization, and dynamics of a mussel bed, Ph.D. dissertation, University of Washington, Seattle, WA, 1979.

Suchanek, T.H., Diversity in natural and artificial mussel bed communities of *Mytilus californianus, Am. Zool.,* 20, 807, 1980.

Suchanek, T.H., The role of disturbance in the evaluation of life history strategies in the intertidal mussels *Mytilus edulis* and *Mytilus californianus, Oecologia (Berlin),* 30, 143, 1981.

Suchanek, T.H., Mussels and their role in structuring the rocky shore, in *The Ecology of Rocky Coasts,* Moore, P.G., Seed, R., Eds., Hodder & Straughton, London, 1985, 70.

Sullivan, M.J., Diatom communities from a Delaware salt marsh, *J. Phycol.,* 11, 384, 1975.

Sullivan, M.L., Daiber, F.C., Responses in production of cord grass, *Spartina alterniflora*, to inorganic nitrogen and phosphorus fertilizer, *Chesapeake Sci.*, 15, 121, 1974.

Sullivan, M.J., Moncreiff, C.A., Primary production of edaphic algal communities in a Mississippi salt marsh, *J. Phycol.*, 24, 49, 1988.

Sullivan, M.J., Moncreiff, C.A., Edaphic algae are an important component of salt marsh food-webs: evidence from multiple stable isotope analysis, *Mar. Ecol. Prog. Ser.*, 10, 307, 1990.

Summers, J.K., McKellar, H.N., Jr., A simulation model of estuarine subsystem coupling and carbon exchange with the sea. I. Model structure, in *State-of-the-Art in Ecological Modelling*, International Society of Ecological Modelling, Copenhagen, 1979, 323.

Summers, J.K., Mckellar, H.N., Jr., The role of forcing functions in an estuarine model of carbon exchange with the sea, *ISEM J.*, 3, 71, 1981a.

Summers, J.K., McKellar, H.N., Jr., A sensitivity analysis of an ecosystem model of estuarine carbon flow, *Ecol. Modelling*, 11, 101, 1981b.

Summers, J.K., McKellar, H.N., Jr., Dame, R.F., Kitchens, W.M., A simulation model of estuarine subsystem coupling and carbon exchange with the sea. II. North Inlet model structure, output and validation, *Ecol. Modelling*, 11, 101, 1980.

Summerson, H.C., Peterson, C.H., Role of predation in organizing benthic communities of a temperate-zone seagrass bed, *Mar. Ecol. Prog. Ser.*, 15, 63, 1984.

Sutcliffe, W.H., Correlations between seasonal river discharge and local landings of American lobster (*Homarus americanus*) and Atlantic halibut (*Hippoglossus hippoglossus*) on the Gulf of St. Lawrence, *J. Fish. Res. Bd. Can.*, 30, 856, 1973.

Sutherland, J.P., Dynamics of high and low populations of the limpet *Acmaea scabra* (Gould), *Ecol. Monogr.*, 40, 169, 1970.

Sutherland, J.P., Multiple stable points in benthic communities, *Am. Nat.*, 118, 859, 1974.

Sutherland, J.P., Functional role of *Schizoporella* and *Styela* in the fouling community at Beaufort, North Carolina, *Ecology*, 59, 257, 1978.

Sutherland, J.P., The fouling community at Beaufort, North Carolina: a study in stability, *Am. Nat.*, 118, 499, 1981.

Sutherland, J.P., Perturbations, resistance and alternative views of the existence of multiple stable points in nature, *Am. Nat.*, 136, 270, 1990.

Sutherland, J.P., Otega, S., Competition conditional on recruitment and temporary escape from predators on a tropical rocky shore, *J. Exp. Mar. Biol. Ecol.*, 95, 155, 1986.

Suzuki, B., Tagawa, H., Biomass of a mangrove forest and a sedge marsh on Ishigaki Island, South Japan, *Jap. J. Ecol.*, 33, 24, 1983.

Swart, D.H., Physical aspects of sandy beaches — a review, in *Sandy Beaches as Ecosystems*, McLachlan, A., Erasmus, T., Eds., Dr. W. Junk Publishers, The Hague, 1983, 5.

Swedmark, B., The interstitial fauna of marine sand, *Biol. Rev.*, 37, 1, 1964.

Swift, M.J., Heal, O.W., Anderson, J.M., *Decomposition in Terrestrial Ecosystems*, University of California Press, Berkeley, CA, 372 pp, 1979.

Swithenbanks, D., Intertidal exposure zones: a way to subdivide the shore, *J. Exp. Mar. Biol. Ecol.*, 62, 69, 1982.

Sze, P., Aspects of the ecology of macrophytic algae in high rockpools at the Isles of Shoals (USA), *Bot. Mar.*, 23, 313, 1980.

Szyrmer, J., Ulanowicz, R.E., Total fluxes in ecosystems, *Ecol. Modelling*, 35, 123, 1987.

Taft, J.L., Elliot, A.J., Taylor, W.R., Box model analyses of Chesapeake Bay ammonium and nitrogen fluxes, in *Estuarien Interactions*, Wiley, M.L., Ed., Academic Press, New York, 78, 113, 1978.

Tagatz, M.E., The biology of the blue crab *Callinectes sapidus* Rathburn, in the St. Johns River, Florida, *U.S. Fish Wildlife Ser. Fish. Bull.*, 67, 614, 1968.

Taghorn, G.L., Optimal foraging by deposit-feeding invertebrates: roles of particle size and organic coating, *Oecologia (Berlin)*, 52, 295, 1982.

Taghorn, G.l., Greene, R.R., Utilization of deposited and suspended particulate matter by bivalve "interface" feeders, *Limnol. Oceanogr.*, 37, 1370, 1992.

Takeda, S., Kurihara, Y., The effects of burrowing *Helice tridens* (Deltann) on the soil of a salt-marsh habitat, *J. Exp. Mar. Biol. Ecol.*, 113, 79, 1987.

Talbot, M.M.B., Bate, G.C., Diel periodicities in cell characteristics of the surf zone diatom *Anaulus birostratus*: their role in the dynamics of cell patches, *Mar. Ecol. Prog. Ser.*, 32, 81, 1986.

Talbot, M.M.B., Bate, G.C., Rip current characteristics and their role in the exchange of water and surf diatoms between the surf zone and nearshore, *Est. Coastal Shelf Sci.*, 25, 707, 1987a.

Talbot, M.M.B., Bate, G.C., The spatial dynamics of surf diatom patches in a medium energy, cuspate beach, *Mar. Biol.*, 30, 459, 1987b.

Talbot, M.M.B., Bate, G.C., The use of false bouyancies by the surf diatom *Anaulus birostratus* in the formation and decay of cell patches, *Est. Coastal Shelf Sci.*, 26, 155, 1988a.

Talbot, M.M.B., Bate, G.C., The response of surf diatom populations to environmental conditions. Changes in the extent of the planktonic fraction and surface patch activity, *Bot. Mar.*, 30, 109, 1988b.

Talbot, M.M.B., Bate, G.C., Distribution patterns of the surf diatom *Anaulus birostratus* in an exposed surf zone, *Est. Coastal Shelf Sci.*, 26, 137, 1988c.

Talbot, M.M.B., Bate, G.C., The relative quantities of live and detrital organic matter in a beach-surf ecosystem, *J. Exp. Mar. Biol. Ecol.*, 121, 255, 1988d.

Talbot, M.M.B., Bate, G.C., Beach morphology dynamics and surf-diatom populations, *J. Exp. Mar. Biol. Ecol.*, 129, 231, 1989.

Talbot, M.M.B., Bate, G.C., Campbell, E.E., A review of the ecology of surf-zone diatoms with special reference to *Anaulus australis*, *Oceanogr. Mar. Biol. Annu. Rev.*, 28, 155, 1989.

Tan, F.C., Strain, P.M., Carbon isotope ratios of particulate organic matter in the Gulf of St. Lawrence, *J. Fish. Res. Bd. Can.*, 36, 678, 1979.

Taylor, D.I., Allanson, B.R., Impacts of dense crab populations on carbon exchange across the surface of a salt marsh, *Mar. Ecol. Prog. Ser.*, 101, 119, 1993.

Taylor, D.I., Allanson, B.R., Organic carbon fluxes between a high marsh and estuary, and the inapplicability of the Outwelling Hypothesis, *Mar. Ecol. Prog. Ser.*, 120, 263, 1995.

Taylor, J.D., A partial food web involving gastropods on a Pacific fringing reef, *J. Exp. Mar. Biol. Ecol.*, 74, 273, 1984.

Taylor, P.R., Littler, M.M., The roles of compensatory mortality, physical disturbance, and substrate retention in the development of a sand-influenced, rocky intertidal community, *Ecology*, 63, 135, 1982.

Taylor, W.R., Light and photosynthesis in intertidal benthic diatoms, *Helg. Wiss. Meersunters.*, 10, 29, 1964.

Teal, J.M., Energy flow in the salt marsh ecosystem of Georgia, *Ecology*, 40, 6, 1962.

Teal, J.M., The ecology of regularly flooded salt marshes of New England: a community profile, *U.S. Fish Wildl. Ser., Rep. No. 85* (7.4), Washington, DC, 61 pp, 1986.

Teal, J.M, Valiela, I., Berlo, D., Nitrogen fixation by rhizosphere and free-living bacteria in salt marsh sediments, *Limnol. Oceanogr.*, 22, 126, 1979.

Teal, J.M., Wieser, W., The distribution and ecology of nematodes in a Georgia salt marsh, *Limnol. Oceanogr.*, 11, 217, 1966.

Teare, M., Price, R., Respiration of the meiobenthic harpacticoid copepod *Tachidus discipes* Gesbretcht, from an estuarine mudflat, *J. Exp. Mar. Biol. Ecol.*, 56, 23, 1969.

Teas, H.J., Ed., *Biology and Ecology of Mangroves*, Dr. W. Junk Publishers, The Hague, 1983.

Tergner, M.J., Dayton, P.K., Population structure, recruitment and mortality of two sea urchins (*Stringylocentrotus franciscanus* and *S. purpuratus)* in a kelp forest, *Mar. Ecol. Prog. Ser.*, 5, 255, 1981.

Tenore, K.R., Growth of the polychaete *Capitella capitata* cultured in different levels of detritus derived from various sources, *Limnol. Oceanogr.*, 22, 936, 1977a.

Tenore, K.R., Utilization of aged detritus derived from different sources by the polychaete, *Capitella capitata*, *Mar. Biol.*, 44, 51, 1977b.

Tenore, K.R., Organic nitrogen and caloric content of detritus. I. Utilization by the deposit-feeding polychaete *Capitella capitata*, *Est. Coastal Shelf Sci.*, 12, 39, 1981.

Tenore, K.R., What controls the availability of detritus derived from vascular plants: nitrogen enrichment or caloric availability?, *Mar. Ecol. Prog. Ser.*, 10, 307, 1983.

Tenore, K.R., Cammen, L., Finlay, S.E.G., Phillips, N., Perspective of research on detritus: do factors controlling the availability of detritus to macro-consumers depend on its source?, *J. Mar. Res.*, 40, 473, 1982.

Tenore, K.R., Chesney, E.J., Jr., The interaction of rate of food supply and population density upon the bioenergetics of the opportunistic polychaete, *Capitella capitata* (type I), *Limnol. Oceanogr.*, 30, 1188, 1985.

Tenore, K. R., Hanson, R. B., Availability of detritus of different types and ages to a polychaete macroconsumer *Capilella capitata, Limnol. Oceanogr.*, 25, 553, 1980.

Tenore, K.R., Harrison, R.B., Availability of detritus of different types and ages to a polychaete macroconsumer, *Capitella capitata*, *Linmol. Oceanogr.*, 25, 553, 1980.

Tenore, K.R., Rice, D.L., A review of trophic factors affecting secondary production of deposit-feeders, in *Marine Benthic Dynamics*, Tenore, K.R., Coull, B.C., Eds., University of South Carolina Press, Columbia, SC, 1980, 325.

Tenore, K.R., Tietjen, J.H., Lee, J.J., Effect of meiofauna on incorporation of aged eelgrass, *Zostera marina* detritus by the polychaete *Nephtys incisa*, *J. Fish. Res. Bd. Can.*, 34, 563, 1977.

Tenore, R.R., Goldman, J.C., Clarner, J.D., The food chain dynamics of the oyster, clam, and mussel in an aquaculture food chain, *J. Exp. Mar. Biol. Ecol.*, 12, 19, 1973.

Thayer, G.W., Phytoplankton production and distribution of nutrients in a shallow unstratified estuarine system near Beaufort, North Carolina, *Chesapeake Sci.*, 12, 240, 1971.

Thayer, G.W., Sediment-mediated biological disturbance and the evolution of the marine benthos, in *Biotic Interactions in Recent and Fossil Benthic Communities*, Tevesz, M.J.S., McCall, P.L., Eds., Plenum Press, New York, 1983, 479.

Thayer, G.W., Adams, S.M., La Croix, M.W., Structural and functional aspects of a recently established *Zostera marina* community, in *Estuarine Research, Vol. 1.*, Cronin, L.E. Ed, Academic Press, New York, 1975, 578.

Thayer, G.W., Kenworthy, W.J., Fonseca, M.S., Ecology of eelgrass meadows of the Atlantic coast: a comunity profile, *U.S. Fish. Wildl. Ser., Publ. No. FWS/OBS-84/02*, 1984.

Thistle, D., Natural physical disturbance and communities of marine soft bottoms, *Mar. Ecol. Prog. Ser.*, 6, 223, 1981.

Thom, B.G., Mangrove ecology and deltaic geomorphology, Tabasco, Mexico, *J. Ecol.*, 55, 301, 1967.

Thom, B.G., Mangrove ecology from a geomorphic viewpoint, in *Proc. Internat. Symp. Biol. Managt. Mangroves*, Walsh, G., Snedaker, S., Teas, H., Eds., Institute of Food and Agricultural Sciences, University of Florida, Gainesville, FL, 1974, 469.

Thom, B.G., Mangrove ecology and deltaic geomorphology, Tabasco, Mexico, *J. Ecol.*, 55, 301, 1981.

Thom, B.G., Mangrove ecology — A geomorphological perspective, in *Mangrove Ecosystems in Australia, Structure, Function and Management*, Clough, B.F., Ed., Australian National University Press, Canberra, 1982, 3.

Thom, B.G., Coastal landforms and geomorphic procsses, in *Mangrove Ecosystem Research Methods*, Snedaker, S.C., Snedaker, J.G., Eds., UNESCO, Paris, 1984, 3.

Thom, R.M., Composition, habitats, seasonal changes and productivity of mangroves in Grays Harbor Estuary, *Estuaries*, 7, 57, 1984.

Thomas, J.P., Influence of the Atamaha River on primary production beyond the mouth of the river, M.Sc. thesis, University of Georgia, Athens, GA, 80 pp, 1966.

Thompson, G.B., Distribution and population dynamics of the limpet *Patella vulgata* L. in Botany Bay, *J. Exp. Mar. Biol. Ecol.*, 45, 173, 1980.

Thompson, J.K., Nichols, F.H., Food availability controls seasonal cycle of growth in *Macoma balthica* (L.) in San Francisco Bay, *J. Exp. Mar. Biol. Ecol.*, 116, 44, 1988.

Thorson, G., Reproduction and larval development of Danish marine bottom invertebrates, with special reference to the planktonic larvae in the Sound (Oresund), *Meddr. Komm. Fisk. og. Havunds., Serie Plankton*, 4, 1, 1946.

Thorson, G., Reproductive and larval ecology of marine bottom invertebrates, *Biol. Rev.*, 25, 1. 1950.

Thorson, G., Bottom communities (sublittoral or shallow shelf), in *Treatise on Marine Ecology & Paleoecology, Vol. 1*, Hedgpeth, J.W., Ed., Memoirs of the Geological Society of America, 67, 1957, 461.

Thursby, G., Harlin, M., Leaf-root interaction in the uptake of ammonia by *Zostera marina*, *Mar. Biol.*, 72, 109, 1984.

Tietjen, J.H., Microbial-meiofaunal relationships: a review, in *Aquatic Microbial Ecology*, Colwell, R.R., Foster, J., Eds., University of Maryland Press, College Park, 1980, 130.

Tiffany, W.J., The tidal migration of *Donax variabilis* Say., *Veliger*, 14, 82, 1972.

Todd, C.D., Reproductive strategies of North Temperate rocky shore invertebrates, in *The Ecology of Rocky Shores*, Moore, P.G., Seed, R., Eds., Hodder & Straughton, London, 1985, 203.

Townsend, C.R., Hughes, R.N., Maximizing net energy returns from foraging, in *Physiological Ecology*, Townsend, C.R., Calow, P. Eds., Sinauer Associates, Sunderland, MA, 1981, 86.

Townsend, C., Lawson, G.W., Preliminary results on factors causing zonation in *Enteromorpha* using a tide simulating apparatus, *J. Exp. Mar. Biol. Ecol.*, 8, 265, 1972.

Trevallion, A., Studies on *Tellina tenuis* da Costa. III. Aspects of general biology and energy flow, *J. Exp. Mar. Biol. Ecol.*, 7, 95, 1971.

Trueblood, P.D., Gallagher, E.D., Gould, D.M., Three stages of seasonal succession in the Savin Hill Cove, *Limnol. Oceanogr.*, 39, 1440, 1994.

Trueman, E.R., The control of burrowing and the migratory behaviour of *Donax denticulatus* (Bivalvia: Tellinacea), *J. Zool., London*, 165, 453, 1971.

Trueman, E.R., *The Locomotion of Soft-Bodied Animals*, Arnold, London, 200 pp, 1975.

Tsuchiya, M., Mass mortality in a population of the mussel *Mytilus edulis* L. on rocky shores, *Bull. Biol. Sta. Asamushi*, 17, 99, 1983.

Tsuchiya, M., Nishihara, M., Islands of *Mytilus* as a habitat for small intertidal animala: effect of island size on community structure, *Mar. Ecol. Prog. Ser.*, 25, 71, 1985.

Tsutsumi, H., Population dynamics of *Capitella capitata* (Polychaeta: Capitellidae) in an organically polluted cove, *Mar. Ecol. Prog. Ser.*, 36, 139, 1987.

Tsutsumi, H., Population persistence of *Capitella* sp. (Polychaeta: Capitellidae) on a mudflat subject to environmental disturbance by organic enrichment, *Mar. Ecol. Prog. Ser.*, 63, 147, 1990.

Tsutsumi, H., Kikuchi, T., Benthic ecology of a small cove with seasonal oxygen depletion caused by organic pollution, *Publ. Amakusa Mar. Biol. Lab.*, 7, 17, 1983.

Turner, J.T., Tester, P.A., Ferguson, R.L., The marine cladoceran *Perilia avirostris* and the 'microbial loop' of pelagic food webs, *Limnol. Oceanogr.*, 30, 1268, 1986.

Turner, R.E., Geographic variations in salt marsh macrophyte production: a review, *Contr. Mar. Sci., Univ. Texas*, 22, 147, 1976.

Turner, R.E., Community plankton respiration in a salt marsh estuary and the importance of macrophyte leacheates, *Limnol. Oceanogr.*, 23, 442, 1978.

Turner, T., Facilitation as a successional mechanism in a rocky intertidal community, *Am. Nat.*, 121, 729, 1983a.

Turner, T., Complexity of early and middle successional states in a rocky intertidal surfgrass community, *Oecologia (Berlin)*, 60, 56, 1983b.

Tussenbrock von, B.T., Seasonal growth and composition of fronds of *Macrocystis pyrifera* in the Falkland Islands, *Mar. Biol.*, 100, 219, 1989.

Twilley, R.R., The exchange of organic carbon in basin mangrove forests in a southwest Florida estuary, *Est. Coastal Shelf Sci.*, 20, 543, 1985.

Twilley, R.R., Coupling of mangroves to the productivity of estuarine and coastal waters, in *Coastal-Offshore Ecosystem Interactions*, Lecture Notes on Coastal and Estuarine Studies, Jansson, B.-O., Springer-Verlag, Berlin, 1986, 155.

Twilley, R.R., Lugo, A.E., Patterson-Zucca, C., Litter production and turnover in basin mangrove forests in southwest Florida, *Ecology*, 67, 670, 1986.

Twilley, R.R., Pozo, M., Gaecia, V.H., Ricero-Monroy, V.H., Zambrano, R., Bodera, A., Litter dynamics in riverine mangrove forests in the Guayas River and estuary, Ecuador, *Oecologia (Berlin)*, 111, 109, 1997.

Twilley, R.R., Snedaker, S.C., Yanez-Aranciba, A., Medina, E., Biodiversity and ecosystem processes in tropical estuaries: persectives in marine ecosystems, in *Functional Roles of Biodiversity. A Global Perspective*, Mooney, H.A., Cushman, J.H., Medina, E., Sala, O.E., Schulze, E.-D., Eds., John Wiley & Sons, New York, 1996, 327.

Ugolini, A., Scarpini, F., Pardi, L., Interaction between solar orientation and landscape visibility in *Talitrus saltator* (Crustacea: Amphipoda), *Mar. Biol.*, 90, 449, 1986.

Ulanowicz, R.E., An hypothesis on the development of natural communities, *J. Theoret. Biol.*, 85, 223, 1980.

Ulanowicz, E.E., Community measures of marine food networks and their possible applications, in *Flows of Energy and Materials in Marine Ecosystems*, Fasham, J.R., Ed., Plenum Press, New York, 1984, 3.

Ulanowicz, R.E., *Growth and Development: Ecosystem Phenomenology*, Springer-Verlag, New York, 1986.

Ulanowicz, R.E., NETWRK4: A package of computer algorithms to analyze ecological flow networks, *UMCEES Ref. No. 82*, University of Maryland, 1987.

Ulanowicz, R.E., A generic simulation model for treating incomplete sets of data, in *Network Analysis in Marine Ecology*: Methods and Applications, Wulff, W., Field, J.G., Mann, K.H., Eds., Springer-Verlag, Belin, 1989, 84.

Ulanowicz, R.E., *Ecology the Antecedent Perspective*, Columbia University Press, New York, 201 pp, 1997.

Ulanowicz, R.E., Kemp, W.M., Towards canonical trophic aggregations, *Am. Nat.*, 114, 871, 1979.

Ulanowicz, R.E., Mann, K.H., Ecosystems under stress, in *Mathematical Models in Biological Oceanography*, Platt, T., Mann, K.H., Ulanowicz, R.E., Eds., The UNESCO Press, Paris, 1987, 133.

Ulanowicz, R.E., Norden, J.S., Symmetrical overhead in flow networks, *Int. J. Systems Sci.*, 21, 429, 1990.

Ulanowicz, R.E., Platt, T., Eds., *Ecosystem Theory for Biological Oceanography*, Can. Bull., Fish. Aquatic Sci., 213, 260 pp, 1985.

Unabia, C.R.C., Hadfield, M.G., Role of bacteria in larval settlement and metamorphosis of the polychaete *Hydroides elegans*, *Mar. Biol.*, 133, 53, 1999.

Underwood, A.J., An experimental evaluation of competition between three species of intertidal prosobranch molluscs, *Oecologia (Berlin)*, 33, 185, 1976.

Underwood, A.J., Movements of intertidal gastropods, *J. Exp. Mar. Biol. Ecol.,* 26, 191, 1977.

Underwood, A.J., A refutation of critical tidal levels as determinants of the structure of intertidal communities on British shores, *Oecologia (Berlin),* 36, 201, 1978a.

Underwood, A.J., The detection of non-random patterns of distribution of species along a gradient, *Oecologia (Berlin),* 36, 317, 1978b.

Underwood, A.J., An experimental evaluation of competition between three species of intertidal prosobranch gastropods, *Oecologia (Berlin),* 33, 185, 1978c.

Underwood, A.J., The ecology of intertidal gastropods, *Adv. Mar. Biol.*, 16, 111, 1979.

Underwood, A.J., The effects of grazing by gastropods and physical factors on the upper limits of distribution of intertidal macroalgae, *Oecologia (Berlin)*, 46, 201, 1980.

Underwood, A. J., Techniques for the analysis of variance in experimental marine biology and ecology, *Oceanogr. Mar. Biol. Ecol., Annu. Rev.,* 19, 513, 1981a.

Underwood, A.J., Structure of a rocky intertidal community in New South Wales: patterns of vertical distribution and seasonal changes, *J. Exp. Mar. Biol. Ecol.,* 51, 57, 1981b.

Underwood, A.J., The vertical ditribution and seasonal abundance of intertidal microalgae on a rocky shore in New South Wales, *J. Exp. Mar. Biol. Ecol.,* 78, 199, 1984a.

Underwood, A.J., Vertical and seasonal patterns in competition for microalgae between intertidal gastropods, *Oecologia (Berlin),* 64, 211, 1984b.

Underwood, A.J., Microalgal food and the growth of the intertidal gastropods *Nerita atramentosa* Reeve and *Bembicium vulgata* L., *J. Exp. Mar. Biol. Ecol.*, 68, 81, 1984c.

Underwood, A.J., Physical factors and biological interaction: the necessity and nature of ecological experiments, in *The Ecology of Rocky Shores*, Moore, P.G., Seed, R., Hodder & Stroughton, London, 1985, 372.

Underwood, A.J., The analysis of competition by field experiments, in *Community Ecology, Patterns and Processes*, Kikkawa, J., Anderson, D.J., Blackwell Scientific Press, Melbourne, 1986, 240.

Underwood, A.J., The logic of ecological experiments: a case history from studies of the distribution of macroalgae on rocky intertidal shores, *J. Mar. Biol. Assoc. U.K.*, 71, 841, 1991.

Underwood, A.J., Competition and marine plant-animal interactions, in *Plant-Animal Interactions in the Marine Benthos* (Systematics Association Special Publication Vol. 46), John, D.M., Hawkins, S.J., Price, J.H., Eds., Clarendon Press, Oxford, 1992, 443.

Underwood, A.J., Rocky intertidal shores, in *Marine Biology,* Hammond, L.S., Synnot, R.N., Eds., Longman Cheshire, Melbourne, 1994, 272.

Underwood, A.J., Chapman, M.G., Multifactorial directions of movements in animals, *J. Exp. Mar. Biol. Ecol.*, 91, 17, 1985.

Underwood, A.J., Chapman, M.G., Experimental analyses of the influence of topography of the substratum on movements and density of the intertidal snail, *Littorina unifasciata*, *J. Exp. Mar. Biol. Ecol.*, 134, 175, 1989.

Underwood, A.J., Denley, E.J., Paradigms, explanations and generalizations in models for the structure of communities on rocky shores, in *Ecological Communities: Conceptual Issues and the Evidence*, Strong, D.R., Simberloff, D., Able, L.G., Thistle, A.B., Eds., Princetown University Press, Princetown, 1990, 141.

Underwood, A.J., Denley, E.J., Moran, M.J., Experimental analysis of the structure and dynamics of mid-shore rocky intertidal communities in New South Wales, *Oecologia (Berlin),* 56, 202, 1983.

Underwood, A. J., Denley, E. J., Paradigms, explanations, and generalizations in models for the structure of low shore rocky intertidal communities, *Oecologia*, 48, 221, 1984.

Underwood, A.J., Fairweather, P.G., Supply-side ecology and benthic marine assemblages, *Trends Ecol. Evol.*, 4, 18, 1989.

Underwood, A.J., Jernakoff, P., Interactions between algae and grazing gastropods in the structure of a low-shore algal community, *Oecologia (Berlin)*, 48, 221, 1981.

Underwood, A.J., Jernakoff, P., The effect of tidal height, wave exposure, seasonality and rock-pools on grazing and the distribution of intertidal macroalgae in New South Wales, *J. Exp. Mar. Biol. Ecol.*, 75, 71, 1984.

Underwood, G.J.C., Paterson, D.M., Seasonal changes in diatom biomass, sediment stability and biogenic stabilization in the Severn Estuary, *J. Mar. Biol. Assoc., U.K.,* 73, 871, 1993.

Vadas, R.L., Preferential feeding-optimization strategy in sea-urchins, *Ecol. Monogr.*, 47, 337, 1977.

Vadas, R.L., Herbiovory, in *Handbook of Phycological Methods. Ecology Field Methods: Macroalgae*, Littler, M.M., Littler, D.S., Eds., Cambridge University Press, Cambridge, 1985, 531.

Vadas, R.L., Elner, R.W., Plant-animal interactions in the northwest Atlantic, in *Plant-Animal Interactions in the Marine Benthos* (Systematics Association Special Volume No. 46), John, D.M., Hawkins, S.J., Price, J.H., Eds., Clarendon Press, Oxford, 1992, 33.

Vadas, R.L., Elner, R.W., Garwood, P.E., Babb, I.G., Experimental evaluation of aggregation behaviour in the sea-urchin *Strongylocentrotus droebachiensis*: a reinterpretation, *Mar. Biol.*, 90, 433, 1986.

Valentine, J.F., Heck, K.L., Jr., Seagrass herbivory: evidence for the continued grazing of marine grasses, *Mar. Ecol. Prog. Ser.,* 176, 291, 1999.

Valiela, I., *Marine Ecological Processes*, 2nd ed., Springer-Verlag, New York, 546 pp, 1995.

Valiela, I., Koumjian, L., Swain, T., Teal, J.M., Hobbie, J.E., Cinnamic acid inhibition of detritus feeding, *Nature (London)*, 280, 55, 1979.

Valiela, I., Teal, J.M., Nutrient limitation in salt marsh vegetation, in *Ecology of Halophytes*, Reimbold, R.J., Queen, W.H., Eds., Academic Press, New York, 1974, 547.

Valiela, I., Teal, J.M., The nitrogen budget of a salt marsh ecosystem, *Nature (London)*, 286, 652, 1979a.

Valiela, I., Teal, J.M., Inputs, outputs and interconversions of nitrogen in a salt marsh ecosystem, in *Ecological Processes in Coastal Environments*, Jeffries, R.L., Davy, A.J., Eds., Blackwell Scientific Publications, Oxford, 1979b, 399.

Valiela, I., Teal, J.M., Allen, S.D., Van Etten, R., Goehringer, D., Volkman, S., Decomposition in salt marsh ecosystem: the phases and major factors affecting disappearance of above-ground organic matter, *J. Exp. Mar. Biol. Ecol.*, 89, 29, 1985.

Valiela, I., Teal, J.M., Deuser, W.G., The nature of growth forms of the salt marsh grass *Spartina alterniflora, Am. Nat.*, 122, 461, 1978.

Valiela, I., Teal, J.M., Persson, N.Y., Production and dynamics of experimentally enriched salt marsh vegetation: below-ground biomass, *Limnol. Oceanogr.*, 2, 245, 1976.

Valiela, I., Teal, J.M., Sass, W., Nutrient retention in salt marsh plots experimentally fertilized with sewage sludge, *Est. Coastal Mar. Sci.*, 1, 262, 1973.

Valiela, I., Wilson, J., Bachsbaum, R., Rietsma, C., Bryanr, D., Foreman, K., Teal, J., Importance of chemical composition of salt marsh litter on decay rates and feeding by detritovores, *Bull. Mar. Sci.*, 35, 261, 1984.

Vandermeer, J.H., Niche theory, *Annu. Rev. Ecol. Syst.*, 3, 107, 1976.

Van Beek, F.A., Antallen groei, produktie en vosd selopname von de zandgrondel (*P. minutus*) wadgrondel (*P. microps*) op het Balgzand, *Internat. NZOI Rapp.*, 1976–1979.

Van Blaricom, G.R., Experimental analysis of structural regulation in a marine sand community exposed to oceanic swell, *Ecol. Monogr.*, 532, 283, 1982.

Van Blaricom, G.R., Estes, T.A., Eds., *The Community Ecology of Sea Otters*, Springer-Verlag, New York, 1988.

Van der Schoot, E., Onderzock naar het aandeel van *Asterias rubens* in de koolstofkringloop van het Grevelingen bekken, Della Institut voor Hydrobiologisch Onderzoek Vereseke, Doctoraalverstag D5-1974.

Van Es, F.B., A preliminary carbon budget for a part of the Ems estuary, *Helgol. Wiss. Meeresunters*, 30, 283, 1977.

Van Es, F.B., Some aspects of the flow of oxygen and organic carbon in the Ems-Dollard estuary, *BOEDE publicalies en verslagen*, 5-1882, 1, 1982.

Van Raalte, C.D., Valiela, I., Teal, J.M., Production of epiphytic salt marsh algae: light and nutrient limitation, *Limnol. Oceanogr.*, 21, 862, 1976.

Van Roon, B., *Water Quality in the Upper Waitemata Harbour and Catchment*, Upper Waitemata Harbour Catchment Study Specialist Report, Auckland Regional Authority, Auckland, 430 pp, 1983.

Van Tamlen, P.G., Early successional mechanisms in the rocky intertidal: the role of direct and indirect interactions, *J. Exp. Mar. Biol. Ecol.*, 122, 39, 1987.

Verity, P.G., Grazing, respiration, excretion, and growth rates in tintinnids, *Limnol. Oceanogr.*, 30, 1268, 1985.

Verity, P.G., Abundance, community composition, size distribution and production rates of tintinnids in Narragansett Bay, Rhode Island, *Est. Coastal Shelf Sci.*, 24, 571, 1987.

Vermeij, G.J., Intraspecific shore-level size gradients in intertidal molluscs, *Ecology*, 53, 693, 1972.

Vermeij, G.J., Morphological patterns in high-intertidal gastropods: adaptive strategies and their limitations, *Mar. Biol.*, 20, 319, 1973.

Vermeij, G.J., *Biogeography and Adaptation Patterns of Marine Life*, Harvard University Press, Cambridge, MA, 1978.

Vermeij, G.J., Gastropod shell growth rate, allometry, and adult size: environmental implications, in *Skeletal Growth of Aquatic Organisms: Biological Records of Environmental Change*, Rhoads, D.C., Lutz, R.A., Eds., Plenum Publishing, New York, 1980, 379.

Verwey, J., On the ecology and distribution of cockles and mussels in the Dutch Wadden Sea, their role in sedimentation and the source of their food supply, with a short review of the feeding behaviour in bivalve molluscs, *Archiv. Neerl. Zool.*, 10, 172, 1952.

Virnstein, R.W., The importance of predation by crabs and fishes on benthic fauna in Chesapeake Bay, *Ecology*, 58, 1199, 1977a.

Virnstein, R.W., Predation on estuarine infauna: response patterns of component species, *Estuaries*, 2, 69, 1977b.

Virnstein, R.W., Predator caging experiments in soft sediments: caution advised, in *Estuarine Interactions Vol. I*, Wiley, M.L., Ed., Academic Press, New York, 1978, 261.

Virnstein, R.W., Nelson, W.G., Lewis, F.G., Howard, R.K., Latitudinal patterns in seagrass fauna: do patterns exist and can they be explained, *Estuaries*, 7, 310, 1984.

Voipio, A., *The Baltic Sea*, Elsevier Publishing, Amsterdam, 418 pp, 1981.

Voller, R.W., Salinity, sediment, exposure and invertebrate macrofaunal distributions on the mudflats of the Avon-Heathcote Estuary, Christchurch, M.Sc. thesis, University of Canterbury, Christchurch, NZ, 51 pp, 1975.

Wadsworth, F., An evaluation of mangroves west of Jobes Bay, Puerto Rico, in *Aguire Power Plant Complex Environmental Report, Appendix C*, Puerto Rico Water Research Authority, San Juan, Puerto Rico, 1972.

Wafar, S., Untawale, A.G., Wafer, M., Litterfall and energy flow in a mangrove ecosystem, *Est. Coastal Shelf Sci.*, 44, 111, 1997.

Wagner, D.R., Seasonal biomass, abundance and distribution of estuarine dependent fishes in the Caminada Bay syatem of Lousiana, Ph.D. dissertation, Louisiana State University, Baton Rouge, LA, 193 pp, 1973.

Wahl, M., Marine epibiosis. I. Fouling and antifouling: some basic aspectes, *Mar. Ecol. Prog. Ser.*, 58, 175, 1989.

Wainwright, P.F., Mann, K.H., Effects of antimicrobial substances on the ability of mysid shrimps *Mysis stenolepis* to digest cellulose, *Mar. Ecol. Prog. Ser.*, 7, 309, 1982.

Waisel, Y., *Biology of Halophytes*, Academic Press, New York, 395 pp, 1972.

Waite, T.D., Mitchell, R., The effect of nutrient fertilization on the benthic alga *Ulva lactuca*, *Bot. Mar.*, 15, 151, 1972.

Waits, E.D., Net primary productivity of an irregularily flooded North Carolina salt marsh, Ph.D. thesis, North Carolina State University, Raleigh, NC, 124 pp, 1967.

Walker, B.H., Biodiversity and ecological redundancy, *Conservation Biol.*, 6, 18, 1991.

Walker, N.A., The carbon species taken up by *Chara*, in *Inorganic Carbon Uptake by Aquatic Photosynthetic Organisms*, Lucas, W.J., Berry, J.A., Eds., American Society of Plant Physiologists, 1985, 31.

Walne, R.R., The influence of current speed, body size and water temperature on the filtration rate of five species of bivalves, *J. Mar. Biol. Assoc. U.K.*, 52, 345, 1972.

Walsh, G.E., Mangroves: a review, in *Ecology of Halophytes*, Reimbold, R.J., Queen, W.H., Eds., Academic Press, New York, 1974, 51.

Walsh, J.J., Utility of species models: a consideration of some possible feedback loops of the Peruvian upwelling ecosystem, in *Estuarine Research Vol. 1.*, Cronin, L.E., Ed., Academic Press, New York, 1975, 617.

Walters, K., Bell, S.S., Significance of copepod emergence to benthic pelagic and phytal linkages in a subtidal seagrass beds, *Mar. Ecol. Prog. Ser.*, 108, 237, 1994.

Wang, Z., Dauvin, J.-C., The suprabenthic crustacean fauna of the infralittoral fine sand communities in the Bay of Seine (Eastern English Channel): composition, swimming activity and diurnal variation, *Cash. Biol. Mar.*, 35, 135, 1994.

Wang, Z., Dauvin, J.-C., Thiébaut, E., Preliminary data on the near-bottom macro-zooplanktonic fauna of the eastern Bay of Seine: faunistic composition, vertical distribution and density variation, *Cash. Biol. Mar.*, 35, 157, 1994.

Ward, L.G., Kemp, W.M., Boynton, W.R., The influence of waves and seagrass communities on suspended particlulates in an estuarine embayment, *Mar. Geol.*, 59, 85, 1984.

Warme, J.E., Graded bedding in the recent sediments of Mogu Lagoon, California, *J. Sedim. Petrol.*, 37, 540, 1967.

Warren, C.E., *Biology and Water Pollution Control,* W.B. Saunders, Philadelphia, 434 pp, 1971.

Warren, C.E., Davies, G.E., Laboratory studies in the feeding, bioenergetics and growth of fish, in *The Biological Basis of Freshwater Fish Production*, Blackwell Scientific Publications, Oxford, 1967, 175.

Warren, L.M., How intertidal polychaetes survive at low tide, in *Proceedings of the 1st International Polychaete Conference*, Hutchings, P.A., Ed., Linnean Society of New South Wales, Sydney, 1984, 238.

Warwick, R.M., The structure and seasonal fluctuation of phytal marine nematode associations on the Isles of Scilly, in *Biology of Benthic Organisms*, Keegan, B.F., Ceidigh, P.O., Boaden, P.J.S., Eds., Pergamon Press, Oxford, 1977, 577.

Warwick, R.M., Survival strategies of meiofauna, in *Feeding and Survival Strategies of Estuarine Organisms*, Jones, N.V., Wolff, W.J., Eds., Plenum Press, New York, 1981, 39.

Warwick, R.M., Davey, J.T., Gee, J.M., George, C.L., Faunistic control of *Enteromorpha* blooms: a field experiment, *J. Exp. Mar. Bio. Ecol.*, 56, 23, 1982.

Warwick, R.M., Joint, I.R., Radford, P.J., Secondary production of the benthos in an estuarine environment, in *Ecological Processes in Coastal Environments,* Jeffries, R.L., Davey, A.J., Eds., Blackwell, Oxford, 1979, 429.

Warwick, R.M., Price, R., Macrofaunal production on an estuarine mudflat, *J. Mar. Biol. Assoc. U.K.*, 55, 1, 1975.

Warwick, R.M., Price, R., Ecological and metabolic studies on free-living nematodes from an estuarine mud-flat, *Est. Coastal Mar. Sci.*, 9, 257, 1979.

Warwick, R.M., Radford, P.J., Analysis of the flow network in an estuarine benthic community, in *Network Analysis in Marine Ecology: Methods and Applications,* Wulff, F., Field, J.G., Mann, K.H., Eds., Springer-Verlag, Berlin, 1989, 220.

Waslenchuk, D.G., Matson, E.A., Zaijac, R.N., Dobbs, F.C., Tramontano, J.M., Geochemistry of burrow waters vented by a bioturbating shrimp in Bermuda sediments, *Mar. Biol.*, 72, 219, 1983.

Watanabe, J., The influence of recruitment, competition, and benthic predation on spatial distribution of three kelp forest gastropods (Trochidae: Tegula), *Ecology,* 65, 920, 1984.

Watson, J.G., Mangrove forests of the Malay Peninsula, *Malay Forest Rec.*, 6, 1, 1928.

Watson, W.S., The role of bacteria in an upwelling system, in *Upwelling Ecosystems*, Boje, R., Tomczak, T., Eds., Springer-Verlag, Berlin, 1978, 139.

Watzin, M.C., The effects of meiofauna on settling macrofauna: meiofauna may structure macrofaunal communities, *Oecologia (Berlin)*, 59, 163, 1983.

Watzin, M.C., Larval settlement into marine soft-sediment systems: interactions with meiofauna, *J. Exp. Mar. Biol. Ecol.*, 98, 65, 1986.

Webb, J.E., Biologically significant properties of submerged marine sands, *Proc. Roy. Soc. Lond., B*, 174, 355, 1969.

Webb, K.L., Conceptual models and processes of nutrient cycles in estuaries, in *Estuaries and Nutrients*, Nielson, B.J., Cronin, L.E., Eds., Humana, NJ, 1981, 25.

Webster, T.J.M., Paranjape, M., Mann, K.H., Sedimentation of organic matter in St. Margarets Bay, Nova Scotia, *J. Fish. Res. Bd. Can.*, 32, 1399, 1975.

Weinberg, J.R., Interactions between functional groups in soft-substrata: do species differences matter?, *J. Exp. Mar. Biol. Ecol.*, 80, 16, 1984.

Wells, R.A., Activity pattern as a mechanism of predator avoidance in two species of acmaeid limpets, *J. Exp. Mar. Biol. Ecol.*, 48, 151, 1980.

Welsh, E.B., Emery, R.M., Matsuda, R.I., Dawson, W.A., The relationship of algal growth in an estuary to hydrographic factors, *Limnol. Oceanogr.*, 21, 1, 1972.

Went, G.E., McLachlan, A., Zonation and biomass of the intertidal macrofauna along a South African sandy beach, *Cash. Biol. Mar.*, 26, 1, 1985.

West, R.J., Larkum, A.W.D., Leaf productivity of the seagrass *Posidonia australis*, in eastern Australian waters, *Fish. Bull.*, 2, 1, 1979.

Westlake, D.F., Comparisons of plant productivity, *Biol. Rev.*, 38, 385, 1963.

Weston, D.P., Quantitative examination of macrobenthic community changes along an organic enrichment gradient, *Mar. Ecol. Prog. Ser.*, 61, 233, 1990.

Wethey, D.S., Sun and shade mediate competition in the barnacles *Balanus* and *Semibalanus* in a field experiment, *Biol. Bull. Mar. Biol. Lab., Woods Hole,* 167, 176, 1984.

Wethey, D.S., Catastrophe, extinction and species diversity: a rocky intertidal example, *Ecology,* 66, 445, 1985.

Wethey, D.S., Ranking of settlement cues by barnacle larvae: influence of settlement contours, *Bull. Mar. Sci.*, 39, 393, 1986.

Wetsteyn, L.P., Kromkamp, J.C., Turbidity and phytoplankton primary production in the Oosterschelde (The Netherlands), during and after a large-scale coastal engineering project (1980-1991), *Hydrobiologica*, 282/283, 61, 1984.

Wetzel, R.L., Carbon resources of a benthic salt marsh invertebrate *Nassarius obsoletus* Say (Mollusca: Nassariidae), in *Estuarine Processes Vol. 2*, Wiley, M., Ed., Academic Press, New York, 1977, 293.

Wetzel, R.G., Detrital, dissolved and particulate organic carbon functions in aquatic systems, *Bull. Mar. Sci.*, 35, 503, 1984.

Wetzel, R.G., Penhale, P.A., Productive ecology of seagrass communities in the lower Chesapeake Bay, *Mar. Technol. Soc. J.*, 17, 22, 1982.

Wharton, W.G., Mann, K.H., Relationship between destructive grazing by the sea urchin *Strongylocentrouus droebachiensis*, and the abundance of American lobster, *Homarus americanus*, on the Atlantic coast of Nova Scotia, *Can. J. Fish. Aquatic Sci.*, 38, 1339, 1981.

Wheeler, W.N., Druehl, L.D., Seasonal growth and productivity of *Macrocystis integrifolia* in British Columbia, Canada, *Mar. Biol.*, 90, 181, 1986.

White, D.C., Analysis of microorganisms in terms of quality and activity in natural environments, in *Microbes in Their Natural Environments*, Slater, J.H., Whittenberg, R., Whimpenny, J.W.P., Eds., London, Cambridge University Press, Cambridge, 1983, 37.

White, D.S., Howes, B.L., Long-term ^{15}N-nitrogen in the vegetated sediment of a New England salt marsh, *Limnol. Oceanogr.*, 39, 1878, 1994.

Whitlach, R.B., Food resource partitioning in the deposit-feeding polychaete *Pectinaria gouldii, Biol. Bull.*, 147, 227, 1974.

Whitlach, R.B., Methods of resource allocation in marine deposit-feeding communities, *Am. Nat.*, 16, 195, 1976.

Whitlach, R.B., Patterns of resource utilization and coexistance in marine intertidal deposit-feeding communities, *J. Mar. Res.*, 38, 743, 1980.

Whitlach, R.B., Johnson, R.G., Methods for staining organic matter in marine sediments, *J. Sed. Petrol.*, 44, 1310, 1974.

Whitman, J.D., Subtidal coexistence: storms, grazing, mutualism, and the zonation of kelps and mussels, *Ecol. Monogr.*, 57, 167, 1987.

Whitney, D.M., Chalmers, A.G., Haines, E.B., Hanson, R.B., Pomeroy, L.R., Skerr, B., The cycles of nitrogen and phosphorus, in *The Ecology of a Salt Marsh*, Pomeroy, L.R., Wiegert, G., Springer-Verlag, New York, 1981, 163.

Whitney, D.E., Woodwell, G.M., Howarth, R.W., Nitrogen fixation in Flax Pond: a Long Island salt marsh, *Limnol. Oceanogr.*, 20, 640, 1975.

Whittaker, R.H., Gradient analysis of vegetation, *Biol. Rev.*, 42, 207, 1967.

Whittaker, R.H., *Communities and Ecosystems*, 2nd ed., Macmillan Publishing Company, New York, 385 pp, 1975.

Whittaker, R.H., Levin, S.A., The role of mozaic phenomena in natural communities, *Theoret. Pop. Biol.*, 12, 117, 1977.

Wiebe, W.J., Christian, R.R., Hansen, J.A., King, G., Sherr, B., Skyring, G., Anaerobic respiration and fermentation, in *The Ecology of a Salt Marsh*, Pomeroy, L.R., Wiegert, R.G., Eds., Springer-Verlag, New York, 1981, 138.

Wiebe, W.J., Smith, D.F., Direct measurement of dissolved organic carbon release by phytoplankton and incorporation by microheterotrophs, *Mar. Biol.*, 42, 213, 1977.

Wiegel, R.L., *Oceanographic Engineering*, Prentice-Hall, New York, 1964.

Wiegert, R.G., Ecological processes characteristic of coastal *Spartina* marshes of the south- eastern USA, in *Ecological Processes in Coastal Environments*, Jeffries, R.L., Davy, A.J., Eds., Blackwell Scientific, Oxford, 1979a, 467.

Wiegert, R.G., Population models: experimental tools for analysis of ecosystems, in *Analysis of Ecological Systems*, Hoen, D., Stairs, G., Mitchell, R., Eds., Ohio State University Press, Columbus, OH, 1979b, 234.

Wiegert, R.C., Christian, R.R., Gallagher, J.L., Hall, J.R., Jones, R.D.H., Wetzel, R., A preliminary ecosystem model of coastal Georgia *Spartina* marsh, in *Estuarine Research Vol. 2*, Cronin, L.E., Ed., Academic Press, New York, 1975, 583.

Wiegert, R.G., Christian, R.R., Wetzel, R.L., A model view of the marsh, in *The Ecology of a Salt Marsh*, Pomeroy, L.R., Wiegert, R.G., Eds., Springer-Verlag, New York, 1981, 183.

Wiegert, R.G., Evans, F.C., Investigations of secondary productivity in grasslands, in *Secondary Productivity of Terrestrial Ecosystems*, Petrusewicz, K., Ed., Polish Academy of Science, Krakow, 1964, 499.

Wiegert, R.G., Pomeroy, L.R., Wiebe, W.J., Ecology of salt marshes: an introduction, in *The Ecology of a Salt Marsh*, Pomeroy, L.R., Wiegert, R.G., Eds., Springer-Verlag, New York, 1981, 3.

Wiegert, R.G., Wetzel, R.L., Simulaton experiments with a fourteen-compartment model of a *Spartina* salt marsh, in *Marsh-Estuarine Systems Simulation*, Dame, R.F., Ed., University of South Carolina Press, Columbia, SC, 1979, 7.

Wieser, W., Die Beziehung zwischen Mundhollengenestaff, Erna hrungsweisse und Vorkommen bie frielebenden marinen Nematoden. Eine okologischmorphologische Studie, *Archiv. Zool.*, 4, 439, 1953.

Wieser, W., The effect of grain size on the distribution of small invertebrates inhabiting the beaches of Puget Sound, *Limnol. Oceanogr.*, 4, 181, 1959.

Wieser, W., Kaniwisher, J., Ecological and physiological studies on marine nematodes from a small salt marsh near Woods Hole, Massachusetts, *Limnol. Oceanogr.*, 6, 262, 1961.

Wildish, D.J., Factors controlling marine and estuarine sublittoral macrofauna, *Helog. Wiss. Meeresunters*, 30, 465, 1977.

Wildish, D.J., Kristmanson, D.D., Importance to mussels of the benthic boundary layer, *Can. J. Fish. Aquatic Sci.*, 41, 1618, 1984.

Wilkinson, M., Survival strategies of attached algae in estuaries, in *Feeding and Survival Strategies of Estuarine Organisms*, Jones, N.V., Wolff, W.J., Eds., Plenum Press, New York, 1981, 29.

Willason, S.W., Factors influencing the distribution and coexistence of *Pachygrapsus crassipes* and *Hemigrapsus oregonensis* (Decapoda: Grapsidae), *Mar. Biol.*, 64, 123, 1981.

Williams, A.H., An analysis of competitive interactions in a patchy backreef environment, *Ecology*, 62, 1107, 1981.

Williams, B.G., Naylor, E., Chatterton, T.D., The activity pattern of New Zealand mud crabs under field and laboratory conditions, *J. Exp. Mar. Biol. Ecol.*, 89, 269, 1985.

Williams, G.B., Substrate specificity in marine invertebrates, Ph.D. thesis, University of Wales, 1965.

Williams, J.A., The effect of dusk and dawn on the locomotory activity on the rhythm of *Talitrus saltator* (Montagu), *J. Exp. Mar. Biol. Ecol.*, 42, 285, 1980.

Williams, J.G., The influence of adults on the settlement of spat of the clam, *Tapes japonica*, *J. Mar. Res.*, 38, 729, 1980.

Williams, P.J.LeB., Biological and chemical aspects of dissolved organic compounds in sea water, in *Chemical Oceanography*, Riley, J.P., Shirrow, G., Eds., Academic Press, London, 1975, 301.

Williams, P.J.LeB., Incorporation of microheterotrophic processes into the classical paradigm of the planktonic food web, *Kiel. Meeresforch.*, 5, 1, 1981.

Williams, R.B., The ecology of diatom populations in a Georgia salt marsh, Ph.D. thesis, Harvard University, Cambridge, MA, 1962.

Williams, R.B., Division rates of salt marsh diatoms in relation to salinity and cell size, *Ecology*, 43, 877, 1964.

Williams, R.B., Annual phytoplankton production in a system of shallow temperate estuaries, in *Some Contemporary Studies in Marine Science,* Barnes, H., Ed., George Allen & Unwin, London, 1966, 699.

Williams, R.B., Murdock, M.B., Phytoplankton production and chlorophyll concentration in the Beaufort Channel, North Carolina, *Limnol. Oceanogr.,* 11, 73, 1966.

Williams, R.B., Murdock, M.B., The potential importance of *Spartina alterniflora* in conveying zinc, manganese and iron into estuarine food chains, in *Proceedings of the Second International Symposium on Radiology,* Nelson, D.J., Evans, F.C., Eds., U.S.A.E.C., Springfield, VA, 1969, 431.

Williams, S.L., McRoy, C.P., Seagrass productivity: the effect of light on carbon uptake, *Aquatic Bot.*, 12, 321, 1982.

Williams, T.M., Wolaver, T.G., Dame, R.F., Spurrier, J.D., The Bly Creek ecosystem study — organic carbon transport within a euryhaline salt marsh basin, North Inlet, South Carolina, *J. Exp. Mar. Biol. Ecol.*, 163, 125, 1992.

Williamson, P., Kendall, M.A., Population age structure and growth of the trochid *Monodonta lineata* determined from shell rings, *J. Mar. Biol. Assoc. U.K.*, 61, 1011, 1981.

Wilson, C.A., Stevenson, L.H., The dynamics of the bacterial population associated with a salt marsh, *J. Exp. Mar. Biol. Ecol.*, 48, 123, 1980.

Wilson, D.P., The influence of the substratum on the metamorphosis of *Noromastus* larvae, *J. Mar. Biol. Assoc. U.K.*, 22, 227, 1937.

Wilson, D.P., The relation of the substratum to the metamorphosis of *Ophelia* larvae, *J. Mar., Biol. Assoc. U.K.*, 27, 723, 1948.

Wilson, D.P., The influence of the nature of the substratum on the metamorphosis of the larvae of marine animals, especially the larvae of *Ophelia birornis* Savigny, *Anneles Inst. Oceanogr. Monaco, N.S.*, 27, 25, 1952.

Wilson, D.P., The role of micro-organisms in the settlement of *Ophelia bicornis* Savigny, *J. Mar. Biol. Assoc. U.K.*, 48, 387, 1955.

Wilson, W.H., Jr., A laboratory investigation of the effect of a terebellid polychaete on the suvivorship of a nereid polychaete larvae, *J. Exp. Mar. Biol. Ecol.*, 46, 73, 1980.

Wilson, W.H., Jr., Sediment-mediated interactions in a densely populated infaunal assemblage: the effects of the polychaete *Abarenicola pacifica*, *J. Mar. Res.*, 39, 735, 1981.

Wilson, W.H., Jr., The role of density dependence in a marine infaunal community, *Ecology*, 64, 295, 1983.

Wiltse, W.I., Effects of *Polinices duplicatus* (Gastropoda: Nticidae) on infaunal community structure at Barnstable Harbor, Massachusetts, USA, *Mar. Biol.*, 56, 301, 1980.

Wiltse, W.L., Foreman, K.H., Teal, J.M., Valiela, I., Effects of predators and food resources on the macrobenthos of salt marsh creeks, *J. Mar. Res.*, 42, 923, 1984.

Winberg, G.C., Rate of metabolism and food requirements of fishes (in Russian), Belorussian State University, Minsk, 251 pp, 1956, (Translation Service Fisheries Research Board of Canada 184).

Winter, J.E., A review on the knowledge of suspension-feeding in lamellibranchiate bivalves, with special reference to artificial aquaculture systems, *Aquaculture*, 13, 1, 1978.

Wisely, B., Observations on the settling behaviour of the larvae of the tubeworm, *Spirorbis borealis* Daudin (Polychaeta), *Aust. J. Mar. Freshw. Res.*, 2, 389, 1960.

Witte, F., de Wilde, P.A.W.J., On the ecological relation between *Nereis diversicolor* and juvenile *Arenicola marina, Neth. J. Sea Res.*, 13, 394, 1979.

Wium-Anderson, S., Borum, J., Biomass and production of eelgrass (*Zstera marina* L.) in the Orensund, Denmark, *Ophelia, Suppl.* 1, 49, 1980.

Wolcott, T.G., Physiological ecology and intertidal zonation of limets (Acmaea): a critical look at "limiting factors, " *Biol. Bull. Mar. Biol. Lab., Woods Hole,* 145, 389, 1973.

Wolff, W.J., A benthic food budget for the Grevelingen Estuary, The Netherlands, and a consideration of the mechanisms causing high benthic secondary production in estuaries, in *Ecology of the Marine Benthos,* Coull, B.C., Ed., University of South Carolina Press, Columbia, SC, 1977, 267.

Wolff, W.J., *Ecology of the Wadden Sea,* Balkema Books, Rotterdam, 1300 pp, in 3 vols, 1980.

Wolff, W.J., Estuarine benthos, in *Estuaries and Enclosed Seas. Ecosystems of the World 26*, Ketchum, B.H., Ed., Elsevier, Amsterdam, 1983, 151.

Wolff, W.J., van Haperen, A.M.M., Sandee, A.J.J., Baptist, H.J.M., Saeijs, H.L.F., The trophic role of birds in the Grevelingen Estuary, The Netherlands, as compared to their role in the saline Lake Grevellingen, in *Proceedings of the 10th European Symposium on Marine Biology. 2. Population Dynamics*, Persoone, G., Jaspers, W., Universal Press, Westeren, 1976, 673.

Wood, J.E.F., Odum, W.E., Zieman, J.C., Influence on the productivity of tropical coastal lagoons, in *United Nations Symposio, Memoir Symposio Internat. Lagunas Costeras, Nov. 1967, Mexico*, UNAM-UNESCO, 1969, 495.

Woodin, S.A., Polychaete abundance patterns in a marine soft-sediment environment. The importance of biological interactions, *Ecol. Monogr.*, 44, 171, 1974.

Woodin, S.A., Adult-larval interactions in dense infaunal assemblages: patterns of abundance, *J. Mar. Res.*, 34, 25, 1976.

Woodin, S.A., Refuges, disturbance, and community structure: a marine soft-bottom example, *Ecology*, 59, 274, 1978.

Woodin, S.A., Disturbance and community structure in a shallow water sand flat, *Ecology*, 62, 1052, 1981.

Woodin, S.A., Recruitment of infauna: positive or negative cues?, *Amer. Zool.*, 31, 797, 1991.

Woodin, S.A., Shallow water benthic ecology: a North American perspective of sedimentary habitats, *Aust. J. Ecol.*, 24, 291, 1999.

Woodin, S.A., Marinelli, R., Biogenic habitat modification in marine sediments: the importance of species composition and activity, *Symp. Zool. Soc. London*, 63, 1991.

Woodin, S.A., Marinelli, R.L., Lincoln, D.E., Biogenic brominated aromatic compounds and recruitment of infauna, *J. Chem. Ecol.*, 19, 517, 1993.

Woodroffe, C.D., *Mangroves of the Upper Waitemata Harbour. Biomass, Productivity and Geomorphology,* Upper Waitemata Harbour Catchment Study, Working Report No. 36, 192 pp, 1982a.

Woodroffe, C.D., Litter production and decomposition in New Zealand mangrove *Avicennia marina* var. *resiniflora, N.Z. J. Mar. Freshw. Res.*, 16, 179, 1982b.

Woodroffe, C.D., Studies of a mangrove basin, Tuff Crater, New Zealand. I. Mangrove biomass and production of detritus, *Est. Coastal Shelf Sci.*, 20, 265, 1985.

Woodroffe, C.D., Mangrove sediments and geomorphology, in *Tropical Mangrove Ecosystems*, Robertson, A.I., Alongi, D.M., Eds., American Geophysical Union, New York, 1992, 7.

Woodwell, G.M., Hall, C.A.S., Whitney, D.E., Houghton, R.A., The Flax Pond ecosystem study: exchanges of inorganic nitrogen between an estuarine marsh and Long Island Sound, *Ecology*, 60, 695, 1979.

Woodwell, G.M., Whitney, D.E., Flax Pond ecosystem study. I. Exchanges of phosphorus between salt marsh and the coastal waters of Long Island Sound, *Mar. Biol.*, 1, 1, 1977.

Woodwell, C.M., Whitney, D.E., Hall, C.A.S., Houghton, R.A., The Flax Pond ecosystem study: exchanges of carbon in water between a salt marsh and Long Island Sound, *Limnol. Oceanogr.*, 22, 833, 1977.

Woolridge, T., Zonation and distribution of the beach mysid, *Gastrosaccus psammodytes*, *J. Zool., London*, 193, 183, 1981.

Wootin, J.T., Indirect effects, prey susceptability, and habitat selection: impacts of birds on limpets and algae, *Ecology,* 73, 981, 1992.

Workman, C., Comparisons of energy partitioning in contrasting age-structure populations of the limpet *Patella vulgata, J. Exp. Mar. Biol. Ecol.*, 68, 81, 1983.

Wright, J.R., Hartnoll, R.G., An energy budget for a population of the limpet *Patella vulgata*, *J. Mar. Biol. Assoc. U.K.*, 61, 627, 1981.

Wright, L.D., Short, A.D., Morphodynamics of beaches and surf zones in Australia, in *Handbook of Coastal Processes and Erosion*, Komar, P.D., Ed., CRC Press, Boca Raton, FL, 1983, 35.

Wright, R.T., Measurement and significance of specific activity in the heterotrophic activity of natural waters, *Applied Envt. Microbiol.*, 36, 297, 1978.

Wright, R.T., Coffin, R.B., Planktonic bacteria in estuaries and coastal waters of northern Massachusetts: spatial and temporal distribution, *Mar. Ecol. Prog. Ser.*, 11, 205, 1983.

Wright, R.T., Coffin, R.B., Measuring mictozooplankton grazing on planktonic bacteria by its impact on bacterial production, *Microbial Ecol.*, 10, 137, 1984.

Wulff, F., Aertebjerg, G., Nicholas, G., Niemi, A., Ciszevski, P., Schultz, S., Kaiser, W., The changing pelagic ecosystem of the Baltic Sea, *Ophelia (Supplement)*, 4, 299, 1986.

Wulff, F., Field, J.G., The importance of different trophic pathways in a nearshore benthic community under upwelling and downwelling conditions, *Mar. Ecol. Prog. Ser.*, 12, 217, 1983.

Wulff, F., Field, J.G., Mann, K.H., Eds., *Network Analysis in Marine Ecology: Methods and Applications*, Springer-Verlag, Berlin, 1989.

Wulff, F., Ulanowicz, R.E., A comparative anatomy of the Baltic Sea and Chesapeake Bay ecosystems, in *Network Analysis of Marine Ecology; Methods and Applications,* Springer-Verlag, Berlin, 1989, 232.

Yamada, S.B., Boulding, E.G., The role of highly mobile crab predators in the intertidal zonation of their gastropod prey, *J. Exp. Mar. Bio. Ecol.*, 204, 59, 1996.

Yamanoto, G., Ecological study in an indicator species of eutrophic waters, Annual Report of the Study Supported by the Grant from the Ministry of Education of Japan, (Grant No. 257219) in 1979, 82 pp, 1980.

Yáñez-Arancibia, A., On studies of fishes in coastal lagoons: Scientific notes, *Ann. Centro. Cienc. del Mar y Limnol., Univ. Nat. Auton, Mexico*, 2, 53, 1975.

Yáñez-Arancibia, A., Observaciones sobre *Mugil curema* Valenciennes, en areas naturales de crianza, Mexico. Alimentacion, madurez, cercimento y relaciones ecologicas, *Ann. Centro. Cienc. del Mar. y Limnol., Univ. Nat. Auton, Mexico*, 3, 92, 1976.

Yáñez-Arancibia, A., Taxonomy, ecology and structure of fish communities in coastal lagoons on the Pacific coasts of Mexico, *Centro. Cienc. de Mar. y Limnol., Univ. Nat. Auton, Mexico, Publ. Esp*, 2, 1-306, 1978.

Yáñez-Arancibia, A., Ed., *Fish Community Ecology in Estuaries and Coastal Lagoons: Towards an Ecosystem Integration*, Editorial Universataria, UNMA-PUAL-ICML, Mexico D.F., 654 pp, 1985.

Yáñez-Arancibia, A., *Ecologia de la Zona Coatera: Analysis de Siete Tropicas*, Editorial ACT, Mexico, D.F., 200 pp, 1986.

Yáñez-Arancibia, A., Sanchez-Gill, M., *Ecologia de les Recuros Demersales Marinos: Fundamentos en Costas Tropicales*, Editorial ACT, Mexico, D.F., 230 pp, 1988.

Yingst, J.Y., Rhoads, D.C., The role of bioturbation in the enhancement of bacterial growth rates in marine sediments, in *Marine Benthic Dynamics*, Tenore, K.R., Coull, B.C., Eds., University of South Carolina Press, Columbia, SC, 1980, 407.

Yool, A.J., Grau, S.M., Hadfield, M.G., Jensen, R.A., Markell, D.A., Norse, D.E., Excess potassium induces larval metamorphosis in four marine invertebrate species, *Biol. Bull.*, 170, 255, 1986.

Young, C.M., Novelty of supply-side ecology, *Science*, 235, 415, 1987.

Young, D.K., Rhoads, D.C., Animal-sediment relations in Cape Cod Bay, Massachusetts. I. A transect study, *Mar. Biol.*, 11, 242, 1971.

Young, D.K., Young, M.W., Regulation of species densities of seagrass-associated macrobenthos: evidence from field experiments in the Indian River estuary, Florida, *J. Mar. Res.*, 36, 569, 1978.

Yule, A.B., Walker, G., Settlement of *Balanus balanoides*: the effect of cyprid antennular secretion, *J. Mar. Biol. Assoc., U.K.*, 65, 707, 1985.

Zajac, R.N., Whitlach, R.B., Responses of estuarine infauna to disturbance. II. Spatial and temporal variation of initial colonization, *Mar. Ecol. Prog. Ser.*, 10, 1, 1982a.

Zajac, R.N., Whitlach, R.B., Responses of estuarine infauna to disturbance. II. Spatial and temporal variation in succession, *Mar. Ecol. Prog. Ser.*, 10, 15, 1982b.

Zedler, J., The ecology of southern California coastal saltmarshes: a community profile, U.S. Fish & Wildlife Service, Report No. FWS/OBS-81/54, Washington, DC, 110 pp, 1982.

Zeitzchel, B., Sediment-water interactions in nutrient dynmics, in *Marine Benthic Dynamics*, Tenore, K.R., Coull, B.C., Eds., University of South Carolina Press, Columbia, SC, 1980, 195.

Zieman, J.C., A study of the growth and decomposition of the seagrass *Thalassia testudinium,* Master's thesis, University of Miami, Miami, FL, 50 pp, 1968.

Zieman, J.C., Quantitative and dynamic aspects of the ecology of turtle grass, *Thalassia testudinium*, in *Estuarine Research, Vol. 1. Chemistry, Biology and the Estuarine System*, Cronin, L.E., Ed., Academic Press, New York, 1975, 541.

Zieman, J.C., Thayer, G.W., Robblee, M.B., Zieman, R.T., Production and export of seagrasses from a tropical bay, in *Proceedings of the Conference on Biological Processes in Coastal and Marsh Systems*, Plenum Press, New York, 1979, 21.

Zieman, J.C., Wetzel, R.G., Productivity in seagrasses, methods and rates, in *Handbook of Seagrass Biology: An Ecosystem Perspective*, Phillips, R.C., McRoy, C.P., Eds., Garland STMP Press, New York, 1980, 87.

Zimmerman, R.C., Kremer, J.N., *In situ* growth and chemical composition of the giant kelp *Macrocystic pyrifera*: The response to temporal changes in ambient nutrient availability, *J. Exp. Mar. Biol. Ecol.*, 27, 277, 1986.

Zimmerman.R.J., Minello, T.J., Densities of *Pinaeus aztecus, Penaeus sertifrons* and other natant macrofauna in a Texas salt marsh, *Estuaries,* 7, 421, 1984.

Zobell, G.E., Factors affecting drift seaweeds on some San Diego beaches, University of California Marine Research Report No. 59(3), 1959.

Zwartz, L., Drent, R.H., Prey depletion and the regulation of predator density: Oystercatchers (*Haemotopus ostralegus*) feeding on mussels (*Mytilus edulis*), in *Feeding and Survival Strategies of Estuarine Organisms,* Jones, N.V., Wolff, W.J., Eds., Plenum Press, New York, 1981, 193.

Index

C

D

E

F

N

O

P

Q

R

S

T

U

V

W

Y

Z